1989 Recommended Dietary Allowances (RDA)

Age (yr)	Energy (cal)	Protein (g)	Vitamin A (µg RE)	Vitamin K (µg)	Iron (mg)	Zinc (mg)	Iodine (µg)
Infants							
0.0–0.5	650	13	375	5	6	5	40
0.5–1.0	850	14	375	10	10	5	50
Children							
1–3	1300	16	400	15	10	10	70
4–6	1800	24	500	20	10	10	90
7–10	2000	28	700	30	10	10	120
Males							
11–14	2500	45	1000	45	12	15	150
15–18	3000	59	1000	65	12	15	150
19–24	2900	58	1000	70	10	15	150
25–50	2900	63	1000	80	10	15	150
51+	2300	63	1000	80	10	15	150
Females							
11–14	2200	46	800	45	15	12	150
15–18	2200	44	800	55	15	12	150
19–24	2200	46	800	60	15	12	150
25–50	2200	50	800	65	15	12	150
51+	1900	50	800	65	10	12	150
Pregnancy	+300	60	800	65	30	15	175
Lactation							
1st 6 mo.	+500	65	1300	65	15	19	200
2nd 6 mo.	+500	62	1200	65	15	16	200

Daily Values (DV) Used on Food Labels[a]

Daily Reference Values (DRVs)[b]

Food Component	Amount
protein[c]	50 g
fat	65 g[d]
saturated fat	20 g
cholesterol	300 mg[e]
total carbohydrate	300 g
fiber	25 g
sodium	2,400 mg
potassium	3,500 mg

Reference Daily Intakes (RDI)

Nutrient	Amount	Nutrient	Amount
Thiamin	1.5 mg	Calcium	1,000 mg
Riboflavin	1.7 mg	Iron	18 mg
Niacin	20 mg	Zinc	15 mg
Biotin	300 µg	Iodine	150 µg
Pantothenic Acid	10 mg	Copper	2 mg
Vitamin B_6	2 mg	Chromium	120 µg
Folate	400 µg[f]	Selenium	70 µg
Vitamin B_{12}	6 µg	Molybdenum	75 µg
Vitamin C	60 mg	Manganese	2 mg
Vitamin A	5,000 IU[g]	Chloride	3,400 mg
Vitamin D	400 IU[g]	Magnesium	400 mg
Vitamin E	30 IU[g]	Phosphorus	1 g
Vitamin K	80 µg		

[a] Based on 2,000 calories a day for adults and children over 4 years old.

[b] Formerly the U.S. RDA, based on National Academy of Sciences' 1968 Recommended Dietary Allowances.

[c] DRV for protein does not apply to certain populations; Reference Daily Intake (RDI) for protein has been established for these groups: children 1 to 4 years: 16 g; infants under 1 year: 14 g; pregnant women: 60 g; nursing mothers: 65 g.

[d] (g) grams

[e] (mg) milligrams

[f] (µg) micrograms

[g] Equivalent values for the three RDI nutrients expressed as IU are: vitamin A, 900 RE (assumes a mixture of 40% retinol and 60% beta-carotene); vitamin D, 10 µg; vitamin E, 20 mg.

Estimated Safe and Adequate Daily Dietary Intakes of Additional Selected Minerals (United States)

	TRACE ELEMENTS[a]			
Age (years)	**Chromium (μg)**	**Molybdenum (μg)**	**Copper (mg)**	**Manganese (mg)**
Infants				
0–0.5	10–40	15–30	0.4–0.6	0.3–0.6
0.5–1	20–60	20–40	0.6–0.7	0.6–1.0
Children				
1–3	20–80	25–50	0.7–1.0	1.0–1.5
4–6	30–120	30–75	1.0–1.5	1.5–2.0
7–10	50–200	50–150	1.0–2.0	2.0–3.0
11+	50–200	75–250	1.5–2.5	2.0–5.0
Adults	50–200	75–250	1.5–3.0	2.0–5.0

[a]Because the toxic levels for many trace elements may be only several times usual intakes, the upper levels for the trace elements given in this table should not be habitually exceeded.

Source: Recommended Dietary Allowances, © 1989 by the National Academy of Sciences, National Academy Press, Washington, D.C.

Estimated Minimum Requirements of Sodium, Chloride, and Potassium

Age (years)	**Sodium[a] (mg)**	**Chloride (mg)**	**Potassium[b] (mg)**
Infants			
0.0–0.5	120	180	500
0.5–1.0	200	300	700
Children			
1	225	350	1,000
2–5	300	500	1,400
6–9	400	600	1,600
Adolescents	500	750	2,000
Adults	500	750	2,000

[a]Sodium requirements are based on estimates for growth and for replacement of obligatory losses. They cover a wide variation of physical activity patterns and climatic exposure but do not provide for large, prolonged losses from the skin through sweat.

[b]Dietary potassium may benefit the prevention and treatment of hypertension and recommendations to include many servings of fruits and vegetables would raise potassium intakes to about 3,500 mg/day.

Source: Recommended Dietary Allowances, © 1989 by the National Academy of Sciences, National Academy Press, Washington, D.C.

Unit Conversions

International Units (IU)

To convert IU to:

RE:[a] from animal sources, divide by 3.33; and from vegetables and fruits, divide by 10.

μg vitamin D: divide by 40 or multiply by 0.025.

mg α-T:[b] multiply by 0.67 if the form of supplement is "natural" (labeled d-α-tocopherol); multiply by 0.45 if labeled dl-α-tocopherol. To convert mg α-TE[c] to mg α-T, multiply by 0.8.

Folate

To convert micrograms of synthetic folate in supplements and enriched foods to Dietary Folate Equivalents (μg DFE):

$$\mu g \text{ synthetic folate} \times 1.7 = \mu g \text{ DFE}$$

For naturally occurring folate, assign each microgram folate a value of 1 μg DFE:

$$\mu g \text{ folate} = \mu g \text{ DFE}$$

Sodium

To convert milligrams of sodium to grams of salt:

$$\text{mg sodium} \div 400 = \text{g of salt}$$

The reverse is also true:

$$\text{g salt} \times 400 = \text{mg sodium}$$

Niacin

1 mg NE (niacin equivalent) = 1 mg niacin = 60 mg dietary tryptophan

[a]Retinol equivalents (vitamin A). [b]Alpha-tocopherol (vitamin E). [c]Alpha-tocopherol equivalents (vitamin E).

Publisher: Peter Marshall
Development Editor: Laura Graham
Editorial Assistant: Keynia Johnson
Marketing Manager: Becky Tollerson
Project Editor: Sandra Craig
Print Buyer: Barbara Britton
Permissions Editor: Joohee Lee

Production: Martha Emry Production Services
Text and Cover Designer: Hespenheide Design
Illustrations: Miyake Illustration, Sandra McMahon, Atherton Customs
Cover Printer: Phoenix Color
Compositor: Parkwood Composition
Printer: R. R. Donnelley/Willard

Printed in the United States of America
1 2 3 4 5 6 7 04 03 02 01 00

Library of Congress Cataloging-in-Publication Data
Boyle, Marie A. (Marie Ann)
Personal nutrition/Marie A. Boyle.—4th ed.
p. cm.
Includes index.
ISBN 0-534-54603-X (pbk.)
1. Nutrition

RA784.B65 2000
613.2—dc21

00-026382

Instructor's Edition ISBN: 0-534-540608-0

Wadsworth/Thomson Learning
10 Davis Drive
Belmont, CA 94002-3098
USA

For more information about our products, contact us:
Thomson Learning Academic Resource Center
1-800-423-0563
http://www.wadsworth.com

International Headquarters
Thomson Learning
International Division
290 Harbor Drive, 2nd Floor
Stamford, CT 06902-7477
USA

UK/Europe/Middle East/South Africa
Thomson Learning
Berkshire House
168-173 High Holborn
London WC1V 7AA
United Kingdom

Asia
Thomson Learning
60 Albert Street, #15-01
Albert Complex
Singapore 189969

Canada
Nelson Thomson Learning
1120 Birchmount Road
Toronto, Ontario M1K 5G4
Canada

FOURTH EDITION

Personal Nutrition

MARIE A. BOYLE
College of St. Elizabeth

Wadsworth
Thomson Learning™

Australia • Canada • Mexico • Singapore • Spain • United Kingdom • United States

In memory of my father, David Michael Boyle, and his Irish-loving ways, and with love and gratitude to my husband, best friend, and willing new recipe critic, Steve Struble

Marie Boyle Struble

About the Author

MARIE A. BOYLE, Ph.D., RD, received her B.A. in Psychology from the University of Maine, her M.S. in nutrition from Florida State University, and, in 1992, her Ph.D. in nutrition and exercise science from Florida State University. She is coauthor of the textbook *Community Nutrition in Action: An Entrepreneurial Approach*. She presently works as Associate Professor and Director of the Didactic Program in Dietetics at the College of Saint Elizabeth in Morristown, NJ. She teaches undergraduate courses in introductory, advanced, lifecycle, and community nutrition. She also teaches Research Methods, Advanced Metabolism of the Micronutrients, Nutrition and Aging, and Complementary and Alternative Medical Therapy in the Graduate nutrition program at the college. She maintains memberships with the American Dietetic Association, New Jersey Dietetic Association, American Public Health Association, and Society for Nutrition Education. She also serves as a reviewer for the American Dietetic Association, *American Journal of Health Promotion, Florida Journal of Public Health,* and as a member of the Osteoporosis Education Coalition of New Jersey.

Contents in Brief

1 *The Art of Understanding Nutrition* 2
Spotlight: Separating Nutrition Fact from Fiction 19

2 *The Pursuit of an Ideal Diet* 26
Spotlight: A Tapestry of Cultures and Cuisines 54

3 *The Carbohydrates: Sugar, Starch, and Fiber* 66
Spotlight: Sweet Talk—Alternatives to Sugar 93

4 *The Lipids: Fats and Oils* 98
Spotlight: Diet and Heart Disease 125

5 *The Proteins and Amino Acids* 134
Spotlight: The Benefits of Soy 159

6 *The Vitamins* 164
Spotlight: Functional Foods—Let Food Be Your Medicine 196

7 *Water and the Minerals* 200
Spotlight: Osteoporosis—The Silent Stalker of the Bones 229

8 *Weight Control* 238
Spotlight: The Eating Disorders 275

9 *Nutrition and Fitness* 282
Spotlight: Alcohol and Nutrition 311

10 *The Life Cycle: Conception Through the Later Years* 318
Spotlight: Nutrition and Cancer Prevention 365

11 *Food Safety and the Global Food Supply* 374
Spotlight: Domestic and World Hunger 405

Appendixes A-1

Contents

1 *The Art of Understanding Nutrition* 2
Nutrition and Health Promotion 5
The Longevity Game Scorecard 8
Nutrition Action: Fast Guide to Eating on the Run 11
Understanding Our Food Choices 14
Availability 14
Income and Food Prices 15
Advertising and the Media 15
Social and Cultural Factors 16
Other Factors That Affect Our Food Choices 17
Spotlight: SEPARATING NUTRITION FACT FROM FICTION 19

2 *The Pursuit of an Ideal Diet* 26
The ABCs of Eating for Health 28
The Nutrients 30
The Energy-Yielding Nutrients 31
Vitamins, Minerals, and Water 31
Nutrient Recommendations 32
The Dietary Reference Intakes (DRI) 33
The RDA for Calories 35
Other Recommendations 35
The Challenge of Dietary Guidelines 36
Tools Used in Diet Planning 38
Food Group Plans 38
Nutrition Action: Grazer's Guide to Smart Snacking 39
Food Labels 42
Exchange Lists 51
Food Composition Tables 52
Rate Your Plate Scorecard 53
Spotlight: A TAPESTRY OF CULTURES AND CUISINES 54
The Savvy Diner: Choosing Healthful International Cuisines 62

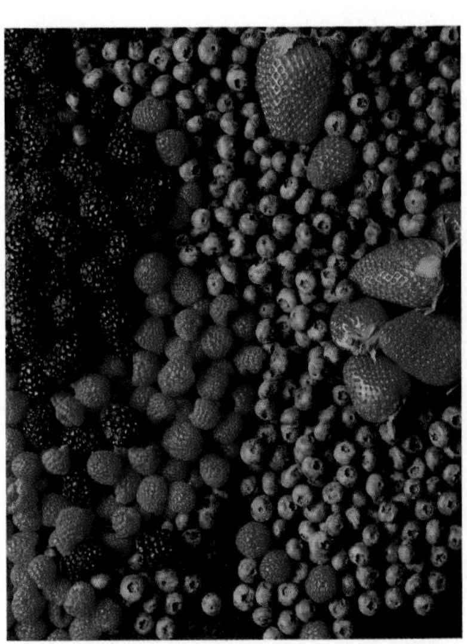

3 *The Carbohydrates: Sugar, Starch, and Fiber* 66

Carbohydrate Basics 68
The Simple Carbohydrates 69
 The Single Sugars: Monosaccharides 69
 The Double Sugars: Disaccharides 69
 Sugar and Health 70
 Keeping Sweetness in the Diet 70
The Complex Carbohydrates: Starch in the Diet 72
Nutrition Action: Keeping Out from Under the Dental Drill 73
 Adding Whole Foods to the Diet 76
The Savvy Diner: Selection Tips for Carbohydrates in the Diet 77
 The Bread Box: Refined, Enriched, and Whole-Grain Breads 80
The Complex Carbohydrates: Fiber in the Diet 81
 The Health Effects of Fiber 81
How the Body Handles Carbohydrates 84
Carbohydrate Consumption Scorecard 86
 Maintaining the Blood Glucose Level 87
 Hypoglycemia 89
 Diabetes 90
Spotlight: SWEET TALK—ALTERNATIVES TO SUGAR 93

4 *The Lipids: Fats and Oils* 98

A Primer on Fats 100
 The Functions of Fat in the Body 100
 The Functions of Fat in Foods 101
 The Terminology of Fat 101
A Closer View of Fats 102
 Saturated Versus Unsaturated Fats 102
 The Essential Fatty Acids 103
 Omega-6 Versus Omega-3 Fatty Acids 103
Characteristics of Fats in Foods 104
Nutrition Action: The Trans Fatty Acid Controversy—Is Butter Better? 106
The Other Members of the Lipid Family: Phospholipids and Sterols 109
How the Body Handles Fat 110
Lipids and Health 111
 "Good" Versus "Bad" Cholesterol 112
 Lowering Blood Cholesterol Levels 113
Fat in the Diet 117
 Understanding Fat Substitutes 118
Scorecard: Rate Your Fats and Health IQ 121
The Savvy Diner: How to Handle Fats in the Diet 122
Spotlight: DIET AND HEART DISEASE 125
Heart Health Scorecard 128

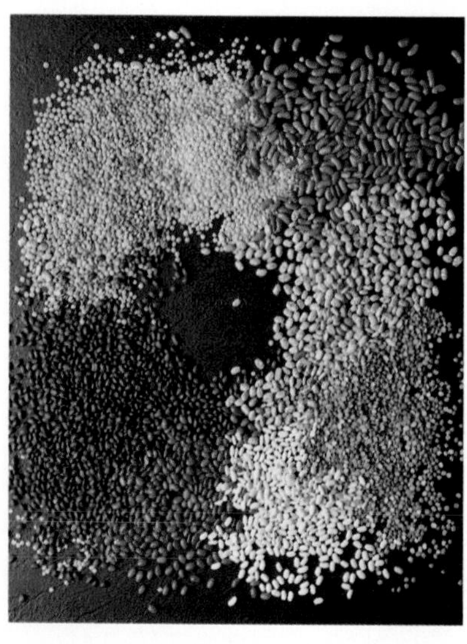

5 *The Proteins and Amino Acids* 134

What Proteins Are Made Of 136
 Essential and Nonessential Amino Acids 136
Nutrition Action: Amino Acid Essentials 137
 Proteins as the Source of Life's Variety 140
 Denaturation of Proteins 140
The Functions of Body Proteins 140
 Growth and Maintenance 140
 Enzymes 141

Hormones 141
Antibodies 142
Fluid Balance 142
Acid-Base Balance 142
Transport Proteins 143
Protein as Energy 143
How the Body Handles Protein 144
Protein Quality of Foods 144
Recommended Protein Intakes 146
Protein and Health 146
Protein-Energy Malnutrition 147
Too Much Protein 148
Protein in the Diet 149
Protein Scorecard 152
The Savvy Diner: Reshape Your Protein Choices for Health 153
The Vegetarian Diet 155
Proteins 155
Vitamins 156
Minerals 157
Health Benefits 158
Spotlight: THE BENEFITS OF SOY 159

6 *The Vitamins* 164
Turning Back the Clock 166
Water-Soluble Vitamins 167
Thiamin 170
Riboflavin 171
Niacin 171
Folate 172
Vitamin B_6 174
Vitamin B_{12} 175
Pantothenic Acid and Biotin 176
Vitamin C 176
Fat-Soluble Vitamins 179
Vitamin A 179
Vitamin D 181
Vitamin E 183
Vitamin K 184
The Savvy Diner: Vitamin Preservation 186
Five a Day Plus Scorecard 188
Nutrition Action: Medicinal Herbs 189
Nonnutrients in Foods: The Phytochemical Superstars 193
Mechanisms of Actions of Phytochemicals 195
How to Optimize Phytochemicals in a Daily Eating Plan 195
Spotlight: FUNCTIONAL FOODS—LET FOOD BE YOUR MEDICINE 196

7 *Water and the Minerals* 200
Water—The Most Essential Nutrient 202
Water in the Diet 203
Keeping Water Safe 203
Bottled Water 205
The Major Minerals 207
Calcium 208
Phosphorus 212

Calcium Sources Scorecard 213
Sulfur and Magnesium 214
Sodium, Potassium, and Chloride 214
Nutrition Action: Diet and Blood Pressure—The Salt Shaker and Beyond 218
The Savvy Diner: Seasoning Foods Without Excess Salt 222
The Trace Minerals 223
Iron 223
Zinc 225
Iodine 226
Fluoride 227
Copper, Manganese, Chromium, Selenium, and Molybdenum 228
Trace Minerals of Uncertain Status 228
Spotlight: OSTEOPOROSIS—THE SILENT STALKER OF THE BONES 229
Scorecard: Do You Have a Bone-Healthy Lifestyle? 236

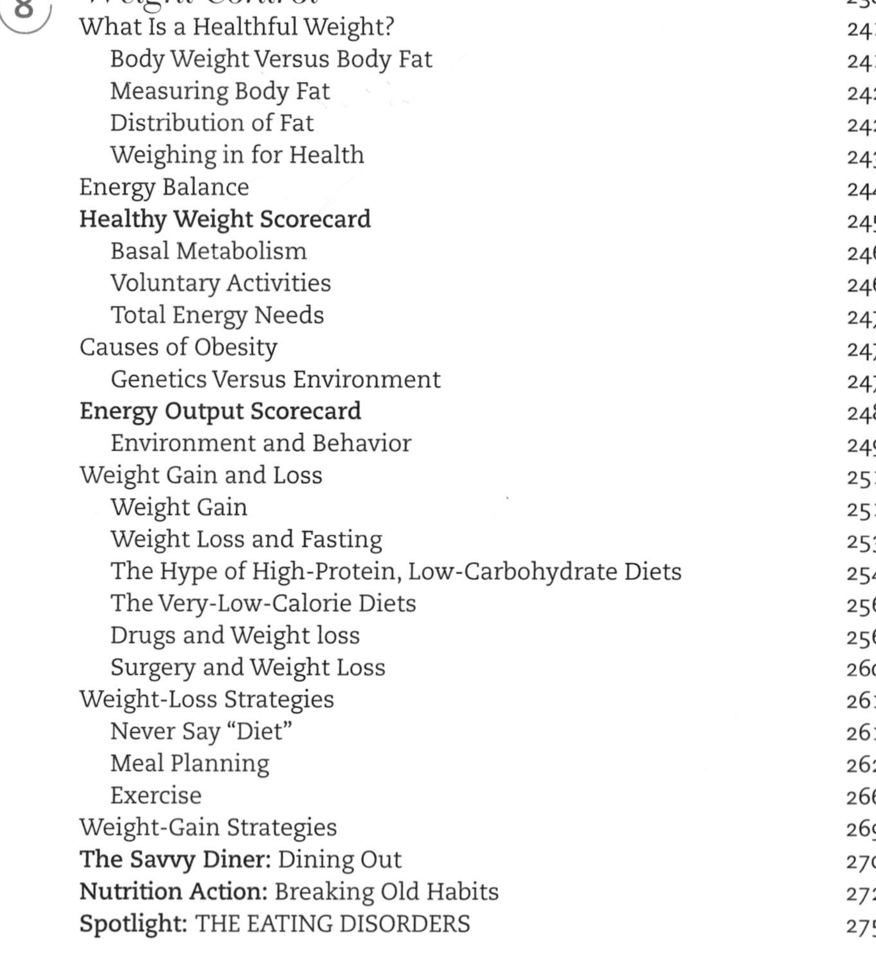

8 *Weight Control* 238
What Is a Healthful Weight? 241
Body Weight Versus Body Fat 241
Measuring Body Fat 242
Distribution of Fat 242
Weighing in for Health 243
Energy Balance 244
Healthy Weight Scorecard 245
Basal Metabolism 246
Voluntary Activities 246
Total Energy Needs 247
Causes of Obesity 247
Genetics Versus Environment 247
Energy Output Scorecard 248
Environment and Behavior 249
Weight Gain and Loss 251
Weight Gain 251
Weight Loss and Fasting 253
The Hype of High-Protein, Low-Carbohydrate Diets 254
The Very-Low-Calorie Diets 256
Drugs and Weight loss 256
Surgery and Weight Loss 260
Weight-Loss Strategies 261
Never Say "Diet" 261
Meal Planning 262
Exercise 266
Weight-Gain Strategies 269
The Savvy Diner: Dining Out 270
Nutrition Action: Breaking Old Habits 272
Spotlight: THE EATING DISORDERS 275

9 *Nutrition and Fitness* 282
Reasons to Exercise 284
Physical Conditioning 286
The Four Components of Fitness 287
Strength 287
Flexibility 289
Muscle Endurance 290
Cardiovascular Endurance 290

Getting Started on Lifetime Fitness 290
Physical Activity Scorecard 291
Energy for Exercise 292
Aerobic and Anaerobic Metabolism 292
Aerobic Exercise—Exercise for the Heart 293
Fuels for Exercise 294
Glucose Use During Exercise 294
Fat Use During Exercise 296
Protein Needs for Fitness 297
Fluid Needs and Exercise 298
Water and Exercise 298
Water and Fluid-Replacement Drinks 299
Vitamins and Minerals for Exercise 300
The Vitamins 301
The Minerals 301
The Bones and Exercise 302
Nutrition Action: Athletes and Supplements—Help or Hype 303
Fitness Quackery Scorecard 307
The Savvy Diner: Food for Fitness 308
Spotlight: ALCOHOL AND NUTRITION 311

10 *The Life Cycle: Conception Through the Later Years* 318

Pregnancy: Nutrition for the Future 320
Nutritional Needs of Pregnant Women 320
Pregnancy Readiness Scorecard 323
Maternal Weight Gain 324
Practices to Avoid 325
Common Nutrition-Related Problems of Pregnancy 326
Adolescent Pregnancy 327
Nutrition Action: Not for Coffee Drinkers Only 328
Nutrition of the Breastfeeding Mother 330
Healthy Infants 330
Milk for the Infant: Breastfeeding 330
Feeding Formula 332
Supplements for the Infant 333
Food for the Infant 334
Nutrition-Related Problems of Infancy 336
Early and Middle Childhood 337
Growth and Nutrient Needs of Children 337
Other Factors That Influence Childhood Nutrition 340
Nutrition-Related Problems of Childhood 341
The Importance of Teen Nutrition 342
Nutrient Needs of Adolescents 343
Nutrition-Related Problems of Adolescents 343
Nutrition Action: Food Allergy—Nothing to Sneeze At 345
Nutrition in Later life 348
Demographic Trends and Aging 348
Aging Scorecard 349
Healthy Adults 350
Aging and Nutrition Status 351
Nutritional Needs and Intakes 351
Nutrition-Related Problems of Older Adults 354
Factors Affecting Nutrition Status of Older Adults 357
Sources of Nutritional Assistance 359

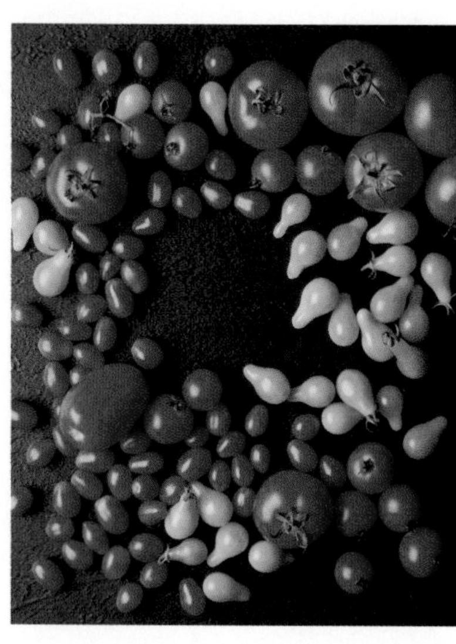

The Savvy Diner: Meals for One 361
Looking Ahead: Growing Seasoned 363
Spotlight: NUTRITION AND CANCER PREVENTION 365

11 *Food Safety and the Global Food Supply* 374

Foodborne Illnesses and the Agents That Cause Them 376
Microbial Agents 378
Natural Toxins 382
Safe Food Storage and Preparation 383
The Savvy Diner: Home Food Safety 387
Food Safety Scorecard 389
Pesticides and Other Chemical Contaminants 391
Chemical Agents 391
Pesticide Residues 393
Nutrition Action: Should You Buy Organically Grown Produce? 396
Food Additives 398
The GRAS List and the Delaney Clause 399
New Technologies on the Horizon 400
Irradiation 400
Genetic Engineering 401
Spotlight: DOMESTIC AND WORLD HUNGER 405

Appendixes

Appendix A
Nutrition Resources A-1
Appendix B
An Introduction to the Human Body A-7
Appendix C
Canadian Dietary Guidelines and Recommendations A-21
Appendix D
Aids to Calculations A-27
Appendix E
Food Exchange Systems A-32
Appendix F
Chapter Notes A-51
Appendix G
Answers to Check Yourself Questions A-65
Appendix H
Table of Food Composition A-68

Glossary G-1
Index I-1

Preface

With this Fourth Edition of *Personal Nutrition*, we continue to develop the vision we had in writing the first edition of this book nearly 15 years ago—that is, to apply basic nutrition concepts to personal everyday life. The text is designed to support the many one- to four-credit introductory nutrition courses available to students today from a variety of majors. This edition reflects the many changes that have taken place in the field of nutrition in recent years, and offers all readers the opportunity to develop practical skills in making decisions regarding their personal nutrition and health. The challenge has been to teach the facts about nutrition as well as how to evaluate them and, most importantly, to motivate readers to apply what they learn in daily life. It is our hope that you will benefit from the many new findings and fundamental information presented in this edition and enjoy its new design and many new figures, photos, and cartoons—each with a clear message of its own.

Nutrition is a subject that is forever changing. Since the last edition was published, we have experienced the growth of the Internet as a resource for learning; we have established new dietary guidelines, nutrient recommendations, national health objectives, and food safety initiatives. Additionally, we have witnessed the emergence of new research findings regarding the aging process, human genetics, and phytochemicals; the marketing of functional foods for disease prevention; the rise of complementary and alternative medicine; and the birth of a new millennium. It is important that you, as a consumer of the latest nutrition information, have the knowledge to evaluate the nutrition issues and controversies that confront you both today and tomorrow. Newspapers are quick to print nutrition breakthroughs, new fad diets appear monthly on the magazine racks, and television advertising extols the wonders of products of questionable value. Nutrition claims bombard us frequently, and we must evaluate and assess them. This Fourth Edition of *Personal Nutrition* continues to provide a sieve through which to separate the valid nutrition information from the rest.

Chapter 1 provides a personal invitation to eat well for optimum health and assists the reader in becoming a sophisticated consumer of new information about nutrition. It also explores the factors that affect food choices, including the media, advertising, and cultural factors. Chapter 2 introduces the basic nutrients the body needs along with the nutrition tools and most recent guidelines needed to help make sound food choices. It provides sample food labels for understanding nutrition information, terminology, and health claims found on labels. Chapter 2 also includes a section on various international and ethnic cuisines that highlights the multicultural heritage of our country. Chapters 3 through 7 present the nutrients and show how they all work together to nourish the body. The chapters on vitamins and minerals spotlight the emerging importance of the antioxidant nutrients and phytochemicals and also feature colorful food photos depicting food sources for individual vitamins and minerals. Chapter 8 discusses weight management issues and includes a new summary table that compares the major weight-loss programs. Chapter 9 addresses the relationships between nutrition and personal fitness. Chapter 10 describes the special nutrition needs and concerns that arise during the various stages of the life cycle from conception through old age. The new Food Guide Pyramids for Young Children and Healthy Aging are included. Chapter 11 addresses consumer concerns about the safety of our food supply and provides a glimpse at some of the problems and advantages of the newer food technologies.

This edition introduces three new features. The first of these—beginning in Chapter 2 and continuing through each of the remaining chapters—is called *The Savvy Diner* and provides practical suggestions for healthy eating. The *Savvy Diners* include tips for choosing healthy ethnic foods, making healthy selections from the Food Guide Pyramid, consuming heart-healthy diets, preserving vitamins in foods, seasoning foods without excess salt, dining out defensively, creating tasteful meals for one, and practicing home food safety. The second new feature you will encounter in every chapter, *Working on the Web,* sends readers to selected Web sites that provide practical experiences and apply the chapter contents to everyday life. These interactive Web activities invite readers to analyze the fast-food meals they eat, evaluate their eating habits and current fitness level, assess their risk for diseases such as diabetes, heart disease, and osteoporosis, and expand their knowledge of food safety and healthy cooking. The final new addition—*Nutrition on the Web*—is located at the end of each chapter and contains an annotated list of World Wide Web (www) addresses that provides links to reliable sources of nutrition and health information from the Internet related to the chapter's topics.

The *Nutrition Action* features that appear in every chapter are magazine-style essays that keep you abreast of current topics important to the nutrition-conscious consumer. The *Nutrition Action* features address topics such as fast food, smart snacking, dental health, the trans fatty acid controversy, amino acid supplements, medicinal herbs, diet and blood pressure, behavior modification for weight management, popular fitness aids and supplements, caffeine, food allergies, and the organic foods industry.

Scorecards are hands-on features included in every chapter. *Scorecards* allow readers to evaluate their own nutrition behaviors and knowledge in many areas. Some of the *Scorecards* assist readers in assessing their longevity, overall diet, fruit and vegetable consumption, calcium intake, weight status, exercise habits, and food safety know-how.

The *Ask Yourself* sections at the beginning of each chapter contain a set of true-false questions designed to provide readers with a preview of the chapter's contents. Answers to the questions appear on the following page. A *Check Yourself* section at the end of each chapter includes review questions designed to test readers' comprehension of the chapter material.

The final special feature of each chapter is the *Spotlight*—many are new to this edition, and the others have been thoroughly updated. Each addresses a common concern people have about nutrition. *Spotlight* topics include nutrition and the media, ethnic cuisines, alternative sweeteners, diet and heart disease, the benefits derived from soyfoods, the emergence of functional foods, osteoporosis, eating disorders, alcohol and nutrition, and nutrition and cancer prevention. The final *Spotlight* covers the many factors that influence nutrition and food insecurity among the people of the world and underscores that the practical suggestions offered throughout this book for attaining the ideals of personal nutrition are the very suggestions that best support the health of the whole earth as well. The *Spotlights* continue in their question and answer format to encourage the reader to ask further questions about nutrition issues. We encourage you to ask us questions, too, in care of the publisher.

The Appendixes have been updated. Appendix A contains an invaluable listing of general nutrition resources, including telephone numbers and Internet Web sites; Appendix B provides a colorfully illustrated introduction to the workings of the human body; Appendix C includes the Canadian Dietary Guidelines and other recommendations; Appendix D presents aids to calculations, including how to calculate the percentage of calories from fat in one's diet; Appendix E provides both the U.S. and Canadian Food Exchange Systems; Appendix F includes the chapter reference notes; Appendix G provides answers to the *Check Yourself* questions at the end of each chapter; and Appendix H includes our Table of Food Composition. The Glossary of terms that follows the Appendixes provides a quick reference to the nutrition terminology defined in the margins of the text and can be used as a review tool.

We welcome you to the fascinating subject of nutrition. We hope that the book speaks to you personally and that you find it practical for your everyday use. We hope, too, that by reading it you may enhance your own personal nutrition and health.

ACKNOWLEDGMENTS

I am grateful to the many individuals who have made contributions to the development of this Fourth Edition of *Personal Nutrition*. I thank my family and friends for their continued support and encouragement throughout this endeavor and countless others. I appreciate the insights and support provided by my colleagues—especially Diane Morris, Ph.D, RD of Mainstream Nutrition and the nutrition faculty at the College of Saint Elizabeth. My thanks to Kathleen Klotzbach-Shimomura of Rutgers University Cooperative Extension for her contributions to Chapters 5 and 6. Thanks also to Kathy Roberts for her assistance with sections of Chapters 10 and 11. I am grateful for the work that Gail Zyla contributed to the second and third editions of this text; her insights are reflected in this new edition, still. Thanks to Bob Geltz and Betty Hands and their staff at ESHA research for creating the food composition table found in Appendix H and the computerized diet analysis program that accompanies this book.

Special thanks go to the editorial team and their staff: Peter Marshall, Publisher; Laura Graham, Development Editor; Martha Emry, Production Editor; and Sandra Craig, Project Editor. Their guidance ensured the highest quality of work throughout all facets of this production. I am especially grateful to Martha Emry, because this text would not appear as it does today without her tireless commitment to excellence. As always, my gratitude goes to Becky Tollerson, Marketing Manager, for her problem-solving skills and fine efforts in marketing this book. My thanks to the many sales representatives who will introduce this new book to its readers. My appreciation goes to other members of the production team: Jim Atherton, artist; Marian Selig, proofreader, and Joohee Lee—for her swift actions in obtaining permissions.

Last, but not least, I owe much to my colleagues who provided expert reviews of the manuscript, not only for their ideas and suggestions, many of which made their way into the text, but also for their continued enthusiasm, support, and interest in *Personal Nutrition*. Thanks to all of you.

Lou Ann Carden
University of Tennessee at Martin

Sharon Davis
University of Nebraska at Kearney

Jeffrey Hampl
Arizona State University

Art Gilbert
University of California at Santa Barbara

Lorrie Miller Kohler
Minneapolis Community and Technical College

Debbie Luffey
University of Pittsburgh

T. C. Proctor
Orange Coast College

Joan Thompson
Weber State University

Anne VanBeber
Texas Christian University

Suzanne Vieira
Johnson and Wales University

Janis White
Sam Houston State University

Marie Boyle Struble

FOURTH EDITION

Personal Nutrition

1

The Art of Understanding Nutrition

CONTENTS

Nutrition and Health Promotion

The Longevity Game Scorecard

Nutrition Action: Fast Guide to Eating on the Run

Understanding Our Food Choices

Spotlight: Separating Nutrition Fact from Fiction

Tell me what you eat,
and I will tell you what you are.

Anthelme Brillat-Savarin
(1755–1826, French politician and gourmet;
author of *Physiology of Taste*)

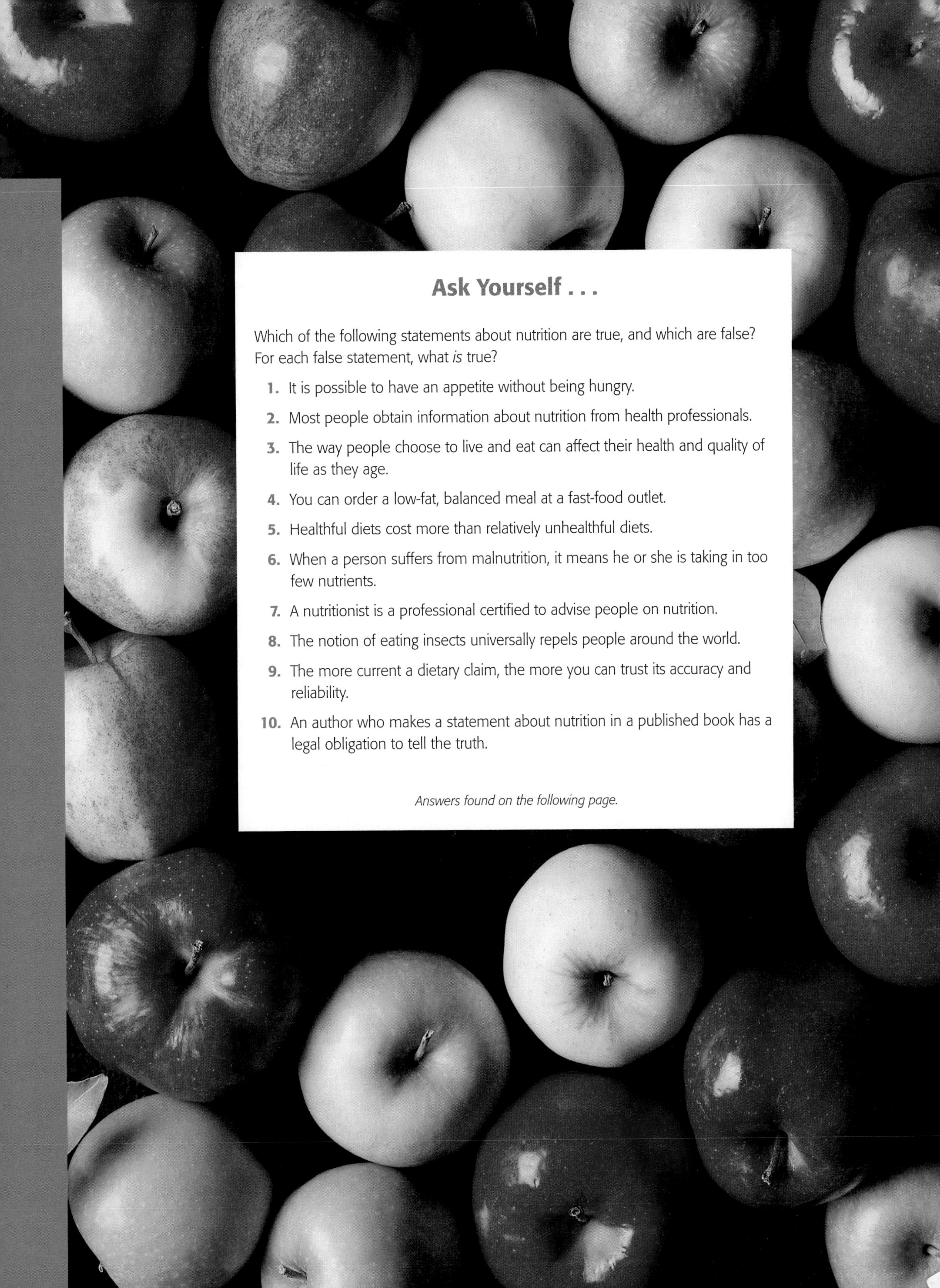

Ask Yourself . . .

Which of the following statements about nutrition are true, and which are false? For each false statement, what *is* true?

1. It is possible to have an appetite without being hungry.
2. Most people obtain information about nutrition from health professionals.
3. The way people choose to live and eat can affect their health and quality of life as they age.
4. You can order a low-fat, balanced meal at a fast-food outlet.
5. Healthful diets cost more than relatively unhealthful diets.
6. When a person suffers from malnutrition, it means he or she is taking in too few nutrients.
7. A nutritionist is a professional certified to advise people on nutrition.
8. The notion of eating insects universally repels people around the world.
9. The more current a dietary claim, the more you can trust its accuracy and reliability.
10. An author who makes a statement about nutrition in a published book has a legal obligation to tell the truth.

Answers found on the following page.

Stroll down the aisle of any supermarket, and you'll see all manner of foods touting such claims as "reduced-fat," "low-calorie," and "fat-free." Flip through the pages of just about any magazine, and you're likely to find advice on how to lose weight. Walk into any gym, and you'll probably hear members discussing the merits of one performance-enhancing food or another. What this all boils down to is that nutrition has become part and parcel of the American lifestyle.

NUTRITION the study of foods, their nutrients and other chemical components, their actions and interactions in the body, and their influence on health and disease.

It wasn't always that way, however. The field of **nutrition** is a relative newcomer on the scientific block. Although Hippocrates recognized diet as a component of health back in 400 B.C., only in the past one hundred years or so have researchers begun to understand that carbohydrates, fats, and proteins are needed for normal growth. The next nutrition breakthrough—the discovery of the first vitamin—occurred in the early 1900s. It wasn't until 1928, when an organization called the American Institute of Nutrition was formed, that nutrition was officially looked upon as a distinct field of study.[1*] It took several more decades before nutrition achieved its current status as one of the most talked about scientific disciplines.

Today we spend billions of dollars each year to investigate the many aspects of nutrition, a science that encompasses the study of not only vitamins, minerals, and the like, but also such diverse subjects as alcohol, caffeine, and pesticides. In addition, nutrition scientists continually expand our understanding of the impact food has on our bodies by examining research in chemistry, physics, biology, biochemistry, immunology, and other nutrition-related fields.

A number of disciplines have made contributions to the study of nutrition. Related fields include psychology, anthropology, geography, agriculture, ethics, economics, sociology, and philosophy.

At the same time that science has shown that to some extent we really are what we eat, many consumers have become more confused than ever about how to translate the steady stream of new findings about nutrition into healthful eating. As Table 1-1 illustrates, people's priorities regarding diet have changed dramatically over the past decade. Each additional nugget of nutrition news that comes along raises new concerns: Is caffeine bad for me? Should I take vitamin supplements? Do diet pills work? Can a sports drink improve my performance? Are pesticides posing a hazard?

Some manufacturers and media outlets feed into the confusion by offering health-conscious consumers unreliable products and misleading dietary advice. For example, in 1970 and 1989, General Nutrition, Inc. (GNC), the largest retailer of nutritional supplements in the United States, was charged with making unsubstantiated claims for a number of nutritional products, including supplements touted as fat melters, muscle builders, and energy boosters. In 1994, however, the U.S. Federal Trade Commission (FTC)—the arm of the government that monitors advertising—alleged that GNC had continued making spurious claims for more than 40 such products, despite FTC orders not to do so. As a result, GNC agreed to pay $2.4 million in penalties—the largest civil penalty for violating an advertising order in the FTC's history.[2]

HEALTH FRAUD conscious deceit practiced for profit, such as the promotion of a false or an unproven product or therapy.

QUACKERY fraud. A quack is a person who practices health fraud.
quack = to boast loudly

Unfortunately, misinformation continues to run rampant in the marketplace. Americans spend more than $30 billion annually on medical and nutritional **health fraud** and **quackery,** up from only $1 billion to $2 billion in the early 1960s.[3] Consider that college athletes alone may spend as much as $400 a month

Ask Yourself Answers: 1. True. **2.** False. Most Americans look first to television for nutrition information, then to magazines, and finally to newspapers. **3.** True. **4.** True. **5.** False. People can save money when they switch from a typical high-fat diet to the grain-based, produce-rich diet recommended by health experts. **6.** False. Malnutrition can be caused either by taking in too few nutrients or by consuming an excess of nutrients. **7.** False. A nutritionist is a person who claims to specialize in the study of nutrition, but some are self-described experts whose training is questionable; a registered dietitian (RD), however, is recognized as a nutrition expert—with training in nutrition, food science, and diet planning. **8.** False. It's true that most people in the United States and Canada find the idea of eating insects repulsive because the practice is not part of the American culture. But people in many other countries, because they have been brought up in cultures in which insects have long been a traditional food, consider dishes prepared with insects a delicacy. **9.** False. If a nutrition claim is too new, it may not have been adequately tested. Findings must be confirmed many times over by experiment and considered in light of other knowledge before they can be translated into recommendations for the public. **10.** False. Authors who write about nutrition are morally but not legally obligated to tell the truth.

*Reference notes for each chapter are in Appendix F.

TABLE 1-1 NATURE OF SHOPPERS' CONCERNS ABOUT DIET

Although consumers have long placed importance on the nutritional profile of their diets, the nature of their concerns has changed over the years. The following figures show how respondents to a national survey answered this question in 1988 and in 1999. **What is it about the nutritional content of what you eat that most concerns you and your family?**

	1988 Total Percentage	1999 Total Percentage	Percentage Point Change
Fat content, low fat	27	50	+23
Cholesterol levels	22	18	−4
Salt content, less salt	26	16	−10
Calories, low calories	14	8	−6
Sugar content, less sugar	20	9	−11
Vitamin/mineral content	21	5	−16
Preservatives	16	6	−10
Getting an overall nutritious diet	11	17	+6
Chemical additives	12	8	−4
Freshness, purity, no spoilage	15	3	−12
As natural as possible, not overly processed	4	2	−2
Desire to be healthy, eat what's good for us	7	3	−4
Ingredients/content	8	1	−7
Fiber content	5	1	−4

Source: The Food Marketing Institute, *Trends in the United States: Consumer Attitudes and the Supermarket*, 1999 edition (Washington, D.C.: The Research Department, Food Marketing Institute).

on nutritional supplements, even though most of the products pitched to serious exercisers are useless and, in some cases, potentially harmful.[4]

To be sure, the widespread interest in nutrition has generated some positive changes in the marketplace. Whereas the sale of low-fat items such as frozen yogurt in fast-food chains was virtually unheard of years ago, those eateries couldn't survive in the current nutrition-conscious environment without offering such healthful fare. (See the Nutrition Action feature later in the chapter for tips on eating healthfully at fast-food outlets.) By the same token, food manufacturers have responded to consumer concerns about diet by developing new technologies, such as the creation of fat substitutes, to provide shoppers with an unprecedented number of choices at the supermarket.

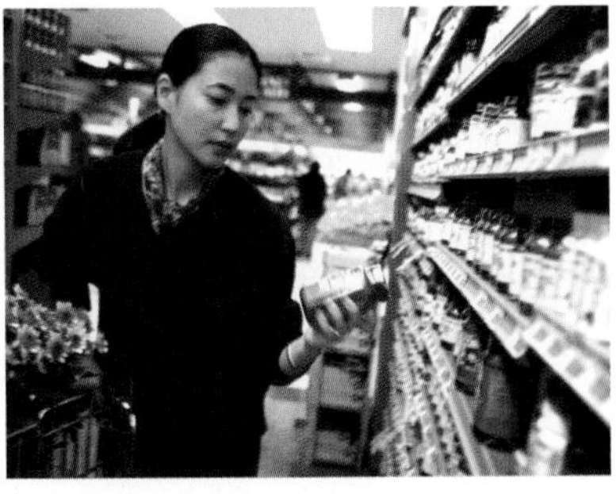

Some stores sell pills and potions touted as fat melters, energy boosters, and muscle builders.

With the amount of nutrition information and the number of food alternatives ever on the rise, choosing a healthful diet can seem like a daunting task. Fortunately, you don't need a degree in nutrition to put the principles of the science to use in your own life. A basic understanding of nutrition can go a long way in helping you protect your health (and your wallet). This book will lay the foundation you need to take the science out of the laboratory and move it into your kitchen, both today and tomorrow. The first step is exploring the current thrust of the field of nutrition.

NUTRITION AND HEALTH PROMOTION

In the past, scientists investigating the role that diet plays in health zeroed in on the consequences of getting too little of one nutrient or another. Until the end of World War II, in fact, nutrition researchers concentrated on eliminating

GOITER (GOY-ter) enlargement of the thyroid gland caused by iodine deficiency.

PELLAGRA (pell-AY-gra) niacin deficiency characterized by diarrhea, inflammation of the skin, and, in severe cases, mental disorders.

MALNUTRITION any condition caused by an excess, deficiency, or imbalance of calories or nutrients.

OVERNUTRITION calorie or nutrient overconsumption severe enough to cause disease or increased risk of disease; a form of malnutrition.

DEGENERATIVE DISEASE chronic disease characterized by deterioration of body organs as a result of misuse and neglect; poor eating habits, smoking, lack of exercise, and other lifestyle habits often contribute to degenerative diseases, including heart disease, cancer, osteoporosis, and diabetes.

deficiency diseases such as **goiter,** a condition in which the thyroid gland swells from lack of the mineral iodine, and **pellagra,** inflammation of the skin caused by deficiency of the B vitamin niacin.

These days, the focus is just the opposite. Deficiency diseases have been virtually eliminated in America because of our country's abundant food supply and the practice of fortifying food with essential nutrients (adding calcium to orange juice, for example). Yet diseases related to **malnutrition** in the form of dietary excess and imbalance run rampant. Four of the ten leading causes of death—heart disease, cancer, stroke, and diabetes—have been linked to diet. Another three are associated with excessive alcohol consumption—accidents, suicide, and liver disease.[5] **Overnutrition** contributes to other ills as well, including high blood pressure and dental disease. Poor dietary habits and a sedentary lifestyle together account for more than 300,000 deaths each year, not to mention hospitalizations, lost time on the job, and poor quality of life among many Americans (see Figure 1-1).[6]

This is not to say that diet is the sole culprit responsible for these conditions. A number of environmental, behavioral, social, and genetic factors work together to determine a person's likelihood of suffering from a **degenerative disease.** For example, diet notwithstanding, someone who smokes, doesn't exercise regularly, and has a parent who suffered a heart attack is more likely to end up with heart disease than a nonsmoker who works out regularly and does not have a close relative with heart disease. The way to alter disease risk is to concentrate on changing the day-to-day habits that can be controlled. The results can be significant.

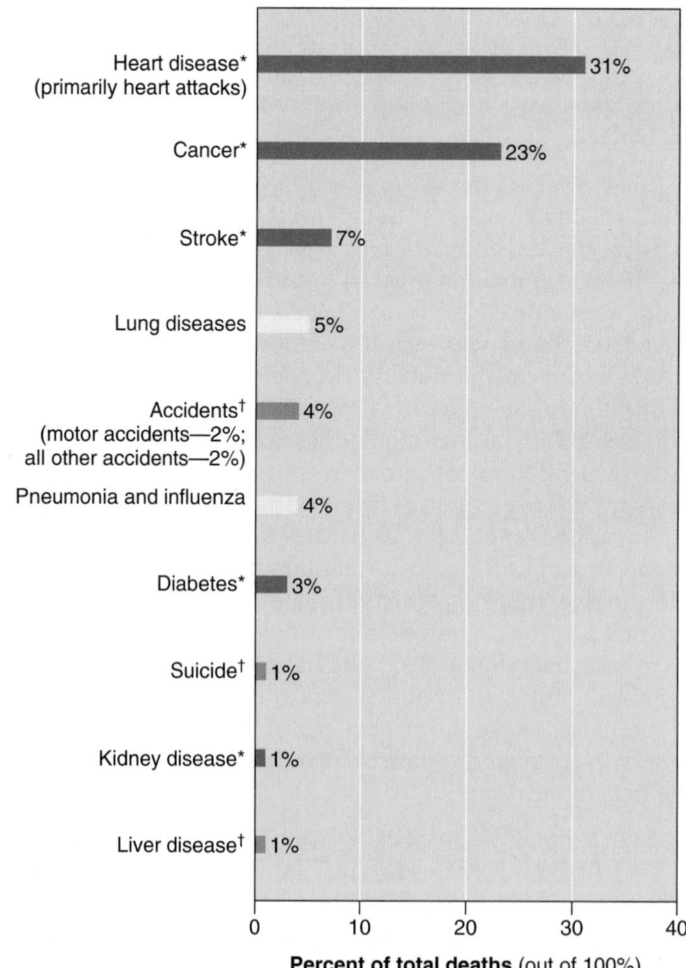

FIGURE 1-1
THE TEN LEADING CAUSES OF DEATH IN THE UNITED STATES

Many of the major killers, such as heart disease and cancer, are influenced by a number of factors, including a person's genetic makeup, eating and exercise habits, and exposure to tobacco.

*Causes of death in which diet plays a part.
†Causes of death in which excessive alcohol consumption plays a part.

Source: Centers for Disease Control and Prevention, *National Vital Statistics Report,* October, 1999.

"You figure it. Everything we eat is 100 percent natural, yet our life expectancy is only 31 years."

Consider that researchers who monitored the habits and health of a group of some 7,000 Californians for nearly two decades were able to pinpoint seven common lifestyle elements associated with optimal quality of life and longevity: avoiding excess alcohol, not smoking, maintaining desirable weight, exercising regularly, sleeping seven to eight hours a night, not snacking between meals, and eating breakfast. In fact, after 20 years, those who had adhered to the healthful habits were only half as likely to have died as those who hadn't. They were also half as likely to have suffered disabilities that interfere with day-to-day living. Granted, the researchers speculated that the last three habits—sleeping seven to eight hours a night, not eating between meals, and eating breakfast—are not necessarily as beneficial as, say, the habit of exercising regularly. Rather, regular eating and sleeping habits are most likely signs that people take the time and have enough control of their lives to take care of their health.[7]

These findings illustrate that you can change the probable length and quality of your life. Since nutrition is involved in at least half of the preceding lifestyle recommendations, it no doubt plays a key role in maintaining good health. This chapter's Scorecard feature, The Longevity Game, further demonstrates the point.

Table 1-2 summarizes ways to improve health with sound nutrition practices. As you read Table 1-2, keep in mind that while everyone can benefit from eating a healthful diet that complies with the guidelines, some people stand to gain more than others. Those who have high blood cholesterol levels, for instance, are already at risk for heart disease, thereby making it especially important for them to eat a low-fat diet and maintain a healthful weight. By the same token, those who have close relatives with, say, diabetes, would do well to keep their weight down and pay particular attention to the other nutrition guidelines that help stave off the condition. (The chapters that follow explain the link between diet and chronic diseases in more detail and offer advice on how to follow each dietary recommendation.)

The exact proportion that dietary factors contribute to each health problem can only be estimated, but some experts speculate that they account for a third or more of all cases of both cancer and heart disease.[8] Moreover, some elements appear to play a more integral role than others in determining disease risk. A high-fat diet, for instance, raises the risk of some types of cancer, heart disease, and obesity, which in turn contributes to a number of other problems, including diabetes and high blood pressure.

THE LONGEVITY GAME SCORECARD

You can't look into a crystal ball to find out how long you will live. But you can get a rough idea of the number of years you're likely to survive based largely on your lifestyle today as well as certain givens, such as your family history. To do so, play the Longevity Game.

Start at the top line—age 76, the average life expectancy for adults in the United States today. For each of the 11 lifestyle areas add or subtract years as instructed. If an area doesn't apply, go on to the next one. If you are not sure of the exact number to add or subtract, make a guess. Don't take the score too seriously, but do pay attention to those areas where you lose years; they could point to habits you might want to change.

START WITH	76
1. Exercise	______
2. Relaxation	______
3. Driving	______
4. Blood pressure	______
5. 65 and working	______
6. Family history	______
7. Smoking	______
8. Drinking	______
9. Gender	______
10. Weight	______
11. Age	______
Your final score:	______

1. Exercise. If your job requires regular, vigorous activity or if you work out each day, add three years. If you don't get much exercise at home, on the job, or at play, subtract three years.
2. Relaxation. If you have a laid-back approach to life (you roll with the punches), add three years. If you're aggressive, hard-driving, or anxious (you suffer from sleepless nights or bite your nails), subtract three years. If you consider yourself unhappy, subtract another year.
3. Driving. Drivers under age 30 who have received traffic tickets in the past year or who have been involved in an accident should subtract four years. Other violations, minus one. If you always wear seatbelts, add a year.
4. Blood pressure. While high blood pressure is a major contributor to common killers—heart attacks and strokes—it can be lowered effectively through drugs and changes in lifestyle. The problem is that rises in blood pressure can't be felt, so many victims don't know they have it and therefore never receive lifesaving treatment. If you *know* your blood pressure, add one year.
5. 65 and working. If you are at the traditional retirement age or older and still working, add three.
6. Family history. If any grandparent has reached age 85, add two; if all grandparents have reached age 80, add six. If a parent died of a stroke or heart attack before age 50, minus four. If a parent or brother or sister has (or had) diabetes since childhood, minus three.
7. Smoking. Cigarette smokers who finish more than two packs a day, minus eight; one or two packs a day, minus six; one-half to one pack, minus three.
8. Drinking. If you drink two cocktails (or beers or glasses of wine) a day, subtract one year. For each additional daily libation, subtract two.
9. Gender. Women live longer than men. Females add three years; males subtract three years.
10. Weight. If you avoid eating fatty foods and don't add salt to your meals, your heart will probably remain healthy longer, entitling you to add two years.

    Now, weigh in: overweight by 50 pounds or more, minus eight; 30 to 40 pounds, minus four; 10 to 29 pounds, minus two.
11. Age. How long you have already lived can help predict how much longer you'll survive. If you're under 30, the jury is still out. But if your age is 30 to 39, plus two; 40 to 49, plus three; 50 to 69, plus four; 70 or over, plus five.

Source: From "The Longevity Game," by Northwestern Mutual Life Insurance Company, with permission.

TABLE 1-2 EATING TO BEAT THE ODDS

Dietary Recommendation	To Help Reduce the Risk of
Fat: Reduce total fat intake to 30% or less of calories. Reduce saturated fat intake to less than 10% of calories and the intake of cholesterol to less than 300 milligrams daily.[a,b]	Some types of cancer, obesity, heart disease, and possibly gallbladder disease.
Weight: Achieve and maintain a desirable weight.[a]	Diabetes, high blood pressure, stroke, cancers (especially breast and uterine), and gallbladder disease.
Carbohydrates and fiber: Increase consumption of fruits, vegetables, legumes, and whole grains.[a,c]	Diabetes, heart disease, and some types of cancer.
Sodium: Limit daily intake of salt (sodium chloride).[d]	High blood pressure and stroke.
Alcohol: Avoid completely or drink only in moderation.[e]	Heart disease, high blood pressure, liver disease, stroke, some forms of cancer, and malformations in babies born to mothers who drink alcohol during pregnancy.
Sugar: Limit consumption of refined sugar to 10% of total calories.	Tooth decay and gum disease.
Calcium: Maintain adequate intake (1,000 to 1,300 milligrams a day).	Osteoporosis, bone fractures, and possibly colon cancer.

[a]Pay particular attention to this guideline if you have glucose intolerance, high blood cholesterol or high triglyceride levels, or high blood pressure.
[b]The intake of fat and cholesterol can be reduced by substituting fish, poultry without skin, lean meats, and low-fat or fat-free dairy products for fatty meats and whole-milk products; by choosing more vegetables, fruits, cereals, and legumes; and by limiting oils, fats, egg yolks, and fried or other fatty foods.
[c]Every day eat five or more servings of a combination of vegetables and fruits, especially green and yellow vegetables and citrus fruits, and six or more servings of a combination of breads, cereals, and legumes.
[d]Limit the use of salt in cooking and avoid adding it to foods at the table. Highly processed, salty, salt-preserved, and salt-pickled foods should be consumed sparingly.
[e]Moderate drinking is defined as no more than one drink a day for the average-sized woman and no more than two drinks a day for the average-sized man. A drink is any alcoholic beverage that delivers ½ ounce of pure ethanol: 5 ounces of wine, 12 ounces of beer, or 1½ ounces of hard liquor (whiskey, scotch, etc.)

Source: Adapted from Committee on Diet and Health, *Diet and Health: Implications for Reducing Chronic Disease Risk,* (Washington, D.C.: National Academy Press, 1989).

The relative importance of certain dietary recommendations was underscored by their appearance in the U.S. Department of Health and Human Services' official strategy for improving the nation's health during the 1990s. Called *Healthy People 2000: National Health Promotion and Disease Prevention Objectives,* the plan of action included a number of health and nutritional goals geared toward increasing the span of healthy life for Americans.[9]

Several goals focused on risk reduction and specified targets for the intakes of nutrients such as fat, saturated fat, and calcium, and foods such as fruits, vegetables, and grain products. Other risk-reduction goals set targets for the prevalence of people who are overweight, the proportion of people who adopt sound dietary practices, and the proportion of people who reduce their use of salt and sodium in foods and at the table. For example:

- Reduce dietary fat consumption among people aged two and older from 36 percent of total calories to an average of not more than 30 percent of calories.*
- Increase the number of servings of fruits and vegetables from two and a half a day to five or more, and increase the daily servings of grain products (bread and cereals) from three to at least six.
- Decrease salt consumption by increasing the proportion of people who avoid using salt at the table from 68 percent to 80 percent and by increasing the number of cooks who season dishes without salt.

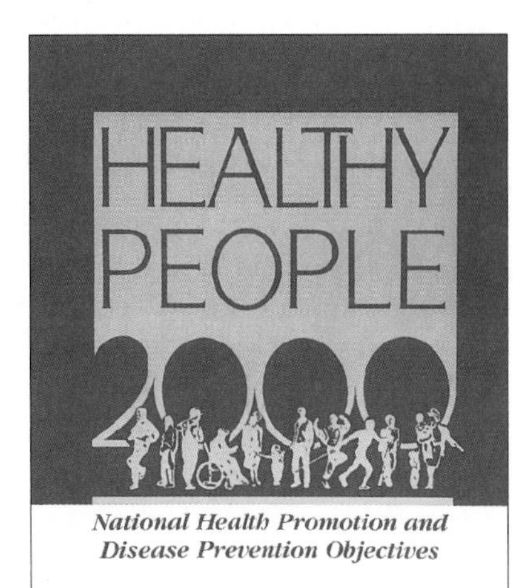

*The comparison figures are based on survey results gathered in the 1980s and serve as rough estimates of the way Americans presently eat.

- Reduce the prevalence of overweight among adults from 26 percent to 20 percent.

Although these are just a few of the goals for nutrition spelled out in *Healthy People 2000,* they represent some of the priorities for maintaining good health. Much of the practical information presented later in this chapter and in those that follow is aimed at guiding you toward developing eating and lifestyle habits that will help you achieve the goals.

How well are Americans meeting the *Healthy People 2000* goals? Since the 1980s, the number of deaths from heart disease, stroke, and certain types of cancer has decreased, but the prevalence of overweight has soared.[10] In fact, overweight increased among all ethnic and age groups of the population. The widespread problem of overweight has occurred despite a slight decrease in dietary fat intake and a modest increase in consumption of fruits, vegetables, and grains. One contributing factor is that people seem to be taking fewer steps to control their weight by adopting sound dietary patterns and being physically active.

A new report has set health objectives for the year 2010 with two broad goals designed to help all Americans achieve their full potential by increasing the quality and years of healthy life and by eliminating health disparities. Table 1-3 lists the nutrition-related objectives considered to be top public health priorities for the present decade. For updates on the *Healthy People* initiative, refer to Nutrition on the Web later in this chapter.

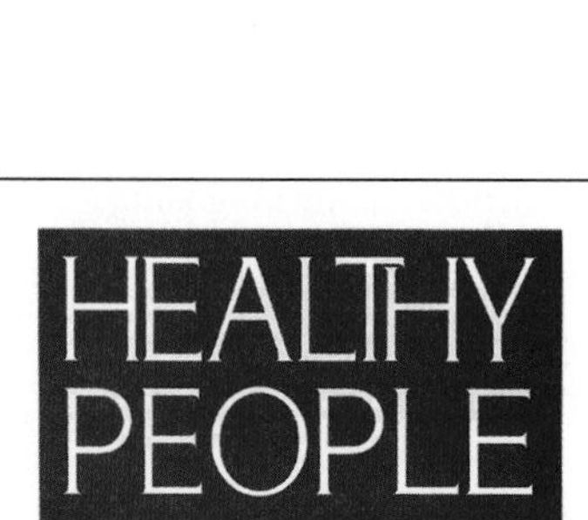

TABLE 1-3 *Healthy People 2010* Nutrition-Related Objectives for the Nation[a]

Disease-Related Objectives

Reduce rates of heart disease, cancer, and diabetes-related deaths.

Reduce prevalence of Type 2 diabetes, osteoporosis, and dental caries.

Nutrition Objectives

Increase prevalence of healthy weight and decrease the prevalence of obesity.

Increase proportion of people who meet the recommended intakes for fat, saturated fat, sodium, and calcium in the diet[b]

Increase intakes of fruits and vegetables, and grain products[c]

Increase proportion of mothers who breastfeed, children whose intake of meals and snacks at school contribute to overall dietary quality, and schools teaching essential nutrition topics.

Reduce rates of growth retardation among low-income children, and iron deficiency in young children, women of childbearing age, and low-income pregnant women.

Food Safety Objectives[d]

Reduce the proportion of infections caused by foodborne pathogens.[e]

Reduce deaths from food allergy.

Increase the proportion of consumers who practice four essential food safety behaviors when handling food: washing hands, avoiding cross contamination, cooking meats thoroughly, and chilling foods promptly.

Reduce occurrences of improper food safety techniques in retail food establishments.

[a]The complete list of *Healthy People 2010 Objectives* may be viewed online at http://www.health.gov/healthypeople/.

[b]Recommendations are: 30% of calories or less from fat, 10% of calories or less from saturated fat, 2,400 mg or less of sodium, and 1,300 mg of calcium for children (ages 9 to 18), 1,000 mg of calcium for adults (ages 19 to 50), and 1,200 mg for adults over 50.

[c]At least five servings a day of fruits and vegetables and at least six servings a day from grains.

[d]See Chapter 11 for more on the topic of food safety.

[e]Pathogens are disease-causing agents such as bacteria or viruses.

Source: Healthy People 2010: National Health Promotion and Disease Prevention Objectives (Washington, D.C.: U.S. Department of Health and Human Services, 2000).

NUTRITION ACTION

Fast Guide to Eating on the Run

Chances are you've stood in line for a burger and fries, a slice of pizza, a taco, or a muffin and coffee at least once this week. You're not alone. Every day one out of two Americans orders food served or prepared outside the home, and much of this food is takeout fare. In fact, fast food is the fastest-growing segment of the restaurant industry.[11]

McDonald's golden arches, Domino's Pizza deliveries, and Burger King's "Have It Your Way" slogan are as much a part of American culture as baseball and apple pie. But while a meal eaten on the run fits easily into a busy schedule, it's not necessarily so simple to work it into the dietary guidelines recommended by major health organizations. Two all-beef patties with special sauce and other requisite trimmings, an order of fries, and a milkshake, for example, can chalk up anywhere from 1,000 to 1,500 calories, more than 1,000 milligrams of sodium, and 65-plus grams of fat—more than the amount in half a stick of butter.

The good news is that in the past several years, McDonald's and other fast-food giants have begun offering such items as low-fat dairy products and salads that can fit more easily into the high-carbohydrate, low-fat diet that experts are recommending for good health. With just a little bit of nutrition know-how, you can combine new alternatives with old favorites to place orders more in keeping with today's dietary guidelines. A regular burger, tossed salad, and cup of low-fat milk, for instance, supply only a third of the calories and a quarter of the fat of a standard double-burger-with-fries-and-a-shake order. (Table 1-4 gives the

"Which came first, Mom, the Chicken McNugget or the Egg McMuffin?"

nutritional breakdown of other fast-food meals.) The following tips can guide you in trimming fat and calories from fast-food meals:

- Select an English muffin, bagel, or toast with margarine for breakfast rather than a Danish, doughnut, or muffin. You'll save upwards of 150 calories and take in about three times less fat. A 3-inch bagel topped with two teaspoons of jelly contains only about a gram of fat, while many takeout muffins contain in the neighborhood of 13 grams of fat.
- Buy breakfast sandwiches or entrees with Canadian bacon or ham instead of sausage. Whereas a 3-inch sausage patty provides 200 calories and 17 grams of fat, two slices of Canadian bacon contribute only 85 calories and 4 grams of fat. Three slices of cured bacon supply about 110 calories and 9 grams of fat.
- Sweeten your pancakes with syrup or spread your toast with jelly or marmalade, but hold the butter. Each tablespoon of butter you spread on your pancakes contributes 100 calories, virtually all of which come from fat. A tablespoon of syrup or jelly, on the other hand, adds only 50 calories and not a trace of fat.
- Hold the mayo. Each dollop of that condiment adds about 100 calories, nearly all of which are fat. Most fast-food sandwiches contain more than just one spoonful of mayo in their toppings and special sauces. The same goes for tartar sauce. Order a fish sandwich without it, and you'll trim at least 70 fat-laden calories (the amount in just one tablespoon) from your meal.
- Opt for a side salad moistened with reduced-calorie dressing with your burger instead of French fries and ketchup. You'll trim at least 150 calories and a good deal of fat from your meal.
- Wash your meal down with low-fat milk instead of a milkshake, thereby cutting the fat and calorie count at least in half.
- Satisfy your sweet tooth with a cup of hot chocolate. Whereas the hot chocolate contains about 100 to 200 calories per cup, sundaes, pies, milkshakes, and other sweets contribute 300 or more calories per serving (and little else in the way of nutrients).
- Don't let the word *chicken* or *fish* fool you. Granted, many health-conscious consumers have heard the advice to choose skinless poultry and fish instead

TABLE 1-4 WHAT'S IN AN ORDER?*

	Calories	Grams of Fat	Milligrams of Sodium
Fat and Calories To Go			
Double burger with sauce	625	40	880
Milkshake	410	10	190
French fries (regular size)	240	15	120
Total	1,275	65	1,190
Fish sandwich with cheese and tartar sauce	495	25	676
Soda (12 ounces)	150	0	15
French fries (regular size)	240	15	120
Total	885	40	811
Chicken nuggets (6)	310	20	700
Apple pie	280	15	400
Coffee with cream	65	5	15
Total	655	40	1,115
Nutrition on the Run			
Single burger	290	13	435
Tossed salad with low-calorie dressing	50	1	445
Low-fat milk (8 ounces)	105	2	125
Total	445	16	1,005
Baked potato (plain)	150	trace	5
Margarine (1 pat)	35	4	45
Tossed salad with low-calorie dressing	50	1	445
Low-fat milk (8 ounces)	105	2	125
Total	340	7	620
Cheese pizza (1 slice)	155	5	455
Tossed salad with low-calorie dressing	50	1	445
Orange juice (8 ounces)	110	0	0
Total	315	6	900

*Figures represent the average nutrient values for similar items from three or more fast-food chains.

Source: Adapted from C. Roberts, Fast-food fare: Consumer guidelines, *New England Journal of Medicine* 321 (1989): 754.

of relatively high-fat red meat. But when it comes to chicken nuggets and fish patties coated with batter and deep fried, it is a different story. Six chicken nuggets, for example, typically contain as many calories (about 300) as an entire burger. What's more, many chicken- and fish-patty sandwiches chalk up as much fat as a pint and a half of ice cream. Even rotisserie-style chicken contains a large amount of fat and calories if you don't remove the skin before eating it.

TABLE 1-5 TOPPING IT OFF

	Calories per Tablespoon	Grams of Fat per Tablespoon
Creamy Italian dressing	70	8
Reduced-calorie Italian dressing	15	2
Imitation bacon bits	30	2
Sunflower seeds	50	4
Chopped egg	15	1
Grated process cheese	25	2
Seasoned croutons	5	trace
Raisins	28	trace

Source: U.S. Department of Agriculture, Human Nutrition Information Service, Home and Garden Bulletin No. 232–11, *Eating Better When Eating Out Using the Dietary Guidelines,* p. 9.

- When ordering a pizza, hold the sausage and pepperoni and ask for mushrooms, green peppers, and onions instead. Pizza is an excellent source of calcium—that bone-building mineral many Americans don't get enough of—as well as protein, carbohydrate, and a number of vitamins and minerals. Two slices of *pepperoni* pizza, however, can easily contain 100 more calories and twice as much fat as the same amount topped with onions, green peppers, and mushrooms.
- Cover your plate with fresh greens, fruits, and vegetables at the salad bar, but go easy on some of the toppings. A tablespoon here and a tablespoon there of "fixings," such as bacon bits or rich dressing, can turn a low-fat, low-calorie meal into a high-fat, high-calorie extravaganza (see Table 1-5).
- Ask for lots of lettuce and tomatoes and less sour cream and guacamole on your nachos and tacos. A tablespoon of either sour cream or guacamole adds about 25 calories to Mexican fare. A few extra chunks of tomato, on the other hand, supply a negligible number of calories, no fat, and a good deal of vitamin C.
- Order a plain baked potato with a pat of margarine. Whereas a potato with margarine has fewer than 200 calories, spuds covered with bacon and cheese, sour cream, or chili and cheese can contain as many calories and as much fat as a double burger.
- Choose frozen yogurt instead of ice cream, but remember that many "mix-ins" rack up additional fat and calories. Two tablespoons of sprinkles, or "Jimmies," contain about 250 calories and 6 grams of fat (see Table 1–6).

TABLE 1-6 FROZEN YOGURT "MIX-INS"

	Calories per Tablespoon	Grams of Fat per Tablespoon
Sprinkles (Jimmies)	122	3
Yogurt-covered raisins	67	2
Chocolate syrup	46	<1
Fudge topping	62	2
Crumbled chocolate chip cookie	66	3
Crumbled Heath Bar	71	4
Reese's Pieces	69	3
m&m's	68	3
Mixed nuts	85	7
Granola	35	2
Fresh strawberries	3	<1

WORKING ON THE WEB

www.cyberdiet.com/profile/profile.html

Sponsor: CyberDiet (in partnership with mediconsult.com)

Description: This Web site provides practical information about adopting a healthy lifestyle. A wide variety of nutrition-related topics are presented in language that is easy to comprehend.

Available Activities: An assessment tool to determine your personal nutrient needs; a comprehensive analysis of fast food items at a variety of popular fast-food restaurants.

Web Work:

1. Complete the Nutritional Profile to determine your recommended daily intakes of nutrients based on your age, gender, weight, height, and activity level.
2. Next, return to CyberDiet's home page and click on Fast Food Quest. Select three of your favorite fast-food categories at the fast-food outlets you prefer. Compare the items selected at the various restaurants for total calories and fat.
3. Using the total daily calorie and fat allowances given for you in the Nutritional Profile exercise, determine which fast-food selections will best keep you within your allowances for total calories and fat on any given day.

Helpful Hints: In the Fast Food Quest, you can compare particular food items (for example, cheese burritos, egg dishes, or hamburgers) at the more than two dozen fast-food outlets listed. To do so, don't choose any restaurants, just choose a food category from the Fast Food Category and Subcategories table. Or, you can compare foods in up to three restaurants: Click on three restaurants in the Fast Food Restaurants table, and then click on the items you wish to have compared in the Fast Food Categories and Subcategories table.

In the Columns to Display box, click on Calories and Fat. Next, click on Display Results. Compare the calorie and fat content of the selected items. This can help you compare similar fast-food items at different outlets and enable you to make more healthful choices.

UNDERSTANDING OUR FOOD CHOICES

HUNGER the physiological need for food.

APPETITE the psychological desire to eat, which is often but not always accompanied by hunger.

The choices you make about what to eat can have a profound impact on your health, both now and in your later years. The healthful eater resists disease and other stresses better than a person with poor dietary habits and is more likely to enjoy an active, vigorous lifestyle for a greater number of years. Even so, the nutritional profile of various foods ranks as only one of many factors that influence your eating habits. Whether you realize it or not, each time you sit down to a meal you bring to the table such factors as your own personal preferences, cultural traditions, and economic considerations. These influences exert as great an impact on your eating habits as does **hunger**—the physiological need for food—and **appetite**—the psychological desire for food, which may arise in response to the sight, smell, or thought of food even when you're not hungry. The following sections examine some of the most influential factors in making food choices.

Many factors influence our food choices, including advertising, early experiences, economics, and cultural traditions.

Availability

Our diets are limited by the types and amounts of food available through the food supply, which in turn is influenced by many forces. Because we have the geographical area, climate, soil conditions, labor, and capital necessary to maintain a large agricultural industry, Americans enjoy what is arguably the most abundant food supply in the world. In addition, unlike many other less wealthy countries, the United States and Canada have the resources needed to import and distribute a wide variety of foods from other countries—everything from kiwi from New Zealand to mangoes from the tropics.

History has shown, however, that when it comes to health, an abundant food supply can be a double-edged sword. Access to many types of foods allows people to choose high-fat diets rich in meats, eggs, and other fatty foods, which can contribute to increased rates of heart disease and other problems. That's one of the reasons why degenerative diseases are sometimes referred to as diseases of affluence.

Income and Food Prices

As most college students know firsthand, the amount of money available to spend on food can mean the difference between ordering pizza every night and resigning yourself to a steady diet of peanut butter and jelly sandwiches. Extremely low incomes can make it difficult for people to buy enough food to meet their minimum nutritional needs, thereby putting them at risk for **undernutrition.**

UNDERNUTRITION severe underconsumption of calories or nutrients leading to disease or increased susceptibility to disease; a form of malnutrition.

A consumer's *perception* of the cost of various foods can also play a role in his or her choices. For example, one barrier that prevents people from adopting healthful eating habits is the belief that it would be too expensive. In fact, 40 percent of consumers who answered one survey said that fruits, vegetables, seafood, and other elements of a low-fat, nutrient-rich diet would strain their budgets.[12] But some research has shown that switching from a high-fat diet to one that is lower in fat can reduce food costs.[13] Just cutting back on the amount of meat and poultry—from which much of the fat in the American diet comes and on which many of our food dollars are spent—goes a long way in trimming food budgets.

Advertising and the Media

Television and radio commercials as well as magazines and newspapers play an extremely powerful role in influencing our food choices and our knowledge of nutrition.[14] Given today's health-conscious environment, food manufacturers are promoting the nutritional merits of their products more than ever before. In fact, in some of the most popular women's magazines, the number of food and beverage ads containing nutrition claims increased by nearly 100 percent between

Numerous factors influence your food choices, including cultural traditions, nutrition knowledge and beliefs, personal food preferences, and other practical considerations.

1975 and 1990.[15] Unfortunately, advertising is not always created with the consumer's best interest in mind. Much of the food advertising that we're exposed to from the earliest ages is aimed at selling products that aren't the optimal choices for regular inclusion in a healthful diet. For example, the great majority of television commercials geared to children and aired on Saturday mornings promote high-fat, sugary foods such as candy and sugar-coated cereals.[16] Yet commercials promoting good nutrition are relatively few and far between.

Along with advertising, the media rank among the most influential sources of diet and nutrition information, which in turn affects our food choices. Consider that most Americans look first to television as a source of nutrition information, then to magazines, newspapers, family or friends, and finally to books.[17] The downside is that the reliability of information delivered by the media varies considerably. The Spotlight at the end of the chapter will help you learn how to evaluate the nutrition information you receive via the media.

See Appendix A for a complete list of resources for reliable nutrition information.

Social and Cultural Factors

The social and cultural groups to which a person belongs have a significant effect on food choices. **Social groups** such as families, friends, and coworkers tend to exert the most influence. The family, particularly the wife/mother, plays one of the most powerful roles in determining our food choices.[18] That makes sense, as the family is both the first social group a person encounters and the one to which he or she typically belongs for the longest period of time. The values, attitudes, and traditions of our family can have a lasting effect on our food choices. Think of the holiday food traditions in your own family and your friends' families. Treasured recipes or rituals surrounding holiday meals are often passed from one generation to the next.

SOCIAL GROUP a group of people, such as a family, who depend on one another and share a set of norms, beliefs, values, and behaviors.

Friends, coworkers, and members of other social networks also influence our food choices and eating behavior. For instance, many weight-loss programs feature group sessions made up of people who are in the same boat so that they can support one another in their efforts to lose weight. Social pressure can also

The act of eating is complex. We derive many benefits—both physical and emotional—when we eat foods.

push us to eat meals we might not choose on our own. For example, as a guest in another country or in a friend's home, your choosing not to partake of the food and drink that's offered might be considered rude. By the same token, it's natural to join your friends on a spontaneous trip for ice cream or pizza even when you're not hungry.

Culture also determines our food choices to a large extent. Many of our eating habits arise from the traditions, belief systems, technologies, values, and norms of the culture in which we live. For example, in the United States and Canada, the idea of eating insects is generally considered repulsive. But many people throughout the rest of the world relish dishes prepared with various bugs, including locust dumplings (northern Africa), red-ant chutney (India), water beetles in shrimp sauce (Laos), and fried caterpillars (South Africa).[19]

CULTURE knowledge, beliefs, customs, laws, morals, art, and literature acquired by members of a society and passed along to succeeding generations.

One of the ways people of different cultures often come together to share their heritage is by sampling each other's traditional foods. American consumers are particularly fortunate in that they don't have to travel far to get a taste of the food of different cultures. **Ethnic cuisine** ranging from Chinese to Mexican to Italian to Indian food has become embedded in American culture. (The Spotlight in Chapter 2 discusses in detail the eating patterns of different cultures.)

ETHNIC CUISINE the traditional foods eaten by the people of a particular culture.

One aspect of culture that affects the food choices of millions of people worldwide is religion. The practice of giving and abstaining from food has long been used by many cultures as a way to show devotion, respect, and love to a supreme being or power. Dietary customs play a major role in the practice of many of the world's major religions. As Table 1-7 shows, many religions specify which foods their followers may eat and how the foods must be prepared.

Other Factors That Affect Our Food Choices

One of the main reasons you choose to eat certain foods is your preferences for certain tastes. Just about everyone enjoys sweet foods, for example, because humans are born with an affinity for sugar.[20] In addition, we usually prefer foods with which we have happy associations—special foods prepared for birthdays

TABLE 1-7 A SAMPLING OF RELIGION-BASED DIETARY PRACTICES

Religion	Dietary Practices
Buddhism	The central tenet of Buddhism is vegetarianism. In the eyes of Buddhists, to eat meat is to destroy the seeds of compassion. All plant foods are considered appropriate to eat except for the "five pungent foods": garlic, leeks, scallions, chives, and onion. These foods are considered unclean and are believed to generate lust when eaten cooked and to induce rage when eaten raw.
Hinduism	Hindus believe that food was created by the Supreme Being for the benefit of humans. Beef is prohibited, and many Hindu followers are vegetarians.
Islam	Islamic food laws prohibit the consumption of "unclean" foods such as swine, animals killed in a manner that prevents their blood from being fully drained from their bodies, carnivorous animals with fangs (such as lions and wolves), birds with sharp claws (including falcons and eagles), and land animals without ears (frogs and snakes, for example).
Judaism	The traditional dietary laws of Judaism prohibit the consumption of swine, carrion eaters, and shellfish and specify other practices, such as the ritual slaughtering of animals. The term *kosher* indicates that the food was prepared according to certain methods. For example, the meat was salted to help remove the blood, or milk and meat were prepared using separate utensils.
Seventh-Day Adventism	Vegetarianism is the foundation of dietary practices for this sect of Christianity. In addition, dietary standards call for abstaining from alcohol and avoiding caffeine. Followers typically eat a wholesome diet consisting of whole grains, fruits, nuts, vegetables, a little milk, and occasional eggs.

Source: Adapted from M. A. Boyle and D. H. Morris, *Community Nutrition in Action* (St. Paul: West Publishing Company, 1994), p. 245.

or holiday gatherings, those given to us by a loved one when we were children, or those eaten by an admired role model. By the same token, intense aversion to certain foods—say, foods you were given when you were sick or foods you were forced to eat—can be strong enough to last a lifetime. Your parents may have taught you to prefer certain foods and pass up others for reasons of their own, without even being aware they were doing so.

Food habits are also intimately tied to deep psychological needs, such as an infant's association of food with a parent's love. Yearnings, cravings, and addictions with profound meaning and significance sometimes surface as food behavior. Some people respond to stress—positive or negative—by eating; others use food to fill a void, such as lack of satisfying personal relationships or fulfilling work.

Some people adopt a certain way of eating or make specific choices based on a larger worldview. For instance, many environmentally conscious people, believing that raising animals for human consumption strains the world's supply of land and water, choose to abstain from meat as much as possible in an effort to preserve the earth's resources. Others may choose to boycott certain manufacturers' items for political reasons, perhaps because they disagree with a company's advertising practices.

The influences on people's eating habits are as many and varied as the individuals themselves. Our food choices reflect our own unique cultural legacies, philosophies, and beliefs. To think about food as nothing more than a source of nutrients would be to deny food's rich symbolism and meaning and would take away from much of the pleasure of breaking bread with friends and family. As you read this book and consider ways to improve your own eating habits, take time to reflect on your own unique background and think about how you can integrate your knowledge of nutrition into your cultural heritage and philosophies.

At many American-style restaurants, you can experience other cultures by sampling from the various ethnic cuisines represented on the menu.

Spotlight
SEPARATING NUTRITION FACT FROM FICTION

You've just watched a television commercial for a vitamin supplement guaranteed to produce a laundry list of benefits, including fewer colds, a better complexion, and a decreased risk of cancer. Should you buy it? You've just read a magazine article with a plan for quick weight loss. Should you believe it? Someone who plays the same sport as you says that improving your diet will help your game. Where do you go for help?

We all find ourselves faced with such decisions at one time or another. It's essential to know how to handle them to protect ourselves from nutrition misinformation. As pointed out earlier, health fraud costs consumers more than $30 billion each year. At the same time, the sale of weight-loss programs and products—not all of them sound—has become a 33-billion-dollar industry.[21] Media attention to "hot" foods and nutritional supplements generally causes spending on those items to soar.

Money down the drain is just one of the problems stemming from misleading dietary information. Although some fraudulent claims about nutrition are harmless and might make for a good laugh, others can lead to tragic consequences. Swallowing false claims about nutritional products has been known to bring about malnutrition, birth defects, mental retardation, and even death in extreme cases. Negative effects from following false claims can come about in two ways. One is that the product in question causes direct harm. Even a seemingly innocuous substance such as vitamin A can cause severe liver damage over time if taken in large enough doses. The other way that using spurious nutritional remedies can cause problems is that such remedies build false hope and may keep a consumer from obtaining sound, scientifically tested medical treatment. A person who relies on a so-called anticancer diet as a cure for the disease, for example, might forego possible lifesaving interventions such as surgery or chemotherapy.

The following questions and answers should help you learn to evaluate the nutrition information you see in the media or on the Internet. It should enable you to develop the skills with which to view nutrition claims with a skeptic's eye or, at the very least, find a qualified professional who can.

Judging by what I've read in newspapers, it seems as if nutritionists are always changing their minds. One week the headlines say to eat high-fiber foods to help prevent cancer and the next week they say that fiber may not, in fact, prevent the disease. Why is there so much controversy?

Part of the confusion stems from the way the media interpret the findings of scientific research. A good case in point is the controversy over whether a high-fiber diet protects against colon cancer—a disease that affects some 130,000 Americans each year. (A detailed discussion of Nutrition and Cancer appears in the Chapter 10 Spotlight.) The fiber and colon cancer connection dates back to the early 1970s when scientists observed that colon cancer was extremely uncommon in areas of the world where the diet consisted largely of unrefined foods and little meat. The researchers theorized that dietary fiber may be protective against colon cancer by binding bile (a chemical substance needed in fat digestion) and speeding the passage of wastes and potentially harmful compounds through the colon. Since then, other studies have suggested that those who eat a high-fiber diet have a lower risk of colon and rectal cancers.[22]

However, in 1998, a flurry of headlines threatened to pull the pedestal out from under the popular fiber theory asking, "Fiber: Is It Still the Right Choice?" A Harvard-based study published in the *New England Journal of Medicine,* one of the most prestigious medical journals, suggested that fiber did nothing to prevent cancer.[23] The 16-year trial of almost 90,000 nurses—called the Nurses' Health Study—found that the nurses who ate low-fiber diets (less than 10 grams daily) were no more likely to develop colon cancer than those eating higher levels of fiber (about 25 grams daily). As a result, the researchers concluded that the study provided no support for the theory that fiber could reduce the risk of colon cancer.

The news was surprising and reinforces the need for research studies to be duplicated. All studies have their limitations, and a number of questions can be raised regarding the conclusions of the Nurses' Health Study. For example, the study relied on participants to recall their eating habits accurately. Is this type of self-reported dietary information reliable? Back in 1980, the nurses were asked for information about their intakes of "dark" breads. However, food labels at that time did not list the fiber content of breads, and some wheat breads on the shelf had similar amounts of fiber found in white bread. Did the nurses mistakenly consider "dark" bread the same as 100 percent whole-wheat bread?

Another question to ask is "What are the optimal levels of fiber intake for colon cancer protection?" Some experts believe that it may take more than 25 grams of fiber a day to show cancer-protective effects, which might explain the lack of effect noted in the Nurses' study.

This fiber story illustrates how news reports based on only one study can leave the public with the impression that scientists can't make up their minds. It seems as if one week scientists are saying that fiber is good, and the next week the word is that fiber doesn't do any good at all. The truth is health experts and major health organizations continue to urge the public to get 20 to 35 grams of fiber a day. (The average fiber intake today is about 13 grams.)

Contrary to what some headlines imply, reputable scientists do not base their dietary recommendations for the public on the findings of one or two studies. Scientists are still conducting research to determine whether fiber does in fact help to prevent colon cancer, and if so, what types of fiber and in what amount. Scientists design their research to test theories, such as the notion that eating a high-fiber diet is associated with lower risk of cancer. Other factors, however, often confound the matter at hand. The study of fiber and colon cancer is complex due to the many other factors linked with colon cancer development, including inactivity, obesity, saturated fat intake, low calcium or folate intakes, and others.

So, should you hang on to your high-fiber cereals and vegetables? Yes, according to The American Institute for Cancer Research. Its analysis of more than 4,500 research studies provides evidence that increased intakes of fiber may be associated with decreased incidence of colon cancer.[24] Most importantly, the health benefits of fruits, vegetables, legumes, and whole grains go beyond the possible protection from colon cancer.[25] Diets rich in fiber from these foods are strongly associated with reduced risks of heart disease, high blood pressure, Type 2 diabetes, and diverticular disease, a condition which can lead to painful inflammation of the large intestine. Fiber also promotes a feeling of fullness after you eat, which can help with weight control.

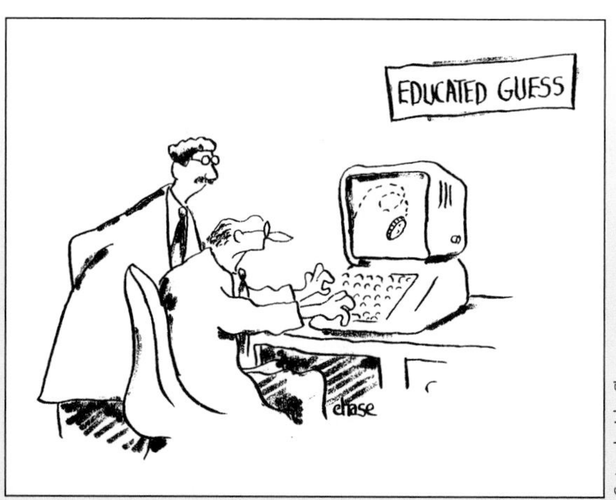

Cartoon by John Chase.

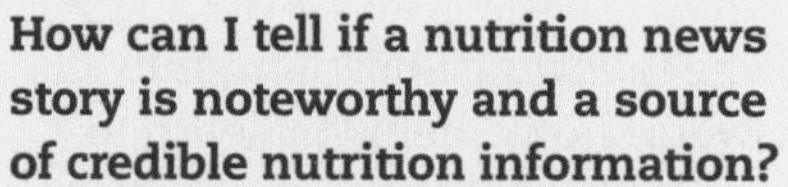

How can I tell if a nutrition news story is noteworthy and a source of credible nutrition information? You can critique the nutrition news you read by asking a series of questions. Consider the following points as a checklist for separating the bogus news stories from those worth your attention:

- *Where is the study published?*

The study being described in the news story should be published in a peer-reviewed journal—a journal that uses experts in the field to review research results. These reviewers serve to point out flaws in the research design and can challenge the researcher's conclusions before the study gets published.

- *How recent is the study?*

The science of nutrition continues to develop. New studies employ state-of-the-art methods and technology and benefit from the scrutiny of experts current in the particular field of study.

- *What research methods were used to obtain the data?*

Are the results from an **epidemiological study** or an **intervention study?** Epidemiological studies examine populations to determine food patterns and health status over time. These population studies are useful in uncovering **correlations** between two factors (for example, whether a high calcium intake early in life reduces the incidence of bone fractures later in life.) However, they are not considered as conclusive as intervention studies. A correlation between two factors may suggest a cause and effect relationship between the factors, but does not prove it.

Intervention studies examine the effects of a specific treatment or intervention on a particular group of subjects and compare the results to a similar group of people not receiving the treatment. An example is a cholesterol-monitoring study in which half the subjects follow dietary advice to lower their blood cholesterol and half do not. Ideally, intervention studies should be randomized and controlled—that is, subjects are assigned to either an **experimental group** or a **control group** based on a random selection process. Each subject has an equal chance of being assigned to either group. The experimental group receives the "treatment" being tested; the control group receives a **placebo** or neutral substance. If possible, neither the researcher nor the participants should know which group the subjects have been assigned to until the end of the experiment. A randomized, controlled study helps to ensure that the study's conclusions are a result of the treatment and minimizes the chances that the results are due to a **placebo effect** or bias on the part of the researcher.

- *What was the size of the study?*

In order to achieve validity—accuracy in results—studies must generally include a sufficiently large number of people. (For example, intervention studies of 50 or more persons.) This reduces the chances that the results are simply a coincidence and helps to generalize the conclusions of the study to a wider audience.

- *Who were the subjects?*

Look for similarities between the subjects in the study and yourself. The more you have in common with the participants (age, diet pattern, gender, etc.), the more pertinent the study results may be for you.

It's wise to be leery of media stories based solely on animal experiments. Although different species may respond in some way for better or for worse to a treatment, the same response will not necessarily occur in humans.

For example, suppose rats became sick when injected with the amount of substance X that a person would normally consume in a year's time. A researcher might correctly conclude that substance X could cause illness. But in a case like this, remember that a rat is a tiny animal with a different physical makeup than a human being. Moreover, the rodent received a huge dose of substance X all at once. Thus, the experiment does not indicate whether small amounts of substance X taken by a person over the course of a year would produce the same harmful effect. A news story that implies otherwise misses the mark.

- *Does a consensus of published studies support the results reported in the news?*

Even if an experiment is carefully designed and carried out perfectly, however, its findings cannot be considered definitive until they have been confirmed by other research. Testing and retesting reduce the possibility that the outcome was simply the result of chance or an error or oversight on the part of the experimenter. Every study should be viewed as preliminary until it becomes just one addition to a significant body of evidence pointing in the same direction.

When making dietary recommendations for the public, experts pool the results of different types of studies, such as analyses of food patterns of groups of people and carefully controlled studies on people in hospitals or clinics. Before drawing any conclusions, they then consider the evidence from all of the research. In fact, the dietary guidelines spelled out in the 1,300-page *Diet and Health* report are based on the results of hundreds of studies. The bottom line is that if you read a report in the newspaper or watch one on television that advises making a dramatic change in your diet or lifestyle based on the results of one study, don't take it to heart. The findings may make for a good story, but they're not worth taking too seriously.

Why doesn't the government do something to prevent the media from delivering misleading nutrition information?

The government lacks the power to do so because the **First Amendment** guarantees freedom of the press. Thus, it is possible for people to express whatever views they like in the media, whether sound, unsound, or even dangerous. This freedom is a cornerstone of the U.S. Constitution, and to deny it would be to deny democracy. Writers cannot be punished by law for publishing misinformation unless it can be proved in court that the information has caused a reader bodily harm.

Fortunately, most professional health groups maintain committees to combat the spread of health and nutrition misinformation. A list of organizations that provide reliable scientific information appears in Appendix A, and any of these organizations can serve as sources for your own inquiries about the authenticity of scientific information in their areas.

Many professional organizations have also banded together to form the National Council for Reliable Health Information—NCRHI (formerly known as the National Council Against Health Fraud). The NCRHI monitors radio, television, and other advertising and investigates complaints. You can contact them online at **www.ncrhi.org**.

Is the Internet a reliable source of nutrition and health information?

Information is rampant on the **Internet.**[26] In a sense, the Internet is information, and the information is continually being revised and created. Internet information exists in many forms (facts, statistics, stories, opinions), is created for many purposes (to entertain, to inform, to persuade, to sell, to influence), and varies in quality from good to bad. One method for determining whether the information found on the Internet is reliable and of good quality is the CARS Checklist. The acronym CARS stands for credibility, accuracy, reasonableness, and support. Each of these is summarized on the following page.[27]

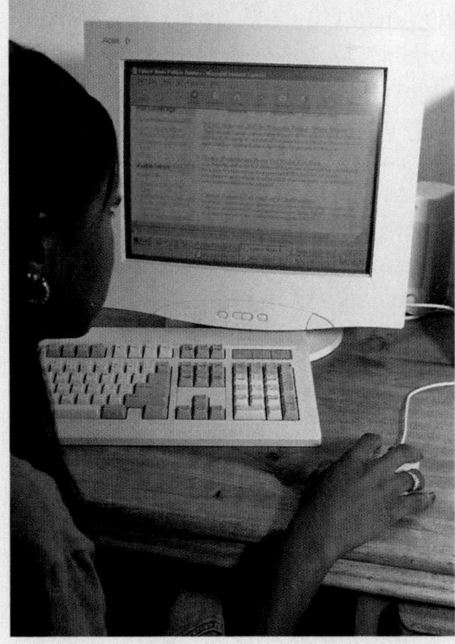

Keep in mind that the information you find on the Internet is only as good as its source.

◗ **Credibility.** Check the credentials of the author (if there is one!) or sponsoring organization. Is the author or organization respected and well known as a source of sound, scientific information? Evidence of a lack of credibility includes no posted author and even the presence of misspelled words or bad grammar. A credible sponsor will use a professional approach to designing the Web site.

◗ **Accuracy.** Check to ensure that the information is current, factual, and comprehensive. If important facts, consequences, or other information are missing, the Web site may not be presenting a complete story. Evidence of a lack of accuracy include no date on the document, the use of sweeping generalizations, and the presence of outdated information. How often is the site updated? Reliable sites are updated regularly, with a date posted on the site. Watch for testimonials masquerading as scientific evidence. This is a common method for promoting questionable products on the Internet.

◗ **Reasonableness.** Evaluate the information for fairness, balance, and consistency. Does the author present a fair, balanced argument supporting his ideas? Are the author's arguments rational? Has she maintained objectivity in discussing the topic? Does the author have an obvious—or hidden—conflict of interest? Evidence of a lack of reasonableness includes gross generalizations ("Foods not grown organically are all toxic and shouldn't be eaten") and outlandish claims ("Kombucha tea will cure cancer and diabetes").

◗ **Support.** Check to see whether supporting documentation is cited for scientific statements. Are there references to legitimate scientific journals and publications? Is it clear where the information came from? An Internet document that fails to show the sources of its information is suspect.

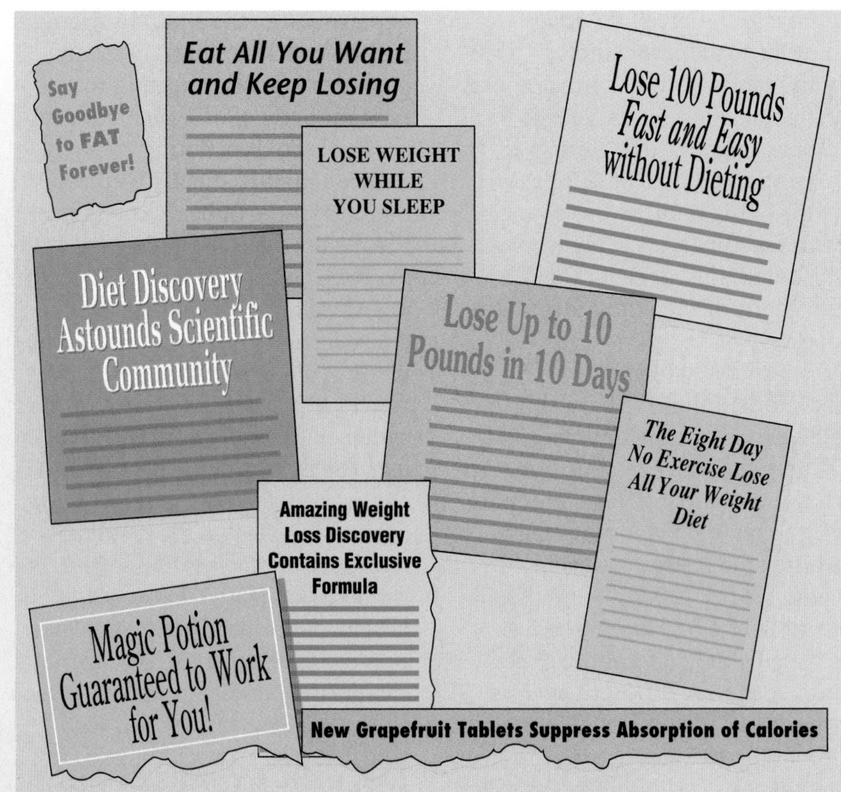

If it sounds too good to be true, it probably is.

Many reputable government agencies and professional organizations host Web sites on the Internet. To help you explore the **World Wide Web,** the publishers of this book maintain a Web site with links to reliable information at **www.wadsworth.com/nutrition.** Another site to check out is the Tufts University *Nutrition Navigator* Web site at **navigator.tufts.edu.** This site reviews and evaluates other nutrition-related Web sites and provides links to dozens of reliable nutrition sites.*

Does the First Amendment also make it legal for companies to say whatever they want about the products they sell?

No. Unlike journalists, purveyors of products are bound by law to make only true statements about their wares. The Food and Drug Administration (FDA) holds the authority to prosecute companies that display false nutrition information on product labels or enclosures, and the Federal Trade Commission (FTC) can take to task manufacturers who make fraudulent or misleading statements in their advertisements. Nevertheless, combating health fraud is an overwhelming job requiring enormous amounts of time and money. As one FDA official has put it, "Quack promoters have learned to stay one step ahead of the laws either by moving from state to state or by changing their corporate names."[28]

*Appendix A provides a list of reliable sources of nutrition information. Many have Web sites.

How can I tell whether a product is bogus?

It's not always easy. Given that many misleading claims are supposedly backed by scientific-sounding statements, it is difficult for even informed consumers to separate fact from fiction. The following red flags can help you spot a quack:

- *The promoter claims that the medical establishment is against him or her and the government won't accept this new "alternative" treatment.*

If the government or medical community doesn't accept a treatment, it's because the treatment hasn't been proven to work. Reputable professionals don't suppress knowledge about fighting disease. On the contrary, they welcome new remedies for illness, provided the treatments have been carefully tested.

- *The promoter uses testimonials and anecdotes from satisfied customers to support claims.*

Valid nutrition information comes from careful experimental research, not from random tales. A few people's reports that the product in question "works every time" are never acceptable as sound scientific evidence.

- *The promoter uses a computer-scored questionnaire for diagnosing "nutrient deficiencies."*

Those computers are programmed to suggest that just about everyone has a deficiency that can be reversed with the supplements the promoter just happens to be selling, regardless of the consumer's symptoms or health.

- *The promoter claims the product will make weight loss easy.*

Unfortunately, there is no simple way to lose weight. In other words, if a claim sounds too good to be true, it probably is.

- *The promoter promises that the product is made with a "secret formula" available only from this one company.*

Legitimate health professionals share their knowledge of proven treatments so that others can benefit from it.

- *The treatment is available only through the back pages of magazines, over the phone, or by mail-order ads in the form of news stories or 30-minute commercials (known as infomercials) in talk-show format.*

Results of studies on credible treatments are reported first in medical journals and administered through a doctor or other health professional. If information about a treatment appears only elsewhere, it probably can't withstand scrutiny.*

If I do buy a product, say, to help me lose weight, but I still need some advice about dieting, should I check with a nutritionist?

To answer that question, first consider the following. About 15 years ago, Charlie Herbert became a professional of the International Academy of Nutrition Consultants. Another member of the household, Sassafras Herbert, met all the requirements for membership in the American Association of Nutrition and Dietary Consultants, a "professional association dedicated to maintaining ethical standards in nutritional and dietary consulting." The only qualification for membership is a 50-dollar fee, regardless of your background (or even your species). Charlie Herbert is a cat, and Sassafras is a poodle. The two obtained their "credentials" with the help of Victor Herbert, M.D., professor of medicine, chairman of the Committee to Strengthen Nutrition at Mount Sinai School of Medicine, New York City, and a leader in combating nutrition fraud. Dr. Herbert had his pets added to the membership rosters of those organizations to demonstrate how easy it is for anyone to get fake nutrition credentials. This is because in some states, the term **nutritionist** is not at present legally defined.

During the past decade, the situation doesn't seem to have improved much. In 1994, a 32-state survey sponsored by the National Council Against Health Fraud found that consumers who turn to the Yellow Pages to find a nutrition counselor have about a 50-50 chance of finding someone with legitimate training. Many so-called practitioners calling themselves nutritionists or nutrition-oriented physicians hold bogus degrees and dispense bad advice and pricey supplements.[29]

Before you pay a fee or follow a nutritionist's advice, inquire about the person's credentials. Some "nutritionists" obtain their diplomas and titles without undergoing the rigorous training required to obtain a legitimate degree in nutrition. Lax state laws make it possible for irresponsible **correspondence schools**—also called **diploma mills**—to grant degrees to unqualified individuals for nothing more than a fee.

Charlie and Sassafras display their professional credentials.

*If you think you've been duped by a quack, write the FDA, Office of Consumer Affairs and Information, HFC-110, 5600 Fishers Lane, Rockville, MD 29857; your state Attorney General's office; the Federal Trade Commission, Correspondence Branch, Sixth and Pennsylvania Avenues, N.W., Washington, DC 20580; and/or the newspaper, magazine, or TV or radio station running the ad. If you ordered the product by mail, call the U.S. Postal Service and ask for a fraud packet. If you suspect that a Web site is promoting misinformation or marketing bogus goods or services on the Internet, e-mail the FDA at otcfraud@cder.fda.gov.

How can I check a nutritionist's credentials?

You can call the institution that the person claims has awarded the degree. To find out about the existence or reputation of an institution of higher learning, you can go to any library and ask for a directory of colleges and universities called *Accredited Institutions of Postsecondary Education*, published by the American Council on Education. Be suspicious of diplomas or degrees issued by institutions that cannot prove that they have **accreditation** from the Council on Education.

Another option is to find out whether the person is a special type of nutritionist, known as a **registered dietitian (RD)**. The RD, an especially meaningful credential, has a standard definition—a professional who has fulfilled coursework required by the American Dietetic Association (ADA), including courses in psychology, business, chemistry, anatomy, physiology, advanced nutrition, medical nutrition therapy, food science, and foodservice administration. In addition, the RD has completed an internship that includes on-the-job training counseling people about diet and has passed a national registration exam. All registered dietitians must keep their credentials current by completing regular continuing education requirements.

You can check on any RD by asking for that person's registration number and calling the ADA's Commission on Dietetic Registration (312-899-0040). If you'd like to contact an RD but don't know where to find one, call the ADA's Consumer Nutrition Hot Line at 1-800-366-1655 for a referral. Or, visit the ADA Web site **(www.eatright.org)**.

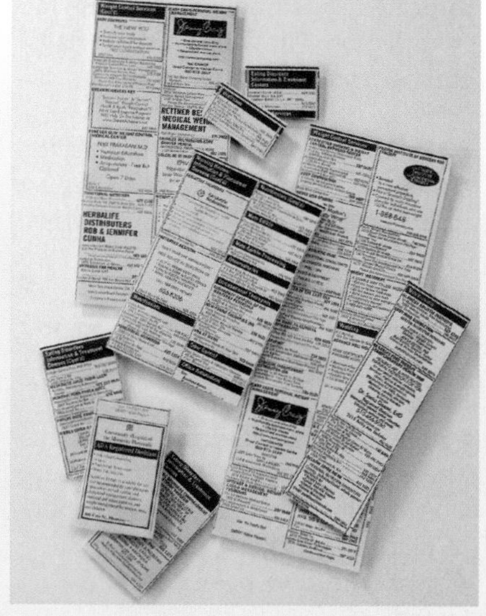

If you're looking for a credible nutritionist, be careful when you "let your fingers do the walking" through the Yellow Pages.

MINIGLOSSARY

ACCREDITATION approval; in the case of hospitals or university departments, approval by a professional organization of the educational program offered. There are phony accrediting agencies; the genuine ones are listed in a directory called *Accredited Institutions of Postsecondary Education.*

CONTROL GROUP a group of individuals with matching characteristics to the group being treated in an intervention study who receive a sham treatment or no treatment at all.

CORRELATION a simultaneous change in two factors, such as a decrease in blood pressure with regular aerobic activity (a direct or positive correlation) or the decrease in incidence of bone fractures with increasing calcium intakes (an inverse or negative correlation).

CORRESPONDENCE SCHOOL a school from which courses can be taken and degrees granted by mail. Schools that are accredited offer respectable courses and degrees.

DIPLOMA MILL a correspondence school that grinds out degrees—sometimes worth no more than the cost of the paper they are printed on—the way a grain mill grinds out flour.

EPIDEMIOLOGICAL STUDY a study of a population to search for possible correlations between nutrition factors and health patterns over time.

EXPERIMENTAL GROUP the participants in a study who receive the real treatment or intervention under investigation.

FIRST AMENDMENT the amendment to the U.S. Constitution that guarantees freedom of the press.

INTERNET a network of millions of computers connected to universities, government agencies, and commercial and nonprofit organizations around the world and linked together to form a "mega-network."

INTERVENTION STUDY a population study examining the effects of a treatment on experimental subjects compared to a control group.

NUTRITIONIST a person who claims to be capable of advising people about their diets. Some nutritionists are registered dietitians, whereas others are self-described experts whose training is questionable.

PLACEBO a sham or neutral treatment given to a control group; an inert, harmless "treatment" that the group's members cannot recognize as different from the real thing, which minimizes the chance that a result of the treatment will appear to have occurred due to the **placebo effect**—the healing effect that the belief in the treatment, rather than the treatment itself, often has.

REGISTERED DIETITIAN (RD) a professional who has graduated from a program of dietetics accredited by the Commission on Accreditation for Dietetics Education (CADE) of the American Dietetic Association (ADA), has served an internship program or the equivalent to gain practical skills, has passed a registration examination, and maintains competencies through continuing education. Some states require licensing for dietitians; that is, they have legislation in place obligating anyone who wants to use the title "dietitian" to receive permission by passing a state examination. Other states do not require dietitians to be licensed. RD (the abbreviation for registered dietitian) is often used to refer to such a professional in the same way M.D. designates a medical doctor.

WORLD WIDE WEB the portion of the Internet that contains sites hosted by government agencies, industry, academic institutions, health organizations, and other groups or individuals. Web sites vary in sophistication from simple displays of text to pages with sophisticated graphics, videos, and sounds. Credible Web sites usually include addresses of related sites—or *links*—that can be accessed simply by clicking on the address, immediately connecting the user to the new site.

Nutrition on the Web

nutrition.wadsworth.com	Go to the *Personal Nutrition* site to check for the latest updates to chapter topics or to access links to related Web pages.
www.healthfinder.gov	Scroll down to Smart Choices for online screening checklists and information on healthy lifestyles; this site offers many links to other reliable health-related sites.
www.eatright.org	The American Dietetic Association home page includes frequently asked questions, nutrition resources, and many reliable links to other food and nutrition sites.
www.5aday.com	Promotes the "5 a Day for Better Health" campaign.
ificinfo.health.org	Nutrition information for consumers and health professionals.
www.thriveonline.com/nutrition/index.html	A general resource for nutrition information.
www.hc-sc.gc.ca/hppb/nutrition	A Canadian health and nutrition information home page.
www.health.gov/healthypeople	Updates on *Healthy People 2010* initiative.
www.navigator.tufts.edu	A search engine for reliable nutrition information.
www.pueblo.gsa.gov	A site for consumer food and nutrition information.
www.nih.gov/health	A search engine from the National Institutes of Health with access to Medline and PubMed databases.
www.ncrhi.org	Information on consumer health and nutrition fraud.

Check Yourself . . .

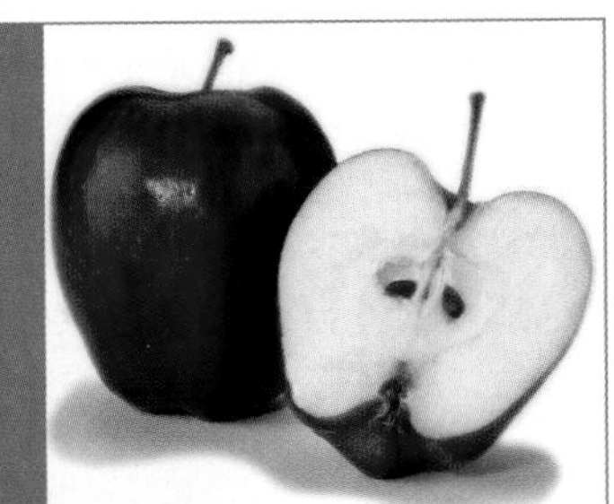

1. Name three diseases associated with overnutrition.
2. Describe the difference between a nutritionist and a registered dietitian.
3. Name four factors that affect your eating habits.
4. Define health fraud.
5. Identify three ways you can cut fat and calories from fast food.
6. Describe three red flags that can help you spot a nutrition quack.
7. Explain how culture influences eating habits.
8. Name three factors besides diet that affect longevity.
9. Describe the role of the Food and Drug Administration and of the Federal Trade Commission in monitoring sellers of nutrition-related products.
10. Name the government initiative that seeks to reduce disease risk and improve the nutrition status of the American population.

Answers to Check Yourself questions are found in Appendix G.

2

The Pursuit of an Ideal Diet

CONTENTS

The ABCs of Eating for Health

The Nutrients

Nutrient Recommendations

The Challenge of Dietary Guidelines

Tools Used in Diet Planning

Nutrition Action: Grazer's Guide to Smart Snacking

Rate Your Plate Scorecard

Spotlight: A Tapestry of Cultures and Cuisines

The Savvy Diner: Choosing Healthful International Cuisines

If the doctors of today will not become the nutritionists of tomorrow, the nutritionists of today will become the doctors of tomorrow.

Thomas Edison
(1847–1931, American inventor)

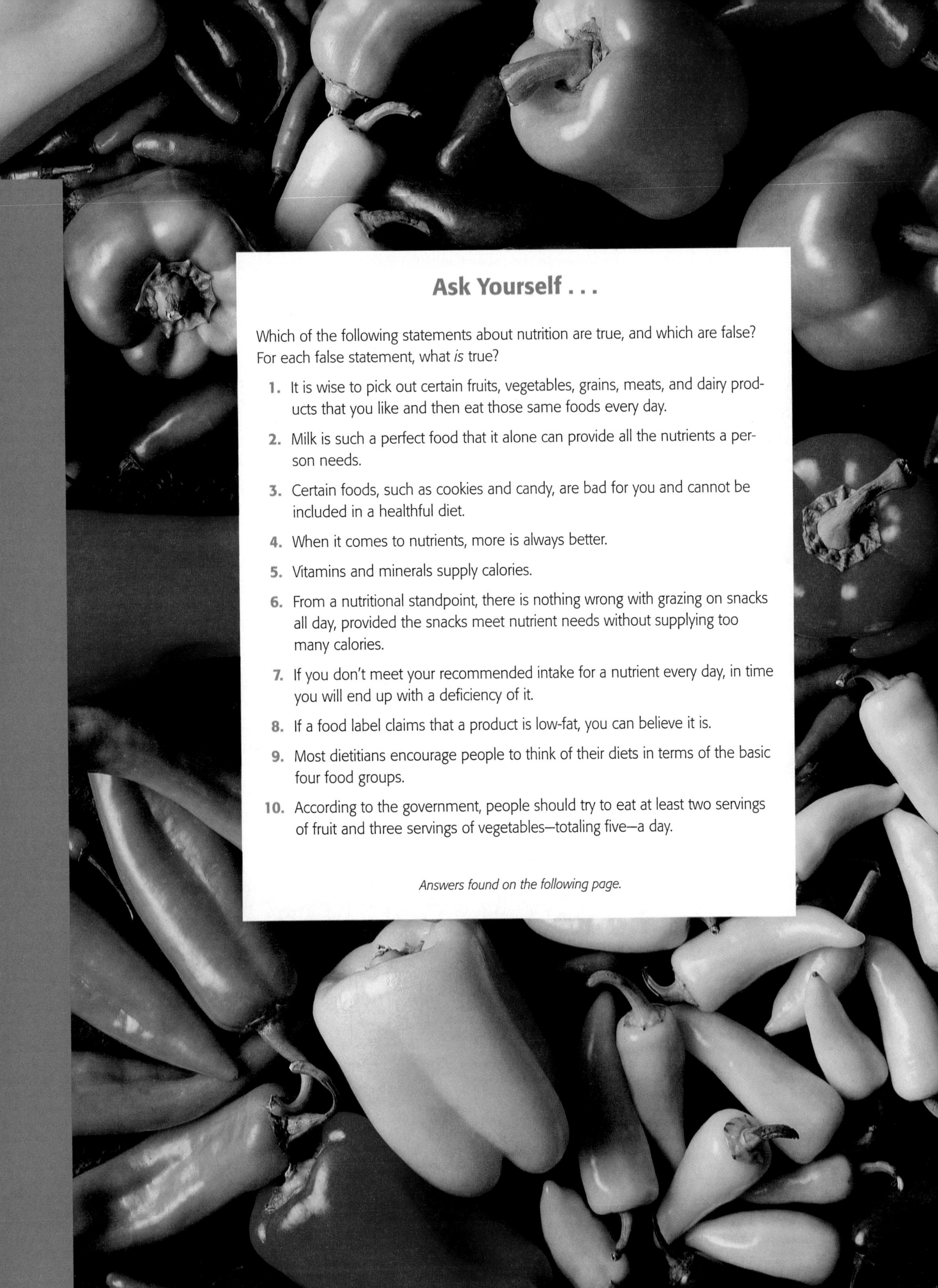

Ask Yourself . . .

Which of the following statements about nutrition are true, and which are false? For each false statement, what *is* true?

1. It is wise to pick out certain fruits, vegetables, grains, meats, and dairy products that you like and then eat those same foods every day.
2. Milk is such a perfect food that it alone can provide all the nutrients a person needs.
3. Certain foods, such as cookies and candy, are bad for you and cannot be included in a healthful diet.
4. When it comes to nutrients, more is always better.
5. Vitamins and minerals supply calories.
6. From a nutritional standpoint, there is nothing wrong with grazing on snacks all day, provided the snacks meet nutrient needs without supplying too many calories.
7. If you don't meet your recommended intake for a nutrient every day, in time you will end up with a deficiency of it.
8. If a food label claims that a product is low-fat, you can believe it is.
9. Most dietitians encourage people to think of their diets in terms of the basic four food groups.
10. According to the government, people should try to eat at least two servings of fruit and three servings of vegetables—totaling five—a day.

Answers found on the following page.

For most people, eating is so habitual that they give hardly any thought to the foods they choose to eat. Yet, as Chapter 1 emphasized, the foods you select can have a profound effect on the quality and possibly even the length of your life. Given all the statistics and government mandates presented so far, however, designing a healthful diet may seem like a complicated matter involving a rigid regimen that excludes certain foods from the diet. Fortunately, that's not the case. The government, as well as many major health organizations, has devised dietary guidelines and tools such as food labels to help you choose the most healthful diet for you. This chapter provides an overview of some of the best guides and tools and shows you how to use them.

As you read the following pages, keep in mind that one of the biggest misconceptions about planning a healthful diet is that some foods, say, carrots and celery sticks, are "good," while others, like cookies and candy, are "bad." People who class foods this way often feel guilty every time they "splurge" on a so-called bad food. The overall diet is what counts, however. A diet consisting of nothing but carrot sticks is just as unhealthful as one made up of only candy bars. The trick is choosing a healthful balance of foods. The ideal diet contains primarily foods that supply adequate nutrients, fiber, and calories without an excess of fat, sugar, sodium, or alcohol.

ADEQUACY characterizes a diet that provides all of the essential nutrients, fiber, and energy (calories) in amounts sufficient to maintain health.

BALANCE a feature of a diet that provides a number of types of foods in balance with one another, such that foods rich in one nutrient do not crowd out of the diet foods that are rich in another nutrient.

CALORIE CONTROL control of consumption of energy (calories); a feature of a sound diet plan.

MODERATION the attribute of a diet that provides no unwanted constituent in excess.

VARIETY a feature of a diet in which different foods are used for the same purposes on different occasions—the opposite of *monotony.*

 THE ABCs OF EATING FOR HEALTH

When you plan a diet for yourself, try to make sure it has certain characteristics: **adequacy** (it will provide enough of the essential nutrients, fiber, and energy—in the form of calories); **balance** (it will not overemphasize any food type or nutrient at the expense of another); **calorie control** (it will supply the amount of energy you need to maintain desirable weight—not more, not less); **moderation** (it will not contain excess amounts of unwanted constituents, such as fat, salt, or sugar); and **variety** (it will be made up of different foods rather than the same meals day

ACCORDING TO THIS TABLE ON MINIMUM DAILY REQUIREMENTS, WE NEED TO WORK ON OUR FRUITS AND VEGETABLES
NUTRITION
...BUT WE'RE DOING GREAT IN FATS AND SUGARS—IN FACT, WE'RE ALL CAUGHT UP UNTIL JANUARY OF THE YEAR 2024...
NUTRITION
ETTA HULME FORT WORTH STAR-TELEGRAM N.E.A. '79

Ask Yourself Answers: **1.** False. It is unwise to eat the same foods day in and day out; your diet will lack variety and probably will not supply all the nutrients your body needs. **2.** False. Milk rates as an excellent source of nutrients such as calcium and protein, but it contains only very small amounts of iron and several other nutrients. **3.** False. Any food can fit into a healthful eating plan. It's the total diet, not individual foods, that can be either good or bad for health. **4.** False. Too much or too little of a nutrient is often equally harmful. **5.** False. Only protein, carbohydrate, and fat supply calories. **6.** True. **7.** False. Even if you don't meet your recommended intake for a nutrient, you still may be consuming a sufficient amount of the nutrient. **8.** True. **9.** False. Dietitians today encourage people to think about eating from the five food groups in the Food Guide Pyramid. **10.** True.

after day). Equally important, it will suit you; that is, it will consist of foods that fit your personality, family and cultural traditions, lifestyle, and budget. At best, it will be a source of both pleasure and good health.

Any nutrient could be used to demonstrate the importance of dietary *adequacy*. Consider iron, an essential nutrient that your body loses daily and must be replaced continually via iron-rich foods. If your diet does not provide adequate iron—that is, it lacks food sources of the mineral—you can develop a condition known as iron-deficiency anemia. If you add iron-rich foods such as meat, fish, poultry, and legumes to your diet, the condition will most likely soon disappear. (More information about iron appears in Chapter 7.)

To appreciate the importance of *balance,* consider a second essential nutrient. Calcium plays a vital role in building a strong frame that can withstand the gradual loss of bone that occurs with age. Thus, adults are advised to consume daily at least two, and preferably more, servings of milk or milk products—the best sources of the bone-building mineral—to meet their calcium needs. Foods that are rich in calcium typically lack iron, however, and vice versa, so you have to balance the two in your diet.

Variety fosters good nutrition.

Balancing the whole diet is a juggling act that, if successful, provides enough, but not too much, of every one of the 40-odd nutrients the body needs for good health. As you will see later in the chapter, you can design a diet that is both adequate and balanced by using food group plans that help you choose from various groups specific amounts of foods that should be eaten each day.

To maintain a desirable weight, energy intakes should not exceed energy needs. *Calorie control* helps ensure a balance between energy we take in from food and energy we expend in activity. A cup and a half of ice cream, for example, has about the same amount of calcium as a cup of milk, but the ice cream may contain more than 500 calories, while the milk may supply only 90. When it comes to iron, a 3-ounce serving of beef pot roast provides the same amount of the mineral as a 3-ounce serving of canned water-packed tuna. But whereas the beef contains 325 calories, the tuna adds only 175 to the diet. The choice of which one to eat depends on personal preferences as well as the calorie content of the other foods in the diet.

Those who are trying to ensure optimal intakes of nutrients without an excess of calories should be sure to include foods that are rich in nutrients (protein, vitamins, and minerals) but relatively low in calories and fat. Such foods are referred to as **nutrient dense.** A baked potato, for example, contains more iron and vitamin C for its calories than French fries. Hence, it is more nutrient dense. Figure 2-1 compares the nutrient density of selected beverages.

NUTRIENT DENSE refers to a food that supplies large amounts of nutrients relative to the number of calories it contains. The higher the level of nutrients and the fewer the number of calories, the more nutrient dense the food is.

Diet planning principles:
- Adequacy—enough of each type of food
- Balance—not too much of any type of food
- Calorie control—not too many or too few calories
- Moderation—not too much fat, salt, or sugar
- Variety—as many different foods as possible

Another characteristic of a healthy diet is *moderation*. In other words, try to eat meals that do not contain excessive amounts of any one nutrient, particularly dietary fat—the culprit linked to a number of chronic diseases. That's not to say that you should choose only foods that supply little or no fat. Such an approach is unrealistic and will only lead to frustration. A more moderate philosophy to adopt is the 80/20 rule: Try to eat low-fat, nutrient-dense foods (and remember to exercise) at least 80 percent of the time, and you're not likely to reverse their benefits to your health if you splurge here and there the remaining 20 percent of the time.

Aside from avoiding the monotony of eating the same foods day after day, we need *variety* in our diet for two reasons. One is that some foods are better sources of nutrients needed in such small amounts that we don't consciously plan diets around them. Another is that a limited diet can supply excess amounts of undesirable substances such as chemical contaminants. Eating many different foods, on the other hand, greatly reduces the likelihood that large amounts of a potential toxin will be consumed.

Research underscores that variety is one of the hallmarks of a healthful diet. Consider that a team of U.S. scientists examined the eating habits of more than 10,000 people in the early 1970s and found that those whose food choices were the least varied were most likely to have died 20 years later. In addition, they

FIGURE 2-1
NUTRIENT DENSITY OF SELECTED BEVERAGES

Understanding the concept of nutrient density will help you locate foods that are high in the amount of nutrients provided per calorie. Compare the nutrient contributions made by the three beverages shown here. The figures show that orange juice and milk are good sources of several essential nutrients, whereas the cola beverage simply provides calories (from sugar).

[a]The Daily Values, used on food labels, are based on 2,000 calories a day for adults and children over 4 years old, and are useful for making comparisons among foods in terms of their nutrient composition.
[b]A good source of this nutrient (more than 10% of the recommended intake).
[c]Certain brands of orange juice are fortified with calcium and provide 30% of the recommended value.

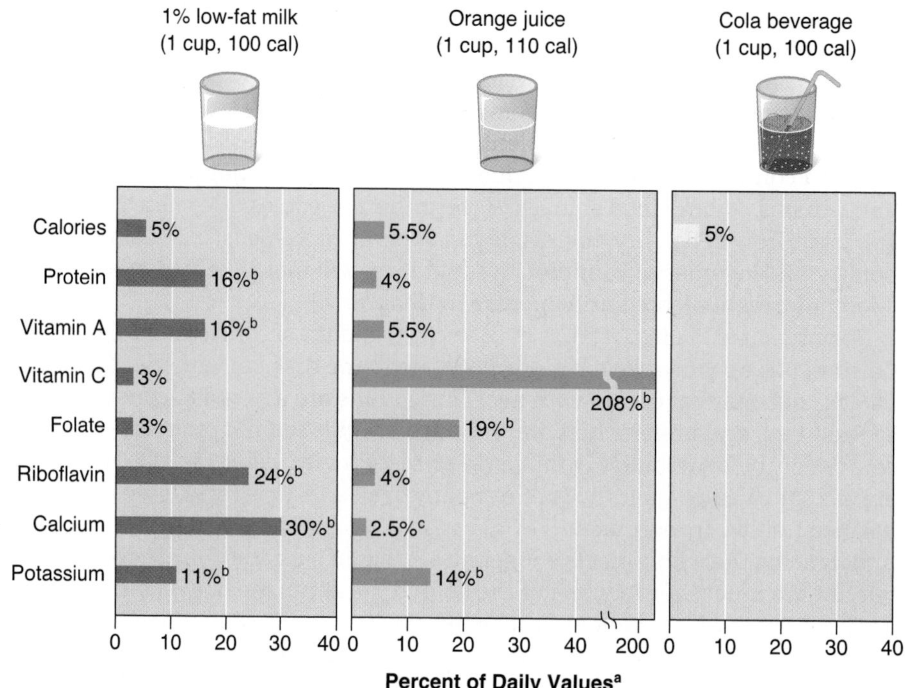

found that many of the people surveyed failed to regularly include items from major food groups in their meals. About 25 percent left out calcium-rich dairy products, and 17 percent went without fruit while 46 percent did not eat vegetables—both of which contain fiber and many essential nutrients.[1] If your diet lacks variety, chances are you're missing out on many nutrients necessary for optimal health. The Japanese, incidentally, recognize variety as such an important part of healthful eating that their dietary guidelines recommend consuming 30 or more different kinds of food every day to achieve a balance of essential nutrients.[2]

THE NUTRIENTS

NUTRIENTS substances obtained from food and used in the body to promote growth, maintenance, and repair.

The six classes of nutrients:
- carbohydrate
- fat
- protein
- vitamins
- minerals
- water

Almost any food you eat is mostly water, and some foods are as high as 99 percent water. The bulk of the solid materials consists of carbohydrate, fat, and protein. If you could remove these materials, you would detect a tiny residue of minerals, vitamins, and other materials. Water, carbohydrate, fat, protein, vitamins, and some of the minerals are **nutrients.** Some of the other materials are not nutrients. There are six classes of nutrients: carbohydrate, fat, protein, vitamins, minerals, and water.

A complete chemical analysis of your body would show that it is made of similar materials in roughly the same proportions as most foods. For example, if you weigh 150 pounds (and that is a desirable weight for you), your body contains about 90 pounds of water and some 30 pounds of fat. The other 30 pounds consist of mostly protein, carbohydrate, and the major minerals of your bones—calcium and phosphorus. Vitamins, other minerals, and incidental extras constitute a fraction of a pound.

ESSENTIAL NUTRIENTS nutrients that must be obtained from food because the body cannot make them for itself.

When referring to nutrients, scientists use the words *essential* and *nonessential* to distinguish between those that the body must obtain from food and those that the body can produce using its own resources. For instance, about 40 nutrients are known to be **essential;** that is, they are compounds that the body cannot make for itself but are indispensable to life processes. Many compounds the body makes for itself are necessary for good health. What distinguishes an essential nutrient is that it must be obtained through foods. In contrast, some of the oils and fats in the body are nonessential because they can be made from any of several different raw materials. Likewise, carbohydrate is nonessential because the body can convert some of the materials in protein into carbohydrate when needed. How can you be sure you're getting all the nutrients you need? The rest of this chapter will help you design a diet that covers all your body's needs.

The Energy-Yielding Nutrients

On being broken down in the body, or digested, three of the nutrients—carbohydrate, protein, and fat—yield the **energy** that the body uses to fuel its various activities. In contrast, vitamins, minerals, and water, once broken down in the body, do not yield energy but perform other tasks, such as maintenance and repair.

ENERGY the capacity to do work, such as moving or heating something.

The energy-yielding nutrients:
- carbohydrate
- fat
- protein

The body uses energy from carbohydrate, fat, and protein to do work or generate heat. This energy is measured in **calories**—familiar to most everyone as markers of how "fattening" foods are. If your body doesn't "burn" the energy you obtained from a food soon after you've eaten it, it stores it, usually as body fat, for use later. If excess amounts of protein, fat, or carbohydrate are eaten fairly regularly, the stored fat builds up over time and leads to obesity. Too much of any food, whether lean meat (a protein-rich food), potatoes (a high-carbohydrate food), or peanuts (a fatty food), can contribute excess calories that result in overweight.

CALORIE the unit used to measure energy. Keep in mind that 1 gram of carbohydrate yields 4 calories, 1 gram of fat yields 9 calories, 1 gram of protein yields 4 calories, and 1 gram of alcohol yields 7 calories.

Only one other compound that people consume provides calories, and that is alcohol. Alcohol is not considered a nutrient, however, because it does not help maintain or repair body tissues the way nutrients do.

Vitamins, Minerals, and Water

Unlike carbohydrate, fat, and protein, **vitamins** and **minerals** do not supply energy, or calories. Instead, they regulate the release of energy and other aspects of metabolism. As Table 2-1 shows, there are 13 vitamins, each with its special

VITAMINS organic, or carbon containing, essential nutrients vital to life and needed in minute amounts.
vita = life
amine = containing nitrogen

MINERALS inorganic compounds, some of which are essential nutrients.

Calorie Value of Carbohydrate, Fat, and Protein

If you know the number of grams of carbohydrate, fat, and protein in a food, you can calculate the number of calories in it. Simply multiply the carbohydrate grams by 4, the fat grams by 9, and the protein grams by 4. Add the totals together to obtain the number of calories. For example, a deluxe fast-food hamburger contains about 45 grams of carbohydrate, 27 grams of protein, and 39 grams of fat:

45 grams carbohydrate × 4 calories	= 180 calories
27 grams protein × 4 calories	= 108 calories
39 grams of fat × 9 calories	= 351 calories
Total	639 calories

The percentage of your total energy intake from carbohydrate, protein, and fat can then be determined by dividing the number of calories from each energy nutrient by the total calories, and then multiplying your answer by 100 to get the percentage:

calories from carbohydrate $= \frac{45 \times 4 \text{ cal/g}}{639} = 0.281 \times 100 = 28\%$

calories from protein $= \frac{27 \times 4 \text{ cal/g}}{639} = 0.168 \times 100 = 17\%$

calories from fat $= \frac{39 \times 9 \text{ cal/g}}{639} = 0.548 \times 100 = 55\%$

Remember that 1 gram is a very small amount. For instance, one teaspoon of sugar weighs roughly 5 grams.

See Appendix D for help with figuring percentages and other calculations.

TABLE 2-1 THE VITAMINS AND MINERALS

The Vitamins			The Minerals			
The water-soluble vitamins:		The fat-soluble vitamins:	The major minerals:		The trace minerals:	
B vitamins		Vitamin A	Calcium	Potassium	Chromium	Manganese
Thiamin	Vitamin B_{12}	Vitamin D	Chloride	Sodium	Copper	Molybdenum
Riboflavin	Folate	Vitamin E	Magnesium	Sulfur	Fluoride	Selenium
Niacin	Biotin	Vitamin K	Phosphorus		Iodine	Zinc
Vitamin B_6	Pantothenic acid				Iron	
Vitamin C						

Note: A number of trace minerals are currently under study to determine possible dietary requirements for humans. These include arsenic, boron, cadmium, cobalt, lead, lithium, nickel, silicon, tin, and vanadium.

Vitamins and minerals—play regulatory roles.
Water—provides the medium for life processes.

roles to play. Vitamins are divided into two classes: water-soluble (the B vitamins and vitamin C) and fat-soluble (vitamins A, D, E, and K). This distinction has many implications for the kinds of foods the different vitamins are found in and the way the body uses them, as you will see in Chapter 6.

The minerals also perform important functions. Some, such as calcium, make up the structure of bones and teeth. Others, including sodium, float about in the body's fluids, where they help regulate crucial bodily functions, such as heartbeat and muscle contractions.

Often neglected but equally vital, water is the medium in which all the body's processes take place. Some 60 percent of your body's weight is water, which carries materials to and from cells and provides the warm, nutrient-rich bath in which cells thrive. It also transports hormonal messages from place to place. When energy-yielding nutrients release "fuel," they break down to water and other simple compounds. Without water, you could live only a few days.

Because each day your body loses water in the form of sweat and urine, you must replace large amounts of it daily—on the order of two to three quarts a day. To be sure, you don't need to drink that much water daily, because the foods and other beverages you consume supply it in abundance.

NUTRIENT RECOMMENDATIONS

At this point, knowing that foods are made of so many different combinations of nutrients, you may be wondering how to determine whether you are eating

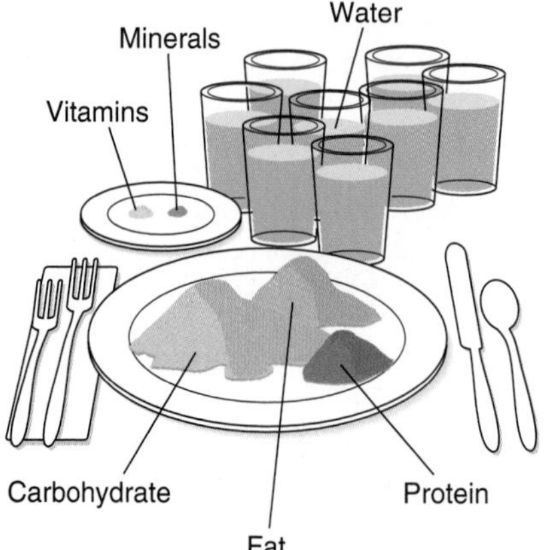

Amounts of Nutrients Eaten Daily
If you could extract and purify the carbohydrate, fat, and protein from your daily diet, they would fill two or three measuring cups, even though the foods they come in weigh much more and occupy much more space. For instance, a half cup of vegetables contains only 10 or so grams of energy-yielding nutrients, the rest being water, fiber, and other noncaloric materials.

the right balance of vitamins and minerals. Obviously, if your diet lacks any of the essential nutrients, you may develop deficiencies. Even if you don't develop a full-blown deficiency disease, if you are less than optimally nourished you may get sick more easily and be at higher risk of certain chronic diseases.

To help prevent such problems and provide a benchmark for people's nutrient needs, the U.S. government has devised the **Dietary Reference Intakes (DRI),** which are used in Canada and the United States.

DIETARY REFERENCE INTAKES (DRI) a set of reference values for nutrients that can be used for planning and assessing diets for healthy populations. DRI are guides for meeting the daily nutritional needs of virtually all healthy people in a specific age and gender group and include the Recommended Dietary Allowances (RDA). Estimated Average Requirements (EAR), Adequate Intakes (AI), and Tolerable Upper Intake Levels (UL).

The Dietary Reference Intakes (DRI)

A committee of nutrition experts selected by the National Academy of Sciences (NAS) sets forth the DRI—a set of daily nutrient standards based on the latest scientific evidence regarding diet and health. The first set, published in 1941, was called the Recommended Dietary Allowances (RDA). The RDA were revised ten times since then. In 1997, the NAS released the first in a series of reports on the DRI, which are intended to expand and update the RDA (see the inside front cover of this book).[3]

The DRI estimate the nutritional requirements of healthy people; those with certain medical problems often have different nutritional needs. In addition, the DRI include separate recommendations for different groups of people. For instance, the committee issues one set of guidelines for children ages 4 through 8, another set for adult men, another for pregnant women, and so on. The DRI encompass four sets of values: Estimated Average Requirements (EAR), RDA, Adequate Intakes (AI), and Tolerable Upper Intake Levels (UL).

To understand how scientists developed the DRI for nutrients, consider the folllowing example. Suppose we were the committee members, and were called upon to set the DRI for nutrient X. First, we would determine the **requirement**—how much of that nutrient the average healthy person needs to prevent a deficiency. To do so, we could review scientific research exploring how the body stores the nutrient, what the consequences of a deficiency might be, what causes depletion of the nutrient, and other factors that affect a person's need for it.

REQUIREMENT the minimum amount of a nutrient that will prevent the development of deficiency symptoms. Requirements differ from the RDA and AI, which include a substantial margin of safety to cover the requirements of different individuals.

In addition, we would consider any scientific evidence linking certain amounts of the nutrient to reduced risk of chronic disease. When considering calcium, for instance, we would take into account not only the amount needed to prevent a serious deficiency in the short term, but also the optimal level that would reduce risk of osteoporosis—the fragile-bone disease that, in the later years of life, can cause fractures.

Our next step would be to decide what amount of nutrient X to recommend. People vary in the amount of a given nutrient they need. One person might need 35 units of nutrient X for optimal health, another 40, and another 70, and so on. Our challenge would be to determine the best amount to recommend for everybody. In doing this, several numbers are calculated as part of the DRI.

First is the **Estimated Average Requirement (EAR),** which is the amount of a nutrient estimated to meet the needs of 50 percent of the people of a particular age and gender—women aged 19 through 30, for example (see Figure 2-2). The EAR is intended as an average daily value and is used in setting the RDA.

ESTIMATED AVERAGE REQUIREMENT (EAR) the average daily intake that meets the estimated nutrient needs of half of the individuals of a specific age and gender.

Once we have determined the EAR, we would use it to set the **Recommended Dietary Allowance (RDA).** The RDA is the amount of nutrient X that would meet the nutritional needs of 97 to 98 percent of the people in a group. Unlike the RDA which were set prior to 1997 and were intended for populations rather than individuals, the new RDA are to be used as goals for individuals to meet. As Figure 2-2 illustrates, the RDA are higher than the EAR.

RECOMMENDED DIETARY ALLOWANCES (RDA) daily dietary intake levels sufficient to meet the nutrient requirements of approximately 98 percent of healthy people.

When the scientific evidence regarding a particular nutrient is not sufficient to allow scientists to set an EAR and RDA, an **Adequate Intake (AI)** is determined instead.* An AI may be used as a rough goal for an individual to meet. The value

ADEQUATE INTAKE (AI) the estimated amount of a nutrient that should be consumed when sufficient scientific evidence is not available to calculate an EAR and RDA.

*This text combines the two sets of nutrient recommendations for individuals (RDA and AI) and refers to them as DRI recommended intakes.

FIGURE 2-2
NUTRIENT REQUIREMENTS VARY FROM PERSON TO PERSON

Each square represents a person with nutrient needs: A, B, and C are Mr. A, Ms. B, and Mr. C. The EAR is the amount of a nutrient estimated to meet the needs of half of the people in a particular age and gender group. The RDA for the nutrient is the amount that covers about 98% of the population.

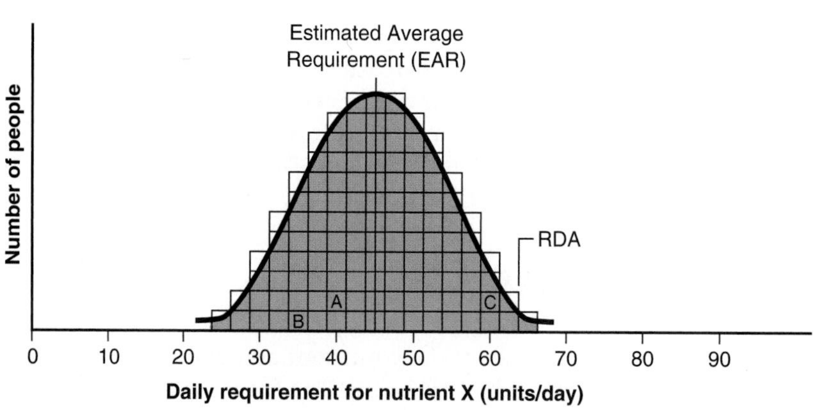

represents scientists' best estimate of the nutritional need of a person of a certain age and gender.

TOLERABLE UPPER INTAKE LEVEL (UL) the maximum amount of a nutrient that is unlikely to pose risk of harm in healthy people. The UL exceeds the RDA and is not intended to be an amount that people should regularly consume.

Finally, the DRI include a **Tolerable Upper Intake Level (UL),** which is the maximum amount of a nutrient that is unlikely to pose risk of harm in healthy people when consumed daily. This figure exceeds the RDA and is not intended as a recommended amount for regular consumption (see Figure 2-3). The UL was included in the DRI because more and more people self-prescribe nutrient supplements in large doses. It is intended to help people determine whether they are taking an excessive amount of a particular vitamin or mineral that may cause harm over time.

One other set of nutrient intake standards—the Daily Values (DV)—are used on food labels. The DV are useful for making comparisons among foods with reference to their nutrient composition. The Daily Values are discussed later in the chapter.

The DRIs are a good example of policy making in action. Compared with the RDAs, the DRIs represent a major shift in thinking about nutrient requirements for humans, from prevention of nutrient deficiencies to prevention of chronic disease. They also herald a new thinking about the role of dietary supplements in achieving good health—such that, "for some individuals at higher risk, use of nutrient supplements may be desirable in order to meet recommended intakes." For example, older women who are at risk of osteoporosis may benefit from dietary calcium supplements to help maintain bone mineral mass. In addition, the development of the UL signals the widespread recognition that there can be a degree of risk with high intakes of nutrients.

People vary in the amount of a given nutrient they need. The challenge of the DRI is to determine the best amount to recommend for everybody.

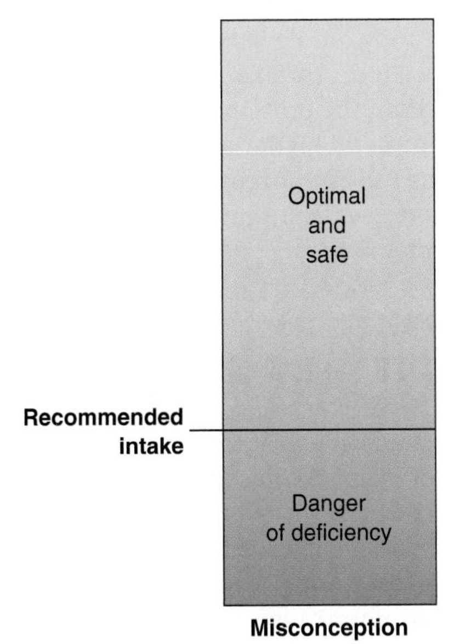

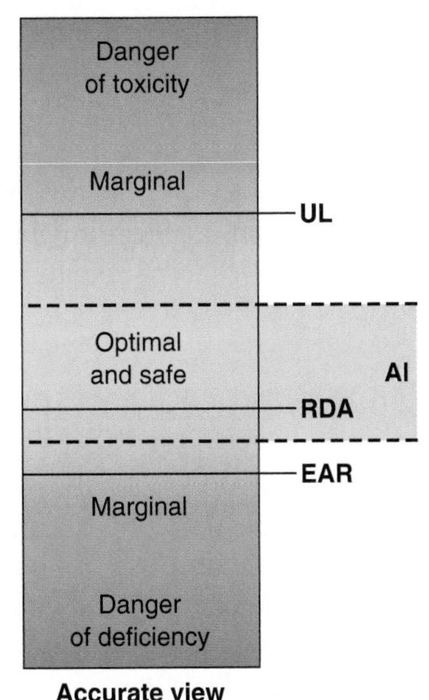

FIGURE 2-3
THE CORRECT VIEW OF THE DRI

People often think that more is better when it comes to nutrients. Too much of a good thing can be dangerous, however, so the RDA and AI fall within an optimal margin of safety, and the UL help people avoid harmful excesses of nutrients. To learn more about these standard nutrient values, go to www.nas.edu or www.nap.edu/readingroom and search for Dietary Reference Intakes.

AI = Adequate Intakes
EAR = Estimated Average Requirements
RDA = Recommended Dietary Allowances
UL = Tolerable Upper Intake Levels

The RDA for Calories

The DRI committee took a different approach when it set allowances for energy, or calories, than it did for the nutrients. That's because although small excesses of protein, vitamins, and most minerals are harmless for the majority of people, excess calories can lead to obesity. Too few calories can cause undesirable weight loss. Thus, the committee set the RDA for calories at the mean for each group, with the stipulation that for most people, an acceptable amount of variation above or below the mean is 20 percent. Figure 2-4 illustrates the difference between the nutrient and the calorie RDA. The RDA for calories, in particular, offer only a very rough estimate of individual calorie needs, which vary tremendously as a result of factors that cannot be accounted for in the RDA, such as a person's ratio of muscle to fat and activity level. If you want to know the number of calories needed to, say, lose or gain weight, you're better off getting a thorough assessment from a health professional than checking the RDA for calories.

Other Recommendations

Different nations and international groups have set forth different sets of standards similar to the DRI. Another widely used set of recommendations comes from two international organizations: the Food and Agriculture Organization (FAO) and the World Health Organization (WHO). The FAO/WHO recommendations

Since 1997, Adequate Intakes (AI) have been set for:
- vitamin D
- pantothenic acid
- biotin
- choline
- calcium
- fluoride

Since 1997, RDA have been set for:
- thiamin
- riboflavin
- niacin
- folate
- vitamin E
- vitamin C
- vitamin B_6
- vitamin B_{12}
- phosphorus
- magnesium
- selenium

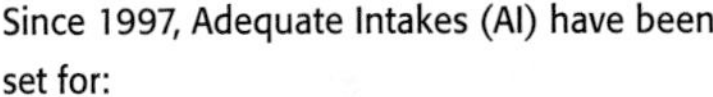

New DRIs are expected for the following nutrients for which a 1989 RDA exists:
- protein
- vitamin A
- vitamin K
- iron
- zinc
- iodine

New DRIs are also to be developed for:
- fiber
- carbohydrate
- fat
- sodium
- potassium
- chloride
- chromium
- copper
- manganese
- molybdenum

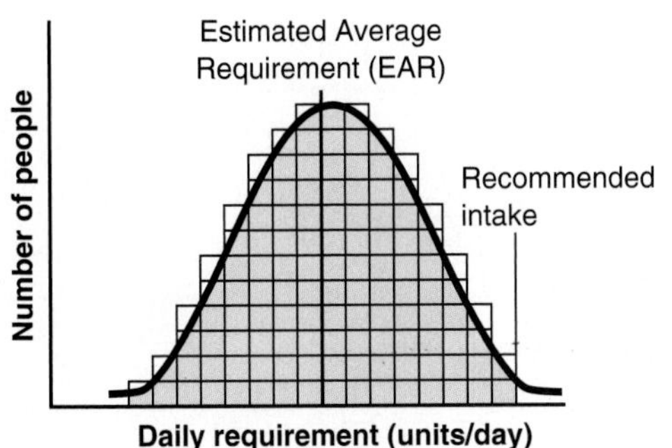

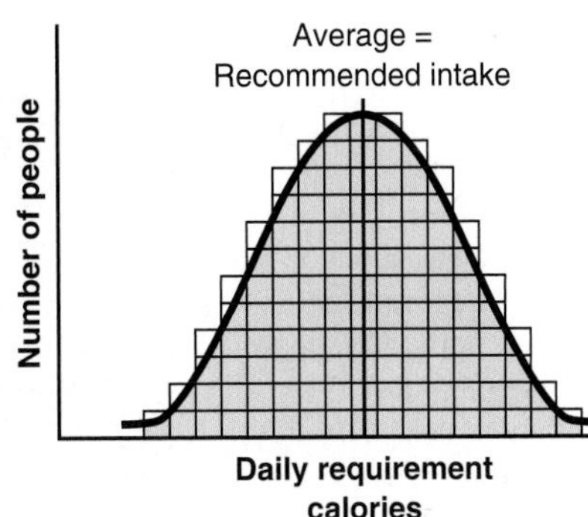

FIGURE 2-4
THE DIFFERENCES BETWEEN THE RECOMMENDATIONS FOR NUTRIENTS AND FOR CALORIES

The recommended nutrient intakes are intended to meet nearly all people's requirements. The recommended calorie intakes, on the other hand, are set at a point at which half the population's requirements will fall below and half will fall above.

TABLE 2-2 GUIDELINES OF THE VOLUNTARY HEALTH GROUPS

American Heart Association

- Balance food intake with physical activity and maintain or reduce weight.
- Eat a variety of foods.
- Limit total fat intake to less than 30% of calories.
- Limit saturated fat intake to to less than 8–10% of calories.
- Limit cholesterol intake to 300 milligrams per day.
- Choose a diet with plenty of vegetables, fruits, and whole-grain products.
- Choose a diet moderate in sugar.
- Use salt and sodium in moderation.
- If you drink, do so in moderation.

American Cancer Society

- Be physically active—achieve and maintain a healthy weight.
- Limit your intake of high fat foods, particularly from animal sources.
- Choose most of the foods you eat from plant sources.
- Limit consumption of alcoholic beverages, if you drink at all.
- Eat foods rich in vitamins A and C.
- Consume salt-cured, salt-pickled, and smoked foods only in moderation.

Source: American Heart Association and American Cancer Society, 1996.

are considered sufficient to address the needs of nearly all people around the world. They differ from the DRI in that they are devised with slightly different priorities and purposes in mind. They assume, for example, that most people's diets contain protein of a lower quality than the protein in the diets of people in the United States. As a result, they recommend higher amounts of that nutrient (Chapter 5 discusses protein quality). The FAO/WHO recommendations also take into consideration the fact that worldwide, people are generally smaller and more physically active than people in the United States.

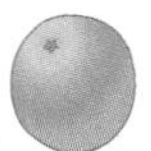

THE CHALLENGE OF DIETARY GUIDELINES

Although the DRI make specific recommendations for protein, vitamin, and mineral consumption, they provide only general guidelines for calorie intake. What's more, they do not address the hazards of nutrient excesses, such as too much fat in the diet. Yet, as Chapter 1 pointed out, health authorities are as concerned today about widespread nutrient excesses among Americans as they used to be about nutrient deficiencies. This is where dietary guidelines come into play. Both the American Heart Association and the American Cancer Society provide suitable recommendations for individuals to follow with ease (listed in Table 2-2). Among the most widely used of the guidelines are the *Dietary Goals for the United States* and the *Dietary Guidelines for Americans* (see Table 2-3). The *Nutrition Recommendations for Canadians* presented in Appendix C are also commonly used.

As you can see, all of these guidelines recommend that people choose diets with low to moderate amounts of fat (particularly saturated fat), sodium, sugar,

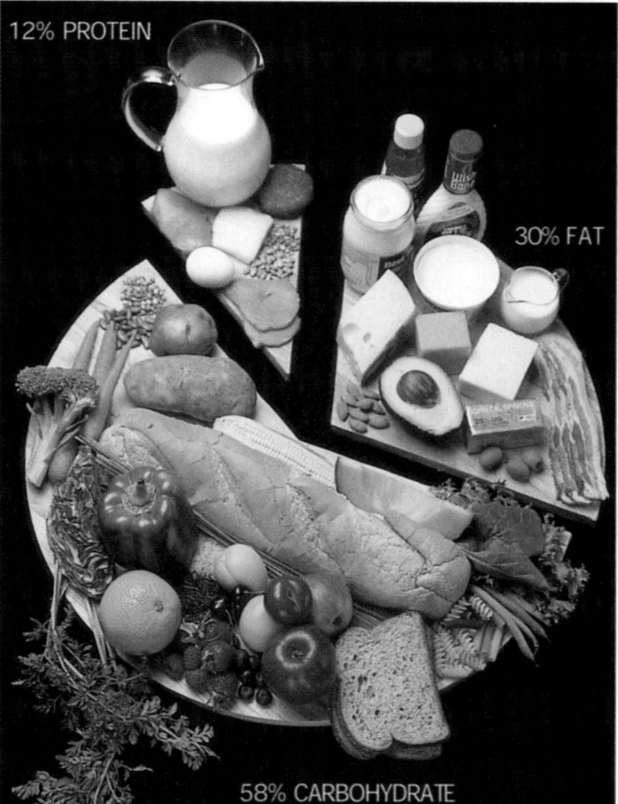

Dietary Goals for the United States
A balanced diet is composed of approximately 12% protein, about 58% carbohydrate (with no more than 10% of this amount from simple sugars), and less than 30% from fat, including the fats from meat, poultry, fish, eggs, milk and milk products, oils, and nuts. Currently, most Americans consume about 17% of their calories from protein, 47% from carbohydrate (with 25% of this amount coming from simple sugars), and 34% of their total calories from fat.

TABLE 2-3 DIETARY GUIDELINES FOR AMERICANS, 2000*

The following ten Dietary Guidelines point the way to good health using three basic messages—the ABCs for health:

Aim for fitness. These two guidelines will help keep you healthy and fit.

- Aim for a healthy weight.
- Be physically active each day.

Build a healthy base. These four guidelines build a base for healthy eating.

- Let the Pyramid guide your food choices.
- Choose a variety of grains daily, especially whole grains.
- Choose a variety of fruits and vegetables daily.
- Keep food safe to eat.

Choose sensibly. These four guidelines help you make sensible choices that promote health.

- Choose a diet that is low in saturated fat and cholesterol and moderate in total fat.
- Choose beverages and foods that limit your intake of sugars.
- Choose and prepare foods with less salt.
- If you drink alcoholic beverages, do so in moderation.

Tips for Following the Dietary Guidelines

Aim for a healthy weight	Choose a lifestyle that combines sensible eating with regular physical activity. Evaluate your current weight by using the Scorecard feature in Chapter 8. If you are overweight, first aim to prevent further weight gain, and then lose weight to improve your health. Choose a healthy assortment of foods that includes vegetables, fruits, whole grains, fat-free milk, and fish, lean meat, poultry, or legumes.
Be physically active each day.	Physical activity and nutrition work together for better health. Engage in 30 minutes or more of moderate physical activity on most, preferably all, days of the week. A moderate physical activity is any activity that requires about as much energy as walking 2 miles in 30 minutes. Choose activities that you enjoy and that you can do regularly. See Chapter 9 for more about the benefits of a physically active lifestyle.
Let the Pyramid guide your food choices.	To get all the nutrients and other beneficial substances you need for health, use the Food Guide Pyramid to help you make healthy food choices. Build your eating pattern on a variety of plant foods, including whole grains, fruits, and vegetables. Also choose moderate amounts of low-fat dairy products and low-fat foods from the Meat, Poultry, Fish, Dry Beans, Eggs, and Nuts group each day. Go easy on foods high in fat or sugars.
Choose a variety of grains daily, especially whole grains.	Eat six or more servings of grain products daily (whole-grain and enriched breads, cereals, pasta, and rice). Include several servings of a variety of whole-grain foods—such as whole wheat, brown rice, oats, and whole corn—every day. Prepare or choose grain products with little added saturated fat and low amounts of added sugars. See Chapter 3 for more about the benefits of whole grains.
Choose a variety of fruits and vegetables daily.	To promote your health, eat a variety of fruits and vegetables—at least two servings of fruits and three servings of vegetables—each day (see Figure 2-6 for serving sizes). Choose fresh, frozen, dried, or canned forms and a variety of colors and kinds. Choose dark-green leafy vegetables, bright orange and red fruits and vegetables, and cooked dried peas and beans often. Enjoy fruits as a naturally sweet end to a meal.
Keep food safe to eat.	Keep things clean: Wash hands and surfaces often. Don't cross-contaminate: Keep raw meat, poultry, eggs, fish, and shellfish away from contact with other foods, surfaces, utensils, or serving plates. Keep hot foods hot (above 140°F). Cook and reheat foods to a safe temperature (see Figure 11-1 in Chapter 11.) Choose pasteurized milk and juices. The risk of contamination is high from rare hamburger, raw eggs, fish (including sushi), clams, and oysters. Keep cold foods cold (below 40°F): refrigerate perishable foods promptly.
Choose a diet that is low in saturated fat and cholesterol and moderate in total fat.	Limit use of animal fats, hard margarines, and partially hydrogenated shortenings. Choose fat-free or low-fat dairy products, legumes, fish, and lean meats and skinless poultry. Trim fats from meats. Use egg yolks and whole eggs in moderation (use egg whites and egg substitutes freely). Limit breaded and deep-fried foods, and foods with creamy sauces. Chapter 4 offers more tips for managing fats in the diet.
Choose beverages and foods that limit your intake of sugars.	Use less sugar, syrup, and honey. Use fewer concentrated sweets such as candy, soft drinks, cakes, cookies, and fruit drinks. Read food labels for sugar content (see Figure 2-9). Don't let soft drinks, or other sweets crowd out other foods you need to maintain health, such as low-fat milk or other good sources of calcium. See Chapter 3 for healthful ways to keep sweetness in the diet.
Choose and prepare foods with less salt.	Use herbs, spices, and fruits to flavor foods. Add little or no salt at the table. Limit salty foods. Read the Nutrition Facts label to compare and help identify foods lower in sodium (see Figure 2-12). Go easy on condiments such as soy sauce, ketchup, mustard, pickles, and olives—they can add a lot of salt to your food. See the Savvy Diner feature in Chapter 7 for more tips for reducing your intake of salt.
If you drink alcoholic beverages, do so in moderation.	Limit alcoholic beverages to one drink per day for women or two drinks per day for men, and take with meals to slow absorption, "One drink" means 12 oz of beer, 5 oz of wine, or 1½ oz of distilled spirits. Pregnant women should not use alcohol. If you drink, do not drive. See the Spotlight feature in Chapter 9.

*These guidelines are intended for healthy children (ages 2 years and older) and adults of any age.

Source: U.S. Department of Agriculture, U.S. Department of Health and Human Services, *Dietary Guidelines for Americans*, 5th ed., 2000.

cholesterol, and alcohol while maintaining a healthy weight and eating plenty of fruits, vegetables, and grains. The goal of the recommendations is to help people decrease their risk of some forms of cancer, heart disease, obesity, diabetes, high blood pressure, stroke, osteoporosis, and liver disease—the so-called **lifestyle diseases.** Following such recommendations certainly makes sense, given the considerable potential health benefits they confer.

LIFESTYLE DISEASES conditions that may be aggravated by modern lifestyles that include too little exercise, poor diets, and excessive drinking and smoking. Lifestyle diseases are also referred to as diseases of affluence.

TOOLS USED IN DIET PLANNING

Although the DRI and the various dietary guidelines provide good frameworks for healthful eating, planning daily menus requires use of other, more specific tools. Dietitians and other nutrition experts often rely on a number of tools that you, too, can use to assess and plan your own diet.

Food Group Plans

One of the most helpful, easy-to-use diet planning tools is the **food group plan,** which separates foods into specific groups and then specifies the number of **servings** from each group to eat each day. One example of such a plan is the Four Food Group Plan, devised in the 1950s and taught to consumers for nearly four decades. Several years ago, however, scientists updated this plan, taking into consideration new nutrition knowledge gained over the years. The revised version, shown in Figure 2-5, is called the Food Guide Pyramid. It contains five food groups rather than four; the tip is not considered to be a major food group because the foods found there provide extra calories and little else in the way of nutrients.

FOOD GROUP PLAN a diet-planning tool, such as the Food Guide Pyramid, that groups foods according to similar origin and nutrient content and then specifies the number of foods from each group that a person should eat.

SERVING the amount of food a person might eat, similar to a helping.

The pyramid was designed to help consumers choose foods that supply a good balance of nutrients, but not too much total fat, saturated fat, sugar, sodium, or alcohol, in keeping with the U.S. government's *Dietary Guidelines for Americans*. The placement of the five food groups on the pyramid emphasizes their role in the diet. The grains that form the base should serve as the foundation of a healthy diet because breads, cereals, rice, and pasta are generally high in carbohydrate

(text continues on page 41)

FIGURE 2-5
EATING FROM THE BOTTOM UP: THE FOOD GUIDE PYRAMID

The pyramid shows the proportions of foods that should make up a healthful diet. Those found in the bottom half—grains, fruits, and vegetables—should make up the bulk of the diet. Those in the top half, including fats and sweets, should be eaten in moderate amounts.
Source: U.S. Department of Agriculture.

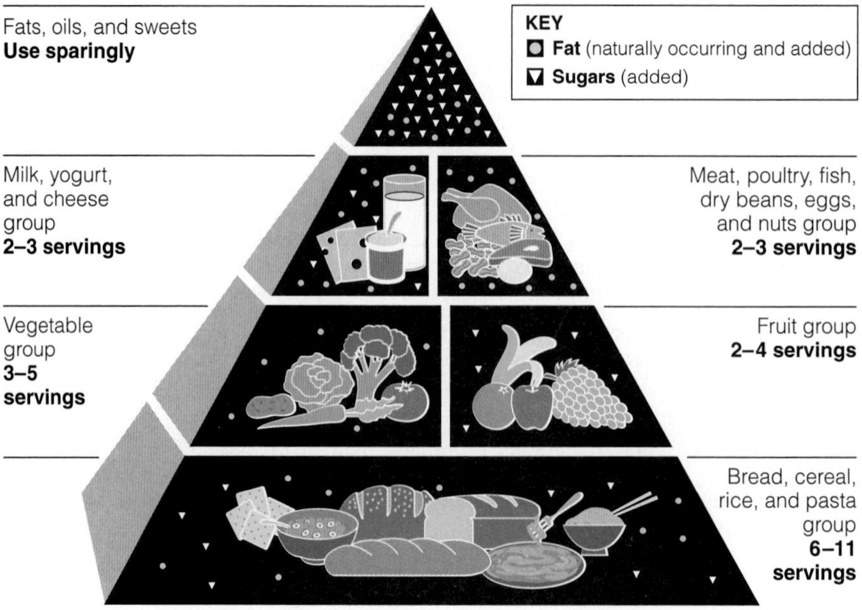

NUTRITION ACTION

Grazer's Guide to Smart Snacking

Government surveys show that just about everyone reaches for a little something between meals. In fact, Americans spend some $16 billion annually on snack foods such as popcorn and potato chips.[4] But while **grazing** is in, many people feel a twinge of guilt now and then about between-meal munching. Perhaps the parental warnings of childhood that snacks can spoil a meal linger in the back of many minds. Nevertheless, nutritious nibbling can make it easier for many people to eat healthfully.

GRAZING eating small amounts of food at intervals throughout the day rather than—or in addition to—eating regular meals.

Snacks can supply essential vitamins, minerals, and calories to the diets of little children, whose small stomachs and appetites often cannot handle larger meals. As for teenagers, who typically don't seem to have the time (or the inclination) to eat regularly as they go to school, baseball practice, music lessons, or engage in other activities, snacks account for upwards of a third of the calories they eat and as much as 20 percent of their vitamin and mineral consumption.[5] Snacks also contribute to the nutritional needs of adults, who may find fitting meals into a busy schedule difficult. Even senior citizens, whose lifestyles tend to be less hectic, can benefit from grazing. That's because lack of activity, certain medications, and isolation can blunt a formerly hearty appetite, making frequent, small meals more desirable than large breakfasts, lunches, and dinners.

At any age, the key to healthful snacking is to choose low-fat, high-fiber, nutrient-rich foods such as fruits, low-fat cheese and yogurt, whole-grain crackers, and plain popcorn. Cakes, cookies, candy, and potato chips are harder to fit into a healthful eating plan because they add fat and calories to the diet and little else in the way of nutrients.

Some snacks that appear nutritious may be deceiving. For example, fruit drinks, mixes, and punches are loaded with sugar and are more similar to soft drinks than to fruit juice. Fruit rolls and bars are nutritionally similar to jams and jelly, not to fresh fruit. Sugar is added, and most of the nutrients in the fruit are lost when it is processed with heat. Granola bars can also be deceiving because, like candy bars, many are loaded with sugar and fat. Even some varieties of microwave popcorn don't rate well as snacks because they contain lots of oil and salt. (You can easily make your own popcorn in the microwave oven without adding an excess of extras.) When you feel like snacking, try some of the alternatives offered in Table 2-4. Use Table 2-5 as a guide to see how some of your favorite snacks rate in terms of calories and fat. In addition, consider the following tips next time you're in the mood to grab a snack.

- Stock your refrigerator and kitchen cupboards with nutritious foods such as fruit juices, low-fat yogurt, fresh fruits and vegetables, plain popcorn, pretzels, whole-grain crackers, and low-fat cheeses so that they are close at hand. Nibblers often reach for a snack just to have something to munch on rather than because of a desire for the food itself. If nutritious choices are easy to get to, chances are that's what you'll eat.
- Carry fresh fruit and crackers and cheese, or even half a sandwich, in your backpack or briefcase so you won't have to resort to buying candy from a vending machine when you get the urge to munch.

TABLE 2-4 SMART SNACKING

Instead of	Try
Fruit drinks	Fruit juice
Soft drinks	Fruit juice concentrate and sparkling water
Milk shakes	Shakes made with fat-free milk and fresh fruit
Potato chips	Pretzels (try low salt)
Chips and sour cream dip	Vegetables and buttermilk dip
Granola bars	Homemade oatmeal cookies (reduce the fat and sugar)
Frosted cake	Angel food cake and fruit
Pie	Fruit cobblers
Cookies	Vanilla wafers, gingersnaps, graham crackers

TABLE 2-5 WHAT'S IN A MUNCHER'S MENU?

Food	Grams of Fat	Calories
Corn chips (½ c)	4	70
Plain popcorn (1 c)	trace	30
Buttered and salted popcorn (1 c)	2	50
Graham cracker squares (2)	1	60
Saltine crackers (4)	1	50
Bread sticks (2)	1	75
Bagel (1 3½" round)	2	200
Bran muffin (1 2½" round)	6	125
Pretzel sticks (10 thin)	trace	10
Carrot and celery sticks (2 each)	trace	5
Tomato juice (6 oz)	trace	30
Potato chips (10)	7	105
French fries (10)	8	160
Apple (small)	trace	60
Banana	1	105
Raisins, 1 small box (about 1½ tbsp)	trace	40
Processed American cheese (1 oz)	9	105
Fat-free milk (1 c)	1	90
Low-fat yogurt with fruit (8 oz)	2	230
Roasted peanuts (¼ c)	18	210
Peanut butter (2 tbsp)	16	190
Frozen yogurt (½ c)	2	105
Ice cream (½ c)	7	135
Chocolate chip cookies (2 2⅓" round)	6	90
Chocolate or vanilla sandwich-type cookies (2)	4	100
Fig bars (2)	2	105
Frosted cream-filled cupcake	4	160
Cake-type doughnut	12	210
Chocolate candy bar (1½ oz)	14	220
Regular cola soft drink (12 oz)	0	160

Source: Adapted from Muncher's Guide in *Making Bag Lunches, Snacks, & Desserts Using the Dietary Guidelines* (U.S. Department of Agriculture Human Nutrition Information Service, Home and Garden Bulletin No. 232-9): pp. 7–8.

- Create your own snacks. Mix together one cup each pretzels, peanuts, raisins, and sunflower seeds to take with you on your next bicycle trip or hike. As a substitute for cream cheese, blend ½ cup drained low-fat yogurt with ½ cup low-fat cottage cheese. Add a bit of chopped pineapple, strawberries, or other fruit and spread this mixture on crackers, bagels, English muffins, or rice cakes.
- Make new versions of old favorites, such as *Frozen Bananas,*[6] *Chili Popcorn, and Mexican Snack Pizzas.*[7]
- Snack with a friend. If you're craving a candy bar or chips and nothing else will do, try splitting a bar or a bag with a friend. That way you'll satisfy your craving without going too far overboard on calories, fat, or salt.
- Try to brush your teeth—or at least rinse your mouth thoroughly—after snacking to prevent tooth decay. (See the Nutrition Action feature in Chapter 3 for a detailed explanation of the role of diet in dental health.)

FROZEN BANANAS

Mix 1 tablespoon peanut butter with ¼ cup evaporated skim milk. Roll a banana, cut in half, in the peanut butter mixture. Then roll the coated banana halves in bran cereal. Place in the freezer until frozen. Makes two banana halves, each of which supplies about 165 calories, 4 grams of fat, and 149 milligrams of sodium.

CHILI POPCORN

Mix 1 quart popped popcorn with 1 tablespoon melted margarine. In a separate bowl, mix 1¼ teaspoons chili powder, ¼ teaspoon cumin, and a dash of garlic powder. Sprinkle seasonings over popcorn and mix well. Makes about four 1-cup servings, each of which contains approximately 50 calories, 3 grams of fat, and 42 milligrams of sodium.

MEXICAN SNACK PIZZAS

Split two English muffins and toast lightly. Mix ¼ cup tomato paste, ¼ cup canned, drained, chopped kidney beans, 1 tablespoon each chopped onion and chopped green pepper, and ½ teaspoon oregano. Spread mixture on muffin halves. Top with ¼ cup shredded part-skim mozzarella cheese and broil until cheese is bubbly (about two minutes). Garnish with ¼ cup shredded lettuce. Makes 4 servings, each of which contains 95 calories, 2 grams of fat, and 300 milligrams of sodium.

The Food Guide Pyramid calls for eating a variety of foods to get the nutrients you need and at the same time the right amount of calories to maintain a healthy weight. Remember to balance the energy consumed with the energy expended in play.

and low in fat. The grains are followed by fruits and vegetables, which supply the vitamins, minerals, and fiber many people's diets lack. The next level suggests eating smaller amounts of dairy products as well as meat, poultry, fish, beans, eggs, and nuts. While foods from these groups provide protein, calcium, iron, zinc, and other nutrients, they often contain large amounts of fat as well as saturated fat. Thus, choosing wisely from these groups goes a long way in limiting the fat content of your diet.

The pyramid isn't the only food group plan. Canada has one of its own—Canada's Food Guide (shown in Appendix C).

Not considered one of the food groups, the tip of the pyramid consists of fats, oils, and sweets—foods such as butter, margarine, salad dressing, oils, candy, soda pop, and other similar items—which typically supply lots of fat and/or calories and few nutrients. As their placement in the tip suggests, these items should be added to the diet sparingly.

Note that alcohol is not included in any portion of the pyramid, but like the foods in the tip, it provides calories and no nutrients to speak of. The *Dietary Guidelines* recommend that consumers have no more than one or two alcoholic drinks a day. A standard drink is a 12-ounce can or bottle of beer, a 5-ounce glass of wine, or a 1½-ounce shot of liquor.

To get an idea of how your current diet compares with the Food Guide Pyramid, try the Rate Your Plate Scorecard on page 53.

Using the pyramid to assess and plan your own diet requires an understanding of what amount of food counts as one serving from the various food groups. For instance, one slice of bread, one half a bagel, and a half cup of pasta each counts as a serving from the bread, cereal, rice, and pasta group. When it comes to vegetables, a half cup of raw or cooked vegetables or one cup of leafy raw vegetables chalks up one serving. Figure 2-6 shows the portions that count as a serving from each of the various food groups, and Figure 2-7 shows where some hard-to-place foods fit on the pyramid. When considering your own diet, you may want to look at some measuring cups to get a good idea of just how much a cup really is. You may be surprised to learn how much you're really eating. In fact, most people tend to over- or underestimate serving sizes.

Note that certain foods within each group are relatively low in fat and added sugars. Since these foods are generally the most nutrient-dense choices within the group, it makes sense to select them the majority of the time. For example, consider the meat, poultry, fish, dry beans, eggs, and nuts group. Since lean cuts of meat, skinless poultry, and fish rank lower in fat than ground beef and chicken with skin, the leaner items should be chosen more often than the high-fat selections. Dry beans, which contain only a trace of fat, are good choices to add to the diet even more often. When it comes to the milk group, low-fat and fat-free dairy products make the best choices.

FIGURE 2-6
WHAT COUNTS AS A SERVING?

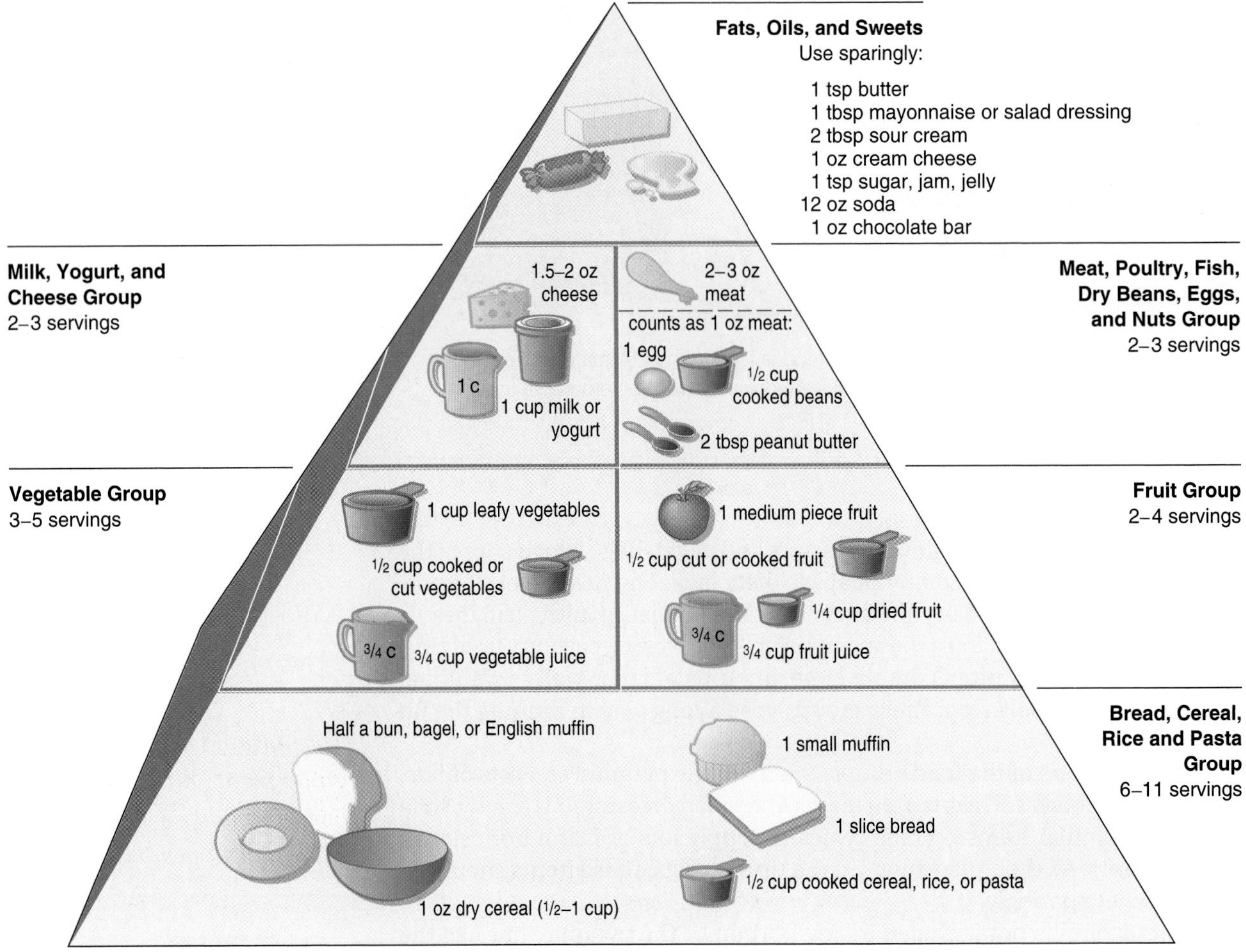

Source: Adapted from the *Pyramid Packet* © 1993, Penn State Nutrition Center, 417 East Calder Way, University Park, PA 16801; 814-865-6323.

Food Labels

In 1990, Congress passed one of the most important pieces of legislation of the twentieth century. Known as the Nutrition Labeling and Education Act, the law called for sweeping changes in the way foods are labeled in the United States. Officials at the U.S. Food and Drug Administration (FDA), the nation's food industry watchdog, spent several years devising regulations aimed at revamping the food label. By May 1994, food manufacturers had to relabel some 300,000 packaged foods sold in American supermarkets.[8]

For consumers, the law ensures that food companies provide the kind of nutrition information that best allows people to select foods that fit into a healthful eating plan. Considering the great variety of packaged foods available, using the food label to understand the nutrients a food supplies or lacks is essential (see Figure 2-8). The label is one of the most important tools you can use to eat healthfully.

By law, all labels must contain the following:

- The name of the food, also known as the statement of identity.
- The name of the manufacturer, packer, or distributor, as well as the firm's city, state, and zip code.

FIGURE 2-7
HARD-TO-PLACE FOODS

The serving sizes that follow the various foods are the equivalent of one serving from the group in which the food has been placed. Certain items in each group, particularly the high-fat choices, may be difficult to fit into a healthful eating plan on a regular basis.

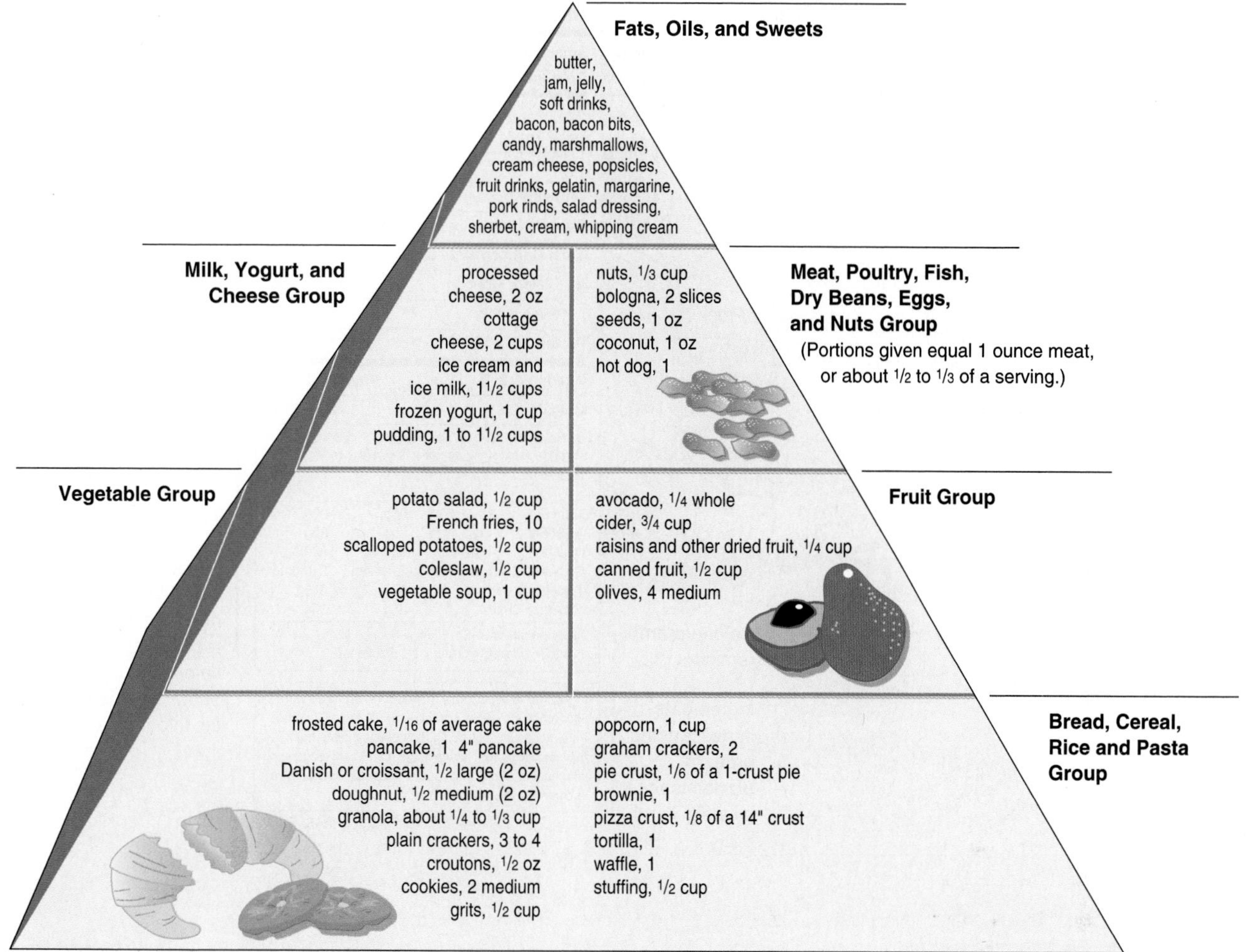

Source: Adapted from the *Pyramid Packet* © 1993, Penn State Nutrition Center, 417 East Calder Way, University Park, PA 16801.

- The net quantity, which tells you how much food is in the container so that you can compare prices. Net quantity has to be stated in both inch or pound units and metric units.
- The **ingredients list,** with items listed in descending order by weight. The first ingredient listed makes up the largest proportion of all the ingredients in the food, the second, the second largest amount, and so on. If the first ingredient in the list is sugar, for example, you know the food contains more sugar than anything else. The list is especially useful in helping people identify ingredients they avoid for health, religious, or other reasons (see Figure 2-9).

INGREDIENTS LIST a listing of the ingredients in a food, with items listed in descending order of predominance by weight. All food labels are required to bear an ingredients list.

- The **Nutrition Facts panel,** unless the package is small—no larger than 12 square inches of surface area, or about the size of a small candy bar or a roll of breath mints; small packages must carry a telephone number or address consumers can contact to obtain nutrition information (see Figure 2-10).

NUTRITION FACTS PANEL a detailed breakdown of the nutritional content of a serving of a food that must appear on virtually all packaged foods sold in the United States.

Nutrition Facts Panel. The Nutrition Facts panel must indicate the amount of certain mandatory nutrients that one serving of the food contains. When you

FIGURE 2-8
ANATOMY OF A FOOD LABEL

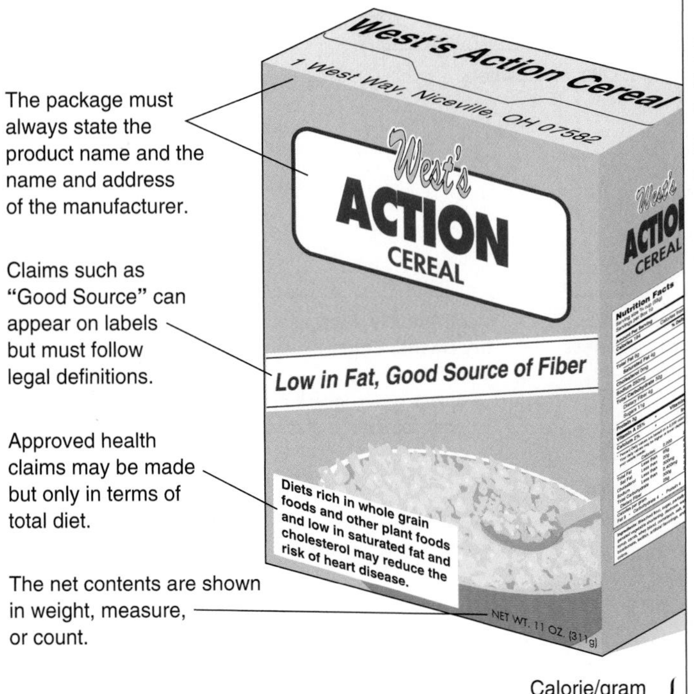

The package must always state the product name and the name and address of the manufacturer.
Claims such as "Good Source" can appear on labels but must follow legal definitions.
Approved health claims may be made but only in terms of total diet.
The net contents are shown in weight, measure, or count.
West's Action Cereal
1 West Way, Niceville, OH 07582
West's
ACTION
CEREAL
Low in Fat, Good Source of Fiber
Diets rich in whole grain foods and other plant foods and low in saturated fat and cholesterol may reduce the risk of heart disease.
NET WT. 11 OZ. (311g)

Nutrition Facts
Serving size 3/4 cup (55g)
Servings per Box 5

Serving size, number of servings per container, and calorie information

Amount Per Serving	
Calories 167	Calories from Fat 27
	% Daily Value*
Total Fat 3g	**5%**
Saturated Fat 1g	**5%**
Cholesterol 0mg	**0%**
Sodium 250mg	**10%**
Total Carbohydrate 32g	**11%**
Dietary Fiber 4g	**16%**
Sugars 11g	
Protein 3g	
Vitamin A 25% •	**Vitamin C** 25%
Calcium 2% •	**Iron** 25%

Quantities of nutrients per serving and percentage of Daily Value for nutrients based on a 2,000 calorie energy intake

*Percent Daily values are based on a 2,000 calorie diet. Your daily values may be higher or lower depending on your calorie needs:

	Calories:	2,000	2,500
Total Fat	Less than	65g	80g
Sat Fat	Less than	20g	25g
Cholesterol	Less than	300mg	300mg
Sodium	Less than	2,400mg	2,400mg
Total Carbohydrate		300g	375g
Dietary Fiber		25g	30g

Calories per gram
Fat 9 • Carbohydrate 4 • Protein 4

Calorie/gram reminder

Reference values

This allows comparison of some values for nutrients in a serving of the food with the needs of a person requiring 2,000 or 2,500 calories per day to show how the product fits into the daily diet.

Ingredients: Whole oats, milled corn, enriched wheat flour (contains niacin, reduced iron, thiamin mononitrate, riboflavin, folic acid), dextrose, maltose, high fructose corn syrup, brown sugar, partially hydrogenated cottonseed oil, coconut oil, walnuts, vitamin C (sodium ascorbate), vitamin A (palmitate), iron.

Ingredients in descending order of predominance by weight

Learn to read food labels to help you achieve a healthful diet.

consider the nutrition information, keep serving sizes in mind. Based on the amount of food most people eat at one time, the FDA has set forth a list of serving sizes for more than 100 food categories. Manufacturers must use these recommended serving sizes on food labels. For instance, the serving size for product X must always be 8 ounces. This ensures that consumers can easily compare one brand of the product to another without going through the calculations that would be necessary if serving sizes varied—say, if another brand's label had a 6-ounce serving size. The nutrient information that must appear on the Nutrition Facts panel is calories, calories from fat, total fat, saturated fat, cholesterol, sodium, total carbohydrate, dietary fiber, sugars, protein, vitamin A, vitamin C, calcium, and iron (in that order).

These nutrients were chosen to appear on the Nutrition Facts panel because they address today's health concerns. Today, many people need to be concerned about getting an excess of certain nutrients, such as fat, rather than too few vitamins and minerals (see Figure 2-11).

The ranking of the required nutrients was spelled out by the FDA to ensure that the label reflects the government's dietary priorities for the public. For example, fat falls near the top of the list because most consumers need to pay closer attention to the amount of fat in their diet. Most people eat too much fat, which raises the risk of developing heart disease, obesity, and cancer—chronic problems suffered by millions of Americans. Protein, on the other hand, appears near the bottom of the label because the amount of protein most people eat does not rate as a major health concern.

FIGURE 2-9
USING THE INGREDIENTS LIST ON FOOD LABELS

Oats 'N' More

Nutrition Facts

Serving Size 1 cup (30g)
Servings Per Container About 14

Amount Per Serving	Cereal	with ½ cup fat-free milk
Calories	110	150
Calories from Fat	15	20
	% Daily Value**	
Total Fat 2g*	**3%**	**3%**
Saturated Fat 0g	**0%**	**3%**
Polyunsaturated Fat 0.5g		
Monounsaturated Fat 0.5g		
Cholesterol 0mg	**0%**	**1%**
Sodium 280mg	**12%**	**15%**
Potassium 95mg	**3%**	**9%**
Total Carbohydrate 22g	**7%**	**9%**
Dietary Fiber 3g	**11%**	**11%**
Soluble Fiber 1g		
Sugars 1g		
Other Carbohydrate 18g		
Protein 3g		
Vitamin A	10%	15%
Vitamin C	10%	10%
Calcium	4%	20%
Iron	45%	45%
Vitamin D	10%	25%
Thiamin	25%	30%
Riboflavin	25%	35%
Niacin	25%	25%
Vitamin B_6	25%	25%
Folic Acid	50%	50%
Vitamin B_{12}	25%	35%
Phosphorus	10%	25%
Magnesium	8%	10%
Zinc	25%	30%
Copper	2%	2%

*Amount in Cereal. A serving of cereal plus fat-free milk provides 2g total fat (0.5g saturated fat, 1g monounsaturated fat), less than 5mg cholesterol, 350mg sodium, 300mg potassium, 28g total carbohydrate (7g sugars) and 7g protein.

**Percent Daily Values are based on a 2,000 calorie diet. Your daily values may be higher or lower depending on your calorie needs:

	Calories:	2,000	2,500
Total Fat	Less than	65g	80g
Sat Fat	Less than	20g	25g
Cholesterol	Less than	300mg	300mg
Sodium	Less than	2,400mg	2,400mg
Potassium		3,500mg	3,500mg
Total Carbohydrate		300g	375g
Dietary Fiber		25g	30g

Ingredients: whole grain oats, (includes the oat bran), modified corn starch, wheat starch, sugar, salt, oat fiber, trisodium phosphate, calcium carbonate, vitamin E (mixed tocopherols) added to preserve freshness. **Vitamins and Minerals:** iron and zinc (mineral nutrients), vitamin C (sodium ascorbate), vitamin B_6 (pyridoxine hydrochloride), riboflavin, thiamin mononitrate, niacinamide. folic acid, vitamin A (palmitate), vitamin B_{12}, vitamin D.

Morning Krisps

Nutrition Facts

Serving Size 3/4 cup (27g/1.0 oz.)
Servings Per Container About 18

Amount Per Serving	Cereal	with ½ cup fat-free milk
Calories	100	140
Calories from Fat	5	5
	% Daily Value**	
Total Fat 0.5g*	**1%**	**1%**
Saturated Fat 0g	**0%**	**0%**
Cholesterol 0mg	**0%**	**0%**
Sodium 50mg	**2%**	**5%**
Potassium 40mg	**1%**	**7%**
Total Carbohydrate 24g	**8%**	**10%**
Dietary Fiber 1g	**4%**	**4%**
Sugars 15g		
Other Carbohydrate 8g		
Protein 2g		
Vitamin A	15%	20%
Vitamin C	25%	25%
Calcium	0%	15%
Iron	10%	10%
Vitamin D	10%	25%
Thiamin	25%	30%
Riboflavin	25%	35%
Niacin	25%	25%
Vitamin B6	25%	25%
Folic Acid	25%	25%
Vitamin B_{12}	25%	35%
Phosphorus	4%	15%
Magnesium	2%	6%
Zinc	2%	6%
Copper	2%	4%

*Amount in Cereal. One half cup of fat-free milk contributes an additional 40 calories. 65mg sodium, 6g total carbohydrate (6g sugars), and 4g protein.

**Percent Daily Values are based on a 2,000 calorie diet. Your daily values may be higher or lower depending on your calorie needs:

	Calories:	2,000	2,500
Total Fat	Less than	65g	80g
Sat Fat	Less than	20g	25g
Cholesterol	Less than	300mg	300mg
Sodium	Less than	2,400mg	2,400mg
Potassium		3,500mg	3,500mg
Total Carbohydrate		300g	375g
Dietary Fiber		25g	30g

Calories per gram: Fat 9 • Carbohydrate 4 • Protein 4

Ingredients: Sugar, wheat, corn syrup, honey, hydrogenated soybean oil, salt, caramel color, soy lecithin, **Vitamins and Iron:** sodium ascorbate (vitamin C), ferric phosphate (iron), niacinamide, pyridoxine hydrochloride (vitamin B_6), riboflavin, vitamin A palmitate, thiamin hydrochloride, BHT (preservative), folic acid, vitamin B_{12}, and vitamin D.

The cereals shown contain sugars. To find out how much, start by checking the carbohydrate section of the Nutrition Facts panel for sugars. This number refers to both the sugars added by the manufacturer and the simple sugars naturally present in the food. Learn to read the ingredients list. Labels list ingredients in order of amount by weight with the greatest amount of an ingredient present in the food listed first. Check labels for sugar terms, in addition to sugar, such as:

- brown sugar
- corn syrup
- dextrose
- fructose
- glucose
- high fructose corn syrup
- honey
- invert sugar
- levulose
- mannitol
- molasses
- sorbitol
- sucrose

Note that only vitamins A and C, iron, and calcium appear on the nutrition panel. Those are the only vitamins and minerals (except for sodium) required to be on food labels, unless a manufacturer makes a nutrition claim about another one. For instance, if a manufacturer says that a cereal is **fortified** with niacin, the amount of niacin in the product must appear on the label.[9] (See Figure 2-12).

FORTIFIED FOOD a food to which manufacturers have added 10 percent or more of the Daily Value for a particular nutrient.

FIGURE 2-10
TYPES OF FOOD LABELS

A container with less than 40 square inches of surface area for nutrition labeling can present fewer facts in the format shown on the can of tuna. A simplified format, shown on the can of cola, is allowed on foods that do not contain significant amounts of nutrients. Packages with less than 12 square inches of surface area, such as small candy bars, need not carry nutrition information, though they must provide a telephone number or address for contacting the company to obtain nutrition information.

Nutrition Facts
Serving Size 1/3 cup (56g)
Servings about 3
Calories 80
Fat Cal. 9
*Percent Daily Values (DV) are based on a 2,000 calorie diet.

Amount/serving	% DV*	Amount/serving	% DV*
Total Fat 1g	2%	**Total Carb.** 0g	0%
Sat. Fat 0g	0%	Fiber 0g	0%
Cholest. 10mg	3%	Sugars 0g	
Sodium 200mg	8%	**Protein** 17g	

Vitamin A 0% • Vitamin C 0% • Calcium 0% • Iron 6%

Nutrition Facts
Serving Size 1 can (360 ml)

Amount Per Serving	
Calories 140	
	% Daily Value*
Total Fat 0g	0%
Sodium 20mg	1%
Total Carbohydrate 36g	12%
Sugars 36g	
Protein 0g	0%

*Percent Daily Values are based on a 2,000 calorie diet.

DAILY VALUES the amount of fat, sodium, fiber, and other nutrients health experts say should make up a healthful diet. The % Daily Values that appear on food labels tell you the percentage of a nutrient that a serving of the food contributes to a healthful diet.

Daily Values. The **Daily Values** for fats, sodium, carbohydrates, and fiber are calculated according to what experts deem a healthful diet for adults should consist of:

- No more than 30 percent of calories as fat
- No more than 10 percent of calories as saturated fat
- Fewer than 300 milligrams of cholesterol
- Fewer than 2,400 milligrams of sodium
- At least 60 percent of calories as carbohydrate
- 25–35 grams of fiber

For instance, since no more than 30 percent of total calories should come from fat, the % Daily Value tells you the percentage of fat that a serving of the food contributes to a 2,000-calorie eater's fat "allowance." A 2,000-calorie diet was chosen as a good point of reference because that's about the amount eaten by most moderately active women, teenage girls, and sedentary men. Of course, more calories may be appropriate for many men, teenage boys, and active women. This is why the nutrition panel also shows Daily Values for a 2,500-calorie diet (refer to the bottom of the Nutrition Facts panel in Figure 2-8).

To understand how the Daily Values for fats, sodium, carbohydrates, and fiber are calculated, let's go through an example. First, look for the grams of total fat and the % Daily Value for the cereal shown in Figure 2-8. The label shows that a serving supplies 3 grams of fat, with a Daily Value of 5 percent. This means that a serving of the cereal contributes 5 percent of the total fat that a person eating 2,000 calories a day should consume.

Now look at the bottom of the Nutrition Facts panel, which indicates that someone eating 2,000 calories a day should take in no more than 65 grams of fat.

Courtesy: Modern Maturity Magazine.

"Henry likes nothing more than to curl up with a good label."

FIGURE 2-11
CHECKING OUT THE FOOD LABEL FOR FAT INFORMATION

% Daily Value compares the total fat, saturated fat, and cholesterol in a serving of the food with a person's daily need. The % Daily Value tells you if the food has large or small amounts of total fat, saturated fat, and cholesterol. Note at the bottom of the Nutrition Facts panel that the Daily Value for total fat is *less than* 65 grams for a person requiring 2,000 calories a day. The reference value of 65 grams for total fat is derived from the recommendation for health that states that 30% *or less* of our calories should come from fat. (For a 2,000 calorie diet, this amount would equal 65 grams—the Daily Value for total fat.) A serving of the French bread pizza shown here provides 6% of the Daily Value for total fat, 10% of the Daily Value for saturated fat, and 3% of the Daily Value for cholesterol; the pepperoni pizza provides 42% of the Daily Value for total fat, 50% of the Daily Value for saturated fat, and 8% of the Daily Value for cholesterol. The overall goal should be to select foods that together do *not* exceed 100% of the Daily Values for fat, saturated fat, and cholesterol over the course of your day for these nutrients.

Papa Solo's French Bread Pizza
French Bread Pizza with Tomato Sauce and Mozzarella Cheese

Nutrition Facts
Serving Size 1 pizza (158 g)
Servings per Container 1

Amount Per Serving	
Calories 310	Calories from Fat 36
	% Daily Value*
Total Fat 4g	**6%**
Saturated Fat 2g	**10%**
Polyunsaturated Fat 0.5g	
Monounsaturated Fat 1g	
Cholesterol 10mg	**3%**
Sodium 470mg	**20%**
Total Carbohydrate 49g	**16%**
Dietary Fiber 6g	**24%**
Sugars 3g	
Protein 20g	
Vitamin A 4% •	**Vitamin C** 0%
Calcium 30% •	**Iron** 15%

* Percent Daily Values are based on a 2,000 calorie diet. Your Daily Values may be higher or lower depending on your calorie needs:

	Calories:	2,000	2,500
Total Fat	Less than	65g	80g
Sat Fat	Less than	20g	25g
Cholesterol	Less than	300mg	300mg
Sodium	Less than	2,400mg	2,400mg
Total Carbohydrate		300g	375g
Dietary Fiber		25g	30g

Calories per gram
Fat 9 • Carbohydrate 4 • Protein 4

Ingredients: French bread (enriched unbleached wheat flour), water, wheat gluten, oat fiber, sugar, soybean oil, nonfat dry milk, mozzarella cheese, tomatoes. Contains 2% or less of each of the following: pizza spice mix (spices, salt, onion and garlic), sugar, coloring (dextrose, cabbage extract, annatto).

MAMA MIA's
Pepperoni Pizza-for-One

Nutrition Facts
Serving Size 1 pizza (191g)
Servings per Container 1

Amount Per Serving	
Calories 520	Calories from Fat 243
	% Daily Value*
Total Fat 27g	**42%**
Saturated Fat 10g	**50%**
Cholesterol 25mg	**8%**
Sodium 1,280mg	**53%**
Total Carbohydrate 53g	**18%**
Dietary Fiber 4g	**16%**
Sugars 6g	
Protein 19g	
Vitamin A 45% •	**Vitamin C** 0%
Calcium 30% •	**Iron** 10%

* Percent Daily Values are based on a 2,000 calorie diet. Your Daily Values may be higher or lower depending on your calorie needs:

	Calories:	2,000	2,500
Total Fat	Less than	65g	80g
Sat Fat	Less than	20g	25g
Cholesterol	Less than	300mg	300mg
Sodium	Less than	2,400mg	2,400mg
Total Carbohydrate		300g	375g
Dietary Fiber		25g	30g

Calories per gram
Fat 9 • Carbohydrate 4 • Protein 4

Ingredients: CRUST: Wheat flour with malted barley flour, water, partially hydrogenated vegetable oil (soybean and/or cottonseed oil) with soy lecithin, soybean oil, yeast, high fructose corn syrup, salt; TOPPING: Part-skim mozzarella cheese substitute, pepperoni (pork and beef, salt, water, dextrose, spices, sodium nitrite), part-skim mozzarella cheese; SAUCE: Tomato puree, water, green peppers, salt, lactose and flavoring, spices, corn oil, xanthum gum.

Calories from fat are now shown on the label to help consumers meet dietary guidelines that recommend people get no more than 30% of their calories from fat.

Total fat refers to all the fat in the food: saturated, monounsaturated, polyunsaturated, and trans fat. Total fat and saturated fat information is required on the label because high intakes of both are linked to high blood cholesterol levels, which in turn are linked to increased risk of heart disease.* Listing the amount of monounsaturated and polyunsaturated fats in the food is voluntary.

Ingredients list—look for terms such as:

- coconut oil
- diglycerides
- hydrogenated oils
- lard
- monoglycerides
- oil
- palmitate
- palm oil
- stearate
- triglycerides
- vegetable shortening

*In December 1999, the FDA announced plans to require manufacturers to soon list the trans fat content in the Nutrition Facts panel on all food products.

Divide 3—the number of fat grams in a serving of the cereal—by 65. Multiply that number by 100 to obtain a percentage. The answer is 5, that is, 5 percent of the total fat.

You can use the % Daily Values to get a good idea of how various foods fit into a healthful diet, regardless of the number of calories you eat. Consider a student who eats only 1,800 calories a day. If she snacks on two servings of potato chips with a Daily Value of 15 percent fat per serving, she's already taken in 30 percent of the fat someone eating 2,000 calories should have in an entire day. Because she eats less than 2,000 calories, the potato chips contribute slightly more than 30 percent. Thus, the 30 percent Daily Value shows that potato chips chalk up a lot of fat for a snack. If she checks the % Daily Value for fat on a label of pretzels, on the other hand, she might see that a serving supplies only about 3 percent. If she eats two servings, she's only up to less than 10 percent of the fat she can have, leaving her much less likely to go overboard on fat throughout the rest of the day. In other words, the % Daily Value column can give you a good idea of how different foods fit into the overall diet.

FIGURE 2-12
CHECKING OUT THE FOOD LABEL FOR VITAMINS AND MINERALS

Nutrition Facts
Serving Size 1 cup (55g)
Servings per Container About 8

Amount Per Serving	Cereal	with 1/2 cup fat-free milk
Calories	220	260
Calories from Fat	35	40
		% Daily Value**
Total Fat 4g*	**6%**	**7%**
Saturated Fat 0.5g	**3%**	**3%**
Cholesterol 0mg	**0%**	**1%**
Sodium 260mg	**11%**	**14%**
Total Carbohydrate 43g	**14%**	**16%**
Dietary Fiber 4g	**16%**	**16%**
Sugars 14g		
Other Carbohydrate 25g		
Protein 5g		
Vitamin A	0%	4%
Vitamin C	15%	15%
Calcium	10%	25%
Iron	25%	25%
Vitamin D	0%	10%
Thiamin	25%	30%
Riboflavin	25%	35%
Niacin	25%	25%
Folate	25%	25%
Phosphorus	15%	25%
Magnesium	10%	15%
Zinc	6%	10%
Copper	8%	8%

* Amount in Cereal. A serving of cereal plus fat-free milk provides 4.5g fat (0.5g saturated) less than 5mg cholesterol, 330mg sodium, 400mg potassium, 49g carbohydrate (20g sugars), and 9g protein.
**Percent Daily Values are based on a 2,000 calorie diet. Your daily values may be higher or lower depending on your calorie needs.

	Calories:	2,000	2,500
Total Fat	Less than	65g	80g
Sat Fat	Less than	20g	25g
Cholesterol	Less than	300mg	300mg
Sodium	Less than	2,400mg	2,400mg
Total Carbohydrate		300g	375g
Dietary Fiber		25g	30g

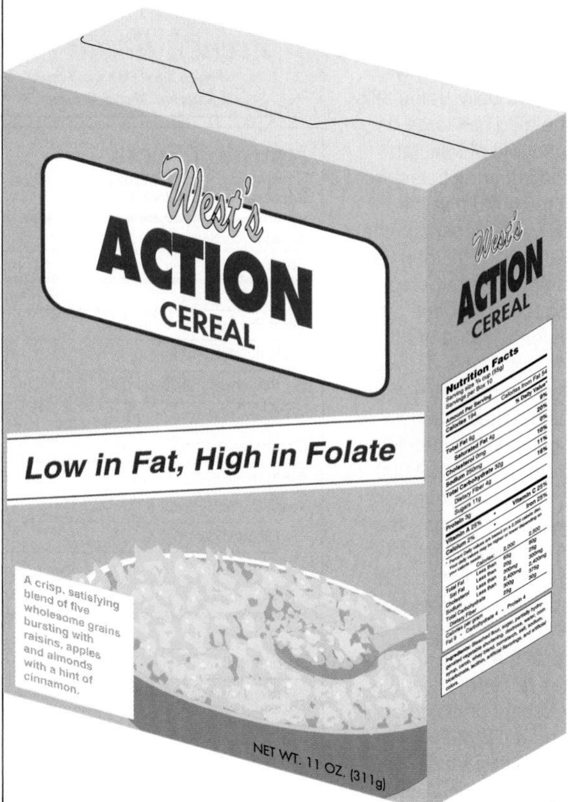

The only vitamins required to appear on the Nutrition Facts panel are vitamins A and C. If a manufacturer makes a nutrition claim about another vitamin, however, the amount of that nutrient in a serving of the product must also be stated on the panel. For instance, the cereal shown is touted as "High in Folate," so the percentage of the recommended intake for folate in a serving of the cereal (25%) is stated on the label. The manufacturer has the option of listing any other nutrients as well.

Watch for health claims for foods or nutrients that help prevent osteoporosis, cancer, heart disease, hypertension, and other conditions. Food products that are good sources of calcium can be labeled with a health claim.

When a label makes a claim about the sodium content of a food product, the label must list the amount of sodium in a serving and adhere to the legal definitions for low-sodium, reduced-sodium, lightly salted, or sodium-free as listed in Table 2-6.

HIGH IN CALCIUM

Regular exercise and a healthy diet with enough calcium help teens and young adult white and Asian women maintain good bone health and may reduce their high risk of osteoporosis later in life.

Westville's
LOWFAT • LOW-SODIUM RASPBERRY
YOGURT

NET WT. 8 oz. (225g)
Distributed by Westville, Inc., Somewhere, USA 12345

Nutrition Facts
Serving size 1 cup (225g)
Servings per Container 1

Amount Per Serving	
Calories 225	Calories from Fat 27
	% Daily Value*
Total Fat 3g	**5%**
Saturated Fat 1.5g	**8%**
Cholesterol 10mg	**3%**
Sodium 120mg	**5%**
Total Carbohydrate 42g	**14%**
Dietary Fiber 0g	**0%**
Sugars 42g	
Protein 9g	**18%**
Vitamin A 2% •	**Vitamin C** 2%
Calcium 30% •	**Iron** 0%

* Percent Daily Values are based on a 2,000 calorie diet. Your Daily Values may be higher or lower depending on your calorie needs.

	Calories:	2,000	2,500
Total Fat	Less than	65g	80g
Sat Fat	Less than	20g	25g
Cholesterol	Less than	300mg	300mg
Sodium	Less than	2,400mg	2,400mg
Total Carbohydrate		300g	375g
Dietary Fiber		25g	30g

Calories per gram
Fat 9 • Carbohydrate 4 • Protein 4

Ingredients: Cultured pasteurized grade A lowfat milk, strawberries, sugar, fat-free milk, modified corn starch, natural flavors, gelatin.

With active cultures including L. Acidophilus.

Information on sodium is required on food labels. It is given as a percent of a Daily Value of 2,400 mg.

Two other minerals—calcium and iron—are required to be listed on the label because they are often low in the American diet. Look for foods that provide 10% or more of these minerals.

Some people find it easiest to bypass the % Daily Values and simply check the grams of total fat a serving of food supplies to see how much it adds to a daily fat tally. Let's say a man eats 2,000 calories a day and therefore should consume no more than 65 grams of fat a day. If he eats a muffin (15 grams of fat) and coffee with cream (10 grams) in the morning, he's up to 25 grams of fat. That means he can have about 40 more grams during the rest of the day to stay within his fat "budget."

Once you determine the maximum number of fat grams you should have in a day, you can use the food label to get a good idea of how many grams of total fat the items you buy add to your daily tally. To figure out your fat allowance, check Table 4-8 on page 120.

You can also use the Daily Values to comparison shop. For example, if you're looking for a high-fiber cereal to increase the amount of fiber in your diet, you can check the % Daily Value for fiber on the labels of several brands of cereal. If a serving of Brand X's cereal has a Daily Value of 16 percent for fiber and Brand Y supplies only 5 percent, Brand X is higher in fiber and the best bet for any fiber seeker, regardless of the number of calories he or she usually consumes.

The % Daily Values for vitamins and minerals are calculated using standard values designed specifically for use on food labels. These values are shown on the inside cover of this book. These standard values for nutrients were created to help manufacturers avoid a stumbling block they face as they label foods. Since manufacturers don't know whether you're an 18-year-old woman or a 30-year-old man, they don't know exactly what your nutritional needs are. You may recall that the DRI include a different set of vitamin and mineral recommendations for each gender and age group.

To help get around the problem, the nutrient recommendations used for vitamins and minerals on labels represent the highest of all the values to ensure that virtually everyone in the population is covered. For most nutrients, the highest recommendation is for an adult man. When it comes to iron, however, the DRI for women is the highest (women require more iron than men), so the women's DRI is used as the standard value on labels.

Nutrient content claims are strictly defined by the FDA.

Nutrient Content Claims. By law, foods carrying terms called **nutrient content claims**—low-fat, low-calorie, light, and so forth—must adhere to specific definitions spelled out by the Food and Drug Administration. For instance, a serving of a food dubbed low-fat must contain no more than 3 grams of fat. An item touted as low-calorie may provide no more than 40 calories per serving. Table 2-6 lists the claims commonly used on food labels and their legal definitions.

NUTRIENT CONTENT CLAIMS claims such as "low-fat" and "low-calorie" used on food labels to help consumers who don't want to scrutinize the Nutrition Facts panel get an idea of a food's nutritional profile. These claims must adhere to specific definitions set forth by the Food and Drug Administration.

Health Claims. A statement linking the nutritional profile of a food to a reduced risk of a particular disease is known as a **health claim.** The FDA has set forth very strict rules governing the use of such health claims. For example, if a food's label bears health claims regarding calcium, a serving of the product must contain at least 20 percent of the Daily Value for calcium, among other restrictions. What's more, the manufacturers are allowed to imply only that the food "may" or "might" reduce risk of disease. They must also note the other factors, such as exercise, that play a role in prevention of the disease. Finally, they must phrase the claim so that the consumer can understand the relationship between the nutrient and the disease.

HEALTH CLAIM a statement on the food label linking the nutritional profile of a food to a reduced risk of a particular disease, such as osteoporosis or cancer. Manufacturers must adhere to strict government guidelines when making such claims.

For example, a health claim on a food low in fat, saturated fat, and cholesterol might read: "While many factors affect heart disease, diets low in saturated fat and cholesterol may reduce the risk of this disease."

The following are the nutrient/disease relationships about which health claims can be made:

- Calcium-rich foods and reduced risk of osteoporosis
- Low-sodium foods and reduced risk of high blood pressure

TABLE 2-6 DEFINITIONS OF NUTRIENT CONTENT CLAIMS

Free means a product contains none or only negligible amounts of fat, saturated fat, cholesterol, sodium, sugar, and/or calories. For instance, "calorie-free" means fewer than 5 calories per serving.

Low indicates the food can be eaten frequently without exceeding dietary guidelines for fat, saturated fat, cholesterol, sodium, and/or calories. More specifically:

low-fat: 3 grams or fewer per serving*
low saturated fat: no more than 1 gram per serving
low sodium: no more than 140 milligrams per serving*
very low sodium: no more than 35 milligrams per serving
low cholesterol: no more than 20 milligrams and no more than 2 grams of saturated fat per serving*
low-calorie: no more than 40 calories per serving*

Lean and **extra lean** describe the fat content of meat, poultry, seafood, and game meats:

lean: fewer than 10 grams of fat, no more than 4.5 grams of saturated fat, and fewer than 95 milligrams of cholesterol per serving (or 100 grams)
extra lean: fewer than 5 grams of fat, fewer than 2 grams of saturated fat, and fewer than 95 milligrams of cholesterol per serving (or 100 grams)

High used when a serving of a food contains 20 percent or more of the Daily Value for a particular nutrient.

Good source indicates that a serving of the food supplies 10 to 19 percent of the Daily Value for a particular nutrient.

Reduced denotes a product that has been nutritionally altered and contains 25 percent less of a nutrient such as fat or calories than the regular, unaltered product. A product cannot be dubbed "reduced," however, if the regular version of the food already meets the requirements for a "low" claim. That is, if a food is "low-fat" to begin with, it cannot be called "reduced" if manufacturers take even more fat out of it.

Less means that a food contains 25 percent less of a nutrient or calories than a comparable food. For example, pretzels containing 25 percent less fat than potato chips carry a "less fat" claim. "Fewer" can be used in the same way.

Light carries several meanings:

First, that a nutritionally altered food contains one-third fewer calories or half the fat as the regular product. If fat supplies 50 percent or more of the calories to begin with, it must be reduced by half to be called "light."

Second, that the sodium content of a low-fat, low-calorie food has been reduced by 50 percent. If the food is not low in fat and calories but the sodium has been decreased by half, it must be labeled "light in sodium."

Third, "light" can be used to describe a food's color and/or texture, as long as the label explains the intent. For example, "light brown sugar."

More means that a serving of the food contains at least 10 percent more of the Daily Value of a particular nutrient than the regular food. The label on calcium-fortified bread can state that the product contains "more calcium" than regular bread.

Percent fat free is an indication of the amount of a food's weight that is fat-free, which can be used only on foods that are low-fat or fat-free to begin with. For instance, a food that weighs 100 grams with 3 grams from fat can be labeled "97 percent fat-free." Note that this term refers to the amount that is fat-free by weight, not calories. If that same food supplies 100 calories, the 3 grams of fat contribute 27 of them (1 gram of fat contains 9 calories). This means that 27 of the 100 calories, or 27 percent of the total calories, come from fat (see Figure 2-13).

Made with oat bran, no tropical oils claims, known as implied claims, are prohibited if they mislead consumers into believing a product supplies (or lacks) significant levels of nutrients. For example, a manufacturer can say a product is "made with oat bran" only if it contains enough oat bran to meet the definition for "good source" of fiber.

Healthy indicates that a food is low in fat and saturated fat; contains no more than 60 milligrams of cholesterol per serving; and provides at least 10 percent of the Daily Value for vitamin A, vitamin C, protein, calcium, iron, or fiber (main dishes must supply at least two of the six nutrients). In addition, the food must meet sodium requirements: no more than 360 milligrams of sodium per serving of individual foods and no more than 480 milligrams per main dish meal.

*On meals and main dish products such as frozen dinners, "low-calorie" can be used if the dish contains no more than 120 calories in 100 grams, or about 3.5 ounces; "low sodium" means the dish supplies no more than 140 milligrams per 100 grams; and "low cholesterol" indicates a maximum of 20 milligrams and 2 grams saturated fat per 100 grams. "Light" means the dish or meal is low-fat or low-calorie.

- Low-fat diet and reduced risk of cancer
- A diet low in saturated fat and cholesterol and reduced risk of heart disease
- High-fiber foods and reduced risk of cancer
- Soluble fiber in fruits, vegetables, and grains and reduced risk of heart disease

- Soluble fiber in oats and psyllium seed husk and reduced risk of heart disease[10]
- Fruit- and vegetable-rich diet and reduced risk of cancer
- Folate-rich foods and reduced risk of neural tube defects[11]
- Sugar alcohols and reduced risk of tooth decay[12]
- Soy protein and reduced risk of heart disease[13]
- Whole-grain foods and reduced risk of heart disease and certain cancers

Exchange Lists

While food group plans provide sufficient detail to help most healthy people plan a good diet, **exchange lists** take meal planning a step further. As their name implies, exchange lists are simply lists of categories of foods, such as fruit, with portions specified in a way that allows the foods to be mixed and matched or exchanged with one another in the diet. For instance, you might strive to eat two servings of fruit each day, and the exchange list shows you that a half cup of orange juice, a small banana, or a small apple each counts as a fruit.

EXCHANGE LISTS lists of foods with portion sizes specified; the foods on a single list are similar with respect to nutrient and calorie content and so can be mixed and matched in the diet.

Portion sizes within groups are determined by considering the calorie, protein, carbohydrate, and fat content of the food. For example, one fruit contains about 60 calories and 15 grams of carbohydrate. One starch, on the other hand, provides about 80 calories, 15 grams of carbohydrate, 3 grams of protein, and a trace of fat. This breakdown makes exchange lists useful tools for people who follow carefully planned diets as a result of a health problem, such as diabetes.

Exchange lists are also useful for people who are following calorie-controlled diets to lose weight. Dietitians sometimes give clients tailor-made diets centered on the exchange lists—say, a 1,500-calorie daily diet that might include eight starches, three vegetables, two fruits, and so forth. A person can take such a framework and use the exchange lists to choose a wide variety of foods that fit

FIGURE 2-13
PERCENT FAT FREE

Percent fat free is an indication of the amount of a food's weight that is fat free, which can be used only on foods that are low-fat or fat free to begin with. For instance, if a food weighs 100 grams and 3 grams are from fat, it can be labeled "97% fat free." Note that this term refers to the amount that is fat free by weight, not calories. If that same food supplies 100 calories, the 3 grams of fat contribute 27 of them (1 gram of fat contains 9 calories). That means 27 of the 100 calories, or 27%, of the total calories come from fat.

Compare labels on packaged deli meats. This package of turkey breast slices states that it is "97% fat free." What this means is that 3% of the product's weight is contributed by fat. If you determine percentage of calories from fat, you will see that this does not mean that only 3% of the calories in a serving come from fat.

Applying the formula for percentage of calories from fat:

$$\frac{\text{14 calories from fat}}{\text{60 total calories}} \times 100 = 23\% \text{ calories from fat}$$

Nutrition Facts
Serving Size 6 slices (54g)
Servings per Container about 3

Amount Per Serving	
Calories 60	60
Calories from Fat	14
	% Daily Value*
Total Fat 1.5g	**2%**
Saturated Fat 0.5g	**3%**
Cholesterol 20mg	**7%**
Sodium 480mg	**20%**
Total Carbohydrate 2g	**1%**
Dietary Fiber 0g	**0%**
Sugars 1g	
Protein 10g	**20%**
Vitamin A 0% •	**Vitamin C** 0%
Calcium 0% •	**Iron** 0%

*Percent Daily Values are based on 2,000 calorie diet.

Ingredients: Turkey breast, water, modified food starch, dextrose, salt, sodium lactate, sodium phosphate, flavorings, celery juice.

TABLE 2-7 THE EXCHANGE LISTS

Carbohydrate Group
- Starch
- Fruit
- Milk
  - Skim (fat-free)
  - Low-fat
  - Whole
- Other carbohydrates
- Vegetable

Meat and Meat Substitute Group
- Very lean
- Lean
- Medium-fat
- High-fat

Fat Group

into the basic eating plan. Table 2-7 shows the seven common exchange lists. Typical portions used in the list are as follows:

Starch: 1 small potato/1 slice bread—80 calories
Fruit: 1 small orange—60 calories
Milk: 1 cup fat-free milk—90 calories
Other carbohydrates: 3 gingersnaps—80 calories (but calories in this list vary)
Vegetable: ½ cup green beans—25 calories
Meat and meat substitutes: 1 ounce lean meat or low-fat cheese—55 calories
Fat: 1 teaspoon butter—45 calories

The complete exchange lists shown in Appendix E can give you a better understanding of the nutritional profile of various foods.

Food Composition Tables

You now have an understanding of the tools you need to set up a healthful eating plan for yourself. You may find useful yet another tool: **food composition tables,** which list the exact number of calories, grams of fat, milligrams of sodium, and other nutrients found in commonly eaten foods. In fact, you may already be familiar with one of the many books of food composition sold in supermarkets and bookstores and geared to consumers.

FOOD COMPOSITION TABLES tables that list the nutrient profile of commonly eaten foods.

Appendix H provides a food composition table that profiles the nutrient content of more than 1,900 different foods. Such tables offer the health-conscious eater a wealth of useful information—everything from the amount of vitamin C in an orange to the number of calories in an order of French fries to the amount of calcium in various cheeses. To be sure, the nutritional content of foods listed in food composition tables varies depending on cooking methods and other factors, and not every table lists every single nutrient. Still, food composition tables give fairly precise estimates of the nutrients in the foods you eat.

Free diet-analysis software programs are available online. See Nutrition on the Web at the end of this chapter.

Computer buffs can take advantage of one of the many software packages containing a database that is, in essence, a food composition table. Simply plug in the foods you eat, and the computer will generate a profile of your daily diet. Dietitians often use such software to analyze both people's diets and recipes. More and more reasonably priced, reliable nutrition-analysis software is becoming widely available.

WORKING ON THE WEB www.intelihealth.com

Sponsor: Intelihealth, with Johns Hopkins University and Aetna US Healthcare

Description: This Web site offers practical advice on how to improve your eating habits and your health.

Available Activities: The site offers information on nutrient needs, weight management, fitness, and much more. You will find clever interactive features that show you how your eating habits compare to the Food Guide Pyramid as well as virtual holiday dinners and special occasion meals (e.g., picnics, ballpark fare) that calculate your calories as you fill your plate. You can also access the USDA database for foods and view the Nutrition Facts panel on any food you choose.

Web Work:

1. Find out the shape of your Food Guide Pyramid.
2. For any areas of your diet needing improvement, list strategies to improve your score.
3. Test you knowledge of serving sizes from the Pyramid.
4. Visit the USDA Food Database and examine the food labels of three foods you frequently purchase at the grocery store. What nutrients do the foods contribute to your diet?

Helpful Hints: From the Intelihealth home page, scroll down and click on Nutrition. Click on What Shape Is Your Pyramid, and specify the number of servings you consume from each food category. When finished, click on Draw My Pyramid and scroll back to the top of the page to view the comments made in each wedge of your pyramid. Go back one page to the What Is the Shape of Your Pyramid home page and click on Fun, Interactive Food Gadgets. Click on Serving Size Surprise and follow the directions given. Finally, from this same Food Gadgets page, click on Food Database and review the Nutrition Facts Panels on your favorite foods.

RATE YOUR PLATE SCORECARD

To see how your diet measures up to the recommendations in the Food Guide Pyramid, follow these steps.

Step 1: Write down everything you ate yesterday, including both meals and snacks. Make note of portion sizes as well.

Step 2: Identify the food group to which each item you ate belongs (refer to Figures 2-5, 2-6, and 2-7 for help).

Step 3: Determine the number of servings from the five food groups that are right for you. The Food Guide Pyramid shows a range of recommended number of servings for each food group. The optimal amount for you depends on the number of calories you need, which in turn is influenced by a number of factors such as your age and activity level. The following chart gives sample daily diets for three different calorie levels to help you get a rough estimate of the type of diet that might work for you. Note that 1,600 calories is a good estimate for sedentary women and some older adults; 2,000 to 2,200 is about right for most children, teenage women, and many sedentary men; and 2,800 is a generous estimate for many teenage boys, active men, and very active women.

	SAMPLE DIETS FOR A DAY		
	1,600 CALORIES	2,000 TO 2,200 CALORIES	2,800 CALORIES
Bread group servings	6	8–9	11
Vegetable group servings	3	4	5
Fruit group servings	2	3	4
Milk group servings	2–3	2–3	2–3
Meat group (ounces)[a]	5	5–6	7

[a]Meat group servings are given in total ounces. That is, five servings from this group means 5 ounces.

Source: Adapted from *USDA's Food Guide Pyramid* (U.S. Department of Agriculture Human Nutrition Information Service, Home and Garden Bulletin Number 249): pp.9, 28–29.

Step 4: Circle the estimated number of servings that are right for you in the left column. In the right column, write down the number of servings you ate yesterday (refer to Figures 2-6 and 2-7 to help determine what counts as a serving). Compare the two columns to see how your diet rates.

	NUMBER OF SERVINGS YOU SHOULD EAT	NUMBER OF SERVINGS YOU ATE
Bread group servings	6 7 8 9 10 11	________
Vegetable group servings	3 4 5	________
Fruit group servings	2 3 4	________
Milk group servings	2 3	________
Meat group (ounces)	5 6 7	________
Fats, oils, and sweets	Use sparingly	________

Step 5: Decide what changes in your eating habits will make your diet more healthful. If your diet is "top heavy," with lots of foods coming from the top of the pyramid rather than the middle and the bottom, make gradual changes, such as eating more fruits and vegetables. The chapters that follow offer tips on how to do so.

A. A. McCourt

"Here's your problem ...You've been reading the Food Pyramid upside down!"

Spotlight
A TAPESTRY OF CULTURES AND CUISINES

American cuisine is often described as nothing more than meat-and-potato meals or burgers, fries, hot dogs, and apple pie. But to define American food in this limited way does the United States an injustice. American eating habits are as diverse as the various ethnic and cultural groups that make up America's people. Throughout American history, immigrant groups—from Poles to Jews to Italians to Irish to Germans to Hispanics to African-Americans to Asians—have had and continue to have a profound effect on the collective American palate.

Certainly, it's not only the Americans who are members of various ethnic groups who enjoy their own traditional fare. Consider that in the past decade, more and more consumers from many different cultural backgrounds have begun to enjoy an ever increasing variety of ethnic foods. In fact, according to the National Restaurant Association, some 90 percent of restaurant menus now include ethnic dishes, and 80 percent use recipes from more than one national cuisine, a 35 percent increase from 1988.[14] What's more, **fusion cuisine**—the merging of elements from different ethnic cuisines to create a new dish—has been gaining in popularity in the past decade. For example, many restaurants are now offering items such as Mexican pizza, which borrows from both Hispanic and Italian food traditions.

As our food writer and critic John Mariani has pointed out:

> *. . . the United States—a stewpot of cultures—has developed a gastronomy more varied . . . than that of any other country in the world. . . . In any major American city one will find restaurants representing a dozen national cuisines, including northern Italian trattorias, bourgeois French bistros, Portuguese seafood houses, Vietnamese and Thai eateries, Chinese dim sum parlors, Japanese sushi bars, and German rathskellers.*[15]

This Spotlight examines some of the more prevalent ethnic and regional food practices to see how they originated and how they fit into a healthful eating plan.

I love to eat Mexican food, but I've heard that it is loaded with fat. Is that true? Do the Mexican people eat a lot of high-fat food?
It's true that many menu selections in Mexican restaurants are loaded with high-fat ingredients such as cheese, ground beef, sour cream, guacamole, and fried tortillas. But most dishes eaten regularly in Mexican-American homes are much simpler. Breakfast, for example, might be tortillas (flat, thin corn or wheat pancakes) served with fried beans, eggs, or cereal, and a beverage. Lunches and dinners often consist of beans and rice, bread or tortillas, meat/sausage (often as part of a stew), a vegetable or lettuce and tomato, and a beverage. The traditional Mexican diet, which draws largely from Spanish and Indian influences, was even simpler, containing mostly vegetables, including beans, squash, and maize (corn). In addition, it often included cactus parts, agave (a plant with spiny-margined leaves and flowers), chili peppers, **amaranth** (a grain), avocado, and **guava** (a sweet, juicy fruit with green or yellow skin and red or yellow flesh).

This emphasis on a diet high in complex carbohydrates, such as rice, and vitamin A- and C-rich fruits and vegetables is a particularly healthful aspect of the traditional Mexican way of eating. Consider that no Mexican meal is complete without salsa, a low-fat condiment consisting of vitamin-rich tomatoes, chilis, and onions. Another plus of the Mexican diet is the frequent use of beans, mostly the pinto variety, which rank as a particularly good source of fiber.

Among the downsides of the Mexican diet, however, is the heavy-handed use of oil and lard. Most foods, even beans and rice, are fried rather than baked or broiled. For example, frijoles refritos (refried beans) are usually fried in lard and contain about 270 calories and 3 grams of fat per cup. Most flour tortillas are also made with lard, sometimes 1 or 2 teaspoons' worth per tortilla. (Corn tortillas, on the other hand, typically contain very little fat.) Another drawback of the Mexican diet is the frequent consumption of high-fat meats, such as **chorizo** (spicy pork or beef sausage) and eggs, which are used in dishes such as **chiles rellenos** (roasted mild green chili pepper stuffed with

cheese, dipped in egg batter, and fried), **burritos** (warm flour tortillas stuffed with a mixture of egg, meat, beans, and/or avocado), and **chilaquiles** (tortilla casserole often made with eggs or meat).

Fortunately, Mexican-food lovers can take advantage of the popular cuisine with a little know-how. Instead of the usual American versions of Mexican fare, such as fried tortillas packed with ground beef and smothered in cheese and sour cream, opt for corn tortillas filled with, say, regular, unfried pinto beans mixed with chopped onion and topped with a sprinkle of shredded cheese and a generous portion of lettuce, salsa, a dollop of fat-free plain yogurt (a low-fat alternative to sour cream), and a garnish of sliced avocado. Or, instead of serving high-fat commercial tortilla chips with salsa, try making "no-fry" chips: immerse several tortillas in warm water, drain quickly, cut into six to eight wedges, and place on a nonstick pan; bake in a 500-degree Fahrenheit oven for three to four minutes, flip, and continue to bake for another minute or two until golden brown.

Adventurous eaters who have access to a wide variety of exotic produce might want to try some of the different fruits and vegetables the Mexican people have enjoyed for decades, such as **jicama** (HE-cah-mah) (yam bean root—a vegetable that is tan outside and white inside, has a mild chestnut flavor, and is always eaten raw).[16] Figure 2-14 shows produce and other foods commonly eaten by Mexican-Americans.

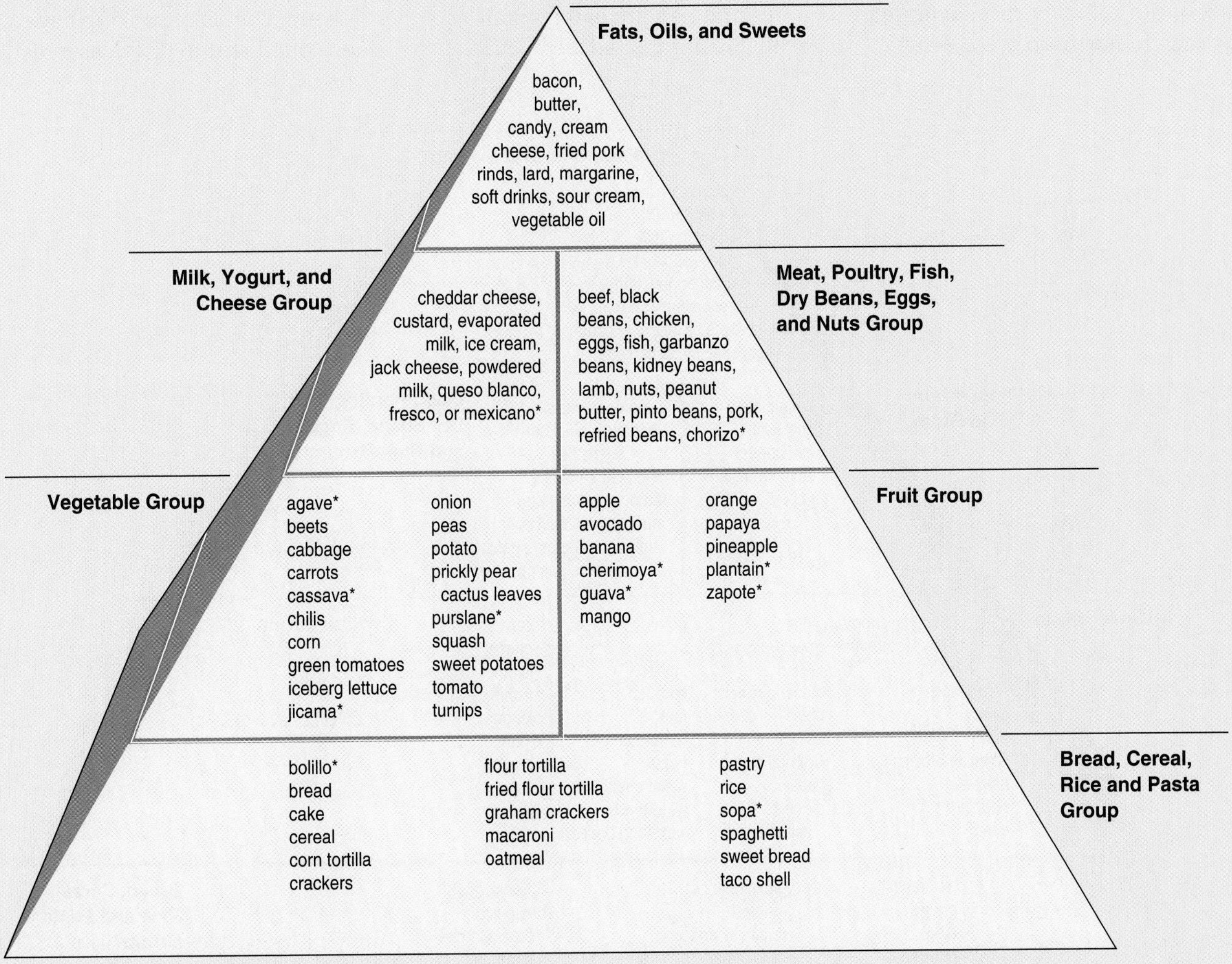

FIGURE 2-14
MEXICAN-AMERICAN FOODS AND THE FOOD GUIDE PYRAMID

*These foods are described in the Miniglossary.

Source: Adapted from the *Pyramid Packet,* © 1993, Penn State Nutrition Center, 417 East Calder Way, University Park, PA 16801.

I always thought chop suey, egg rolls, and fortune cookies were traditional Chinese foods, but a Chinese-American friend of mine says they aren't. Is she right?
Yes, it's true that chop suey and the like are American inventions. The traditional Chinese diet consists of much simpler, lower-fat dishes. Overall, about 80 percent of calories in traditional Chinese fare comes from grains, legumes, and vegetables, while the other 20 percent comes from animal meats, fruits, and fat. In southern China, rice can be easily produced and provides the bulk of the complex carbohydrate in the diet. In northern areas, where wheat grows readily, noodles, dumplings, and steamed buns are staples. In all regions of the country, fruits and vegetables are eaten in abundance. Typically, they are not eaten raw but rather are steamed, added to soups, or stir-fried in peanut or corn oil or, less frequently, lard. In addition, vegetables are often salted, pickled, and dried, a practice resulting from lack of the facilities needed to refrigerate and transport fresh produce across the country. As for meat and other protein foods, pork is considered the staple meat, though poultry, eggs, lamb, and fish are eaten when available. Tofu, or soybean curd, is another staple often added to stir-fries or fermented into sauces. Dairy foods, such as milk and cheese, have never been part of the Chinese diet.

Traditional Chinese meals follow basically the same pattern. Breakfast might be rice congee (rice gruel containing bits of meat), a salty side dish such as pickles, and tea. Lunch typically consists of soup, rice, and mixed dishes made with vegetables and fish, meat, or poultry, and dinner is usually a larger version of lunch, occasionally followed by fresh fruit.

Four schools of cooking have developed within China as a result

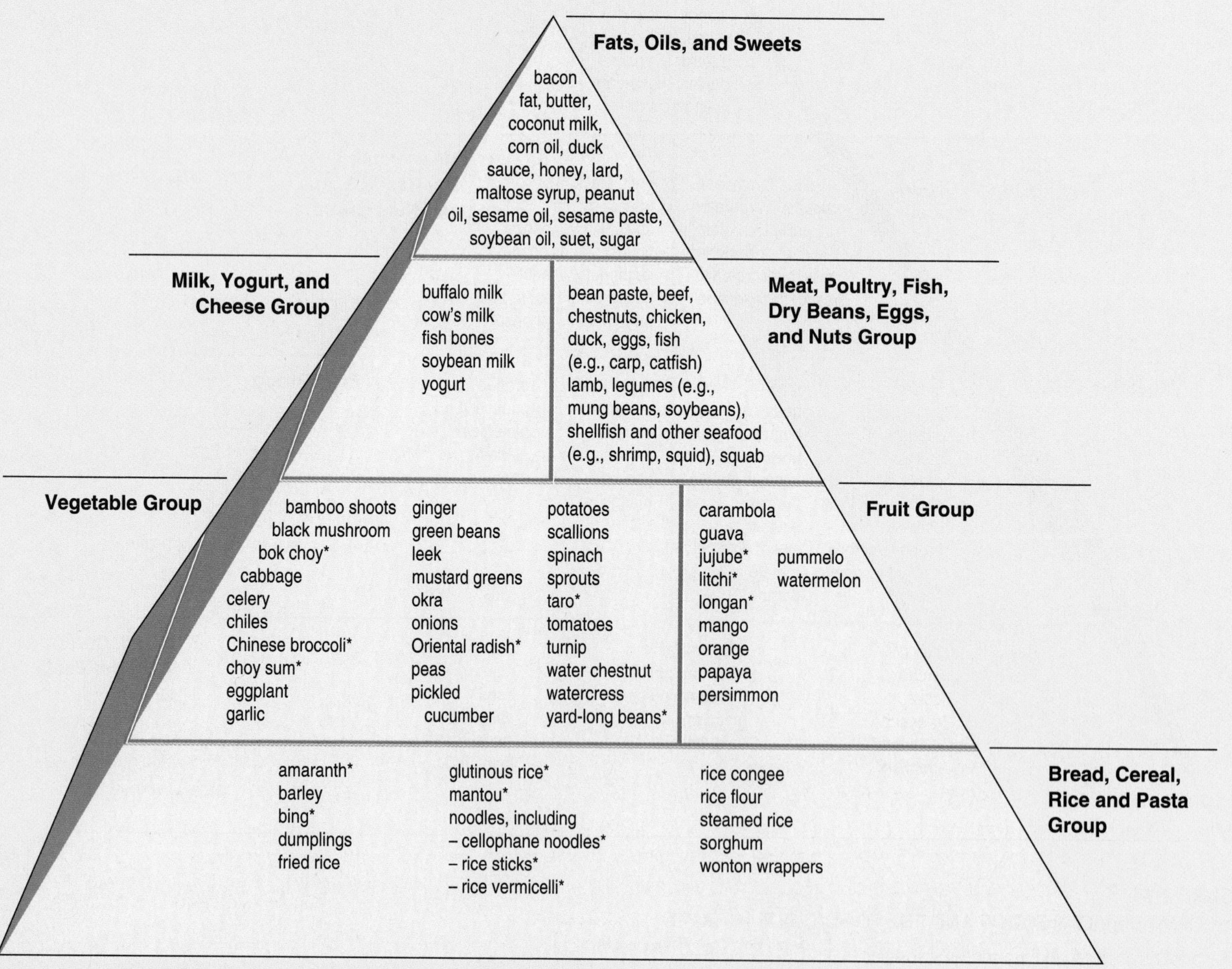

FIGURE 2-15
A CHINESE-AMERICAN FOOD GUIDE PYRAMID

Source: Adapted from *Pyramid Packet,* © 1993, Penn State Nutrition Center, 417 East Calder Way, University Park, PA 16801.

of differences in climate, food production, religion, and custom: Peking, Shanghai, Szechwan or Hunan, and Cantonese. Peking cooking, which comes from the north and northeast part of China, is distinguished by use of garlic, leeks, and scallions and has given rise to familiar dishes such as Peking duck and spring rolls. Shanghai cuisine, from the east and coastal areas, is characterized by "red-cooking"—braising foods with large amounts of soy sauce and sugar. Pickled and salted vegetables are also frequently served with meat. In the west and central part of China, people favor the Szechwan or Hunan cooking style, in which chili peppers and hot pepper sauces are added liberally to dishes, and the food tends to be spicy and oily. The cuisine of southern China, known as Cantonese cooking, is the style Americans know best. Because foods are usually steamed or stir-fried and chicken broth is often used as a cooking medium, Cantonese food tends to be the least fatty. One popular example of Cantonese food is **dim sum,** steamed or fried dumplings stuffed with pork, shrimp, beef, sweet paste, or preserves and steamed or fried.[17]

The regional differences in cooking styles and eating habits among people living in rural China make the country fertile ground for scientific research into the effects of diet on disease. Another reason rural China is such an ideal place to conduct research is that most people residing in rural China today spend their entire lives within the vicinity of the community in which they were born. In addition, eating habits are dictated by climate and environment because the country has little or no means of transporting foods from one region to another. Thus, scientists have been able to carefully study eating patterns and disease rates in an attempt to see how the two are related. One of the largest such studies was led by T. Colin Campbell, Ph.D., of Cornell University. Begun in 1983, the research was carried out in 130 villages located in 65 counties of rural China and included 6,500 adults. Although the data accumulated in this study are enormous and analyses are still underway, the findings suggest that the traditional, plant-based Chinese diet is associated with low rates of many of the chronic diseases that plague Americans, such as heart disease and some types of cancer.[18]

To be sure, Chinese food served in American Chinese restaurants is a far cry from the type eaten day in and day out by the rural Chinese people. Many Chinese restaurant meals are swimming in oil and contain much more meat and poultry and fewer vegetables than "real" Chinese food. Consider that a typical American Chinese meal might include won ton soup, barbecued spareribs, chicken lo mein, and fried rice—chalking up some 1,400 calories and more than 80 grams of fat. A typical meal eaten in rural China, on the other hand, would likely contain a heaping portion of rice, along with fiber- and nutrient-rich vegetables and less than an ounce of meat and fish and would thereby contain only a fraction of the fat.[19] Americans who want to enjoy both the flavor and the health benefits of traditional Chinese cuisine can "stretch" one of the many, delicious vegetable-based dishes with relatively large portions of rice and go easy on deep-fried appetizers such as egg rolls. People who enjoy cooking can also follow the Chinese people's lead and make low-fat, high-carbohydrate stir-fries with lots of vegetables, little oil, and small amounts of meat, poultry, or seafood. People who are sodium-conscious may also want to go easy on soy sauce, which contains large amounts of the mineral, or try some of the reduced-sodium soy sauces on the market. Figure 2-15 shows Chinese foods placed on a food pyramid.

How does Italian food rate? I've heard that the olive oil in Italian pasta dishes is good for health. Much of the Italian food served in the United States runs very high in fat and calories. Rich cream sauces in dishes such as fettucine alfredo, an abundance of cheese and sausage in entrees, including meat-topped pizza and lasagna, and liberal use of olive oil to prepare all manner of pasta and other Italian fare can all chalk up extraordinary amounts of fat and calories.

With a little modification, the Italian food so many Americans love can easily fit into a high-carbohydrate, low-fat diet. For example, substituting vegetables for sausage and pepperoni on pizza and in pasta sauces and lasagna reduces fat content considerably. Using reduced-fat cheeses, such as part-skim mozzarella and ricotta and low-fat cottage cheese, as substitutes for high-fat versions called for in many Italian recipes also helps skim some of the fat.

When it comes to olive oil, a staple in traditional Italian cuisine, it's true that using it instead of butter is preferable from a health standpoint. For reasons that will be explained in detail in Chapter 4, olive oil and other vegetable oils are less likely than certain other types

of fat, such as butter and lard, to boost blood cholesterol levels. Some scientists even believe that people can eat large amounts of fat—in the neighborhood of 35 to 40 percent of total calories—without raising their risk of heart disease, as long as the predominant fat is olive oil. This line of thinking stems from research conducted in southern Italy and the Greek island of Crete, as well as southern regions of other European nations, including France; North African areas such as Morocco and Tunisia; and Middle Eastern countries such as Israel and Syria. Collectively known as the Mediterranean region, these countries share an overall dietary pattern that includes an abundance of fruits and vegetables, breads, and other grains, beans, nuts, and seeds; low to moderate amounts of cheese, yogurt, fish, and poultry; small amounts of red meat; moderate consumption of wine; and liberal use of olive oil. Scientists are fascinated with this region because the people living there historically have enjoyed long lives and low rates of chronic diseases. That was particularly true in Crete during the 1950s and 1960s. At that time, researchers found that residents of the island shared one of the lowest rates of heart disease and cancer ever recorded and one of the longest life expectancies. That held true despite the fact that their diets contained nearly 40 percent of calories as fat—well above the 30 percent limit recommended by major U.S. health organizations.[20]

With this historical perspective in mind, some scientists recommend that Americans adopt a "Mediterranean diet" that includes lots of fruits, vegetables, and grains; little meat and other flesh food; moderate amounts of wine; and generous amounts of olive oil (see Figure 2-16). This advice has sparked a good deal of controversy within the scientific community, however. Many nutrition experts question the wisdom of recommending such a diet in America, where excess fat from olive oil and other sources might contribute to the epidemic of obesity in this country. In addition, many experts point out that diet is not the only lifestyle factor that may have promoted the long, healthy lives of the people of Crete and Italy several decades ago. For example, the residents of this region farmed the land and thereby engaged in far more physical activity than the average American, a factor that probably played a role in their health and longevity. In addition, unlike many modern-day Americans, they enjoyed the social support of extended networks of family and friends, another part of their lifestyle that probably contributed to their well-being. Given these caveats, major U.S. health organizations, including The American Dietetic Association, the American Heart Association, and the U.S. Department of Health and Human Services, maintain that Americans should keep the amount of total fat—whether in the form of olive oil or anything else—to no more than 30 percent of total calories.

Our advice is that you consider adopting the aspects of the Mediterranean diet virtually all nutrition experts advocate: eat an abundance of produce, grains, and

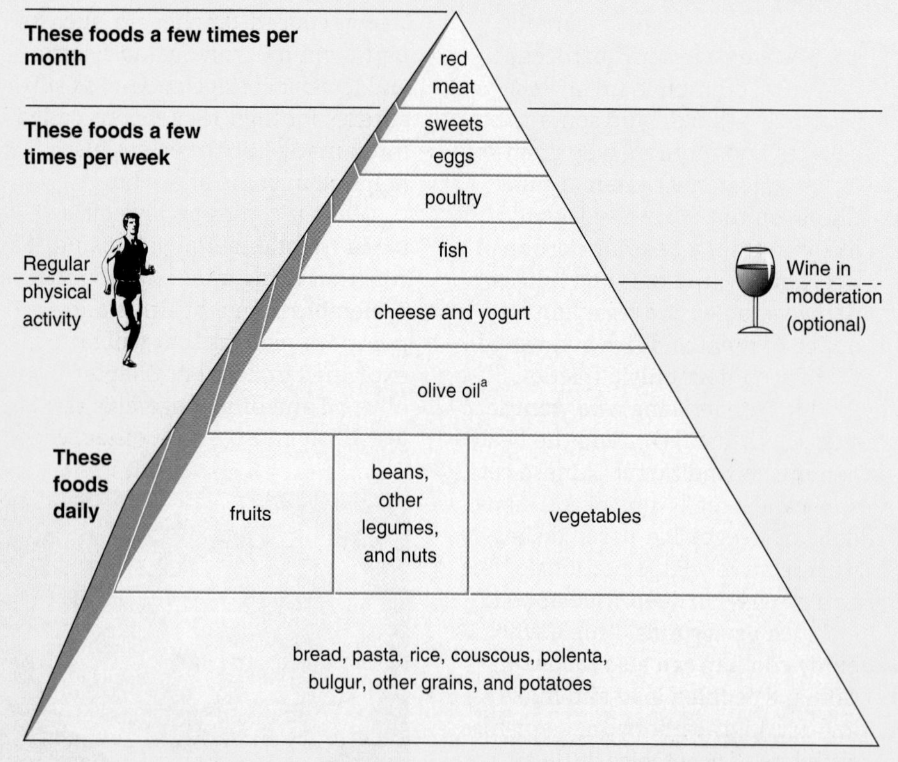

**FIGURE 2-16
MEDITERRANEAN FOOD GUIDE PYRAMID**

[a]Other oils rich in monounsaturated fats, such as canola or peanut oil, can be substituted for olive oil. People who are watching their weight should limit their oil consumption.

Source: © 1994 Oldways Preservation & Exchange Trust.

legumes combined with moderate amounts of dairy products, and relatively smaller amounts of meats, poultry, and fish.

What about "soul food"? Do most African-Americans eat lots of it? Soul food is a term that was coined in the mid-1960s to promote ethnic pride and solidarity among African-Americans.[21] But the origins of soul food date back to a much earlier time in history. Black-eyed peas, grits (coarsely ground cornmeal), collard greens, okra, and other soul foods evolved from the traditional diet of West African slaves living in the South. When West Africans were brought to the United States to work the fields, their dietary habits revolved around the foods provided by slave owners. Corn was commonly given as a staple, and the slaves prepared it in many forms, such as grits, cornmeal pudding, and **hominy** (hulled, dried corn kernels with certain parts removed). Salt pork was also frequently a staple supplied to the slaves, so pork fat was used to fry and flavor greens, breads, stews, and other foods. In addition, some owners allowed their slaves to grow vegetables in small plots. Okra and black-eyed peas, two West African favorites, were introduced in the United States by the slaves who farmed such plots, and American

vegetables, including cabbage, collard and mustard greens, sweet potatoes, and turnips, were often grown as well. Other southern favorites, such as fried chicken and fried catfish, were made popular by the slaves who worked as cooks in the homes of slave owners.

After emancipation, the food-eating patterns of the African-

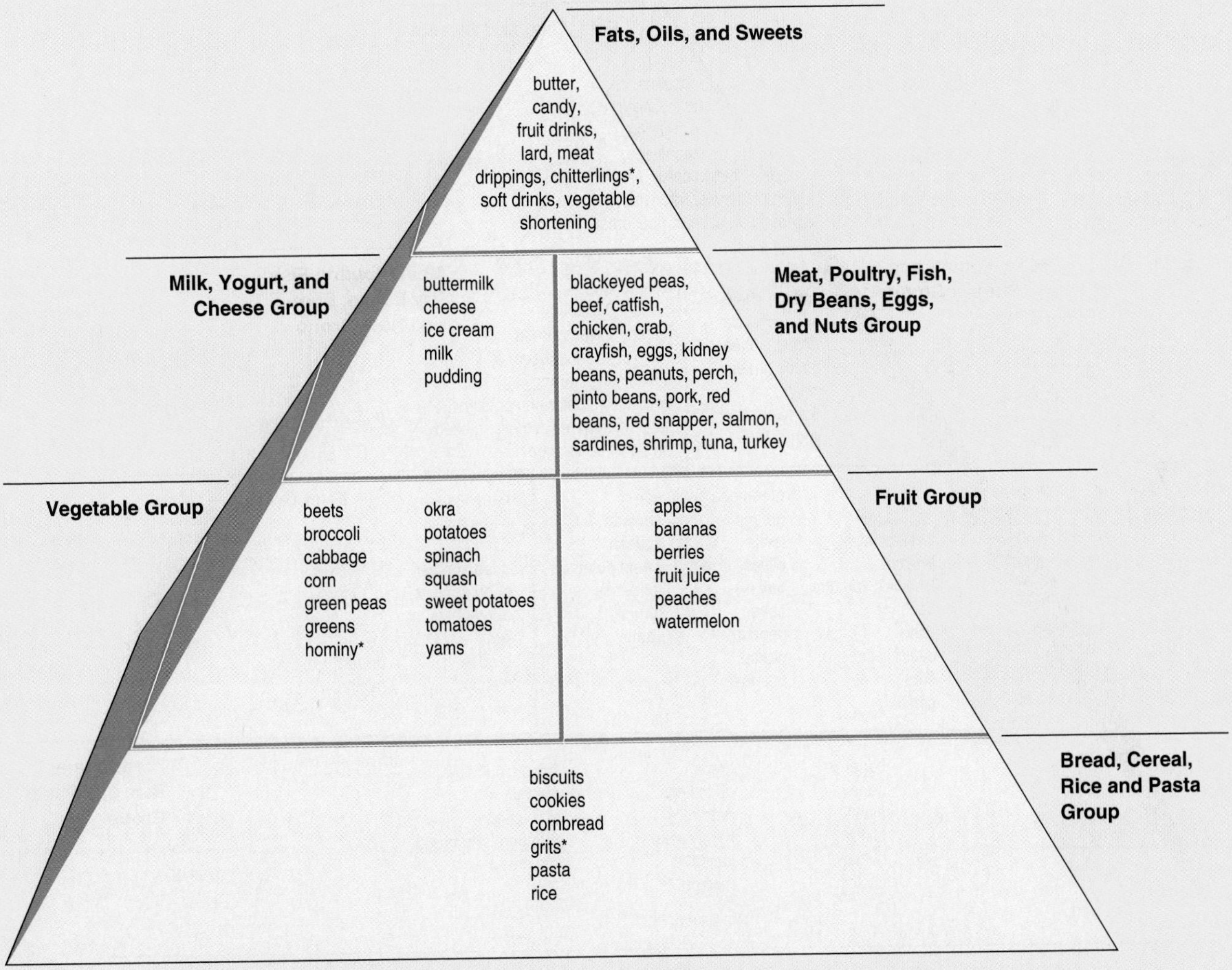

Fats, Oils, and Sweets

butter, candy, fruit drinks, lard, meat drippings, chitterlings*, soft drinks, vegetable shortening

Milk, Yogurt, and Cheese Group

buttermilk
cheese
ice cream
milk
pudding

Meat, Poultry, Fish, Dry Beans, Eggs, and Nuts Group

blackeyed peas, beef, catfish, chicken, crab, crayfish, eggs, kidney beans, peanuts, perch, pinto beans, pork, red beans, red snapper, salmon, sardines, shrimp, tuna, turkey

Vegetable Group

beets
broccoli
cabbage
corn
green peas
greens
hominy*
okra
potatoes
spinach
squash
sweet potatoes
tomatoes
yams

Fruit Group

apples
bananas
berries
fruit juice
peaches
watermelon

Bread, Cereal, Rice and Pasta Group

biscuits
cookies
cornbread
grits*
pasta
rice

FIGURE 2-17
TRADITIONAL AFRICAN-AMERICAN FOODS AND THE FOOD GUIDE PYRAMID

Source: Adapted from *Pyramid Packet,* © 1993, Penn State Nutrition Center, 417 East Calder Way, University Park, PA 16801.

Americans did not change significantly and represent much of what we think of as southern cuisine today. The underpinnings of this eating style, which is very high in fat, remain corn-based dishes, greens, pork, and pork products such as **chitterlings** (chitlins—pig intestines) and ham hocks. The food habits of African-Americans around the country tend to be influenced more by economic status and geographic location than by heritage, although soul food remains a symbol of identity and heritage for many African-Americans.[22] Figure 2-17 shows traditional African-American foods in pyramid form.

I see lots of foods marked kosher in the supermarket, especially during Jewish holidays such as Passover. Are kosher foods better for health than regular items? Do other religious groups eat special foods?

As pointed out in Chapter 1 (refer to Table 1-7 on page 17), food is part of the symbolism and traditions of many major religions. In the predominantly Christian United States, however, most people's eating habits are not dictated by religion to a large extent.

Nevertheless, during the past few decades "Jewish" foods have been growing in popularity among Americans of all different religious backgrounds. For instance, bagels are one of the fastest growing breakfast menu items in the United States.[23]

Most of the traditional Jewish foods eaten in America come from a particular group known as Ashke-

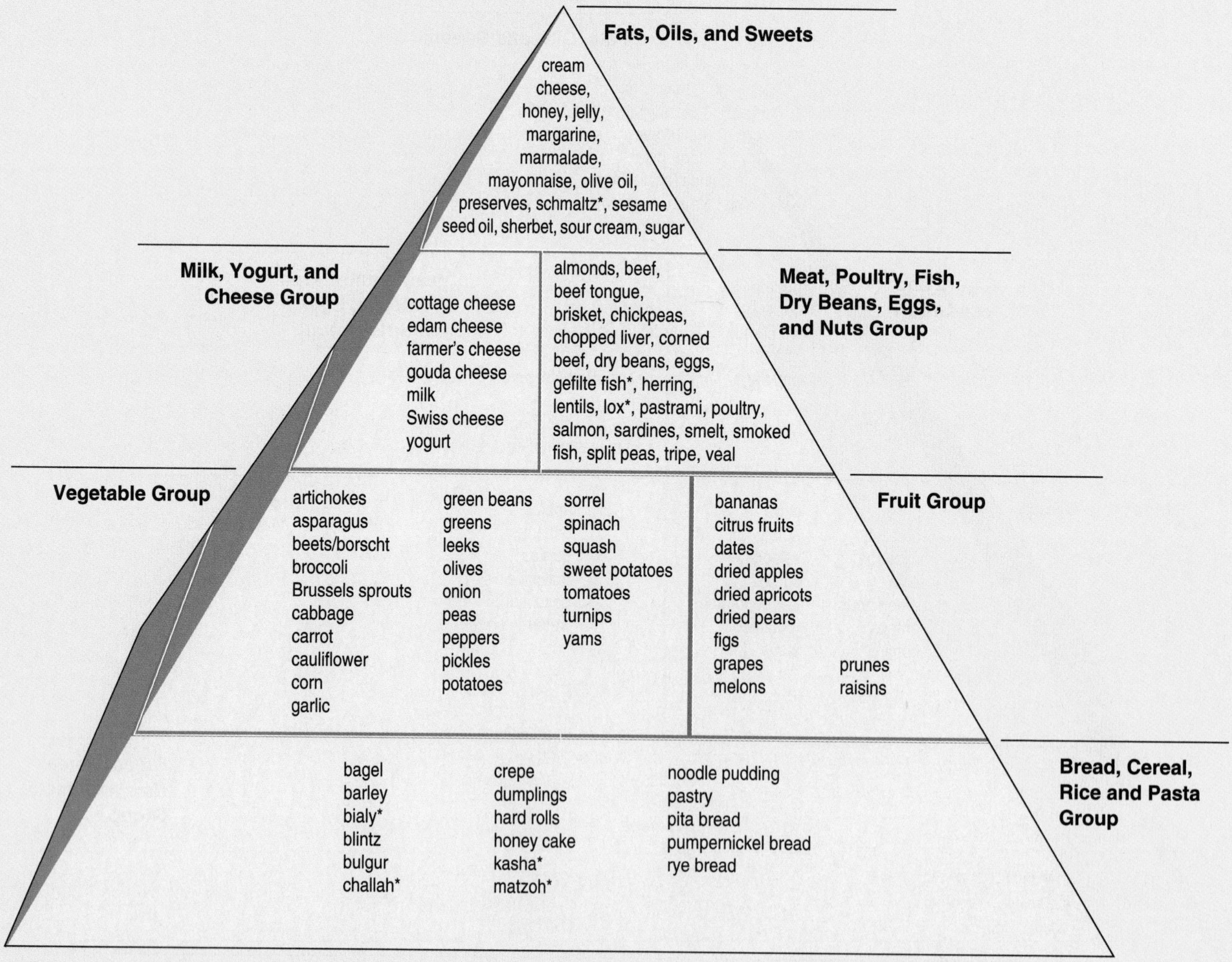

FIGURE 2-18
FITTING JEWISH-AMERICAN FOODS INTO THE FOOD GUIDE PYRAMID

Source: Adapted from *Pyramid Packet,* © 1993, Penn State Nutrition Center, 417 East Calder Way, University Park, PA 16801.

nazic Jews—Jews from central and eastern European countries such as Russia, Germany, Poland, and Romania. (The other major group is Sephardic Jews, who come mainly from Spain and Portugal.) Although few Jewish people in the United States strictly abide by all the dietary laws Judaism prescribes, many adhere to at least some of the rules of kashrut, biblical ordinances specifying which foods are **kosher,** or fit to eat.

Most people assume that the laws of kashrut were set forth to protect the health of the Jewish people. For example, a popular misconception is that kashrut forbids the consumption of pork products because eating undercooked pork can cause serious illness. In truth, however, Jewish dietary laws are considered divine commandments set forth to maintain spiritual, not physical, health. Foods labeled as kosher are not necessarily more healthful than their unmarked counterparts. Instead, the designation kosher indicates that a food has been prepared in accordance with the basic tenets of kashrut. For instance, one principle of kashrut is separation of milk and meat products, meaning that an item containing, say, both ground beef and cheese would not be kosher. Another tenet is selection of appropriate meat, poultry, and seafood items: Only animals with cloven hooves who chew their cud are allowed—cattle, sheep, goats, and deer; chicken, turkey, goose, pheasant, and duck can be kosher, but birds of prey are not; and seafood with both fins and scales can be kosher, while shellfish is forbidden. Finally, as a result of the salting process used to prepare kosher animal foods, many traditional Jewish foods are high in sodium. Herring, smoked fish, canned beef, tongue, corned beef, and other deli-style meats are examples. Other traditional Jewish foods, many of which are high in fat, include **schmaltz** (chicken fat), **knishes** (potato pastry filled with ground meat or potato), cream cheese, and chopped liver.[24] Figure 2-18 shows how some traditional Jewish foods fit into the Food Guide Pyramid.

MINIGLOSSARY OF FOODS

FUSION CUISINE a term used to describe food that combines the elements of two or more cuisines—say, European and Oriental—to create a new one.

MEXICAN–AMERICAN

AGAVE a plant with spiny-margined leaves and flower

BOLILLO a roll-like bread often used instead of tortillas or to make sandwiches

BURRITOS warm flour tortillas stuffed with a mixture of egg, meat, beans, and/or avocado

CASSAVA a starchy root that is never eaten raw because it must be cooked to eliminate its bitter smell

CHERIMOYA a fruit with a rough green outer skin and sherbetlike flesh

CHILAQUILES tortilla casserole often made with eggs or meat

CHILES RELLENOS roasted mild green chili pepper stuffed with cheese, dipped in egg batter, and fried

CHORIZO spicy beef or pork sausage

GUAVA a sweet juicy fruit with green or yellow skin and red or yellow flesh

JICAMA a crisp, bean root vegetable that is tan outside and white inside and is always eaten raw; jicama is as popular in Mexico as the potato is in the United States

PLANTAIN a greenish, starchy banana; because it is starchy even when ripe, it is never eaten raw and is usually pan-fried

QUESO BLANCO, FRESCO, OR **MEXICANO** soft white cheese made of part-skim milk

SOPA rice or pasta that is fried and cooked in consomme

PURSLANE leafy vegetable that can be used in salads or cooked like spinach

ZAPOTE an apple-size fruit with green skin and black flesh

CHINESE-AMERICAN

AMARANTH a golden-colored grain

BING thin pancakes

BOK CHOY a vegetable with broad, white or greenish-white stalks and dark green leaves; also called Chinese chard

CELLOPHANE NOODLES thin, translucent noodles made from mung beans

CHINESE BROCCOLI a green leafy vegetable often stir-fried; also called Chinese kale

CHOY SUM a bright-green vegetable commonly stir-fried; also called field mustard or Chinese flowering cabbage

DIM SUM steamed or fried dumplings stuffed with pork, shrimp, beef, sweet paste, or preserves and steamed or fried

GLUTINOUS RICE short-grained, opaque, white rice that turns sticky when cooked

JUJUBE Chinese date

LITCHI small, round fruits with orange-red skin and opaque, white flesh; also called litchee or lychee

LONGAN a small, round fruit with smooth brown skin and clear pulp

MANTOU steamed bread

ORIENTAL RADISH large, cylindrically shaped vegetables with smooth skin; also called daikon

RICE STICKS flat, opaque, wide noodles made from rice flour

RICE VERMICELLI thin, white noodles made from rice flour

TARO a starchy vegetable with brown, hairy skin and a pink-purple interior

YARD-LONG BEANS thin, tender string beans that grow to as long as 18 inches

AFRICAN-AMERICAN

GRITS coarsely ground cornmeal

HOMINY hulled, dried corn kernels

CHITTERLINGS (chitlins) pig intestine

JEWISH-AMERICAN

BIALY a flat breakfast roll that is softer than a bagel

CHALLAH an egg-containing yeast bread, often braided, and served on the Sabbath and holidays

GELFILTE FISH a chopped fish mixture often made with pike and whitefish as well as matzoh crumbs, eggs, and seasonings

KASHA cracked buckwheat, barley, millet, or wheat that is served as a cooked cereal or potato substitute

KNISH a potato pastry filled with ground meat, potato, or kasha

KOSHER fit, proper, or in accordance with religious law

LOX smoked salmon

MATZOH a crackerlike bread eaten most often at Passover

SCHMALTZ chicken fat

Choosing Healthful International Cuisines

From Pre-Columbian times until today, Mexicans have inadvertently added calcium to their diet as they prepared corn for making tortillas. By soaking the corn in slaked limewater, it became more digestible as well as calcium rich.[25]

Every culture has its own typical foods and ways of combining foods into meals. The challenge for healthful ethnic dining is in learning to choose foods for good health without sacrificing good taste. The following tips guide you through a variety of international cuisines with this challenge in mind.

Chinese

Many Chinese dishes are based on an abundance of rice and vegetables, small portions of meat, and low-fat preparation methods, making them quite healthful. Here are a few things to look for when you buy or prepare Chinese food.

LIMIT

1. Foods prepared with whole eggs, such as lobster sauce, egg drop soup, and egg foo yung.
2. Fried foods, such as egg rolls, fried noodles, fried rice, "crispy" meat, poultry, or fish, and fried dumplings.
3. Fatty foods, such as duck, goose, poultry with skin, and spareribs.
4. Regular soy sauce, duck sauce, plum sauce, and MSG—they are high in sodium.

TRY INSTEAD

1. Dishes made with low-fat, cholesterol-free egg whites.
2. Foods cooked using the stir-fry method with a small amount of unsaturated oil, such as chicken with broccoli and shrimp with Chinese vegetables, or by steaming, such as dim-sum selections and steamed dumplings.
3. Lean cuts of meat, skinless poultry, all types of fish, and shellfish—also try bean curd (tofu) to replace or enhance meat.
4. Reduced-sodium soy sauce. Also, most Chinese restaurants will prepare food without MSG, if requested.

Japanese

Most Japanese cuisine fits easily into a healthful lifestyle with its emphasis on fresh fish and vegetables prepared with little fat. The basis for much of its flavor, however, is soy sauce (shoyu). This salty condiment may be a problem if you are watching your sodium intake.

LIMIT

1. Deep-fried foods, such as fish, meat, or vegetable tempura.
2. Bottled prepared sauces, such as soy sauce, teriyaki, or tempura. They can be high in sodium.
3. Commercial miso soups.

TRY INSTEAD

1. Steamed or grilled (kushiyaki) vegetable, fish, shellfish, or skinless poultry dishes with a little tempura dipping sauce on the side.
2. Reducing the amount of bottled sauce you use, switching to a reduced-sodium soy sauce, or preparing daikon sauce. To make this sauce, grate fresh daikon radish and add lemon juice, chopped green onions, and a little reduced-sodium soy sauce.
3. Preparing miso soup at home with a lower-salt miso paste.

African-American

The traditional African-American diet has copious amounts of healthful vegetables and whole-grain foods. Many dishes, though, also call for fatty meats and a fair amount of fat added during preparation. Try some of these alternatives.

LIMIT

1. Fried foods, such as fried chicken, fried fish, or beef livers with onions.
2. High-fat cuts of meat, such as pig feet, chitterlings (chitlins) and hogmaws, pig tails, spareribs, sausage, regular bacon, oxtails, goat, and ham hocks.
3. Vegetables and beans prepared with fat back, neck bones, salt pork, ham hocks, or coconut oil.
4. Hominy grits with butter, baked macaroni and cheese, and home fries.
5. Ice cream, frosted cake, and fruit cobbler made with lard or shortening.

TRY INSTEAD

1. Baked, broiled, or stewed chicken (no skin); baked, broiled, or steamed fish. Eat liver (chicken or beef) only occasionally and bake or broil instead of frying.
2. Lower-fat meats, such as skinless chicken or turkey, fish, lean stew beef, beef tenderloin, center-cut ham, Canadian bacon, pork tenderloin, and pork loin chops.
3. Vegetables and beans prepared with turkey parts (no skin) and selected herbs for seasoning.
4. Hominy grits with a little margarine, baked macaroni and low-fat cheese (5 or fewer grams of fat per ounce), or baked or boiled potatoes.
5. Low-fat frozen yogurt, ice milk, fruit ice, angel food cake with fruit instead of icing, or fruit cobbler made with a small amount of liquid vegetable oil or margarine.

Hispanic

Traditionally, Hispanic foods are hearty fare, relying heavily on rice, meat, tortillas, beans, salted fish, and fats. By making some simple ingredient and preparation changes, traditional Hispanic cuisine can be more healthful as well.

LIMIT

1. Corn chips (fried), pork skins (*chicharron*), and guacamole.
2. Butter, lard, shortening, coconut oil, palm oil, and salt pork.
3. Prepared sauces, such as *sofrito* and *mojo criollo.*
4. Fried foods, such as fried tortillas, fried vegetable fritters (*alcapurrias*), banana slices (*tostones*), fried pork or chicken with skin, and refried beans.
5. High-fat meats and salad dishes, such as roast pork *(lechon asado)*, shrimp, tuna, and lobster salads made with mayonnaise, *picadillo* with ground beef, and chicken stew (*pollo asopao*).
6. Higher-fat ground beef-filled tacos, burritos, and enchiladas.
7. Malt beer and egg yolk (*malta*), coffee with cream or evaporated whole milk, and blender drinks with whole milk (*batidas*).

TRY INSTEAD

1. Baked tortillas, popcorn, regular salsa, or sauce made from low-fat or plain yogurt with tabasco and coriander.
2. Small amounts of corn, olive, safflower, canola, sunflower, soybean, cottonseed oils, or margarines/spreads made from these oils.
3. Homemade *sofrito* with a little liquid oil instead of lard. For *mojo criollo,* use garlic powder and seasonings, but no meat drippings.
4. Baked tortillas, boiled dumplings (*pasteles*), boiled bananas (*tuineo, platano*), baked or broiled meat or chicken without skin, and boiled beans.
5. Well-trimmed roast pork, salads with reduced-calorie mayonnaise or low-fat or fat-free plain yogurt, *picadillo* with lean ground turkey, or chicken stew with skinless chicken.
6. Lean beef-, chicken-, or bean-filled soft burritos, enchiladas, or tacos.
7. Plain malt beer, juice, coffee with fat-free or low-fat milk, and blender drinks with fat-free yogurt or fat-free milk.

Source: Adapted from Ethnic and Healthful, *Nutrition Counselor,* Vol. VII, 1990.

Nutrition on the Web

nutrition.wadsworth.com	Go to the *Personal Nutrition* site to check for the latest updates to chapter topics or to access links to related Web sites.
www.nap.edu/readingroom	Look here for updates about the new Dietary Reference Intakes (DRIs).
www.nal.usda.gov/fnic/consumer/index.html	Go to It's All About You for practical tips for following the Dietary Guidelines.
www.nal.usda.gov/fnic/Fpyr/pyramid.html	Scroll down and click on the line for even more Food Guide Pyramids to view a variety of Ethnic/Cultural/Special Audience Pyramids.
www.oldwayspt.org/html/p_latin.htm	Latin American, Asian, Mediterranean, and vegetarian Pyramids.
www.nal.usda.gov/fnic/foodcomp	Click on Search the Database for free food analysis. Just type in the food you want to analyze, and get a breakdown of its calories, fat, fiber, protein, vitamins, and minerals.
www.ag.uiuc.edu/~food-lab/nat	A free diet analysis program developed at the University of Illinois—Urbana/Champaign. Allows anyone to analyze the foods they eat for various nutrients.
www.nal.usda.gov	The Food and Nutrition Information Center's Web site contains resources for comparing nutrient analysis software packages. Click on Publications/Databases. Scroll down to Food and Nutrition Software and Multimedia Programs.
vm.cfsan.fda.gov/label.html	Useful facts about food labels; updates on label health claims.
www.healthfinder.gov/searchoptions/topicsaz.htm	Search for food label information.
www.eatethnic.com	Useful facts and fun trivia about ethnic foods, international holiday traditions, religious diets, regional customs, recipes. Be sure to take the Cultural Foods Quiz.

Check Yourself . . .

1. Name two characteristics of a healthful diet.
2. Describe the concept of nutrient density.
3. Identify the five food groups shown in the Food Guide Pyramid and the approximate number of servings you need from each group.
4. Name six of the nutrients that must appear on virtually all food labels (excluding small labels and other exceptions).
5. Explain how the % Daily Value column on food labels can be used.
6. Name the ten Dietary Guidelines for Americans.
7. Explain the limitations of using the RDA for calories to evaluate individual diets.
8. List three examples of relatively low-fat snacks.
9. Describe the order in which ingredients are listed on food labels.
10. List the three calorie-yielding nutrients.

Answers to Check Yourself questions are found in Appendix G.

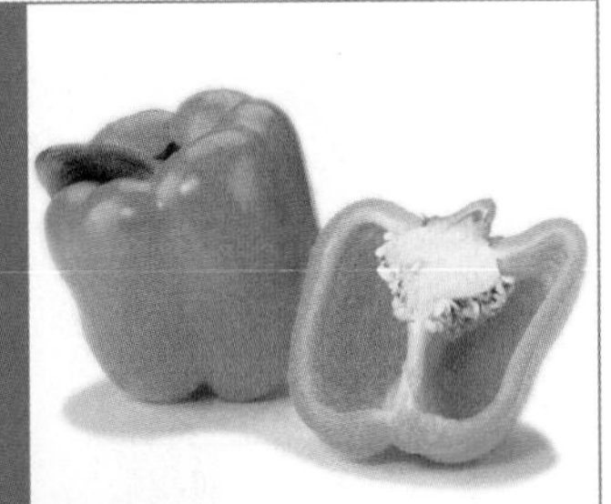

3

The Carbohydrates: Sugar, Starch, and Fiber

CONTENTS

Carbohydrate Basics

The Simple Carbohydrates

The Complex Carbohydrates: Starch in the Diet

Nutrition Action: Keeping Out from Under the Dental Drill

The Savvy Diner: Selection Tips for Carbohydrates in the Diet

The Complex Carbohydrates: Fiber in the Diet

How the Body Handles Carbohydrates

Carbohydrate Consumption Scorecard

Spotlight: Sweet Talk—Alternatives to Sugar

Rabbit said, "Honey or condensed milk with your bread?" [Pooh] was so excited that he said, "Both," and then, so as not to seem greedy, he added, "But don't bother about the bread, please."

from *Winnie the Pooh*, A.A. Milne (1882–1956, children's book author)

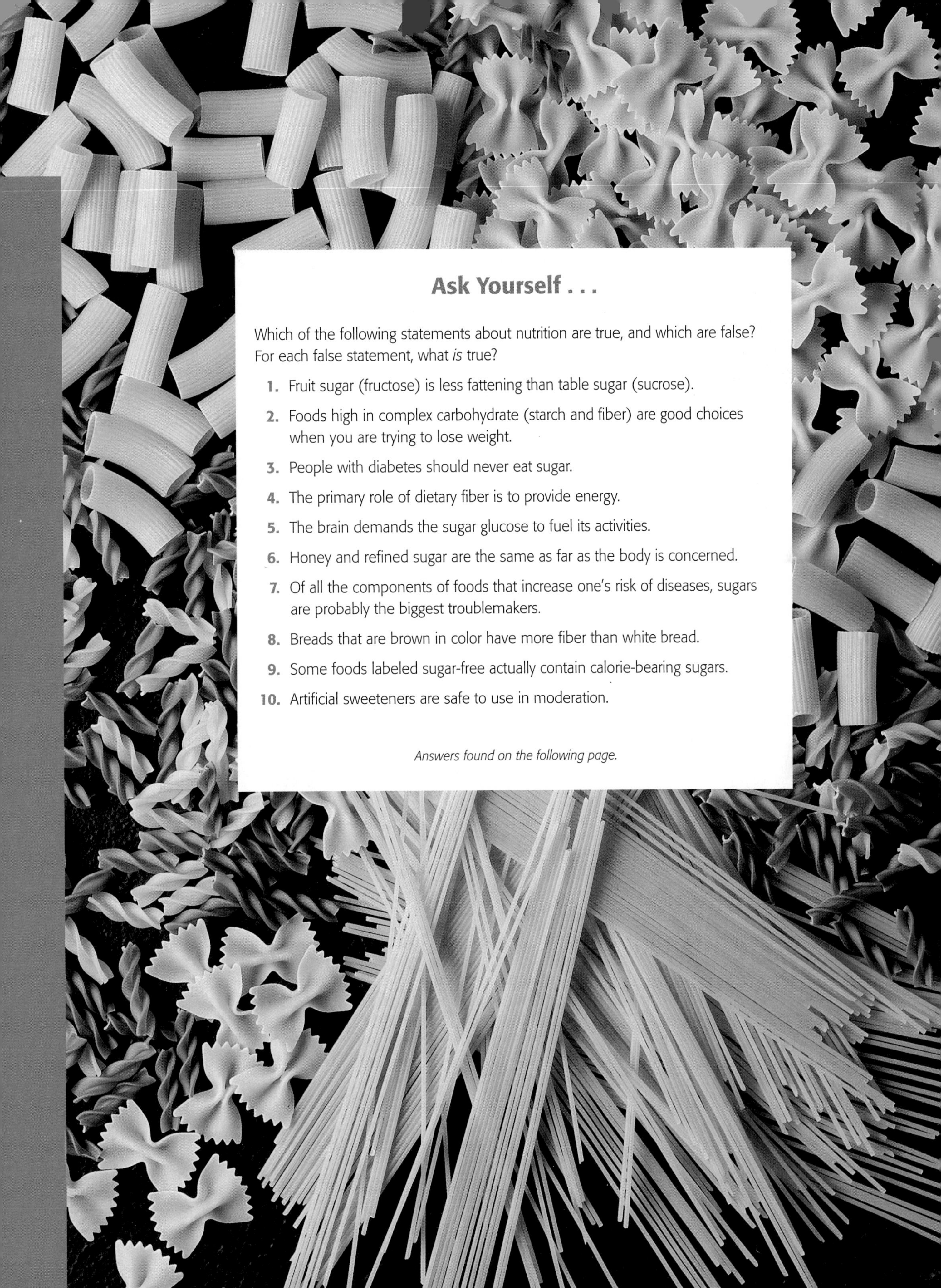

Ask Yourself . . .

Which of the following statements about nutrition are true, and which are false? For each false statement, what *is* true?

1. Fruit sugar (fructose) is less fattening than table sugar (sucrose).
2. Foods high in complex carbohydrate (starch and fiber) are good choices when you are trying to lose weight.
3. People with diabetes should never eat sugar.
4. The primary role of dietary fiber is to provide energy.
5. The brain demands the sugar glucose to fuel its activities.
6. Honey and refined sugar are the same as far as the body is concerned.
7. Of all the components of foods that increase one's risk of diseases, sugars are probably the biggest troublemakers.
8. Breads that are brown in color have more fiber than white bread.
9. Some foods labeled sugar-free actually contain calorie-bearing sugars.
10. Artificial sweeteners are safe to use in moderation.

Answers found on the following page.

Once upon a time, bread, potatoes, pasta, and other starchy foods were placed on the dieter's list of most-fattening or "illegal" foods. This unattractive image doubtless comes from the practice of serving many carbohydrate-rich foods laden with fat—potatoes with sour cream and butter, vegetables or pasta with rich cream sauces, toast with butter, and salads with fat-rich dressings. People who need to lose weight must limit high-calorie foods, but they are ill-advised to try to avoid all **carbohydrate.** It is the fat, not the carbohydrate, that raises the calorie count the most.

CARBOHYDRATES compounds made of single sugars or multiples of them and composed of carbon, hydrogen, and oxygen atoms.
carbo = carbon (C)
hydrate = water (H_2O)

This chapter invites you to learn to distinguish between certain carbohydrates, such as starch and fiber, and others, such as concentrated sugars. You will learn to choose your calories by the company they keep (see Table 3-1).

CARBOHYDRATE BASICS

The primary role of carbohydrates is to provide the body with energy (calories), and for certain body systems (for example, the brain and the nervous system), carbohydrates are the preferred energy source. Carbohydrates are the ideal fuel for the body. There are only two alternative calorie sources: protein and fat. Protein-rich foods are usually expensive and provide no advantage over carbohydrates when used to provide fuel for the body. Fat-rich foods might be less expensive, but fat cannot be used efficiently as fuel by the brain and nerves, and diets high in fat are associated with many chronic diseases. Thus, of all alternative food-energy sources, carbohydrates are preferred; they provide most of the day's energy for most of the world's people.

TABLE 3-1 CHOOSING CARBOHYDRATES BY THE COMPANY THEY KEEP

Food	Calories	Grams of Fat
Toast (2 slices)		
with margarine (2 tsp)	188	9
with low-sugar jelly/fruit spread (2 tsp)	139	2
Potato (medium)		
with margarine (1 tsp) and sour cream (1 tbsp)	287	10
with yogurt cheese (2 tbsp)*	223	0
Bagel (medium)		
with regular cream cheese (2 tbsp)	263	11
with fat-free cream cheese (2 tbsp)	188	1
Pasta (1 cup)		
with Alfredo sauce (⅓ c)	390	20
with tomato and mushroom sauce (⅓ c)	232	4

*To make yogurt cheese: Drain nonfat plain yogurt through cheesecloth to thicken to consistency of cream cheese. Add herbs for flavor. Or, try the *fat-free* sour cream now on the market.
Source: Adapted from *Environmental Nutrition* Vol. 17, February 1994, p.2.

Carbohydrates are divided into two categories: complex carbohydrates and simple carbohydrates. **Complex carbohydrates** include starch and fiber. Starches make up a large part of the world's food supply—mostly as grains. Consider such staples as wheat, rice, and corn, which are rich sources of starch. Fiber is found abundantly in plants, especially in the outer portions of cereal grains, and in fruits, legumes, and most vegetables. **Simple carbohydrates** include naturally occurring sugars in fresh fruits and some vegetables and in milk and milk products and added sugars in concentrated form, as in honey, corn syrup, or sugar in the sugar bowl. All of these carbohydrates have characteristics in common, but they are of different merit nutritionally. Table 3-2 introduces the different types of carbohydrates.

COMPLEX CARBOHYDRATES long chains of sugars (glucose) arranged as starch or fiber. Also called polysaccharides.
poly = many
saccharides = sugar unit

SIMPLE CARBOHYDRATES (sugars) the single sugars (monosaccharides) and the pairs of sugars (disaccharides) linked together.

Ask Yourself Answers: 1. False. Fructose and sucrose are equally fattening because they have the same number of calories per gram. 2. True. 3. False. People with diabetes need to watch the total carbohydrate in their diets, but they can choose foods with sugar as a small portion of that total. 4. False. Although certain fibers may provide negligible calories to the diet, fiber's primary role is in providing bulk for the digestive tract. 5. True. 6. True. 7. False. Of all the things in foods associated with risk of diseases, fat is by far the biggest troublemaker. 8. Not always. Being brown in color does not always mean the bread is high in fiber. The color can come from molasses or caramel. To be high in fiber, the label must say *whole-grain* or *whole-wheat*, not simply *wheat* bread. *Whole-grain* flour should be listed first in the ingredients list. 9. True. 10. True.

TABLE 3-2 CATEGORIES AND SOURCES OF CARBOHYDRATE

Carbohydrate Type	Common Names	Examples of Food Sources
Monosaccharides		
Glucose ●	Dextrose, blood sugar	Fruits, sweeteners
Fructose ■	Fruit sugar, levulose	Fruits, honey, high-fructose corn syrup
Galactose ▲	—	Part of lactose, found in milk
Disaccharides		
● + ■ Sucrose (glucose + fructose)	Table sugar	Beet and cane sugar, fruit, most sweets
● + ▲ Lactose (glucose + galactose)	Milk sugar	Milk and milk products
● + ● Maltose (glucose + glucose)	Malt sugar	Sprouted seeds
Polysaccharides		
Starches ●-●-●-●-●-●-●	Dextrins	Potatoes, legumes, corn, wheat, rye, and other grains
Dietary fiber	Roughage, bulk:	Whole grains, legumes, fruits, vegetables
Insoluble fibers	cellulose, hemicellulose	Wheat products, brown rice, vegetables, legumes, seeds
Soluble fibers	pectins, gums, mucilages, some hemicelluloses	Oat products, barley, legumes, fruits, vegetables, seeds

THE SIMPLE CARBOHYDRATES

Carbohydrate-rich foods are obtained almost exclusively from plants. Milk is the only animal-derived food that contains significant amounts of carbohydrate. All carbohydrates are composed of single sugars, alone or in various combinations.

The Single Sugars: Monosaccharides

All of the carbohydrates are made of simple sugars, and all carbohydrates but fiber can quickly be converted to **glucose** in the body. Green plants make glucose from carbon dioxide and water through a process known as photosynthesis in the presence of chlorophyll and sunlight, as illustrated in the margin.

Glucose is not a very sweet sugar, but plants can rearrange its atoms to form another sugar, **fructose,** which is sweet to the taste. Fructose is found mostly in fruits, in honey, and as part of another sugar—table sugar. Glucose and fructose are the most common single sugars in nature.

The Double Sugars: Disaccharides

Some sugars are double sugars, made by bonding two single sugars together. When glucose and fructose are bonded together, they form **sucrose,** or table sugar, the product most people refer to when they use the term *sugar.* The sweet taste of sucrose comes primarily from the fructose in its structure. It occurs naturally in many fruits and vegetables. Sugar cane and sugar beets are two sources from which sucrose is purified and granulated to various extents to provide the brown, white, and powdered sugars available in the supermarket. Sucrose is one of the two most caloric ingredients of candy, cakes, pastries, frostings, cookies, presweetened ready-to-eat cereals, and other concentrated sweets. (The other major calorie contributor is fat, discussed in Chapter 4.)

Another double sugar, **maltose,** consists of two glucose units. It occurs in sprouting seeds and arises during the digestion of starch in the human body. The malt found in beer contains maltose. Enzymes used in the brewing process break down the long chains of starch in barley and wheat into maltose units.

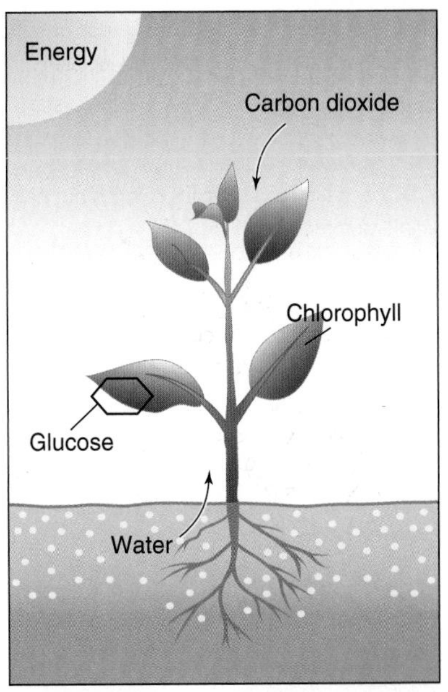

Photosynthesis: In the presence of chlorophyll and the energy of the sun, plants make glucose. Water, absorbed by the plant's roots, and carbon dioxide, absorbed through the plant's leaves, combine to form a molecule of glucose.

GLUCOSE (GLOO-koce) the building block of carbohydrate; a single sugar used in both plant and animal tissues as quick-energy. A single sugar is known as a **monosaccharide.**
mono = one

FRUCTOSE (FROOK-toce) fruit sugar—the sweetest of the single sugars. Another single sugar, **galactose** (ga-LACK-toce), occurs bonded to glucose in the sugar of milk. A double sugar is known as a **disaccharide.**
di = two

SUCROSE (SOO-crose) a double sugar composed of glucose and fructose.

MALTOSE a double sugar composed of two glucose units.

LACTOSE a double sugar composed of glucose and galactose; commonly known as milk sugar.

ENZYMES protein catalysts. A catalyst facilitates a chemical reaction without itself being altered in the process. (Proteins are discussed in Chapter 5; digestive enzymes in Appendix B.)

LACTOSE INTOLERANCE inability to digest lactose as a result of a lack of the necessary enzyme lactase. Symptoms include nausea, abdominal pain, diarrhea, or excessive gas that occurs anywhere from 15 minutes to a couple of hours after consuming milk or milk products. Chapter 7 provides the lactose contents of selected dairy products.

EMPTY-CALORIE FOODS a phrase used to indicate that a food supplies calories but negligible nutrients.

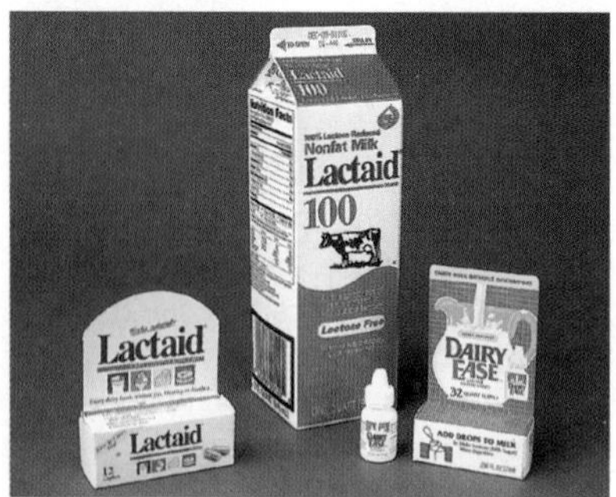

Lactose-reduced milk, enzyme solutions containing lactase for treating dairy products, and lactase tablets can help reduce the symptoms of lactose intolerance.

Finally, there is **lactose,** the major sugar in milk, a double sugar made by mammals from galactose and glucose units. A human baby is born with the digestive **enzymes** necessary to split lactose into its two simple sugars—glucose and galactose—so that they can be absorbed. Lactose facilitates the absorption of calcium and promotes the growth of beneficial bacteria in the intestines. Breastmilk and infant formula, which contain lactose, are ideal foods for babies because they provide a simple, easily digested carbohydrate to meet an infant's energy needs.

When you eat a food containing lactose, the enzyme lactase in your small intestine first splits the double sugar into single sugars so that they can enter your bloodstream. Many people can lose the ability to digest lactose during or after childhood. Thereafter, the person may experience nausea, bloating, abdominal pain or cramping, diarrhea, or excessive gas after drinking milk or eating lactose-containing products because the intestinal bacteria will use the lactose for energy, producing gas and other products that irritate the intestine. This condition—**lactose intolerance**—is inherited by about 75 percent of the world's people. It is most common in African-Americans, Mediterranean peoples, Native Americans, and Asians, and less common in people of northern European origin.[1] It also can develop temporarily in anyone who is malnourished or sick, making the avoidance of milk and milk products temporarily necessary.

Many people with lactose intolerance are able to consume small amounts of lactose without symptoms.[2] For them, lower-lactose foods such as yogurt, acidophilus milk, aged cheeses, cottage cheese, or specially prepared milk products that have been treated with an enzyme to reduce lactose may be tolerated.

Sugar and Health

Sugar is often in the headlines and has been accused of contributing to a host of human ills such as tooth decay, obesity, diabetes, heart disease, hyperactive behavior in children, and even criminal behavior. Nevertheless, research studies have not shown a direct link between sugar and any of these conditions, except tooth decay (see the Nutrition Action feature starting on page 73).[3] However, eating a lot of sugar could mean that you are eating an inadequate amount of foods containing essential nutrients. Conversely, if you eat a lot of sugar without eating less of other foods, you might be getting too many calories. Excess calories from any energy nutrient, even protein, are stored as body fat. Evidence from population studies in many countries shows that obesity rates rise as sugar consumption increases. One reason may be that many sugary foods, such as candy bars, are also high in fat.

There is no reason to believe that moderate consumption of sugar (5 percent to 10 percent of calories) is dangerous to a healthy human being. The dietary goal to limit sugar to less than 10 percent of total calories does not apply to *all* sugars in the diet. The diluted *naturally occurring* sugars found in milk and fruits should not be confused with concentrated refined sugars, such as table sugar, honey, and corn syrup. These concentrated sweets should be limited, so as not to displace needed nutrients.

When people learn that fruit's energy comes from simple sugars, they may think that eating fruit is the same as consuming concentrated sweets such as candy or soft drinks. However, fruits differ from candy and soft drinks in important ways. Their sugars are diluted in large volumes of water, packaged in fiber, and mixed with many vitamins and minerals needed by the body (see Table 3-3). In contrast, concentrated sweets such as honey and table sugar are merely—as the popular phrase calls them—**empty-calorie foods.** How much sugar *do* you eat? Try the Carbohydrate Consumption Scorecard on page 86, which checks your diet for sugar and fiber, to find out.

Keeping Sweetness in the Diet

The taste of sweetness is a pleasure; the liking for it is innate. However, the *Dietary Guidelines for Americans* recommends that people choose beverages and foods that

TABLE 3-3 SAMPLE NUTRIENTS IN FRUITS AND SUGARS

	Size of 100-calorie portion	Carbohydrate (g)	Fiber (g)	Vitamin A (RE)	Vitamin C (mg)	Folate (mg)	Potassium (mg)
Fruits							
Apricots	6	24	6	548	20	18	622
Apple	1 large	25	5	11	12	6	244
Cantaloupe	½ 5″ melon	23	2	889	116	47	853
Orange	1¼ c sections	26	5	48	120	68	408
Pineapple	1¼ c chunks	24	2.5	4	30	20	219
Strawberries	2 c	20	6	8	164	50	478
Watermelon	1¼ c chunks	22	2	112	30	6	352
Sugars							
Cola beverage	8 oz	26	0	0	0	0	3
Honey	1½ tbsp	26	0	0	0	<1	22
Jelly	2 tbsp	26	0	0	<1	0	24
Sugar, white	2 tbsp	24	0	0	0	0	0
Sugar, brown	3 tbsp	24	0	0	0	0	90
Recommended Daily Values		—	25	1,000	60	400	3,500

limit their intake of sugars. To help with this task while still catering to the sweet tooth, consider the following pointers:

- Use less of all sugars, including white sugar, brown sugar, honey, jelly, and syrups.
- Eat less of foods containing large amounts of sugars, such as soft drinks, fruit drinks, candy, ice cream, cakes, and cookies.
- Select fresh fruits or fruits canned without sugar or in fruit juice rather than heavy syrup to satisfy your urge for sweets.
- Read ingredient lists on food labels for clues on sugar content—if the word *sucrose, glucose, maltose, dextrose, fructose,* or *syrup* appears first, the food contains a large amount of sugar (see Figure 2-9 on page 45). Use sparingly foods in which these sweeteners are among the first three ingredients listed. Table 3-4 shows the amounts of sugar in some common products.
- Remember, for dental health, how frequently you eat sugar is as important as—and perhaps more important than—how much sugar you eat at one time (see the Nutrition Action feature on page 73).

Alternatives to sweet desserts might be whole-grain crackers, low-fat cheese, and yogurt. Snacks for children could include fruits, vegetables, string cheese, popcorn, homemade fruit juice popsicles, and other wholesome foods. Here are some other suggestions:

- Substitute fruit *juices* or plain water for fruit drinks, regular soft drinks, and punches that contain considerable amounts of sugar.
- Buy *unsweetened* cereals so that you can control the amount of sugar added. Many cereals are presweetened. Check the Nutrition Facts panel for the grams of sugar present. Many list sugar first or second among their ingredients.
- Experiment with reducing the sugar in your favorite recipes. Some recipes taste just the same even after a 25 percent to 50 percent reduction in sugar content. Others taste different, but just as—if not more—delicious.

Honey and sucrose contain the same monosaccharides, but in sucrose they are linked together. Compared with honey or sugar, fruit is more nutrient dense.

TABLE 3-4 SUGAR IN SELECTED FOODS

Food	Teaspoons of Sugar Per Serving
Fruit drink, ade (12 oz)	12
Chocolate shake (10 oz)	9
Cola (12 oz)	8
Jellybeans (10)	7
Yogurt, fruit flavored (1 c)	7
Apple pie (⅙ of pie)	6
Cake, frosted (1/16 of cake)	6
Angel food cake (1/12 of cake)	5
Applesauce, sweetened (½ c)	5
Fig bars (1)	5
Fudge (1 oz)	5
Sherbet (½ c)	5
Cereal, sweetened (Sugar Pops, ¾ c)	4
Doughnuts, glazed (1)	4
Fruit, canned in heavy syrup (½ c)	4
Gelatin dessert (½ c)	4
Chocolate milk, 2% (1 c)	3
Ice cream, ice milk, or frozen yogurt (½ c)	3
Syrup or honey (1 tbsp)	3
Dairy creamer (1 tbsp)	2
Doughnuts, plain (1)	2
Catsup (1 tbsp)	1
Chewing gum (2 sticks)	1
Cookie, Oreo type (1)	1
Sugar, jam, or jelly (1 tsp)	1

- The sweet spices—allspice, anise, cardamom, cinnamon, cloves, fennel, ginger, and nutmeg—can replace substantial sugar in recipes. Use half as much sugar and increase one and a half times the amount of spice the recipe calls for. Increasing the amount of extracts like vanilla can enhance sweetness, too. Experiment with other extracts like maple, coconut, banana, and chocolate; add chopped dried fruit to baked goods for extra sweetness and nutrients as you decrease sugar.

Still another alternative is to use sugary foods that convey nutrients as well as calories. Examples: rather than sugar cookies, serve oatmeal cookies; rather than brownies, eat apricot bars; and rather than table sugar as a topping, add raisins and banana slices, which are really very sweet.

Keep in mind that sugar is delicious and that you can use it with discretion, but use it in moderation. The person with nutrition sense and a taste for sweets can artfully combine the two by using sugar with creative imagination to enhance the flavors of nutritious foods.

You may wonder whether using artificial sweeteners to reduce some of the total sugar in your diet is a safe and recommended strategy. The Spotlight feature "Sweet Talk," starting on page 93, will help you decide.

THE COMPLEX CARBOHYDRATES: STARCH IN THE DIET

STARCH a plant polysaccharide composed of hundreds of glucose molecules, digestible by human beings.

STAPLE GRAIN a grain used frequently or daily in the diet—for example, corn (in Mexico) or rice (in Asia).

All starchy foods are plant foods. **Starch** is made up of many glucose units bonded together—3,000 or so in each molecule of starch. Seeds such as grains, peas, and beans are the richest starch source. Most societies have a primary or **staple grain**

(text continues on page 75)

NUTRITION ACTION

Keeping Out from Under the Dental Drill

For years, conventional wisdom has held that staying away from candy, cookies, sugary soda pop, and the like is the most important line of dietary defense against cavities. Even Aristotle asked in about 350 B.C., "Why do figs when they are sweet, produce damage to the teeth?"[4]

These days, dentists advise that cutting down on the amount of sugar eaten is not the only way to prevent **dental caries.** Every bit as—if not more—important to consider is whether a food clings to the teeth and lingers in the mouth as well as when and how often it's eaten. It has to do with the mouth's level of acid. Each time you bite into a food, the bacteria that live in your mouth feed on the sugar in it and release an acid that eats away at tooth enamel. When a large number of bacteria living in a film referred to as **dental plaque** produce enough acid to dissolve a "hole" in the enamel over a period of time, the result is a cavity.

DENTAL CARIES decay of the teeth, or cavities.
caries = rottenness

DENTAL PLAQUE a colorless film, consisting of bacteria and their by-products, that is constantly forming on the teeth.

While consuming sugary items such as soft drinks boosts acid production in the mouth, so does munching on starchy foods such as crackers or pretzels. This is because enzymes in the saliva can break the carbohydrate in the cracker or pretzel into the simple sugars that bacteria feast on. If a crumb or two from a starchy food gets caught between the teeth, it might provide enough carbohydrate for bacteria to feed on for hours, thereby prolonging the teeth's exposure to acid. In fact, some research indicates that even high-sugar foods such as chocolate bars and hot fudge sundaes are less likely to contribute to cavity formation than stickier, starchier items that are likely to linger in the mouth, such as potato chips and crackers.[5]

To be sure, the carbohydrate content and stickiness of a food are only two of the many factors that influence the food's effect on teeth. Another is how often you eat the food. Each time you eat a carbohydrate-containing food, your teeth are bathed in acids for about 20 minutes. Thus, the more often you eat, the longer your teeth are exposed to harmful acids.[6]

Yet another factor is what you eat along with the food. Starchy, sugary foods tend to be less harmful to the teeth when eaten with a meal than when consumed alone. One reason may be that the mouth makes more saliva during a full meal. That's crucial, because even though saliva helps make sugar for bacteria, it also clears food particles from the mouth and neutralizes destructive acids before they can dissolve the teeth. Consider that one study at the University of Iowa's College of Dentistry found that after people nibbled on foods with a strong tendency to produce acid, such as raisins and chocolate bars, and then chewed sugarless gum, within ten minutes the gum helped stimulate the release of enough saliva to neutralize the acid flow that the sweets had caused.[7] In addition, some research suggests that, like sugarless chewing gum, such foods as cheese and peanuts help fight acid attacks stimulated by eating carbohydrate-rich foods.[8]

"Just pull my sweet tooth."

Of course, while saliva helps fight cavities, the best way to ensure good dental health is to brush your teeth as soon as possible after eating, or at least swish the mouth with water after a meal to help rinse the teeth and dislodge stuck particles. Flossing daily is also important because flossing rids the mouth not only of food but also of plaque before it becomes so widespread that it produces tooth-threatening levels of acid. Regular visits to the dentist play a role in keeping dental health up to snuff. In addition, drinking water containing fluoride, a mineral that helps strengthen tooth enamel, can prevent cavities. People whose drinking water does not supply fluoride should check with a dentist about the need for fluoride supplements.

Along with practicing good dental hygiene, eating a balanced diet helps keep the mouth healthy. One reason is that if the diet lacks essential nutrients, mouth tissues can become compromised, leaving them particularly vulnerable to infection. In fact, some experts believe that **periodontal disease** is especially severe among people who have poor diets.[9] Another reason to eat an adequate diet, particularly for children, is that it helps to ensure proper development of the teeth.[10] Similarly, eating well should rank as a high priority for a pregnant woman, whose unborn baby's teeth, among other things, start forming after just six weeks and begin to harden between the first and second trimester of pregnancy.

PERIODONTAL DISEASE inflammation or degeneration of the tissues that surround and support the teeth.

Incidentally, parents should never allow an infant to sleep with a bottle filled with sweetened liquids, fruit juices, milk, or formula. Little ones allowed to do so often develop what is known as **nursing bottle syndrome.** As a baby sucks on a bottle, the tongue pushes outward slightly and covers the lower teeth. If the infant falls asleep with the bottle in the mouth, the liquid bathes the teeth, particularly the upper teeth not protected by the tongue, thereby literally soaking them in cavity-causing carbohydrates for hours at a time.

NURSING BOTTLE SYNDROME (also called baby bottle tooth decay) decay of all the upper and sometimes the back lower teeth that occurs in infants given carbohydrate-containing liquids when they sleep. The syndrome can also develop in babies given bottles of liquid to carry around and sip all day.

For optimal dental health, the American Dental Association (ADA) recommends:[11]

- Eat a balanced diet.
- Keep snacking to a minimum, if possible. The ADA recognizes that some people, such as people with diabetes, may require snacks. For others, however, the ADA suggests limiting the number of snacks if brushing the teeth is not possible shortly after eating them.
- Eat sweets with meals rather than between them.
- Brush and floss thoroughly each day to remove dental plaque.
- Use an ADA-accepted fluoride toothpaste and mouth rinse and talk to your dentist about the need for supplemental fluoride.
- Visit a dentist regularly.
- Do not allow infants to sleep with bottles in their mouth that contain sweetened liquids, fruit juices, milk, or formula.

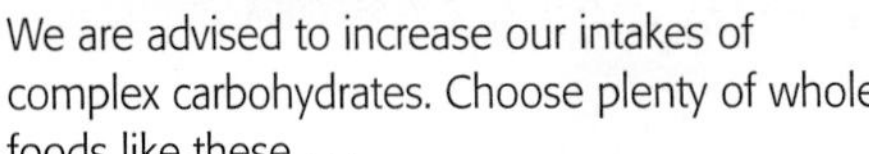

We are advised to increase our intakes of complex carbohydrates. Choose plenty of whole foods like these . . .

. . . and fewer foods like these—foods that no longer resemble their original farm-grown products.

that provides most of the people's food energy. In many Asian nations, the staple grain is rice. In Canada, the United States, and Europe, the staple grain is wheat. If you consider all the food products made from wheat—bread (and other baked goods made from wheat flour), cereals, and pasta—you will realize how all-pervasive this grain is in the food supply. The staple grains of other peoples include corn, millet, rye, barley, and oats.

A second important source of starch is the legume family, including such dried beans and peas as butter beans, kidney beans, pinto beans, navy beans, black-eyed peas, chick-peas (garbanzo beans), lentils, and soybeans. These vegetables are about 40 percent starch by weight and contain abundant protein. Root vegetables (such as yams) and tubers (such as potatoes) are other sources of starch that are important in many societies. Table 3-5 shows the types of carbohydrates found in foods from the Food Guide Pyramid.

Starch can be broken down during food processing to shorter chains of glucose units known as dextrins. The word dextrins sometimes appears on food labels, because dextrins can be used as thickening agents in foods.

TABLE 3-5 TYPES OF CARBOHYDRATES FOUND IN SELECTED FOODS IN THE FOOD GUIDE PYRAMID

	Sugar	Starch	Fiber
Bread, Cereal, Rice, and Pasta Group			
Bread, cooked grains	✔	✔	✔
Vegetable Group			
Corn, peas	✔	✔	✔
Green beans	✔		✔
Potato	✔	✔	✔
Tomato	✔		✔
Fruit Group			
Apple, banana	✔ (mostly fructose)		✔
Orange juice	✔ (mostly fructose)		
Milk, Yogurt, and Cheese Group			
Milk	✔ (lactose)		
Meat, Poultry, Fish, Dry Beans, Eggs, and Nuts Groups			
Meat, poultry, fish, eggs			
Legumes	✔	✔	✔
Sugar	✔ (sucrose)		

FIGURE 3-1
COMPARING PYRAMIDS: ACTUAL CONSUMPTION VERSUS RECOMMENDED SERVINGS OF CARBOHYDRATE IN THE DIET

While total carbohydrate in the diet is supposed to increase, sugar as a percentage of the total carbohydrate is supposed to decrease. The figure shows the actual servings being consumed from the Food Guide Pyramid compared to the recommended number of servings.

*Data from U.S. Department of Agriculture, Agricultural Research Service, Continuing Survey of Food Intakes by Individuals, 1997.

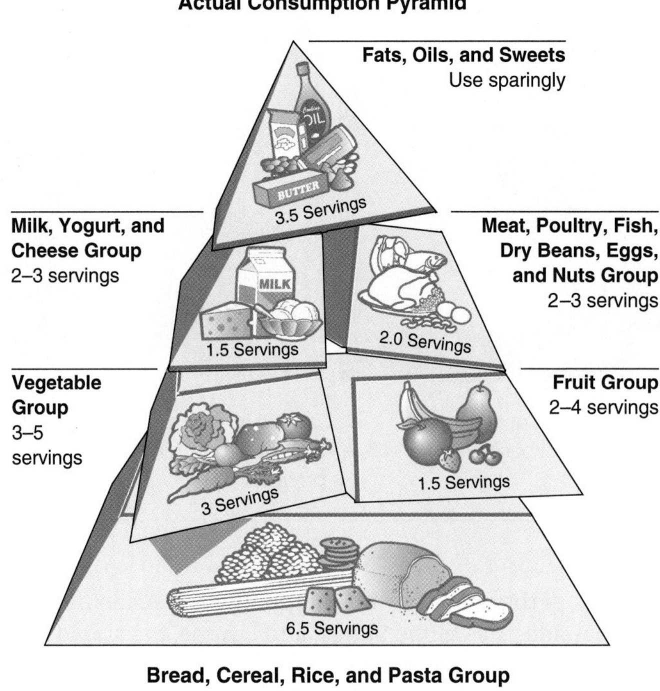

Complex carbohydrates are thought to be our most valuable energy nutrient. The *Dietary Guidelines for Americans* includes the following suggestions:

- Choose a variety of grains daily, especially whole grains.
- Choose a variety of fruits and vegetables daily.

Figure 3-1 compares the number of servings of carbohydrates Americans actually consume with the recommended number of servings from the Food Guide Pyramid.[12] As you can see, the shape of the pyramid actually consumed is somewhat top-heavy, with liberal amounts of servings from the *Fats, Oils, and Sweets Group*, and is rather lean with servings from the groups offering complex carbohydrates—notably the *Bread, Cereal, Rice, and Pasta Group, Vegetable Group,* and *Fruit Group.* The Savvy Diner feature on the following page presents pointers on how to build a more stable pyramid.

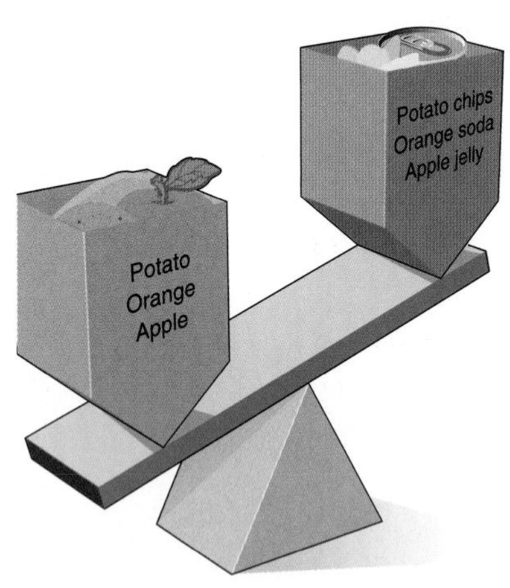

WHOLE FOOD a food that is altered as little as possible from the plant or animal tissue from which it was taken—such as beets, milk, oats, potatoes, or apples. A whole food has the nutritional advantage over its empty-calorie processed forms.

Adding Whole Foods to the Diet

As it happens, most of the nutrition goals and guidelines offered to date have had one point in common. They tend to favor a return to a more **whole-food,** plant-based diet. Generally speaking, the more a food resembles the original, farm-grown product, the more nutritious it is likely to be. During processing, some nutrients may be lost, and often nutrient-poor additions such as sugar, salt, and fat are made. For example, a potato contains 20 milligrams of vitamin C, the same number of calories in French fries contain only about 7 milligrams, and the same number of calories in potato chips contain only 2 milligrams of vitamin C.

The health benefits to be expected from a change to fewer processed and more whole-food, plant-based products are many. A diet lower in servings from the *Fats, Oils, and Sweets Group* and higher in foods containing complex carbohydrates will almost necessarily be lower in fat and calories and higher in fiber. Working together, these factors might be expected to bring about or contribute to lower

(text continues on page 80)

Selection Tips for Carbohydrates in the Diet

Not long after potatoes were brought from the Americas to Europe 500 years ago, they were banned for fear of causing leprosy, as well as sold at high prices as a cure for impotence. By the 1700s, the Irish were among the first to recognize their value as an important source of nourishment.[13]

The Food Guide Pyramid illustrates the goal of healthy eating as a shift away from a diet based on high-protein, higher-fat foods to one that uses more of the complex carbohydrates found in whole grains, vegetables, and fruits. This shift is reflected visually by the lower portions of the pyramid, which encourage you to eat proportionally more servings of grain products, vegetables, and fruit. This dining feature provides tips for selecting carbohydrates for your diet—at the supermarket, in the kitchen, and at the table—while using the Food Guide Pyramid as your guide.

At the Supermarket

BREAD, CEREAL, RICE, AND PASTA GROUP

One serving equals 1 slice of bread; half an English muffin, bun, or bagel, a 6-inch tortilla, 1 small muffin; ½ cup of cooked cereal, rice, or pasta; or 1 ounce (about 1 cup) dry cereal.

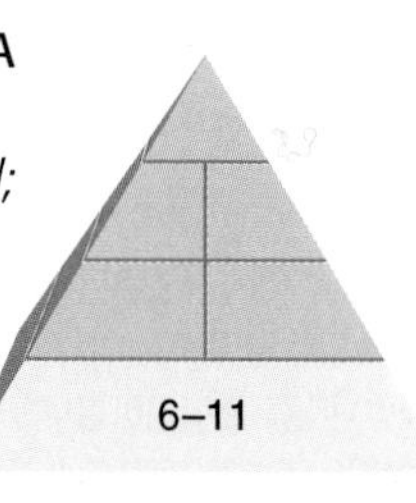

- Choose most often those foods that contain little fat (less than 3 grams of fat per serving): bagels, breadsticks, low-fat crackers (for example, animal crackers, matzoh crackers, melba toast, oyster crackers, saltines, whole-wheat crackers), low-fat cookies (for example, fig bars, gingersnaps, graham crackers, vanilla wafers), angel food cake, tortillas (not fried), couscous, English muffins, enriched breads, farina, grits, pancakes, pastas, popcorn (air-popped), pretzels, ready-to-eat cereals, rice, rice cakes, taco shells, whole grains (wheat, barley, oats, bulgur, quinoa, millet, rye).
- Select sparingly those foods that are higher in fat (more than 3 to 5 grams of fat per serving): biscuits, brownies, cakes, cheese curls, chow mein noodles, cookies, corn bread, corn chips, croissants, croutons, cupcakes, doughnuts, fried rice, granola, pastry, pizza crust, popcorn (commercially popped), presweetened cereals, stuffing, toaster pastry, tortillas (fried), waffles.
- Choose whole-grain breads, cereals, and snack foods. Buy products that list a whole grain or whole-wheat or other whole-grain flour as the first ingredient. If the label states "wheat flour" or "enriched wheat flour" the product is not whole grain. Check the food label for fiber content.
- Choose whole fruits instead of juice. Look for nonwaxed varieties of fruits and vegetables so that you can leave the skin on.
- Bring home a variety of fiber-rich snacks such as popcorn, whole-grain crackers and low-fat muffins, fresh fruits and vegetables, and dried fruits.

VEGETABLE GROUP

One serving equals 1 cup of raw leafy vegetables, ½ cup of other cooked or raw vegetables, or ¾ cup of vegetable juice.

3–5

- Choose a variety of vegetables; different types of vegetables provide different nutrients. Most vegetables (less than 2 grams of fat per serving) make good choices.
- Select brightly colored (red, orange, yellow) and dark green leafy vegetables several times a week—they are especially good sources of vitamins, minerals, and other beneficial substances.
- Purchase higher-fat items (more than 3 grams of fat per serving) less often: coleslaw, French fries, fried vegetables, hash browns, mashed potatoes, potato chips, potato salad, scalloped potatoes.

FRUIT GROUP

One serving equals 1 medium piece of fruit; ½ cup of chopped, cooked, or canned fruit; ¼ cup dried fruit; or ¾ cup fruit juice.

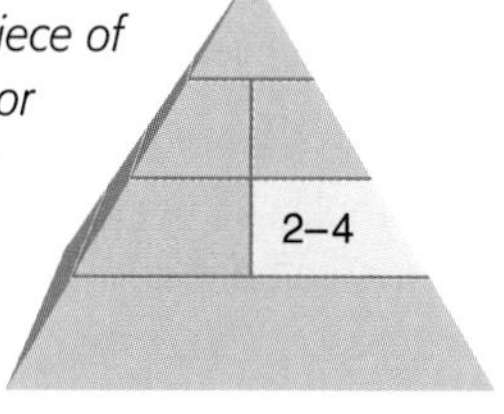

- Choose whole, fresh fruits most often—they are higher in fiber than juice.

Continued

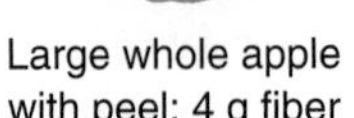

Large whole apple with peel: 4 g fiber | Applesauce, 1/2 cup: 1.5 g fiber | Apple juice, 3/4 cup: 0.2 g fiber

- Select canned fruits packed in juice or water.
- Look for frozen fruits without added sugar or syrup.
- Choose juices that are 100 percent fruit juice. Punches, ades, and most fruit drinks, contain only a little juice and lots of added sugars.
- Select olives and avocados less often—these are higher in fat.

MILK, CHEESE, AND YOGURT GROUP

The milk group provides carbohydrate as lactose, a naturally occurring simple sugar. One serving equals 1 cup of milk or yogurt, 1½ ounces natural cheese, or 2 ounces processed cheese.

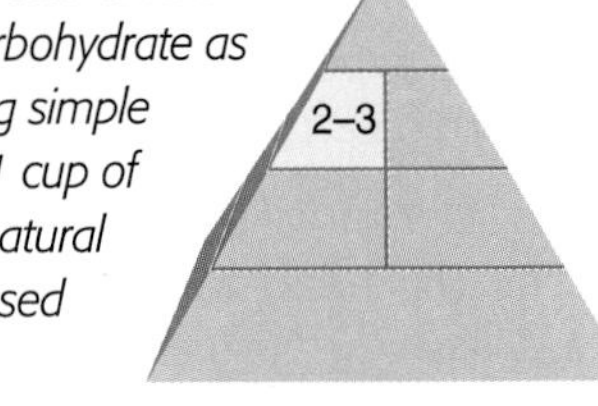

- Items in this group vary in fat content, check the food label and select the lower-fat items most often.

MEAT, POULTRY, FISH, DRY BEANS, EGGS, AND NUTS GROUP

Foods in this group provide almost no carbohydrate with the exception of dry beans and nuts. A serving equals ½ cup of cooked dry beans, ⅓ cup nuts or ¼ cup seeds, or 2 tablespoons of peanut butter.

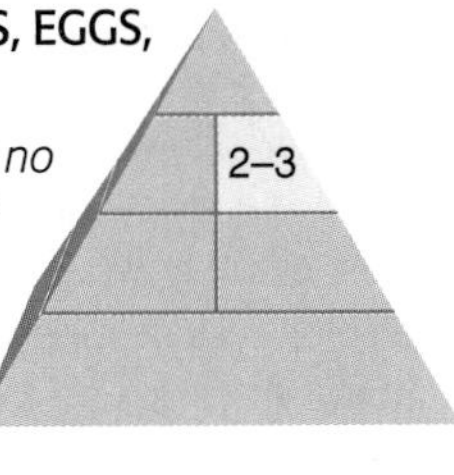

- Plan to include legumes (such as kidney beans, garbanzo beans, lentils, lima beans) in meals three or more times a week—they are especially good sources of low-fat protein, fiber, vitamins, and minerals.
- Go easy on the use of nuts and seeds—although they contain a little starch and fiber, they are high in fat.

FATS, OILS, AND SWEETS GROUP

The tiny tip of the pyramid represents foods to be used sparingly.

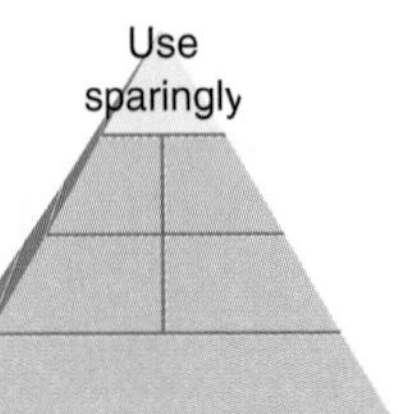

- Choose fewer foods that are high in simple sugars—candy, dessert items, soft drinks.
- Check the food labels of fat-free and reduced-fat items—these are often high in sugars.

In the Kitchen

- Complex carbohydrates are naturally low in fat. Keep them that way by going easy on the fats you add in baking and cooking—butter, margarine, gravies, dressings.
- Whenever possible, substitute whole-grain flour for all-purpose flour when baking.
- Use oats in place of flour in crumb toppings for fruit crisps. Using a blender, grind oats into flour, which you can then use to replace one-third or more of the all-purpose flour called for in recipes. Try toasted oats as a replacement for bread crumbs on top of casseroles, cooked vegetables, or fish fillets; add cinnamon to the toasted oats and sprinkle over fresh fruit and yogurt.
- Add fresh bananas or berries to muffin and pancake batters.
- Use unpeeled vegetables in soups and stews, enhance tossed green salads with high-fiber additions such as shredded carrots, sunflower seeds, roasted soybeans, and apple slices.
- Experiment with legumes: toss split peas, garbanzo beans, kidney beans, or other varieties of cooked dry beans into salads, soups, stews, pastas, tacos, or burritos; one-half cup of cooked kidney beans in a bowl of chili adds about 8 grams of dietary fiber.

At the Table

- Start your day with a high-fiber selection: a warm bowl of oatmeal with fresh fruit, bran cereals or shredded wheat with sliced fruit, banana bread, low-fat apple bran muffin, whole-grain English muffin.
- Go easy on the fats and sugars you add as spreads, seasonings, or toppings—butter, margarine, sour cream, gravy, salad dressing, sugar, jams, jellies.
- Eat at least five servings of fruits and vegetables each day. Although a variety of fruit and vegetable selections are preferred, you can meet this recommendation for five servings by eating 1½ cups of steamed broccoli and 1½ cups of grapefruit juice.
- Increase consumption of a variety of fiber-rich foods, including fruit, vegetables, legumes, and whole-grain cereals, breads, and pastas.

It's not the potatoes
that are fattening;
it's the butter, sour cream,
or gravy that they
put on us!
BUTTER
SOUR
CREAM

- Check the food label for names of sugars used in foods: brown sugar, caramel, confectioner's sugar (powdered sugar), corn syrup or sweeteners, glucose (also called dextrose), fructose (also called levulose), fruit juice concentrate, high-fructose corn syrup, honey, invert sugar, lactose, maltose, malt syrup, maple syrup, molasses, polydextrose, sucrose, sugar, syrup, turbinado sugar.

- Naturally sweet foods such as fruit can satisfy your sweet tooth.
- For a fun dessert, put a whole ripe banana on a cookie sheet (leave the peel on) and bake at 350 degrees for 20 minutes. Split baked fruit with knife; sprinkle with cinnamon or nutmeg.
- Eat fiber-rich snacks—popcorn, fresh fruits, raw vegetables with low-fat bean dip, dried fruits, whole-grain pita wedges with high-fiber salsas (see recipe).
- Track your fiber intake: For every serving of whole-grain food, count 2 grams of fiber; for each serving of refined-grain products, count 1 gram of fiber; for every serving of fruits or vegetables, count 2 grams of fiber; for every serving of cooked dry beans, count 5 grams; for high-fiber breakfast cereals count 3 to 5 grams. Note that an intake of three servings of whole grains, three servings of refined-grain foods, five servings of fruits and vegetables, one serving of legumes, and a high-fiber breakfast cereal meets the recommended intake for fiber.
- Remember to add high-fiber foods to the diet gradually. Be sure to consume adequate fluids to allow your body to adjust.

Fresh Tomato, Corn, and Black Bean Salsa

2 large tomatoes, cut into small chunks
1½ cups fresh or frozen corn kernels (steam fresh corn for 5 minutes and remove from cobs; thaw frozen corn before using)
1 15-oz can black beans, rinsed and drained (or 1 ¾ cup cooked black beans)
½ cup red onion, chopped fine
1–2 tbsp minced fresh cilantro
4 cloves garlic, minced
juice of two limes
2 tsp ground cumin
½ tsp salt

Combine all ingredients and refrigerate for two or more hours to allow flavors to blend.

Yield: about 4 cups; Serving size: 2 tablespoons; Calories per serving: 20; Fat: <1 gram

Serve with pita bread wedges, reduced-fat crackers, or baguette slices; or make a vegetarian sandwich by spooning salsa into a warm flour tortilla and rolling it up.

rates of the so-called lifestyle diseases—including obesity, heart disease, diabetes, cancer, and osteoporosis.

Foods containing complex carbohydrates are usually lower in calories per portion than protein-rich foods because the latter often include considerable fat. A diet high in complex carbohydrates might be more beneficial for weight loss than a diet of comparable calories that is high in fat. Researchers report that altering the composition of the diet to achieve a higher carbohydrate-to-fat ratio may actually decrease the incidence of obesity.[14] One reason is that switching to a high complex carbohydrate (high-fiber) diet tends to make you feel full faster, which may reduce caloric intake.

The Bread Box: Refined, Enriched, and Whole-Grain Breads

For many people, grains supply much of the carbohydrate, or at least most of the starch, in a day's meals. Because grains have such a primary place in the diet, be sure that the grains you choose—wheat, rice, oats, or corn—contribute the nutrients you need. Learn the meanings of the words associated with the flours that make up the grain products you use—**refined, enriched, fortified,** and **whole grain.** This discussion of the nutritional differences between different breads provides an example of an important principle of nutrition: foods far removed from the original state of wholeness may be lacking in significant nutrients.

The part of the wheat plant that is made into flour and then into bread and other baked goods is the kernel. The wheat kernel (a whole grain) has four main parts. The **germ** is the part that grows into a wheat plant, and it contains concentrated food to support the new life. It is especially rich in protein, vitamins, and minerals. The **endosperm** is the soft, white inside portion of the kernel containing starch and protein. The **bran,** a protective coating around the kernel similar in function to the shell of a nut, is also rich in nutrients and fiber. The **husk,** commonly called chaff, is unusable for most purposes except for animal feed.

In earlier times, people milled wheat by grinding it between two stones, then sifting out the inedible chaff but retaining the nutrient-rich bran and germ as well as the endosperm. Improved milling machinery made it possible to remove the dark, heavy germ and bran as well, leaving a whiter, smoother-textured flour. People came to look on this flour as more desirable than the crunchy, dark-brown, "old-fashioned" flour but at first were unaware of the nutrition implications. Bread eaters suffered a tragic loss of needed nutrients in turning to white bread. A U.S. survey done in 1936 revealed that many people were suffering from deficiencies of the nutrients iron, thiamin, riboflavin, and niacin, which they had formerly received from bread. The Enrichment Act of 1942 standardized the return of these four lost nutrients to commercial flour. This legislation was amended in 1996 to include folic acid, a form of the B vitamin—folate—considered important in the prevention of certain birth defects. Since 1998, most grain products in the United States are fortified with folic acid. This doesn't make a single slice of bread "rich" in these added nutrients, but people who eat several or many slices of bread a day obtain significantly more of them than they would from unenriched white bread.

Today, you can assume that almost all breads, grains such as rice, wheat products such as macaroni and spaghetti, and cereals, both cooked and ready-to-eat, have been enriched. Figure 3-2 shows that although enrichment makes refined bread comparable to whole-grain bread with respect to the enrichment nutrients, it does not do so with respect to other important nutrients. Therefore, although the enrichment of flour and other cereal products does improve them, it doesn't improve them enough: Whole-grain products are still preferred over enriched products. If bread is a staple food in your diet—that is, if you eat it every day—you would be well advised to learn to like the hearty flavor of whole-grain bread.

REFINED refers to the process by which the coarse parts of food products are removed. For example, the refining of wheat into flour involves removing three of the four parts of the kernel—the chaff, the bran, and the germ—leaving only the endosperm.

ENRICHED refers to a process by which the B vitamins thiamin, riboflavin, niacin, folic acid, and the mineral iron are added to refined grains and grain products at levels specified by law.

FORTIFIED FOODS foods to which nutrients have been added. Typically, commonly eaten foods are chosen for fortification with added nutrients to help prevent a deficiency of a nutrient (iodized salt, milk with vitamin D) or to reduce the risk of chronic disease (juices with added calcium).

WHOLE GRAIN refers to a grain that is milled in its entirety (all but the husk), not refined. Whole grains include wheat, corn, rice, rye, oats, amaranth, barley, buckwheat, sorghum, and millet; two others—bulgur and couscous—are processed from wheat grains.

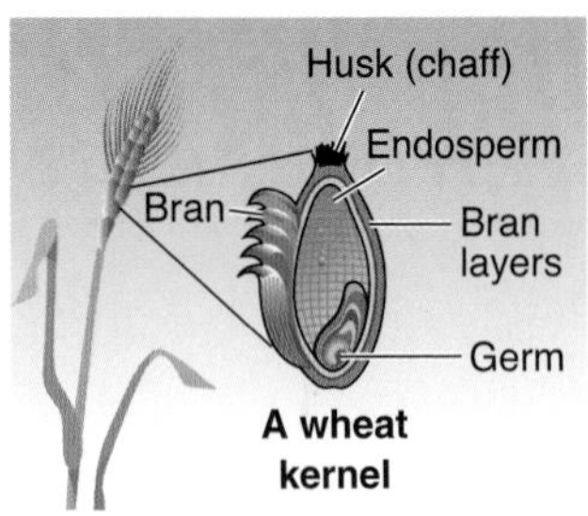

GERM the nutrient-rich and fat-dense inner part of a whole grain.

ENDOSPERM the bulk of the edible part of a grain; contains starch grains embedded in a protein matrix.

BRAN the fibrous protective covering of a whole grain and the chief source of fiber.

HUSK the outer, inedible covering of a grain.

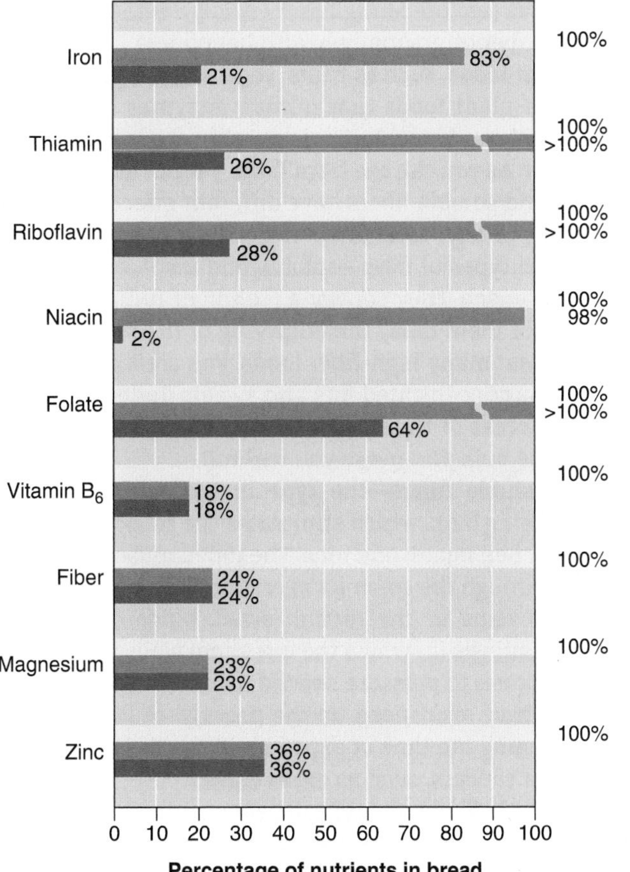

FIGURE 3-2
NUTRIENTS IN WHOLE-GRAIN, ENRICHED WHITE, AND UNENRICHED WHITE BREADS

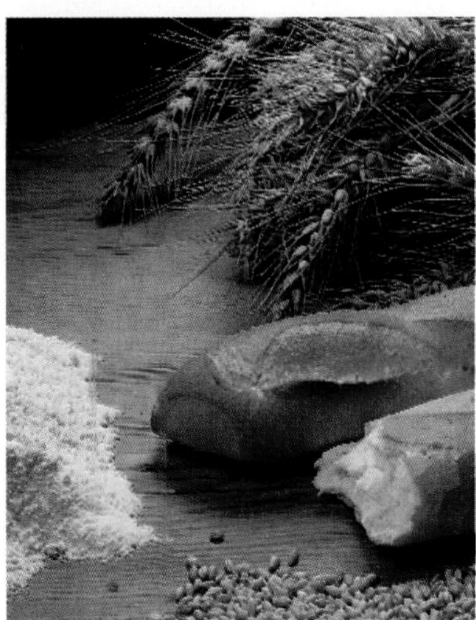

The nutrients present in the wheat plant at harvest are not always present in the wheat products you eat.

THE COMPLEX CARBOHYDRATES: FIBER IN THE DIET

The **fibers** of a plant form the supporting structures of the plant's leaves, stems, and seeds. Most fibers are polysaccharides, just as starch is, but with different bonds between the glucose units— bonds that cannot be broken by human digestive enzymes. The term *fiber* is used by almost everyone as if it represented a single entity. It was known generations ago as roughage. However, there are many compounds, mostly carbohydrates, that make up fiber.* Such compounds are familiar as the strings of celery, the skins of corn kernels, and the membranes separating the segments in citrus fruits. Isolated from plants, they may be used to thicken jelly (citrus pectin), to keep salad dressing from separating (guar gum), to provide bulk (wheat and other brans), and to exert other effects on texture and consistency.

FIBERS the indigestible residues of food, composed mostly of polysaccharides. Thus fibers are the nonstarch polysaccharides in foods. The term *dietary fiber* refers to the fiber that resists human digestive enzymes. The best known of the fibers are **cellulose, hemicellulose, pectin,** and **gums.**

The bonds that hold the units of fiber together cannot be broken by human digestive enzymes, but some can be broken by the bacteria that reside in the human digestive tract. Therefore, we may obtain a trace amount of glucose and some related products from fiber molecules. Fibers exert important effects on people's health, as described in the following section.

The Health Effects of Fiber

Many health experts are encouraging consumers to eat more fiber. According to recent evidence, inadequate levels of fiber in the diet are associated with several

* The woody material of heavy stems and bark is the noncarbohydrate lignin, classed by some as a fiber.

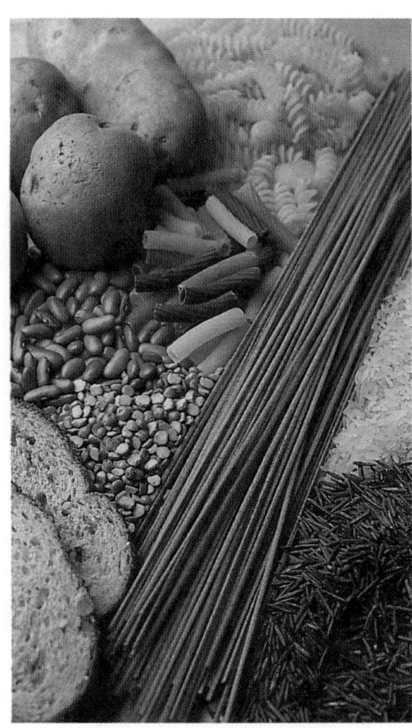

Foods such as potatoes, dried beans and peas, rice and whole-grain breads, cereals, and pastas are especially nutritious because of their starch, fiber, vitamin, and mineral content—and because they are virtually fat and cholesterol free.

diseases (see the Miniglossary below), while consumption of recommended levels of fiber offers many health benefits.[15] Recall that dietary fiber is found only in plant foods, such as fruits, vegetables, legumes, and whole grains, and it is the part of plant foods that human enzymes cannot digest. Fiber has two forms: insoluble and soluble. Table 3-6 shows the health benefits from these two types of fiber. As you can see from Table 3-6, not all fibers are created equal. Since insoluble and soluble fibers have different effects in the body, it is important to eat a variety of high-fiber foods to get both types.

Both types of fiber—soluble and insoluble—can help with weight control. In the stomach, they convey a feeling of fullness because they absorb water, and some of them delay the emptying of the stomach so that you feel fuller longer. If you eat many high-fiber foods, you are likely to eat fewer empty-calorie foods such as concentrated fats and sweets. Indeed, producers of some diet aids base the success of their products on the ability of certain fibers in the products to provide bulk and make you feel full.

Insoluble fibers—the type in wheat bran—hold water in the colon, thus increasing bulk, which stimulates the muscles of the digestive tract so that they retain their health and tone. The toned muscles can more easily move waste products through the colon for excretion. This prevents **constipation, hemorrhoids** (in which veins in the rectum swell, bulge out, become weak, and bleed), and **diverticulosis** (in which the intestinal walls become weak and bulge out in places in response to pressure needed to excrete waste when bulk is inadequate). These fibers may also speed up the passage of food through the digestive tract, thus shortening the time of exposure of the tissue to agents in food that might cause certain cancers, such as **colon cancer.**[16]

Soluble fibers, the type in beans and oats, are credited with reducing the risks of heart and artery disease—**atherosclerosis**—by lowering the level of cholesterol in the blood. It appears that the products of bacterial digestion of soluble fiber in the colon are absorbed into the body and may inhibit the body's production of cholesterol, as well as enhance the clearance of cholesterol from the blood.[17] Cholesterol levels may also decrease if food sources of soluble fiber (for example, barley, lentils, peas, beans, oat bran, or psyllium-enriched cereal) are used

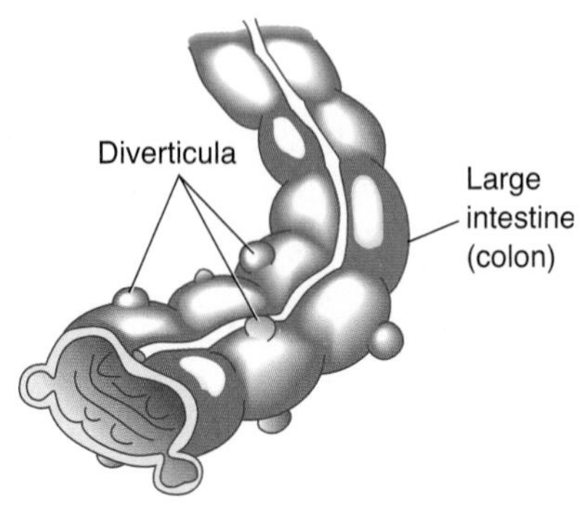

Diverticulosis
The outpocketings of intestinal linings that balloon through the weakened intestinal wall muscles are known as diverticula.

MINIGLOSSARY OF DISEASES and Conditions Associated with Lack of Fiber in the Diet

APPENDICITIS inflammation and/or infection of the appendix, a sac protruding from the large intestine.

ATHEROSCLEROSIS (ATH-er-oh-scler-OH-sis) a type of cardiovascular disease; the most common kind of hardening of the arteries characterized by the formation of fatty deposits, or plaques, in their inner walls.

COLON CANCER cancer of the large intestine (colon), the terminal portion of the digestive tract (see Appendix B).

CONSTIPATION hardness and dryness of bowel movements associated with discomfort in passing them.

DIABETES a disorder characterized by insufficiency or relative ineffectiveness of insulin that renders a person unable to regulate the blood glucose level normally.

DIVERTICULOSIS (dye-ver-tic-you-LOCE-iss) outpocketings of weakened areas of the intestinal wall, like blowouts in a tire, that can rupture, causing dangerous infections.

HEMORRHOIDS (HEM-or-oids) swollen, hardened (varicose) veins in the rectum, usually caused by the pressure resulting from constipation.

OBESITY body weight high enough above normal weight to constitute a health hazard (see also Chapter 8).

TABLE 3-6 HEALTH BENEFITS AND SOURCES OF DIETARY FIBER

Foods Rich in Insoluble Fiber		Foods Rich in Soluble Fiber	
apples	pears	apples	green beans
bananas	peaches	apricots	green peas
beets	plums	bananas	legumes
brown rice	rice bran	barley	oat bran
cabbage family	seeds	berries	oatmeal
cauliflower	skins/peels of fruits and vegetables	black-eyed peas	pears
corn bran	strawberries	broccoli	potatoes
green beans	tomatoes	cabbage	prunes
green peas	wheat bran	carrots	rye
legumes	whole-grain breads and cereals	cherries	seeds
mature vegetables		citrus fruits	sweet potatoes
nuts		corn	zucchini
		grapes	

Health Problems	Fiber Type	Possible Health Benefits
Obesity	Soluble Insoluble	Replaces calories from fat, provides satiety, and prolongs eating time because of chewiness of food
Digestive tract disorders: Constipation Diverticulosis Hemorrhoids	Insoluble	Provides bulk and aids intestinal motility; binds bile acids
Colon cancer	Insoluble	Speeds transit time through intestines and may protect against prolonged exposure to carcinogens.*
Diabetes	Soluble	May improve blood sugar tolerance by delaying glucose absorption
Heart disease	Soluble	May lower blood cholesterol by slowing absorption of cholesterol and binding bile

*This effect is based on epidemiologic studies and is usually observed along with a reduced-fat intake; it is unclear whether the protection comes from fiber or from other components that accompany fiber in foods—such as vitamins (e.g., folate), minerals (e.g., calcium), or phytochemicals.

as part of a low-fat, low-cholesterol diet.[18] Certain insoluble fibers also bind cholesterol compounds and carry them out of the body with the feces so that the whole body content of cholesterol is lowered.

Soluble fibers also improve the body's handling of glucose, even in people with diabetes, perhaps by slowing the digestion or absorption rate of carbohydrates.[19] Blood glucose levels therefore stay moderate, helping to prevent symptoms of diabetes or hypoglycemia. The list of fiber's contributions to human health, therefore, is impressive.

When people choose high-fiber foods in hopes of receiving some of these benefits, they must choose with care. Wheat bran, which is composed mostly of cellulose, has no cholesterol-lowering effect, whereas oat bran and the fibers of legumes, carrots, apples, and grapefruits do lower blood cholesterol. On the other hand, the fiber of wheat bran in whole-wheat bread is one of the most effective stool-softening fibers that help prevent constipation and hemorrhoids. If a single practical conclusion were to be drawn from what is known about fiber, it would have to be that all whole plant foods seem to contain many kinds of fibers and so can be expected to have the whole range of effects previously mentioned. To obtain the greatest benefits from fiber, therefore, you have to eat a variety of foods that contain it rather than take doses of purified fiber such as bran from a single source.

The wholesale addition of purified fiber to foods is probably ill-advised because it can be taken so easily to extremes. Taking only one isolated type of

INSOLUBLE FIBER includes the fiber types called cellulose, hemicellulose, and lignin; insoluble fibers do not dissolve in water.

SOLUBLE FIBER includes the fiber types called pectin, gums, mucilages, some hemicelluloses, and algal substances (e.g., carageenan); soluble fibers either dissolve or swell when placed in water. Psyllium seed husk is an ingredient in certain cereals and bulk-forming laxatives and contains both soluble and insoluble properties.

Enjoy the hearty flavor of whole grains.

fiber deprives the taker of the benefits the other types of fiber provide. On the other hand, if you add a variety of whole grains, legumes, nuts, fruits, and vegetables to your diet, you get the various types of fiber you need, together with a package of benefits—water, minerals, vitamins, the energy nutrients, and other beneficial substances. Fiber out of context is similar to sugar out of context: It can be viewed as nutrient-empty fiber, just as concentrated sugar is sometimes seen as nutrient-empty calories.

Undoubtedly, including fiber in a daily meal plan has benefits—but how much is enough? Even fiber has potential to cause harm if taken in excess. Since fiber carries water out of the body, taking too much can cause dehydration and intestinal discomfort. Iron is mainly absorbed early during digestion, and because fiber speeds the movement of foods through the digestive system, it may limit the opportunity for the absorption of iron and other nutrients. Binders in some fibers link chemically with minerals such as calcium and zinc, making them unavailable for absorption by carrying them out of the body (more about this in Chapter 7). Too much bulk from the diet could reduce the total amount of food consumed and cause deficiencies of both nutrients and energy. The malnourished, the elderly, and children, because they eat small amounts of food anyway, are especially vulnerable to these concerns.

Major health organizations agree that about 20 to 35 grams of dietary fiber daily is a desirable intake.[20] The diet can supply that amount, given ample choices of whole foods (see Table 3-7). The diet does have to be high in fruits, vegetables, legumes, and grains, and moderate in meats, fats, and concentrated sugars—the same recommendations made in the *Dietary Guidelines*.

 HOW THE BODY HANDLES CARBOHYDRATES

Just as glucose is the original unit from which the variety of carbohydrate foods are made, so is glucose the basic carbohydrate unit that each cell of the body uses. Cells cannot use lactose, sucrose, or starch—they require glucose. The task of the digestive system, then, is to disassemble the double sugars and starch to single sugars and to absorb these monosaccharides into the blood. The liver converts to glucose those carbohydrates that are not already in the form of glucose so that they can be transported to the cells. The cells can then store this glucose, use it for energy, or convert it to fat.

The first digestive enzymes to work on starch are those in the saliva; they begin taking the starch apart, and enzymes in the intestines continue digestive action (see Figure 3-3). The enzymes release the individual glucose units, which are absorbed across the intestinal wall into the blood (see Appendix B). One to four hours after a meal, all the starch has been digested and absorbed and is circulating to the cells as glucose.

Cooking facilitates the digestive process by spreading out the tightly packaged chains of glucose so that during digestion the digestive enzymes can break the chains down into glucose units for absorption.

TABLE 3-7 LOOKING FOR FIBER IN THE FOOD GUIDE PYRAMID*

Food Group	Fiber Content: High (5–10 g per serving)	Medium (2–4 g per serving)	Low (<1 g per serving)
Bread, Cereal, Rice, and Pasta	bran cereals, ½ c Shredded Wheat, 1 c	oatmeal, 1 c whole-grain flakes, 1 c (e.g., Wheaties, Total) puffed wheat, 1½ c whole-wheat bread, 1 slice rye bread, 1 slice whole-grain crackers, 5 popped popcorn, 2 c brown rice, ½ c barley, ½ c	corn flakes, 1 c Special K, 1 c Rice Krispies, 1 c Cheerios, 1 c white rice, ½ c pasta, ½ c
Vegetable†	legumes	broccoli Brussels sprouts cabbage carrots corn eggplant green beans potato, with skin, 1 medium spinach tomatoes, 1 large	asparagus cauliflower celery lettuce onions peppers pickle, 1 large potato, without skin, 1 medium
Fruit‡		apple, 1 small apricot banana, 1 small berries, 1 c cantaloupe, ½ melon cherries, 16 large orange peach pear prunes, 2 raisins, ¼ c	canned fruit, 1 c juices, 1 c
Milk, Yogurt, and Cheese	0	0	0
Meat, Poultry, Fish, Dry Beans, Eggs, and Nuts	legumes, cooked or canned, ½ c	nuts and seeds, 1 oz peanut butter, 2 tbsp	

*Appendix H lists the grams of dietary fiber in more than 1,900 foods.
†The serving size for vegetables is 1/2 cup, unless otherwise noted.
‡The serving size for fruits is one medium piece, unless otherwise noted.

CARBOHYDRATE CONSUMPTION SCORECARD

RATING YOUR DIET: HOW SWEET IS IT?

Now that you are aware of some of the sources of added sugars, let's take a look at *your* diet. Check the box that most closely describes your eating habits to see how the foods you choose affect the amount of added sugars in your diet.

How often do you	SELDOM OR NEVER	1 OR 2 TIMES A WEEK	3 TO 5 TIMES A WEEK	ALMOST DAILY
1. Drink soft drinks, sweetened fruit drinks, or punches?	☐	☐	☐	☐
2. Choose sweet desserts and snacks, such as cakes, pies, cookies, and ice cream?	☐	☐	☐	☐
3. Use canned or frozen fruits packed in heavy syrup or add sugar to fresh fruit?	☐	☐	☐	☐
4. Eat candy?	☐	☐	☐	☐
5. Add sugar to coffee or tea?	☐	☐	☐	☐
6. Use jam, jelly, or honey on bread or rolls?	☐	☐	☐	☐

HOW DID YOU DO? The more often you choose the items listed above, the higher your diet is likely to be in sugars. You may need to cut back on sugar-containing foods, especially those you checked as "3 to 5 times a week" or more. This does not mean eliminating these foods from your diet. You can moderate your intake of sugars by choosing foods that are high in sugar less often and by eating smaller portions.

CHECK YOUR DIET FOR FIBER

To reap the benefits of fiber, healthy adults need between 20 and 35 grams of fiber a day. Most Americans get less than 14 grams a day. To figure how much fiber you consume in a day, write down the number of servings you eat in a typical day for each of the food categories below. Next, multiply by the factor shown (this number represents the average amount of fiber from a serving in that food group). Then add up the total amount of fiber consumed. Check the scoring to determine how you compare to the recommended goal of 20 to 35 grams per day.

FOOD GROUP	NUMBER OF SERVINGS		AMOUNT (g)
Vegetables (½ c cooked; 1 c raw)	______ × 2 g	=	______
Fruits (1 medium; ½ c cut; ¼ c dried)	______ × 2 g	=	______
Dried beans, lentils, split peas (½ c cooked)	______ × 8 g	=	______
Nuts, seeds (¼ c; 2 tbsp peanut butter)	______ × 2 g	=	______
Whole grains (1 slice bread; ½ c rice, pasta; ½ bun, bagel, muffin)	______ × 2 g	=	______
Refined grains* (1 slice bread; ½ c rice, pasta; ½ bun, bagel, muffin)	______ × 1 g	=	______
†Breakfast cereals (1 oz)	______ × grams	=	______
	Total	=	______

*Refined grains refer to products made with white or wheat flour (not whole-wheat flour); white rice, or other processed grains.
†Check the Nutrition Facts panel on the cereal you eat to determine the number of grams of fiber in a 1-ounce serving.

SCORING

20–35 Good News! Your fiber intake meets the recommended goal.

15–20 You consume more fiber than the average American. Another serving from the fruit, vegetable, and whole-grain categories would help bring your fiber intake into the recommended goal area.

10–15 Your intake is similar to the average American intake. Consider adding a serving of legumes and additional servings of fruits, vegetables, and whole grains to your daily diet.

0–10 Your fiber intake is too low. Try getting at least five servings of fruits and vegetables, along with three servings of whole-grain foods each day. Add a high fiber cereal to your morning routine and consider adding legumes to some of your weekly meals.

If the blood delivers more glucose than the cells need, the liver and muscles take up the surplus to build the polysaccharide **glycogen.** The muscles hold two-thirds of the body's total store of this carbohydrate and use it during exercise. (Chapter 9 explores the relationship between glycogen and exercise and offers tips on how to make the most of glycogen stores.) The liver stores the other one-third, making it available to maintain the blood glucose level.

GLYCOGEN (GLY-co-gen) a polysaccharide composed of chains of glucose, manufactured in the body and stored in liver and muscle. As a storage form of glucose, liver glycogen can be broken down by the liver to maintain a constant blood glucose level when carbohydrate intake is inadequate.

After the glycogen stores are full and the cells' immediate energy needs are met, the body takes a third path for using carbohydrates. Say you have eaten dinner that includes enough carbohydrates to fill your glycogen stores. Now, you watch a movie, eat popcorn, and drink a cola. Your digestive tract is delivering glucose from the popcorn and soda to your liver; your liver breaks these extra energy compounds into small fragments and puts them together into the more permanent energy-storage compound—fat. The fat is then released from the liver, carried to the fatty tissues of the body, and deposited there. (Fat cells, too, can utilize excess glucose to make fat for storage.) Unlike the liver cells, though, which can store only about half a day's worth of glucose as glycogen, the fat cells can store unlimited quantities of fat.

This section refers to many body organs. To review them, turn to Appendix B.

Maintaining the Blood Glucose Level

The brain and nervous system are sensitive to the concentration of glucose in the blood. Normal blood glucose levels are important for a feeling of well-being.

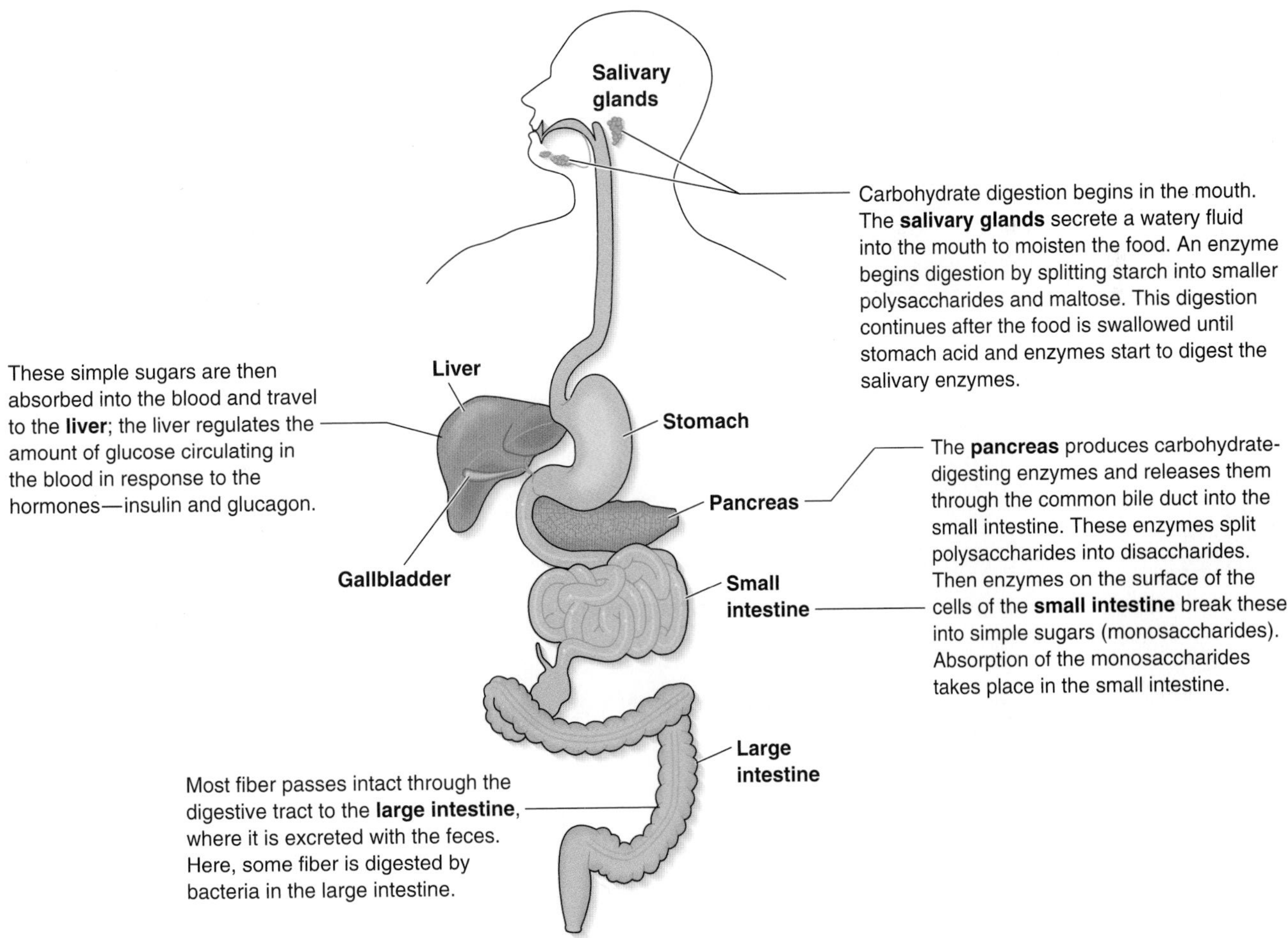

Salivary glands
Carbohydrate digestion begins in the mouth. The **salivary glands** secrete a watery fluid into the mouth to moisten the food. An enzyme begins digestion by splitting starch into smaller polysaccharides and maltose. This digestion continues after the food is swallowed until stomach acid and enzymes start to digest the salivary enzymes.
Liver
These simple sugars are then absorbed into the blood and travel to the **liver**; the liver regulates the amount of glucose circulating in the blood in response to the hormones—insulin and glucagon.
Stomach
Pancreas
The **pancreas** produces carbohydrate-digesting enzymes and releases them through the common bile duct into the small intestine. These enzymes split polysaccharides into disaccharides. Then enzymes on the surface of the cells of the **small intestine** break these into simple sugars (monosaccharides). Absorption of the monosaccharides takes place in the small intestine.
Gallbladder
Small intestine
Large intestine
Most fiber passes intact through the digestive tract to the **large intestine**, where it is excreted with the feces. Here, some fiber is digested by bacteria in the large intestine.

FIGURE 3-3
A SUMMARY OF CARBOHYDRATE DIGESTION AND ABSORPTION

When your blood glucose level becomes too high, you get sleepy; when the concentration falls too low, you get weak and shaky. Only when blood glucose is within the normal range can you feel energetic and alert. The person who wants to feel energetic and alert all day should make the effort to eat so as to maintain blood glucose levels in the normal range to fuel the critical work of the brain and nervous system.

The maintenance of a normal blood glucose level depends on two safeguards, as shown in Figure 3-4. When the level gets too high—for example, immediately after a meal—it can be corrected by siphoning off the excess into liver and muscle glycogen and into body fat. When it gets too low—for example, following an overnight fast—it can be replenished by drawing on liver glycogen stores.

INSULIN a hormone secreted by the pancreas in response to high blood glucose levels; it assists cells in drawing glucose from the blood.

When the blood glucose level rises, the body adjusts by storing the excess. The first organ to detect the excess glucose is the pancreas, which releases the hormone **insulin** in response. Most of the body's cells respond to insulin by taking up glucose from the blood to make glycogen or fat. Thus, the blood glucose level is quickly brought back down to normal as the body stores the excess. Insulin's opposing hormone, released by the pancreas when blood glucose is too low, is

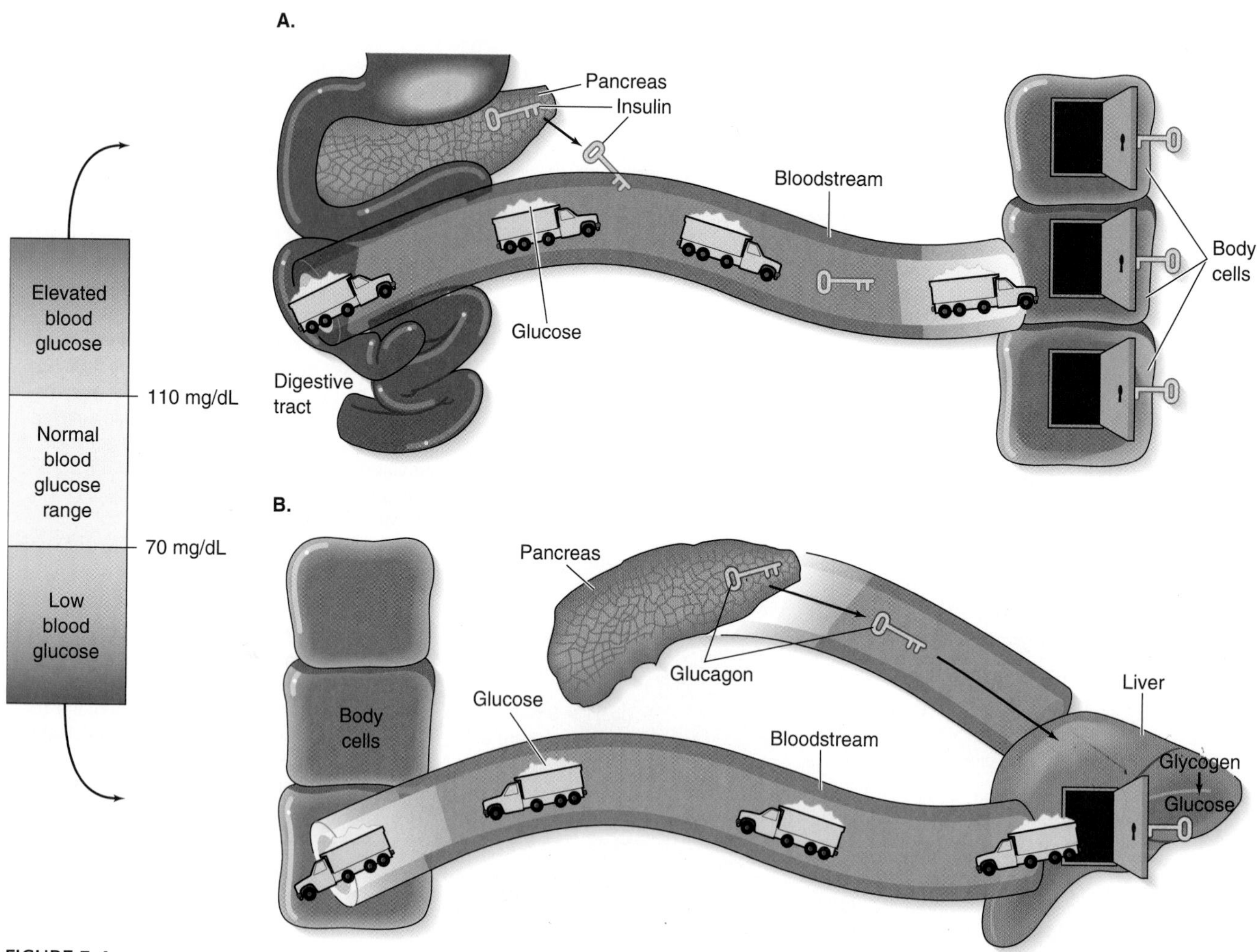

FIGURE 3-4
BLOOD GLUCOSE REGULATION

A. When a person eats, blood glucose rises. High blood glucose (sugar) stimulates the pancreas to release insulin. Insulin serves as a key for entrance of blood glucose into cells. Liver and muscle cells store the glucose as glycogen. Excess glucose can also be stored as fat.
B. Later, when blood glucose is low, the pancreas releases glucagon, which serves as the key for the liver to break down stored glycogen to glucose and release it into the blood to raise blood glucose levels.

glucagon, which draws forth glucose from storage, making it available to supply energy. Insulin and glucagon both work to maintain the concentration of glucose in the blood within the normal range—neither too high nor too low.

GLUCAGON (GLUE-cuh-gon) a hormone released by the pancreas that signals the liver to release glucose into the bloodstream.

Obviously, when the blood glucose level falls and stores are depleted, a meal or a snack can replenish the supply. An appropriate choice is to eat a balanced meal containing foods that offer carbohydrate, protein, and fat:

- The carbohydrates in the meal provide a quick source of glucose.
- The protein in the meal stimulates glucagon secretion, which opposes insulin and prevents it from storing glucose too quickly.
- The soluble fibers and fat in the meal slow down digestion so that a steady stream of glucose is received rather than a sudden flood.[21]

By these standards, a bowl of whole-grain cereal and fresh fruit with low-fat milk for breakfast (offering complex carbohydrate, protein, and fat) is a good choice. Eating well-spaced, carefully chosen meals that provide the balance of protein, carbohydrates, and fat recommended in the *Dietary Guidelines* can prevent rapid rises and falls in the blood glucose level.

Researchers rank carbohydrate foods according to how fast each is digested into glucose and how much the food causes blood glucose to rise.* In general, legumes produce the most even blood glucose response, dairy products next, and fruits and cereals next; pure sugar produces the greatest rise in the blood glucose level.[22] The effect of a food on the blood glucose level is important to people with abnormalities of blood glucose regulation, notably **diabetes** and **hypoglycemia.** Rapid swings in blood glucose can affect the performance of both an athlete during an endurance event and an office employee following lunch. People are wise to eat meals that contain complex carbohydrate, an adequate amount of protein, and a little fat.

DIABETES (dye-uh-BEET-eez) a disorder (technically termed *diabetes mellitus*) characterized by insufficiency or relative ineffectiveness of insulin, which renders a person unable to regulate the blood glucose level normally.

HYPOGLYCEMIA (HIGH-po-gligh-SEEM-ee-uh) an abnormally low blood glucose concentration—below 60 to 70 mg/100 ml.

KETOSIS abnormal amounts of ketone bodies in the blood and urine; ketone bodies are produced from the incomplete breakdown of fat when glucose is unavailable for the brain and nerve cells.

Hypoglycemia

Suppose the blood glucose falls, the glycogen reserves are exhausted, and you do *not* eat. Gradually, your body will shift into a fasting state, breaking down its muscle to provide amino acids to the liver. The liver converts some of these into glucose to fuel the brain. The fat released from the muscle cells is used to fuel other cells. (Chapter 8 describes this state—**ketosis**—in more detail. Most times, the transition is smooth and is not noticeable. But there may be times when your blood glucose level falls rapidly or below what is normal for you, and you may experience symptoms of glucose deprivation to the brain: irritability, weakness, and dizziness. Your muscles become weak, shaky, and trembling, and your heart races in an attempt to speed more fuel to your brain. These symptoms disappear once calories are consumed. While *true* hypoglycemia is rare, the symptoms of low blood sugar may occur in people who are not eating often enough to keep their blood sugar levels from dropping. The treatment consists of eating balanced meals and learning to eat more frequently throughout the day (see the sample menu that follows on page 91).

People who suspect they might have *true* hypoglycemia should consult a physician for diagnosis and treatment. It is a serious condition in which abnormal amounts of insulin are secreted, perhaps because of a pancreatic tumor or other health problem. As a result, the person's blood glucose is constantly too low. This kind of hypoglycemia causes symptoms such as headache, confusion, fatigue, nervousness, sweating, or unconsciousness. A person with hypoglycemia

*The effect of food on a person's blood glucose and insulin response is called the *glycemic effect*—how fast and how high the blood glucose rises and how quickly the body responds by bringing it back to normal. A *glycemic index* ranks foods on the basis of the extent to which the foods raise the blood glucose level as compared with pure glucose or white bread.

is told to avoid alcohol and snacks high in simple sugars and usually benefits from eating six small, evenly spaced meals a day that include a variety of complex carbohydrates and lean protein foods. These foods will increase blood sugar levels slowly and more gradually than highly refined carbohydrates.

Diabetes

HYPERGLYCEMIA an abnormally high blood glucose concentration, often a symptom of diabetes.

Knowing how the blood glucose level is maintained, you can appreciate the problem of the person with diabetes whose insulin response is slow or ineffective. Most adults with diabetes fall into this category (see Table 3-8).[23] Even if the blood glucose level rises too high (**hyperglycemia**), glucose still fails to get into cells and so stays too high for an abnormally long time. The kidneys may respond by shifting some glucose into the urine so that it can be excreted. Short-term effects of hyperglycemia may include thirst, frequent urination, weakness, lack of ability to concentrate, hunger, and blurred vision. People with diabetes must be careful to eat regularly scheduled, balanced meals—providing a constant, steady, moderate flow of glucose to the bloodstream—so the insulin response can keep up.

Research indicates that many people with diabetes actually do best on a diet that is high in complex carbohydrate-rich foods—as high as is recommended for any healthy person. The starch and protein in these foods help to regulate the blood glucose level, as already described.

The most common type of diabetes—Type 2 diabetes is usually diagnosed in people over 45 and results from genetic factors, too little insulin, or cellular resistance to insulin.[24] The incidence of Type 2 diabetes is rising, mostly because the U.S. population is aging, sedentary, and gaining excess weight. Obesity is considered a major risk factor for this type of diabetes, especially when the excess fat is carried around the abdomen. Usually, eating a healthful diet, exercising, and losing weight can normalize blood glucose levels for people with Type 2 diabetes. For others, oral medications and sometimes insulin shots are necessary.[25]

Recent studies show that Type 2 diabetes, which normally occurs only in adults, is now affecting an increasing number of children and adolescents. Obesity in children and teens seems to play a major role in the early development of Type 2 diabetes.

People with Type 2 diabetes—those who make insulin but whose cells resist responding to it—tend to become obese, storing more fat than normal and being constantly hungry. This is because the liver takes up from the blood the glucose

TABLE 3-8 DISTINCTIONS BETWEEN THE TWO MAJOR TYPES OF DIABETES

	Type 1 Diabetes	Type 2 Diabetes
Incidence	5–10% of cases	90–95% of cases
Age of onset	< 20	> 40
Insulin deficiency	Yes, pancreas unable to make insulin to meet needs	In some cases there may be insufficient insulin, or cells may be unresponsive to insulin
Risk factors	Genetic predisposition plus environmental factor (for example, viral infection)	Genetic predisposition plus obesity (especially central-type obesity), family history
Treatment	Insulin injections, diet, and exercise	Weight loss, diet, and exercise; oral drugs or insulin sometimes needed

A SAMPLE MEAL PLAN for Keeping Blood Sugar Levels on an Even Keel

Breakfast
Spread ¼ cup low-fat ricotta cheese on 2 halves of a small cinnamon raisin bagel. Top with 2 tbsp raisins and sprinkle with cinnamon. Heat under broiler for one to two minutes.

Midmorning Snack
Mix 1 sliced fresh peach with 1 cup low-fat plain yogurt and top with ¼ cup low-fat granola.

Lunch
Stuff a whole wheat pita with lettuce, ¾ cup Zesty Vegetable Tuna (see recipe) and top with shredded lettuce.
1 cup fresh melon cubes.

Midafternoon Snack
Spread 2 tbsp of hummus over 4 slices of whole grain melba toast.

Dinner
¾ cup Quick Kidney Bean Salad (see recipe)
3 oz grilled chicken breast
1 cup steamed broccoli florets
1 medium baked sweet potato
2 tsp butter

Evening Snack
Combine ¾ cup fat-free milk, 1 frozen peeled banana, and 1 tbsp peanut butter and blend for one to two minutes until smooth.

Zesty Vegetable Tuna
1 6-oz can water-packed tuna
½ cup chopped green pepper
½ cup chopped red pepper
½ cup shredded carrots
3 tbsp light mayonnaise
2 tbsp dijon mustard
3 tbsp balsamic vinegar
Combine last three ingredients, then add remaining ingredients.
Makes 3 servings.

Quick Kidney Bean Salad
1 16-oz can kidney beans, rinsed and drained
½ cup chopped celery
½ cup chopped onion
⅓ cup orange juice
2 tbsp cider vinegar
Mix all ingredients in a bowl and refrigerate until chilled.
Makes 4 servings.

Source: Adapted from Getting even with low blood sugar, *Health Wise* 5 (1999): 6.

that the cells cannot take up, converts the glucose to fat, and then ships it out to the fat cells for storage. The fat cells respond—slowly, to be sure—but ultimately they store all the fat that is sent to them. Due to the insulin resistance of body cells, insulin does not move glucose into cells effectively, and the message that energy fuels are coming in from food is delayed, causing the person with diabetes to be constantly hungry. Unfortunately, the larger the fat cells become, the more resistant they may be to insulin, thus making the diabetes worse. People with diabetes in their family are urged not to gain excess weight, for it is likely to precipitate the onset of the disease.

Persons with the less common type of diabetes—Type 1—produce no (or very little) insulin and require insulin injections to control their blood sugar levels. Type 1 diabetes is more likely to be diagnosed during childhood and has been linked to a possible viral or allergic reaction in susceptible persons that causes the destruction of pancreatic cells that produce insulin. The person who does not produce any insulin is likely to experience a sudden onset of the disease. Such a person may lose weight rapidly because without insulin, the cells cannot store glucose or fat. Thus, the two types of diabetes have opposite effects on body weight—one making the person fat; the other, thin. To determine your own risk for diabetes, take the test in the margin.

Recently, the American Diabetes Association eased its restriction on sugar use in the diets of people with diabetes because sucrose-containing foods do not seem to exert any higher of a glycemic response than do some starchy foods.[26] Any foods containing sucrose must be used in place of other carbohydrate-containing foods and should not exceed 5 percent to 10 percent of the total carbohydrate calories.

WHAT IS YOUR RISK FOR DIABETES?

1. I have given birth to a baby weighing more than nine pounds.
   Yes = 1. No = 0 ______
2. I have a sister or a brother with diabetes.
   Yes = 1. No = 0 ______
3. I have a parent with diabetes.
   Yes = 1. No = 0 ______
4. My weight is equal to or above that listed in the At-Risk Weight Chart below.
   Yes = 5. No = 0 ______
5. I am under 65 years of age *and* I get little or no exercise during a typical day.
   Yes = 5. No = 0 ______
6. I am between 45 and 64 years of age.
   Yes = 5. No = 0 ______
7. I am 65 years or older.
   Yes = 9. No = 0 ______

Total: ______

AT-RISK WEIGHT CHART

If you weigh the same or more than the weight listed here for your height (BMI >27) you may be at risk for diabetes.

Height, without shoes (feet/inches)	Weight, without clothing (pounds)
4′10″	128
4′11″	133
5′0″	138
5′1″	143
5′2″	147
5′3″	152
5′4″	157
5′5″	162
5′6″	167
5′7″	172
5′8″	177
5′9″	183
5′10″	188
5′11″	193
6′0″	198
6′1″	204
6′2″	210
6′3″	218
6′4″	221

SCORING

3–9 points: Chances are you are at low risk for diabetes now. Maintain a healthy weight, eat nutritious low-fat meals, and exercise regularly to keep risk low.

10 or more points: You are at high risk for diabetes. See your doctor to be tested.

Source: Reprinted with permission, © 2000, American Diabetes Association. For more information about diabetes, call 1-800-DIABETES.

WORKING ON THE WEB

www.diabetes.org

Sponsor: American Diabetes Association (ADA)

Description: This Web site provides practical information to consumers about dining out, sugars and artificial sweeteners, alcohol, and weight loss. A wide variety of nutrition-related topics are posted from *Diabetes Forecast,* a consumer magazine. The site also features guides for health professionals for diagnosis and screening, and nutrition recommendations for people with diabetes.

Available Activities: The site provides an interactive quiz to test your knowledge about nutrition and diabetes; access to diabetes-related news and events in your local area; information to help you learn about smart food shopping for diabetes; recipes and cooking information; abstracts of articles from both consumer and professional diabetes journals; research updates; nutrition games; a chance to shop in the ADA bookstore; plus free Medline searches through TopicDoc.

Web Work:

1. Test your knowledge about diabetes and nutrition. Complete the Feed Your Brain Quiz. After you receive your score, review the correct answers for any questions you missed.
2. Review the tips provided for Healthy Restaurant Eating.
3. Visit The Webb Cooks to learn more about healthy cooking.

Helpful Hints: From the ADA home page, scroll down and click on Nutrition. On the Nutrition home page, click on Feed Your Brain to take the diabetes nutrition quiz. When you finish with the quiz, return to the Nutrition home page and click on Healthy Restaurant Eating. Finally, return to the site's home page and click on The Webb Cooks. Scroll through the list of articles provided (e.g., One Pot Wonders or Take a Wok on the Wild Side) and review those of interest to you. Check out the recipes, too!

Spotlight

SWEET TALK—ALTERNATIVES TO SUGAR

It all started back in 1879—the year that a substance called saccharin was first discovered and found to be able to sweeten foods without adding calories. Since that time, the sale of **artificial sweeteners** for use in tabletop sweeteners, such as Sweet 'N Low and Equal, and in artificially sweetened soft drinks, yogurt, and numerous other products has soared.[27] The major artificial sweeteners on the market today are **acesulfame-K, aspartame, saccharin,** and **sucralose.** But despite the ever-growing popularity of such sweeteners, doubts about their safety have stirred a good deal of controversy over the years. In addition, confusion about the characteristics of the various alternatives to sugar (see Table 3-9) and

TABLE 3-9 HOW SWEET IT IS

Name (Trade Name)	Sweeteners Compared with Sucrose	Typical Uses	Characteristics*
Sweeteners approved for use in the United States			
Acesulfame K (Sunette/Sweet One)	200 times sweeter	Tabletop sweetener, puddings, gelatins, chewing gum, candies, baked goods, desserts, diet drink mixes	Stable in high temperatures; soluble in water. A new blend of Ace-K and aspartame synergizes the sweetening power of these sweeteners.
Aspartame† (NutraSweet/Equal)	180 times sweeter	General purpose sweetener in foods, beverages, chewing gum; tabletop sweetener	Loses sweetness at high temperatures and may lose sweetness over time. New forms can increase its sweetening power in cooking and baking.
Saccharin (Sweet'N Low)	300 times sweeter	Diet soft drinks, tabletop sweetener	Stable at high temperatures.
Sucralose (Splenda)	600 times sweeter	Soft drinks, jams, frozen desserts, dairy products, baked goods, chewing gum, salad dressings, syrups, tabletop sweetener	Stable at various temperatures.
Sweeteners for which U.S. approval is pending			
Alitame (Novasweet)	2,000 times sweeter	Baked goods, beverages, frozen desserts, tabletop sweetener	Stable at high temperatures and in both acidic and nonacidic foods; highly soluble in water.
Cyclamate‡	30 times sweeter	Tabletop sweetener, baked goods	Stable at high temperatures; long shelf life.

*In addition to the characteristics listed, all of the sweeteners have a synergistic effect when combined with other sweeteners. In other words, together they enhance each other's sweetness, yielding a combined sweetness greater than the sum of all the substances' sweetness.

†The NutraSweet Company held the patent on aspartame from 1981 to 1992. NutraSweet is still the most common trade name for aspartame, but others may become common in the future.

‡Cyclamate was banned in the United States in 1970 because studies suggested that it may cause cancer in rats. The validity of the studies has been questioned, however. Currently, a petition is before the FDA to reapprove cyclamate for use in the United States. In Canada, cyclamate has been approved for use in tabletop sweeteners and as a sweetening agent in drugs.

Source: Position of the American Dietetic Association: Use of nutritive and nonnutritive sweeteners, *Journal of the American Dietetic Association* 98 (1998): 580–587.

Aspartame and other sugar substitutes are used to sweeten a wide variety of products.

their role in managing problems such as obesity, diabetes, and tooth decay prevail among the millions of Americans who use them. The following discussion about sugar substitutes will help put matters into perspective.

Is it true that small amounts of saccharin cause cancer?

This assumption has never been proven. It's true that some studies of saccharin, the sweetener in Sweet 'N Low, have found that it can cause bladder cancer in laboratory rats. A human, however, would have to drink 850 cans of soft drinks a day to take in a dose equivalent to what the rats in those studies were given. Moreover, some research suggests that while very high doses of saccharin may promote bladder cancer in rats, the sweetener does not have the same effect on mice, hamsters, monkeys, or humans.[28] In addition, investigations of large groups of people have yet to establish any clear-cut link between saccharin consumption and the risk of cancer. One study conducted by the National Cancer Institute and involving more than 9,000 men and women showed no association between saccharin use and bladder cancer.[29] Thus, it appears that the health risks, if any, posed by saccharin are minuscule at most. Those fearful of getting cancer would be better off quitting smoking or reducing the amount of fat in their diet than worrying about putting Sweet 'N Low in their coffee now and then. As the American Medical Association's Council on Scientific Affairs has stated, "In humans, available evidence indicates that the use of artificial sweeteners, including saccharin, is not associated with an increased risk of bladder cancer."[30]

Does aspartame cause headaches?

Not in most people. While the U.S. Food and Drug Administration has received numerous complaints from consumers who claim to suffer from headaches, nausea, anxiety, and other symptoms after consuming aspartame-containing foods or beverages, scientific studies have never confirmed that the sweetener is truly the culprit. Consider that scientists at Duke University who conducted carefully controlled research into the matter concluded that tablets containing the amount of aspartame in about four liters of diet soft drinks were no more likely to prompt headaches in people than **placebo** pills administered for the sake of comparison.[31] Furthermore, numerous careful scientific investigations carried out both before and after aspartame first appeared on the market have indicated that the product brings about adverse health effects only in a small group of people with a rare metabolic disorder known as **phenylketonuria,** or PKU. People born with PKU must carefully control their consumption of phenylalanine—one of the two amino acids that make up aspartame—to prevent health problems as severe as mental retardation.The American Medical Association's Council on Scientific Affairs, the Centers for Disease Control and Prevention, the Food and Drug Administration, The American Dietetic Association, and numerous other health organizations all consider the sweetener safe for use except by people with PKU.[32] Table 3-10 shows how to determine the amount of aspartame in your diet.

Can the use of artificial sweeteners bring about weight loss?

If only it were that simple. Obviously, because artificially sweetened foods typically contain fewer calories than their sugar-sweetened counterparts, substituting a low-calorie alternative for a sugar-laden one can help a person who is trying to lose weight enjoy sweet-tasting foods and save calories. But simply *adding* foods sweetened with sugar substitutes to the diet will not do the trick. Moreover, eating artificially sweetened foods to have an excuse to splurge on a high-calorie food defeats the purpose. A person who drinks diet soda pop to justify eating an ice cream sundae later is reaping little, if any, benefit. In other words, it's the way in which you fit artificial sweeteners into the rest of your diet that counts. The American Dietetic Association takes the position that individuals who desire to lose weight may choose to use artificial sweeteners, but should do so as part of a sensible weight-management program that includes a nutritious diet and regular physical activity. (See Chapter 8 for healthy weight management guidelines).

On the flip side, some consumers may be under the impression that artificial sweeteners bolster the appetite. A report that aspartame sends mixed signals to the brain and thereby increases appetite as well as food consumption spawned the notion that artificial sweeteners can interfere with weight loss.[33] That report was based on comparisons of the effects of drinking plain water and drinking aspartame- and sugar-sweetened water, however, rather than consuming actual foods or beverages that are typically sweetened with aspartame. Moreover, the researchers did not measure how much food people ate after drinking the liquids. Thus, the study did not show how consuming aspartame-sweetened foods influences appetite and eating behavior in the real world.

Since that time, about a dozen investigations into the relationship between appetite and foods and beverages sweetened with the substances have indicated that aspartame either lessens, or doesn't affect, feelings of hunger or the amount of food ultimately eaten.[34] So for now, at least, it appears that aspartame's impact on appetite need not be of concern to those trying to lose weight.

I have heard that foods sweetened with sugar substitutes do not contribute to tooth decay. Is this true?
Not necessarily. As pointed out earlier, any carbohydrate-containing food, be it sugar sweetened or otherwise, can promote tooth decay. When you eat a sugary food, the millions of bacteria lurking on the surfaces of and between your teeth feast on the sugar and, in the process, release an acid that eats away at tooth enamel. Once inside the mouth, carbohydrates can be devoured by cavity-causing bacteria because enzymes in the saliva break the complex carbohydrates into simple sugars, which bacteria thrive on.

Should people with diabetes eat foods sweetened only with sugar substitutes?
That depends. It is often assumed that because one of the hallmarks of diabetes is high blood sugar (glucose), sugar from foods is the major culprit behind high blood sugar and should therefore be off limits for people with diabetes. But it's not that simple. The total amount of carbohydrate, including both simple and complex, exerts the most influence on blood glucose levels. What's more, many other factors, including the amount of fat and fiber in a food, affect the body's blood glucose response to it. That's

TABLE 3-10 CHECKING THE ASPARTAME IN YOUR DIET

The Food and Drug Administration has set forth an "acceptable daily intake" of 50 milligrams of aspartame per kilogram (2.2 pounds) of body weight. Most people, however, take in much less—on the order of fewer than 5 milligrams per kilogram daily. To reach the FDA's limit, consider that a 150-pound adult would have to consume about 19 12-ounce cans of diet soda pop or 97 packets of Equal, and a 40-pound child would have to consume four 12-ounce cans of diet soft drinks or 24 packets of Equal. To see how much aspartame you're getting in your diet, check the following numbers for the average amount of aspartame found in typical artificially sweetened foods.

Aspartame-sweetened Food	Milligrams of Aspartame
1 packet Equal tabletop sweetener	37
12 ounces diet soda pop	180–225
8 ounces fruit drink made from powder	100
4 ounces gelatin dessert	80
6 ounces hot chocolate	50
4 ounces pudding	25
1 cup cereal	45
8 ounces iced tea made with instant, sweetened mix	80
12 ounces wine cooler	89
8 ounces yogurt	80
4 ounces frozen dairy dessert	47

Sources: Position of the American Dietetic Association: Use of nutritive and nonnutritive sweeteners, *Journal of The American Dietetic Association* 98 (1998): 584, Approximate aspartame content in Food and Drug Administration—approved categories, The NutraSweet Company.

MINIGLOSSARY OF SWEETENERS

ACESULFAME-K (AY-see-sul-fame) a derivative of acetoacetic acid approved for use in the United States in 1988. Since it is not metabolized by the body, acesulfame K does not contribute calories and is excreted from the body unchanged. It is currently approved for use in more than 70 countries and found in more than 100 international products, including chewing gum, gelatins, nondairy creamers, powdered drink mixes, and puddings.

ALTERNATIVE SWEETENERS nutritive (calorie-containing) sweeteners such as fructose, sorbitol, mannitol, and xylitol.

ARTIFICIAL SWEETENERS nonnutritive sugar replacements such as acesulfame-K, aspartame, saccharin, and sucralose.

ASPARTAME a dipeptide (see Chapter 5) containing the amino acids aspartic acid and phenylalanine and used in the United States and Canada since 1981. Although it is digested as protein and supplies calories, it is so sweet that only small amounts, which contribute negligible calories, are needed to sweeten foods. Thus, it is classified as a nonnutritive sweetener. Often sold under the trade name NutraSweet, aspartame is blended with lactose and an anticaking agent and sold commercially as Equal.

PHENYLKETONURIA an inborn error of metabolism, detectable at birth, in which the body lacks the enzyme needed to convert the amino acid phenylalanine to the amino acid tyrosine. As a result, derivatives of phenylalanine accumulate in the blood and tissues, where they can cause severe damage, including mental retardation.

PLACEBO (plah-SEE-bo) a sham treatment given to a control group; an inert, harmless "treatment" that the group's members cannot recognize as different from the real thing. This will minimize the chance that an effect of the treatment will appear to have occurred due to the **placebo effect**—the healing effect that the belief in the treatment, rather than the treatment itself, often has.

SACCHARIN a zero-calorie sweetener discovered in 1879 and used in the United States since the turn of the century. A possible link to bladder cancer has led to saccharin's being banned as a food additive in Canada, although it is available there as a tabletop sweetener; it is used with a warning label in the United States and is the sweetening agent in Sweet 'N Low.

SUCRALOSE a nonnutritive sweetener approved in 1998 as a tabletop sweetener and for use in a variety of desserts, confections, and nonalcoholic beverages. Sucralose is a noncaloric, heat-stable sweetener derived from a chlorinated form of sugar. Although sucralose is made from sugar, the body does not recognize it as sugar, and the sucralose molecule is excreted in the urine essentially unchanged.

SUGAR ALCOHOLS (MALTITOL, MANNITOL, SORBITOL, ISOMALT, XYLITOL) can be derived from fruits or commercially produced from dextrose; absorbed more slowly and metabolized differently than other sugars in the human body. The sugar alcohols are not readily used by ordinary mouth bacteria and therefore are associated with less cavity formation. Although the sugar alcohols are used as sugar substitutes, they do add calories (about 1.5 to 3 calories per gram) to a food product. They are found in a wide variety of chewing gums, candies, and dietetic foods. Sorbitol and mannitol can have a laxative effect in some people.

why people with diabetes do not have to limit themselves to only artificially sweetened foods. Sugar-containing items can be incorporated into a carefully designed eating plan.

Another reason that blanket statements about the use of sugar substitutes are not made for people with diabetes is that the various types of sweeteners behave differently in the body. A group of sugar substitutes known as **alternative sweeteners** (fructose, sorbitol, mannitol, and xylitol), for instance, contain calories but used to be frequently recommended for people with diabetes because the body generally absorbs them more slowly than it does table sugar. Most, however, have side effects that detract from their desirability. Some research suggests, for example, that large amounts of fructose may contribute to rises in blood cholesterol levels, making fructose a poor choice for the many people whose diabetes goes hand in hand with heart disease. By the same token, some people experience diarrhea after consuming large amounts of sorbitol or mannitol, and those sweeteners' overall effect on blood glucose control is insignificant, according to the American Diabetes Association. Products with sorbitol and mannitol may carry the following warning on the label because high intakes increase the risk of malabsorption: "excess consumption may have a laxative effect." Thus, the role sugar substitutes play in the overall diet is a decision that should be made individually with the help of a dietitian and a physician.[35]

Isn't it true that sugar-free chewing gum doesn't contain any calories?

Not so, if the sugar-free chewing gum is sweetened with certain alternative sweeteners, such as xylitol and sorbitol, also known as **sugar alcohols.** While sugar alcohols impart a sweet taste and supply calories (about 8 per stick), unlike sucrose, they do not promote tooth decay. Xylitol may actually inhibit the production of tooth-damaging acid by the caries-producing bacteria in the mouth and prevent them from adhering to the teeth. For this reason, the FDA authorizes use of the health claim on food labels that sugar alcohols do not promote tooth decay.

Some manufacturers sweeten chewing gum with artificial sweeteners or a combination of sweeteners (for example, both xylitol and aspartame). Chewing gum sweetened with artificial sweeteners contain negligible calories and also do not promote tooth decay.

Nutrition on the Web

nutrition.wadsworth.com	Go to the *Personal Nutrition* site to check for the latest updates to chapter topics or to access links to related Web sites.
www.healthfinder.gov/searchoptions/topicsaz.htm	Search for information on lactose intolerance, tooth decay, diabetes, and artificial sweeteners.
ificinfo.health.org	Search for information on sugars, sweeteners, and fiber.
www.cdc.gov/health/diabetes.htm	Provides guidelines and updates of information for people with diabetes.
www.diabetes.org	The American Diabetes Association's site with consumer information for people with diabetes.
www.nlm.nih.gov	Free access to National Library of Medicine's Medline for information searches on a variety of health-related topics.
www.wheatfoods.org	Information about the benefits of eating a high-carbohydrate diet.
www.eatright.org	The American Dietetic Association's site with position papers and resources on carbohydrate topics.
www.flaxcouncil.ca	Look for information about the benefits of fiber from flaxseed.

Check Yourself . . .

1. Which carbohydrates are described as simple and in which foods are they found?
2. Which carbohydrates are described as complex and in which foods are they found?
3. Why are carbohydrates the ideal fuel for the body?
4. What are the current dietary recommendations regarding the intake of complex carbohydrates and sugar?
5. State whether people with diabetes should eat foods sweetened only with sugar substitutes instead of sugar and give a rationale.
6. Why are enriched products not always as nutritious as whole-grain products?
7. List the health effects of fiber.
8. Suggest ways for increasing the fiber content of one's diet.
9. Summarize the digestion and absorption of carbohydrates in the body.
10. Describe how the body maintains blood glucose concentrations.

Answers to Check Yourself questions are found in Appendix G.

4

The Lipids: Fats and Oils

CONTENTS

A Primer on Fats

A Closer View of Fats

Characteristics of Fats in Foods

Nutrition Action: The Trans Fatty Acid Controversy—Is Butter Better?

The Other Members of the Lipid Family: Phospholipids and Sterols

How the Body Handles Fat

Lipids and Health

Fat in the Diet

Scorecard: Rate Your Fats and Health IQ

The Savvy Diner: How to Handle Fats in the Diet

Spotlight: Diet and Heart Disease

Heart Health Scorecard

The ultimate reality of nutrition rests with the chemistry of the food we eat and its effects on the processes of life.

R.M. Deutsch
(1928–1988, nutrition author and educator)

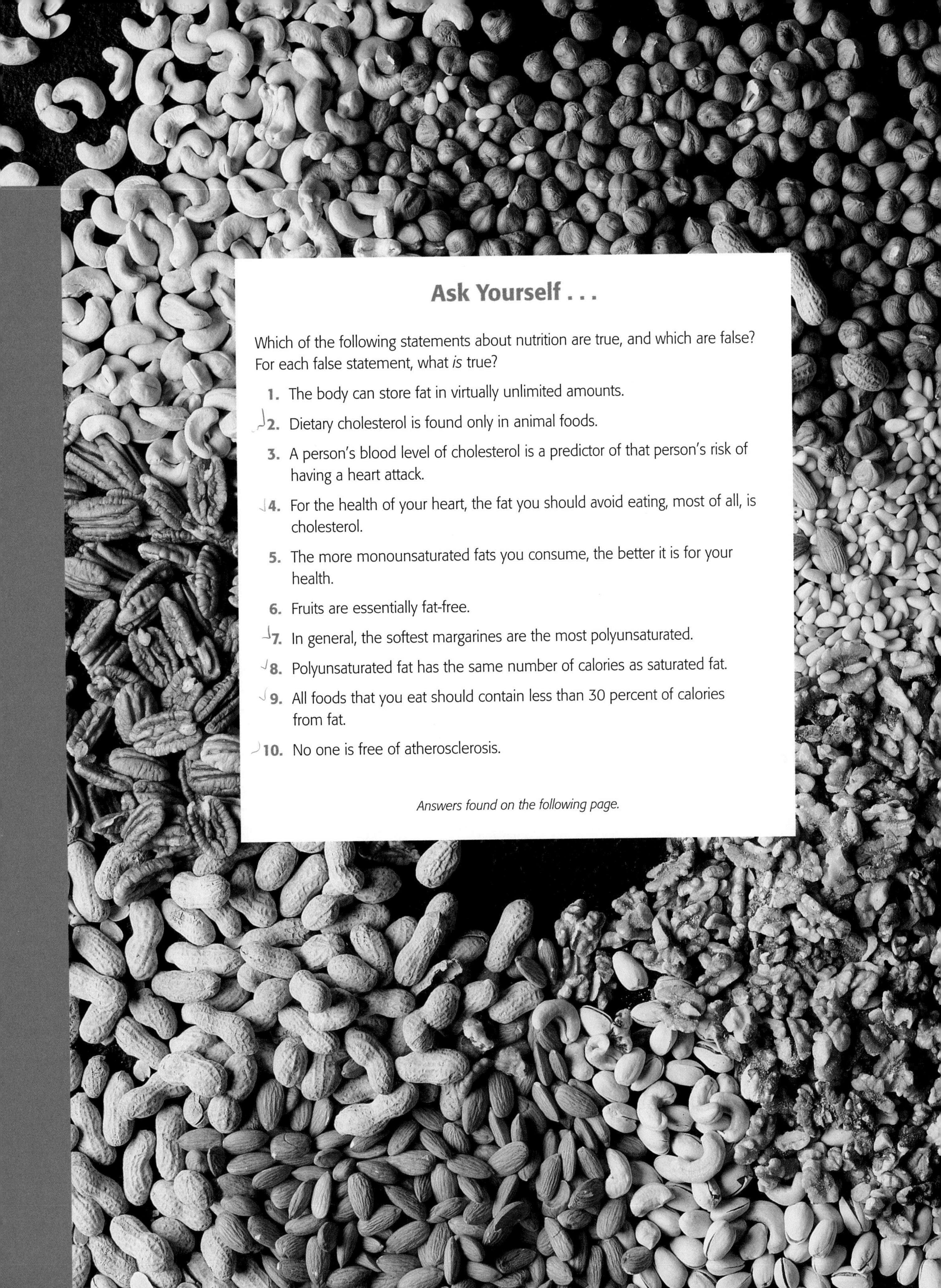

Ask Yourself . . .

Which of the following statements about nutrition are true, and which are false? For each false statement, what *is* true?

1. The body can store fat in virtually unlimited amounts.
2. Dietary cholesterol is found only in animal foods.
3. A person's blood level of cholesterol is a predictor of that person's risk of having a heart attack.
4. For the health of your heart, the fat you should avoid eating, most of all, is cholesterol.
5. The more monounsaturated fats you consume, the better it is for your health.
6. Fruits are essentially fat-free.
7. In general, the softest margarines are the most polyunsaturated.
8. Polyunsaturated fat has the same number of calories as saturated fat.
9. All foods that you eat should contain less than 30 percent of calories from fat.
10. No one is free of atherosclerosis.

Answers found on the following page.

While you and your friend are visiting a health fair at a local shopping mall, you learn that your blood cholesterol level is high. Your friend, a registered dietitian, urges you to see your health-care provider to request a blood test to determine the level of triglycerides and the ratio of "good" to "bad" cholesterol in your blood. As you are leaving the mall, you notice a bookstore display for a new diet book that urges you to cut fat intake to reduce cancer risk. Then you stop at the window of a health-food store. Your friend points to the freshly ground peanut butter, mentioning to you that it is not hydrogenated, while you stare at the bottles of antioxidant and fish oil supplements that line the wall. Later, during the evening news, your attention is drawn to a television commercial featuring a loving couple promising to use heart-healthy spreads on their morning muffins. "What does all of this mean?" you ask.

LIPIDS a family of compounds that includes triglycerides (fats and oils), phospholipids (lecithin), and sterols (cholesterol).

FATS lipids that are solid at room temperature.

OILS lipids that are liquid at normal room temperature.

Actually you may know more than you think you do about the **lipids,** more commonly called **fats** and **oils.** The most obvious dietary sources of fat are oil, butter, margarine, and shortening. Other food sources that provide fat to the diet are meats, nuts, mayonnaise, salad dressings, eggs, bacon, gravy, cheese, ice cream, and whole milk. You may know that egg yolk and liver are high in cholesterol, and you probably know that the cholesterol in the body is in some way related to heart disease. However, you may be confused by the many terms related to the fat in your diet. This chapter explores the terminology of fat and describes how fats can both contribute to health and detract from it.

A PRIMER ON FATS

Most people have the impression that fat is bad for them, and it might come as a surprise that fats are valuable. More than valuable, some are absolutely essential, and some fats must be present in the diet for you to maintain good health. Even if you wanted to, it would be impossible to remove all the fat from your diet, because at least a trace of fat is found in almost all foods.

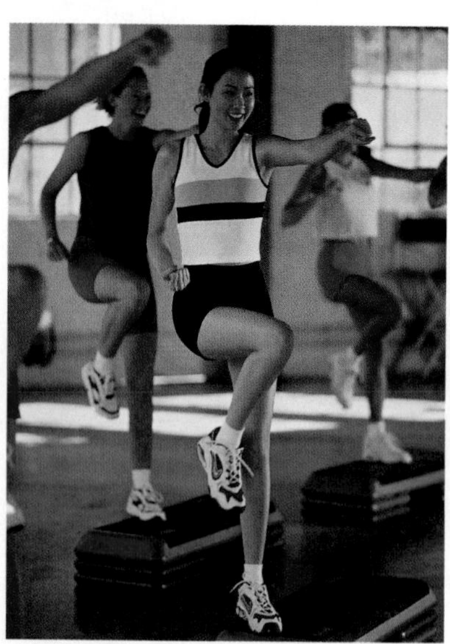

Muscles derive most of the fuel they need for work from body fat.

The Functions of Fat in the Body

Fat is the body's chief storage form for the energy (or calories) from food eaten in excess of immediate need. Fats provide most of the energy needed to perform much of the body's work, and especially muscular work.

Fat serves as an energy reserve. Whenever you eat, you store some fat, and within a few hours after a meal, you take the fat out of storage and use it for energy until the next meal. Thus, both glucose and fat are stored after meals, and both are released later when needed as fuel for the cells' work. However, whereas excess carbohydrate and protein can be converted to fat, they cannot be made from fat. Fat can serve only as an energy fuel for cells equipped to use it.

The body has scanty reserves of carbohydrate and virtually no protein to spare, but it can store fat in practically unlimited amounts. A pound of body fat is worth 3,500 calories, and a person's body can easily carry 30 to 50 pounds of fat without appearing fat at all.

Both fats and oils are found in your body, and both help to keep your body healthy. Fat is important to all your body's cells as part of their cell membranes. Natural oils in the skin provide a radiant complexion; in the scalp they help to nourish the hair and make it glossy. Fat insulates the body and cushions the body's vital organs. It serves as a shock absorber. The fat blanket under the skin

Ask Yourself Answers: 1. True. **2.** True. **3.** True. **4.** False. For the health of your heart, the fat you should avoid eating, most of all, is saturated fat. **5.** False. Monounsaturated fats such as olive or canola oil are good for the health of your heart, but like all fats, they should be consumed in moderation. **6.** True. **7.** True. **8.** True. **9.** False. Your *total* diet, not individual foods, should contain 30 percent or less of calories from fat. **10.** True.

also provides insulation from extremes of temperature, thus achieving internal climate control. Table 4-1 summarizes the major functions of fats in the body.

The Functions of Fat in Foods

Fat is a nutrient found in many foods. As the most concentrated source of calories, fat contains more than twice as many calories, ounce for ounce, as protein or carbohydrate. High-fat foods may therefore deliver many *unneeded* calories in only a few bites to the person who is not expending much physical energy. Fats in foods also provide **satiety** by slowing the rate at which the stomach empties. This is the reason you feel fuller longer after eating meals that include fat.

Fat is important for another reason. Some essential nutrients are soluble in fat and therefore are found mainly in foods that contain it. These nutrients are the essential fatty acids (to be described shortly) and the fat-soluble vitamins—A, D, E, and K (described in Chapter 6). Fat also carries many dissolved compounds that give foods their aroma and flavor. This accounts for the aromatic smells associated with foods that are being fried, such as onions or French fries. It also helps explain why a plain doughnut is more flavorful than a plain roll—it is higher in fat. Table 4-2 summarizes the functions of fats in foods.

The Terminology of Fat

About 95 percent of the lipids in foods and in the human body are **triglycerides.** Other members of the lipid family are the **phospholipids** (of which lecithin is one) and the **sterols** (**cholesterol** is the best known of these). The blood lipid profile refers to a test analyzed by a medical laboratory that reveals the amounts of various lipids (especially triglycerides and cholesterol) found in the blood and the carriers (such as LDL and HDL, described later) in which they are found. The results of this test tell much about a person's risk of heart disease, or cardiovascular disease (CVD). The blood cholesterol level is especially telling, and it bears on the question of whether people should avoid foods containing fat, those containing cholesterol, or both.*

Two kinds of lipids in foods are:

- triglycerides (commonly called fat)
- cholesterol

Similarly, two kinds of lipids in the blood are:

- triglycerides
- cholesterol

A person's *blood* level of cholesterol is considered to be a predictor of that person's likelihood of suffering a heart attack or stroke. The fat in *food* that contributes most to a high blood cholesterol level is triglycerides (especially those that contain *saturated* fat), *not cholesterol.* People often fail to understand this point, and the question arises again and again: "Should I eat cholesterol?" When told, "It doesn't matter much," the questioner often jumps to the wrong conclusion—the conclusion that cholesterol doesn't matter. It does matter. High *blood* cholesterol is an indicator of risk for CVD, but the main *food* factor associated with it is a high *saturated fat intake.*† One more distinction must be made clear about

*Blood, plasma, and serum cholesterol all refer to about the same thing; this book uses the term blood cholesterol. Plasma is simply blood with the cells removed; serum has the clotting factors also removed. The concentration of cholesterol is not much altered by these treatments.

†A few individuals have a hereditary inability to clear from their blood the cholesterol they have eaten and absorbed. This condition is rare but well-known in medical circles, because the study of it led to the discovery of how cholesterol is transported in the body. People with hereditary high blood cholesterol levels must refrain from eating cholesterol in foods. The vast majority of people can eat eggs and other cholesterol-containing foods in moderation without fear of incurring high blood cholesterol levels.

TABLE 4-1 THE ROLE OF FATS IN THE BODY

Fats in the Body:

- Provide a concentrated source of energy
- Serve as an energy reserve
- Form the major components of cell membranes
- Nourish skin and hair
- Insulate the body from extremes of temperature
- Cushion the vital organs to protect them from shock

TABLE 4-2 THE ROLE OF FATS IN FOODS

Fats in Foods:

- Provide calories (9 calories per gram)
- Provide satiety
- Carry fat-soluble vitamins and essential fatty acids
- Contribute aroma and flavor

SATIETY the feeling of fullness or satisfaction that people feel after meals.

TRIGLYCERIDES (try-GLISS-er-ides) the major class of dietary lipids, including fats and oils. A triglyceride is made up of three units known as fatty acids and one unit called glycerol.

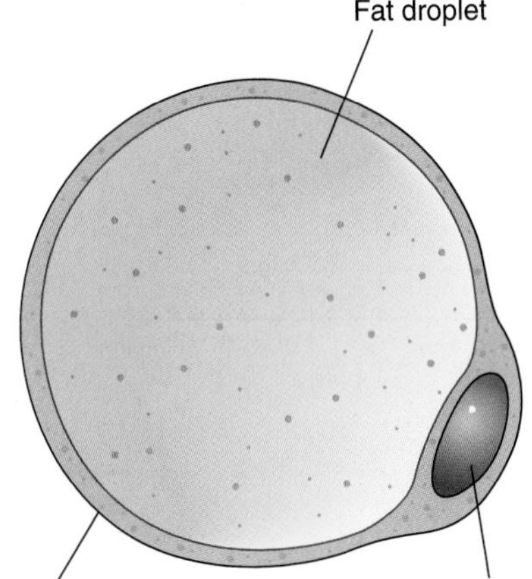

A Fat Cell
Within the fat cell, lipid is stored in a droplet. This droplet can greatly enlarge, and the fat cell membrane will grow to accommodate its swollen contents. (More about fat cells and obesity in Chapter 8.)

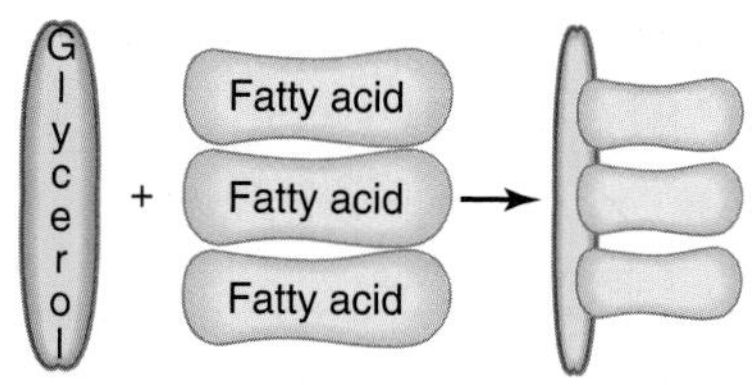

Glycerol + 3 Fatty acids → **Triglyceride**

GLYCEROL (GLISS-er-all) an organic compound that serves as the backbone for triglycerides.

PHOSPHOLIPIDS (FOSS-foh-LIP-ids) one of the three main classes of lipids; a lipid similar to a triglyceride but containing phosphorus.

STEROLS (STEER-alls) one of the three main classes of lipids; a lipid with a structure similar to that of cholesterol.

CHOLESTEROL (koh-LESS-ter-all) one of the sterols, manufactured in the body for a variety of purposes.

FATTY ACIDS basic units of fat composed of chains of carbon atoms with an acid group at one end and hydrogen atoms attached all along their length.

fats on the plate: they come in two varieties, saturated and unsaturated, and the saturated type is strongly implicated in raising blood cholesterol.

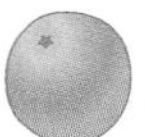

A CLOSER VIEW OF FATS

When the energy from any of the energy-yielding nutrients is to be stored as fat, it is first broken into fragments—small molecules made of carbon, hydrogen, and oxygen. These fragments are then linked together into chains known as **fatty acids**—the major building blocks of triglycerides, the chief form of fat found in the body.

Saturated Versus Unsaturated Fats

Fatty acids differ from one another in two ways: in chain length and in degree of saturation. Chain length refers to the number of carbons that are hooked together in the fatty acid.* Chain length is significant because it affects solubility of the fat in water—the short-chain fatty acids are somewhat soluble in water. Milk, butter, and cheese are rich in the short-chain fatty acids; vegetable oils and red meat contain triglycerides with long-chain fatty acids.

Of more significance than chain length is the degree of *saturation,* mentioned earlier in relation to heart disease. Saturation refers to the chemical structure—specifically to the number of hydrogens the fatty acid chain is holding. If every available bond from the carbons is holding a hydrogen, we say the chain is a **saturated fatty acid**—filled to capacity, or saturated, with hydrogen (see Figure 4-1).

*Short-chain fatty acids contain 4 to 6 carbons; medium-chain fatty acids contain 8 to 10 carbons; long-chain fatty acids contain 12 or more carbons.

FIGURE 4-1
THE TYPES OF FATTY ACIDS

*The omega number (e.g., omega-6, omega-3) designates the location of the first double bond, counting from the far end (the methyl [CH_3] end) of the fatty acid. Linoleic acid and linolenic acid are *essential* fatty acids.

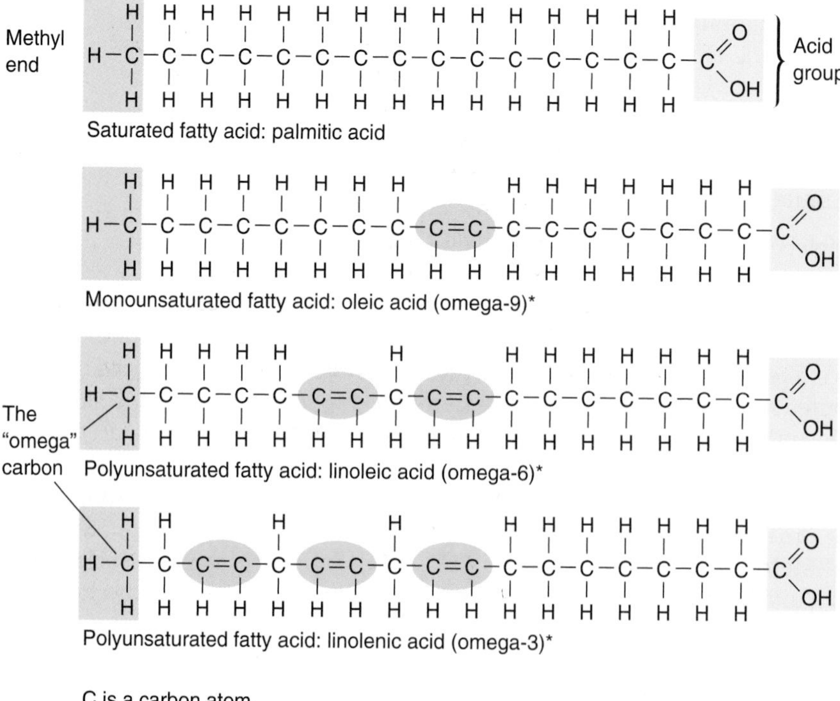

C is a carbon atom.
H is a hydrogen atom.
– is a single bond.
= is a double bond.
C=C is a point of unsaturation.

Sometimes, especially in the fatty acids in plants and fish, there is a place in the chain where hydrogens are missing—an "empty spot," or **point of unsaturation** (see Figure 4-1). A chain that possesses a point of unsaturation is an **unsaturated fatty acid.** If there is one point of unsaturation, it is a **monounsaturated fatty acid.** If there are two or more points of unsaturation, it is a **polyunsaturated fatty acid.**

The Essential Fatty Acids

The human body can synthesize all the fatty acids it needs from carbohydrate, fat, or protein, except two—**linoleic acid** and **linolenic acid.** These two cannot be made from other substances in the body or from each other, and they must be supplied by the diet; they are, therefore, **essential fatty acids.** These fatty acids are polyunsaturated fatty acids, widely distributed in the diet, especially in plant and fish oils. To be sure, they are readily stored in the adult body, making deficiencies unlikely. Still, deficiency symptoms can appear in the person deprived of these acids—a characteristic skin rash and, in children, poor growth.

Omega-6 Versus Omega-3 Fatty Acids

One further classification system for unsaturated fatty acids classifies the fatty acid as either an **omega-6** or an **omega-3** fatty acid (see Figure 4-1). Of the two essential fatty acids, linoleic acid is an omega-6 fatty acid, related to a whole series of others. Linolenic acid is an omega-3 fatty acid, with a similar family of its own.

Of interest in relation to dietary fat are findings on the omega-3 fatty acids found in fish oils, which offer a protective effect on health. Interest in fish oils was first kindled when someone thought to ask why the Eskimos of Greenland, who eat a diet very high in fat, have such a low rate of heart disease. The trail led to the abundance of fish they eat, then to the oils in those fish, and finally to the omega-3 fatty acids—EPA and DHA—in the oils. Now scientists are unraveling the mystery of what those fatty acids do.[1]

The omega-3 fatty acids—found in highest amounts in fish that live in cold waters—have a profound effect on the synthesis of certain hormonelike compounds that play many regulatory roles in the body. They affect a number of body functions, including the formation of blood clots, the raising and lowering of blood pressure and blood lipid levels, the immune response, and the inflammatory response to injury and infection.* The types of these compounds made in the body determine the degree of vulnerability to certain diseases, including cardiovascular disease and cancer.[2] For example, a study from the Netherlands reported that two or three fish meals a week in place of meat are all it takes to exert a significant positive effect.[3] The fatty acids in the fish appear to favorably alter the body's blood-clotting balance by reducing the ability of the blood to clot. These omega-3 fatty acids provide protection against heart disease, possibly by producing substances that dilate the arteries, lower the blood pressure, and help to dissolve blood clots.[4]

People who eat large amounts of fish tend to have lower blood cholesterol and triglyceride levels and slower clot-forming rates.[5] More than 70 studies have now documented other connections as well. Diets high in omega-3 fatty acids seem to bring about enhanced defenses against cancer (via the immune response) and reduced inflammation in arthritis and asthma sufferers. The Eskimos apparently did well using fish for food. Perhaps we, too, could benefit from eating fish in abundance.

*The hormonelike compounds referred to here are the **eicosanoids** (eye-COSS-uh-noyds)—compounds that regulate blood pressure, clotting, and other body functions. Names of classes of eicosanoids are prostaglandins, thromboxanes, prostacyclins, and leukotrienes.

SATURATED FATTY ACID a fatty acid carrying the maximum possible number of hydrogen atoms (having no points of unsaturation). Saturated fats are found in animal foods like meat, poultry, and full-fat dairy products, and in tropical oils such as palm and coconut.

POINT OF UNSATURATION a site in a molecule where the bonding is such that additional hydrogen atoms can easily be added.

UNSATURATED FATTY ACID a fatty acid with one or more points of unsaturation. Unsaturated fats are found in foods from both plant and animal sources. Unsaturated fatty acids are further divided into monounsaturated fatty acids and polyunsaturated fatty acids.

MONOUNSATURATED FATTY ACID a fatty acid containing one point of unsaturation, found mostly in vegetable oils such as olive, canola, and peanut.

POLYUNSATURATED FATTY ACID (sometimes abbreviated PUFA) a fatty acid in which two or more points of unsaturation occur, found in nuts and vegetable oils such as safflower, sunflower, and soybean, and in fatty fish.

LINOLEIC (lin-oh-LAY-ic) **ACID, LINOLENIC** (lin-oh-LEN-ic) **ACID** polyunsaturated fatty acids, essential for human beings.

ESSENTIAL FATTY ACID a fatty acid that cannot be synthesized in the body in amounts sufficient to meet physiological need.

The Essential Fatty Acids

Polyunsaturated Fatty Acids

Omega-6
Linoleic acid
Arachidonic acid

Omega-3
Linolenic acid
(DHA) (EPA)

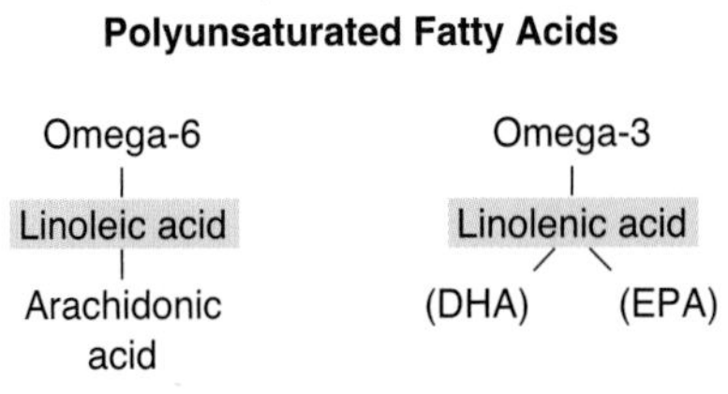

The fatty acids in fish oils include eicosapentaenoic (EYE-kossa-PENTA-ee-NOH-ic) acid (EPA) and docosahexaenoic (DOE-cosa-HEXA-ee-NOH-ic) acid (DHA), both of which are omega-3 fatty acids.

OMEGA the last letter—the far end—of the Greek alphabet (ω), used by chemists to refer to the position of the end-most double bond in a fatty acid. The **omega-6** fatty acids have their end-most double bonds after the sixth carbon in the chain; the **omega-3** acids, after the third.

SEAFOOD ESPECIALLY HIGH IN OMEGA-3 FATTY ACIDS (AT LEAST 1 GRAM OF OMEGA-3 FATTY ACIDS IN 3.5 OUNCES OF FISH): anchovies, Atlantic bluefish, herring, mackerel, sablefish, salmon (all varieties), sardines, sturgeon, tuna (white albacore—canned in water—or bluefin), and whitefish.

SEAFOOD MODERATELY HIGH IN OMEGA-3 FATTY ACIDS (0.5–0.9 GRAM IN 3.5 OUNCES OF FISH): pompano, rainbow trout, shark, smelt, striped bass, and swordfish.

OTHER SEAFOOD SOURCES OF OMEGA-3 FATTY ACIDS (0.5 GRAM OR LESS IN 3.5 OUNCES OF FISH): catfish, clams, cod, crab, crayfish, flounder, grouper, haddock, hake, halibut, lobster, mahi mahi, mullet, ocean perch, orange roughy, oysters, pike, pollock, rockfish, scallops, sea bass, sea trout, shrimp, snapper, and whiting.

OTHER SOURCES OF OMEGA-3 FATTY ACIDS (LINOLENIC ACID) BESIDES FISH AND SEAFOOD: canola oil, flaxseed, leafy green vegetables, soyfoods, wheat germ, and walnuts.

Current western intakes of omega-3 fats are well below the levels that experts consider optimal.[6] The optimal ratio of omega-6 to omega-3 fatty acids is believed to be about 5 to 1 (omega-6 to omega-3), but in modern western diets the ratio has shifted toward the omega-6 fatty acids, leading to a ratio of more than 25 to 1. Health experts now advise that you decrease consumption of foods rich in omega-6 fatty acids, such as vegetable oils (corn, safflower, sesame, sunflower), and increase the intake of omega-3 fatty acids—those found in fish, canola oil, flaxseed, and soyfoods—to achieve a more healthful balance between the two types of fats.

If our diets need modification, it is not by self-prescribing supplements of fish oils. Many hazards may be associated with the taking of such supplements. One reason is that they could unbalance the diet too far; you do need omega-6, too, remember. Too much fish oil might make a person susceptible to stroke or hemorrhage because of its effect of prolonging bleeding and clotting time. Fish oils—made from fish livers—may contain toxic amounts of fat-soluble vitamins and pesticide residues. These and other risks do not normally accompany the eating of a variety of fish. Furthermore, fish is both a good source of high-quality protein and relatively low in calories. The purified oils contain just oil, and oil is high in calories. If you tried to take enough fish oil to match the Eskimos' intake, you would need 300 to 500 calories a day—enough to gain about 25 pounds a year. Instead, substitute two or three fish meals a week for meals based on other protein-rich foods, and you may improve your health more safely.

CHARACTERISTICS OF FATS IN FOODS

The amount of unsaturated fatty acids in a fat affects the temperature at which the fat melts. The more unsaturated a fat, the more liquid it is at room temperature. In contrast, the more saturated a fat (the more hydrogen it has), the firmer it is. Thus, of three fats—beef fat, chicken fat, and corn oil—beef fat is the most saturated and the hardest; chicken fat is less saturated and somewhat soft; and corn oil, which is the most unsaturated, is a liquid at room temperature (and the only one of the three that comes from a plant). If your health-care provider tells you to use monounsaturated fats or polyunsaturated fats, you can usually judge which ones to choose by their hardness at room temperature. Fats and oils contain mixtures of saturated, monounsaturated, and polyunsaturated fatty acids; the fatty acids that predominate determine whether the fat is solid or liquid at room temperature. Figure 4-2 compares the most common fats and oils with regard to their percentages of the various types of fat.

The more unsaturated a fat, the more liquid it is at room temperature. The more polyunsaturated the fat is, the sooner it melts.

Because fats differ chemically, they behave differently in foods. To control their characteristics, food manufacturers sometimes alter them—and sometimes the alterations have health consequences, as described in the Nutrition Action feature on page 106.

Fats in common use. The vegetable oil is polyunsaturated; the olive oil is monounsaturated; the butter is largely saturated; and the margarine, no doubt, is partially hydrogenated.

Points of unsaturation in fatty acids are like weak spots in that they are vulnerable to attack by oxygen. When the unsaturated

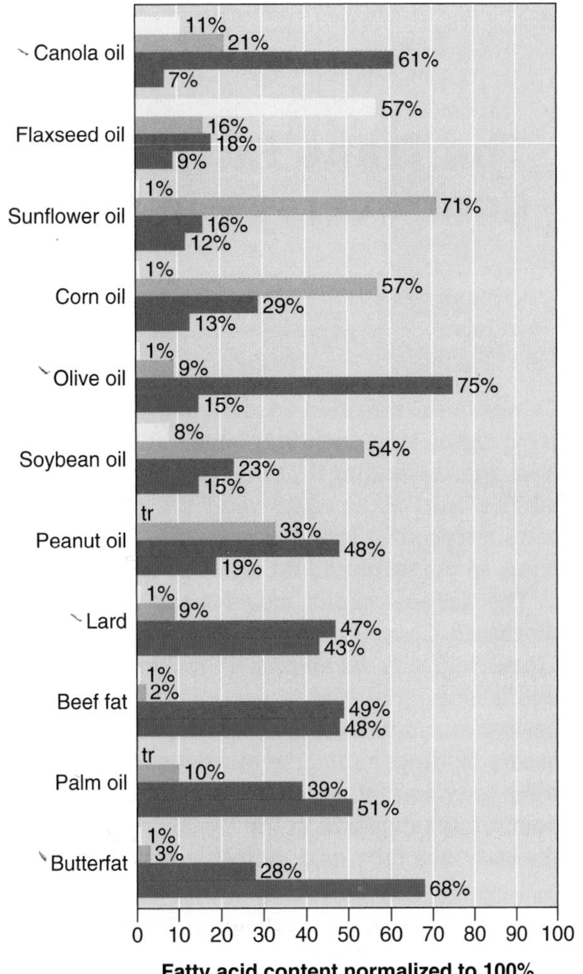

FIGURE 4-2
A COMPARISON OF SATURATED AND UNSATURATED FATTY ACIDS IN DIETARY FATS AND OILS

points react with oxygen, the oils become rancid. This is why unprocessed oils should be stored in tightly covered containers. If stored for long periods, they need refrigeration to prevent spoilage.

One way to prevent spoilage of oils containing unsaturated fatty acids is to change them chemically by **hydrogenation,** but this causes them to lose their unsaturated character and the health benefits that go with it. When food producers want to use a polyunsaturated oil such as corn oil to make a spreadable margarine, they hydrogenate the oil. Hydrogen is forced into the oil, some of the unsaturated fatty acids accept the hydrogen, and the oil becomes harder. The spreadable margarine that results is more saturated than the original oil but not as saturated as butter.

A second way to prevent spoilage of oils is to add a chemical that will compete for the oxygen and thus protect the oil. Such an additive is called an **antioxidant.** Examples are the well-known additives BHA and BHT listed on the labels of many processed foods, such as breakfast cereals and snack foods, and the natural antioxidants vitamin C and vitamin E. A third alternative, already mentioned, is to keep the product refrigerated.

Another way that the food industry alters natural fats and oils is by adding an **emulsifier** to a food product to allow fats and water to mix and remain mixed in a food product. Mayonnaise, margarines, salad dressings, and cake mixes often list an emulsifier on their labels. Mono- and diglycerides are good emulsifiers, as is the emulsifier found in egg yolk—lecithin.

HYDROGENATION (high-droh-gen-AY-shun) the process of adding hydrogen to unsaturated fat to make it more solid and more resistant to chemical change.

ANTIOXIDANT (anti-OX-ih-dant) a compound that protects other compounds from oxygen by itself reacting with oxygen.

EMULSIFIER a substance that mixes with both fat and water and can break fat globules into small droplets, thereby suspending fat in water.

NUTRITION ACTION

The Trans Fatty Acid Controversy—Is Butter Better?

Conventional nutrition wisdom has long held that margarine is better than butter to use on food, including bagels and other breads. But in recent years, rumors have been spreading that margarine may not be the most healthful choice after all. The heart of the controversy is a growing body of research suggesting that a certain type of fat found in margarine may be as likely to boost blood cholesterol levels as the saturated fat found in butter.[7]

TRANS FATTY ACID a type of fatty acid created when an unsaturated fat is hydrogenated. Found primarily in margarines, shortenings, commercial frying fats, and baked goods, trans fatty acids have been implicated in research as culprits in heart disease.

The alleged culprit, called **trans fatty acid,** is formed when margarine is processed. Consider that to make margarine, manufacturers take a highly unsaturated vegetable oil and partially hydrogenate (add hydrogen to) it; hydrogenation is what gives the spread its relatively solid consistency and helps protect against rancidity. During the hydrogenation process, however, a chemical "fluke" occurs. Hydrogenating the oil creates a new chemical configuration, known as a trans fatty acid, in which hydrogen atoms of the fat lie on opposite sides of the point of unsaturation in the carbon chain (see Figure 4-3). (You may recall that the body of a fatty acid molecule consists of a chain of carbon atoms to which hydrogen is attached, as shown in Figure 4-1 on page 102.)

Because trans fatty acids are formed whenever oils are hydrogenated, margarine isn't the only food on the market that contains trans fatty acids. Shortenings, baked goods, certain brands of peanut butter, commercial frying fats used in fast-food outlets to cook French fries and other fried items, and any other foods that list "partially hydrogenated vegetable oil" on their label are among the foods containing trans fatty acids.

Although nutrition experts have long advised that using margarine as a spread is preferable to using butter and that partially hydrogenated oils are the more healthful alternative to, say, lard, studies conducted during the past several years caught the attention of the media and sparked numerous news stories challenging that advice. The research suggests that high levels of trans fatty acids

FIGURE 4-3
TYPES OF UNSATURATED FATTY ACIDS: CIS VERSUS TRANS

Unsaturated fatty acids are either in *cis* form or in *trans* form, depending upon the way in which the hydrogen atoms are attached to the points of unsaturation in the carbon chain. If the hydrogen atoms are attached to the same side of the points of unsaturation, the arrangement is called cis. If the hydrogen atoms are attached to different sides, the arrangement is called trans.

```
 H  H  H  H          H  H     H
 |  |  |  |          |  |     |
-C--C==C--C-        -C--C==C--C-
 |        |          |     |  |
 H        H          H     H  H

    cis                 trans
```

cis = same
trans = across

JEFF STAHLER reprinted by permission of Newspaper Enterprise Association, Inc.

may raise blood levels of "bad" LDL-cholesterol nearly as much as saturated fats. What's more, the research suggests that in large amounts, trans fatty acids lower "good" HDL-cholesterol in the blood.[8] One large-scale study published in 1993 even indicated that women who ate a diet high in trans fatty acid-containing foods, particularly margarine, were more likely to suffer heart disease than their counterparts who ate relatively little margarine and other trans fatty acid-containing foods. The study prompted a flurry of headlines and news reports questioning whether margarine was a healthful alternative after all.[9] In 1997, the same group of researchers again demonstrated a link between heart disease and trans fatty acids with the results of a well controlled study of longer duration.[10]

Fueling the controversy even further was an analysis by researchers at Harvard University suggesting that more than 30,000 American deaths each year are attributable to trans fatty acids. The authors of the report went so far as to call for government legislation mandating that manufacturers include trans fatty acid amounts on food labels and phase out the use of partially hydrogenated oils in the United States.[11]

In November 1999, the Food and Drug Administration announced plans to require manufacturers to include specific information about trans fats in the Nutrition Facts panel on all food products. Under the proposed ruling, products containing trans fats will carry an asterisk on the line for saturated fats, directing consumers to a footnote about trans fat content. Manufacturers can label a product "trans fat free" if the food contains less than 0.5 grams of trans fat and less than 0.5 grams of saturated fat per serving.[12] It is expected to take two years before consumers will see the new information on food packages.

Many major health organizations, including the American Heart Association, The American Dietetic Association, and the American Institute of Nutrition point out that the trans fatty acid research does not indicate that consumers should switch back from eating margarine to eating butter, which contains an excess of saturated fat. That's because trans fatty acids account for only about 5 to 8 percent of the total calories in a typical American diet, whereas saturated fat comprises some 12 percent of calories. Thus, the long-standing recommendation to keep to a minimum consumption of saturated fat, which a comprehensive body of research has clearly shown raises blood cholesterol levels, far outweighs concerns about trans fatty acids.

To be sure, continuing research may shed more light on the health effects of trans fatty acids. And while the evidence that has come to light doesn't warrant going back to butter, it does underscore the value of keeping the amount of *total* dietary fat to a minimum. The foods that contain large amounts of trans fatty acids—French fries, corn chips, deep-fat fried doughnuts—are also high in total fat (see Table 4-3). Thus,

TABLE 4-3 **TUNING INTO TRANS FATTY ACIDS**

Trans fatty acids occur naturally in meat, poultry, and dairy products. They show up in packaged and fast foods to which hydrogenated vegetable oils have been added. Note that high levels of trans fatty acids go hand in hand with high-fat foods. Thus, if you eat a low-fat diet, you will consume minimal amounts of trans fatty acids.

Food	Grams of Trans Fatty Acids*	Grams of Total Fat
Animal Foods		
5 oz beef	0.9	27.7
1 tsp butter	0.1	4.1
5 oz chicken	0.1	10.4
5 oz pork	0.1	21.7
Vegetable Fats		
1 tsp vegetable shortening	0.63	4.3
1 tsp stick margarine	0.62	4
1 tsp soft margarine	0.27	4
1 tsp vegetable oil	0.02	4.5
Packaged Foods and Fast Foods		
4 oz deep-fat fried French fries	5.5	18.5
1 cake doughnut	3.19	10.8
1 oz corn chips	1.42	9.5
1 piece yellow cake with chocolate frosting	1.04	11.2
1 slice apple pie	1	13.8
1 oatmeal raisin cookie	0.86	3.3
1 slice cheese pizza	0.13	6.1
1 oz potato chips	0.11	9.8
1 blueberry muffin	0.09	3.7

*Figures represent the average values derived from several brands or varieties of the foods.

Source: Adapted from L. Litin and F. Sacks, Trans-fatty-acid content of common foods, *New England Journal of Medicine* 329 (1994): 1969–1970.

health-conscious eaters who want to hedge their bets against the possibility that trans fatty acids contribute to high blood cholesterol levels should stick with the same principles of low-fat eating that have been outlined throughout the book. They should also pay particular attention to the following:

- Shop for margarine containing no more than two grams of saturated fat per tablespoon and with liquid vegetable oil as the first ingredient. Choose soft tub or liquid forms over stick; "hard" stick margarines typically contain the most trans fatty acids. Look for the canola- or olive-oil based margarines containing zero trans fat.
- Limit total daily consumption of fats and oils to no more than five to eight teaspoons to keep both total fat and trans fatty acid consumption to a minimum.[13]
- Cook and bake with a vegetable oil, such as canola, olive, or corn oil, instead of butter, shortening, or margarine whenever possible.
- Check ingredients lists for "partially hydrogenated vegetable oil." The higher on the list it appears, the more of both total fat and trans fatty acids the product probably contains, and the more difficult it is to fit into a healthful diet. As an alternative, look for similar products made with unhydrogenated oils and smaller amounts of total fat.
- Remember that deep-fat fried foods such as French fries, doughnuts, chicken nuggets, and fried seafood contain excessive amounts of total fat as well as trans fatty acids and should be eaten in moderation.

In contrast to the margarines containing trans fatty acids are certain new brands of margarine that contain **stanol esters**—compounds derived from plants that may actually reduce blood cholesterol when consumed as part of a low-fat diet.[14] These compounds block the absorption of dietary cholesterol from the intestine. Margarines containing these compounds, however, typically cost three or four times more than regular margarine and the safety of their long-term daily use is not known.

STANOL ESTERS members of the sterol class of lipids, derived from plants, and capable of reducing blood cholesterol levels when eaten as part of a low-fat diet. The Spotlight feature on Functional Foods in Chapter 6 provides more information about stanol esters added to margarine.

Certain brands of margarine are made from unhydrogenated oils and contain negligible or no trans fatty acids. Check the package label to find margarine made this way.

THE OTHER MEMBERS OF THE LIPID FAMILY: PHOSPHOLIPIDS AND STEROLS

Lecithin and other phospholipids are important components of cell membranes. Because of the way the phospholipids are constructed, they can serve as emulsifiers in the body—joining with both water and fat so that they can help fats travel back and forth across the lipid-containing membranes of cells into the watery fluids on both sides. Due to its role as an emulsifier, lecithin is often listed in the ingredients list on food labels as a food additive. Food manufacturers add lecithin to foods such as margarine, chocolate, salad dressings, and frozen desserts to keep the fats dispersed with the other ingredients.

Magical properties are sometimes attributed to lecithin, and health-food advertisers try to persuade people to supplement their diets with it. But lecithin is widespread in food and is also made by the liver in abundant quantities and therefore most people's diets contain adequate amounts.

Cholesterol is found only in animal foods and is also made in the body, where it is an important compound with many functions. It is a part of **bile,** which is necessary in the digestion of fats. It is the starting material from which the sex hormones and many other hormones are made. In the skin, one of its derivatives is made into vitamin D with the help of sunlight. It is an important lipid in the structure of brain and nerve cells. In fact, cholesterol is a part of every cell. But while it is widespread in the body and necessary to the body's function, it also is the major component of the plaque that narrows the arteries in the killer disease atherosclerosis (see the Spotlight at the end of the chapter for more on heart disease). Figure 4-4 lists the cholesterol content of selected foods.

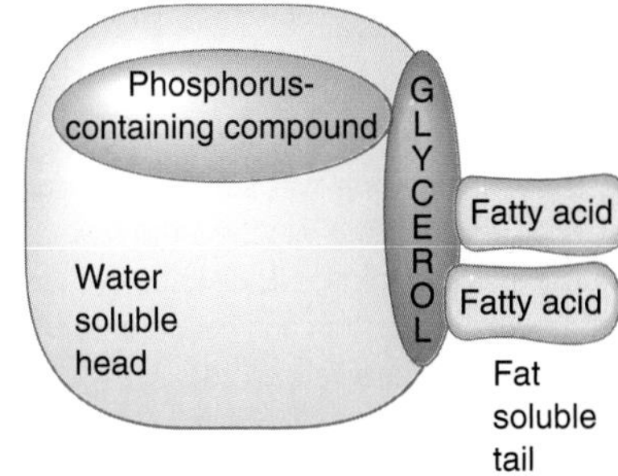

A Phospholipid: Lecithin
This phospholipid (lecithin) consists of a water-soluble head and a fat-soluble tail; thus, a phospholipid is a phosphorus-containing fat.

Cholesterold can be:
- Incorporated as an integral part of the structure of cell membranes
- Used to make bile for digestion
- Used to make sex hormones (estrogen and testosterone)
- Made into vitamin D
- Deposited in the artery walls leading to plaque buildup and heart disease

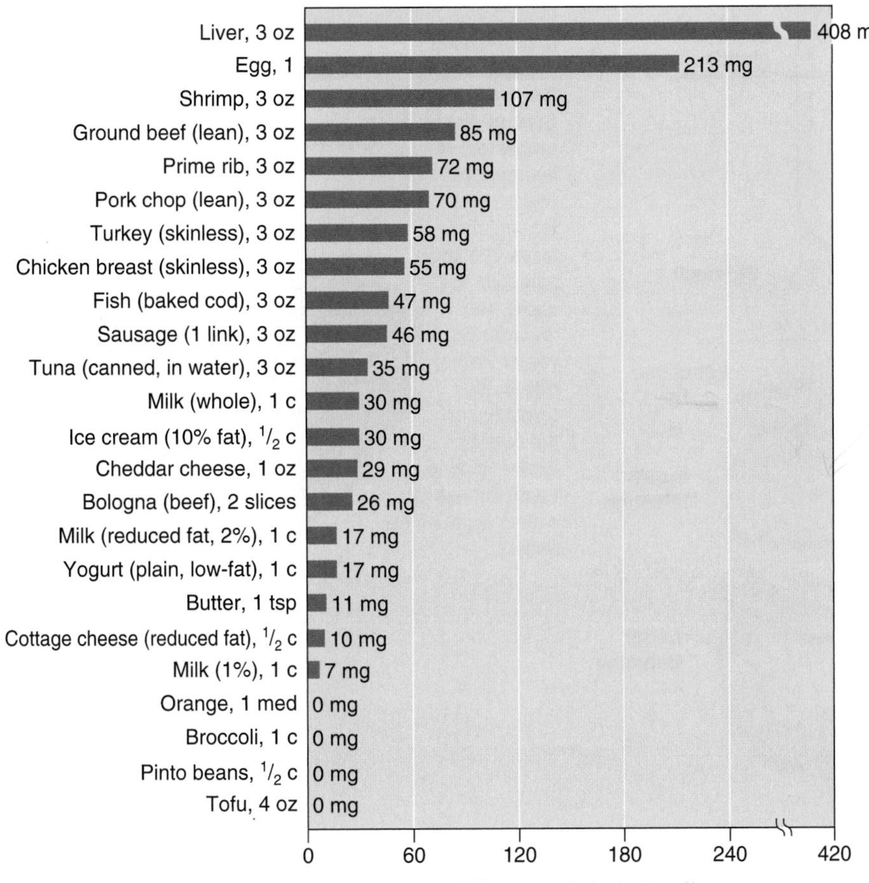

FIGURE 4-4
CHOLESTEROL CONTENT OF SELECTED FOODS

*Daily Value is 300 mg.

HOW THE BODY HANDLES FAT

BILE a mixture of compounds, including cholesterol, made by the liver, stored in the gallbladder, and secreted into the small intestine. Bile emulsifies lipids to ready them for enzymatic digestion and helps transport them into the intestinal wall cells.

MONOGLYCERIDE (mon-oh-GLISS-er-ide) a glycerol molecule with one fatty acid attached to it. A *diglyceride* is a glycerol molecule with 2 fatty acids attached to it.

LYMPH (LIMF) the body fluid that transports the products of fat digestion toward the heart and eventually drains back into the bloodstream; lymph consists of the same components as blood with the exception of red blood cells.

A summary of lipid digestion is shown in Figure 4-5. After digestion in the upper small intestine, the products of fat digestion—fatty acids, glycerol, and **monoglycerides**—must enter the bloodstream if they are to be of use to the body's cells. The shortest free fatty acids pass by simple diffusion into the cells that line the intestine. Because these short-chain fatty acids are somewhat water soluble, they can, without any further processing, enter the body's capillaries. Like the products of carbohydrate digestion, the short-chain fatty acids are transported from these capillaries through collecting veins to the capillaries of the liver. The liver cells pick them up and convert them to other substances the body needs. The glycerol follows the same path as the short-chain fatty acids because it, too, is water soluble.

The larger products of fat digestion (long-chain fatty acids, cholesterol, and phospholipids) are insoluble in water, a difficulty that must be overcome. The body's fluids—**lymph** and blood—are watery and will not accept these larger molecules as they are. The longer-chain fatty acids do pass into the intestinal cells, but there they reconnect with glycerol or with monoglycerides, forming new triglycerides. Then the cells package them for transport before releasing them into the lymph system (see Figure 4-5, part B).

The cells allow triglycerides and other lipids to form and combine with special proteins to make **chylomicrons,** one of the four types of **lipoproteins** found

A. Digestion of Fat

Mouth
Some hard fats begin to melt as they reach body temperature.

Stomach
The stomach's churning action mixes fat with water and acid. A stomach enzyme accesses and breaks apart a small amount of fat. Fat is last to leave the stomach.

Small Intestine
Once in the small intestine, fat encounters **bile**, an emulsifier made in the liver. The gallbladder, a storage organ, squirts bile into the contents of the small intestine to blend the fat with the watery digestive secretions.

Pancreas
Fat-digesting enzymes from the pancreas (pancreatic lipase) enter the small intestine. The enzymes can attack fat only after emulsification by bile. They break down the triglycerides to fatty acids, glycerol, and monoglycerides.

Large intestine
Some fat and cholesterol, trapped in fiber, are carried out of the body with other wastes.

Mouth
Salivary glands
Liver
Stomach
Pancreas
Gallbladder
Small intestine
Large intestine

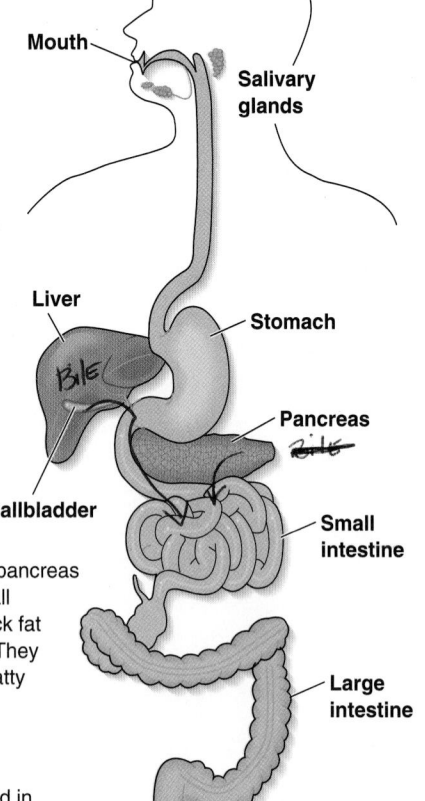

B. Absorption of Fat
Most of the newly digested fats are absorbed into lymph as part of a special package—the chylomicron.
A chylomicron (lipoprotein) contains an interior of triglycerides and cholesterol surrounded by phospholipids.
Proteins cover the structure. Such an arrangement of hydrophobic (water-fearing) molecules (the fatty acids) on the inside and hydrophilic (water-loving) molecules (proteins) on the outside allows lipids to travel through the watery fluids of the body.

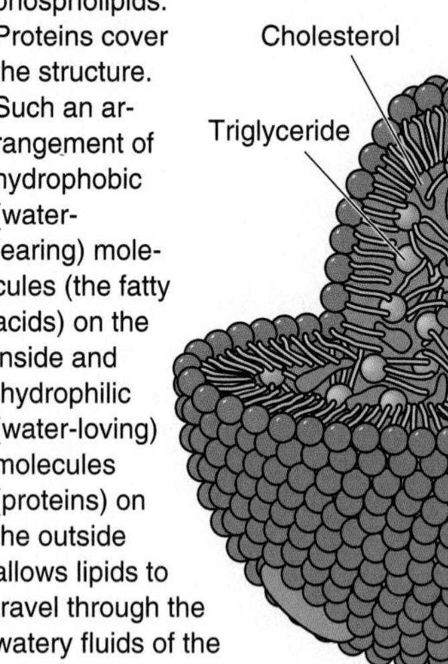

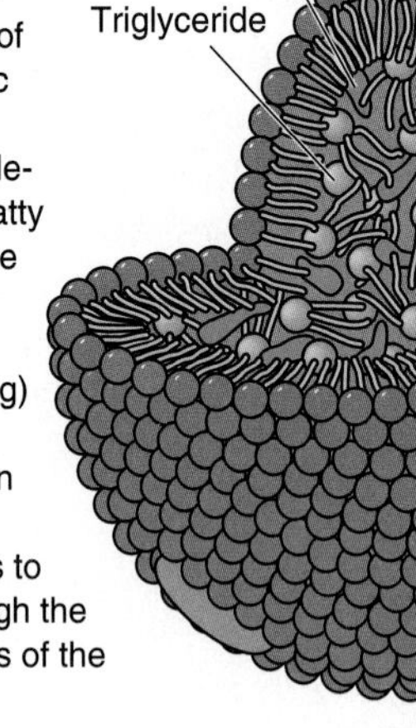

FIGURE 4-5
A SUMMARY OF LIPID DIGESTION AND ABSORPTION

in the blood. Within the body, the larger fats always travel in lipoproteins. In this ingenious configuration, the water-soluble proteins enable the fats to travel in the watery body fluids. That way, when the tissues of the body need energy from fat, they can extract what they need from lipoproteins. The remnants that remain are picked up by the liver, which dismantles them and reuses their parts. The characteristics of the four types of lipoproteins circulating in the blood are shown in Figure 4-6.

Lipoproteins are very much in the news these days. In fact, the health-care provider who measures your blood lipid profile is interested not only in the types of fats in your blood (triglycerides and cholesterol) but also in the lipoproteins that carry them. One distinction among types of lipoproteins is of great importance because it has implications for the health of the heart and blood vessels—the distinction between low-density lipoproteins (**LDL**) and high-density lipoproteins (**HDL**). The more protein in the lipoprotein molecule, the higher the density.

This description of the intestine's processing of fat has omitted the few other dietary fats, such as phospholipids and cholesterol, that may have entered the body in food. These fats enter the circulation the same way the triglycerides do and travel packaged in chylomicrons. After transport, they end up in the liver as part of the chylomicron remnants.

CHYLOMICRON (KIGH-loh-MY-cron) a type of lipoprotein that transports newly digested fat from the intestine through lymph and blood.

LIPOPROTEINS (LIP-oh-PRO-teens) clusters of lipids associated with protein that serve as transport vehicles for lipids in blood and lymph. The four main types of lipoproteins are **CHYLOMICRONS, VLDL, LDL** and **HDL.**

VLDL (very-low-density lipoprotein) carries fats packaged or made by the liver to various tissues in the body.

LDL (low-density lipoprotein) carries cholesterol (much of it synthesized in the liver) to body cells. A high blood cholesterol level usually reflects high LDL.

HDL (high-density lipoprotein) carries cholesterol in the blood back to the liver for recycling or disposal.

LIPIDS AND HEALTH

The question of what kind of fat to include in the diet can be puzzling. Research has potentially linked the fat in the diet to several diseases, including certain types of cancer, heart disease, arthritis, and gallbladder disease. For both cancer and heart disease, the most important strategy is to lower the total fat content of the diet.

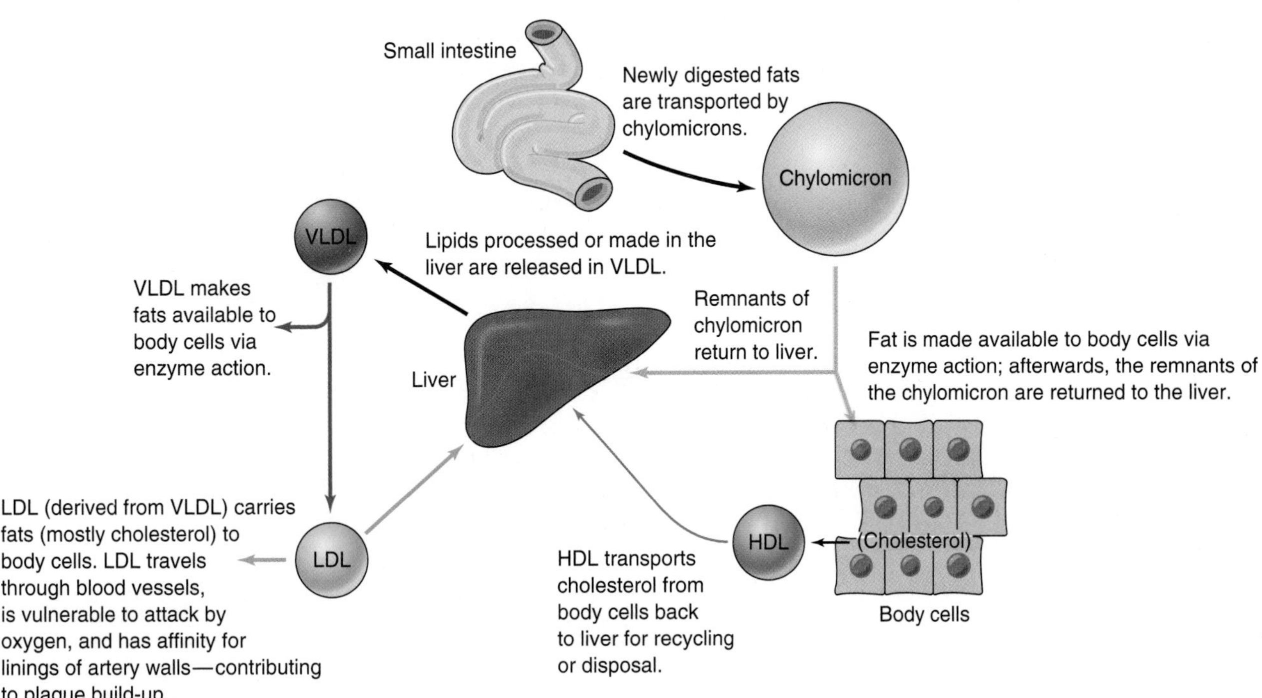

FIGURE 4-6
FUNCTIONS AND INTERACTIONS OF LIPOPROTEINS

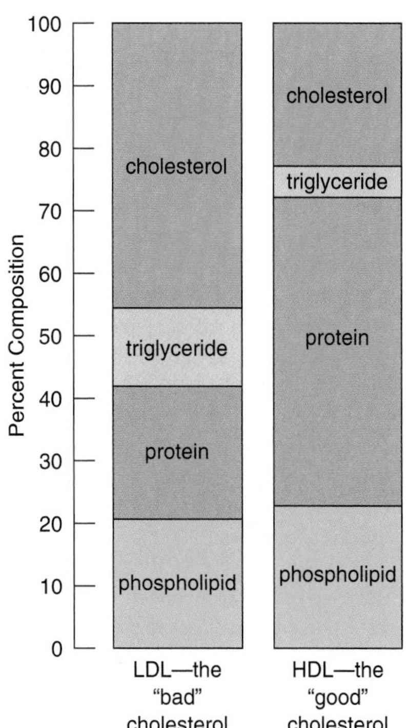

"Good" Versus "Bad" Cholesterol

A silent, symptomless risk factor for heart disease is much talked about but little understood: elevated blood cholesterol levels. Blood cholesterol levels may be high for any of a number of reasons. Some people inherit tendencies to make too much cholesterol or to fail to destroy it on schedule. Others have high blood cholesterol for any or all of the following lifestyle reasons: eating too much fat or too much saturated fat, exercising too little, or carrying too much weight. The blood lipid profile mentioned earlier can give you an idea of your standing as to this risk factor. Figure 4-7 shows how to interpret your blood cholesterol level.[15]

The underlying cause of heart disease is atherosclerosis—the narrowing of the arteries caused by a buildup of cholesterol-containing plaque in the arterial walls (this chapter's Spotlight further discusses atherosclerosis and heart disease). The initiating step in the process of atherosclerosis is some form of injury or inflammation in the artery wall.[16] High blood pressure, high blood cholesterol levels, and cigarette smoke are potential sources of injury, as are other causes (see Table 4-4). Raised LDL concentrations in the blood are a sign of high heart attack risk because LDLs in the blood tend to deposit cholesterol in the arteries.

Researchers now theorize that LDL-cholesterol is damaging to the artery walls once it has been oxidized. Circulating LDL-cholesterol is more likely to settle along the linings of the artery walls after it first reacts with an unstable form of oxygen to become **oxidized LDL-cholesterol** (o-LDL) (see Figure 4-8 on page 114).

Researchers believe that scavenger cells from the immune system known as macrophages ingest more and more of the o-LDL particles and eventually become **foam cells**—so called because of their resemblance to seafoam. These foam cells eventually burst and deposit their accumulated cholesterol as debris in the arterial wall, leading to the development of fatty streaks—the precursors of plaque. Thus, scientists speculate that the oxidized form of LDL catalyzes the process of atherosclerosis in the artery walls by attracting these macrophages to the arterial area.

One of the key steps in the development of atherosclerosis is the accumulation of o-LDL in the walls of the artery. The more LDLs in the circulating blood, the greater the chance for oxidation to occur. This leaves more o-LDL available for ingestion by the scavenger cells and more debris left behind in the arterial wall. In theory then, steps to reduce the total amount of LDL circulating in the blood (reducing saturated fat in the diet) or steps to prevent the oxidation of LDL-

OXIDIZED LDL-CHOLESTEROL (o-LDL) the cholesterol in LDLs that is attacked by reactive oxygen molecules inside the walls of the arteries; o-LDL is taken up by scavenger cells and deposited in plaque.

FOAM CELLS cells from the immune system containing scavenged oxidized LDL-cholesterol that are thought to initiate arterial plaque formation.

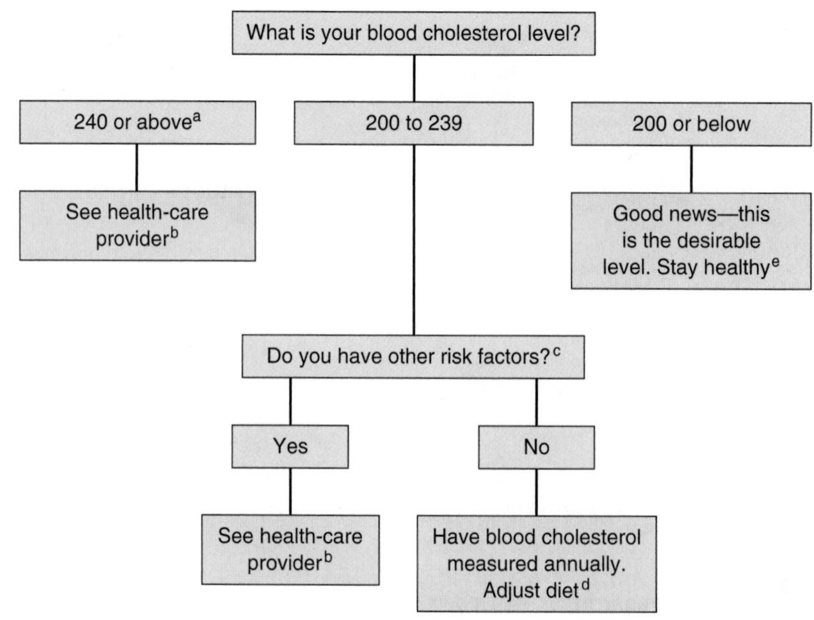

FIGURE 4-7
WHAT IS YOUR CHOLESTEROL LEVEL AND RISK FOR HEART DISEASE?

A diet high in fiber and complex carbohydrates and low in fats and cholesterol plus regular exercise can help you achieve heart-healthy cholesterol levels.

[a]Blood cholesterol is measured in milligrams per deciliter (mg/dL) of blood.
[b]Ask to have your low-density-lipoprotein (LDL) level determined. If LDL is high, your healthcare provider may suggest a diet and/or drug therapy.
[c]Risk factors: See Table 4-4.
[d]Reduce total fat, saturated fat, and cholesterol in the diet.
[e]A reading of 200 mg/dL or below is recommended. Stay heart-healthy with a nutritious diet and exercise.

TABLE 4-4 RISK FACTORS FOR HEART DISEASE

Risk Factors*
- Age: Men 45 or over
  Women 55 or over or premature menopause without estrogen replacement therapy
- High blood cholesterol (greater than 200 mg/dL)
- High LDL-cholesterol (greater than 130 mg/dL)*
- Low HDL-cholesterol (less than 35 mg/dl)
- Cigarette smoking*
- Hypertension (high blood pressure)*
- Lack of exercise*
- Obesity*
- Diabetes
- Diet high in total fat and saturated fat
- Gender (males are at higher risk)
- Family history of heart disease before the age of 55
- Circulation disorders of blood vessels to the legs, arms, and brain

Protective Factors
- High HDL-cholesterol level (≥ 60 mg/dl)
- Regular physical activity
- Low-fat, heart-healthy diet

*There are five *major* risk factors for heart disease: High LDL-cholesterol, smoking, high blood pressure, lack of exercise, and obesity.

Source: Adapted from Expert Panel on Detection, Evaluation, and Treatment of High Blood Cholesterol in Adults, *Journal of the American Medical Association* 269 (1993): 3015–3023, and R. H. Eckel and R. M. Krauss, American Heart Association call to action: Obesity as a major risk factor for coronary heart disease, *Circulation* 97 (1998): 2099–2100.

LDL, the "bad" cholesterol

HDL, the "good" cholesterol

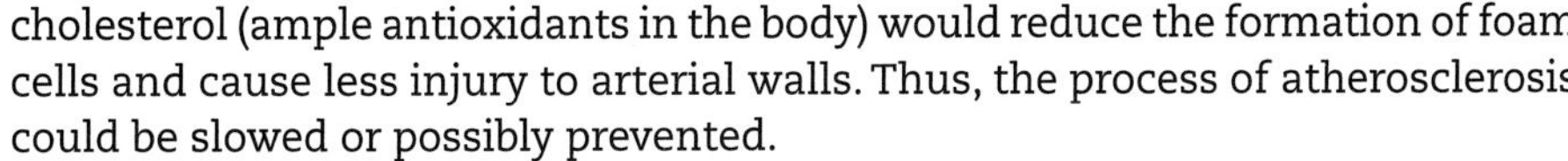

cholesterol (ample antioxidants in the body) would reduce the formation of foam cells and cause less injury to arterial walls. Thus, the process of atherosclerosis could be slowed or possibly prevented.

Most blood cholesterol is carried in LDL and correlates *directly* with heart disease risk, but some is carried in HDL and correlates *inversely* with risk. In fact, the most potent single predictor of heart attack risk may be the HDL level. Recent research indicates that an acceptable total blood cholesterol reading of 200 mg/dl or below may not be protective against heart disease if the HDL level is low. Raised HDL concentrations relative to LDL represent cholesterol on its way out of the arteries back to the liver—and a reduced risk of heart attack. The Spotlight feature at the end of the chapter gives further tips on how to raise your HDL level and lower your LDL level.

Some cases of elevated blood cholesterol do not respond to changes in lifestyle. In such cases, cholesterol-lowering drugs might be prescribed. However, for many, a few simple changes in diet can improve cholesterol readings, as discussed next.

Lowering Blood Cholesterol Levels

Among the most influential diet-related factors that raise blood cholesterol levels are total fat intake, saturated fat intake, and obesity.[17] As it turns out, the changes in diet that reduce blood cholesterol concentrations mostly do so by reducing LDL-cholesterol. Dietary modifications that help lower LDL include substituting highly monounsaturated fats (canola, olive, and peanut oils) and highly polyunsaturated fats (vegetable oils and fish oils—see Table 4-5) for saturated

Saturated fats and cholesterol contribute to high blood cholesterol levels.

FIGURE 4-8
ATHEROSCLEROSIS

As LDL particles penetrate the walls of the arteries, they become oxidized-LDL and next are scavenged by the body's white blood cells. These foam cells are then deposited into the lining of the artery wall. This process, known as atherosclerosis, causes plaque deposits to enlarge, artery walls to lose elasticity, and the passage through the artery to narrow.

*Early injury may be the result of smoking, high blood pressure, elevated blood cholesterol, elevated homocysteine level, diabetes, genetics, or possibly a bacterial or viral infection.

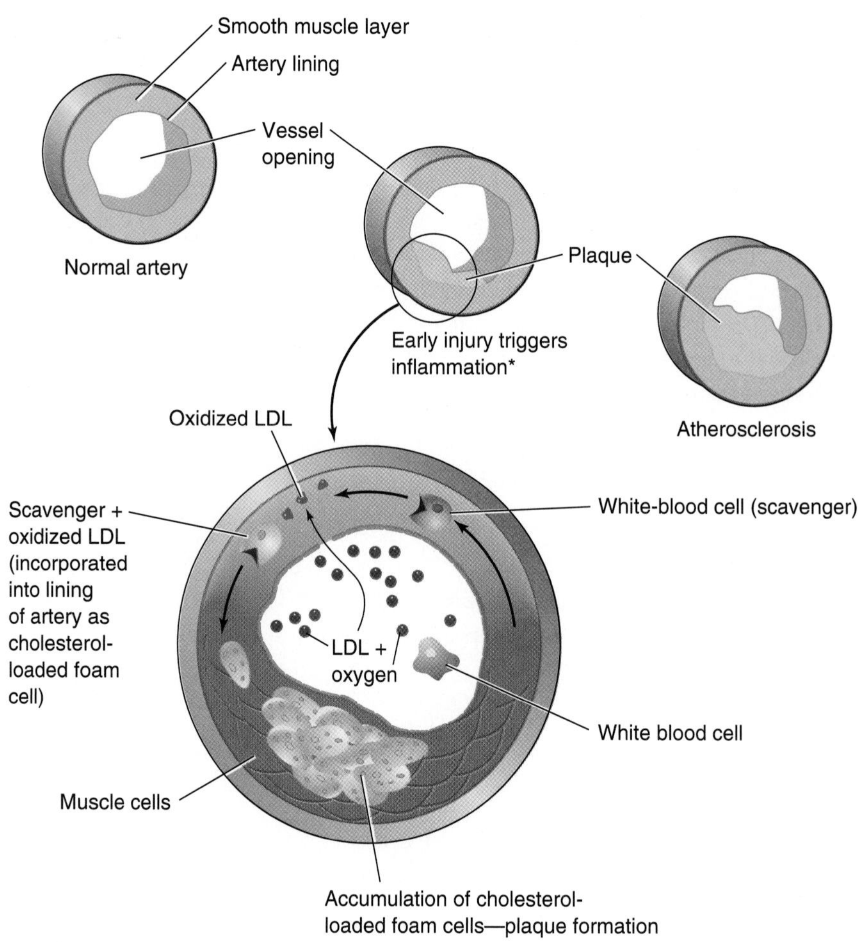

fats.[18] As for dietary cholesterol itself, it raises LDL levels slightly for some people, depending on the amount consumed and on the body's ability to compensate by making less.

The current recommendations for diet, based on these findings, include the following:

- Eat no more than 30 percent of calories as fat.
- Eat no more than 8 percent to 10 percent of calories as saturated fat.
- Eat no more than 10 percent of calories as polyunsaturated fats.
- Monounsaturated fats (see Table 4-5) should make up 10 percent to 15 percent of total calories.
- Limit daily cholesterol intake to no more than 300 milligrams.

Researchers continue to debate which type of fat makes the best replacement for saturated fat in the diet.[19] For years, polyunsaturated fats were recommended as substitutes for saturated fats in the diet until evidence suggested that PUFAs play a role in the oxidation of LDL-cholesterol. A diet high in polyunsaturated fats (corn or vegetable oil) may increase the oxidation of LDL-cholesterol and therefore increase the risk of heart disease. Since the polyunsaturated fat (being transported on the LDL carrier) has several points of unsaturation, it is vulnerable to attack by oxygen or other oxidizing agents.[20] Monounsaturated fats (olive and canola oil) are more stable than the polyunsaturated fats because they have only one point of unsaturation.

Researchers have shown that the substitution of monounsaturated for saturated fatty acids in the diet brings about a significant decrease in blood levels of

TABLE 4-5 THE EFFECTS OF VARIOUS KINDS OF FAT ON BLOOD LIPIDS

Type of Fat*	Dietary Sources	Effects on Blood Lipids
Saturated Fat	All animal meats, beef tallow, butter, cheese, chocolate, cocoa butter, coconut oil, cream, hydrogenated oils, lard, palm oil, stick margarine, shortening, whole milk	Increases total cholesterol Increases LDL-cholesterol
Polyunsaturated Fat	Almonds, corn oil, cottonseed oil, filberts, fish, liquid/soft margarine, mayonnaise, pecans, safflower oil, sesame oil, soybean oil, sunflower oil, walnuts	If used to *replace* saturated fat in the diet, polyunsaturated fat may: Decrease total cholesterol Decrease LDL-cholesterol Decrease HDL-cholesterol
Monounsaturated Fat	Avocados, canola oil, cashews, olive oil, olives, peanut butter, peanut oil, peanuts, poultry	If used to *replace* saturated fat in the diet, monounsaturated fat may: Decrease total cholesterol Decrease LDL-cholesterol without decreasing HDL-cholesterol
Omega-3 Fat	Canola oil, flaxseed, ocean fish (salmon, mackerel, tuna), shellfish, some vegetables (spinach, broccoli, lettuce), soyfoods, walnuts, wheat germ	If used to *replace* saturated fat in the diet, omega-3 fat may: Decrease total cholesterol Decrease LDL-cholesterol Increase HDL-cholesterol Decrease triglycerides
Trans Fat	Margarine (hard stick), cake, cookies, doughnuts, crackers, chips, meat and dairy products, peanut butter (hydrogenated), shortening	Increases total cholesterol Increases LDL-cholesterol

*All fats, whether classified as mainly saturated fat, monounsaturated fat, or polyunsaturated fat, contain mixtures of saturated and unsaturated fats and provide the same number of calories: 9 calories per gram.

LDL-cholesterol, without the decrease in HDL-cholesterol seen when polyunsaturated fats (at intakes greater than 10 percent of calories) are substituted for saturated fats.[21] Some researchers warn, however, that diets rich in any type of fat can be calorically dense and could worsen the problem of obesity in the United States and Canada.

Still, researchers continue to investigate whether the high consumption of olive oil, typical in the Mediterranean region, may be partly responsible for the lower risk of heart disease seen among the people of Greece, France, Italy, and other Mediterranean countries. Other factors undoubtedly contribute to the lower incidence of heart disease in the Mediterranean region as implied in the following anecdote:[22]

> He's a shepherd or small farmer, a beekeeper or fisherman, or a tender of olives or vines. He walks to work daily and labors in the soft light of his Greek Isle. His midday, main meal is of eggplant, with large mushrooms, crisp vegetables, and country bread dipped in golden olive oil. Once a week there is a bit of lamb. Once a week there is chicken. Twice a week there is fish fresh from the sea. Other meals are hot dishes of legumes seasoned with meats and condiments. The main dish is followed by a tangy salad, then by dates, Turkish sweets, nuts, or fresh fruits. A sharp local wine completes the meal.

In understanding the paradox of how a high-fat diet (more than 30 percent of calories from mostly monounsaturated fats) can coexist with a low incidence of heart disease, it is important to note that people from the Mediterranean region in general are more physically active, have the social and emotional support

Nuts are rich in many nutrients and other beneficial substances but are also high in fat. Two whole walnuts or ten large peanuts contain the same amount of fat as is found in a teaspoon of butter or margarine (5 grams of fat and 45 calories).

For more about the dietary customs and health status of the people living in the Mediterranean region, refer to the Spotlight feature in Chapter 2.

WORKING ON THE WEB

www.mayohealth.org

Sponsor: Mayo Clinic

Description: A credible source for general health information on a variety of topics, practical tips for reducing the amount of fat in your favorite recipes, and tools to assess your food and nutrition knowledge.

Available Activities: The site provides an interactive quiz section in each of its "Centers" which allows you to test your nutrition and health knowledge on a variety of topics, with new quizzes added frequently. A Virtual Cookbook includes heart-healthy makeovers of favorite recipes and allows you to submit recipes of your own for modification. The site also provides reference articles on numerous health-related topics, headline news information, and useful links to related health sites.

Web Work:

1. Test your knowledge about nutrition by completing one of the quizzes posted in the Nutrition Center.
2. Explore the Virtual Cookbook for examples of how to make your favorite recipes more heart healthy.
3. Assess your basic knowledge of dietary cholesterol and heart disease by completing the Cholesterol Quiz—Part I.
4. Test your knowledge about blood cholesterol and heart disease by completing the Cholesterol Quiz—Part II.

Helpful Hints: From the home page, scroll to Centers and click on Nutrition. Click on Quizzes, followed by the quiz of your choice, and answer the questions. Review the answers and information provided at the end of the quiz. Return to the Nutrition Center home page and click on Virtual Cookbook. Browse through the modified recipes, and select one of interest to you. Return to the site's home page and under Center, click on Heart. From this page you will first click on Quizzes, followed by Cholesterol Quiz—Part I and Cholesterol Quiz—Part II. At the end of each quiz, review the answers and scroll through the related topics listed under For More Information.

found in extended networks of family and friends, eat more fruits, vegetables, legumes, and nuts and less meat and animal fat, use monounsaturated fat (found in vegetable oils such as olive or canola) more than saturated fat, and consume more of each day's calories earlier in the day than people in the United States and Canada. Any one of these factors may be significant for the lower incidence of heart disease in the region. A healthful eating pattern (with a Mediterranean twist) for most people in the United States and Canada is one that is lower in total fat and higher in complex carbohydrates and fiber (fruits, vegetables, whole grain breads and other grains, and legumes).

In the past few years, attention has focused on reducing heart disease risk by both reducing contributory factors in the diet—notably, saturated fat—and increasing the intake of protective factors, such as the antioxidant vitamins. Antioxidants—beta-carotene, vitamins C and E, and the mineral selenium—act to strengthen the body's natural defenses against cell damage by blocking the potentially damaging **free radicals** that arise as a part of numerous normal cell activities. Free radicals—the chemical compounds that oxidize LDL-cholesterol—become a problem when there are too many of them.

FREE RADICALS highly toxic compounds created in the body as a result of chemical reactions that involve oxygen. Environmental pollutants such as cigarette smoke and ozone also prompt the formation of free radicals.

Antioxidants in the body may reduce the amount of LDL-cholesterol that becomes oxidized because the antioxidants (particularly vitamin E) can neutralize the highly reactive oxygen before it gets a chance to oxidize the LDL particle, thereby lessening the buildup of plaque in the artery walls.[23] Scientists are now exploring the potential abilities of all the antioxidant substances found in foods to protect against heart disease.[24] Indeed, the role of the antioxidants in protecting against heart disease is a rapidly growing area of research.[25] Important questions remain unanswered: What is the optimal level of vitamin E (or beta-carotene or vitamin C) intake that confers the greatest benefit in reducing risk of heart disease? Also, can this amount be obtained from the diet, or will an antioxidant supplement be recommended?

Chapter 6 discusses the role of antioxidants and phytochemicals in reducing risks for chronic diseases such as heart disease and cancer.

While waiting for the results of this ongoing research, two suggestions can be made for the time being:

1. Keep blood cholesterol at or below the recommended levels. For every 1 percent reduction in a high blood cholesterol level, there is a 2 percent to 3 percent reduction in the risk of heart attack. Substitute monounsaturated fats

and omega-3 fats for saturated fats in the diet because both lower LDL-cholesterol levels in the blood.

2. Eat generous amounts of fruits and vegetables—at least five servings a day—as a source of antioxidants and other protective compounds in the diet.

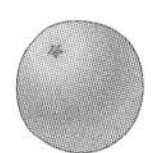

FAT IN THE DIET

The remainder of this chapter will help you to apply what you have learned about fats—that is, how to choose foods that supply enough but not too much of the right kinds of fat to support optimal health and provide pleasure in eating. To start, you must know where the fats are located in the Food Guide Pyramid. Three groups—the *Fats, Oils, and Sweets Group;* the *Meat, Poultry, Fish, Dry Beans, Eggs, and Nuts Group;* and the *Milk, Yogurt, and Cheese Group*—have traditionally accounted for about nine-tenths of the fat in the U.S. diet. Currently, the average American diet includes about 34 percent of its calories from fat with about 12 percent of calories from saturated fat.[26] Dietary guidelines recommend that *total* fat not exceed 30 percent of the day's total calories, and saturated fat contribute less than 8 to 10 percent. Most of the fat, especially saturated fat, in our diet comes from animal products (see Figure 4-9).

"Fat on the plate" includes visible fats and oils, such as butter, the oil in salad dressing, and the fat you trim from a steak. It also refers to some you cannot see, such as the fat that marbles a steak or that is hidden in such foods as nuts, cheese, biscuits, crackers, doughnuts, cookies, muffins, avocados, olives, fried foods, and chocolate.

Meats probably conceal most of the fat that people unwittingly consume. Many people, when choosing a serving of meat (or certain meat alternates), don't realize that they are electing to eat a large amount of fat (see Table 4-6). Recently, some animal breeders have begun producing beef and pork that is lower in fat. This is a help to those people who choose lean cuts—they get less fat in the same

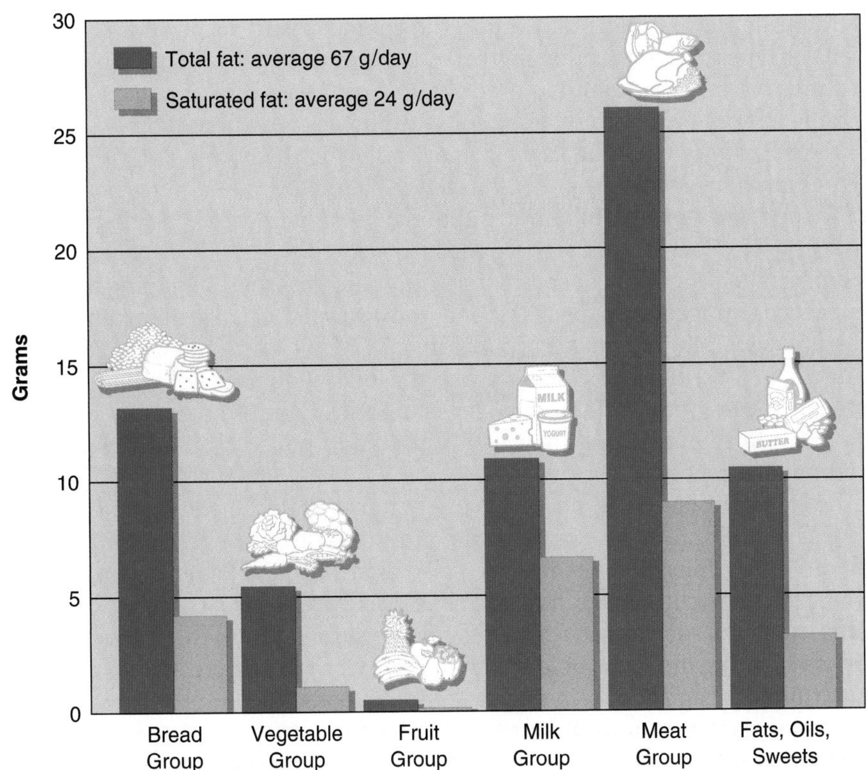

FIGURE 4-9
FAT AND SATURATED FAT CONTRIBUTED BY THE FOOD PYRAMID GROUPS IN THE U.S. DIET

Source: *Eating in America Today: A Dietary Pattern and Intake Report/Edition II (Eat II)*, 1994, p.13.

A 3-ounce portion of lean beef, chicken, or fish has roughly the dimensions of a deck of playing cards. Three ounces is also the approximate size of the palm of the average woman's hand.

TABLE 4-6 COMPARE THE SATURATED FAT CONTENT OF FISH, POULTRY, AND BEEF

	Grams Per 3-oz Serving	
	Total fat	Saturated fat
Turkey breast, roasted (without skin)	1	<1
Halibut, baked	2.5	<1
Chicken breast, roasted (without skin)	3	1
Beef, top round, broiled	4	1.5
Beef, sirloin, broiled	6	2
Chicken breast, roasted (with skin)	7	2
Turkey, ground	11	3
Beef, T-bone steak, broiled	10	4
Beef, ground lean	15	6

TYPES OF FAT-REPLACER INGREDIENTS

Carbohydrate-based:
Carrageenan, fruit purees, gelatin, gels derived from cellulose or starch, guar gum, xanthum gum, maltodextrins made from corn, corn starch (Stellar), polydextrose, Oatrim (made from oat fiber), and Z-trim (a modified form of insoluble fiber)

Protein-based:
Whey protein concentrate (Dairy-Lo), Microparticulated protein products (Simplesse,® K-Blazer) made from whey, or milk and egg white protein

Fat-based:
Mono- and diglycerides; Caprenin—a substitute for cocoa butter in candy—and Salatrim—found in reduced fat baking chips—both containing long-chain fatty acids, which are partially absorbed, and short-chain fatty acids—providing 5 calories per gram; Olestra (noncaloric artificial fat made from fatty acids and sucrose)

CARRAGEENAN a seaweed derivative used by food manufacturers to add "body" to numerous products, including ice cream, frozen yogurt, and salad dressings.

quantity of meat. When selecting beef or pork, look for the words loin or round on the label—these words represent lean cuts from which the fat can be trimmed. The Scorecard on page 121 will help you rate your own dietary selections for fat.

Understanding Fat Substitutes

Fat-free ice cream, cookies, cakes, and salad dressings have long been the stuff that dieters' dreams are made of. The food industry has now made such fatless fare a reality. With more and more low-fat and fat-free items introduced daily into supermarkets across the country, sales of fat-free products are soaring.

The boom in both low-fat and fat-free products has to do with the country's expanding health consciousness. Although most Americans have heard the warnings that fat can contribute to heart disease, cancer, and obesity, many people find low-fat diets particularly unpalatable since fat adds a desirable flavor and texture—known as "mouth feel"—to foods. With innovations in food chemistry, however, the food industry can concoct new recipes or ingredients that yield low-fat or nonfat products that retain the characteristic flavor and mouth feel of fat. The types of fat-reduction ingredients currently in use are listed in the margin. Entenmann's, the manufacturer of a line of fat-free cakes and cookies, for example, makes those products by substituting egg whites for whole eggs and fat-free milk for whole milk as well as by removing the butter from its recipes. The end product: sweets that contain fewer calories and 4 to 5 fewer grams of fat per serving than the company's traditional desserts.[27] Table 4-7 compares regular foods with low-fat and fat-free foods.

Similarly, McDonald's cooked reduced-fat burger patties by adding an ingredient called **carrageenan** to beef. Derived from seaweed, carrageenan helps retain moisture, which in turn makes up for the loss of juiciness that accompanies a reduction in fat. Dubbed McLean Deluxe, the low-fat quarter-pound burger contained 11 fewer grams of fat and 99 fewer calories than the company's Quarter Pounder.[28] In February, 1996, however, McDonald's pulled McLean Deluxe from its menu due to poor sales.

Another technique manufacturers use to replace fat is to add starches, gums, and gels to their products. For example, Kraft uses cellulose gel, a complex carbohydrate, as a filler to make fat-free salad dressings and Sealtest nonfat dairy dessert.[29] Other companies add starches and gums—also complex carbohydrates—to items such as sauces and yogurts to skim fat from those foods. That's because starches and gums hold water and impart a smooth creamy texture similar to that of fat and add form and structure to foods. These substitutes cannot, however, replace the fat used for cooking and frying.[30]

Yet another more innovative approach the food industry has taken to provide fat-free fare is the development of fat substitutes. In 1990, a substance called

TABLE 4-7 FAT-FREE FARE: BEFORE AND AFTER

Crackers/Cookies/Cakes	Grams of Fat	Calories
Better Cheddars Snack Crackers (22 crackers)	8	150
Reduced Fat Better Cheddars Snack Crackers (24)	6	140
Wheatsworth Stoned Ground Wheat Crackers (5)	3.5	80
SnackWell's Fat Free Wheat Crackers (5)	0	60
Fig Newtons (2)	2.5	110
Fat Free Fig Newtons (2)	0	100
Marshmallow Puffs Fudge Cookies (1)	4	90
SnackWell's Devil's Food Cookie Cakes (1)	0	50
Oreo cookies (3)	7	160
Reduced Fat Oreo cookies (3)	5	140
Hostess Chocolate Cup Cakes (1)	5	170
Hostess Chocolate Cup Cakes Light (1)	1.5	120
Entenmann's All Butter Pound Loaf (⅙ loaf)	10	220
Entenmann's Golden Loaf Cake (fat-free) (⅛ loaf)	0	130
Dairy Products		
Regular Swiss cheese, average (1 slice)	8	107
Kraft Singles Swiss Flavor (1.3 slice)	3	67
Weight Watchers Fat Free Swiss (1.3 slice)	0	40
Regular vanilla ice cream (½ cup)	7	133
Weight Watchers Oh! So Very Vanilla (½ cup)	2.5	120
Healthy Choice Premium Low Fat Ice Cream vanilla (½ cup)	2	100
Breakstone Sour Cream (2 tbsp)	5	60
Breakstone Fat-Free Sour Cream (2 tbsp)	0	35
Condiments		
Hidden Valley Ranch salad dressing (2 tbsp)	14	140
Hidden Valley Ranch Light salad dressing (2 tbsp)	7	80
Hidden Valley Fat Free Ranch salad dressing (2 tbsp)	0	45
Kraft Mayonnaise (1 tbsp)	11	100
Kraft Light Mayonnaise (1 tbsp)	5	50
Kraft Free Mayonnaise (1 tbsp)	0	10
Meats		
Ground beef, regular (4 oz)	22	331
Ground beef, extra lean (4 oz)	18	300
Swift 95 Supreme Extra Lean ground beef (4 oz)	5	140
Oscar Mayer Wiener (1 link)	13	150
Oscar Mayer Light Wiener (1 link)	9	110
Oscar Mayer Free Hot Dog (1 link)	0	40

Source: Manufacturer's information, product labels, and Reduced-fat foods: Dieter's dream or marketer's ploy? *Consumer Reports on Health,* July 1995, p. 79.

"Figby Foods has reduced its fat content by another gram! This is war, gentlemen!"

Simplesse® became the first such product to gain the approval of the Food and Drug Administration. Six years in the making, Simplesse® is a mixture of food proteins such as egg white, whey, and milk protein that are cooked and blended to form tiny round particles that trap water. Inside the mouth, the particles roll over one another, and the tongue perceives them as a creamy, smooth liquid similar to fat.

The FDA allows use of Simplesse® in foods including cheese, baked goods, ice creams, frozen desserts, mayonnaise, salad dressings, yogurts, sour cream, and butter. Foods made with the fat substitute contain considerably less fat and fewer calories than their traditional fat-containing counterparts.[31] The reason is that while fat contains 9 calories per gram, Simplesse® supplies only 1 to 2. Simplesse®

SIMPLESSE® the trade name for a protein-based, low-calorie artificial fat, approved by the FDA for use in foods such as frozen desserts; cannot be used for frying or baking.

OLESTRA an artificial fat derived from vegetable oils and sugar combined in such a way that the body cannot break them down. Sold under the brand name Olean®, olestra does not contribute calories to food. It can, however, prevent absorption of some nutrients. Thus, the FDA requires all products made with it to bear this warning: "This Product Contains Olestra. Olestra may cause abdominal cramping and loose stools. Olestra inhibits the absorption of some vitamins and nutrients. Vitamins A, D, E, and K have been added."

FIGURE 4-10
OLESTRA: A FAT-FREE FAT

Compare the structure of olestra to that of a triglyceride. Whereas triglycerides have 3 fatty acids attached to a glycerol core, olestra has up to 6, 7, or 8 fatty acids on a sucrose core. The extra fatty acids make it too large to be digested or absorbed by the body, which is why it adds no fat or calories to the diet. Therefore, a 1-ounce bag of potato chips which typically has 10 grams of fat and 150 calories, has 0 grams of fat and 60 calories when Olean® is added in place of fat.

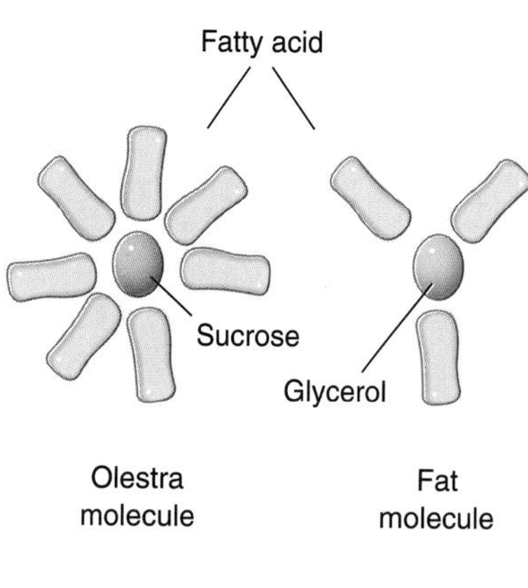

cannot be used for frying or baking, since heat causes it to gel and lose its creaminess. The sample menu shown below illustrates the role of fat-reducers in the reduction of fat and calories in the diet.

Another artificial fat recently approved for use in salty snacks and crackers is **olestra** (sold under the brand name Olean®). Created by Procter & Gamble, the olestra molecule resembles a triglyceride but is structured in a way that prevents its breakdown by digestive enzymes in the body, thereby allowing it to pass through the digestive tract completely unabsorbed (see Figure 4-10). Many scientists have a number of concerns about olestra, however. One is that it interferes with the absorption of fat-soluble vitamins as well as beta-carotene and other nutrients. Another is that large amounts of the product can cause abdominal cramping and loose stools. Because of these issues, the FDA requires any product containing olestra to bear a warning label.[32]

Certainly, the growing number of fat-free foods and fat substitutes provides consumers with viable alternatives to fattier fare. Nevertheless, many experts view the fat-free boom with skepticism. Although reduced-fat foods can help lower the overall fat content of the diet, they *do* contain calories, and they are not a replacement for a healthful diet rich in whole grains and fresh fruits and vegetables. Nor are they likely to become the panacea that will prevent problems such as heart disease and obesity. Fit into a low-fat diet, however, they can help consumers reach and adhere to the goal of taking in no more than 30 percent of total calories as fat (see Table 4-8).[33]

TABLE 4-8

HOW TO DETERMINE TOTAL DAILY FAT ALLOWANCE

We are advised to limit our total daily fat intake to 30 percent or less of our total daily calories. Every single food item does not need to be low in fat if you balance higher fat items with lower fat foods throughout the day.

Use this table to find your total daily fat allowance, based on your daily calorie intake, to keep your fat grams at or below 30 percent of total calories.

Calories	Fat
1400 calories	47 grams
1600 calories	53 grams
1800 calories	60 grams
2000 calories	65 grams
2200 calories	73 grams
2400 calories	80 grams
2800 calories	93 grams
3000 calories	100 grams

Example: If you are eating 1,500 calories per day, multiply 1,500 by 30 percent to determine the maximum number of calories that should come from fat in one day (1,500 × .30 = 450 calories from fat). Since 1 gram of fat provides 9 calories, divide the calories from fat by 9 to see how many grams of fat you should have per day (450 ÷ 9 = 50 grams fat per day):

Total daily calories × .30 = total fat calories

Total fat calories ÷ 9 = total fat grams

The Role of Fat Replacers in Fat and Calorie Reduction

This sample menu shows the fat and calorie difference foods that contain fat replacers can make.

Regular Lunch

	Calories	Fat (grams)
Bread, 2 slices	130	2
American cheese, 1 oz	105	9
Bologna, 2 oz	180	17
Mayonnaise, 1 tbsp	100	11
Banana	105	0
Chocolate cookies, 2	140	6
	760	45

Fat-Replaced Lunch

	Calories	Fat (grams)
Bread, 2 slices	130	2
Reduced-fat cheese product, 1 oz	75	4
Fat-free bologna, 2 oz	40	0
Low-fat mayonnaise/dressing, 1 tbsp	25	1
Banana	105	0
Reduced-fat chocolate cookies, 2	120	3
	495	10

Source: Adapted from P. Kurtzweil, Taking the fat out of food, FDA Consumer, July–August, 1996.

RATE YOUR FATS AND HEALTH IQ SCORECARD

DO YOU?	OFTEN	SOMETIMES	RARELY
Trim or drain the fat from meats, remove the skin from chicken, and serve fish-based meals?	10	5	1
Eat a variety of fresh, frozen, or canned fruits and vegetables?	10	5	1
Eat high-fat foods such as bacon, sausage, regular franks, and luncheon meat several times a week?	1	5	10
Limit whole eggs or egg yolks to four per week?	10	5	1
Read food labels when shopping?	10	5	1
Choose low-fat or fat-free milk, yogurt, cheese, and sour cream?	10	5	1
Bake, rather than fry, foods?	10	5	1
Maintain a healthy weight?	10	5	1
Think "eating right" and make trade-offs when eating out?	10	5	1
Choose doughnuts, croissants, or sweet rolls for breakfast?	1	5	10
Choose reduced-fat or fat-free products when available?	10	5	1
Routinely add margarine, butter, salad dressing, and sauces to foods?	1	5	10
Balance a high-fat dinner by choosing low-fat foods for breakfast and lunch?	10	5	1
Plan exercise (walking, running, swimming, bike riding) into your schedule three to four times a week?	10	5	1
TOTAL			

SCORING

113–150 You practice heart-healthy habits. Keep up the good work.

75–112 Not a bad score, but there's room for improvement.

5–74 Too low a score! Learn more about the relationships between fats and health.

Large potato with 1 tablespoon butter and 1 tablespoon sour cream (14 grams fat, 350 calories).

Large potato with 2 tablespoons fat-free sour cream or yogurt seasoned with chives (less than 1 gram fat, 235 calories).

Source: Adapted from American Dietetic Association and Mosby Great Performance, *Healthy Eating For the Whole Family* (St. Louis, MO: Mosby–Year Book, Inc. 1995), p. 5.

How to Handle Fats in the Diet

Chocolate, which first came from the Americas, was considered a gift from the gods in pre–Columbian times—and a form of currency. In the 1500s, Native Americans served it to honor their European guests, the explorers who came to what is now Mexico.[34]

It is not the food itself but how you prepare it that often determines the total fat (and calories) in a food. Compare the fat in 3 ounces of broiled chicken (3 grams fat, 141 calories) versus 3 ounces of fried chicken (18 grams fat, 364 calories). What makes the difference in calories? Fat. This feature suggests a variety of ways to manage a healthy level of fat in your diet.

At the Grocery Store

Foods you purchase in the grocery store can tell you much about their fat content—if you take the time to read their labels.

- Read manufacturers' labels to determine both the amounts and the types of fat contained in foods.
- Choose low-fat dairy products, including fat-free, 1 percent, and 2 percent milk, low-fat cheeses, and low-fat or fat-free yogurt.
- Choose lean meats, fish, chicken, and turkey.
- Choose the low-fat and lean varieties of processed meats (sausages, luncheon meats, bacon, and frankfurters).
- Buy bread and other baked goods with added flaxseed.
- Look for omega-3-enriched eggs in the dairy case (eggs obtained from hens fed flaxseed).
- Select water-packed canned fish rather than oil-packed varieties.
- Choose foods naturally rich in the antioxidant nutrients and other protective compounds by eating five or more servings of fruits and vegetables *every* day.
- Include soyfoods such as tofu, soy powder, soy milk, miso, tempeh, and soy cheese.

In the Kitchen

- Use canola oil and olive oil for baking and cooking.
- Try reducing the fat in recipes a little at time, and notice that you can do so and still get a good-tasting product. If you reduce fat by one tablespoon of butter or oil, you lower the total fat in the product by about 12 grams and at least 100 calories. Or, to reduce the fat even further, try substituting unsweetened applesauce or fruit purees for the oil called for in your favorite cake, quickbread, or cookie recipe.
- Serve no more than 3-ounce portions of lean meats and poultry, adding them to stews, soups, stir-fry recipes, or pasta.
- Prepare lean cuts of meat; remove visible fat from meat; remove skin from poultry.
- Incorporate plant-based protein sources in your diet, such as dried peas, soyfoods and other legumes.
- Eat fried foods sparingly; rather **bake, braise, broil, steam, poach,** and **sauté** (see Miniglossary of Cooking Terms). Cook meats on a rack so that the invisible fat can drain off.
- Experiment with some new low-fat, low-cholesterol recipes.
- Use low-fat or fat-free dairy products. Learn to substitute: low-fat or fat-free milk for whole milk, low-fat or fat-free yogurt for sour cream, low-fat Neufchatel cheese for cream cheese, and part-skim ricotta and mozzarella cheese for whole-milk varieties.
- For flavor in sauces and dressings, experiment using herbs and spices, onions or garlic, salsa, ginger, lemon juice, plain, fat-free yogurt with lemon juice, mustard, or butter-flavored granules instead of butter, margarine, or oil.
- Use nonstick sprays rather than fat to coat pans.
- If you are sauteing a vegetable such as onion in butter, margarine, or oil, try reducing the amount used and substituting water, fat-free vegetable or chicken broth, or wine in its place.

- Prepare broths, soups, and stews ahead of time, refrigerate them, and then skim off hardened fat from the surface. this also gives the flavors time to blend and develop.
- Use low-fat or fat-free milk when making cream soups.
- Learn to substitute. Table 4-11 on page 131 offers some ways to reduce both fat and calories. More suggestions to help you lighten your favorite recipes can be found under the Recipe Modification heading in this feature.

At the Table

- Eat more fresh fruits and vegetables, whole-grain breads and cereals, potatoes, rice, noodles, dried beans, peas, and lentils.
- Try using fruit jams instead of butter on your bagel or toast. Most fruit butters and jams contain half the calories per teaspoon of butter.
- Limit your intake of butter, cream, margarine, vegetable shortenings, coconut and palm oils, half-and-half, sour cream, and mayonnaise.
- Use mostly monounsaturated vegetable oils such as canola or olive oils; use the low-fat or fat-free varieties of margarine, mayonnaise, and salad dressings.
- Try brushing a small amount of olive oil instead of butter or margarine on breads warm from the toaster or oven.
- When you use margarine, butter, or cream cheese, try the whipped varieties, which contain half the calories of the regular types.
- Eat fewer high-fat desserts (such as cheesecake, ice cream, brownies, pies, pastries, and butter-and-cream-frosted cakes).

MINIGLOSSARY OF COOKING TERMS

BAKE to cook in an oven surrounded by heat.

BRAISE to cook by browning in fat and then simmering in a covered container with a little liquid.

BROIL to cook quickly over or under a direct source of intense heat, allowing fats to drip away.

POACH to cook foods (fish, an egg without its shell, etc.) in liquid such as water, wine, juice, or bouillon near the boiling point.

SAUTÉ (saw-TAY, the French word for stir-fry) to cook in a pan using little fat; foods are stirred frequently to prevent sticking.

STEAM to cook foods suspended over boiling water.

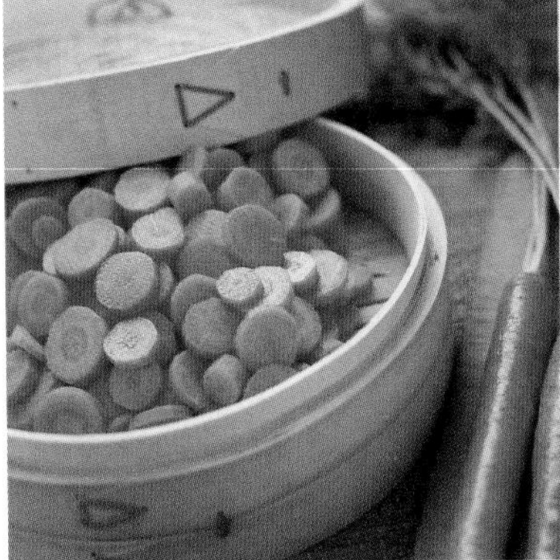

Bake, broil, poach, or steam.

Season with herbs and spices.

Buy low-fat foods.

Continued

Recipe Modification

Use this substitution information to modify your own favorite recipes. They'll be more healthful but still look and taste as good as your originals!

INSTEAD OF	SUBSTITUTE	RESULT
1 whole egg	2 whipped egg whites	Less fat, less cholesterol, fewer calories
Whole milk	Fat-free milk	Less fat, less cholesterol, fewer calories
Evaporated milk	Evaporated nonfat or low-fat milk	Less fat, less cholesterol, fewer calories
Sugar (baking)	Use ⅓ less than called for in baking, add a small amount of vanilla, cinnamon or nutmeg	Fewer calories
Oil	Use a vegetable oil spray for preventing foods from sticking	Less fat, fewer calories
Cake frosting	Sprinkled powdered sugar	Less sugar, less fat, fewer calories
Heavy cream (which does not need to be whipped)	Evaporated nonfat or low-fat milk	Less fat, less cholesterol, fewer calories
Sour cream	Reduced-fat or fat-free sour cream or plain yogurt	Less fat, less cholesterol, fewer calories
Solid shortening (baking)	Vegetable oil, margarine	Less saturated fat
Solid shortening (stir-frying)	Peanut oil	Less saturated fat
Mayonnaise	Reduced-fat or fat-free mayonnaise or plain yogurt (mixed with 1 tbsp mayo per cup yogurt)	Less fat, less cholesterol, fewer calories
Salt	Use ½ the salt called for in recipes (try seasoning with herbs and other spices) except yeast breads	Less sodium
Cream cheese	Reduced-fat or fat-free cream cheese	Less fat, fewer calories
Regular cheese	Cheese made from part-skim milk (mozzarella, Swiss lace, farmer) or reduced-fat cheeses	Less fat, less cholesterol, fewer calories
Salad dressing	Fat-free or oil-free salad dressings or reduced-calorie dressing	Less fat, less cholesterol, fewer calories
White sauce (made with cream and butter)	Use low-fat milk and cornstarch or flour by blending the starch into cold liquid to eliminate the need for fat	Less fat, less cholesterol, fewer calories

TRY A LOWER-FAT GUACAMOLE Blend 1 medium avocado with 1 cup low-fat (1 percent) cottage cheese; add 1 tbsp lime juice, 1 tsp chives, and ¼ tsp red pepper flakes. Per ¼ cup serving: 60 calories and 4 grams of fat—about half the fat of traditional guacamole.

TRY OVEN-BAKED FRIES (3 grams fat per serving) instead of traditional French fries (12 grams fat per serving): Place 4 medium russet potatoes (cut into wedges) into bowl and sprinkle with 1 tbsp vegetable oil, ¼ tsp black pepper, and a pinch of salt (optional); toss gently to combine; arrange potatoes in single layer on nonstick baking sheet sprayed with vegetable cooking spray. Bake in oven (425° F) for 35 minutes or until lightly browned. Serve with salsa or nonfat yogurt mixed with fresh herbs. *Makes 4 servings.*

TRY A HEALTHY TUNA SALAD Mix together one 6½-ounce can of drained, water-packed tuna with one 19-ounce can of your favorite cooked bean (cannellini, kidney, or black), rinsed and drained. Add 1 small, thinly sliced red onion, ½ green bell pepper—diced, ¼ tsp ground black pepper, and ⅓ cup prepared fat-free Italian dressing. Toss to combine and chill before serving. *Makes 6 ½-cup servings.*

Spotlight
DIET AND HEART DISEASE

More than half the people who die in the United States each year die of heart and blood vessel disease. The underlying condition that contributes to most of these deaths is atherosclerosis, which leads to closure of the arteries that feed the heart and brain and thus to heart attacks and strokes. In terms of direct health care costs, lost wages, and lost productivity, heart disease costs the United States more than $60 billion a year. There is little wonder, then, that much effort has been focused on preventing it.

The twin demons that lead to most forms of heart disease are atherosclerosis and hypertension. Atherosclerosis, the subject of this Spotlight, is the common form of hardening of the arteries; hypertension (discussed in the Nutrition Action feature of Chapter 7) is high blood pressure; and each aggravates the other.

How can I know whether I have atherosclerosis?
No one is free of atherosclerosis. The question is not whether you have it but how far advanced it is and what you can do to retard or reverse it. As mentioned in the chapter, atherosclerosis usually begins with the accumulation of soft mounds of lipid, known as plaques, along the inner walls of the arteries, especially at the branch points. These plaques gradually enlarge, making the artery walls lose their elasticity and narrowing the passage through them. Most people have well-developed plaques by the time they are 30.

Normally, the arteries expand with each heartbeat to accommodate the pulses of blood that flow through them. Because arteries hardened and narrowed by plaques cannot expand, the blood pressure rises. The increased pressure puts a strain on the heart and further damages the artery wall. At damaged points, plaques are especially likely to form; thus, the development of atherosclerosis is a self-accelerating process.

Hypertension makes atherosclerosis worse. A stiffened artery, already strained by each pulse of blood surging through it, is stressed even more if the internal pressure is high. Injured places develop more frequently, and plaques grow faster.

Also, atherosclerosis makes hypertension worse. By hardening the arteries, it makes the arteries unable to expand with each beat of the heart, so the pressure rises instead. This leads to further hardening of the arteries, as already explained. Hardened arteries also fail to let blood flow freely through the body's blood pressure-sensing organs—the kidneys—which respond as if the blood pressure were too low and raise it further.

How can I slow the process down?
Learn your risk factors, and control the ones you can control. Among the many factors linked to heart disease are smoking, gender (being male), age (men older than 45 and women older than 55), postmenopausal status in women, heredity, diabetes, high blood pressure, lack of exercise, obesity, high blood cholesterol level, and low HDL-cholesterol level (see

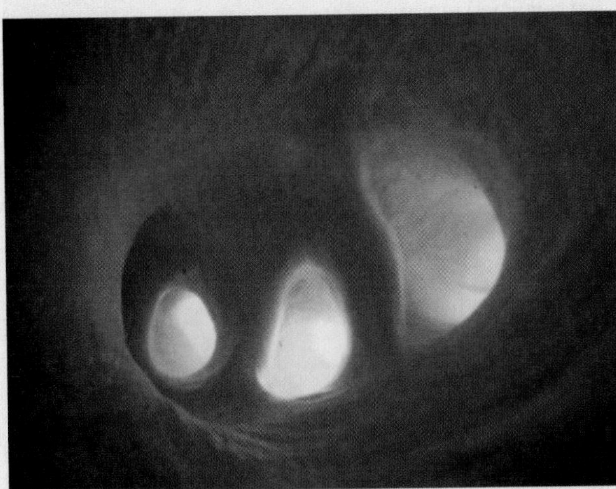

A normal artery provides open passage for blood to circulate.

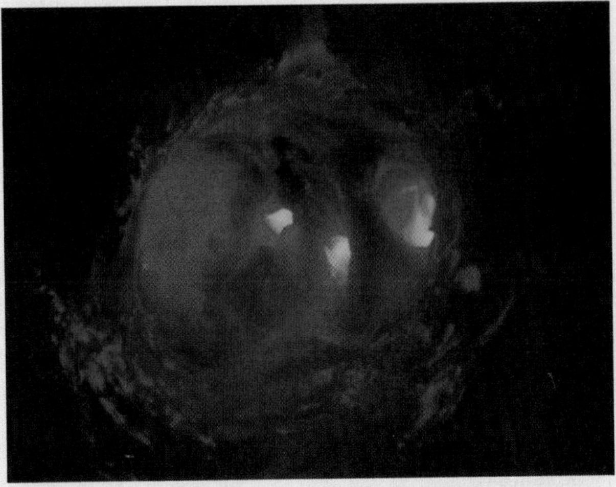

Plaques along an artery narrow the passage and obstruct blood flow.

Table 4-9). Some of the risk factors are powerful predictors of heart disease. If you have none of them, the statistical likelihood of your developing heart disease may be only 1 in 100. If you have three major ones, the chance may rise to over 1 in 20. The five factors that have emerged as the most powerful predictors of risk are high LDL cholesterol, high blood pressure, smoking, obesity, and lack of exercise.[35]

The accompanying Heart Health Scorecard shows one way of calculating your risk score based on present knowledge. Such a quiz not only helps you look at what your risks may be but also points out areas that you can change to reduce your risk.

What can I do to improve my risk score?
Obviously, some risk factors cannot be altered. Being born male or inheriting a predisposition to develop high blood cholesterol or high blood pressure are factors beyond your control. Even so, you can make conscious choices that may reduce your risk of developing heart disease. Let's examine one of the two major risk factors related to diet—high blood cholesterol—with an eye toward learning which dietary and lifestyle changes will help reduce your heart disease risk.

To what extent does a high blood cholesterol level raise the risk of developing heart disease?
The likelihood of a person's developing or dying from heart disease increases as blood cholesterol level rises. Figure 4-11 presents this relationship graphically: The number of deaths from heart disease increases steadily among those with elevated blood cholesterol levels, particularly when the level rises above 200 milligrams per deciliter. Individuals with a blood cholesterol level in the neighborhood of 300 milligrams per deciliter run four times the risk of dying from heart disease as those whose cholesterol level is lower than 200 milligrams per deciliter. Having a low HDL-cholesterol value is now also recognized as being a risk factor for heart disease. A low HDL value is defined as one below

TABLE 4-9 LEADING RISK FACTORS FOR HEART DISEASE

Heart disease rarely develops from a single risk factor. Clusters of risk factors usually occur, and a small increase in one risk factor, such as blood pressure, becomes more critical when combined with other risk factors. Luckily, a similar pattern happens in reverse. Even moderate changes in one risk factor can decrease several others at the same time. For example, a weight loss of just 5 to 10 pounds can reduce blood pressure in overweight people. Or, becoming physically active can lower blood pressure, increase HDL ("good" cholesterol), and help control weight.

	Risk Factors You Can Change
Risk Factor	**How to Minimize the Risk**
High blood cholesterol (especially LDL-cholesterol)	Limit intake of cholesterol, saturated fat, and trans fat. Increase your intake of soluble fiber, soyfoods, and omega-3 fats.
High blood pressure	Control high blood pressure with medication and a heart-healthy diet. Maintain a healthy weight. Losing just 5 to 10 pounds may lower your blood pressure.
Cigarette smoking	Stop smoking. Nicotine constricts blood vessels and forces your heart to work harder. Carbon monoxide reduces oxygen in blood and damages the lining of blood vessels.
Diabetes	Maintain proper weight. Losing excess weight helps control blood sugar level. Eat high-fiber foods. Limit saturated fat and sugar. Get regular exercise.
Lack of exercise	Get at least 30 minutes of moderate-paced physical activity on most days of the week. See Chapter 9 for more about exercise.
Obesity	Maintain a healthy weight and exercise. Being only 10 percent overweight increases heart disease risk. See Chapter 8 for more about weight management.
Unhealthy diet	Keep total fat intake to less than 30 percent of daily calories and saturated fat to under 8–10 percent. Substitute olive and canola oils for saturated fat. Increase fiber intake to 20 to 35 grams a day by eating cereal grains, legumes, fruits, and vegetables. Eat five to nine servings of fruits and vegetables a day to receive the beneficial antioxidants and phytochemicals they contain.
Stress	Get regular exercise. Avoid excessive caffeine and alcohol. Practice relaxation techniques. Maintain good social relationships.
	Risk Factors You Can't Change
Age	Men over age 45 and women over age 55 are at increased risk.
Gender	Men are at higher risk. Estrogen may protect women before menopause.
Genetics	Increased risk if you have a father or brother under age 55 or a mother or sister under 65 who had heart disease.

35 milligrams per deciliter. A combined effort to lower LDL-cholesterol and raise HDL-cholesterol delivers a double punch in the fight against heart disease.

Reducing high blood cholesterol levels, particularly the "bad" LDL-cholesterol level, is thus an important strategy toward lowering the risk of heart disease. This is especially true for persons having one or more of the other major heart disease risk factors, such as smoking. Figure 4-7 on page 112 and Table 4-10 on page 129 provide information for determining what to do if your blood cholesterol level is high and whether you might need to seek treatment.

What sorts of dietary changes should I make to reduce my total cholesterol and LDL-cholesterol? Probably the most significant dietary change you can make is to reduce the amount of fat that you eat, particularly saturated fat, since high intakes of these dietary constituents are related to high blood levels of LDL-cholesterol.[36] The goals to work toward are to reduce your intake of total fat to 30 percent of total calories or less and reduce your intake of saturated fat to 8–10 percent of total calories or less. Many Americans are on the right track toward reducing their intakes of fat and saturated fat and lowering their blood LDL-cholesterol levels.

In addition to reducing fat intakes, people can make other dietary changes that might also help lower LDL-cholesterol. Dietary fiber, for example, may confer benefits. People on high-fiber diets have been shown to excrete more cholesterol and fat than those on low-fiber diets. A 5 percent decrease in total cholesterol can be achieved with a 5- to 10-gram increase in soluble fiber intake.[37] One reason is that the high-fiber diet decreases food's transit time through the digestive tract, allowing less time for cholesterol to be absorbed. When cholesterol from the diet is thus reduced, the body must turn to its own supply for making necessary body compounds. Diets high in fiber are typically low in fat and cholesterol—another advantage to emphasizing fiber.

The various dietary fibers have varying effects on blood cholesterol. Rolled oats, oat bran, and psyllium-fortified cereals have favorable effects on blood cholesterol, whereas wheat bran does not appear effective.[38] Apples, pears, peaches, oranges, and grapes are good sources of pectin, another type of cholesterol-lowering fiber.

"Excuse me, but which is it that practically kills you—polysaturated or polyunsaturated!"

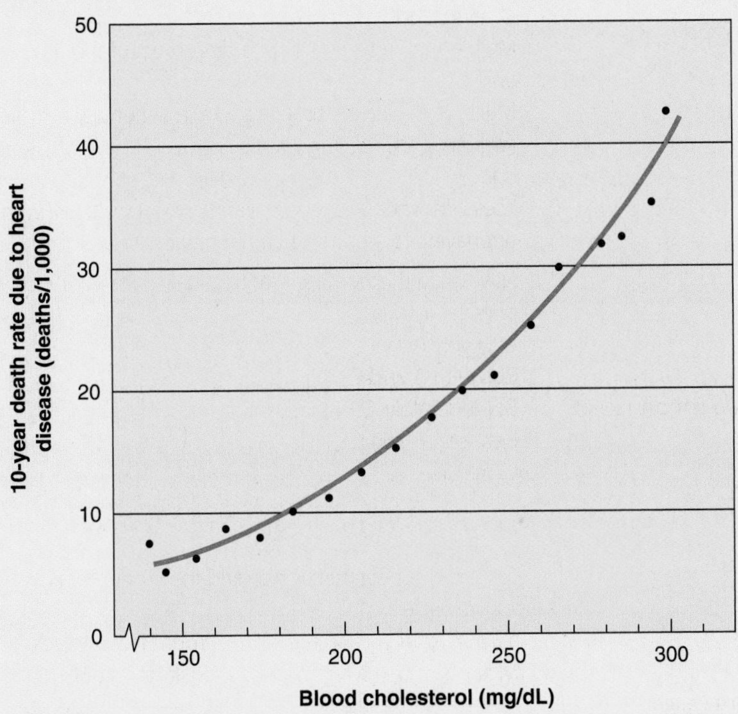

FIGURE 4-11
RELATIONSHIP BETWEEN BLOOD CHOLESTEROL LEVEL AND DEATH RATE FROM HEART DISEASE

Persons with blood cholesterol levels of 300 mg/dL are four times as likely to die from heart disease as those with blood cholesterol levels below 200 mg/dL.

Source: Data from 361,662 men screened for the Multiple Risk Factor Intervention Trial (MRFIT). Adapted from National Cholesterol Education Program, *Report of the Expert Panel on Population Strategies for Blood Cholesterol Reduction,* Bethesda, Md.: U.S. Department of Health and Human Services, Public Health Service, National Institutes of Health, National Heart, Lung, and Blood Institute, NIH Publication No. 90-3046, November 1990, p. 7.

HEART HEALTH SCORECARD

Do you know your heart disease risk score? Respond to the statements below, and score yourself as directed. Be aware that a high-risk score does not mean you *will* develop heart disease, but it should warn you of the possibility. Consult your physician if you have questions about your score results.

In each category, circle the number next to the statement that's most true for you.

CIGARETTE SMOKING

I never smoked or stopped smoking three or more years ago.	1
I don't smoke but live and/or work with smokers.	2
I stopped smoking within the past three years.	3
I smoke regularly.	4
I smoke regularly and live and/or work with other smokers.	5

TOTAL BLOOD CHOLESTEROL

Use the number from your most recent blood cholesterol measurement:

Less than 160	1
160–199	2
Don't know	3
200–239	4
240 or higher	5

HDL CHOLESTEROL

Use the number from your most recent HDL cholesterol measurement:

Over 60	1
56–60	2
Don't know	3
35–55	4
Less than 35	5

SYSTOLIC BLOOD PRESSURE

Use the first (highest) number from your most recent blood pressure measurement:

Less than 120	1
120–139	2
Don't know	3
140–159	4
160 or higher	5

EXCESS BODY WEIGHT

I am within 10 pounds of my desirable weight.	1
I am 10–20 pounds above my desirable weight.	2
I am 21–30 pounds above my desirable weight.	3
I am 31–50 pounds above my desirable weight.	4
I am more than 50 pounds above my desirable weight.	5

PHYSICAL ACTIVITY

Determine which statements best describe your usual level of physical activity:

A: HIGHLY ACTIVE

My job requires very hard physical labor (such as digging or loading heavy objects) at least four hours a day

or

I do vigorous activities (jogging, cycling, swimming, etc.) at least three times a week for 30 minutes or more.

or

I do at least one hour of moderate activity such as brisk walking at least four days a week.

B: MODERATELY ACTIVE

My job requires that I walk, lift, carry, or do other moderately hard work for several hours a day (day-care worker, stock clerk, or busboy/waitress)

or

I spend much of my leisure time doing moderate activities (dancing, gardening, walking, or housework).

C: INACTIVE

My job requires that I sit at a desk most of the day

and

Much of my leisure time is spent in sedentary activities (watching TV, reading, etc.)

and

I seldom work up a sweat, and I cannot walk fast without having to stop to catch my breath.

Now circle the number that best describes your level of physical activity:

A: Highly Active	1
Between A and B	2
B: Moderately Active	3
Between B and C	4
C: Inactive	5

SCORING YOUR HEART ATTACK RISK

To learn your estimated risk, add the six numbers you've circled.

If Your Total Score Is:	Your Heart Attack Risk Is:
6–13	Low
14–22	Moderate
23–30	High

Source: American Heart Association

Should I also reduce my cholesterol intake?

The *Dietary Guidelines* recommend that we limit the cholesterol in our daily diets to 300 milligrams. On average, men consume 400 to 500 milligrams of cholesterol and women consume about 300 milligrams daily. The average for American men is above that consumed in countries with little heart disease.

Some experts say that all adults should cut their cholesterol intake; others say only those medically identified as at risk for heart disease should do so. The question remains open, but most people who develop heart-healthy eating habits, such as decreasing saturated fat, tend to lower their cholesterol intake along with their fat intake.

What factors determine my HDL-cholesterol level?

One is gender. Women have higher HDL-cholesterol levels than men. However, heart disease is the major cause of death among women after menopause, when low levels of estrogen lead to a decrease in HDL levels along with an increase in LDL levels.[39] Another factor, interestingly, seems to be smoking habits. Nonsmokers have uniformly higher HDL levels than do smokers. Still another factor is weight reduction for those who are overweight.[40]

If there are dietary factors of any significance, one may be the use of some fish in the diet.[41] Another may be the use of soyfoods and foods containing soluble fibers that lower LDL levels. By far the most powerful influence on HDL levels is not a dietary factor at all—it is regular exercise. The discovery that exercise raises HDL levels has given great impetus to the physical fitness movement—and especially to the popularity of running and walking. The earliest reports were of raised HDL levels in long-distance runners, and the continuing study of this elite group has repeatedly demonstrated that running does indeed elevate HDL. People do not, however, have to become competitive athletes to raise their HDL—moderate exercise, such as walking, may both lower LDL levels and raise HDL levels if the activity is consistently pursued for long enough periods. Evidently, then, almost all people are capable of exercising enough to reap this and many other benefits (see Chapter 9).

TABLE 4-10 STANDARDS FOR BLOOD CHOLESTEROL LEVELS AND RISK OF HEART DISEASE

Evaluating Blood Lipid Levels

Total Cholesterol Levels	Classification
Less than 200 mg/dL	Desirable
200 to 239 mg/dL	Borderline high
240 mg/dL or higher	High
HDL-Cholesterol Levels	
Less than 35 mg/dL	Increased risk for CHD (Coronary Heart Disease)
60 mg/dL or higher	Decreased risk for CHD
LDL-to-HDL Ratio	
Men: greater than 5.0 indicates risk	
Women: greater than 4.5 indicates risk	
Triglyceride Levels	
Less than 200 mg/dL	Desirable
200 to 400 mg/dL	Borderline high
400 to 1,000 mg/dL	High
1,000 mg/dL or higher	Very High

Treating LDL-Cholesterol Levels

Risk Status	Target/Recommendations for LDL-Cholesterol
Low-risk individuals[a]	Less than 160 mg/dL Begin dietary therapy at 160 mg/dL; consider drug therapy at 190 mg/dL in men over 45 and post-menopausal women. Begin drug therapy at 220 mg/dL in younger men and women.
High-risk individuals[b]	Less than 130 mg/dL Begin dietary therapy at 130 mg/dL; consider drug therapy at 160 mg/dL.
Very high risk individuals[c]	Less than 100 mg/dL Begin dietary therapy at 100 mg/dL; consider drug therapy if dietary therapy fails to reduce LDL-cholesterol levels below 100 mg/dL.

[a]Low-risk individuals are those with one or no risk factors for heart disease.
[b]High-risk individuals are those who have at least two risk factors for CHD. (Men over age 45 and postmenopausal women are considered to have one risk factor.)
[c]Very-high-risk individuals are those with known heart disease (history of heart attack or previous bypass surgery or angioplasty).

What is homocysteine and how does it contribute to heart disease?

Recently, a substance called homocysteine (pronounced ho-mo-SIS-teen) has been linked to heart disease. The body naturally produces homocysteine when it breaks down protein. The B vitamins—folate, B_6, and B_{12}—then convert homocysteine into other amino acids that the body requires. However, when dietary intakes of these B vitamins are low, blood levels of homocysteine rise. Scientists believe that high levels of homocysteine cause damage to the linings of arteries and accelerate the formation of blood clots. Research shows that people with high blood levels of homocysteine have significantly more heart attacks and strokes.[42]

The good news is that as a risk factor for heart disease, high homocysteine levels may be preventable and reversible by consuming generous amounts of the B vitamins. Good sources of folate include orange juice; fortified breads, cereals, and other grain products; green leafy vegetables; and legumes. Vitamin B_6 is found in meat, chicken, fish, whole grains, and legumes. Vitamin B_{12} is found in fortified cereals, fortified soy milk, and all animal foods. (See Chapter 6 for more information about these B vitamins.)

What about alcohol? I've heard that moderate alcohol intake can be beneficial to heart disease.

The evidence with regards to alcohol and blood lipids points to a possible protective mechanism between moderate alcohol consumption and heart disease risk factors.[43] Investigators have reported that the consumption of moderate amounts of alcohol appears to raise HDL levels.[44] Moderate drinking is defined as no more than one drink a day for most women, and no more than two drinks a day for most men (see the Spotlight feature in Chapter 9). A strong association between moderate alcohol consumption and low rates of heart disease was first observed in France and then in other wine-drinking areas of the Mediterranean. Researchers have since identified antioxidants and other compounds that decrease blood clotting in red wine, which may help explain part of the "French paradox"—or how a region with high fat intakes could have such low rates of heart disease. Researchers are quick to point out, however, that other factors certainly may contribute to the paradox. For example, the French consume about 57 percent of their day's calories before 2 P.M., whereas most Americans have only consumed about 38 percent of their calories by that time.[45] Also, the French mostly consume their wine with meals and eat more fruits and vegetables and leaner cuts of meat.

Scientists warn that caution is needed in recommending moderate alcohol consumption to the public. Many people need to refrain from alcohol consumption altogether (pregnant women, recovering alcoholics, people under age 21, persons on certain medications, and those intending to drive a vehicle).[46] Keep in mind that alcohol can have profoundly negative effects on the body and is associated with many disease states (see the Chapter 9 Spotlight feature).

What about garlic? I've heard that garlic pills and whole cloves help lower blood cholesterol.

A sizeable body of evidence from around the world suggests that garlic may help lower blood cholesterol, reduce blood pressure, and help arteries remain elastic.[47] Although the research is promising, some studies have shown conflicting results regarding the effect of garlic on blood cholesterol. Further studies are needed before scientists can come to any conclusions about the usefulness of fresh garlic or garlic pills, which vary a great deal in composition, in lowering blood cholesterol levels. See Chapter 6 for more about the phytochemical benefits derived from garlic and its relatives (leeks, onions, chives, scallions, and shallots).

How do I translate these recommendations into a healthful eating pattern?

Most people have a difficult time translating the dietary recommendations into actual meal patterns. What foods should you eat, for example, to achieve the goal of reducing your fat intake to 30 percent or less of total calories? It's virtually impossible to know exactly what your fat intake is without having your daily food intake analyzed using a computerized program. Even so, some general tips will help you make the heart-healthy food choices needed to achieve your goals (see Table 4-11). If you were to take all of the steps suggested, you would:

- Choose healthful portions of skinless poultry, lean meat, and fish, especially omega-3-fatty-acid-rich fish such as mackerel, salmon, and canned albacore tuna.
- Choose fat-free or low-fat dairy products, such as fat-free milk or low-fat or fat-free yogurt.
- Consume abundant legumes of many varieties, including soybeans, kidney beans, and lentils.*
- Eat generous quantities of fiber-rich, antioxidant-rich fruits and vegetables, including many raw ones.[48]
- Eat whole grains (oats, wheat, corn, rice, pasta) often.
- Limit your use of foods particularly rich in saturated fat, such as fatty red meats. Trim away all visible fat from meats before cooking.
- Adopt low-fat cooking methods, such as broiling, roasting, steaming, braising, and stir-frying.
- Become a savvy supermarket shopper—learn to read food labels to help you choose low-fat and fat-free food products.
- Consume alcohol only in moderation, if at all.

It seems that the factors affecting the health of the heart are all tangled together. The exact relationships among them have not yet been worked out; we don't know which causes what, but all evidence

*For a discussion of the benefits of soy in prevention of heart disease, see the Spotlight in Chapter 5.

points to the same general recommendations. For good health and to avoid heart disease, stop smoking; reduce blood pressure and weight, if necessary; eat a balanced, adequate, and varied diet; reduce total fat intake, especially saturated fat; and exercise regularly.

Attention to emotional health is also important in reducing the risk of heart disease. Both love and affection seem to affect the heart. People with many social ties appear to develop less heart disease than people with few or none. Married men have less heart disease than single men, and pet owners (even owners of pet fish) have lower blood pressure than do people without pets. Clearly, the mystery of heart disease, like all the great human mysteries, involves the mind and spirit as well as the body. So nourish yourself in all ways—not just physically.

Should children, like their parents, eat low-fat diets for heart health?

The current Dietary Guidelines and experts representing 42 major U.S. health and professional organizations recommended that children aged 2 and older eat diets containing no more than 30 percent of total calories from saturated fat and no more than 300 milligrams of cholesterol daily. The reason is that hardening of the arteries often begins in childhood. However, from birth to 2 years of age, a child's fat consumption should not be restricted, because fat is a concentrated source of the calories needed to ensure proper physical development.

The panel also advised that children or teens should get their blood cholesterol measured if they have one parent with a high blood cholesterol level. For children and adolescents, a total of 200 milligrams per deciliter or more is considered high, 170 to 199 is borderline high, and less than 170 is acceptable. In addition, children born to families with a history of premature heart disease should have both total blood cholesterol and HDL-cholesterol checked.[49]

Enjoy a variety of low-fat foods for the health of your heart.

TABLE 4-11 HOW TO CUT BACK ON FAT IN YOUR DIET

Substitute This . . . Food	Fat (g)	With This . . . Food	Fat (g)	And Save Fat/Calories Fat (g)	Calories
Whole milk, 8 oz	8	Fat-free milk, 8 oz	Trace	8	72
Sour cream, 4 oz	22	Low-fat yogurt, 4 oz	2	20	180
Ice cream, 1 c	14	Sorbet, 1 c	0	14	126
Chicken breast, with skin, 3 oz	7	Chicken breast, skinless, 3 oz	3	4	36
Hamburger, 3 oz	17	Lean hamburger, 3 oz	10	7	63
Biscuits, 2 dinner	6	Bread, 2 slices	2	4	36
Butter/margarine, 1 tbsp	12	Parmesan cheese, 1 tbsp	2	10	90
Mayonnaise, 1 tbsp	11	Fat-free mayonnaise, 1 tbsp	0	11	99
Asparagus, 1 c, with hollandaise sauce	18	Asparagus, 1 c, with lemon	0	18	162
French fries, 15	12	Baked potato with 1 pat butter	4	8	72
Fried egg, 1	9	Broiled or poached egg, 1	6	3	27
Bacon, 1 oz	14	Canadian bacon, 1 oz	2	12	108
Round steak, 8 oz	30	Round steak, 4 oz	15	15	135
Apple pie, 1 slice	18	Apple crisp, 1 portion	8	10	90
Potato chips, 2 oz (1 small bag)	24	Unbuttered popcorn, 3 c	2	22	198
Danish pastry, 1	15	English muffin with 1 tbsp jam	1	14	126

Nutrition on the Web

nutrition.wadsworth.com — Go to the *Personal Nutrition* site to check for the latest updates to chapter topics or to access links to related Web sites.

www.eatright.org/cgi/search.cgi — Search for information on heart-healthy diets, fat and cholesterol in foods, and review the ADA position paper on fat replacers.

www.mayohealth.org — Provides practical health and nutrition information to consumers, current articles on a variety of nutrition topics, a Virtual Cookbook with lower-fat versions of family-favorite recipes, and a quiz to test your knowledge of fats.

www.fda.gov — Visit Foods and search for information on Olestra and other fat substitutes and updates on labeling of trans fats in foods.

ificinfo.health.org — Provides scientific-based information about managing fat, trans fat, cholesterol, and fat replacers in your diet.

www.cnn.com/HEALTH — The CNN Health page posts research updates, food preparation tips, and fat and cholesterol information for a healthy diet.

www.healthfinder.gov/searchoptions/topicsaz.htm — This site presents a wide assortment of health and nutrition information from government agencies. Use this search engine to locate dietary fat and cholesterol information.

www.nhlbi.nih.gov/chd — The National Heart, Lung, and Blood Institute provides consumers and health professionals with practical information for cardiovascular health. This site explains how heart disease develops, provides tips for reducing your cholesterol level and risk for heart disease. Visit the CyberKitchen to determine how well you estimate portion sizes, and go to Create a Diet to see how your diet compares to the current dietary recommendations for a healthy heart.

www.americanheart.org — The American Heart Association provides practical advice about following a healthy lifestyle, a risk assessment tool, a quiz to test your knowledge about diet, exercise, and heart disease, recipes, and useful links to related sites.

www.heartinfo.org — Provides information on healthy shopping and cooking, eating healthy away from home, general nutrition tips, heart disease, high blood pressure, supplements and heart disease, and weight management.

www.caloriecontrol.org — Search for information on fat replacers and reduced-fat products, low-fat recipes, and note the discussion on "Calories Still Count."

www.flaxcouncil.ca — The Flax Council of Canada gives useful information on the health benefits of flaxseed, soluble fiber, and omega-3 fatty acids. The site posts research updates, fact sheets, and recipes.

www.hsf.ca/ — The Heart and Stroke Foundation of Canada offers a tutorial on heart-healthy food choices, an assessment of your risk for heart disease, and a quiz to test your heart health knowledge.

www.healthyfridge.org — This site provides information on heart disease, practical shopping tips for stocking a healthy refrigerator, heart-healthy recipes, and a quiz to test your saturated fat IQ.

www.ncbi.nlm.nih.gov/PubMed — A search engine to help you locate information from current scientific articles on any topic related to fat.

Check Yourself . . .

1. What are the functions of fat in the diet?
2. What are the functions of fat in the body?
3. Give an example of a food source of (a) a saturated fat, (b) a monounsaturated fat, (c) a polyunsaturated fat, (d) a trans fat, (e) cholesterol, and (f) an omega-3 fat.
4. What does hydrogenation do to fats?
5. How do the four lipoproteins differ in function from one another in the body?
6. How do the levels of HDL and LDL in the blood relate to risk of heart disease?
7. Describe the process of fat digestion in the body.
8. What are the current dietary recommendations regarding fat and cholesterol intake?
9. List four risk factors for heart disease.
10. What steps are recommended for raising one's HDL-cholesterol level in the blood?

Answers to Check Yourself questions are found in Appendix G.

5

The Proteins and Amino Acids

CONTENTS

What Proteins Are Made Of

Nutrition Action: Amino Acid Essentials

The Functions of Body Proteins

How the Body Handles Protein

Protein Quality of Foods

Recommended Protein Intakes

Protein and Health

Protein Scorecard

The Savvy Diner: Reshape Your Protein Choices for Health

The Vegetarian Diet

Spotlight: The Benefits of Soy

The amino acids of proteins are the raw materials of heredity, the keys to life chemistry, handed from generation to generation.

R. M. Deutsch
(1928–1988, nutrition author and educator)

Ask Yourself . . .

Which of the following statements about nutrition are true, and which are false? For each false statement, what *is* true?

1. Protein eaten in excess of need is stored intact in the body, as is fat, so that it can be used when a person's diet falls short of supplying the day's need for essential proteins.
2. No new living tissue can be built without protein.
3. Whenever cells are lost, protein is lost.
4. All enzymes and hormones are made of protein.
5. When antibodies enter the body, they produce illness.
6. When a person doesn't eat enough food to meet the body's energy needs, the body devours its own protein tissue.
7. Once the body has assembled its proteins into body structures, it never lets go of them.
8. Milk protein is the standard against which the quality of other proteins is usually measured.
9. It is impossible to consume too much protein.
10. People who eat no meat have to eat a lot of special foods to get enough protein.

Answers found on the following page.

PROTEINS compounds—composed of atoms of carbon, hydrogen, oxygen, and nitrogen—arranged as strands of amino acids. Some amino acids also contain atoms of sulfur.

An Amino Acid: glycine

Amine group H—N—C—C—OH Acid group (with H atoms on N and C, and =O on the acid carbon)

Side group

The **proteins** are perhaps the most highly respected of the three energy nutrients, and the roles they play in the body are far more varied than those of carbohydrate or fat. First named 150 years ago after the Greek word *proteios* ("of prime importance"), proteins have revealed countless secrets about the ways living processes take place, and they account for many nutrition concerns. How do we grow? How do our bodies replace the materials they lose? How does blood clot? What makes us able to become immune to diseases we have been exposed to? The answers to these and many other such questions arise to a great extent from an understanding of the nature of the proteins.

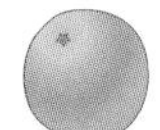

WHAT PROTEINS ARE MADE OF

AMINO (a-MEEN-o) ACIDS building blocks of protein; each is a compound with an amine group at one end, an acid group at the other, and a distinctive side chain.

AMINE (a-MEEN) GROUP the nitrogen-containing portion of an amino acid.

To appreciate the many vital functions of proteins, we must understand their structure. One key difference from carbohydrate and fat, which contain only carbon, hydrogen, and oxygen atoms, is that proteins contain nitrogen atoms. These nitrogen atoms give the name *amino* ("nitrogen containing") to the **amino acids** of which protein is made. Another key difference is that in contrast to the carbohydrates—whose repeating units, glucose molecules, are identical—the amino acids in a strand of protein are different from one another.

All amino acids have the same, simple chemical backbone with an **amine group** (the nitrogen-containing part) at one end and an acid group at the other end. The differences between the amino acids depend on a distinctive structure—the chemical side chain—that is attached to the backbone. Twenty amino acids with 20 different side chains make up most of the proteins of living tissue.

The side chains vary in complexity from a single hydrogen atom like that on glycine to a complex ring structure like that on phenylalanine. Not only do these structures differ in composition, size, and shape, they also differ in electrical charge. Some are negative, some are positive, and some have no charge. These side chains help to determine the shapes and behaviors of the larger protein molecules that the amino acids make up.

ESSENTIAL AMINO ACIDS amino acids that cannot be synthesized by the body or that cannot be synthesized in amounts sufficient to meet physiological need.

THE NINE ESSENTIAL AMINO ACIDS FOR HUMAN ADULTS THAT MUST BE OBTAINED FROM THE DIET:

tryptophan	lysine
valine	phenylalanine
threonine	methionine
isoleucine	histidine
leucine	

THE NONESSENTIAL AMINO ACIDS—ALSO IMPORTANT IN NUTRITION:

alanine	glutamine
arginine	glycine
asparagine	proline
aspartic acid	serine
cysteine	tyrosine
glutamic acid	

Essential and Nonessential Amino Acids

The body can make about half of the amino acids (known as nonessential amino acids) for itself, given the needed parts: nitrogen to form the amine group, along with backbone fragments derived from carbohydrate or fat. But there are some amino acids that the healthy body cannot make. These are known as **essential amino acids.** If the diet does not supply them, the body cannot make the proteins it needs to do its work. The indispensability of the essential amino acids makes it necessary for people to eat protein food sources every day.

The distinction between essential and nonessential amino acids is not quite as clear-cut as the list in the margin makes it appear.* Histidine often appears not to be essential, perhaps because the diet supplies it in abundance.[1] Arginine, under some conditions, may be synthesized too slowly to fully meet the human need for it.[2] States of illness can interfere with amino acid transformations in the body and so make other amino acids essential for certain individuals.[3]

Ask Yourself Answers: **1.** False. Protein eaten in excess of need is not stored in the body, as is fat, so it has to be eaten every day if it is not to become depleted. **2.** True. **3.** True. **4.** False. All enzymes, but not all hormones, are made of protein. **5.** False. Antibodies protect the body from illness caused by antigens. **6.** True. **7.** False. Your body loses protein every day. **8.** False. Egg white protein, not milk protein, is the standard against which the quality of other proteins is usually measured. **9.** False. It is possible to consume too much protein. **10.** False. People who eat no meat can easily get enough protein without eating a lot of special foods.

*Cysteine and tyrosine normally are not essential because the body makes them from methionine and phenylalanine. However, if there are not enough of these precursors from which to make them, they have to be supplied in the diet. Likewise, glutamine is considered a conditionally essential amino acid during critical illness associated with inflammation and injury.

NUTRITION ACTION

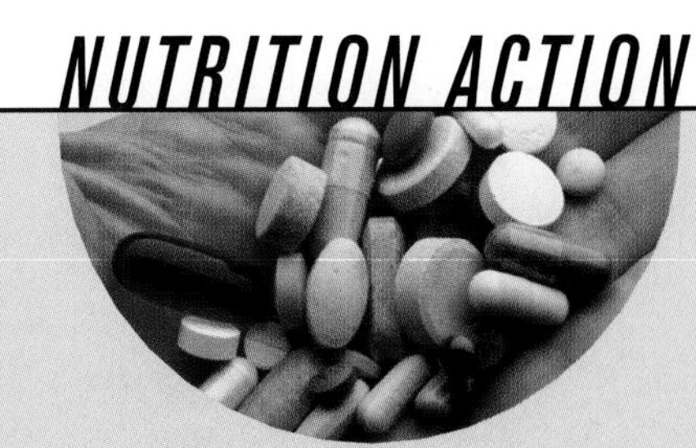

Amino Acid Essentials

Dr. Gerald Gleich had a hunch. Back in 1989, the blood disorder expert from Minnesota's Mayo Clinic received telephone calls from several doctors baffled over the cause of a bizarre set of symptoms, including muscle pain, mouth sores, and abnormally high levels of a certain type of white blood cell in the blood of a handful of patients. But Dr. Gleich identified a suspect: a dietary supplement called **L-tryptophan,** which all of the victims had been taking before their symptoms appeared. One of the essential amino acids, L-tryptophan had been sold in supermarkets, drug stores, health food stores, and other retail outlets and had been self-prescribed for everything from sleeping problems to depression to stress to premenstrual syndrome to drug addiction.

L-TRYPTOPHAN an essential amino acid that has been sold in tablets, capsules, and powders as a dietary supplement.

Dr. Gleich contacted the Centers for Disease Control and Prevention, the Atlanta-based public-health watchdog, sparking a national investigation that confirmed his suspicions within two weeks and prompted the Food and Drug Administration to ban the supplement shortly thereafter.[4] Nevertheless, by 1992, more than 1,500 consumers had reported similar symptoms as a result of popping L-tryptophan supplements, and 38 deaths had been linked to the product.[5] The cause of the illness, dubbed **eosinophilia-myalgia syndrome (EMS),** was later suspected to be an impurity that contaminated a particular manufacturer's L-tryptophan supplements, though some experts theorize that the amino acid itself caused EMS in some people.[6]

EOSINOPHILIA-MYALGIA (ee-o-sin-o-FIL-ia my-AL-jia) **SYNDROME (EMS)** a disease characterized primarily by a high level of eosinophils, a type of white blood cell, as well as myalgia—that is, muscle pain and weakness.

The case of the contaminated tryptophan brought to the forefront concerns about the safety of amino acid supplements. Granted, the Food and Drug Administration had long cautioned that excesses of one or more of those building blocks of protein can be toxic and create imbalances of other amino acids in the body. That's why, in 1974, the FDA removed amino acids from its list of substances that are generally recognized as safe. (The only uses of amino acids approved by the FDA are (1) their addition to foods to improve the value of an item's protein and (2) their use in the making of special "medical foods" designed to be taken under a physician's supervision by people who require special protein formulations because of health problems such as kidney or liver disease.)[7]

Still, as a result of the eosinophilia-myalgia syndrome scare, the FDA asked a panel of experts to review the safety of amino acids currently sold in the marketplace. The panel's conclusions underscore the need for better regulation of these supplements. For instance, the panel found that the labels of most amino acid supplements failed to carry vital information including suggested doses, shelf life, and contraindications for use of the product. In addition, the panel identified certain groups of people who may be at particularly high risk of suffering health problems as a result of swallowing amino acid supplements. Children and teenagers, for example, may not grow properly if they take amino acid pills or powders. That's because young, underdeveloped bodies may metabolize amino acids differently than adult bodies, possibly leaving young people more vulnerable to harmful effects of excess amino acids. Other groups at a particularly high risk of harm as a result of long-term use of amino acid supplements:[8]

- Infants, children, and teenagers
- Women of childbearing age

We cannot force extra protein into our muscles just by eating more of it—the way to make muscles grow is to make them work (see Chapter 9).

- Pregnant and breastfeeding women
- Elderly men and women
- People with medical conditions that alter the body's ability to metabolize amino acids
- People who regularly consume low amounts of protein
- Smokers
- People with medical problems, especially if they take medications

Safety hazards aren't the only problems that the panel identified. The panel also found that much of the advertising and product label information regarding the effectiveness of the amino acid supplements is questionable and based on anecdotes rather than grounded in careful, scientific research. In fact, the lack of data to support the usefulness of many amino acid supplements has been a source of concern for years. Consider that L-tryptophan has never been shown to relieve depression, insomnia, stress, premenstrual syndrome, or any other ills its manufacturers claim it treats.

The same goes for arginine, an amino acid that has been touted as "causing weight loss overnight" by stimulating secretion of a substance called human growth hormone, which in turn supposedly promotes weight loss. Although it's true that arginine can prompt the release of the hormone, it will do so only when people take in whopping doses of it that are unlikely to be found in supplements. And even if a person were to take enough arginine to prompt a surge of the hormone into the body, he or she wouldn't automatically shed pounds; human growth hormone has not been found to cause weight loss. Thus, claims that arginine "burns fat" are spurious, at best. An FDA advisory panel on over-the-counter weight-loss products investigated arginine along with 11 other amino acids touted as diet aids—namely, cystine, histidine, isoleucine, leucine, L-lysine, methionine, phenylalanine, threonine, tryptophan, tyrosine, and valine—and found no basis for the claims about the effectiveness of any of them in controlling weight.

Another popular amino acid supplement about which overblown claims are often made is lysine. Some research indicated that lysine could help treat herpes infections, a suggestion that has been used to hawk the substance. That evidence was based on poorly controlled studies, however, which carefully controlled research failed to confirm. Thus, lysine is not recommended by reputable professionals as a treatment for herpes.[9]

Another substance—the hormone melatonin—is derived from tryptophan in the body. Melatonin is currently marketed as a sleep aid. The body typically makes larger levels of melatonin at night and lower levels in the day. Short-term studies of the substance show that low doses (0.1 to 0.3 milligrams) may relieve insomnia in older adults who sometimes have low levels of melatonin in their blood. However, more studies are needed to test its effectiveness and long-term safety.[10]

Athletes rank as prime targets of amino acid supplement manufacturers, whose products promote false hopes of benefits.

In addition to people suffering from health problems, athletes rank as prime targets of amino acid supplement manufacturers. Pick up any copy of one of the bodybuilding magazines, and you're likely to see ads for supplements packed with "free-form," "predigested," and "peptide-bond" amino acids touted as optimum sources of protein for athletes. Scientific-sounding names notwithstanding, such products have never been shown to increase muscle size or enhance athletic prowess. Consider that one comparison of Marine officer candidates given protein supplements with another set of trainees who received a placebo indicated that the groups performed equally well before, during, and after the program, regardless of supplement use or lack thereof.[11] Refer to the Spotlight feature in Chapter 1 for some tips on how to spot fraudulent nutritional products.

The latest group to begin swallowing fraudulent claims about amino acids is young professionals who down "smart drinks," made with amino acids such as phenylalanine, choline, taurine, and L-cysteine and touted as intelligence, energy, and memory boosters. Smart drinks started to become popular in the early 1980s, when the baby boom generation began coming of age and looking for ways to make career strides. But while the efficacy of these so-called intellect enhancers has never been proven, the side effects of taking smart drinks are well-known: Phenylalanine can make an otherwise healthy person irritable or lead to insomnia, choline can cause gastrointestinal illness, and L-cysteine can cause health problems for people who take antidepressants, to name a few such side effects.[12]

Health risks aside, consumers should note that special protein supplements command a high price. Foods, on the other hand, supply ample amounts of protein at a fraction of the cost. One glass of milk, a serving of rice and beans, or a 3-ounce portion of chicken, for example, provides a generous helping of all nine essential amino acids for less than half the price of a dose of most amino acid tablets, liquids, or powders. That's why it's easy for most Americans to eat a diet that supplies the body with more than enough protein and amino acids. Table 5-1 compares costs of protein supplements with costs of high-protein foods.

Finally, consumers should be wary of companies that sell tests that measure the amino acid content of the blood and urine. Costing in the neighborhood of $150, these tests are then used to help sell concoctions designed to replace the "missing" amino acids. That practice, however, is completely fraudulent; amino acid levels in the blood and urine vary greatly from day to day and have no bearing on the body's supply or requirement for amino acids.[13]

TABLE 5-1 THE HIGH COST OF PROTEIN SUPPLEMENTS

Food or Supplement	Amount Needed to Obtain About 15 Grams of Protein	Cost*
Fat-free milk	2 cups	$0.30
Tuna (light, canned in water)	2 ounces	0.30
Twinlab Amino Fuel Anabolic Liquid Amino Acid Concentrate	3 tablespoons	1.25
Twinlab Amino Fuel Tablets	15 tablets	2.00
Hot Stuff Fitness Enhancing Nutritional Powder	About 2 tablespoons	1.50

*Based on 1999 prices in New York metropolitan area stores.

Source: Manufacturers' information.

Proteins as the Source of Life's Variety

PROTEIN SYNTHESIS the process by which cells assemble amino acids into proteins. Each individual is unique because of minute differences in the ways his or her body proteins are made. The instructions for making every protein in a person's body are transmitted in the genetic information the person receives at conception.

PEPTIDE BOND a bond that connects one amino acid with another.

In the first step of **protein synthesis,** each amino acid is hooked to the next. A bond, called a **peptide bond,** is formed between the amino end of one and the acid end of the next. Proteins are made of many amino acid units, from several dozen to many hundred.

A strand of protein is not a straight, but a tangled, chain. The amino acids at different places along the strand are attracted to one another, and this attraction causes the strand to coil into a shape similar to that of a metal spring. Not only does the strand of amino acids form a long coil, but the coil tangles, forming a globular structure.

The charged amino acids are attracted to water, and in the body fluids they orient themselves on the outside of the globular structure. The neutral amino acids are repelled by water and are attracted to one another; they tuck themselves into the center, away from the body fluid. All these interactions among the amino acids and the surrounding fluid result in the unique architecture of each type of protein. Additional steps may be needed for the protein to become functional. A mineral or a vitamin may be needed to complete the unit and activate it, or several proteins may gather to form a functioning group.

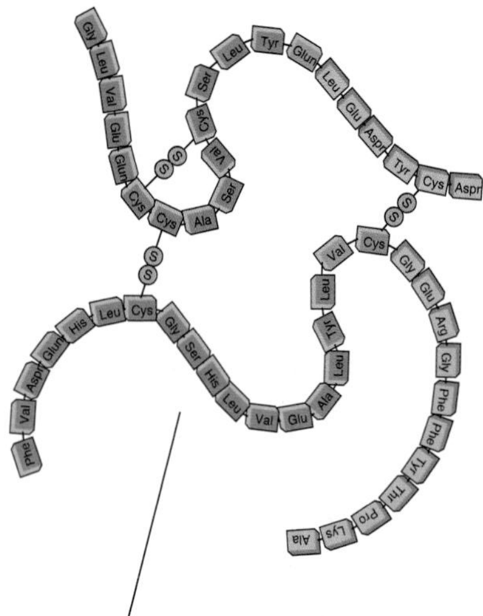

Amino acids are linked together with peptide bonds to form strands of protein. Note how the amino acids are linked together in this molecule of human insulin. The sulfur groups (S) on two cysteine (cys) molecules can bond together, creating a "sulfur-bridge" between the two protein strands.

The differing shapes of proteins enable them to perform different tasks in the body. In proteins that give strength and elasticity to body parts, several springs of amino acids coil together and form ropelike fibers. Other proteins, like those in the blood, do not have such structural strength but are water soluble, with a globular shape like a ball of steel wool. Some are hollow balls that can carry and store minerals in their interiors. Still others provide support to tissues. Some—the enzymes—act on other substances to change them chemically.

Denaturation of Proteins

DENATURATION the change in shape of a protein brought about by heat, alcohol, acids, bases, or other agents. Many well-known poisons are salts of heavy metals such as mercury and silver; these salts alter the structure of proteins wherever they touch them.

Proteins can undergo **denaturation,** resulting in distortion of shape by heat, alcohol, acids, bases, or the salts of heavy metals. The denaturation of a protein is the first step in the protein's breakdown. Denaturation is useful to the body in digestion. During the digestion of a food protein, an early step is denaturation by the stomach acid, which opens up the protein's structure, permitting digestive enzymes to cleave the peptide bonds (see Figure 5-1 on page 145). Denaturation can also occur during food preparation. For example, cooking an egg denatures the proteins of the egg and makes the egg firmer. Perhaps more important, cooking denatures two raw-egg proteins that bind the B vitamin biotin and the mineral iron, and another that slows the digestion of other proteins. Cooking eggs liberates biotin and iron and aids in protein digestion.

THE FUNCTIONS OF BODY PROTEINS

No new living tissue can be built without protein, for protein is part of every cell. About 20 percent of our total body weight is protein. Proteins come in many forms: enzymes, antibodies, hormones, transport vehicles, oxygen carriers, tendons and ligaments, scars, the cores of bones and teeth, the filaments of hair, the materials of nails, and more (see Table 5-2). A few of the many vital functions of proteins are described here to show why they have rightfully earned their position of importance in nutrition.

Cooking an egg denatures its protein.

Growth and Maintenance

One function of dietary protein is to ensure the availability of amino acids to build the proteins of new tissue. The new tissue may be found in an embryo; in a growing child; in the blood that replaces that which has been lost in burns, hemor-

TABLE 5-2 THE FUNCTIONS OF PROTEINS IN THE BODY

- *Growth and Maintenance.* Proteins provide building materials—amino acids—for growth and repair of body tissues.
- *Enzymes.* Proteins facilitate numerous chemical reactions in the body; all enzymes are proteins.
- *Hormones.* Some proteins act as chemical messengers—regulating body processes; not all hormones are proteins.
- *Antibodies.* Proteins assist the body in maintaining its resistance to disease by acting against foreign disease-causing substances.
- *Fluid balance.* Proteins help regulate the quantity of fluids in body compartments.
- *Acid-Base Balance.* Proteins act as buffers to maintain the normal acid and base concentrations in body fluids.
- *Transportation.* Proteins move needed nutrients and other substances into and out of cells and around the body.
- *Body structures.* Proteins form vital parts of most body structures such as skin, nails, hair, membranes, muscles, teeth, bones, organs, ligaments, and tendons.
- *Energy.* Protein can be used to provide calories (4 calories per gram) to help meet the body's energy needs.

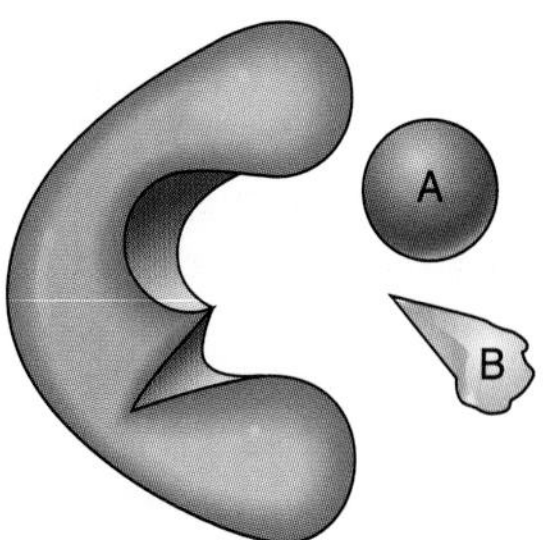

Enzyme plus two compounds, A and B

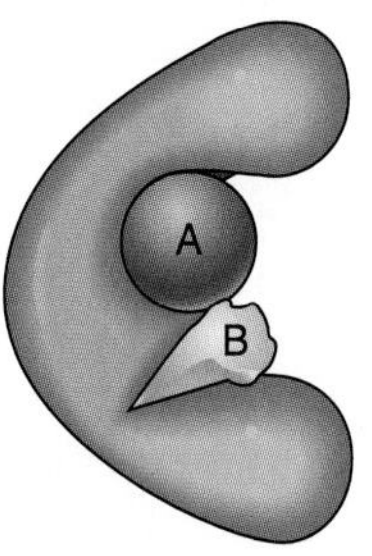

Enzyme complexed with A and B

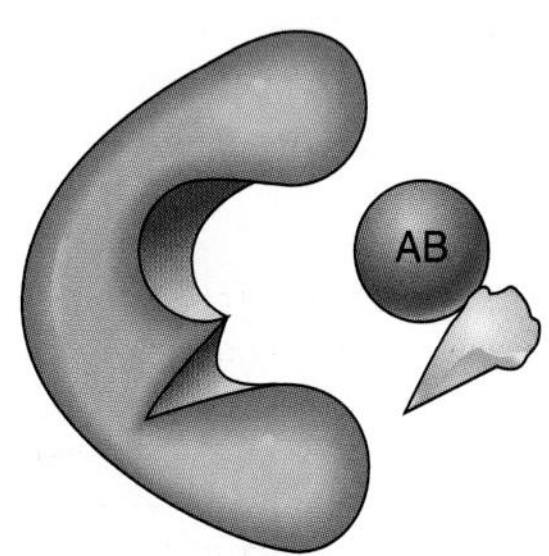

Enzyme plus new compound AB

Enzyme Action
Each enzyme facilitates a specific chemical reaction.

rhage, or surgery; in the scar tissue that heals wounds; or in new hair and nails. Not so obvious is the protein that helps replace wornout cells. The cells that line the digestive tract live for about three days and are constantly being shed and excreted. You have probably observed that the cells of your skin die, rub off, and are replaced from underneath. For this new growth, amino acids must constantly be resupplied by food.

Enzymes

All enzymes are proteins, and they are among the most important proteins formed in living cells. Enzymes are catalysts—biological spark plugs—that help chemical reactions take place. There are thousands of enzymes inside a single cell, each type facilitating a specific chemical reaction. Enzymes are involved in such processes as the digestion of food, the release of energy from the body's stored energy supplies, and tissue growth and repair.

A mystery that has been partially explained is how an enzyme can be specific for a particular reaction. The surface of the enzyme is contoured so that the enzyme can recognize the substances it works on and ignore others. The surface provides a site that attracts one or more specific chemical compounds and promotes a specific chemical reaction. For example, two substances might become attached to the enzyme and then to each other. The newly formed product is then expelled by the enzyme into the fluid of the cell. Enzymes are the hands-on workers in the production and processing of all substances needed by the body.

Hormones

Similar to the enzymes in the profoundness of their effects are the **hormones.** However, these molecules differ from the enzymes. For one thing, not all of them are made of protein. For another, they don't catalyze chemical reactions directly but rather are messengers that elicit the appropriate responses to maintain a normal environment in the body. Hormones regulate overall body conditions, such as the blood glucose level (the hormones insulin and glucagon) and the metabolic rate (thyroid hormone).

HORMONES chemical messengers. Hormones are secreted by a variety of glands in the body in response to altered conditions. Each affects one or more target tissues or organs and elicits specific responses to restore normal conditions.

Antibodies

Of all the great variety of proteins in living organisms, the **antibodies** best demonstrate that proteins are specific for one organism. Antibodies are formed in response to the presence of **antigens** (foreign proteins or other large molecules) that invade the body. The foreign protein may be part of a bacterium, a virus, or a toxin, or it may be present in food that causes allergy. The body, after recognizing that it has been invaded, manufactures antibodies, which inactivate the foreign substance. Without sufficient protein to make antibodies, the body cannot maintain its resistance to disease.

An optimal diet helps to provide strength and support to the body's immune system.

ANTIBODIES large proteins of the blood and body fluids, produced by one type of immune cell in response to invasion of the body by unfamiliar molecules (mostly foreign proteins). Antibodies inactivate the foreign substances and so protect the body. The foreign substances are called antigens.

IMMUNITY specific disease resistance derived from the immune system's memory of prior exposure to specific disease agents and its ability to mount a swift response against them.

OPPORTUNISTIC INFECTIONS infections produced by organisms that do not affect people whose immune systems are working normally. An example is the unusual form of pneumonia caused by *Pneumocystis carinii*, often seen in individuals with AIDS.

ACQUIRED IMMUNE DEFICIENCY SYNDROME (AIDS) an immune system disorder caused by the human immunodeficiency virus (HIV). Its attack on the individual's immune cells (T-cells) results in a decreased ability to fight foreign organisms, thus increasing the individual's susceptibility to a variety of opportunistic infections. AIDS is transmitted to a person through direct contact of the person's body fluids with contaminated body fluids. It is most often transmitted through sexual intercourse, contaminated needles, contaminated blood products, or from mother to infant during pregnancy or lactation. Since the early 1980s, AIDS has become a major public health problem.

One of the most fascinating aspects of this response is that each antibody is designed specifically to destroy one foreign substance. An antibody that has been manufactured to combat one strain of flu virus would be of no help in protecting a person against another strain. Once the body has learned to make a particular antibody, it never forgets, and the next time it encounters that same foreign substance, it will be equipped to destroy it even more rapidly. In other words, it develops an **immunity.** This is the principle underlying the vaccines and antitoxins that have nearly eradicated most childhood diseases in the Western world.

Clearly, malnutrition injures the immune system. Without adequate protein in the diet, the immune system will not be able to make its specialized cells and other tools to function optimally. Often protein deficiency and immune incompetence appear together. For this reason, measles in a malnourished child can be fatal. Protein deficiency also can put the malnourished person at risk for increased incidence of **opportunistic infections**—as is the case with many of those diagnosed with **acquired immune deficiency syndrome (AIDS).** Such infections are often the cause of death in the AIDS patient. While adequate protein cannot prevent infection with the AIDS virus, nutrition intervention can be important in preventing the weight loss and malnutrition seen in people with AIDS.

Many other nutrients besides proteins participate in conferring immunity, and many factors besides the antibodies are involved.[14] The immune system is extraordinarily sensitive to nutrition, and almost any nutrient deficit can impair its efficiency and reduce resistance to disease.

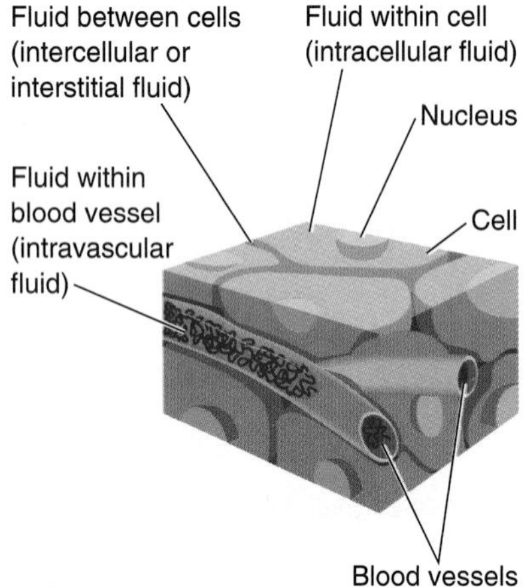

Shown here are the fluids within and surrounding a cell. Body proteins help hold fluid within cells, tissues, and blood vessels.

Fluid Balance

Proteins help regulate the quantity of fluids in the compartments of the body to maintain the **fluid balance.** To remain alive, a cell must contain a constant amount of fluid. Too much might cause it to rupture, and too little would make it unable to function. Although water can diffuse freely into and out of the cell, proteins cannot—and proteins attract water. By maintaining a store of internal proteins, the cell retains the fluid it needs (it also uses minerals this way). Similarly, the cells secrete proteins (and minerals) into the spaces between them to keep the fluid volume constant in those spaces. The proteins secreted into the blood cannot cross the blood vessel walls, and thus they help to maintain the blood volume in the same way.

Acid-Base Balance

Normal processes of the body continually produce **acids** and their opposite, **bases,** which must be carried by the blood to the organs of excretion. The blood

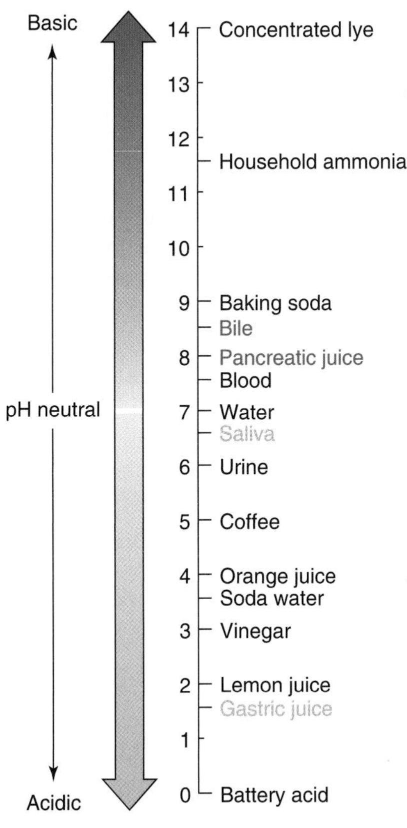

pH Values of Selected Fluids
A fluid's acidity or alkalinity is measured in pH units.

must do this without allowing its own **acid-base balance** to be affected. To accomplish this, some proteins act as **buffers** to maintain the blood's normal **pH.** They pick up hydrogens when there are too many in the blood (the more hydrogen, the more concentrated the acid). Likewise, protein buffers release hydrogens again when there are too few in the blood. The secret is that the negatively charged side chains of the amino acids can accommodate additional hydrogens (which are positively charged) when necessary.

The acid-base balance of the blood is one of the most accurately controlled conditions in the body. If it changes too much, the dangerous condition **acidosis** or the opposite, basic condition **alkalosis** can cause coma or death. The hazards of these conditions are a result of their effect on proteins. When the proteins' buffering capacity is exceeded—for example, when proteins have taken on board or released all the acid hydrogens they can—additional acid or base deranges their structure by pulling them out of shape; that is, it denatures them. Knowing how indispensable the structures of proteins are to their functions and how vital their functions are to life, you can imagine how many body processes would be halted by such a disturbance.

FLUID BALANCE distribution of fluid among body compartments.

ACIDS compounds that release hydrogens in a watery solution; acids have a low pH.

BASES compounds that accept hydrogens from solutions; bases have a high pH.

ACID-BASE BALANCE equilibrium between acid and base concentrations in the body fluids.

BUFFERS compounds that help keep a solution's acidity (amount of acid) or alkalinity (amount of base) constant.

PH the concentration of hydrogen ions. The lower the pH, the stronger the acid; pH 2 is a strong acid; pH 7 is neutral; and a pH above 7 is alkaline.

ACIDOSIS (a-sih-DOSE-sis) blood acidity above normal, indicating excess acid.

ALKALOSIS (al-kah-LOH-sis) blood alkalinity above normal.

Transport Proteins

A specific group of the body's proteins specializes in moving nutrients and other molecules into and out of cells. Some of these act as pumps—picking up compounds on one side of the membrane and depositing them on the other—and thereby decide what substances the cell will take up or release. One such pump is the "sodium-potassium pump," which resides in the cell membrane and acts as a revolving door—picking up potassium from outside the cell and depositing it inside the cell, and picking up sodium from within the cell and depositing it outside the cell as necessary. The protein machinery of cell membranes can be switched on or off in response to the body's needs. Often hormones do the switching with a marvelous precision.

Other transport proteins move about in the body fluids, carrying nutrients and other molecules from one organ to another. Those that carry lipids in the lipoproteins are an example. Special proteins also can carry fat-soluble vitamins, water-soluble vitamins, and minerals. As a result, a protein deficiency can cause a vitamin A deficiency or a deficiency of whatever other nutrient is in need of a transport protein in order to reach its destination in the body.

This sampling of the major roles proteins play in the body should serve to illustrate their versatility, uniqueness, and importance. All the body's tissues and organs—muscles, bones, blood, skin, and nerves—are made largely of proteins. No wonder proteins are said to be the primary material of life.

Protein as Energy

Only protein can perform all the functions previously described, but it will be sacrificed to provide needed energy if insufficient fat and carbohydrate are

UREA (yoo-REE-uh) the principal nitrogen-excretion product of metabolism, generated mostly by the removal of amine groups from unneeded amino acids or from those amino acids being sacrificed to a need for energy.

PROTEIN-SPARING a description of the effect of carbohydrate and fat, which, by being available to yield energy, allow amino acids to be used to build body proteins.

DIPEPTIDES (dye-PEP-tides) protein fragments two amino acids long. A peptide is a strand of amino acids.

TRIPEPTIDES (try-PEP-tides) protein fragments three amino acids long.

eaten. The body's number one need is for energy. All other needs have a lower priority.

When amino acids are degraded for energy, their amine groups are usually incorporated by the liver into **urea** and sent to the kidney for excretion in the urine. The remaining components are carbon, hydrogen, and oxygen, which are available for immediate energy use by the body.

Only if the **protein-sparing** calories from carbohydrate and fat are sufficient to power the cells will the amino acids be used for their most important function—making proteins. Thus, energy deficiency (starvation) is always accompanied by the symptoms of protein deficiency.

If amino acids are oversupplied, the body has no place to store them. It will remove and excrete their amine groups and then convert the fragments that remain to glucose and glycogen, or to fat, for energy storage. Amino acids are not stored in the body except in the sense that they are present in proteins in all the tissues. When there is a great shortage of amino acids, such tissues as the blood, muscle, and skin have to be broken down so that their amino acids can be used to maintain the heart, lungs, and brain.

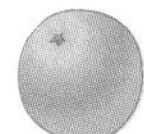

HOW THE BODY HANDLES PROTEIN

When a person eats a food protein, whether from cereals, vegetables, meats, or dairy products, the digestive system breaks the protein down and delivers the separated amino acids to the body cells. The cells then put the amino acids together in the order necessary to produce the particular proteins they need.

To review the digestive and absorptive systems relevant to the body's handling of protein, turn to Appendix B.

The stomach initiates protein digestion (see Figure 5-1). By the time proteins slip into the small intestine, they are already broken into different-sized pieces—some single amino acids and many strands of two, three, or more amino acids—**dipeptides, tripeptides,** and longer chains. Digestion continues until almost all pieces of protein are broken into dipeptides, tripeptides, and more free amino acids. Absorption of amino acids takes place all along the small intestine. As for dipeptides and tripeptides, the cells that line the small intestine capture them on their surfaces, split them into amino acids on the cell surfaces, absorb them, and then release them into the bloodstream.

Once they are circulating in the bloodstream, the amino acids are available to be taken up by any cell of the body. The cells can then make proteins, either for their own use or for secretion into the circulatory system for other uses.

If a *non*essential amino acid (that is, one the body can make for itself) is unavailable for a growing protein strand, the cell will make it and continue attaching amino acids to the strand. If, however, an essential amino acid (one the body cannot make) is missing, the building of the protein will halt. The cell cannot hold partially completed proteins to complete them later, for example, the next day. Rather, it has to dismantle the partial structures and return surplus amino acids to the circulation, making them available to other cells. If other cells do not soon pick up these amino acids and insert them into protein, the liver will remove their amine groups for the kidney to excrete. Other cells will then use the remaining fragments for other purposes. Whatever need prompted the calling for that particular protein will not be met.

Both meals shown here supply an adequate assortment of amino acids needed for health.

PROTEIN QUALITY OF FOODS

The role of protein in food, as already mentioned, is not to provide body proteins directly but to supply the amino acids from which the body can make its own proteins. Since body cells cannot store amino acids for future use, it follows that all the essential amino acids must be eaten as part of a balanced diet. To make body protein, then, a cell must have all the needed amino acids available. Three important characteristics of dietary protein, therefore, are (1) that it should sup-

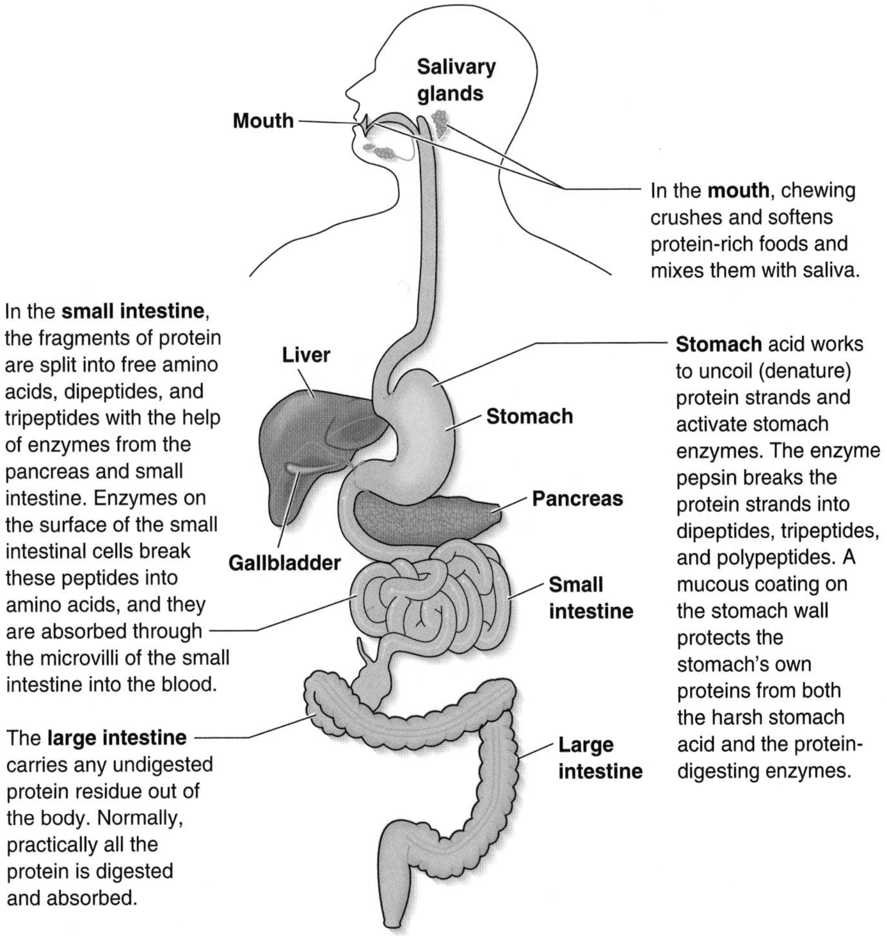

FIGURE 5-1
A SUMMARY OF PROTEIN DIGESTION AND ABSORPTION

Salivary glands
Mouth
In the **mouth**, chewing crushes and softens protein-rich foods and mixes them with saliva.
In the **small intestine**, the fragments of protein are split into free amino acids, dipeptides, and tripeptides with the help of enzymes from the pancreas and small intestine. Enzymes on the surface of the small intestinal cells break these peptides into amino acids, and they are absorbed through the microvilli of the small intestine into the blood.
Liver
Stomach
Stomach acid works to uncoil (denature) protein strands and activate stomach enzymes. The enzyme pepsin breaks the protein strands into dipeptides, tripeptides, and polypeptides. A mucous coating on the stomach wall protects the stomach's own proteins from both the harsh stomach acid and the protein-digesting enzymes.
Pancreas
Gallbladder
Small intestine
The **large intestine** carries any undigested protein residue out of the body. Normally, practically all the protein is digested and absorbed.
Large intestine

ply at least the nine essential amino acids, (2) that it should supply enough other amino acids to make nitrogen available for the synthesis of whatever nonessential amino acids the cell may need to make, and (3) that it should be accompanied by enough food energy (preferably from carbohydrate and fat) to prevent sacrifice of its amino acids for energy. This presents no problem to people who regularly eat **complete proteins,** such as those of meat, fish, poultry, cheese, eggs, milk, or many soybean products, as part of balanced meals.[15] The proteins of these foods contain ample amounts of all the essential amino acids relative to our bodies' need for them, and the rest of the diet provides protein-sparing energy and needed vitamins and minerals. An equally sound choice is to eat two or more **incomplete protein** foods from plants, each of which supplies the **limiting amino acid** in the other—also, of course, as part of a balanced diet. The *quality* of plant proteins (legumes, grains, and vegetables) having different *limiting* amino acids can therefore be improved by combining different sources of plant proteins, either during a meal or over the course of a day, so that sufficient amounts of all the essential amino acids become available for protein synthesis. This strategy—using **complementary proteins**—is shown in Figure 5-2. Note that by combining a grain (whole-wheat bread) that is low in lysine with a legume (peanut butter) that is low in methionine, the limiting amino acid disappears.

A person in good health can be expected to use dietary protein efficiently. However, malnutrition or infection can seriously impair digestion (by reducing enzyme secretion), absorption (by causing degeneration of the absorptive surface of the small intestine or losses from diarrhea), and the cells' use of protein (by forcing amino acids to meet other needs). In addition, infections cause the stepped-up production of antibodies made of protein. Malnutrition or infection can greatly increase protein needs while making it hard to meet them.[16]

COMPLETE PROTEINS proteins containing all the essential amino acids in the right proportion relative to need. The *quality* of a food protein is judged by the proportions of essential amino acids that it contains relative to our needs. Animal proteins are the highest in quality.

INCOMPLETE PROTEIN a protein lacking or low in one or more of the essential amino acids.

LIMITING AMINO ACID a term given to the essential amino acid in shortest supply (relative to the body's need) in a food protein; it therefore *limits* the body's ability to make its own proteins.

COMPLEMENTARY PROTEINS two or more food proteins whose amino acid assortments complement each other in such a way that the essential amino acids limited in or missing from each are supplied by the others.

FIGURE 5-2
HOW TWO PLANT PROTEINS COMBINE TO YIELD A COMPLETE PROTEIN

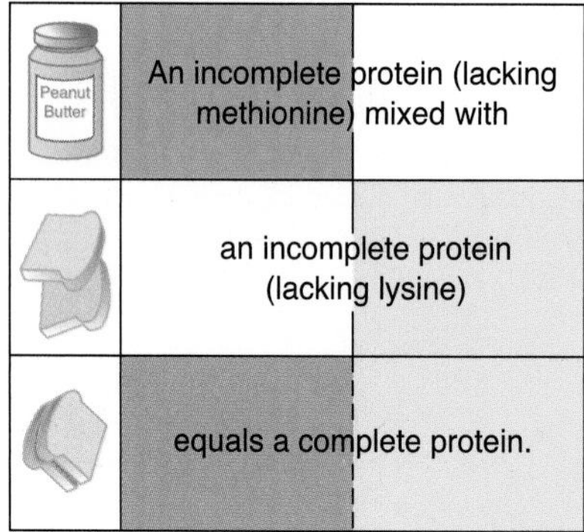

People usually eat many foods containing protein. Each food has its own characteristic amino acid balance, and together, a mixture of foods almost invariably supplies plenty of each individual amino acid. However, when food energy intake is limited, this is not the case (as discussed in the section titled "Protein-Energy Malnutrition," later in the chapter). Also, even if food energy intake is abundant, if the selection of foods available is severely limited (where, for example, a single food such as potatoes or rice provides 90 percent of the calories), protein intake may not be adequate. The primary food source of protein must be checked, for its protein quality is of great importance.

Researchers have studied many different individual foods as protein sources and have developed many different methods of evaluating their **protein quality.** In general, amino acids from animal proteins are the best absorbed (over 90 percent). Those from legumes follow (about 80 percent), and those from grains and other plant foods vary (from 60 percent to 90 percent).

When amino acids are wasted, their amine groups (which contain their nitrogen) cannot be stored. Therefore, the efficiency of a protein can be assessed experimentally by measuring the net loss of nitrogen from the body. The higher the amount of nitrogen retained, the higher the quality of the protein. This is the basis for determination of the **biological value (BV)** of proteins. A high-quality protein by this standard is egg white protein, which has been designated the **reference protein** and given a score of 100. Other proteins are compared with it.* The best guarantee of amino acid adequacy is to eat a variety of food containing protein in the presence of adequate amounts of vitamins, minerals, and energy from carbohydrate and fat.

PROTEIN QUALITY a measure of the essential amino acid content of a protein relative to the essential amino acid needs of the body.

BIOLOGICAL VALUE (BV) a measure of protein quality, assessed by determining how well a given food or food mixture supports nitrogen retention.

REFERENCE PROTEIN egg white protein, the standard with which other proteins are compared to determine protein quality.

RECOMMENDED PROTEIN INTAKES

Recommended protein intakes can be stated in one of two ways—as a percentage of total calories or as an absolute number (grams per day). It is recommended that protein provide about 12 percent of total caloric intake.

The recommended intake for protein is stated in grams per day. The recommended protein allowance for a healthy adult is 0.8 gram per kilogram (or 2.2 pounds) of desirable body weight per day.

The recommendation for protein uses the desirable, not the actual, weight for a given height because the desirable weight is proportional to the *lean* body mass of the average person. Lean body mass determines protein need. If you gain weight, your fat tissue increases in mass, but fat tissue is composed largely of fat and, as mentioned, does not require much protein for maintenance.

The recommendations for protein are based on the assumption that the protein eaten will be a combination of plant and animal proteins, that it will be consumed with adequate calories from carbohydrate and fat, and that other nutrients in the diet will be adequate. These protein recommendations apply only to healthy individuals with no unusual metabolic need for protein.

To calculate the percentage of calories you derive from protein:

1. Use your total calories as the denominator (example: 1,900 cal).
2. Multiply your protein *grams* by 4 cal/g to obtain calories from protein as the numerator (example: 70 g protein × 4 cal/g = 280 cal).
3. Divide to obtain a decimal, multiply by 100, and round off (example: 280/1,900 × 100 = 15% cal from protein).

To figure your recommended protein intake (RDA):

1. Find the desirable weight for a person your height (see inside back cover). Assume this weight is appropriate for you.
2. Change pounds to kilograms (divide pounds by 2.2; one kilogram = 2.2 pounds).
3. Multiply kilograms by 0.8 g/kg.

Example (for a 5′8″ male):

1. Desirable weight: about 150 lb.
2. 150 lb ÷ 2.2 lb = 68 kg (rounded off).
3. 68 kg × 0.8 g/kg = 54 g protein (rounded off).

PROTEIN AND HEALTH

With all the attention that has been paid to the health effects of starch, sugars, fibers, fats, oils, and cholesterol, protein has been slighted. Protein deficiency effects are well-known because together with energy deficiency, they are the world's main form of malnutrition. But the health effects of too much protein—and particularly the effects of proteins of different kinds—are far less well-

*Another method of evaluating the protein quality of foods is the protein digestibility-corrected amino acid score (PDCAAS) which takes into account both the proportion of amino acids that a food provides and the relative digestibility of the protein. The PDCAAS is used in determining the protein values listed on food labels.

known. The following sections discuss protein deficiency, excess protein, and types of protein.

Protein-Energy Malnutrition

Protein deficiency and energy deficiency go hand in hand so often that public health officials have given a nickname to the pair: **protein-energy malnutrition (PEM).** The two diseases and their symptoms overlap all along the spectrum, but the extremes have names of their own. Protein deficiency is **kwashiorkor,** and energy deficiency is **marasmus.**[17]

PROTEIN-ENERGY MALNUTRITION (PEM), also called **PROTEIN-CALORIE MALNUTRITION (PCM)** the world's most widespread malnutrition problem, including both kwashiorkor and marasmus as well as the states in which they overlap.

KWASHIORKOR (kwash-ee-OR-core) a deficiency disease caused by inadequate protein in the presence of adequate food energy.

MARASMUS (ma-RAZ-mus) an energy-deficiency disease; starvation.

Kwashiorkor is the Ghanaian name for "the evil spirit that infects the first child when the second child is born." In countries where kwashiorkor is prevalent, parents customarily give their newly weaned children watery cereal rather than the food eaten by the rest of the family. The child has been receiving the mother's breast milk, which contains high-quality protein designed beautifully to support growth. Suddenly the child receives only a weak drink with scant protein of very low quality. It is not surprising that the just-weaned child sickens when the new baby arrives.

The child who has been banished from its mother's breast faces this threat to life by engaging in as little activity as possible. Apathy is one of the earliest signs of protein deprivation. The body is collecting all its forces to meet the crisis and so cuts down on any expenditure of protein not needed for the heart, lungs, and brain. As the apathy increases, the child doesn't even cry for food. All growth ceases; the child is no larger at age 4 than at age 2. New hair grows without the protein pigment that gives hair its color. The skin also loses its color, and open sores fail to heal. Digestive enzymes are in short supply, the digestive tract lining deteriorates, and absorption fails. The child can't assimilate what little food is eaten. Proteins and hormones that previously kept the fluids correctly distributed among the compartments of the body now are diminished, so that fluid leaks out of the blood (**edema**) and accumulates in the belly and legs. Blood proteins, including hemoglobin, are not synthesized, so the child becomes anemic; this increases the child's weakness and apathy. The kwashiorkor victim often develops a fatty liver, caused by a lack of the protein carriers that transport fat out of the liver. Antibodies to fight off invading bacteria are degraded to provide amino acids for other uses; the child becomes an easy target for any infection. Then **dysentery,** an infection of the digestive tract that causes diarrhea, further depletes the body of nutrients, especially minerals. Measles, which might make a healthy child sick for a week or two, kills the kwashiorkor child within two or three days. If the condition is caught in time, the starving child's life may be saved by careful nutrition therapy.

EDEMA (eh-DEEM-uh) swelling of body tissue caused by leakage of fluid from the blood vessels, seen in (among other conditions) protein deficiency.

DYSENTERY (DISS-en-terry) an infection of the digestive tract that causes diarrhea.

Children with marasmus suffer symptoms similar to those of children with kwashiorkor, since both conditions cause loss of body protein tissue, but there are differences between the two conditions. Kwashiorkor children retain some of their stores of body fat (because they are still consuming calories), accumulate fat in their livers (because they can't make protein to carry it away), and develop edema (from protein deficiency). Marasmic children experience **ketosis** to conserve body protein, whereas kwashiorkor children do not, because they are receiving some carbohydrate.

KETOSIS (kee-TOE-sis) an adaptation of the body to prolonged (several days') fasting or carbohydrate restriction: body fat is converted to ketones, which can be used as fuel for some brain cells. (More about ketosis in Chapter 8.)

A marasmic child looks like a wizened little old person—just skin and bones. The child is often sick because his or her resistance to disease is low. All the muscles are wasted, including the heart muscle, and the heart is weak. Metabolism is so slow that body temperature is subnormal. There is little or no fat under the skin to insulate against cold. The experience of hospital workers with victims of this disease is that the victims' primary need is to be wrapped up and kept warm. Marasmic patients also need love because they often have been deprived of parental attention as well as food.

Unlike the kwashiorkor child, who is fed milk until weaning, the marasmic child may have been neglected from early infancy. The disease occurs most

Kwashiorkor. These children have the characteristic edema and swollen belly often seen with kwashiorkor.

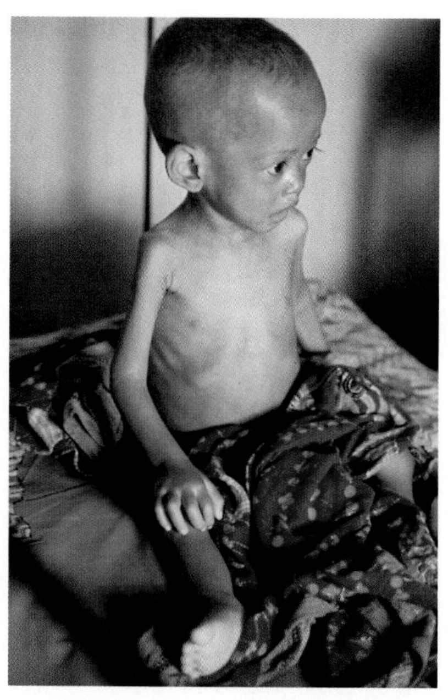

Marasmus. This child is suffering from the extreme emaciation of marasmus.

commonly in children from 6 months to 18 months of age in all the overpopulated city slums of the world. Since the brain normally grows to almost its full adult size within the first two years of life, marasmus impairs brain development and so may have a permanent effect on a child's learning ability.

PEM is prevalent in Africa, Central America, South America, and Asia. Cases have also been reported on American Indian reservations and in the inner cities and impoverished rural areas of the United States.[18] PEM has also been recognized in many undernourished hospital patients, including those with anorexia nervosa, AIDS, cancer, and other wasting conditions. The extent and severity of malnutrition worldwide is a political and economic problem and is discussed further in the Spotlight feature in Chapter 11.

Too Much Protein

Many of the world's people struggle to obtain enough food and enough protein to keep themselves alive, but in the developed countries, where protein is abundant, the problems of protein *excess* can be seen. Animals fed high-protein diets experience a protein overload effect, seen in the enlargement of their livers and kidneys. In human beings, diets high in animal protein necessitate higher intakes of calcium as well, because such diets promote calcium excretion.[19] In persons with chronically low calcium intakes, high protein intakes may increase the risk for osteoporosis. Excess protein may also create an increased demand for vitamin B_6 in the diet, so that the body can utilize the protein.

Animals experimentally fed high-protein diets similar to those typically eaten in the United States experience loss of zinc from their tissues as they age. Increased zinc excretion is also seen in pregnant women and infants on protein supplements. The use of such supplements during pregnancy may do more harm than good, even to undernourished women. In infants, the use of protein supplements has been linked to deficits in cognitive development.

High dietary protein also increases the tendency to obesity, a finding in direct contrast to the popular belief that high-protein diets cause people to "burn off fat." Protein-rich foods are often high in saturated fat, cholesterol, and calories.

The higher a person's intake of such protein-rich foods as meat and milk, the more likely it is that fruits, vegetables, and grains will be crowded out of the diet, making it inadequate in other nutrients.

Although protein is essential to health, the body converts extra protein to energy or to glucose, which gets stored as body fat when energy needs are met. Despite the flood of new protein-packed snack bars and other products in the marketplace, there are evidently no benefits to be gained from consuming excess protein, and the recommended upper limit for protein intake is no more than twice the recommended amount when calorie intake is adequate. Note the qualification "when calorie intake is adequate" in the preceding statement. Remember that your recommended protein intake can be stated as a percentage of calories in the diet or as a specific number of grams of dietary protein. The recommended protein intake for a 150-pound person is roughly 55 grams, or about 12 percent of their daily caloric intake. Fifty-five grams of protein is equal to 220 calories and equals 11 percent of a 2,000 calorie intake—a reasonable calorie intake for a 150-pound active person. If this person was to drastically reduce their caloric intake—to, say, 800 calories a day—then 220 calories from protein is suddenly 28 percent of the total, yet it is still this person's recommended intake for protein, and a reasonable intake. It is the caloric intake that is unreasonable in this example. Similarly, if the person eats too many calories—say 4,000—this protein intake represents only 6 percent of the total caloric intake, yet it is *still* a reasonable intake. It is the caloric intake that may be unreasonable.

Be careful when judging protein intakes as a percentage of calories. Always ask what the absolute number of grams is, too, and compare it with the recommended protein intake in grams. Recommendations stated as a percentage of calories are useful only when food energy intakes are within reason.

Protein in the Diet

Misconceived notions abound regarding protein in the diet; the most obvious of these is that more is better. American women eat about 60 to 65 grams of protein a day, notably higher than the recommended 45 to 50 grams a day; men average about 100 grams a day when young and drop to about 75 grams as older adults, still considerably higher than their recommended intake of 60 to 65 grams. Moreover, about 70 percent of this protein comes from animal and dairy products (see Figure 5-3). Saturated fats supply half or more of the calories in some animal protein foods. You could better balance your food choices by

Foods that supply protein in abundance are shown here in the *Milk, Yogurt, and Cheese Group* and the *Meat, Poultry, Fish, Dry Beans, Eggs, and Nuts Group* of the Food Guide Pyramid (top two photos). Servings of foods from the *Vegetable Group* and the *Bread, Cereal, Rice and Pasta Group* can also contribute protein to the diet (bottom two photos).

FIGURE 5-3
PROTEIN CONTRIBUTED BY THE FOOD GUIDE PYRAMID GROUPS IN THE AVERAGE AMERICAN DIET*

*The average protein consumption in the United States is 67.5 grams of protein per day, of which 37 grams (55%) come from the Meat Group. The Milk and Bread Groups are the next two largest contributors, providing 36% of the total daily protein.

Source: Adapted from *Eating in America Today,* Edition II, © 1994 by the National Live Stock and Meat Board, p. 17.

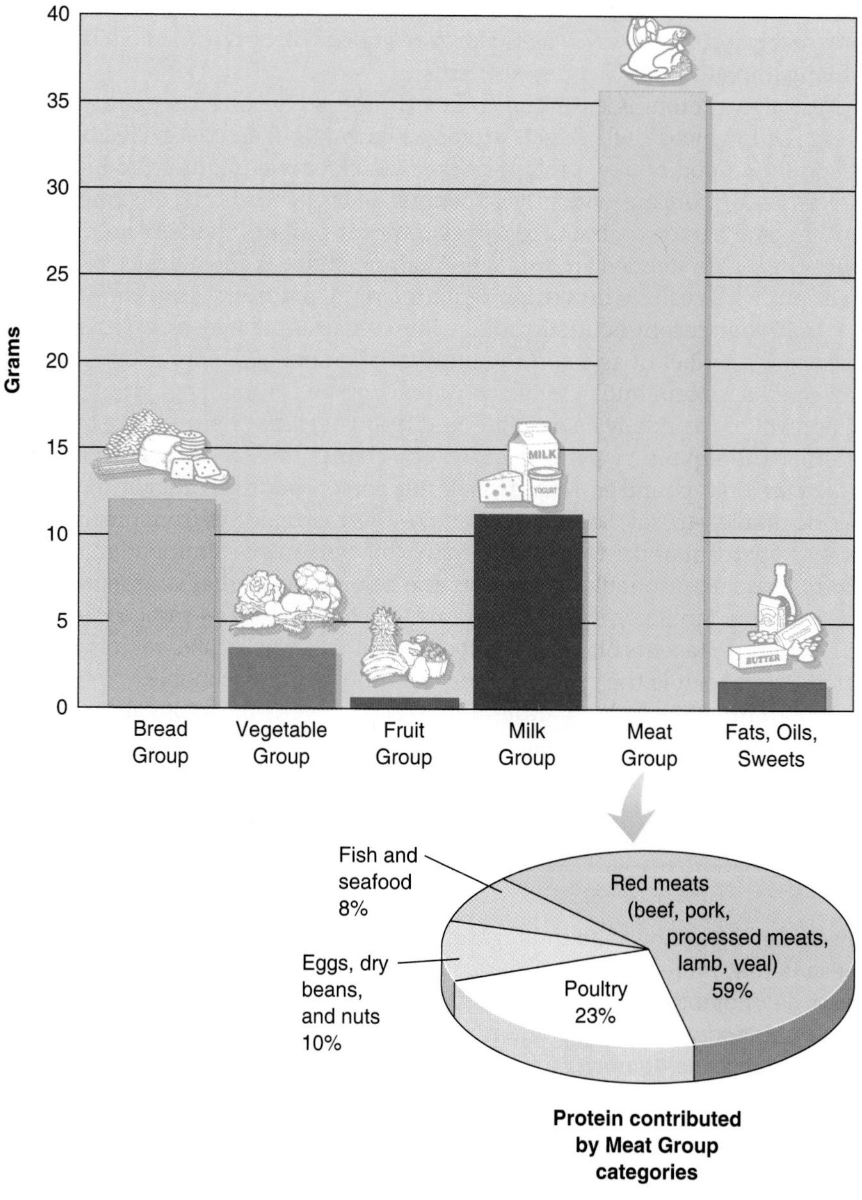

selecting one-third or less of your protein from animal sources and the rest from plants.

Foods that supply protein in abundance can be found in the *Milk, Yogurt, and Cheese Group* and the *Meat, Poultry, Fish, Dry Beans, Eggs, and Nuts Group* of the Food Guide Pyramid. Servings of vegetables and grain products from the *Vegetable Group* and the *Bread, Cereal, Rice, and Pasta Group* can also contribute significant amounts of protein to the diet. As the vegetarian knows, one can easily design a perfectly acceptable diet around plant foods alone by choosing an appropriate variety.

LEGUMES (leg-GYOOMS) plants of the bean and pea family having roots with nodules that contain bacteria that can trap nitrogen from the air in the soil and make it into compounds that become part of the seed. The seeds are rich in high-quality protein compared with those of most other plant foods.

Many interesting sources of protein are available. Try adding **legumes** to your meals. (For a variety of legumes used in cooking, see the accompanying Miniglossary.) This chapter's Spotlight reviews the benefits of soy products, an excellent source of protein, in the diet.

Adequate protein in the diet is easy to obtain. A breakfast of one egg, two slices of wheat bread, and a glass of milk provides close to 20 grams of protein. This meets about a third of an average man's recommended protein intake of 60 grams and about half of a woman's recommendation for protein of 45 grams. The following Scorecard shows you how to estimate your own protein intake.

Legumes include such plants as the soybean, kidney bean, garbanzo bean, black bean, lentil, garden pea, black-eyed pea, and lima bean.

MINIGLOSSARY OF LEGUMES

BLACK, CUBAN, OR TURTLE BEANS These medium-size black-skinned ovals have a rich, sweet taste. They are best served in Mexican and Latin American dishes or thick soups and stews.

BLACK-EYED PEAS These are small and oval shaped, creamy white with a black spot. They have a vegetable flavor with mealy texture. Use in salads with rice and greens.

GARBANZO BEANS OR CHICK-PEAS These legumes are large, round, and tan colored. They have a nutty flavor and crunchy texture. Use in soups and stews and puréed for dips.

GREAT NORTHERN BEANS This variety is medium white and kidney shaped. Enjoy the delicate flavor and firm texture in salads, soups, and main dishes.

KIDNEY BEANS These familiar beans are large, red, and kidney shaped (the white variety is called cannellini). They have a bland taste and soft texture but tough skins. Use in chili, bean stews, and Mexican dishes for red; Italian dishes for white.

LENTILS These legumes are small, flat, and round. Usually brown colored, lentils also can be green, pink, or red. They have a mild taste with firm texture. Best used when combined with grains or vegetables in salads, soups, or stews.

LIMA OR BUTTER BEANS Limas are soft and mealy in texture. They are flat, oval shaped, and white tinged with green. The smaller variety has a milder taste. Use in soups and stews.

PINTO BEANS These medium ovals are mottled beige and brown with an earthy flavor. They are most often used in Mexican dishes, such as refried beans, stews, or dips.

RED BEANS This versatile bean is a medium-size, dark red oval. The taste and texture are similar to kidney beans. Use in soups and stews, and serve with rice.

SOYBEANS You can find these creamy white ovals in numerous food products, such as tofu, flour, grits, and milk. They have a firm texture and bland flavor. See the Spotlight feature later in this chapter for a discussion of the possible health benefits derived from soyfoods.

SPLIT PEAS Green or yellow, these small halved peas supply an earthy flavor with mealy texture. They are best used in soups and with rice or grains.

WHITE NAVY BEANS These beans are small, white ovals and are best used in soups and stews and as baked beans.

*All the legumes in this guide except lentils and split peas require at least 1 hour of soaking. After soaking, rinse and cover with fresh water. Then bring to a boil; simmer for 30 to 60 minutes or until soft. Canned varieties are available, too.

Source: Adapted from K. Mangum, *Life's Simple Pleasures: Fine Vegetarian Cooking for Sharing and Celebration* (Boise, ID: Pacific Press Publishing, 1990), p. 149.

WORKING ON THE WEB

www.phys.com

Sponsor: CondeNet

Description: This Web site provides practical information about nutrition, fitness, weight loss, eating well during pregnancy, snacking, and food safety.

Available Activities: The self-analysis section allows you to analyze your eating patterns, calculate your healthy weight, body fat percentage and nutrient needs, analyze your overall eating habits, and assess your health risks. You can also look up information in an encyclopedia and access articles from the *Tufts University Health and Nutrition Letter* from this site.

Web Work:

1. Calculate the recommended amounts of fat, protein, and carbohydrate for your diet. Keep these daily totals in mind when you compare food labels at the grocery store.
2. Play the Wheel of Portion game to test your knowledge about serving sizes from the Food Guide Pyramid.
3. Visit the Portion Teller to determine what counts as a serving of your favorite foods.
4. Bonus question: What six common objects can be used to help you understand what a serving of food looks like?

Helpful Hints: From the PHYS home page, scroll down and click on Nutrition. Next, click on Self-Analysis followed by PHYS Calculators. Scroll down and click on Fat Need, Protein Need, and Carbohydrate Need to determine your nutrient needs. Next, go back one page to the Self-Analysis home page and click on Wheel of Portion. When you finish the last question, click on What Now? Scroll down to view the recommended serving sizes from the groups of the Food Guide Pyramid. Finally, click on The Portion Teller. Click on your favorite foods from the list provided to learn more about serving sizes.

PROTEIN SCORECARD

ESTIMATE YOUR PROTEIN INTAKE

The average American consumes much more than his or her recommended protein intake. How do you compare? First, figure your recommended protein intake (divide your weight in pounds by 2.2 and then multiply by 0.8). Next, write down everything you ate and drank yesterday. Using the values given below, estimate the grams of protein you ate from both animal and plant sources. If an item is not listed here, use Appendix H to determine the amount of protein it contains. How close are you to your recommended protein intake? What percentage of your protein comes from animal versus plant sources?

Recommended Protein Intake: _________ grams

PROTEIN FOODS	AMOUNT	GRAMS OF PROTEIN IN 1 SERVING	GRAMS OF PROTEIN IN YOUR TYPICAL DIET
Animal Sources			
Hard cheese (e.g., cheddar)	1 oz	7	_______
Cottage cheese	½ c	14	_______
Milk	1 c	9	_______
Yogurt	1 c	12	_______
Egg	1 large	7	_______
Poultry	3 oz	21	_______
Ground beef, lean	3 oz	24	_______
Beef steak, lean	3 oz	26	_______
Pork chop, lean	3 oz	20	_______
Other	_____	_____	_______
		Animal Proteins Subtotal (grams):	_______
Plant Sources			
Vegetables	½ c	2	_______
Legumes, cooked	½ c	8	_______
Tofu	4 oz	9	_______
Cereals	1 c	2–6	_______
Bread	1 slice	2	_______
Tortilla	1	2	_______
Rice	½ c	3	_______
Pasta	½ c	3	_______
Peanut butter	1 tbsp	4	_______
Nuts	2 tbsp	3	_______
Seeds	2 tbsp	3	_______
Other	_____	_____	_______
		Plant Proteins Subtotal (grams):	_______

Day's Total Protein: _________ grams

Reshape Your Protein Choices for Health

The ancient inhabitants of South America liked to eat a kind of paste made from peanuts. But modern peanut butter came into being around 1890 as the bright idea of a St. Louis physician, who thought it would be a good health food for elderly people. It was not linked with jelly until the 1920s.[20]

Many health organizations now recommend a diet that emphasizes vegetables, fruits, legumes, and whole grains in order to protect against cancer, heart disease, stroke, diabetes, and obesity.[21] The key to getting enough, but not too much, protein seems to be to use a variety of plant-based foods in ample quantities and to de-emphasize meats.[22] Consider the following pointers.

At the Grocery Store

- Dairy foods are excellent sources of protein, calcium, and other important nutrients. Look for fat-free and low-fat varieties for the recommended two to three servings each day. Look for low-fat cheeses that have less than 5 grams of fat per ounce, such as part-skim or fat-free ricotta or mozzarella, farmer's cheese, feta cheese, string cheese, or other reduced-calorie cheeses. These choices are lower in saturated fat and calories than their full-fat counterparts.
- Meat, chicken, and fish all provide excellent protein, as well as iron, zinc, and vitamin B_{12}. We are advised to choose low-fat varieties. Look for the leanest meats:

  Flank steak, round steak, sirloin, tenderloin, or extra-lean ground beef

  Lean ham, Canadian bacon, pork tenderloin, and center-loin pork chops

  Chicken, turkey, or game hens without the skin; fresh ground turkey breast or chicken breast meat
- At the deli counter, select items with less than 1 gram of fat per ounce, such as lean ham, turkey or chicken breast, and lean roast beef.
- Fish and shellfish—fresh, frozen, or canned in water—make excellent protein choices and are low in fat. Experts recommend you eat at least two fish meals per week.
- Bring home legumes—they are good sources of protein, fiber, folate and minerals. Choose dried or canned varieties. A ½-cup serving counts as 1 ounce of meat.
- Nuts and nut butters are good sources of protein, but are also high in fat. Look for fresh ground varieties. A 2-tablespoon serving counts as 1 ounce of meat.
- Eggs are another excellent source of protein. Because they are high in cholesterol, we are advised to eat no more than 4 egg yolks a week. Look for egg substitutes and substitute ¼ cup for one whole egg in recipes. One egg counts as 1 ounce of meat.

In the Kitchen

- Small meat portions tend to work best mixed into dishes with lots of vegetables and grains; try stir-fries, pastas, soups and stews, burritos, casseroles, and main-dish salads. For example, cook a large pot of soup, stew, or chili. Minimize the amount of meat you use and load it up with vegetables (fresh or frozen) and cooked beans (kidney, black, cannellini, or other favorite beans).
- Go meatless one or more days each week. Experiment with new recipes for vegetables, grains, or legumes from health-minded cookbooks or magazines. Try one new recipe each week.
- Take a fresh look at your favorite recipes. Try to use less meat and add more vegetables, grains, and legumes. Instead of a chicken and broccoli stir-fry, try stir-fried veggies with a little chicken.
- For quick, colorful, meals rich in nutrients and flavor, try various combinations of stir-fried vegetables on beds of steamed brown rice, whole-grain bulgur, or couscous. You can use the stir-fry sauce below with many of your favorite stir-fry combinations.

Simple Stir-Fry Sauce

1 low-sodium bouillon cube dissolved in ¾ cup water
1 tbsp cornstarch
1 tsp sugar
¼ tsp ground ginger
2 tsp reduced-sodium soy sauce

Mix all ingredients together in a bowl. Add to your stir fry at the end of cooking, stirring for 2 to 3 minutes or until the cornstarch turns translucent.

In the Lunch Box

- Get out of the peanut butter and jelly rut by filling sandwiches with water-packed tuna mixed with mandarin oranges, bean sprouts, and a bit of plain low-fat yogurt; chopped, cooked, skinless chicken combined with raw sliced vegetables and a little French dressing; cooked, mashed dried beans seasoned with chopped onion, garlic powder, rosemary, thyme, and pepper; or low-fat cottage cheese flavored with drained, chopped pineapple and a dash of cinnamon.
- Take a thermos filled with chili, vegetable soup, or a milk-based soup, such as cream of tomato, prepared with nonfat milk instead of a sandwich. Try cold lunches such as low-fat or nonfat yogurt and fruit, brown rice with cubes of skinless poultry or lean meat, or cooked pasta tossed with raw vegetables, low-fat cheese, and a bit of Italian dressing.

At the Table

- When dining out, choose an ethnic restaurant with plant-based entrées on the menu. Consider Spanish paella, Asian stir-fries, Moroccan stew, Italian pasta primavera, Indian curries, or French ratatouille as your entrée. Or try Chinese, Vietnamese, or Thai take-out with lots of rice and vegetables.
- Keep portion size in mind. The Food Guide Pyramid recommends 5 to 7 ounces a day from the meat, poultry, fish, dry beans, eggs, and nuts group. A cooked 3-ounce serving of meat, fish, or poultry is about the size of a deck of cards.
- Choose entrées that are steamed, poached, broiled, roasted or baked, rather than fried.
- Make whole grains, vegetables, and legumes the main event of your meals. At least two-thirds of your meal should come from these plant-based foods and one-third or less from lean meat, poultry, fish, or low-fat dairy products.

THE VEGETARIAN DIET

More and more people are following vegetarian diets. Their reasons for becoming vegetarian vary widely.[23] Some have health reasons while others have religious or ethical reasons. Some believe that vegetarianism is ecologically sound, and others that it is less costly than the meat-eating alternative. In addition to the traditional types of vegetarians (see Table 5-3), there are people who eat seafood but not other meats, and those who include chicken and other poultry but not red meat. Whatever the particular reasons for choosing a vegetarian diet, the vegetarian needs to be aware of the nutrition and health implications of it.[24]

Important goals for any diet planner include the following:

- To obtain neither too few nor too many calories—that is, to maintain a healthful weight.
- To obtain adequate quantities of complete protein.
- To obtain the needed vitamins and minerals.

The vegetarian can use the special *Vegetarian Food Pyramid* (see Figure 5-4) to balance his or her diet.

Well-planned, plant-based meals consisting of a variety of whole grains, legumes, nuts, vegetables, fruits, and for some vegetarians, eggs and dairy products, can offer sound nutrition and health benefits to vegetarians and nonvegetarians alike.

Proteins

The vegetarian needs adequate amounts of all the essential amino acids. Proteins from animals contain ample amounts of the essential amino acids, so the lacto-ovo vegetarian can get a head start on meeting protein needs by drinking two cups of milk daily or by consuming the equivalent in milk products in the day's diet.

Adequate amounts of amino acids can be obtained from a plant-based diet when a varied diet is routinely consumed on a daily basis. Mixtures of proteins from unrefined grains, vegetables, legumes, seeds, and nuts eaten over the course of a day complement one another in their amino acid profiles so that deficits in

TABLE 5-3 TYPES OF VEGETARIANS

Semi-Vegetarian	Some but not all groups of animal-derived products, such as meat, poultry, fish, seafood, eggs, milk, and milk products, included in this diet.
Lactovegetarian	Milk and milk products included in this diet, but meat, poultry, fish, seafood, and eggs excluded. *possible limiting nutrient: iron*
Lacto-Ovovegetarian	Milk and milk products and eggs included in this diet, but meat, poultry, fish, and seafood excluded. *possible limiting nutrient: iron*
Ovovegetarian	Eggs included in this diet, but milk and milk products, meat, poultry, fish, and seafood excluded. *possible limiting nutrients: iron, vitamin D, calcium, riboflavin*
Strict-Vegetarian/Vegan	All animal-derived foods, including meat, poultry, fish, seafood, eggs, milk, and milk products excluded from this diet. *possible limiting nutrients: iron, vitamin D, calcium, riboflavin, vitamin B_{12}, high-quality protein*

FIGURE 5-4
THE VEGETARIAN FOOD PYRAMID

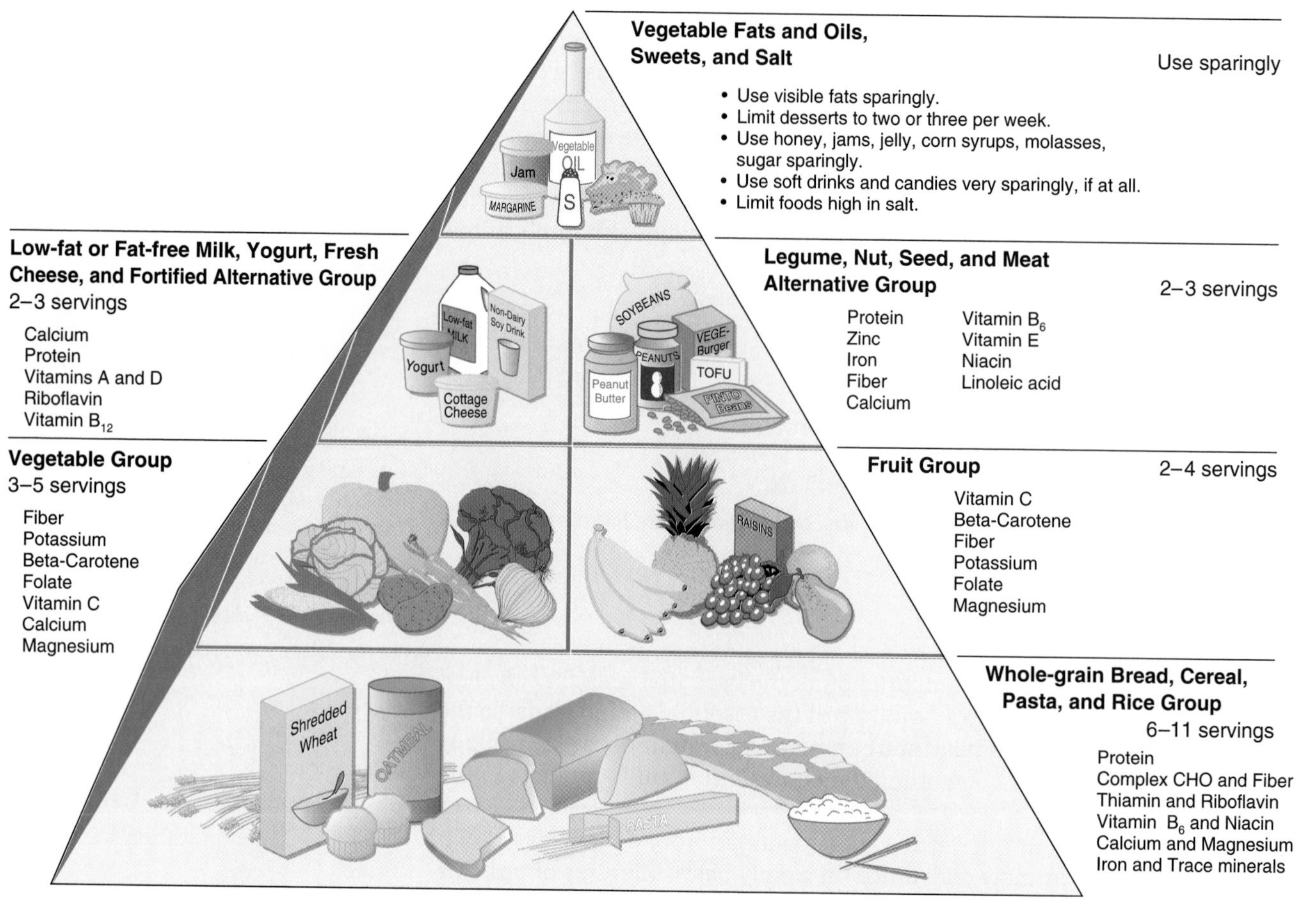

Source: Adapted from Seventh-Day Adventist Dietetic Association, 2100 Douglas Blvd., Roseville, CA 95661, the Health Connection (800-548-8700). Used by permission.

one are made up by another.[25] Table 5-4 gives examples of how such mixtures of foods can be combined to form complete proteins.

Vitamins

The lacto-ovo vegetarian diet can be adequate in all vitamins, but several vitamins may be a problem for the vegan. One such vitamin is vitamin B_{12}, which doesn't occur naturally in plant foods but is available in fortified foods, such as breakfast cereals or **nutritional yeast** grown in a vitamin B_{12}-enriched environment. The vegan needs a reliable B_{12} source, such as vitamin B_{12}-fortified soy milk, breakfast cereals, or **meat replacements.** Some vegetarians use seaweeds, fermented soy, and other products in the belief that they provide vitamin B_{12} in adequate amounts, but these products are not currently recommended as reliable sources. A pregnant or lactating woman who is eating a vegan diet should be aware that her infant can develop a vitamin B_{12} deficiency that can damage the baby's nervous system, even if the mother remains healthy. Since large amounts of vitamin B_{12} are stored in the body, it may take years for a deficiency to develop. Vegan diets are not generally recommended for infants and young children.

NUTRITIONAL YEAST a fortified food supplement containing B vitamins, iron, and protein that can be used to improve the quality of a vegetarian diet.

MEAT REPLACEMENTS textured vegetable protein products formulated to look and taste like meat, fish, or poultry. Many of these are designed to match the known nutrient contents of animal protein foods.

Another vitamin of concern is vitamin D.[26] The milk drinker is protected, provided the milk is fortified with vitamin D, but there is no practical source of vitamin D in plant foods. Fortified margarines, soy milk, and breakfast cereals can

TABLE 5-4 COMPLEMENTARY PROTEIN COMBINATIONS THAT PROVIDE HIGH-QUALITY PROTEIN

Combine			Examples
Cereal grains	+	Legumes	
Barley		Dried beans	Bean taco
Bulgur		Dried lentils	Chili and cornbread
Oats		Dried peas	Lentils or beans and rice
Rice		Peanuts	Peanut butter sandwich
Whole-grain breads			
Pasta			
Cornmeal			
Legumes (or Grains)	+	Seeds and nuts	
Dried beans		Sesame seeds	Hummus (chick-pea and sesame paste)
Dried lentils		Sunflower seeds	Split pea soup and sesame crackers
Dried peas		Walnuts	Noodles with sesame seeds
Peanuts		Cashews	
		Nut butters	

Examples

Hummus and bread

Corn and black-eyed peas

Peanut butter and wheat bread

Bean burrito

supply some vitamin D. Regular exposure to the sun can help prevent a deficiency too.

Riboflavin, another B vitamin obtained from milk, is present in the diet of the vegan who eats ample servings of dark greens, whole and enriched grains, mushrooms, legumes, nuts, and seeds. The vegan who doesn't eat these foods, however, may not meet riboflavin needs.

Minerals

Iron and zinc need special attention in the diets of all vegetarians.[27] Whole-grain products, soyfoods, other legumes, dried fruit, nuts, and seeds are important sources of iron in the vegetarian diet. The iron in these foods, however, is not as easily absorbed by the body as that in meat. Because the vitamin C in fruits and vegetables can triple iron absorption from other foods eaten at the same meal, vegetarian meals should be rich in foods offering vitamin C.

Zinc is widespread in plant foods, but its availability may be hindered by the fibers and other binders found in fruits, vegetables, and whole grains. Vegetarians are advised to eat varied diets that include wheat germ, legumes, nuts, seeds, and whole-grain products. Milk, yogurt, and cheese provide zinc to the lactovegetarian as well.

Special efforts are necessary to meet the calcium needs of the vegan.[28] Whereas the milk-drinking vegetarian is protected from calcium deficiency, the vegan must find other sources of calcium. Some good sources of calcium are *regular* servings of calcium-fortified breakfast cereals, flours, and juices; legumes; firm-style tofu; other soyfoods, including calcium-fortified soy milk; dried figs, some nuts, such as almonds; certain seeds, such as sesame seeds; and some vegetables, such as broccoli, collard greens, kale, mustard greens, turnip greens, okra, rutabaga, and Chinese cabbage (bok choy). The choices should be varied, because the absorption of calcium from some of these foods is hindered by binders in them. The strict vegetarian is urged to use *calcium-fortified* soy milk in *ample quantities, regularly.*[29]

Health Benefits

Vegetarian protein foods are higher in fiber, richer in certain vitamins and minerals, and lower in fat as compared to meats.[30] Vegetarians can enjoy a nutritious diet very low in fat, provided that they eat in moderation high-fat foods such as margarine, oil, cheese, sour cream, and nuts. Table 5-5 offers tips for nutritious, easy-to-fix vegetarian meals and snacks.

Studies have found that people with vegetarian or near-vegetarian traditions, such as the Seventh-Day Adventists and the Chinese, have lower rates of heart disease, cancer, and obesity than those consuming the typical North American diet.[31] Informed vegetarians are more likely to be at the desired weights for their heights and to have lower blood cholesterol levels, lower blood pressure, lower rates of certain types of cancer, better digestive function, and better health in other ways.[32] Even compared with people who are health conscious, vegetarians experience fewer deaths from cardiovascular disease. Often vegetarianism goes with a healthful lifestyle (no smoking, abstinence from alcohol, emphasis on supportive family life, and so forth), so it is unlikely that dietary practices *alone* account for all the aspects of improved health. However, they may contribute to it.

TABLE 5-5 EASY-TO-PREPARE VEGETARIAN MEALS AND SNACKS

Breakfast

- Cold cereal (preferably iron-enriched, as noted on the label); eat with fat-free or low-fat milk, yogurt, or soy milk.
- Hot cereals: add fresh fruit slices and yogurt and sprinkle with cinnamon.
- Toast, bagels: top with low-fat cheese, low-fat cottage cheese, or 1 to 2 tbsp of peanut butter.

Snacks

- Assorted fresh fruits and vegetables with yogurt dip.
- Low-fat cheese or peanut butter on rice cakes or crackers.
- Fat-free or low-fat yogurt.
- English muffin pizza with part-skim mozzarella cheese.
- Hummus with pita bread wedges and crisp vegetables.*

Lunch and Dinner

- Salads: add tofu, chick-peas, three-bean salad, kidney beans, low-fat cottage cheese, sunflower seeds, and hard-cooked egg.
- Salad dressings: add salad seasonings to plain yogurt or blenderized tofu.
- Pasta: add diced tofu and/or canned kidney beans to tomato sauce; top with grated part-skim mozzarella.
- Baked potato: top it with canned beans, steamed vegetables, or low-fat cheese.
- Hearty soups: enjoy lentil, split pea, bean, and minestrone soups, either homemade or canned.
- Vegetarian pizza: top with nonfat or low-fat cheeses and lots of vegetables.

*To make hummus: Blend 1 cup cooked chick-peas, 2 tablespoons tahini (sesame seed paste), 2 tablespoons lemon juice, 1 minced clove garlic, and 1/4 cup chopped fresh parsley in food processor until smooth. Chill and serve with pita bread wedges, crackers, or crisp vegetables. (Makes 2 servings.)

Spotlight
THE BENEFITS OF SOY

Soybeans have a long rich history in the Eastern World cuisine. The early Chinese recognized the importance of this food. They called it *Ta Tau* which means "greater bean." According to Chinese tradition, soybeans were named as one of the five most sacred crops by the emperor who reigned 5000 years ago.

Not too long ago in the Western world, soybeans were fodder for livestock. During the 1970s, many people turned to a vegetarian style of eating, often as a form of protest, but more frequently as a way to adopt a healthier lifestyle. Soy protein became the meat substitute of choice. Twenty-five years later, soybeans are the ideal functional food, a food that has the potential to reduce the risk of disease.

Soyfoods are currently a hot area of research. Recent findings show that substances such as phytoestrogens and isoflavones, found in soybeans, can lower cholesterol and help prevent disease. Numerous studies attest to the role soyfoods may play in reducing risk for certain forms of cancer, heart disease, and osteoporosis and in controlling diabetes and easing a woman's transition through menopause. Is it any wonder that over 5000 years ago the soybean was called the "greater bean!"

What is soy?

Soybeans are legumes, members of the same plant family that includes other beans, peas, and lentils. Among edible legumes, however, the soybean is somewhat unusual. Compared to beans, peas, and lentils, soybeans are relatively low in carbohydrates. They are, however, high in fiber.

Among plant foods, legumes are high in protein. What distinguishes the soybean from its cousins, however, is the nature of the protein: soybeans supply all of the essential amino acids needed for health. The amino acid pattern of soy protein is essentially equivalent in quality to that of meat, milk, and egg protein. Soybeans are the only vegetable food that contains complete protein. What emerges from this nutritional analysis of the soybean is the image of "balance." Soybeans are a food that basically can stand alone, give or take a few vitamins and minerals. Add some vegetables to the beans, and a high-quality, nutrient-dense meal is created!

You have mentioned isoflavones, what are they?

Isoflavones are a type of phytoestrogen, compounds that have a weak, estrogenic activity. There are many types of phytoestrogenic compounds available in edible plants. Foods made from soybeans have varying amounts of the isoflavones, depending on how they are processed (see Table 5-6). Foods such as tofu, soy milk, soy flour, and soy nuts have higher isoflavone concentrations than foods made with a combination of soy and grains. Soy sauce and soybean oil have virtually no isoflavones.[33]

Research in several areas of health care has shown that consumption of soyfoods may play a role in lowering risk for disease. Soy isoflavones are being studied intensively to clarify the physiological effects they exert. In some cases, the research has shown that the isoflavones may be one of the key factors in soybeans that have disease-fighting potential.

It is important to keep in mind that our knowledge of the long-term effects of isoflavones is based on their content in soyfoods. These foods have been consumed for

MINIGLOSSARY

WHAT FOODS CONTAIN SOY?

GREEN SOYBEANS (Edamame) These large soybeans are harvested when the beans are still green and sweet and can be served as a snack or a main vegetable dish, after boiling in water for 15 to 20 minutes. They are high in protein and fiber and contain no cholesterol. Edamame is more often found in Asian and natural food stores, shelled, or still in the pod.[34]

HYDROLYZED VEGETABLE PROTEIN (HVP) Hydrolyzed vegetable protein (HVP) is a protein obtained from any vegetable, including soybeans. The protein is broken down into amino acids by a chemical process called acid hydrolysis. HVP is a flavor enhancer that can be used in soups, broths, sauces, gravies, flavoring and spice blends, canned and frozen vegetables, and meats and poultry.

MEAT ALTERNATIVES Meat alternatives made from soybeans contain soy protein or tofu and other ingredients mixed together to simulate various kinds of meat. These meat alternatives are sold as frozen, canned, or dried foods.

MISO Miso is a rich, salty condiment that characterizes the essence of Japanese cooking. The Japanese make miso soup and use miso to flavor a variety of foods. A smooth paste, miso is made from soybeans and a grain such as rice, plus salt and a mold culture, and then aged in cedar vats for one to three years. Miso should be refrigerated. Use miso to flavor soups, sauces, dressings, marinades, and pâtés.

NONDAIRY SOY FROZEN DESSERT Nondairy frozen desserts are made from soy milk or soy yogurt. Soy ice cream is one of the most popular desserts made from soybeans and can be found in many grocery stores.

SOY CHEESE Soy cheese is made from soy milk. Its creamy texture makes it an easy substitute for sour cream or cream cheese. It can be found in a variety of flavors in natural food stores.

SOY FLOUR Soy flour is made from roasted soybeans ground into a fine powder. Soy flour gives a protein boost to recipes. Soy flour is gluten-free so yeast-raised breads made with soy flour are more dense in texture. Replace one-quarter to one-third wheat flour with soy flour in recipes for muffins, cakes, cookies, pancakes, and quick breads.

SOY PROTEIN, TEXTURIZED Texturized soy protein usually refers to products made from texturized soy flour. Texturized soy flour is made by running defatted soy flour through an extrusion cooker, which allows for many different forms and sizes. When hydrated, it has a chewy texture. It is widely used as a meat extender.

SOY YOGURT Soy yogurt is made from soy milk. Its creamy texture makes it an easy substitute for sour cream or cream cheese. Soy yogurt can be found in a variety of flavors in natural food stores.

SOYBEANS As soybeans mature in the pod they ripen into a hard, dry bean. Most soybeans are yellow. However, there are also brown and black varieties. Whole soybeans can be cooked and used in sauces, stews, and soups.

SOY MILK, SOY BEVERAGES Soybeans that are soaked, ground fine, and strained, produce a fluid called soybean milk. Soy milk is an excellent source of high quality protein and B vitamins. Look for calcium-fortified varieties.

SOY-NUT BUTTER Made from roasted, whole soy nuts, which are then crushed and blended with soy oil and other ingredients, soy-nut butter has a slightly nutty taste, significantly less fat than peanut butter, and provides many other nutritional benefits.

SOY NUTS Roasted soy nuts are whole soybeans that have been soaked in water and then baked until browned. Soy nuts can be found in a variety of flavors, including chocolate-covered. High in protein and isoflavones, soy nuts are similar in texture and flavor to peanuts. Try sprinkling some on salads.

TEMPEH Tempeh, a traditional Indonesian food, is a chunky, tender soybean cake. Whole soybeans, sometimes mixed with another grain such as rice or millet, are fermented into a rich cake of soybeans with a smoky nutty flavor. Tempeh can be marinated and grilled and added to soups, casseroles, or chili.

TOFU AND TOFU PRODUCTS Tofu, also known as soybean curd, is a soft cheeselike food made by curdling fresh hot soy milk with a coagulant. Tofu is a bland product that easily absorbs the flavors of other ingredients with which it is cooked. Tofu is rich in high-quality protein and B vitamins and is low in sodium. Firm tofu is dense and solid and can be cubed and served in soups, stir fried, or grilled. Firm tofu is higher in protein, fat, and calcium than other forms of tofu. Silken tofu is a creamy product and can be used as a replacement for sour cream in many dip recipes.

hundreds of years, and are known to be safe. It is still best to obtain isoflavones by enjoying a variety of soyfoods.

The new information about soy seems promising. What are the potential health benefits of adding soyfoods to my diet?

SOY AND HEART DISEASE

High blood cholesterol is a major risk factor for heart disease.[35] There is a great deal of evidence that soy protein helps lower blood cholesterol levels.[36] Replacing animal protein with soy protein in the diet lowers total and LDL cholesterol levels in people with high cholesterol.[37] A meta-analysis of 38 research studies concluded that soy protein lowers total and LDL cholesterol and triglycerides, without lowering HDL cholesterol in people with high cholesterol.[38] In these studies, the average consumption of soy protein was 47 grams per day. The greatest decreases in blood cholesterol were seen in those with the highest starting levels. Even adding soy protein to an omnivorous diet has been shown to produce this effect.[39] The newly approved health claim for food labels states that as little as 25 grams of soy protein per day may be enough to lower cholesterol levels. Refer to Table 5-6 for the protein content of selected soyfoods.

SOY AND OSTEOPOROSIS

Soybeans and soyfoods may help prevent and treat osteoporosis, a disease that weakens bones and often results in bone fractures. As women age, it becomes more important than ever to maintain adequate levels of calcium in bones. Soyfoods such as fortified soy milk, texturized soy protein, and tofu made with calcium salt are all good calcium sources.

Isoflavones found in soy protein may also play an important role in protecting bones.[40] A breakthrough study at the University of Illinois at

Urbana concluded that consuming soybean isoflavones can increase bone mineral content and bone density. As little as 40 grams of soy protein, consumed each day for six months, led to positive results in a test group of postmenopausal women.[41] Forty grams of soy protein can be found in 2 ounces of soy protein isolate (see Table 5-6).

TABLE 5-6 PROTEIN AND ISOFLAVONE CONTENT OF SELECTED SOYFOODS

	Serving Size	Protein* (g)	Isoflavones† (mg)
Green soy beans, edamame	½ c	11	50
Soy beans, roasted	½ c	30	110
Miso	2 tbsp	4	15
Soy milk	1 c	7	24
Soy flour, roasted	¼ c	7	42
Tempeh	½ c	16	36
Tofu	½ c	6	24
Texturized soy protein (TVP), dry	¼ c	6	29
Soy burger (check label)	3 oz	15–17	38–55
Soy protein isolate, dry	1 oz	23	28
Soy protein concentrate (alcohol extracted)	1 oz	17	4‡

*Soy protein data (rounded to whole numbers) from USDA nutrient Database for Standard Reference, Release 12.

†Isoflavone data (rounded to whole numbers) from USDA–Iowa State University database on the Isoflavone Content of Foods, 1999.

‡In order to isolate soy protein from defatted soybeans, the carbohydrates must be removed by using a solvent, which can remove some or all of the isoflavone content. Since isoflavones are soluble in alcohol, much of the isoflavone content is lost if alcohol or repeated water washings are used in the extraction process. Soy milk, soy flour, tofu, tempeh, and soy protein isolate are not prepared with alcohol or repeated water extraction and therefore have a higher isoflavone content than soy protein concentrate.

SOY AND MENOPAUSE

The hormonal changes that occur during menopause can cause a variety of symptoms and increase risk for heart disease and osteoporosis.[42] Soyfoods that contain phytoestrogens are being studied for their possible efficacy in decreasing the negative effects of menopause. Fluctuating levels of estrogen can cause hot flashes, night sweats, insomnia, vaginal dryness, or headaches. Hormone replacement therapy (HRT) is commonly prescribed to help prevent the negative health effects of menopause. However, many women do not want to take HRT because of the possible increased risk for breast cancer. Scientists are now investigating the question: Can soyfoods provide the same kinds of health benefits as HRT, without the risks?

In women who are producing little estrogen, phytoestrogens may produce enough estrogenic activity to relieve symptoms such as hot flashes. A recent study found that women who were fed 45 grams of soy flour per day had a 40 percent reduction in the incidence of hot flashes.[43] From an epidemiological point of view, it is interesting to note that in Japan, where soy consumption is very high, menopause symptoms of any kind are rarely reported. In addition, bones tend to be stronger in Asia, and broken hips and spinal fractures are less common.

Soy contains phytoestrogens in the form of isoflavones, genistein, and daidzein. These are known to have weak estrogenic effects when consumed by animals and humans.[44] Researchers continue to study the physiological effects of the isoflavones to find out whether they can serve some of the same functions as estrogen, and thereby decrease the health risks associated with menopause.

SOY AND CANCER

One out of every four deaths in the United States is due to cancer. Epidemiological studies show that populations that consume a typical Asian diet have lower incidences of breast, prostate, and colon cancers than those consuming a Western diet.[45] The Asian diet includes mostly plant foods, including legumes, fruits, and vegetables, and is low in fat. The Japanese have the highest consumption of soyfoods. Japan has a very low incidence of hormone-dependent cancers. The mortality rate from breast and prostate cancers in Japan is about one-fourth that of the United States.[46] There is evidence that suggests that the difference in cancer rates is not due to genetics, but rather to diet. Migration studies have shown that when Asians move to the United States and adopt a Western diet, they ultimately have the same cancer incidence as Americans.[47] Other long-term studies have noted an inverse association between regular consumption of miso soup and breast cancer risk in premenopausal women.[48] In Hawaii, a long-term study of 8,000 men of Japanese ancestry showed that men who ate tofu daily were only one-third as likely to get prostate cancer as those who ate tofu once a week or less.[49]

Soybeans contain five classes of compounds, which have been identified as anticarcinogens.[50] Most of these compounds can be found in many different plant foods, but soy is the only significant dietary source of isoflavones. Soy isoflavones, especially genistein, have been the subject of a tremendous amount of cancer research.

SOY AND DIABETES

Another interesting benefit of soyfoods is its effect on glucose control. Recently, scientists have become interested in the role of soyfoods in regulating diabetes. Because soybeans are a complex carbohydrate and also have a low glycemic index, they are an ideal food to help in regulating blood glucose levels in people with diabetes.

How much soy should I be eating on a daily basis to receive these health benefits?

Science has not yet established a recommended daily amount of soy to be consumed to achieve all of the health benefits mentioned here. In many cases, just one serving of soy per day may help improve your health. Although the new health claim for food labels states that 25 grams of soy protein each day may lower risk of heart disease, no recommendation for daily isoflavone intake has yet been made. In Asian countries, where people typically consume 25 to 40 milligrams of isoflavones a day, the incidence of osteoporosis, heart disease, and certain cancers is low. Still, many questions remain about how soy acts in the body and how much is needed for benefits at various stages of the lifecycle. A balanced diet that includes soy is recommended, but loading up on one food, nutrient, or phytochemical is not advisable. Here are some quick tips for adding some soy to your diet.

QUICK AND EASY SOY RECIPES

STRAWBERRY-BANANA FROSTY

3 cups plain or vanilla soy milk
1 ripe banana
1 cup strawberries

Blend in blender until smooth.

Makes 6 servings.

MULTI-GRAIN APPLE PANCAKES

¼ cup yellow corn meal
½ cup rolled oats
½ cup unbleached flour
¼ cup soy flour
1 tsp cinnamon
1 tbsp baking powder
1½ cups plain soy milk
½ cup applesauce

In a large bowl, combine the rolled oats, corn meal, unbleached flour, soy flour, cinnamon, and baking powder. Add the soy milk, and blend with a few swift strokes. Fold in the applesauce. Pour ¼ cup of the batter on a hot nonstick griddle or pan. Cook for about 2 minutes or until bubbles appear on the surface. Flip the pancake and cook for another minute or until heated through. Serve the pancakes with maple syrup, fruit spread, or applesauce.

Makes 12 pancakes.

SOY NUT TRAIL MIX

2 cups roasted soy nuts
1 cup oat-ring cereal
2 cups mini-wheat squares cereal
1 cup raisins
1 cup dried cranberries
½ cup dried cherries

Mix all ingredients in large bowl or container. Keep tightly closed in container or zippered plastic bag.

Yield: 7 cups

Source: "Soyfoods Cookbook" www.soyfoods.com/recipes, © 1999 Indiana Soybean Board. Reprinted by permission.

SOY MILK IN THE MORNING

- Pour soy milk over breakfast cereal once or twice a week.
- Cook oatmeal or cream of wheat in soy milk. You might try vanilla flavored soy milk.
- Make pancakes or French toast with vanilla soy milk.
- Use soy milk to make hot chocolate.

STIR-FRIED TOFU

- Add firm tofu chunks in place of meat or poultry in stir-fries, fajitas, or shish-ka-bobs.

MEATLESS HAMBURGERS

- Try a "garden" burger or soy burger. Check food labels; not all are made with soy.
- Replace ground beef on nachos, pizzas, or in spaghetti sauce with crumbled soy burgers.

DELIGHTFUL TOFU DIPS

- Add tofu to the blender with your favorite seasoning packet, such as "ranch," onion soup mix, or taco seasoning, and serve with tortilla chips or fresh vegetables.

QUICK BREAD WITH SOY FLOUR

- Add soy flour to quick bread recipes. It adds moisture and a soy protein boost. Just replace one-fourth of the total flour with soy flour in recipes for quick breads, muffins, or cakes.

SOYNUT BUTTER

- Enjoy on bagels, breads, English muffins, or carrot and celery sticks.

Nutrition on the Web

nutrition.wadsworth.com	Go to the *Personal Nutrition* site to check for the latest updates to chapter topics or to access links to related Web sites.
www.eatright.org	Search for information about protein in foods; view the ADA Position paper on Vegetarian Diets.
www.fda.gov	Go to Foods and search for information on vegetarian diets.
www.nemsn.org	The National EMS Network provides information on eosinophilia-myalgia syndrome (EMS), tryptophan and melatonin.
www.who.org	Search this site for more information about protein-energy malnutrition worldwide.
www.nal.usda.gov/fnic/Fpyr/pyramid.html	View the Food Guide Pyramid for Vegetarians and a variety of pyramids from other countries at this site. Provides helpful links to other sites.
ww.vrg.org	The Vegetarian Resource Group provides information on vegetarianism, vegetarian books and recipes, links to related sites, and hosts a fun Vegetarian Quiz to test your knowledge.
www.navigator.tufts.edu	Use this search engine to locate credible information about vegetarian diets.
www.ncbi.nlm.nih.gov/PubMed	A search engine to help you locate information from current scientific articles on any topic related to protein.
www.soyfoods.com	This U.S. Soyfoods Directory Web site is an essential resource for anyone interested in learning more about soyfoods. The site includes a searchable database, recipes, research information about the health benefits of soyfoods, and a free monthly email newsletter, plus useful links to other related sites.
talksoy.com/info.htm	The United Soybean Board provides information about soyfoods and answers questions about soyfoods.

Check Yourself . . .

1. Name the element that appears exclusively in protein.
2. What will happen to protein synthesis if an essential amino acid is not in the diet?
3. On what does the quality of protein depend?
4. What is the recommendation regarding the percentage of calories in the diet that should come from protein?
5. Identify three roles of protein in the body.
6. What are the risks associated with using amino acid supplements?
7. Describe the concept of complementary proteins.
8. Can a vegetarian diet meet protein needs?
9. Which nutrients may be lacking in the strict vegetarian diet?
10. What are the possible health benefits derived from soyfoods in the diet?

Answers to Check Yourself questions are found in Appendix G.

The Vitamins

CONTENTS

Turning Back the Clock

Water-Soluble Vitamins

Fat-Soluble Vitamins

The Savvy Diner: Vitamin Preservation

Five a Day Plus Scorecard

Nutrition Action: Medicinal Herbs

Nonnutrients in Foods: The Phytochemical Superstars

Spotlight: Functional Foods—Let Food Be Your Medicine

In France old Crainquebille sold leeks from a cart, leeks called "the asparagus of the poor." Now asparagus sells for the asking, almost, in California markets, and broccoli, that strong age-old green, leaps from its lowly pot to the Ritz's copper saucepan.

Who determines, and for what strange reasons, the social status of a vegetable?

M. F. K. Fisher
(1908–1992, U.S. food writer)

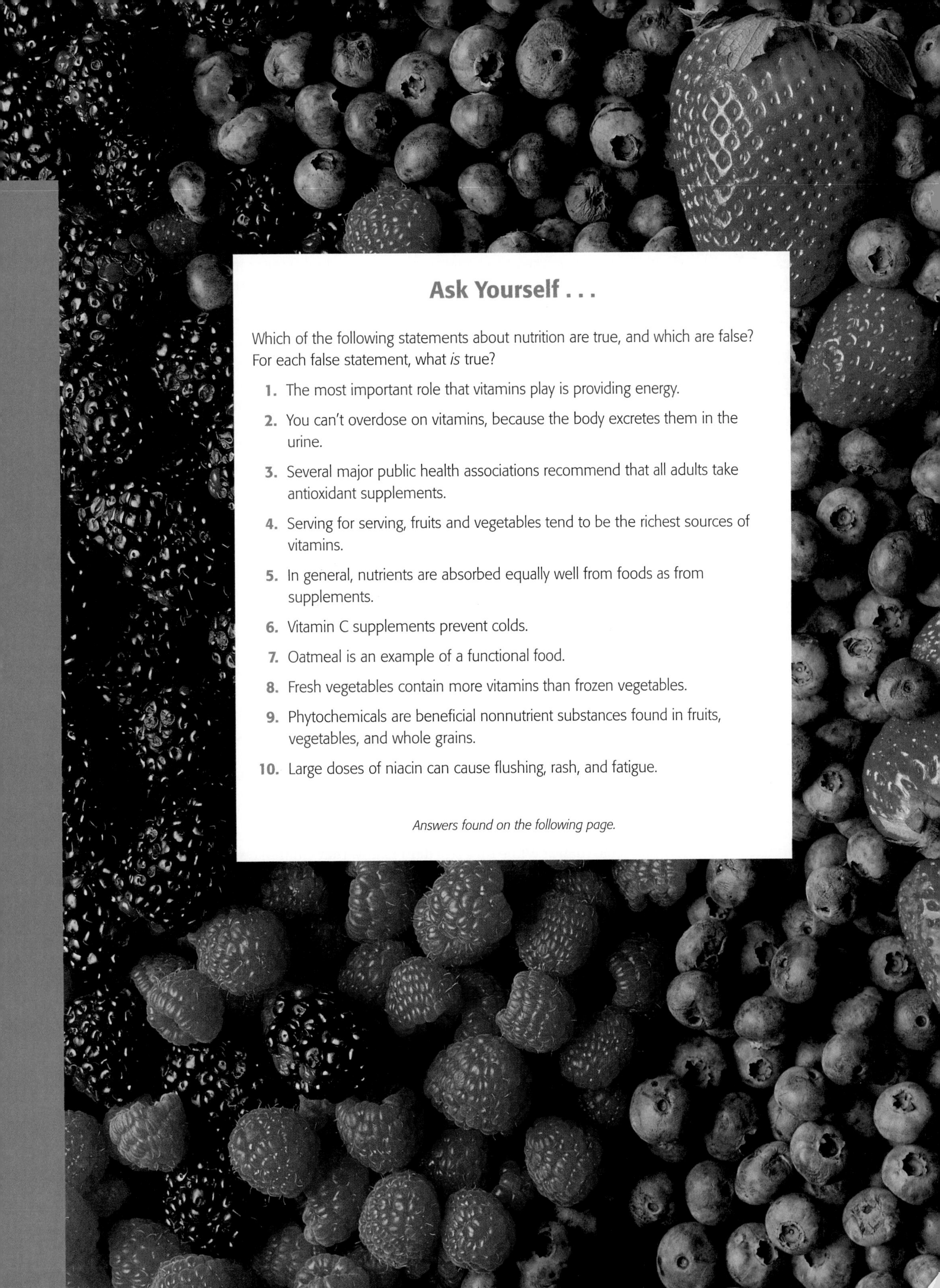

Ask Yourself . . .

Which of the following statements about nutrition are true, and which are false? For each false statement, what *is* true?

1. The most important role that vitamins play is providing energy.
2. You can't overdose on vitamins, because the body excretes them in the urine.
3. Several major public health associations recommend that all adults take antioxidant supplements.
4. Serving for serving, fruits and vegetables tend to be the richest sources of vitamins.
5. In general, nutrients are absorbed equally well from foods as from supplements.
6. Vitamin C supplements prevent colds.
7. Oatmeal is an example of a functional food.
8. Fresh vegetables contain more vitamins than frozen vegetables.
9. Phytochemicals are beneficial nonnutrient substances found in fruits, vegetables, and whole grains.
10. Large doses of niacin can cause flushing, rash, and fatigue.

Answers found on the following page.

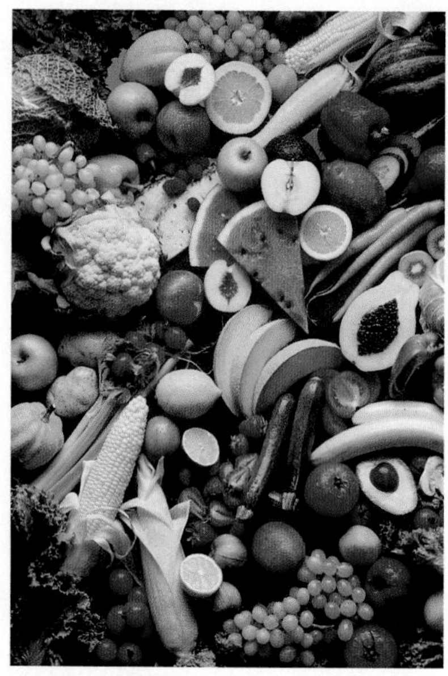

The more varied the kinds of fruits and vegetables you eat, the better nourished you are likely to be.

About a century ago, scientists ushered in a new era in the science of nutrition: the discovery of vitamins. They quickly realized that these substances, found in minute amounts in foods, were just as essential to health as fats, carbohydrates, and proteins. A diet lacking in one could cause a barrage of symptoms and, ultimately, death. Knowledge of the vital role played by vitamins quickly advanced, and today life-threatening vitamin deficiencies are rare in developed countries such as the United States.

Still, the vitamin research that has been conducted during the past decade or so has marked the beginning of yet another chapter in the annals of nutrition. Throughout the past ten years, more and more scientists have been investigating the possibility that large doses of certain vitamins will help stave off chronic diseases such as cancer and heart disease, problems that rank as major killers today. In fact, the study of vitamins, particularly a class known as the antioxidant vitamins, is one of the hottest, most widely publicized areas in nutrition research today. In addition, the pros and cons of taking vitamin supplements are the subject of heated debate among the scientific community. To help you sort through the steady stream of controversy regarding vitamins, this chapter explores the history, roles, and current thrust of research of the various vitamins and offers practical advice on how to incorporate the information into decisions about your own lifestyle.

TURNING BACK THE CLOCK

SCURVY the vitamin C deficiency disease characterized by bleeding gums, tooth loss, and even death in severe cases.

Many of the vitamin-deficiency diseases that have been virtually eliminated today were first recognized in Greek and Roman times and ultimately led to the discovery of vitamins centuries later. One of the most prevalent was **scurvy,** a disease characterized by bleeding gums, tooth loss, and even death due to lack of vitamin C. The scourge of armies, sailors, and other travelers forced to do without vitamin C-rich foods for weeks on end, scurvy was recognized by Hippocrates, a Greek physician heralded today as the father of medicine.[1]

A cure for the disease was not recorded until the sixteenth century, however, when a beverage made of spruce needles or oranges and lemons was recommended. In 1753, a British physician named James Lind published a famous report recommending consumption of herbs, lettuce, endive, watercress, and summer fruits to prevent scurvy. By the early 1800s, sailors in the British navy had been dubbed "limeys" because they were required to drink lemon or lime juice daily.[2] While they still didn't know that vitamin C was the real antidote, they did recognize that certain foods prevented and cured the illness.

RICKETS a disease that occurs in children as a result of vitamin D deficiency and that is characterized by abnormal growth of bone, which in turn leads to bowed legs and an outward-bowed chest.

PELLAGRA (pell-AY-gra) niacin deficiency characterized by diarrhea, inflammation of the skin, and, in severe cases, mental disorders and death.

Similarly, a deficiency disease called **rickets** dates back to Roman times, when children frequently suffered skeletal deformities as a result of a lack of vitamin D. By the 1600s, rickets was known as the English disease because it afflicted so many English children. Some 200 years later, cod liver oil was finally recognized as a cure for the disease; no one knew at the time, however, that the "magic" ingredient in the oil was vitamin D.

Another deficiency disease, called **pellagra,** was not recognized until 1730, when a Spanish physician named Gaspar Casal first described the crusty, dry,

Ask Yourself Answers: **1.** False. Vitamins do not provide energy, though they do play roles in energy-yielding reactions in the body. **2.** False. Excess doses of all of the vitamins can be toxic. **3.** False. No major health organization recommends that all adults take antioxidant pills. **4.** True. **5.** False. In general, nutrients are absorbed best from foods, because they are accompanied by other ingredients that facilitate their absorption. **6.** False. Vitamin C has never been proved to prevent colds; at best, it may reduce the severity of cold symptoms. **7.** True. Oatmeal, oat bran, and whole-oat products contain a soluble fiber shown to reduce cholesterol levels when eaten as part of a heart-healthy diet. **8.** False. Fresh vegetables do not necessarily contain more vitamins than their frozen counterparts, depending on such factors as how the fresh vegetable has been stored and how long since it has been harvested. **9.** True. **10.** True.

scabby, blackish patches of skin symptomatic of the disease. In Italy, the disease was named pellagra, from the Italian *pelle agra,* meaning sour skin. Called *mal de la rosa* in Spanish, pellagra was thought to be incurable until Dr. Casal noticed that the people who developed the disease were typically poor and had inadequate diets made up of mostly corn and little meat.

By the nineteenth century, physicians had recognized that certain foods prevented or cured pellagra and other deficiency diseases. But they still hadn't determined exactly what it was in the various foods that worked as a remedy. By the middle of the nineteenth century, however, the science of chemistry had advanced to a point at which foods could be analyzed. Chemists had determined that foods consisted of fats, proteins, and carbohydrates along with minerals and water, and they assumed that they had identified all the nutritionally significant compounds.

Then, in the early twentieth century, scientists detected minute amounts of other substances that they found were essential in preventing disease and maintaining health. The substances were dubbed *vitamines,* a term coined in 1912 by a scientist named Dr. Casimir Funk to indicate that these substances were *vital* for survival and that they contained nitrogen—that is, they were *amines.* (The *e* was later dropped when scientists discovered that some of the vitamins were not amines.)

Over the next few decades, scientists identified the various **vitamins,** established their chemical formulas, and determined their functions in the body. They also measured the amount of vitamins in various foods and determined human and animal requirements for the compounds. Knowledge of vitamins constantly evolves as scientists continue to study their actions in the human body.

Today, scientists recognize that vitamins are potent compounds that perform many tasks in the body that promote growth and reproduction and maintain health and life. Vitamins constantly work to keep your nerves and skin healthy; build bone, teeth, and blood; and heal wounds, among other things. While they do not provide calories, they are essential to helping the body make use of the calories consumed via foods.

Vitamins fall into two categories: those that dissolve in water, or water-soluble, and those that dissolve in fat, or fat-soluble. To date, scientists have identified 13 vitamins, each with its own special roles to play (see Table 6-1). As Figure 6-1 shows, each of the major food groups in the Food Guide Pyramid supplies a number of vitamins. Eating plans that exclude entire food groups, or fail to include the minimum number of servings from each of the groups, may lead to vitamin deficiencies over time.

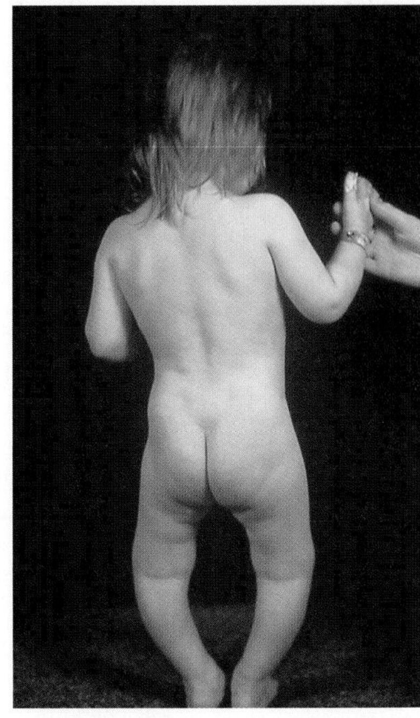

This child has the bowed legs characteristic of rickets. Worldwide, rickets afflicts many children who live in poverty and do not have access to sunlight or adequate foods containing vitamin D. (See vitamin D discussion starting on page 181.)

VITAMIN a potent, indispensable compound that performs various bodily functions that promote growth and reproduction and maintain health. Vitamins are **organic,** meaning that they contain or are related to carbon compounds. Contrary to popular belief, vitamins do not supply calories.

ORGANIC of, related to, or containing carbon compounds.

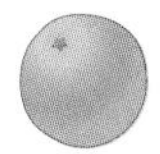

WATER-SOLUBLE VITAMINS

There are nine water-soluble vitamins: eight B vitamins and vitamin C. Found in the watery compartments of foods, such as the juice of an orange, these vitamins are distributed into water-filled compartments of the body, including the fluid that surrounds the spinal cord. The body excretes water-soluble vitamins if the blood levels rise too high. As a result, they rarely reach toxic levels in the body. This is not to say, however, that excess levels cannot cause problems, at least in some people.

In the body, water-soluble vitamins act as **coenzymes**—that is, they assist enzymes in doing their metabolic work within the body. (You may recall from Chapter 5 that enzymes are proteins that act as catalysts that help to boost chemical reactions in the body, as described on page 141.)

COENZYMES enzyme helpers; small molecules that interact with enzymes and enable them to do their work. Many coenzymes are made from water-soluble vitamins.

In foods, the water-soluble vitamins are relatively fragile. Although large proportions of them are naturally present in many foods, they can be washed out or destroyed during food storage, processing, and preparation. These effects are spelled out in detail in the Savvy Diner feature later in this chapter.

TABLE 6-1 A GUIDE TO THE VITAMINS

Vitamin (Chemical Name)	Best Sources	Chief Roles	Deficiency Symptoms	Toxicity Symptoms
Water-soluble Vitamins				
Thiamin	Meat, pork, liver, fish, poultry, whole-grain and enriched breads, cereals, pasta, nuts, legumes, wheat germ, oats	Helps enzymes release energy from carbohydrate; supports normal appetite and nervous system function	Beriberi: edema, heart irregularity, mental confusion, muscle weakness, low morale, impaired growth	None reported
Riboflavin	Milk, leafy green vegetables, yogurt, cottage cheese, liver, meat, whole-grain or enriched breads and cereals	Helps enzymes release energy from carbohydrate, fat, and protein; promotes healthy skin and normal vision	Eye problems, skin disorders around nose and mouth	None reported
Niacin	Meat, eggs, poultry, fish, milk, whole-grain and enriched breads and cereals, nuts, legumes, peanuts	Helps enzymes release energy from energy nutrients; promotes health of skin, nerves, and digestive system	Pellagra: skin rash on parts exposed to sun, loss of appetite, dizziness, weakness, irritability, fatigue, mental confusion, indigestion	Flushing, nausea, headaches, cramps, ulcer irritation, heartburn, abnormal liver function, rapid heartbeat with doses above 500 mg per day; UL equals 35 mg/day from fortified foods or supplements*
Vitamin B_6 (pyridoxine)	Meat, poultry, fish, shellfish, legumes, whole-grain products, green leafy vegetables, fruits	Protein and fat metabolism; formation of antibodies and red blood cells; helps convert tryptophan to niacin	Nervous disorders, skin rash, muscle weakness, anemia, convulsions, kidney stones	Depression, fatigue, irritability, headaches, numbness, damage to nerves, difficulty walking; UL equals 100 mg/day
Folate (folacin, folic acid)	Green leafy vegetables, liver, legumes, seeds, citrus fruits, melons, enriched breads and grain products	Red blood cell formation; protein metabolism; new cell division	Anemia, heartburn, diarrhea, smooth tongue, depression, poor growth, neural tube defects, increased risk of heart disease, stroke, and certain cancers	Diarrhea, insomnia, irritability, may mask a vitamin B_{12} deficiency; UL equals 1 mg/day
Vitamin B_{12} (cobalamin)	Animal products: meat, fish, poultry, shellfish, milk, cheese, eggs; fortified cereals	Helps maintain nerve cells; red blood cell formation; synthesis of genetic material	Anemia, smooth tongue, fatigue, nerve degeneration progressing to paralysis	None reported
Pantothenic acid	Widespread in foods	Coenzyme in energy metabolism	Rare; sleep disturbances, nausea, fatigue	None reported
Biotin	Widespread in foods	Coenzyme in energy metabolism; fat synthesis; glycogen formation	Loss of appetite, nausea, depression, muscle pain, weakness, fatigue, rash	None reported

*UL = Tolerable Upper Intake Level

TABLE 6-1 A GUIDE TO THE VITAMINS—*Continued*

Vitamin (Chemical Name)	Best Sources	Chief Roles	Deficiency Symptoms	Toxicity Symptoms
Water-soluble Vitamins				
Vitamin C (ascorbic acid)	Citrus fruits, cabbage-type vegetables, tomatoes, potatoes, dark green vegetables, peppers, lettuce, cantaloupe, strawberries, mangos, papayas	Synthesis of collagen (helps heal wounds, maintains bone and teeth, strengthens blood vessels); antioxidant; strengthens resistance to infection; helps body absorb iron	Scurvy: anemia, depression, frequent infections, bleeding gums, loosened teeth, pinpoint hemorrhages, muscle degeneration, rough skin, bone fragility, poor wound healing, hysteria	Intakes of more than 1 g per day may cause nausea, abdominal cramps, diarrhea, and increased risk for kidney stones; UL equals 2,000 mg/day
Fat-soluble Vitamins				
Vitamin A	*Retinol:* fortified milk and margarine, cream, cheese, butter, eggs, liver *Beta-carotene:* Spinach and other dark leafy greens, broccoli, deep orange fruits (apricots, peaches, cantaloupe), and vegetables (squash, carrots, sweet potatoes, pumpkin)	Vision; growth and repair of body tissues; maintenance of mucous membranes; reproduction; bone and tooth formation; immunity; hormone synthesis; antioxidant (in the form of beta-carotene only)	Night blindness, rough skin, susceptibility to infection, impaired bone growth, abnormal tooth and jaw alignment, eye problems leading to blindness, impaired growth	Red blood cell breakage, nosebleeds, abdominal cramps, nausea, diarrhea, weight loss, blurred vision, irritability, loss of appetite, bone pain, dry skin, rashes, hair loss, cessation of menstruation, liver disease, birth defects
Vitamin D (cholecalciferol)	Self-synthesis with sunlight; fortified milk, fortified margarine, eggs, liver, fish	Calcium and phosphorus metabolism (bone and tooth formation); aids body's absorption of calcium	Rickets in children; osteomalacia in adults; abnormal growth, joint pain, soft bones	Deposits of calcium in organs such as the kidneys, liver, or heart, mental retardation, abnormal bone growth. UL equals 50 µg/day or 2,000 IU
Vitamin E	Vegetable oils, green leafy vegetables, wheat germ, whole-grain products, liver, egg yolk, salad dressings, mayonnaise, margarine, nuts, seeds	Protects red blood cells; antioxidant (protects fat-soluble vitamins); stabilization of cell membranes	Muscle wasting, weakness, red blood cell breakage, anemia, hemorrhaging	Doses over 800 IU/day may increase bleeding (blood clotting time); UL equals 1,000 mg/day
Vitamin K	Bacterial synthesis in digestive tract, liver, green leafy and cabbage-type vegetables, milk, grain products	Synthesis of blood-clotting proteins and a blood protein that regulates blood calcium	Hemorrhaging, decreased calcium in bones	Interference with anticlotting medication; synthetic forms may cause jaundice

FIGURE 6-1
GOOD SOURCES OF VITAMINS IN THE FOOD GUIDE PYRAMID

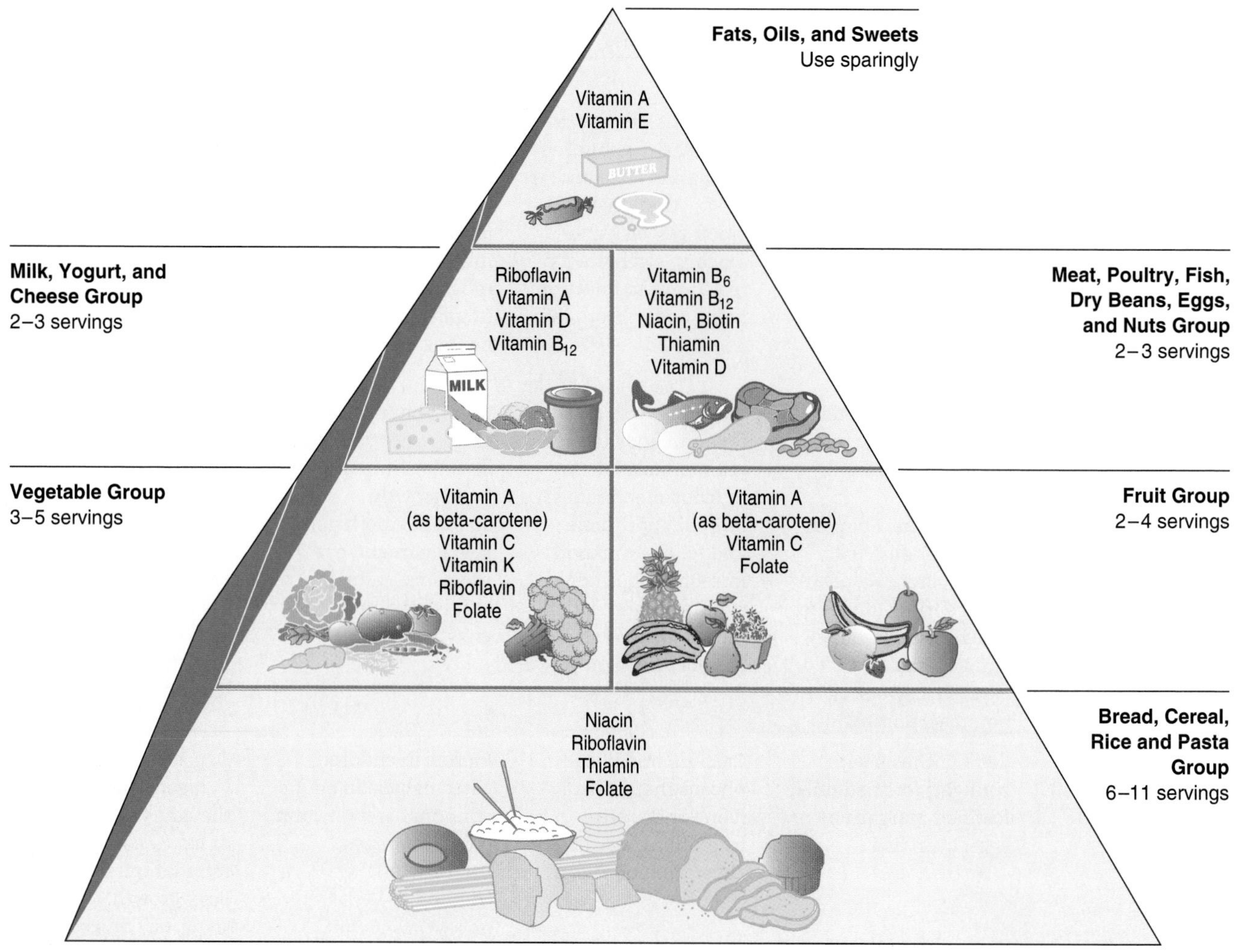

TABLE 6-2 THIAMIN IN FOODS

(mg)	Sources
0.98	Pork chop (3 oz)
0.41	Sunflower seeds (2 tbsp)
0.23	Watermelon (1 slice)
0.23	Green peas (½ c)
0.22	Baked potato (1)
0.21	Black beans (½ c)
0.21	Peanuts (⅓ c)
0.17	Black-eyed peas (½ c)
0.13	Oatmeal, cooked (½ c)
0.11	Sirloin steak (3 oz)
0.11	Orange (1)
0.11	Wheat bread (1 slice)

BERIBERI the thiamin deficiency disease, characterized by irregular heartbeat, paralysis, and extreme wasting of muscle tissue.

Thiamin

One of the B vitamins, thiamin acts primarily as a coenzyme in reactions that release energy from carbohydrate. It also plays a crucial role in processes involving the nerves. So vital is thiamin to the functioning of the entire body that a deficiency affects the nerves, muscles, heart, and other organs. A severe deficiency, called **beriberi,** causes extreme wasting and loss of muscle tissue, swelling all over the body, enlargement of the heart, irregular heartbeat, and paralysis. Ultimately, the victim dies from heart failure. A mild thiamin deficiency, on the other hand, often mimics other conditions and typically manifests itself as vague, general symptoms such as stomachaches, headaches, fatigue, restlessness, sleep disturbances, chest pains, fevers, personality changes (aggressiveness and hostility), and neurosis.

Thiamin is found in a wide variety of foods, and virtually no single food will supply your daily needs in a single serving the way, say, an orange provides a plentiful supply of vitamin C. But people who eat a balanced diet that follows the framework of the Food Guide Pyramid typically take in plenty of thiamin. As Table 6-2 shows, thiamin is found in a variety of meats, legumes, fruits, and vegetables, as well as in all enriched and whole-grain products.

*Dietary Reference Intakes (DRI) for all age groups are listed on the inside front cover.

Riboflavin

Like thiamin, the B vitamin called riboflavin acts as a coenzyme in energy-releasing reactions in the body. In addition, riboflavin helps to prepare fatty acids and amino acids for breakdown. Deficiencies of the vitamin, which are rare, are characterized by severe skin problems, including painful cracks at the corners of the mouth; a red, swollen tongue; and teary or bloodshot eyes.

Table 6-3 shows the riboflavin content of foods. Milk and dairy products contribute a good deal of the riboflavin in most people's diets. Meats are another good source, as are dark green vegetables such as broccoli. Leafy green vegetables and whole-grain or enriched bread and cereal products also supply a generous amount of riboflavin in most people's diets.

Note that riboflavin can be destroyed by the ultraviolet rays of the sun or fluorescent lamps. That's why milk is usually sold in protective cardboard or opaque plastic containers rather than in transparent glass bottles.

TABLE 6-3 RIBOFLAVIN IN FOODS

(mg)	Sources
0.52	Low-fat yogurt (1 c)
0.41	Fat-free milk (1 c)
0.37	Almonds (⅓ c)
0.24	Pork chop (3 oz)
0.23	Ricotta cheese (½ c)
0.23	Sirloin steak (3 oz)
0.21	Beet greens, cooked (½ c)
0.21	Poached egg (1)
0.20	Ground beef (3 oz)
0.17	Spinach, cooked (½ c)
0.17	Cheddar cheese (1.5 oz)
0.16	Turkey (3 oz)
0.11	Asparagus, cooked (½ c)
0.10	Strawberries (1 c)
0.08	Wheat bread (1 slice)

Niacin

Like thiamin and riboflavin, the B vitamin niacin is part of a coenzyme vital to obtaining energy. Without niacin to form this coenzyme, energy-yielding reactions come to a halt. Over time, a deficiency of niacin leads to the disease pellagra, characterized by diarrhea, dermatitis, and, in severe cases, dementia—a progressive mental deterioration resulting in delirium, mania, or depression, and eventually death.

While niacin deficiency can be prevented by eating a diet rich in niacin itself, consuming plenty of protein also staves off the problem. That's because the essential amino acid tryptophan, which is a component of protein, can be converted to niacin in the body. In fact, 60 milligrams of tryptophan yield one milligram of niacin. Thus, the DRI for niacin is expressed in niacin equivalents (NEs)—that is, the amount of niacin present in food, including the amount that can be theoretically made from the tryptophan in the food.

Milk, eggs, meat, poultry, and fish contribute most of the niacin equivalents consumed by most people, followed by enriched breads and cereals. Table 6-4 shows the niacin content of some common foods.

Diet aside, in recent years, niacin has been increasingly used as a druglike supplement to help lower cholesterol. Doses ranging from 10 to 15 times the RDA have been shown to reduce "bad" LDL-cholesterol and raise "good" HDL-cholesterol. Large doses of a form of niacin may also prove effective in preventing or delaying the onset of Type 1 diabetes.[3] The hitch, however, is that such high doses of niacin can lead to side effects such as nausea, flushing of the skin, rash, fatigue, and liver damage. Because of the side effects, many experts argue that

TABLE 6-4 NIACIN IN FOODS

(mg NE)	Sources
10.80	Chicken breast (½)
8.10	Tuna (3 oz)
6.57	Peanuts (⅓ c)
6.05	Halibut (3 oz)
5.08	Ground beef (3 oz)
4.63	Turkey (3 oz)
3.31	Baked potato (1)
3.29	Sirloin steak (3 oz)
1.85	Flounder/sole (3 oz)
1.53	Cantaloupe (½)
1.49	Brown rice, cooked (½ c)
1.13	Wheat bread (1 slice)
0.97	Asparagus, cooked (½ c)
0.89	Broccoli, cooked (½ c)
0.86	Peach (1)

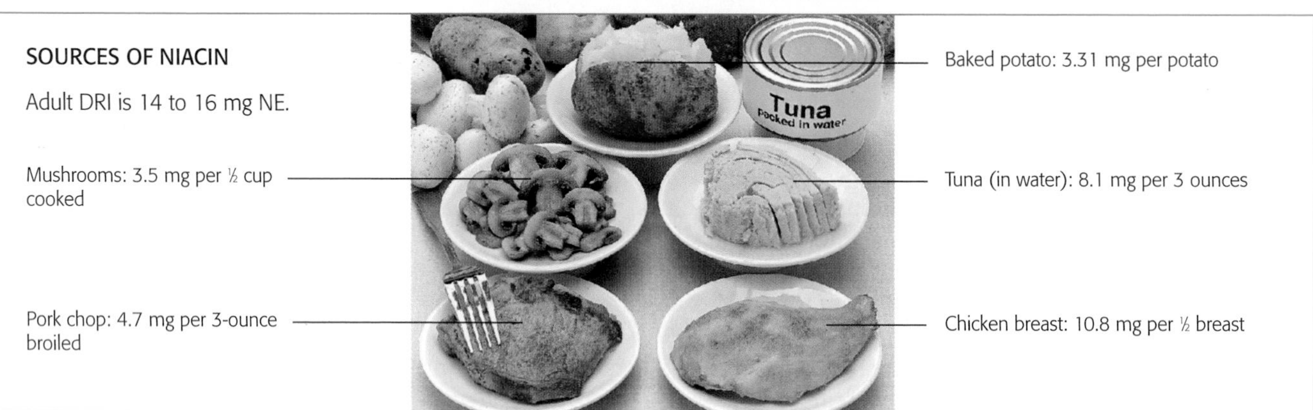

niacin pills should be sold not as over-the-counter dietary supplements but rather as drugs prescribed and taken only while under a physician's supervision.

Folate

ANEMIA any condition in which the blood is unable to deliver oxygen to the cells of the body. Examples include a shortage or abnormality of the red blood cells. Many nutrient deficiencies and diseases can cause anemia.

Folate (also called folic acid or folacin) is a coenzyme with many functions in the body. It is particularly important in the synthesis of DNA and the formation of red blood cells. A deficiency makes the red blood cells misshapen and unable to carry sufficient oxygen to all the body's other cells, thereby causing a certain kind of **anemia.** Thus, folate deficiency results in a kind of generalized malaise with many symptoms, including fatigue, diarrhea, irritability, forgetfulness, lack of appetite, and headache. Folate deficiency can easily be confused with general ill health, depressed mood, and senility in the elderly. Folate deficiency may also elevate a person's risk for certain cancers—notably colon cancer, and cervical cancer in women.[4]

Folate deficiency tends to occur in people who eat few fresh vegetables, because folate is easily lost when foods are overcooked, canned, dehydrated, or otherwise processed. In addition, people who are growing rapidly run a high risk of folate deficiency because folate is needed to promote the rapid multiplication of cells that occurs during growth. That's why, for example, the need for folate increases during pregnancy, when large amounts of the vitamin are needed to support the growth of the fetus.*

*Recommended folate intakes are stated in micrograms (μg) DFE. A microgram is a thousandth of a milligram. DFE stands for Dietary Folate Equivalent, a unit of measure expressing the amount of folate available to the body from naturally occurring food sources. The measure accounts for the differences in absorption between food folate and the more absorbable synthetic folate added to foods and supplements. Appendix D offers a conversion factor for calculating DFE.

Folate plays a crucial role in a healthy pregnancy. A growing body of evidence indicates that consuming a generous amount of folate reduces the risk of bearing a baby with a type of birth defect called **neural tube defect.** Afflicting some 2,500 infants born in the United States each year, neural tube defects include spina bifida—the incomplete closing of the bony casing around the spinal cord that causes partial paralysis—and anencephaly—a condition in which major parts of the brain are missing. All women of childbearing age are advised to consume the recommended amount of folate because adequate levels of the nutrient must be ingested before and during the first few weeks of pregnancy, the period during which the neural tube of the embryo is closing but when most women are not aware that they are pregnant.

TABLE 6-5 FOLATE IN FOODS

(µg)	Sources
127	Asparagus, cooked (½ c)
120	Black-eyed peas (½ c)
113	Spinach, cooked (½ c)
108	Fresh spinach (1 c)
100	Oatmeal, instant, fortified (½ c)
85	Turnip greens, cooked (½ c)
82	Enriched cereal (¾ c)
76	Romaine (1 c)
71	Peanuts (⅓ c)
65	Kidney beans (½ c)
55	Lima beans (½ c)
47	Cantaloupe (¼)
41	Sunflower seeds (2 tbsp)
40	Orange (1)
39	Broccoli, cooked (½ c)
27	Cauliflower, cooked (½ c)
19	Tofu (soybean curd) (½ c)
14	Whole-wheat bread (1 slice)
13	Fat-free milk (1 c)
13	Strawberries (½ c)
8	Sirloin steak (3 oz)

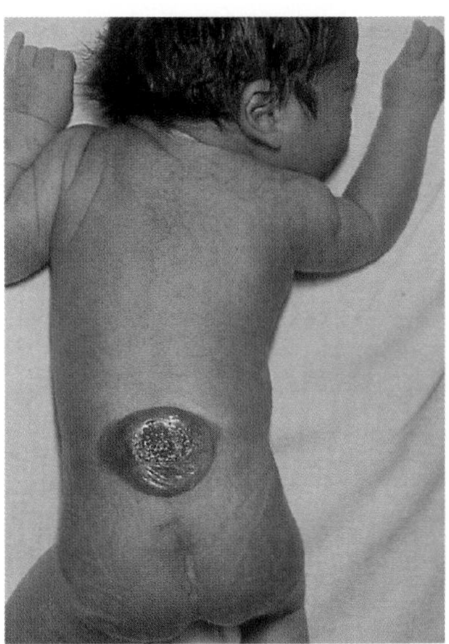

This child has spina bifida—a birth defect characterized by the incomplete closing of the casing around the spinal cord. Afflicting some 2,500 U.S. infants born each year, the problem causes partial paralysis.

Although folate is abundant in vegetables, legumes, and seeds, as shown in Table 6-5, most women typically consume foods that provide only about half the 400-microgram amount recommended. For this reason, the Food and Drug Administration (FDA) has mandated that grain products be fortified with folate to improve folate intakes in the United States.[5]* Labels on fortified products may make the health claim that "adequate intake of folate has been shown to reduce the risk of neural tube defects." A person can obtain the recommended 400 micrograms of folate by increasing consumption of foods naturally rich in folate (orange juice and green vegetables) and foods fortified with folate (breakfast cereals, breads, rice, or pasta), or by taking a folic acid supplement daily.[6] Since high levels of blood folate can mask a true vitamin B_{12} deficiency, total folate intake should not exceed 1 milligram daily.[7]

Low intakes of three B vitamins—folate, vitamin B_{12}, and vitamin B_6—are linked with increased risk of fatal heart disease in both men and women.[8] People with low blood levels of these B vitamins tend to have high blood levels of the protein-related compound **homocysteine.**[9] High levels of homocysteine seem

NEURAL TUBE DEFECTS malformations of the brain and/or spinal cord during embryonic development.

HOMOCYSTEINE (ho-mo-SIS-teen) a chemical that appears to be toxic to the blood vessels of the heart. High blood levels of homocysteine have been associated with low blood levels of vitamin B_{12}, vitamin B_6, and folate.

*Since 1998, breads, flour, corn grits, cornmeal, farina, macaroni, rice, and noodles are fortified with 1.4 milligrams of folate per 100 grams of food.

to enhance blood clot formation and damage to arterial walls and raise the risk of suffering a heart attack or stroke as much as fourfold.[10] In addition, scientists suspect that high homocysteine levels may not only be damaging to the cardiovascular system but may also be toxic for brain tissue and impair cognitive ability.[11] The B vitamins serve to clear homocysteine from the blood and prevent its toxic buildup. The research linking a vitamin B-poor diet with increased homocysteine levels highlights the importance of consuming generous amounts of these nutrients.

Vitamin B_6

Like the other B vitamins, vitamin B_6 functions as a coenzyme and is an indispensable cog in the body's machinery. For example, vitamin B_6 plays many roles in protein metabolism. In fact, a person's requirement for vitamin B_6 is proportional to protein intakes. Because vitamin B_6 performs this and so many other tasks, a deficiency causes a multitude of symptoms, including weakness, irritability, and insomnia. Low levels of vitamin B_6 may also weaken the body's immune response and increase a person's risk for heart disease. Vitamin B_6 is found in meats, vegetables, and whole-grain cereals, and true vitamin B_6 deficiencies are rare, occurring in some people who eat inadequate diets and whose nutrient needs are higher than usual because of pregnancy, alcohol abuse, some diseases, use of certain prescription drugs, and other unusual circumstances. Table 6-6 shows the vitamin B_6 content of various foods.

TABLE 6-6 VITAMIN B_6 IN FOODS

(mg)	Sources
0.70	Baked potato (1)
0.69	Watermelon (1″ by 10″ slice)
0.68	Banana (1)
0.51	Chicken breast (3 oz)
0.42	Figs, dried (10)
0.35	Pork chop (3 oz)
0.34	Sirloin steak (3 oz)
0.31	Cantaloupe (½)
0.30	Tuna (3 oz)
0.26	Ground beef (3 oz)
0.22	Spinach, cooked (½ c)
0.20	Flounder/sole (3 oz)
0.20	Soybeans (½ c)
0.19	Salmon (3 oz)
0.15	Navy beans (½ c)
0.14	Brown rice, cooked (½ c)
0.14	Sunflower seeds (2 tbsp)
0.11	Asparagus, cooked (½ c)
0.11	Broccoli, cooked (½ c)
0.10	Fat-free milk (1 c)
0.07	Zucchini, cooked (½ c)

Vitamin B_6 is also widely reputed as a cure for **premenstrual syndrome (PMS).** Some people have claimed that a deficiency of the vitamin goes hand in hand with imbalances of hormones, particularly estrogen, which cause the depression, mood swings, and other symptoms characteristic of PMS. Although this theory has never been proven to be scientifically sound, women have taken **megadoses** of B_6—as much as 2,000 times the RDA in some cases—in an effort to treat PMS. However, a number of these women began to experience symptoms associated with

PREMENSTRUAL SYNDROME (PMS) a cluster of physical, emotional, and psychological symptoms that some women experience seven to ten days before menstruating. Symptoms can include acne, anxiety, food cravings (especially for sweets), back pain, breast tenderness, cramps, depression, fatigue, headaches, irritability, moodiness, water retention, and weight gain. Because a clear-cut treatment for the symptoms of PMS has not been identified, women who suffer from the problem rank as prime targets for unproved nutritional remedies for the condition.

MEGADOSE a dose of ten or more times the amount normally recommended. An overdose is an amount high enough to cause toxicity symptoms. Megadoses taken over a long period often result in an overdose.

damage to the nervous system, such as loss of sensation in the hands and mouth.[12] Granted, not everyone is likely to suffer toxicity symptoms as a result of swallowing megadoses of vitamin B_6, because excess amounts are excreted in the urine. But the problems seen in women who take these megadoses underscore the potential hazards of taking megadoses of any vitamin or nutritional supplement.

TABLE 6-7 VITAMIN B_{12} IN FOODS

(μg)	Sources
16.50	Chicken liver (3 oz)
7.60	Sardines (3 oz)
2.50	Tuna (3 oz)
2.01	Ground beef (3 oz)
1.60	Cottage cheese (1 c)
1.39	Plain nonfat yogurt (1 c)
1.27	Shrimp (3 oz)
1.18	Haddock (3 oz)
0.93	Fat-free milk (1 c)
0.50	Egg (1)
0.35	Cheddar cheese (1.5 oz)
0.29	Chicken breast (½)

Vitamin B_{12}

Vitamin B_{12} maintains the sheaths that surround and protect nerve fibers. The nutrient also works closely with folate, enabling it to manufacture red blood cells. When a B_{12} deficiency is present, folate is unable to do its work building red blood cells. As a result, a person suffering a lack of vitamin B_{12} ends up with the same sort of anemia seen in people with a folate deficiency and characterized by large, immature red blood cells. While extra folate will clear up the anemia, it will not take care of the other problems resulting from a B_{12} deficiency, namely a creeping paralysis of the nerves and muscles that can cause permanent nerve damage if left untreated. Thus, because excess folate can clear up the blood problems that signal an otherwise hard-to-diagnose vitamin B_{12} deficiency, the amount of folate in over-the-counter supplements is limited by law to an amount that is too low to cover up a B_{12} deficiency.

To be sure, dietary deficiencies of vitamin B_{12} are not likely to occur among people who eat animal foods such as meat, milk, cheese, and eggs, all of which supply generous amounts of the nutrient (see Table 6-7). Strict vegetarians who eschew meat, eggs, and dairy products, however, need to find alternative sources of the nutrient, such as vitamin B_{12}-fortified soy beverages, fortified cereals, or B_{12} supplements.

Several other groups of people are also at high risk for vitamin B_{12} deficiency, not because of a lack of the vitamin in their diets but because of physical conditions that hamper their body's ability to make use of the nutrient. One such group is people who inherit a genetic defect that leaves the body unable to make a compound known as **intrinsic factor.** Produced in the stomach, intrinsic factor enables the body to absorb and make use of vitamin B_{12}; without the compound, vitamin B_{12} deficiency develops. In this instance or in the case of stomach damage that interferes with the production of intrinsic factor, people must get vitamin B_{12} injections.

INTRINSIC FACTOR a compound made in the stomach that is necessary for the body's absorption of vitamin B_{12}.

Another group likely to experience vitamin B_{12} deficiencies is the elderly. An estimated 20 percent of seniors in their sixties and 40 percent in their eighties develop **atrophic gastritis,** an age-related condition characterized by the stomach's inability to produce enough acid, which in turn hampers the body's ability to use vitamin B_{12}. In severe cases, the condition also limits the stomach's ability to make intrinsic factor. Vitamin B_{12} deficiencies resulting from atrophic gastritis appear to be easily treated with vitamin B_{12} supplements or injections.[13]

ATROPHIC GASTRITIS an age-related condition characterized by the stomach's inability to produce enough acid, which in turn leads to vitamin B_{12} deficiencies.

SOURCES OF BIOTIN AND PANTOTHENIC ACID*

Adult DRI:
Biotin: 30 μg
Pantothenic acid: 5 mg

*Information concerning biotin and pantothenic acid in foods is incomplete: deficiencies are rare.

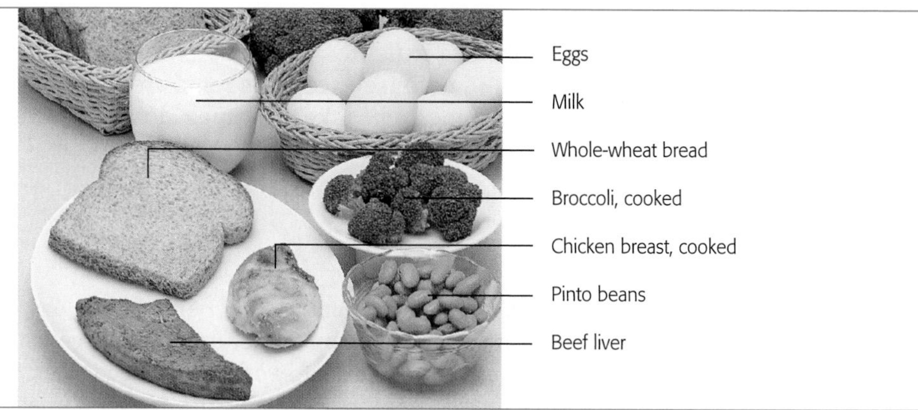

Pantothenic Acid and Biotin

Two other B vitamins—pantothenic acid and biotin—are needed for the synthesis of coenzymes that are active in a multitude of body systems. Biotin is also required for cell growth, synthesis of DNA (the genetic "blueprint" present in every cell), and maintenance of blood glucose (sugar) levels. Because both pantothenic acid and biotin are widespread in foods, people who eat a varied diet are not at risk for deficiencies.

Vitamin C

COLLAGEN the characteristic protein of connective tissue.
kolla = glue
gennan = to produce

Vitamin C is required for the production and maintenance of **collagen,** the protein foundation material for the body's connective tissue, including bones, teeth, skin, and tendons. Vitamin C has also been touted as a nutrient that can help fight stress. And it's true that in times of stress, the body uses more vitamin C than usual because the vitamin is involved in the release of stress hormones. Still, the amount of extra vitamin C used as a result of, say, on-the-job deadline stress or the stress of ending a significant relationship, is minuscule and is more than accounted for by a diet that regularly includes vitamin C-rich foods (see Table 6-8).

ANTIOXIDANT a substance, such as a vitamin, that is "anti-oxygen"—that is, it helps to prevent damage done to the body as a result of chemical reactions that involve the use of oxygen.

Vitamin C also boosts the body's ability to fight infections, and a growing body of research suggests that it may protect against heart disease and certain types of cancer. Vitamin C's potential role as a chronic-disease fighter stems from its workings as an **antioxidant.**

"A little vitamin C ought to clear that up in no time."

"What do you have that's rich in antioxidants?

TABLE 6-8 VITAMIN C IN FOODS

(mg)	Sources
187	Papaya (1)
113	Cantaloupe (½)
93	Orange juice, fresh (¾ c)
71	Grapefruit juice (¾ c)
70	Orange (1)
67	Green pepper (1)
57	Mango (1)
48	Broccoli, cooked (½ c)
43	Pink/red grapefruit (½)
42	Brussels sprouts, ckd (½ c)
42	Strawberries (½ c)
27	Cauliflower, cooked (½ c)
26	Baked potato (1)
22	Bok choy, cooked (½ c)
22	Cabbage, raw (1 c)
22	Tomato, fresh (1)
16	Raspberries (½ c)
12	Pineapple, fresh (½ c)
10	Asparagus, cooked (½ c)
9	Spinach, cooked (½ c)

As their name suggests, antioxidants are "anti-oxygen"—they fight oxygen, in a manner of speaking. Consider that certain chemical reactions occur in the body that involve the use of oxygen. While these reactions are essential to the body's ability to function, they lead to the creation of highly toxic compounds called **free radicals.** Environmental pollutants such as cigarette smoke and ozone also prompt the formation of free radicals. Left unchecked, these compounds can cause severe cell injury and ultimately may contribute to the development of chronic diseases such as cancer and heart disease.

FREE RADICALS highly toxic compounds created in the body as a result of chemical reactions that involve oxygen. Environmental pollutants such as cigarette smoke and ozone also prompt the formation of free radicals.

Fortunately, the body has a built-in defense system to protect against potential damage from free radicals. That defense system makes use of the antioxidant nutrients—vitamin C as well as vitamin E and the carotenoids (discussed later with vitamin A). In addition, the body manufactures certain enzymes, one of which contains the mineral selenium, that help to fight free radicals.

The antioxidants all work in one way or another to squelch free radicals before they injure the body (see Figure 6-2). Vitamin C helps to stop free radicals in their tracks, working with vitamin E to block damaging chain reactions that appear to promote heart disease and cancer. In addition, vitamin C is a powerful scavenger of environmental air pollutants. In fact, the National Academy of Sciences advises smokers to consume an additional 35 milligrams of vitamin C a day as compared to nonsmokers. The more smoke a person inhales, the more free radicals that are produced and the more vitamin C that is needed to fight them.

FIGURE 6-2
THE ANTIOXIDANTS VERSUS FREE RADICALS IN THE BODY

A. Free radicals—unstable oxygen molecules—can be formed from sunlight, in cigarette smoke and environmental pollution, and as a result of many normal chemical reactions involving oxygen in the body. These free radicals attack healthy molecules in the body in hopes of stealing an electron to help stabilize themselves, which in turn can cause cell and tissue damage to the body. Free radical activity can cause damage to the body's enzymes, cell membranes, nuclear DNA, and result in the formation of oxidized LDL-cholesterol in the arteries. **B.** The antioxidant team includes vitamin C, vitamin E, the carotenoids (for example, beta-carotene, lutein, lycopene), selenium (a trace mineral), and many naturally occurring nonnutrients—called phytochemicals—found in fruits, vegetables, legumes, and whole grains. The antioxidants prevent free radicals from attacking cells and causing damage by neutralizing the free radicals and converting them back into stable oxygen molecules.

Source: Adapted from *Antioxidant Nutrients* (Mount Olive, NJ: BASF Corporation, 1997).

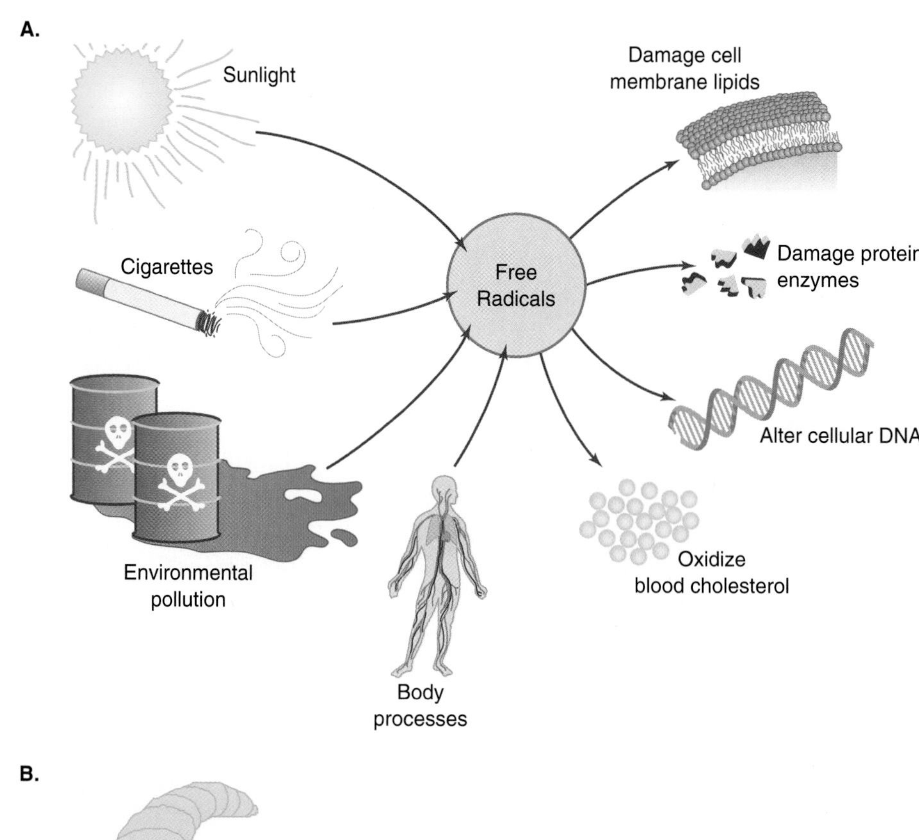

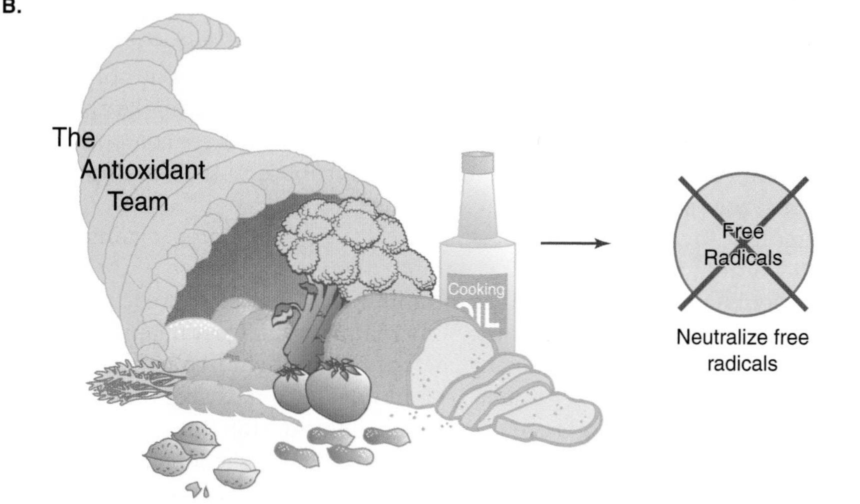

Of course, vitamin C is most famous for its long-standing notoriety as a cure for the common cold. Ever since the publication of the controversial book *Vitamin C and the Common Cold* by the award-winning scientist Linus Pauling, millions of Americans have followed Dr. Pauling's advice and swallowed megadoses of vitamin C—sometimes exceeding 30 times the RDA for the nutrient.[14] Despite the popularity of vitamin C as a cold remedy, however, many carefully controlled studies have shown that it plays an insignificant, if indeed it plays any, role in preventing colds. At best, the nutrient may slightly reduce the severity of cold symptoms in some people. To be sure, many people swear by vitamin C, and it may be that their belief in the nutrient is so strong that they experience a placebo effect as a result of their faith in its curative powers.

Vitamin C-rich foods are widely available in the United States and include not only oranges and other citrus fruits but also broccoli, Brussels sprouts, cantaloupe, and strawberries. A single serving of any of those foods provides more

WORKING ON THE WEB **www.produceoasis.com**

Sponsor: P-O-P Interactive, Inc.

Description: Produce OASIS (Online Advice and Shopper Information System) provides shoppers with information that makes selecting fresh fruits and vegetables easier, faster, more informative, and fun.

Available Activities: You can access information on any item in the produce department. You can find nutritional information by serving size, tips for selecting various fruits and vegetables, recipes designed for busy lifestyles, information on how to use herbs in cooking, and fun facts and trivia about all your favorite fruits and vegetables.

Web Work:

1. Make a list of your favorite three fruits and three vegetables.
2. Go to Fruits and discover the nutrients and calories that you obtain from a serving of each of your favorite fruits.
3. Go to Vegetables and discover the nutrients and calories that you obtain from a serving of each of your favorite vegetables.
4. Go to Produce Nutrition and review the information provided. Identify fruits and vegetables that are excellent sources of the following nutrients: vitamin A (as beta-carotene), riboflavin, folate, vitamin C, calcium, iron, and potassium.
5. Note the Tips & Trivia information provided on this Web site. Find the answers to the following questions:
   a. What did it mean in ancient Greece when a man tossed an apple to a woman?
   b. What vegetable was used to make black dye for teeth?
   c. What is the real name of the fruit that Americans call cantaloupe?
   d. What vegetable is associated with the name Chicago?

Helpful Hints: From the Produce Oasis home page, click on Fruits. When you finish finding the nutrient content of your favorite fruits, click on Tips & Trivia for more information about the fruits. Next, return to the home page and do the same for your favorite vegetables, by clicking on Vegetables, followed by Tips & Trivia. Finally, return to the home page and click on Produce Nutrition. Click on each of the nutrients listed to identify the fruits and vegetables that are excellent sources of each. From this page you can also click on Herbs, for advice on cooking and seasoning with herbs, or Alphabetical Search to locate specific information on any item located in the produce department.

than half the DRI for the vitamin. Potatoes also contribute significant amounts of vitamin C to the American diet because they are eaten so often. Note that vitamins C and A are the only vitamins required to appear on the Nutrition Facts panel of the food label (see Figure 2-12 on page 48).

Vitamin C is widespread in the food supply. Still, deficiencies arise both in infants not given a source of vitamin C and in children and the elderly, due to inadequate consumption of fruits and vegetables.

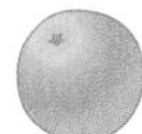

FAT-SOLUBLE VITAMINS

The four fat-soluble vitamins—A, D, E, and K—are generally found in the fats of foods and are absorbed from the digestive tract with the aid of fats in the diet and bile produced by the liver. Any disorder that interferes with fat digestion or absorption can precipitate a deficiency of the fat-soluble vitamins. Once in the bloodstream, these vitamins are escorted by protein carriers because they are insoluble in water. Since they are stored in the liver and in body fat, you need not consume them daily unless your intakes are typically marginal. It is possible for megadoses of the fat-soluble vitamins to build up to toxic levels in the body and cause undesirable side effects.

Vitamin A

Vitamin A has the distinction of being the first fat-soluble vitamin to be identified. It is one of the most versatile vitamins, playing roles in several important body processes.

The best known function of vitamin A is in vision. For a person to see, light reaching the eye must be transformed into nerve impulses that the brain

SOURCES OF VITAMIN A AND BETA-CAROTENE

Adult RDA is 800 to 1,000 RE.

Sweet potato: 1,936 RE[b] per ½ cup cooked

Carrots: 1,584 RE[b] per ½ cup cooked

Fortified milk: 150 RE[a] per cup

Beef liver: 9,124 RE[a] per 3 ounces

Apricots: 274 RE[b] per 3 fresh apricots

Spinach: 739 RE[b] per ½ cup cooked

[a]Preformed vitamin A.
[b]Beta-carotene.

PIGMENT a molecule capable of absorbing certain wavelengths of light. Pigments in the eye permit us to perceive different colors.

RETINA (RET-in-uh) the paper-thin layer of light-sensitive cells lining the back of the inside of the eye.

RETINAL (RET-in-al) one of the active forms of vitamin A that functions in the pigments of the eye. Other active forms of vitamin A include retinol.

NIGHT BLINDNESS slow recovery of vision following flashes of bright light at night; an early symptom of vitamin A deficiency.

EPITHELIAL (ep-ih-THEE-lee-ul) **TISSUE** those cells that form the outer surface of the body and line the body cavities and the principal passageways leading to the exterior. Examples include the cornea, digestive tract lining, respiratory tract lining, and skin. The epithelial cells produce mucus to protect these tissues from bacteria and other potentially harmful substances. Without this mucus, infections become more likely.

PREFORMED VITAMIN A vitamin A in its active form.

BETA-CAROTENE an orange pigment found in plants that is converted into vitamin A inside the body. Beta-carotene is also an antioxidant.

PRECURSOR a compound that can be converted into another compound. For example, beta-carotene is a precursor of vitamin A.
pre = before
cursor = runner, forerunner

interprets in producing visual images. The transformers are molecules of **pigment** in the cells of the **retina,** a paper-thin tissue lining the back of the eye. A portion of each pigment molecule is **retinal,** a compound the body can synthesize only if vitamin A is supplied by the diet in some form. Thus, when vitamin A is deficient, vision is impaired. Specifically, the eye has difficulty in adapting to changing light levels. A flash of bright light at night (after the eye has adapted to darkness) will be followed by a prolonged spell of **night blindness.** Because night blindness is easy to diagnose, it aids in the identification of vitamin A deficiency. Night blindness is only a symptom, however, and may indicate a condition other than vitamin A deficiency.

Vitamin A serves other roles in the body. It helps to maintain healthy skin and **epithelial tissue**—the cells (called epithelial cells) lining such body cavities as the small intestine. It is also involved in the production of sperm, the normal development of fetuses, the immune response, hearing, taste, and growth.

Up to a year's supply of vitamin A can be stored in the body, 90 percent of it in the liver. If you stop eating good food sources of vitamin A, deficiency symptoms will not begin to appear until after your stores are depleted. Then, however, the consequences are profound and include blindness and reduced resistance to infection. While vitamin A deficiency is rarely seen in developed countries such as the United States and Canada, it is one of the most serious public health problems in developing countries, where millions of children suffer from blindness, infections, and the other consequences of vitamin A deficiencies.

Vitamin A toxicity, on the other hand, is not nearly as widespread as deficiency. Nevertheless, it can lead to severe health consequences, including joint pain, dryness of skin, hair loss, irritability, fatigue, headaches, weakness, nausea, and liver damage. That's why it's especially important not to take megadoses of this nutrient.

Although toxicity poses a hazard to people who take supplements of **preformed vitamin A,** toxicity poses virtually no risk to people who obtain vitamin A from foods in the form of **beta-carotene,** an orange plant pigment that is a vitamin A **precursor.** Inside the body, beta-carotene is converted into vitamin A, but this happens so slowly that excess amounts are not stored as vitamin A, but rather are stored in fat deposits.

Beta-carotene is a member of the **carotenoid** family of pigments. The carotenoids possess antioxidant properties and work with vitamins C and E in the body to protect against free radical damage that leads to diseases of the respiratory tract, such as lung cancer, as well as other chronic conditions. Certain carotenoids with antioxidant properties found in dark green leafy vegetables such as spinach, kale, collard greens, and Swiss chard may help prevent **age-related macular degeneration** as well as lower the risk of cataracts.[15] These carotenoids work by filtering out harmful light rays that could cause free-radical damage to the eye.

Because the body uses both the preformed vitamin A and the beta-carotene in foods to make **retinol,** the amount of vitamin A that comes from foods is usually expressed in **retinol equivalents (RE)**—a measure of the amount of retinol the body will derive from the food. Table 6-9 shows the amount of vitamin A, expressed in RE, that comes from various foods. Note that plant foods provide RE from beta-carotene, whereas animal foods such as milk and cheese provide RE via preformed vitamin A.

TABLE 6-9 VITAMIN A IN FOODS

(RE)	Sources
9,124	Beef liver (3 oz)*
2,486	Sweet potato (1)
2,024	Carrot, fresh (1)
860	Cantaloupe (½)
805	Mango (1)
739	Spinach, cooked (½ c)
718	Butternut squash (½ c)
396	Turnip greens, cooked (½ c)
253	Apricot halves, dried (10)
219	Bok choy, cooked (½ c)
178	Watermelon (1 slice)
149	Fat-free milk (1 c)*
146	Romaine (1 c)
129	Cheddar cheese (1.5 oz)*
109	Broccoli, cooked (½ c)
89	Tomatoes, cooked (½ c)
85	Papaya (1)
76	Tomato, fresh (1)
54	Flounder/sole(3 oz)*
49	Asparagus, cooked (½ c)

*Preformed vitamin A. The rest of the items on the chart derive vitamin A from beta-carotene.

CAROTENOIDS (kah-ROT-eh-noyds) a group of pigments (yellow, orange, and red) found in foods. See the discussion on phytochemicals later in the chapter for more about this family of compounds.

AGE-RELATED MACULAR DEGENERATION oxidative damage to the central portion of the eye—called the macula—that allows you to focus and see details clearly (peripheral vision remains unimpaired).

RETINOL one of the active forms of vitamin A.

RETINOL EQUIVALENTS (RE) a measure of the amount of retinol the body will derive from a food containing preformed vitamin A or beta-carotene. Note that some tables list vitamin A in terms of *International Units (IU)*. See Appendix D for methods of converting from one measure to another.

CHLOROPHYLL the green pigment of plants that traps energy from sunlight and uses this energy in photosynthesis (the synthesis of carbohydrate by green plants).

The major sources of vitamin A (in the form of beta-carotene) are almost all brightly colored in hues of green, yellow, orange, and red. Any plant food with significant vitamin A activity must have some color, since beta-carotene is a rich, deep yellow, almost orange color. (Preformed vitamin A is pale yellow.) The dark green leafy vegetables contain large amounts of the green pigment **chlorophyll,** which masks the carotene in them.

In the United States, about half of the vitamin A consumed in foods comes from fruits and vegetables, and about half of that comes from *dark* leafy greens, such as broccoli and spinach and *rich* yellow or *deep* orange vegetables, such as winter squash, carrots, and sweet potatoes. The other half of the vitamin A comes from milk, cheese, butter, and other dairy products, eggs, and a few meats, such as liver. When whole milk is processed to produce fat-free milk, the vitamin is lost along with the fat that is skimmed. (Remember that vitamin A is found in the fat.) In the United States and Canada, fat-free milk is fortified with the nutrient to compensate for its loss. Likewise, margarine is fortified to provide the amount of vitamin A typically found in butter. (Milk and margarine are also fortified with vitamin D.)

One of the best and easiest ways of ensuring that you meet your vitamin A needs is to consume generous amounts of a variety of dark green and deep orange vegetables and fruits. Because these foods provide such an abundance of beta-carotene, along with other nutrients, most dietary guidelines advise eating at least five servings of fruits and vegetables daily, including at least one dark green or deep orange item every other day.

Vitamin D

Vitamin D is a member of a large bone-making and bone-maintenance team composed of several nutrients and other compounds, including vitamin C and vitamin K; hormones; the protein collagen; and the minerals calcium, phosphorus, magnesium, and fluoride. Vitamin D's special role involves assisting in the absorption of dietary calcium and helping to make calcium and phosphorus available in the blood that bathes the bones so that these minerals can be deposited as the bones harden. In addition, vitamin D acts very much like a hormone—a compound manufactured by one organ of the body that affects another. Indeed, vitamin D exerts an influence on a number of organs, including the kidneys and the intestines.

The precursor of vitamin D is made from cholesterol in the liver. This is one of the body's many good uses for cholesterol.

Another particularly unique feature of vitamin D is that the body can synthesize it with the help of sunlight, regardless of dietary consumption of the

Sun exposure should always be moderate. It takes a minimum amount of exposure—5 to 15 minutes on the hands, arms, and face—for the body to meet vitamin D needs. The rest of the time, body parts exposed to ultraviolet rays should be protected with a formula containing a sun protection factor of 15 or more to protect against skin cancer.

TABLE 6-10 VITAMIN D IN FOODS

(μg)	Sources
3.0	Shrimp (3 oz)
2.5	Fat-free milk (1 c)
1.3	Corn flakes (1 cup)
0.9	Cod liver oil (1 tbsp)
0.6	Egg (1)
0.5	Margarine (1 tsp)
0.3	Hot dog (1)

nutrient. Vitamin D is commonly called the sunshine vitamin because the liver makes a vitamin D precursor, which is converted to vitamin D with the help of the sun's ultraviolet rays. The liver alters the molecule, and the kidney alters it further to produce the active form of the vitamin. This is why diseases affecting either the liver or the kidneys, which in turn upset vitamin production, may ultimately lead to bone deterioration.

Because the body can make vitamin D with the help of sunlight, you can meet your needs for the nutrient either via sun exposure or through diet. However, significant amounts of the nutrient come from only a few animal foods—notably eggs, liver, and some fish (see Table 6-10). And even in these, the vitamin D content varies greatly.[16]

Although vitamin D is not prevalent in the diet, most adults, especially those living in sunny, southern regions, need not worry about the vitamin D content of the foods they eat because their bodies are getting plenty of the nutrient as a result of sun exposure. Sun exposure of the face, hands, and arms for just 5 to 15 minutes several times a week is usually all it takes. However, people who live in northern parts of the country—above an imaginary line drawn between Boston and the Oregon-California border—are not exposed from November through February to enough ultraviolet rays from the sun to synthesize vitamin D. The same holds true for housebound or institutionalized elderly people, who not only get outside less often than younger people but also tend to be much less efficient at producing vitamin D via the skin/sun.[17] For these people, eating vitamin D-rich foods, such as fortified milk, fatty fish (including sardines, herring, mackerel, and swordfish), eggs, and some fortified cereals, is particularly important.

As discussed earlier in the chapter, children who fail to get enough vitamin D characteristically develop bowed legs, which are often the most obvious sign of the deficiency disease rickets (refer to the photograph on page 167). In adults, vitamin D deficiency causes **osteomalacia,** most often in women whose diets lack calcium, who get little exposure to the sun, and who go through several closely spaced pregnancies and prolonged periods of breastfeeding. Osteomalacia causes the bones, particularly the leg bones and spine, to become soft, porous, and weak.

Vitamin D is the sunshine vitamin. A 15-minute walk, on a clear summer day, two to three times per week, can help supply much of your needed vitamin D.

Although vitamin D deficiency depresses calcium absorption, resulting in low blood calcium levels and abnormal bone development, an excess of vitamin D does just the opposite. It increases calcium absorption, causing abnormally high concentrations of the mineral in the blood, which in turn tend to be deposited in the soft tissues. This is especially likely to happen in the kidneys, where calcium-containing stones may form as a result. Thus, vitamin D supplements should be taken only on the advice of a physician.

OSTEOMALACIA (os-tee-o-mal-AY-shuh) the disease resulting from vitamin D deficiency in adults. (Its counterpart in children is called rickets.) Osteomalacia is characterized by bowed legs and a curved spine.

Vitamin E

Vitamin E is known as a vitamin in search of a disease.[18] That's because vitamin E is widespread in the food supply, and deficiencies of the nutrient are rare. The great majority of the nutrient in the diet comes from vegetable oils and products such as margarine, salad dressings, and shortenings (animal fats such as butter and lard contain negligible amounts of the nutrient). Soybean, cottonseed, corn, and safflower oils contain generous amounts of vitamin E, as do nuts and seeds. Smaller amounts come from fruits, vegetables, grains, and other foods. Wheat germ, for example, is an excellent source of vitamin E. Table 6-11 lists sources of vitamin E.

TABLE 6-11 VITAMIN E IN FOODS

(mg)	Sources
10.0	Sunflower seeds (1 oz)
6.0	Safflower oil (1 tbsp)
6.0	Wheat germ (2 tbsp)
3.0	Corn oil (1 tbsp)
3.0	Peanut butter (2 tbsp)
3.0	Peanuts (1 oz)
2.9	Canola oil (1 tbsp)
2.32	Avocado (1)
2.32	Mango (1)
1.67	Olive oil (1 tbsp)
1.60	Peanut oil (1 tbsp)
1.0	Shrimp (3 oz)

Despite the rarity of deficiency, however, vitamin E is one of the most popular vitamin supplements. For decades, people have swallowed all manner of extravagant claims reputing the power of the nutrient to improve athletic performance; increase sexual potency and performance; and prevent graying of the hair, wrinkling of the skin, development of age spots, and other signs of aging, to name just some of the claims. Vitamin E has never been proven to be a panacea for any of those problems. Vitamin E supplements are also often touted as a remedy for nighttime leg cramps. Again, the evidence to support that claim is limited, so the nutrient should not be self-prescribed as a treatment for the condition.[19]

A much more tangible link between vitamin E and heart disease and other chronic diseases is supported by accumulating scientific research. Because vitamin E performs a key role as an antioxidant in the body, scientists suspect it is involved in protecting the membranes of the lungs, heart, and other organs against damage from pollutants and other environmental hazards. Scientists believe that vitamin E residing in the fatty cell membranes that surround cells, acts as a scavenger of free radicals that enter the area.[20] When vitamin E is absent, the free radicals can attack the cell and start a chemical chain reaction that damages the cell membrane, making it leaky and ultimately causing it to break down completely. (A similar reaction can occur in fatty foods such as vegetable oils, causing rancidity.) A good deal of evidence suggests that vitamin E may protect against heart disease because it may thwart the free radicals that would otherwise damage the walls of blood vessels and contribute to hardening of the arteries.[21]

The cell membranes harbor vitamin E, and a deficiency of the nutrient causes those membranes, particularly the red blood cells, to rupture and cause a type of anemia. Although extremely rare in healthy adults, this scenario sometimes occurs in premature infants who are born before vitamin E is transferred to them from their mothers. Other groups of people who run the risk of deficiency include

those who cannot absorb fats as a result of diseases and those with certain blood disorders.

Vitamin E toxicity appears to be rare, occurring only in people who take extremely high doses. Suspected symptoms include alteration of the body's blood-clotting mechanisms and interference with the function of vitamin K.

Vitamin K

K stands for the Danish word koagulation ("coagulation" or "clotting").

The key function of vitamin K is its role in the blood-clotting system of the body, where its presence can mean the difference between life and death. It is essential for the synthesis of at least 4 of the 13 proteins involved, along with calcium, in making the blood clot. When *any* of these blood-clotting factors is absent, blood cannot clot, leaving a person vulnerable to excessive bleeding upon injury. Vitamin K works with vitamin D in synthesizing a bone protein that helps to regulate the calcium levels in the blood.

Accumulating evidence supports an active role for vitamin K in maintenance of bone health. Low levels of vitamin K in the blood have been associated with low bone mineral density, and researchers have noted a lower risk of hip fracture in older women with high intakes of vitamin K than in those with low intakes of the vitamin.[22]

INTESTINAL FLORA the normal bacterial inhabitants of the digestive tract.
flora = plant inhabitants

Vitamin K can be synthesized by the **intestinal flora**—the bacteria that reside in the digestive tract. In addition, as Table 6-12 shows, many foods supply ample amounts of the vitamin, green leafy vegetables and members of the cabbage family in particular.

Because vitamin K is obtained both in the diet and via the intestinal bacteria, deficiencies are rare and occur only under unusual circumstances. Taking antibiotics for an extended period of time, for instance, could kill some of the intestinal bacteria and thereby prompt a deficiency.

Newborn babies are the one group that is commonly susceptible to a vitamin K deficiency because a baby's digestive tract is free of bacteria until birth. After birth, the infant's intestinal tract gradually becomes populated with bacteria, but this happens over time. What's more, the formula or breast milk fed to the baby generally doesn't contain adequate amounts of vitamin K. Thus, newborns are given a dose of vitamin K to prevent the possibility of a life-threatening hemorrhage in the case of injury.

Vitamin K toxicity is rare, but it can occur when supplemental doses are taken. One group of adults who often have to keep an eye on the amount of vitamin K in their diet is people taking drugs designed to prevent the blood from clotting and causing, say, a stroke. People taking such medications (known as anticoagulants) are advised to keep their consumption of vitamin K fairly constant from day to day, because large fluctuations can limit the effectiveness of the anticlotting drugs.[23]

TABLE 6-12 VITAMIN K IN FOODS

(μg)	Sources
364	Turnip greens, raw (1 c)
148	Spinach, raw (1 c)
104	Cabbage, raw (1 c)
96	Cauliflower, raw (½ c)
89	Beef liver (3 oz)
58	Broccoli, raw (½ c)
52	Chick-peas (½ c)
25	Egg (1)
10	Nonfat milk (1 c)
10	Strawberries (½ c)

CHOOSING A VITAMIN/MINERAL SUPPLEMENT

For years, health experts have been saying that most healthy people can meet their vitamin and mineral needs with a balanced diet. Nevertheless, about half of young adults and college students pop vitamin and mineral pills. And Americans overall spend some $13 billion annually on pills, powders, liquids, and other vitamin and mineral supplements, often in the mistaken belief that such preparations will ensure proper nutrition, help reduce stress, decrease fatigue, and increase pep and energy.[24]

But before you buy a supplement, remember that most major health organizations—from the American Dietetic Association to the American Medical Association to the American Academy of Pediatrics—essentially agree that healthy children and adults should be able to get all the nutrients they need by eating a variety of foods. However, those organizations and other experts say that taking a multivitamin/mineral supplement, under the guidance of a physician or dietitian, may be in order for these particular groups of people:

- People following very-low-calorie diets
- People with certain diseases or those taking medications that interfere with appetite, digestion, absorption, or excretion of nutrients
- Strict vegetarians, whose diets may fall short in vitamin B_{12}, vitamin D, calcium, iron, and zinc
- Women who are pregnant or breastfeeding, phases that bolster the need for nutrients, including iron and folate
- Women with excessive menstrual bleeding, who may need iron supplements
- Women during their childbearing years who do not consume folate-rich or folate-fortified foods may need more folate in their diets to prevent neural tube defects in infants
- Anyone with lactose intolerance or who does not consume milk or other dairy products needs a source of calcium; those with inadequate exposure to sunlight may also need vitamin D
- Elderly people who may have difficulty choosing an adequate diet, chewing problems, or a reduced ability to absorb and metabolize certain nutrients (see Chapter 10)
- People who are recovering from surgery, burn injuries, or other illnesses that increase nutrient needs
- People with heart disease or who are at risk for heart disease and consume diets inadequate in antioxidant nutrients (vitamins C and E) and the B vitamins (folate, vitamin B_6, and vitamin B_{12})
- People with chronic diseases of the digestive tract or other conditions that lead to poor eating habits or deplete nutrient stores
- People with alcohol or other drug addictions are likely to have a shortage of vitamins and minerals in their diets

If you decide to start taking a supplement, keep the following points in mind when choosing one:

- Remember that price is not an indication of quality. Many products sold at major retail chains and drugstores are just as high in quality as pricier versions sold in health food stores.
- Look for a product that meets high standards for manufacturing. One way to do this is to check the label to see whether the product meets USP standards—manufacturing practices set forth by the U.S. Pharmacopeia, the organization that establishes drug standards. The organization's standards require that a supplement be able to disintegrate and dissolve thoroughly in the stomach within a certain period of time, thereby increasing the chances that the nutrients inside are absorbed and used by the body.
- Look for a bottle or package that carries an expiration date. If it doesn't, you run the risk of buying a product that has been sitting on a shelf for an indefinite period of time. After a while, the product may lose its potency.
- Look for a supplement that contains both vitamins and minerals, with no more than 100 percent to 150 percent of the recommended Daily Values for each. For the most part, nutrients work in concert with one another, promoting the body's ability to make use of them. Products that include a balanced mix of vitamins and minerals are the best bet for most people.
- Steer clear of products containing extraneous substances such as PABA, hesperidin, inositol, and bee pollen. These nonvitamin substances have never been proved essential to humans and only add to the price of the supplement.
- Be wary of spurious claims, such as statements that suggest that expensive "natural" nutrients are preferable to synthetic versions. As far as the body is concerned, natural and synthetic nutrients are one in the same. An exception to this rule may be vitamin E, which is more readily absorbed in its natural form.
- Buy products sold in childproof bottles or packages if you have children around. Vitamins and minerals, especially iron, can be highly toxic to children. Every year, tens of thousands of children swallow excess vitamin/mineral supplements, and iron-tablet overdoses alone are one of the top causes of accidental death in youngsters.
- Be wary of taking multivitamins that also contain herbs. Although herbal products are considered to be dietary supplements, the unregulated herbal industry of today is a buyer-beware market. The Nutrition Action that follows on page 189 discusses regulatory issues regarding herbal remedies and includes a list of proposed effects and potential adverse reactions for a wide variety of herbal supplements.

Vitamin Preservation

Tomatoes were considered poisonous when first brought from the Americas to Europe 500 years ago. For a long while it was thought that the "red berries" were more appropriate for deterring ants and mosquitoes than for eating.[25]

Buying vitamin-rich fruits and vegetables at the market is one of the first steps to obtaining plenty of those all-important nutrients in your diet. The next step is storing and cooking those foods in ways that minimize the loss of vitamins that can occur as a result of improper storage and preparation. To help get the most from the produce you buy, use the following tips on fruit and vegetable storage and preparation.

- Shop for produce at least once a week. The longer fruits and vegetables stay in your refrigerator before eating, the more nutrients that are likely to be lost.
- Store fruits and vegetables (other than bananas and potatoes) in your refrigerator rather than in a fruit bowl or on the kitchen counter. Chilling slows the metabolic rate of the cells of a fruit or vegetable, which in turn causes the cells to use less of the item's own nutrient supply. Thus, chilling prevents nutrient depletion.[26]
- Place potatoes in a basket or burlap or paper bag and store in a cool, dark place such as a cellar (the darkness will prevent the potatoes from greening). If you put potatoes in the refrigerator, do not keep them there for more than a day or two. Chilling causes the starch in the potato to turn to sugar, thereby altering its flavor and texture.[27]
- Store fruits and vegetables whole, peeling and cutting only what you need immediately before cooking or eating whenever possible. Once you cut into the skin of an item and expose it to air, vitamin loss begins. That's because the oxygen in the air causes a chemical reaction that "rusts" the produce (picture how brown an apple slice looks after an hour or two). After slicing, the vitamin C content of oranges, grapefruits, tomatoes, and strawberries begins to decline. If you do have leftover, cut produce, wrap it tightly in airtight plastic or store it in an airtight container inside the refrigerator. In addition, keep fruit juice in containers with tight-fitting lids; opened cans or bottles of fruit juice are susceptible to rapid vitamin loss.
- In the refrigerator, keep whole fruits and vegetables in perforated plastic bags. The perforated plastic allows some airflow while helping to maintain a moist environment, preventing produce from drying out. Thus, the bag creates an optimal environment for the preservation of the food's nutrients. Consider that water loss not only decreases the appeal of a fruit or vegetable but also may go hand in hand with nutrient loss.[28]
- Store frozen fruits and vegetables in a freezer kept at 0 degrees Fahrenheit (−17.7 degrees Centigrade) or less to ensure that they are solidly frozen and therefore retain their nutrients. To be sure, since the freezing process itself destroys few nutrients, the nutrient content of frozen foods is similar to that of fresh foods, provided they are stored properly. In fact, properly stored frozen foods often contain more nutrients by the time they reach the table than fresh fruits and vegetables that have been sitting in the supermarket for a day or two. Still, foods stored at temperatures higher than 0 degrees Fahrenheit may appear to be frozen on the exterior, but much of the interior may actually be partially thawed. In this unfrozen state, vitamin loss can occur quickly. To ensure that your freezer is sufficiently cold, buy a freezer thermometer and make sure its temperature is at or below 0 degrees Fahrenheit.
- Try to eat frozen vegetables within a month or two of purchase. Although they will last in your freezer indefinitely, nutrient losses occur over time even in properly stored vegetables. For example, frozen vegetables such as snap beans, broccoli, cauliflower, and spinach can lose 60 percent to 75 percent of their vitamin C in a 0-degree Fahrenheit freezer over the course of a year.[29]
- If you prepare fresh vegetables for freezing at home, dip them in boiling water for a minute or two. This process, known as blanching, deactivates enzymes that would otherwise deplete the food's nutrients. Immediately dunk the vegetables in cold water to cool, place them in plastic containers with tight-fitting lids, and freeze.
- When it comes to freezing fruit, note that most varieties can be sliced and frozen in containers or freezer bags without any special preparation. However, light-colored

varieties such as apples, bananas, pears, and peaches sometimes turn brown in the freezer. To help decrease the chances of discoloration, such fruits should be dipped in orange or lemon juice before freezing.[30]

- Don't throw away the liquid from canned vegetables. The liquid bathing the cut, canned vegetables leaches water-soluble vitamins as well as minerals such as calcium, iron, and phosphorus. Instead of getting rid of this nutrient-containing liquid, known as pot liquor, use it to make soups, cook rice, or moisten casseroles. The process used to can foods, incidentally, destroys some of the food's vitamins, such as thiamin, because during the canning process, when the food is heated to high temperatures to kill food-spoiling bacteria and yeast, heat-sensitive vitamins are also destroyed. This doesn't mean that canned foods have no place in the diet. Many vitamins hold up quite well under heat: riboflavin, niacin, vitamin D, and the vitamin A precursor, beta-carotene. Despite some nutrient losses, canned foods rank as convenient pantry stockers that provide essential nutrients year-round.[31]

- Cook vegetables in the least amount of water and for the shortest period of time possible. Water-soluble vitamins readily dissolve into cooking water, and heat destroys some as well. To minimize such losses, steam vegetables over water or boil or sauté them in very small amounts of water. Better yet, cook vegetables in a microwave oven for optimal nutrient retention. Because microwave cooking requires no water and a short heating period, few nutrient losses occur; the same goes for stir-frying.

- Plan to eat at least five servings of fruits and vegetables a day, regardless of the form in which you buy them. Remember that since the priority is to consume lots of fruits and vegetables, look for ways to do so that are realistic for your lifestyle and budget. In the best of all possible worlds, everyone would eat only vegetables purchased and prepared with optimal vitamin retention in mind. But that's not always possible. If you can't make it to the market every few days to buy fresh produce, don't give up fruits and vegetables altogether. Instead, keep canned, dried, and frozen alternatives on hand. To evaluate your intake of fruits and vegetables, take the 5 a Day challenge in the Scorecard that follows.

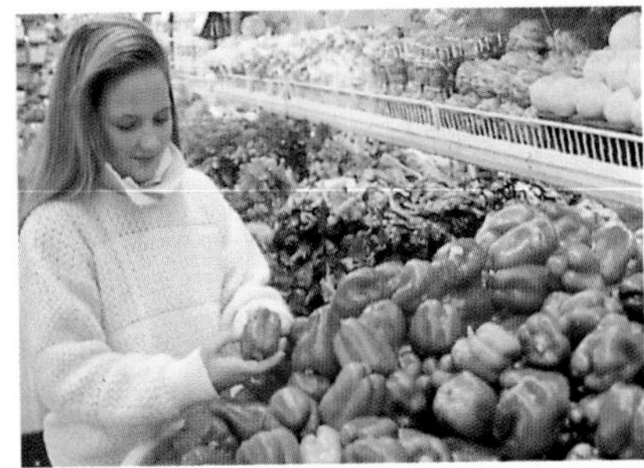

Once you've taken the time to select fresh, vitamin-rich vegetables, be sure to handle them with care when you get them home.

Wrap cut produce tightly to preserve its vitamin content.

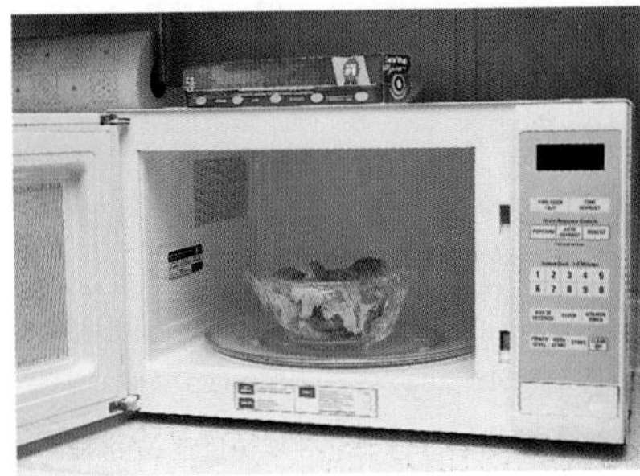

To preserve nutrients, don't boil vegetables. Instead, microwave cook them or steam them over small amounts of water.

5 A DAY PLUS SCORECARD

ARE YOU REAPING THE POWER OF PRODUCE?

The message is simple: *Five to nine servings of fruits and vegetables* every *day.* Most Americans, however, find this simple bit of nutritional advice challenging to put into practice. Overwhelming evidence points to the health benefits derived from diets rich in fruits and vegetables, including an enhanced immune system and reduced risk for many chronic conditions, such as heart disease, high blood pressure, certain types of cancer (e.g., prostate, stomach, colon), and age-related vision loss. To achieve the maximum benefits from the vitamins, minerals, antioxidants, and phytochemicals available in produce, you should aim for a minimum of five servings every day. Be adventurous: Select from as wide a variety of fruits and vegetables as possible. Be sure to frequently include members of the "superstar" categories of fruits and vegetables listed below. Evaluate the Five a Day goal in your own diet by answering the following questions and putting a check mark in the appropriate number of boxes.

TAKE THE 5 A DAY CHALLENGE

	NUMBER OF SERVINGS (EACH BOX REPRESENTS ONE SERVING)*				
I start the day with a serving of fruit at breakfast.	☐	☐	☐	☐	☐
I snack on fruits or vegetables during the day.	☐	☐	☐	☐	☐
I include a serving of fruit or a vegetable at lunch.	☐	☐	☐	☐	☐
I eat one or more vegetable servings at dinner.	☐	☐	☐	☐	☐
I choose berries, melon, pears, apples, or other fruit for a sweet dessert.	☐	☐	☐	☐	☐
Total Boxes Checked: ______					

EVALUATE YOUR INTAKE OF THE SUPERSTAR FRUITS AND VEGETABLES

	Daily	Occasionally	Seldom	Never
I eat citrus fruits (oranges, grapefruits, etc.).	☐	☐	☐	☐
I eat berries (strawberries, blueberries, etc.), kiwi, or grapes.	☐	☐	☐	☐
I eat cruciferous vegetables such as cabbage, broccoli, cauliflower, Brussels sprouts, and kale.	☐	☐	☐	☐
I eat dark green leafy vegetables such as spinach, turnip greens, collard greens, mustard greens, or beet greens.	☐	☐	☐	☐
I eat deep yellow, orange, and red fruits and vegetables such as sweet potatoes, carrots, winter squash, pumpkin, cantaloupe, nectarines, peaches, apricots, mangos, papayas, tomatoes, red peppers, and watermelon.	☐	☐	☐	☐

SCORING

Less than five boxes checked:

Ouch! You're missing out on the health benefits from produce. Take the 5 a Day challenge.

At least five boxes checked:

Good! You're getting the *minimum* number of servings of fruits and vegetables for a healthy diet. Be sure to select a variety of fruits and vegetables and to include servings from the superstar groups as well.

Five to nine boxes (or more) checked:

Excellent! You're getting the recommended number of servings for fruits and vegetables in a healthy diet. Be sure to select a variety of fruits and vegetables and to include servings from the superstar groups as well.

*A serving of fruit equals one medium piece of fruit, ½ cup cut or cooked fruit, ¼ cup dried fruit, or ¾ cup fruit juice. A serving of vegetables equals 1 cup leafy vegetables, ½ cup cooked or cut vegetables, or ¾ cup vegetable juice.

NUTRITION ACTION

Medicinal Herbs

Throughout human history, people have relied on herbal medicines; in fact, the use of herbs and medicinal plants for any number of ailments is believed to be a universal phenomenon. The World Health Organization estimates that about 80 percent of the world's population depends on traditional medicine, involving the use of herbs, for primary health care.[32] It is only with the development of twentieth-century Western medicine that synthetic chemicals found their place in the medical system. Yet, even in a modern pharmacy in the United States today, over 25 percent of medicines are extracted from plants or are synthetic copies or derivatives of plant chemicals.[33]

In the United States, plant medicines composed of whole plants (crude drugs) or complex extracts are sold as dietary supplements because natural (or herbal) medicines are not economically viable candidates for drug research and development. Most botanicals contain one or several relatively dilute compounds, and for this reason, they tend to have milder actions than the more concentrated single chemicals found in most drugs. Therefore, herbal medicines usually take longer to act than regular medicinal products and few herbs have the potency of a prescription.

Pharmaceutical companies are less willing to spend the millions of dollars to fund research on plants that grow in the wild and therefore cannot be patented, and most herb manufacturers don't have the funds to support large research studies. Despite this, herbal products are becoming increasingly popular in the United States, with annual growth rates in natural food stores as high as 60 percent to 80 percent for medicinal herbs in bulk, capsules, extracts, tinctures, tablets, and teas for medicinal use.[34] Since 1990, the use of herbal medicines in the United States has increased by 380 percent.[35] In 1997, American consumers spent approximately 5.1 billion dollars on herbal supplements, and it is anticipated that by the year 2010 sales of herbal products will be in the range of 25 billion dollars annually.[36]

What is driving the trend toward increased use of herbal products? Primarily the growth is consumer driven. Most consumers learn of herbal products through the media either in magazines, television or radio commercials, or by word of mouth from others. Other factors responsible for this trend are an interest in returning to a more natural lifestyle; dissatisfaction with the current state of Western health care; the unwanted side effects of prescription drugs; the spiraling cost and disarray of managed health care; aging "baby boomers" who want a better quality of health; a strong interest in alternative and complementary therapies; and finally, a whole arena of sales and marketing campaigns, often making use of famous personalities to market herbal products to consumers. Several of the major pharmaceutical companies are now marketing a line of herbal products that consumers can purchase right in the grocery store. See Table 6-13 for a list of herbs thought to be effective and those that should be avoided.

Many herbal product users believe that herbal medicines are the "natural" way to good health. However, natural is not always synonymous with safe. There are no regulations that oversee the manufacture and marketing of herbal supplements. And when it comes to botanicals, several species may look identical, but one may, in fact, be toxic. If the person collecting the herbs is not entirely knowledgeable, there is a danger that the toxic herb may be mixed with the medicinal

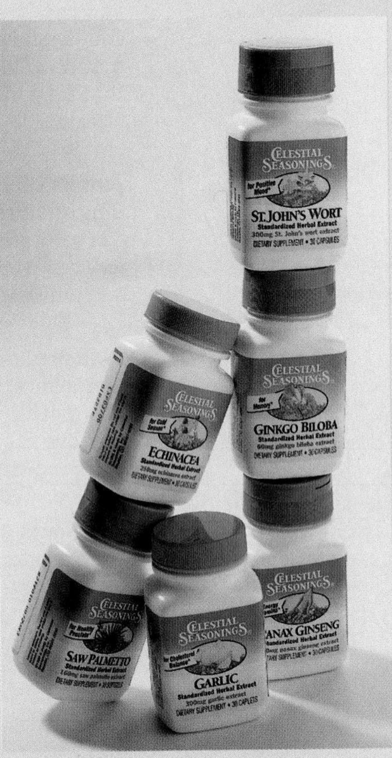

Research is currently underway in the United States to test the safety and efficacy of a few of the most popular herbs on the market today.

TABLE 6-13 THE GARDEN OF HERBAL REMEDIES

These are herbs thought to be effective.*

Herb	Why It Is Used	How It Works	Cautions
 Black Cohosh	Reduces symptoms of premenstrual syndrome, painful menstruation, and the hot flashes associated with menopause.	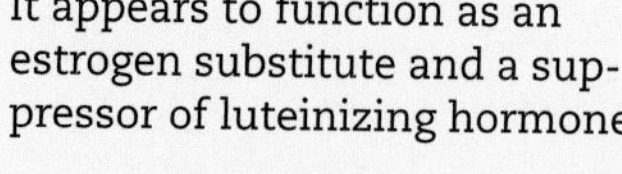 It appears to function as an estrogen substitute and a suppressor of luteinizing hormone.	Occasional stomach pain or intestinal discomfort. Since no long-term studies have been done, use of black cohosh should be limited to six months.
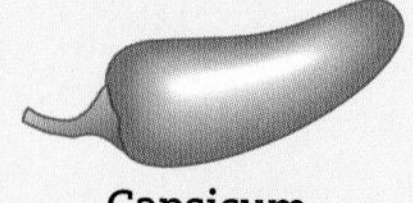 Capsicum	Best known as the hot red peppers cayenne and chili. Applied topically for chronic pain from conditions such as shingles and trigeminal neuralgia.	The active ingredient, capsaicin, works as a counterirritant and decreases sensitivity to pain by depleting substance P, a neurotransmitter that facilitates the transmission of pain impulses to the spinal cord.	Overuse can result in a prolonged insensitivity to pain. More concentrated products can cause a burning sensation. Users must avoid contact with eyes, genitals and other mucous membranes.
 Echinacea	These species of purple cornflower appear to shorten the intensity and duration of colds and flus, may help to control urinary tract infections, and, when applied topically, speed the healing of wounds.	Though echinacea lacks direct antibiotic activity, it helps the body muster up its own defenses against invading microorganisms.	Experts warn against using echinacea for more than eight weeks at a time, and against its use by people with autoimmune diseases like multiple sclerosis and rheumatoid arthritis or by those who are infected with H.I.V.
 Feverfew	The dried leaves of this plant have been shown to reduce the frequency and severity of migraine as well as its frequently associated symptoms of nausea and vomiting.	The active ingredient, parthenolide, appears to act on the blood vessels of the brain making them less reactive to certain compounds.	Most commercial preparations recommend doses that are much too high. 250 micrograms a day of parthenolide—or 125 milligrams of the herb—is an adequate dose.
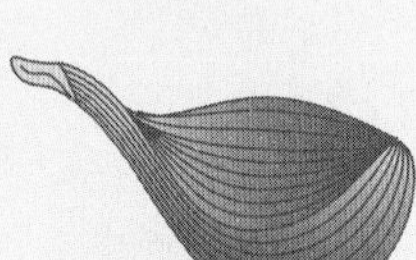 Garlic	Active against viruses, fungi, and parasites. It may also lower cholesterol and inhibit the formation of blood clots, actions that might help to prevent heart attacks.	When fresh garlic is crushed, enzymes convert alliin to allicin, a potent antibiotic. Garlic tablets and capsules containing alliin and the enzyme can be absorbed when dissolved in the intestines and not the stomach.	More than five cloves of garlic a day can result in heartburn, flatulence and other gastrointestinal problems. People taking anticoagulants should be cautious about taking garlic.
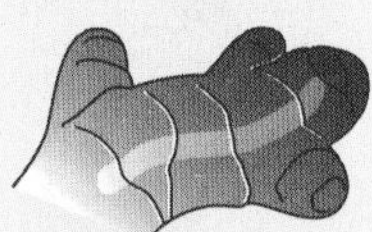 Ginger	A time-honored remedy for settling an upset stomach, ginger has been shown in clinical studies to prevent motion sickness and nausea following surgery.	Components in the aromatic oil and resin of ginger have been found to strengthen the heart and to promote secretion of saliva and gastric juices.	May prolong postoperative bleeding, aggravate gallstones and cause heartburn. There is also debate about its safety when used to treat morning sickness.
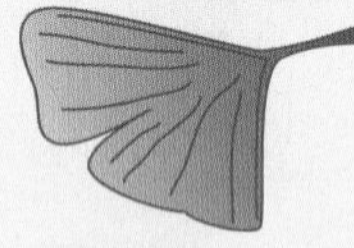 Ginkgo	Used medicinally in China for hundreds of years, Ginkgo biloba was recently reported to improve short-term memory and concentration in people with early Alzheimer's disease.	It appears to work by increasing the brain's tolerance for low levels of oxygen and by enhancing blood flow to the brain and extremities.	Possible side effects include indigestion, headache and allergic skin reactions.
 Kava	This South Pacific herb, also known as kava-kava, has a growing coterie of enthusiasts who use it to reduce anxiety, stress and restlessness and to facilitate sleep without developing problems of tolerance, dependence, and withdrawal.	The active ingredients in the root and stem act as muscle relaxants and anticonvulsants.	Should not be used with alcohol or other depressants, by people who are clinically depressed or by women who are pregnant or nursing. Kava should also not be taken for longer than three months without medical supervision.
 Milk Thistle	One of the few herbs that has been demonstrated to protect the liver against toxins. Also encourages regeneration of new liver cells.	The seeds contain a compound called silymarin, which helps liver cells keep out toxins and may promote formation of new liver cells.	When used as capsules containing 200 milligrams of concentrated extract (140 milligrams of silymarin), no harmful effects have been reported.

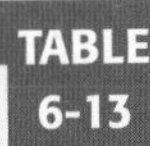

TABLE 6-13 THE GARDEN OF HERBAL REMEDIES—*Continued*

These are herbs thought to be effective.*

Herb	Why It Is Used	How It Works	Cautions
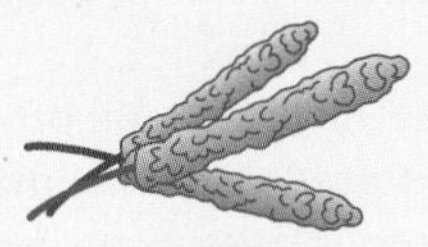 **Psyllium**	A laxative that doesn't undermine the natural action of the gut. It may reduce the risk of colorectal cancer and reduce blood levels of cholesterol.	The seeds of this herb have husks that are filled with mucilage, a fiber that swells with water in the intestines to add bulk and lubrication to the stool.	Increase in flatulence in some people, especially if a lot is consumed.
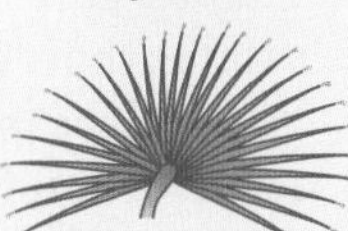 **Saw Palmetto**	Studies on patients with an enlarged prostate have shown that extracts of this palm tree can reduce urinary symptoms even though the gland may not shrink.	Nonhormonal chemicals in saw palmetto appear to work through their antiandrogen and anti-inflammatory activity.	Some experts are concerned that those taking the herb may have inaccurate P.S.A. readings, used as an early warning sign of prostate cancer.
 **St John's Wort**	Reports attest to this herb's ability to relieve mild depression. It may also have sedating and antianxiety activity.	Though products are standardized for hypericin, the chemicals responsible for the herb's activity have not yet been identified.	Based on the sun-induced toxicity of hypericin in animals, people are advised to avoid exposure to bright sunlight.
 **Valerian**	Perhaps best characterized as a mild tranquilizer.	Parts of this plant have antianxiety effects making it useful in treating nervousness and insomnia.	Long-term use can cause headache, restlessness, sleeplessness and heart function disorders.

These herbs should be avoided.

Herb	Why It Is Used	Reasons for Caution	Herb	Why It Is Used	Reasons for Caution
Aconite	Pain, rheumatism, headaches	Numerous poisonings in China	**Germander**	Digestion, fever	Stimulant can cause heart problems
Belladonna	Spasms, gastrointestinal pain	Contains three toxic alkaloids, including atropine	**Kombucha Tea**	AIDS, insomnia, acne	Can cause liver damage, intestinal problems and death
Blue Cohosh	Menstrual ailments, worms	Can induce labor	**Lobelia**	Mood booster	Can cause rapid heartbeat, coma and death
Borage	Coughs, diuretic, mood booster	May contain liver toxins and carcinogens	**Pennyroyal**	Stimulant, gastric distress	Liver damage, convulsions, and deaths
Broom	Intoxicant, diuretic, heart problems	May slow heart rhythm; contains toxic alkaloids	**Poke Root**	Emetic, rheumatism	Extremely toxic; low blood pressure, respiratory depression
Chaparral	Arthritis, cancer, pain, colds	Can cause severe hepatitis, liver failure	**Sassafras**	Stimulant, sweat producers, syphilis	Contains the carcinogen safrole. Banned from use in food
Comfrey	Cuts, bruises, ulcers	Contains toxins linked to liver disease and death	**Scullcap**	Tranquilizer	Can cause liver damage
Ephedra	Stimulant, decongestant	Contains cardiac toxins resulting in dozens of deaths	**Wormwood**	Tonic, digestion	Can cause convulsions, loss of consciousness and hallucinations

*In general, much more controlled research is needed to examine the potential active components, efficacy, and long-term safety of the herbs listed here.

Source: Copyright © 1999 by The New York Times Co. Reprinted with permission.

herb. It has happened. A further concern is that consumers have no way of knowing whether the product they are purchasing has an "effective" amount of the active compound.

In 1994, Congress passed the Dietary Supplement Health Education Act (DSHEA), which severely restricted the FDA's authority over virtually any product labeled "supplement" so long as the product made no claim to affect a disease. DSHEA allowed herbal medicines to be marketed without prior approval from FDA. What DSHEA did allow manufacturers to state on a label are how it affects a structure or function of the body, such as the claim that the herbal product can "support," "promote," or "maintain" health. DSHEA states that a product cannot claim that it affects disease, and a manufacturer cannot state on the label that the herbal product will "prevent," "treat," "diagnose," "mitigate," or "cure" disease. A disclaimer must always be included on the label, which states that "This product has not been evaluated by the Food and Drug Administration. This product is not intended to diagnose, treat, cure, or prevent any disease." Therefore, herbal products are not obliged to meet any standards of effectiveness or safety that have been established for other medicines, which require extensive laboratory and clinical trials before approval. Today a supplement is presumed safe until the FDA receives well-documented reports of adverse reactions. Consumers and professionals can call the FDA Med Alert hot line at 800-332-1088 to report adverse effects.

Consumers and public interest groups have been lobbying for the regulation of herbal supplements. DSHEA did give the FDA the power to require that supplement makers follow "good manufacturing practices." This would specify standards for sanitation, but not necessarily for efficacy or purity. The FDA has not yet mandated these practices, but it is considering it. Additionally, the supplement industry is hoping to put into place a system whereby they will monitor themselves. In this system, the National Nutritional Foods Association will randomly test members' products to determine if they match label claims. Hopefully, this movement within the industry, along with increasing consumer demand, will spur manufacturers to comply with guidelines for producing a "safe and effective" product. Pharmaceutical and herbal companies are also working to standardize product quality so consumers can feel secure that they are getting consistent amounts of the active compounds to be found in herbal products.

In October 1998, Congress established The National Center for Complementary and Alternative Medicine (NCCAM), formerly the Office of Alternative Medicine, a division of the National Institutes of Health. The center will be devoted to conducting and supporting basic and applied research and training, and will disseminate information on complementary and alternative medicine to practitioners and the public. Some of the herbs that are currently undergoing research here in the United States are garlic, St. John's wort, Ginkgo biloba, saw palmetto, echinacea, hawthorn, and cranberry.

Until we have further research studies from which to evaluate the safety and efficacy of herbal medicines, physicians, health professionals, and consumers need to continue to seek valid information and further education from reliable sources. Use the following guidelines for choosing and using herbal medicines.

- Be informed; seek out unbiased, scientific sources.
- Inform your physician, especially if taking prescribed medications.
- Do not exceed recommended doses, or use for prolonged periods.
- Know the benefits and risks as well as potential side effects.
- Never use herbal medicines if pregnant or nursing.
- Avoid prolonged storage. Herbs do not maintain their potency for as long as over-the-counter drugs.
- Call your physician or the FDA Med Alert number if you experience side effects.

To find out more about herbal medicines, some recommended publications and Web sites are:

- American Botanical Council, *The Complete German Commission E Monographs: Therapeutic Guide to Herbal Medicines*
- S. Foster and V. E. Tyler, *Honest Herbal: A Sensible Guide to the Use of Herbs and Related Products,* 1999
- J. E. Robbers and V. E. Tyler, *Herbs of Choice: The Therapeutic Use of Phytomedicinals,* 1999
- *101 Medicinal Plants* by S. Foster, 1999
- Medical Economics, *Physicians' Desk Reference for Herbal Medicines,* 1998
- HerbalGram, a peer-reviewed journal, from the American Botanical Council
- American Botanical Council: www.herbalgram.org
- Herb Research Foundation: www.herbs.org
- U.S. Food and Drug Administration: www.fda.gov
- National Institutes of Health/NCCAM: nccam.nih.gov
- U.S. Pharmacopeia: www.usp.org

NONNUTRIENTS IN FOODS: THE PHYTOCHEMICAL SUPERSTARS

One of the newest and most exciting areas of nutrition research today focuses on a class of substances in plant foods called **phytochemicals** (*phyto* is the Greek word for plant), nonnutritive substances in plants that possess health-protective benefits. Phytochemicals are the compounds that give plants their brilliant colors—for example, lycopene, a pigment that makes tomatoes red and watermelon pink—and distinctive aromas—for example, the allium compounds that give us garlic breath. These natural compounds also protect plants from the ravages of overexposure to sunlight and other environmental threats and insects.

PHYTOCHEMICALS (FIGH-toe-CHEM-icals) physiologically active compounds found in plants that are not nutrients but that appear to help promote health and reduce risk for cancer, heart disease, and other conditions.
phyto = plant

Most recently, the term phytochemicals has been popularized to refer in particular to plant chemicals that may affect health and prevent disease. Of particular interest are phytochemicals in edible plants, including fruits and vegetables, grains, legumes, herbs, and seeds. See Figure 6-3 for a sampling of the phytochemicals that have been identified, as well as common food sources and beneficial effects and attributes of each phytochemical group. In the classic sense, the naturally occurring phytochemicals are not vitamins, minerals, or nutrients; they do not provide energy or building materials. However, ongoing research shows that phytochemicals might perform important functions by acting as powerful antioxidants, decreasing blood pressure and cholesterol, preventing cataracts, reducing menopause symptoms, and preventing osteoporosis. Though cause and effect have not been firmly established, many of these compounds are currently under investigation for their roles in blocking the formation of some cancers. Scientists who study them also say that phytochemicals may have the potential to slow the aging process, boost immune function, prevent, slow, or even reverse certain cancers; and strengthen our hearts and circulatory systems.

An individual fruit or vegetable can contain many (50 or more) of the thousands of known phytochemicals, but one, or only a select few, are usually present in a large amount. For example, garlic contains more than 160 identified compounds. When a clove of garlic is cut or crushed it produces sulfur compounds, such as *allicin*, which scientists believe is partly responsible for the health benefits that have been attributed to garlic. Finding the specific chemical in a food that offers the disease-protecting potential is not easy, however. Clinical studies in which people eat foods rich in certain phytochemicals are currently underway. If health benefits, such as protection against cancer, are observed, then the active ingredient in the food must be determined. Is it the phytochemical, a vitamin, fiber, or the low-fat, low-calorie content of the increased fruit or vegetable diet that is responsible for the health benefit? Based on current research, any or a combination of these factors might be responsible.

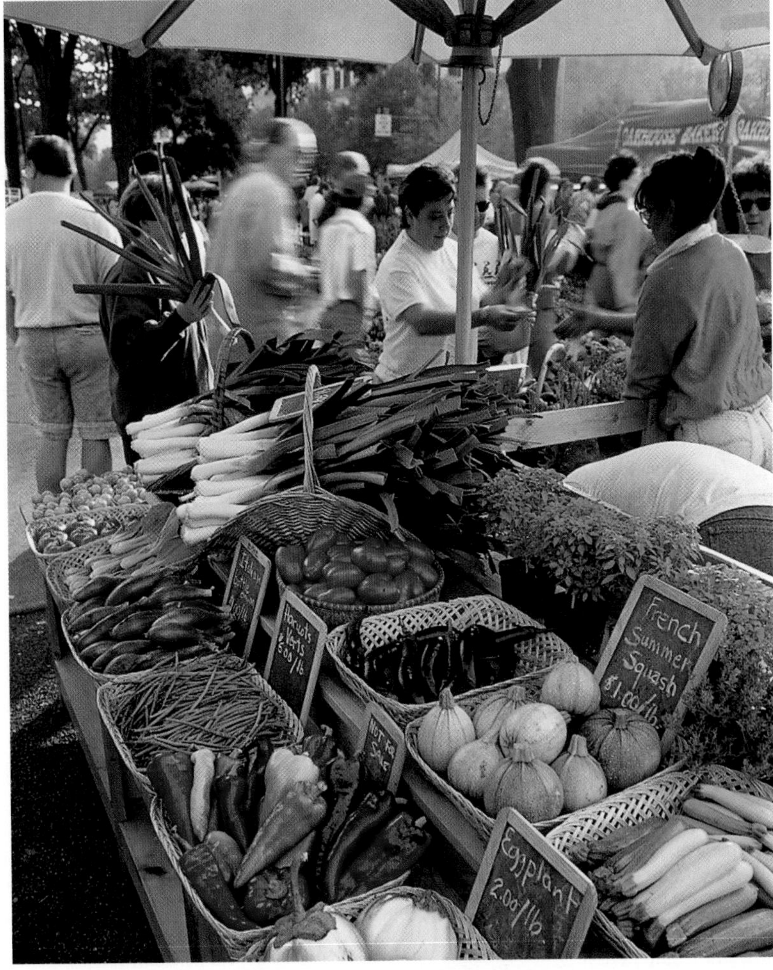

Visit your local farmers' market or produce aisle for a cornucopia of health-promoting compounds: carotenoids in carrots, spinach, and tomatoes; sulphoraphane in broccoli and bok choy, capsaicin in chili peppers. To find the farmers' market nearest you, log on to the USDA Web site at www.ams.usda.gov/farmersmarkets or consult the National Farmers' Markets Directory, available at no cost from the USDA (202-720-8317).

FIGURE 6-3
A SAMPLING OF PHYTOCHEMICALS: CLASSIFICATIONS, FOOD SOURCES AND POSSIBLE HEALTH BENEFITS

Phytochemicals are found in all fruits and vegetables and may help prevent heart disease, many types of cancer, and other degenerative conditions.

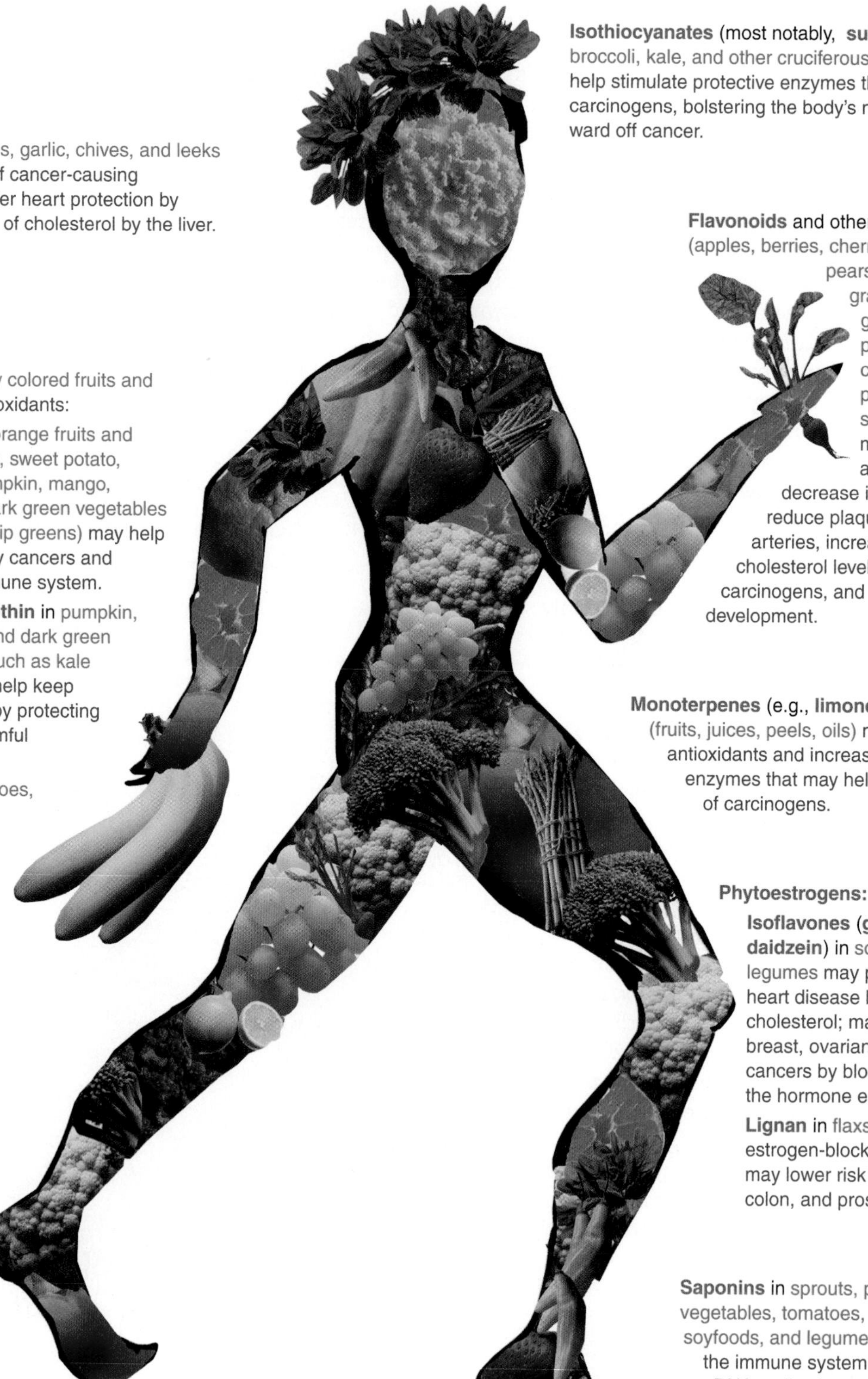

Isothiocyanates (most notably, **sulphoraphane**) in broccoli, kale, and other cruciferous vegetables may help stimulate protective enzymes that detoxify carcinogens, bolstering the body's natural ability to ward off cancer.

Allyl sulfides in onions, garlic, chives, and leeks may block the action of cancer-causing chemicals and may offer heart protection by decreasing production of cholesterol by the liver.

Flavonoids and other **phenols*** in fruits (apples, berries, cherries, citrus, grapes, pears, prunes), whole grains, nuts, black or green tea, eggplant, potato peel, red cabbage, celery, peppers, onions, soy, and red wine may act as antioxidants, decrease inflammation, reduce plaque buildup in arteries, increase HDL-cholesterol levels, deactivate carcinogens, and inhibit cancer development.

Carotenoids in deeply colored fruits and vegetables act as antioxidants:

Beta-carotene in orange fruits and vegetables (carrots, sweet potato, winter squash, pumpkin, mango, cantaloupe) and dark green vegetables (spinach, kale, turnip greens) may help reduce risk of many cancers and strengthen the immune system.

Lutein and **zeaxanthin** in pumpkin, summer squash, and dark green leafy vegetables, such as kale and spinach, may help keep your eyes healthy by protecting the retina from harmful radiation.

Lycopene in tomatoes, tomato products, watermelon, red grapefruit, and red peppers may help reduce risk of prostate and other cancers.

Monoterpenes (e.g., **limonene**) in citrus (fruits, juices, peels, oils) may act as antioxidants and increase production of enzymes that may help the body dispose of carcinogens.

Phytoestrogens:

Isoflavones (genistein and **daidzein)** in soyfoods and other legumes may protect against heart disease by lowering blood cholesterol; may lower risk of breast, ovarian, and other cancers by blocking the action of the hormone estrogen.

Lignan in flaxseeds exhibits estrogen-blocking activity and may lower risk of breast, ovarian, colon, and prostate cancer.

Indoles in cruciferous vegetables such as broccoli, kale, cauliflower, cabbage, turnip, and Brussels sprouts stimulate enzymes that make the hormone estrogen less effective, possibly reducing breast cancer risk.

Saponins in sprouts, potatoes, green vegetables, tomatoes, nuts, grains, soyfoods, and legumes may strengthen the immune system and interfere with DNA replication, preventing cancer cells from multiplying.

*Flavonoids are a subset of phenols. Flavonoids of interest include anthocyanins, catechins, ferulic acid, flavones, glycosides, quercetin, resveratrol, rutin, tangeretin, and nobiletin. The larger class of phenolic phytochemicals includes phenols, ellagic acid, capsaicin (in chili peppers), coumarin, curcumin, and others.

Source: Adapted from Food and Health Communications, Inc. Copyright 1998, www.foodandhealth.com. Used with permission.

Mechanisms of Actions of Phytochemicals

Many foods contain numerous phytochemicals, each one acting on one or several mechanisms. Some have antioxidant properties (protecting against harmful cell damage), others have anticancer properties (preventing initiation and promotion of cancer), and some have antiestrogen properties (blocking the action of estrogen and lowering the risk of some cancers).

See the Spotlight feature in Chapter 10 for a discussion on nutrition's role in cancer prevention.

Different phytochemicals have different modes of action, and an individual phytochemical may exhibit more than one mechanism of action. A phytochemical may influence one or more stages of cancer, from initiation to promotion and progression to a malignant tumor. Phytochemicals act by both *direct* and *indirect* mechanisms. They may act directly to *inhibit* enzymes that activate carcinogens or to *induce* enzymes that detoxify carcinogens. They may act indirectly by stimulating the immune response or scavenging free radicals to prevent DNA damage.

Although many phytochemicals act on cells to suppress cancer development, they may also help protect against other diseases, notably heart disease. Some phytochemicals may influence blood pressure and blood clotting, while others reduce the synthesis and absorption of cholesterol. Certain pigments (especially carotenoids) in plant foods may protect the eye against free radical damage, and thus prevent or postpone macular degeneration, which can lead to blindness in older adults. One of the most widely studied groups of phytochemicals, the isoflavones found in soy products, is currently being researched for its potential role in preventing diseases such as osteoporosis, heart disease, various cancers (mainly prostate, breast, and ovarian cancers) and alleviating symptoms associated with menopause.

See the Spotlight feature in Chapter 5 for more about the benefits of soy.

How to Optimize Phytochemicals in a Daily Eating Plan

Research indicates that pure extracts of phytochemicals in supplements are less effective than phytochemicals in whole foods. Some phytochemicals might not be metabolized in pure form, and some might not function by themselves, so they have less protective power when ingested as concentrated extracts, such as in pills.

We know that the absorption, metabolism, and distribution of some nutrients are dependent upon the presence of other nutrients. Likewise, it appears that the absorption, metabolism, distribution, and function of phytochemicals are also dependent on the combination of a phytochemical with other phytochemicals or other substances that occur naturally in food. It is not necessarily an individual phytochemical that provides protection against disease, but rather the combination of the phytochemical with other phytochemicals or food components.

Remember when Mom used to say "eat your vegetables?" This childhood advice is now supported by scientific evidence. Scientists believe that this advice would go a long way toward improving Americans' health because of the phytochemicals and antioxidant vitamins found in plant foods.

Until more is known about phytochemicals and how they function, it is best to follow the recommendations of the Food Guide Pyramid and consume at least the minimum of three vegetables and two fruits per day along with a variety of whole grains, soyfoods, other legumes, nuts, and seeds. In addition, consider the following tips:

- Consume many differently colored fruits and vegetables. For color variety, select at least three differently colored fruits and vegetables a day. The red pigment in tomatoes has different bioactive properties than the orange pigments in carrots, melon, or squash.
- Put fruit and sliced vegetables in easy-to-reach places, such as sliced vegetables in the refrigerator and fresh fruit on the table.
- Keep frozen and canned fruits and vegetables on hand to add to soups, salads, or rice dishes.

Spotlight

FUNCTIONAL FOODS—LET FOOD BE YOUR MEDICINE

Lycopene in your tomato sauce? Beta-carotene in your soup?

Today, one of the hottest areas in food science and nutrition policy is **functional foods.** Nutritionists, food scientists, food marketers, and others are exploring how today's traditional foods, and perhaps new formulations, may open doors to a healthier tomorrow.[37]

That food is intimately linked to optimal health is not a novel concept. "Let food be your medicine and medicine be your food" was a tenet espoused by Hippocrates in approximately 400 B.C.[38] Almost 2500 years later, as we begin the twenty-first century, this philosophy is once more of utmost importance, as it is the "food as medicine" philosophy that underpins the paradigm of functional foods. In this Spotlight, we will look at the role of functional foods and disease prevention, promising benefits as well as regulatory issues surrounding functional foods, and, most of all, how we can incorporate functional foods into our daily eating plan so we can reap the benefits.

MINIGLOSSARY

FUNCTIONAL FOOD a general term for foods that provide an *additional* physiological or psychological benefit beyond that of meeting basic nutritional needs. Also called *medical foods.*

DESIGNER FOODS foods "fortified" with phytochemicals or plants bred to contain high levels of phytochemicals; also known as "future foods." Genetic engineering of foods—also called *biotechnology*—is discussed in Chapter 11.

NUTRACEUTICALS a term without any legal or scientific meaning, but used to refer to foods, nutrients, phytochemicals, or dietary supplements believed to have medicinal attributes. The term is sometimes used to extend the credibility of known nutrients to a wide group of substances which have, to date, scientifically unproven effects.

What Are Functional Foods?

From the Western perspective, the concept of functional foods is still evolving. The Food and Drug Administration (FDA) defines foods as "articles used primarily for taste, aroma, or nutritive value."[39] In comparison, functional foods are foods that provide an additional physiological benefit that may prevent disease or promote health.[40]

A plethora of terms have been used interchangeably to describe foods for disease prevention and health promotion, most notably designer foods and nutraceuticals. **Designer foods** was coined in 1989 to describe foods that naturally contain or are enriched with nonnutritive, biologically active chemical components of plants (for example, phytochemicals) that are effective in reducing cancer risks.[41] The term **nutraceuticals** has been popularized by the Foundation for Innovation in Medicine and refers to "any substance that may be considered a food or part of a food and provides medical or health benefits, including the prevention and treatment of disease."[42] More recently, the Institute of Medicine of the U.S. National Academy of Sciences has defined *functional foods* as those that "encompass potentially healthful products," including "any modified food or food ingredient that may provide a health benefit beyond the traditional nutrients it contains."[43]

Exactly what foods or food components are creating excitement in the area of disease prevention?

Functional foods are not really a new concept; grocery stores are already filled with numerous foods that would meet the definition of functional foods. One of the first was calcium-fortified orange juice, which gave people who don't like milk as much bone-healthy calcium per glass as milk. We now have grain products fortified with folate to prevent birth defects and possibly heart disease. If you go for a glass of cranberry juice at the first sign of a urinary tract infection, then that juice is a functional food. Many cereal grains, fruits, and vegetables, are touted for their potential for cancer prevention benefits as well as cardiovascular protection. The current research on lycopene, present abundantly in tomatoes, ruby red grapefruit and red peppers, and reduced risk of prostate cancer is very exciting.

The Quaker Oats Man continues to smile as the cholesterol-lowering effects of the soluble fiber beta-glucan in oats are well documented, and omega-3 fatty acids in flaxseed oil and fish also appear to reduce the risk of heart disease. Research indicates that organosulfur compounds in garlic may have it all—cancer-fighting, cholesterol-lowering, antibiotic, and antihypertensive properties—which far outweigh the herb's ability to cause bad breath. Probiotics (friendly bacteria) in fermented milk products such as yogurt also appear to have multiple effects. Probiotics have been associated with reducing the risk of colon cancer, lowering cholesterol, and out-competing potentially disease-causing bacteria in the gastrointestinal tract. Soy protein has been the focus of intense research efforts because of its ability to lower cholesterol and possibly reduce the risk of cancer, osteoporosis, and symptoms associated with menopause.[44]

Two manufacturers have recently developed a margarine spread that helps lower cholesterol, according to peer-reviewed clinical studies sponsored by the manufacturers. The key ingredients in these spreads are plant stanol esters. *Benecol* uses a stanol ester derived from pine trees and *Take Control* uses one made from soy. The FDA has recently given the manufacturers approval for their label to state that the products "promote healthy cholesterol levels."

There is still so much that is unknown when it comes to functional foods. Dr. Gary Beecher, research chemist with USDA's Agriculture Research Service, indicated the need for the USDA and others to conduct additional research to clarify the role of phytochemicals in health promotion. This will also provide plant physiologists and human nutritionists with vital information needed to develop an enhanced food supply and better target dietary recommendations for the general public.[45]

Visit your nearest grocery store for a wide assortment of functional foods.

Exactly what is driving this increasing interest in functional foods today?

Currently, there are several factors responsible for the prevalence of functional foods in society today: scientific advances, consumer demand, an aging population, increasing health care costs, technical advances in the food industry, and a changing regulatory environment. Consumers want good health; they no longer view food as merely a means of providing sustenance or preventing classic nutrient-deficiency diseases. Food is now viewed as a "miracle medicine." More than ever, consumers are looking at food and food ingredients and their role in disease prevention. Many consumers are using a variety of functional foods to lower cholesterol or reduce the risk of heart disease. They are using foods such as garlic, omega-3 fatty acids, flaxseed, oat products, and folate-enriched foods. Consumers are looking for foods that might reduce their use of drugs and other medical therapy.

We have observed major technical advances in the food industry. For years the food industry focused on "taking out the bad stuff" (fat, sodium, cholesterol, calories) and restoring or enhancing favorable nutrients (calcium, folate, fiber). We may soon see the day where we have tomatoes with extra lycopene, sulforaphane-enriched broccoli, and sweet potatoes with extra carotene.

What are the regulatory issues regarding functional foods?

Do you remember the oat bran craze of several years ago, when everything from cereal to doughnuts contained "healthful oat bran"? When new "healthful" foods hit the market manufacturers jump on the bandwagon, often at the consumer's expense. Sometimes these new foods create confusion on the part of consumers. Can you imagine a soup containing the herb St. John's wort to "give your mood a natural lift" or a tea drink containing echinacea to "fight off a cold." Foods like these are available in the market. Despite the excitement of the potential health benefits of functional foods, experts are still cautious about how functional foods should be regulated. Because of the many different types of products that come under the umbrella of "natural products with health benefits," there is confusion about what these products should be called, how they should be regulated, whether or not they raise safety concerns, and whether or not health claims should be permitted. The functional food confusion centers on two issues.

- *What health claims companies can legitimately make.* The FDA strictly limits which ingredients have enough scientific proof to claim they help prevent or treat disease and only allows such a claim in a food that is healthy overall. But the law allows more vague health claims of how a food supports bodily "structure or function."

Some companies interpret that to mean they can say a food "promotes a healthy heart" without showing proof, as long as they do not say it actually "reduces the risk of heart disease."

- *How to regulate foods with added dietary supplements such as herbs or amino acids.* The FDA must declare new foods safe before Americans buy them. However, Congress allows dietary supplements to be sold without any FDA finding that they are safe or effective. To give you a good example, the FDA did set one boundary on a functional food, Benecol, the margarine spread mentioned earlier. FDA ordered McNeil Pharmaceuticals to prove that its much-hyped Benecol cholesterol-lowering margarine was safe before it began selling. McNeil argued that Benecol was not a new food, just a dietary supplement—it is made from an ingredient in trees—and thus did not require FDA approval. The FDA cited a law that says dietary supplements cannot masquerade as foods. Benecol looks and tastes like regular margarine and is sold in supermarkets next to the butter!

In the United States, functional foods do not have a separate regulatory category so they must fit into an existing category. "The primary determinant of regulatory category is intended use" says Walter Glinsman, an advisor to the FDA. "Functional foods could be considered conventional foods, special dietary supplements, or medical foods used by physicians to manage disease. Functional foods will be judged in terms of their safe use and suitability for health-related claims."[46] Although functional foods hold much promise, experts agree that scientific evidence is still unfolding. As more research becomes available, the regulatory environment will offer more clarity for consumers.

EASY PHYTOCHEMICAL-RICH SOUP

MINESTRONE

Four antioxidant-rich veggies make this soup a real winner.

1 16-ounce package frozen broccoli, cauliflower, and carrot blend
2 15-ounce cans stewed tomatoes
2 14½-ounce cans broth (beef, vegetable, or poultry)
1 15-ounce can of great northern beans
2 ounces uncooked vermicelli (break into 2-inch pieces)
Grated Parmesan cheese

In a large saucepan, combine vegetables, tomatoes, broth, beans, and pasta; bring to a boil. Reduce heat; cover and simmer 6 to 8 minutes or until vegetables and pasta are tender. Sprinkle with Parmesan cheese.

Makes 4 to 6 (1½ cup) servings.

Nutrition information per serving: 210 calories and 2 grams fat.

How can I incorporate functional foods into my diet to receive the most health benefits?

Very easily! You can start by incorporating the Food Guide Pyramid into your daily food planning if you are not already using it. Begin with the grain group—consume several servings of whole-grain breads, muffins, and cereals, especially oatmeal. Grind flaxseed in a coffee grinder and add it to your cereal. Experiment with new grains like quinoa and amaranth.

Consume at least nine servings of vegetables and fruits daily. Be sure to include a variety of vegetables and fruits, including dark orange vegetables and fruits such as sweet potatoes, carrots, winter squash, apricots, cantaloupes, and peaches. Add dark green leafy vegetables such as kale, mustard greens, collards, and spinach, along with broccoli and Brussels sprouts. Include tomatoes as often as possible and add to that garlic and onions. Eat plenty of purple vegetables such as eggplant and red cabbage.

Fruits are a vital powerhouse of the antioxidant nutrients. Blueberries top the list for being one of the highest in antioxidants and fiber too. Be sure to include citrus fruits such as oranges and grapefruit regularly, as they contain many compounds that have vital health benefits. Red wine and red grapes are the source of resveratrol, the subject of many current research projects that are studying the compound and its potential for heart health.

Soybeans and soyfoods are loaded with health-promoting compounds. Try experimenting with soy milk, tofu, miso, and tempeh and include them regularly in your diet.

Salmon and other fatty fish (sardines, mackerel, herring, tuna) contain omega-3 fatty acids which are shown to offer heart protection. Try selecting two to three fish meals per week.

Yogurt, cheese, and milk products go a long way to offer the calcium needed for healthy bones; be sure to have at least three servings daily. Many dairy products are functional foods. The live bacteria used to ferment milk products such as yogurts and yogurt-based drinks enhance intestinal flora.

Last, but not least, top your meal off with a cup of green or black tea, a source of phenolic compounds, which many researchers are convinced protect against cancer, heart disease, and stroke. As you can see, the best advice is to consume a variety of foods that contain both known beneficial compounds and those still awaiting discovery.

Nutrition on the Web

nutrition.wadsworth.com	Go to the *Personal Nutrition* site to check for the latest updates to chapter topics or to access links to related Web sites.
www.thriveonline.com/eats/vitamins/guide.index.html	A general guide to the vitamins.
www.healthfinder.gov/searchoptions/topicsaz.htm	Search for reliable consumer information on individual vitamins.
www.nas.edu	Search for updates regarding new Dietary Reference Intakes.
www.nal.usda.gov/fnic	Search here for individual vitamins, food composition, and vitamin-related topics.
bookman.com.au/vitamins	A general information guide to the vitamins.
www.navigator.tufts.edu	Use this search engine for information about phytochemicals and functional foods.
www.nlm.nih.gov	Free access to national Library of Medicine's Medline for information searches on a variety of health-related topics.
www.eatright.org	The American Dietetic Association's site with position papers on vitamin supplements, functional foods, and many resources.
www.5aday.com	Information about the National 5 a Day program.
vm.cfsan.fda.gov/~dms/supplmnt.html	FDA Center for Food Safety and Applied Nutrition; search for supplements.
www.herbs.org	Information on herbs from the Herb Research Foundation.
nccam.nih.gov	National Center for Complementary and Alternative Medicine.

Check Yourself . . .

1. Name three water-soluble vitamins and their key functions in the body.
2. Explain the difference between beta-carotene and preformed vitamin A.
3. Name three groups of people who might need a multivitamin/mineral supplement.
4. Compare the nutrient losses likely to occur when cooking vegetables via boiling, steaming, and microwave cooking.
5. Explain why people need not obtain all their vitamin D from food.
6. Describe vitamin A's role in vision.
7. Explain why women of childbearing age are advised to consume generous amounts of folate.
8. Define phytochemical.
9. Name three reasons why major health organizations advise that everyone consume a minimum of five servings of fruits and vegetables a day.
10. Name three things to look for when purchasing a vitamin supplement.

Answers to Check Yourself questions are found in Appendix G.

7

Water and the Minerals

CONTENTS

Water—The Most Essential Nutrient

The Major Minerals

Calcium Sources Scorecard

Nutrition Action: Diet and Blood Pressure—The Salt Shaker and Beyond

The Savvy Diner: Seasoning Foods Without Excess Salt

The Trace Minerals

Spotlight: Osteoporosis—The Silent Stalker of the Bones

Scorecard: Do You Have a Bone-Healthy Lifestyle?

The pleasure of eating . . . is of all times, all ages, all conditions. . . . Because it may be enjoyed with other enjoyments, and even console us for their absence. . . . Because its impressions are more durable and more dependent on our will. . . . Because in eating we experience a certain indescribably keen sensation of pleasure, by what we eat we repair the losses we have sustained, and prolong life.

Anthelme Brillat-Savarin (1755–1826, French politician and gourmet; author of *Physiology of Taste*)

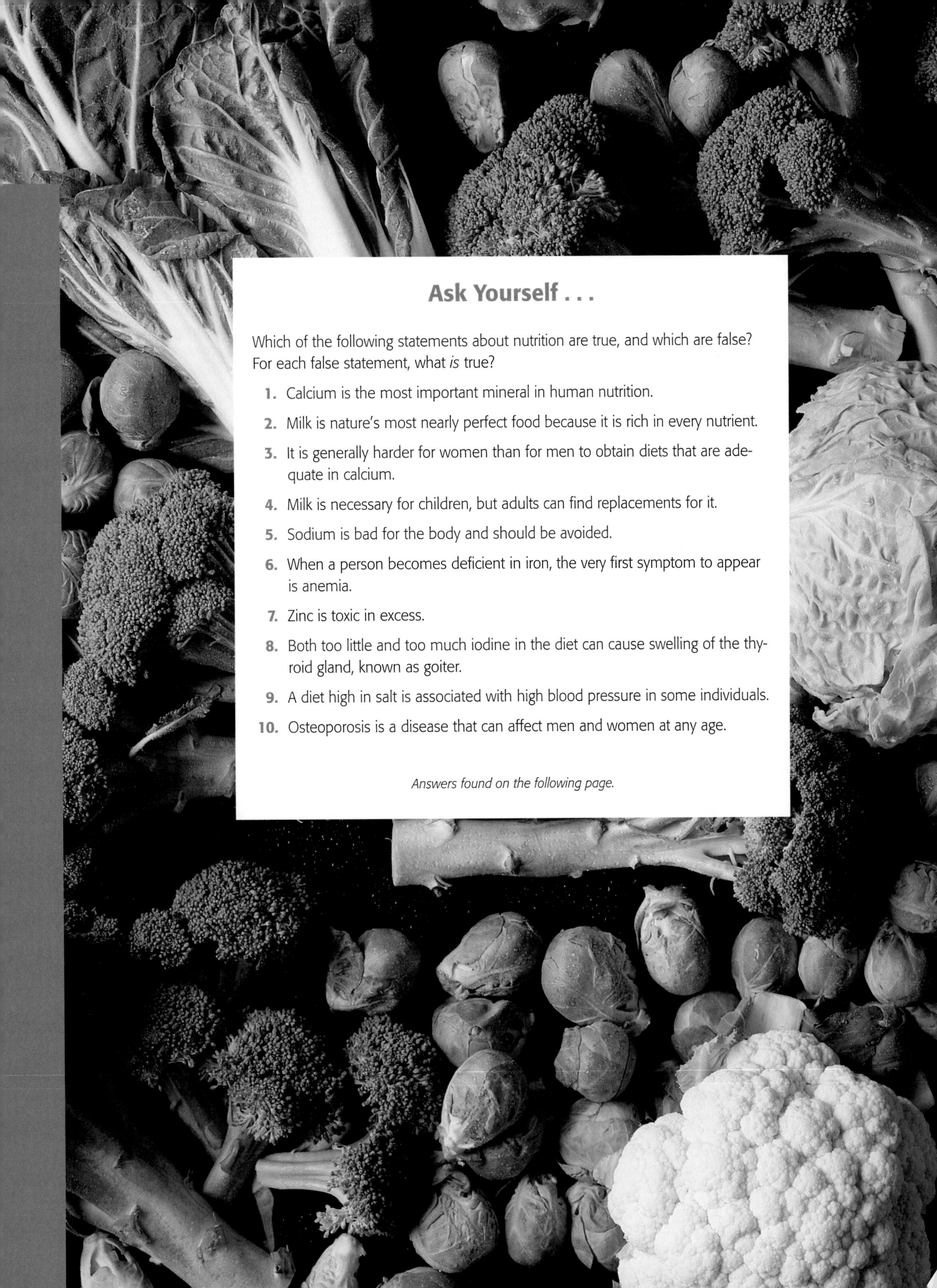

Ask Yourself . . .

Which of the following statements about nutrition are true, and which are false? For each false statement, what *is* true?

1. Calcium is the most important mineral in human nutrition.
2. Milk is nature's most nearly perfect food because it is rich in every nutrient.
3. It is generally harder for women than for men to obtain diets that are adequate in calcium.
4. Milk is necessary for children, but adults can find replacements for it.
5. Sodium is bad for the body and should be avoided.
6. When a person becomes deficient in iron, the very first symptom to appear is anemia.
7. Zinc is toxic in excess.
8. Both too little and too much iodine in the diet can cause swelling of the thyroid gland, known as goiter.
9. A diet high in salt is associated with high blood pressure in some individuals.
10. Osteoporosis is a disease that can affect men and women at any age.

Answers found on the following page.

For hundreds, if not thousands, of years, the physical and chemical properties of minerals such as gold, silver, lead, and copper, were known to metallurgists and alchemists. Even so, the role of some minerals in biological processes was recognized only within the past few hundred years. The discovery of iron in blood, for example, occurred in 1713. The identification of calcium in bone was made in 1771.[1] Not until the late nineteenth century was the role of minerals in human nutrition fully appreciated.

The previous chapter described the water-soluble and fat-soluble vitamins, their biological roles, food sources, and human requirements. This chapter discusses the minerals known to be important in human nutrition. In some respects, minerals are similar to vitamins. Like the vitamins, minerals do not themselves contribute energy (calories) to the diet. Of those known to be important in human nutrition, most minerals have diverse functions within the body and work with enzymes to facilitate chemical reactions. As with the vitamins, most minerals are required in the diet in very small amounts.

MINERALS small, naturally occurring, inorganic, chemical elements; the minerals serve as structural components and in many vital processes in the body.

INORGANIC being or composed of matter other than plant or animal.

In other respects, minerals are different from vitamins. Whereas vitamins are organic compounds, **minerals** are **inorganic** compounds that occur naturally in the earth's crust. And unlike the vitamins, some minerals (such as calcium) contribute to the building of body structures (such as bone).

As with other areas of research in human nutrition, there are many unanswered questions related to mineral metabolism. Scientists are studying the biochemical functions of minerals, the mechanisms by which they activate enzymes, the factors that control their blood and tissue levels, and the ways in which the composition of the diet affects the body's ability to make use of these dietary components. The many complex metabolic interactions of minerals make this a challenging area of research.

Although we can survive for months or even years without some vitamins and minerals, we can last only a few days without water. This chapter begins with a discussion of water—the most essential nutrient of all.

FIGURE 7-1
WATER—THE NUMBER ONE NUTRIENT

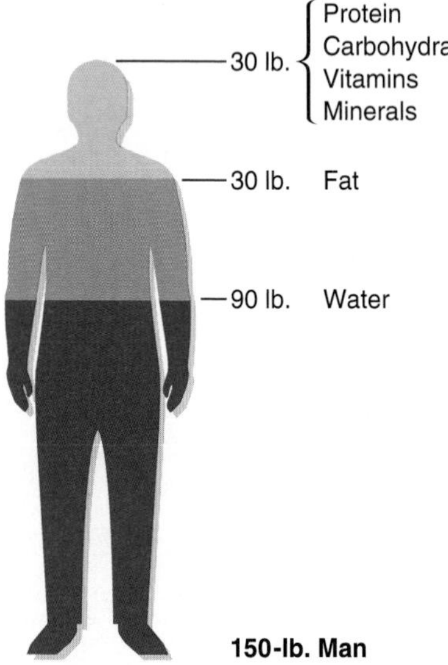

Source: C. Lecos, Water: The Number One Nutrient, *FDA Consumer,* November 1983.

 WATER—THE MOST ESSENTIAL NUTRIENT

We often take it for granted, yet water is by far the nutrient most needed by the body. A combination of hydrogen and oxygen, water makes up part of every cell, tissue, and organ in the body and accounts for about 60 percent of body weight (see Figure 7-1), even contributing to body parts thought of as "dry." Bone is more than 20 percent water, for instance; muscle is 75 percent water; and teeth are about 10 percent water.

Inside the body, water performs many tasks vital to life (see Table 7-1). It helps transport the nutrients needed to nourish the cells, for example. The blood is a river of water that flows through the arteries, capillaries, and veins, bringing each cell the exact substance and particles it requires. The same river carries away waste products formed during the reactions that take place in the cells. In addition, water acts as a shock absorber in joints and around the spinal cord, lubricates the digestive tract as well as all the tissues moistened with mucus, and surrounds and cushions an unborn child. Moreover, water plays a key role in maintaining body temperature. When water is changed from a liquid to a gas, a great deal of heat is used. Thus, when sweat evaporates, heat is released, leaving the body cooler.

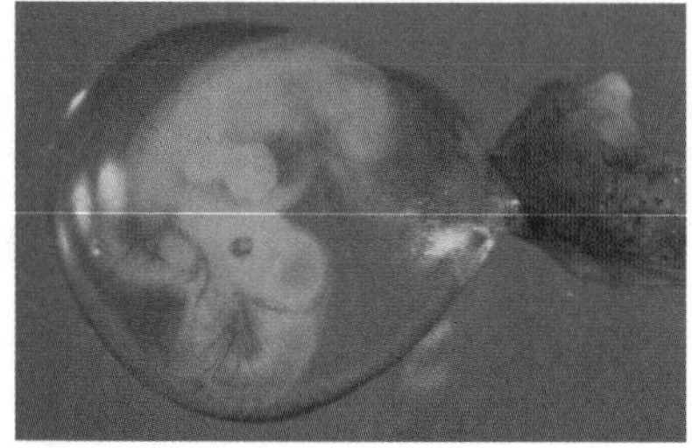

Life begins in water.

Ask Yourself Answers: **1.** False. No one essential mineral is more important than any other. **2.** False. Milk is an excellent food, but it is poor in several nutrients, including iron. **3.** True. **4.** False. Strictly speaking, milk is not absolutely necessary in anyone's diet, but its nutrients are hard to obtain from other foods, and it is recommended for both children and adults. **5.** False. Sodium is an essential nutrient, but excesses should be avoided. **6.** False. When a person becomes deficient in iron, one of the last symptoms to appear is anemia; fatigue and weakness appear first. **7.** True. **8.** True. **9.** True. **10.** True.

Water in the Diet

Adults consume and excrete some one and a half to three quarts of water a day (see Figure 7-2). Although most of the water we take in comes from juice, milk, soft drinks, and other beverages, including tap water, foods also add considerable amounts of water to the diet. Water makes up 85 to 95 percent of fruits and vegetables, for instance.[2]

Just as the sources of water in our diet vary, so too does the water we drink "straight." The makeup of water differs depending on where it comes from and how it is processed, variations that can have significant health implications. One of the most basic distinctions, hard versus soft water, is based on the concentrations of three minerals: calcium, magnesium, and sodium. **Hard water** usually comes from shallow ground and contains relatively high levels of minerals, primarily calcium and magnesium. **Soft water,** on the other hand, generally flows from deep in the earth and has a higher concentration of sodium.

Although your water utility company can tell you whether your water is soft or hard, you can probably distinguish between the two based on your own experience. Soft water helps soap lather better than hard water and leaves less of a ring on the bathtub. Hard water, to the contrary, doesn't clean clothes as thoroughly as soft water and leaves a residue of rocklike crystals on the inside of the teakettle over time. That's why many consumers prefer soft to hard water.

From a health standpoint, however, hard water seems to be the better alternative. One reason is that the excess sodium carried in soft water, even in small amounts, adds more of the mineral to our already sodium-laden diets. More importantly, soft water dissolves potentially toxic substances such as lead from pipes. Therefore, people who install water softeners in their homes for the purpose of getting cleaner laundry and better mileage from soap would do well to connect them only to their hot water lines for washing and bathing and use cold, hard water for drinking and cooking.

TABLE 7-1 THE FUNCTIONS OF WATER IN THE BODY

- Transports nutrients
- Carries away waste
- Moistens eyes, mouth, and nose
- Hydrates skin
- Ensures adequate blood volume
- Forms main component of body fluids
- Participates in many chemical reactions
- Helps maintain normal body temperature
- Acts as a lubricant around joints
- Serves as a shock absorber inside the spinal cord and amniotic sac surrounding a fetus

HARD WATER water with a high concentration of minerals such as calcium and magnesium.

SOFT WATER water containing a high sodium concentration.

Keeping Water Safe

In addition to having varying concentrations of minerals, water taken from the earth contains different levels of bacteria, microorganisms, and heavy metals such as lead. To ensure that the water that flows from the tap is safe to drink, the Environmental Protection Agency (EPA), the arm of the government responsible for monitoring municipal water supplies, sets limits for potential contaminants such as mercury, nitrate, and silver in drinking water. By law, the public must be notified by a water utility company within 24 hours of discovering any potentially dangerous contaminants in drinking water. The EPA also mandates that tap

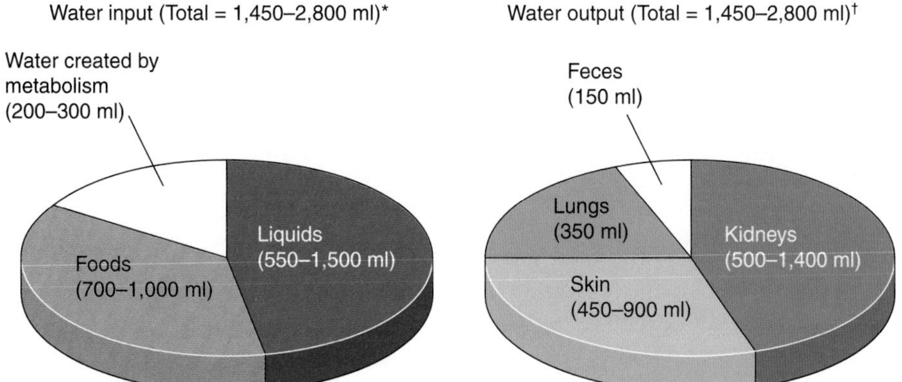

Source: Adapted from S. Sizer and E. Whitney, *Nutrition: Concepts and Controversies,* 8th ed. (Belmont, CA: Wadsworth/Thomson Learning, 1999), p. 267. Information on fluid requirements from S. Kleiner, Water: An essential but overlooked nutrient, *Journal of the American Dietetic Association* 99 (1999): 200.

FIGURE 7-2
WATER BALANCE IN THE BODY

Water enters the body in liquids and foods, and some water is created in the body as a by-product of metabolic processes. Water leaves the body through the evaporation of sweat, in the moisture of exhaled breath, in the urine, and in the feces.

*This amount equals 1½ to 3 quarts (1 oz equals approximately 30 ml).
†Adults are advised to consume 1.0 to 1.5 ml of fluid for each calorie expended. For instance, if you require 2,000 calories a day, you need approximately 2 quarts of fluid.

2,000 cal × 1 ml/cal = 2,000 ml
2,000 ml ÷ 30 ml/1 oz = 67 ml
67 ml ÷ 32 oz/qt = 2 qt

water be disinfected if bacteria levels run high in order to prevent the spread of waterborne diseases such as typhoid and dysentery. Such precautionary measures go a long way in keeping our water supply one of the safest in the world.[3]

One potential contaminant is a parasite called *Cryptosporidium*. Found in lakes and rivers that have come into contact with sewage or animal waste, *Cryptosporidium* has emerged as a health threat to vulnerable people because it is highly resistant to chlorine and other disinfectants used in municipal water supplies. In healthy people, the parasite can cause diarrhea and other flu-like symptoms that typically subside within a week to ten days. In people with weak immune systems, however, the parasite leads to severe, long-lasting gastrointestinal problems that can even cause death.

Because of the *Cryptosporidium* hazard, in 1995 the Environmental Protection Agency and the Centers for Disease Control and Prevention recommended special precautions for people with severely weakened immune systems—those with HIV infection, cancer and transplant patients taking immunosuppressive drugs, and people born with a weakened immune system. The public health groups advised those groups to talk to a health-care provider about bringing tap water intended for drinking to a boil for one minute, which will destroy any *Cryptosporidium* present, buying a special water filter system,* or drinking distilled bottled water or bottled water that has been filtered by a technique called reverse osmosis.[4]

Another potential health threat over which the EPA has little control is the level of lead that comes out of your faucet. Although the EPA has put a ceiling on the concentration of lead that may be in public water supplies, once the water leaves, say, a reservoir, unhealthy high levels of the metal may be dissolved into it (see Figure 7-3). That's because it may flow through pipes made of lead or joined by lead solder, which then can leach the metal into the water as it passes through. Granted, the government banned the use of lead-containing plumbing systems back in 1986, but dwellings built before that time may not have lead-free pipes.

The issue has generated a good deal of publicity and concern of late. Certainly attention to the matter is warranted given that once lead accumulates in the body, it begins to damage the nerves, kidneys, and liver along with the cardiovascular, reproductive, immunologic, and gastrointestinal systems. The metal is especially toxic to children and fetuses, in whom it can cause neurologic problems as severe as brain damage.

Luckily, lead levels can usually be kept to a minimum simply by "flushing" the tap, that is, letting the water run until it becomes as cold as possible, thereby ridding it of water that has been sitting in pipes and dissolving lead for any length

*For more information, visit the EPA Drinking Water Homepage at www.epa.gov. To obtain a list of suitable filters, contact NSF International at 1-800-673-8010 and ask for a list of filters certified under a regulation called Standard 53, or visit their Web site at www.nsf.org.

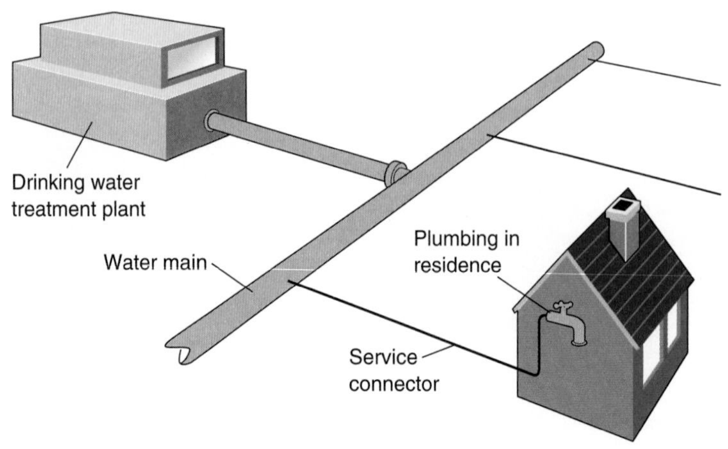

FIGURE 7-3
LEAD IN DRINKING WATER

Lead usually gets into water after the water leaves the local drinking-water treatment plant and makes its way through lead-containing plumbing systems.

of time. Pipes should be flushed with cold water only, because cold water is less likely to dissolve lead as it flows through them than warm or hot water. To be sure, in some instances, flushing is not enough to reduce harmful lead levels. Some homes or apartments may need special water treatment systems, the necessity of which can only be determined by subjecting tap water to a lead test. Most tests cost between $20 and $100; names of certified laboratories which analyze water can be obtained from a local branch of the EPA, or by calling the EPA's Safe Drinking Water Hotline (800-426-4791).[5]

Bottled Water

Americans drink nearly three times as much bottled water today as a decade ago, spending about $4 billion a year on it.[6] Although the reasons for the trend are many, bottled water's perceived health benefits fall near the top of the list. Surveys have found that about 25 percent of bottled water drinkers choose the beverage for health and safety reasons; another quarter believe it is pure and free of contaminants.[7]

Regardless of its pristine image, bottled water is not necessarily any purer or more healthful than what flows right out of the tap. Consider that the Food and Drug Administration (FDA), the bottled-water industry watchdog, does not require that bottled water meet higher standards for quality, such as the maximum level of contaminants, than public water supplies regulated by the EPA. For the most part, the FDA simply follows EPA's regulatory lead. Granted, bottled water is often filtered to remove chemicals such as chlorine that may impart a certain taste. But that doesn't make it any safer. In fact, about 25 to 40 percent of bottled water comes from the same municipal water supplies as tap water.[8] Furthermore, some bottled waters do not contain any or enough of the fluoride needed to fight cavities.[9] The only way to determine whether a certain water contains the mineral is to check with the company that bottles it.

MINIGLOSSARY OF BOTTLED WATERS

ARTESIAN WATER OR ARTESIAN WELL WATER water drawn from a well that taps a confined water-bearing rock or rock formation.

GROUND WATER water that comes from an underground body of water that does not come into contact with any surface water.

MINERAL WATER water that is drawn from an underground source and that contains at least 250 parts per million of dissolved solids. If the water contains between 250 and 500 parts per million total dissolved solids, the statement "low mineral content" must appear. If it contains more than 1,500 parts per million, the statement "high mineral content" must appear. If a cup of the water contains at least 20 milligrams of calcium, 0.36 milligrams of iron, or 5 milligrams of sodium, the product must carry nutrition labeling.

PURIFIED WATER (also know as **DEMINERALIZED WATER, DISTILLED WATER, DEIONIZED WATER, OR REVERSE OSMOSIS WATER)** water from which all the minerals have been removed, thereby eliminating the possibility that the minerals might corrode, say, a steam iron.

SPARKLING BOTTLED WATER water whose carbon dioxide (the ingredient that makes soda pop bubbly) is naturally present. That is, carbonation is not added from an outside source.

SPRING WATER water derived from an underground formation from which water flows naturally to the surface of the earth and to which minerals have not been added or taken away. It may be collected either at the spring itself or through a hole tapping the underground formation feeding the spring.

WELL WATER water derived from a rock formation by way of a hole bored, drilled, or otherwise constructed in the ground.

FROM A COMMUNITY WATER SYSTEM OR FROM A MUNICIPAL SOURCE statement that must appear on bottles containing water derived from a municipal water supply. The phrase must conspicuously precede or follow the name of the brand.

SELTZER* tap water injected with carbon dioxide and containing no added salts.

CLUB SODA* artificially carbonated water containing added salts and minerals.

TONIC WATER* artificially carbonated water with added sugar and/or high-fructose corn syrup, sodium, and quinine.

*These are not considered bottled water in government parlance. The FDA defines bottled water as water that is sealed in bottles or other containers and is intended for human consumption, excluding soda, seltzer, flavored, and vended water products.

This is not to say that bottled water is necessarily any better or worse, from a health standpoint, than tap water. It's certainly preferable to tap water for those who like its taste. The problem is that many consumers pay 300 to 1,200 times more per gallon for bottled than tap water because they think bottled water is the more healthful of the two. Bottlers add to the confusion by sprinkling terms such as "pure," "crystal pure," and "premium" on labels illustrated with pictures of glaciers, mountain streams, and waterfalls, even when the water inside comes from a public reservoir. However, the FDA has set forth regulations mandating clear labeling of bottled waters. The miniglossary of bottled waters on page 205 explains what some of the terms used on bottles actually mean.

TABLE 7-2 A GUIDE TO THE MINERALS

Mineral	Best Sources	Chief Roles	Deficiency Symptoms	Toxicity Symptoms
Major Minerals				
Calcium	Milk and milk products, small fish (with bones), tofu, certain green vegetables, legumes, fortified juices	Principal mineral of bones and teeth; involved in muscle contraction and relaxation, nerve function, blood clotting, blood pressure	Stunted growth in children; bone loss (osteoporosis) in adults	Excess calcium is usually excreted except in hormonal imbalance states; Tolerable Upper Intake Level (UL) is 2,500 mg
Phosphorus	Meat, poultry, fish, dairy products, soft drinks, processed foods	Part of every cell; involved in acid-base balance and energy transfer	Muscle weakness and bone pain (rarely seen)	May cause calcium excretion; UL is 4,000 mg
Magnesium	Nuts, legumes, whole grains, dark green vegetables, seafoods, chocolate, cocoa	Involved in bone mineralization, protein synthesis, enzyme action, normal muscular contraction, nerve transmission	Weakness, confusion, depressed pancreatic hormone secretion, growth failure, behavioral disturbances, muscle spasms	Excess intakes (from overuse of laxatives) has caused low blood pressure, lack of coordination, coma, and death; UL is 350 mg
Sodium	Salt, soy sauce; processed foods: cured, canned, pickled, and many boxed foods	Helps maintain normal fluid and acid-base balance; nerve impulse transmission	Muscle cramps, mental apathy, loss of appetite	High blood pressure (in salt-sensitive persons)
Chloride	Salt, soy sauce; processed foods	Part of hydrochloric acid found in the stomach, necessary for proper digestion, fluid balance	Growth failure in children, muscle cramps, mental apathy, loss of appetite	Normally harmless (the gas chlorine is a poison but evaporates from water); disturbed acid-base balance; vomiting
Potassium	All whole foods: meats, milk, fruits, vegetables, grains, legumes	Facilitates many reactions, including protein synthesis, fluid balance, nerve transmission, and contraction of muscles	Muscle weakness, paralysis, confusion; can cause death; accompanies dehydration	Causes muscular weakness; triggers vomiting; if given into a vein, can stop the heart
Sulfur	All protein-containing foods	Component of certain amino acids; part of biotin, thiamin, and insulin	None known; protein deficiency would occur first	Would occur only if sulfur amino acids were eaten in excess; this (in animals) depresses growth

THE MAJOR MINERALS

Minerals are traditionally divided into two large classes: the **major minerals** and the **trace minerals.** The distinction between them is that the major minerals occur in relatively large quantities in the body and are needed in the daily diet in relatively large amounts—on the order of a gram or so each. The trace minerals occur in the body in minute quantities and are needed in smaller amounts in the daily diet. Table 7-2 lists the minerals known to be essential in human nutrition and Figure 7-4 offers tips for locating the minerals in the Food Guide Pyramid. The discussions that follow focus on the minerals that are of particular

MAJOR MINERAL an essential mineral nutrient found in the human body in amounts greater than 5 grams.

TRACE MINERAL an essential mineral nutrient found in the human body in amounts less than 5 grams.

TABLE 7-2 A GUIDE TO THE MINERALS—*Continued*

Mineral	Best Sources	Chief Roles	Deficiency Symptoms	Toxicity Symptoms
Trace Minerals				
Iodine	Iodized salt, seafood, bread	Part of thyroxine, which regulates metabolism	Goiter, cretinism	Depressed thyroid activity
Iron	Red meats, fish, poultry, shellfish, eggs, legumes, dried fruits, fortified cereals	Hemoglobin formation; part of myoglobin; energy utilization	Anemia: weakness, pallor, headaches, reduced immunity, inability to concentrate, cold intolerance	Iron overload: infections, liver injury, acidosis, shock
Zinc	Protein-containing foods: meats, fish, shellfish, poultry, grains, vegetables	Part of many enzymes; present in insulin; involved in making genetic material and proteins, immunity, vitamin A transport, taste, wound healing, making sperm, fetal development	Growth failure in children, delayed development of sexual organs, loss of taste, poor wound healing	Fever, nausea, vomiting, diarrhea, kidney failure
Copper	Meats, drinking water	Helps make hemoglobin; part of several enzymes	Anemia, bone changes (rare in human beings)	Nausea, vomiting, diarrhea
Fluoride	Drinking water (if naturally fluoride containing or fluoridated), tea, seafood	Formation of bones and teeth; helps make teeth resistant to decay and bones resistant to mineral loss	Susceptibility to tooth decay	Fluorosis (discoloration of teeth); nausea, vomiting, diarrhea, UL is 10 mg
Selenium	Seafood, meats, grains, vegetables (depending on soil conditions)	Helps protect body compounds from oxidation; works with vitamin E	Fragile red blood cells, cataracts, growth failure, heart damage	Nausea, abdominal pain; nail and hair changes; liver and nerve damage; UL equals 400 µg
Chromium	Meats, unrefined foods, vegetable oils	Associated with insulin and required for the release of energy from glucose	Abnormal glucose metabolism	Occupational exposures damage skin and kidneys
Molybdenum	Legumes, cereals, organ meats	Facilitates, with enzymes, many cell processes	Unknown	Enzyme inhibition
Manganese	Widely distributed in foods	Facilitates, with enzymes, many cell processes	In animals: poor growth, nervous system disorders, abnormal reproduction	Poisoning, nervous system disorders
Cobalt	Meats, milk, and milk products	As part of vitamin B_{12}, involved in nerve function and blood formation	Unknown except in vitamin B_{12} deficiency	Unknown as a nutrition disorder

interest in human nutrition, primarily because people are known to suffer deficiencies of them.

Calcium

Calcium is the most abundant mineral in the body. Ninety-nine percent of the body's calcium is stored in the bones, which play two important roles. First, they support and protect the body's soft tissues. Second, they serve as a calcium bank, providing calcium to the body fluids whenever the supply is running low.

Although only a small part (about 1 percent) of the body's calcium is in the fluids, circulating calcium is vital to life. Calcium is required for the transmission of nerve impulses. It is essential for muscle contraction and thus helps maintain the heartbeat. It appears to be essential for the integrity of cell membranes and for the maintenance of normal blood pressure. Calcium must also be present if blood clotting is to occur, and it is a **cofactor** for several enzymes.

COFACTOR a mineral element that, like a coenzyme, works with an enzyme to facilitate a chemical reaction.

Everyone knows that children need calcium daily to support the growth of their bones and teeth, but not everyone is aware of adults' needs for daily intakes of calcium. Abundant evidence now supports the importance of calcium for adults, especially women, who need about as much calcium in their later years

FIGURE 7-4
GOOD SOURCES OF MINERALS IN THE FOOD GUIDE PYRAMID

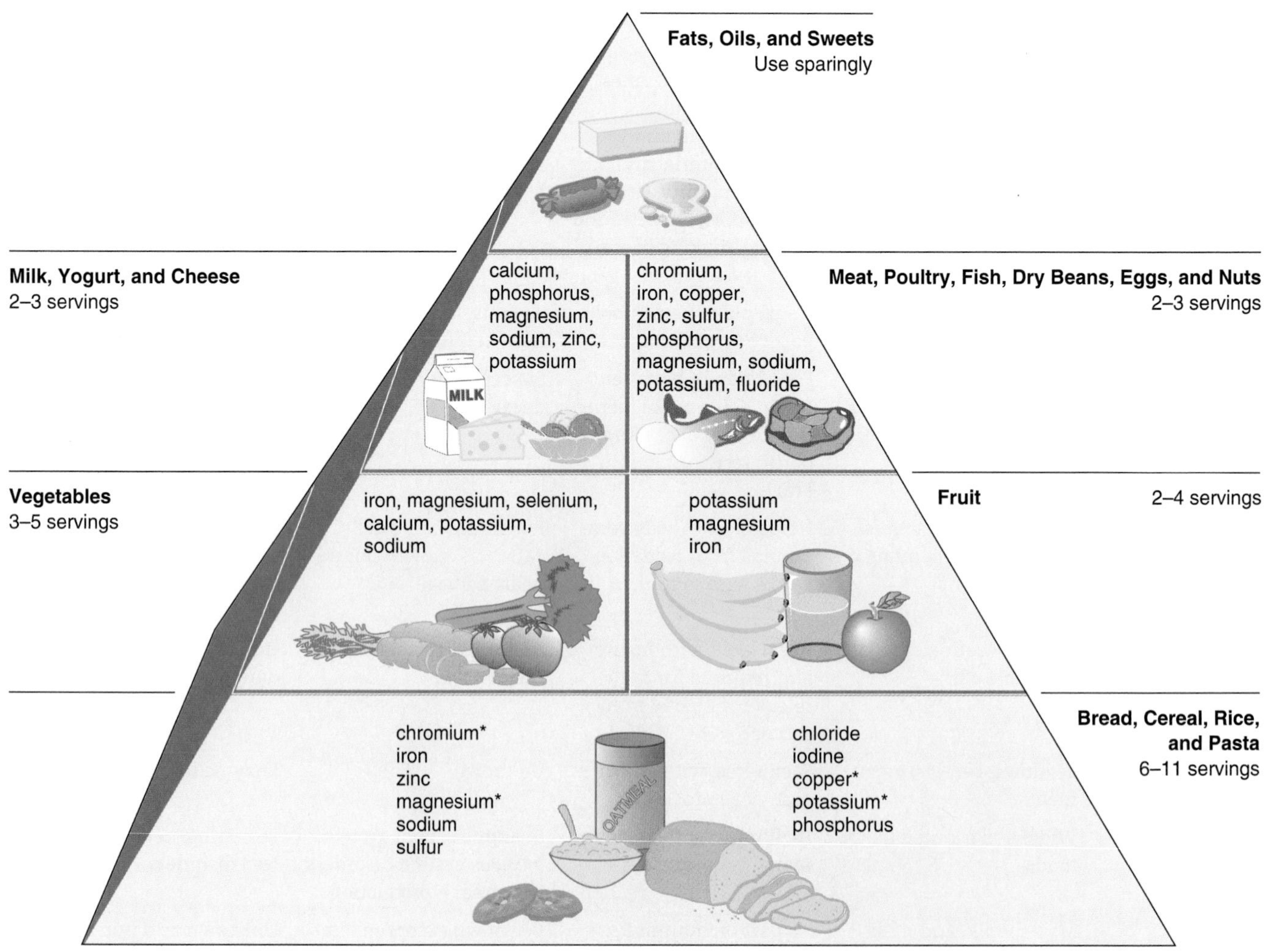

*In whole-grain choices.

as they did when they were adolescents.[10] A deficit of calcium during the growing years and in adulthood contributes to gradual bone loss, **osteoporosis,** which can totally cripple a person in later life.

Other nutrients are also important to bone growth and maintenance. Fluoride and vitamin D deficiencies, like calcium deficiencies, can cause loss of bone density. So can heredity, abnormal hormone levels, alcohol (even in moderate use), prescription medications, other drugs, and lack of exercise (especially weight-bearing exercises), but dietary calcium is one of the most important factors. This chapter's Spotlight feature discusses osteoporosis and its possible causes and prevention.

Table 7-3 shows that calcium appears almost exclusively in three classes of foods: milk and milk products, green vegetables such as broccoli, kale, bok choy, collards, and turnip greens, and a few fish and shellfish. Milk and milk products typically contain the most calcium per serving. Many greens are also good choices, but a complication enters in—absorption. It is not clear to what extent calcium is absorbed from certain green vegetables—notably, spinach and Swiss chard—while calcium is known to be very well absorbed from milk.[11] Milk contains both vitamin D and lactose, which both enhance calcium absorption and promote bone health. Milk and milk products also normally supply about 40 percent of people's intake of riboflavin. Figure 7-5 shows the current recommendations for calcium intakes.

Some foods contain **binders** that combine chemically with calcium and other minerals such as iron and zinc to prevent their absorption, carrying them out of the body with other wastes. For example, **phytic acid** renders the calcium, iron, zinc, and magnesium in certain foods less available than they might be otherwise; **oxalic acid** also binds calcium and iron. Phytic acid is found in oatmeal and other whole-grain cereals; oxalic acid is found in beet greens, rhubarb, and spinach, among other foods. These binders seem to depress absorption of the calcium present in the same food as the binder but not of calcium in other calcium-containing foods consumed at the same time. Since fiber in general seems to hinder calcium absorption, the higher the diet is in fiber (especially wheat bran), the higher it should be in calcium. This fact doesn't diminish the overall value of high-fiber foods; such foods are nutritious for many reasons, but they are not as useful as calcium sources.

Protein also affects calcium status by affecting excretion, not absorption. The higher the diet is in protein, the greater the amount of calcium excreted. This is why people in the United States and Canada are told to ingest more calcium than people in countries whose protein intakes are lower.

Alternative Sources of Calcium. Milk and milk products need not be taken as such; there are ways to include them in other foods. Yogurt is an acceptable substitute for regular milk. Puddings, custards, and baked goods can be prepared in such a

Guess what, Mommy—you were right! My teacher says milk is good for us."

OSTEOPOROSIS (OSS-tee-oh-pore-OH-sis) also known as adult bone loss; a disease in which the bones become porous and fragile.
osteo = bones
poros = porous

BINDERS in foods, chemical compounds that can combine with nutrients (especially minerals) to form complexes the body cannot absorb. Examples of such binders are **phytic** (FIGHT-ic) **acid** and **oxalic** (ox-AL-ic) **acid.**

SOURCES OF CALCIUM
Adult DRI*:
19 to 50 years: 1,000 mg
51 years and older: 1,200 mg
Broccoli: 47 mg per ½ cup cooked
Sardines: 325 mg per 3 ounces
Milk: 316 mg per cup
Pork and beans: 77 mg per ½ cup
Cheddar cheese: 306 mg per 1½ ounces
Almonds: 50 mg per 2 tablespoons

*DRI for all age groups are listed on the inside front cover.

TABLE 7-3 CALCIUM IN FOODS

(mg)	Sources
413	Yogurt, plain, low-fat (1 c)
408	Swiss cheese (1.5 oz)
350	Orange juice, calcium-fortified (1 c)
348	American cheese (2 oz)
345	Yogurt with fruit, low-fat (1 c)
325	Sardines (with bones) (3 oz)
316	Fat-free milk (1 c)
306	Cheddar cheese (1.5 oz)
300	Romano/Parmesan cheese (1 oz)
275	Shrimp (3 oz)
250	Pizza (1 slice)
248	Frozen yogurt (1 c)
200	Soy milk, calcium-fortified (1 c)
197	Turnip greens, cooked (1 c)
186	Cream soup (1 c)
182	Salmon (with bones) (3 oz)
180	Kale, cooked (1 c)
179	Collard greens, cooked (½ c)
165	Instant oatmeal, (1 packet)
160	Bok choy (1 c)
154	Cottage cheese, low-fat (1 c)
144	Pudding, chocolate (½ c)
130*	Tofu (½ c)
126	Almonds (⅓ c)
94	Broccoli (1 c)
87	Ice cream (½ c)
52	Tortilla, corn (1)
50	Dried beans, cooked (½ c)
20–40	Bread (1 slice)

*The calcium content of tofu varies depending on processing methods. Look for tofu processed with calcium salts.

Got milk? If not, be sure to consume the recommended amount of calcium—the amount found in three to four glasses of milk—from the many alternate sources of calcium available on the market today (see Table 7-3 and Table 7-4).

way that they also contain appreciable amounts of milk. Powdered nonfat milk, which is an excellent and inexpensive source of protein, calcium, and other nutrients, can be added to many foods (such as cookies, soups, casseroles, and meatloaf) during preparation. Nonfat yogurt fortified with extra milk solids is another excellent calcium source. Similar to milk and milk products in calcium richness are small fish such as Atlantic sardines or canned, pink salmon with soft edible bones.

A number of calcium-fortified foods are available, including calcium-fortified juices, fruit drinks, rice, breads, cereals, waffles, snack bars, and candy (see Table 7-4). Many times, these foods are fortified to match or exceed the amount of calcium in a cup of milk; for example, an 8-ounce serving of fortified orange juice provides about 350 milligrams of calcium. Remember to check the Nutrition Facts panel on food labels for the amount of calcium contained in the foods you eat.

The word *daily* should be stressed with respect to food sources of calcium. Because of the body's limited ability to absorb calcium, it cannot handle massive doses periodically but needs frequent opportunities to take in small amounts. To evaluate your own diet for sources of calcium, see the calcium sources scorecard on page 213.

Milk Substitutes. Some people have **milk allergy** or **lactose intolerance** and can't drink milk. For them, calcium-rich substitutes must be found. Among the possible substitutes for persons with milk allergy are boiled milk, milk of goats or other species, calcium-fortified soy milk, nondairy foods containing the nutrients of milk, and calcium supplements. People with lactose intolerance can choose enzyme-treated milk, calcium-fortified soy milk, small amounts of milk

FIGURE 7-5
CALCIUM RECOMMENDATIONS

Age (years)	Calcium needed (mg) per day	Number of milk/milk product servings per day (or the equivalent)
1–3	500	2
4–8	800	3
9–18	1,300	4
19–50	1,000	3
51+	1,200	4

TABLE 7-4 A SAMPLING OF FORTIFIED SOURCES OF CALCIUM

Product	Calcium* (milligrams)	Comments
Calcium-Fortified Fat-Free Lactaid 100 Milk (1 c)†	500	67% more calcium than regular milk. Lactose-free
VIACTIV Soft Calcium Chews, Mead Johnson (1 piece)	500	Chocolate-flavored candy (other flavors available)
Edy's/Dreyer's Fat-Free Frozen Yogurt—all flavors (½ c)†	450	50% more calcium than 1 cup of milk
Farmland Dairies Special Request Skim Plus Milk (1 c)†	405	34% more calcium than regular milk; also higher in protein
Health Valley Fat-Free Soy Moo Nondairy Soy Drink (1 c)	400	Vitamin D also added. Good source of soy protein (7 grams); lactose-free
Tropicana Pure Premium Orange Juice with Calcium (1 c)‡	350	Good way for non-milk-drinkers to get easily absorbed calcium
Swiss Miss Fat-Free Hot Cocoa Mix (1 envelope)†	300	Contains aspartame; regular Swiss Miss is low in calcium
White Wave Silk Dairyless Soy Milk (1 c)	300	Good source of soy protein
Rice Dream Original Enriched Rice Drink (1 c)	300	Made with brown rice; good alternative to dairy and soy milks
Kellogg's Eggo Homestyle Frozen Waffles (2 pieces)	300	Contains sodium (480 mg) and fat (8 grams); low-fat Eggo waffles do not contain added calcium
Power Bar—all flavors (1 bar)	300	Watch the calories (about 230 each)
Weight Watchers Fat-Free Hot Cocoa Mix (1 package)†	250	Contains aspartame; for added calcium boost, make with fat-free milk
Starbucks Frappuccino Coffee Drink—all flavors 1 (9.5-oz bottle)†	220	High in sugar (more than 7 teaspoons)
Soyco Veggy Singles (1 slice)	200	Cheese substitute made from tofu
Soyco American Flavor "Rice Slice"	200	Alternative to dairy and soy cheeses
Light n' Lively "Twice the Calcium" Fat-Free Cottage Cheese (½ c)†	200	Contains sodium (440 mg); low-fat version contains 380 mg sodium
EdenSoy Original Extra Soy Beverage (1 c)	200	Made from soybeans; vitamins D and B_{12} added
Kellogg's Nutri-Grain Low-Fat Cereal Bars—all flavors (1 bar)	200	Contains more than 3 teaspoons sugar
Kix Cereal (1⅓ c)	150	Contains sodium (270 mg)

*Calcium sources are ranked by calcium content, from most to least. Serving size is as specified on product labels. Compare to milk at 300 mg. calcium per 8-oz serving.
†Calcium source is partly or all dairy.
‡Similar calcium-fortified juices are also available from Minute Maid, Florida's Natural, and store brands.

Source: Adapted with permission from EN's Guide to Calcium in Unexpected Places, *Environmental Nutrition*, May 1999. © Copyright, 1999 by Environmental Nutrition, Inc., 52 Riverside Drive, New York, NY 10024. For subscription information: 800-529-5384.

products such as plain yogurt and aged cheese, as well as nondairy foods containing calcium, or calcium supplements.

In theory, it should be easy to choose the appropriate milk substitute. If the person is allergic to milk, in theory, the milk protein is the offending substance, and a substitute with altered or different proteins must be found—such as soy milk. If the person is intolerant to lactose, a lactose-free substitute is needed—such as enzyme-treated milk. It is often difficult, however, to determine why someone tolerates milk poorly. Both the kinds and the amounts of milk any given person can tolerate can be determined only by experimenting. The selection of a substitute may have to proceed by trial and error.

The treatment of milk with enzymes to digest its lactose offers a possible solution to the problem of lactose intolerance. An enzyme preparation (for example, LactAid) can be purchased over the counter and mixed with the milk before it's served. Or, you can take lactase tablets just before eating dairy products. Another alternative is low-lactose or lactose-free milk. Fermented dairy products, such

MILK ALLERGY the most common food allergy; caused by the protein in raw milk. Milk allergy is sometimes overcome by cooking the milk to denature the protein; it is sometimes alleviated by an abstinence from and a gradual reintroduction to milk.

LACTOSE INTOLERANCE as described in Chapter 3, an inherited or acquired inability to digest lactose as a result of a failure to produce the enzyme lactase. Lactose intolerance is prevalent in the majority of adult human populations.

TABLE 7-5 LACTOSE CONTENTS OF SELECTED DAIRY PRODUCTS

Dairy Product	Lactose (grams)
Whole milk (1 c)	11.0
2% low-fat milk (1 c)	9.0–13.0
Fat-free milk (1 c)	12.0–14.0
Chocolate milk (1 c)	10.0–12.0
Lactose-reduced low-fat milk (1 c)	3.3
Buttermilk (1 c)	9.0–11.0
Low-fat yogurt (1 c)*	11.0–15.0
Cheese (1 oz)	
Blue, cream, Parmesan, Colby	0.7–0.8
Camembert, Limburger	0.1
Cheddar, Gouda	0.4–0.6
Processed American	0.5
Processed Swiss	0.4–0.6
Cottage cheese, low-fat (1 c)	7.0–8.0
Butter (2 pats)	0.1
Ice cream/ice milk (1 c)	9.0–10.0

*Choose yogurt with "active cultures"—they help to digest the lactose found in yogurt.

Source: Adapted from K. Meister, *Much ado about milk*, American Council on Science and Health, July, 1993, p. 6.

as aged cheeses, offer the same nutrients as milk but contain bacteria that use the lactose as an energy source to do their work.[12] Table 7-5 shows the lactose contents of fermented dairy products compared with the lactose content of milk.

Phosphorus

Phosphorus is second to calcium in abundance in the body. About 85 percent of it is found combined with calcium in the crystals of the bones and teeth as calcium phosphate, the chief compound that gives them strength and rigidity. Phosphorus is also a part of DNA and RNA, the genetic code material present in every cell. Phosphorus is thus necessary for all growth because DNA and RNA provide the instructions for new cells to be formed.

Phosphorus plays many key roles in the cells' use of energy nutrients. Many enzymes and the B vitamins become active only when a phosphate group is attached. The B vitamins, you will recall, play a major role in energy metabolism. Phosphorus is critical in energy exchange.

Some lipids (phospholipids) contain phosphorus as part of their structure. They help to transport other lipids in the blood; they also form part of the structure of cell membranes, where they affect the transport of nutrients and wastes into and out of the cells.

TABLE 7-6 PHOSPHORUS IN FOODS

(mg)	Sources
422	American cheese (2 oz)
341	Cottage cheese (1 c)
325	Yogurt (1 c)
280	Salmon (canned) (3 oz)
242	Pork (3 oz)
235	Fat-free milk (1 c)
208	Sirloin steak (3 oz)
186	Turkey (3 oz)
174	Peanuts (⅓ c)
161	Hamburger (3 oz)
149	Shredded wheat (1 c)
143	Navy beans (½ c)
139	Tuna (3 oz)
127	Sunflower seeds (2 tbsp)
115	Potato (1)
104	Peanut butter (2 tbsp)
67	Corn (½ c)
51	Cola (12 oz)
46	Broccoli (½ c)
46	Wheat bread (1 slice)
45	Diet cola (12 oz)

Animal protein is the best source of phosphorus because phosphorus is so abundant in the energetic cells of animals. People who eat large amounts of animal protein have high phosphorus intakes. The intakes of phosphorus are also greater in people who regularly consume carbonated beverages because of their phosphoric acid content. Soft drink consumption is now estimated to be about 40 gallons per person each year. When soft drinks replace milk in the diet, fracture risks in both girls and women increase.[13]

The new recommended intake of phosphorus is lower than that for calcium, a level believed to provide a sufficient intake of phosphorus and ensure adequate absorption and retention of calcium. Higher intakes of phosphorus can interfere with the absorption of calcium. People need not make a special effort to eat foods containing phosphorus, since phosphorus is present in virtually all foods (see Table 7-6).

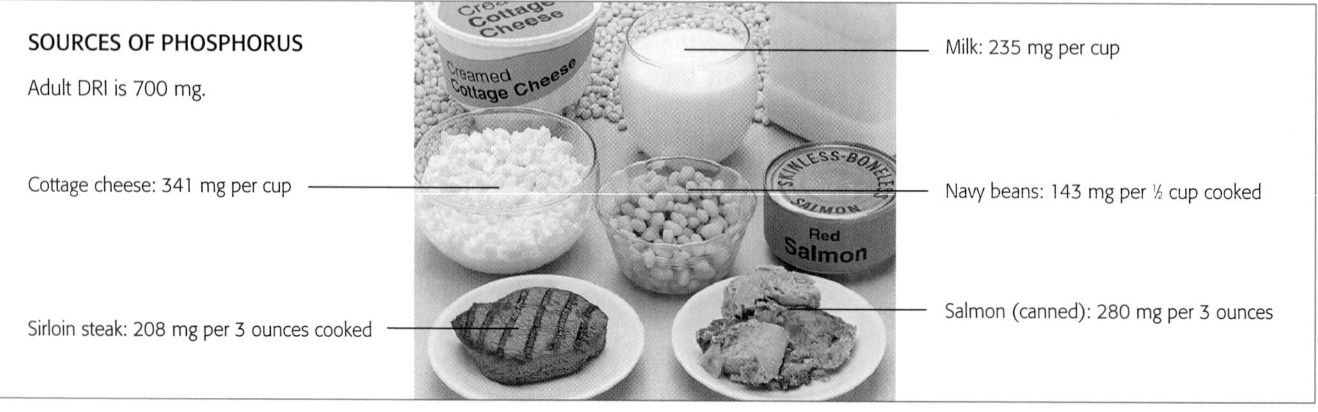

SOURCES OF PHOSPHORUS

Adult DRI is 700 mg.

Milk: 235 mg per cup

Cottage cheese: 341 mg per cup

Navy beans: 143 mg per ½ cup cooked

Sirloin steak: 208 mg per 3 ounces cooked

Salmon (canned): 280 mg per 3 ounces

CALCIUM SOURCES SCORECARD

We need at least 1,000 milligrams of calcium daily. Answer these questions by considering what you ate yesterday.

1. Did you drink milk (nonfat, low-fat, or whole) yesterday? If so, give yourself 3 points for every 8-ounce glass (1 cup). ________
2. Did you eat yogurt? Give yourself 4 points for each 8-ounce serving. ________
3. Did you eat (1 cup) calcium-fortified cereal with ½ cup of milk? Give yourself 4 points for every serving. ________
4. Did you eat 1 cup other type of cereal with ½ cup of milk? Give yourself 2 points for every serving. ________
5. Did you drink juice that is fortified with calcium? For every 6-ounce serving give yourself 2 points. ________
6. Did you eat canned salmon with bones or tofu (that's been processed with calcium) yesterday? Give yourself 3 points for each 3-ounce portion eaten (or ½ cup tofu). ________
7. Did you eat cheese yesterday? For every 1 ounce eaten, give yourself 2 points. ________
8. Did you eat cottage cheese? For each ½-cup serving give yourself 1 point. ________
9. Did you eat broccoli, kale, collards, or bok choy? For every 1 cup, raw or cooked, give yourself 1 point. ________
10. Did you have ice cream, pudding, or frozen yogurt yesterday? For a 1-cup serving give yourself 1 point. ________

Now add up all your points. *Total Points:*________________

Multiply your total points x 100:________

This gives you an idea of how many milligrams of calcium you are getting each day.

Source: Adapted from *The Calcium Connection,* © 1994 Continental Baking Company.

WORKING ON THE WEB

www.dairycouncilofca.org

Sponsor: Dairy Council of California

Description: Provides online learning activities for children and adults to evaluate their calcium intake as well as the overall adequacy of their diet.

Available Activities: The site provides information about making healthy lifestyle changes, such as finding time for breakfast, making healthy lunches, and including adequate calcium in the diet. A Personal Fitness Planner, Food Pyramid Game for children, Calcium Quiz, calcium-rich recipes, and links to related sites are included.

Web Work:

1. Make a list of what you ate or drank yesterday for breakfast, lunch, dinner, and snacks.
2. Assess your calcium status by completing the Calcium Quiz.
3. Review your results and compare your calcium intake to the recommended calcium intake for your age group. If there was a gap between your calcium intake and the recommended intake, consider adding to your diet some of the many excellent sources of calcium listed.
4. Go to the Kids & Family page and scroll through the general tips on making healthy breakfasts, lunches, and dinners with a focus on calcium.
5. Finally, retake the Calcium Quiz in a week or so to see how your intake has improved.

Helpful Hints: From the site's home page, scroll down and click on Activities. Click on Calcium Quiz and fill in your gender and age. Next, enter the number of servings you ate yesterday of the foods listed. At the end, click on Check My Calcium Intake to see how your intake compares to current recommendations. Note the tips given for improving your calcium intake. When you finish, click on the Kids & Family icon from the home page, followed by Apple of My Eye and review the tips provided.

SOURCES OF MAGNESIUM

Adult DRI (19 to 30 years) is 310 to 400 mg.

TABLE 7-7 MAGNESIUM IN FOODS

(mg)	Sources
140	Almonds (⅓ c)
126	Tofu (½ c)
119	Cashews (⅓ c)
95	Raisin bran (1 c)
85	Peanuts (⅓ c)
81	Oysters (steamed) (3 oz)
75	Spinach, cooked (½ c)
60	Black beans (½ c)
54	Baked potato (1)
46	Soy milk (1 c)
45	Avocado (½ c)
45	Black-eyed peas (½ c)
43	Yogurt, plain (1 c)
42	Brown rice, cooked (½ c)
40	Lima beans (½ c)
33	Dried figs (¼ c)
28	Fat-free milk (1 c)
27	Chicken (3 oz)
21	Sunflower seeds (2 tbsp)
21	Hamburger (3 oz)
20	Pork (3 oz)
17	Milk chocolate (1 oz)

Sulfur and Magnesium

Sulfur is present in some amino acids and in all proteins. Its most important role is in helping strands of protein to assume and hold a particular shape, thus enabling them to do their specific jobs, such as enzyme work. Skin, hair, and nails contain some of the body's more rigid proteins, and these have a high sulfur content. There is no recommended intake for sulfur, and no deficiencies are known. Only a person who lacks dietary protein to the point of severe deficiency will lack the sulfur-containing amino acids.

Magnesium acts in all the cells of the muscles, heart, liver, and other soft tissues, where it forms part of the protein-making machinery and is necessary for the release of energy. Magnesium also helps to relax muscles after contraction and promotes resistance to tooth decay by helping to hold calcium in tooth enamel. Bone magnesium seems to be a reservoir to ensure that some will be on hand for vital reactions regardless of recent dietary intake. Areas of the country that have hard water—higher in magnesium and calcium—have lower rates of death from cardiovascular disease. A deficiency of magnesium may be related to sudden death from heart failure and to high blood pressure.[14] A dietary deficiency of magnesium is not likely but may occur as a result of vomiting, diarrhea, alcohol abuse, or protein malnutrition; in people who have been fed incomplete fluids into a vein for too long; or in people using diuretics. Good food sources of magnesium include nuts, legumes, cereal grains, dark green vegetables, seafoods, chocolate, and cocoa (see Table 7-7).

Sodium, Potassium, and Chloride

Special conditions are needed to regulate the amounts of water inside and outside the cells so that the cells do not collapse from water leaving them or swell up under the stress of too much water entering them. The cells cannot manage this by pumping water across their membranes, because water slips back and forth freely. However, they can pump minerals across their membranes, and these minerals attract the water to come along with them. This is how the cells maintain water balance. Minerals are used for this purpose in a special form: as **ions** or **electrolytes.** In this form, as single, electrically charged particles, the minerals play many roles, including helping to maintain water balance and acid-base balance.

IONS (EYE-ons) electrically charged particles, such as sodium (positively charged) and chloride (negatively charged).

ELECTROLYTES compounds that partially dissociate in water to form ions; examples are sodium, potassium, and chloride.

Sodium, potassium, and chloride are examples of electrolytes—dissolved substances in blood and body fluids that carry electric charges. Sodium is the chief positively charged ion used to maintain the volume of fluid outside cells; potassium is the chief positively charged ion inside body cells. Chloride is the major negatively charged ion of the fluids outside the cells, where it is found mostly in association with sodium. Electrolytes influence the distribution of fluids among the various body compartments. As an example of how electrolytes affect fluid volume, let's consider the hypothetical case of a man given one tablespoon of salt (sodium chloride) to eat and no water (NOT something we recommend trying!). The salt would become distributed in the space outside of his body's cells and would be excluded from his cells. To counterbalance this sudden change in electrolyte concentration, water would move from the cells into the space outside of the cells. The result would be an increase in the man's fluid volume outside of his cells and a decrease in the fluid volume inside his cells. Under normal circumstances, many factors work together to keep the fluid volume fairly constant inside and outside of cells.

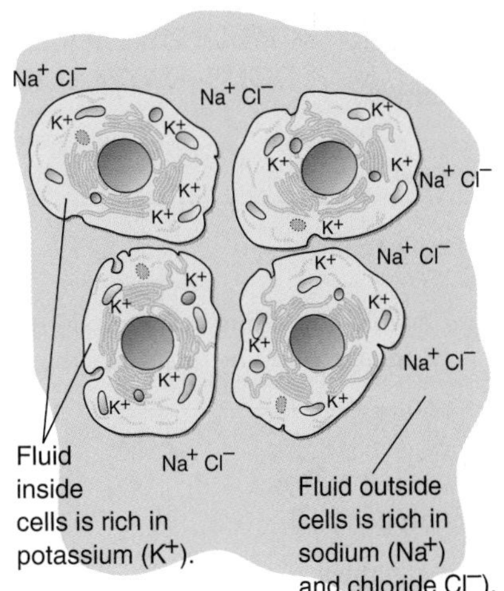
Na^+ Cl^-
K^+
Fluid inside cells is rich in potassium (K^+).
Fluid outside cells is rich in sodium (Na^+) and chloride Cl^-).

Electrolytes
Sodium, potassium, and chloride are examples of body electrolytes. Potassium, which is usually found in the fluids inside the cells, carries a positive charge. Sodium and chloride are usually found in the fluids outside the cells; sodium carries a positive charge, whereas chloride carries a negative charge.

Electrolytes also provide the environment in which the cells' work takes place—work such as nerve-to-nerve communication, heartbeats, and contraction of muscles. When a person's body loses fluid—whether it be sweat, blood, or urine—the person also loses electrolytes. The concentrations of electrolytes are crucial to the life-sustaining activities of the vital organs. When large amounts of body fluid are lost, as in heat stroke, infant diarrhea, or injury, their replacement is a task for a medical team.

About 40 percent of the body's water weight is inside the cells, and about 15 percent bathes the outsides of the cells. The remainder is in the blood vessels.

Sodium. Sodium is part of sodium chloride, ordinary table **salt**, a food seasoning and preservative. The minimum sodium requirement for U.S. adults is estimated to be 500 milligrams. Since sodium is so abundant in the U.S. diet, however, a Daily Value of 2,400 milligrams or less is suggested because the use of highly salted foods can contribute to high blood pressure (**hypertension**) in those who are genetically susceptible. (A discussion of the causes and prevention of hypertension appears in the Nutrition Action feature starting on page 218.)

SALT a pair of charged mineral particles, such as sodium (Na+) and chloride (Cl–), that associate together. In water, they dissociate and help to carry electric current—that is, they become electrolytes.

HYPERTENSION sustained high blood pressure.
hyper = too much
tension = pressure

If some members of your family have high blood pressure, you are advised to curtail your sodium consumption and make sure that your potassium and calcium intakes are ample. Table 7-8 shows a sampling of the sodium content of common foods and reveals that generally, the more processed a food is, the more sodium and the less potassium it contains. Whole, unprocessed foods, on the other hand, tend to be high in potassium and low in sodium.

Persons who wish to avoid salt need to know that what they pour from the salt shaker may be only a sixth of the total salt they consume. Up to 75 percent of the salt in the diet is that added to foods by food processors. Processed foods don't always taste salty. This makes eating something of a guessing game; remember to check food labels for sodium content. The serious sodium avoider must eat fresh rather than processed foods, or look for reduced-sodium processed foods. Reducing your salt intake doesn't have to be unpleasant. Many sodium-free flavoring agents such as lemons, onions, chilies, curries, and other spices excite the taste buds and enhance the flavors of traditional foods.

Potassium. Potassium is critical to maintaining the heartbeat. The sudden deaths that occur during fasting, severe diarrhea, or severe vomiting are thought to be due to heart failure caused by potassium loss. As the principal positively charged ion inside body cells, potassium plays a major role in maintaining water balance and cell integrity.

People who exercise lose fluids and must replace them to avoid dehydration. Chapter 9 discusses heat stroke, fluid needs during and after exercise, and popular sports drinks.

When sodium is lost with water from the body, the ultimate damage comes when potassium moves out of the cells with cell water and is excreted. This is especially dangerous because potassium deficiency affects the brain cells,

See the Savvy Diner feature on page 222 for a list of foods that are high in sodium and for suggested ways to reduce salt intake.

TABLE 7-8 WHOLE UNPROCESSED FOODS VERSUS PROCESSED FOODS—POTASSIUM AND SODIUM CONTENT

Food	Potassium (mg)*	Sodium (mg)*	Potassium-to-Sodium Ratio
Milk Foods			
Milk, 1 c	370	120	3:1
Chocolate pudding (homemade), 1 c	506	274	2:1
Chocolate pudding (instant), 1 c	488	834	1:2
Meats			
Beef roast, 3 oz	254	54	5:1
Corned beef (canned), 3 oz	116	855	1:7
Chipped beef, 3 oz	377	2,946	1:8
Vegetables			
Corn (cooked), 1 c	242	8	30:1
Creamed corn (canned), 1 c	390	572	1:2
Cornflakes, 1 c	25	300	1:12
Fruits			
Peaches (fresh), 1	193	0	171:1
Peaches (canned), 1 c	241	16	15:1
Peach pie, 1 piece	131	253	1:2
Grains			
Whole-wheat flour, 1 c	486	6	81:1
Whole-wheat bread, 1 slice	71	148	1:2
Wheat crackers, 4	17	69	1:4

*Data are taken from Appendix H.

SOURCES OF SODIUM Daily Value recommended for sodium is no more than 2,400 mg.

20 mg (1 small potato) 120 mg (1 c)

50 mg (3 oz) 3 mg (1 ear) 60 mg (1 egg) 50 mg (½ c)

Many whole, unprocessed foods are low in sodium. These foods contribute less than 10% of the sodium in the U.S. diet.

2,300 mg (1 tsp) 100 mg (per shake: 23 shakes per tsp)

The salt added during cooking or at the table contributes about 15% of the sodium in the U.S. diet.

400 mg (½ c) 830 mg (1 pickle) 1,410 mg (1 dinner) 725 mg (1 small cheeseburger)

900 mg (½ c) 1,470 mg (1 fast-food breakfast biscuit) 200 mg (½ c) 960 mg (3 oz) 400 mg (1-oz slice)

Most of the sodium (about 75%) in our diet is added by food manufacturers to processed foods such as these.

SOURCES OF POTASSIUM

Estimated minimum requirement for adults: 2,000 mg/day.

making the victim unaware of the need for water. Adults are warned not to take **diuretics,** except under the direction of a physician, because some of them cause potassium excretion. Physicians prescribing such diuretics will tell their patients to eat potassium-rich foods to compensate for the losses and, depending on the diuretic, may also advise a lowered sodium intake.

DIURETICS (dye-you-RET-ics) medications causing increased water excretion.
dia = through
ouron = urine

The relationship of potassium and sodium in maintaining the blood pressure is not entirely clear. Abundant evidence supports the simple view that the two minerals have opposite effects. In any case, it is clear that increasing the potassium in the diet can promote sodium excretion under most circumstances and thereby lower the blood pressure.[15] A lifelong intake of foods low in sodium and high in potassium protects against hypertension and is thought to play a role in the low blood pressure seen in vegetarians.[16]

A dietary deficiency of potassium is unlikely, but high-sodium, highly processed diets low in fresh fruits and vegetables can make it a possibility. Whole foods of all kinds, including fruits, vegetables, grains, meats, fish, and poultry, are among the richest sources of potassium. Potassium is also abundant in milk. Table 7-9 shows the potassium content of foods.

Some people have medical reasons for needing potassium supplementation, but these people need to be medically supervised. For example, people on medically-supervised, very-low-calorie weight-loss diets may be advised to take a potassium supplement. A physician must monitor the potassium status of these people and order supplements commensurate with the degree of depletion.

Potassium supplements should never be self-prescribed. Potassium toxicity from potassium in supplement form is a greater concern than potassium deficiency. The body protects itself from this eventuality as best it can. If you consume more than you need, the kidneys accelerate their excretion and so maintain control. Should their limit be exceeded (if you ingest too much potassium too fast), a vomiting reflex is triggered. However, if the digestive tract is bypassed and potassium is injected into a vein at a high rate, the heart can stop.

Chloride. The negative ion, chloride, accompanies sodium in the fluids outside the cells. Chloride can move freely across membranes and so is also found inside the cells in association with potassium. In the blood, chloride helps in maintaining the acid-base balance. In the stomach, the chloride ion is part of hydrochloric acid, which maintains the strong acidity of the stomach—needed for protein digestion. Nearly all dietary chloride comes from sodium chloride or salt.

TABLE 7-9 POTASSIUM IN FOODS

(mg)	Sources
529	Yogurt (1 c)
477	Baked potato (1)
467	Banana (1 medium)
419	Spinach, cooked (½ c)
400	Pinto beans (½ c)
398	Lima beans (½ c)
383	Fat-free milk (1 c)
355	Orange juice (¾ c)
329	Kidney beans (½ c)
319	Salmon (3 oz)
315	Bok choy, cooked (1/2 c)
299	Pork (3 oz)
297	Hamburger (3 oz)
273	Tomato (1 medium)
272	Raisins (¼ c)
254	Raisin bran (1 c)
237	Chicken (3 oz)
232	Carrots (½ c)
228	Broccoli (½ c)
213	Cantaloupe (¼ melon)

NUTRITION ACTION

Diet and Blood Pressure—The Salt Shaker and Beyond

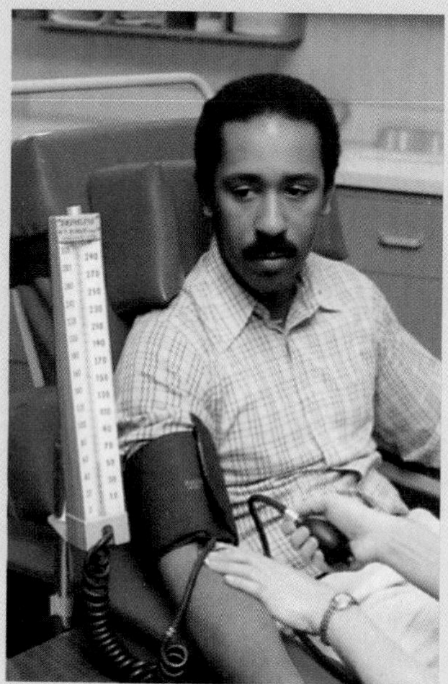

The most effective single measure you can take to protect yourself against high blood pressure is to know what your blood pressure is.

Think "diet" and "blood pressure," and the first thing that comes to mind is salt. Small wonder, given that experts have long emphasized that a diet high in salt—or more specifically, the sodium it contains—contributes to a public health problem known as hypertension (or high blood pressure—see Figure 7-6).[17] Nevertheless, a number of other, albeit less notorious, dietary factors play a role in the unhealthful rise in blood pressure that afflicts some 50 million Americans.

Obesity, for instance, inarguably ranks as the number one dietary culprit linked to hypertension.[18] That's because excess weight forces the heart to work harder to supply the extra pounds of tissue with blood. High blood pressure occurs about twice as often among obese people as it does in thinner people. Fortunately, losing as little as 5 percent of the excess weight has been shown, in some cases, to lower blood pressure to a point at which taking antihypertensive medications may become unnecessary.[19] In other words, a little weight loss, regardless of sodium intake, can be enough to get a high blood pressure level back under control.

Another dietary factor that may lead to high blood pressure (not to mention poor compliance with drug regimens and other treatments for the problem) is drinking excessive amounts of alcohol. In fact, consuming more than a couple of ounces of alcohol daily has been blamed for leading to 5 percent to 11 percent of cases of hypertension among men.[20] The Joint National Committee on Detection, Evaluation, and Treatment of High Blood Pressure thus recommends that to control high blood pressure, those who drink should limit themselves to no more than 1 ounce of alcohol a day—an amount found in 2 ounces of 100-proof whiskey, an 8-ounce glass of wine, or two 12-ounce cans of beer.[21]

In addition to recommending that hypertensive people lose weight and drink alcohol in moderation, the Committee advises those with hypertension to limit the amount of sodium in their diet. Whether such advice is appropriate for the public at large, however, has been widely debated; some experts argue against the necessity of doing so in all but a few select cases. The controversy revolves around a growing body of evidence suggesting that not everyone's blood pressure rises as a result of eating a high-sodium diet. Only certain people, presumably because of their genetic makeup, appear to be **salt sensitive,** meaning that their blood pressure rises in proportion to the amount of sodium they consume. For them, of course, keeping dietary sodium to a minimum can help lower elevated blood pressure.

SALT SENSITIVE the tendency for blood pressure to rise in proportion to salt consumption that certain people seem to have from birth.

Risk factors for high blood pressure:

- Obesity
- Family history of high blood pressure
- Race (African Americans are more likely than whites to develop hypertension)
- Age (in the United States, blood pressure tends to rise with age)
- Excess alcohol consumption
- Sedentary lifestyle

The problem is that it is difficult to identify who is salt sensitive and who is not. Thus, some experts argue that a call to the general public to reduce sodium consumption is unwarranted because it recommends that even people who are not salt sensitive avoid the seasoning. On the other side of the fence are experts and organizations, the National Research Council among them, whose position is that because lowering the amount of sodium in the diet does not pose any health hazard and may be beneficial to some people, it is still prudent for the public as a whole to keep sodium intake in check.[22] And for those who take medication to control their blood pressure, the argument in favor of adhering to a low-

FIGURE 7-6
ARE YOUR NUMBERS UP?

One of the characteristics of hypertension is that it has been called a "silent killer" that cannot be felt and may go undetected for years. That's why it is crucial to have your blood pressure checked on a regular basis. Diagnosis of hypertension requires at least two elevated readings. The first of the two numbers in a blood pressure reading—systolic pressure—represents the force exerted by the heart as it contracts to pump blood throughout the body, as measured by the number of millimeters that the pressure pushes a column of mercury up a tube. The second number—diastolic pressure—is a measure of the pressure the blood exerts on artery walls between heart beats. A reading of below 120 over 80 millimeters of mercury is considered "optimal" for adults 18 years and older. Measurements under 130 over 85 are classified as "normal," but even slight elevations of blood pressure above the optimal level are unhealthy. Beyond those levels, the risks of heart attacks and strokes rise in direct proportion to increasing blood pressure.*

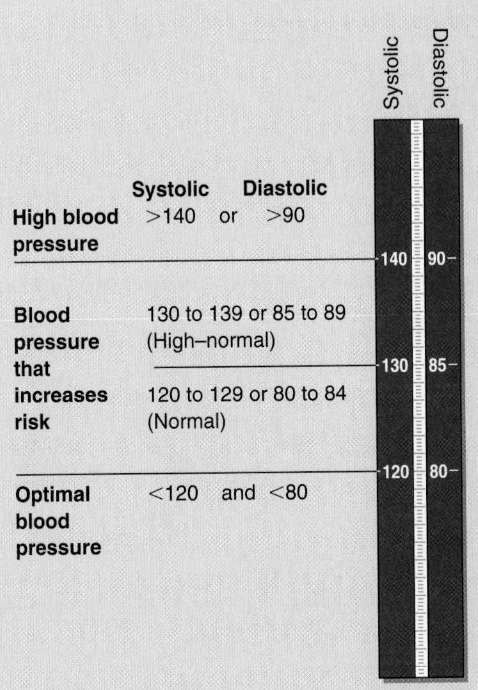

*The categories are for adults not taking high blood pressure medications. If your systolic and diastolic numbers fall into different categories, your overall status depends on the higher category.

Source: National Institutes of Health, National Heart, Lung, and Blood Institute, *The DASH Diet* (Washington, D.C.: U.S. Department of Health and Human Services, February, 1999).

sodium diet is stronger: Eating too much salt can limit the effectiveness of some antihypertensive medications.

In any case, it is clear that modest changes in lifestyle can add up to significant gains in controlling high blood pressure.[23] In addition, it appears that attention to a healthful lifestyle can help to prevent the problem from arising in the first place. Consider that researchers at Northwestern University Medical School in Chicago put one group of adults prone to elevations in blood pressure on a preventive program that included losing weight, lowering sodium intake, cutting down on alcohol, and exercising more and compared them to a similar group left to their own devices. The researchers found that over the course of five years, only one in eleven persons who changed their lifestyle wound up with full-blown hypertension. But of those who didn't alter their habits, one in five became hypertensive.[24]

Along with the factors just explained, scientists are exploring other nutrients that may influence blood pressure.[25] Some researchers believe, for instance, that it is not dietary sodium per se that alters blood pressure, but rather the ratio between it and potassium, sodium's partner in regulating the body's water balance. In other words, the more potassium and less sodium in the diet, the greater the likelihood that the body will maintain a normal blood pressure. While the evidence is not strong enough to imply that people with high blood pressure need to take potassium supplements (such a practice can be dangerous), it's worth keeping in mind, particularly because diets rich in high-potassium foods such as fresh fruits and vegetables tend to go hand in hand with low sodium consumption.[26]

In addition, some studies indicate that people who eat high-calcium diets are less likely to have high blood pressure; conversely, people with inadequate calcium intakes are more likely to have hypertension.[27] It is recommended, therefore, that people with hypertension or at risk for hypertension eat a diet that contains two to three servings daily of calcium-rich foods such as milk or calcium-fortified orange juice.

Another nutrient under investigation for contributing to hypertension is chloride, the mineral found along with sodium in salt. (Salt is about 40 percent sodium and 60 percent chloride.) Some evidence has come to light that dietary sodium raises blood pressure only when accompanied by chloride. In other words, it may not be sodium in and of itself, but rather salt, that contributes to elevations in blood pressure. In the United States, incidentally, about 95 percent of sodium consumed comes from salt.[28]

Finally, some evidence suggests that eating a diet that includes monounsaturated or polyunsaturated fat and is low in saturated fat helps to lower blood pressure. Though the relationship remains tenuous, it certainly makes sense for those with high blood pressure to eat such a diet if for no other reason than to help prevent heart disease, a condition that hypertensives run a high risk of developing (refer to the Spotlight feature in Chapter 4).[29]

An Eating Plan to Reduce High Blood Pressure

A recent landmark study called DASH—Dietary Approaches to Stop Hypertension—examined the effects of overall diet on 459 adult participants with normal to high blood pressures. After just eight weeks, those people with mild hypertension who were consuming the "DASH Diet"—a diet high in fruits, vegetables, whole grains, and low-fat dairy products and low in total fat, saturated fat, and cholesterol saw their blood pressures decrease.[30] Figure 7-7 illustrates the DASH Diet in pyramid form and a sample DASH Diet meal pattern is included.

Previous research that altered nutrient intakes with supplements (for example, of calcium or magnesium) found inconclusive effects on blood pressure. In contrast, the DASH study designed an overall diet, without supplements or drugs, to lower blood pressure. The results of the DASH study echo the recommendations for an overall healthful diet: Choose low-fat dairy products, smaller portions of meat, plenty of fruits and vegetables, and ample servings of high-fiber whole-grain products. Such an eating pattern may reduce risk not only of hypertension, but also for heart disease (low in fat, saturated fat, and cholesterol), diabetes (high in complex carbohydrates and fiber), osteoporosis (adequate in calcium), and cancer (lots of fruits and vegetables). The take-home message from the DASH clinical trials seems to be that if you want to help prevent a rise in blood pressure as you age, and reduce your risk of developing high blood pressure, you can do so by following these six tips:

1. Adopt a low-fat eating pattern rich in fruits, vegetables, legumes, and low-fat dairy products—similar to the DASH Diet.
2. Maintain a normal weight. Lose weight if you're overweight; even losing just a few pounds can reduce blood pressure if you're overweight.
3. Keep your sodium intake at recommended levels—not more than 2,400 milligrams a day.
4. Pursue a moderately active lifestyle: Walk, swim, jog, cycle, or do other aerobic activities at least three times a week for 30 minutes or more.
5. If you drink, use moderation—no more than one drink a day for women, and no more than two drinks a day for men.
6. Don't smoke. Cigarette smoking raises blood pressure and seriously increases risk for heart disease.

The Savvy Diner feature that follows offers you tips in designing your own healthful diet with a focus on seasoning foods without excess salt.

FIGURE 7-7
THE DASH DIET PYRAMID

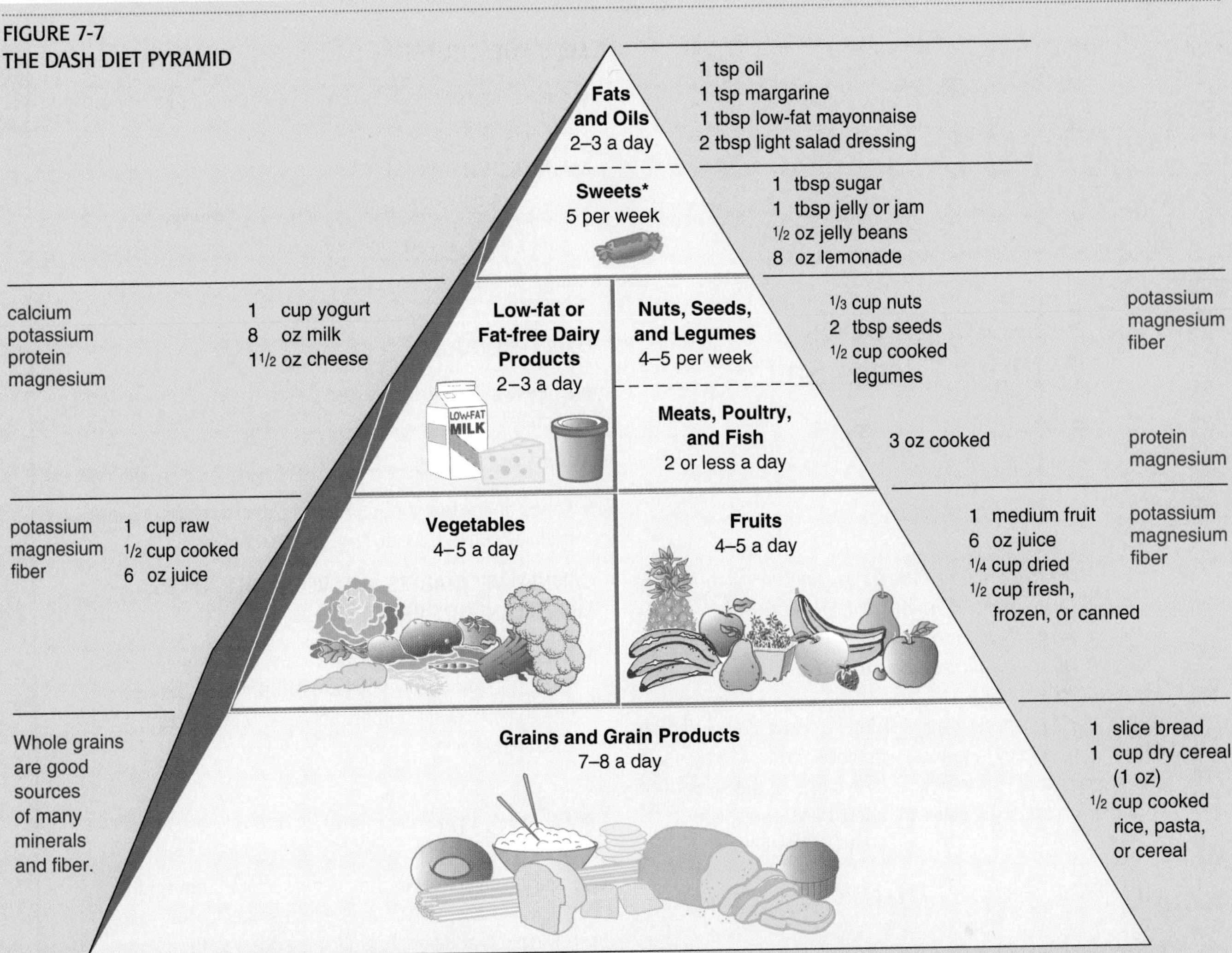

*Sweets should be low in fat (for example, jelly, jam, sugar, maple syrup, fruit-flavored gelatin, jelly beans, hard candy, sorbet, fruit ices).

Sample DASH Menu: The DASH eating pattern is lower in fat, saturated fat, cholesterol, and sodium, and higher in complex carbohydrates, potassium, magnesium, and calcium than the typical American diet. The meal plan shown contains approximately 2,000 calories.

Breakfast	***Serving***
lowfat milk (1 cup)	1 dairy
cornflakes (1 cup)	2 grain
medium banana (1)	1 fruit
whole wheat bread (1 slice)	1 grain
jelly (1 tbsp)	1 sweet

Lunch	***Serving***
chicken taco	
corn tortilla (1)	1 grain
chicken (3 oz)	1 meat
low-fat cheddar cheese (1½ oz)	1 dairy
lettuce (½ cup)	½ vegetable
tomato (¼ cup)	½ vegetable
Mexican rice (½ cup)	1 grain
apple (1)	1 fruit

Dinner	***Serving***
tossed salad (1 cup)	1 vegetable
Italian dressing, low-fat (2 tbsp)	1 fat
meatloaf (3 oz)	1 meat
medium baked potato (1)	2 vegetables
soft margarine (1 tsp)	1 fat
turnip greens (½ cup)	1 vegetable
whole wheat dinner roll (1)	1 grain
honeydew melon (½ cup)	1 fruit
fat free milk (1 cup)	1 dairy

Snack	***Serving***
peanuts (⅓ cup)	1 nuts
raisins (¼ cup)	1 fruit
pretzels (1 oz or ¾ cup)	1 grain

Source: A Clinical Trial of Dietary Patterns on Blood Pressure, *New England Journal of Medicine,* 336 (1997): 1117–1124.

Seasoning Foods Without Excess Salt

To the ancient Greeks, parsley was thought to promote appetite and good humor. Perhaps that's why, even today, a sprig of parsley brings appeal to a meal.[31]

You've decided to cut your salt intake. How do you know which foods to buy? How can you cook foods with less salt and good flavor? Fortunately, cutting some of the salt out of your diet isn't difficult. Here are a few basic principles.

At the Supermarket

- Read labels! Look for the key words *salt* and *sodium* in the ingredients list. These are signals that the foods contain added sodium. These sodium-containing ingredients are used in food processing: baking powder, baking soda, disodium phosphate, monosodium glutamate (MSG), salt, sodium benzoate, sodium hydroxide, sodium nitrite, sodium propionate, and sodium sulfite.
- Choose reduced-sodium products when possible. Many commercially prepared products are available in sodium-reduced versions. For example, most supermarkets carry reduced-sodium soy sauce, canned tuna, soups, and canned vegetables.
- Buy fresh, natural foods more frequently than processed foods, which tend to be high in salt. Here is a list of processed and convenience foods that are typically high in salt:

MEAT, POULTRY, FISH

- Cured meats: ham, bacon, luncheon meats; Sausages; Fish: commercially frozen, prebreaded; Canned shellfish; Fish: canned in oil or brine; Hot dogs

MEAT SUBSTITUTES

- Salted nuts or seeds; Canned beans and peas; Soy protein products

MAIN DISH ITEMS

- Pizza, lasagna, macaroni and cheese; Commercially prepared main course foods (e.g., frozen dinners); Tacos, enchiladas, burritos; Canned or dehydrated soups or chowders

DAIRY PRODUCTS

- Cheeses; Buttermilk; Instant cocoa mixes; Dutch process cocoa

BREAD PRODUCTS

- Salted snack foods (e.g., crackers, potato chips); Commercially baked goods (e.g., cakes, cookies)

CANNED VEGETABLES AND VEGETABLE JUICES; MISCELLANEOUS PRODUCTS

- Catsup, chili sauce; Bouillon cubes; Salted gravies and sauces; Commercial salad dressings; Olives, pickles, pickle relish; Meat tenderizers (e.g., MSG, Accent); Seasoning salts

At Home

- Put the salt shaker in the kitchen cabinet, not on the table, and use it only on rare occasions. It's surprising how much additional sodium is added to foods with just a few shakes of the salt shaker. Here's a handy guide for the sodium content of common kitchen measures of salt:

  ¼ tsp of salt = 575 mg of sodium
  ½ tsp of salt = 1,150 mg of sodium
  1 tsp of salt = 2,300 mg of sodium

- Flavor your foods with seasonings and spices, most of which are virtually salt and sodium free. Try some of the seasonings listed in your favorite dishes:

allspice	curry powder	onion powder
basil	dill	paprika
bay leaves	garlic powder	parsley
caraway seeds	ginger	rosemary
chives	lemon juice	sage
cilantro	mustard (dry)	thyme
cinnamon	nutmeg	tumeric

- Salt substitutes can be used as a means of reducing sodium intake. They usually consist of a mineral base other than sodium and are compounded to give a taste sensation similar to sodium chloride. Or, try one of the many salt-free blends of herbs and spices (e.g., Mrs. Dash).

THE TRACE MINERALS

If you could extract all of the trace minerals from the body, you would obtain only a bit of dust, hardly enough to fill a teaspoon. As tiny as their quantities are, though, each of the trace minerals performs several vital roles for which no substitute will do. A deficiency of any of them may be fatal, and an excess of many is equally deadly.

HEMOGLOBIN (HEEM-oh-globe-in) the oxygen-carrying protein of the blood; found in the red blood cells.

IRON-DEFICIENCY ANEMIA a reduction of the number and size of red blood cells and a loss of their color because of iron deficiency.

Iron

Iron is the body's oxygen carrier. Bound into the protein **hemoglobin** in the red blood cells, it helps transport oxygen from lungs to tissues and so permits the release of energy from fuels to do the cells' work. When the iron supply is too low, **iron-deficiency anemia** occurs, characterized by weakness, tiredness, apathy, headaches, increased sensitivity to cold, and a paleness that reflects the reduction in the number and size of the red blood cells. A person with this anemia can do very little muscular work without disabling fatigue but can replenish iron status by eating iron-rich foods (see Table 7-10).

It is difficult to convey the extent and severity of iron deficiency among the world's people. Iron deficiency occurs in as many as half of all persons in some settings, even in developed countries—most predictably in the inner city and among the rural poor. People begin to feel iron deficiency's impact, without knowing it, long before anemia is diagnosed. They don't appear to have an obvious deficiency disease; they just appear unmotivated and apathetic. Because they work and play less, they are less physically fit. Prevalence rates for iron-deficiency anemia in developed countries range from 5 percent to 20 percent. In the United States, iron deficiency remains relatively prevalent among toddlers, adolescent girls, and women of childbearing age.[32] If this one worldwide malnutrition problem could be alleviated, millions of people would benefit.[33]

The cause of iron deficiency is usually malnutrition—that is, inadequate intake, either from limited access to food or from high consumption of foods low in iron. Among causes other than malnutrition, blood loss is the primary one, caused in many countries by parasitic infections of the digestive tract.

The usual Western mixed diet provides only about 5 to 6 milligrams of iron in every 1,000 calories. The recommended daily intake for an adult man is 10 milligrams. Because most men easily eat more than 2,000 calories, they can meet their iron needs without special effort.

The situation for women is different. Women may have normal blood cell counts or hemoglobin levels and yet may need more iron because their body stores may be depleted, a factor that doesn't show up in standard tests. Because most women typically eat less food than men, their iron intakes are lower. And

TABLE 7-10 IRON IN FOODS

(mg)	Sources
23.80	Clams, steamed (3 oz)
8.90	Raisin bran (1 c)
6.60	Tofu (½ c)
3.70	Enriched Cereals (¾ c)
3.13	Beef pot roast (3 oz)
2.90	Sirloin steak (3 oz)
2.75	Baked potato (1)
2.65	Shrimp (3 oz)
2.48	Sardines (3 oz)
2.40	Spinach, cooked (½ c)
2.35	Hamburger (3 oz)
2.30	Navy beans (½ c)
2.25	Lima beans (½ c)
2.25	Prune juice (¾ c)
2.15	Black-eyed peas (½ c)
2.00	Swiss chard (½ c)
1.61	Kidney beans (½ c)
1.59	Oatmeal, cooked (1 c)
1.30	Tuna (3 oz)
1.30	Dried figs (¼ c)
1.26	Green peas (½ c)
0.94	Wheat bread (1 slice)
0.82	Apricot halves, dried (5)
0.76	Raisins (¼ c)
0.65	Broccoli, cooked (½ c)

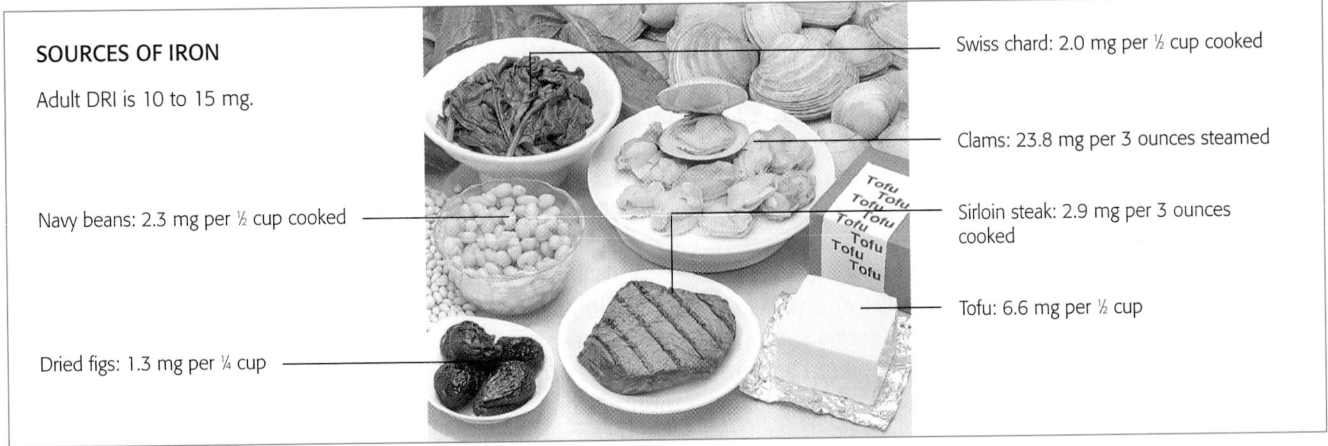

You can combine foods to achieve maximum iron absorption—the heme iron in the meat and the vitamin C in the tomatoes in this chili help you absorb the nonheme iron from the beans.

HEME (HEEM) **IRON** the iron-holding part of the hemoglobin protein, found in meat, fish, and poultry. About 40 percent of the iron in meat, fish, and poultry is bound into heme. Meat, fish, and poultry also contain a factor (MFP factor) other than heme that promotes the absorption of iron, even of the iron from other foods eaten at the same time as the meat.

NONHEME IRON the iron found in plant foods.

CONTAMINATION IRON iron found in foods as the result of contamination by inorganic iron salts from iron cookware, iron-containing soils, and the like.

IRON OVERLOAD a condition in which the body contains more iron than it needs or can handle; excess iron is toxic and can damage the liver. The most common cause of iron overload is the genetic disorder hemochromatosis.

Iron cookware adds supplemental iron to foods.

because women menstruate, their iron losses are greater. These two factors may put women much closer to the borderline of deficiency.

The recommended intake for a woman before menopause is 15 milligrams per day; pregnant women need 30 milligrams daily. Because women typically consume fewer than 2,000 calories per day, they understandably have trouble achieving adequate iron intakes. A woman who wants to meet her iron needs from foods must increase the iron-to-calorie ratio of her diet so that she will receive about double the average amount of iron—about 10 milligrams per 1,000 calories. This means she must emphasize iron-rich foods in her daily diet.

Table 7–10 shows the amounts of iron in foods. Meats, fish, and poultry are superior sources on a per-serving basis. People must select foods carefully to obtain enough iron because it is present in such small quantities in most foods. The best meat sources are liver, red meats, poultry, fish, oysters, and clams. Among the grains, whole grains and enriched and fortified breads and cereals are best, and dried beans are a good source. Some fruits and vegetables contain appreciable amounts of iron, as shown in the table. Foods in the milk group are notoriously poor iron sources.

Ways to Enhance Iron Absorption. Iron occurs in two forms in foods—as **heme iron,** bound into the iron-carrying proteins such as hemoglobin in meats, poultry, and fish, and as **nonheme iron** in both plant and animal foods. Heme iron is much more reliably absorbed than is nonheme iron. Nonheme iron's absorption is affected by many factors, including the amount of vitamin C consumed with meals.[34] Vitamin C promotes iron absorption and can triple the amount of nonheme iron absorbed from foods eaten at the same meal.

Contamination iron—that is, iron obtained from cookware or soil—can also increase iron intake significantly.[35] Consumers who cook their foods in iron cookware can contribute to their iron intake. For example, the iron content of a half-cup of spaghetti sauce simmered in a glass dish is 3 milligrams, but it is 87 milligrams when the sauce is cooked in an iron skillet. Similarly, the reason why dried apricots and raisins contain more iron than the fresh fruit is that they are dried in iron pans. Admittedly, this form of iron is not as well absorbed as the iron from meat, but every little bit helps.

Some food components interfere with iron absorption. Phytic acid is one example; it occurs in some fruits, vegetables, and whole grains, as mentioned earlier, as well as in nonherbal tea. Other examples are the tannins, which occur in black teas, coffee, cola drinks, chocolate, and red wines. Fiber also can reduce iron absorption because it speeds up the transit of materials through the intestine.

Iron Toxicity. Large amounts of iron can be toxic to the body. **Iron overload** is a condition in which the body absorbs excessive amounts of iron and is more common in men than in women. As a result, tissue damage occurs, especially in organs that store iron such as the liver. Infections are more likely to occur because bacteria thrive on iron-rich blood.

Researchers are currently investigating a possible link between excess iron stores in the body and increased risk of chronic conditions such as heart disease.[36] Researchers in Finland found that in a three-year study of 2,000 healthy men, the risk of heart attack was twice as great for men with the highest levels of stored iron in their bodies.[37] Although iron is an important essential nutrient in the body, it can also act as a powerful oxidizing agent in reactions that produce free radicals in the body. As noted in Chapter 4, free radicals can initiate the changes in LDL-cholesterol that eventually damage artery walls and lead to heart disease. Much more research is needed in this area before the relationship between iron and heart disease can be clarified. Until all the answers are in, avoid taking extra supplemental iron unless you have been diagnosed as deficient. Be sure to keep iron supplements out of the reach of children, too, who can fatally overdose on such tablets.

Zinc

Zinc is found in every cell of the body and plays a major role with more than 50 enzymes that regulate cell multiplication and growth, normal metabolism of protein, carbohydrate, fat, and alcohol, and the disposal of damaging free radicals. Zinc is associated with the hormone insulin, which regulates the body's fuel supply. It is involved in the utilization of vitamin A, taste perception, thyroid function, wound healing, the synthesis of sperm, and the development of sexual organs and bone.

Most recently, zinc has become known for its role in promoting a healthy immune system.[38] A flurry of zinc lozenges have appeared on the market following the results of a study seeking an antidote for the common cold.[39] People who used zinc gluconate lozenges every 2 hours starting within 24 hours of a cold's onset reduced the length of their colds by three days. However, other studies have not reported similar results, and whether zinc lozenges are any more effective than placebos remains to be determined. Furthermore, the long-term effects of such supplemental use of zinc is unknown.

Zinc deficiencies were first reported in the 1960s in the Middle East, where studies on adolescent boys revealed severe growth retardation and delayed sexual maturation—symptoms responsive to zinc supplementation. The native diets were typically low in animal protein and zinc and high in fiber and other compounds that bind minerals. The researchers learned that the binders were carrying zinc out of the boys' bodies, thus causing the deficiency.

Since then, cases of zinc deficiency have been discovered closer to home.[40] Zinc deficiency can cause night blindness, hair loss, poor appetite, susceptibility to infection, delayed healings of cuts or abrasions, decreased taste and smell sensitivity, and poor growth in children. Because zinc is lost from the body daily in much the same way as protein is, it must be replenished daily.

People who are building new tissue have the highest zinc needs—infants, children, teenagers, and pregnant women. The pregnant teenager is at particular risk because she needs zinc for her own growth as well as for her fetus's growth. Pregnant vegetarians are at risk, too, because their diets are high in fiber and zinc-binding factors. Dieters also need to be reminded that very-low-calorie diets cause not only a low zinc intake but also a loss of zinc from body tissues as they break down to release fuel.

Zinc is a relatively nontoxic element. However, it can be toxic if consumed in large enough quantities. Consumption of high levels of zinc can cause a host of symptoms, including vomiting, diarrhea, fever, and exhaustion.[41] The hazards of overconsumption are greatest when consumers dose themselves with supplements. Excess supplemental zinc can cause imbalances of both copper and iron in the body. Chronic consumption of zinc exceeding 15 milligrams per day is not recommended without close medical supervision.

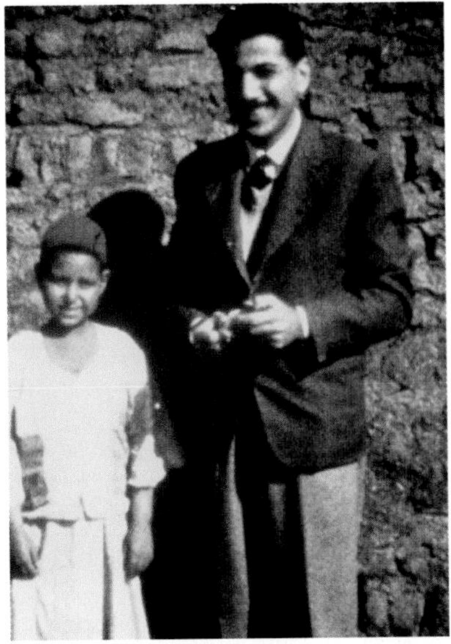

The Egyptian boy in the picture is 17 years old but is only 4 feet tall, the height of an average 7-year-old in the United States. His genitalia are like those of a 6-year-old. The retardation, known as dwarfism, is rightly ascribed to zinc deficiency because it is partially reversible when zinc is restored to the diet.

SOURCES OF ZINC

Adult DRI is 12 to 15 mg.

TABLE 7-11 ZINC IN FOODS

(mg)	Sources
28.00	Oysters, cooked (3 oz)
6.5	Crabmeat (3 oz)
5.6	Sirloin steak (3 oz)
5.27	Ground beef (3 oz)
4.65	Beef pot roast (3 oz)
4.0	Soybeans, dry-roasted (½ c)
3.1	Enriched cereal (¾ c)
2.20	Yogurt (1 c)
1.73	Turkey (3 oz)
1.66	Swiss cheese (1.5 oz)
1.61	Peanuts (⅓ c)
1.34	Shrimp (3 oz)
1.32	Cheddar cheese (1.5 oz)
1.10	Black-eyed peas (½ c)
1.00	Black beans (½ c)
1.00	Tofu (½ c)
0.98	Fat-free milk (1 c)
0.75	Green peas (½ c)
0.70	Kidney beans (½ c)
0.68	Spinach, cooked (½ c)
0.55	Whole-wheat bread (1 slice)

Table 7-11 shows the zinc amounts in foods. An average 1,500-calorie diet provides about 6.3 milligrams of zinc per day, or about 40 percent to 50 percent of the recommended intake. Zinc is highest in foods of high protein content, such as shellfish (especially oysters), meats, and liver. As a rule of thumb, two servings a day of animal protein will provide most of the zinc a healthy person needs. Whole-grain products are good sources of zinc if large quantities are eaten (the phytate in grains does not inhibit the absorption of zinc in people consuming ordinary diets). Cow's milk protein (casein) binds zinc avidly and seems to prevent its absorption somewhat; infants absorb zinc better from human breast milk. Fresh and canned vegetables vary in zinc content, depending on the soil in which they are grown. The zinc content of cooking water varies from region to region as well.

Iodine

Iodine occurs in the body in an infinitesimal quantity, but its principal role in human nutrition is well-known. It is part of the thyroid hormones, which regulate body temperature, metabolic rate, reproduction, and growth. The hormones enter every cell of the body to control the rate at which the cells use oxygen and release energy.

When the iodine level of the blood is low, the thyroid gland may enlarge until it causes swelling in the throat area—a condition called **goiter.** Goiter is estimated to affect 200 million people the world over.

In addition to causing sluggishness and weight gain, an iodine deficiency can have serious effects on fetal development. Severe thyroid undersecretion of a woman during pregnancy causes the extreme and irreversible mental and physical retardation of the child known as **cretinism.** A person with this condition has mental retardation and a face and body with many abnormalities. Much of the mental retardation associated with cretinism can be averted by early diagnosis and treatment of the mother's iodine deficiency.

GOITER (GOY-ter) enlargement of the thyroid gland caused by iodine deficiency.

CRETINISM (CREE-tin-ism) severe mental and physical retardation of an infant caused by iodine deficiency during pregnancy.

The amount of iodine in foods reflects the amount present in the soil in which plants are grown or on which animals graze. Soil iodine is greatest along the coastal regions. In the United States, in areas where the soil is low in iodine (most notably in the plains states and around the Great Lakes and St. Lawrence River areas and in the Willamette Valley of Oregon), widespread goiter and cretinism appeared in the local people during the 1930s. Iodized salt was introduced as a preventive measure, and these scourges disappeared.

By the 1970s, a dramatic increase in iodine intakes had occurred—well above the recommended 150 micrograms a day.[42] The toxic level at which detectable harm results is thought to be only a few times higher than the average consumption levels at that time. Excessive intakes of iodine can cause an enlargement of the thyroid gland resembling a goiter; in infants it can be so severe as to block the airways and cause suffocation. Most of the excess iodine came from iodates (dough conditioners used in the baking industry) and from milk produced by cows exposed to iodine-containing medications and from disinfectants used during the milk treatment process.

At the beginning of the new millennium, new data show a surprising decline in iodine intake during the last two decades of the twentieth century, especially in women of reproductive age.[43] Although the average American dietary iodine intake is considered sufficient, some 15 percent of women may be iodine deficient. The recent decrease in iodine intake may be attributed to several factors, including the replacement of iodine with bromine salts as the dough conditioner in commercial bread production, reduced intake of iodine-rich egg yolks for cholesterol concerns, reduced use of iodized table salt for high blood pressure concerns, the dairy industry's effort to reduce iodine in milk, and the increasing use of noniodized salt in manufactured foods.[44] The sudden emergence of this problem points to a need for continued surveillance of the food supply to prevent the effects associated with an iodine deficiency, including miscarriages, goiter, and mental retardation in babies of iodine-deficient mothers.

People sometimes ask whether they should be sure to buy the *iodized* variety of salt in the grocery store to ensure adequate iodine intake. Although most consumers now have access to fruits and vegetables grown in coastal areas rich in iodine, health experts state the importance of using iodized salt to maintain an adequate iodine intake.

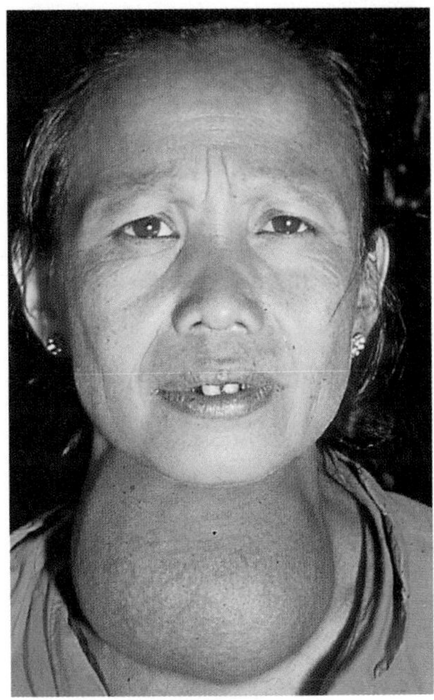

In iodine deficiency, the thyroid gland enlarges—a condition known as simple goiter.

Fluoride

Only a trace of fluoride occurs in the human body, but studies have demonstrated that where diets are high in fluoride, the crystalline deposits in teeth and bones are larger and more perfectly formed than where diets are low in it. Fluoride not only protects children's teeth from decay but also makes the bones of older people resistant to adult bone loss (osteoporosis). Its continuous presence in body fluids is desirable.

Drinking water is the usual source of fluoride. Where fluoride is lacking in the water supply, the incidence of dental decay is very high. Fluoridation of community water where needed to raise its fluoride concentration to one part per million (ppm) is thus an important public health measure (see Figure 7-8).

In some communities, the natural fluoride concentration in water is high (2 to 8 ppm), and children's teeth develop with mottled enamel. This condition, called **fluorosis,** may not be harmful (in fact, these children's teeth may be extraordinarily decay resistant), but it violates the prejudice that teeth should be white. True toxicity from fluoride overdoses can occur, but usually only after years of chronic daily intakes of 20 to 80 times the amounts normally consumed from fluoridated water.

FLUOROSIS (floor-OH-sis) discoloration of the teeth from ingestion of too much fluoride during tooth development.

Although drinking fluoridated water ranks as one of the most effective ways to prevent dental cavities, some 40 percent of Americans continue to drink water not treated with the mineral, putting themselves at a high risk of developing

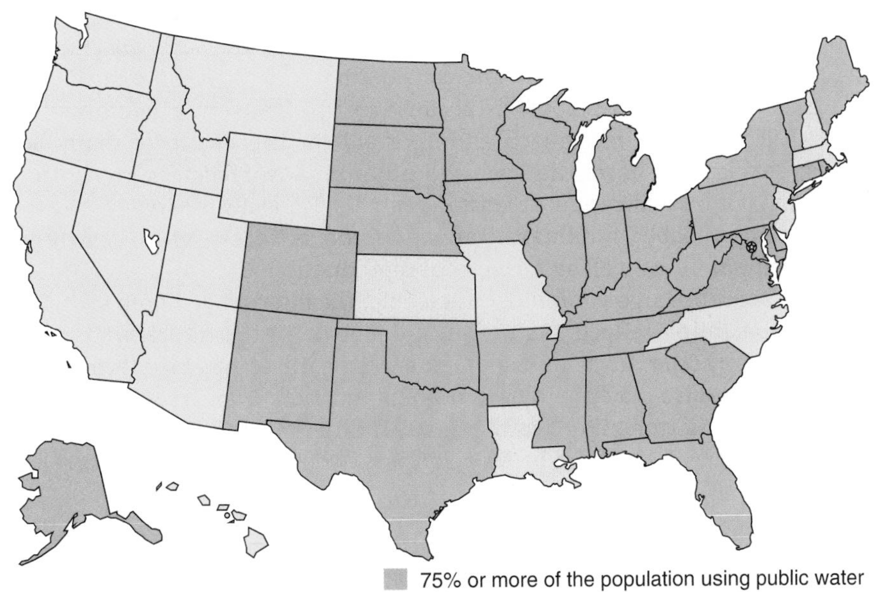

FIGURE 7-8
FLUORIDATION IN THE UNITED STATES

Source: Fluoridation Census Summary, U.S. Department of Health and Human Services, Public Health Service, Centers for Disease Control and Prevention.

tooth decay. Part of the problem is that the practice has been the subject of bitter controversy ever since it was introduced in 1945. Antifluoridation groups claim that drinking fluoride-containing water can cause everything from cancer to birth defects, despite reports to the contrary issued by major health organizations, including the National Institute of Dental Health and the American Dietetic Association.[45] In fact, a 1993 report from the National Research Council—based on hundreds of studies and the most comprehensive investigation of the health effects of fluoride to date—concluded that current levels of fluoride in drinking water are not associated with cancer, kidney disease, stomach and intestinal problems, infertility, birth defects, genetic mutations, or any other health problems for which fluoride has been blamed. However, fluoride in water combined with excess fluoride from toothpaste, mouth rinses, supplements, and other sources can lead to fluorosis, the irreversible condition mentioned earlier which only occurs during tooth development. For this reason, children living in areas with fluoridated water should not be given fluoride supplements unless prescribed by a physician.[46]

Alternatives to water fluoridation:
- Fluoride toothpastes
- Fluoride treatments for teeth
- Fluoride tablets and drops

Copper, Manganese, Chromium, Selenium, and Molybdenum

See Chapter 6 for more discussion of the role of the antioxidants in health and disease.

Several trace minerals have been found to have important roles in a variety of metabolic and physiologic processes. Copper, for example, is involved in making red blood cells, manufacturing collagen, healing wounds, and maintaining the sheaths around nerve fibers. Chromium works closely with the hormone insulin to help the cells take up glucose and break it down for energy.[47] Selenium functions as part of an antioxidant enzyme and can substitute for vitamin E in some of that vitamin's antioxidant activities. Research is currently under way to investigate a possible role for selenium in protecting against the development of prostate and other forms of cancer.[48] Manganese and molybdenum both function as working parts of several enzymes. (Refer to Table 7-2 on page 207 for a brief description and common food sources of each of these minerals.)

Trace Minerals of Uncertain Status

None of the trace minerals has been known for very long, and some are extremely recent newcomers. Some researchers consider boron to be one factor that contributes to the risk of osteoporosis because of its effects on calcium metabolism.[49] Nickel is now recognized as important for the health of many body tissues; deficiencies harm the liver and other organs. Silicon is known to be involved in bone calcification, at least in animals. Tin is necessary for growth in animals and probably in humans. Vanadium, too, is necessary for growth and bone development as well as for normal reproduction. Cobalt is recognized as the mineral in the large vitamin B_{12} molecule; the alternative name for this vitamin—cobalamin—reflects the presence of cobalt. In the future, we may discover that many other trace minerals—for example, silver, mercury, lead, barium, and cadmium—also play key roles in human nutrition. Even arsenic, famous as the poisonous instrument of death in many murder mysteries and known to be a carcinogen, may turn out to be an essential nutrient in tiny quantities.

Spotlight

OSTEOPOROSIS—THE SILENT STALKER OF THE BONES

- *As she waited in a grocery checkout line, Nancy Miller Friesen felt a stab of pain in her left hip. The 40-year-old Fort Worth woman had always been healthy and active—suddenly she could barely stand. When she went to her family doctor, she learned that several of her vertebrae had fractured. The radiologist studying her x-rays thought he was looking at the bones of a 70-year-old woman.*
- *Lila Rubin had just teed off at a New Jersey golf course. Putting down her club, the robust 63-year-old felt a searing back pain. When Rubin met with her orthopedic surgeon, she was told she'd suffered a serious compression fracture.*
- *Debra Epstein of New Haven, Connecticut, wasn't worried when her physician diagnosed a mild curvature of her upper spine seven years ago. She was only 31 and in otherwise excellent health. Then last year, her right wrist began aching and her back would tire easily. After tests, she learned that at age 37, she had the brittle bones of a woman twice her age.*

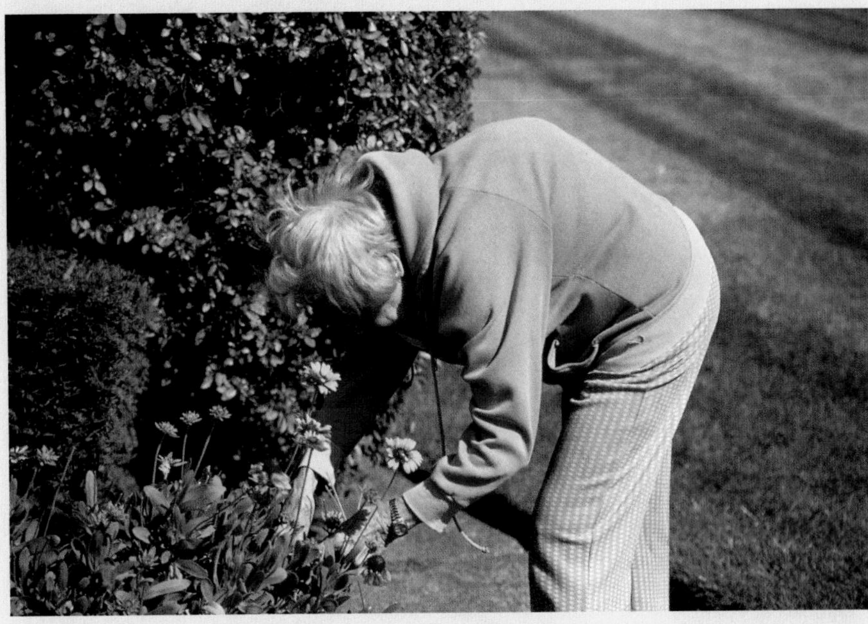

Osteoporosis strikes the bones. This 75-year-old woman has lost height—not because her legs have grown shorter but because vertebrae in her back have collapsed. These fractures occur on the front side of the spinal bones resulting in a hunched over posture.

The three women just described share the potentially crippling and sometimes deadly condition: osteoporosis.[50] Osteoporosis threatens the integrity of the skeleton in some 28 million adults in the United States. One out of two women and one out of eight men over 50 will suffer an osteoporosis-related fracture in their lifetime. Osteoporosis is more prevalent in women than in men after age 50 for several reasons:[51]

- Women generally have less bone mass than men.
- Women typically have lower calcium intakes than men.
- Women more often use weight-loss diets, which tend to be low in calcium and lead to bone loss.[52]
- Bone loss begins earlier in women because of women's different hormonal make-up, and the loss is accelerated at **menopause,** when their protective **estrogen** secretion declines.
- Pregnancy and lactation decrease the calcium reserves in bones whenever calcium intake is inadequate.
- Women live longer than men, and bone loss continues with aging.

What is bone loss?

Adult bone loss affects the entire skeleton, but it occurs first in the pelvis and the spine. Unfortunately, osteoporosis establishes itself silently in its victims' lives. Symptoms do not usually occur until late in life. The disease often results in fractures of the hip, spine, and wrists—over 1.5 million such fractures a year—that occur with very little or no stress or pressure (see Figure 7-9).[53] Often, it first becomes apparent when someone's hip suddenly gives way. People say, "She fell and broke her hip." The fact of the matter is that often the hip was so fragile that it was already broken and caused the fall. Fewer than 50 percent of all women with hip fractures return to previous activity levels—many are confined to wheelchairs for the remainder of their lives. Approximately 20 percent of these women die of complications within the first year after hip fracture.

Bones are made up of a complex matrix based on the protein collagen, into which the crystals of the

FIGURE 7-9
BONE LOSS AND MOST COMMON TYPES OF BONE FRACTURES IN WOMEN

A total of 1.5 million fractures occur each year including 250,000 fractures at sites not shown here.

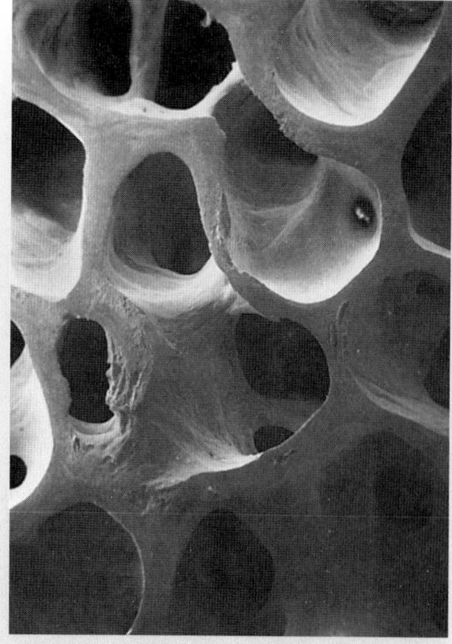

Electron micrograph of healthy trabecular bone.

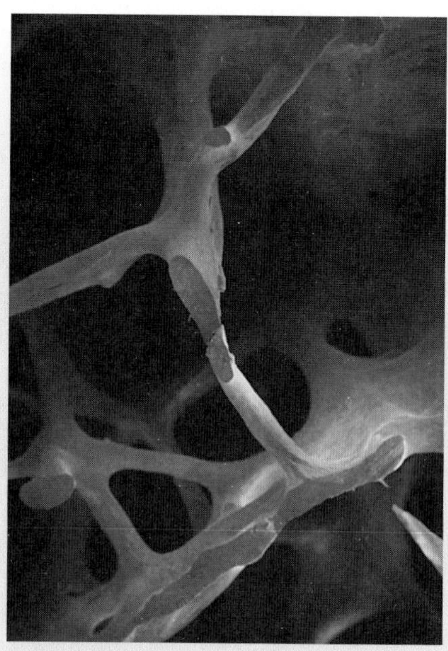

Electron micrograph of healthy trabecular bone affected by osteoporosis.

Spinal Vertebrae Fracture
- More likely at ages 55 to 75 years
- Fracture can occur from bending, lifting, or spontaneously
- Bone weakens and collapses leading to loss of height and chronic back pain
- Fractures occur on front side of vertebrae, and a hunched-over posture results

700,000

250,000

300,000

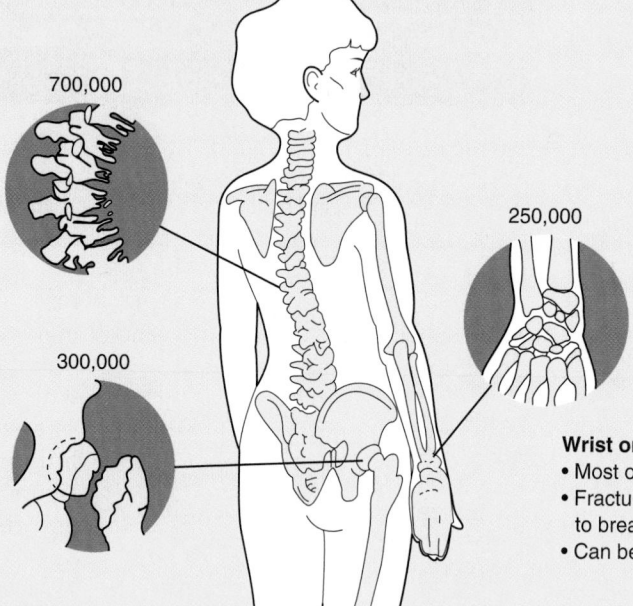

Hip Fracture
- Most serious type of fracture
- 300,000 per year
- Most occur at 70 years or older
- May lead to serious complications
- 50% become institutionalized

Wrist or Forearm Fracture
- Most occur at 50 years or older
- Fracture often occurs when hands are used to break a fall
- Can be early warning sign for osteoporosis

bone minerals—principally calcium and phosphorus—are deposited. Bone tissue has two forms: **trabecular bone** and **cortical bone.** The thick ivorylike outer portion of a bone is the cortical bone. Cortical bone provides a covering for the inner trabecular bone—a lacy network of calcium-containing crystals, almost spongelike in appearance (see Figure 7-10). When calcium intakes are low, hormones call first upon the trabecular bone to release calcium into the blood for use by the rest of the body whenever levels fall too low.* Over time, the lacy network of bones becomes less

*Vitamin D and parathyroid hormone circulate in the blood whenever the calcium concentration in the blood falls too low. These hormones cause the release of calcium from the bones, a decreased excretion of calcium from the kidneys, and an increased absorption of dietary calcium from the intestines. The hormone calcitonin is released from the thyroid gland when levels of blood calcium get too high; it acts to stop the release of calcium from bone and to slow intestinal absorption of the mineral.

dense and fragile as calcium deposits are withdrawn.

Although bones undergo remodeling throughout life—adding and losing bone minerals—the total amount of bone mass in the human body reaches a peak by about age 30.[54] Afterwards, bones lose strength and density as bone minerals are lost. The principal determinant of bone health is peak bone mass. Think of bone mass as money in the bank: The larger the savings account, the easier it will be to withstand some loss of bone as one ages. To attain a healthful peak bone mass, it is necessary to have an optimal intake of calcium during the years of bone growth—all through the growing years and on into the years of young adulthood.[55]

Can osteoporosis be prevented? Although there is no cure for osteoporosis, the good news is that you can take steps to minimize your risk. First, determine your risk by considering the following list of factors that play a role in osteoporosis:

- *Age*. During childhood, adolescence, and the young adult years, the cells that build bone (**osteoblasts**) form more bone than the bone-dismantling cells (**osteoclasts**) take away. With aging, the bone-building cells become less active, while the bone-dismantlers continue to work—causing bone tissue and strength to decline.

- *Gender*. Osteoporosis is four times more common in women. Men achieve a higher peak bone mass than women do, and the rate of age-related bone loss is lower in men than in women.

- *Age-related decline in hormones*. Hormones are important in bone health for both men and women. Menopause deprives women of the protective effects of estrogen. Estrogen improves calcium absorption from the intestines and reduces excretion of the mineral by the kidneys. Bone loss accelerates in women for the five to ten years following menopause. The earlier that menopause occurs, the greater a woman's risk of osteoporosis. Women who have ceased menstruating (a reliable indicator of reduced estrogen levels) comprise the largest segment of people with osteoporosis. Menopause occurs naturally in women typically between the ages of 45 and 55, but it can occur earlier in life with the removal of diseased ovaries. Estrogen replacement therapy slows the calcium losses from the bones of such women.[56] Risk of osteoporosis increases in men with an age-related decline in testosterone; low testosterone levels increase risk of fractures in men.[57]

- *Abnormal absence of menstrual periods (estrogen deficiency)*. Menstruation can also cease in women who overexercise and are underweight, a condition called athletic amenorrhea (discussed in Chapter 9).[58] For the same reason, eating disorders (anorexia nervosa and bulimia) increase the risk of osteoporosis later in life.[59]

- *Family history*. The greater the history of skeletal fractures of the hips and vertebrae among elderly relatives, the greater your risk for osteoporosis.

- *Race and ethnic background*. Those of British, Northern European, Chinese, Japanese, or Mexican-American background or Hispanic people from Central and South America are at the highest risk; African-American women tend to have denser bones and are at a lower risk.[60] Undoubtedly, environmental factors—such as calcium intakes, physical activity, smoking, body weight, and alcohol intake—can influence the ultimate outcome of one's genetic heritage.

- *Body build*. The smaller the frame and the thinner the person, the greater the risk. Petite women have less bone to lose than larger-boned women.

- *Sedentary lifestyle*. Inactivity leads to bone loss.[61]

- *Smoking and alcohol*. Smoking and excessive alcohol intake increase the risk. People who smoke tend to have lower body weights than nonsmokers. Some women who smoke have lower levels of estrogen in their blood and experience an earlier onset of menopause compared to nonsmokers.[62] Alcohol interferes with the absorption of calcium and may increase excretion of calcium from the kidneys. People who abuse alcohol also tend to have poor nutrition status, including low intakes of calcium.

- *Medical conditions*. Certain medical conditions including Type 1 diabetes, thyroid disorders, rheumatoid arthritis, asthma, seizures, and organ transplantation are associated with increased risk, primarily because of the drugs used to treat them. Prolonged use of medications such as excessive thyroid hormone, anti-seizure drugs, and anti-inflammatory drugs—such as prednisone—used to treat asthma, arthritis, and some cancers may reduce calcium absorption, impair bone formation, and accelerate bone loss.

- *A bone-healthy diet—calcium, vitamin D, and other nutrients*. Calcium intake early in life affects the attainment of peak bone mass, achieved by about age 30, which may be the most important determinant of the risk of a fracture in later life.[63] Hence, ensuring an adequate calcium intake throughout life appears to be a sensible strategy. Adequate intake of Vitamin D is required for the absorption of calcium. Many older adults, who typically have lower intakes of vitamin D and reduced ability to synthesize the vitamin from sunshine, absorb less calcium as they age.

The calcium intake of a sizeable portion of the U.S. population, especially female adolescents and

FIGURE 7-10
A BONE'S LIFE

A. Bone is living tissue that is continuously remodeling itself. Bone contains cells embedded in a hard mineralized matrix. The osteoblast cells make new bone; the osteoclast cells break down old or damaged bone.

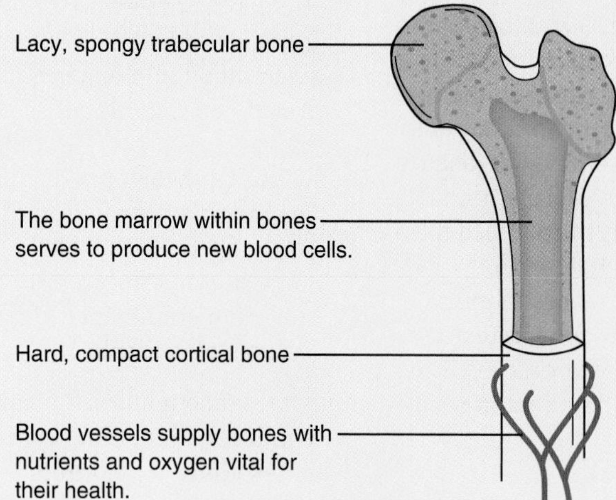
Lacy, spongy trabecular bone
The bone marrow within bones serves to produce new blood cells.
Hard, compact cortical bone
Blood vessels supply bones with nutrients and oxygen vital for their health.

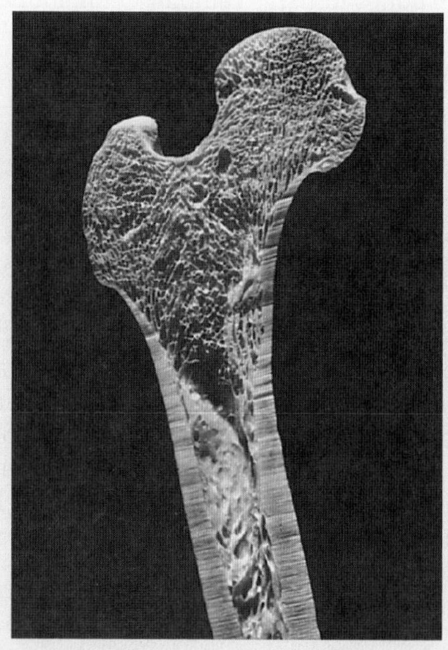

Cross section of bone. The lacy structural elements are trabeculae (tra-BECK-you-le), which can be drawn on to replenish blood calcium.

B. Peak bone mass occurs at approximately 30 years of age. Afterwards, bone loss starts to outpace bone deposition; at menopause there is a surge of calcium out of the bones.

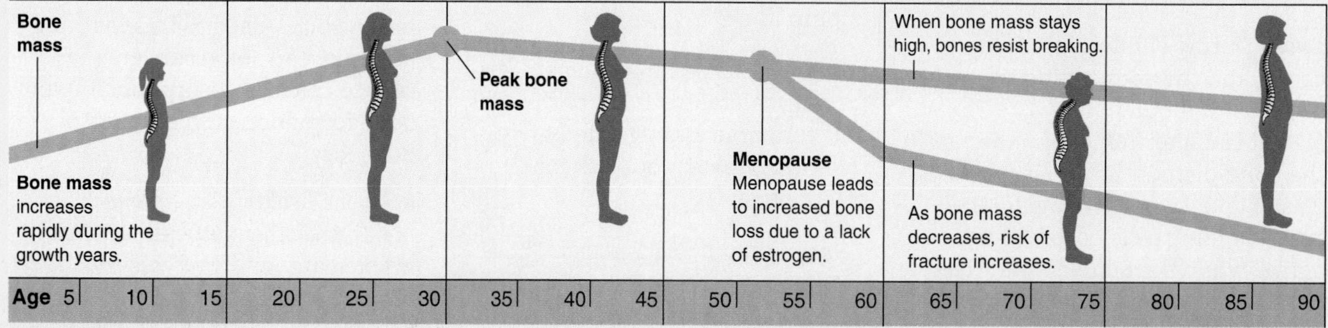
Bone mass
Bone mass increases rapidly during the growth years.
Peak bone mass
Menopause
Menopause leads to increased bone loss due to a lack of estrogen.
When bone mass stays high, bones resist breaking.
As bone mass decreases, risk of fracture increases.
Age 5 10 15 20 25 30 35 40 45 50 55 60 65 70 75 80 85 90

Source of peak bone mass timeline: Adapted from The Stay Well Company, San Bruno, CA. © 1998.

female adults, is estimated to be below the recommended intake for calcium (see Figure 7-11). Between the ages of 18 and 30 years—the time of the attainment of peak bone mass—more than two-thirds of all American women consume less calcium than they need. One explanation is the obsession—especially among adolescent girls—with dieting. Consumption of diet sodas has increased dramatically over the past three decades, displacing milk as an accompanying beverage to meals.[64]

Other nutrients also play a role in bone health.[65] For example, high intakes of phosphorus, protein, vitamin A (as retinol), and sodium may adversely affect calcium in the body.[66] Adequate intakes of vitamins C and K are needed to maintain healthy bone. Low vitamin K intakes have recently been associated with increased risk of hip fractures in the elderly.[67] Potassium, magnesium, and fruit and vegetable intakes are also associated with greater bone density in older adults.[68] The many phytochemicals found in fruits, vegetables, and legumes also offer protective benefits to bone (as discussed in Chapter 6).

Should I consider taking calcium supplements if I am unable to

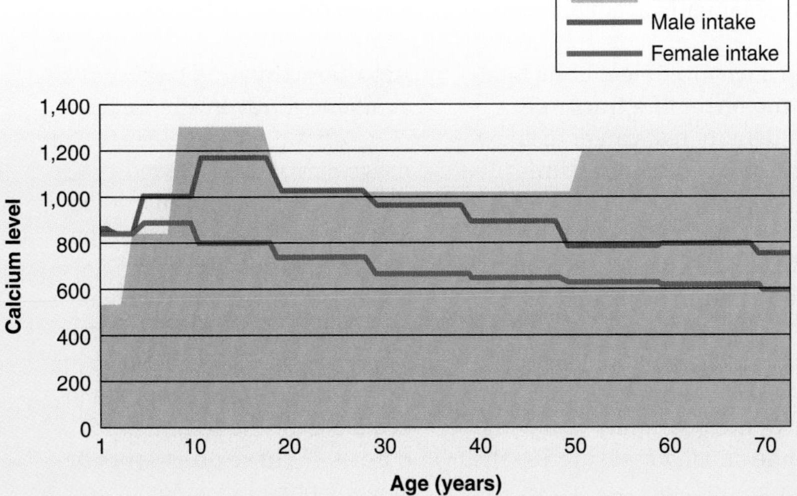

FIGURE 7-11
THE CALCIUM GAP: DAILY CALCIUM INTAKES IN THE UNITED STATES COMPARED TO RECOMMENDED INTAKES

For the most part, young children meet their recommended calcium intakes. In contrast, according to data from the most recent Health and Nutrition Examination Survey, 81 percent of girls and 48 percent of boys fail to get enough calcium in their diets after age 11. Likewise, the majority of adult women of any age and adult men after age 50 do not meet their recommended calcium intakes.

meet the recommended intake for calcium using foods? If so, what should I consider in choosing a supplement?
Although it is preferable to meet calcium needs by consuming calcium-rich foods, calcium supplements are an alternative way to obtain calcium.[69] Supplements may be needed by those who either cannot or will not adjust their diets to get enough calcium from food. Supplements have the advantage of being easy to take, but they have the disadvantage of not offering other nutrients, such as the thiamin, riboflavin, niacin, potassium, phosphorus, magnesium, zinc, vitamin A, and vitamin D found naturally in milk.

Choosing a calcium supplement can also be confusing. Calcium comes in a number of different salts, and the cost of a one-month supply of 1,000 milligrams per day can vary widely. The organic salts include calcium citrate, calcium citrate-malate, and calcium lactate; the inorganic salts include calcium carbonate and calcium phosphate. Calcium carbonate is the least expensive and contains the most calcium per pill, but an organic salt, such as calcium citrate, is better absorbed, especially by older people whose secretion of stomach acid—a factor known to enhance calcium absorption—tends to be reduced.[70] A strategy to overcome this problem is to take calcium carbonate two or three times a day in divided doses, with meals, to improve absorption. Or, try a chewable tablet, or take the calcium carbonate pill with a glass of orange juice to aid its dissolving. Since regular vitamin-mineral pills contain only small amounts of calcium, read the label carefully. Note, too, that many calcium supplements come with added nutrients, such as vitamin D and magnesium. You may not need these additional nutrients if you are consuming a nutritious diet, taking a multivitamin-mineral pill, or eating highly fortified cereals.

Keep in mind that a cup of milk contains about 300 milligrams of calcium. If you are using

Soft drinks often replace milk as a beverage in the diet of teenagers, accounting for about 10 percent of their caloric intake.

supplements to replace the calcium found in milk, then choose a supplement that contains at least 300 milligrams of calcium. Since calcium is best absorbed in doses of 500 milligrams or less, consuming supplements with more than 500 milligrams of calcium per pill is not advantageous. For example, if you need to take 1,000 milligrams of calcium in supplement form, choose a 500 milligram tablet and take one in the morning and one in the evening. A generic brand is fine to use—just check the label to see how much elemental calcium and which form of calcium it contains. Avoid calcium supplements derived from dolomite, oyster shell, or bonemeal, as some have been found to be contaminated with lead.[71] Table 7-12 offers tips for selecting from among the many popular calcium supplements available today.

Is there a way to detect bone loss before the onset of a fracture?
A **bone density** test can detect low bone density before a fracture occurs, or predict your chances of fracturing a bone in the future. Since 1998, bone density screening has been covered by Medicare for all women over age 65. The National Osteoporosis Foundation encourages women to discuss with their physician the possibility of having their bone density measured at the onset of menopause, when bone loss sharply accelerates, or if they have other medical conditions that increase their risk for osteoporosis. The more risk factors you have for osteoporosis, the more you need to evaluate your bone density as you age.

Are there any drug treatments available to reverse bone loss?
Yes. Among the bone-conserving drugs currently approved by the Food and Drug Administration are **estrogen replacement therapy, designer estrogens,** and **bisphosphonates** for osteoporosis prevention (see the accompanying Miniglossary). These options and the hormone, **calcitonin,** are also available for the treatment of osteoporosis.[72] These drugs work by slowing bone loss, allowing bones to slowly rebuild and increase bone density, and by reducing the risk of fractures by as much as 50 percent. To be effective, these treatments should be accompanied by the rec-

TABLE 7-12 TIPS FOR SELECTING AND USING A CALCIUM SUPPLEMENT

Supplement	Calcium (milligrams/pill)	Advantages	Possible Problems[a]	How to Take[b]
Calcium carbonate		Least expensive and most available		Take with meals
Caltrate, One A Day, Os-Cal,[c] Tums,[d] GNC, Solgar	600, 500, 500, 500, 400, 600			
Calcium citrate		Easily absorbed, regardless of the amount of stomach acid		Take on an empty stomach
Citracal	315			
Solgar	250			
Calcium lactate			Contains less calcium per pill	
Twin Lab	100			
Calcium phosphate		May be less likely to cause constipation		Take with meals
Posture-D	600			
Your Life[c]	250			

[a]Calcium with other supplements:
- If you also take an iron supplement, take it at a different time of day. Each mineral is better absorbed alone.
- Very high calcium intakes may limit zinc absorption—a concern for the many Americans who already get too little zinc. High intakes may also increase the risk of kidney stones, especially when taken on an empty stomach. The safe, upper limit for *total* calcium (from foods and supplements) is 2,500 milligrams per day.
- Although vitamin D is necessary for calcium absorption, excessive amounts may be harmful. Supplemental vitamin D in the amount of 400 to 600 IU per day should be sufficient for most individuals.

[b]Calcium with medications:
- Calcium should be taken separately from some medications such as Fosamax or else the drug's effectiveness can be diminished. Ask your doctor or pharmacist if you have any questions about food-drug interactions.

[c]Calcium source is oyster shells.

[d]Chewable.

Source: Adapted from How to Choose and Use a Calcium Supplement, *Nutrition in Clinical Care* 1(2) (1998): 100.

ommended intakes of calcium (1,200 mg for women over 50 years of age) and vitamin D (10 micrograms for adults 51 to 70 years), as well as regular weight-bearing exercise.[73]

MINIGLOSSARY

BISPHOSPHONATES drugs that decrease the risk of fractures by acting on the bone-dismantling cells (osteoclasts) and inhibiting their resorption of bone tissue; an example is alendronate (Fosamax). The side effects of long-term use have yet to be determined.

BONE DENSITY a measure of bone strength that reflects the degree of bone mineralization. The bone density test compares your bone density to that of a healthy young adult. Bone density tests—using a dual beam of low-level X-rays—take a snapshot of bone density in the spine, wrist, and hip. Simpler, less precise tests use ultrasound to measure bone density in the wrist or heel; these tests can be done in a physician's office and may help identify individuals in need of more precise testing.

CALCITONIN a hormone used as a drug to decrease the rate of bone loss in osteoporosis; administered as a nasal spray (Miacalcin) or by injection, calcitonin works by inhibiting the bone resorption activity of osteoclasts.

CORTICAL BONE the dense outer ivorylike layer of bone that provides an exterior shell over trabecular bone.

DESIGNER ESTROGENS (Selective Estrogen Receptor Modulators—SERMs) drugs that act on *estrogen receptors* in osteoblasts to promote an increase in bone mass; an example is raloxifene (Evista). Unlike estrogen, SERMs have little effect on reproductive tissues of the breast or uterus. *Estrogen receptors* are cellular molecules that bind to estrogen, selective estrogen receptor modulators, or phytoestrogens and deliver these compounds to the nucleus of the cell. Phytoestrogens—estrogen-like compounds found in soyfoods—may act as SERMs (see the Spotlight in Chapter 5).

ESTROGEN a major female hormone—important in connection with nutrition because it maintains calcium balance and because its secretion abruptly declines at menopause.

ESTROGEN REPLACEMENT THERAPY (ERT) administration of estrogen to replace the natural hormone that declines with menopause. Since ERT may increase the risk of uterine and breast cancer, *hormone replacement therapy*—the administration of a combination of estrogen with the hormone, progesterone—is often used. The combination of hormones lowers the risk of cancer.

MENOPAUSE the time of life at which a woman's menstrual cycle ceases, usually at about 45 to 50 years of age.

OSTEOBLAST a bone-building cell; responsible for formation of bone.

OSTEOCLAST a bone-destroying cell; responsible for resorption and removal of bone.

PEAK BONE MASS the highest bone density achieved for an individual—accumulated over the first three decades of life; typically occurs by 30 years of age. After age 30, bone resorption slowly begins to exceed bone formation.

TRABECULAR (tra-BECK-you-lar) **BONE** the lacy inner network of calcium-containing crystals—spongelike in appearance—that supports the bone's structure.

What can be done to prevent the debilitating effects of osteoporosis?

Not surprisingly, the same recommendations have been made before for lowering risks for other chronic diseases: Eat a nutritious diet and exercise *throughout life*. Consider the following steps for building bone mass and lowering your risk for osteoporosis:[74]

- Maximize peak bone mass and be vigilant about keeping the bones well supplied with calcium. Consume the recommended amount of calcium (and vitamin D) for your age group (refer back to Figure 7-5 on page 210). According to researcher Robert Heaney, only a few simple changes can add an extra 1,000 milligrams of calcium to an ordinary diet: Put a couple of heaping teaspoons of nonfat dried milk powder in each cup of coffee or tea, drink calcium-fortified orange juice, and eat one cup of low-fat yogurt, a dark green leafy vegetable, and a 1-inch cube of hard cheese.[75]
- Consume alcohol only in moderation, if at all, and avoid cigarettes altogether.
- Exercise regularly, since exercise can reduce the risk of developing osteoporosis by making bones stronger and increasing their ability to absorb calcium.[76] Incorporate regular sessions of weight-bearing exercise, such as brisk walking, weight training, stair climbing, rope jumping, dancing, hiking, tennis, or aerobic dance classes, into your weekly schedule to slow and possibly even prevent bone loss.[77] Swimming and cycling are less effective at building bone mass, since they put less weight on the bones. Exercises designed to strengthen the back muscles and improve posture are also recommended, since they will aid the skeleton in bearing its burden of weight. Increased muscle strength and flexibility will also improve balance and help prevent debilitating falls from occurring.[78]
- For women at or nearing menopause, talk to your health-care provider about the need for bone density testing, hormone replacement therapy, or other treatments to slow bone loss after menopause.

Taking these steps to lower your risk of osteoporosis or to treat it if you already have it, improves the chances of enjoying a quality life in your later years. Determine your own risk for osteoporosis using the Scorecard, which examines your current lifestyle for osteoporosis risk factors, that follows.

DO YOU HAVE A BONE-HEALTHY LIFESTYLE?

SCORECARD

No matter what your age, your food and lifestyle choices make a difference to your bone health. How are you "banking" on bones? Give yourself 10 points for each *Always,* 5 points for each *Sometimes,* and zero for each *Never.*

DO YOU . . . ?	ALWAYS	SOMETIMES	NEVER
1. Eat three or more servings of milk, yogurt, cheese, or other calcium-rich foods?	☐	☐	☐
2. Drink vitamin D-fortified milk or get moderate exposure to sunlight?	☐	☐	☐
3. Keep your body moving—with at least 30 minutes of weight-bearing activity, at least three times weekly?	☐	☐	☐
4. Consume enough calories to maintain a healthy weight (not too thin)?	☐	☐	☐
5. Avoid cigarettes?	☐	☐	☐
6. Go easy on alcoholic beverages?	☐	☐	☐
SCORE	___ +	___ +	___ = ___

SCORING

A perfect 60

You are banking on beautiful bones as a lifelong investment. Toast to a perfect score with an ice-cold glass of milk!

35–55

For healthier bones, you'd be wise to improve your investment strategies. Otherwise, there may be future penalties.

30 or less

Your calcium withdrawals may exceed your deposits. Make changes now before your window of opportunity closes.

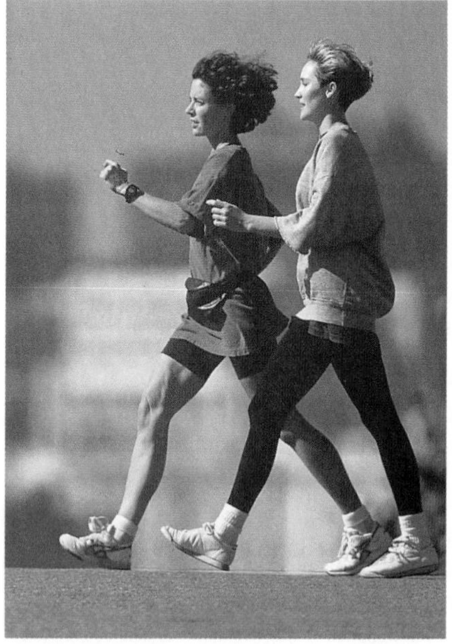

Incorporate regular sessions of weight-bearing exercise, such as brisk walking, into your weekly schedule to slow and possibly even prevent bone loss (for more about exercise, see Chapter 9).

Source: Banking on Beautiful Bones: A Lifelong Commitment to Calcium (Rosemont, Il: National Dairy Council, 1998). Reprinted by permission.

Nutrition on the Web

nutrition.wadsworth.com	Go to the *Personal Nutrition* site to check for the latest updates to chapter topics or to access links to related Web sites.
www.thriveonline.com/eats/vitamins/guide.index.html	A general guide for vitamins and minerals.
www.healthfinder.gov/searchoptions/topicsaz.htm	Search for information on water and the minerals.
www.nas.edu	Search for updates regarding new Dietary Reference Intakes.
www.nal.usda.gov/fnic	Search here for mineral-related topics.
www.nlm.nih.gov	Free access to National Library of Medicine's Medline for information searches on a variety of health-related topics.
www.eatright.org	American Dietetic Association's resources on mineral topics.
www.nationaldairycouncil.org	Search here for the latest in calcium and nutrition research.
www.whymilk.com	Check out the milk tips, recipes, and contests for consumers.
www.osteo.org	Fact sheets on many osteoporosis-related topics.
www.nof.org	Information from the National Osteoporosis Foundation.
www.nhlbi.nih.gov/health/public/heart/index.htm	Information on the DASH diet, recipes, menus, and other resources for people with high blood pressure.
dash.bwh.harvard.edu	The official DASH Diet site provides sample menus and tips.
www.navigator.tufts.edu	Use this search engine to find information about water and health, osteoporosis, anemia, and hypertension.
www.cdc.gov	Information about water safety from the Centers for Disease Control and Prevention.

Check Yourself . . .

1. List three functions of water in the body.
2. State the major functions of chloride, potassium, magnesium, and sulfur in the body.
3. Discuss the dietary factors that may affect blood pressure and describe the lifestyle changes recommended for reducing blood pressure.
4. What do the terms *major* and *trace* mean when referring to the minerals?
5. Name the bone minerals.
6. State the function of iodine in the body and describe the two conditions that can result from an iodine deficiency.
7. List factors that enhance iron absorption; factors that inhibit iron absorption.
8. What are the deficiency symptoms of zinc in children? In adults?
9. Define osteoporosis and list the major risk factors associated with it.
10. When would the use of a calcium supplement be recommended; how would you go about selecting one?

Answers to Check Yourself questions are found in Appendix G.

8 Weight Control

CONTENTS

What Is a Healthful Weight?

Energy Balance

Healthy Weight Scorecard

Causes of Obesity

Energy Output Scorecard

Weight Gain and Loss

Weight-Loss Strategies

Weight-Gain Strategies

The Savvy Diner: Dining Out

Nutrition Action: Breaking Old Habits

Spotlight: The Eating Disorders

Before you begin a thing remind yourself that difficulties and delays quite impossible to foresee are ahead. . . . You can see only one thing clearly, and that is your goal. Form a mental vision of that and cling to it through thick and thin.

K. Norris

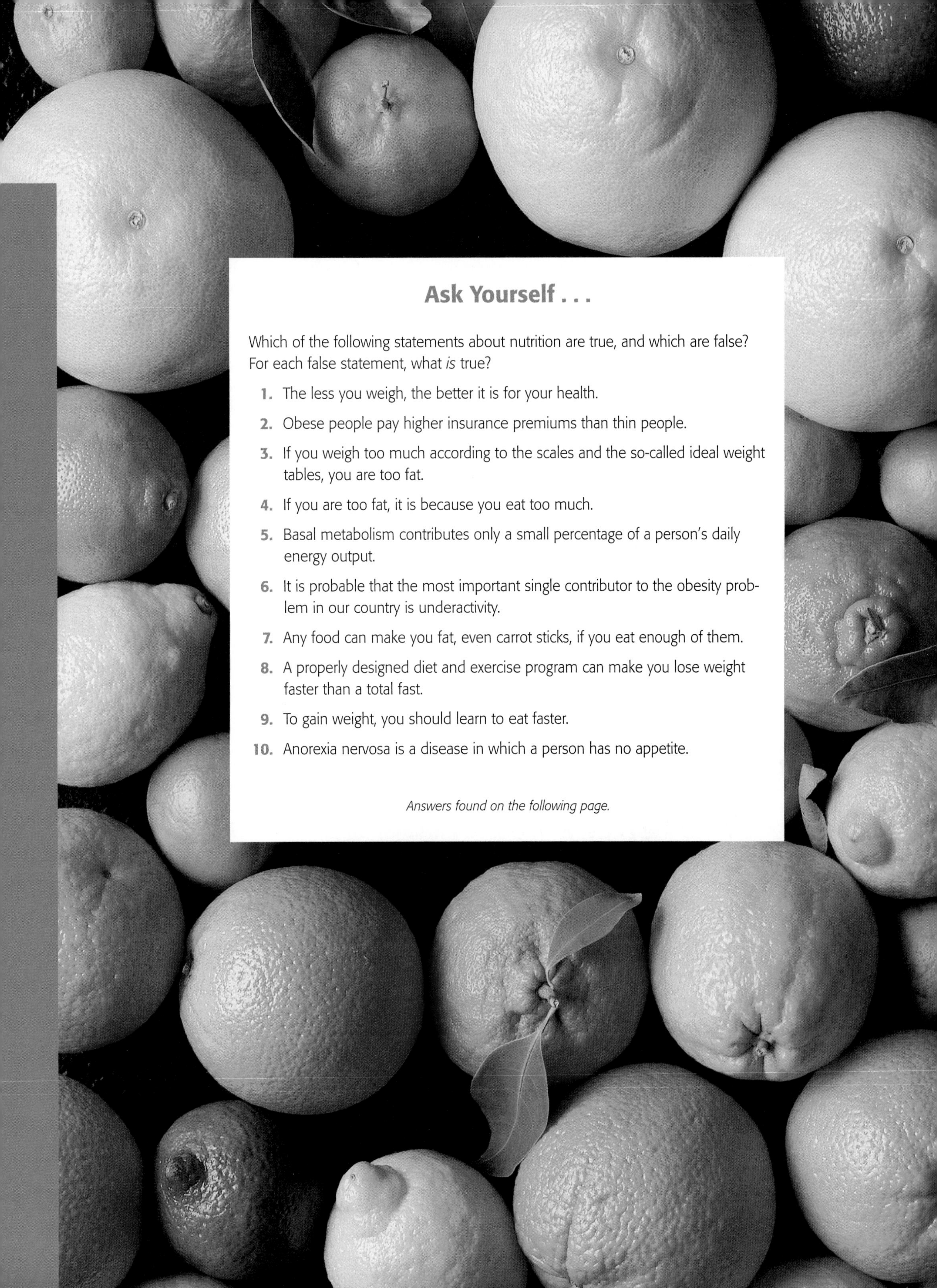

Ask Yourself . . .

Which of the following statements about nutrition are true, and which are false? For each false statement, what *is* true?

1. The less you weigh, the better it is for your health.
2. Obese people pay higher insurance premiums than thin people.
3. If you weigh too much according to the scales and the so-called ideal weight tables, you are too fat.
4. If you are too fat, it is because you eat too much.
5. Basal metabolism contributes only a small percentage of a person's daily energy output.
6. It is probable that the most important single contributor to the obesity problem in our country is underactivity.
7. Any food can make you fat, even carrot sticks, if you eat enough of them.
8. A properly designed diet and exercise program can make you lose weight faster than a total fast.
9. To gain weight, you should learn to eat faster.
10. Anorexia nervosa is a disease in which a person has no appetite.

Answers found on the following page.

OVERWEIGHT conventionally defined as weight between 10 percent and 20 percent above the desirable weight for height, or a body mass index (BMI) of 25.0 through 29.9 (see page 243).

UNDERWEIGHT weight 10 percent or more below the desirable weight for height, or a BMI of less than 18.5.

OBESITY conventionally defined as weight 20 percent or more above the desirable weight for height, or a BMI of 30 or greater.

How much should you weigh? At one time, you could look up the answer in the so-called ideal weight tables. Defining **overweight** and **underweight** was simple: If a person's weight was 10 percent or more above the ideal weight, the person was overweight; 20 percent or more meant the person was **obese.** If the person's weight was 10 percent or more below the ideal weight, the person was underweight. Now the term *ideal weight* is seldom used, and defining overweight and underweight is no longer simple. Regardless of how you define it, obesity is described by the World Health Organization as "an escalating epidemic" and one of the greatest neglected public health problems of the new millenium.[1] Obesity is a disease with multiple health risks, resulting in some 300,000 deaths each year—ranking second only to smoking as a cause of preventable death.[2]

Currently, an estimated 107 million U.S. adults (55 percent) and 11 percent of children and adolescents are either overweight or obese—exceeding their healthy weight range (see Table 8-1).[3] The annual costs of overweight are staggering: more than $33 billion spent on weight-reduction products and services, and another $68 billion spent on health care for medical complications associated with obesity.[4] The *Healthy People 2010* objectives mandate a reversal in the rising trend toward obesity, as shown in Figure 8-1 on page 242.

Clearly, such a thing exists for each individual as a weight too low or a weight too high to be healthful. The too-thin person has minimal body fat stores and will be at a disadvantage in situations where energy reserves might be needed, such as a prolonged period of physiological stress or injury. Other problems include menstrual irregularity, infertility, and osteoporosis.

The physical risks of obesity are greater for some people than for others, depending on inherited susceptibilities to conditions such as high blood pressure, high blood cholesterol, and diabetes.[5] High blood pressure is made worse by weight gain and can often be normalized merely by weight loss. Diabetes can be precipitated in genetically susceptible people if they become overweight. If any of these conditions run in your family, you are urgently in need of a sensible weight-management program to prevent obesity.

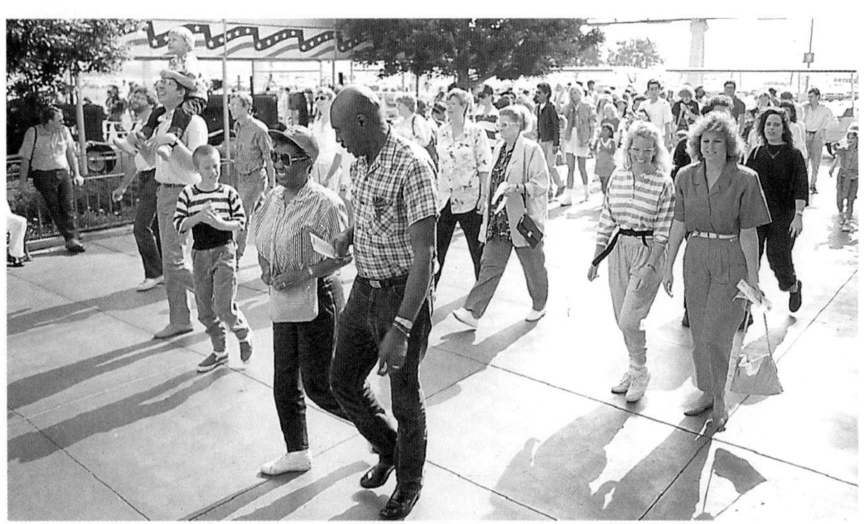

Bodies come in many shapes and sizes. Which are healthy?

Obesity also increases the risk of heart disease because excess fat pads crowd the heart muscle and the lungs within the body cavity. These fat pads encumber the heart as it beats, making it work hard to deliver oxygen and nutrients. The lungs, too, cannot expand fully, which limits the oxygen intake of each breath, causing the heart to work even harder to pump the needed amount of oxygen to the other body parts. Furthermore, since each extra pound of fat tissue demands to be fed by way of miles of capillaries, the heart labors to pump its blood through a network of blood vessels vastly larger than that of a thin person. Even a healthy heart is strained by excess fatness. If a diseased heart finds itself in this bind, a sudden burst of exercise

Ask Yourself Answers: **1.** False. Being thin is good for health only to a point; being too thin is as risky as being too fat. **2.** True. **3.** False. A high weight according to the scales and the so-called ideal weight tables may reflect heavy bones and muscles rather than excess fatness. **4.** False. If you are too fat, it could be because you exercise too little. **5.** False. Basal metabolism contributes about 60 percent or more of the average person's daily energy output. **6.** True. **7.** True. **8.** True. **9.** True. **10.** False. Anorexia nervosa is misnamed; people with anorexia nervosa are constantly hungry but control their hunger.

may be more of a workload than it can handle. Chapter 9 discusses precautions relevant to exercise.

Gallbladder disease, too, can be brought on in susceptible people merely by excess weight.[6] Similarly, obesity increases a woman's risk of developing breast cancer. Table 8-2 shows other conditions brought on or made worse by obesity.

Besides being at risk for these health hazards, millions of obese people throughout much of their lives incur risks from ill-advised, misguided dieting. Some fad diets are more hazardous to health than is obesity itself. Many of the claims, treatments, devices, and gadgets for losing weight can be described as simply ineffective to truly dangerous to your health.[7] Over the centuries, "magical" weight-loss plans have been offered time and again, the success of which is in their popularity, not in their achievements.

Some people can be obese and suffer none of the risks of these physical health hazards, but there is one disadvantage of obesity that no one in our society quite escapes. Obesity in many parts of North America is a social and economic handicap.[8] Obese individuals suffer from discrimination in many areas, including social relationships and the job market. Psychologically, a body size that embarrasses or shames a person confers severe disadvantages. For people who cannot talk freely about their own self-doubt, a less-than-ideal body image becomes a private anguish. For people who perceive themselves as fat in a society that prizes thinness, real or imagined obesity can thrust them into withdrawal, shame, humiliation, and isolation.

How thin, then, is too thin—and how fat is too fat? Because the term *ideal weight* cannot be defined, the following section discusses the concept of a healthful weight.

WHAT IS A HEALTHFUL WEIGHT?

The problems of defining a healthful weight are many. Consider the question, Healthful for what? For long-distance runners, every unneeded pound is a disadvantage; the lowest amount of body fat that doesn't compromise hormonal balance and fuel availability is desirable. Weight matters less for swimmers, and fat contributes to their buoyancy and insulates them against the cold. In the case of swimmers, to a point, more is desirable. Dancers and models may value thinness so highly that to attain it, they compromise their health.

Some societies value fatness, equating it with prosperity; others value thinness to the point of obsession (our own being an example). This chapter first asks what range of weights is compatible with wellness and long life and then suggests that personal preferences dictate the choice of a weight within that range.

Body Weight Versus Body Fat

The question of what weight is healthful is harder to answer than you might at first think. To think merely in terms of weight oversimplifies the issue of body fatness and health. Two people of the same sex, age, and height may both weigh the same, yet one may be too fat and the other too thin. The difference lies in their body composition. One may have small, light bones and minimally developed muscles, while the other has big, heavy bones and well-developed muscles. The first person could have too much body fat and the second person too little. For example, football players, body builders, and other athletes may weigh in as overweight for their height according to the weight tables. However, they probably will have less body fat than the amount that poses a risk to health. Likewise, many sedentary persons may weigh in as normal weight but be overly fat. This comparison points to the need to define obesity in terms of people's body fatness rather than in terms of their weight.

The health risks of obesity refer to people who are overfat. On the average, men having over 25 percent body fat and women having over 30 percent body

TABLE 8-1 WEIGHT FOR HEIGHT STANDARDS

Healthy Weight Ranges

Height*	Weight (lb)† Midpoint	Range
4′10″	105	91–119
4′11″	109	94–124
5′0″	112	97–128
5′1″	116	101–132
5′2″	120	104–137
5′3″	124	107–141
5′4″	128	111–146
5′5″	132	114–150
5′6″	136	118–155
5′7″	140	121–160
5′8″	144	125–164
5′9″	149	129–169
5′10″	153	132–174
5′11″	157	136–179
6′0″	162	140–184
6′1″	166	144–189
6′2″	171	148–195
6′3″	176	152–200
6′4″	180	156–205

BMI Cut-off Weights for Overweight and Obesity

Height	BMI of 25 overweight (lb)	BMI of 30 obese (lb)
4′11″	124	148
5′0″	128	153
5′1″	132	158
5′2″	136	164
5′3″	141	169
5′4″	145	174
5′5″	150	180
5′6″	155	186
5′7″	159	191
5′8″	164	197
5′9″	169	203
5′10″	174	207
5′11″	179	215
6′0″	184	221
6′1″	189	227
6′2″	194	233
6′3″	200	240
6′4″	205	246

*The higher weights in the ranges generally apply to men, who tend to have more muscle and bone; the lower weights more often apply to women, who have less muscle and bone.
†Without shoes or clothes.

FIGURE 8-1
THE PREVALENCE OF HEALTHY WEIGHT AND OBESITY AMONG U.S. ADULTS AND THE *HEALTHY PEOPLE 2010* OBJECTIVES

Currently, only 42% of adults are at a healthy weight, whereas 23% of adults are identified as obese.

*Healthy weight is defined as having a Body Mass Index (BMI) from 18.5 through 24.9.
†Obesity is defined as having a BMI at or above 30.
Data Source: National Health and Nutrition Examination Survey (NHANES III), 1988–1994.

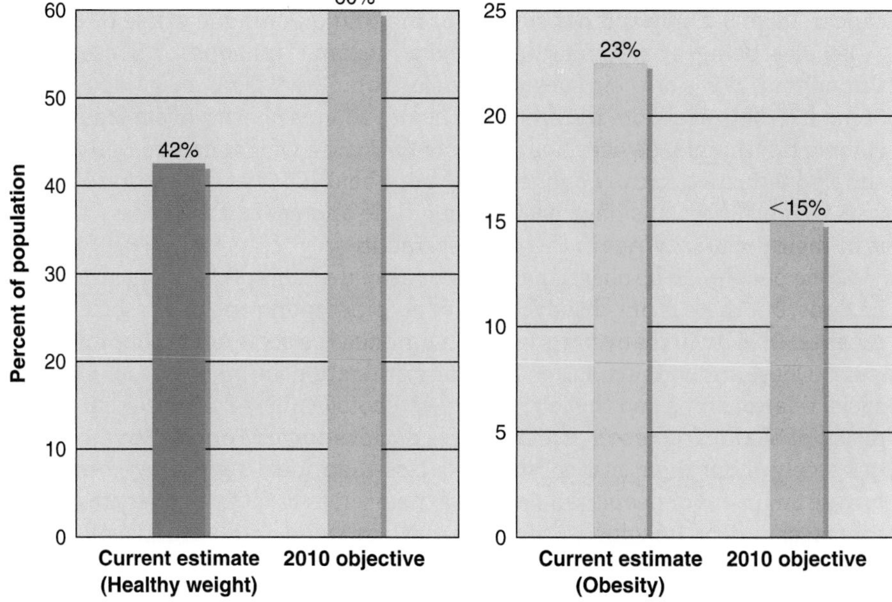

TABLE 8-2 PROBLEMS ASSOCIATED WITH OBESITY

- Abdominal hernias
- Accidents
- Certain cancers
  In men: colon, rectum, prostate
  In women: breast, uterus, cervix, ovaries, colon
- Complications during pregnancy
- Complications after surgery
- Decreased longevity
- Decreased quality of life
- Depression
- Diabetes (Type 2)
- Gallbladder and liver disease
- Gout
- Heart disease
- High blood cholesterol levels
- Hypertension
- Injury to weight-bearing joints
- Poor self-esteem
- Osteoarthritis (knees, hips, lower spine)
- Respiratory problems
- Sleep disturbances
- Varicose veins

fat are considered obese. More desirable measures are 12 percent to 20 percent body fat for most men and 20 percent to 28 percent body fat for most women.

Measuring Body Fat

Body fatness is hard to measure. One very accurate way is to obtain a measure of the body's density—that is, weight divided by volume. Weight is easy enough; just step on an accurate scale. But to obtain the body's volume, you have to immerse the whole body in a tank of water. Not many health professionals have space in their offices for the equipment needed for this procedure, known as hydrostatic weighing or **underwater weighing.** Most employ the **skinfold test,** using a caliper—a pinching device that measures the thickness of a fold of fat in such areas as the back of the arm, below the shoulder blade, and the side of the waist. About 50 percent of the body's fat lies beneath the skin, and its thickness at these locations can be compared with standard tables to give a fair approximation of total body fat, at least for most people.

An alternative method for assessing body fatness is **bioelectrical impedance,** in which electrodes are attached to a person's hand and foot. This method provides a measure of how much fat a person has by measuring the speed at which electrical current is conducted through the body (from the ankle to the wrist). Since fat is a poor conductor of electricity, the more fat one has, the more resistance this current encounters in the body.[9]

Distribution of Fat

Not everyone carries his or her body fat distributed in the same way. To complicate matters still further, the distribution itself turns out to have health implications. Excess fat around the middle—**central obesity**—is associated with increased health hazards.[10] Body types are compared as being either apple-shaped or pear-shaped. People who store most of their excess fat around the abdomen (typically men) are at a greater risk for developing diabetes, hypertension, elevated levels of blood cholesterol, and heart disease than are people who

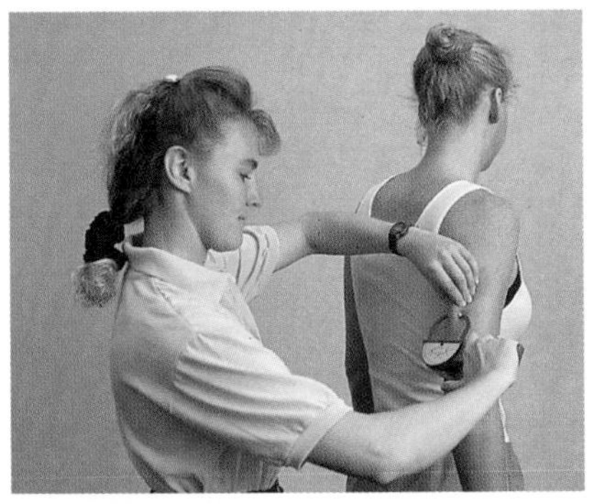

The fatfold test gives a fair approximation of total body fat.

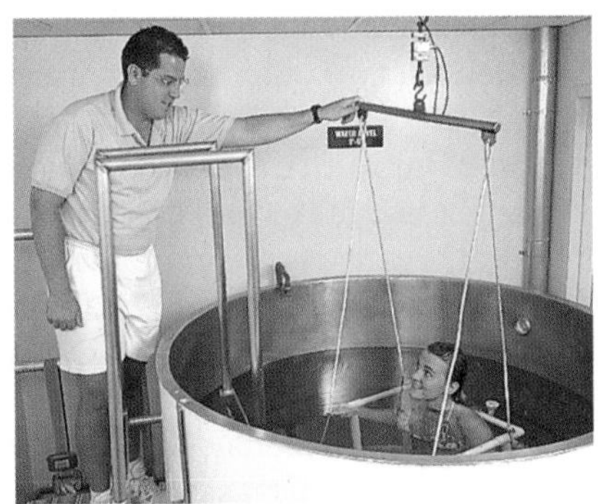

Hydrostatic (underwater) weighing.

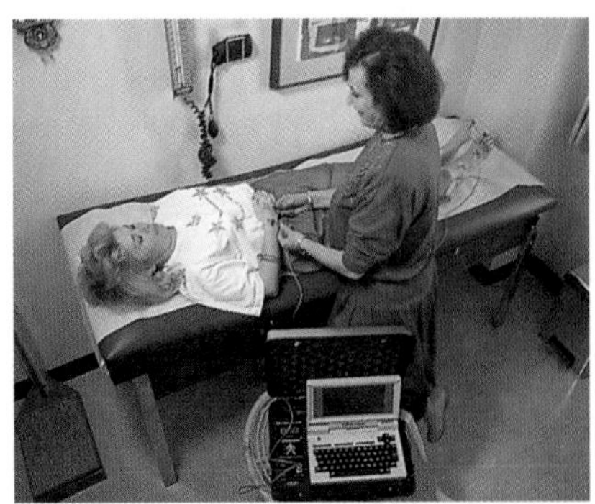

Electrical impedance measurement of body fat content.

store excess fat elsewhere on the body—notably, hips, thighs, and buttocks (typically women).[11] A simple calculation of a person's waist circumference can be used to assess abdominal fat, as discussed on the following page.

Weighing in for Health

Obesity presents one of the most serious health risks that people face—and one that people should, theoretically, be able to control. In response to the rising epidemic of excessive weights, new guidelines for the assessment and treatment of overweight and obesity were developed.[12] A person's health risk is dependent on three factors: body weight, amount and location of body fat, and current health status.

The first measure is the **body mass index (BMI),** which is an index of your weight in relation to your height. Overweight is defined as a BMI of 25 to 29.9, and obesity is defined as a BMI of 30 or above (refer to Table 8-1 on page 241 for body weights that equate with these BMI values).[13] Keep in mind that BMI does not account for location of fat in the body, and a muscular person with a low

UNDERWATER WEIGHING (hydrostatic weighing) a measure of density and volume; the less a person weighs underwater compared to the person's out-of-water weight, the greater the proportion of body fat (fat is less dense or more buoyant than lean tissue).

SKINFOLD TEST a clinical test of body fatness in which the thickness of a fold of skin on the back of the arm (the triceps), below the shoulder blade (subscapular), or in other areas is measured with an instrument called a caliper. Obesity is defined by triceps skinfold thickness equal to or greater than 18–19 mm in adult men or 25–26 mm in women.

Central obesity, characterized by an "apple-shaped" body with large abdominal-type fat stores, is a strong risk factor for Type 2 diabetes, heart disease, and other problems.

A muscular person, such as this body builder, often has a low percentage of body fat but a high BMI.

BIOELECTRICAL IMPEDANCE estimation of body fat content made by measuring how quickly electrical current is conducted through the body.

CENTRAL OBESITY excess fat on the abdomen and around the trunk. Peripheral obesity is excess fat on the arms, thighs, hips, and buttocks.

BODY MASS INDEX an index of a person's weight in relation to height which correlates with totoal body fat content.

- BMI < 18.5 = underweight
- BMI 18.5 to 24.9 = normal weight
- BMI 25 to 29.9 = overweight
- BMI ≥ 30 = obesity
- BMI ≥ 40 = severe obesity

WAIST CIRCUMFERENCE a measure used to assess a person's abdominal (visceral) fat; excess fat in the abdomen increases a person's risk for obesity-related health problems.

Waist circumference denoting risk of obesity-related health problems:

	Increased risk	Substantially Increased Risk
Men	> 33″	> 40″
Women	> 32″	> 35″

People storing excess fat in the chest and stomach areas (apple-shaped) are at a higher risk for diabetes, heart disease, and hypertension than people storing excess fat in the hips, thighs, and buttocks (pear-shaped).

percentage of body fat may have a high BMI. In consideration of this shortcoming of the BMI, two other factors are evaluated.

First is the **waist circumference,** which provides information about the distribution of fat in the abdomen.* Excess fat in the abdomen is a greater health risk than excess fat in the hips and thighs. The extra abdominal fat crowds the abdominal organs and its proximity to the liver means that when metabolized, abdominal fat can raise blood cholesterol levels and lower the body's sensitivity to insulin. Disease risk rises significantly with a waist circumference of over 35 inches in women and over 40 inches in men.[14]

The last factor to consider is the presence of weight-related health problems and risk factors for diseases. These may include family health history, heart disease, Type 2 diabetes, high blood cholesterol, high blood pressure, cigarette smoking, osteoarthritis, gallstones, or sleep apnea (irregular breathing during sleep). According to the new treatment guidelines, an initial goal for treatment of overweight and obese people with risk factors is to reduce body weight by about 10 percent at a rate of about 1 to 2 pounds per week.[15] For overweight individuals, losing as little as 5 to 10 percent of their body weight may improve many of the problems linked to being overweight, such as diabetes and high blood pressure.[16] The accompanying Healthy Weight Scorecard helps you make sense of these guidelines and evaluate your own weight status.

ENERGY BALANCE

Suppose you decide that you are too fat or too thin. You got that way by having an unbalanced energy budget—that is, by eating either more or less food energy than you spent. Fatness and thinness are reflections of excessive or deficient energy stores. You store extra energy as fat only if you eat *more* food energy in a day than you use to fuel your metabolic and other activities. Similarly, you lose stored fat only if you eat *less* food energy in a day than you use as fuel. A day's energy balance can be stated like this:

Change in energy stores = Energy in − Energy out.

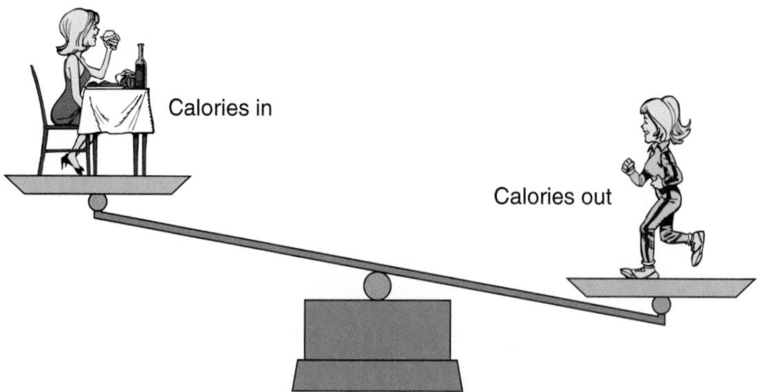

The balance between energy in and energy out determines whether a person stores or uses body fat.

You know about the "energy in" side of this equation. An apple brings you 100 calories; a candy bar, 290 calories. You probably also know that for each 3,500 calories you eat in excess of need, you store one pound of body fat.

As for the "energy out" side, the body spends energy in two major ways: to fuel its metabolic activities and to fuel its muscle activities. You can change your

*The *waist-to-hip ratio (WHR)* was previously used to assess abdominal fat (a ratio above 0.80 for women and above 0.95 for men suggests an excess of abdominal fat).

HEALTHY WEIGHT SCORECARD

A wide range of weights is compatible with good health. Within this range, the definition of desirable or healthful weight is up to the individual, depending on such factors as family history, occupation, physical and recreational activities, and personal preferences. To determine if your weight is a healthful weight for you:

1. *Calculate your body mass index (BMI).* Find your height and weight in the following chart.* Your BMI is at the top of the column that contains your weight. Record your BMI here: ________

	19	20	21	22	23	24	25	26	27	28	29	30	35	40
HEIGHT	WEIGHT (POUNDS)													
5′0″	97	102	107	112	118	123	128	133	138	143	148	153	179	204
5′1″	100	106	111	116	122	127	132	137	143	148	153	158	185	211
5′2″	104	109	115	120	126	131	136	142	147	153	158	164	191	218
5′3″	107	113	118	124	130	135	141	146	152	158	163	169	197	225
5′4″	110	116	122	128	134	140	145	151	157	163	169	174	204	232
5′5″	114	120	126	132	138	144	150	156	162	168	174	180	210	240
5′6″	118	124	130	136	142	148	155	161	167	173	179	186	216	247
5′7″	121	127	134	140	146	153	159	166	172	178	185	191	223	255
5′8″	125	131	138	144	151	158	164	171	177	184	190	197	230	262
5′9″	128	135	142	149	155	162	169	176	182	189	196	203	236	270
5′10″	132	139	146	153	160	167	174	181	188	195	202	207	243	278
5′11″	136	143	150	157	165	172	179	186	193	200	208	215	250	286
6′0″	140	147	154	162	169	177	184	191	199	206	213	221	258	294
6′1″	144	151	159	166	174	182	189	197	204	212	219	227	265	302
6′2″	148	155	163	171	179	186	194	202	210	218	225	233	272	311
	Healthy Weight						Overweight					Obese		

Source: National Heart, Lung and Blood Institute

2. Determine if your fat distribution is associated with health risks. Measure your waist circumference by placing a tape measure around your waist just above your belly button. Record your waist in inches here. ________
3. Is your weight affecting your health? Do you have any of these weight-related health problems or risk factors?
   - Heart disease
   - Type 2 diabetes
   - High blood pressure
   - High LDL cholesterol
   - Low HDL cholesterol
   - Cigarette smoking
   - Osteoarthritis
   - Recurrent gallstones
   - Sleep disturbances
4. How does your current weight measure up to these considerations?
   - If your BMI is acceptable for good health and if your waist measurement is not high, you will want to maintain this weight. If you need to lose weight or gain weight, consider the tips offered throughout this chapter for healthfully changing your weight.
   - You should consider losing weight if:
     Your BMI is 30 or greater
     Your BMI is 25 to 29 *and* you have two or more weight-related health problems or risk factors
     Your waist circumference exceeds 40 inches (for men) or 35 inches (for women) *and* you have two or more weight-related health problems or risk factors.
   - Weight loss is optional for you if your BMI is 25 to 29 and you do *not* have two or more weight-related health problems (particularly if your BMI is under 27 or you have large muscles and bones).

*Or you can use this formula to calculate your BMI:

$$\frac{\text{weight (in pounds)}}{\text{height}^2 \text{ (in inches)}} \times 705$$

Example: A 5′8″ person weighing 145 pounds has a BMI of 22

$$\text{BMI} = \frac{145}{68^2} \times 705$$

$$\text{BMI} = \frac{145}{4624} \times 705$$

BMI = 22 (rounded)

Weight Loss: calories eaten are less than calories burned.
Weight Gain: calories eaten are greater than calories burned.

activity level to spend more energy in a day, and if you do so consistently, your metabolic activities also will ultimately speed up somewhat. The following three sections discuss the body's needs for energy.

Basal Metabolism

About 60 percent or more of the energy the average person spends goes to support the ongoing metabolic work of the body's cells, the **basal metabolism.**[17] This is the work that goes on all the time, without conscious awareness. The beating of the heart, the inhaling and exhaling of air, the maintenance of body temperature, and the sending of nerve and hormonal messages to direct these activities are the basal processes that maintain life. Basal metabolic needs are surprisingly large. A person whose total energy expenditure amounts to 2,000 calories per day spends as many as 1,200 to 1,400 of them to support basal metabolism.

BASAL METABOLISM the sum total of all the chemical activities of the cells necessary to sustain life, exclusive of voluntary activities—that is, the ongoing activities of the cells when the body is at rest.

BASAL METABOLIC RATE (BMR) the rate at which the body spends energy to support its basal metabolism. The BMR accounts for the largest component of a person's daily energy (calorie) needs.

The **basal metabolic rate** (often abbreviated **BMR**) is influenced by a number of factors (see Table 8-3). In general, the younger a person is, the higher the basal metabolic rate, partly because of the increased activity of cells undergoing division. The BMR is most pronounced during the growth spurts that take place during infancy, puberty, and pregnancy. Body composition also influences metabolic rate. Muscle tissue is highly active even when it is resting, whereas fat tissue is comparatively inactive. The more lean tissue in a body, the higher the BMR; the more fat tissue, the lower the BMR. Lean body mass decreases with age, and it is estimated that the BMR decreases by about 5 percent every ten years after age 50.

Gender correlates roughly with body composition. Men generally have a faster metabolic rate than women, and researchers believe that this is because of men's greater percentage of lean tissue. (A woman athlete has a greater percentage of lean tissue than a sedentary man of the same weight and so would have a higher metabolic rate.) Fever increases the energy needs of cells, whose increased activities to generate heat and fight off infection speed up the metabolic rate.

Fasting and constant malnutrition lower the metabolic rate because of the loss of lean tissue and the slowdown of activities the body can't afford to support fully. This slowing of metabolism seems to be a protective mechanism to conserve energy when there is a shortage, and it hampers weight loss in a person who fasts or undertakes a very-low-calorie diet.

Some hormones—the stress hormones, for example—influence metabolism. They increase the energy demands of every cell and thus raise the metabolic rate. This raised metabolism partly accounts for the weight loss sometimes seen in people experiencing extreme stress in their life, although other factors, such as upset digestion and loss of appetite, also enter in.

The activity of the thyroid gland also influences the basal metabolic rate. The less thyroid hormone secreted, the lower the energy requirement for maintenance of basal functions.

TABLE 8-3 FACTORS THAT INFLUENCE THE BASAL METABOLIC RATE

Factors That Increase BMR:
- Fever
- Growth (higher in children and pregnant women)
- Height (higher in tall, thin people)
- Male gender (more lean tissue)
- Muscle mass (the more lean tissue, the higher the BMR)
- Stress
- Thyroid hormone

Factors That Decrease BMR:
- Age (slows down with age)
- Reduced energy intake (fasting, starvation, low-calorie diets)
- Sleep (BMR is lowest when sleeping)

Voluntary Activities

Muscular activity does not make as big a contribution as basal metabolism does to most people's energy outputs. On the average, it amounts to only about 30 percent of the total. But unlike basal metabolism, which cannot be changed immediately, physical activity can be changed at will. If you want to tinker with your energy balance, this is the component—on the output side—that you can alter significantly in the short term. If you increase it consistently, you will also ultimately increase the energy your body spends on metabolic activity because you will have an increase in lean body mass.

The energy spent on physical activity is the energy spent moving the body's skeletal muscles—the muscles of the arms, back, abdomen, legs, and so forth—and the extra energy spent to speed up the heartbeat and respiration rate as needed. The number of calories spent depends on three factors: (1) the amount of muscle mass required, (2) the amount of weight being moved, and (3) the amount of time the activity takes. Thus, an activity involving both the arms and the legs requires more calories than an activity of the same intensity involving only the legs; an activity performed by a heavier person requires more energy than the same activity performed by a lighter person; and an activity performed for 40 minutes requires twice as much energy as that same activity performed for 20 minutes.

As disheartening as it may be for a college student to discover, mental activity requires little energy, even though it may be tiring. Studying for an exam may be hard work, but it won't burn body fat. People who are very, very busy—writing letters, making phone calls, riding in their cars from place to place—may wonder why they tend to gain weight, because they think of themselves as active people. They may be socially or intellectually active, but such activity involves few muscles and therefore little energy expenditure.

Body composition influences metabolic rate. Weight training can help shift your body composition toward more lean tissue, thereby speeding up your metabolism. See Chapter 9 for guidelines on how to get started.

Total Energy Needs

A typical breakdown of the total energy spent by a lightly active person (for example, a student who walks back and forth to classes) might look like this:

Energy for basal metabolism:	1,400 calories
Energy for physical activity:	560 calories
	Total: 1,960 calories

The first component is larger, and you cannot change it much. You can, however, change the second component—physical activity—and so use more calories. If you want to increase your basal metabolic output, make exercise a daily habit. Your body composition will gradually change, and your basal energy output will pick up the pace as well. You can figure out roughly how much energy you spend in a day by using the Energy Output Scorecard that follows.

In summary, the amount of fat stored in a person's body depends on the balance between the total food energy the person has taken in and the total energy the person has expended. Later in the chapter you will learn how to alter both—with diet and exercise—to regulate body weight. But first: Why do so many people have excessive fat stores?

CAUSES OF OBESITY

Some people eat more than they need or exercise less than they should to maintain their body weight, and they get fat. Some eat less or exercise more, and they get thin. Perhaps most amazingly, some people—most people, in fact—eat exactly what they need and stay at the same weight year after year. A single extra pat of butter each day would make them gain five pounds in a year, but if they overeat by that much one day, they undereat by the same amount the next. How do they do this—and, in contrast, why do some people fail to maintain their weight?

Genetics Versus Environment

In general, two schools of thought address the problem of obesity's causes.[18] One attributes it to inside-the-body causes; the other, to environmental factors. One popular inside-the-body theory is the so-called **set-point theory.** Noting that many people who lose weight on reducing diets subsequently return to their original weight, some researchers have suggested that the body "wants" to maintain a certain amount of fat and regulates eating behaviors and hormonal actions to defend its "set point." The theory implies that science should search inside obese people to find the causes of their problems—perhaps in their hunger-regulating mechanisms.

SET-POINT THEORY the theory that the body tends to maintain a certain weight by adjusting hunger, appetite, and food energy intake on the one hand and metabolism (energy output) on the other so that a person's conscious efforts to alter weight may be foiled.

ENERGY OUTPUT SCORECARD

A. First figure the energy you spend on BMR.* It's about 10 calories per pound of your body weight if you are a woman or 11 calories per pound if you are a man (men generally have more muscle than women): ________ cal

B. Now figure the energy you spend on activities:

If you are not very active (sedentary)—you sit down most of the day and drive or ride whenever possible—take 25 percent to 40 percent (men) or 25 percent to 35 percent (women) of your BMR calories in A.

If you are lightly active—you sit most of the day, but you move around two to four hours of the day as a teacher might—take 50 percent to 70 percent (men) or 40 percent to 60 percent (women) of A.

If you are moderately active—you do some amount of regular exercise four or five times a week or your job requires some physical labor—take 65 percent to 80 percent (men) or 50 percent to 70 percent (women) of A.

If you are very physically active—your job requires much physical labor or you are physically active four or more hours each day—and you do little standing or sitting, take 90 percent to 120 percent (men) or 80 percent to 100 percent (women) of A. ________ cal

C. Total energy need = A + B = ________ cal

*Note: Another way to estimate the energy spent on basal metabolism is to use the factor (for men) 1.0 cal/kg/hour (or for women 0.9 cal/kg/hr) for a 24-hour period. Then add an increment of this amount depending on how physically active you are.

Recently, researchers identified a gene—named *ob* (for *obese*)—that appears to produce a hormone called *leptin,* after the Greek word for slender, that seems to tell the body to stop eating when it is released from fat cells.[19] Researchers report that as body fat stores increase, blood leptin increases. The brain responds by decreasing appetite and increasing energy expenditure. Likewise, when body fat stores decrease, blood leptin decreases, and the brain responds by stimulating appetite and decreasing energy expenditure. Mice with a defective form of the gene fail to produce leptin and can weigh as much as three times more than normal mice. Overweight people, too, may have a defective form of this gene (or may be unresponsive to its hormone), which fails to give an accurate report of the size of the fat cells to the brain, thus making the set point too high—resulting in weight gain.[20] Much more research is needed to clarify this mechanism.

Given a wide variety of tempting foods, these laboratory animals become obese, just as human beings do.

The other point of view is that obesity is environmentally determined. Proponents of this view hold that people overeat or underexercise because they are pushed to do so by factors in their surroundings—foremost among them, the availability of a multitude of delectable foods and the lack of opportunities for vigorous physical activity. The two views are not mutually exclusive, and they may both be operating, even within the same person.

Perhaps some people have inherited or learned a way of resisting external stimuli to eat, while others have not. Of interest in this connection is the report of a classic experiment with "cafeteria rats." Ordinary rats fed regular rat chow are of normal weight (for rats), but if those very same rats are offered free access to a wide variety of tempting, rich, highly palatable foods, they greatly overeat and become obese. Similarly, one study found a positive correlation between overfatness and a diet offering a wide variety of snacks and sweets.[21] This is the basis of the **external cue theory**—the theory that, at least in some people, the internal regulatory systems are easily overridden by environmental influences. A question to ponder: Does this make obesity hereditary, environmental, or both?

EXTERNAL CUE THEORY the theory that some people eat in response to such external factors as the presence of food or the time of day rather than to such internal factors as hunger.

It seems likely that both hereditary and environmental factors influence obesity in human beings. The tendency to obesity is probably inherited, but the environment is probably influential in the sense that it can prevent or permit the development of obesity when the potential is there.

The complexity of the situation is reflected in observations of the ways obesity develops from infancy on. Some overweight infants become overweight adults, but most grow out of their obesity in childhood. An overweight child, however, is more likely to remain overweight into adulthood.[22] Researchers propose the **fat cell theory**—that childhood obesity is persistent because early overfeeding (during the growing years) may cause fat cells to increase abnormally in *number* (**hyperplastic obesity**). The number of fat cells is thought to become fixed by adulthood; afterwards, a gain or loss of weight either increases or diminishes the *size* of the fat cells (**hypertrophic obesity**). Unfortunately, persons with greater than normal numbers of fat cells are least likely to lose weight successfully. Some researchers suggest that the body triggers hunger signals when the fat stored in these cells begins to decrease. Since fat cells increase in number during childhood, prevention of obesity is critical during the growing years.

FAT CELL THEORY states that during the growing years, fat cells respond to overfeeding by producing additional fat cells (**hyperplastic obesity**); the number of fat cells eventually becomes fixed, and overfeeding from this point on causes the body to enlarge existing fat cells (**hypertrophic obesity**). Hypertropic obesity is the more common type and is usually seen in adults.

Additionally, fat cells of obese people contain higher levels of the enzyme, **lipoprotein lipase (LPL),** which determines the rate at which adipose cells store fat. The larger the fat cell (and the greater the number of fat cells), the more LPL and the more easily the body can pull triglycerides into fat cells for storage. Unfortunately, LPL activity rises further with weight loss, enhancing the body's ability to regain the lost weight.[23] A question still to be answered is whether some people develop obesity because their fat cells contain an abnormal amount of LPL from birth.

LIPOPROTEIN LIPASE (LPL) an enzyme located on the surfaces of fat cells that enables the cell to convert blood triglycerides into fatty acids and glycerol to be pulled into the cell for reassembly and storage as body fat.

The theory that a hereditary, inside-the-body basis for obesity may exist is supported by the existence of animal strains that are genetically fat. Such animals tend to be fat in any environment—that is, they are fat regardless of the kind or variety of food offered. In humans, studies have shown that identical twins—whether raised together or apart, tend to have similar weight-gain patterns. Also, twins raised by adoptive parents tend to have body shapes similar to their biological parents. Not all studies confirm this, however.[24] Moreover, pairs of twins purposefully overfed in clinical experiments tend to respond similarly to the extra calories. Some sets of twins gain considerable amounts of weight when fed a certain number of calories, whereas other pairs put on relatively few pounds even on the same diet.[25]

Environment and Behavior

In human beings, learning plays an important role. Although we have genetically inborn instincts, superimposed on these are learnings from our early childhood experiences, and depending on our environments, the two may differ. Thus, **hunger** is a drive programmed into us by our heredity, but learned responses to **appetite** can teach us to ignore or overrespond to our hunger. In contrast, appetite is more influenced by learning. Another way to say this is to say that hunger is physiological, whereas appetite is psychological, and the two don't always coincide. We have all experienced appetite without hunger: "I'm not hungry, but I'd love to have some." We also often experience the reverse, hunger without appetite: "I know I'm hungry, but I don't feel like eating." Hunger is a negative experience (and we may eat to avoid it); appetite is positive.

HUNGER the physiological drive to find and eat food, experienced as an unpleasant sensation.

APPETITE the psychological desire to find and eat food, experienced as a pleasant sensation, often in the absence of hunger.

The ways people respond to hunger and appetite determine whether they eat too much, too little, or just enough to maintain their weight. A third factor enters in, too: **satiety,** which signals that it is time to stop eating. One view holds that eating behavior is turned on all the time, except when the satiety signal turns it

SATIETY the feeling of fullness or satisfaction that people feel following a meal.

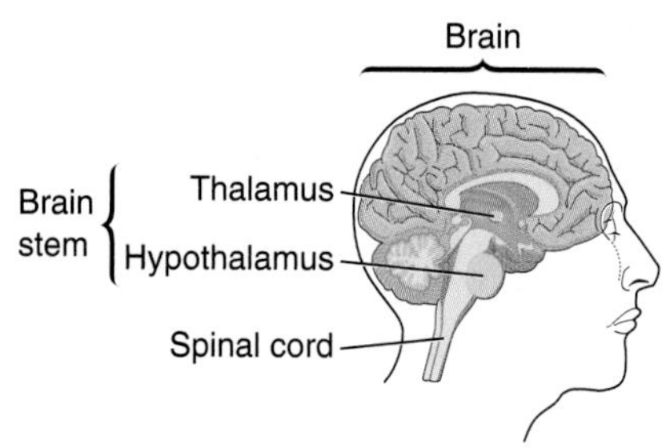

Researchers believe that the hypothalamus controls the sensations of hunger and satiety.

HYPOTHALAMUS (high-poh-THALL-ah-mus) a part of the brain that senses a variety of conditions in the blood, such as temperature, salt content, and glucose content, and then signals other parts of the brain or body to change those conditions when necessary.

AROUSAL as used in this context, heightened activity of certain brain centers associated with excitement and anxiety.

It is probable that the most important single contributor to the obesity problem in our country is underactivity.

off. But much remains to be learned about the ways in which hunger, appetite, and satiety regulate food intake.

In human physiology, research is beginning to find possible answers to what regulates food behavior. The stomach's nerves perceive stretching, and you stop eating when your stomach feels stretched full. Blood glucose level is thought to be involved: You get hungry when your blood glucose level falls—or perhaps when your liver glycogen is beginning to be exhausted. Blood lipids, and possibly amino acids and other molecules, also play a role. When you eat, you secrete hormones to regulate digestive activity; these hormones may also convey the message to the brain that it is time to stop eating.

This brings us to the question, Where in the brain are these messages received (whatever they are)? One brain area stands out as a regulator for food behavior—the **hypothalamus.** The hypothalamus is a center that communicates with both the hormonal and the nervous systems. It integrates many kinds of signals received from the rest of the body, including information about the blood's temperature, sodium content, and glucose content. We know it is important in regulating eating because damage to the hypothalamus produces derangements in eating behavior and body weight—in some cases causing severe weight loss; in others, vast overeating. In the person with a normal hypothalamus, however, appropriate eating behavior seems to be a response not to a single signal arriving at some one location in the hypothalamus but to a whole host of signals. Somehow these many inputs become integrated into a final common path—the act of eating.

A person who eats inappropriately may have established a habitual behavior pattern that wrongly links many different stimuli to the act of eating. In this connection, the study of behavior offers insight into the problem of overeating by viewing it as a conditioned response to a variety of stimuli. Sometimes eating behavior tends to get turned on by the wrong triggers. A crying child with a skinned knee who is offered a lollipop to help soothe the hurt may learn to associate food with comfort and so seek food inappropriately when experiencing emotional pain later in life.

Eating behavior, then, may be a response not only to hunger or appetite but also to complex human sensations such as yearning, craving, addiction, or compulsion. For an emotionally insecure person, eating may be less threatening than calling a friend when lonely and risking rejection. Often, eating is used to relieve boredom or to ward off depression. Some people respond to anxiety—or, in fact, to any kind of **arousal**—by eating. Significantly, however, if they are able to give a name to their aroused condition and thereby gain a feeling that they have some control over it, they are not as likely to overeat.

Stress may also directly promote the accumulation of body fat. The stress hormones favor the breakdown of energy stores (glycogen and fat) to glucose and fatty acids, which can be used to fuel the muscular activity of fight or flight. If a person fails to use the fuel in physical exertion, however, the body cannot turn these fragments back into glycogen. It has no alternative but to convert them to fat. Each time glucose is pulled out of storage in response to stress and then transformed into fat, the lowered glucose level or exhausted glycogen will signal hunger, and the person will eat again soon after.

Stress eating may appear in different patterns. Some people eat excessively at night, while others characteristically binge during emotional crises. The overly thin often react oppositely. Stress causes them to reject food and thus become thinner. It is not yet known why these behaviors occur, but research continues.

The many possible causes of obesity mentioned so far all relate to the input side of the energy equation. What about output? It is probable that the most important single contributor to the obesity problem in our country is underactivity.[26] The control of hunger and appetite works well in active people and fails only when activity falls below a certain minimum level. Some obese people eat less than lean people, but they are so extraordinarily inactive that they still manage to store surplus calories. Some people move more efficiently than others, too. Two people of the same age, height, and weight might use different amounts of calories walking five miles because of the different ways in which they move their muscles.

No two people are alike, either physically or psychologically. No doubt, the causes of obesity are as varied as the obese people themselves. Many causes may contribute to the problem in a single person. Given this complexity, it is obvious that there is no panacea. The top priority should be prevention, but where prevention has failed, the treatment of obesity must involve a simultaneous attack on many fronts.[27]

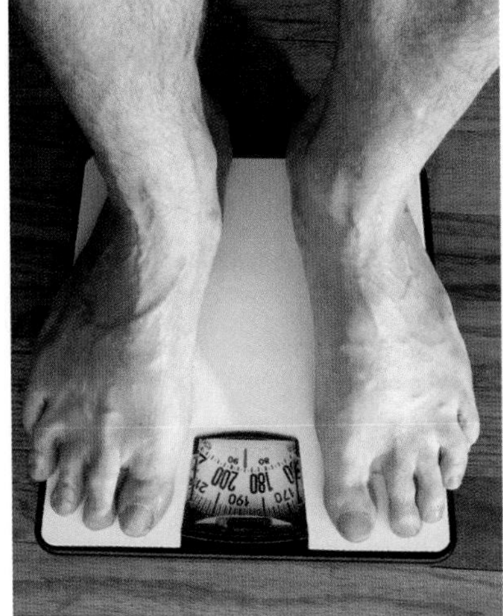

The loss (or gain) of a pound does not always reflect the loss (or gain) of body fat.

WEIGHT GAIN AND LOSS

When you step on the scale and note that you weigh a pound more or less than you did the last time you weighed yourself, this doesn't necessarily mean you have gained or lost body fat. Changes in body weight reflect shifts in many different materials—not only fat but also fluid, bone minerals, and lean tissues such as muscles. This means that the loss of a pound does not always reflect the loss of fat. Similarly, the gain of a pound may not reflect gained fat; some of it may reflect gained muscle and bone and an overall shift toward a leaner body type. Because it is so important for people concerned with weight control to realize this, this section discusses the changes that take place with gains and losses of weight.

A healthy man or woman about 5 feet 10 inches tall who weighs 150 pounds carries about 90 of those pounds as water and 30 as fat. The other 30 pounds are the so-called lean tissues: muscles; organs such as the heart, brain, and liver; and the bones of the skeleton.* Stripped of water and fat, then, the person weighs only 30 pounds. This lean tissue is the body's vital machinery that maintains health and life. When a person who is too fat seeks to lose weight, it should be fat—not this precious lean tissue—that is lost. And for someone who wants to gain weight, it is desirable to gain weight in proportion—lean *and* fat, not just fat.

Weight is gained or lost in different body tissues, depending on how a person goes about it. Most quick-weight-loss diet schemes promote large losses of fluid that create large, temporary changes in the scale's weight with little or no real loss of body fat. The rest of this chapter underscores this distinction, and a later section on weight-loss strategy stresses exercise as a means of supporting lean tissue during weight loss.

Weight Gain

When you eat more calories than you need, where does this excess go in your body? The energy nutrients—carbohydrate, fat, and protein—contribute to body stores as follows:

- Carbohydrate is broken down to glucose for absorption. Inside the body, glucose may be built up to glycogen or converted to fat and stored as such.
- Fat is broken down to its component parts (including fatty acids) for absorption. Inside the body, these components are most easily converted to storage fat.
- Protein, too, is broken down to its basic units (amino acids) for absorption. Inside the body, these units may be used to replace body proteins. Those

*For a healthy woman or man 5 feet tall who weighs 100 pounds, the comparable figures would be 60 pounds of water, 20 pounds of fat, and 20 pounds of lean tissue.

amino acids that are not used cannot be stored as protein for later use. They lose their nitrogen and are converted to fat.

Notice in Figure 8-2 that although three kinds of materials enter the body, they are stored for later use in only two forms: glycogen and fat. Also notice that when protein is stored in the form of fat, it cannot be recovered later as protein. The amino acids lose their nitrogen—the nitrogen is actually excreted in the urine. It does not matter whether you are eating hamburgers, brownies, or carrot sticks; if you eat enough of the food, the excess will be turned to fat within hours. (On the other hand, as Chapter 9 will make clear, a judicious program of eating well and exercising will help build muscle.)

Of the three energy nutrients, fat from food is especially easy for the body to store as fat tissue. This implies, then, that the calories from fat may be more fattening than those from carbohydrate or protein. Researchers have shown in both animals and humans that subjects who ate higher fat diets had higher body fat contents than those who ate diets high in carbohydrates and lower in fat, even though the two diets were similar in terms of total calories.[28] This suggests that obesity more easily develops from eating a high-fat diet.[29] If you choose to overeat, therefore, there may be some advantage to overeating carbohydrate-rich

FIGURE 8-2
FEASTING AND FASTING

In A, the person is storing energy. In B, the person is drawing on stored energy. In C, the person is in ketosis.

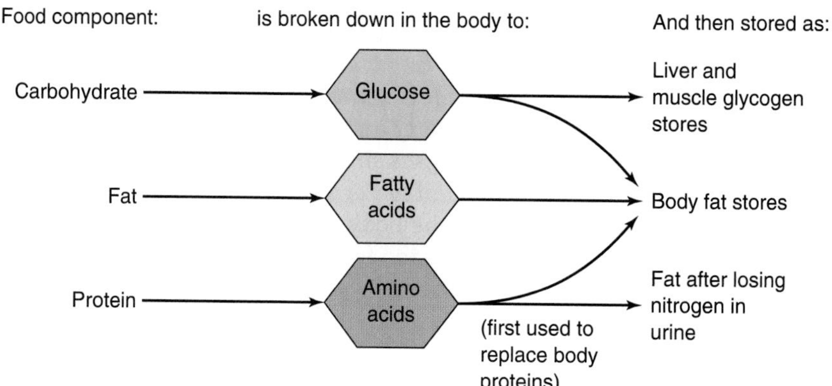

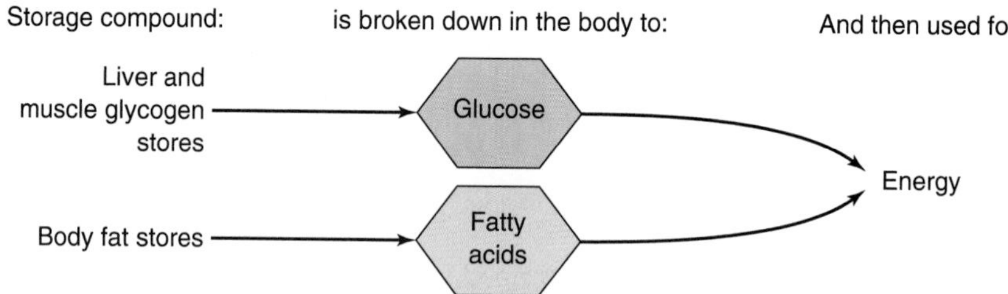

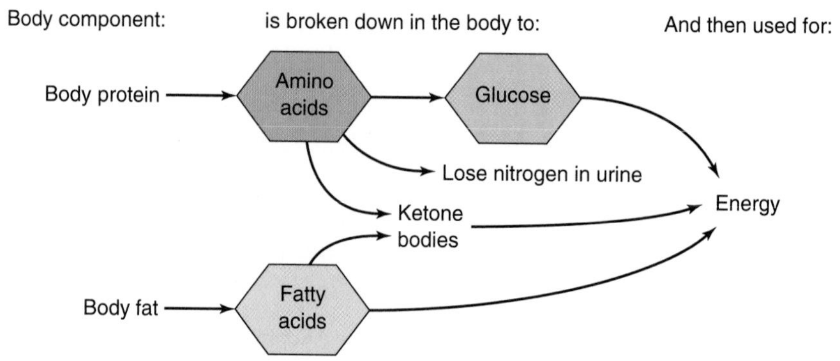

foods such as vegetables or legumes than overeating fat-rich butter or sour cream—you may deposit less fat.

Weight Loss and Fasting

When the tables are turned and you stop eating altogether, your body has to draw on its stored supplies of nutrients to keep going. Nothing is wrong with this; in fact, it is a great advantage to you that you can eat periodically, store fuel, and then use up that fuel between meals. The between-meal interval is ideally about 4 to 6 hours—about the length of time it takes to use up most of the available liver glycogen—or 12 to 14 hours at night, when body systems are slowed down and the need for energy is lower. If a person doesn't eat for, say, three whole days or a week, the body makes one adjustment after another.

The first adjustment is to use the liver's glycogen. (The muscles' glycogen is reserved for the muscles' own use—and they are using it.) The liver's glycogen, remember, is the body's source of blood glucose to fuel the brain's and nerves' activities. Ordinarily, the brain and nerves can use no other fuel. But after about a day, the primary supply is gone. Where, then, does the body turn to keep its nervous system going? Whatever it has to do, it will do, for the nervous system runs the body, and when it stops, the body dies.

An obvious alternative source of energy would be the abundant fat stores most people carry. At first, these are of no use to the nervous system. The muscles and other organs use fat as fuel, but the nervous system ordinarily cannot. Nor can the body convert this fat to glucose, because it possesses no enzymes to do so. It does, however, possess enzymes to convert protein to glucose.

As the fast continues, the body turns to its own lean tissues to keep up the supply of glucose (see Figure 8-2). One reason why people lose weight so dramatically within the first three days of a fast is that they are devouring their own protein tissues as fuel. Since protein contains only half as many calories per pound as fat, it disappears twice as fast. Also, with each pound of body protein, three or four pounds of associated water are lost. As you will see in a moment, the same reasons account for the rapid weight loss seen in the early stages of a low-carbohydrate diet.

If the body were to continue to consume itself at this rate, death would ensue within about ten days. After all, the liver, the heart and skeletal muscles, the lung tissue, and the blood—all vital tissues—are being burned as fuel. (In fact, fasting or starving people remain alive only until their body fat is gone or until half their lean tissue is gone, whichever comes first.) But now the body plays its last ace.

Source: The St. Petersburg Times. Reprinted with permission.

KETOSIS (kee-TOE-sis) an adaptation of the body to prolonged (several days') fasting or carbohydrate restriction: body fat is converted to ketones, which can be used as fuel for some brain cells.

It begins converting fat stores into a form it can use to help feed the nervous system and so forestall the end. This is known as **ketosis.**

Ketosis is an adaptation to prolonged fasting or carbohydrate deprivation. Instead of breaking down fat molecules all the way to carbon dioxide and water, as it normally does, the body takes partially broken down fat fragments, combines them into ketone bodies (compounds that are normally rare in the blood), and lets them circulate in the bloodstream. The advantage is that about half of the brain's cells can use these compounds for energy. Thus, indirectly, the nervous system begins to feed on the body's fat stores. This reduces the nervous system's need for glucose. It spares the muscle and other lean tissue from being devoured so quickly and prolongs the starving person's life. Because of ketosis, an initially healthy person totally deprived of food can live for as long as six to eight weeks.

Fasting has been practiced as a periodic discipline by respected, wise people in many cultures. However, ketosis may be harmful to the body by upsetting the acid-base balance of the blood. For the person who merely wants to lose weight, then, fasting is not the best way. For one thing, even in ketosis, the body's lean tissue continues to be lost at a rapid rate to supply glucose to those nervous system cells that cannot use ketones as fuel. For another, the body becomes conservative during a fast and slows its metabolism so as to lose as little energy as it possibly can. A well-designed low-calorie diet, accompanied by the appropriate exercise program, has actually been observed to promote the same rate of *weight* loss as, and a faster rate of *fat* loss than, a total fast. Just how to design a low-calorie diet is the subject of a later section. But first the low-carbohydrate diet—an example of how *not* to design a diet—deserves attention.

The Hype of High-Protein, Low-Carbohydrate Diets

To identify fraudulent low-carbohydrate diets for what they are, learn to add up the carbohydrate grams they supply: fewer than 100 to 125 grams a day is inadequate.

People are attracted to high-protein, low-carbohydrate diets because of the dramatic weight loss that occurs within the first few days. Such people would be disillusioned if they realized that the major part of this weight loss is a loss of body protein, along with quantities of water and important minerals.

The low-carbohydrate diet is designed to make a person go into the potentially harmful condition called ketosis. The sales pitch is that "you'll never feel hungry" and that "you'll lose weight fast—faster than you would on any ordinary diet." Both claims are true, but knowledgeable consumers see through them. They know that the loss of appetite is common to any low-calorie diet. To the fast weight loss, they say, "Yes, but what kind of weight loss: water and lean tissue,or fat?"

The body responds to a low-carbohydrate diet as it does to a fast. It is receiving protein and fat (on a fast it draws on its own protein and fat), but it has used up its stored glycogen. It therefore turns to protein to make the needed glucose. Why should you give your body protein if it will only convert that protein to glucose? And why *not* give it carbohydrate if that is the very material it needs? Carbohydrate will sustain it and allow it to use up its stored fat at the maximal rate.

Protein, then, is inefficient fuel for a carbohydrate-deprived body. It has another disadvantage as well. On being converted to glucose, protein loses its nitrogen, and that nitrogen has to be excreted. This puts a burden on the kidneys. Advising the low-carbohydrate dieter to drink sufficient water is intended to prevent kidney damage that may result from the large amounts of nitrogen-containing waste materials and ketone bodies circulating in the blood. Other unpleasant side effects associated with continued use of a low-carbohydrate (low-calorie) diet include constipation, nausea, dehydration, headaches, and fatigue. What is achieved by quick-weight loss dieting is loss of lean tissue, glycogen, bone minerals, and fluids—all materials vital to healthy body functioning. In contrast, weight loss of not more than one to two pounds per week achieved by a balanced diet and exercise will encourage fat loss and minimize muscle loss.

The low-carbohydrate diet is perhaps the most resilient of popular diets and comes in many disguises with many different names. In the 1960s and 1970s there were the Drinking Man's Diet, the Air Force Diet, Dr. Stillman's Quick Weight-Loss Diet, the Complete Scarsdale Medical Diet, and Dr. Atkins' Diet Revolution. Today, high-protein, low-carbohydrate diets are making a comeback with diets such as Protein Power, Enter the Zone, Sugar Busters, and Dr. Atkins' New Diet Revolution. Following the diets will inevitably lead to weight loss only because they provide so few calories. Typically, these diets are also low in calcium and dietary fiber and are not recommended for long-term weight loss. When the eating pattern recommended in the popular high-protein diet books is followed in amounts to meet calorie needs for maintenance, the diets are excessively high in protein and fat.[30]

Bringing out new diet books and products every year is a profitable business, and it will continue to be successful as long as people are deceived by the initial rapid weight loss into thinking that the diets work. Before adopting any new diet plan, compare it to the guidelines presented in Table 8-4.[31]

The caloric distribution of eating patterns recommended by high-protein diets differs significantly from the current recommendations for a healthful diet:

	High-Protein Regimen	Current Recommendations
Carbohydrate	40% (or less)	55% to 58%
Protein	30%	12% to 15%
Fat	30% (or more)	< 30%

TABLE 8-4 GUIDELINES FOR EVALUATING WEIGHT MANAGEMENT PROGRAMS

Seek programs that provide:

- Information about program costs, format, potential risks, qualifications of professional staff, expected outcomes, expected time frame for reaching weight goal, and duration of maintenance.
- Qualified experts in nutrition, exercise, and behavior change.
- Physician-evaluated screenings of weight-loss candidates with medical conditions that may make weight loss risky.
- Reasonable goals for weight loss.
- Three-pronged approach to weight management: nutritious eating plan, exercise, and behavior modification.
- Maintenance program for at least two years of follow-up.

Beware of weight-loss programs that:

- Promise or imply dramatic, rapid weight loss (i.e., substantially more than 1% of total body weight per week).
- Promote diets that are extremely low in calories (i.e., below 800 cal/day; 1,200-cal/day diets are preferred) unless under the supervision of competent medical experts.
- Do not encourage permanent, realistic lifestyle changes, including regular exercise and the behavioral aspects of eating wherein food may be used as a coping device (i.e., programs should focus on changing the causes of overweight rather than simply on the overweight itself).
- Misrepresent salespeople as "counselors" supposedly qualified to give guidance in nutrition and/or general health. Even if adequately trained, such counselors would still be objectionable because of the obvious conflict of interest that exists when providers profit directly from products they recommend and sell.
- Require large sums of money at the start or require that clients sign contracts for expensive, long-term programs. Such practices too often have been abused as salespeople focus attention upon signing up new people rather than delivering continuing, satisfactory service.
- Promote unproven or spurious weight-loss aids, such as chromium picolinate, human chorionic gonadotropin (HCG) hormone, starch blockers, diuretics, sauna belts, body wraps, passive exercise, ear stapling, acupuncture, electric muscle stimulating (EMS) devices, spirulina, amino acid supplements (e.g., arginine or ornithine), and glucomannan.

The Very-Low-Calorie Diets

Very-low-calorie diets (VLCDs) are mostly powdered formulas available by prescription, are usually medically supervised, and provide fewer than 800 calories. VLCDs consist of about 50 grams of carbohydrate (not enough to spare protein), and high-quality protein equivalent to about twice the recommended daily intake. Supplements of vitamins and minerals are provided.

For particular individuals, especially the morbidly obese, these diets may provide some benefit in that they promote rapid weight loss and free the individual from having to make decisions regarding food intake—dieters simply drink the prescribed formula. Because of the health risks that may accompany these diets, it is recommended that a VLCD be undertaken only when other more traditional approaches have failed and always under the supervision of a health team that includes both a physician and a registered dietitian. A significant loss of heart muscle can occur and lead to sudden death from heart attack if the client loses weight too rapidly on a VLCD.[32] Other risks are listed in Table 8-5. The more valid VLCD programs include exercise, nutrition education, behavior modification, and support groups to improve their long-term effectiveness.[33] Table 8-6 evaluates popular weight-management programs.

Drugs and Weight Loss

The search is on to find a safe and effective drug solution to the problem of obesity. The ideal drug needs to be safe, free of undesirable side effects and abuse potential, and effective at reducing body fat. Additionally, as with the drug treatment of other chronic disorders (for example, high blood pressure), the ideal drug should be safe and effective for long-term use in the treatment of obesity. To foster long-term success with weight loss, any such drug treatment should be combined with lifestyle changes, including exercising and consuming a healthy diet.[34] Until recently, mostly appetite–suppressant drugs were available, either as prescription or over–the–counter drugs.

Stimulant drugs such as the amphetamines Dexedrine and Benzedrine can suppress appetite and thereby cause a drop in food intake. However, their effects are usually short-lived, and a person's appetite returns to normal after a few weeks. The drugs can be addictive and leave the dieter with another problem—how to get off them without gaining more weight. Other side effects include nervousness, headache, dizziness, weakness, fatigue, and insomnia.[35]

Another class of drugs, chemically similar to the amphetamines, but generally nonaddicitve as drugs, are the agents that enhance or stimulate the release

TABLE 8-5 RISKS ASSOCIATED WITH THE VERY-LOW-CALORIE DIETS

- Blood sugar imbalance
- Cold intolerance
- Constipation
- Decreased basal metabolic rate
- Dehydration
- Diarrhea
- Emotional problems
- Fatigue/weakness
- Gallstones and kidney stones
- Headaches
- Heart irregularity
- Ketosis
- Kidney infection
- Loss of lean body tissue
- Menstrual irregularity
- Mineral and electrolyte imbalances
- Sleeplessness
- Sudden death

Source: Adapted from Position of The American Dietetic Association: Very-low-calorie weight loss diets, *Journal of the American Dietetic Association,* May 1990, p. 722.

of the brain chemical (neurotransmitter), serotonin. Upon its release, serotonin acts to curb the appetite and thus reduces food intake. Examples include the two drugs recently withdrawn from the market: dexfenfluramine (Redux) and fenfluramine (Pondimin). Either of the pair was often prescribed in combination with a third drug—the appetite-curbing stimulant, phentermine—and referred to as "fen-phen." These agents were effective at increasing the production of serotonin in the brain and thus suppressing appetite. However, after only one year on the market, the FDA persuaded the manufacturers of Redux and Pondimin to pull the drugs off the market when it was discovered that patients taking the drugs were at increased risk of developing potentially fatal heart valve damage and elevated blood pressure in the lungs (pulmonary hypertension).[36] Although both drugs had been approved by the FDA as safe weight-loss agents, the drugs were never approved for use in the ways that many physicians had prescribed them: for long-term periods, by people who were not obese, or in combination with other drugs. The FDA advises that anyone who took Pondimin or Redux or fen-phen should have an ultrasound imaging test of the heart, called an echocardiogram (ECG), to check for heart valve damage. Phentermine (Adipex-P or Fastin), which raises the level of norepinephrine, signals satiety, and acts as a stimulant to increase the rate at which you burn calories is still available for use.

In 1998, following the removal of Pondimin and Redux from the market, the FDA approved sibutramine (Meridia), another appetite-suppressing drug. Meridia enhances the effects of serotonin by slowing the body's breakdown of the serotonin it naturally produces. Potential side effects include dry mouth, insomnia, and elevated blood pressure. People who are not obese or anyone with high blood pressure are cautioned not to use the drug.

A different class of weight-loss drugs are the lipase inhibitors. Approved for use in 1999, Orlistat (Xenical) acts by inhibiting the enzymes—gastric and pancreatic lipase—needed for fat digestion.[37] Xenical binds to the enzymes, making them unavailable to digest dietary fat. Since fat absorption is reduced by as much as 30 percent, people often lose weight. Side effects include bloating, gas, and anal leakage of undigested fats in people who do not adhere to a low-fat diet. Since Xenical may block the absorption of fat-soluble vitamins, users are advised to take a multivitamin supplement. The drug is intended only for the obese person—with a BMI greater than 30.

Only two over-the-counter (OTC) appetite-suppressant ingredients are currently approved for sale by FDA without prescription. The first—**phenylpropanolamine (PPA) hydrochloride**—is found in some cold remedies and in weight-loss products such as Dexatrim, AcuTrim, and Appedrine. PPA is a chemical relative of the amphetamines, and researchers report an extra half pound of weight loss for users of this drug compared to a placebo. Misuse of the drug (for example, taking larger than recommended doses) can produce dizziness, high blood pressure, heart palpitations, and sleeplessness. The second approved OTC ingredient is the anesthetic—**benzocaine**—usually found in gum or candy form. Benzocaine acts by numbing the taste buds and other sensory signals, thereby reducing the desire for food. The safety and effectiveness of both of these drugs as well as a host of new weight-loss drugs is a matter of ongoing research and review by the Food and Drug Administration (FDA).

PHENYLPROPANOLAMINE HYDROCHLORIDE (PPA) a stimulant of the central nervous system available in over–the–counter weight-loss products used to suppress appetite.

BENZOCAINE an anesthetic found in gum or candy form that numbs the taste buds and reduces the desire for food.

Fiber pills increase bulk in the stomach and could ideally lead to satiety, but they can also cause intestinal gas and blockage. Moreover, the fiber typically found in diet pills does not seem to be effective at producing satiety.[38]

Diuretics, or water pills, do nothing to solve an overweight problem, although they may bring about the loss of a few pounds on the scale for half a day and cause dehydration. Numerous other diet aids on the market include products with mysterious sounding ingredients, such as spirulina, inositol, chromium picolinate, ginseng, and numerous other herbs. Manufacturers suggest such products can aid in weight loss, but their ingredients often serve as little more than fillers. To date, none have proven effective in aiding weight loss. For

TABLE 8-6 WEIGHT MANAGEMENT PROGRAMS COMPARED

With a balanced perspective on foods and a sense of what is important in diet planning and what is not, you can evaluate the many different available diets and decide which might be best for you. Here's a summary of the questions you might ask, followed by a comparison of several weight management programs.

1. Is this a diet you could live with indefinitely?
2. What is the recommended rate of weight loss?
3. Does the program take individual differences into account to determine caloric needs?
4. To what extent does the plan educate the client about nutrition, behavior modification, and the importance of exercise?
5. Does the program put you in contact with professionals such as physicians or registered dietitians?
6. Does the program offer a maintenance plan once the weight is lost?
7. What is the nature of the advertisements and endorsements?
8. How much does the program cost (can you afford it)?

Program*	Overall Approach	
Diet Center 800-333-2581 www.dietcenterworldwide.com	Personalized diet and exercise program, emphasizing healthy body composition. *Exclusively You* option based on supermarket foods; prepackaged cuisine optional. Minimum daily calorie level: 1,200. Vitamin and fiber supplements provided during reducing phase. *Concept 1000* option provides 1,000 daily calories from three meal replacements (shakes or bars) and one regular meal. Body composition analysis provided.	**Pros:** Emphasizes body composition, not pounds, as a measure of health. Choice of two diet plans; only *Concept 1000* requires Diet Center foods. **Cons:** Expensive. No professional guidance. No group support.
Health Management Resources 617-357-9876 www.yourbetterhealth.com	Makes use of a very-low-calorie diet (VLCD) consisting of fortified, high-protein liquid meal replacements (520–800 calories/day) under medical supervision. *Healthy Solutions* plan (1,000–1,600 calories/day) combines meal replacements with regular foods, including optional prepackaged HMR entrees and five servings of fruits and vegetables daily. Mandatory weekly 90-minute group meetings (60 minutes during maintenance). Dieters assigned personal coaches for weekly meetings or calls. Receive health risk appraisal.	**Pros:** Few eating decisions. Emphasizes exercise. Supervised by health professionals. **Cons:** Expensive if not covered by insurance. Requires prepackaged food and strong commitment to exercise. May be difficult to transition to regular foods. Side effects of VLCD include intolerance to cold, constipation, dizziness, dry skin, and headaches.
Jenny Craig 619-812-7000 www.jennycraig.com	*ABC (About Better Choices)* program relies on *Jenny Craig's Cuisine* plus additional supermarket foods. 1,000 to 2,600 calories daily. Optional, weekly *Options Lifestyle* classes and one-to-one counseling. After losing half of goal weight, clients given option to transition to regular foods.	**Pros:** Little food preparation required. Plans available for vegetarians, people with diabetes, breastfeeding moms. **Cons:** Must rely on prepackaged foods, making dining out and socializing difficult. Lacks professional guidance at client level. Limited maintenance options.
Nutri/System 215-442-5300 www.nutrisystem.com	Diet based mostly on Nutri/System's prepackaged foods. Reducing diet averages minimum of 1,200 calories/day for women; 1,500 for men. Maintenance diet based on optional purchase of prepackaged fare. Nutri/System's multivitamin/mineral supplement recommended for all dieters, but not included in price. Mostly one-to-one weekly counseling with weigh-in; some centers offer group classes.	**Pros:** Few eating decisions. **Cons:** Expensive. Weak on lifestyle education component. Little contact with health professionals.

TABLE 8-6 WEIGHT MANAGEMENT PROGRAMS COMPARED—*Continued*

Program*	Overall Approach	
Optifast 800-662-2540 www.optifast.com	Medically supervised program of fortified liquid meal replacements or prepackaged foods, eventually including regular foods. Dieters assigned one of three plans: 800, 950 or 1,200 calories daily. Mandatory weekly sessions promote positive eating behaviors. One-to-one counseling available.	**Pros:** Close contact with health professionals. Beneficial for people with serious health problems who need low calorie level to promote quick weight loss. Few eating decisions. **Cons:** Expensive if not covered by insurance. Must rely on Optifast products during much of reducing phase. May be difficult to transition from liquid diet to regular food.
Registered Dietitian Consultation 800-366-1655 for free referral to local RD www.eatright.org	Provides a personalized approach to weight control that takes into consideration your individual needs, including medical history, family situation, eating and exercise habits and preferences, travel and dining-out routines, and budget.	**Pros:** Eating prescription adapted to your lifestyle and medical history. Appropriate for any age group and entire families. **Cons:** Expensive if not covered by insurance.
Shapedown Pediatric Obesity Program 800-322-5487 www.shapedown.com	Program (designed by *The Solution* staff) addresses weight-loss needs of children and young adults (ages 6 to 20 years) with an emphasis on balanced eating, aerobic exercise, and positive behavior change. Over 600 sites nationally provide group and individual counseling in ten-week program. Minimum of 1,200 calories recommended to achieve 1-pound weight loss per week.	**Pros:** Qualified program delivery team includes dietitian, social worker, and psychologist with training in pediatric obesity. **Cons:** Some participants may benefit from programs of longer duration to achieve long-lasting behavior changes.
The Solution 415-457-3331 www.weightsolution.com	Program offers family-systems approach to weight loss with a focus on developing skills for the mind, body, and spirit. Small "community" of members meet for 12 sessions addressing body pride and making positive changes in eating and exercise habits. Between-session support provided by members. No set calorie limit given; available at 150 sites nationwide.	**Pros:** Qualified program delivery team includes program-trained registered dietitian and licensed mental health professionals. Slow weight loss of 1 pound per week recommended. **Cons:** Some participants may benefit from more structured eating plan with defined calorie limits.
TOPS (Take Off Pounds Sensibly) 800-932-8677 www.tops.org	Nonprofit organization whose members meet weekly in groups. Requires members to submit weight goals and diets in written form from health professionals. Provides peer support. Holds periodic contests and recognition programs for weight loss.	**Pros:** Inexpensive form of group support. No purchases required. Has potential for long-term participation. **Cons:** Focuses on weight loss as chief measure of success. Must weigh in weekly. Groups vary widely in approach. Program lacks professional guidance.
Weight Watchers 800-651-6000 www.weightwatchers.com	Emphasizes calorie-controlled, high-fiber eating and healthful lifestyle habits. 1-2-3 *Success* program assigns members daily food point allotment, which averages 1,250–1,500 calories/day for women. Weekly group meetings with mandatory weigh-in. Need to lose at least 5 pounds to join.	**Pros:** Flexible, easy-to-use program offering group support. Plans available for vegetarians, teens, and breastfeeding moms. **Cons:** Lacks professional guidance at client level. No personalized counseling. Weekly weigh-ins.

*Listed alphabetically.

Source: Adapted from E. M. Ward, *Environmental Nutrition*, January 1998. © Copyright 1998 by Environmental Nutrition, Inc., 52 Riverside Drive, New York, N.Y. 10024.

For more about herbs thought to be effective and those that should be avoided, see the Nutrition Action feature in Chapter 6.

example, manufacturers of products that include chromium picolinate praise its druglike abilities to reduce fat, build lean tissue, suppress appetite, and increase metabolism and imply that our diets lack chromium. However, chromium picolinate has not been approved for weight loss by the FDA, and its claims are not backed by scientific data.[39]

The popularity of herbal supplements for weight loss is on the rise, partly because consumers have the mistaken notion that herbs are a "natural" alternative to prescription drugs for weight loss. Thus far, herbal preparations in the United States are produced and marketed without regulations to ensure their safety or effectiveness, and some have been linked to side effects ranging from nausea and headaches to heart attacks and death. For example, the Chinese herb ephedra—commonly called ma huang—contains ephedrine, a stimulant that mimics the action of the drug phentermine. Ephedrine can cause tremors, insomnia, severe headaches, high blood pressure, heart attacks, and stroke. Some herbal preparations combine ma huang with other substances (for example, caffeine-containing guarana), to enhance ephedrine's effects. Herbal fen-phen combines ma huang with extracts of St. John's wort, an antidepressant-like herb that mimics the calming effect of fenfluramine. However, such combinations increase the risk of stroke and death and should be avoided. More than 800 reports of illness and dozens of deaths are linked to ma huang use. For that reason, Canada bans ma huang and the FDA has issued warnings against its use.

Other herbal preparations sold as dieter's tea contain senna, rhubarb root, aloe, cascara, or buckthorn. The "tea" has a laxative effect and can cause dehydration, diarrhea, nausea, fainting, and in some cases, even death.[40]

Surgery and Weight Loss

Sheer desperation prompts some obese people to request surgery. One operation—intestinal bypass surgery—involves removing or disconnecting a portion of the small intestine to reduce absorption. After bypass surgery, the person can continue overeating but will absorb fewer calories. Dangerous side effects from this procedure are many, including liver failure, massive and frequent diarrhea, urinary stones, intestinal infections, and malnutrition. Such surgery is now seldom performed because results have been so disappointing.

GASTROPLASTY surgery on the stomach (also called stomach stapling) that reduces its volume to less than 2 ounces (the size of a shot glass) to prevent overeating.

Another more common operation—**gastroplasty**—involves stapling the stomach to make it smaller, thus forcing the person to eat less (see Figure 8-3). Nau-

FIGURE 8-3
SURGICAL PROCEDURES FOR OBESITY

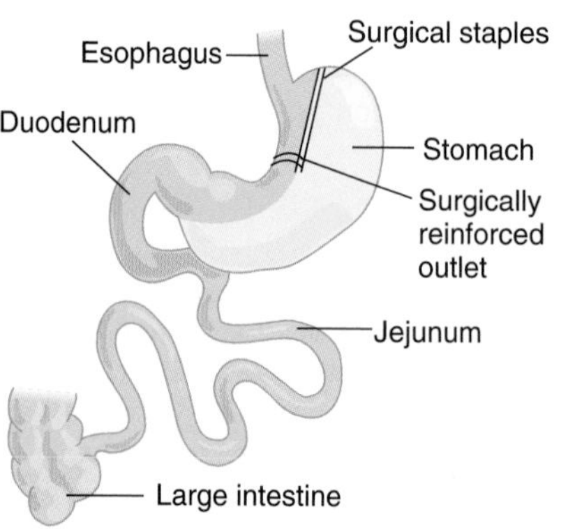

In vertical-banded gastroplasty, a small pouch is constructed at the top of the stomach that both limits the amount of food that the stomach can hold and restricts the passage of food to the small intestine (duodenum).

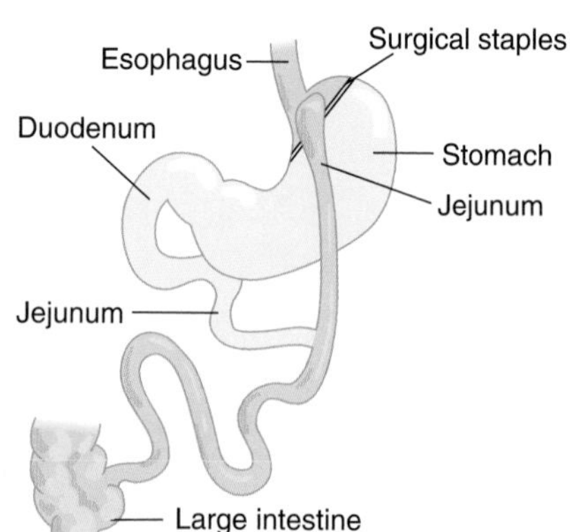

In gastric bypass operations, the stomach is stapled closed near the top, leaving only a small pouch that is connected directly to the middle portion of the small intestine (jejunum).

sea and vomiting occur if the person continues to overeat following the procedure. Still, although the theory is pleasingly simple, stapling involves hazards in practice: stomach tissue is damaged, scars are formed, staples pull loose. The person contemplating such surgery should think long and hard before submitting to it.

Another approach involves cosmetic surgery. One such procedure is **liposuction,** in which the surgeon uses a small hollow tube to suction out fatty tissue from beneath the skin. People who wish to remove the fat from a particular area can elect this procedure, which sometimes brings pleasing results but sometimes produces a figure in which one part of the body is disproportionately thin relative to the others.

LIPOSUCTION a type of surgery (also called lipectomy) that vacuums out fat cells that have accumulated, typically in the buttocks and thighs. If the person continues to eat more calories than are expended through physical activity, fat will return to the fat cells that remain in those regions.

Surgery is appropriate in some instances.[41] Cosmetic surgery can minimize disfigurements, improve self-confidence, and ease the way toward concentration on life issues more important than external appearances. But after surgery, the same person resides within the skin as before. A changed appearance does not guarantee changed eating habits, a better personality, reduced interpersonal conflicts, or any other improvements in the quality of one's life.

WEIGHT-LOSS STRATEGIES

Given that so many approaches are likely to fail, what weight-loss strategies work? How can a person lose weight safely and permanently? The secret is a sensible (not to say *easy*) three-pronged approach involving healthful eating habits, exercise, and behavior change.[42] Such an approach takes tremendous dedication, especially at first, for a person whose habits have all promoted obesity to learn and make habits of the hundred or so new behaviors necessary to promote a healthful weight. Even the most effective weight-loss programs reveal a grim pattern: Those who complete weight-loss programs lose about 10 percent of their body weight, only to regain two-thirds of it back within one year and almost all of it back within five years.[43] But when people succeed, they typically do so because they have employed many of the techniques described in this chapter.

Never Say "Diet"

Say the word *diet*, and most people think "starvation," "deprivation," "hunger," and something to go "on" and "off." But this kind of thinking can be unrealistic and self-defeating. Strict, temporary diets—and the attitudes that go with them—don't work for the great majority of overweight people, as history has borne out. In fact, while Americans spend more than $33 billion a year on programs and products aimed at weight loss, obesity rates are higher than ever.[44]

The problem with going on a rigid diet with a goal of, say, a 15-pound weight loss in three weeks is that it's a quick fix—the dieter attempts to gain a temporary solution to what is typically a chronic problem. Often, the dieter tries to rigidly restrict eating by, for example, skipping meals or eating salads all day. But trying not to eat is like trying not to breathe, as one group of psychologists put it.[45] After a while, the body and mind rebel and, like a person gasping for air, the dieter loses control and binges. In fact, experts believe that rigid diets play a strong role in the development of binge-eating problems.[46] As a result of the binge, the person feels that he has failed, gives up the diet altogether, eats more to make himself feel better, puts on some weight, feels even worse, gains even more weight, decides to try another restrictive diet, and begins the whole cycle all over again.

The solution, say many experts, is not to focus on losing a certain amount of weight within a set period of time. A more healthful alternative is to gradually develop habits that you can live with permanently and that will help you shed pounds and keep them off over the long run. Instead of measuring your success by the needle on the scale, gauge your progress by the strides you make in adopting good eating and exercise habits as well as healthful attitudes about yourself and your body.

Although most people recognize the importance of the eating and exercise habits discussed in this chapter in helping to achieve and maintain a healthy weight, few recognize the attitude problems that often go hand in hand with restrictive dieting and can stand in the way of long-term weight loss. One of the most common is "all-or-nothing," or "on-or-off," thinking. People with all-or-nothing attitudes tend to view the world as either black or white, right or wrong, good or bad, and so forth. When it comes to food, the attitude translates into good or bad, on or off limits, diet or junk food.[47] Often, a person who thinks this way sets the stage for failure by trying to live up to extremely rigid, unrealistic goals: "Ice cream is bad, so I must never eat it," "I will jog three miles every day," or "I will never order anything but a salad when I go out to eat."

Typically, a person who has an all-or-nothing attitude might go several days without, say, ice cream, which she views as "bad junk food," and then splurge one day on a double-dip cone. Instead of recognizing that ice cream isn't "bad" and that indulging in some every now and then won't undo all her previous efforts, she tells herself, "I've blown my diet completely—I have no willpower." Consequently, she feels guilty and depressed, and because she's "blown it" anyway, she decides she might as well eat another pint of ice cream and some cookies to make herself feel better. If she had looked at her situation with a different attitude—"I've been eating lots of fruits and vegetables and bicycling several times a week, so one ice cream cone won't hurt me"—she may have avoided a bout of despair and binging.

Another type of attitude that can thwart efforts to achieve a healthy weight is the "lookist" attitude—that is, the notion that weight and appearance are the determinants of a person's worth and happiness.[48] Just about everyone holds lookist attitudes in one form or another and wishes that he could change something about his appearance in the hopes that it would make life more enjoyable. But in our "thin-is-in" society, overweight people suffer tremendous prejudice and tend to be painfully self-conscious about their bodies. As a result, they may mistakenly believe that all their dreams will come true once they lose weight and their problems will be lost with the pounds. By pinning all their hopes for happiness on losing weight, they put themselves under enormous pressure to shed pounds. In addition, people who hold this type of attitude often put off living in the present because they are so caught up in fantasizing about what they think life will be like once they lose weight.

Using weight and appearance as a measure of self-worth and happiness can be extremely destructive, however. Consider that the person who does so may desperately try to lose weight with restrictive, unrealistic diets or unhealthfully strenuous exercise regimens. Each "slip-up" whittles away at the person's self-esteem, which in turn may lead to feelings of rejection, depression, and social isolation, which in turn may prompt a binge, and so forth. And even a person who drops a desired number of pounds may then realize that she still has many of the same problems as before, which can lead to depression and loss of self-esteem.

The way around lookist thinking is to try to focus on health—both mental and physical—rather than appearance. Table 8-7 lists some common characteristics of the obese self (which holds lookist attitudes) and the healthy self. Although it can be difficult to overcome society's prejudices about weight, striving to adopt the ideals of the healthy self, regardless of your weight, can be a major step in helping you take care of your mental and physical health.

Oh please, scale, say that I'm down to 110, so I'll look really good and I'll meet someone, and we'll fall in love and get married and I'll be happy for the rest of my life...
SIPRESS

K. D. Brownell, *The LEARN Program for Weight Control,* 1994. Reproduced with permission. All rights reserved. Dallas: American Health Publishing Company.

Meal Planning

The way a particular person loses weight is a highly individual matter. Two weight-loss plans may both be successful and yet have little or nothing in com-

TABLE 8-7 THE OBESE SELF VERSUS THE HEALTHY SELF

The Obese Self	The Healthy Self
Uses appearance as measure of self-worth	Uses caring for others/self as measure of self-worth
Is socially isolated	Is socially involved
Rejects self	Accepts self
Goes on and off extreme diets	Eats healthfully
Rarely exercises (or overdoes it and quits)	Exercises regularly and sensibly
Focuses on mistakes; feels like a failure	Views mistakes as learning experiences and recognizes that nobody is perfect
Uses obesity as excuse for failures in life	Doesn't allow obesity to interfere with life
Focuses on past	Focuses on present
Allows negative emotions to reduce self-control	Uses positive attitude to enhance self-control
Focuses on appearance	Focuses on health
Wants to improve appearance to get love/affection	Wants to be healthy to participate fully in loving relationships

Source: Adapted from J. P. Foreyt and G. K. Goodrick, *Living Without Dieting* (New York: Warner Books, 1992), p. 55.

mon. To emphasize the personal nature of weight-loss plans, the following sections are written in terms of advice to "you." This is intended not to put you under pressure to take it personally but to give you the illusion of listening in on a conversation in which a person with, say, anywhere from 10 to 200 pounds to lose is being competently counseled by someone familiar with the techniques known to be effective. Notes in the margin highlight the principles involved.

No particular diet is magical, and no particular food must be either included or avoided. Since you are the one who will have to live with the eating plan, you had better be involved in its planning. Don't think of it as a diet you are going "on"—because then you may be tempted to go "off" it. Lifestyle changes can be called successful only if the pounds do not return. Think of it as an eating plan that you will adopt for life. The diet must consist of foods that you like or can learn to like, that are available to you, and that are within your means.

For the person wanting to lose weight, a deficit of 500 calories a day for seven days (3,500 calories a week) is enough to lose a pound of body fat a week. If you were to spend an extra 250 calories a day in some form of exercise, you could increase this energy deficit.

Choose a calorie level you can live with. The 10-calorie rule will enable you to lose a pound or two a week while supporting your basal metabolism: Allow 10 calories a day for each pound of your present body weight. As you lose weight, you can gradually adjust calories downward to keep losing at this rate. Thus, a person who starts at 220 pounds should eat 2,200 calories a day at first; one who starts at 150 should eat 1,500 calories.

Put nutritional adequacy high on your list of priorities. This is a way of putting yourself first. "I like me, and I'm going to take good care of me" is the attitude to adopt. This means including foods that are rich in valuable nutrients—vegetables and fruits; whole-grain breads and cereals; and a reasonable amount of protein-rich foods such as lean meats, skinless poultry, fish, eggs, legumes, low-fat cheeses, and fat-free milk. Within these categories, learn what foods you like, and eat them often. If you resolve to include a certain number of servings of food from each of these groups each day, you may be so busy making sure you get what you need that you will have little time or appetite left for high-calorie or empty-calorie foods. Researchers have shown that reducing the intake of fat alone can promote significant weight loss, especially when coupled with a high complex-carbohydrate diet.[49]

EATING PLAN STRATEGIES

1. Get involved personally.
2. Adopt a realistic plan, and then keep track of calories.
3. Make the eating plan adequate.
4. Emphasize high nutrient density.
5. Individualize. Eat foods you like.
6. Stress dos, not don'ts.
7. Eat regular meals.
8. Take a positive view of yourself.
9. Visualize a changed future self.
10. Take well-spaced weighings to avoid discouragement.

To burn 250 extra calories a day (equal to half a pound of fat loss per week), add one of these activities to your daily routine: Walk (briskly) for 45 minutes. Bike (moderate pace) for 36 minutes. Swim (fast) for 23 minutes. Run (moderate pace) for 18 minutes.

FIGURE 8-4
SAMPLE BALANCED WEIGHT-LOSS DIETS USING THE FOOD GUIDE PYRAMID[a]

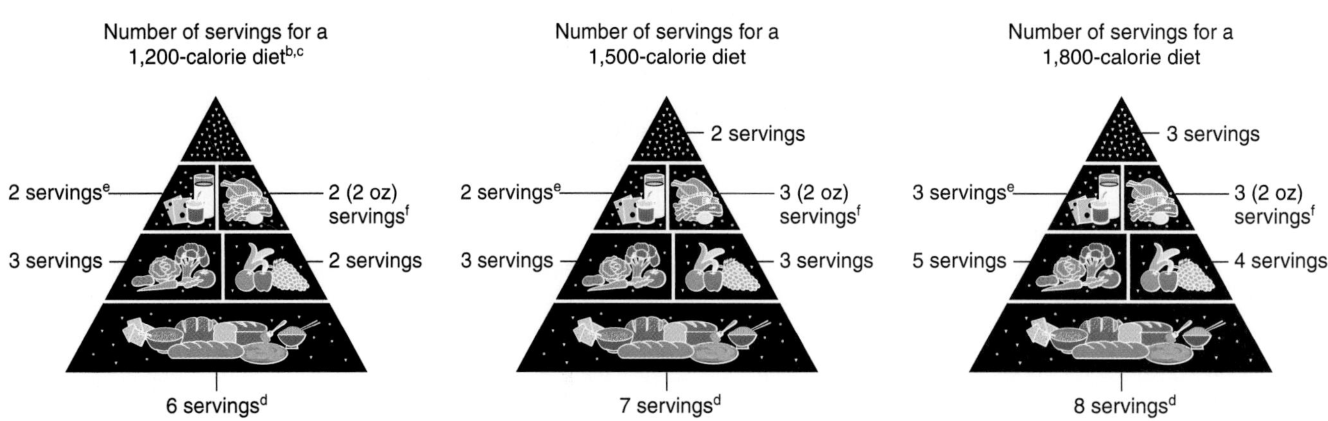

[a]Assumes no alcohol intake.
[b]Not recommended for pregnant or lactating women, children (depending on age), or those who have special dietary needs. At or below this low level of calorie intake, it may not be possible to obtain recommended amounts of all nutrients from foods; therefore, it is important to make careful food choices, and the need for dietary supplements should be evaluated.
[c]This plan allows up to 1 teaspoon of added sugar and 5 grams (1 teaspoon) of added fat.
[d]For maximum nutritional value, make whole-grain, high-fiber choices.
[e]Choose fat-free milk products.
[f]Select lean meat and use cooking methods that do not require added fat.

A small amount of fat should be included in each meal to make it satisfying and keep you from getting hungry again too soon. You don't have to use pure fat—butter, margarine, or oil. Rather, most of the fat should come from protein-rich foods, such as lean meats, eggs, poultry, fish, and low-fat cheeses. Add any pure fat with extra caution. A slip of the butter knife adds more calories than a slip of the sugar spoon. Keep concentrated sweets to a minimum, and let your carbohydrate come from starchy foods. Figure 8-4 shows three suitable patterns for weight-loss eating plans. Table 8-8 offers more tips on cutting back on fat and calories and the Savvy Diner feature on page 270 gives suggestions for defensive dining.

If you include alcohol or other empty-calorie items in your eating plan, limit it to no more than 150 calories a day. Budget this amount into your chosen calorie level, and reconcile yourself to a slower rate of weight loss.

Three meals a day is standard for our society, but no law says you shouldn't have four or five meals—only be sure that they are smaller, of course. What is

TABLE 8-8 HEALTHFUL EATING TIPS: CUTTING BACK ON FAT AND CALORIES

- Watch out for second helpings of higher calorie foods.
- Choose low-calorie versions of foods you like.
- Go easy on foods that are high in fat or sugar.
- Limit alcoholic beverages.
- Roast, broil, boil, steam, or poach foods rather than fry them.
- Select lean cuts of meat and trim visible fat.
- Eat poultry and fish without skin.
- Use spices and herbs instead of sauces, butter, or other fats.
- Consume low-fat or fat-free dairy products.
- Try fresh fruit for dessert or baked products made with less fat and sugar (e.g., angel food cake).
- Use alternatives to foods as rewards (e.g., long walks, relaxing baths, a visit with a friend, a hobby, gardening, a good book).

important is to eat regularly and, if at all possible, to eat before you become very hungry.

Keep a record of what you have eaten each day for at least a week or two until your habits are automatic. Resume record keeping whenever you need to.

At first it may seem as if you have to spend all your waking hours thinking about and planning your meals. Such a massive effort is always required when a new skill is being learned. But after about three weeks, it will be much easier. Your new eating pattern will become a habit. Some of the characteristics of successful dieters are listed in the margin.[50]

Do not weigh yourself more than once a week. Although 3,500 calories roughly equals a pound of body fat, there is no simple relationship between calorie balance and weight loss over short intervals. Gains or losses of a pound or more in a matter of days reverse themselves quickly; the smoothed-out average is what is real. Don't expect to lose continuously as fast as you did at first. A sizable water loss is common during the first week, but it will not happen again.

Many dieters experience a temporary plateau after about three weeks—not because they are slipping but because they have gained water weight temporarily while they are still losing body fat. The fat they are hoping to lose must be used for energy. To use it, the body must combine it with oxygen (oxidize it) to make carbon dioxide and water. These compounds are heavier than the fat they are made from because oxygen has been added to them.* The carbon dioxide will be exhaled quickly, but the water stays in the body for a longer time. The water takes a while to leave the cells, then makes its way into the spaces between them, and finally enters the bloodstream. Only after the water has arrived in the blood do the kidneys "see" it and send it to the bladder for excretion. Meanwhile, the dieter has a weight gain, but one day the plateau will break. The signal that this is happening is frequent urination.

If you have been working out lately, successive weighings may show an occasional gain when you expect a loss. This may reflect a welcome development: the gain of lean body mass—just what you want if you want to be healthy. In fact, weight loss without exercise can have a negative effect on body composition. No doubt you've heard someone say as a joke, "I've lost 200 pounds, but I've never been more than 20 pounds overweight." This person expects to diet, lose weight, regain the weight, and diet again throughout life, without exercising.

This pattern of losing, gaining, and then losing weight again may set the stage for a lifetime of weight fluctuations that may have a lasting impact on the makeup of the body and the way it burns calories. Dubbed the *yo-yo effect,* the up-and-down movement of the needle on the scales may cause the body to accumulate a greater percentage of fat and less lean muscle with each round of dieting.[51]

PROFILE OF SUCCESSFUL DIETERS

They know their weight (they check their scale weight weekly).

They are motivated from within.

They know what they eat (they keep records).

They engage in regular exercise.

They have social support (relationships/groups).

They control their intake of alcohol, fat, and sugar.

They follow an individualized diet plan—one that they can enjoy permanently.

They lose no more than a pound or two a week.

They set reasonable goals—such as losing just 10 percent of their weight as a start.

They accept an occasional lapse rather than aim for perfection.

They view weight management as a long-term commitment.

*Water weight accumulates during fat oxidation because one fatty acid weighing 284 units leaves behind water weighing 324 units—14% more.

WORKING ON THE WEB www.cyberdiet.com

Sponsor: Timi Gustafson, RD and Cynthia Fink

Description: A credible source for information on nutrition, weight loss, and how to adopt a healthy lifestyle. Practical tips on getting started on weight loss and exercise, with lots of fun, easy-to-use tools to assess your health risks.

Available Activities: The site provides an interactive section that allows you to determine your nutrition profile with calorie and nutrient needs. Additional tools calculate body mass index, body fat distribution, and target heart rate. The Activity Calculator lists how many calories you would spend doing dozens of activities—everything from aerobics to white-water rafting. Using the Daily Food Planner, you can develop a personalized eating plan, complete with recipes and shopping lists. Tips for recipe makeovers and healthy dining at a variety of ethnic and fast food restaurants are included.

Web Work:

1. Construct a personalized eating plan, based on your calorie and nutrient needs.
2. Calculate your BMI and body fat distribution to assess your risk for weight-related problems.
3. Calculate the number of minutes you would need to spend doing exercises that you enjoy in order to burn 150 calories a day in moderate activity.
4. Learn healthy cooking tips for eating at home: Choose a favorite recipe and compare the fat and calorie content in the before and after versions of the recipe.
5. Learn how to make healthy choices while dining out: Review the tips offered for breakfast and deli items and for your favorite ethnic or fast-food restaurant.

Helpful Hints: From the Cyberdiet home page, scroll down and click on self-assessment to calculate your nutrition profile, BMI, and body fat distribution. Next, go back to the home page and scroll down to Resource Center and click on Healthy Heart. From the Eating Smart for a Healthy Heart page, scroll down to Cooking In, and click on Recipe Makeovers to view modified recipes. When finished, click on Return to Eating Smart for a Healthy Heart. Scroll down to Dining Out and click on Restaurant Choices to review tips for eating at a variety of ethnic and fast-food restaurants.

If you cut back drastically the number of calories you consume without exercising for, say, a month or two, each week you'll probably lose a great deal of weight in the form of not only fat but also lean muscle. The problem is that if you later put the weight back on, the regained pounds may be primarily fat—not muscle. Thus, you may end up weighing the same as you did when you began the diet, but your body will now be composed of more fat and less muscle. Because muscle burns more calories just to sustain itself than fat does, your body will need fewer calories to maintain its weight than it did before. If you go back to eating the way you used to, chances are you will gain even more weight.

Weight loss—even modest weight loss—can also result in a depletion of bone mineral density, according to a University of Pittsburgh study of women.[52] Declines in fat loss with weight loss may lower estrogen levels, a hormone necessary for healthy bone maintenance. The study emphasizes the importance of including adequate calcium intakes and exercise in any weight-loss regimen, particularly such weight-bearing exercises as walking, jogging, stair climbing, and weight lifting.

In addition to the risk of altering the body's overall composition during repeated bouts of weight loss, consider that each unsuccessful attempt to keep weight off is often viewed by the dieter as a personal failure, which can erode the person's self-esteem and trigger painful feelings of guilt and depression.[53] Despite these possible consequences of repeated bouts of dieting, a recent Task Force on the Prevention and Treatment of Obesity states that "obese individuals should not allow concerns about weight cycling to deter them from efforts to control their body weight," since evidence of the adverse effects of weight cycling in humans is insufficient.[54] As the Nutrition Action feature on page 272 points out, a more healthful alternative to crash dieting is to gradually develop healthful, permanent lifestyle habits that will help you to lose weight and keep it off.

Exercise

The physical contributions exercise makes to a weight management program are threefold: exercise increases one's calorie expenditure, it alters body composition in a desirable direction, and it alters metabolism.[55] It also offers the psychological benefits of looking and feeling healthy and the increased self-esteem that accompanies these benefits, which can enhance the motivation to maintain a healthful lifestyle for the long run.

Compared with lean tissue, fat tissue is relatively inactive metabolically. Metabolic activity burns calories—lots of them. Thus, the more lean tissue you develop, the faster your metabolism becomes, the more calories you spend, and the more you can afford to eat. This brings you both pleasure and nutrients. Exercise, by shifting body composition toward more lean tissue, speeds up the metabolism *permanently*—that is, for as long as you keep your body conditioned. Furthermore, the more muscle and lean tissue you have, the more fat you will burn—all day long, even when you are resting.

The next chapter offers many pointers about becoming fit, but a few notes on strategy are in order here. For one thing, keep in mind that if exercise is to help with weight loss, it must be active exercise—voluntary moving of muscles. Being moved passively, such as by a machine at a health spa or by a massage, does not increase calorie expenditure. The more muscles you move, the more calories you spend (see Table 8-9).

People sometimes ask about spot reducing. Can you lose fat in particular locations? Unfortunately, muscles don't "own" the fat that surrounds them. Since all body fat is shared by all the muscles and organs, spot-reducing exercises that work only the flabby parts won't help reduce the fat located there. There is some good news, though: Tightening muscles in trouble spots by way of a balanced, all-over exercise program may improve the appearance of the fatty areas.

Another thing to keep in mind is that the number of calories spent in an activity depends more upon how much a person weighs than on how fast the person can do the exercise (see Table 8-9). For example, a person who weighs 125 pounds burns off 85 calories by running a 12-minute mile. That same person, walking a mile in 15 minutes, burns almost the same amount—86 calories. Similarly, a 200-pound person spends 136 calories on the 12-minute mile, and a similar amount—138 calories—on the 15-minute walk. The rule seems to be that you don't have

Examples of easy-paced to moderate exercise:

Walking at about 3 miles per hour.

Bicycling on level ground at about 9 miles per hour.

You can tell that the exercise is moderate if you are breathing a little faster than normal but can still easily carry on a conversation.

STRATEGIES FOR USING EXERCISE FOR WEIGHT CONTROL

1. Make it active exercise; move your muscles.
2. Think in terms of quantity, not speed.
3. Exercise informally, in daily routines.

Exercise is essential for weight control.

TABLE 8-9 CALORIES SPENT DURING VARIOUS ACTIVITIES

To calculate the exercise calories you expend per hour, find your "exercise" in the left column and your "weight" in the right column. In the place where they intersect is the figure indicating the calories burned *per hour*. For example, if you aerobic dance for 1 hour and weigh 125 pounds, you will expend 285 calories.

Weight (in pounds): Exercise	110	125	150	175	200
	Calories expended per hour				
Aerobic Dancing	250	285	340	395	450
Archery	225	255	305	360	410
Baseball	225	255	305	360	410
Basketball	415	470	565	660	750
Bowling	180	205	245	285	325
Calisthenics (vigorous)	225	255	305	360	410
Cross-country skiing					
moderately hilly	595	675	810	945	1080
indoor machine (11 mph)	330	375	450	525	600
Cycling					
outdoor (5.5 mph)	195	220	260	305	350
outdoor (9.4 mph)	300	340	410	475	545
outdoor racing (19 mph)	505	575	690	805	920
Schwinn Aerodyne	510	580	695	810	925
stationary (moderate tension)	330	375	450	525	600
Golf					
with cart (90–120 minutes)	145	165	200	230	265
no cart (90–120 minutes)	185	210	255	295	340
Handball/Squash	635	725	870	1015	1155
Hiking 4 mph, 20-lb pack	355	405	490	570	650
Horseback Riding	225	255	305	360	410
Ice Skating	275	300	350	390	425
Nordic Ski Machine					
heavy (18 mph)	1100	1250	1500	1750	2000
medium (11 mph)	330	375	450	525	600
light (6 mph)	225	255	305	360	410
Racquetball	550	625	750	875	1000
Roller Skating/Blading	275	300	350	390	425
Rope Skipping (100 skips/min)	560	640	765	895	1020
Rowing (sculling or machine)	620	705	845	990	1130
Running (Jogging)					
6 min/mile (10 mph)	755	860	1030	1200	1375
7 min/mile (8.5 mph)	685	780	935	1090	1245
8 min/mile (7.5 mph)	625	710	850	990	1135

Weight (in pounds): Exercise	110	125	150	175	200
	Calories expended per hour				
Running (Jogging) (*cont.*)					
9 min/mile (6.5 mph)	580	660	790	920	1050
10 min/mile (6 mph)	535	605	730	850	970
11 min/mile (5.5 mph)	470	530	640	745	850
12 min/mile (5 mph)	375	425	510	600	680
Scuba Diving	355	405	490	570	650
Snow Skiing—Downhill	300	340	410	480	545
Softball	225	255	305	360	410
Stair Climbing (moderate)	515	600	750	850	960
Stairmaster (machine)	595	675	810	945	1080
Step Aerobics—120 steps/min	550	625	750	875	1000
Swimming					
45 min/mile	385	435	525	610	700
60 min/mile	300	335	405	475	540
Table Tennis (moderate)	200	225	270	315	360
Tennis					
doubles	225	255	305	360	410
singles	325	370	445	520	600
Treadmill					
12 min/mile	375	425	510	600	680
13.5 min/mile	330	375	450	525	600
Volleyball					
competitive	435	495	595	700	800
recreational	165	185	225	260	300
Walking/Race Walking					
12 min/mile (5 mph)	435	495	595	700	800
Walk/Jog Combination					
13.5 min/mile (4.5 mph)	330	375	450	525	600
Walking					
15 min/mile (4 mph)	300	345	415	480	550
17 min/mile (3.5 mph)	250	285	345	400	450
20 min/mile (3 mph)	225	255	310	360	410
30 min/mile (2 mph)	145	165	200	230	265
Weight Training/Lifting (Light)	270	310	370	430	500

Soutce: G. Kostas, M.P.H., RD, *The Balancing Act Nutrition and Weight Guide*, Dallas, TX, 1993.

to work fast to use up calories effectively. If you choose to walk rather than run the distance, you will use up about the same energy; it will just take you longer.

You may have heard the suggestion to incorporate more exercise into your daily schedule in many simple, small-scale ways. Park the car at the far end of the parking lot; use the stairs instead of the elevator; do a round of sit-ups before you get up in the morning. If you also incorporate both regular aerobic exercise and strength-training into your schedule, your heart and lungs as well as your skeletal muscles will be fit (see Chapter 9).

WEIGHT-GAIN STRATEGIES

It is as hard for a person who tends to be thin to gain a pound as it is for a person who tends to be fat to lose one. The person who wants to gain weight is faced with some of the same challenges as the one who wants to lose weight—learning new habits, learning to like new foods, and establishing discipline related to meals and mealtimes. But there are major differences.

Add extra calories to the milk or juice you drink by tossing in a frozen banana or other fruit and whipping up a smoothie in the blender.

Knowing that vigorous physical activity costs calories, an active person may wonder whether it is advisable to curtail activity. The answer is no, not unless the underweight condition is so extreme as to be life-threatening. The healthful way to gain weight is to build yourself up by patient and consistent training while eating nutritious foods containing enough extra calories to support the weight gain. If you add extra snacks of high-calorie nutritious foods, such as a peanut butter sandwich with a glass of milk before bedtime, and a healthful midafternoon fruit smoothie or milkshake, you can eat 700 to 800 extra calories a day and achieve a healthful weight gain of 1 to 1½ pounds per week. Choose calorie-dense snacks, such as fruit yogurt, granola, dried fruits, apple or grape juice, nuts, sunflower seeds, fig bars, and baked potatoes topped with low-fat cheese.

A person wanting to gain weight often has to learn to eat different foods. No matter how many helpings of carrots you consume, you won't gain weight very fast because carrots simply don't offer enough calories. The person who can't eat much volume is encouraged to use calorie-dense foods at meals (the very ones the dieter is trying to stay away from). Choose a milkshake instead of milk, peanut butter instead of lean meat, avocado instead of cucumber, blueberry bran muffin instead of whole-wheat bread. When you do eat carrots, dip them in hummus; use creamy dressings on salads, yogurt on fruit, cottage cheese on potatoes, and the like. Because fat contains twice as many calories per teaspoon as sugar, it adds calories without adding bulk. For heart health, be sure to choose the "good" fats, found in foods such as avocados, olives, peanuts, pistachios, and other nuts and seeds, and limit your intake of saturated fats.

Try eating more food at each meal. If you normally eat one sandwich for lunch, try eating an extra sandwich.

Eat more frequently. Make three sandwiches in the morning, and eat them between classes in addition to the day's three regular meals. Spend more time eating each meal: If you fill up fast, eat the highest-calorie items first. Don't start with soup or salad; start with the main course. Always finish with dessert. Many an underweight person has simply been too busy (for months) to eat enough to gain or maintain weight. These strategies will help you change this behavior pattern.

Whether you need to gain, lose, or maintain weight, attention to what you eat can pay off in long-term wellness benefits. This chapter has emphasized the relationship of food and eating to body weight and has come to the same conclusion reached earlier in the book: To support wellness, you should eat regular, balanced meals composed of a wide variety of foods you enjoy.

Dining Out

Pizza Yankee . . . The story goes that a pizza parlor waitress asked Yogi Berra whether he wanted his pizza cut into six or eight slices. "Six, please," he said. "No way am I hungry enough to eat eight."[56]

Dining out can be a pleasant time to relax, socialize, taste the foods of different cultures, and provide a break from your usual schedule, and it does not have to involve overindulgence in high-calorie, low-nutrient foods. With practice, you can maximize the benefits of eating out by adopting some of the following strategies:

- Take time to examine your options before selecting foods at buffets, cafeterias, or food courts. Otherwise, the array of choices can seem overwhelming.
- Take the edge off hunger by starting with a broth-based soup, small salad, fruit, or light appetizer.
- Make specific requests: "I'd like to have this sandwich with only half the cheese and just a teaspoon of mayonnaise served on the side."
- Ask for salad dressings, sauces, or gravies *on the side.* This puts you in control of the portions. Ask if reduced-fat or fat-free salad dressings are available.
- Request fresh fruit, sorbet, or low-fat frozen yogurt for dessert. It will give you the sweetness you want without the calories of fat and sugar. If you choose to splurge on your favorite chocolate-raspberry torte, enjoy it by sharing it with a friend. Afterward, get back to your healthful habits of eating well and exercising.
- At fast-food windows, ask that they hold the sauce on your burger and give you extra tomato and lettuce or other vegetables for your sandwiches or tacos. For more tips on selecting foods from fast-food menus, see the Nutrition Action feature in Chapter 1. The table that follows offers additional tips on ordering at a variety of quick-service restaurants.
- Order mineral water or club soda with a twist, alcoholic drinks with just the mix and not the alcohol (e.g., Bloody Marys without the vodka), or a fruit juice. Remember, alcohol supplies empty calories.
- Be a menu sleuth: The following terms typically describe high-fat items: au gratin, Alfredo, batter-dipped, breaded, bearnaise, carbonara, creamy, crispy, croquette, deep-fried, flaky, fritters, parmigiana, tempura, with gravy, or with hollandaise sauce. Order red pasta sauces rather than white.

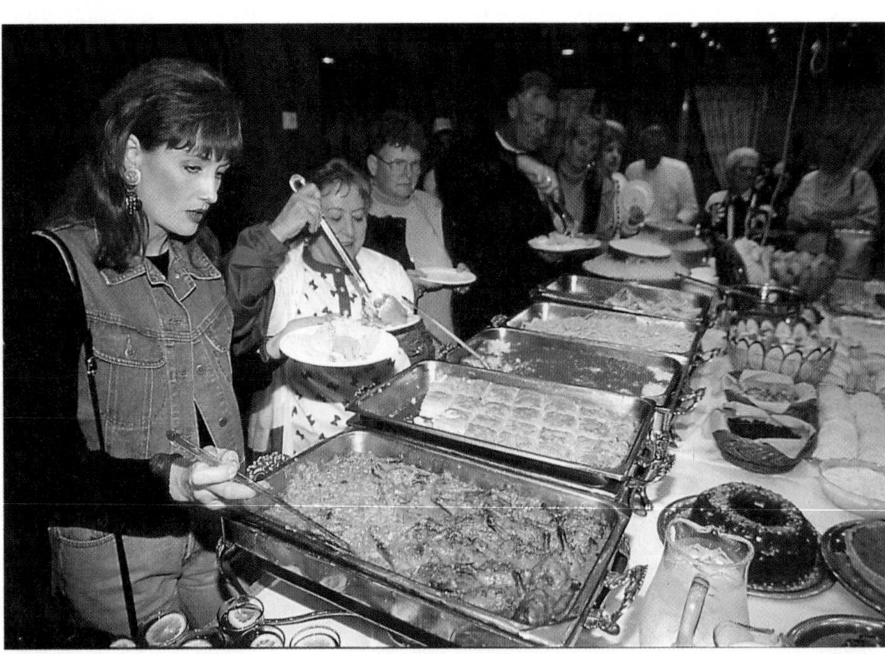

- When ordering, inquire about preparation. Ask for entrées to be broiled, baked, grilled, poached, steamed, or roasted, with only a minimal amount of fat (or none at all).
- Downsize your order. Many restaurants serve portions that are three or four times the recommended serving size. Request a doggie bag so you can take half the amount home. Or, consider sharing the entrée with a friend and ordering a salad to go with it. You can also order à la carte: one or two nutritious low-fat appetizers such as steamed shrimp or raw vegetables and a low-fat dip along with a salad and bread instead of a huge main meal.
- Request a plain baked potato or one with plain fat-free yogurt and chives instead of butter and sour cream. Order your vegetables steamed or stir-fried without added butter or sauces.
- Request a vegetable-based sauce for your pasta rather than the traditional meat sauces.
- Order a special omelet. Instead of the usual three-egg version, ask for one made with egg substitutes, or with one egg plus extra egg whites with no salt, filled with plenty of vegetables.
- Watch out for the all-you-can-eat restaurants and those that specialize in fried foods. Overindulgence is easy at these places.
- Finally, take time to enjoy your meal, along with the company and conversation. Eating slowly gives your body time to digest the food and feel satisfied.

Eating Well On the Run*

SMART CHOICES	HEALTHY TIPS
Sandwich Shop: Fresh sliced veggies in a pita with low-fat dressing, cup of minestrone soup, turkey breast sandwich with mustard, lettuce, tomato, fresh fruit	For that big deli sandwich, order two extra slices of bread, remove half the filling, assemble an extra sandwich, wrap in foil, and refrigerate for the next day.
Rotisserie Chicken: Chicken breast (remove skin), steamed vegetables, mashed sweet potatoes, tossed salad, fruit salad	Select plain rolls instead of cornbread or biscuits.
Fast Food: Grilled chicken breast sandwich (no sauce), single hamburger without cheese, grilled chicken salad, garden salad, low-fat or nonfat yogurt, fat-free muffin, cereal, low-fat milk	Bring a piece of fresh fruit to round out your meal. Keep salads healthy with reduced-fat or fat-free dressings.
Salad Bars: Broth-based soups, fresh bread or breadsticks, fresh greens, chopped veggies, beans, low-fat dressing, fresh fruit salad	Avoid marinated beans and oily pasta salads.
Asian Take-Out: Wonton soup, pho (Vietnamese noodle soup), hot and sour soup, steamed vegetable dumplings, vegetable mixtures over steamed rice or noodles	Request that vegetables be steamed or stir-fried with as little oil as possible. Soy sauce is high in sodium, so use sparingly.
Pizza Night: Choose flavorful, low-fat toppings such as peppers, onions, sliced tomatoes, spinach, broccoli or mushrooms	Ask for your pizza with less cheese.

**Source:* Adapted from *Healthy Eating Away from Home* (Washington, D.C.: American Institute for Cancer Research, 1996).

Breaking Old Habits

ELEMENTS OF BEHAVIOR CHANGE[58]

1. *Precontemplation:* You need to change, but you're not yet ready to accept that fact.
2. *Contemplation:* You want to change, but you're not sure how.
3. *Preparation:* You gain knowledge to set up a plan of action for change.
4. *Action:* You jump in and "just do it."
5. *Maintenance:* You work on sticking to your plan of action.
6. *Termination:* You have achieved lasting change and experience few, if any, temptations or relapses.

"Just do it," urge the makers of Nike sportswear on billboards, in magazines, and wherever else they promote their products, implying that adhering to a fitness program is just a matter of donning sneakers and heading for the track or gym. But as most of us know from experience, sticking to a regular exercise program for the first time or changing eating habits isn't always easy. The majority of people who successfully quit smoking make three or four attempts before they finally kick the habit. And most people make the same New Year's resolution at least five years in a row before they stick to it permanently.[57] That's because breaking old patterns of behavior and developing new ones involves going through a number of stages before reaching the point at which you're able to change for good.

First, you must be aware that you could change. "I could lose some weight and try to become more physically fit," you might tell yourself. Then you must be inspired to *want* to change, which in turn will help to motivate you to find out how to do so, say, by talking to a registered dietitian about strategies for cutting back on fat and calories in your diet and by reading about various forms of exercise. After you've gone through those steps, you will be ready to take action—that is, to "just do it." Once you've taken these initial steps, you need to maintain your behavior change by creating a game plan that will help you handle the inevitable slips and lapses that will occur over time. After you've done so and have stuck with your plan for a good deal of time, you're home free.

BEHAVIOR MODIFICATION a process developed by psychologists for helping people make lasting behavior changes.

Of course, as you go through those steps, making lasting changes in behavior poses ongoing challenges. That's where a process known as **behavior modification** can help. Developed by psychologists to help people change their habits, behavior modification uses techniques similar to mapping out a game plan.

To start, identify your goals. You might decide, for example, that you want to lower the fat and calorie content of your diet. Once you've chosen your goals, record your present pattern of behavior along with the reasons behind it. Keep a food diary by writing down everything you eat for five days as well as how you feel at the time you eat (see Table 8-10 for a sample food diary). This technique

TABLE 8-10 SAMPLE ENTRIES IN FOOD DIARY

Food Eaten	Time and Place	With Whom	Mood	Other Activity
1 large blueberry muffin 1 cup coffee with 1 oz cream and 1 tsp sugar	7:30 a.m., donut shop	Alone	Stressed—late for class	Skimming notes for upcoming class
1 hot dog with 2 tbsp ketchup 1 small bag potato chips 12-oz can diet soda	12:45 p.m., cafeteria	2 friends	Relaxed	Talking with friends
1 candy bar	2:35 p.m., bedroom	Alone	Angry about argument with significant other	Watching TV

can help you understand why you behave the way you do in certain situations. You might find, for instance, that every time you are angry at your significant other or get a bad grade you eat a candy bar or two to comfort yourself. Or perhaps without even thinking about it, you nibble on whatever happens to be handy while watching television.

Once you've identified your goals and current patterns of behavior, determine some new strategies that will help you meet these goals and set yourself some rewards for sticking to them. If you talk to a friend instead of heading to the kitchen cupboard or vending machine when you're angry, for instance, you'll save hundreds of calories and probably experience the relief of getting your feelings off your chest. After you've pinpointed behaviors you want to alter, commit yourself to make the changes and try to envision the healthier person you'll become as a result.

The next step is to divide the behavioral goals into small portions that can be achieved one at a time. Although you may have come up with 20 behaviors that you want to change to meet your overall goals, trying to accomplish all 20 in a week or two is asking a great deal of yourself and may be setting the stage for disappointment and failure. Instead, pick one or two, see whether you can stick to them for, say, two weeks, and then reward yourself for doing so before going on to try more.

As you meet your goals, remind yourself of the progress you're making as well as the benefits you're gaining—weight loss, increased energy, better health. Give yourself a tangible reward, such as a night out at the movies or a new item of clothing. This step is key because behavior research indicates that once you have instituted a change in your life, you will maintain it only by positive reinforcement. Along with rewarding yourself, feel free to modify your goals if you think that those you initially set were too difficult or too simple to achieve, and evaluate your progress on a regular basis.

While you're working on modifying your behavior, it is critical to keep in mind that you're only human and there will be days when you slip and revert back to an old pattern of behavior. Instead of dwelling on it or berating yourself, realize that relapses are par for the course, forgive yourself, and simply move on. Keep in mind that even though you may feel a setback, you're still moving forward if you learn from your mistakes and get right back on track again. That's why some experts prefer the term *recycle* to *relapse;* people who make a mistake and then learn from it can reuse, or recycle, their new knowledge.

In addition, set priorities to ensure that you adopt one or two behaviors that you make it a point to stick to all the time. Furthermore, be aware of how other areas of your life are affecting your readiness to make changes and meet your goals. Many people waste a great deal of time and energy trying to change when the timing isn't right for them. Someone who is going through the breakup of a long-term relationship or having difficulties at work or school, for example, may

BEHAVIOR MODIFICATION STEPS

1. Identify the goal.
2. Record your present behavior pattern. Identify the reasons you practice these behaviors.
3. Identify the behaviors that will lead to the goal and the rewards of those behaviors.
4. Commit yourself to changing. Face what you'll have to give up or change to make the desired behavior a reality. Envision your changed future self.
5. Plan. Divide the behavior into manageable portions. Set small, achievable goals and plan periodic rewards.
6. Try out the plan. Modify the plan, if necessary, in ways that will help you succeed.
7. Evaluate your progress on a regular basis.

Continued motivation:

- Persist long enough to experience the rewards, such as improved self-image and enhanced self-esteem.
- Remember the price of the old behavior.
- Keep in mind where you started.
- Tune in to the benefits of the new behavior.

be under too much stress to deal with making a change in behavior on top of everything else. Realizing that you have limits and waiting until the timing is right can stave off additional frustration and guilt brought about by feeling like a failure if you aren't able to change.

Another point to keep in mind is that you might need to try new behaviors more than once before you begin to like them. Dietitians sometimes use a "rule of three" in counseling. Try fat-free milk once, and you may not like it. Try it a second time, and you may be able to tolerate drinking it, even though it doesn't taste as good as whole milk. Try it a third time, and you may conclude that although it's not as good as whole milk, you don't really mind drinking it. Then you'll be able to maintain your new habit more easily.

Strategies for Changing the Way You Eat

This section covers the steps designed specifically to help with weight control. (Chapter 9 includes strategies for modifying exercise behavior.) To begin with, keep a record of your present eating behavior that you can compare with your future progress and use to determine situations that trigger your reaching for food. In this way, you can see how far you've come in changing your eating habits.

Next, try to identify cues that prompt you to eat when you're not hungry, such as watching television, talking on the telephone, and walking by a convenience store or vending machine. Then resolve to stop responding to those cues, and try to eat only in one place in a certain room, such as at the table in the dining room. In addition, try the following tips to eliminate the temptation to eat when you really don't want to:

- Don't buy hard-to-resist foods such as cakes, cookies, and ice cream.
- Don't shop when you're hungry and thereby more likely to buy tempting foods.
- Don't leave large amounts of food within easy reach. Keep serving dishes off the table, for instance, and put the cookie jar out of sight.
- Make small portions of food look large by spreading food out and putting it on smaller plates. Garnish empty space with low-calorie vegetables.
- Try to eat regular meals and snacks instead of skipping them, thereby reducing the likelihood of becoming uncomfortably hungry and then overeating later.
- Ask your family and friends to encourage you to eat a healthful diet and not to criticize you if you splurge occasionally.

Third, make it easy for yourself to eat the way you want to eat.

- Keep a variety of nutritious foods such as fruits and vegetables readily available.
- If you like to eat between meals, plan snacks that fit easily into your diet, such as low-fat yogurt and fruit.
- Prepare attractive meals for yourself.

Fourth, take a look at and then, if necessary, alter the manner in which you eat. If you eat quickly, slow down by chewing food thoroughly, pausing between bites, putting down your utensils and swallowing before reloading your fork or spoon. Eating slowly will give your body a chance to feel full and satisfied.

Finally, reward yourself for positive behaviors.

- Plan to buy something new to wear or do something you enjoy, such as attending a favorite sporting event or a concert, each time you meet your goals.

A RECIPE FOR SUCCESS

- **Be Realistic.**
  Make small changes over time in what you eat and your level of activity. Aim for losing no more than 1 to 2 pounds per week. Even a small amount of weight loss can bring major health benefits.
- **Be Adventurous.**
  Expand your tastes to enjoy a wide variety of foods and activities. Get fewer calories by going "large" on flavor and small on portions.
- **Be Flexible.**
  You don't have to give up your favorite foods to manage your weight. Just balance what you eat and what you do over several days.
- **Be Sensible.**
  Enjoy all foods, just don't overdo it. Slow down when you're eating and listen to your body's signals to know when you're hungry or full.
- **Be Active.**
  It's not *what* you do—it's whether you enjoy what you're doing so you can do it more often!

Source: Adapted from Weighing In On Health, *Food and Nutrition News* 70 (1998): 6.

Spotlight
THE EATING DISORDERS

The relentless pursuit of thinness and fear of being fat are a haunting nightmare that drives millions of American teens and adults to starve, vomit, or purge.[59] The illness of **anorexia nervosa**—self-starvation—has been recognized as a psychiatric syndrome since the 1870s. Its companion disease **bulimia nervosa**—gorging on food and then purging—was not recognized as a separate eating disorder until the 1960s and 1970s. Some researchers suspect that a complex interplay between environmental, social, and perhaps genetic factors triggers the disorder in victims, mostly women. Others question whether or not the fear of fatness isn't a mask for underlying emotional problems. Experts speculate that focusing on the body diverts these people's attention away from and suppresses the painful emotion of anger, feelings of low self-esteem, the inability to express their feelings, or poor family relationships. The acute focus on the body develops into an intense fear of fatness, a characteristic intrinsically linked with food.[60] As a result, food is seen only as a source of body fat and so becomes carefully controlled. But why food? Why not use some other method of coping with stress?

Enter the societal link. In a society where thinness is equated with material success and even self-worth, especially for women, becoming thin appears to be the yellow brick road that leads to happiness. Unfortunately, as victims of eating disorders come to learn, practicing self-starvation or gorging and purging leads instead to physical and emotional pain.

What conditions are referred to by the term *eating disorder?*

The term **eating disorder** comprises a wide spectrum of conditions including **anorexia nervosa, bulimia nervosa,** and **binge-eating disorder.**[61] Although the various conditions differ in their origin and consequences, they appear to have similarities among them—all of the conditions exhibit an excessive preoccupation with body weight, a fear of body fatness, and a distorted body image. Many times the person with an eating disorder falls short of the diagnostic criteria shown in Table 8-11 for anorexia nervosa or bulimia nervosa. Some of these people are described as having an **unspecified eating disorder** and can include people who:[62]

- Meet all the criteria for anorexia nervosa except irregular menses.
- Meet all the criteria for anorexia nervosa except that their weight remains within a normal range.
- Meet all the criteria for bulimia nervosa except that their binges are less frequent.
- Have recurrent episodes of binge eating but do not compensate using the methods of those with bulimia nervosa, a condition known as binge-eating disorder (see Table 8-12 on page 277).[63]

What is the difference between an eating disorder and disordered eating?

Disordered eating occurs when you eat (or don't eat) because of an external stimulus rather than an internal one. For example, an external stimulus might be, "I'm lonely," "I'm angry," or "I'm anxious," and "these chocolate cookies will distract me and help me feel better." An internal stimulus is a physical sensation—a message from your brain regarding your current state of hunger or satiety, such as "my stomach is growling," or "I feel full." People with eating disorders spend more time in "external eating" than in responding to internal hunger cues.[64] Many people eat for reasons other than hunger from time to time, but if this type of behavior controls our thoughts and interferes with our everyday routines, it becomes an eating disorder.

What are the symptoms of anorexia nervosa?

Anorexics deprive themselves of food except for controlled amounts of very-low-calorie foods such as

Looking in the mirror, a woman with anorexia nervosa distorts body image and overestimates body fatness.

TABLE 8-11 DIAGNOSIS OF EATING DISORDERS

Anorexia Nervosa	Bulimia Nervosa
A person with anorexia nervosa demonstrates the following: 1. Refusal to maintain body weight at or above a minimal normal weight for age and height. 2. Intense fear of weight gain, or becoming fat, even though underweight. 3. Distorted body image, undue influence of body weight or shape on self-evaluation, or denial of the seriousness of the current low body weight. 4. In females past puberty, amenorrhea, that is, the absence of at least three consecutive menstrual cycles. Two types: *Restricting type:* During the episode of anorexia nervosa, the person does not regularly engage in binge eating or purging behavior (i.e., self-induced vomiting or the misuse of laxatives or diuretics). *Binge eating/purging type:* During the episode of anorexia nervosa, the person regularly engages in binge eating or purging behavior (i.e., self-induced vomiting or the misuse of laxatives or diuretics).	A person with bulimia nervosa demonstrates the following: 1. Recurrent episodes of binge eating characterized by: a. eating in a discrete period of time an amount of food that is definitely larger than most people would eat during a similar period of time and under similar circumstances. b. a sense of lack of control over eating during the episode. 2. Recurrent compensatory behavior to prevent weight gain, such as self-induced vomiting; misuse of laxatives, diuretics, or other medications; fasting; or excessive exercise. 3. Binge eating and compensatory behaviors that occur, on an average, at least twice a week for three months. 4. Self-evaluation unduly influenced by body shape and weight. 5. The disturbance does not occur exclusively during episodes of anorexia nervosa. Two types: *Purging type:* the person regularly engages in self-induced vomiting or the misuse of laxatives or diuretics. *Nonpurging type:* the person uses other behaviors, such as fasting or excessive exercise, but does not regularly engage in self-induced vomiting or the misuse of laxatives or diuretics.

Source: Adapted from American Psychiatric Association, *Diagnostic and Statistical Manual of Mental Disorders* (DSM-IV) (Washington, D.C.: American Psychological Association, 1994).

unbuttered toast or popcorn, apples, and green beans. And even these foods are painstakingly limited. After three to four days of eating very small amounts of food, hunger pangs subside. Once their appetite is suppressed, anorexics report feeling quite energetic, as if on a high, making the strict fast easier to stick to. But the body needs fuel to run. To compensate for the lack of fuel from food, the body turns inward for its fuel and begins to slowly destroy muscle and fat tissue for energy. The following are some of the physical symptoms associated with starvation seen in anorexics:[65]

- Wasting of the whole body, including muscle tissue and bones
- Arresting of sexual development and stopping of menstruation due to loss of body fat
- Drying and yellowing of the skin from an accumulation of a stored vitamin A compound released from body fat
- Intolerance to cold weather due to loss of subcutaneous fat
- Growth of hair on the body, perhaps in response to a decrease in body temperature
- Loss of health and texture of hair
- Pain on touch
- Lowered blood pressure and metabolic rate
- Anemia
- Severe sleep disturbance[66]
- Depression, possibly related to changes in neurotransmitter function in the brain[67]

Simultaneously, distorted psychological symptoms develop. Looking in the mirror, the anorexic does not see the emaciated body others see but continues to see someone who is too fat.[68] A preoccupation with death develops, accompanied by a frantic pursuit of physical fitness by means of stringent exercise routines. The person deals with parents and family in a manipulative way so as to become the center of attention. Diet becomes so all-engrossing that the person may be quite isolated socially except from friends who stick close by and

TABLE 8-12 CRITERIA FOR DIAGNOSIS OF BINGE-EATING DISORDER

A person with a binge-eating disorder demonstrates the following:

1. Recurrent episodes of binge eating. An episode of binge eating is characterized by both of the following:
   a. Eating, in a discrete period of time (e.g. within any two-hour period) an amount of food that is definitely larger than most people would eat in a similar period of time under similar circumstances.
   b. A sense of lack of control over eating during the episode (e.g., a feeling that one cannot stop eating or control what or how much one is eating).
2. Binge-eating episodes are associated with at least three of the following:
   a. Eating much more rapidly than normal.
   b. Eating until feeling uncomfortably full.
   c. Eating large amounts of food when not feeling physically hungry.
   d. Eating alone because of being embarrassed by how much one is eating.
   e. Feeling disgusted with oneself, depressed, or very guilty after overeating.
3. The binge eating causes marked distress.
4. The binge eating occurs, on average, at least twice a week for six months.
5. The binge eating is not associated with the regular use of inappropriate compensatory behaviors (e.g., purging, fasting, excessive exercise) and does not occur exclusively during the course of anorexia nervosa or bulimia nervosa.

Source: Adapted from American Psychiatric Association, *Diagnostic and Statistical Manual of Mental Disorders*, 4th ed. (Washington, D.C.: American Psychiatric Association, 1994).

worry without knowing how to help.

By this time, the anorexic has reached absolute minimum body weight—for example, 65 to 70 pounds for a woman of average height. The person is on the verge of incurring permanent brain damage and chronic debilitation or death. The National Association of Anorexia Nervosa and Associated Disorders (ANAD) estimates that of those with severe eating disorders, 6 percent die, usually because major organs—heart and kidneys—fail.

What are the symptoms of bulimia?

Unlike anorexics, bulimics don't shrink to skeletonlike proportions. They usually are of healthy body weight or even slightly overweight. Bulimics also follow rigid rules of dietary restraint, but their routine is not as rigorous as the anorexic's starvation routine, and occasionally, they break their rigid rules.

Quickly, and usually privately, bulimics gorge on foods that are often sweet, starchy, and high in fat or calories and require little chewing. The binge ends when it would hurt to eat any more, when they are interrupted, or when they go to sleep or induce vomiting to expel the food just eaten.

Bulimics often feel controlled by the vicious circle that develops: anxiety about being "fat" leads to rigid dietary restraint. Mounting hunger from not eating and an increased preoccupation with food lead to a break in the rules—binging. After gorging, an intense fear of fatness overtakes the person, who then vomits to get rid of the food and release the fear of becoming "fatter." Feelings of guilt and shame follow the purge, building a new level of anxiety over the body, and the cycle begins again. Each binge reinforces the idea that additional rigidity of dietary restraint is required to prevent weight gain. Excessive use of laxatives or diuretics or bouts of vigorous exercise replace vomiting in some cases.

Binge eating is seldom life-threatening, but at the extreme, it can be physically damaging, causing lacerations of the stomach, tearing or irritation of the esophagus (in those who vomit frequently), dental caries (from acidic vomit attacking the teeth), electrolyte imbalances and malnutrition (in vomiters and laxative takers).

Bulimics also suffer from distorted body image.[69] They see themselves as fat and needing to restrict food, even though they usually have a healthy body weight. Bulimics prefer a body size somewhat smaller than normal.

How are anorexia nervosa and bulimia treated?

There are several philosophies regarding the treatment of eating disorders. The four major approaches include individual psychotherapy, hospitalization, family

For many people with anorexia nervosa, a full day's diet may consist of no more than three or four items.

For many people with bulimia, guilt, depression, and self-condemnation follow a binge-eating episode.

TABLE 8-13 HOW THE EATING DISORDERS COMPARE

	Anorexia Nervosa	Bulimia Nervosa	Binge-Eating Disorder
Estimated prevalence*	Up to 0.5–1.0%	Up to 1.0%	Up to 2.0%
Male versus female incidence	5–10% male versus 90–95% female	5–10% male versus 90–95% female	Unknown
Typical age of onset	Early to middle adolescence	Late teens to early twenties	Any age, but not usually recognized until adulthood
Weight	Extremely thin and emaciated; <85% of desirable body weight	Near-desirable body weight, but often has weight fluctuations	Usually overweight or obese
Self-esteem	Low	Low	Low
Depression	Common	Common	Common
Substance abuse	Rare	Common	Rare
Rate of weight loss	Rapid	Repeatedly loses and gains weight or chronically diets without losing weight	Repeatedly loses and gains weight or chronically diets without losing weight
Past dieting	Yes	Yes	Yes

*Determining accurate statistics is difficult because physicians are not required to report eating disorders to a health agency and because people who have eating disorders tend to deny that they have a problem and are very secretive about their behaviors.

Source: J. Kirby, for the American Dietetic Association *Dieting for Dummies* (Foster City, CA: IDG Books Worldwide, 1998), pp. 67–68.

therapy, and behavior modification therapy.[70] Some therapists use more than one approach. Most treatment methods focus on identifying the societal and environmental pressures that triggered the eating disorder and on exposing the emotions masked by it. An interdisciplinary team made up of a psychologist, a social worker, a family therapy counselor, and a dietitian work with the family and the patient to reestablish emotional and nutritional health.[71] Length of treatment varies from two months to two to three years, depending on the patient's readiness for change and the type of treatment. American researcher Hilde Bruche, M.D., focuses on the problems of low self-esteem, guilt, anxiety, depression, and a sense of helplessness. In addition, family therapy focuses on changing patterns of family interaction.

Normal nutrition must be restored in the anorexic. After some progress is made in counseling, the dietitian can help the patient gain a new understanding of a healthful eating pattern, and clear up some earlier misconceptions about food and nutrition.[72]

Are there any early warning signs to watch for regarding eating disorders?

Table 8-13 shows the similarities and differences among the eating disorders. Families and friends can be alerted to several possible warning signs of eating disorders (see Table 8-14). Severe dieting often precedes the illness. Anorexics develop an exaggerated interest in food but at the same time deny their hunger and stop eating. A distorted body image makes them feel fat even as weight loss continues. The anorexic begins to have sleep problems, shows unusual devotion to schoolwork, and often undertakes a program of unrelenting exercise. Bulimics may binge and self-induce vomiting or use excessive amounts of laxatives. Reduced food intake usually causes sufficient weight loss to stop menstrual cycles in women. People with eating disorders were usually good children who did not indulge in rebellion. Not all anorexics and bulimics exhibit all symptoms. Early detection is vital. Use the Eating Attitudes Quiz on page 280 to see if your own eating attitudes and behaviors are within a normal range.

Is there anything that can be done to prevent eating disorders?

Yes. Consider the following suggestions for enabling people to remain healthy and for preventing the occurrence of eating disorders altogether.[73]

- Discourage restrictive dieting and meal skipping. Severe dieting may be a precursor of eating disorders, especially in adolescent girls. Model a lifestyle of healthy eating and exercise; never encourage unhealthy "quick-fix" weight loss.
- Promote fitness and a healthy body rather than thinness and numbers on a scale.
- Help teens understand and accept the normal physiological changes in body composition and weight that occur with puberty—changes which may be interpreted as "getting fat."
- Don't plant unhealthy seeds. Avoid assigning good/bad attributes to food. Don't offer food as a reward. Don't imply that the shape or size of your body has

MINIGLOSSARY

ANOREXIA NERVOSA literally "nervous lack of appetite," a disorder (usually seen in teenage girls) involving self-starvation to the extreme.
an = without
orexis = appetite

BINGE-EATING DISORDER an eating disorder characterized by uncontrolled chronic episodes of overeating (compulsive overeating) without other symptoms of eating disorders. Typically, the episodes of binge eating occur at least twice a week on average for a period of six months or more.

BULIMIA NERVOSA, BULIMAREXIA (byoo-LEE-me-uh, byoo-lee-ma-REX-ee-uh) binge eating (literally, "eating like an ox"). Combined with an intense fear of becoming fat and sometimes followed by self-induced vomiting or the taking of laxatives.
buli = ox

DISORDERED EATING eating food as an outlet for emotional stress rather than in response to internal physiological cues.

EATING DISORDER general term for several conditions (anorexia nervosa, bulimia nervosa, binge-eating disorder) that exhibit an excessive preoccupation with body weight, a fear of body fatness, and a distorted body image.

UNSPECIFIED EATING DISORDERS some people suffer from unspecified eating disorders; that is, they exhibit some but not all of the criteria for specific eating disorders.

anything to do with self-worth ("I'd be happy if I was thin").

- Be sensitive and careful about making weight-related comments or recommendations regarding weight. Even playful teasing about physical appearance may trigger a long-term struggle with disordered eating or eating disorders.
- Encourage normal expression of the basic emotions: joy, anger, fear, sadness, loneliness, guilt, and shame.

What can I do when I am worried that someone I know has an eating disorder?[74]
If you or others have observed behaviors in a friend, roommate, or family member that are suggestive of one of the eating disorders, consider the following points:

- Make a plan to approach the person in a private place when there is no immediate stress and time to talk.
- Present in a caring but straightforward way what you have observed and what your concerns are. Tell her or him that you are worried and want to help.
- Give the person time to talk and encourage them to verbalize their feelings. Listen carefully; accept what is said in a nonjudgmental manner.
- Do not argue about whether there is or is not a problem—power struggles are not helpful. Perhaps you can say, "I hear what you are saying and I hope you are right that this is not a problem. But I am still very worried about what I have seen and heard, and that is not going to go away."
- Provide information about resources for treatment. Offer to go with the person and wait while they have their first appointment with a counselor, physician, or dietitian.
- If you are concerned that the eating disorder is severe or life-threatening, enlist the help of a counseling center staff member, or a relative, friend, or roommate of the person *before* you intervene. Present a united and supportive front with others.
- If the person denies the problem, becomes angry, or refuses

TABLE 8-14 WARNING SIGNS OF AN EATING DISORDER

- Preoccupation with weight, food, calories, and dieting, to the extent that it consistently intrudes on conversations and interferes with other activities.
- Excessive exercise—despite weather, fatigue, illness, and injury
- Dramatic weight loss; arrested growth in children
- Anxiety about being fat which does not diminish with weight loss
- Reluctance to be weighed
- Inability to gain weight
- Difficulty eating in social situations
- Deceptive or secretive behavior
- Alternating periods of severely restrictive dieting and overeating
- Evidence of binge-eating or consumption of large quantities of food inconsistent with the person's weight
- Evidence of use of laxatives, diuretics, or agents that cause vomiting (emetics)
- Extreme concern about appearance as a defining feature of self-esteem (e.g., I am "thin and good," or "gross and bad")
- Social withdrawal or avoidance of activities because of weight and shape concerns
- Absence from school or work
- Depression
- Disrupted menstruation
- Constipation or diarrhea
- Paleness or lightheadedness not accounted for by other medical conditions
- Susceptibility to fractures

Sources: Adapted from Eating Disorders Awareness and Prevention, Inc. materials and D. Rosen, Dieting Disorder, *New England Journal of Medicine*, April 8, 1999.

treatment, understand that this is often part of the illness. Besides, they have a right to refuse treatment (unless their life is in danger). You may feel helpless, angry, and frustrated with them. You might say, "I know you can refuse to go for help, but that will not stop me from worrying about you or caring about you. I may bring this up again to you later, and maybe we can talk more about it then." Follow through on that—and on any other promise you make.

- Do not try to be a hero or a rescuer; you will probably be resented. Eating disorders are stubborn problems, and treatment is most effective when the person is truly ready for it. You may have planted a seed that helps them get ready.

For more information regarding anorexia nervosa, bulimia, and associated disorders, log on to the following Web sites:

- www.anred.com
  Anorexia Nervosa and Related Eating Disorders.
- www.edap.org
  Eating Disorders Awareness and Prevention, Inc.
- www.aabainc.org
  The American Anorexia Bulimia Association, Inc.
- www.gurze.com
  Gurze Books' Eating Disorders Resource Catalog.
- www.nedic.on.ca
  Canadian National Eating Disorder Information Centre.
- Other resources are listed in Appendix A.

EATING ATTITUDES QUIZ

Answer the following questions to evaluate your own eating attitudes and behaviors. Do they fall within a normal range? Use these responses:

A = always S = sometimes
U = usually R = rarely
O = often N = never

_________ 1. I am terrified about being overweight.
_________ 2. I avoid eating when I am hungry.
_________ 3. I find myself preoccupied with food.
_________ 4. I have gone on eating binges where I feel that I may not be able to stop.
_________ 5. I cut my food into very small pieces.
_________ 6. I am aware of the calorie content of the foods I eat.
_________ 7. I particularly avoid foods with a high carbohydrate content.
_________ 8. I feel that others would prefer that I ate more.
_________ 9. I vomit after I have eaten.
_________ 10. I feel extremely guilty after eating.
_________ 11. I am preoccupied with a desire to be thinner.
_________ 12. I think about burning up calories when I exercise.
_________ 13. Other people think I am too thin.
_________ 14. I am preoccupied with the thought of having fat on my body.
_________ 15. I take longer than other people to eat my meals.
_________ 16. I avoid foods with sugar in them.
_________ 17. I eat diet foods.
_________ 18. I feel that food controls my life.
_________ 19. I display self-control around food.
_________ 20. I feel that others pressure me to eat.
_________ 21. I give too much time and thought to food.
_________ 22. I feel uncomfortable after eating sweets.
_________ 23. I engage in dieting behavior.
_________ 24. I like my stomach to be empty.
_________ 25. I enjoy trying new rich foods.
_________ 26. I have the impulse to vomit after meals.

Scoring:

Never = 3
Rarely = 2
Sometimes = 1
Always, usually, and often = 0

A total score under 20 points may indicate abnormal eating behavior and the risk of having or developing an eating disorder. Share your results with a health professional for further evaluation.

Source: J. A. McSherry, Progress in the diagnosis of anorexia nervosa, *Journal of the Royal Society of Health* 106 (1986): 8–9; *Eating Attitudes Test* and scoring developed by Dr. P. Garfinkel.

Nutrition on the Web

nutrition.wadsworth.com	Go to the *Personal Nutrition* site to check for the latest updates to chapter topics or to access links to related Web sites.
www.eatright.org	Find useful links to weight-management resources.
www.nal.usda.gov/fnic/foodcomp	Look up calories in foods and beverages.
www.healthfinder.gov/searchoptions/topicsaz.htm	Search for information on obesity on this U.S. Government site.
www.shapeup.org	Find information on body composition and physical activity.
www.niddk.nih.gov/health/nutrit/nutrit.htm	Information, resources, and links for weight-management.
www.niddk.nih.gov/health/nutrit/win.htm	Resources for numerous weight-related topics.
www.healthyweight.net	The Healthy Weight Journal site.
www.navigator.tufts.edu	A search engine for locating reliable sites for weight control.
www.quackwatch.com	Evaluate weight-loss gimmicks, products, and promotions.
www.fda.gov/cder	Information on drugs for weight loss from the Center for Drug Evaluation and Research.
www.ncbi.nlm.nih.gov/PubMed/	Use this search engine to locate information on obesity and related topics
www.hugs.com	HUGS International, Inc. offers books, tapes, and training guides for teen and adult eating disorder programs and nondieting approach to weight management.

Check Yourself . . .

1. What is the primary factor that determines BMR (basal metabolic rate)?
2. What does the skinfold test tell us?
3. What are the possible risk factors associated with obesity?
4. What is the most likely cause of obesity in this country?
5. List the four components recommended for a successful weight-loss program.
6. What is the recommended rate for weekly weight loss (in pounds)?
7. List three examples of behavior modification techniques.
8. What is central obesity, and what are the risks associated with it?
9. What is the measure of a successful weight-loss program?
10. What are the characteristics of anorexia nervosa and bulimia?

Answers to Check Yourself questions are found in Appendix G.

Nutrition and Fitness

CONTENTS

Reasons to Exercise

The Four Components of Fitness

Getting Started on Lifetime Fitness

Physical Activity Scorecard

Energy for Exercise

Fuels for Exercise

Protein Needs for Fitness

Fluid Needs and Exercise

Vitamins and Minerals for Exercise

Nutrition Action: Athletes and Supplements—Help or Hype?

Fitness Quackery Scorecard

The Savvy Diner: Food for Fitness

Spotlight: Alcohol and Nutrition

Get health. No labor, effort, nor exercise that can gain it must be grudged.

R. W. Emerson
(1803–1882, American essayist and poet)

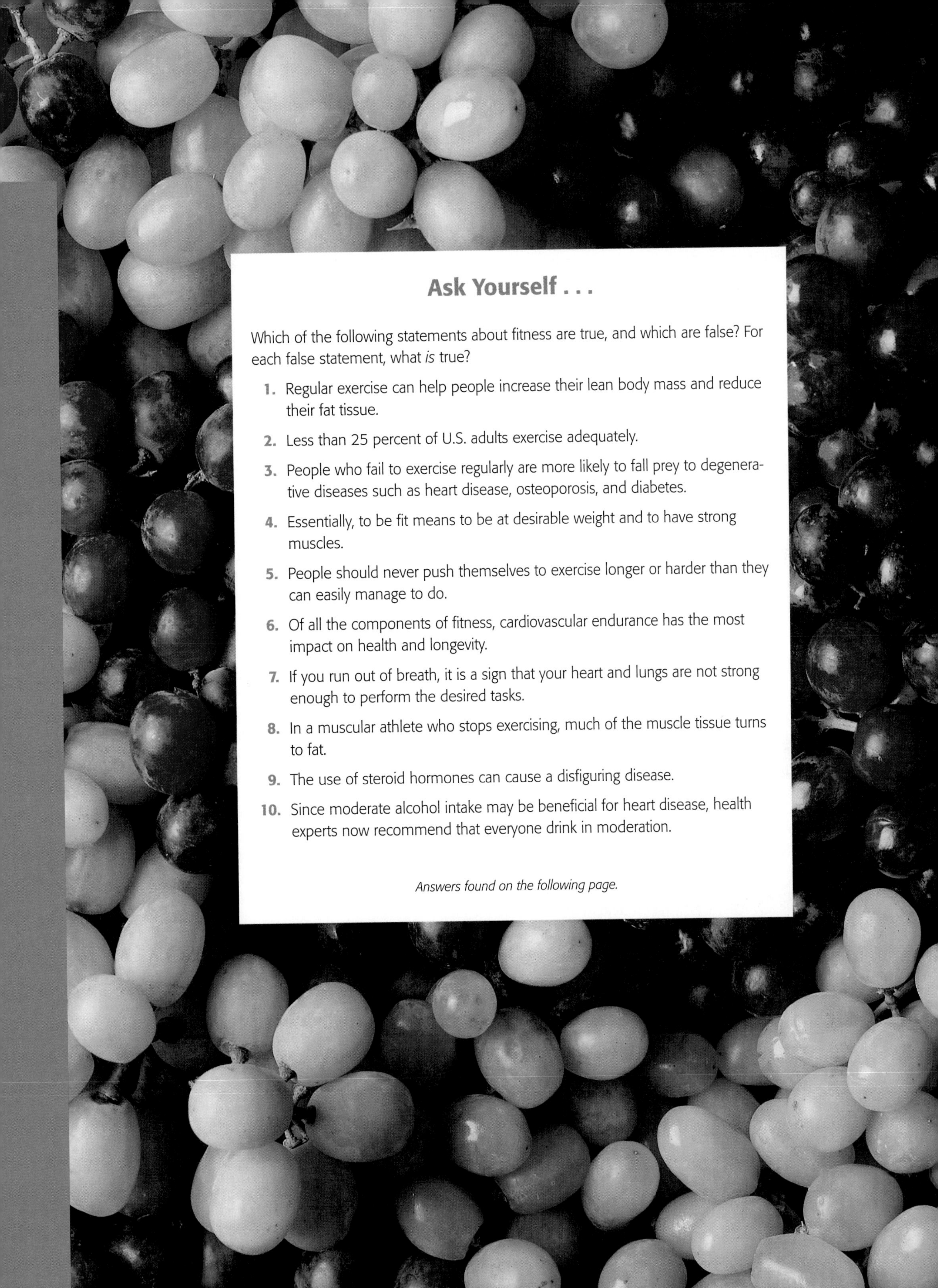

Ask Yourself . . .

Which of the following statements about fitness are true, and which are false? For each false statement, what *is* true?

1. Regular exercise can help people increase their lean body mass and reduce their fat tissue.
2. Less than 25 percent of U.S. adults exercise adequately.
3. People who fail to exercise regularly are more likely to fall prey to degenerative diseases such as heart disease, osteoporosis, and diabetes.
4. Essentially, to be fit means to be at desirable weight and to have strong muscles.
5. People should never push themselves to exercise longer or harder than they can easily manage to do.
6. Of all the components of fitness, cardiovascular endurance has the most impact on health and longevity.
7. If you run out of breath, it is a sign that your heart and lungs are not strong enough to perform the desired tasks.
8. In a muscular athlete who stops exercising, much of the muscle tissue turns to fat.
9. The use of steroid hormones can cause a disfiguring disease.
10. Since moderate alcohol intake may be beneficial for heart disease, health experts now recommend that everyone drink in moderation.

Answers found on the following page.

Fitness is in! Never before has our culture been so focused on fitness. That's good news, because people who are fit not only look good but also feel good. They can perform the activities they must and still have energy left over to do the things they want to do. In a general sense, then, fitness is a state of physical well-being that lets you lead a higher quality of life.

In addition, mounting evidence suggests that our bodies need regular, moderate exercise that gets our hearts beating and forces our muscles to work harder than they usually do to stay healthy. Physiologically speaking, overall fitness is a balance between different body systems. With respect to the joints, flexibility is important. With respect to muscles, strength and endurance are important. Endurance is also important to the heart and lungs (this type of endurance is called cardiovascular endurance, discussed later in the chapter).

This chapter illustrates the effects of nutrition on fitness—and vice versa, for a two-way relationship exists. Optimal nutrition contributes to athletic performance; and conversely, regular exercise contributes to a person's ability to use and store nutrients optimally. The two together are indispensable to a high quality of life; **fitness,** like good nutrition, is an essential component of good health.

FITNESS the body's ability to meet physical demands, composed of four components: flexibility, strength, muscle endurance, and cardiovascular endurance.

Benefits of physical activity include:

- Increased self-confidence
- Easier weight control
- More energy
- Less stress and anxiety
- Improved sleep
- Enhanced immunity
- Lowered risk of heart disease
- Lowered risk of certain cancers
- Stronger bones
- Lowered risk of diabetes
- Lowered risk of high blood pressure
- Increased quality of life
- Increased independence in life's later years

REASONS TO EXERCISE

The benefits of regular exercise make up an impressive list, which is growing longer as new discoveries are made. Through regular exercise, people can gain energy and confidence; increase their lean body tissue; reduce their body fatness; improve the health of their skin and their muscle tone; improve their sleeping habits; reduce their risk of heart disease, diabetes, and certain cancers; reduce their blood cholesterol levels; reduce their blood pressure; build bones that remain strong into old age; enhance their immunity; and live a more enjoyable, perhaps even longer, life.[1]

No one can promise that you will receive all of these benefits if you exercise, but almost everyone who exercises reaps at least some of them. Despite evidence of the benefits, only 22 percent of U.S. adults are adequately active (see Figure 9-1).[2] Perhaps this is because an unworked body requires some time to adapt—and creating a new habit takes time. However, once a routine of physical activity is established, the felt benefits far outweigh any inconveniences. People may begin a fitness program to trim fat or add muscle, but soon they are pleasantly surprised to find that they have more energy, feel less tense, sleep better, and feel healthier, and so they keep it up.

People can get in the habit of doing things that are good for them, just as they can get hooked on things that are not good for them, such as alcohol and other drugs, including nicotine. You can cultivate the habit of engaging in regular physical activity if you:

Choose an activity that you can do without a great deal of mental effort.

Choose an activity that doesn't depend on other people's presence—that is, one you can do alone when a partner is not available.

Believe that it has physical, mental, or spiritual value for you.

Enjoy the activity or some aspect of it (even just enjoying the knowledge that you do it).

Being fit is more than being free of disease; it is feeling full of vitality and enthusiasm for life.

Ask Yourself Answers: **1.** True. **2.** True. **3.** True. **4.** False. To be fit means not only to be at desirable weight and to have strong muscles but also to be flexible and, most importantly, to have muscular and cardiovascular endurance. **5.** False. The overload principle states that people *should* push themselves to exercise longer or harder than they can easily manage to do—although not, of course, to the point of strain. **6.** True. **7.** False. If you run out of breath, it is not a sign that your heart and lungs are weak but a sign that you are going into oxygen debt. **8.** False. Muscle tissue does not turn to fat, but in a muscular athlete who stops exercising, muscle tissue is lost and fat is gained. **9.** True. **10.** False. Although drinking in moderation may lower risk for heart disease among men over age 45 and women over age 55, moderate consumption provides little, if any, health benefit for younger people.

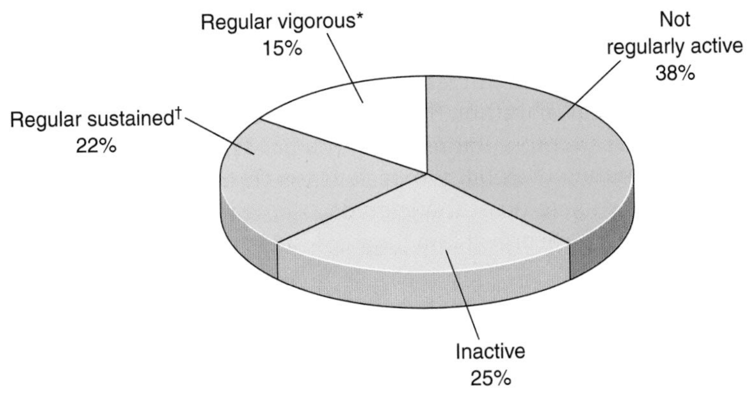

Source: Physical Activity and Health: A Report of the Surgeon General, 1996.

FIGURE 9-1
ADULTS AND PHYSICAL ACTIVITY

Physical activity need not be strenuous to achieve health benefits. Moderate amounts of physical activity are recommended for people of all ages. However, at present, only about 22% of adults engage regularly (5 times a week for at least 30 minutes) in sustained physical activity of any intensity during leisure time. About 25% of adults report no physical activity at all.

*Regular vigorous activity: three times a week for at least 20 minutes.
†Regular sustained activity (of any intensity): five times a week for at least 30 minutes.

Can experience progress through doing it.

Can do the activity without criticizing yourself.[3]

The notion that a certain minimum daily average amount of activity is indispensable to health is just now reaching public consciousness. A consequence of sedentary living is accelerated development of the diseases associated with a sedentary lifestyle—cardiovascular disease, some forms of cancer, stroke, diabetes, osteoporosis, and hypertension.[4] The American College of Sports Medicine recommends that we spend a total of at *least* 30 minutes on most days of the week engaged in any one of numerous forms of physical activities (see Table 9-1). *Moderate* physical activity means 30 minutes of brisk walking (2 miles in 30 minutes), dancing, bicycling—any activity that will expend approximately 150 to 200 calories a day. Activity durations will vary according to the intensity of the activity, as shown in the margin.[5] This 30 minutes can be accumulated in relatively brief sessions of activity—just mix and match your preferred activities in periods as short as 8 to 10 minutes that total 30 minutes by the end of the day. For example, walk your dog for 12 minutes in the morning, enjoy a 10-minute bike ride through the neighborhood after classes, and take an 8-minute walk after dinner. The new guidelines stress the value of moderate activity and suggest that the total *amount* of activity is more important than the manner in which it is carried out.

"Now all we need is fake exercise."

Examples of moderate physical activity:
- Gardening for 30–45 minutes
- Raking leaves for 30 minutes
- Walking 2 miles in 30 minutes (15 min/mile)
- Wheeling self in wheelchair for 30–40 minutes
- Pushing a stroller 1½ miles in 30 minutes
- Running 1½ miles in 15 minutes (10 min/mile)
- Stairwalking for 15 minutes
- Playing touch football for 30–45 minutes
- Basketball (shooting baskets) for 30 minutes
- Bicycling 5 miles in 30 minutes
- Dancing fast (social) for 30 minutes
- Swimming laps for 20 minutes
- Water aerobics for 30 minutes
- Jumping rope for 15 minutes

TABLE 9-1 EXERCISE GUIDELINES FOR DEVELOPING AND MAINTAINING FITNESS

- *Frequency:* 3 to 5 days per week; preferably most days*
- *Duration:* 30 minutes†
- *Intensity:* Any moderate physical activity that uses large muscle groups
- *Flexibility training:* Stretching of the major muscle groups two to three times per week for 10 to 20 minutes
- *Strength training:* At least two times per week

**Healthy People 2010* Objective: Increase the proportion of adults who engage regularly, preferably daily, in moderate physical activity for at least 30 minutes per day.
†Brief sessions of activity during the course of the day can be almost as beneficial as one 30-minute session. However, for additional cardiovascular benefits, 20 to 60 minutes of sustained aerobic activity (brisk walking, jogging, cycling) at least three times a week is necessary.

Source: Adapted from American College of Sports Medicine, Position stand: The recommended quantity and quality of exercise for developing and maintaining cardiorespiratory and muscular fitness, and flexibility in healthy adults, *Medicine and Science in Sports and Exercise* 30 (1998): 975–991.

People's bodies are shaped by what they do.

Fitness is not restricted to the seasoned athlete. With a basic understanding of the concept of total fitness and a personal commitment to a physically active lifestyle, anyone can become fit (see Figure 9-2). To be fit, you don't have to be able to finish the local marathon, nor do you have to develop the muscles of a Mr. Universe or Miss Olympia. Rather, what you need is a reasonable weight (refer to Chapter 8) and enough flexibility, muscle strength, muscle endurance, and cardiovascular endurance to meet the everyday demands that life places on you, plus some to spare. The Scorecard on page 291 helps you evaluate your own level of physical activity.

Physical Conditioning

OVERLOAD an extra physical demand placed on the body. A principle of training is that for a body system to improve, its workload must be increased by increments over time.

HYPERTROPHY an increase in size in response to use.

It seems obvious that people's bodies are shaped by what they do, but this fact is often overlooked. Physical conditioning refers to a planned program of exercise directed toward improving the function of a particular body system. Placing regular, physical demand on the body and forcing the body to do more than it usually does will cause it to adapt and function more efficiently. This is called **overload.** Muscles respond to the overload of exercise by gaining strength and ability to endure. Strength gains may not be visible in all cases. But in some, such as with male bodybuilders, muscles increase in strength and size, a response called **hypertrophy.** The converse is also true: muscles, if not called on to per-

If you are generally *inactive*, increase daily activities at the base of the Activity Pyramid by:

- taking the stairs
- hiding the TV remote control
- making extra trips around the house or yard
- stretching while standing in line
- walking whenever you can

If your activity routines are *sporadic* (active in summer, but not in the winter), become consistent with activity by increasing activity in the middle of the Pyramid by:

- finding activities you enjoy
- planning activities in your day
- setting realistic goals

If your physical activity is *consistent* (at least 5 times per week), think about the long term as you move throughout the Pyramid by:

- changing your routine if you start to get bored
- exploring new activities

Sit Sparingly: watch TV; play computer games.

2–3 Times per Week Enjoy Leisure Activities: golf; bowling; yardwork.

2–3 Times per Week Stretch/Strengthen: curl-ups; push-ups; weight training.

3–5 Times per Week Do Aerobic Activities: swimming; biking; long wallks.

3–5 Times per Week Enjoy Recreational Sports: basketball; tennis; racquetball.

Everyday Make Extra Steps in Your Day: take the stairs instead of the elevator; walk or ride your bike instead of getting a ride.

FIGURE 9-2
THE ACTIVITY PYRAMID

Physical activity helps you feel and perform at your best. The key to sticking with a regular exercise program is to choose activities that you enjoy. Be flexible. Remember that you don't always have to spend 30 consecutive minutes engaged in physical activity—a few minutes here and a few minutes there add to the benefits of moderate exercise. Keep in mind that *any* activity is good (for example, "Just do something"), but *more* is better. The goal is to have a *physically active* lifestyle.

Source: Adapted from J. Norstrom, Ten Tips to Healthy Eating and Physical Activity for You, © 1995 The American Dietetic Association National Center for Nutrition and Dietetics, International Food Information Council, and President's Council on Physical Fitness and Sports.

form, decrease in size, a response called **atrophy.** For example, an arm that is in a cast for six weeks will gradually become smaller in size because the arm muscles are not being used.

ATROPHY a decrease in size in response to disuse.

Runners often have well-developed strong legs; a tennis player may have one arm that is stronger than the other. Swimmers usually develop in a balanced way—all limbs, chest, back, and so forth are called on to perform and so develop uniformly. This doesn't mean that everyone should give up tennis and running and take up swimming; it only means that a variety of exercises will produce the most uniform overall fitness. This is why people are told to use different muscle groups in their exercise from day to day.

The overload principle applies equally to all aspects of fitness: flexibility, muscle strength, muscle endurance, and cardiovascular endurance. It also applies to the skeleton. To develop a strong, dense skeletal system, you must start by demanding that the bones bear slightly more stress (through weight-bearing exercises such as walking, running, weightlifting, or aerobic dance) than they are used to. The bones respond by depositing more minerals. Eventually, a maximum is reached. People can develop an amazing fitness level by progressively increasing overload.

You can apply overload in several ways. You can do the activity more often—that is, increase its **frequency;** you can do the activity more strenuously—that is, increase its **intensity;** or you can do it for longer periods of **time**—that is, increase its duration. All three strategies work well, and you can pick one or a combination, depending on your fitness goals.

For *fitness*, remember the *FIT* principle:

F—FREQUENCY number of exercise sessions per week; at least three to five sessions per week are recommended.

I—INTENSITY how hard you exercise (for example, the degree of exertion while exercising); it is recommended that you exercise at 60% to 85% of your maximum heart rate per minute—known as your target heart rate.

T—TIME duration or length of time that you exercise with your heart rate elevated into your target heart rate zone (the minimum amount is thought to be 20 to 30 minutes per session).

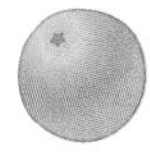

THE FOUR COMPONENTS OF FITNESS

Strength

Strength is the ability of the muscles to work against resistance: pulling yourself up and out of a swimming pool, carrying a backpack full of large books, or opening a jar of pickles. The purpose of strength training is to build well-toned muscles that let you accomplish daily activities at work and during recreation as well as to prevent injury. As muscles get stronger, individual fibers thicken and enlarge. Our ability to respond to strength training continues to a very old age.[6] The connective tissues making up muscles, tendons, and ligaments also strengthen and become more efficient at using energy. The benefit: Strong muscles, tendons, and ligaments play a key role in preventing injury. For example, strong quadriceps—muscles on the front of the thigh—stabilize your knee as you bike. Strong calf muscles and ankle ligaments decrease the risk of an ankle sprain when walking briskly or jogging.

STRENGTH the ability of muscles to work against resistance.

Many of today's mechanical aids invented to make life easier rob us of the opportunity to develop strength—for example, the strength we would gain from chopping firewood instead of turning up a thermostat. Today, we must put forth conscious effort to develop strength. The kind of equipment you use is largely a matter of availability and personal preference. If you belong to a health club, you probably have access to weight-training equipment. If not, you can use equipment you have at home, such as plastic gallon milk jugs filled with water or sand or store-bought ankle weights and dumbbells (see Figure 9-3). No matter what method you choose, safety in strength training is essential. Get proper instruction. Before you set up a strength-training program, consult an exercise physiologist or physical therapist or enroll in a college class.

The American College of Sports Medicine now recommends strength training as an integral part of a good overall exercise program.[7] To reap the benefits of strength training, consider the following tips:

- Aim for two to three sessions a week.
- Choose an intensity—the amount of weight you lift—that allows you to

FIGURE 9-3
EXERCISES FOR ADDED STRENGTH

Choose a chair that has a firm seat and back and no arms. The seat should be high enough so that when you sit in it your feet barely touch the floor. It should also be long enough to reach the back of your knees. And the back of the chair should be high enough for you to hold onto while standing behind it.

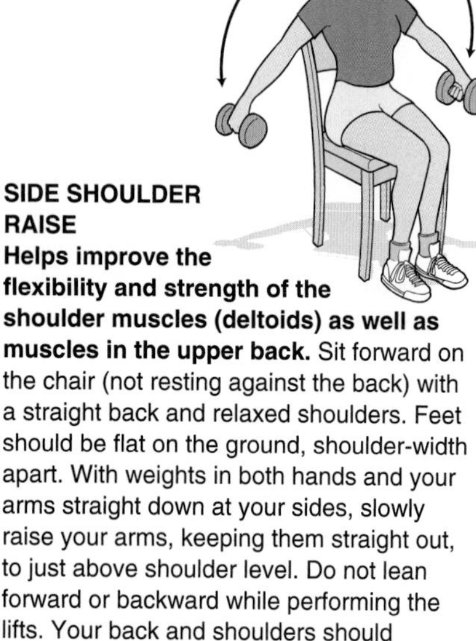

SIDE SHOULDER RAISE
Helps improve the flexibility and strength of the shoulder muscles (deltoids) as well as muscles in the upper back. Sit forward on the chair (not resting against the back) with a straight back and relaxed shoulders. Feet should be flat on the ground, shoulder-width apart. With weights in both hands and your arms straight down at your sides, slowly raise your arms, keeping them straight out, to just above shoulder level. Do not lean forward or backward while performing the lifts. Your back and shoulders should remain fixed (but not hunched) throughout.

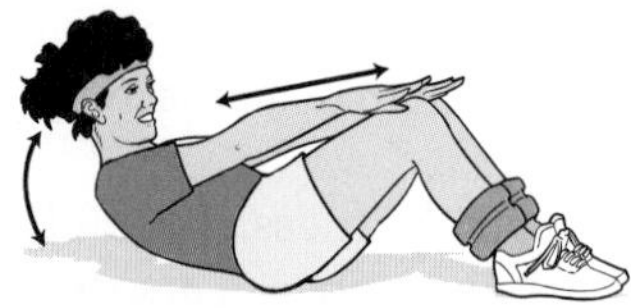

BENT-KNEE SITUP Works the abdominal muscles. Lie on your back with your knees bent and your feet flat on the floor, about 12 inches apart. Keep your palms down on your thighs. Gently tuck in your chin and lift your shoulders off the floor while sliding your palms up toward your knees. Stop about halfway up, at the point where it would be a struggle to continue. Hold for a moment, then go slowly back down. When it becomes too easy to do 8 to 12 repetitions at a time, increase the number of repetitions.

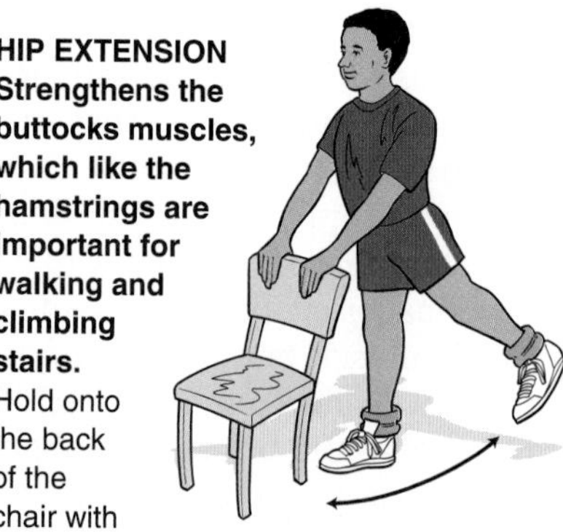

HIP EXTENSION
Strengthens the buttocks muscles, which like the hamstrings are important for walking and climbing stairs. Hold onto the back of the chair with weights around the ankles and bend forward about 45 degrees at the waist. Lock the knee and lift one leg straight out behind you as high as possible without moving your upper body. (The movement should be smooth and controlled.) Slowly lower your leg to the starting position. Alternate with the other leg until you have completed 8 to 12 repetitions on *both* sides.

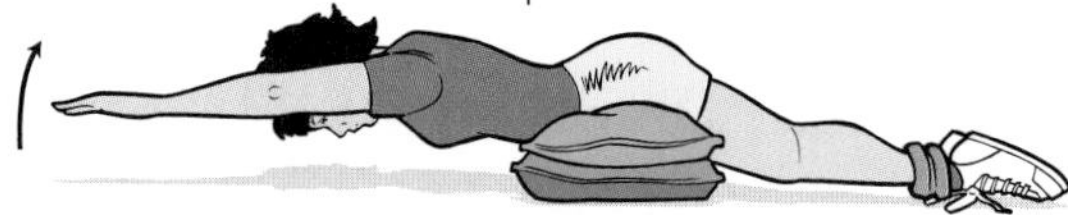

BACK EXTENSION Works the lower back muscles. Lie on your stomach with two pillows under your pelvis (hips). Leave ankle weights on to anchor your feet to the ground. With your arms straight out in front of your head, slowly lift your back 4 to 5 inches off the floor (straight, not arching). Hold, then slowly lower your back. When it becomes too easy to do 8 to 12 repetitions at a time, increase the number of repetitions.

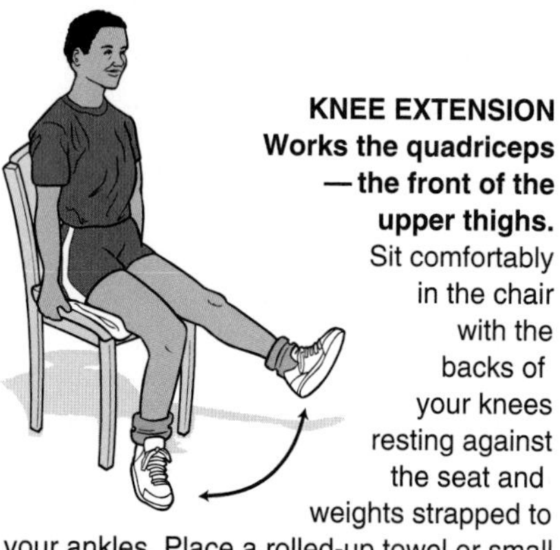

KNEE EXTENSION
Works the quadriceps —the front of the upper thighs. Sit comfortably in the chair with the backs of your knees resting against the seat and weights strapped to your ankles. Place a rolled-up towel or small cushion under your knees to lift them slightly so that just the balls of your feet touch the floor. Extend one leg out in front of you until your leg is as straight as possible. Do not grip the chair as you lift, but you may gently hold onto the seat of the chair to help stabilize you. Slowly lower your leg until your foot is resting on the floor. Alternate legs from lift to lift. One set equals 8 to 12 repetitions on *both* sides.

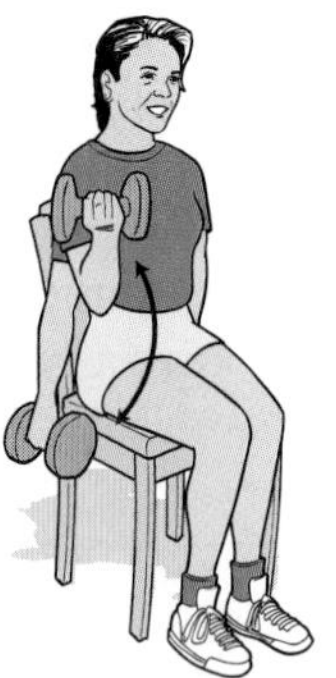

BICEPS CURL
Strengthens the biceps muscles at the front of the upper arm, needed for carrying and lifting. Sit straight in the chair with feet shoulder-width apart. Place your lifting arm straight down to the side of the chair with your palm facing your body. Bring the hand with the weight three-quarters of the way up toward your shoulder (bend from the elbow). While lifting, slowly rotate your hand around so that at the end of your lift, your palm is facing your shoulder. Slowly lower the hand. After lifting 8 to 12 times with one arm, proceed to the other. It takes 8 to12 repetitions on *both* sides to equal one set.

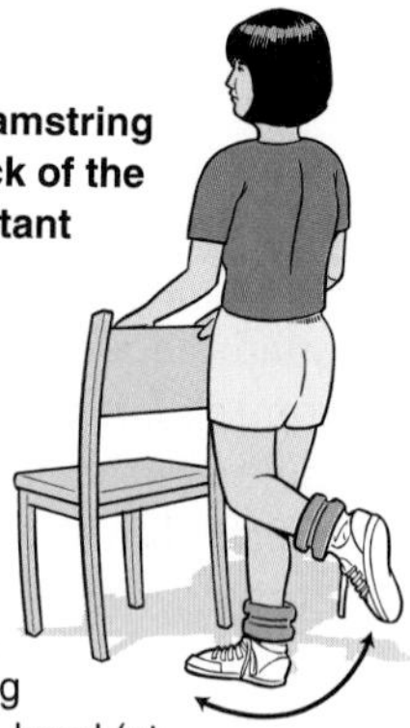

KNEE FLEXION
Strengthens the hamstring muscles at the back of the upper thigh, important for walking and climbing stairs. Stand behind the chair with weights strapped to your ankles. Place one foot slightly farther back than the other. Then, without moving your upper leg at all, bend (at the knee) the leg that is slightly back so that your heel comes as close to the back of your thigh as possible. Slowly lower your leg to the starting position. Do the same with the other leg. Eight to 12 repetitions on *both* sides make a set.

Source: Reprinted with permission from *Tufts University Diet & Nutrition Letter* 13(1), March 1995, p. 5.

perform 8 to 12 repetitions; adjust the amount of weight upwards as your strength improves so that you are still able to perform in the 8 to 12 repetitions range.

- Perform each repetition slowly—allowing six to nine seconds per repetition.
- Remember to breathe! Inhale before you lift, exhale as you lift, and breathe in slowly as you lower the weight to the starting position; pause and breathe fully between repetitions.
- Warm up before lifting weights and cool down afterwards—you can begin and end each session with several repetitions without using weights or by doing some stretching exercises.
- Allow your muscles time to recover between sessions; do not strength train the same muscle groups on consecutive days.

Because many women mistakenly think they will develop enlarged muscles, which they view as undesirable and unattractive, they don't participate in strength-training programs. Testosterone—a male hormone—is responsible for muscle overdevelopment. Women only have 10 percent of the amount of testosterone that men do. Because of their lower testosterone levels, it is unlikely that women would develop bulging or overdeveloped muscles as a result of two or three weekly sessions of weight training. What will occur, though, is improved muscle strength, shape, and tone.[8] Strength training also helps with weight loss by increasing lean muscle mass and thus increasing a person's resting metabolic rate.[9]

Flexibility

Keeping your muscles and joints pliable is critical for developing a fit body. A flexible body can move as it was designed to move and will bend rather than tear or break in response to sudden stress. **Flexibility** (range of motion) depends upon the condition and interrelationships of bones, ligaments, muscles, and tendons.

Stretching exercises improve flexibility by increasing muscle and tendon elasticity and length. Stretching should be done slowly—called **static stretches.** When you feel a slight strain in the muscle, hold the position for 10 to 30 seconds. Bouncy, rapid stretches—known as **ballistic stretches**—can cause minute tears in the muscle and also set up a reaction in the muscle that makes it resist the stretch. Avoid painful stretches. They are clearly excessive.

Flexibility tends to decrease as you age but improves in response to stretching, and it can be maintained in most people by doing frequent stretching exercises. While regular exercise will increase muscle tone and strength, some types of exercise can also make muscles stiffer—making them more prone to strain and injury. For example, jogging and dancing can reduce flexibility of the lower back, front of thighs, and calves. Joggers and dancers should emphasize flexibility exercises for these areas. When beginning an exercise program, stretching is especially critical for previously sedentary people whose muscles are short and tight.

Stretching routines are commonly done as part of a warm-up routine before exercise. Low-intensity preliminary exercise allows your heart—also a muscle—to slowly accelerate and make adjustments in bloodflow and oxygen supply, preparing for the work it is about to perform. Using calisthenics as a warm-up, such as walking, marching in place, or doing some other moderate rhythmic activity, prepares your heart muscle for action. After your light warm-up, stretch the muscles that you will be using in your main exercise activity. Waiting to stretch until after your warm-up allows blood to move into the muscles, making them easier to stretch. Doing stretches after your exercise session gives your heart a chance to gradually slow its pace. It also allows you to lengthen those

FLEXIBILITY the ability to bend or extend without injury; flexibility depends on the elasticity of the muscles, tendons, and ligaments and on the condition of the joints.

STATIC STRETCHES stretches that lengthen tissues without injury; characterized by long-lasting, painless, pleasurable stretches.

BALLISTIC STRETCHES stretches characterized by short, choppy, sometimes painful movements that often pull connective tissues beyond their elastic limits.

ENDURANCE the ability to sustain an effort for a long time. One type, **muscle endurance,** is the ability of a muscle to contract repeatedly within a given time without becoming exhausted. Another type, **cardiovascular endurance,** is the ability of the cardiovascular system to sustain effort over a period of time.

muscles that have become tight and tense from the exercise. You can make greater gains in flexibility by stretching after your workout because muscles are warm and easier to stretch.

Muscle Endurance

Muscle endurance, the third component of fitness, is the power of a muscle to keep on going for long periods. Your muscle endurance influences your performance in the last set of a tennis match, your swing on the eighteenth hole of a golf game, or your ability to pedal during the last 10 miles of a 100-mile bike tour. Endurance of certain muscles can be tested by the number of situps or pushups you can accomplish in a certain period of time. But remember, these tests evaluate only the abdominal and upper arm muscles.

To develop cardiovascular fitness, choose an activity that:

- Is steady and constant.
- Uses large muscle groups, such as the arms, legs, and buttocks. If you move 50 percent of your muscle mass, the activity is aerobic.
- Is uninterrupted, lasts for more than 20 minutes, and is done at least three times a week.

Cardiovascular Endurance

Another realm in which endurance is important is the length of time that you can keep going with an elevated heart rate—that is, how long your heart can endure a given demand. This kind of endurance is **cardiovascular endurance.** The heart is a muscle, and it, like your other muscles, can respond to repeated demands by becoming larger and stronger.

Exercises that promote cardiovascular endurance are the best for making short-term fitness gains and long-term health improvements as well as for weight control. The best exercises to develop cardiovascular endurance are those that repetitively use large muscle groups—arms and legs—and that last for a continuous 20 to 60 minutes. Examples include brisk walking, aerobic dance, running, cycling, cross-country skiing, and rowing. The American College of Sports Medicine recommends that people participate in cardiovascular conditioning activities at least three times a week for a continuous 20 to 60 minutes.[10]

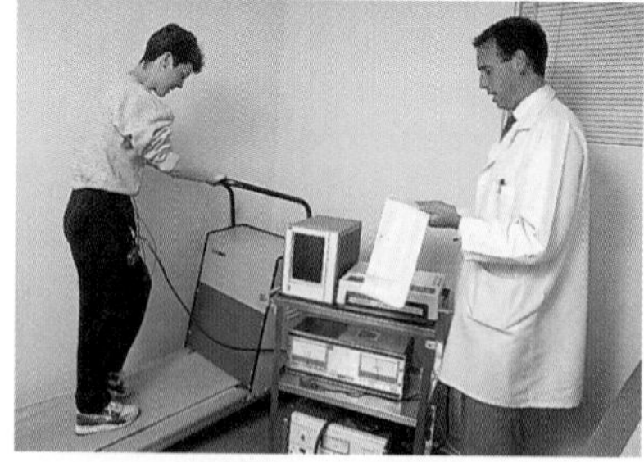

The exercise stress test measures heart function during exercise.

EXERCISE STRESS TEST a test that monitors heart function during exercise to detect abnormalities that may not show up under ordinary conditions; exercise physiologists and trained physicians or health-care professionals can administer the test.

GETTING STARTED ON LIFETIME FITNESS

For total fitness, an exercise program that incorporates strength training, stretching, and cardiovascular endurance activity is best. The more active you are, the more fit you are likely to be. Tests that measure your ability to perform various physical activities reveal more, and if you were to have an **exercise stress test** or other such measurement taken by a professional, you would obtain an accurate estimate of your fitness.

In proceeding with a fitness program, it is important to keep your own goals in mind, since they can carry you through periods of discouragement and help you to choose the activities that best meet your needs. Keep in mind that fitness builds slowly, and so activity should increase gradually. Don't rush things by taking on too much, too soon. View your exercise time as a lifelong commitment.

PHYSICAL ACTIVITY SCORECARD

How physically active are you? For each question answered yes, give yourself the number of points indicated. Then total your points to determine your score.

A. VIGOROUS EXERCISE ROUTINES

1. I participate in active recreational sports such as tennis or racquetball for an hour or more:
   a. about once a week (*2 points*)
   b. about twice a week (*4 points*)
   c. three times a week (*6 points*)
   d. four times a week (*8 points*)
   e. not at all (*0 points*)
2. I participate in vigorous fitness activities like aerobic dancing, roller blading, jogging, or swimming (at least 20 minutes each session):
   a. about once a week (*3 points*)
   b. about twice a week (*6 points*)
   c. three times a week (*9 points*)
   d. four times a week (*12 points*)
   e. not at all (*0 points*)

B. OTHER EXERCISE ROUTINES

3. At least two times a week, I work out with weights for at least ten minutes:
   a. two sessions a week (*2 points*)
   b. three sessions a week (*3 points*)
   c. four or more sessions a week (*4 points*)
   d. not at all (*0 points*)
4. At least two times a week, I perform floor workouts (sit-ups, push-ups) for at least ten minutes:
   a. two sessions a week (*2 points*)
   b. three sessions a week (*3 points*)
   c. four or more sessions a week (*4 points*)
   d. not at all (*0 points*)
5. At least two times a week, I participate in yoga or perform stretching exercises for at least ten minutes:
   a. two sessions a week (*2 points*)
   b. three sessions a week (*3 points*)
   c. four or more sessions a week (*4 points*)
   d. not at all (*0 points*)

C. OCCUPATION AND DAILY ACTIVITIES

6. I walk to and from school, work, and shopping (½ mile or more each way), two or three times a week or more. (*1 point*)
7. I climb stairs rather than using elevators or escalators, every other day or more. (*1 point*)
8. My school, job, or household routine involves physical activity that fits the following description:
   a. Most of my day is spent in desk work or light physical activity. (*0 points*)
   b. Most of my day is spent in farm activities, moderate physical activity, brisk walking, or comparable activities. (*4 points*)
   c. My typical day includes several hours of heavy physical activity (shoveling, lifting, etc.). (*2 points per day*)

D. LEISURE ACTIVITIES

9. I do several hours of gardening, lawn work, or similar hobby work each week. (*1 point*)
10. At least once a week I dance vigorously (folk or line dancing) for an hour or more. (*1 point*)
11. In season, I play 9 to 18 holes of golf at least once a week, and I do not use a power cart. (*2 points*)
12. I walk for exercise or recreation:
    a. one to two hours a week (*1 point*)
    b. three to four hours a week (*2 points*)
    c. five hours or more a week (*3 points*)
    d. not at all (*0 points*)
13. In *addition* to the above, I engage in other forms of physical activity:
    a. one to two hours a week (*1 point*)
    b. three to four hours a week (*2 points*)
    c. five hours or more a week (*3 points*)

SCORING

Record your point scores here.

Category	Score
A. Vigorous Exercise Routines	______
B. Other Exercise Routines	______
C. Occupation and Daily Activities	______
D. Leisure Activities	______
Total:	______

Evaluation of total score (circle one).

- Inactive (0 to 5 points).
- Moderately active (6 to 11 points).
- Active (12 to 20 points).
- Very active (21 points or over).

If your score categorized you as inactive or only moderately active, think of activities that you could realistically engage in on a regular basis to raise your score to "active" (12 points).

Source: Adapted with permission of Russell Pate (University of South Carolina, Human Performance Laboratory).

An "apparently healthy" individual has no more than one of the following risk factors:

- Sedentary lifestyle
- Age (Men > 45 years of age; women > 55 years of age)
- Family history of heart disease
- Cigarette smoking
- High blood pressure
- High blood cholesterol (> 200 mg/dL)
- Diabetes

A few cautions on getting started: If you are an apparently healthy male older than 45 years of age or an apparently healthy female older than 55 years of age, the American College of Sports Medicine recommends that you have a medical examination and diagnostic exercise test before you start a *vigorous* exercise program. Beginning a *moderate* program, such as walking, however, would not require the physician's exam.[11]

For most people, physical activity should not pose any problem or hazard. However, medical advice concerning suitable type of activity is needed for anyone with two or more of the risk factors shown in the margin, or anyone diagnosed with cardiac or other known diseases.

Aerobic exercises:

Aerobic dancing	Roller skating
Bench stepping	Rope jumping
Bicycling	Rowing
Cross-country skiing	Running
Fast walking	Speed skating
Jogging	Stair climbing
Mini-trampoline jumping	Swimming
Roller blading	Treadmill walking or running

AEROBIC requiring oxygen.

ANAEROBIC not requiring oxygen.

ENERGY FOR EXERCISE

Your body runs on water, oxygen, and food—primarily carbohydrate and fat. The chemical reactions that use these substances to make energy are called metabolism. Your body has two interrelated energy-producing systems—one dependent on oxygen—**aerobic** metabolism—and the other able to function without oxygen—**anaerobic** metabolism. An understanding of how the two systems work is important because it explains why you choose certain exercises over others to strengthen your heart, why you eat what you do, and what factors influence your performance during sporting events.

Aerobic and Anaerobic Metabolism

At rest, your muscles burn mostly fat and some carbohydrate for energy. During exercise, though, the amounts the muscles use depend on an interplay between fuel availability and oxygen availability. To an exercising muscle, oxygen is everything. With ample oxygen, muscles can extract all available energy from carbohydrate and fat by means of aerobic metabolism. During moderate exercise, your lungs and circulatory system have no trouble keeping up with the muscles' need for oxygen. You breathe deeply and easily, and your heart beats steadily—the exercise is aerobic. But the heart and lungs can supply only so much oxygen so fast.

When the muscles' exertion is great enough that their energy demand outstrips their oxygen supply, they must also rely on anaerobic metabolism to make energy. Since the anaerobic metabolic pathway can burn only carbohydrate for fuel, it draws heavily on your limited body stores of carbohydrate. Nevertheless, this system does provide an immediate source of energy without relying on oxygen. Because of this system, you can dash out of the way of an oncoming car or sprint ahead of your competitor at the finish line. Unfortunately, this energy-yielding system is extremely inefficient. Only 5 percent of carbohydrate's energy-producing potential is harnessed by this pathway.[12]

Make exercise a habit: Choose an activity you enjoy.

Because the anaerobic metabolic pathway only partially burns your carbohydrate, it litters your muscle with lactic acid—partly broken down portions of glucose. The buildup of lactic acid causes burning pain in the muscles and can lead to muscle exhaustion within seconds if it is not drained away. A strategy for dealing with lactic acid buildup is to relax the muscles at every opportunity so that the circulating blood can carry it away and bring oxygen to support aerobic metabolism. Fortunately, lactic acid is not a waste product. When oxygen reaches your muscles, lactic acid is ushered to your liver, which converts it back to glucose.

Neither the aerobic nor the anaerobic metabolic pathway functions exclusively to supply energy to your body. The two work together, complementing and supporting each other. Keep in mind, however, that carbohydrate is absolutely essential for exercise. Without it, your muscles can't perform. You want to exercise aerobically because muscles burn fat and extract energy from carbohydrate more efficiently in the presence of oxygen, thereby conserving your body's limited store of carbohydrate. Thus, you want to exercise at an intensity that allows your heart and lungs to keep pace with your working muscles' oxygen needs.

Sports:

Sports add fun to your exercise routine, but because of frequent starts and stops, they don't allow for a continuous 20-minute bout of aerobic exercise. Aerobic training, however, will help you perform these sports:

Baseball	Ice hockey
Basketball	Lacrosse
Downhill skiing	Racquetball
Fencing	Soccer
Football	Squash
Frisbee	Softball
Golf	Tennis
Handball	Volleyball
Horseback riding	

Aerobic Exercise—Exercise for the Heart

To meet your body's increased oxygen needs during aerobic exercise, your heart must pump oxygen-rich blood to muscles at a faster pace than normal. This increased demand on the heart makes the heart stronger and increases its endurance. In addition, aerobic exercise improves the endurance of the lungs and the muscles along the arteries and in the walls of the digestive tract and, of course, the muscles directly involved in the activity. These all-over improvements are called **cardiovascular conditioning** or the **training effect.** In cardiovascular conditioning, the total blood volume increases so that the blood can carry more oxygen. The heart muscle becomes stronger and larger. Since each beat of the heart pumps more blood, it needs to pump less often. The muscles that work the lungs gain strength and endurance, and breathing becomes more efficient. Circulation through the body's arteries and veins improves. Blood moves easily, and the blood pressure falls. Muscles throughout the body become firmer. Figure 9-4 shows the major relationships between the heart, circulatory system, and lungs.

CARDIOVASCULAR CONDITIONING or **TRAINING EFFECT** the effect of regular exercise on the cardiovascular system—including improvements in heart, lung, and muscle function and increased blood volume.

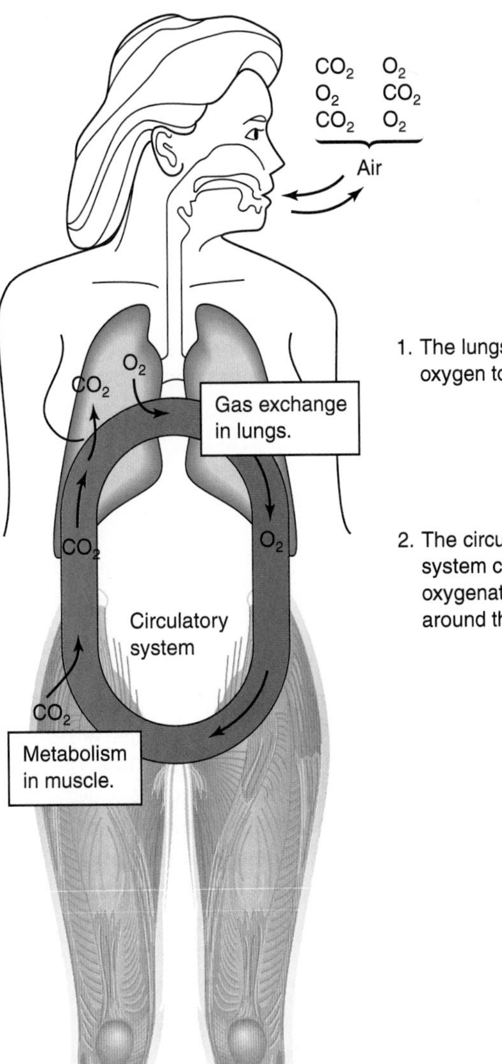
CO_2 O_2
O_2 CO_2
CO_2 O_2
Air
1. The lungs deliver oxygen to the blood.
2. The circulatory system carries the oxygenated blood around the body.
3. The muscles and other tissues remove oxygen from the blood and release carbon dioxide into it.
4. The blood carries the carbon dioxide back to the lungs.
5. The lungs remove carbon dioxide, making the blood ready to be reloaded with oxygen.
Gas exchange in lungs.
Circulatory system
Metabolism in muscle.
O_2
CO_2

FIGURE 9-4
DELIVERY OF OXYGEN BY THE HEART AND LUNGS TO THE MUSCLES

The more fit a muscle is, the more oxygen it draws from the blood. That oxygen is drawn from the lungs, so the person with more fit muscles extracts more oxygen from the inhaled air than a person with less fit muscles. The cardiovascular system responds to the demand for oxygen by building up its capacity to deliver oxygen. Researchers can measure cardiovascular fitness by measuring the amount of oxygen a person consumes per minute while working out, a measure called VO_2 max.

To take your pulse and monitor your heart rate during exercise, lightly press your middle and index fingers on the carotid artery, next to your trachea (windpipe) as shown here. Count your pulse for ten seconds, and then multiply by six to give beats per minute. Another convenient place to take your pulse is the radial artery (on the thumb side of the wrist).

TARGET HEART RATE the heartbeat rate that will achieve a cardiovascular conditioning effect for a given person—fast enough to push the heart but not so fast as to strain it.

To make these gains in cardiovascular conditioning, you must work up to a point where you can continuously exercise aerobically for 20 minutes or longer. This means you must elevate your heart rate (pulse). This heart rate—called your **target heart rate**—must be considerably faster than the resting rate to push (overload) the heart but not so fast as to strain it.

An informal pulse check can give you some indication of how conditioned your heart is to start with. As a rule of thumb, the average resting pulse rate for adults is around 70 beats per minute, but the rate can be higher or lower. Active people can have resting pulse rates of 50 or even lower.

For cardiovascular conditioning, your target heart rate can be calculated from your age. The older you are, the lower your maximum heart rate. As your heart gets stronger, more intense exercise will be required to reach the same target rate. For example, at first, walking at a pace of three miles per hour may cause you to reach your target heart rate. After six to eight weeks of walking at this pace, you may notice you no longer reach your target heart rate. That's because your heart is stronger. It now needs more of a challenge to beat faster. Increasing the intensity of your workout by walking faster can provide this challenge.

To calculate your target heart rate range, take the following steps:

1. *Estimate your maximum heart rate (MHR).* Subtract your age from 220. This provides an estimate of the absolute maximum heart rate possible for a person your age. You should never exercise at this rate, of course.
2. *Determine your target heart rate range.* Multiply your MHR by 60 percent and 85 percent to find your upper and lower limits.

When you can work out at your target heart rate for 20 to 30 minutes, you know that you have arrived at your cardiovascular fitness goal.

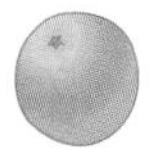

FUELS FOR EXERCISE

Your energy-producing pathways require oxygen and the two muscle fuels: glucose and fatty acids. As Figure 9-4 showed, the oxygen comes from the lungs, which pass it to the blood, which carries it to the muscles. Your muscles, and to some extent your liver, supply carbohydrate to your muscles from their carbohydrate supply (see Figure 9-5). The fatty acids come mostly from fat inside the muscles but partly from fat that is released from the body's fat stores, and the blood delivers these fatty acids to the muscles.

Glucose Use During Exercise

Glucose comes from carbohydrate-rich foods—breads, pasta, rice, legumes, fruits, vegetables, milk, and yogurt. Your body stores glucose in your liver and muscles as glycogen, a long chain of glucose molecules linked together.

During exercise, the body supplies glucose to the muscles from the stores of glycogen in the liver and in the muscles themselves. The longer the exercise lasts or the more intense it is, the more glucose a person uses. Recall that exercise done at an intensity that outstrips the ability of the heart and lungs to supply oxygen to working muscles relies primarily on glucose for fuel. Thus, activities such as sprinting quickly deplete the body's stores of glycogen. Other activities, such as jogging or brisk walking, where the body can meet the muscles' oxygen demands, are more conservative of glycogen. Nonetheless, joggers and walkers still use it, and eventually they can run out of it.

When a person begins exercising, for the first 20 minutes or so, about one-fifth of the body's total glycogen store is rapidly used.[13] If exercise continues beyond 20 minutes, glycogen use slows down (see Figure 9-6 on page 296). To conserve the remaining glycogen supply, the body begins to rely more on fat for fuel. At some point, if exercise continues long enough, glycogen will run out almost

FIGURE 9-5
THE USE OF GLYCOGEN AND BODY FAT FOR ENERGY DURING EXERCISE

Training can increase the amount of glycogen a muscle can conserve during exercise. The more fit a muscle is, the more fat it can burn for energy when oxygen is present—sparing the valuable glycogen.

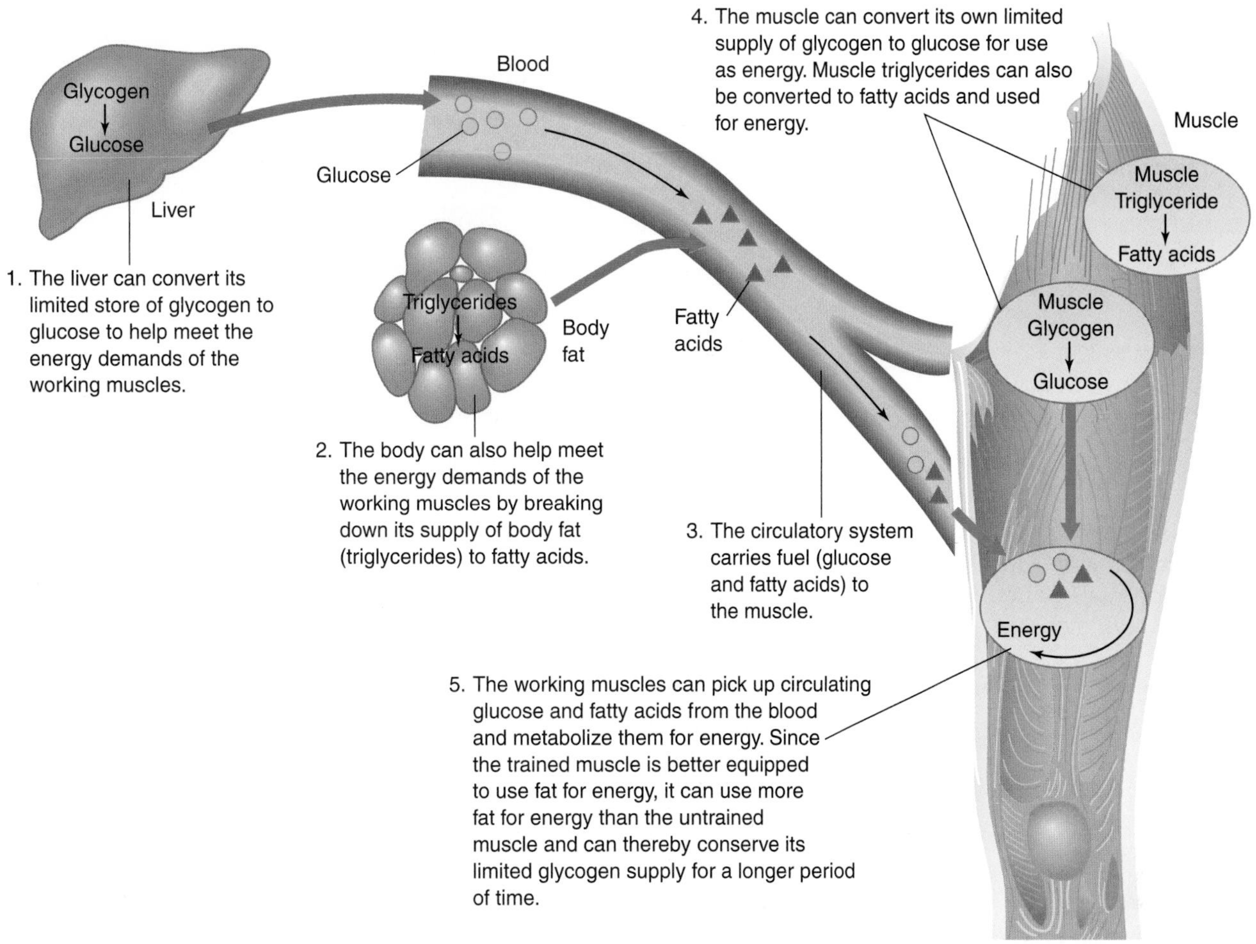

Glycogen
↓
Glucose
Liver

1. The liver can convert its limited store of glycogen to glucose to help meet the energy demands of the working muscles.

Blood
Glucose

Triglycerides
↓
Fatty acids
Body fat

2. The body can also help meet the energy demands of the working muscles by breaking down its supply of body fat (triglycerides) to fatty acids.

Fatty acids

3. The circulatory system carries fuel (glucose and fatty acids) to the muscle.

4. The muscle can convert its own limited supply of glycogen to glucose for use as energy. Muscle triglycerides can also be converted to fatty acids and used for energy.

Muscle
Muscle Triglyceride
↓
Fatty acids
Muscle Glycogen
↓
Glucose
Energy

5. The working muscles can pick up circulating glucose and fatty acids from the blood and metabolize them for energy. Since the trained muscle is better equipped to use fat for energy, it can use more fat for energy than the untrained muscle and can thereby conserve its limited glycogen supply for a longer period of time.

completely. People who run out of muscle glycogen during an event (for example, before the finish line in a marathon) "hit the wall"—they have to slow down their pace since muscle glycogen is no longer available to fuel their activity. Exercise can continue for a short time after that, only because the liver scrambles to produce the minimum amount of glucose needed to briefly forestall body shutdown. When blood sugars dip too low, the nervous system function comes almost to a halt, making exercise difficult, if not impossible, although there is still plenty of fat left to burn.

Another factor that influences how much glycogen a person uses during exercise is how well trained the person is to do the particular exercise. When first attempting an activity, a person uses more glucose than a trained athlete. This is because the muscles can quickly and easily extract energy from glucose. Extracting energy from fat takes longer and requires that the muscle cells contain abundant fat-burning enzymes. Untrained muscle cells must rely heavily on the quick energy source, glucose; with training, the muscles adapt, packing their cells with more fat-burning enzymes. As a result, trained muscles use more fat and conserve their glucose.

The amount of glycogen present in the muscles before exercise also influences glycogen use. Following the diet prescription in Figure 9-7 will provide athletes'

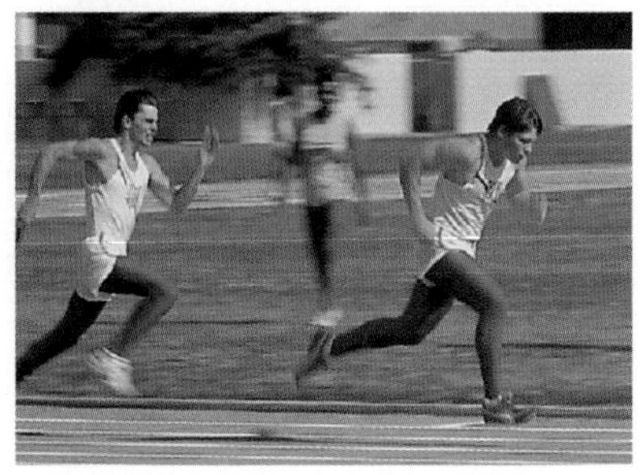

Anaerobic exercise. Glucose is the principal source of energy for activities of high intensity.

FIGURE 9-6
FUEL USE AND DURATION/INTENSITY OF EXERCISE

For most people, fat isn't used much as a fuel for exercise until you've been working out aerobically for at least 20 minutes, and it is not used as a primary fuel until after two hours.

*The more moderate the intensity of activity (brisk walking, jogging, aerobic dancing), and the longer the duration, the greater the use of fat for fuel.
†The higher the intensity of activty (sprinting, hurdles, rowing), the greater the use of carbohydrate for fuel.

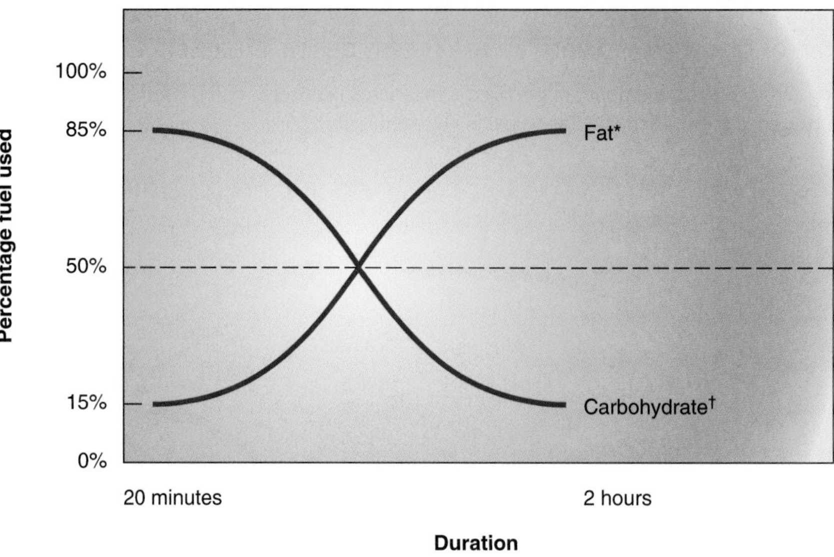

muscles with enough glycogen to support exercise. For the casual exerciser who participates in low- to moderate-intensity activities (walking, bicycling, dancing), the activity is rarely sufficient to totally deplete glycogen stores. Carbohydrate loading—a practice endurance athletes follow to trick their muscles into storing extra glycogen—may not be beneficial for people who exercise less than 90 minutes per workout at a low intensity, although competitive athletes who exercise at a high intensity for more than 90 minutes at a time may benefit from carbohydrate loading. Muscles typically have enough glycogen to fuel one-and-a-half to two-hour bouts of activity.

An athlete who follows the glycogen-loading technique in preparation for an upcoming event will first exercise intensely without restricting carbohydrates, then gradually cut back on exercise the week before the competition, rest completely the day before, and eat a very-high-carbohydrate diet.

Endurance athletes who follow this plan can keep going longer than their competitors without ill effects.[14] In a hot climate, extra glycogen offers an additional advantage. As glycogen breaks down, it releases water, which helps to meet the athlete's fluid needs.

For people who exercise less than 90 minutes per session, such extremes in glycogen storage are unnecessary. Glycogen isn't likely to run out in short exercise sessions, no matter how intense. All that is necessary to provide consistently full glycogen stores for workouts is to eat a balanced diet based on the Food Guide Pyramid and allow muscle groups to recover fully before working them again. Full recovery of glycogen stores takes from 24 to 48 hours. This doesn't mean you can work out only every other day. It means you should vary your exercise routine from day to day to work different groups of muscles on different days.

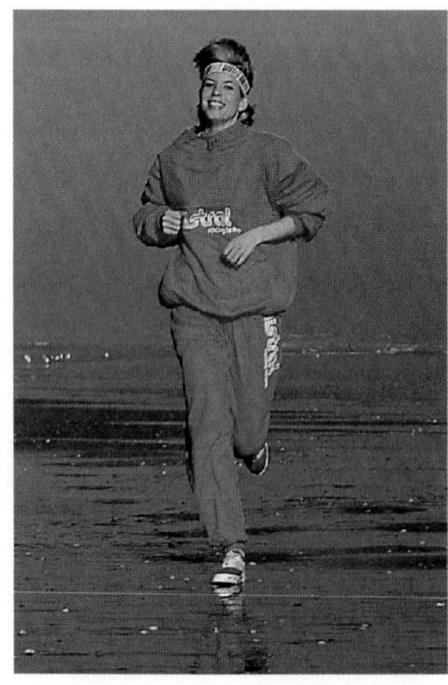

Aerobic exercise. Fats are the main source of energy for activities of low- to moderate-intensity.

Fat Use During Exercise

When you exercise, the fat your muscles burn comes from the fatty deposits all over the body, especially from those with the greatest amounts of fat to spare. That is why physically fit people look trim all over—they reduce their fat stores all over the body, not just those overlaying the working muscles.

A person who is of desirable body weight may store 25 to 30 pounds of body fat but only about one pound of carbohydrate. Although your supply of fat is almost unlimited, the ability of your muscles to use fat for energy is not.

Recall that for a working muscle to burn fat, oxygen must be present. If you work out at a rate that allows your heart to supply ample oxygen to working mus-

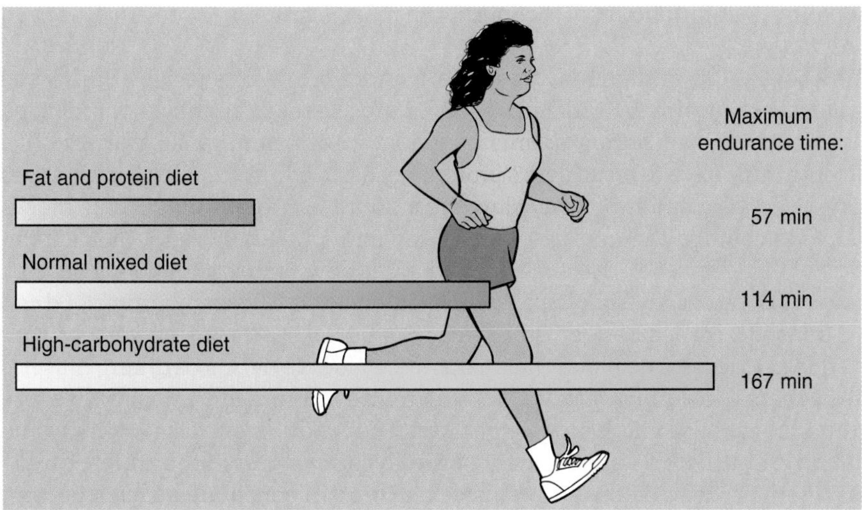

FIGURE 9-7
THE EFFECT OF DIET ON PHYSICAL ENDURANCE

A high-carbohydrate diet can increase an athlete's endurance. In this study, the fat and protein diet provided 94 percent of calories from fat and 6 percent from protein; the normal mixed diet provided 55 percent of calories from carbohydrate; and the high-carbohydrate diet provided 83 percent of calories from carbohydrate.

cles, the muscles will draw heavily on fat stores for fuel. When exercise intensity outstrips your oxygen supply, fat still contributes as much energy as ever, but glucose pitches in and is burned by the anaerobic pathway. Thus, the percentage of the total energy supplied by fat declines. Your breathing rate can signal which fuel is providing most of the energy. A rule of thumb for gauging exercise intensity for aerobic workouts is this: If you can't talk normally, you are incurring oxygen debt and are burning more glucose than fat; if you can sing, you aren't getting a cardiovascular workout or burning much of anything (so speed up).

Athletic training also controls the amount of fat used during a bout of exercise. Exercise training improves the body's ability to deliver fat to working muscles, and trained muscles have an increased ability to use the fat.

Much attention has been focused on the type of fuel used for varying exercise intensities and duration.[15] Research shows that when athletes exercise at a moderate intensity, they initially use more carbohydrate than fat for fuel.[16] Gradually, as exercise continues for more than 20 minutes, the fuel ratio shifts, and the athletes use more fat. For athletes participating in endurance sports, such as marathon runners and long-distance cyclists, who want to conserve their limited supply of carbohydrate, switching to a fat-burning energy system is crucial.

People who participate in endurance events know to build up their reserves of muscle glycogen before an event, so that they do not run out of glycogen—"hit the wall"—before the finish line.

PROTEIN NEEDS FOR FITNESS

Fit people have more muscle than fat; exercise involves muscles; muscles are made largely of protein. It would seem logical, then, that to become or stay fit, an athlete might need more protein. Although it's true that fat and glucose are the primary fuels for working muscles, 5 percent to 10 percent of energy needs of weightlifters and athletes

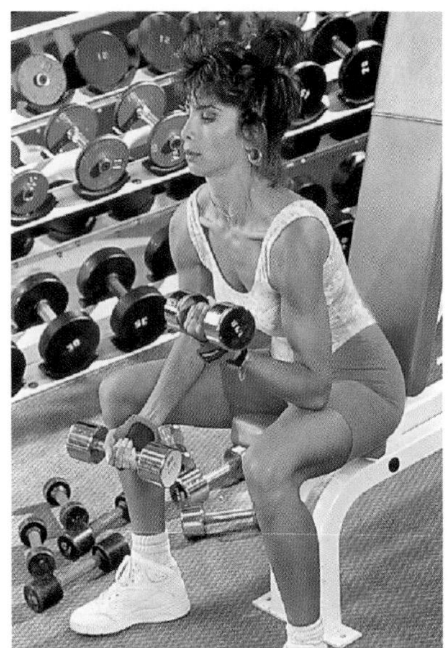

Muscles grow in response to work, not to eating protein.

competing in endurance sports comes from muscle protein. So, do athletes need more protein?

The body of an athlete may use slightly more protein, especially during the initial stages of training. Initial increases in muscle mass, numbers of red blood cells to carry oxygen, and amounts of aerobic enzymes in muscles to use fuel efficiently may elevate an athlete's protein needs. In addition, hormonal changes during exercise can temporarily slow the amount of protein the muscle makes and can encourage the muscle to break down its protein stores.[17] This combination of using fewer amino acids and breaking down muscle, thereby releasing amino acids, builds up a pool of available amino acids. The circulating blood can then transport some liberated amino acids to the liver, where some can be converted into glucose. The blood then ushers the glucose back to working muscles to feed them. How much protein an athlete uses for fuel during hard exercise (endurance exercise and heavy weightlifting) depends on exercise intensity and duration, the athlete's fitness level, and the glycogen stores in the athlete's muscles. When glycogen stores are well stocked, protein contributes only 5 percent of fuel needs.

Although muscle protein breakdown dominates during exercise, muscle growth escalates after exercise. Muscles use the available amino acids to repair and build. The net effect of these changes is muscle protein buildup. Training enhances muscle protein buildup after exercise.

The American Dietetic Association recommends that athletes consume 1 to 1.5 grams of protein per kilogram of desirable body weight.[18] Some athletes involved in prolonged heavy endurance training may need more than 1.5 grams of protein per kilogram.[19] The recommendation of one gram per kilogram is not much more than the 0.8 gram per kilogram recommended for the general population. This is nothing for most athletes to be concerned about, given that the average American eats twice the protein needed.

You might ask whether eating even more protein would help build muscle. Muscles won't respond to excess protein by helplessly accepting it. They respond to the hormones that regulate them and to the demands put upon them. So the way to make muscle cells grow is to put a demand on them—that is, to make them work. They will respond by taking up nutrients—amino acids included—so that they can grow.

 FLUID NEEDS AND EXERCISE

Replenishing fluid lost during exercise is easily accomplished by drinking fluid before, during, and after exercise. Yet many athletes and fitness enthusiasts either don't drink enough or don't drink at all.[20] Ignoring body fluid needs can hinder performance and increase risk of heat-related injury.[21] This section reviews both why your body needs fluids during exercise and how much and what to drink.

Plan to drink fluids before, during, and after exercise.

Water and Exercise

The water in your blood—known as plasma volume, or just plasma—serves a similar function as the water in the radiator of your car. It continually circulates throughout your body, picking up the tremendous amount of heat generated by working muscles. The plasma then transports this heat to your skin, through which the heat is expelled from the body primarily by evaporation of sweat. Think of sweating as your body's air-conditioning system. As sweat evaporates from your skin, it expels large amounts of heat, helping to keep your body cool. However, sweating works only when the sweat is evaporated from your skin. If the sweat simply rolls down your face or down your back, body heat is not released. On a humid day, the air is already saturated with water, which impairs evaporation of sweat from your skin. Hot, humid days, then, are doubly dangerous: You continue to sweat and lose precious body water, but your body temperature doesn't fall. You need to pay particular attention to your fluid needs on humid days.

Sweat is the primary way your body loses water during exercise. How much sweat you lose depends on the intensity and duration of the activity. The more intense the exercise, the more heat you generate and the more sweat you will lose. If you don't replace the water you lose from sweat, your plasma volume decreases. In an attempt to maintain plasma volume, your body will pull water from your muscles and organs. As water is pulled from muscles, cramps may occur, along with premature fatigue and a noticeable decline in performance.

Lower plasma levels also force your heart to beat faster. Since the heart pumps less blood with each beat, it must beat more often to supply oxygen to your muscles. Finally, with less plasma available to transport the heat to your skin, the heat builds, and your body's internal temperature continues to rise. All these changes force your body to work at a higher intensity level, leading to early exhaustion. A water loss equal to 2 percent of body weight can reduce muscular work capacity by 20 percent to 30 percent.

The recommended amount of fluid sufficient to prevent dehydration and **heat stroke** can be quite a bit. Athletes can lose two or more quarts of fluid during every hour of heavy exercise and must rehydrate before, during, and after exercise to replace the lost fluid. Even casual exercisers must drink some fluids while exercising. Thirst is unreliable as an indicator of how much to drink—it signals too late, after fluid stores are depleted.[22] Table 9-2 presents one schedule of hydration before, during, and after exercise. To know how much water is needed to replenish fluid losses after a workout, weigh yourself before and after—the difference is all water. One pound equals roughly two cups of fluid.

HEAT STROKE an acute and dangerous reaction to heat buildup in the body, requiring emergency medical attention; also called *sun stroke.*

Signs of heat stroke include:
- Very high body temperature (104° Fahrenheit or higher)
- Hot, dry red skin
- Sudden cessation of sweating
- Deep breathing and fast pulse
- Blurred vision
- Confusion, delirium, hallucinations
- Convulsions
- Loss of consciousness

To prevent heat stroke, drink plenty of fluid before, during, and after exercise; avoid overexercising in hot weather; and stop exercising at any sign of heat exhaustion. Signs of heat exhaustion include:
- Cool, clammy, pale skin
- Dizziness
- Dry mouth
- Fatigue/weakness
- Headache
- Muscle cramps
- Nausea
- Sweating
- Weak and rapid pulse

Water and Fluid-Replacement Drinks

For fitness enthusiasts, the choice between water and a sports drink (a properly balanced carbohydrate-electrolyte fluid-replacement drink) is more a matter of personal taste and desired performance abilities. But for endurance events (continuous exercise for longer than 60 minutes), mounting evidence indicates that consuming a properly balanced sports drink during exercise enhances energy status and endurance and maintains plasma volume levels better than drinking water does.[23]

How the body manages water and carbohydrate use during exercise determines how well it performs. Sports drinks are designed to enhance the body's use of carbohydrate and water (see Table 9-3). The carbohydrate in a sports beverage serves three purposes during exercise: (1) it becomes an energy source

TABLE 9-2 SCHEDULE OF HYDRATION BEFORE, DURING, AND AFTER EXERCISE*

When to Drink	Amount of Fluid
2 hours before exercise	About 3 c
15 to 30 minutes before exercise	About 2 c
Every 15 minutes during exercise	1 c (about 1 qt per hour)
After exercise	2 c fluid for each pound of body weight lost

*These guidelines are for exercise lasting less than 1 hour. During intense exercise lasting more than 1 hour, the consumption of approximately 1 liter of sports drink per hour (containing 6% to 7% carbohydrate or 60 to 70 grams carbohydrate per liter) is recommended to maintain oxidation of carbohydrates and delay fatigue.

Source: Adapted from American College of Sports Medicine. Position stand on exercise and fluid replacement, *Medicine and Science in Sports and Exercise* 28 (1996): i–vii.

TABLE 9-3 **FLUID-REPLACEMENT DRINKS**

Sports Drink	Calories/Cup	Carbohydrate Percentage	Carbohydrate Type	Sodium
All Sport	75	8.5%	High fructose corn syrup	55 mg
Exceed	70	7.0%	Glucose polymer, fructose	50 mg
Gatorade Thirst Quencher	70	6.0%	Sucrose, glucose	110 mg
Hydra Fuel	70	7.0%	Glucose polymer, fructose, glucose	25 mg
10-K Thirst Quencher	60	6.5%	Sucrose, fructose, and glucose	54 mg
PowerAde	70	7.0%	High fructose corn syrup	55 mg

for working muscles, (2) it helps maintain blood glucose at an optimum level, and (3) it helps increase the rate of water absorption from the small intestine, helping to better maintain plasma volume. In addition, the drink can supply water and minerals lost from sweating.[24]

There are many factors to consider when choosing a sports drink. The ideal beverage should leave the digestive tract rapidly and enter circulation, where it is needed. Carbohydrate solutions don't all empty from the stomach at the same rate. The drink should contain at least 6 percent but no more than 10 percent carbohydrate by volume. Drinks containing more than 10 percent carbohydrate, such as sodas, fruit juice, Kool-Aid types of drinks, and some sports drinks, take longer to absorb. They also can cause cramps, nausea, bloating, and diarrhea. Drinks with less than 6 percent carbohydrate may not offer an endurance-enhancing effect. Drinks using a blend of glucose polymers—short chains of carbohydrate—and fructose leave the stomach at the same rate as water, speeding the availability of the carbohydrate and water to working muscles.[25]

Sodium is another ingredient to which attention should be paid. Since most people eat enough salt in their regular diet to replace the sodium they lose during exercise, it's not essential that the fluid-replacement drink provide large amounts of sodium. In fact, too much sodium can delay muscles' receipt of water.

Research shows that about 50 milligrams of sodium per cup will help stimulate water absorption from your gut. Other studies have found that people who drink a beverage with some sodium drink more of it. If the drink tastes good, athletes and exercisers will want to drink it and so meet their fluid needs.[26]

There are several practical considerations to keep in mind when using a fluid replacement drink. The drink should be consumed cold, in 6-ounce to 8-ounce portions, and frequently (every 10 to 15 minutes *once exercise begins*). Cooled beverages leave the digestive tract quickly to supply water to muscles and cool the body. Competitive athletes should establish fluid consumption patterns during practice. An athletic event is not the time to try a new product or employ a new system of drinking.

For people who are exercising to lose weight, however, drinking a quart of a sports drink to meet fluid needs may supply the amount of calories they expended during a 40-minute aerobic class or in 30 minutes of continuous swimming or biking. A better choice might be plain water or diluted juice.

As for salt loss, when the exercise is over, eating regular food can make it up. Salt tablets are not recommended because they increase potassium losses and can irritate the stomach.

Sports drinks can enhance energy status during endurance events.

VITAMINS AND MINERALS FOR EXERCISE

Your muscles burn food and oxygen to make energy. How well they burn these fuels depends, however, on your supply of vitamins and minerals. Without small amounts of these potent substances, your muscles' ability to work is compromised.

The Vitamins

Vitamins are the links and regulators of energy-producing and muscle-building pathways. Without them, your muscles' ability to convert food energy to body energy is hindered and muscle protein formation is slowed. Table 9-4 shows a few vitamins and minerals and their exercise-supporting functions.

The B vitamins are of special interest to athletes and exercisers because they govern the energy-producing reactions of metabolism. Needs for these vitamins increase proportionally with energy expenditure. A person who expends 4,000 calories per day needs twice as much of the B vitamins as someone who expends 2,000 calories.

TABLE 9-4 EXERCISE-RELATED FUNCTIONS OF VITAMINS AND MINERALS

Vitamin or Mineral	Function
Thiamin, riboflavin, pantothenic acid, niacin, magnesium	Energy-releasing reactions
Vitamin B_6, zinc	Building of muscle protein
Folate, vitamin B_{12}, copper	Building of red blood cells to carry oxygen
Biotin	Fat and glycogen synthesis
Vitamin C	Collagen formation for joint and other tissue integrity; antioxidant ability may reduce oxidative tissue damage
Vitamin E	Protect cell membranes from oxidative damage
Iron	Transport of oxygen in blood and in muscle tissue
Calcium, vitamin D, vitamin A, phosphorus	Building of bone structure; muscle contractions; nerve transmissions
Sodium, potassium, chloride	Maintenance of fluid balance; transmission of nerve impulses for muscle contraction
Chromium	Assistance in insulin's glucose-storage function
Magnesium	Cardiac and other muscle contraction

Note: This is just a sampling. All vitamins and minerals play indispensable roles in exercise.

Many athletes have been falsely led to believe that they can enhance energy production by taking supplements of B vitamins. No experimental evidence exists to support this theory.[27] Once the system that uses B vitamins is full, generally the extra vitamins will be washed away through the urine. A well-balanced diet that meets athletes' energy needs and that features complex carbohydrate-rich foods will ensure B vitamin intakes proportional to energy intake.

Researchers are presently studying the protective effects of antioxidants on recovery from exercise and performance. Since the body uses oxygen at a higher rate during exercise, the generation of free radicals and the potential for exercise-induced tissue damage increase in the body. Although more research is needed, preliminary studies support a role for the antioxidant nutrients—particularly vitamin E—in enhancing recovery from exercise by reducing exercise-induced oxidative injury.[28] The evidence for any role of the antioxidants in improving performance is inconsistent thus far.[29]

The Minerals

Iron is a core component of the body's oxygen taxi service: hemoglobin and myoglobin. A lack of oxygen compromises the muscles' ability to perform. Iron deficiency has not been reported to be a problem for fitness enthusiasts who exercise moderately.[30] Male and female endurance athletes, though, may be prone to developing mild iron deficiency, diagnosed by low blood ferritin levels, a measure of the body's store of iron. Menstruating female athletes are at particular risk—growth and menstruation combined with strenuous training can take a toll on a woman's iron stores.[31]

A combination of factors increases an athlete's chances of depleting his or her iron stores. Inadequate dietary intakes of iron-rich foods combined with iron losses aggravated by physical activity compromise iron status. Physical activity may cause increased iron losses in sweat, feces, and urine, plus increased destruction of red blood cells that occurs during exercise. Chapter 7 contains numerous suggestions for obtaining sufficient iron from foods.

Sometimes iron deficiencies can be corrected only with iron supplements. If you are concerned about your iron level, see a physician. Iron supplements should not be taken without medical supervision. High iron intakes can induce deficiencies of trace minerals, such as copper and zinc, and produce an iron overload in some people.

SPORTS ANEMIA a temporary condition of low blood hemoglobin level, associated with the early stage of athletic training.

An apparent anemia—sometimes called **"sports anemia"**—also can occur in athletes that reflects no reduction in the blood's iron supply, but rather an increase in the blood plasma volume.[32] This occurs because athletic training causes the kidneys to conserve sodium and water. In other words, the extra blood volume dilutes the concentration of iron, thereby making it seem as if the blood does not contain enough of the mineral. Sports anemia is considered a temporary state and probably reflects a normal adaptation to physical training.

The Bones and Exercise

STRESS FRACTURE bone damage or breakage caused by stress on bone surfaces during exercise.

AMENORRHEA cessation of menstruation associated with strenuous athletic training.

Bones absorb great stresses during exercise, and like the muscles, they respond by growing thicker and stronger. Weight-bearing exercises—running, walking, dancing, rope skipping, or activities such as strength training in which significant muscular force can be generated against the long bones of the body—encourage bone development. A bone not strong enough to withstand the strain placed on it by athletic exertion can break in what has become known as a **stress fracture.** When a person suffers such a break, there are three probable causes. One is unbalanced muscle development, which allows strong muscles to pull against the bone opposed only by weaker, undeveloped muscles, thereby leaving the bone susceptible to fractures. Another is bone weakness caused by inadequate calcium intake. A possible third cause, which occurs in women, is reduced estrogen concentration, which leads to bone mineral loss and therefore to fragile bones in women who have ceased menstruating.

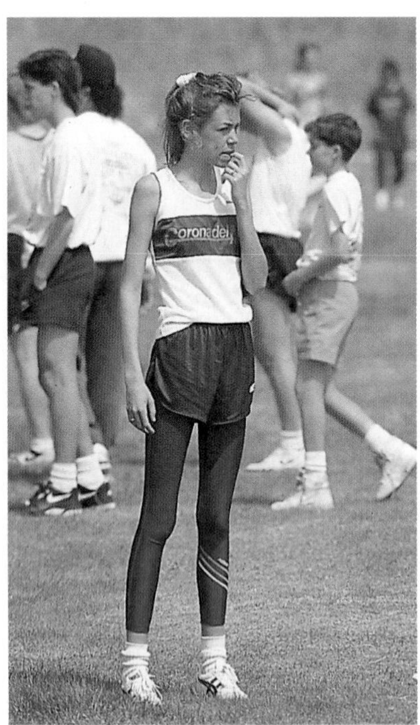

The condition characterized by the potentially fatal combination of disordered eating, amenorrhea, and low bone density is referred to as the **female athlete triad.**

Balanced muscle development can protect the bones from undue stresses. Each set of muscles pulling against bone should be kept in check by an equally strong set of opposing muscles. You should not work a set of muscles in training without working the opposing muscles. So if you work your back and leg muscles a lot (by jogging or walking, for example), work your abdominal muscles too (do situps). Bones, like muscles, take time to develop strength. Giving bones and muscles plenty of time to build up to one level of performance before moving up to the next level can also prevent stress fractures. Eating an adequate amount of calcium throughout life may be one of the primary defenses against developing weak bones.

Some women who exercise strenuously cease to menstruate, a condition called **amenorrhea.** Such women have lower than normal amounts of estrogen, a hormone essential for maintaining the integrity of the bones. With low estrogen levels, the mineral structures of the bones are rapidly dismantled, weakening the skeleton. Women who have athletic amenorrhea are at risk for stress fractures now and adult bone loss later in life.[33] To reverse the condition, they should not stop exercising altogether, for reasonable amounts of exercise may be a key defense against bone depletion. They should, however, seek evaluation from a health-care provider who specializes in sports medicine to find the cause of and receive treatment for their amenorrhea.

Eating disorders are sometimes related to athletic amenorrhea, and a logical part of diagnosis is to look carefully at the woman's diet for adequacy. It could be that a diet too low in calories, coupled with low body fat stores and strenuous exercise, sets the stage for amenorrhea to develop. In such cases, calcium intakes between 1,000 and 1,300 milligrams per day may help to protect the bones somewhat.

NUTRITION ACTION

Athletes and Supplements—Help or Hype?

Competitors in the ancient Greek Olympiad reportedly used mushrooms and herbs.[34] Since then, virtually every food has at one time or another been touted as the "magic bullet" that will enhance performance. Athletes have been known to swallow everything from bee pollen to brewer's yeast to kelp to wheat germ in their quest to gain the competitive edge.

Nutritional Supplements as Ergogenic Aids

Seductive as the idea of using pills and potions to achieve peak performance may be, the scientific evidence to support claims that special supplements will make an athlete run farther or jump higher is sorely lacking.[35] Most so-called **ergogenic aids,** that is, substances that increase the ability to exercise harder, are costly versions of vitamins, minerals, sugar, and other substances provided easily by a balanced diet.[36] Table 9-5 describes some of the many substances currently promoted as ergogenic aids.

ERGOGENIC AIDS anything that helps to increase the capacity to work or exercise.
ergo = work
genic = give rise to

Take bee pollen, a mixture of protein, carbohydrate, a bit of fat, and a few vitamins and minerals.[37] Though touted by one manufacturer as a "natural and balanced source of extra energy" appreciated by "athletes worldwide," it has been tested at Louisiana State University among both runners and swimmers and found to confer no benefit whatsoever on an athlete's training or performance abilities.[38] The same goes for chromium picolinate, touted to increase lean body mass and delay fatigue due to its role in glucose utilization. Although chromium

An endless array of ergogenic aids are marketed to athletes and other sports enthusiasts. Although big on claims, few are based on scientific evidence.

TABLE 9-5 A SAMPLING OF POPULAR ERGOGENIC AIDS[a,b]

Ergogenic Aids with Unproven Claims

Amino acids (e.g., arginine, ornithine, glycine) nonessential amino acids falsely promoted to increase muscle mass and strength by stimulating growth hormone and insulin. Individual amino acids do not significantly increase muscle mass or growth hormone. Weight lifting and endurance training do.

Anabolic steroids synthetic male hormones (related to testosterone) that stimulate growth of body tissues, with many adverse effects as listed on the following page.

Bee pollen mixture of bee saliva, plant nectar, and pollen touted falsely to enhance athletic performance. May cause allergic reactions in people with a sensitivity to bee stings and honey allergies.

Carnitine a compound synthesized in the body from two amino acids (lysine and methionine) and required in fat metabolism. Falsely touted to increase the use of fatty acids and spare glycogen during exercise, delay fatigue, and decrease body fat. The body produces sufficient amounts on its own. No evidence that supplementation in healthy people improves energy or enhances fat loss.

Chromium picolinate Chromium is an essential component of the glucose tolerance factor, which facilitates the action of insulin in the body. Picolinate is a natural derivative of the amino acid tryptophan. Falsely promoted to increase muscle mass, decrease body fat, enhance energy, and promote weight loss. Choose instead, a diet rich in whole, unprocessed foods.

Coenzyme Q10 a lipid made by the body and used by cells in energy metabolism; falsely touted to increase exercise performance and stamina in athletes; potential antioxidant role. May increase oxygen use and stamina in heart disease patients, but no significant effect seen in healthy athletes.

DHEA (Dehydroepiandrosterone) a precursor of the hormones, testosterone and estrogen; falsely promoted to increase production of testosterone, build muscle, burn fat, and delay the effects of aging.[c] Long-term effects unknown, self-supplementation not recommended.

Ginseng a collective term used to describe several species of plants, belonging to the genus *Panax,* containing bioactive compounds in their roots. Falsely touted to enhance exercise endurance and boost energy. A lack of well-controlled research has yielded inconclusive evidence for the benefits of ginseng. The potential for adverse drug-herb or herb-herb interactions with ginseng exists.[a]

Pyruvate a 3-carbon compound derived from the breakdown of glucose for energy in the body. Falsely promoted to increase fat burning and endurance. Side effects include intestinal gas and diarrhea, which could interfere with performance.

Ergogenic Aids with Some (Not All) Scientific Support for Claims

Creatine A nitrogen-containing substance made by the body which combines with phosphate to form the high-energy compound, creatine phosphate (CP). CP is stored and used by muscle for ATP production. Some (not all) studies show that creatine may increase CP content in muscles and improve short-term (< 30 seconds) strenuous exercise performance (e.g., sprinting, weight lifting). Long-term effects unknown.

Caffeine a stimulant which increases blood levels of epinephrine; promoted for improved endurance and utilization of fatty acids during exercise. Consuming 2 to 3 cups of coffee (equal to 3 to 6 milligrams of caffeine per kilogram body weight) 1 hour before exercise may improve endurance performance. High caffeine consumption may cause dehydration, headache, nausea, muscle tremors, and fast heart rate.

HMB (beta-hydroxy-beta-methylbutyrate) a metabolite of the branched-chain amino acid leucine and promoted to increase muscle mass and strength by preventing muscle damage or speeding up muscle repair during resistance training. More research on long-term safety and effectiveness is needed.

Phosphate salt a salt with claims for improved endurance. Found to increase a substance in red blood cells (diphosphoglycerate) and enhance the cell's ability to deliver oxygen to muscle cells and reduce levels of disabling lactic acid in elite athletes. However, more research on safety and efficacy of phosphate loading is needed. Excess can cause loss of bone calcium.

Sodium bicarbonate baking soda is touted to buffer lactic acid in the body and thereby reduce pain and improve maximal-level anaerobic performance. May cause diarrhea in users due to the high sodium load; effects of repeated ingestions unknown.

[a]See also Table 6-13 on page 190 for information on herbal supplements.
[b]For more information, see E. Coleman, *Eating for Endurance* (Palo Alto, CA: Bull Publishing Co., 1997); S. A. Sarubin, *The Health Professional's Guide to Popular Dietary Supplements* (Chicago, IL: The American Dietetic Association, 2,000).
[c]See also Table 10-11 on page 355 for more about anti-aging supplements.

is necessary for muscle function by transferring glucose from the blood to the muscle cells, true chromium deficiencies are rare to nonexistent. To date, there is no evidence that chromium supplements improve athletic performance in healthy individuals without a chromium deficiency.[39]

Surveys of athletes have found that between 53 percent and 80 percent use a vitamin or mineral supplement, although no evidence exists that doing so improves performance.[40] More than 40 years of research has provided no strong evidence that popping vitamins and minerals increases energy or athletic prowess of adequately nourished people. Except for iron, vitamin and mineral deficiencies are rare among athletes. Since most athletes eat more food than nonathletes eat to meet their increased energy demands, they usually get the additional vitamins, minerals, and other beneficial substances they need, provided they eat a well-balanced diet.

Of course, when athletes firmly hold that one or another ergogenic aid does indeed improve performance, convincing them otherwise can be extremely difficult. One reason is that the profound belief that a substance will help can actually produce a psychological benefit, known as the **placebo effect.**

PLACEBO EFFECT an improvement in a person's sense of well-being or physical health in response to the use of a placebo (a substance having no medicinal properties or medicinal effects).

ANABOLIC STEROIDS synthetic male hormones with a chemical structure similar to that of cholesterol; such hormones have wide-ranging effects on body functioning.

Anabolic Steroids: Use and Abuse

Psychological impact aside, pill popping may be harmful in some cases. Taking large doses of vitamin A, for example, can cause liver damage over time. Swallowing supplements known as **anabolic steroids**—synthetic hormones that appear to help build muscle—can be even more dangerous.

A particularly popular practice among weightlifters and bodybuilders, steroid abuse often begins around the age of 18 years.[42] But while steroids may help to increase muscular size and strength in some people, they can bring about numerous side effects, including acne, liver abnormalities, temporary infertility, and offensive outbursts, often referred to as "roid rages."

Among adults, many effects of steroid use are reversible. Unfortunately, adolescents aren't so lucky. Several studies show that adolescent steroid users may suffer serious consequences of premature skeletal maturation, decreased spermatogenesis, and elevated risk of injury.[43]

Along with being unhealthful, steroid use is considered unethical by domestic and international sports organizations such as the American College of Sports Medicine and the International Olympic Committee. As a case in point, track star Ben Johnson lost his Olympic gold medal for the 100-meter sprint in 1988 after officials discovered he had been using steroids.[44]

Side effects of steroids:[41]

- acne
- anxiety
- blood clots
- blood poisoning
- cancer
- diarrhea
- dizziness
- fatigue
- heart disease
- hypertension
- irreversible baldness in women
- jaundice
- kidney damage
- liver damage
- male pattern baldness
- mood swings
- nausea
- oily skin
- prostate enlargement
- psychotic depression
- shrunken testicles
- sterility (reversible)
- stroke
- stunted growth in adolescents
- swelling of feet or lower legs
- yellowing of the eyes or skin

Amino Acid Supplements and the Athlete

Many athletes also swallow amino acid supplements in hopes of building larger muscles. These supplements are advertised as "predigested protein" and so supposedly are easily absorbed and readily available to encourage muscle growth brought on by training. Let's look at the facts.

Your intestinal tract is better prepared to handle protein in its complex form of di- and tripeptides (refer to Chapter 5). You absorb 85 percent to 99 percent of the animal protein you eat and 90 percent of protein from vegetable sources. Amino acid solutions, on the other hand, will draw water into your gut, and their digestion can cause cramping and diarrhea.

In addition, your body can't store extra amino acids, whether they come from food you eat or from supplements. Your body converts the excess into fat. This conversion of amino acids to fat generates urea, which increases your body's need for water. Both diarrhea and increased urination of urea can lead to dehydration, impeding training and performance.

There is no benefit to rapid absorption of amino acids. It takes your body hours, not minutes, to rebuild muscle protein damaged by exercise. And as mentioned before, most athletes already eat more protein than they need.

Caffeine, Alcohol, and Performance

Some athletes look to other dietary means besides pill popping to promote athletic prowess. For instance, many exercisers drink caffeine-containing beverages such as coffee, tea, or cola to enhance performance and endurance. Caffeine apparently stimulates the release of fats into the blood that the body can then use instead of glycogen as a source of energy. Thus, the glycogen is "spared," or saved, for later use, and the amount of time an exerciser can endure physical activity before running out of fuel is prolonged.

The glycogen-sparing effect of caffeine, however, is beneficial only for athletes who exercise for more than one and a half to two hours at a time. As we said earlier, the muscles generally store enough glycogen to fuel as many as 90 minutes of activity. Moreover, even endurance athletes can experience certain downsides to consuming caffeine. A diuretic, the drug promotes frequent urination and fluid loss that can lead to dehydration. In addition, caffeine can induce rapid heart rate and jitters, which can interfere with performance. Athletes would also do well to remember that caffeine is a drug that neither the American College of Sports Medicine nor the International Olympic Committee condones for use among athletes.

Along with caffeinated beverages, alcoholic drinks are often touted as choice fluids for athletes. Beer, for example, is sometimes portrayed as the perfect carbohydrate-containing complement to both before- and after-competition meals. Despite such images, alcoholic drinks rank as poor sources of fluid and energy for several reasons. For one, alcohol is a diuretic that can bring about fluid loss and dehydration. More importantly, the amount of alcohol in just one beer or glass of wine depresses the nervous system, thereby slowing an athlete's reaction time and interfering with reflexes and coordination. Also, one can of beer provides only 50 carbohydrate calories. The rest of the calories come from alcohol, which must be metabolized by your liver, not your muscles. The American College of Sports Medicine and the American Dietetic Association both conclude that use of alcohol hinders performance. (The Spotlight feature later in the chapter presents a detailed explanation of how alcohol affects the body.)

The special supplements and drinks discussed here are just a sampling of the many "magic" pills and potions promoted to athletes. If you're in doubt about a particular product you see boasted as an ergogenic aid, use the Fitness Quackery Scorecard to separate legitimate claims from bogus ones.

FITNESS QUACKERY SCORECARD

Here are a few questions to ask yourself about any health gimmick, product, or device that you see advertised or that someone tries to sell you. Every yes answer gets one point.

1. Is the promised action of the product based on magical thinking? ("Eat all you want and lose weight." "Develop a trim body with no exercise.")
2. Does the promotion claim that "doctors agree" or "research has determined," without clarification? (Which doctors? What research?)
3. Does the promotion state that hormones, drugs, or nutrient doses useful in correcting an abnormal condition are needed to make the normal, healthy person more fit?
4. Does the promoter use scare tactics to pressure you into buying the product? ("It's the only one available without poisons.")
5. Is the product being sold or promoted by a crusading organization, a faith-healing group, or a self-styled health adviser who has no acceptable credentials?
6. Is the product advertised as having a multitude of different beneficial effects? ("Makes bigger muscles; gives that pumped-up feeling, improves digestion, coordination, and breathing.")
7. Does the sponsor claim persecution by the medical community and the government because they do not accept this wonderful discovery?
8. Is the product available only from the sponsor by mail order and with payment in advance?
9. Does the promoter use many case histories or testimonials from grateful users?
10. Is the product a special or "secret" formula not available from any other source?

SCORING

Even 1 point is a point against the claimant: it's your warning signal that you are dealing with misinformation. Three or more points is a sure sign of quackery.

A nutritious diet and regular physical activity enhance performance far better than the products touted as magic bullets for gaining the competitive edge.

Food for Fitness

Barley has been grown as a staple food since prehistoric times. Because it was considered to be a mild grain, athletes in ancient Greece trained on barley mush.[45]

The best nutrition prescription for peak performance is a well-balanced diet. Although no one eating plan meets every athlete's needs, certain fundamental components are common to all well-balanced diets. For athletes, the diet should account for increased energy needs, vitamin/mineral needs, the relative efficiency of various foods as fuels, and current knowledge about long-term health. An eating plan that supplies 60 percent of calories from complex carbohydrate, 15 percent of calories from protein, and 25 percent of calories from fat will enable both athletes and fitness enthusiasts to supply muscles with a proper fuel mix and maintain health.[46] Two critical nutrition periods for the athlete are the training diet and the precompetition diet.

Planning the Diet

- Athletes should consume a significant proportion of their day's calories before beginning physical activities. This will assure that muscle fuel needs are met during exercise.
- A diet rich in complex carbohydrate and low in fat not only provides the best balance of nutrients for health but also supports physical activity best. The table on page 310 shows some sample balanced eating plans for athletes who wish to increase their carbohydrate intake along with their calories.
- Choose foods to provide nutrients as well as calories—extra milk for calcium and riboflavin, many vegetables for B vitamins, meat or alternates for iron and other vitamins and minerals, and whole grains for magnesium and chromium. The photos on the next page provide examples of high-carbohydrate meals for the athlete.
- An athlete may be able to eat more food by consuming it in six or eight meals each day rather than in three or four meals. Large snacks of milkshakes, dried fruits, peanut butter sandwiches, or cheese and crackers can add substantial calories and nutrients.
- A homemade milkshake made with low-fat milk, ice milk, malted milk powder, and half a cup of fresh or frozen fruit can add 300 or more calories to the day's intake. High-carbohydrate liquid nutritional supplements also are available on the market.

The Pregame Meal

- The best choices for the meal before a competitive event are foods that are high in carbohydrate and low in fat, protein, and fiber. Fat and protein slow the stomach's emptying, and the protein's waste products generated during metabolism require that too much water be excreted with them.
- Fiber is not desirable right before physical exertion because it stays in the digestive tract too long and attracts water out of the blood.
- A high-carbohydrate meal will support blood glucose levels during competition. Olympic training tables are laden with foods such as breads, whole-grain cereals, pasta, rice, potatoes, and fruit juices.
- For pregame meals and snacks, choose: grape juice, apricot nectar, pineapple juice, jello, sherbet, popsicles, raisins, apricots, figs, dates, jams and toast, pancakes with syrup, honey, pasta, baked white or sweet potatoes, steamed vegetables, low-fat frozen yogurt, angel food or sponge cake with fresh fruit, gingersnaps, and graham crackers.
- Stay away from higher-fat foods such as meats, cheese, nuts, gravies, cream, French fries, muffins, croissants, biscuits, butter, potato chips, pies, and ice cream.
- Include plenty of fluids—two or more 8-ounce glasses of water or juice per meal—to ensure adequate hydration.
- Any meal should be finished a good two to four hours before the event because digestion requires routing the blood supply to the digestive tract to pick up nutrients. By the time the contest begins, the circulating blood should be freed from that task and should be available instead to carry oxygen and fuel to the muscles.

SAMPLE MEALS FOR HIGH-CARBOHYDRATE INTAKES FOR THE EXERCISE ENTHUSIAST AT TWO CALORIE LEVELS

Breakfast
- 1 c coffee
- 8 oz low-fat milk
- 2 pieces whole-wheat toast
- 4 tsp jelly
- ½ c strawberries
- ½ c orange juice
- 1 c oatmeal and raisins with 2 tsp brown sugar

Morning Snack
- 4 tsp trail mix

Lunch
- 12 oz iced tea with sugar
- 1 orange
- 1 banana
- 2 beef and bean burritos

Afternoon Snack
- A smoothie made from
  - 12 oz nonfat milk
  - 1 frozen banana
- 1 apple
- 4 rye wafers with 1 oz low-fat cheese

Dinner
- 8 oz low-fat milk
- 1 c sherbet
- 1 c spinach salad with 1 tbsp dressing
- 1 dinner roll with 2 tsp butter
- ¼ tomato
- 1 c broccoli
- 4 oz salmon
- ¾ c noodles with parsley and 2 tsp butter

Total Calories: 3,119
61% cal from carbohydrates, 24% cal from fat, 15% cal from protein

Breakfast
- 1 c coffee
- ½ c strawberries
- 8 oz nonfat milk
- 1 c oatmeal and raisins

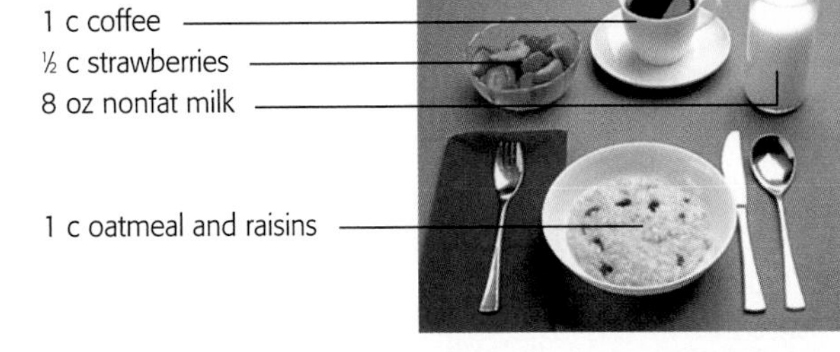

Morning Snack
- 4 tsp trail mix

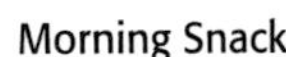

Lunch
- 12 oz iced tea with sugar
- 1 orange
- 1 beef and bean burrito

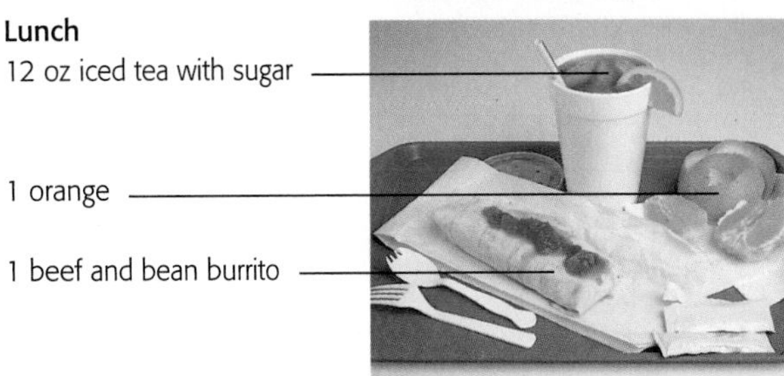

Dinner
- ½ c sherbet
- 1 c spinach salad with 1 tbsp dressing
- 8 oz nonfat milk
- ¼ tomato
- 1 c broccoli
- ½ c noodles with parsley and 2 tsp butter
- 4 oz salmon

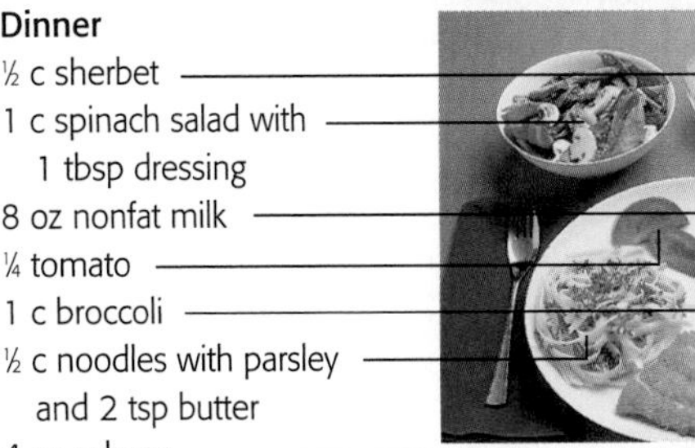

Total Calories: 1,759
57% cal from carbohydrates, 24% cal from fat, 19% cal from protein

- Liquid nutritional products can be consumed closer to the event (one to two hours) because they are emptied relatively rapidly from the stomach.[47] In addition, they supply an effective way to assure hydration.

In summary, the diet that supports performance is similar to the diet recommended for good health. Both for health promotion and for fitness, the diet should consist mostly of whole, minimally processed foods low in fat, sugar, and salt and high in complex carbohydrates and vitamins and minerals.

HIGH-CARBOHYDRATE EATING PATTERNS FOR VARIOUS ENERGY LEVELS

Use the number of portions indicated to arrive at the specified energy levels.*

FOOD	CALORIE LEVEL					
GROUP	1,500	2,000	2,500	3,000	3,500	4,000
Milk	3	3	4	4	4	4
Fruit	5	6	7	9	10	12
Vegetable	3	3	3	5	6	7
Grain	7	11	16	18	20	24
Fat†	2	3	5	6	8	10
Meat‡	5	5	5	5	6	6
Percentage of carbohydrate:	58%	58%	63%	64%	60%	62%

*Refer to Chapter 2, Figure 2-6 on page 42 for portion sizes.
†A serving of fat is equivalent to 1 tsp butter, margarine, or oil.
‡Meat portions are given as total ounces of meat; a typical serving includes 2 to 3 ounces.
Note: These plans supply 58% to 64% of calories as carbohydrate and less than 30% as fat. To increase the carbohydrate content to over 64%, substitute ½-cup servings of legumes for the meats. People who cannot eat these quantities of whole foods may have to replace some of them with milkshakes or liquid-meal replacements to meet their energy needs.

WORKING ON THE WEB

www.shapeup.org

Sponsor: Former U.S. Surgeon General C. Everett Koop in partnership with industry and nonprofit organizations

Description: Practical advice on fitness and nutrition for every level.

Available Activities: Provides handy tools for assessing your readiness to incorporate physical activity into your life and for rating your activity and fitness levels. You will find information about the benefits of becoming more active, a Physical Activity IQ Test, and definitions for physical activity buzzwords. In addition, you will learn different ways for adding physical activity to your life by charting your progress and overcoming barriers. Finally, you can learn what to eat to maximize your workout and receive feedback from experts.

Web Work:

1. Assess your current activity and fitness levels: Take the Par Q test and Activity Level Test.
2. Take the Physical Activity IQ Test.
3. Learn about what to eat and what to take along with you to maximize your workout.

Helpful Hints: From the Shape Up America home page, scroll down and click on Fitness Center. Click on Assessment and follow the directions to rate your current activity and fitness levels. Next, from the home page, click on Information/Exercise IQ Test. Fill in the best answer to each question in the Physical Activity IQ Test, and then check your answers in order to improve your fitness IQ. Return to the Fitness Center home page and click on Improvement to learn ways to improve your physical activity level. Finally, from the Fitness Center page, click on Nutrition to learn more about the recommended diet for optimal performance.

Spotlight
ALCOHOL AND NUTRITION

One of the many drugs— and usually the most common one—that enters people's lives as they arrive at their teen and adult years is alcohol. The National Institute on Alcohol Abuse and Alcoholism has estimated that about two-thirds of all adults in the United States drink alcohol. Of these, 33 percent are classified as light drinkers (three drinks or less per week), about 24 percent are moderate drinkers (two drinks or less per day), and 9 percent are heavy drinkers (five or more drinks per day). Some 15.3 million people in the United States meet the criteria for alcohol abuse, alcohol addiction, or both.[48]

Even though people in the 20- to 40-year-old age group have the greatest alcohol consumption, many teenagers are regular alcohol users.[49] One survey showed that two-thirds of the high school seniors interviewed reported that they had consumed alcohol within the past month and 5 percent indicated that they drank alcohol every day. In addition, a third of U.S. college students are categorized as **binge drinkers.**[50]

Alcohol abuse and alcohol addiction are recognized as major public health problems in the United States today.[51] An alcohol-related traffic accident occurs about every 30 minutes.[52] Approximately 100,000 excess deaths per year are associated with alcohol-related causes—including traffic accidents, homicides, suicides, cirrhosis, and hemorrhagic stroke—while the annual health cost to the nation associated with alcohol abuse and dependence is more than $100 billion.[53] Whether an individual is a social drinker, a **problem drinker,** or an alcoholic, the common element is managing the physiological, psychological, and social consequences of alcohol consumption. The difference between the problem drinker and the alcoholic is generally one of degree, with the alcoholic experiencing more difficulties in terms of illness, dependence on alcohol, loss-of-control symptoms, and disruption of family and work life than the problem drinker. As we will see, alcohol has a profound effect on all aspects of metabolism and health.

Alcohol abuse refers to problem drinkers with patterns of alcohol use that result in health problems, social problems, or both. *Alcohol addiction* (sometimes called *dependence*) refers to the disease—**alcoholism**—that is characterized by abnormal alcohol-seeking behaviors and a lack of control over drinking. Many of the harmful health effects of alcohol consumption are seen in both problem drinkers and those with alcoholism.[54] However, the disease—alcoholism—has four distinguishing features:[55]

- *Tolerance*—more and more alcohol is needed to produce the desired effects.
- *Physical dependence*—when the person refrains from using alcohol, a characteristic withdrawal syndrome appears that can be relieved by consuming alcohol.
- *Impaired control over drinking*—the person cannot regulate alcohol intake once a drink has been taken.
- *Discomfort of abstinence*—strong "cravings" for alcohol pervade the person addicted to alcohol during periods of abstinence and can lead to relapse.

Why doesn't everyone who drinks alcohol become an alcoholic?

As the concept grew that alcoholism is a disease and not a mental illness or a lack of self-discipline, it also became apparent that alcoholism tends to run in families. Studies have shown that the natural children of alcoholics are three to four times more likely to be alcoholic themselves than are the biological children of nonalcoholic parents. This observation tends to hold true whether the child was raised by its natural parent or by an adoptive parent.[56] Research is underway to understand this process in greater detail and develop biochemical markers that will identify individuals who are at risk of becoming alcoholic.

How does alcohol affect the body?

Alcohol (specifically, ethanol, the form found in alcoholic beverages) affects every organ, but the most dramatic evidence of its disruptive behavior appears in the liver—the only organ whose cells can oxidize alcohol for fuel to any great extent. All other cells are affected by the presence of alcohol but can do practically nothing about getting rid of it. Ethanol, like the other alcohols, is toxic—but less so than some. Sufficiently diluted and taken in small enough doses, it produces **euphoria**—an effect that people seek—not with zero risk but with a low enough risk (if the doses are low enough) to be tolerable. Used to achieve this effect, alcohol is a **drug**—that is, a substance that can modify one or more of the body's functions. Like all drugs, alcohol offers some benefits; it also has tremendous abuse potential.

What are the benefits offered by alcohol?

The beneficial effects of alcohol have long been appreciated and praised. Wine, beer, and other fermented beverages have given pleasure and relaxation to people for more than 5,000 years. Only recently have we begun to learn exactly how the ethanol molecule acts in our bodies, but people have always known that it affects their mood, sensations, and behavior. Because it alters mood, alcohol has many uses. Taken in moderation, alcohol relaxes people, reduces their inhibitions, and encourages desirable social interactions. The regular and moderate consumption of alcohol, especially wine, by elderly people results in better sleep patterns, improved appetite and mood, a lower heart rate, and reduced anxiety levels.[57] Alcohol may even lower an individual's susceptibility to heart disease, possibly by increasing the blood HDL level and preventing blood clot formation (see the Spotlight feature in Chapter 4).[58]

What do you mean by *moderation?*

We can't name an exact amount of alcohol per day that would be appropriate for everyone, because people differ in their tolerance levels. Still, authorities have attempted to set a limit that is appropriate for most healthy people.[59] Moderate drinking is defined as *no more* than one **drink** a day for the average-sized woman, and *no more* than two drinks a day for the average-sized man. A drink is any alcoholic beverage that delivers ½ ounce of *pure ethanol:* 5 ounces of wine, 10 ounces of wine cooler, 12 ounces of beer, 1½ ounces of hard liquor (whiskey, gin, scotch, vodka, or rum). Research findings had suggested that these "moderate" levels of intake were associated with reduced mortality (compared to abstainers), whereas larger intakes were associated with increased mortality. However, more recent evidence suggests that the protective benefits of alcohol occur at lower intakes—one to nine drinks per *week* for men and one to three drinks per *week* for women.[60] Doubtless some people could consume slightly more; others could definitely not handle nearly so much without significant risk.

Since moderate alcohol intake may be beneficial to heart disease, are scientists now recommending that everyone drink in moderation?

No. Researchers warn that caution is needed in recommending moderate alcohol consumption to the public.[61] Many people need to refrain from alcohol consumption altogether (pregnant women, recovering alcoholics, people under age 21, persons on certain medications, and those intending to drive a vehicle). Alcohol can have profoundly negative effects on the body, and it is associated with five of the ten leading causes of death in this country—namely certain types of cancer, cirrhosis of the liver, motor vehicle and other accidents, suicides, and homicides.

In addition, the impact (positive or negative) of alcohol use varies with age and gender.[62] For example, the leading cause of death for middle-aged men is heart disease. For this group, there may be an advantage to moderate alcohol use, since the evidence with regards to alcohol and blood lipids points to a possible protective mechanism between moderate alcohol consumption and heart disease risk factors.[63] For younger men, however, the leading causes of death are accidents or violence—and both are often alcohol-related. An increase in alcohol consumption among this group is not encouraged. For premenopausal-aged women, the risk of death from breast cancer is greater than that for heart disease. Some researchers now point to a possible association between alcohol use (one, two, or three drinks per day) and breast cancer.[64] Therefore, any protection against heart disease derived from moderate consumption of alcohol in these women would be negated by the deleterious effects of the alcohol on breast cancer.

How does the body handle alcohol?

From the moment alcohol enters the body in a beverage, it is treated as if it has special privileges. Foods sit around in the stomach for a while, but not alcohol. The tiny alcohol molecules need no digestion; they can diffuse as soon as they arrive, right through the walls of the stomach, and they reach the brain within a minute.

A small amount of alcohol can be metabolized by the enzyme **alcohol dehydrogenase** housed within the cells lining the stomach.[65] The extent to which the stomach cells metabolize this alcohol differs between women and men. The observation that "women can't seem to hold their liquor" compared with men is explained in part by a woman's lower alcohol dehydrogenase activity in the stomach. Women's inability to handle this alcohol holds true even when adjustments are made for the differences in body size between women and men.[66]

What does the alcohol do once it arrives at the liver?

Although a small amount of alcohol can be metabolized by stomach cells, liver cells are the only cells in the body that can make enough alcohol dehydrogenase to oxidize alcohol at an appreciable rate. However, there is a limit to the amount of alcohol anyone can process in a given time. This limit is set by the number of molecules of the alcohol-processing enzymes in the liver. If more molecules of alcohol arrive at the liver cells than the enzymes can handle, the extra alcohol must wait. It enters the general circulation and moves on past the liver. From the liver, it is carried to all parts of the body, circulating again and again through the liver until enzymes are available to convert it to **acetaldehyde.** The rate at which the liver enzymes can work limits the rate of the body's handling of alcohol.

How fast do the liver enzymes work?
That depends on many individual factors. For example, the amount of enzymes present depends on when you last ate a meal. Fasting causes degradation of the enzyme within the cells and can reduce the rate of alcohol metabolism by half. Drinking on an empty stomach thus not only lets the drinker feel the effects more promptly but also brings about higher blood alcohol levels for longer periods of time and increases the effect of alcohol in anesthetizing the brain.

What if you drink too fast?
If you drink so fast that your liver enzymes can't keep up with the load, metabolic products of alcohol's metabolism accumulate in the blood and body organs. One is acid—dangerous for all body processes. Another is acetaldehyde, already mentioned, which is toxic to the brain and other organs. Another is fat, which accumulates in the liver itself and can't be moved out until the body is free of alcohol.

Alcohol also causes loss of body water by excretion, and that leads to thirst. Many people have observed the increase in urination that accompanies drinking, but they may not realize that they can easily get into a vicious cycle as a result. Loss of body water leads to thirst. Thirst leads to more drinking—but drinking of what? The only fluid that will relieve dehydration is water, but the thirsty person welcomes any cold fluid—even concentrated alcohol—that relieves the dry mouth associated with thirst. If a person tries to quench thirst with concentrated alcoholic beverages, the thirst only becomes worse.

What are the nutritional consequences of drinking alcohol regularly?
If you are a light drinker in good health and otherwise well nourished, the occasional consumption of alcohol will probably have little effect on your nutritional status. The biggest risk to you will come most likely from the additional calories provided by alcohol; these extra calories may contribute to unwanted weight gain (see Table 9-6).[67] This is not to say that alcohol has *no effect* on your nutritional status, for it does. Alcohol causes fundamental changes in metabolism that occur whenever you consume alcohol. The extent to which alcohol affects your nutritional status depends on how much alcohol you consume and your current nutritional and health status.

If you drink excessively on a regular basis, your nutritional status will become compromised. Protein deficiency can develop, both from the depression of protein synthesis in the cells and, in the drinker who substitutes alcohol for food, from poor diet.

Alcohol affects every tissue's metabolism of nutrients in other ways. Stomach cells become inflamed and vulnerable to ulcer formation. Intestinal cells fail to absorb thiamin, folate, and vitamin B_{12}. Liver cells lose efficiency in activating vitamin D, and they alter their production and excretion of bile. Rod cells in the retina, which normally process vitamin A alcohol (retinol) to the form needed in vision (retinal), find themselves processing drinking alcohol instead. The kidney excretes increased quantities of magnesium, calcium, potassium, zinc, and folate.

Acetaldehyde interferes with metabolism, too. It dislodges vitamin B_6 from its protective binding protein so that it is destroyed, causing a vitamin B_6 deficiency and thereby lowered production of red blood cells.

Do alcoholics have special nutritional problems?
Definitely. Some alcoholics tend to substitute alcohol for food and can consume more than 50 percent of their calories from alcohol. Because of the metabolic derangements that occur with alcohol consumption, the chronic alcoholic tends to develop hyperlipidemia (a high blood triglyceride level), a **fatty liver** that ultimately leads to **cirrhosis** of the liver, and impaired kidney function. Over time, a high level of

TABLE 9-6 CALORIES IN ALCOHOLIC BEVERAGES AND MIXERS

Beverage	Amount (oz)	Calories
Beer	12	150
Light beer	12	80–130
Gin, rum, vodka, whiskey (80 proof)	1½	100
Dessert wine	3½	140
Table wine (red, rosé, white)	3½	85
Wine cooler	12	170
Liqueurs	1½	155–185
Tonic, ginger ale	12	125
Cola	12	150
Fruit juice	8	110
Club soda, plain seltzer, diet soda pop	12	0

alcohol intake may also damage the lining of the stomach, produce low blood sugar and high blood concentrations of uric acid, and cause metabolic bone disease. Even fairly young men in their thirties and forties can develop osteoporosis from consuming excessive alcohol over the course of 10 to 15 years.[68] In addition, alcoholics have an increased risk of hypertension, stroke, and a variety of cancers, particularly cancers of the tongue, mouth, pharynx, esophagus, larynx, and liver, the latter occurring because of cirrhosis.

I've heard that alcohol and other drugs interact in a dangerous way. Is this true?

Yes, they do. The liver's special treatment of alcohol is reflected in its altered handling of other drugs as well as nutrients. The liver possesses an enzyme system in addition to the enzyme alcohol dehydrogenase that metabolizes *both* alcohol *and* other drugs (any compounds that have certain chemical features in common). Called the **MEOS (microsomal ethanol-oxidizing system),** this system handles only about one-fifth of the total alcohol a person consumes, but the MEOS enlarges if repeatedly exposed to alcohol. This may not make the drinker able to handle much more alcohol at a time than before, because the total alcohol-metabolizing ability of the MEOS is small, but the effect on the ability to metabolize other drugs is considerable.

When the MEOS enlarges, it makes the body able to metabolize other drugs much faster than before. This can make it confusing and tricky to work out the correct doses of medications. The physician who prescribes sedatives to be taken every four hours, for example, assumes that the MEOS will dispose of the drug at a certain predicted rate. Well and good; but in a client who is a heavy drinker, the MEOS is adapted to metabolizing large quantities of alcohol. It therefore metabolizes the other drug extra fast. The drug's effects wear off unexpectedly fast, leaving the client undersedated. Imagine a surgeon's alarm if a client wakes up on the table during an operation! A skilled anesthesiologist always asks clients about their drinking patterns before putting them to sleep.

An enlarged MEOS will oxidize drugs *faster* than expected, but only as long as there is no alcohol in the system. If the person drinks and uses the drug at the same time, the drug will be metabolized more *slowly* and so will be much more potent. Since the MEOS is busy disposing of alcohol, the other drug can't be handled until later, and the dose may build up to where it greatly oversedates, or even kills, the user.

What does alcohol do to the brain?

Alcohol acts as a **narcotic.** It was used for centuries as an anesthetic because of its ability to deaden pain. But it wasn't a very good anesthetic because one could never be sure how much a person would need and how much would be a lethal dose. As new, more predictable anesthetics were discovered, they quickly replaced alcohol. However, alcohol continues to be used today as a kind of anesthetic on social occasions, to help people relax or to relieve anxiety. People think that alcohol is a stimulant because it seems to make them lively and uninhibited at first. Actually, though, the way it does this is by sedating *inhibitory* nerves, which are more numerous than excitatory nerves. Ultimately, alcohol acts as a depressant because it affects all the nerve cells.

When alcohol flows to the brain, it first sedates the frontal lobe, the reasoning part. As the alcohol molecules diffuse into the cells of this lobe, they interfere with reasoning and judgment. If the drinker drinks faster than the rate at which the liver can oxidize the alcohol, the speech center of the brain becomes narcotized, and the area that governs reasoning becomes more incapacitated. Later, the cells of the brain responsible for large-muscle control are affected; at this point, people under the influence stagger or weave when they try to walk. Finally, the conscious brain is completely subdued, and the person passes out. Now, luckily, the person can drink no more. This is fortunate because a higher dose's anesthetic effect could reach the deepest brain centers that control breathing and heartbeat, and the person could die. Table 9-7 shows the blood alcohol levels that correspond with progressively greater intoxication.

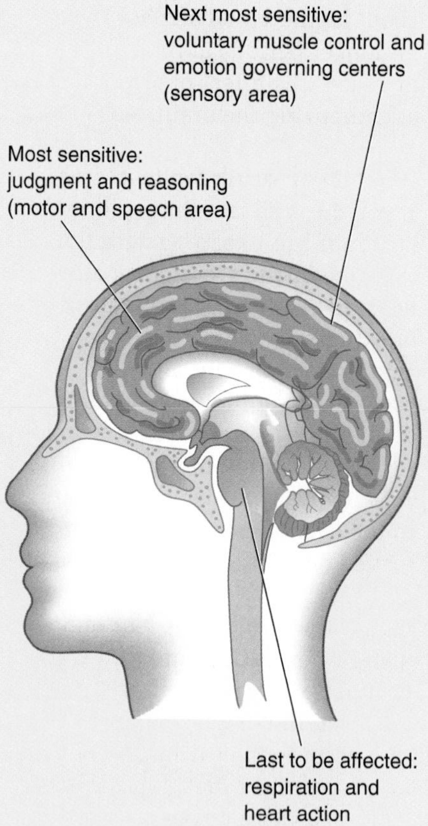

Alcohol is rightly termed an anesthetic because it puts brain centers to sleep in order: first the cortex, then the emotion-governing centers, then the centers that govern muscular control, and finally the deep centers that control respiration and heartbeat.

Why, then, don't people die from drinking alcohol?

Since the brain centers respond to alcohol in the order just described, people usually pass out before they can drink enough to kill them. It is possible, though, to drink fast enough that the effects of alcohol continue to accelerate after one has fallen asleep. The deaths that take place during drinking contests are attributed to this effect. The drinker drinks fast enough before passing out to receive a lethal dose.

The lack of glucose for the brain's function and the length of time needed to clear the blood of alcohol account for some diverse consequences of drinking. Responsible aircraft pilots know that they must allow 24 hours for their bodies to clear alcohol completely, and they refuse to fly any sooner. Major airlines enforce this rule. Women who might become pregnant are warned to abstain from the use of alcohol. Alcohol consumption during pregnancy can cause fetal alcohol syndrome—a condition of irreversible mental and physical retardation of the fetus.[69] One of the effects of an acute dose in experimental animals is to collapse the umbilical cord temporarily, depriving the developing fetus of oxygen. This can occur even before a woman knows that she is pregnant.

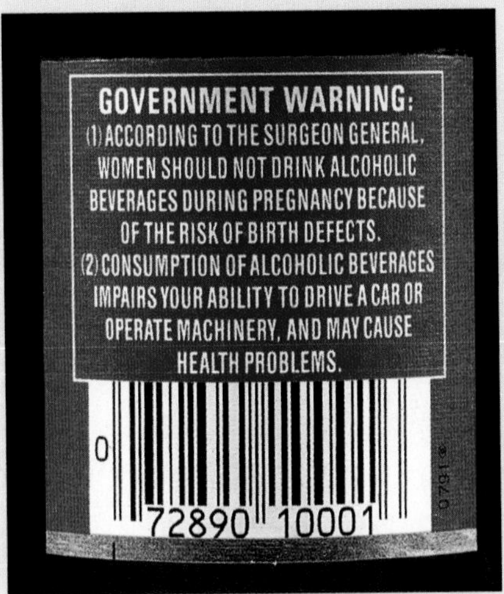

GOVERNMENT WARNING:
(1) ACCORDING TO THE SURGEON GENERAL, WOMEN SHOULD NOT DRINK ALCOHOLIC BEVERAGES DURING PREGNANCY BECAUSE OF THE RISK OF BIRTH DEFECTS.
(2) CONSUMPTION OF ALCOHOLIC BEVERAGES IMPAIRS YOUR ABILITY TO DRIVE A CAR OR OPERATE MACHINERY, AND MAY CAUSE HEALTH PROBLEMS.
0
72890 10001

TABLE 9-7 ALCOHOL DOSES AND BRAIN RESPONSES

Number of Drinks*	Blood Alcohol Level (%)†	Brain Response
2	0.05	Judgment impaired.
4	0.10‡	Emotional control impaired; impaired ability to operate a vehicle.
6	0.15	Muscle coordination and reflexes impaired.
8	0.20	Vision impaired.
12	0.30	Drunk, totally out of control.
14	0.35	Stupor.
More than 14	0.40–0.60	Total loss of consciousness; finally, death.

*Taken within an hour or so.
†Percentage of blood alcohol in a 150-pound person.
‡The legal limit for intoxication according to most states' highway safety ordinances.

MINIGLOSSARY

ACETALDEHYDE (ass-et-AL-duh-hide) a substance to which drinking alcohol (ethanol) is metabolized.

ALCOHOL DEHYDROGENASE a liver enzyme that converts ethanol to acetaldehyde. The MEOS also oxidizes alcohol.

ALCOHOLISM a dependency on alcohol marked by compulsive uncontrollable drinking with negative effects on physical health, family relationships, and social health.

BINGE DRINKER a person who drinks 4 or more drinks in a short period.

CIRRHOSIS (seer-OH-sis) advanced liver disease, often associated with alcoholism, in which liver cells have died and hardened and have permanently lost their function.

DRINK a dose of any alcoholic beverage that delivers one-half ounce of pure ethanol:
5 ounces of wine
12 ounces of beer
1½ ounces of hard liquor (whiskey, gin, rum, or vodka)

DRUGS substances that can modify one or more of the body's functions.

EUPHORIA (you-FORE-ee-uh) a feeling of great well-being that people often seek through the use of drugs such as alcohol.
eu = good
phoria = bearing

FATTY LIVER an early stage of liver disease seen in several conditions (kwashiorkor, alcoholic liver disease), characterized by accumulation of fat in the liver cells.

MEOS (MICROSOMAL ETHANOL-OXIDIZING SYSTEM) a system of enzymes in the liver that oxidize not only alcohol but also several classes of drugs. (The **microsomes** are tiny particles of membranes with associated enzymes that can be collected from broken-up cells.)
micro = tiny
soma = body

NARCOTIC (nar-KOT-ic) any drug that dulls the senses, induces sleep, and becomes addictive with prolonged use.

PROBLEM DRINKER (or alcohol abuser) a person who experiences psychological, social, family, employment, or school problems because of alcohol. Problem drinkers often binge drink and turn to alcohol when facing problems or making decisions.

Suppose I want to drink but not get drunk. How can I go about drinking in moderation?
If you want to drink socially, you should drink slowly, and you should sip, not gulp, your drinks. You should eat available snacks or a light meal along with the alcoholic beverages. With this strategy, the alcohol molecules should dribble slowly enough into the liver cells, allowing the enzymes to handle the load. Spacing of drinks is important, too. It takes about an hour and a half to metabolize one drink, depending on your body size, previous drinking experience, how recently you have eaten, and how you are feeling at the time.

Finally, you might elect to dilute your alcoholic beverages over the course of your drinking period. For example, after you've drunk half a glass of wine, fill up the glass with a non-cola soft drink to make a wine cooler. If you choose a low-calorie carbonated drink, you eliminate extra calories and alcohol. Or choose a low-alcohol beer or wine cooler.

TABLE 9-8 DIAGNOSTIC SIGNS OF ALCOHOLISM

The presence of three or more of these conditions is required to make a diagnosis.

- Tolerance—higher intakes of alcohol are needed to achieve intoxication.
- Withdrawal symptoms—anxiety, agitation, increased blood pressure, or seizures.
- Impaired control—the person intends to have one or two drinks, but drinks many more instead, or the person tries to quit drinking, but fails.
- Disinterest—a loss of interest in social, family, job, or school activities.
- Time—a great deal of time is spent drinking alcohol or recovering from excess drinking.
- Impaired ability—at work, school, or home—due to intoxication or withdrawal symptoms.
- Personal problems—the person continues drinking despite medical, legal, psychological, family, employment, or school problems.

Source: Adapted from *Diagnostic and Statistical Manual of Mental Disorders*, 4th ed. (Washington, D.C.: American Psychiatric Association, 1994).

What can I do to help a friend who has drunk too much? Does it help to walk my friend around?
No, don't wear yourself out walking your friend around the block (and never allow your friend to drive a vehicle while under the influence of alcohol). The muscles have to work harder, but since they can't metabolize alcohol, they can't help clear it from the blood. Time is the only thing that will do the job; each person has a particular level of the enzyme alcohol dehydrogenase, and it clears the blood at a steady rate. This is not true for most nutrients. If you bring in more of a nutrient, generally the body can promptly step up the rate at which it metabolizes that nutrient. But not with alcohol.

Nor will it help to give your friend a cup of coffee. Caffeine is a stimulant, but it won't speed up the metabolism of alcohol. The police say ruefully, "If you give a cup of coffee to a drunk, you'll just have a wide-awake drunk on your hands."

How can you know whether you or a friend has or is developing a problem with alcohol?
Sometimes, a person's judgment may tell him that he should limit himself to two drinks at a party, but the first drink may take his judgment away, so he has many more. The failure to stop drinking as planned, on repeated occasions, is a danger sign that indicates that the person should not drink at all. The diagnostic signs of alcoholism are listed in Table 9-8.

If you answer yes to one or more of the following questions, it is suggestive of an alcohol problem and you may want to seek a professional evaluation:[70]

- Have you ever felt you ought to cut down on drinking?
- Have people annoyed you by criticizing your drinking?
- Have you ever felt bad or guilty about your drinking?
- Have you ever had a drink first thing in the morning to steady your nerves or get rid of a hangover?

Early identification and intervention—involving abstinence from alcohol—are critical to the successful treatment of alcoholism. If you think your own drinking patterns are not normal or beyond moderate, if your drinking has caused problems in your life, or if the alcohol abuse of a friend or relative is of concern to you, contact the National Clearinghouse for Alcohol and Drug Information (1-800-729-6686) for more information, (www.health.org). Additional resources are listed in Appendix A.

Nutrition on the Web

nutrition.wadsworth.com	Go to the *Personal Nutrition* site to check for the latest updates to chapter topics or to access links to related Web sites.
www.cdc.gov/nccdphp/sgr/sgr.htm	Look here for the Surgeon General's Report on Physical Activity.
www.whitehouse.gov/WH/PCPFS/html/fitnet.html	Information from the President's Council on Fitness and Sports.
www.ama-assn.org/insight/gen_hlth/fitness/fitness.htm	The American Medical Association's Health Insight Web page for fitness basics, interactive assessments, and nutrition advice.
www.sportsci.org	Web page of Sportscience News.
www.acsm.org	The American College of Sports Medicine Web site.
www.shapeup.org/fitness	Provides practical advice on fitness and nutrition.
www.physsportsmed.com	Web site of the *Physician and Sportsmedicine Journal*.
www.cdc.gov/nccdphp/dnpa/	Provides links to many nutrition and physical activity sites.
www.nutrifit.org	SCAN: Sports, Cardiovascular, and Wellness Nutritionists—a dietetic practice group of the American Dietetic Association.
www.gssiweb.com	The Gatorade Sports Science Institute provides updates on exercise science and sports drinks.
www.nal.usda.gov/fnic/etext/fnic.html	See the sport nutrition link for nutrition and exercise information.
www.ncrhi.org	The National Council for Reliable Health Information Web site offers current information on ergogenic aids and nutrition fads.
www.quackwatch.com	Quackwatch: Your Guide to Health Fraud and Quackery.
www.navigator.tufts.edu	Provides links to nutrition and fitness-related sites.
www.ncbi.nlm.nih.gov/PubMed/	A search engine to help you locate information about exercise.
www.health.org	The National Clearinghouse for alcohol and drug information.

Check Yourself . . .

1. Name and describe the four components of fitness. Which has top priority?
2. What are the benefits of exercise?
3. Compare aerobic versus anaerobic exercise. Give examples.
4. Define target heart rate and calculate your own target heart rate.
5. What is the recommended prescription for cardiovascular fitness?
6. What is the training effect of cardiovascular exercise?
7. Define anabolic steroids; what are the potential side effects associated with steroid use?
8. How much protein is recommended for the diet of the athlete?
9. What is the best fluid replacement for exercise?
10. Describe the components of a healthful diet for athletic performance.

Answers to Check Yourself questions are found in Appendix G.

10

The Life Cycle: Conception Through the Later Years

CONTENTS

Pregnancy: Nutrition for the Future

Pregnancy Readiness Scorecard

Nutrition Action: Not for Coffee Drinkers Only

Healthy Infants

Early and Middle Childhood

The Importance of Teen Nutrition

Nutrition Action: Food Allergy—Nothing to Sneeze At

Nutrition in Later Life

Aging Scorecard

The Savvy Diner: Meals for One

Looking Ahead: Growing Seasoned

Spotlight: Nutrition and Cancer Prevention

How far you go in life depends on your being tender with the young, compassionate with the aged, sympathetic with the striving, and tolerant of the weak and the strong. Because someday in life you will have been all of these.

George Washington Carver
(1864–1943, American botanist)

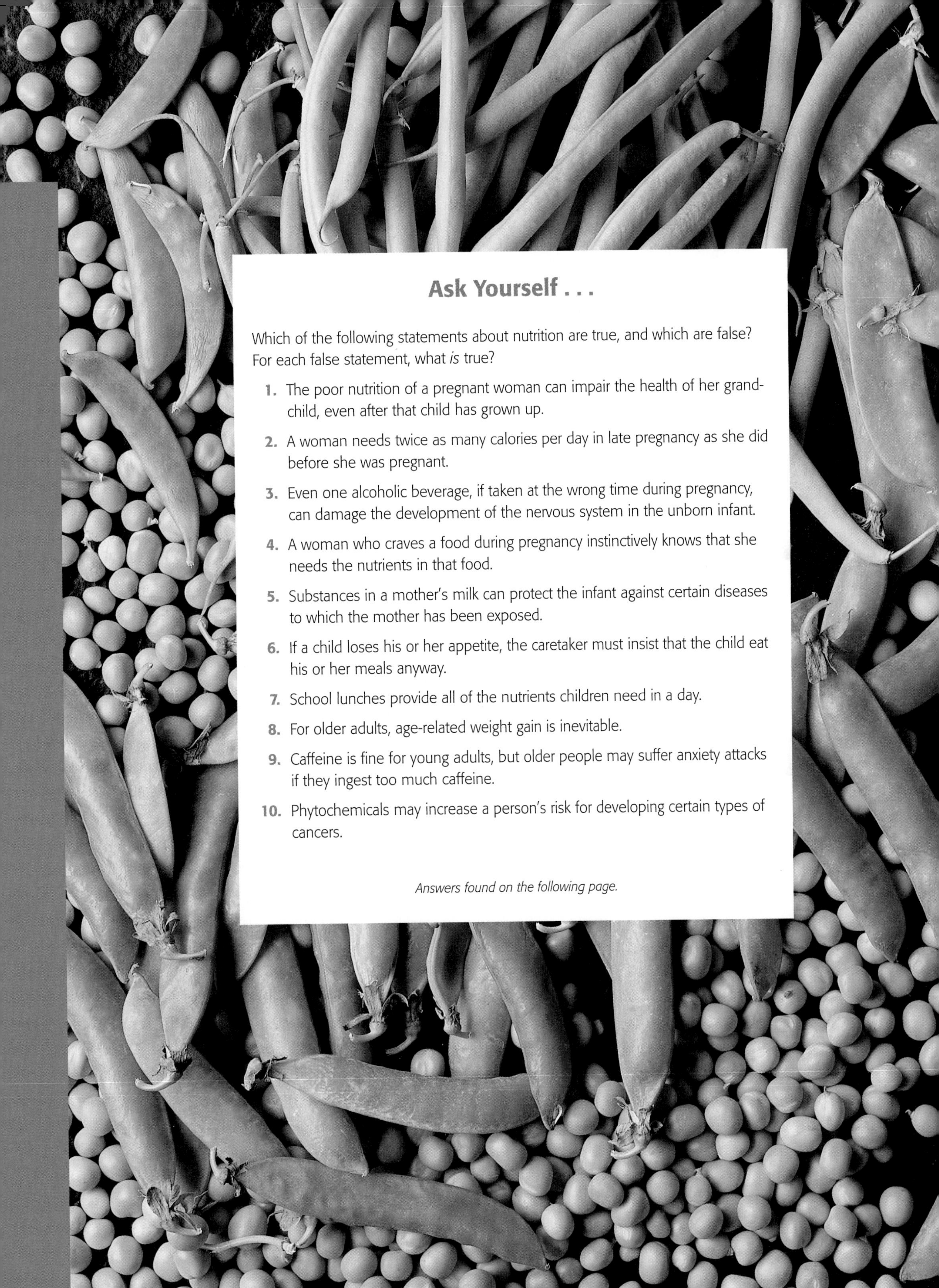

Ask Yourself . . .

Which of the following statements about nutrition are true, and which are false? For each false statement, what *is* true?

1. The poor nutrition of a pregnant woman can impair the health of her grandchild, even after that child has grown up.
2. A woman needs twice as many calories per day in late pregnancy as she did before she was pregnant.
3. Even one alcoholic beverage, if taken at the wrong time during pregnancy, can damage the development of the nervous system in the unborn infant.
4. A woman who craves a food during pregnancy instinctively knows that she needs the nutrients in that food.
5. Substances in a mother's milk can protect the infant against certain diseases to which the mother has been exposed.
6. If a child loses his or her appetite, the caretaker must insist that the child eat his or her meals anyway.
7. School lunches provide all of the nutrients children need in a day.
8. For older adults, age-related weight gain is inevitable.
9. Caffeine is fine for young adults, but older people may suffer anxiety attacks if they ingest too much caffeine.
10. Phytochemicals may increase a person's risk for developing certain types of cancers.

Answers found on the following page.

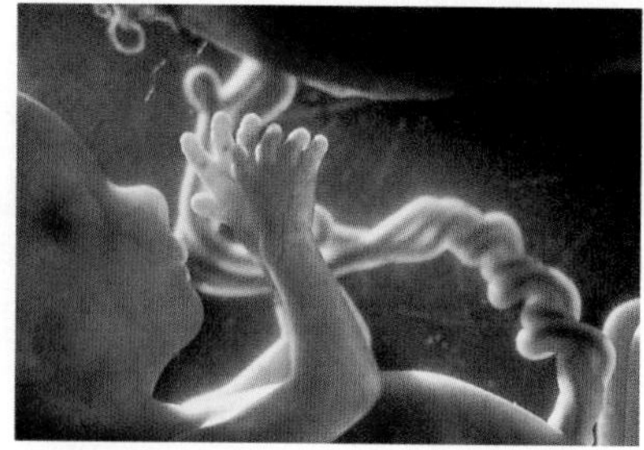

Notice the umbilical cord connecting this 16-week-old fetus with the **placenta.** The placenta is the organ inside the uterus in which maternal blood vessels lie side by side with fetal blood vessels entering it through the umbilical cord. This close association between the two circulatory systems permits the mother's bloodstream to deliver nutrients and oxygen to the fetus and to carry away fetal waste products.

By the time you are 65 years old, you will have eaten about 100,000 pounds of food. Each bite may or may not have brought with it the nutrients you needed. The impact of the food you have eaten, together with your lifestyle habits, accumulates over a lifetime, and people who have lived and eaten differently all their lives are in widely different states of health by the time they reach 65.

Nutrition shares with other lifestyle factors the responsibility for maintaining good health. The complete prescription for good health presented in Chapter 1 reads as follows: avoid excess alcohol, don't smoke, maintain a desirable weight, exercise regularly, get regular sleep, and eat nutritious, regular meals. Nutrition is represented by three of the six items on this list—one-half of the total. A person who abides by good health habits can expect to delay the onset of even minimal disability by several years, compared with a person who abides by few or none of them (see Figure 10-1).[1] If you subscribe to the view that your job is to accept the things you can't control and control the things you can, your nutritional health falls into the second category and deserves your conscientious attention. This chapter follows people through the life cycle and attends to their special nutritional needs at each stage.

PREGNANCY: NUTRITION FOR THE FUTURE

The only way nutrients can reach the developing fetus in the uterus is through the **placenta,** the special organ that grows inside the uterus to support the new life. If the mother's nutrient stores are inadequate early in pregnancy when the placenta is developing, the fetus will develop poorly, no matter how well the mother eats later. After getting such a poor start on life, a female child may grow up poorly equipped to support a normal pregnancy, and she, too, may bear a poorly developed infant. Thus, the poor nutrition of a woman during her early pregnancy can impair the health of her *grandchild.*

PRENATAL prior to birth.

POSTNATAL after birth.

Infants born of malnourished mothers are more likely than healthy women's infants to become ill, to have birth defects, and to suffer retarded mental or physical development. This remains true even if they later receive abundant, nourishing food. Malnutrition in the **prenatal** and early **postnatal** periods also affects learning ability and behavior. Clearly, it is critical to provide the best nutrition at the early stages of life.

Nutritional risk factors in pregnancy:

- Age 15 or under
- Unwanted pregnancy
- Many pregnancies close together (depletes nutrient stores)
- History of poor pregnancy outcome
- Poverty
- Lack of access to health care
- Low education level
- Inadequate diet (e.g., due to food faddism or dieting)
- Cigarette smoking
- Alcohol or drug abuse
- Chronic disease requiring special diet (e.g., diabetes)
- Underweight or overweight
- Insufficient or excessive weight gain in pregnancy
- Carrying twins or triplets

A number of factors contribute to maternal and infant health.[2] Genetic, environmental, and behavioral factors affect risk and the outcome of pregnancy. A woman's nutrition prior to and throughout pregnancy is crucial both to her health and to the growth, development, and health of the infant she conceives. Ideally, a woman will start pregnancy at a healthful weight, with filled nutrient stores, and with the firmly established habit of eating a balanced and varied diet. The Pregnancy Readiness Scorecard permits women to evaluate their nutritional readiness for pregnancy and to identify the eating habits that might need improvement.

Nutritional Needs of Pregnant Women

For most women, nutrient needs during pregnancy and lactation are higher than at any other time in their adult life and are greater for certain nutrients than for

Ask Yourself Answers: **1.** True. **2.** False. A woman needs only 15 percent more calories per day during pregnancy than she did before. **3.** True. **4.** False. A woman's cravings during pregnancy do not seem to reflect real physiological needs. **5.** True. **6.** False. A child's appetite regulates food intake to meet need; caretakers should not force food on children because this will only create conflict. **7.** False. School lunch is designed to provide one-third of the nutrients schoolchildren need in a day. **8.** False. Although age-related weight gain is a fact of life for many people, it is not inevitable; consuming low-fat meals within your calorie allowance and making physical activity part of your daily routine can help. **9.** False. Too much caffeine can cause anxiety attacks in individuals of any age, even children. **10.** False. Phytochemicals found in fruits, vegetables, and other foods may decrease one's cancer risk.

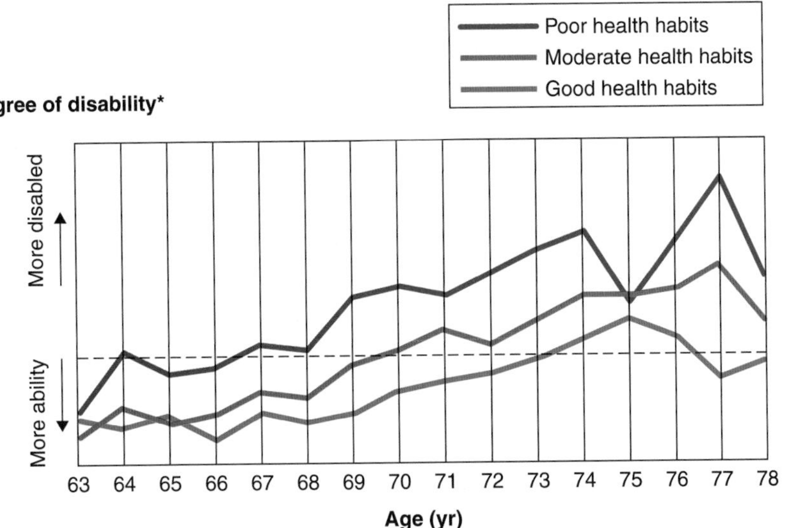

FIGURE 10-1
HEALTHY AGING

In a study of more than 1,700 people, those who smoked the least, maintained a healthy weight, and exercised regularly not only lived longer but postponed disability. As shown here, people with the best health habits delayed the onset of even minimal disability—to about age 73, compared with age 66 for those with the poorest health habits.

*The horizontal line represents minimal disability, defined as having some difficulty performing the everyday tasks of daily living (such as bathing, dressing, eating, walking, toileting, and getting outside).

Source: New England Journal of Medicine, April 9, 1998, pp. 1035–1041.

others (see Figure 10-2). Notice that although nutrient needs are much higher than usual, energy needs are not. An increase of only 15 percent of maintenance calories is recommended to support the metabolic demands of pregnancy and fetal development. The recommendation is an additional 300 calories per day during the second and third **trimesters.**

Nearly all nutrients are recommended in increased amounts during pregnancy and lactation.[3] The nutrient needs of pregnancy are best met by the routine intake of a variety of foods (see Table 10-1 on page 324). The nutrients deserving special attention in the diets of pregnant women include protein, folate, iron, zinc, and calcium, as well as vitamins known to be toxic in excess amounts.[4]

The recommended intake for protein is 60 grams —an additional 10 to 16 grams per day over nonpregnant requirements. Most women are already eating enough protein to cover the increased demand of pregnancy.

The pregnant woman's recommended folate intake is 50 percent greater than that of the nonpregnant woman due to the large increase in her blood volume and rapid growth of the fetus. Certain studies have shown that folate supplements given around the time of conception reduce the recurrence of **neural tube defects,** such as spina bifida, in the infants of women who previously have had such births.[5] To lower the risk of neural tube birth defects, women are advised to get the recommended amounts of folate—especially *before* becoming pregnant (400 micrograms) and during the first trimester of pregnancy (600 micrograms).[6]

As of 1999, all refined grain products (bread, cereal, cornmeal, farina, flour, grits, pasta, and rice) are fortified with folate. A woman can obtain the recommended amounts of folate by eating a breakfast cereal fortified with 100 percent of the daily value of folate, increasing her consumption of foods fortified with folate (for example, cereal, bread, pasta, or rice) and foods naturally rich in folate—green leafy vegetables, legumes, and citrus fruits. If the woman's dietary intake of folate is low, a 300 microgram folate supplement is recommended.*[7]

The body conserves iron even more than usual during pregnancy. Menstruation ceases, and absorption of iron increases up to threefold. However, the developing fetus draws on its mother's iron stores to create stores of its own to carry it through the first three to six months of life. This drain on the mother's supply

TRIMESTER one-third of the normal duration of pregnancy; the first trimester is 0 to 13 weeks, the second is 13 to 26 weeks, and the third trimester is 26 to 40 weeks.

NEURAL TUBE DEFECTS include any of a number of birth defects in the orderly formation of the neural tube during early gestation. Both the brain and the spinal cord develop from the neural tube; defects result in various central nervous system disorders. The two main types are *spina bifida* (incomplete closure of the bony casing around the spinal cord) and *anencephaly* (a partially or completely missing brain). Since the neural tube closes before the sixth week of pregnancy, women are advised to consume adequate folate from as early as three months prior to conception.

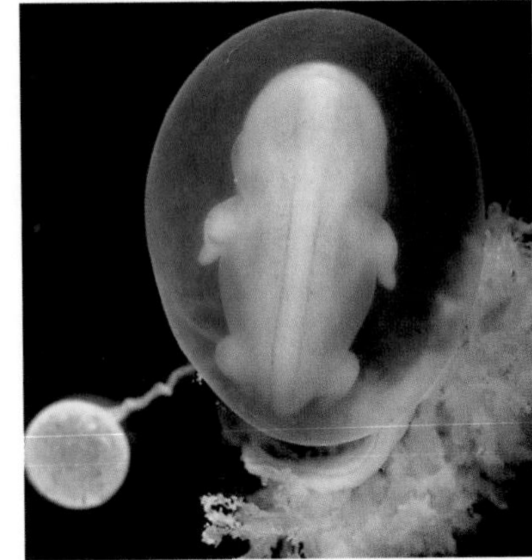

The neural tube (outlined by the delicate red arteries) has successfully closed after only six weeks of pregnancy.

*Folic acid—the synthetic form of folate used to fortify foods and found in folate supplements—is better absorbed by the body than the folate found naturally in fruits and vegetables. Women are advised to choose a variety of foods naturally high in folate together with foods that have been fortified with folic acid. Because high intakes of folate can mask a vitamin B_{12} deficiency, folate intakes should not exceed 1 milligram per day.

FIGURE 10-2
COMPARISON OF NUTRIENT NEEDS OF NONPREGNANT, PREGNANT, AND LACTATING WOMEN*

*For actual values, turn to the table on the inside front cover.

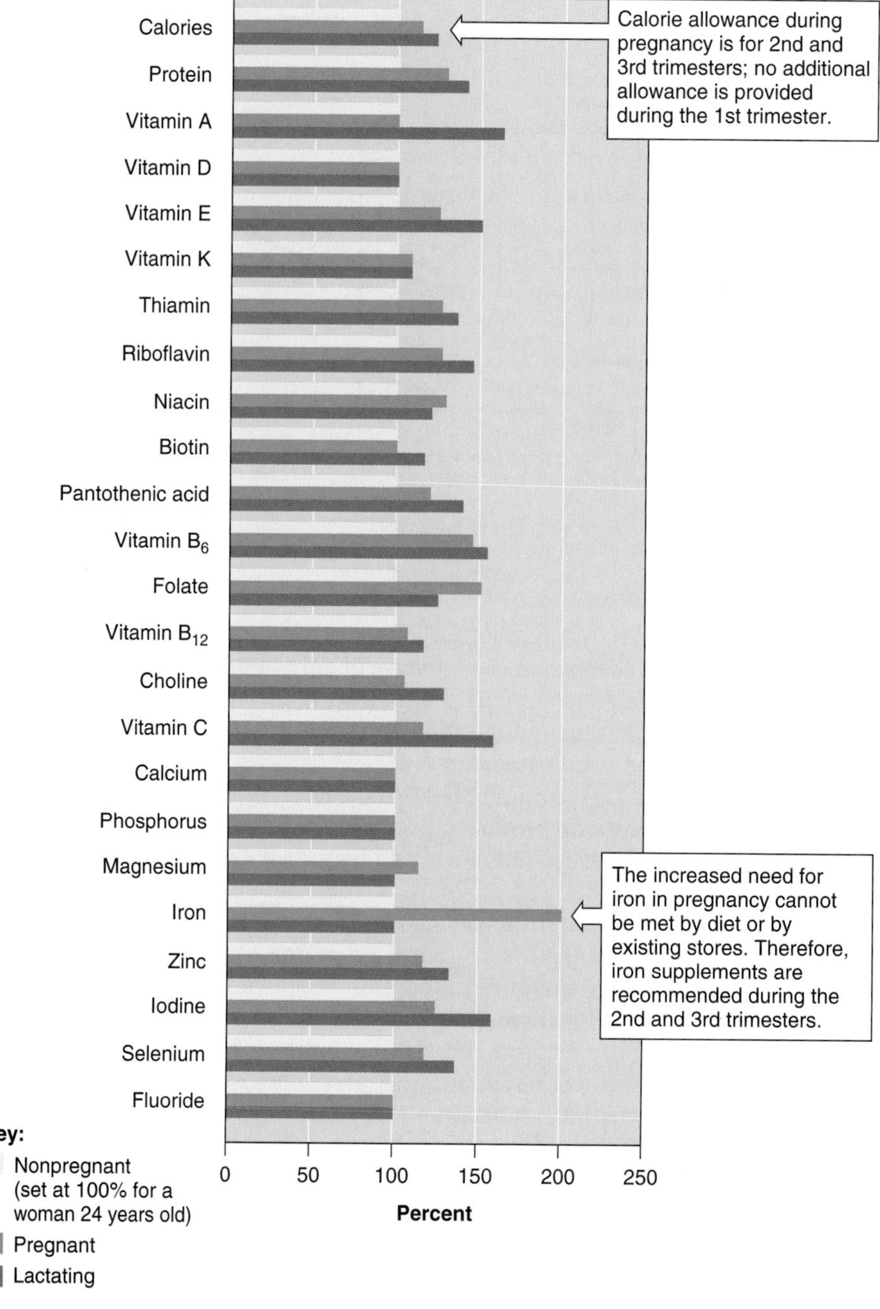

Foods containing iron:
- Red meat, fish, and other meat
- Dried fruits
- Legumes
- Whole-grain and fortified breads and cereals
- Dark green vegetables

Foods containing zinc:
- Lean meats and shellfish
- Legumes
- Fortified cereals
- Nuts and seeds
- Milk and milk products

can precipitate a deficiency; furthermore, the mother loses blood when she gives birth.

The recommended intake for iron during pregnancy is an additional 15 milligrams per day to meet maternal and fetal needs. Iron deficiency is a common problem among nonpregnant women, and as a result, many women begin pregnancy with diminished iron stores. For this reason, an iron supplement of 30 milligrams ferrous iron daily during the second and third trimesters is recommended. To facilitate absorption from the supplement, iron should be taken between meals with vitamin C-rich fruit juices or at bedtime. Since supplemental doses of iron greater than 30 milligrams can interfere with zinc absorption, pregnant women are encouraged to include good sources of zinc in their daily diets.

The DRI for calcium during pregnancy is 1,300 milligrams for teens and 1,000 milligrams for adults over 18 years of age. Intestinal absorption of calcium doubles early in pregnancy, and the mineral is stored in the mother's bones. Later,

PREGNANCY READINESS SCORECARD

Are you nutritionally ready for pregnancy? Score each question as shown. A score of 21 is perfect; scores below 3 per question identify areas that need improvement.

1. My body weight is desirable for my height, according to the standards given in this book:
   - **a.** right on target (3 *points*)
   - **b.** within 10% (2 *points*)
   - **c.** 10% to 20% above (1 *point*)
   - **d.** more than 20% above or 20% below (0 *points*)
2. I drink milk or use milk substitutes every day:
   - **a.** equivalent of 3 cups or more a day (3 *points*)
   - **b.** about 2 cups a day (2 *points*)
   - **c.** about 1 cup a day (1 *point*)
   - **d.** no milk or milk substitutes (0 *points*)
3. I eat vegetables daily:
   - **a.** five servings a day (3 *points*)
   - **b.** four servings a day (2 *points*)
   - **c.** three servings a day (1 *point*)
   - **d.** two or fewer servings a day (0 *points*)
4. I eat fruits daily:
   - **a.** four servings a day (3 *points*)
   - **b.** three servings a day (2 *points*)
   - **c.** two servings a day (1 *point*)
   - **d.** one or fewer servings a day (0 *points*)
5. I eat folate-rich foods, such as green leafy vegetables, orange juice, cantaloupe, legumes, and fortified grain products, daily:
   - **a.** three to four servings, or enough to provide 400 μg daily (3 *points*)
   - **b.** two to three servings, or enough to meet half the current DRI (1 *point*)
   - **c.** one or fewer servings, or less than half the current DRI (0 *points*)
6. I eat iron-rich foods, such as meats or legumes, daily:
   - **a.** two servings, or enough to meet the recommended intake (3 *points*)
   - **b.** one serving, or enough to meet about half the recommended intake (1 *point*)
   - **c.** less than one serving, or less than half the recommended intake (0 *points*)
7. I am physically fit because I have a well-established habit of exercising daily or every other day, and I will be able to continue exercising during pregnancy:
   - **a.** I am as fit as I can be (3 *points*)
   - **b.** I am fairly fit (2 *points*)
   - **c.** I am not fit (0 *points*)

during the last trimester of pregnancy when fetal skeletal growth is maximum and teeth are being formed, the fetus draws approximately 300 milligrams per day from the maternal blood supply.[8] Dairy products are recommended because they are also sources of vitamin D and riboflavin. This is particularly important for women under 25 years of age whose bone mineral density is still increasing. For the woman who normally consumes less than 600 milligrams of calcium a day, a 600-milligram supplement of calcium per day during pregnancy is recommended.

Routine supplementation with vitamins during pregnancy is not advised, and excess intakes of certain vitamins, notably vitamins A and D, can cause fetal malformations.[9] Supplements should not contain more than one to two times the recommended levels. More than 10,000 International Units (more than two times the recommended intake) of vitamin A taken per day early in pregnancy may cause malformations of the newborn, for example.[10]

Nutrient supplementation may be appropriate in certain circumstances, however. For example, supplements of vitamin D (10 micrograms per day) and vitamin B_{12} (2 micrograms per day) may be recommended for vegans, or a multivitamin-mineral supplement (providing 100 percent of the DRI) beginning in the second trimester for women who do not ordinarily consume an adequate diet or are in high-risk categories, such as women carrying more than one fetus, heavy cigarette smokers, and alcohol and other drug abusers.[11]

Foods containing calcium:

- Milk (4 cups per day will supply 1,200 mg of calcium)
- Other dairy products (yogurt, cheese)
- Green leafy vegetables (kale, broccoli, collard greens)
- Legumes
- Fortified soy milk
- Certain brands of tofu
- Fortified juice

TABLE 10-1 FOOD GUIDE FOR PREGNANT AND LACTATING WOMEN

	Number of Servings*	
Food	*Nonpregnant Woman*	*Pregnant or Lactating Woman*†
Breads/cereals	6 to 11	7 to 11 (7+)
Vegetables	3 to 5	4 to 5 (5+)
Fruits	2 to 4	3 to 4 (4+)
Meat/meat alternates	2 to 3	3 (3+)
Milk/milk products	2	3 to 4 (4+)

*Refer to Figures 2-5 and 2-6 in Chapter 2 for a summary of foods in each group and serving sizes.
†Numbers in parentheses indicate numbers of servings recommended for the pregnant teenager.

Because calorie needs increase less than nutrient needs, the pregnant woman must select foods of high nutrient density. A woman who already eats well can simply increase her servings of nutritious foods to meet her increasing nutrient needs. For most women, appropriate choices include such foods as fat-free milk, cottage cheese, lean meats, legumes, eggs, dark green vegetables, and whole-grain or fortified breads and cereals along with generous quantities of vitamin C-rich foods (refer to Table 10-1).

If the pregnant woman does not receive adequate nourishment and does not gain the recommended amount of weight, she might give birth to a baby of **low birthweight (LBW).** Not all small babies are unhealthy, but birthweight and length of gestation are the primary indicators of an infant's future health status. A low-birthweight baby is more likely than a normal-weight baby to experience complications during delivery and has a statistically greater chance of having physical and mental birth defects, developing diseases, and dying early in life.

LOW BIRTHWEIGHT (LBW) a birthweight of 5½ lb (2,500 g) or less, used as a predictor of poor health in the newborn and as a probable indicator of poor nutrition status of the mother during and/or before pregnancy. Normal birthweight for a full-term baby is 6½ to 8¾ lb (about 3,000 to 4,000 g). LBW infants are of two different types. Some are premature (they are born early). Others have suffered growth failure in the uterus; they may or may not be born early, but they are small.

Low birthweight in full-term infants is a major contributing factor to infant mortality.[12] Clearly, a key to reducing infant mortality is reducing the incidence of low-birthweight babies. A host of factors need to be addressed to do so: low socioeconomic status, the lack of access to health care, inadequate prenatal care, poor nutrition, low level of educational achievement, and unhealthful habits such as smoking, drinking, and other drug use.[13]

Maternal Weight Gain

Normal weight gain and adequate nutrition support the health of the mother and the development of the fetus. The recommendations for weight gain take into account a mother's prepregnancy weight for height or body mass index (BMI), as shown in Table 10-2. A woman who begins pregnancy at a healthful weight should gain between 25 and 35 pounds. Women pregnant with twins need to gain 35 to 45 pounds. An underweight woman needs to gain between 28 and 40 pounds; an obese woman, between 16 and 25 pounds. Weight gains at the upper end of the range are recommended for pregnant teenagers because of their increased risk of low weight gains and delivery of low-birthweight infants.

TABLE 10-2 RECOMMENDED WEIGHT GAIN FOR PREGNANT WOMEN

Weight Category*	Recommended Gain (lbs)
Underweight	28–40
Normal weight	25–35
Overweight	15–25
Obesity	≥ 15

*Underweight is defined as BMI < 18.5; normal weight as BMI 18.5 to 24.9; overweight as BMI 25.0 to 29.9; and obesity as BMI ≥ 30.0.

Source: Adapted from Food and Nutrition Board, *Nutrition During Pregnancy* (Washington, D.C.: National Academy Press, 1990).

Low weight gain in pregnancy is associated with increased risk of delivering a low-birthweight infant; such infants have high mortality rates. Excessive weight gain in pregnancy increases the risk of complications during labor and delivery as well as postpartum obesity.[14] Obese women also have an increased risk for complications during pregnancy, including hypertension and gestational diabetes.

The infant at birth will weigh only about 6½ to 8 pounds, but the body tissues the mother builds (blood, blood vessels, muscle, fat stores, and others) to provide a healthful environment for the fetus's development weigh more than 20 pounds (see Table 10-3). Weight gain should be lowest during the first trimester—two to four pounds for the entire trimester—followed by a steady gain of about a pound per week thereafter. If a woman gains more than the recommended amount of weight early in pregnancy, however, she should not try to diet in the last weeks. Dieting during pregnancy is not recommended. A *sudden* large weight gain may indicate the onset of pregnancy-induced hypertension (discussed later). A woman experiencing this type of weight gain should see her health-care provider.

Some of the weight a woman gains in pregnancy is lost at delivery. For the woman who has gained the recommended 25 to 35 pounds, the remainder is generally lost within a few months as her blood volume returns to normal and she loses the fluids she has accumulated.

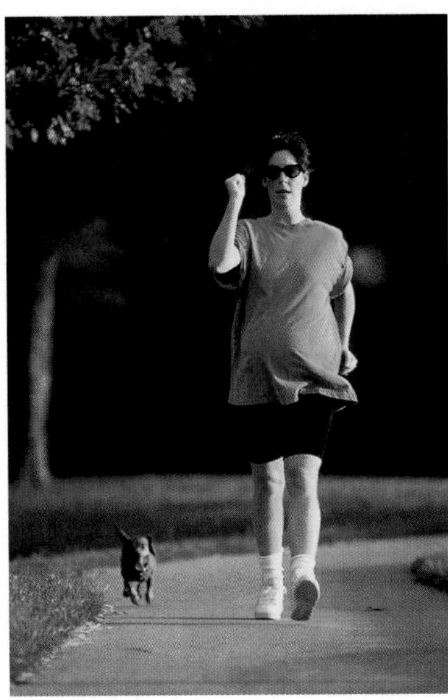

For most women, the surest way to have a healthy baby is to follow a healthy lifestyle: Get early prenatal care, eat a well-balanced diet, exercise regularly with your doctor's permission, and avoid cigarettes, alcohol, and other drugs.

TABLE 10-3 AN EXAMPLE OF THE PREGNANT WOMAN'S WEIGHT GAIN

Development	Weight Gain (lb)
Infant at birth	7–8
Placenta	1
Increase in mother's blood volume to supply placenta	4
Increase in mother's fluid volume	4
Increase in size of mother's uterus and the muscles to support it	2
Increase in size of mother's breasts	2
Fluid to surround infant in amniotic sac	2
Mother's fat stores (varies)	3–12
Total	25–35

Note: The pattern of gain should be about a pound a month for the first three months and a pound a week thereafter. Different patterns of weight gain are suggested for underweight, normal-weight, and overweight women.

Practices to Avoid

Optimal pregnancy outcome is influenced by maternal nutrient intake but can also be affected by maternal use of nonfood substances, excess caffeine, low-calorie diets, megadoses of certain vitamins, tobacco, alcohol, and illicit drugs. **Pica** refers to the craving of nonfood items having little or no nutritional value. Pica of pregnancy typically involves the consumption of dirt, clay, or laundry starch, but episodes of pica have included compulsive ingestion of such things as ice, paper, and coffee grounds.[15] The medical consequences of pica can include malnutrition as nonfood items replace nutritious foods in the diet, obesity from overconsumption of items such as starch, the ingestion of toxic compounds, or intestinal obstruction by consuming large amounts of clay or starch.

PICA the craving of nonfood items such as clay, ice, and laundry starch. Pica does not appear to be limited to any particular geographic area, race, sex, culture, or social status.

The Food and Drug Administration has advised pregnant women to avoid unnecessary consumption of caffeine because of animal studies suggesting that it causes birth defects.[16] Studies in humans have generally failed to show that caffeine use has a negative effect on pregnancy outcome, although limited evidence has shown that moderate to heavy use may contribute to lower infant birthweight.[17] Women who choose to use caffeine during pregnancy are generally advised to do so in moderation (the equivalent of one cup of coffee or two 12-ounce cola beverages a day) if at all (see the Nutrition Action feature starting on page 328).

Some practices are truly harmful, and their potential impact on pregnancy outcome is too great to risk. Low-carbohydrate or low-calorie diets that cause ketosis deprive the fetus's brain of needed glucose and cause congenital deformity. Most serious may be the invisible effects. For example, carbohydrate metabolism may be rendered permanently defective, or the infant's brain may be permanently damaged. Protein deprivation can cause children's height and head circumference to diminish markedly and irreversibly.

Another harmful maternal practice is smoking, which restricts the blood supply to the growing fetus, thereby limiting the delivery of nutrients and removal of wastes. Smoking stunts growth, thus increasing the risk of premature delivery, low infant birthweight, retarded development, and spontaneous abortions (fetal deaths).[18] Smoking is responsible for 20 percent to 30 percent of all

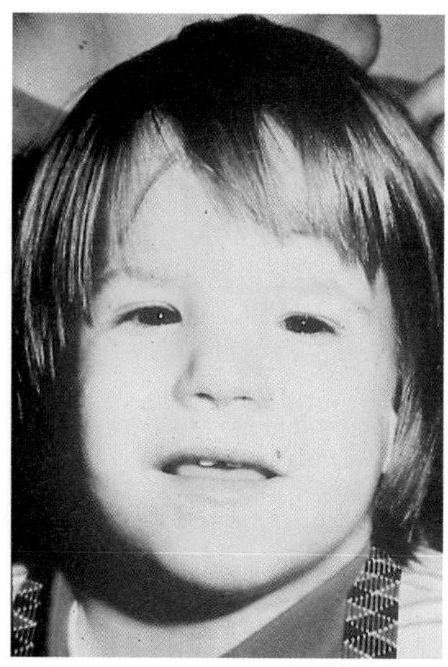

These facial traits (low nasal bridge, short eyelid opening, small head circumference, undeveloped groove in center of upper lip) are typical of fetal alcohol syndrome, caused by drinking alcohol during pregnancy. Irreversible abnormalities of the brain and other organs accompany these facial features.

FETAL ALCOHOL SYNDROME (FAS) the cluster of symptoms seen in an infant or child whose mother consumed excess alcohol during pregnancy, including retarded growth, impaired development of the central nervous system, and facial malformations. A lesser condition—called *fetal alcohol effect* (*FAE*)—causes learning impairment and other more subtle abnormalities in infants exposed to alcohol during pregnancy.

low-birthweight deliveries in the United States.[19] Sudden infant death syndrome (SIDS) has also been linked to a mother's smoking during pregnancy.[20]

During the past 20 years, research has confirmed that consumption of alcohol adversely affects fetal development.[21] Even as few as one or two drinks daily can cause **fetal alcohol syndrome (FAS)**—irreversible brain damage and mental and physical retardation in the fetus. The most severe impact of maternal drinking is likely to occur in the first month, before the woman even is sure she is pregnant. This preventable condition (FAS) is estimated to occur in approximately three infants per 1,000 live births and is the leading known cause of mental retardation in the United States.[22] Birth defects, low birthweight, and spontaneous abortions occur more often in pregnancies of women who drink even as little as 2 ounces of alcohol daily during pregnancy. Accumulating evidence that even one drink may be too much has led the American Academy of Pediatrics to take the position that women should stop drinking as soon as they *plan* to become pregnant.[23]

Drugs other than alcohol taken during pregnancy can also cause birth defects. A particularly dramatic example is the acne medication Accutane (isotretinoin), which causes major deformities during fetal development. Pregnant women should avoid taking all drugs except on the advice of their physician.

Common Nutrition-Related Problems of Pregnancy

Common physical problems in pregnancy include morning sickness and, later, constipation. The nausea of morning sickness seems unavoidable because it arises from the hormonal changes taking place early in pregnancy, but it can sometimes be alleviated. Suggested strategies to alleviate the nausea and vomiting of morning sickness include:

- Eat small, frequent meals alternating dry and fluid feedings.
- Before getting up in the morning, eat saltines, hard candies, or other dry starchy foods.
- Eat as soon as you feel hungry.
- Avoid fatty foods.
- Avoid highly seasoned foods.
- Avoid any specific food (especially foods with strong odors) causing nausea or vomiting.[24]

Later, as the hormones of pregnancy alter her muscle tone and the growing fetus crowds her intestinal organs, an expectant mother may complain of constipation. A high-fiber diet, plentiful fluid intake, and regular exercise will help to relieve this condition.

Women's cravings during pregnancy do not seem to reflect real physiological needs. If a woman craves pickles and chocolate sauce at two o'clock in the morning, for example, it is probably not because she lacks a combination of nutrients uniquely supplied by these foods. The woman is, however, expressing a need as real and as important as her need for nutrients—the need for support, understanding, and love. More serious problems needing control during pregnancy include hypertension and diabetes, described briefly in the following paragraphs.

Hypertension in Pregnancy. Ideally, a woman with preexisting hypertension has her blood pressure under control before becoming pregnant. Otherwise, she may have an increased risk of delivering a low-birthweight baby. Some women develop a *transient hypertension of pregnancy* during the second half of their pregnancy. Usually, this is a mild form of hypertension with no adverse effects on pregnancy outcome, and blood pressure returns to normal shortly after the baby is born. Sometimes, however, high blood pressure in a pregnant woman signals the onset of **pregnancy-induced hypertension.** Preeclampsia and eclampsia are hypertensive conditions induced by pregnancy. Preeclampsia is characterized by high blood pressure, protein in the urine, and generalized edema that may cause sudden, large weight gain from retained water. Fluid retention alone, which is quite common in pregnant women, is not sufficient to diagnose preeclampsia.[25] Warning signs of **preeclampsia** include severe and constant headaches; sudden weight gain (1 lb/day); swelling of face, hands, and feet; dizziness; and blurred vision. **Eclampsia,** the most severe form of this pregnancy-induced hypertension (PIH), is characterized by convulsions that may lead to coma. PIH can retard fetal growth and cause the placenta to separate from the uterus, resulting in stillbirth. Both conditions present serious health risks to mother and fetus and demand careful medical treatment.

PREGNANCY-INDUCED HYPERTENSION (PIH) high blood pressure that develops during the second half of pregnancy.

PREECLAMPSIA a condition characterized by hypertension, fluid retention, and protein in the urine.

ECLAMPSIA a severe extension of preeclampsia characterized by convulsions.

Diabetes. Infants born to women with diabetes are at greater risk for prematurity, congenital defects, excessively high birthweight, and respiratory distress syndrome.[26] Metabolic control of diabetes before and throughout pregnancy is critical. In some women, pregnancy can alter carbohydrate metabolism and precipitate a condition known as **gestational diabetes.** The abnormal blood glucose levels seen during pregnancy return to normal after pregnancy for about two-thirds of women diagnosed with the condition. Risk factors include age 25 or older, previous history of gestational diabetes, obesity, and a family history of diabetes.

GESTATIONAL DIABETES the appearance of abnormal glucose tolerance during pregnancy, with a return to normal following pregnancy.

Some women with gestational diabetes have the classic symptoms of diabetes—increased thirst, hunger, urination, weakness—but other women have no warning signs of the condition. For this reason, pregnant women at risk for developing gestational diabetes are screened for the condition between the twenty-fourth and twenty-eighth weeks of gestation.[27]

Women meeting one or more of the following criteria are screened for gestational diabetes:

- ≥ 25 years of age
- < 25 years of age and obese
- Family history of diabetes
- Member of an ethnic or racial group with a high prevalence of diabetes (Hispanic, Native American, Asian, African-American, or Pacific Islander)

Adolescent Pregnancy

More than a million teenagers become pregnant in the United States each year—one out of every five babies is born to a teenager—and more than a tenth of these mothers are under age 15. The complexity of social, emotional, and physical factors makes teen pregnancy one of the most challenging situations for meeting nutritional needs. According to a position paper from the American Dietetic Association, pregnant adolescents are nutritionally at risk and require intervention early in and throughout pregnancy.[28] Medical and nutritional risks are particularly high when the teenager is within two years of menarche (usually 15 years of age or younger).[29] Risks include higher rates among pregnant teens than in older women of pregnancy-induced hypertension, iron-deficiency anemia, premature birth, stillbirths, low-birthweight infants, and prolonged labor.

Pregnancy places adolescent girls, who are already at risk for nutrition problems, at even greater risk because of the increased energy and nutrient demands of pregnancy. To support the needs of both mother and infant, adolescents are encouraged to strive for pregnancy weight gains at the upper end of the ranges recommended for pregnant women (refer to Table 10-2). Those who gain between 30 and 35 pounds during pregnancy have lower risks of delivering low-birthweight infants.[30] Adequate nutrition can substantially improve the course and outcome of adolescent pregnancy.[31]

Emphasis on preparing young girls for future pregnancy is needed in local schools and public health programs. A model program for giving nutritional help to teenage mothers is, among others, the WIC (Women's, Infants' and Children's) program, a federally funded program that provides nutrition education and low-cost nutritious foods to low-income pregnant women, mothers, and their children.

NUTRITION ACTION

Not for Coffee Drinkers Only

In 1657, when merchants first introduced Londoners to a Middle Eastern brew known as coffee, they boasted it to be a "wholesome and physical drink," an elixir of health suitable for treating colds, coughs, gout, and many other ills.[32] Modern-day coffee drinkers have heard otherwise. During the past decade, the "coffee generation" has been subject to a barrage of reports linking coffee and other **caffeine**-containing products such as colas and tea with more than 100 diseases. Fingers have repeatedly pointed at caffeine as the culprit behind breast disease, cancer, heart disease, birth defects, and high blood pressure, to name just a few.

CAFFEINE a type of compound, called a methylxanthine, found in coffee beans, cola nuts, cocoa beans, and tea leaves. A central nervous system stimulant, caffeine's effects include increasing the heart rate, boosting urine production, and raising the metabolic rate.

Despite the brouhaha over the substance, the jury is still out as to whether caffeine is truly to blame. Scientists have yet to confirm long-standing suspicions that caffeine contributes to any health problems other than jitteriness. One reason for all the controversy is that much of the evidence linking the substance with different diseases has been clouded by a number of issues. Some studies do not measure sources of caffeine other than coffee and tea, such as soft drinks, chocolate, and certain medications. In addition, the amount of caffeine and other substances in coffee or tea can vary considerably depending on how the beverage is brewed, a fact most studies fail to take into account.

At least eight out of ten Americans consume caffeine, the most widely used behaviorally active drug in the world.

Consider the widely debated question of whether drinking coffee raises blood cholesterol. Granted, a great deal of strong evidence suggests that the beverage does contribute to high blood cholesterol and therefore to heart disease. The hitch is that most of the evidence comes from Scandinavia, where coffee is boiled rather than brewed in automatic drip coffeemakers or electric percolators. Subsequent research has found that whereas boiled coffee appears to boost blood cholesterol levels, filtered coffee does not, most likely because substances in boiled coffee other than caffeine may be the cholesterol-raising culprits.[33]

Contrary to popular belief, black coffee will not sober up a person who has had too many beers or other alcoholic drinks.

Another issue that most research has not filtered out is that although coffee drinking may not contribute to ill health in and of itself, it seems to be part of a lifestyle that does. After questioning some 2,600 men and women about their health habits, a group of Boston-based researchers found that women who opted for decaffeinated coffee exclusively were more likely than regular-coffee drinkers to, among other things, eat vegetables frequently and exercise regularly. Male decaf drinkers also tended to have adopted more healthful habits, such as eating low-fat diets, than did their regular-coffee-drinking counterparts. Thus, it may not be the coffee but rather the poor health habits that often go along with it that contribute to health problems.[34]

None of this is to say that caffeine is necessarily good for people. As anyone who can't get going in the morning without a cup of coffee or can of caffeinated soda is well aware, consuming caffeine day in and day out can be habit forming. In fact, some researchers have identified a condition called **caffeine dependence syndrome,** characterized by at least three of the four following criteria: withdrawal symptoms such as headache and fatigue; caffeine consumption despite knowledge that it may be causing harm; repeated, unsuccessful attempts to cut back on caffeine; and tolerance to caffeine.[35]

CAFFEINE DEPENDENCE SYNDROME dependence on caffeine characterized by at least three of the four following criteria: withdrawal symptoms such as headache and fatigue; caffeine consumption despite knowledge that it may be causing harm; repeated, unsuccessful attempts to cut back on caffeine; and tolerance to caffeine.

Along with people who are dependent on caffeine, people with certain medical conditions would do well to consume caffeine in moderation or avoid it completely. Pregnant women, for example, should limit the amount of caffeine they take in. Although the substance has never been proven to cause birth defects, it does cross the placenta and enter the fetus, where large amounts can affect the unborn baby's heart rate and breathing.[36] Some research also suggests that the amount of caffeine in three or four cups of coffee could raise the risk of suffering a miscarriage, perhaps by decreasing bloodflow through the placenta.[37] In addition, those with ulcers should steer clear of caffeinated *and* decaffeinated coffee, both of which stimulate the secretion of acid, which can irritate the stomach's lining.

For a healthy person, drinking one or two cups of coffee, tea, or cola a day does not seem to pose any hazard. Only those who are particularly sensitive to caffeine and suffer symptoms such as headaches, nervousness, and insomnia after consuming it really need to consider avoiding—or at least cutting back on—it. (See Table 10-4 to figure out how much caffeine you're taking in each day.)

TABLE 10-4 CAFFEINE COUNTDOWN

Drinks and Foods	Average (mg)
Coffee (6-oz cup)	
Brewed	103
Instant	57
Decaffeinated, brewed or instant	2
Espresso (1-oz cup)	40
Tea (6-oz cup)	
Brewed, black, steeped for 3 minutes	40
Instant iced tea (12-oz glass)	30
Soft drinks (12-oz can)	
Dr. Pepper	41
Colas:	
Regular	38–46
Diet	36–50
Caffeine free	0
Mountain Dew, Mello Yello	52–54
7-Up	46
Cocoa beverage (6-oz cup)	3–5
Chocolate milk beverage (8 oz)	5–8
Milk chocolate candy (1 oz)	7–18
Dark chocolate, semisweet (1 oz)	21
Baker's chocolate (1 oz)	25
Chocolate-flavored syrup (1 oz)	5
Drugs*	**Average (mg)**
Pain relievers (standard dose)	
Excedrin Extra Strength	130
Bayer Select Headache Pain Relief Formula	130
Maximum Strength Multi-Symptom Formula Midol	120
Stimulants	
NoDoz, Vivarin	200

*Because products change, contact the manufacturer for an update on products you use regularly.

Sources: J. A. T. Pennington, *Bowes & Church's Food Values of Portions Commonly Used, 16th ed.* (Philadelphia: J. B. Lippincott Company, 1994), pp. 381–383; *Physicians' Desk Reference for Nonprescription Drugs, 15th ed.* (Montvale, N.J.: Medical Economics Data Production Company, 1994), pp. 522, 526, 721, 733, 738.

If you drink coffee or a can or two of cola every day and decide to quit, make sure you do it gradually. Even moderate caffeine users who try to stop cold turkey often suffer from withdrawal symptoms, such as splitting headaches, fatigue, moodiness, and nausea. Try instead to cut back gradually by, say, no more than a cup or so every couple of days. You can do that by substituting decaffeinated coffee for some of the regular roast in your morning brew and gradually using more and more decaf and less regular coffee. Likewise, you can drink a glass of decaffeinated rather than caffeinated soda here and there until you've weaned yourself off the caffeinated version.

Chapter 9 discusses caffeine's effect on physical performance.

Smokers, incidentally, metabolize the caffeine in their bloodstreams more quickly than nonsmokers, thereby needing more of the substance to obtain its stimulating effect. When smokers try to give up cigarettes, however, the caffeine stays in the bloodstream longer, giving them a stronger than usual dose and possibly causing jitteriness and irritability on top of the jitters already occurring as a result of giving up nicotine. Thus, smokers who kick the cigarette habit may want to try to cut back on caffeine at the same time to avoid that problem.[38]

Nutrition of the Breastfeeding Mother

Adequate nutrition of the mother makes a highly significant contribution to successful lactation; without it, lactation is likely to falter or fail. The mother should continue to eat high-quality foods until the end of her pregnancy, not attempt to restrict her weight gain unduly, and plan to enjoy ample food and fluid at frequent intervals.

A nursing mother produces 30 ounces of milk a day, on the average, with wide variations possible. Current recommendations suggest that 500 calories to support this milk production come from added food and that the rest come from the stores of fat the mother's body has accumulated during pregnancy for this purpose. (Table 10-1 on page 324 shows a food pattern that will meet the lactating woman's nutrient needs.)

The period of lactation is the natural time for a woman to lose the extra body fat she accumulated during pregnancy. Once lactation has been established, if her choice of foods is judicious, a nursing mother can tolerate a calorie deficit and a gradual loss of weight (1 pound per week) without any effect on her milk output. Fat can only be mobilized slowly, however, and too large an energy deficit will inhibit lactation. On the other hand, if a mother does not breastfeed, she may not as easily lose the fat she gained during pregnancy.

HEALTHY INFANTS

The growth of infants directly reflects their nutritional well-being and is the major indicator of their nutrition status. A baby grows faster during the first year of life than ever again, doubling its birthweight during the first four to six months, and tripling its birthweight by the end of the first year (see Figure 10-3). Adequate nutrition during infancy is critical to support this rapid rate of growth and development. Clearly, from the point of view of nutrition, the first year is the most important year of a person's life. This section provides an overview of nutrient requirements, current recommendations for feeding healthy infants, and the relationship between infant feeding and selected pediatric nutrition issues.

Milk for the Infant: Breastfeeding

Toward the end of her pregnancy, a woman needs to decide whether to breastfeed her baby or not. If she plans to breastfeed, she should begin to prepare so that she can get started smoothly. It is wise, ahead of time, to read a handbook on breastfeeding, talk with other women who have successfully breastfed their babies, and seek family and medical support. Ideally, before the baby is born, the mother and a partner can attend classes together to obtain basic information regarding breastfeeding.

Breastfeeding has both emotional and physical health advantages.[39] Emotional bonding is facilitated by many events and behaviors of mother and infant during the early months and years; one of the first can be breastfeeding.

During the first two or three days of lactation, the breasts produce **colostrum,** a premilk substance containing antibodies and white cells from the mother's blood. Because it contains immunity factors, colostrum helps to protect the newborn infant from those infections against which the mother has developed an immunity—precisely those in the environment against which the infant needs protection. Entering the infant's body with the milk, these antibodies inactivate bacteria within the digestive tract, where they could otherwise cause intestinal infections.

Breast milk also contains antibodies, although not as much as colostrum. Colostrum and breast milk both contain the **bifidus factor** that favors the growth of the "friendly" bacteria *Lactobacillus bifidus* in the infant's digestive tract so that

COLOSTRUM (co-LAHS-trum) a milklike secretion from the breast, rich in protective factors, present during the first day or so after delivery and before milk appears.

BIFIDUS FACTOR (BIFF-id-us) a factor in colostrum and breast milk that favors the growth in the infant's intestinal tract of the "friendly" bacteria *Lactobacillus bifidus* so that other, less desirable intestinal inhabitants will not flourish.

other, harmful bacteria cannot grow there. Breast milk also contains the powerful antibacterial agent **lactoferrin,** as well as other factors, including several enzymes, several hormones (including thyroid hormone and prostaglandins), and lipids that help to protect the infant against infection. Breast-fed infants have lower rates of hospital admissions, ear infections, diarrhea, rashes, allergies, and other health problems than bottle-fed infants.[40]

Breast milk is tailor-made to meet the nutrient needs of the young infant. It offers its carbohydrate in the easy-to-assimilate form of lactose; its fat contains a generous proportion of the essential fatty acid linoleic acid; and its protein, alpha-lactalbumin, is one that the infant can easily digest. With the exception of vitamin D, its vitamin contents are ample. As for minerals, calcium, phosphorus, and magnesium are present in amounts appropriate for the rate of growth expected in a human infant, and breast milk is low in sodium. Its iron is highly absorbable, and the presence of a zinc-binding protein favors the absorption of the zinc it contains.

Breastfeeding provides other benefits as well. It protects a newborn against allergy development during the vulnerable first few weeks of life, the act of suckling favors a baby's normal tooth and jaw alignment, and breastfed babies are less likely to be obese because they are less likely to be overfed. A woman who wants to breastfeed can derive justification and satisfaction from all these advantages.

These attributes, along with the convenience and lower cost of breastfeeding, have led many organizations and medical experts to encourage breastfeeding for all normal full-term infants. The American Academy of Pediatrics recommends that infants receive breast milk for the first 12 months of life.[41] Despite the health benefits, however, the incidence of breastfeeding has declined since the mid-1980s.[42] Approximately 60 percent of mothers currently initiate breastfeeding—far lower than the *Healthy People 2000* goal, which was "to increase to at least 75 percent the proportion of mothers who breastfeed their babies." Only 22 percent of mothers, however, are still breastfeeding after six months. Analysis of data from a survey of mothers indicates that

FIGURE 10-3
GROWTH AND DEVELOPMENTAL CHARACTERISTICS FROM BIRTH THROUGH ONE YEAR

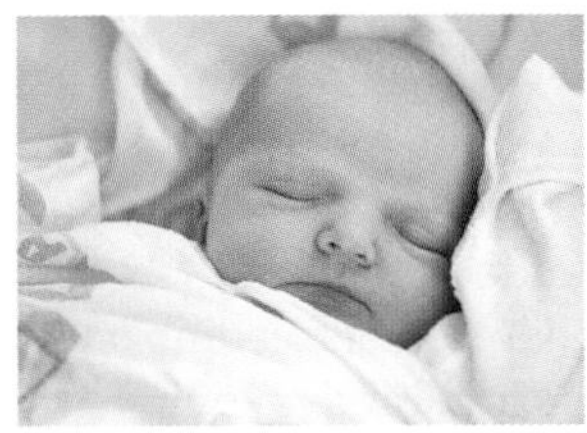

First Days of Life
Generally weighs from 7 to 9 pounds; length is 19 to 21 inches. Head is relatively large and has soft spot on top. Startles and sneezes easily. Jaw may tremble. May hiccup and spit up.

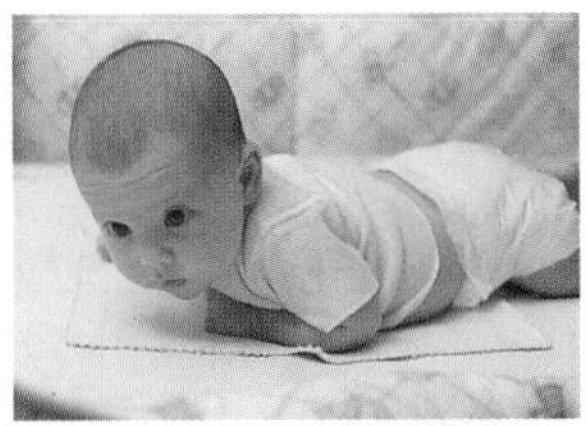

One Month
Has regained weight lost after birth and more. Lifts head briefly when placed on stomach. Whole body moves when infant is touched or lifted. Eats every two to four hours.

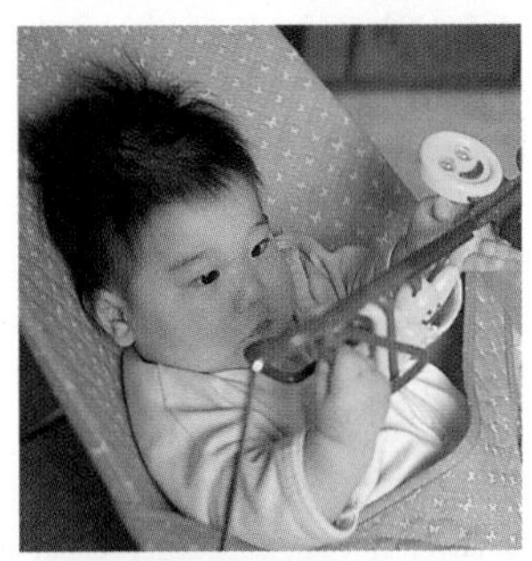

Four Months
Weight nearly doubled. Has grown three to four inches. Follows objects with eyes. Reaches toward objects with both hands; plays with fingers; puts fingers and objects into mouth. Holds head up steadily though back needs support. Attempts to roll over. Awake longer at feeding time. Eats seven or eight times per day. Sleeps six to seven hours at night.

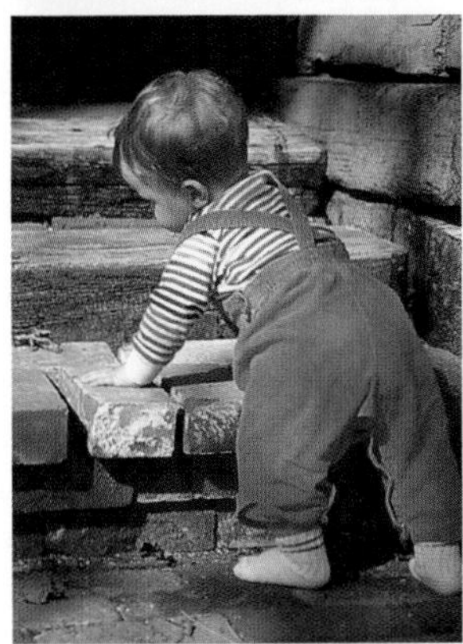

Eight Months
Gains in weight and height are less rapid, appetite has decreased. Rolls over; stands up with help; sits up; hitches self along the floor. Reaches for, grasps, and examines objects with hands, eyes, and mouth. Has one or two teeth. Takes two naps a day. Can feed self from bottle.

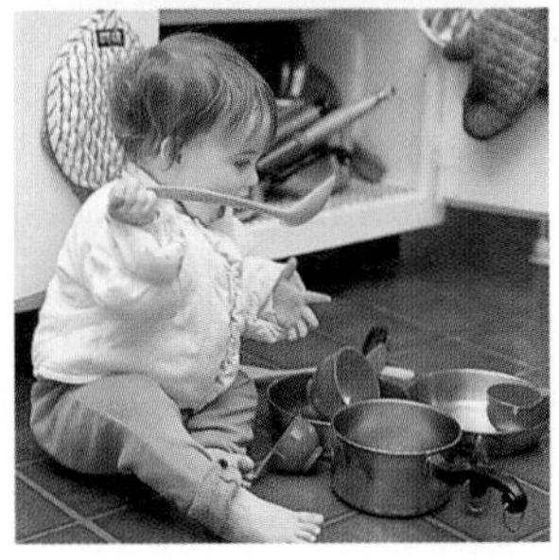

Twelve Months
Usually has tripled birthweight and increased length by 50%. Grasps and releases objects with fingers. Holds spoon, but uses it poorly.

Source: Reprinted with permission from J. Brown, *Nutrition Now* (St. Paul, MN: West Publishing Company, 1994) p. 29-20.

LACTOFERRIN (lak-toe-FERR-in) a factor in breast milk that binds and helps absorb iron and keeps it from supporting the growth of the infant's intestinal bacteria.

breastfeeding rates continue to be the highest among women who are older, are well-educated, are relatively affluent, and/or live in the western United States. Among those least likely to breastfeed are women who are low-income, are black, are under age 20, and/or live in the southeastern United States.[43]

A number of barriers to achieving the nation's health objective for increasing the incidence of breastfeeding have been noted. They include lack of knowledge, an absence of work policies and facilities that support lactating women (for example, extended maternity leave, part-time employment, facilities for pumping breast milk or breastfeeding, and on-site child care), the portrayal of bottle feeding rather than breastfeeding as the norm in the American society, the lack of breastfeeding incentives, and lack of support for low-income women.

Under most circumstances, a woman can freely choose to feed her baby breast milk or formula, knowing that the two modes of feeding are beneficial to the infant. However, if the infant is premature or if other factors act to the baby's disadvantage, breastfeeding becomes the preferred choice. The composition of the milk from a premature infant's mother is ideal for the premature infant. Even when separation prevents the mother from breastfeeding, she can express (pump) her milk to be given to the infant by bottle. Breastfeeding manuals show how to use manual massage or breast pumps to obtain milk. If the mother chooses not to breastfeed, a premature baby can, however, be successfully nourished on special formula for premature infants.

If a woman has a communicable disease such as tuberculosis or hepatitis or if she must take a medication that is secreted in breast milk and is known to affect the infant, she must not breastfeed. Drug addicts—including alcohol abusers—are capable of ingesting such high doses of their drug that their infants can become addicts by way of breast milk. In such cases, breastfeeding is also contraindicated.

Since the human immunodeficiency virus (HIV), responsible for AIDS, can be passed to an infant through breast milk, mothers who have been infected with HIV should not breastfeed. Sometimes, however, the nutritional and immunologic benefits of breast milk are considered to outweigh the risks of HIV transmission through breastfeeding. Such is the case in many developing countries, where lack of safe drinking water increases the risk of diarrhea and disease when formula feeding is used. In these situations, the World Health Organization encourages mothers to breastfeed irrespective of HIV infection.[44]

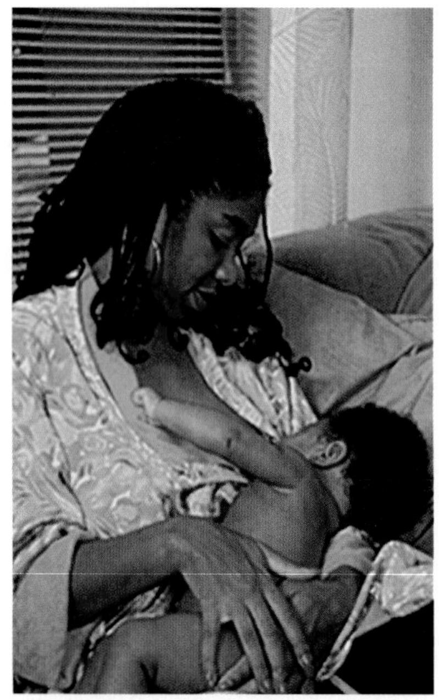

Breast milk is a very special substance.

Most prescription drugs do not reach nursing infants in sufficiently large quantities to affect them adversely. As a precaution, a nursing mother should consult with the prescribing physician before taking any drug. Minimal use of alcohol is compatible with breastfeeding. Smoking between feedings is permissible, although it is important not to expose the infant to secondhand smoke in the air. Coffee drinking is fine in moderation (two to three cups of coffee a day), as is the eating of foods such as garlic and spices. A particular food might affect the baby's liking for the mother's milk; this is a matter that requires individual detective work. (Examples are chocolate for some babies, excess caffeine for others, and foods that cause gas in the mother for still others.) If a woman has an ordinary cold, she can go on nursing without worry. The infant may well catch it from her anyway but may actually be less susceptible than a bottle-fed baby would be, thanks to the immunologic protection offered by breast milk.

A woman sometimes hesitates to breastfeed because she has heard that environmental contaminants may enter her milk and harm her baby. The decision whether to breastfeed on this basis might best be made after consulting with a physician or dietitian familiar with the local environment or with the state health department.

Feeding Formula

Like the breastfeeding mother, the mother who offers formula to her baby has reasons for making her choice, and her feelings should be honored. Infant formulas are manufactured to approximate the nutrient composition of breast milk. The immunologic protection of breast milk, however, cannot be duplicated, but the high level of preventive medical care (vaccinations) and public health measures achieved in the developed countries, especially in the United States and Canada, make this consideration less important than it was in the past. Safety and sanitation can be achieved with either mode of feeding by the educated mother whose drinking water supply is reliable.

One of the major advantages of formula feeding is that gained by the mother whose attempts at breastfeeding have met with frustration. Formula provides adequate nourishment for the infant, and a mother can choose this alternative with confidence. Other aspects of formula feeding include the following:

- The parents can see that the baby is getting enough milk during feedings.
- The mother can offer similar closeness, warmth, and stimulation during feedings as the breastfeeding mother does.
- Other family members can get close to the baby and develop a warm relationship in feeding sessions.

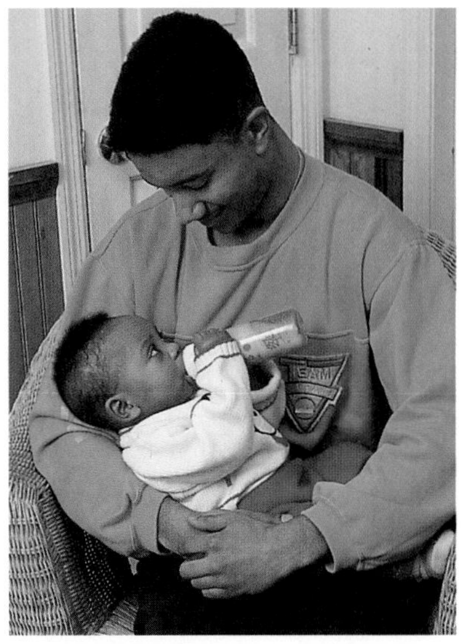

Formula feeding allows other family members to enjoy feeding the infant.

Many mothers breastfeed at first and then wean the baby within the first one to six months. When a woman chooses to wean her infant during the first twelve months of life, it is imperative that she shift to *infant formula,* not to plain cow's milk of any kind—whole or low-fat. Only formula contains enough iron (to name but one of many factors) to support normal development in the baby's first months of life. National and international standards have been set for the nutrient content of infant formulas.

For infants with special problems, many variations of infant formulas are available. Special formulas based on soy protein are available for infants allergic to milk protein, and formulas with the lactose replaced can be used for infants with lactose intolerance.

Cow's milk, both whole and reduced-fat, is not recommended during the first year of life, according to the American Academy of Pediatrics.[45] Cow's milk is an inappropriate replacement for breast milk or infant formula because it provides insufficient vitamin C and iron and excessive sodium and protein. Feeding cow's milk to infants may increase the risk of iron-deficiency anemia and cow's milk protein allergy.

Supplements for the Infant

Breast milk or formula and the infant's own internal stores will meet most nutrient needs for the first four to six months. Thereafter, the introduction of properly chosen juices and foods will normally keep up with the infant's changing requirements. At four to six months, infants require additional iron, preferably in the form of iron-fortified cereal.

Breast milk does not provide enough vitamin D for the infant, and vitamin D deficiency causes impaired bone mineralization in children. Manufacturers fortify infant formulas with vitamin D, but due to the low concentration of vitamin D in breast milk, pediatricians routinely prescribe a vitamin D supplement for breastfed infants whose mothers are vitamin D deficient or those who do not receive adequate exposure to sunlight.

If the water supply is deficient in fluoride, supplemental fluoride may also be needed by the breastfed infant after six months of age. The pediatrician should prescribe it, if appropriate. Fluoride does not appear to be secreted into breast milk even if the mother's fluoride supply is ample.

For the formula-fed infant, the makeup of the formula determines what further supplementation may be necessary. The pediatrician is the expert to consult on individual needs. In a baby's first six months, the choice of formula is important because whatever is chosen must supply the nutrients of human milk in similar forms and proportions.

Food for the Infant

The infant's rapid growth and metabolism demand an adequate supply of all essential nutrients.[46] Because of their small size, infants need smaller total amounts of the nutrients than adults do, but when comparisons are made based on body weight, infants need over twice as much of many of the nutrients. Figure 10-4 compares a five-month-old baby's needs with those of an adult man. As you can see, some of the differences are extraordinary. After six months, calorie needs increase less rapidly as the growth rate begins to slow down, but some of the energy saved by slower growth is spent on increased activity.

The most important nutrient of all—for infants as for everyone—is the one easiest to forget: water. The younger a child, the greater the percentage of the body weight is water and the more rapid the turnover. Since proportionately more of the infant's body water than the adult's is between the cells and in the vascular spaces, this water is easy to lose. Conditions that cause fluid loss, such as vomiting, diarrhea, sweating, or normal urinary loss without replacement, can rapidly propel an infant into life-threatening dehydration. Fluid and electrolyte imbalances caused by diarrhea and infection kill more of the world's children than any disease or disaster. Because infants can only cry and cannot tell you what they are crying for, it is important to remember that they may need fluid and to let them drink it until their thirst is quenched.

Iron is the nutrient hardest to provide for infants after weaning from breast milk or formula because its concentration in milk, the infant's major food, is low. By the end of the first year, half or more of all infants are receiving less than the recommended intake for iron. Iron may be the nutrient most needing attention in infant nutrition.

FIGURE 10-4
RECOMMENDED INTAKES OF NUTRIENTS FOR AN INFANT AND AN ADULT COMPARED ON THE BASIS OF BODY WEIGHT

Source: Adapted with permission from *Understanding Nutrition,* 8th ed., Copyright 1999 by Wadsworth Publishing Co. All rights reserved.

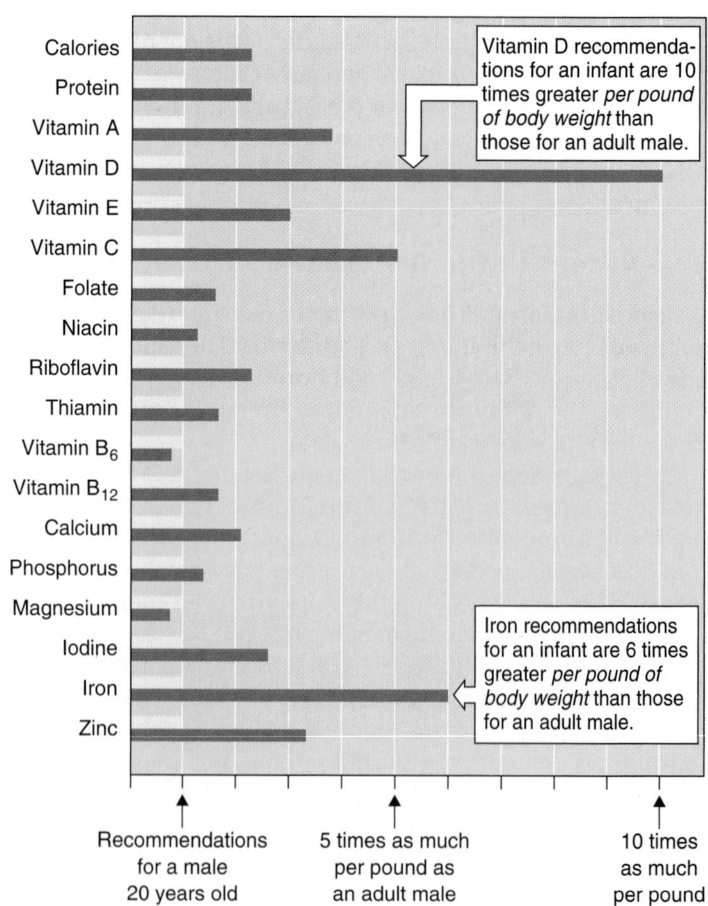

Solid foods may normally be added to a breastfed baby's diet when the baby is between four and six months old. Likewise, a baby who is formula fed might be started on solid foods between four and six months, depending on readiness. The following are indicators of readiness:

- The infant is six months old.
- The infant is developmentally ready—he or she can sit upright with support and can control head movements.

Solids should not be introduced too early because infants are more likely to develop allergies to them in the early months. But all babies are different, and the program of additions should depend on the individual baby, not on any rigid schedule. Table 10-5 presents a suggested sequence for feeding infants.

The addition of foods to a baby's diet should be governed by three considerations: the baby's nutrient needs, the baby's physical readiness to handle different forms of foods, and the need to detect and control allergic reactions. Nutrients needed early are iron and vitamin C. Juices and fruits that contain vitamin C are usually among the first foods introduced. Since a baby's stored iron supply from before birth runs out after the birthweight doubles, formula with iron, iron-fortified cereals, and, later, meat or meat alternates such as legumes are recommended.

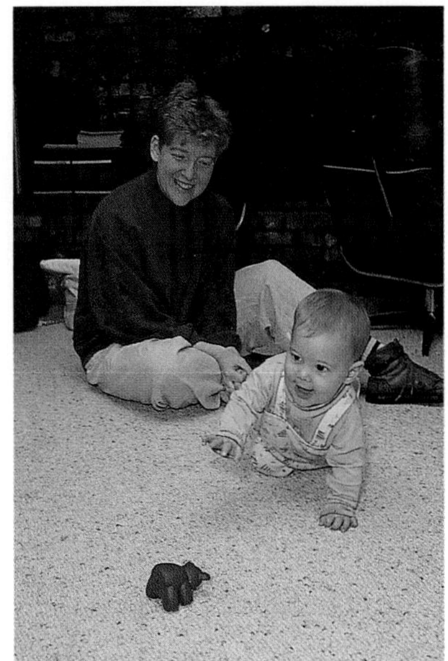

Energy saved by slower growth is spent in increased activity.

It has been suggested that the early introduction of sweet fruits to a baby's diet might favor the baby's developing a preference for sweets and lessen the baby's liking for vegetables introduced later. To prevent this, the order can perhaps be vegetables first, fruits later. As for other types of sweets (soda pop, candy, rich pies, and cakes), there is little room in the baby's diet for these empty-calorie foods. In contrast, naturally sweetened fruits supply not only calories, but also needed nutrients to support normal growth and development.

Physical readiness to handle foods develops in many small steps. For example, the ability to swallow solid food develops at around four to six months, and experience with solid food at that time helps to develop swallowing ability by desensitizing the gag reflex. Later still, when a baby can sit up, can handle finger

TABLE 10-5 FIRST FOODS FOR THE INFANT

Age (Months)	Addition
4 to 6	Iron-fortified rice cereal, followed by other cereals (for iron; baby can swallow and can digest starch now)* Pureed vegetables and/or fruits and their juices, one by one (perhaps vegetables before fruits so that the baby will learn to like their less sweet flavors)†
6 to 8	Mashed vegetables and fruits and their juices, infant breads and crackers
8 to 10	Protein foods (soft cheeses, yogurt, tofu, mashed cooked beans, finely chopped meat, fish, chicken, egg yolk), toast, teething crackers (for emerging teeth), soft-cooked vegetables and fruit
10 to 12	Whole egg (allergies are less likely now), whole milk (at one year)

*Mix with breast milk, formula, or water. Later, other cereals can be introduced, but they should still be iron-fortified varieties.

†All baby juices are fortified with vitamin C. Orange juice may cause allergies; apple juice may be a better juice to feed first. Dilute juices with water and offer in a cup to prevent nursing bottle syndrome.

Source: Adapted from Committee on Nutrition, American Academy of Pediatrics, *Pediatric Nutrition Handbook*, 4th ed., ed. R. E. Kleinman (Elk Grove Village, IL: American Academy of Pediatrics, 1998), pp. 89–91.

Let infants handle food as they become ready.

foods, and is teething, hard crackers and other hard finger foods may be introduced. Such foods promote the development of manual dexterity and control of the jaw muscles. (An infant can choke on these foods, however, so an adult should keep a watchful eye on the learning process.)

Some parents want to feed solids at an earlier age on the theory that "stuffing the baby" at bedtime promotes sleeping through the night. There is no proof for this theory. On the average, babies start to sleep through the night at about the same age regardless of when solid foods are introduced. By three months, 75 percent are sleeping through the night whether or not they are receiving any solid foods.

As for the choice of foods, baby foods commercially prepared in the United States and Canada are safe, and except for mixed dinners and heavily sweetened desserts, are generally nutritious, and of high quality. In response to consumer demand, baby food companies have removed much of the added salt and sugar that many of their products contained in the past. Baby foods generally have high nutrient density, except for mixed dinners (which contain little meat) and desserts (which are heavily sweetened).

An alternative for parents who want the baby to have family foods is to "blenderize" a small portion of the table food at each meal. This necessitates cooking without salt, however, since foods that adults prepare for themselves often contain much more salt than commercial baby foods. The adults can salt their own food, if desired, after the baby's portion has been taken. Canned vegetables are inappropriate for infants; their sodium content is often too high. It is also important to take precautions against food poisoning. Honey should *never* be fed to infants because of the risk of botulism. Babies and even young children have difficulty swallowing certain foods—popcorn, whole grapes, bite-size hot dogs, and nuts, for instance. An infant can easily choke on these foods; it is not worth the risk to give such foods to infants.

At one year of age, the obvious food to supply most of the nutrients the baby needs is still milk; two to three and a half cups a day are now sufficient. Infants under two years should drink whole milk and not low-fat or fat-free milk; they need the fat and vitamins A and D of fortified whole milk until two years of age. The other foods—meat, iron-fortified cereal, enriched or whole-grain bread, fruits, and vegetables—should be supplied in variety and in amounts sufficient to round out total calorie needs. Ideally, the one-year-old is sitting at the table and eating many of the same foods everyone else eats. A meal plan that meets the requirements for the one-year-old is shown in Table 10-6.

The wise parent of a one-year-old offers nutrition and love together. Both promote growth. It is literally true that "feeding with love" produces better growth in both weight and height of children than feeding the same food in an emotionally negative climate. It also promotes better brain development. The formation of nerve-to-nerve connections in the brain depends both on nutrients and on environmental stimulation.

The person feeding a one-year-old should keep in mind that the baby is also developing eating habits that will persist throughout life. Mealtimes should be relaxed and leisurely. Children should learn to eat slowly, pause and enjoy their table companions, and stop eating when they are full. The "clean your plate" dictum should be stamped out for all time, and in its place, parents who wish to avoid waste should learn to serve smaller portions or teach their children to serve themselves as much as they truly want to eat. Physical activity should be encouraged on a daily basis to promote strong skeletal and muscular development and to establish habits that will undergird good health throughout life.

TABLE 10-6 MEAL PLAN FOR A ONE-YEAR-OLD

Breakfast
½ c whole milk
½ c iron-fortified cereal
1–2 tbsp fruit

Snack
½ c yogurt
Teething crackers
1–2 tbsp fruit

Lunch
1 c whole milk
2 to 3 tbsp vegetables
1 egg or 1 oz chopped meat or well-cooked mashed legumes
½ c noodles

Snack
½ c whole milk
½ slice toast
1 tbsp peanut butter

Supper
1 c whole milk
2 oz chopped meat or well-cooked, mashed legumes
½ c potato or rice
2 to 3 tbsp vegetables
2 to 3 tbsp fruit

Nutrition-Related Problems of Infancy

Iron deficiency and food allergies are two of the most significant nutrition-related problems of infants.

Iron Deficiency. Iron deficiency remains a prevalent nutritional problem in infancy, although it has declined in recent years in large part because of the increasing use of iron-fortified formulas. The use of cow's milk earlier than recommended in infancy can cause iron deficiency as a result of its poor iron content and the potential to cause gastrointestinal blood loss in susceptible infants.[47] Other factors contributing to iron deficiency in infancy include breastfeeding for more than six months without providing supplemental iron, intake of infant formula not fortified with iron, the infant's rapid rate of growth, low birthweight, and low socioeconomic status.[48] To prevent iron deficiency, the American Academy of Pediatrics recommends that infants be fed breast milk or iron-fortified formula for the first year of life, with appropriate foods added between the ages of four and six months as shown in Table 10–5.

Food Allergies. Genetics is probably the most significant factor affecting an infant's susceptibility to food allergies. At-risk infants can be identified by means of careful skin testing and by a family history. Breast milk is recommended for those infants allergic to cow's milk protein and is preferable to soy or goat's milk formulas, since infants are sometimes allergic to these proteins as well. To reduce the risk of food sensitivity or allergic reactions to other foods, new foods should be introduced one at a time to facilitate prompt detection of allergies. For example, when cereals are introduced, try rice cereal first for five to seven days; it causes allergy least often. Try wheat cereal last; it is the most common offender. If a cereal causes irritability from skin rash, digestive upset, or respiratory discomfort, discontinue its use before going on to the next food. About nine times out of ten, the allergy won't be evident immediately but will manifest itself in vague symptoms occurring up to five days after the offending food is eaten, so it isn't easy to detect. Several days should elapse between the introduction of each new food to allow time for clinical symptoms to appear, so that the offending food may be identified. The Nutrition Action feature on page 345 provides more information about food allergies and intolerances in children and adults.

EARLY AND MIDDLE CHILDHOOD

Childhood is a critical time in human development. Children typically grow taller by two to three inches and heavier by five or more pounds each year between the age of one and adolescence. They master fine motor skills (including those related to eating and drinking), become increasingly independent, and learn to express themselves appropriately. Nutrition plays a critical role in the development and growth of children. This section describes the nutrient requirements of children and the primary nutritional problems of this population.

Growth and Nutrient Needs of Children

After age one, a child's growth rate slows, but the body continues to change dramatically. At one, most babies have just learned to stand and toddle; by two, they can take long strides with solid confidence and are learning to run, jump, and climb. The internal change that makes these new accomplishments possible is the accumulation of a larger mass and greater density of bone and muscle tissue. The changes are obvious in Figure 10-5.

Children generally become leaner between the ages of six months and six years, after which time there is a gradual increase in fat thickness in both males and females until puberty is reached. Females have a greater body fat content than males at all stages of development.[49] The energy requirements of children are determined by their individual basal metabolic rates, activity patterns, and rates of growth. Toddlers (ages one to three years) need about 1,300 calories per day. By the age of 10, children need about 2,000 calories per day. Appetite

FIGURE 10-5
ONE-YEAR-OLD AND TWO-YEAR-OLD

The two-year-old has lost much of the baby fat; the muscles (especially in the back, buttocks, and legs) have firmed and strengthened, and the leg bones have lengthened and increased in density.

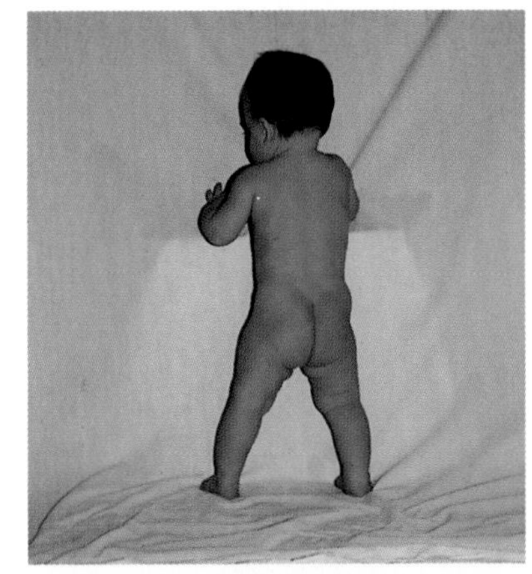

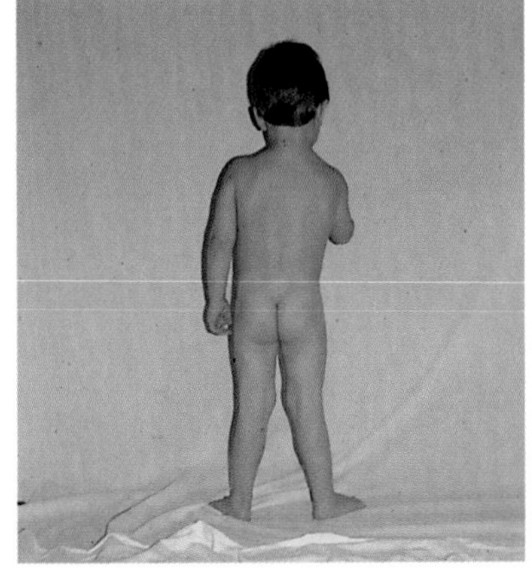

FIGURE 10-6
FOOD GUIDE PYRAMID
FOR YOUNG CHILDREN

Source: USDA Center for Nutrition and Policy Promotion, March 1999, Program AID 1649.

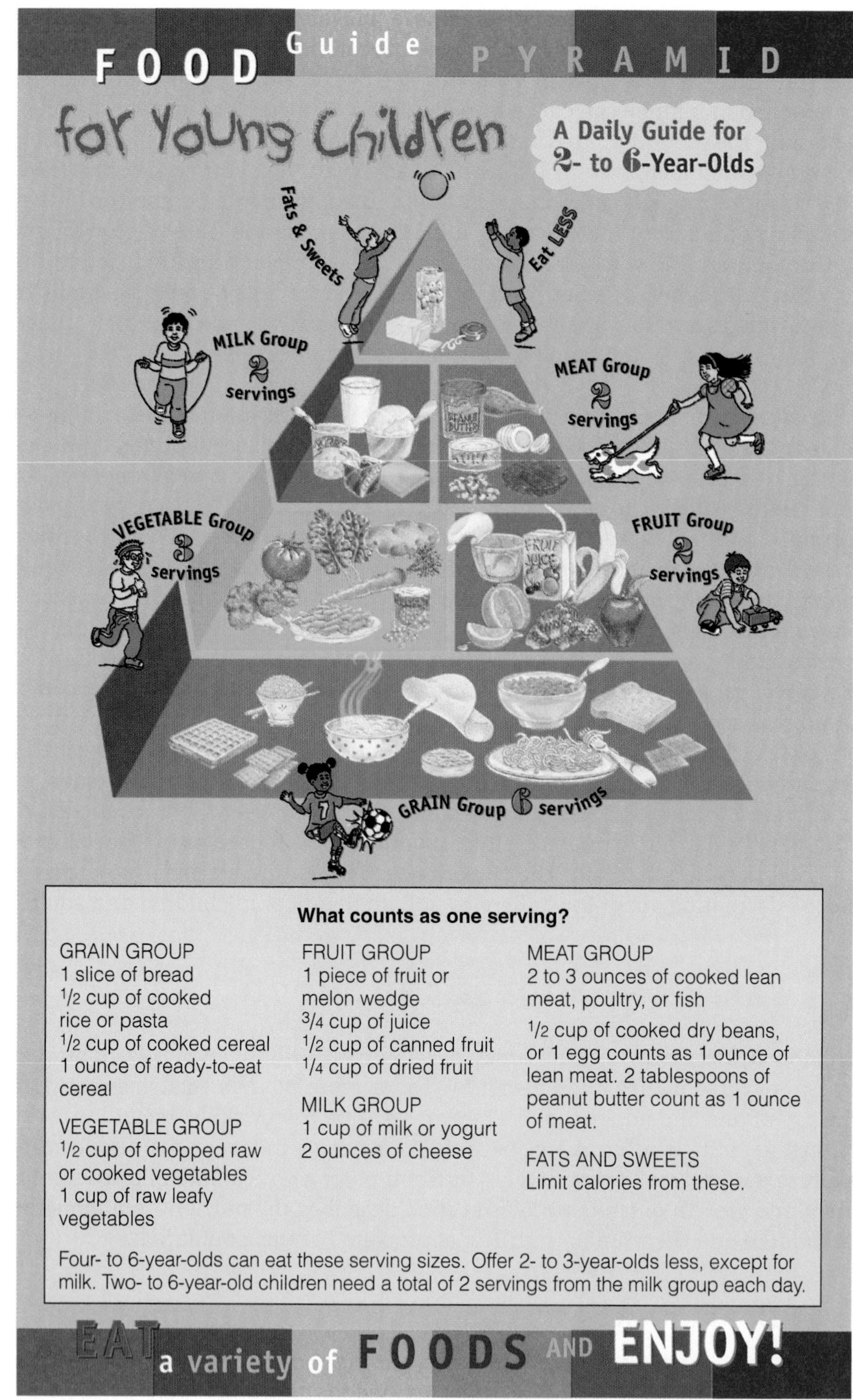

decreases markedly around the age of one year, in line with the great reduction in growth rate. Thereafter, the appetite fluctuates; a child will need and demand much more food during periods of rapid growth than during slow periods. To provide the gradually increasing needs for all nutrients during the growing years, the Food Guide Pyramid recommends a balance among milk and milk products, meats and meat alternates, fruits, vegetables, and breads and cereals (see Figure 10-6).

After the crucial first year, there is still much a parent can do to foster the development of healthful eating habits. Table 10-7 offers tips to make feeding

TABLE 10-7 STRATEGIES TO FOSTER HEALTHFUL EATING HABITS AND HAPPY MEALTIMES

These tips may make feeding time easier and more relaxing for both parent and child:

- Schedule regular meals and snacks for toddlers since they require frequent feeding to ensure adequate intake of calories and nutrients.
- Always try to offer at least one food that the child likes.
- Remain calm if the child leaves an entire meal untouched.
- Do not be concerned about short food jags, stretches of time when the child wants the same food over and over. If this behavior continues for a long period and eliminates entire food groups, consult with your pediatrician or registered dietitian.
- Teach and reinforce good table manners.
- Allow the child to eat slowly.

Offer healthy food in a relaxed manner and children will eat what they need. Try these suggestions for healthful snacking:

- Stock up on carbohydrates—bagels, pretzels, whole-grain breads and rolls, low-fat crackers, and English muffins.
- Keep plenty of washed and cut raw vegetables in the refrigerator. Team up with a yogurt or bean dip.
- Have an assortment of bread spreads handy, such as peanut butter, smooth cottage cheese, reduced-fat cream cheese, jams, jellies, and preserves.
- Top frozen waffles and pancakes with fresh fruit for a tasty and refreshing snack.
- Serve canned soups—they are easy for a child to make for an after-school winter warmer.
- Make your own frozen juice pops in an ice cube tray, or freeze grapes, strawberries, or chunks of banana, pineapple, or melon.
- Bake your own oatmeal cookies. You can cut the shortening (margarine, butter, oil) and sugar in many recipes by at least one-third without affecting quality.
- Put together a large batch of cereal, pretzel, and nut mix. Divide into individual plastic bags.
- Serve healthy party foods—try serving carrot cake cupcakes, bobbing for apples, making fresh fruit kabobs.
- Prepare air-popped popcorn and sprinkle with grated parmesan cheese.

Source: Adapted from M. G. Hermann, *The ABCs of Children's Nutrition* (Chicago: The American Dietetic Association, 1991), pp. 3–7.

times enjoyable for both parent and child. The goal is to teach children to like nutritious foods in all food categories.

Candy, cola, and other concentrated sweets must be limited in a child's diet if the needed nutrients are to be supplied. If such foods are permitted in large quantities, there are only three possible outcomes: nutrient deficiencies, obesity, or both. A child can't be expected to choose nutritious foods on the basis of taste alone; the preference for sweets is innate. On the other hand, an active child can enjoy the higher-calorie nutritious foods in each category: ice cream or pudding in the milk group and whole-grain cookies and crackers in the bread group. These foods, made from milk and grain, carry valuable nutrients and encourage a child to learn, appropriately, that eating is fun.

Children sometimes seem to lose their appetites for a while; this is nothing to worry about. The perfection of appetite regulation in children of normal weight guarantees that such children's calorie intakes will be right for each stage of growth.[50] As long as the food energy they do consume is from nutritious foods, they are well provided for. An overzealous parent, unaware that a one-year-old is supposed to slow down, may begin a lifelong conflict over food by trying to force more food on the child than the child feels like eating. Keep in mind that parents are responsible only for *what* the child is given to eat and where and when it is presented. The child is responsible for *how much,* if anything, is eaten.[51]

Nutrient intake recommendations for children are given on the inside front cover.

Other Factors That Influence Childhood Nutrition

While parents are doing what they can to establish favorable eating behaviors during the transition from infancy to childhood, children in preschool or grade school are encountering foods prepared and served by outsiders. The U.S. government funds several programs to provide nutritious, high-quality meals for children at school. School lunches are designed to meet certain requirements. They must include specified servings of milk, protein-rich foods (meat, cheese, eggs, legumes, or peanut butter), vegetables, fruits, and bread or other grain foods (see Table 10-8). They are intended to follow the recommendations of the *Dietary Guidelines for Americans* and to provide at least a third of the recommended intakes for protein, vitamins A and C, iron, calcium, and calories.[52]

Children growing up today need not only to be fed well in the interest of their growth and development but also to learn enough about nutrition to become able to make adaptive food choices when the choices become theirs to make. It is desirable for children to learn to like nutritious foods in all of the food groups. With one exception, this liking usually develops naturally. The exception is vegetables, which young children sometimes dislike and refuse. Even a tiny serving of spinach, cooked carrots, or squash may elicit an expression of disgust. Since most youngsters need to eat more vegetables, parents need to know how to make them appealing to children. Children prefer vegetables that are slightly undercooked and crunchy, attractive in color and shape, and easy to eat. They should be warm, not hot, because a child's mouth is much more sensitive than an adult's. The flavor should be mild (a child has more taste buds), and smooth foods like mashed potatoes or pea soup should have no lumps in them (a child wonders, with some disgust, what the lumps might be).

Little children like to eat at little tables and to be served little portions of food. They also love to eat with other children and have been observed to stay at the table longer and eat much more when in the company of their peers. A bright, unhurried atmosphere free of conflict is also conducive to good appetite.

TABLE 10-8 SCHOOL LUNCH PATTERNS FOR DIFFERENT AGES

	Preschool (Age)		Grade School Through High School (Grade)		
Food Group	*1 to 2*	*3 to 4*	*k to 3*	*4 to 6*	*7 to 12*
Meat or meat alternate					
1 serving:					
Lean meat, poultry, or fish	1 oz	1½ oz	1½ oz	2 oz	3 oz
Cheese	1 oz	1½ oz	1½ oz	2 oz	3 oz
Large egg(s)	½	¾	¾	1	1½
Cooked dry beans or peas	¼ c	⅜ c	⅜ c	½ c	¾ c
Peanut butter	2 tbsp	3 tbsp	3 tbsp	4 tbsp	6 tbsp
Peanuts, soy nuts, tree nuts, seeds*	½ oz	¾ oz	¾ oz	1 oz	1½ oz
Vegetable and/or fruit					
2 or more servings, both to total	½ c	½ c	½ c	¾ c	¾ c
Bread or bread alternate					
Servings†	5 per week	8 per week	8 per week	8 per week	10 per week
Milk					
1 serving of fluid milk	¾ c	¾ c	1 c	1 c	1 c

*Can be used to meet up to one-half serving of meat, but must be accompanied by other meat/meat alternate in the meal.
†A serving is 1 slice bread; 1 biscuit, roll, or muffin; 1/2 cup cooked rice, pasta, or cereal grain.

Source: U.S. Department of Agriculture, *Food Program Facts—National School Lunch Program*, 1999.

Ideally, each meal is preceded, not followed, by the activity the child looks forward to the most. A number of schools have discovered that children eat a much better lunch if recess occurs before rather than after the meal. With recess after, children are likely to hurry out to play, leaving food on their plates that they were hungry for and would otherwise have eaten. Before sitting down to eat, small children should be helped to wash their hands and faces to decrease likelihood of contaminating the food with bacteria.

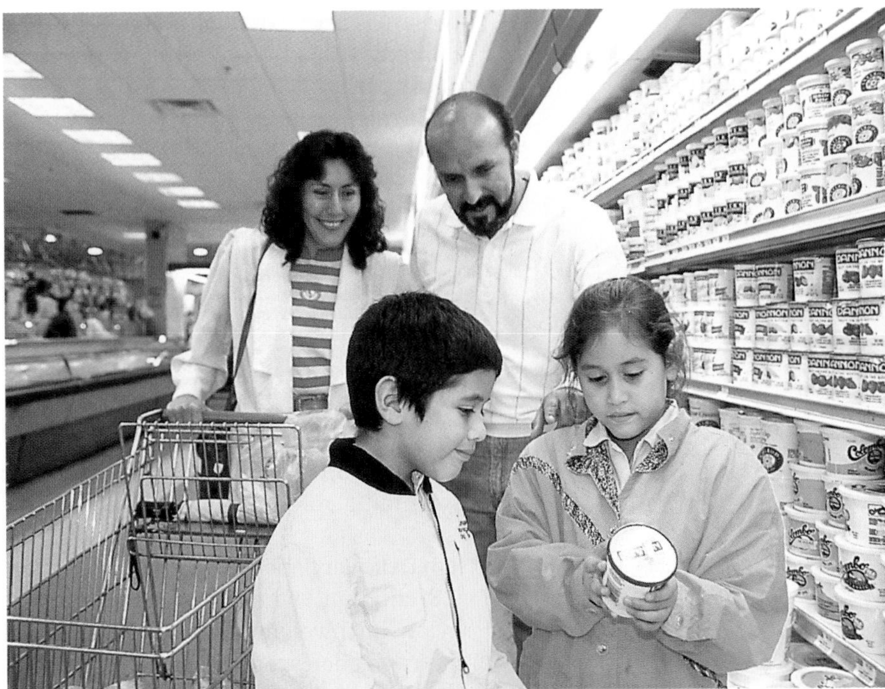

Make learning about nutrition fun. Read food labels together at the grocery store, checking for the supply of vitamins and minerals and the amounts of sugar, salt, and fat.

Many little children, both boys and girls, enjoy helping in the kitchen. Their participation provides many opportunities to encourage good food habits. Vegetables are pretty, especially when fresh, and provide opportunities to learn about color, about growing things and their seeds, about shapes and textures—all of which are fascinating to young children. Measuring, stirring, decorating, cutting, and arranging vegetables are skills even a very small child can practice with enjoyment and pride.

When introducing new foods at the table, parents are advised to offer them one at a time—and only a small amount at first. Whenever possible, the new food should be presented at the beginning of the meal, when the child is hungry. If the child is cross, irritable, or feeling sick, don't insist, but withdraw the new food and try it again a few days later. Remember, parents have inclinations and dislikes to which they feel entitled; children should be accorded the same privilege.

Never make an issue of food acceptance; a power struggle almost invariably results in a permanently closed mind on the child's part. Rather, let children participate in the planning and preparation of meals. If the beginnings are right, children will grow up with positive feelings toward themselves and the ways they relate to food. Remember, too, that parents can act as good role models—enjoying healthful meals and keeping in good physical shape themselves.

Nutrition-Related Problems of Childhood

Although most children in the United States are well nourished, malnutrition does occur in this population. Undernutrition, which occurs when children do not eat enough food or energy, is a problem for some children in the United States, especially those from low-income and migrant families or certain ethnic and racial minority groups (for example, African-Americans, Asians).[53] Children in foster care, many of whom live in poverty, and young homeless children are also at risk for undernutrition.[54] More than 20 percent of U.S. children are considered poor.[55] The most common nutrition-related problems occurring among U.S. children include obesity, iron-deficiency anemia, and high blood cholesterol levels.

Obesity. The problem of overnutrition is one of the most widespread nutrition disorders among children in the United States. The prevalence of obesity among 6- to 11-year-old children has doubled over the past two decades.[56] Childhood

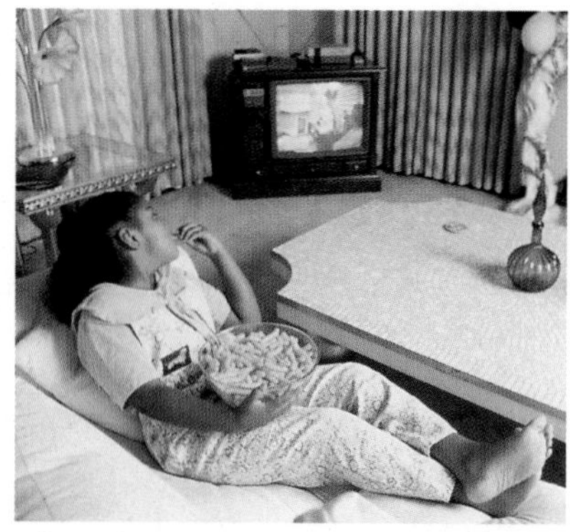

Television can have adverse effects on children by influencing their selection of foods, snacking habits, and activity levels.

obesity is associated with high blood insulin levels, high triglyceride levels, and reduced HDL-cholesterol concentrations and is considered a risk factor for obesity in adulthood.[57] According to recent studies, second to prior obesity, the strongest predictor of subsequent obesity in children is television viewing.[58]

Children also hear a great deal about foods via the television set. Many authorities are concerned that television commercials may have a less than desirable impact on children. It is estimated that the average child sees more than 21,000 commercials a year, of which approximately half are for foods high in fat, sugar, and salt.[59] Hundreds of millions of dollars are spent in the effort to sell these foods to children.

The more time children spend watching television, the less time they have for physical activity. Research results show that children who watch television more often had lower activity levels and were less likely to participate in organized sports or community activities.[60] Parents can encourage physical activity in children to help prevent overweight by participating in activities such as cycling, swimming, rollerblading, skating, or hiking with their children.[61]

Iron-Deficiency Anemia. Of all nutritional disorders other than obesity found in U.S. children, the most common is iron-deficiency anemia. It is most prevalent in low-birthweight infants, babies from six months to two years of age, and children and adolescents from low-income families.[62] To ensure adequate iron nutrition, parents should offer an abundance of such iron-rich foods as lean meats, fish, poultry, eggs, and legumes. Grain products should be whole-grain or enriched only. Milk, beneficial as it is, is a poor iron source; dairy products should be consumed only in the amounts needed to ensure optimal calcium intakes.

High Blood Cholesterol. Considerable evidence exists that atherosclerosis begins in childhood and that this process is related to high blood cholesterol levels. Children and adolescents in the United States have higher blood cholesterol levels and higher dietary intakes of saturated fat and cholesterol than children in other countries.[63] A panel of experts representing 42 major U.S. health and professional organizations has recommended that children age two and older eat diets containing no more than 30 percent of total calories from fat, less than 10 percent of total calories from saturated fat, and less than 300 milligrams of cholesterol daily. However, from birth to two years of age, a child's fat consumption should not be restricted because fat is a concentrated source of the calories needed to ensure proper physical development.

The panel also advised that youngsters and teens should have their blood cholesterol measured if they have one parent with a high blood cholesterol level. For children and adolescents, a total of 200 milligrams per deciliter of blood or more is considered high; 170 to 199 is borderline high; and less than 170 is acceptable. In addition, youngsters born to families with a history of premature heart disease should have both total blood cholesterol and HDL-cholesterol checked.[64]

THE IMPORTANCE OF TEEN NUTRITION

Adolescence is a time of change. Between the ages of about 10 and 18 years in girls and 12 and 20 years in boys, marked changes take place in physical, intellectual, and emotional growth and development. The maturation process is initiated and controlled by a variety of hormones, including, among others, growth hormone, prolactin, estrogen, testosterone, and the thyroid hormones. Many aspects of the maturation process are influenced by dietary intake and nutritional status. This section reviews the nutrient requirements and special nutrition-related problems of teenagers.

Nutrient Needs of Adolescents

Consult the DRI table shown on the inside front cover for the nutrient intake recommendations for adolescents.

The dramatic changes in body composition and the rate of growth that occur during early adolescence give rise to the familiar phrase "the adolescent growth spurt." The magnitude of these changes is such that the linear growth increments during adolescence can contribute about 15 percent to 25 percent of adult stature. The rate of weight gain can contribute anywhere from 40 percent to 50 percent of the adult body mass. This remarkable growth rate requires adequate intakes of energy and nutrients.[65]

No universally applicable formula exists for expressing the calorie demands of adolescents. The individual teenager's energy need is influenced by body size, activity levels, and biological factors affecting growth.

Rule of thumb for teens:
For 5 ft, consider 100 lb a reasonable weight for girls (110 lb for boys).
For each inch over 5 ft, add 5 lb.
For each year under 25 (down to 18), subtract 1 lb.

There is tremendous variation in the rates and patterns of individual teenagers in terms of growth.[66] The only way to be sure teenagers are growing normally is to compare their heights and weights with previous measures taken at intervals and note whether reasonably smooth progress is being made. Teenagers who want to know what they should weigh should be reassured that any of a wide range of weights is considered normal at this time in life. A rule of thumb can be applied to teenage girls (see margin); the result can be considered a weight to aim at, but weights well in excess of these are normal, too. Teenage boys can be told that when they have finished growing, they should expect to weigh what the adult charts show, but that while they are growing, it is not unusual for their weights to be quite different from the adult standards.

Nutrition-Related Problems of Adolescents

Most adolescents in the United States are perceived as "healthy." However, many U.S. adolescents experience a variety of health and nutrition problems, some related to their risk-seeking behaviors and an inability to deal with abstract notions such as "good health" and the link between current behaviors and long-term health. In general, the nutritional health of U.S. teenagers is better today than ever before. Overt nutrient deficiencies, with the exception of iron deficiency, are not the public health problems they once were. Specific nutrition-related problems among U.S. adolescents include undernutrition, obesity, iron-deficiency anemia, low dietary calcium intakes, high blood cholesterol levels, dental caries, and eating disorders.

Undernutrition. Some groups of adolescents are at risk for reduced energy and food intakes. Adolescents from low-income families and those who have run away from home or abuse alcohol or other drugs are at risk nutritionally. Approximately 16 percent of youth ages 14 to 21 are estimated to be living below the poverty line. African-American and Hispanic teenagers are nearly three times more likely to live in poverty than are white youth. Chronic dieters are also at risk. Thirteen percent of the 17,354 females in grades 7 through 12 who were interviewed in the Minnesota Adolescent Survey reported being chronic dieters, defined as always on a diet or having been on a diet for 10 of the previous 12 months.[67]

Successful nutrition education activities for teenagers focus on their interests and the relationship of good eating and physical activity to health.

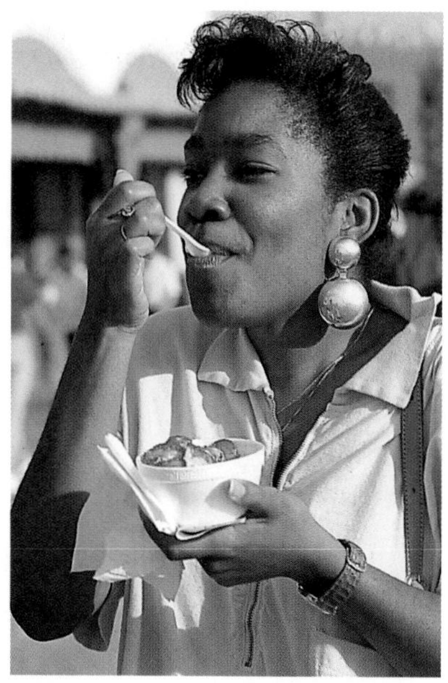

Nutritious snacks can supply added nutrients to the active teenager's diet.

Obesity. Eleven percent of adolescents are overweight or obese and the prevalence of overweight among teens has more than doubled in the past 30 years.[68] Given that childhood obesity can lead to adult obesity and that approximately 55 percent of the adult population is overweight or obese, obesity among teenagers is cause for concern. Genetic susceptibility to obesity, lifestyle, family eating patterns, excessive consumption of soft drinks, lack of positive role models, and physical inactivity all contribute to overweight and obesity in this population.[69] Obese teenagers are also at risk for hypertension and diabetes mellitus, among other disorders.[70]

Iron-Deficiency Anemia. Iron needs increase during adolescence, especially for females as they start to menstruate. In boys, the requirements for absorbed iron increase because of an expanding blood volume and rise in hemoglobin concentration that accompanies the development of larger muscles and sexual maturation. (After the adolescent growth spurt, the need for iron falls off.[71]) Whereas most males have an adequate iron intake during adolescence, many females between 12 and 22 years of age have iron intakes below the current recommended intake. Girls typically consume fewer total calories and less meat than boys do.

Low Calcium Intakes. Another problem nutrient during the teen years is calcium. Low intakes of calcium-rich foods during adolescence compromise peak bone mass development and increase risk of osteoporosis later in life. Adolescents need a minimum of 1,300 milligrams of calcium each day in order to achieve an optimal peak bone mass and healthy bones. However, many teens, especially teenage girls, have calcium intakes below the recommended intake.[72] Adolescent girls following low-calorie diets and drinking diet soft drinks in place of milk are at particular risk for age-related bone loss and subsequent fractures. See the Spotlight feature in Chapter 7 for more about calcium and osteoporosis.

High Blood Cholesterol Levels. Teenagers have many of the same risk factors for high blood cholesterol as adults: family history of coronary heart disease; diets high in total fat, saturated fat, and cholesterol; hypertension; low activity levels; and smoking. Although the process of atherosclerosis is not completely understood, it is believed that the fatty streaks that develop in young people progress to the fibrous plaques of adulthood.[73] For more information about diet and heart disease, see the Spotlight feature in Chapter 4.

Dental Caries. Dental caries are a significant public health problem among teenagers, with about 78 percent of adolescents having one or more caries in permanent or primary teeth.[74] Dental caries are more prevalent among adolescents than among children. A National Dental Research Survey found that only 22 percent of 15-year-olds were caries free, compared with 56 percent of 10-year-olds.[75] Fortunately, the incidence of dental caries has decreased by as much as 30 percent to 50 percent over the past two decades, partly because of fluoridation of public drinking water, improved dental hygiene, and the use of fluoride in toothpastes and mouthwashes. Some population subgroups, such as American Indians, Alaskan Natives, African-Americans, and Hispanics, are at greater risk of dental caries than other groups because they lack access to or do not avail themselves of dental services.

See the Spotlight feature in Chapter 8 for more information about eating disorders.

Eating Disorders. Eating disorders have become serious health problems in recent years. The most common eating disorders are anorexia nervosa and bulimia nervosa. A constellation of individual, familial, sociocultural, and biological factors contribute to these disorders, which threaten physical health and psychological well-being.

Some individuals are more predisposed to developing an eating disorder than others. For example, about 90 percent of people with eating disorders are female.

(text continues on page 348)

NUTRITION ACTION

Food Allergy—Nothing to Sneeze At

In December 1995, a 33-year-old woman with peanut allergy read the contents of the label of a container of split pea soup. Peanuts were not listed. She ate only part of the soup and within minutes experienced a life-threatening reaction and was rushed to the emergency room for treatment. Following the incident, the split pea soup was analyzed and found to contain peanut flour as a component of the "flavoring" ingredients. The soup manufacturer subsequently discontinued using peanut flour in the product.[76]

The woman in the story belongs to the estimated 1 percent to 2 percent of the adult population in the United States suffering from food allergies. Interestingly, although a relatively small percentage actually have food allergies, nearly 25 percent of people "think" they do and develop **food aversions.**[77] What is the explanation? The answer may be in understanding the difference between **food allergy** and **food intolerance.** Although the physical response to a food allergy and food intolerance may be very similar, the difference between the two is whether or not the immune system is involved in the reaction.

FOOD AVERSION a strong desire to avoid a particular food.

FOOD ALLERGY an adverse reaction to an otherwise harmless substance that involves the body's immune system.

FOOD INTOLERANCE a general term for any adverse reaction to a food or food component that does not involve the body's immune system.

ADVERSE REACTION an unusual response to food, including food allergies and food intolerances.

Food Intolerance Versus Food Allergy

Food intolerance is far more common than food allergy. A food intolerance is an **adverse reaction** to food that *does not* involve the immune system. Lactose intolerance is one example of a food intolerance. A person with lactose intolerance lacks the enzyme needed to digest lactose. When that person eats milk products, symptoms such as gas, cramps, and bloating can occur. Gluten intolerance is caused by an intolerance to gluten, a protein found in wheat, oats, barley, and rye. People with gluten intolerance must avoid products prepared with these grains. Some people may develop a sensitivity to various other agents in a food. Tyramine, found in cheese or red wine, can induce a headache if you are susceptible. Others may have a sensitivity to certain food additives such as monosodium glutamate (MSG), sulfites, or coloring agents. The physical reaction to these agents can include hives, rashes, nasal congestion or asthma.

A food allergy, on the other hand, is an abnormal response to a food triggered by the immune system. The allergic reaction involves three main components: **food allergens,** immunoglobulin E (IgE), and mast cells. Food allergens are the fragments of food that are responsible for the allergic reaction. They consist of proteins from the food that are not broken down during the digestive process which then cross the gastrointestinal lining to enter the bloodstream. IgE is a type of protein called an **antibody** that circulates through the blood. When allergic people eat certain foods, their immune system reacts to the food allergen by making IgE that is specific to that food. Once released, the IgE antibody attaches to a cell found in all body tissues called the mast cell. The mast cells are specialized cells of the immune system that serve as the storehouse for various chemical substances, including **histamine.** Mast cells are found throughout all tissues, but are especially common in the areas of the body that are typical sites of allergic reaction: the nose and throat, lungs, skin, and gastrointestinal tract. When an allergic reaction occurs, the food allergen interacts with the IgE on the surface of the mast cells which triggers those cells to release histamine. Depending upon

FOOD ALLERGEN a substance in food—usually a protein—that is seen by the body as harmful and causes the immune system to mount an allergic reaction.

ANTIBODY a protein made by the body to destroy foreign substances (allergens).

HISTAMINE a substance released by cells of the immune system during an allergic reaction to an antigen, causing inflammation, itching, hives, dilation of blood vessels, and a drop in blood pressure.

ANAPHYLAXIS (an-ah-fa-LAX-is) a potentially fatal reaction to a food allergen causing reduced oxygen supply to the heart and other body tissues. Symptoms include difficulty breathing, low blood pressure, pale skin, a weak, rapid pulse, and loss of consciousness.

CROSS-REACTION the reaction of one antigen with antibodies developed against another antigen.

the tissue in which the histamine is released, these chemicals will cause a person to have various symptoms of a food allergy (see Figure 10-7). The most severe allergic reaction is **anaphylaxis.** This potentially fatal condition occurs when several parts of the body experience food allergy reactions at the same time. Signs of anaphylaxis include difficulty breathing, swelling of the mouth and throat, a drop in blood pressure, and loss of consciousness. The reaction can occur in a few seconds or minutes, and without immediate medical attention, death may result. The foods most associated with anaphylactic reactions include peanuts, tree nuts (for example, walnuts, cashews), eggs, and shellfish.

According to the Food Allergy Network, eight foods cause 90 percent of all allergic reactions: egg, fish, milk, peanuts, shellfish, soy, tree nuts, and wheat.[78] In adults, the most common foods to cause allergic reactions include shellfish, peanuts, tree nuts, fish, and egg. In children, the problem foods are typically egg, milk, and peanuts. **Cross-reaction** is a concern for someone diagnosed with a food allergy. For instance, if someone has a history of allergic reaction to shrimp, testing may show that person is also allergic to other shellfish, such as crab and lobster. Table 10-9 lists the most common foods that provoke food allergy as well as possible cross-reacting foods.

Children and Food Allergy

The prevalence of food allergy is greatest in the first few years of life, with up to 6 percent of children younger than 3 years experiencing allergic responses.[79] Increased susceptibility of infants to food allergic reactions is believed to be the result of their immature immune system. The immature digestive system may also allow more intact allergen proteins to enter the bloodstream. Cow's milk and soy are the most common allergens for infants and produce reactions such as hives, bloating, and diarrhea. Breastfeeding is recommended for the first 12 months to avoid early exposure to cow's milk or soy, thus avoiding allergy. Breast milk also contains several powerful agents that stimulate the development and maintenance of the digestive tract, which further reduces the risk of food allergy. Early exposure to certain foods can also play a role in development of food allergy in children. Delaying introduction of foods such as cow's milk, egg, soy, peanuts,

FIGURE 10-7 COMMON SITES FOR ALLERGIC REACTIONS

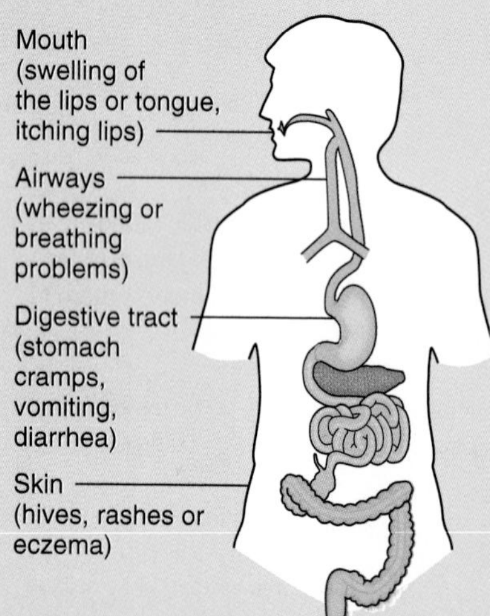

Source: U.S. Food and Drug Administration, *FDA Consumer,* May 1994.

The top eight foods causing adverse reactions in some individuals: milk, eggs, peanuts, nuts, fish, shellfish, soy, and wheat.

nuts, wheat, fish, and shellfish may delay the onset of food allergies by allowing the digestive system and immune system to further develop before being exposed to these known allergens. Young children often lose their sensitivity to most of the common allergenic foods in a few years. Unfortunately, sensitivity to certain foods, such as peanuts, tree nuts, fish, and shellfish, is rarely lost, and sensitivity persists into adulthood.[80]

Diagnosis and Management of Food Allergy

Because both food intolerance and food allergy present very similar symptoms, diagnosing a food allergy involves determining if the reaction is mediated by the immune system. A complete physical is conducted to rule out the possibility of any underlying physical condition which may be producing the symptoms. The physician takes a detailed case history and considers the type and timing of symptoms as well as suspected offending food. Individuals may be asked to keep a food diary and record all physical symptoms. Once food allergy is identified as a likely cause of symptoms, confirmation of the diagnosis and positive identification of the allergen involves various tests. A prick skin test (PST) is a simple test done in the doctor's office. The doctor puts a drop of the substance being tested on the forearm of the patient and pricks it with a needle, which allows a tiny amount of the substance to enter the skin. If the patient is allergic, a small bump will occur at the site in approximately 15 minutes. The radioallergosorbent test (RAST) requires a blood sample. Laboratory tests are done with specific foods to see if the patient has IgE antibodies to those foods. An elimination diet may be recommended by the physician. In an elimination diet the person does not eat the food suspected of causing the allergy for a period of about two to four weeks. If the allergic symptoms improve, a diagnosis can be made.

The only treatment for food allergy is avoidance of the offending food. Although this may appear to be a simple task, it is often quite difficult. People with food allergies must develop a skill for reading food labels. Foods may not necessarily be listed by their common name on the label. For instance, cow's milk protein is present in ingredients such as casein, whey, lactalbumin, caramel color, and nougat. The allergy-causing food may also be included in a general term, such as "flavors" or "spices." Hidden ingredients are also a concern—egg white may be used to glaze pretzels, or peanut butter may be used to seal the end of egg rolls. Cross-contamination is a problem with food allergies. When a food comes in contact with another food, trace amounts of each food mix with the other. This can occur in the processing plant if different foods are processed on the same equipment. It can also occur in the home—a knife inserted into the peanut butter jar and then used in the jelly will contaminate the jelly with peanut protein. The emerging field of biotechnology presents an additional challenge through genetic engineering (see Chapter 11). In genetic engineering, a gene from one species can be "spliced" into another species. The FDA requires that any food product of biotechnology that contains protein from a food that is known as a common allergen be properly labeled.

Currently, no drugs are available to cure food allergy. Some medications, such as antihistamines and corticosteroids, are used to treat the symptoms of food allergy. Individuals with a history of anaphylactic reactions to food allergens typically carry self-injecting syringes of epinephrine to use in case of accidental exposure to the allergen. Research is now underway to develop vaccines for combating anaphylactic allergies.[81]

Many questions remain to be answered in the field of food allergy. As scientists gain insight into the reactions of the immune system to various components of allergens, we can expect advances in the diagnosis and treatment of this disorder. As the mystery of DNA is unraveled it is conceivable that food allergy can be prevented. Until that time, however, there are many organizations that can help with additional information on food allergy and food intolerance, as listed in the margin.

TABLE 10-9

Common Foods Provoking Food Allergy

Food	Cross-Reacting Foods
Cows' milk*	Goats' milk, ewes' milk
Hens' eggs*	Eggs from other birds
Peanuts*	Soybeans, green beans, green peas
Soybeans	Peanuts, green beans, green peas
Cod	Mackerel, herring
Shrimp	Other crustaceans
Wheat	Other grains, mostly rye

Patients allergic to pollen may have cross reactions with hazelnuts, green apples, peaches, almonds, kiwis, tomatoes, and potatoes (birch pollen) or wheat, rye, and corn (grass pollen).

*Most common allergic foods for children.

Source: Adapted from C. Bindslev-Jensen, ABCs of Allergies: Food allergy, *British Medical Journal* 316 (1998): 7140.

The Food Allergy Network
800-929-4040
www.foodallergy.org

American Academy of Allergy, Asthma, and Immunology
414-272-6071
www.aaaai.org

American Dietetic Association
800-877-1600
www.eatright.org

Asthma and Allergy Foundation of America
800-7-ASTHMA

Most are Caucasian, with few cases seen among blacks and other minority groups. Most individuals who develop eating disorders are adolescents or young adults who typically began experiencing food-related and self-image problems between the ages of 14 and 30 years.

Because these syndromes are surrounded by secrecy, their prevalence is not known with certainty, although it has increased dramatically within the past two decades. Estimates of the prevalence in the general population range from 1 percent for anorexia nervosa to 1 to 5 percent for bulimia nervosa. These two types of eating disorders are also sometimes seen in adolescent athletes, many of whom compete in sports such as gymnastics, wrestling, distance running, diving, horse racing, and swimming, which demand a rigid control of body weight.[82]

NUTRITION IN LATER LIFE

Technically, the life span is the oldest documented age to which a member of a given species is known to have survived. For humans, the maximum life span is probably between 120 and 125 years.

It is easy to get the impression from mortality statistics that people are living longer and longer lives, but this is not the case. Certainly the *average* age at death (life expectancy) has changed dramatically. In the United States, the life expectancy at birth is now 73 years for men and 79 years for women.[83] On the other hand, the *maximum* age at which people die—that is the *maximum life span*—has changed very little. It seems that the aging phenomenon cuts off life at a rather fixed point in time.[84]

To what extent is aging inevitable? Apparently, aging is an inevitable, natural process programmed into our genes at conception. Nevertheless, we can adopt lifestyle habits, such as consuming a healthful diet, exercising, and paying attention to our work and recreational environments, that will slow the process within the natural limits set by heredity. Clearly, good nutrition can retard and ease the aging process in many significant ways. However, no potions, foods, or pills will prolong youth. People who claim to have found the fountain of youth have been selling its waters for centuries, but products advertised to prevent aging profit only the sellers, not the buyers.

One approach to the prevention of aging has been to study other cultures in the hope of finding an extremely long-lived race of people and then learning their secrets of long life. The views of the experts can best be summed up by saying that disease can *shorten* people's lives and poor nutrition practices make diseases more likely to occur. Thus, by postponing and slowing disease processes, optimal nutrition can help to prolong life up to the maximum life span—but cannot extend it further.[85] This section focuses on the nutrient requirements of older adults, the diseases that seem to come with age, their risk factors, and the relevance of nutrition to them. Test your own knowledge of the aging process with the Aging Scorecard found on the following page.

FIGURE 10-8
THE AGING OF THE POPULATION

In 1900, 4% of the U.S. population were over 65 years of age; in 1990, 12.7% were over 65; by 2030, 22% will have reached age 65.

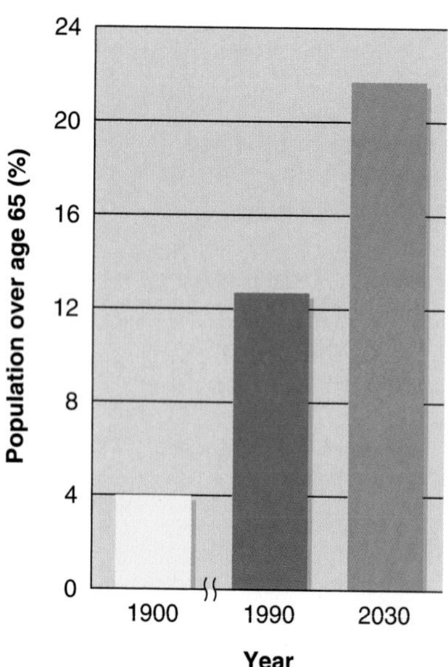

Source: Data from R. Chernoff, *Geriatric Nutrition: The Health Professional's Handbook*, 2nd ed. (Gaithersburg, MD: Aspen Publishers, 1999), p. 2.

Demographic Trends and Aging

The number of elderly (aged 65 years and older) in the United States will double by 2030 to more than 65 million people. In 1988, the elderly accounted for 12.4 percent of the population, and this proportion is expected to rise to approximately 14 percent in 2010 and increase to nearly 22 percent by 2030. Nearly 12 percent of the population will be over age 74 by 2030. The increased growth in the elderly population in the United States is illustrated in Figure 10-8.[86]

Both the baby boom that took place between 1946 and 1964 and improved life expectancy are important contributors to the growing elderly population in the United States. Baby boomers will increase the numbers of the older middle-aged (ages 46 through 63) until 2010, when they will begin to swell the ranks of the retired population.

Improved life expectancy has resulted from better prenatal and postnatal care and improved means of combating disease in older adults. For example, the

AGING SCORECARD

TRUE	FALSE	
☐	☐	1. Everyone becomes "senile" sooner or later if he or she lives long enough.
☐	☐	2. American families have by and large abandoned their older members.
☐	☐	3. Depression is a serious problem for older people.
☐	☐	4. The numbers of older people are growing.
☐	☐	5. The vast majority of older people are self-sufficient.
☐	☐	6. Mental confusion is an inevitable, incurable consequence of old age.
☐	☐	7. Intelligence declines with age.
☐	☐	8. Sexual urges and activity normally cease around age 55 to 60.
☐	☐	9. If a person has been smoking for 30 or 40 years, it does no good to quit.
☐	☐	10. Older people should stop exercising and rest.
☐	☐	11. As you grow older, you need more calories, but fewer vitamins and minerals, to stay healthy.
☐	☐	12. Only children need to be concerned about calcium for strong bones and teeth.
☐	☐	13. Extremes of heat and cold can be particularly dangerous to older people.
☐	☐	14. Many older people are hurt in accidents that could have been prevented.
☐	☐	15. More men than women survive to old age.
☐	☐	16. Deaths from stroke and heart disease are declining.
☐	☐	17. Older people on the average take more medications than younger people.
☐	☐	18. Snake oil salesmen are as common today as they were on the frontier.
☐	☐	19. Personality changes with age, just like hair color and skin texture.
☐	☐	20. Sight changes with age.

ANSWERS

1. False. Even among those who live to be 80 or older, only 20 percent to 25 percent develop Alzheimer's disease or some other incurable form of brain disease. "Senility" is a meaningless term that should be discarded.
2. False. The American family is still the number one caretaker of older Americans. Most older people live close to their children and see them often; many live with their spouses. In all, 8 out of 10 men and 6 out of 10 women live in family settings.
3. True. Depression, loss of self-esteem, loneliness, and anxiety can become more common as older people face retirement, the deaths of relatives and friends, and other such crises—often at the same time. Fortunately, depression is treatable.
4. True. Today, 12 percent of the U.S. population are 65 or older. By the year 2030, 22 percent of the population will be over 65 years of age. The fastest growing group are those over age 85.
5. True. Only 6 percent of people over age 65 live in nursing homes; the rest are basically healthy and self-sufficient.
6. False. Mental confusion and serious forgetfulness in old age can be caused by Alzheimer's disease or other conditions that cause incurable damage to the brain, but some 100 other problems can cause the same symptoms. A minor head injury, a high fever, poor nutrition, adverse drug reactions, and depression can all be treated, and the confusion will be cured.
7. False. Intelligence per se does not decline without reason. Most people maintain their intellect or improve as they grow older.
8. False. Most older people can lead an active, satisfying sex life.
9. False. Stopping smoking at any age not only reduces the risk of cancer and heart disease but also leads to healthier lungs.
10. False. Many older people enjoy—and benefit from—exercises such as walking, swimming, bicycle riding, and strength training. Exercise at any age can help to strengthen the heart and lungs and lower blood pressure. See your physician before beginning a new exercise program.
11. False. Older people need fewer calories (due to lower basal metabolic rates) but the same amounts of most vitamins and minerals as younger people. Some authorities recommend higher intakes of certain nutrients (calcium, vitamin D, antioxidants, B vitamins) for older adults.
12. False. Adequate intake of calcium as part of a bone-healthy diet is needed throughout life. This is particularly true for women, whose risk of osteoporosis increases after menopause.
13. True. The body's thermostat tends to function less efficiently with age, and the older person's body may be less able to adapt to heat or cold.
14. True. Falls are the most common cause of injuries among the elderly. Good safety precautions, including proper lighting, nonskid carpets, and living areas free of obstacles, can help to prevent serious accidents.
15. False. Women tend to outlive men by an average of six years. There are 150 women for every 100 men over age 65, and nearly 250 women for every 100 men over 85.
16. True. Fewer men and women are dying of stroke or heart disease. This has been a major factor in the increase in life expectancy.
17. True. The elderly consume 25 percent of all medications and as a result have many more problems with adverse drug reactions.
18. True. Medical quackery is a 10-billion-dollar business in the United States. People of all ages are commonly duped into "quick cures" for aging, arthritis, and cancer.
19. False. Personality doesn't change with age. Therefore, old people can't all be described as rigid and cantankerous. You are what you are for as long as you live, unless you *choose* to make a change (for example, to be more outgoing or more flexible).
20. True. Changes in vision become more common with age, but any change in vision, regardless of age, is related to a specific disease. If you are having problems with your vision, see your doctor.

Source: Adapted from *What Is Your Aging I.Q.?* U.S. Department of Health and Human Services, Public Health Service, National Institutes of Health (Washington, D.C.: U.S. Government Printing Office).

death rate from heart disease began to decline in the 1960s and continues to fall today. Over half of the drop is attributed to a decline in smoking and fewer people with high blood pressure or high blood cholesterol.

Healthy Adults

An individual's current health profile is substantially determined by behavioral risk factors. The leading causes of death for adults ages 25 through 64 are cancer, heart disease, stroke, injuries, chronic lung disease, and liver disease, all of which have been associated with behavioral risk factors. To what extent can nutrition prevent or retard the development of these diseases?

Many of the health problems associated with the later years are preventable or can be controlled.[87] For example, changing certain risk behaviors into healthy ones can improve the quality of life for older persons and lessen their risk of disability. For example, incorporating exercise and a balanced, low-fat diet into one's lifestyle can contribute to weight loss and to controlling three important risk factors for heart disease: high fat intake, overweight, and a sedentary lifestyle. Figure 10-9 illustrates how the same set of dietary recommendations can both promote general health and help to prevent a broad spectrum of chronic diseases.[88] As Figure 10-9 shows, these basic dietary recommendations reduce the risk of a variety of chronic diseases or their complications.

Figure 10-10 puts nutrition (a factor you *can* control) in perspective with respect to heredity (a factor you can't control). It illustrates the point that some diseases are much more responsive to nutrition than others and that some are not responsive at all. At one extreme are diseases that can be completely cured by supplying missing nutrients, and at the other extreme are certain genetic, or inherited, diseases that are completely unaltered by nutrition. Most fall in between, being influenced by inherited susceptibility but responsive to dietary manipulations that help to normalize metabolism and counteract the disease process. Thus, diabetes may be controlled by means of a diet low in fat and high in complex carbohydrate; arthritis may be somewhat relieved by weight reduction, and cardiovascular disease may respond favorably to a diet low in saturated fat and cholesterol.

FIGURE 10-9
CONSISTENCY OF RECOMMENDATIONS FOR REDUCING THE RISK OF CHRONIC DISEASES OR THEIR COMPLICATIONS

Change diet → / Reduce risk ↓	Reduce fats	Control calories	Increase starch[a] and fiber	Reduce sodium	Control alcohol	Increase antioxidants[b]
Heart disease	●	●	●	●		●
Cancer	●	●	●	●[d]	●	●
Stroke	●	●		●	●	●
Diabetes	●	●	●			●
Gastrointestinal diseases[c]	●	●	●		●	

[a]Starch refers to complex carbohydrates provided by fruits, vegetables, and whole-grain products.
[b]Aim for a minimum of 5 to 9 servings of fruits and vegetables daily.
[c]Gastrointestinal diseases affected by dietary factors are primarily gallbladder disease (fat and calories), diverticular disease (fiber), and cirrhosis of the liver (alcohol).
[d]As used for pickling and preserving foods.

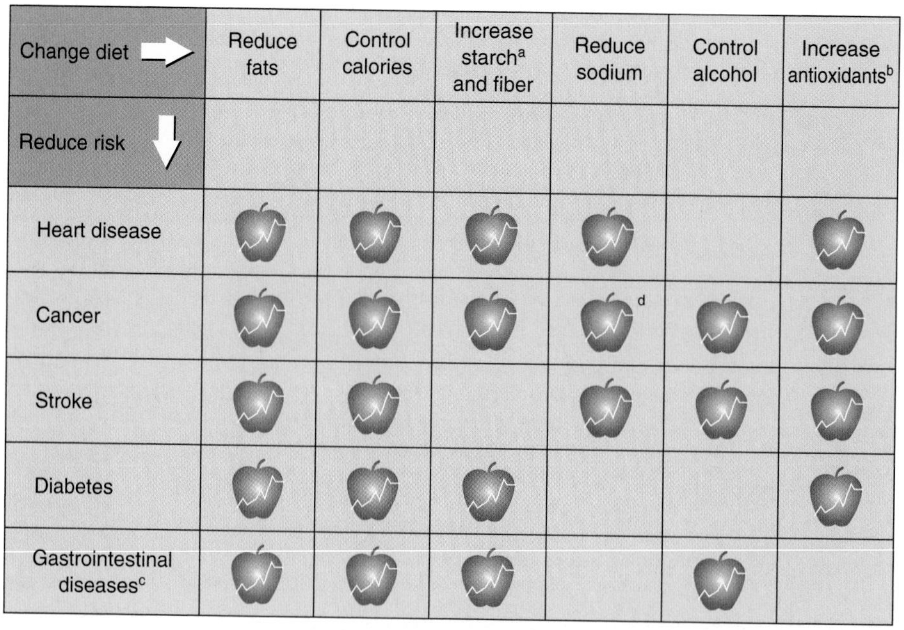

Source: J. M. McGinnis and M. Nestle, The Surgeon General's Report on Nutrition and Health: Policy implications and implementation strategies, *American Journal of Clinical Nutrition* 49 (1989): 26. © American Society for Clinical Nutrition; AHA Conference Proceedings, Unified Dietary Recommendations, Summary of a Scientific Conference on Preventive Nutrition: Pediatrics to Geriatrics, *Circulation* 100 (1999): 450.

Nutrition-unresponsive (genetic) diseases

Sickle-cell anemia	
Adult bone loss (osteoporosis) Cancer	
Arthritis Diabetes Hypertension Heart disease	
Anemias Vitamin deficiencies Mineral deficiencies Toxicities Low birthweight Poor resistance to disease	

Nutrition-responsive diseases

FIGURE 10-10
NUTRITION AND DISEASE

Not all diseases are equally influenced by diet. Some are purely hereditary, such as sickle-cell anemia. Some may be inherited (or the tendency to develop them may be inherited) but may be influenced by diet, such as some forms of diabetes. Some are purely dietary, such as the vitamin and mineral deficiency diseases. Some authorities are concerned that the offering of dietary advice to the public may seem to exaggerate the power of nutrition in preventing disease. Nutrition alone is certainly not enough to prevent many diseases, but it helps.

Aging and Nutrition Status

Growing old is often associated with frailty, sickness, and a loss of vitality. Although chronic illness and associated disabilities are experienced by the aging in our society, there is great heterogeneity among this population—that is, older people vary greatly in their social, economic, and lifestyle situations, functional capacity, and physical conditions.[89] Each person ages at a different rate, sometimes making chronologic age different from biologic age. Most older persons live at home, are fully independent, and have lives of good quality.[90] Only 5 percent to 6 percent of older adults live in nursing homes.[91] Older persons who have problems with the **activities of daily living (ADL)** are known as the frail elderly. Because they depend on others to perform these essential activities, they are likely to be at risk for malnutrition.[92]

Any form of physical activity—even a 10-minute walk around the block—is better than nothing.

Nutritional Needs and Intakes

Many of the nutrient needs of the elderly are the same as for younger persons, but some special considerations deserve emphasis. Calorie needs often decline with age because of a decrease in basal metabolism related to loss of lean tissue and a decrease in physical activity. Beginning at age 51, the recommended energy intake is lower for adults, assuming about a 5 percent reduction in energy output per decade. Given their limited energy allowances, older adults are advised to select mostly nutrient-dense foods.

On one side of the energy budget, energy is taken in; on the other side, it is spent. If you are motivated to maintain your good health into the later years, you should plan regular exercise into your days. Ideally, the exercise should be intense enough to increase the heartbeat and respiration rate and to prevent the atrophy of all muscles (not only the heart) that otherwise takes place. Many older people believe that they can't participate in strenuous exercise, but studies have shown that they can do more than they think they can. Any exercise—even a ten-minute walk each day—is better than none, and with persistence, one can achieve great improvement. Modest endurance training can improve cardiovascular and respiratory function and promote good muscle tone while controlling the accumulation of body fat.

The older person who has never worked out strenuously before may be encouraged to learn that the trainability of older people does not depend on their physical prowess in their youth. Improvement during training is due not only to improvement in muscles but also to the increased bloodflow to the brain engendered by the exercise. A major benefit is that with an increased energy output, a person can afford to eat more food and so obtain more nutrients.

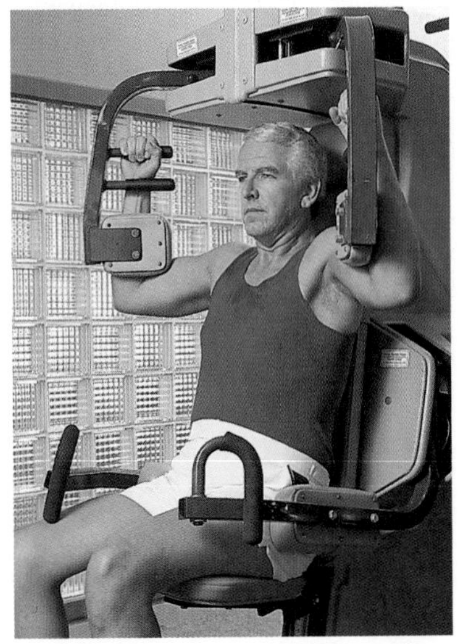

Strength training promotes healthy bones and muscles at any age.

ACTIVITIES OF DAILY LIVING (ADL) include bathing, dressing, grooming, transferring from bed to chair, going to the bathroom, and feeding oneself.

Although caloric needs may decrease with age, the need for certain nutrients such as calcium, vitamin D, vitamin C, vitamin B_{12}, and vitamin B_6 may actually increase with the effects of aging (see Table 10-10).[93] For example, as many as 30 percent of persons older than 50 years may experience reduced stomach acidity, which can interfere with their ability to absorb vitamin B_{12}, calcium, and iron effectively from foods. In addition, there may be increased needs for vitamin D in older adults who have low intakes of fortified dairy products, because the skin becomes less efficient at making vitamin D when exposed to sunlight.[94]

Perhaps the most important nutrient of all is *water.* Older adults need to be reminded to take in fluids because they are likely to be somewhat insensitive to their own thirst signals.[95] They should consume six to eight glasses of fluid a day.

TABLE 10-10 THE IMPACT OF AGING ON NUTRIENT NEEDS

Nutritional Concern	Age-Related Finding	Implications for Healthy Aging
Energy	Energy needs decrease due to decline in lean tissue.	Get regular physical activity, including strength training exercises (see Chapter 9).
Water	Reduced sense of thirst; increased urine output.	To prevent dehydration, drink six to eight glasses of fluid a day.
Fiber	Constipation is a common complaint due to low fiber and fluid intakes and lack of exercise.	Consume 20 to 35 grams of fiber from a variety of sources such as fresh fruits, vegetables, legumes, and whole-grain products; get adequate amounts of fluid and exercise.
Vitamin D	Intakes may be low; skin becomes less able to synthesize vitamin from sunlight; less exposure to sunlight.	Choose fortified milk and cereals, salmon and other fatty fish; get daily exposure to sunlight if possible; or use a supplement (400 to 600 IU per day).
Calcium	Intakes typically low; reduced gastric acidity impairs absorption, reduced bone mass.	To offset bone loss, choose calcium-rich foods (low-fat or fat-free milk, yogurt, cheese), fortified sources of calcium (orange and other juices), or take a calcium supplement (see the Spotlight feature in Chapter 7 for guidelines); get regular weight-bearing exercise.
Iron	Anemia less common after menopause; reduced gastric acidity may impair absorption; chronic blood loss or low energy intakes may increase risk of deficiency.	Consume adequate calories from a variety of nutrient-dense foods with a source of vitamin C to enhance absorption of dietary iron.
Protein, vitamin E, vitamin B_6, zinc	Intakes may be low; decline in immune function.	Meet nutrient needs from low-fat servings of meat, fish, poultry, and legumes; use nuts in moderation; choose at least six servings of whole-grain products.
Antioxidants: vitamin C, vitamin E, carotenoids (beta-carotene, lutein)	Decline in vision (cataracts, macular degeneration); increased oxidative stress to body tissues.	Eat at least five servings of fruits and vegetables a day; choose regular servings of green leafy vegetables, brightly colored fruits and vegetables, and citrus fruits; Choose whole-grain products; use nuts in moderation.
Vitamin A	Decline in liver's uptake of vitamin A; increased absorption may lower the aging body's requirement.	Avoid supplements of vitamin A.
Vitamin B_{12}	Intakes may be low; reduced gastric acidity impairs absorption of food-bound vitamin B_{12}.	Choose foods fortified with vitamin B_{12} (e.g., breakfast cereals), or take a supplement that contains vitamin B_{12}.
Folate, vitamin B_6, vitamin B_{12}	Intakes may be low—causing elevated blood levels of homocysteine and increased risk of heart disease; use of certain medications may impair B-vitamin status.	Consume citrus and other fresh fruits and green leafy vegetables, legumes, whole grains, and foods fortified with folate and vitamin B_{12} (e.g., breakfast cereals).

The new, narrower food guide pyramid for adults over age 70 shown in Figure 10-11 shows a recommended eating pattern that reflects the lower caloric needs of most healthy older adults and emphasizes the need for adequate fluid intake. Clearly, an adequate intake of nutrients and fiber from a variety of foods throughout life, together with moderate intakes of calories and fat, and regular physical activity helps immensely to promote good health in the later years.

The recommended nutrient intakes for older adults are shown inside the front cover.

Some health practitioners recommend nutritional supplements for older adults, particularly for chronic conditions such as osteoporosis, arthritis, or anemia. However, advertisers often target older adults by recommending supple-

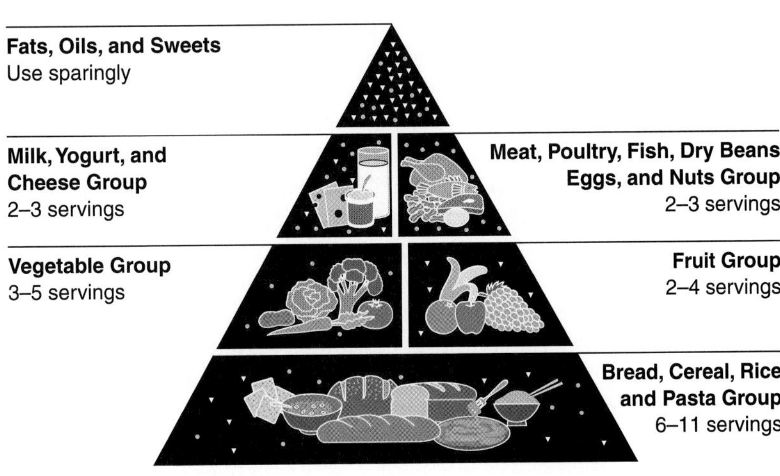

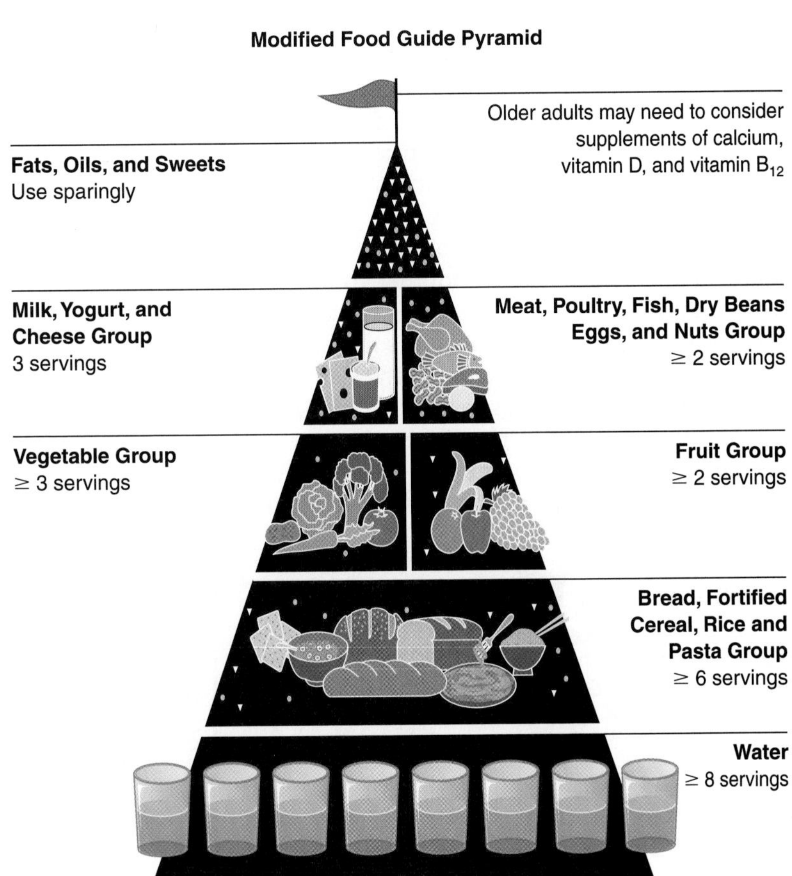

FIGURE 10-11
HEALTHY AGING

A new food guide pyramid for older Americans emphasizes nutrient-dense foods, fewer calories, and lots of fluids. To meet nutritional needs, the diets of older adults should include:

- At least three servings of low-fat dairy products or suitable alternatives
- Two or more servings from the meat group that feature dried beans (rich in fiber), fish, eggs, and lean cuts of meat and poultry
- At least five servings of fruits and vegetables featuring foods that are richly colored (dark green, orange, red, or yellow)
- At least six servings of nutrient-dense, fiber-rich whole grains and fortified cereals
- At least eight glasses of water per day to prevent dehydration

Sources: U.S. Departments of Agriculture and Health and Human Services; Modified food guide pyramid for people over 70 years of age, *Journal of Nutrition,* 129 (1999): 751-753, American Society for Nutritional Science.

The evidence for using dietary supplements to slow the aging process is not as strong as the evidence supporting a diet rich in fruits, vegetables, and other nutrient-dense foods along with a physically active lifestyle.

ments to reduce the effects of aging and increase longevity (see Table 10-11). Such advertising makes the elderly vulnerable to exploitation by quacks, when their money is better spent on nutritious foods. Surveys designed to find out what kinds of nutrient supplements older people are using tend to support the view that in many cases, older people are wasting their money.[96]

Nutrition-Related Problems of Older Adults

Although aging is not completely understood, we know that it involves progressive changes in every body tissue and organ: the brain, heart, lungs, digestive tract, and bones (see Table 10-12 on page 357). After age 35, functional capacity declines in almost every organ system. Such changes affect nutrition status. Some, including oral problems, interfere with nutrient intake; others affect absorption, storage, and utilization of nutrients; and still others increase the excretion of and need for specific nutrients.[97] Examples of various conditions associated with aging that can affect nutritional status include sensory impairments, altered endocrine, gastrointestinal, and cardiovascular functions, and changes in the renal and musculoskeletal systems. Both genetic and environmental factors contribute to these declines. Many of the changes are inevitable, but a healthful lifestyle that combines regular physical activity with adequate intakes of all essential nutrients can forestall degeneration and improve the quality of life into the later years.

As a person gets older, the chances of suffering a chronic illness or functional impairment become greater. Among the diseases that befall some people in later life are heart disease, hypertension, cancer, diverticulosis, osteoporosis, dementia, diabetes, and gum disease. More than 35 percent of people over age 65 have high blood pressure, and approximately 30 percent have heart disease. Chronic conditions contributing to **disability** include arthritis, heart disease, strokes, disorders of vision and hearing, nutritional deficiencies, and oral-dental problems (see Figure 10-12 on page 357). Dementia (especially Alzheimer's disease) is a major contributor to disability and nursing home placement for those over age 75.[98] As noted in Table 10-13 on page 358, malnutrition can occur secondary to these conditions, many of which require special diets that can further compromise nutrition status in the older adult.

DISABILITY any restriction on or impairment in performing an activity in the manner or within the range considered normal for a human being.

Polypharmacy, or the use of multiple drugs, is problematic for many older adults. The average older person receives more than 13 prescriptions a year and may take as many as six drugs at a time.[99] Cardiac drugs are most widely used by the elderly, followed by drugs to treat arthritis, psychiatric disorders, and respiratory and gastrointestinal conditions. Long-term use of a variety of drugs increases the risk of drug-nutrient interactions. Individuals with impaired nutrition status and poor dietary intakes are at the highest risk.

POLYPHARMACY the taking of three or more medications regularly; occurs in one-third of those over 65 years.

Steve Dickenson, Copley News.

TABLE 10-11 ANTI-AGING SUPPLEMENTS:* FOUNTAIN OF YOUTH OR ALL WET?

Supplement	Physiological Role	Anti-Aging Claims Under Investigation	Comments
Vitamin C	Acts as an antioxidant by scavenging free radical compounds, preventing them from attacking healthy tissue.	Associated with detoxifying environmental cancer-causing agents. May block formation of nitrosamines in foods and digestive tract and lower risk for stomach cancer. Possible role in prevention of heart disease.	Vitamin C is found in many food sources: cantaloupe, citrus fruits, green vegetables, peppers, strawberries, and tomatoes. Conflicting results from studies using vitamin C supplementation versus fruit and vegetable consumption may indicate that vitamin C plus other substances in food, such as phytochemicals, are responsible for the beneficial effects. High fruit and vegetable intake has been consistently associated with protective benefits in conditions such as macular degeneration, visual loss, cataracts, and certain cancers. More controlled research to determine whether vitamin C supplements offer a health-protective effect is needed.
Vitamin E	Acts as an antioxidant; particularly effective in protecting cell membranes from oxidative damage and in preventing the oxidation of LDL cholesterol inside the arteries.	May play a role in protecting against cardiovascular, Alzheimer's, and Parkinson's diseases. Studies have shown improvement in immune function with vitamin E supplementation.	The American Heart Association recognizes the benefits of vitamin E supplementation in reducing risk of cardiovascular disease and other conditions as promising, but not proven. Canola, safflower, and sunflower oil are the primary sources of vitamin E in the American diet. Extremely high doses of vitamin E may interfere with the blood-clotting action of vitamin K and intensify the effect of drugs used to oppose blood clotting.
Beta-carotene	One of a large group of orange pigments (carotenoids) found in plants; a precursor for vitamin A. Beta-carotene may help to minimize the damage caused by free radicals.	Beta-carotene has been associated with reduction in risk for certain types of cancers.	Conflicting research results raise concerns about beta-carotene supplementation. Two studies showed an *increased* risk of lung cancer in smokers who took beta-carotene supplements. A diet which includes dark green, yellow, orange, and red plant foods will provide several types of carotenoids and has been associated with reducing the risk of some cancers.
Coenzyme Q 10	A vitamin-like substance which plays a role in production of energy in cells. It serves as an antoixodant to protect the mitochondria from free radical damage.	May be beneficial in treatment of brain-degenerative disorders such as Parkinson's and Alzheimer's diseases. CoQ10 may be useful in treating congestive heart failure.	Preliminary studies involving Coenzyme Q 10 as a therapy for congestive heart failure seem promising. Studies for other uses are limited and claims that it can slow aging warrant further substantiation. It is found in dietary sources such as meats, poultry, seafood, and some fruits (oranges) and vegetables (broccoli).
DHEA	Dehydroepiandrosterone (DHEA) is a hormone which is converted by the body into the sex hormones estrogen and testosterone.	Age-related decline in DHEA levels has led to speculation that it may be involved in the aging process. It has been hypothesized that low levels of DHEA may be associated with decline in muscle mass, loss of sleep,	Of the research that has been done, there are no firm conclusions about the role of DHEA in the prevention of age-related disorders. More studies need to be done in this area. Supplementation with DHEA raises the estrogen levels in women and testosterone levels in men. It is unknown whether this increase may lead to the development of ovarian, prostate, and other types of cancer. Some

continued

TABLE 10-11 ANTI-AGING SUPPLEMENTS*—*Continued*

Supplement	Physiological Role	Anti-Aging Claims Under Investigation	Comments
DHEA—*continued*		diminished feelings of well-being and may occur in diseases such as cancer, atherosclerosis, hypertension, diabetes, osteoporosis, and Alzheimer's.	liver damage has been reported in users of DHEA. Long-term safety and efficacy of DHEA is unknown; self-supplementation is not recommended.
Melatonin	A hormone made from tryptophan and produced by the pineal gland in the brain to regulate the body's daily rhythm or sleep-wake cycles.	Melatonin is marketed as a sleep aid for insomnia and jet lag and anti-aging remedy. Other proposed benefits include stimulation of the immune system and cancer prevention.	More controlled studies on humans are required to support initial claims of possible health benefits to warrant supplementation. Little information is available regarding possible long-term effects of the hormone. Users may feel groggy or mildly depressed, and taken improperly can interrupt sleep cycle.
Human Growth Hormone (GH)	A hormone that stimulates body cells to divide and increase in size; major targets are bones and skeletal muscles. Also affects protein, carbohydrate and lipid metabolism.	Age-related decline in GH levels are associated with decline in lean body mass and muscle strength. Studies indicate administration of GH increases lean body mass. However, functional ability of muscle does not improve in relation to increased lean body mass.	Side effects of GH treatment limit usefulness as an agent for increasing muscle protein. Commonly reported side effects include swelling, fluid retention and carpal tunnel syndrome.
Ginkgo Biloba	Extract from the leaves of the ginkgo tree which is commonly used in Europe as a treatment for dementia. Sold as a licensed drug in France and Germany.	Researchers postulate that ability to expand blood vessels, as well as antioxidant properties of phytochemicals (flavonoids) found in the plant, play a role in improved mental functioning.	Commonly used in Europe and integral part of Chinese herbalist practice for centuries. A study on large U.S. populations is currently under way. Side effects include stomach upset, headache or allergic skin reaction. When taken with anticoagulants such as aspirin or warfarin, serious bleeding can occur. Not recommened for pregnant women.
Glucosamine and Chondroitin Sulfate	Both compounds are synthesized in the body and found in cartilage. The body uses glucosamine to build compounds that provide the building blocks or shape of cartilage; chondroitin sulfate is part of a cartilage-building protein that gives cartilage its elasticity.	May relieve pain, improve mobility, and help repair and maintain cartilage in people with osteoarthritis.	A recent Belgian study showed promise that glucosamine may be able to halt progression of osteoarthritis in the knee. More longer-term studies are needed, however. A large, multicenter U.S. study on glucosamine is currently under way. People with allergies to shellfish (glucosamine is made from the shells of certain shellfish) or taking prescription medications should seek the supervision of their physicians. Until more is known, seek weight-loss to relieve stress on joints and exercises to improve flexibility and strength.

**Note:* The largely unregulated supplement industry is a buyer-beware market. Sometimes, supplements do not contain what is stated on the label. If you choose to use supplements, select brands from reliable manufacturers (see the guidelines for selecting supplements given in Chapter 6 on page 185).

TABLE 10-12 BIOLOGIC FUNCTION CHANGES BETWEEN THE AGES OF 30 AND 70

Biologic Function	Change
Work capacity of heart (%)	↓25–30
Maximum heart rate	↓24
Blood pressure (mm Hg)	
Systolic	↑10–40
Diastolic	↑ 5–10
Maximum breathing capacity (%)	↓40–50
Basal metabolic rate (%)	↓ 8–12
Musculature (%)	
Muscle mass	↓25–30
Hand grip strength	↓25–30
Nerve conduction velocity (%)	↓10–15
Flexibility (%)	↓20–30
Bone (%)	
Women	↓25–30
Men	↓15–20
Kidney function (%)	↓30–50

Source: Adapted from R. R. Watson, *Handbook of Nutrition in the Aged,* 2nd ed. (Boca Raton, FL: CRC Press, 1994), pp. 11–73.

Factors Affecting Nutrition Status of Older Adults

Individually or in combination, the social, economic, psychological, cultural, and environmental factors associated with aging may interact with the physiological changes and further affect nutrition status in older adults. These interactions are illustrated in Figure 10-13 on page 359.

Up to one-quarter of all elderly patients and one-half of all hospitalized elderly may be suffering from malnutrition.[100] In addition, a national survey showed that one-third of noninstitutionalized Americans over the age of 65 live alone, 45 percent take multiple prescription drugs that can interfere with appetite and nutrient absorption, 30 percent skip meals almost daily, and 25 percent have annual

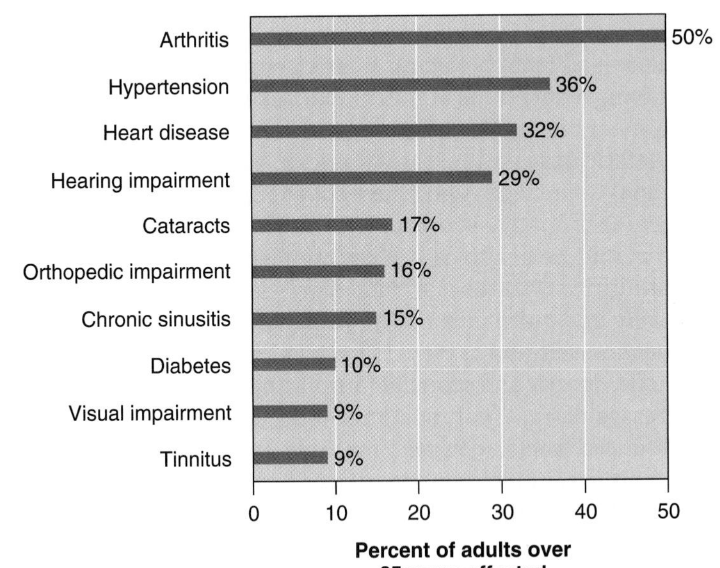

FIGURE 10-12 COMMON CAUSES OF DISABILITY IN OLDER PERSONS

Source: U.S. Department of Health and Human Services, *A Profile of Older Americans: 1999* (Washington, D.C.: U.S. Department of Health and Human Services, 1999).

TABLE 10-13 MALNUTRITION THAT IS SECONDARY TO DISEASE OR PHYSIOLOGIC STATE

Disease or Condition	Effects on Nutrition Status
Atherosclerosis	May increase difficulties in regulating fluid balances if caused by congestive heart failure. If the individual is incapacitated, calorie needs decrease.
Cancer	Weight loss, lack of appetite, and secondary malnutrition are common.
Dental and oral disease	May alter the ability to chew and thus reduce dietary intake. Increased likelihood of choking and aspiration.
Depression and dementia	Increased or decreased food intakes are common. A person with dementia may have decreased ability to get food, or the appetite may be very small or very great. Judgment and balance in meal planning are generally absent.
Diabetes mellitus (Type 1)	If untreated, increased risk of undernutrition results; increased risk of other diet-related diseases.
Diabetes mellitus (Type 2)	Increased risk of other diet-related diseases; weight loss is needed if obesity is present.
End-stage kidney disease	Alters fluid and electrolyte needs. Infections and low-grade fever may increase energy output and weight loss.
Gastrointestinal disorders	Increased risk of malabsorption of nutrients and consequent undernutrition.
High blood pressure	Weight gain may exacerbate high blood pressure.
Osteoarthritis	Makes motion difficult, including those activities related to purchasing, serving, eating, and cleaning up after meals. Predisposes people to a sedentary lifestyle and may give rise to obesity. Drug-nutrient interactions are common.
Osteoporosis	Limits the ability to purchase and prepare food if mobility is affected. If severe scoliosis is present, the appetite may be altered.
Smoking	May alter weight status. Alters serum levels of some nutrients, such as ascorbic acid and beta-carotene. Chronic smoking gives rise to emphysema, which makes it difficult to eat because of breathing problems.
Stroke	May alter abilities in the cognitive and motor realms related to food and eating. If the individual is incapacitated, his or her energy needs decrease.

Source: Adapted from Institute of Medicine, *The Second Fifty Years: Promoting Health and Preventing Disability* (Washington, D.C.: National Academy Press, 1992), pp. 168–69.

incomes under $10,000—all factors placing elderly people at potential for nutritional risk.[101] Identifying older adults at nutritional risk is an important first step in maintaining quality of life and functional status.

The American Dietetic Association, the American Academy of Family Physicians, and the National Council on Aging have collaborated on an effort—called the Nutrition Screening Initiative—to promote nutrition screening and early intervention as part of routine health care. A key premise to the Nutrition Screening Initiative is that nutrition status is a "vital sign"—as vital to health assessment as blood pressure and pulse rate.[102] The Nutrition Screening Initiative has developed a ten-question self-assessment "Checklist" (see Figure 10-14 on page 360) to help individuals identify and score factors placing them at nutritional risk. This checklist addresses disease, eating status, tooth loss or mouth pain, economic hardship, reduced social contact, multiple medications, involuntary weight loss or gain, and need for assistance with self-care. The word *DETERMINE* is used as a mnemonic device with the checklist. Each letter in DETERMINE begins a word or phrase that describes a risk factor.[103]

The DETERMINE signs of malnutrition in the elderly:

- **D**isease
- **E**ating poorly
- **T**ooth loss or oral pain
- **E**conomic hardship
- **R**educed social contact
- **M**ultiple medications
- **I**nvoluntary weight loss or gain
- **N**eed of assistance with self-care
- **E**lderly person older than 80 years

Financial resources, living arrangements, and a social support network, including availability of caregivers, can also directly affect a person's nutrition

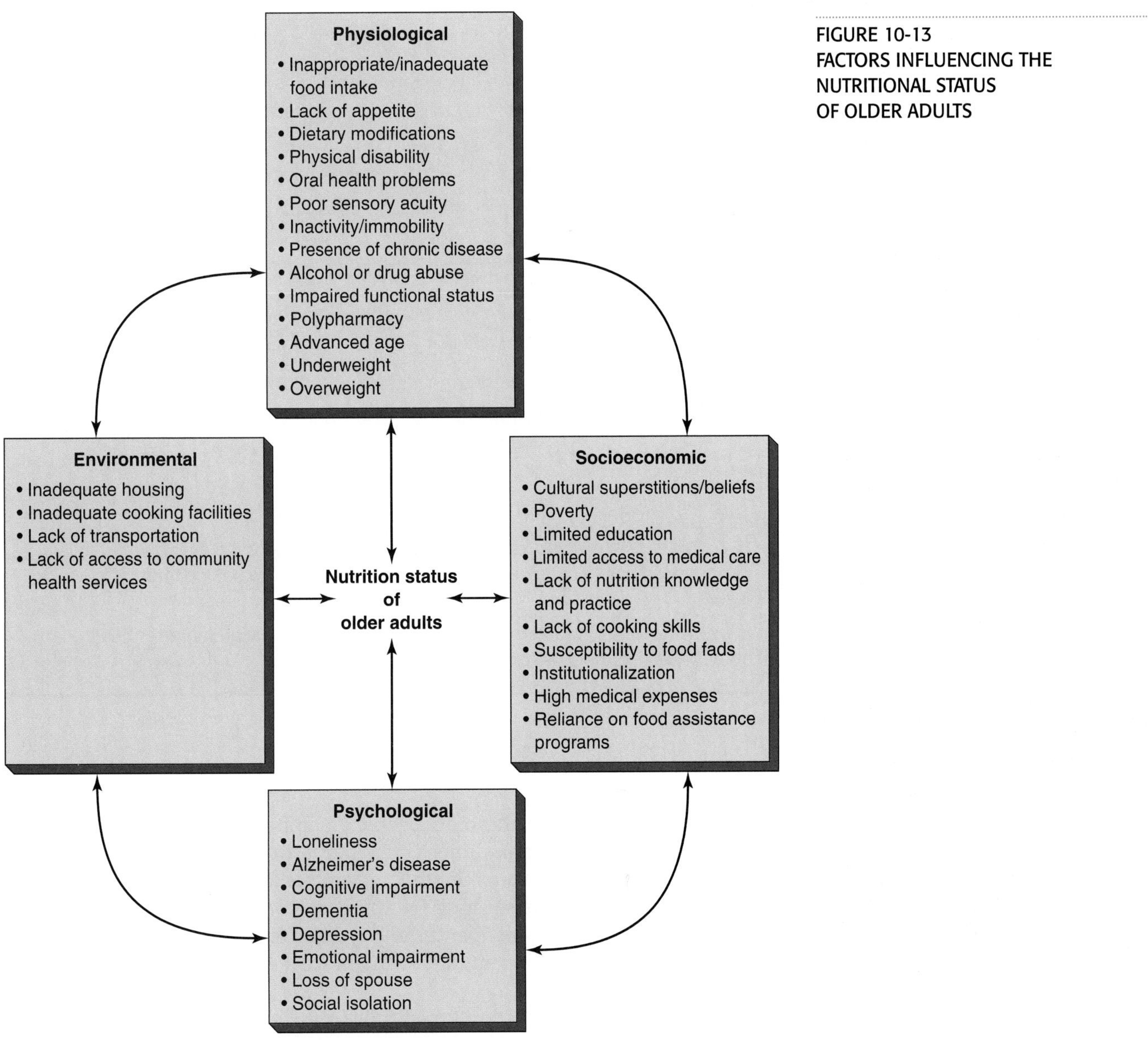

Physiological
• Inappropriate/inadequate food intake
• Lack of appetite
• Dietary modifications
• Physical disability
• Oral health problems
• Poor sensory acuity
• Inactivity/immobility
• Presence of chronic disease
• Alcohol or drug abuse
• Impaired functional status
• Polypharmacy
• Advanced age
• Underweight
• Overweight
Environmental
• Inadequate housing
• Inadequate cooking facilities
• Lack of transportation
• Lack of access to community health services
Nutrition status of older adults
Socioeconomic
• Cultural superstitions/beliefs
• Poverty
• Limited education
• Limited access to medical care
• Lack of nutrition knowledge and practice
• Lack of cooking skills
• Susceptibility to food fads
• Institutionalization
• High medical expenses
• Reliance on food assistance programs
Psychological
• Loneliness
• Alzheimer's disease
• Cognitive impairment
• Dementia
• Depression
• Emotional impairment
• Loss of spouse
• Social isolation

FIGURE 10-13
FACTORS INFLUENCING THE NUTRITIONAL STATUS OF OLDER ADULTS

status. Poverty and social isolation particularly impair the nutrition status of many older adults, as noted perceptively by a professor of psychiatry:[104]

> It is not what the older person eats but with whom that will be the deciding factor in proper care for him. The oft-repeated complaint of the older patient that he has little incentive to prepare food for only himself is not merely a statement of fact but also a rebuke to the questioner for failing to perceive his isolation and aloneness and to realize that food . . . for one's self lacks the condiment of another's presence which can transform the simplest fare to the ceremonial act with all its shared meaning.

Sources of Nutritional Assistance

In response to the socioeconomic problems—low income, inadequate facilities for preparing food, lack of transportation, and inability to afford dental care, among others—that trouble many older adults and may lead to malnutrition, federal, state, and local agencies have mandated nutrition programs for the elderly.

The Food Stamp program enables people who qualify to obtain paper coupons or electronic benefits in the form of a "credit" card with which to buy food. In

FIGURE 10-14
CHECKLIST TO DETERMINE YOUR NUTRITIONAL HEALTH

The warning signs of poor nutritional health are often overlooked. To see whether you (or people you know) are at nutritional risk, take this simple quiz. Read the statements at right. Circle the number in the "yes" column for those that apply. To find your total nutritional score, add up all of the numbers you circled.

Source: Nutrition Screening Initiative.

	Yes
I have an illness or condition that makes me eat different kinds and/or amounts of food.	2
I eat fewer than 2 meals per day.	3
I eat few fruits or vegetables and use few milk products.	2
I have 3 or more drinks of beer, liquor, or wine almost every day.	2
I have tooth or mouth problems that make it hard for me to eat.	2
I don't always have enough money to buy the food I need.	4
I eat alone most of the time.	1
I take 3 or more different prescribed or over-the-counter drugs a day.	1
Without wanting to, I have lost or gained 10 pounds in the past 6 months.	2
I am not always physically able to shop, cook, and/or feed myself.	2
Total:	

Score:

0–2: Good! Recheck your score in six months.

3–5: Moderate nutritional risk. Visit your local office on aging, senior nutrition program, senior citizens center, or health department for tips on improving eating habits.

6 or more: High nutritional risk. See your doctor, dietitian, or other health-care professional for help in improving your nutrition status.

many areas, food banks enable older people on limited incomes to buy good food for less money. A food bank project buys industry "irregulars"—products that have been mislabeled, underweighted, redesigned, or mispackaged and would therefore ordinarily be thrown away. Nothing is wrong with this food; the industry can credit it as a donation, and the buyer (often a food-preparing site) can obtain the food for a small handling fee and make it available at a greatly reduced price.

The federal Elderly Nutrition Program is intended to improve older people's nutrition status and enable them to avoid medical problems, continue living in communities of their own choice, and stay out of institutions. Its specific goals are to provide the following:

- Low-cost, nutritious meals.
- Opportunities for social interaction.
- Nutrition education and shopping assistance.
- Counseling and referral to other social and rehabilitation services.
- Transportation services.

The current Title III legislation makes one hot noon meal available five days a week, supplying a third of the recommended nutrient intakes. There is no cost for meals, but participants sometimes make voluntary contributions.

A part of Title III is the congregate meal program. Administrators try to select sites for congregate meals so as to feed as many of the eligible elderly as possible. The congregate meal sites are often community centers, senior citizen centers, religious facilities, schools, extended care facilities, or elderly housing complexes. Volunteers may also deliver meals to those who are homebound either permanently or temporarily; these efforts are known as the Home-Delivered Meals Program. The home-delivery program ensures nutrition, but its

Meals for One

Eating a food you haven't tasted before prolongs your life by 75 days, according to an old Japanese saying.[105] Perhaps finding someone to share that food with might extend your life even more.

Planning nourishing meals for one that offer variety and optimal nutrition can be particularly challenging. The following tips help to solve some of the problems that singles of any age face concerning buying and preparing foods.

- The first step to simple meal preparation is your grocery list. Keep your cupboards and refrigerator stocked with healthy foods that you like and that can be prepared quickly and easily. Keep a supply of: milk, eggs, bread, tortillas, pita bread, canned beans, jars of spaghetti sauce, rice, pasta or noodles, potatoes, onions, canned soups and broth, margarine, cooking oil, and frozen vegetables.
- Keep fresh fruits or vegetables on hand for easy-to-grab snacking.
- Buy only what you will use: The individual-sized containers may be expensive, but it is also expensive to let the unused portion of a large-sized container spoil before you are ready to use it. Don't be timid about asking the grocer to break up a package of meat, eggs, fresh fruits, or vegetables and wrap a smaller quantity for you.
- Buy a loaf of bread and store half in the freezer. The freezer keeps it fresher than the refrigerator.
- Think up a variety of ways to use a vegetable when you buy it in a large quantity. For example, you can divide a head of broccoli into thirds. Cook one-third and eat it as a hot vegetable. Put one-third in a soup, and use one-third as an appetizer or in a salad. Buy large quantities of frozen vegetables if you have sufficient freezer space. You can take out the exact amount you need at mealtime.
- Buy large packages of meat such as pork chops, ground meat, or chicken when they are on special sale. Divide the package into individual servings and freeze them separately.
- Get the maximum nutritive value out of what you cook. Steam vegetables over water rather than boiling them. Broil meats rather than frying them. Experiment with stir-fried foods. A variety of vegetables and meats can be enjoyed this way; inexpensive vegetables such as cabbage and celery are delicious when crisp-cooked in a little oil with soy sauce and lemon or ginger added. Cooked, leftover vegetables can be dropped in at the last minute.
- Get the maximum value out of the time you spend cooking. Cook for several meals at a time. Roast a turkey breast, or skinless, boneless chicken breasts, and use half for dinner and the rest for lunches—in sandwiches, tortillas, or salads.
- Make twice as much as you need of a recipe that takes time to prepare: spaghetti sauce, a casserole, vegetable pie, chili, or meat loaf. Label and store the extra servings in the freezer. Be sure to date these so that you will use the oldest first.
- Leftover chili is a delicious topping for baked potatoes, rice, or tortillas. Or spoon it into steamed bell peppers and top with a little grated cheddar cheese.
- Pasta and sauce keeps well—up to three months in the freezer. Pasta sauce can also be used as pizza sauce; spread it on ready-made pizza crust or French rolls and top with low-fat cheese and vegetables.
- Make mixtures of leftovers you have on hand: a thick stew prepared from leftover green beans, carrots, cauliflower, broccoli, and any beans or meat with some added fresh herbs, onion, pepper, celery, and potatoes makes a complete and balanced meal.
- Set aside a place in your kitchen for rows of shelf staple items that you can't buy in single-serving quantities—rice, tapioca, lentils or other dried beans, flour, cornmeal, dry nonfat milk, pasta, or cereal, to name only a few possibilities. The jars make an attractive display and will remind you of possibilities for variety in your menus.
- Cook for yourself with the idea that you are cooking for special guests. Invite guests when you can and make enough food so that you will have enough for a later meal.

WORKING ON THE WEB www.thriveonline.com

Sponsor: Oxygen Media

Description: A credible source of practical nutrition and health information for all age groups on a variety of topics. Includes practical advice from a registered dietitian for following a healthy diet, losing weight, packing healthy school lunches, feeding adolescent athletes, and meeting the nutrient needs of older adults.

Available Activities: The site allows you to determine your body mass index, assess your eating style, improve your diet, locate recipes for delicious low-fat meals, and ask for expert advice on losing weight, low-fat cooking, and food safety. You can submit your favorite recipe for a heart-healthy makeover. Available quizzes assess your knowledge on assorted topics. A list of organizations and hotlines providing information on over 100 health-related conditions is included.

Web Work:

1. Assess how well you eat when you are away from home or away from your usual routine by taking the Food Traps Quiz.
2. Learn how to downsize your cooking when making nutritious meals for one or two.
3. See what a recipe makeover can do to the calorie and fat content of popular meals.
4. Have fun selecting one new recipe to try this week from your favorite cuisine.

Helpful Hints: From the home page, click on Nutrition. Scroll down to Tools for You and click on Food Traps Quiz. Answer the questions and click on Submit. Review the comments provided to help you avoid your food traps in the future. Return to the Nutrition home page and scroll down to click on 50+ Nutrition. Click on Downsizing Dinnertime for tips on creating quick and healthy meals for one or two people. Return to the Nutrition home page and scroll down to Nutrition and click on Recipe Makeovers. Choose three recipes from the Makeover Archive and compare the calorie and fat content of the before and after versions of each recipe. Note the changes made in the lighter version. Finally, return to the Nutrition home page and scroll down to Cuisine and click on Everyday Meals followed by Kitchen of 1000 Pleasures. Select a new recipe to try from one of the cuisines listed. Note that you can specify the course (appetizer, salad, entreé) and method of preparation (quick, grill, make-ahead).

recipients miss out on the social benefit of the congregate meal sites. Every effort is made to persuade recipients to attend the shared meals if they can.

Evaluations of the programs for congregate and home-delivered meals generally show that the programs improve the dietary intake and nutrition status of their clients.[106] Participants generally have greater diversity in their diets and higher intakes of essential nutrients and are less likely to report food insecurity than nonparticipants. Other benefits come as a result of screening and the referrals generated by such programs. Additional benefits are derived from the activities associated with the congregate meals services: diet counseling, exercise, adult education, and other classes and activities. Participants benefit, too, from the opportunity for improved socialization.

Sometimes assistance programs are not enough to meet the needs of older people, some of whom require institutional care. A variety of options exist. A familiar alternative is the nursing home, which may be necessary for people who need constant medical care.

The relative inquiring into retirement centers or nursing homes should ask the director or dietitian some questions about the food service. Does the resident have a choice in the selection of food? How often are the menus repeated (is the cycle monotonous)? How often are fresh fruits and vegetables served? Does the staff keep track of each person's weight? Does good communication exist between the nursing staff and the dietitian so that the dietitian will know whether someone is not eating? Is the resident encouraged and helped to go to the dining room to eat so that some socializing will occur? Is the dining room attractive? Does someone help those who can't manage feeding themselves? Are religious dietary restrictions honored? Other questions that the investigator will want to ask have to do with the general atmosphere of the retirement center or nursing home in recognition of the effect of social climate on a person's appetite. A facility that views residents as people, not as patients, gets a mark in its favor.

In the nursing home, the dietitian, nutritionist, or nurse responsible for the residents' care should keep in mind the special needs associated with the residents' time in life. The average age of a nursing home resident is 81, and many residents have problems or habits that can affect their nutrition status:

- At least one chronic disease.
- Constipation or incontinence.
- Confusion due to the change in environment or dementia.
- Poor eyesight or hearing.
- Ill-fitting or missing dentures.
- Inability to feed themselves because of arthritis or stroke.
- Psychological problems, especially depression.
- Anorexia and loss of interest in eating.
- Lack of opportunity to socialize at mealtimes.
- Long-established food preferences.
- Slowed reactions (seeing, holding utensils, chewing, swallowing).

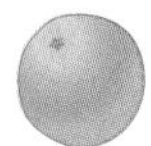

LOOKING AHEAD: GROWING SEASONED

As a nation, we tend to value the future more than the present, putting off enjoying today so that we will have money, prestige, or time to have fun tomorrow. The elderly feel this loss of future. The present is their time for leisure and enjoyment, but they often have no experience in using leisure time.

The solution is to begin to prepare for old age early in life, both psychologically and nutritionally. Preparation for this period should, of course, include financial planning, but other lifelong habits should be developed as well (see Figure 10-15). Each adult needs to learn to reach out to others to forestall the loneliness that will otherwise ensue. Adults need to develop some skills or activities—volunteer work with organizations, reading, games, hobbies, or intellectual pursuits—that they can continue into their later years and will give meaning to their lives. Every adult needs to develop the habit of adjusting to change, especially when it comes without consent, so that it will not be seen as a loss of control over his or her life. The goal is to arrive at maturity with as healthy a mind and body as possible; this means cultivating good nutrition status and maintaining a program of daily exercise.

In general, the ability of the elderly to function well varies from person to person and depends on several factors. The following "life advantages" seem to contribute to good physical and mental health in later years:[107]

- Genetic potential for extended longevity. Some persons seem to have inherited a reduced susceptibility to degenerative diseases.
- A continued desire for new knowledge and new experiences. Some studies suggest that active minds, ever involved in learning new things, may be more resistant to decline.
- Socialization, intimacy, and family integrity. Older persons thrive in situations in which love, understanding, shared responsibility, and mutual respect are nurtured.

FIGURE 10-15
THE AGING WELL PYRAMID

The time to prepare for old age is early in life. Practice the items found at the base of the pyramid to achieve an optimal sense of well-being. Use the inner four compartments of the pyramid to create a balance among all aspects of your life: nutrition, physical activity, social health, and emotional well-being. Use the top of the pyramid to manage everyday stresses such as traffic gridlock, exams, and work deadlines.

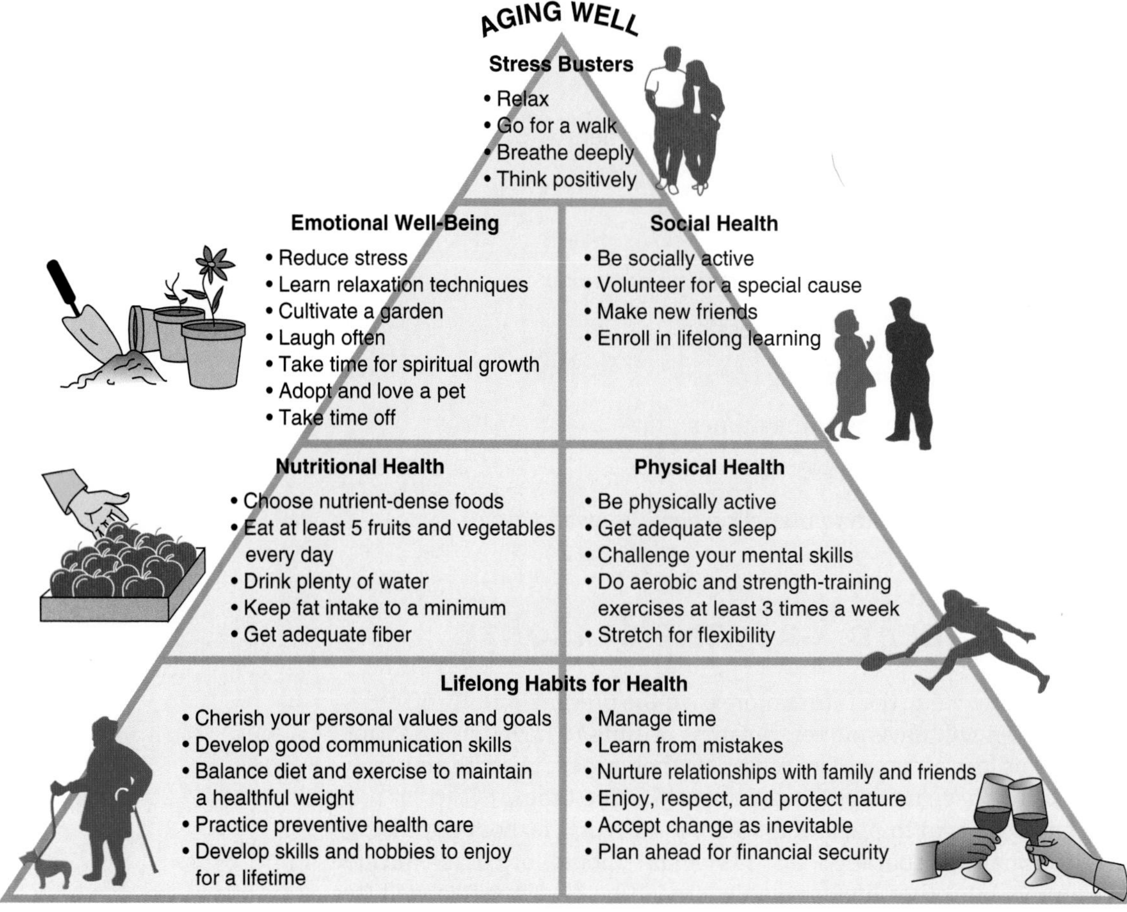

- Adherence to a prudent diet while avoiding excesses of food energy, fat, cholesterol, and sodium. A prudent diet with adequate intakes of all essential nutrients has a positive impact on health and weight management.
- Avoidance of substance abuse.
- Acceptable living arrangements.
- Financial independence.
- Access to health care—to include a family physician, health clinic, public health nursing service providing home health care, dentist, podiatrist, physical therapist, pharmacist, and registered dietitian.

Everyone knows older people who have maintained many contacts—through relatives, friends, church, synagogue, or fraternal orders—and have not allowed themselves to drift into isolation. Upon analysis, you will find that their favorable environment came through a lifetime of effort. These people spent their entire lives reaching out to others and practicing the art of weaving others into their own lives. Likewise, a lifetime of effort is required for good nutrition status in the later years. A person who has eaten a wide variety of foods, stayed trim, and remained physically active will be best able to withstand the assaults of change.

Spotlight
NUTRITION AND CANCER PREVENTION

You probably know someone who has cancer or who has recovered from it or who has died of it. After heart disease, cancer is our most prevalent disease and can be expected, in one form or another, to affect one out of every three Americans living today. Given what is known now about the link between diet and cancer, you are well advised to learn about this connection. Unlike so many factors in our environment, the food you choose to eat is a factor you can control to a great extent. This discussion attempts to answer questions about the connection between diet and cancer.

How is diet associated with cancer?
Numerous studies conducted in both laboratory animals and humans over the past two decades have shown that many connections exist between diet and cancer.[108] Constituents in foods may be responsible for starting the cancer (a process called *initiation*) or for speeding its development (a two-step process that includes tumor *promotion* and *progression*), or they may protect against cancer (see Figure 10-16). Also, for the person who has cancer, diet can make a crucial difference to recovery by helping to restore body weight and improve nutritional status.

Not all studies have shown a firm relationship between cancer and food and nutrient intake. In particular, some epidemiologic studies—that is, studies of disease rates and food patterns of groups of people—have failed to demonstrate such a link. Where a positive association has been found, caution should be used in interpreting data linking dietary components with cancer or other chronic diseases. Remember, an increase in one component of the diet can cause increases or decreases in others. If a close correlation is shown between cancer and, say, the consumption of animal protein by a human population, how can we be sure that the critical factor is the animal protein? It may be increased fat consumption, because fat goes with animal protein in foods. Or the cancer may occur because of what is crowded out: vitamins, minerals, fiber, or nonnutrients contained in the missing fruits, vegetables, legumes, and whole grains.

These issues must be considered when examining the results of studies describing a connection between diet and cancer. Our diets are complex and diverse, making it difficult to separate the effect of a single dietary component from the hundreds of other constituents found in foods. In addition, the difficulty of evaluating the diet-cancer link is compounded by the fact that many cancers take up to 20 years to develop. Thus, assessing a cancer patient's diet today is not as helpful as knowing how that patient ate 10, 20, or even 30 years before the cancer was diagnosed. Finally, our eating pattern is only one of many factors that contribute to the development of cancer (see Figure 10-17 on page 368).

Is nutrition related to cancer causation the way other environmental factors are—such as smoking or air pollution?
Yes. The National Cancer Institute estimates that more than 85 percent of all cancers are associated with lifestyle and environmental factors, including nutrition. Lifestyle factors can usually be controlled by the individual and include tobacco use, diet, exercise, consumption of alcohol, exposure to sunlight, patterns of sexual behavior, and personal hygiene. Individuals typically have little control over environmental factors, such as the exposure to carcinogens in the workplace, radiation during medical and dental procedures, or contaminants (either naturally occurring or artificially created) present in the soil, air, and water. Of course, we have no control over the genetic factors that contribute to the development of cancer.

Nutrition is a lifestyle factor that may account for as much as 35 percent of all cancer deaths.[109] Thus, researchers have taken the study of nutrition and cancer seriously. They are attempting to discover what dietary differences exist between people who do and do not get cancer. In the process, they are trying to identify the various dietary factors that may contribute to or protect against many different types of cancer, including cancer of the esophagus, stomach, liver, pancreas, colon, rectum, breast, ovary, prostate, and lung (see Table 10-14 on page 369).

What is some of the evidence linking diet and cancer?
Studies of the eating habits of different population groups provide some of the knowledge we have about the diet-cancer connection. Particularly telling is research involving Seventh-Day Adventists, a group of people with a remarkably low death rate from cancers of all kinds. This religious group has rules against smoking and using alcohol, discourages the use of hot condiments and spices, and encourages a meatless diet. After cancers linked

FIGURE 10-16
HOW CANCER DEVELOPS

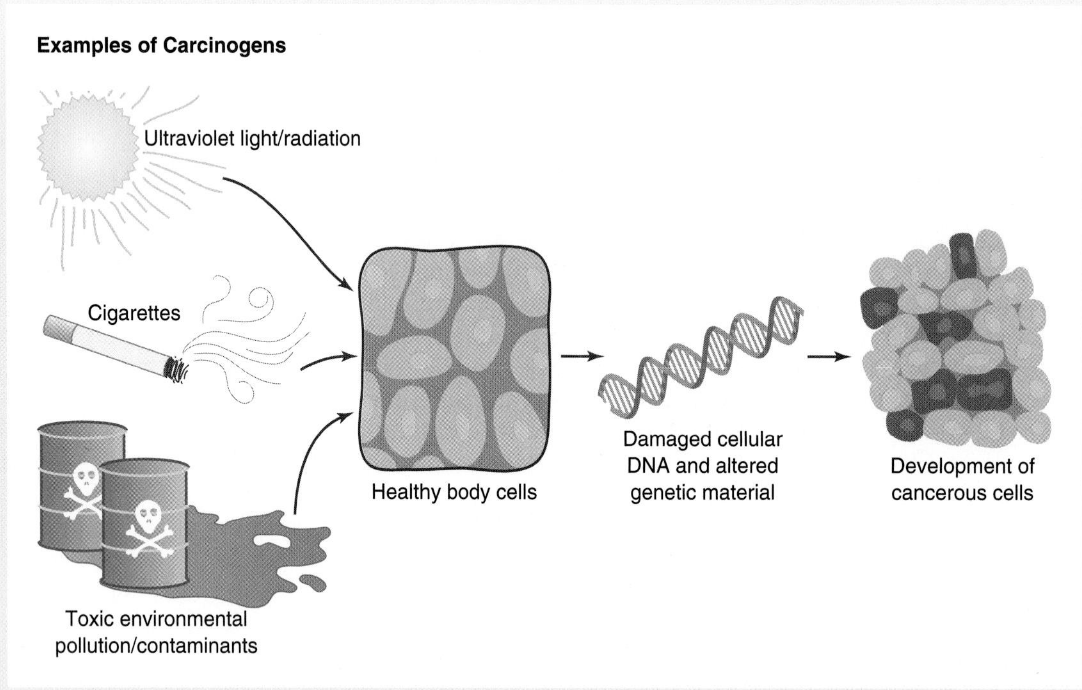

A. *Cancer initiation.* Initiation by a carcinogen causes cancerous alterations in previously healthy body cells.

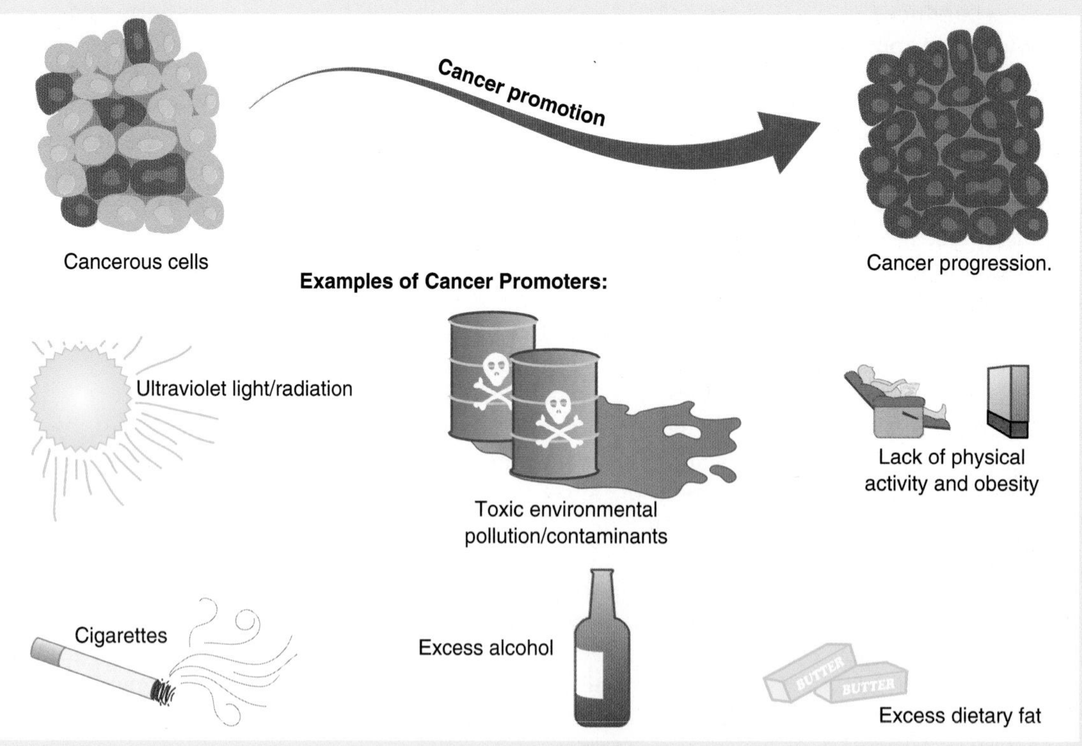

B. *Cancer promotion.* Cancer promoters enhance the growth of abnormal cancerous cells.

FIGURE 10-16, *continued*
HOW CANCER DEVELOPS

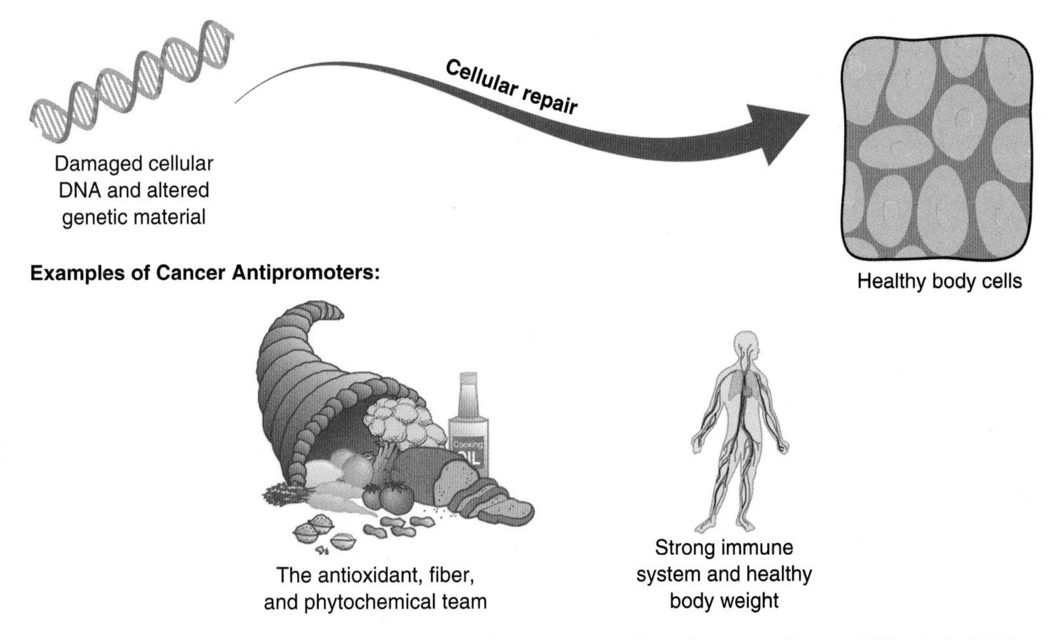

C. *Cellular repair.* Cancer antipromoters squelch free radical damage and enhance the body's ability to repair damaged DNA strands.

to smoking and alcohol are discounted, these people still have a cancer mortality rate about one-half to two-thirds that of the rest of the population. This may be due to a low consumption of meat (and therefore *fat*) and to a high consumption of fruits, vegetables, and cereal grains.

When it comes to colon cancer, studies lend weight to the theory that colon cancer is associated with indicators of affluence, such as a high-fat diet rich in animal protein.[110] In one study, people with colon cancer were compared with a carefully matched set of people without cancer. Those with cancer were found to have a strikingly higher consumption of meat, especially red meat.[111] Another study showed fiber consumption to be lower in colon cancer victims than in comparable people who did not have cancer. However, another study showed colon cancer victims to be eating about the same amount of fiber as people without cancer, indicating that the exact role of fiber in colon cancer prevention is yet to be determined.[112] Furthermore, among U.S. women between the ages of 34 and 59, the risk of colon cancer was related to their intake of animal (but not vegetable) fat, according to researchers at Harvard Medical School.[113] These examples are just a few of the many studies that have led researchers to the view that a diet high in total fat, especially animal fat, and possibly low in fiber is associated with cancer of the colon.[114] Other results suggest that beta-carotene, vitamin C, calcium, and folate or possibly other components of the foods that contain them, can help reduce the risk of colon cancer.[115]

So, diet is associated with the development of colon cancer. What about breast cancer?

The case of the connection between diet and breast cancer is a good example of the difficulty of interpreting study results. Laboratory animals, particularly rodents, consistently develop mammary or breast tumors when fed diets high in either vegetable oils (omega-6 fatty acids) or animal fats.[116] Despite the strength of this link in the animal model, similar results have not been seen consistently in human studies.[117]

A group of researchers in Athens, Greece, who studied 120 women with breast cancer and 120 women without cancer, have reported no association between breast cancer and the consumption of fats and oils.[118] By comparison, a research group in China reported that women with a high intake of fat and calories and a low intake of vegetables and dietary fiber had an increased risk of breast cancer.[119] As you can see from these apparently conflicting results, the role of dietary fat in breast cancer development remains unclear. Most studies

FIGURE 10-17
THE IMPACT OF GENETIC, ENVIRONMENTAL, AND LIFESTYLE RISK FACTORS ON CANCER

Source: American Institute for Cancer Research, *Stopping Cancer Before It Starts: The American Institute for Cancer Research's Program for Cancer Prevention* (New York: Golden Books, 1999), p. 10.

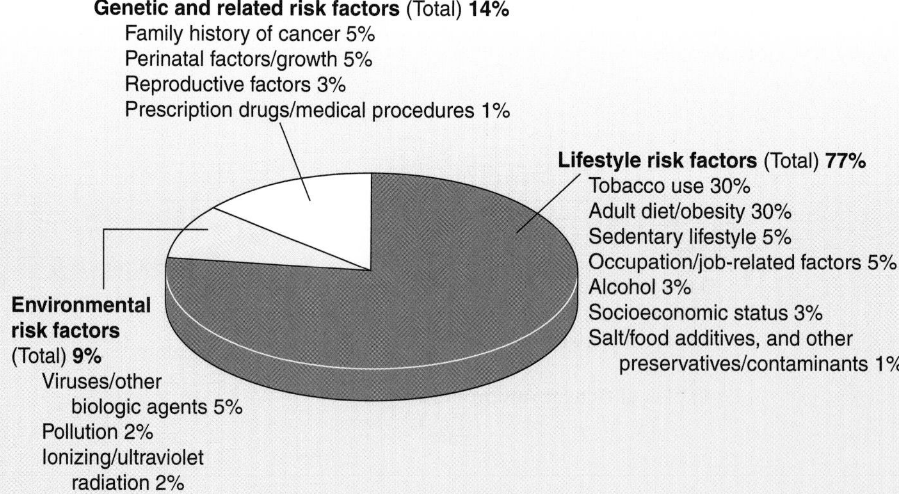

of human populations fail to confirm an association between dietary fat and breast cancer. Laboratory studies, however, show that eating a low-fat diet can reduce estrogen levels and theoretically may reduce the risk of breast cancer. Many ongoing studies are further examining the role of dietary fat and breast cancer.

Of course, excess fat in the diet can contribute to overweight, and recent studies have indicated that women who have gained excess weight during adulthood may have an increased risk of breast cancer.[120] Health experts advise us to limit weight gain during adulthood to no more than ten pounds and to get regular exercise, keeping body mass index between 18.5 and 25.0 (refer to Chapter 8). In fact, getting regular exercise throughout the reproductive years may significantly lower breast cancer risk. Body fat is involved in the production of estrogen. Therefore, by discouraging the accumulation of excess body fat (especially around the abdomen), exercise may decrease the total amount of estrogen a woman is exposed to over her lifetime.

What should consumers do? The American Cancer Society and the National Cancer Institute have reviewed the evidence independently and have pointed out to consumers the following specific concerns:[121]

- *Total calorie intake.* Studies on animals show that reduced food intakes reduce cancer incidence at any age, but the evidence is less clear for human beings. Obesity, however, does increase risks for some cancers in both animals and people (refer to Chapter 8).[122]
- *Fat.* Both animal studies and population studies support the view that high-fat intakes increase the incidence of cancers of the ovaries, colon, and prostate.
- *Protein.* High protein intakes may be associated with increased risks of certain kinds of cancer, but the evidence is not yet firm enough to permit a definitive statement.
- *Carbohydrate.* There is little evidence that carbohydrates as such play a role in cancer development.
- *Beta-carotene.* Inadequate intakes of beta-carotene correlate with a high incidence of cancers of the lung, bladder, and larynx; by inference, adequate intakes may help protect against these cancers.[123]
- *Vitamin C.* Vitamin C may help prevent the formation of cancer-causing agents and thereby protect against cancers of the esophagus and stomach.[124]
- *Cruciferous vegetables.* The consumption of cruciferous vegetables is associated with a reduced incidence of cancer at several sites.[125] Cruciferous vegetables, a group of vegetables named for their cross-shaped blossoms, have been shown to protect against cancer in laboratory animals. Examples of cruciferous vegetables are cauliflower, cabbage, Brussels sprouts, broccoli, kohlrabi, and rutabagas.
- *Other vitamins and minerals.* Bits of evidence suggest that other nutri-

Cruciferous vegetables, such as cauliflower, broccoli, and Brussels sprouts, contain nutrients and nonnutrients that protect against cancer.

ents may protect against certain types of cancer, but no firm conclusions can yet be made. The effect of the antioxidant nutrients—vitamin C, beta-carotene, vitamin E, and selenium—on cancer risk is an area of active research.[126] The trace mineral, selenium, used in the body's production of its own antioxidants, is believed to have a protective effect against cancer of the esophagus, stomach, colon, and rectum.[127] Likewise, vitamin E may protect against cancer, particularly cancer of the gastrointestinal tract.[128]

- *Calcium.* Low calcium intakes have been associated with increased colon cancer. Conversely, people consuming more calcium tend to develop less colon cancer.[129]
- *Fiber.* Fiber might help to protect against some cancers by, for example, speeding up the passage of all materials through the colon so that its walls are not exposed for long to cancer-causing substances.

TABLE 10-14 DIET AND CANCER

Cancer Site	Associated with These Dietary Risk Factors
Breast	High intakes of calories and alcohol; obesity, low fruit and vegetable intake
Colon or rectum	High-fat diets (particularly saturated fat) and alcohol; low-fiber diets
Esophagus	Excessive alcohol intakes; low intakes of vitamins and minerals
Lung	Low fruit and vegetable intake
Ovary and endometrium	High-fat diets; obesity; low fruit and vegetable intake
Prostate	High-fat diets may promote tumor growth
Pancreas	Low fruit and vegetable intake
Stomach	Regular consumption of smoked foods and foods cured with salt or nitrite compounds; low fruit and vegetable intake
Liver	Ingestion of *aflatoxin*-contaminated grains; regular consumption of smoked foods and foods cured with salt or nitrite compounds; alcohol abuse
Mouth and throat	Excessive alcohol intake; low fruit and vegetable intake

Sources: Food, Nutrition, and Prevention of Cancer: A Global Perspective (Washington, D.C.: American Institute for Cancer Research, 1997); and J. H. Weisburger, Nutritional approach to cancer prevention with emphasis on vitamins, antioxidants, and carotenoids, *American Journal of Clinical Nutrition* 53 (1991): 226S–237S; R. G. Ziegler, Vegetables, fruits, and carotenoids and the risk of cancer, *American Journal of Clinical Nutrition* 53 (1991): 251S–259S.

Fat is linked to certain cancers, and fiber is associated with cancer prevention. Do vegetarians have a lower incidence of those cancers?
Yes, they do. The Seventh-Day Adventists have already been mentioned. Vegetarian women also have less breast cancer than do women who eat meat.

A number of studies have examined the relationship between cancer and vegetable consumption. Many of them have shown that people with colon cancer eat vegetables less frequently than do others; one study revealed that colon cancer victims specifically consumed less cabbage, broccoli, and Brussels sprouts than did people free of cancer. Similarly, comparisons of stomach cancer victims' diets with those of carefully matched people without cancer showed lower consumption of vegetables in the cancer group—in one case, vegetables in general; in another, fresh vegetables; in others, lettuce and other fresh greens or vegetables containing vitamin C. Some of the suspects for the causation of stomach cancer are the chemicals known as nitrosamines, produced in the stomach and intestines from nitrites found in foods. Vegetables may help in cancer prevention by contributing vitamin C, which inhibits the conversion of nitrites to nitrosamines.[130]

Another healthy aspect of many plant-based diets is the use of soyfoods. A number of studies conducted in China and Japan have shown that consumers of tofu, soy milk, and other soyfoods have lowered cancer risk than those who rarely consume these foods (see the Spotlight feature in Chapter 5). The data suggest that consuming one or two servings of soy a day (for example, 1 cup of soy milk or 4 ounces of tofu) may reduce risk of breast, prostate, lung, and colon cancers.[131]

What about alcohol. Is it connected with cancer?
Yes. Environmental causes of head, neck, and esophageal cancer have been studied, and the major factor appears to be the combination of alcohol and tobacco use. However, dietary factors have turned up, pointing to a low intake of fruits and raw vegetables, specifically of the fruits and vegetables that contribute the orange pigment beta-carotene (which converts to vitamin A in the body) and the vitamin riboflavin. Beta-carotene—noted for giving carrots, winter squash, sweet potatoes, apricots, cantaloupe, and other fruits and vegetables their familiar colors—and its relatives may also be important in reducing the risk of skin cancer.

Researchers are currently probing the possible association of alcohol consumption with increased risk of breast cancer.[132] Although some studies have shown a modest but significant increase in risk of breast cancer for women who were classified as heavy drinkers (more than 40 grams of alcohol or more than 4 drinks per day), other studies have found no relationship.[133] More research is needed to clarify this issue. However, if a causal relationship between relatively heavy drinking and increased risk of breast cancer is confirmed, such evidence would lend support to the societal benefits derived from not consuming excessive amounts of alcohol.

What is the association between beta-carotene and other carotenoids and cancer?
Among the known actions of vitamin A (beta-carotene) are the important roles it plays in maintaining immune function. A strong immune system may be able to prevent cancers from gaining control even after they have gotten started in the body.

The carotenoids are a family of powerful antioxidants found in dark yellow, orange, and red fruits and vegetables, and deep green vegetables. They are potent free-radical fighters which may play important roles in preventing cancer. The family includes alpha- and beta-carotene, lycopene, lutein, zeaxanthin, and others. Alpha- and beta-carotene appear to protect against the progression of cancer, whereas other carotenoids may be more protective at earlier stages of cancer development. Carotenoids may work against cancer by boosting the immune system and supporting the enzymes that detoxify carcinogens. Beta-carotene and lycopene may be effective in preventing damaged cells from proliferating and becoming malignant. Some scientists now believe lycopene—found in red fruits and vegetables such as tomatoes, tomato products, red peppers, and watermelon—is the most powerful of the antioxidant carotenoids. Research is currently under way to examine a link between lycopene and lowered risk of prostate and other cancers.[134] Meanwhile, we are advised to consume five to nine servings of a variety of fruits and vegetables each day. Be sure to include deep-green and brightly colored fruits and vegetables every day. By doing so, your ability to repair free-radical damage in the body is likely to be enhanced by the carotenoids or possibly by something else in these plant foods (see Chapter 6 for more about the benefits of produce in your diet).

Would that "something" in the vegetables be a vitamin?
Not necessarily. Both fruits and vegetables appear to have a protective effect beyond those already discussed for beta-carotene, vitamin C, and fiber. Researchers have identified substances known as *phytochemicals*—naturally occurring plant compounds—in fruits and vegetables that may play a role in decreasing cancer risk and strengthening the immune system.[135] (Refer to Figure 6-3 on page 194 in Chapter 6 for a sampling of these compounds.) These vegetables also contain folate, a vitamin, which is involved in cell multiplication, and may prove to play an important role in cervical cancer prevention.[136] The effects of the members of the cabbage family may be due to their containing substances known as indoles—a type of phytochemical—which may act by inducing an enzyme in the host that destroys cancer-causing agents.

Choose from a variety of fruits and vegetables rich in fiber, vitamins, minerals, antioxidants, and phytochemicals to help protect against cancer.

Do these findings have any implications for the way a person should eat right now, today?
Although clearly there is still much to learn, many experts believe that enough is known to take the first preventive steps. An exhaustive review of some 4,500 previous studies on cancer has led scientists at The American Institute for Cancer Research to release a comprehensive set of recommendations for cancer prevention.[137] To put these recommendations into action, consider these simple steps:

- Choose a diet rich in a variety of plant-based foods.
- Eat plenty of vegetables and fruit.
- Maintain a healthy weight and be physically active.
- Drink alcohol only in moderation, if at all.
- Select foods low in fat and salt.
- Prepare and store food safely.
- Do not use tobacco in any form.

Your lifestyle choices regarding diet, exercise, and smoking can be powerful tools in reducing your risk for cancer (see Figure 10-18). Taken together, this advice agrees with most recommendations made to the public for helping prevent heart disease, diabetes, and many other ills, as well as cancer.

Public efforts are now under way to reduce the major risk factors for cancer.[138] The National Cancer Institute (NCI) designed its *National 5 a Day for Better Health* program to increase per capita fruit and

FIGURE 10-18
FACTORS THAT DECREASE OR INCREASE CANCER RISK

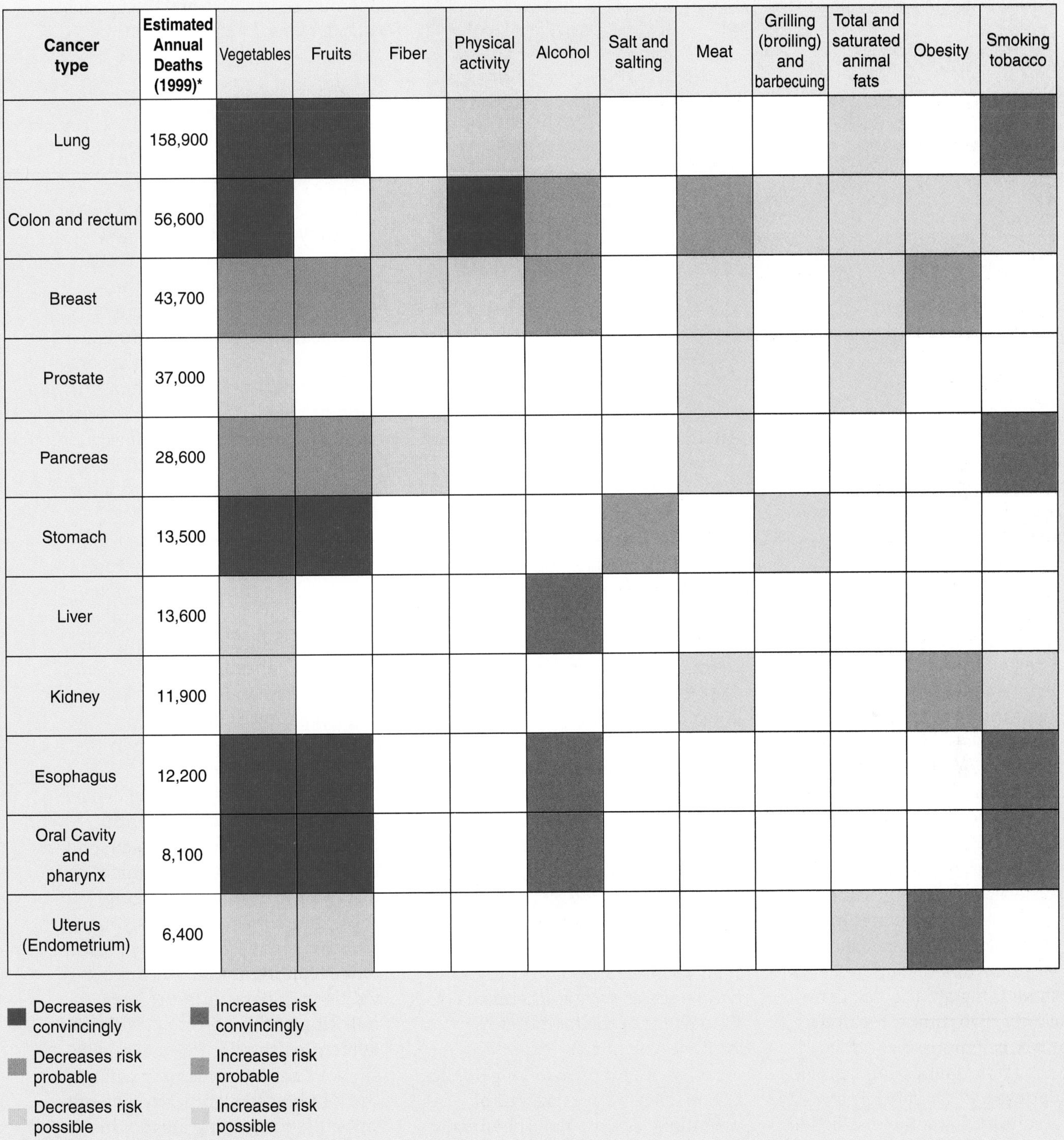

Cancer type	Estimated Annual Deaths (1999)*	Vegetables	Fruits	Fiber	Physical activity	Alcohol	Salt and salting	Meat	Grilling (broiling) and barbecuing	Total and saturated animal fats	Obesity	Smoking tobacco
Lung	158,900											
Colon and rectum	56,600											
Breast	43,700											
Prostate	37,000											
Pancreas	28,600											
Stomach	13,500											
Liver	13,600											
Kidney	11,900											
Esophagus	12,200											
Oral Cavity and pharynx	8,100											
Uterus (Endometrium)	6,400											

Decreases risk convincingly — Increases risk convincingly
Decreases risk probable — Increases risk probable
Decreases risk possible — Increases risk possible

*From: Cancer Facts and Figures, 1999; available at www.cancer.org.

Source: Adapted from World Cancer Research Fund and American Institute for Cancer Research, *Food, Nutrition, and the Prevention of Cancer: A Global Perspective* (Washington, D.C.: American Institute for Cancer Research, 1997), pp. 506–507.

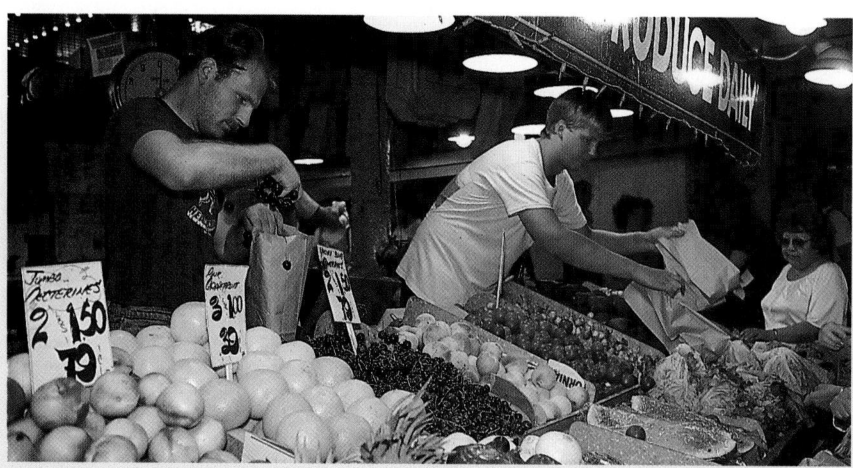

The National Cancer Institute's *National 5 a Day for Better Health* program urges consumers to eat five or more servings of fruits and vegetables every day for better health.

vegetable consumption. The NCI's 5 a Day program promotes a simple nutrition message: *Eat five or more servings of fruits and vegetables every day for better health.*[139]

Does it make a difference whether I choose to take supplements instead of eating vegetables?
Yes. Fiber, vitamin C, beta-carotene, riboflavin, indoles, and other phytochemicals found in *foods* appear to have preventive effects on cancer development. And because it is obvious that researchers do not have *the* answer—just many partial answers—it is best to stick with foods. Supplements may not contain some as yet unidentified components found in foods that may help to protect against cancer. Also, with vitamin/mineral supplementation, there is always the risk of an excessively high intake; recall that vitamin A, selenium, and other vitamins and minerals can be toxic at high doses. *It is best to rely on foods.* Fruits and vegetables will add vitamins, minerals, and fiber to your diet—and color and flavor as well.

Are there any anti-cancer benefits derived from drinking tea?
Scientists note that people in Asia, where large quantities of tea—both green and black—are consumed, have a lower incidence of esophageal, colon, stomach, and other cancers. Laboratory studies suggest that polyphenols—phytochemicals having potent antioxidant properties—found in tea may help protect against these cancers by blocking the formation of carcinogenic compounds in the body. However, the implications for humans are not yet known.

Does cooking food at high temperatures increase the risk of cancer?
Some research suggests that frying or charcoal-broiling meats—especially fatty meats—at very high temperatures creates chemicals on the surface of the food that may increase cancer risk. Preserving meats by methods involving smoke also increases their content of potentially carcinogenic chemicals. We are advised to eat grilled, smoked, or cured foods only occasionally. When you do, consume them with fruits and vegetables that contain protective antioxidant and phytochemical factors. Choose techniques such as braising, poaching, stewing, baking or roasting instead.

Why should I vary my diet?
The recommendation to eat a varied diet is based on an important cancer-prevention strategy—dilution. The standard advice to eat a variety of foods takes on new meaning in this quote: "The wider the variety of food intake, the greater the number of different chemical substances consumed, and the less is the chance that any one chemical will reach a hazardous level in the diet."[140] In other words, whenever you add new foods to the diet, you are diluting whatever is in one food with what is in another.

The variety principle has traditionally meant to eat foods from each of the various food groups. This principle needs to be applied within each of the food groups as well. Don't alternate just between corn and potatoes. Select different vegetables each time you go to the store—broccoli, peas, green beans, squash, and many others.

Although there are many cancer-causing factors that you cannot control, you can decide which food habits you will keep and which ones you will change. By using these guidelines in making your choices, you will have every reason to feel confident that you are providing your body with the best nutrition at the lowest possible risk. Remember that in the final analysis, your risk of developing cancer can be reduced significantly by not smoking, consuming alcohol in moderation if at all, and by adopting healthful eating and exercise habits.

Nutrition on the Web

nutrition.wadsworth.com	Go to the *Personal Nutrition* site to check for the latest updates to chapter topics or to access links to related Web sites.
www.navigator.tufts.edu	Click on Lifecycle for descriptions and ratings of related Web sites.
www.mayohealth.org	Information on pregnancy from the Mayo Clinic Children's Center.
www.modimes.org	Information on birth defects from the March of Dimes.
www.nofas.org	A site for information about fetal alcohol syndrome.
www.lalecheleague.org	Information about breastfeeding available from La Leche League.
www.kidsfood.org	Lots of fun and creative nutrition-related ideas for children.
www.aap.org	Web site of the American Academy of Pediatrics.
ificinfo.health.org/index3.htm	Information on promoting healthy lifestyles for children.
www.kidshealth.org	Information on health promotion for children.
www.eatright.org	The American Dietetic Association offers information on nutrition and pregnancy and the nutrient needs of children and older adults.
www.healthfinder.gov/searchoptions/topicsaz.htm	Search for nutrition and lifecycle-related topics.
www.aarp.org	Resources from the American Association for Retired Persons.
www.nih.gov/nia	Information on aging from the National Institute on Aging.
www.fns.usda.gov/fns	Facts about the U.S. food assistance programs.
www.cancer.org	Information about cancer from the American Cancer Society.
www.aicr.org	Information and research updates about cancer.
www.nal.usda.gov/fnic	Click on chapter topics—including cancer.

Check Yourself . . .

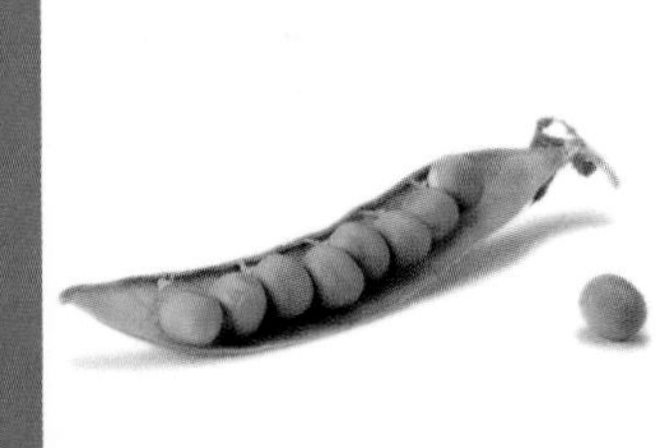

1. State which nutrients are needed in increased amounts in pregnancy.
2. Name the most potent indicator of an infant's future health status.
3. List practices that should be eliminated during pregnancy and provide a reason why.
4. Name three advantages of breastfeeding derived by the infant.
5. Describe the recommended infant feeding practices—when to introduce solid foods and beverages.
6. State food habits to be encouraged in children to avoid nutrient deficiencies, obesity, and dental caries.
7. Describe three of the nutrition-related problems common among teenagers.
8. List physiological changes associated with aging.
9. Compare and contrast energy and nutrient requirements of older adults with those of young adults.
10. List nutrients thought to be protective of cancer formation and nutrients associated with increased risk of developing cancer.

Answers to Check Yourself questions are found in Appendix G.

11

Food Safety and the Global Food Supply

CONTENTS

Foodborne Illnesses and the Agents That Cause Them

Safe Food Storage and Preparation

The Savvy Diner: Home Food Safety

Food Safety Scorecard

Pesticides and Other Chemical Contaminants

Nutrition Action: Should You Buy Organically Grown Produce?

Food Additives

New Technologies on the Horizon

Spotlight: Domestic and World Hunger

The role of the infinitely small is infinitely large.

Louis Pasteur
(1822–1895, French chemist and microbiologist who developed the pasteurization process)

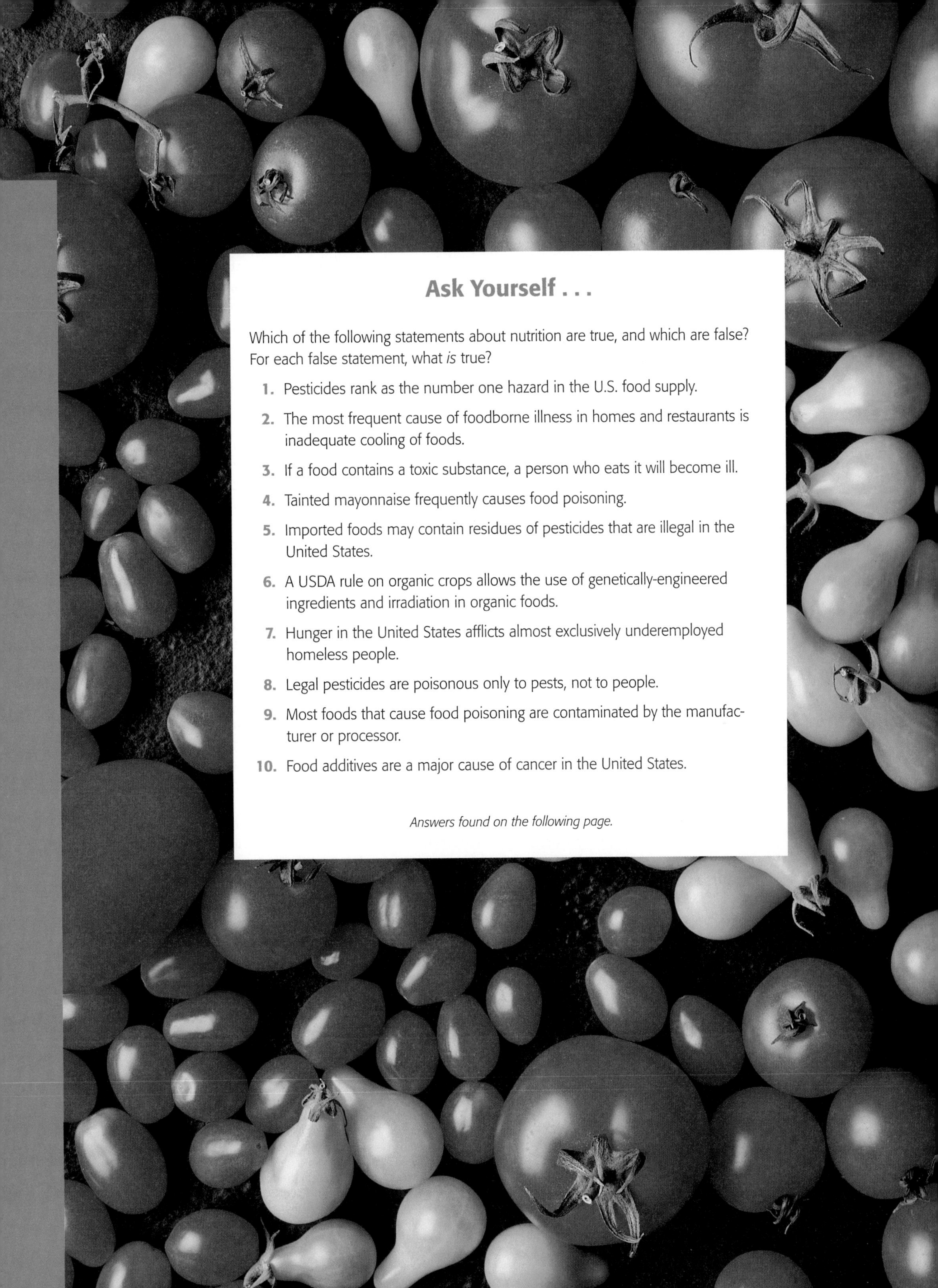

Ask Yourself . . .

Which of the following statements about nutrition are true, and which are false? For each false statement, what *is* true?

1. Pesticides rank as the number one hazard in the U.S. food supply.
2. The most frequent cause of foodborne illness in homes and restaurants is inadequate cooling of foods.
3. If a food contains a toxic substance, a person who eats it will become ill.
4. Tainted mayonnaise frequently causes food poisoning.
5. Imported foods may contain residues of pesticides that are illegal in the United States.
6. A USDA rule on organic crops allows the use of genetically-engineered ingredients and irradiation in organic foods.
7. Hunger in the United States afflicts almost exclusively underemployed homeless people.
8. Legal pesticides are poisonous only to pests, not to people.
9. Most foods that cause food poisoning are contaminated by the manufacturer or processor.
10. Food additives are a major cause of cancer in the United States.

Answers found on the following page.

So far this book has dealt primarily with the nutrients and how your body handles them. This chapter takes the study of nutrition one step further and examines some nonnutrient components of food—bacteria, additives, and pesticides, to name a few—and how they affect the food supply. In addition, the chapter takes a global view of the science of nutrition, looking at how foodways in different countries influence each other as well as how some of our food habits affect the environment.

What sorts of additives do foods contain, and what are the effects of those additives? Are foods ever contaminated? How can you reduce your risk of food poisoning? What potential do new food technologies, such as irradiation and genetic engineering, hold for our lives and for the health of the environment? This chapter addresses these and other questions.

FOODBORNE ILLNESSES AND THE AGENTS THAT CAUSE THEM

FOODBORNE ILLNESS OR FOOD POISONING illness occurring as a result of eating food contaminated with disease-producing microorganisms, such as bacteria, viruses, or parasites, or toxic substances such as environmental pollutants.

Diarrhea, nausea, abdominal pain, or vomiting without fever or upper respiratory distress is often taken to be flu, but people who experience such symptoms are highly likely to be suffering from food poisoning. If you suspect you've eaten tainted food, especially from a restaurant or other place where many people may have been infected, be sure to alert your local health department.

PASTEURIZATION the process of sterilizing food via heat treatment.

North America has the safest, most plentiful food supply in the world, thanks to the concerted efforts of food suppliers, food processors, and federal, state, and local governments, all of which are concerned with food safety. North Americans also enjoy the most diverse food supply in the world, consisting of an incredible array of fresh and processed foods. Most American supermarkets stock a variety of foods from other countries—cookies and crackers from Denmark and the United Kingdom, Belgian chocolates, Italian cheeses, beef and veal products from New Zealand and Australia, goose liver paté from France, and specialty foods from Japan, Mexico, and even China. Our international bent can be seen in the produce section as well, where exotic star fruit, papaya, and mango from overseas markets are widely available. Still, the diverse mix of fresh and processed foods coming from local, national, and international markets underscores the need to understand where the culprits behind foodborne illnesses originate.

Foodborne illness, or **food poisoning,** is one of the greatest concerns of public health experts and the food industry. Each year, as many as 76 million Americans experience foodborne illness, and an estimated 5,000 deaths are linked to tainted foods.[1] Incredible as these figures are, they probably represent only a fraction of the whole picture. Many mild cases of food poisoning are never reported for a number of reasons: The victims pass off the symptoms as flu and do not seek medical attention, the illness is misdiagnosed as another problem with similar symptoms, the victim fails to recognize food as the source of the illness, or the physician doesn't report the food poisoning to local health agencies.

That more people aren't afflicted with foodborne illness is surprising because disease-producing microorganisms proliferate in our environment. Consider that all raw foods contain microbes, and foods often pick up more during production, processing, packaging, transport, storage, or preparation. That's why the food industry uses different types of control measures to limit the risk of foodborne illness to consumers. Destruction or inactivation of bacteria or their spores is accomplished through the use of heat treatments such as **pasteurization** and canning. Freezing, dehydrating, and refrigerating food also halts or slows down

Ask Yourself Answers: **1.** False. The greatest hazard present in the U.S. food supply today is not pesticides but bacteria, viruses, mold, and other microorganisms that cause food poisoning. **2.** True. **3.** False. If a food contains a toxic substance, a person who eats it *may* become ill, depending on whether enough of the toxin to cause illness is present. **4.** False. Mayonnaise rarely carries high levels of the bacteria that cause illness. **5.** True. **6.** False. A March 2000 USDA proposed rule on organic crops rejects the use of genetically-engineered ingredients and irradiation in organic foods. **7.** False. Hunger in the United States affects not only the unemployed homeless but also homeowners and the working poor. **8.** False. Legal pesticides can be poisonous to people, animals, and plants, depending on the amount of the pesticide. **9.** False. Most cases of food poisoning are the result of improper handling of food *after* it leaves the processor or manufacturer. **10.** False. Food additives, which pose minimal health risks, are not a major cause of cancer.

Americans enjoy the safest, most diverse food supply in the world.

bacterial growth. In addition, special packaging techniques and antimicrobial preservatives help to control food-related pathogens.

What causes most cases of foodborne illness? Many people believe that chemical additives and pesticide residues added during the growing and processing of food pose the greatest risk. But contrary to popular belief, most foodborne disease is caused by mishandling of food, either in food-service establishments such as restaurants or in homes. In fact, food eaten in restaurants, cafeterias, and other food-service establishments accounts for about two-thirds of all reported outbreaks of foodborne illness, and food eaten at home, about a quarter.[2] Most cases of foodborne disease are caused by faulty handling, cooking, and storing of food long after it has left the manufacturer or processor. Table 11-1 ranks the most common food hazards.

Experts classify foodborne diseases into two types: intoxications and infections. **Food intoxications** occur when a chemical or toxin transmitted by way of food causes the body to malfunction. An example of food intoxication is the vomiting, nausea, abdominal cramping, sweating, and chills that result from eating food contaminated with a strain of bacteria called *Staphylococcus aureus*. This bacterium produces what is known as an **enterotoxin,** a toxin that causes severe gastrointestinal distress. Other types of bacteria can produce **neurotoxins,** or toxins that afflict the nervous system. Food intoxication can also be caused by eating food that has been contaminated with a chemical, such as lead or some other heavy metal.

Foodborne infections, on the other hand, occur as a result of eating a food that contains living microorganisms, such as bacteria, viruses, or parasites, capable

FOOD INTOXICATION illness caused by eating food that contains a harmful toxin.

ENTEROTOXIN a toxic compound, produced by microorganisms, that harms the gastrointestinal tract.
entero = intestine

NEUROTOXIN a poisonous compound that disrupts the nervous system.
neuro = nerve

FOODBORNE INFECTION illness caused by eating a food containing bacteria or other microorganisms capable of growing and thriving in a person's tissues.

TABLE 11-1 RANK OF AREAS OF CONCERN IN THE FOOD SUPPLY

Although most people think that chemicals such as pesticides and additives rank as the most dangerous, illness-producing substances in our food supply, the real hazard comes from naturally occurring bacteria and other microbes.

Most dangerous:	1. Microbial contamination
	2. Naturally occurring toxicants
	3. Environmental contaminants, such as metals
	4. Pesticide residues
	5. Food processing and nutrients in foods
Least dangerous:	6. Food additives

Source: D. O. Cliver, *Eating Safely* (New York, NY: American Council on Science and Health, 1993), p. 3.

of multiplying and thriving in the body. Ingested in large amounts, the microorganisms can wreak havoc in the digestive tract or other areas of the body. An example of this type of foodborne illness is infection with *Vibrio* bacteria, which often reside in raw seafood such as oysters and clams. Inside the body, the bacteria settle in quickly and cause abrupt onset of chills, fever, or prostration.

The following section provides an overview of the various agents that can cause a foodborne illness—be it an intoxication or infection.

Microbial Agents

When we scan a restaurant menu or reach for an egg or glass of milk from the refrigerator at home, we aren't usually thinking about the microorganisms that might be lurking in the foods or their potential to cause illness. We tend to assume that the foods and beverages we consume are safe to eat or drink. Granted, most of the time we don't need to worry. Typically, a food must harbor thousands of microorganisms before it causes nausea, diarrhea, cramps, or other symptoms of foodborne illness. What's more, a healthy body can usually defend itself against small amounts of the "bad bugs."

But when proper food-handling procedures are not followed carefully, the risk of food poisoning from bacteria or other microbial agents soars. Mishandled food, such as items cooked or stored improperly, provides the perfect medium in which microorganisms can flourish. In children, the elderly, people whose immune systems are compromised, or other vulnerable people, it might only take small amounts of the offending microorganisms to cause trouble. For these fragile people, proper handling of food rates as a particularly high priority.

"I need 148 get-well cards."

In the United States, the many microorganisms responsible for most cases of foodborne illness range from *Campylobacter* to *Salmonella* to *Clostridium botulinum* to *E. coli* 0157:H7. Table 11-2 summarizes the common food sources of these and other microbial agents responsible for most outbreaks of foodborne illness.

Of the microbial pathogens, *Campylobacter jejuni* ranks as the most prevalent pathogen and one of the leading causes of foodborne illness in the United States, and it appears to be growing more widespread.[3] Symptoms of the illness it causes, called campylobacteriosis, usually begin 2 to 5 days after eating the contaminated food and last for 7 to 10 days. Many cases of campylobacteriosis probably go unreported because the illness's characteristic diarrhea, abdominal pain, and fever mimic flulike gastrointestinal ills.

In the past, egg products were a major source of salmonellosis—an illness caused by *Salmonella* bacteria. However, this is no longer the case

WORKING ON THE WEB www.foodsafety.gov

Sponsor: USDA's Cooperative State Research, Education, and Extension Service and the University of Wisconsin's Food Research Institute

Description: A gateway to government information on all aspects of food safety and the National Food Safety Initiative.

Available Activities: The site provides information about the major food pathogens, practical tips for reducing risk of foodborne illness, a guide to reporting illness and product complaints, food safety news releases, and links to educational and training resources and government and state agencies offering food safety information.

Web Work:

1. Choose an organism to learn more about from those listed in Table 11-2.
2. Describe the nature of the illness caused by this organism and its associated foods.
3. Describe typical outbreaks of foodborne illness caused by this organism, including any populations at high risk for contracting the illness.
4. Bonus question: List the ten *least* wanted foodborne pathogens.

Helpful Hints: From the home page, scroll down and click on Foodborne Pathogens. Next, click on The Bad Bug Book. From this page select an organism of interest to you. Click on the organism and scroll down to review the information about nature of disease and associated foods. Next, scroll down to Selected Outbreaks (or Typical Outbreaks) and make a list of recent outbreaks, noting the problematic foods associated with the outbreaks. Finally, return to the Foodborne Pathogens page and scroll down and click on Ten Least Wanted Foodborne Pathogens. Make a list of them and note the photos provided for each of the organisms.

as a result of mandatory pasteurization of eggs used in the commercial preparation of ice cream, baked goods, and other egg-containing food products. Raw eggs, however, have been implicated in outbreaks of salmonellosis, as has raw, unpasteurized milk. (Because of this concern, the sale of raw milk is banned in most states.) Today, most outbreaks of salmonellosis are caused by faulty handling of raw meat and poultry, unpasteurized juice, raw sprouts, and mangos.

Honey has been found to contain dormant bacterial spores, which can awaken in the human body to produce botulism. In adults, this is not a hazard, but infants under one year of age should never be fed honey. Honey has been implicated in several cases of sudden infant death.

Staphylococcus aureus is another strain of bacteria responsible for many of the reported cases of foodborne illnesses that occur in the United States. The bacterium, found in the nose and throat and on the skin of most people, can be transmitted to food when an individual with an infected wound or boil or a respiratory infection handles food improperly. The bacterium itself isn't directly responsible for the illness, however. Rather, it produces staphylococcal enterotoxin, which causes food poisoning within one half hour to eight hours of eating a contaminated food. The foods typically implicated in S. *aureus* intoxication include meat and poultry products, egg products, tuna, potato and macaroni salads, and cream-filled pastries. Proper food-handling and sanitation procedures, discussed in the Savvy Diner feature beginning on page 387, help to prevent food poisoning by this bacterium as well as from *Campylobacter* and *Salmonella*.

Another particularly deadly foodborne pathogen is *Clostridium botulinum*, the agent that causes the nausea, vomiting, dizziness, and muscle paralysis known as botulism. This severe illness results from eating food in which the bacterium has flourished and produced a neurotoxin. The toxin binds irreversibly to nerve endings and causes paralysis, which makes swallowing and breathing difficult. Botulism typically develops within 4 to 36 hours of eating the contaminated food. Because it can be fatal, botulism warrants immediate medical attention.

Handle raw meat and poultry with care and cook it thoroughly to destroy any bacteria present. Place it on a clean plate when it is cooked.

Spores of *C. botulinum* are ubiquitous, having been detected in everything from shellfish to fruits and vegetables to honey to corn syrup. But the *botulinum* bacteria only produce the deadly neurotoxin if they are in a warm, oxygen-free, low-acid environment. That's why improperly

TABLE 11-2 MICROBIAL FOOD AGENTS: ORGANISMS THAT CAN BUG YOU

Disease and Organism That Causes It	Source of Illness	Usual Onset After Eating	Symptoms
Bacteria			
Botulism (*Botulinum* toxin produced by *Clostridium botulinum* bacteria)	Spores of these bacteria are widespread. But these bacteria produce toxin only in an anerobic (oxygenless) environment of little acidity. Found in low-acid canned foods such as corn, green beans, soups, beets, asparagus, mushrooms, tuna, and liver paté. Also in luncheon meats, ham, sausage, stuffed eggplant, lobster, smoked and salted fish and homemade, unrefrigerated garlic-in-oil preparations.	4–36 hours	Neurotoxic symptoms, including double vision, inability to swallow, speech difficulty, and progressive paralysis of the respiratory system. **Get medical help immediately. Botulism can be fatal.**
Campylobacteriosis (*Campylobacter jejuni*)	Raw poultry, meat, and unpasteurized milk.	2–5 days	Diarrhea, abdominal cramping, fever, and sometimes bloody stools. Lasts 7–10 days.
Listeriosis (*Listeria monocytogenes*)	Found in soft cheese, unpasteurized milk, imported seafood products, frozen cooked crab meat, cooked shrimp, and cooked surimi (imitation shellfish). *Listeria* bacteria resist heat, salt, nitrite, and acidity better than many other microorganisms.	48–72 hours (though symptoms can strike 7–30 days after eating)	Fever, headache, nausea, and vomiting.
Perfringens food poisoning (*Clostridium perfringens*)	In most instances, caused by failure to keep food hot. A few organisms are often present after cooking and multiply to toxic levels during cooldown and storage of prepared foods. Meats and meat products are the foods most frequently implicated.	8–12 hours	Abdominal pain and diarrhea and sometimes nausea and vomiting. Symptoms last a day or less and are usually mild.
Salmonellosis (*Salmonella* bacteria)	Raw or undercooked eggs, meats, poultry, milk and other dairy products, shrimp, frog legs, yeast, coconut, pasta, and chocolate are most frequently involved.	6–48 hours	Nausea, abdominal cramps, diarrhea, fever, and headache.
E. coli 0157:H7 infection (Toxin released by *Escherichia coli* 0157:H7)	Undercooked hamburger and roast beef, raw milk, raw apple cider, contaminated water, mayonnaise, and vegetables grown in cow manure.	12–72 hours	Abdominal cramps, vomiting, nausea, watery diarrhea that often turns bloody, low-grade fever, and, in severe cases, kidney failure, strokes, and seizures.
Shigellosis (*Shigella* bacteria)	Found in milk and dairy products, poultry, and potato salad. Food becomes contaminated when a human carrier does not wash hands and then handles liquid or moist food that is not cooked thoroughly. Organisms multiply in food left at room temperature.	1–7 days	Abdominal cramps, diarrhea, fever, sometimes vomiting, and blood, pus, or mucus in stools.

sterilized low-acid canned goods are the most common culprits in botulism. The Savvy Diner feature on page 387 outlines some safety precautions for avoiding *C. botulinum* in canned goods. Note that trendy garlic-in-oil preparations can also harbor *C. botulinum*. Whereas manufacturers of such mixtures must add antibacterial agents to their preparations, people who make their own should be sure to store them in the refrigerator; covered, bottled garlic-in-oil left at room temperature provides the perfect warm, oxygen-free, low-acid environment for the toxin-releasing spores.[4]

TABLE 11-2 MICROBIAL FOOD AGENTS: ORGANISMS THAT CAN BUG YOU—*Continued*

Disease and Organism That Causes It	Source of Illness	Usual Onset After Eating	Symptoms
Staphylococcal food poisoning (Staphylococcal enterotoxin produced by *Staphylococcus aureus* bacteria)	Toxin produced when food contaminated with the bacteria is left too long at room temperature. Meats, poultry, egg products, tuna, potato and macaroni salads, and cream-filled pastries are good environments for these bacteria to produce toxin.	30 minutes–8 hours	Diarrhea, vomiting, nausea, abdominal pain, cramps, and prostration. Lasts 24–48 hours. Rarely fatal.
Vibrio infection (*Vibrio vulnificus*)	The bacteria live in coastal waters and can infect humans either through open wounds or through consumption of contaminated seafood. The bacteria are most numerous in warm weather.	Abrupt	Chills, fever, and/or prostration. At high risk are people with liver conditions, low gastric (stomach) acid, and weakened immune systems.
Parasite			
Trichinosis (*Trichinella spiralis*)	Raw or undercooked pork or wild game.	8–15 days	Nausea, vomiting, abdominal pain, diarrhea, fever; after two weeks, muscle pain, fluid retention, weight loss, and fever.
Protozoa			
Amebiasis (*Entamoeba histolytica*)	Exist in the intestinal tract of humans and are expelled in feces. Polluted water and vegetables grown in polluted soil spread the infection.	3–10 days	Severe crampy pain, tenderness over the colon or liver, loose morning stools, recurrent diarrhea, loss of weight, fatigue and sometimes anemia.
Giardiasis (*Giardia lamblia*)	Most frequently associated with consumption of contaminated water. May be transmitted by uncooked foods that become contaminated while growing or after cooking by infected food handlers. Cool, moist conditions favor organism's survival.	5–25 days	Sudden onset of explosive watery stools, abdominal cramps, anorexia, nausea, and vomiting.
Virus			
(Hepatitis A virus)	Mollusks (oysters, clams, mussels, scallops, and cockles) become carriers when their beds are polluted by untreated sewage. Raw shellfish are especially potent carriers, although cooking does not always kill the virus.	15–60 days	Begins with malaise, appetite loss, nausea, vomiting, and fever. After 3–10 days, patient develops jaundice with darkened urine. Severe cases can cause liver damage and death.

Sources: Adapted from A. Hecht, *The Unwelcome Dinner Guest: Preventing Food-borne Illness* (Washington, D.C.: U.S. Government Printing Office, DHHS Publication No. (FDA) 91-2244; U.S. Department of Agriculture Food Safety and Inspection Service, E. coli 0157:H7 at a glance, *Food News for Consumers* 10 (1993): 5.

Another especially virulent pathogen that is emerging as a major public health concern is *Escherichia coli* 0157:H7.* First recognized as a cause of foodborne illness in 1982, *E. coli* garnered national attention about a decade later, when it caused a major outbreak of foodborne illness in the northwest part of the United States. Undercooked, contaminated hamburgers sold at a major fast-food chain

**Escherichia coli* 0157:H7 is a particular strain of *E. coli* bacteria. When *E. coli* is mentioned throughout the rest of the chapter, it refers to the 0157:H7 strain.

prompted the 1993 scare, which ultimately led to more than 500 reported cases of illness, some 150 hospitalizations, and three deaths. The outbreak sparked a national debate about the safety of the U.S. meat and poultry supply.[5]

Found in the intestinal tracts of mammals, *E. coli* is usually transmitted via animal feces. It appears to be proliferating in the food supply and poses special concern because it is so dangerous. Unlike most other illness-producing microorganisms, *E. coli* need not be present in large numbers to make a person sick; just a few bacteria seem to do the trick. Once ingested, the bacteria clings to the intestinal wall, where it releases an enterotoxin that causes abdominal pain, watery diarrhea that often turns bloody, and, in vulnerable groups such as children and the elderly, serious complications, including kidney failure.

Most *E. coli* outbreaks have been linked to undercooked beef, particularly hamburger. Fresh apple cider, presumably made from apples exposed to tainted animal manure, has also been implicated in some outbreaks.[6] As with all illness-causing microorganisms, the best defense against *E. coli* is careful food handling.

AFLATOXIN a poisonous toxin produced by molds.

Mold is another potential microbial food contaminant. Certain molds produce poisonous compounds called **aflatoxins.** The compounds are powerful liver toxins in animals, are known to be carcinogenic in some species, and can be lethal if consumed in large doses. Aflatoxins have been found on peanuts, wheat, corn, meat pies, dry beans, and even refrigerated and frozen pastries. Molds that produce the aflatoxins typically flourish when foods such as corn and peanuts get wet and are then stored in a warm place, such as a grain silo or railroad boxcar. Controlling aflatoxins in food is difficult, but it is given high priority by the food industry and regulatory agencies.

To be sure, not all microorganisms are bad. A mold called *Penicillium roquefortii* imparts the special, pungent flavor of Roquefort cheese. Another mold strain, *P. camembertii,* lends flavor to Camembert cheese. Likewise, a strain of mold called *Bacteria aceti* causes the alcohol in wine and hard cider to turn to vinegar.[7] What's more, yogurt owes its existence to active cultures of bacteria added during processing, and a new area of research—called probiotics—suggests that the "good bugs" in the yogurt may help fight "bad bugs" that cause yeast infections and other ills.[8]

Natural Toxins

TOXICANTS poisons, that is, agents that cause physical harm or death when present in large amounts.

Most people assume that products derived from plants are safe because they are "natural." But natural food **toxicants** in plants—especially herbs—are sometimes to blame for poisonings. Even familiar foods that are generally safe sometimes harbor potential toxicants. For example, potatoes contain a substance called solanine, a powerful inhibitor of nerve impulses. The green substance accumulates just beneath the vegetable's skin, usually in harmless amounts. When potatoes are exposed to light, however, they sometimes develop excess solanine. The green area and about one-half inch of potato around it should be removed before cooking.

When plants are transformed into powders and potions, their components become more concentrated, as do their potentially harmful effects. Nevertheless, people often assume that because they come from plants, these substances must be safe. Unlike the chemical composition of standard drugs, however, that of herbal products is not regulated by the government. This lack of safety standards for herbal pills, powders, teas, and other potions, which have become increasingly popular in recent years, ranks as a major concern among public health officials. The potential risks are amplified when an herb is mislabeled, misidentified, or mixed with another potentially toxic substance.

Consider that in 1993, several children and adults suffered life-threatening respiratory problems and liver malfunction as a result of taking a Chinese herb called Jin Bu Huan. When investigators looked into the matter, they found, among other problems, that the plant from which the product had been derived was misidentified and that the product carried false and misleading medical claims.[9]

Along the same lines, in 1995, the FDA issued a warning about a product containing caffeine and a plant derivative called ma huang—an amphetamine-like substance often used in weight-loss concoctions. Together, the two substances make a deadly combination that has been linked to more than 100 injuries and several deaths.[10]

These are just a few of the herb-related problems that continue to surface. They highlight the need to be cautious about using herbs of any sort. Adults should not take large amounts of a particular herbal product or take more than one at a time without consulting a competent professional regarding the product's various effects. Pregnant and breastfeeding women should avoid using herbs because they can expose a fetus or an infant to a toxic dose. In addition, people with any type of liver disease or condition should be wary of herbal products; liver failure is one of the hallmarks of an herbal overdose, because toxins accumulate in that organ.

Numerous herbs have been implicated in liver failure and other health problems. Here are just a few:[11]

- chaparral
- Jin Bu Huan
- ma huang
- germander
- comfrey
- mistletoe
- skullcap
- margosa oil
- maté tea
- Gordolobo yerba tea
- pennyroyal (squawmint) oil

See Chapter 6 for more information regarding herbal remedies.

Like plants, other types of food, notably seafood, often harbor natural toxicants. For example, a type of fish called puffers, long considered a delicacy in Japan, serves as host to a potent poison—tetrodotoxin—which doesn't harm the fish but can be lethal to people and other animals. Tetrodotoxin, which is 275 times deadlier than cyanide, works by blocking nerve impulses; over a period of several hours, it will eventually close down a person's entire nervous system. The toxin is so deadly that sale of puffers is illegal in the United States.[12]

Puffer poisoning is just one example of a foodborne disease traced to eating a toxic sea creature. Others also exist. For instance, scombroid fish poisoning—which involves the scombroid fish family, including tuna, bonito, mackerel, and skipjack—results from ingesting a toxin produced by the action of bacteria on the dark meat in the fish. Fish contaminated with the scombroid toxin have a honeycombed look and a sharp, peppery taste. Another type, called ciguatera fish poisoning, results from eating certain fish that inhabit warm waters near coral reefs: grouper, snapper, amberjack, and barracuda. The poisoning brings about symptoms such as nausea, diarrhea, headache, face pain, and nerve problems. These fish ingest the toxin as a result of eating marine plants that contain it.

For additional information about seafood safety and other food safety topics, call the Center for Food Safety and Applied Nutrition Outreach and Information Center Hotline: 800-332-4010, or visit their web site: www.cfsan.fda.gov.

Finally, paralytic shellfish poisoning can occur after eating mollusks (clams, mussels, oysters, and scallops) contaminated with marine algae that produce a neurotoxin. The mollusks themselves don't become ill, but people eating them do, experiencing such symptoms as nausea, vomiting, cramps, and muscle weakness. In each of these cases, the toxin is not destroyed by heat. Fish can also harbor viruses, including hepatitis viruses, worms, and other parasites. This is why it is especially important to buy seafood from reputable vendors and to handle it with care.

SAFE FOOD STORAGE AND PREPARATION

Commercially prepared, canned, or packaged food is usually free of harmful microbial agents when it leaves its manufacturer. When a batch of food is contaminated, batch numbering ensures that it can be recalled quickly and the public can be forewarned. When it comes to tampering, the chances are slim that this would occur in a grocery store food. To protect against this unlikely event, however, carefully inspect the seals and wrappers of packages. Jars should be firmly sealed (many have safety buttons—areas of the lid designed to pop up once opened). Packages should be free of holes or tears. A broken seal or mangled package is not providing protection against microorganisms or other contaminants. Likewise, canned goods should be free of dents and cracks or bulges, which can indicate contamination with *Clostridium botulinum*.

Common food safety mistakes:

- Excessive store-to-refrigerator lag time
- Not washing hands before food handling
- Unclean equipment or utensils
- Countertop thawing
- Room-temperature marinating
- Using same spoon to stir and taste
- Cross-contamination:
  - Using same board or knife to cut raw meat and vegetables or fruits
  - Using same plate for raw and cooked foods
- Inadequate cooking or reheating
- Failure to keep hot foods at high temperature
- Improper cooling: leftovers and "doggie bags" left out

Most cases of food poisoning occur in the home or the restaurant and are caused by improper storage or handling. Those that arise from kitchen mistakes can be avoided by doing three simple things: keep cold food cold; keep your hands, the utensils, and the kitchen clean; and keep hot food hot. Most bacteria

Never thaw food on a kitchen counter. Bacteria can flourish at room temperature.

flourish in warm environments, whereas heat kills them and chilly temperatures halt their growth. That's why you can keep bacteria in check by paying attention to proper storage and cooking temperatures.

The first step, keeping cold food cold, starts when you leave the grocery store. If you are running errands in a car, shop last so that the groceries do not stay in the car too long, especially in hot weather. Immediately upon arrival at home, pack foods into a refrigerator set at 40 degrees Fahrenheit or a freezer kept at 0 degrees Fahrenheit, and be sure not to leave them in the refrigerator too long (see Figure 11-1). Place packages of raw meat, poultry, or fish on a plate before refrigerating or store in plastic storage bags to prevent bacteria-containing juices from dripping onto other foods. Thaw frozen food in the refrigerator or microwave oven—not on the kitchen counter. Since bacteria can multiply at room temperature, they can thrive in the relatively warm, exterior of a food before the interior has thawed.

Along with keeping foods properly chilled, keeping the kitchen clean prevents contamination of otherwise wholesome foods. Before you handle food, wash your hands in warm, soapy water. In addition, be sure to wash your hands after touching meat, poultry, or fish to prevent the spread of any bacteria that your hands have picked up. Likewise, keep countertops and all kitchen equipment clean with soap and warm water. Because bacteria love to nestle down in the fibers of kitchen cloths, sponges, and wooden cutting boards, take particular care to keep such items clean. In addition, wash wooden and plastic cutting boards in the dishwasher if possible; the hot temperatures in the dishwasher are especially effective at killing bacteria.

Keeping hot food hot requires cooking food thoroughly to ensure that the heat destroys any bacteria present. See Figure 11-1 for proper cooking temperatures. After cooking, foods must be kept hot until serving to prevent bacterial growth. Never leave perishable food at room temperature for more than two hours. Before refrigerating large quantities of hot foods, such as a pot of chili or a large casserole, divide it up and place it in shallow containers to allow easy cooling. Otherwise the inside of the pot may stay warm for a dangerously long time, even in the refrigerator.

CROSS-CONTAMINATION the inadvertent transfer of bacteria from one food to another that occurs, for instance, by chopping vegetables on the same cutting board used to skin poultry.

Meat, poultry, and fish require special handling because they often harbor high levels of bacteria. In addition, they provide a moist, nutritious environment—just right for microbial growth. Wash areas that come into contact with such foods after handling to prevent **cross-contamination.** For instance, after marinating raw meat in a dish, don't put the meat back in the same dish after cooking it, and don't use the marinade unless it has been cooked thoroughly. Wash the dish in hot, soapy water before reusing it, or the bacteria inevitably left in the dish from the raw meat can contaminate and grow in the cooked product or other food—a classic example of cross-contamination. Similarly, wash a cutting board (and your hands) after, say, skinning chicken on it. If you don't, and you chop raw vegetables for a salad on the contaminated board, the vegetables could pick up the bacteria the poultry left behind; since the salad won't be heated, the bacteria won't be killed.

Especially susceptible to bacterial contamination is ground meat. Consider that steaks and roasts are not as risky because bacteria usually settle on the outside of the cuts and are easily destroyed when the outside is heated. But when meat is ground, the bacteria are spread throughout, so more thorough cooking is needed to kill the bacteria in the middle of, say, hamburger patties or meat loaf. To decrease your risk of eating contaminated ground beef, order burgers well done and cook them so that the juices run clear and not a trace of pink is left on the inside. For a meat loaf, use a thermometer to test the internal temperature. Be especially careful when cooking ground meat in the microwave oven, because sometimes that appliance cooks foods unevenly if the foods are not handled properly.

MAD COW DISEASE (bovine spongiform encephalopathy or BSE) a rare and fatal degenerative disease first diagnosed in 1986 in cattle in the United Kingdom. The bovine disease may be passed to humans who eat the meat of infected animals and may lead to death due to brain and nerve damage.

Recently, outbreaks of **mad cow disease** in the United Kingdom received exaggerated media coverage, raising concerns among American consumers regarding the safety of consuming beef. The bovine disease may also be linked with

FIGURE 11-1
TAKE CONTROL OF HOME FOOD SAFETY

24-hour bug? Or something you ate? Very often what seems like the flu may be foodborne illness, commonly called food poisoning. Unfortunately, mishandling of food at home is a leading cause of foodborne illness. Consider the following four simple actions to take control of food safety in your kitchen.

1. Wash Hands and Surfaces Often

Proper hand washing may eliminate nearly half of all cases of foodborne illness and significantly reduce the spread of the common cold and flu.

- Hands should be washed in warm, soapy water before preparing foods and after handling raw meat, poultry and seafood.
- Sing two choruses of "Happy Birthday" while you lather up—cleaning your hands for 20 seconds.

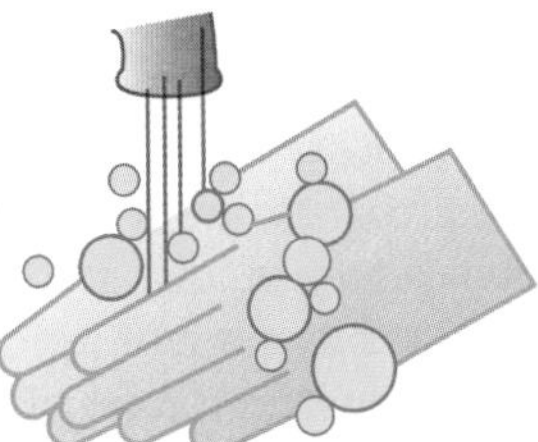

... And don't forget surfaces:

- Keep kitchen surfaces such as appliances, countertops, cutting boards and utensils clean with hot, soapy water.
- A smelly dishcloth, towel, or sponge is a sure sign that unsafe bacterial growth is lurking nearby. Bacteria live and grow in damp conditions.
  - Wash dishcloths and towels often in the hot cycle of your washing machine.
  - Disinfect sponges in a chlorine bleach solution.
  - Replace worn sponges frequently.

2. Keep Raw Meats and Ready-to-Eat Foods Separate

Never work with raw meat or poultry on the same surface that you use for other foods without thoroughly cleaning the surface after you've finished. Otherwise bacteria that the meat or poultry left behind can contaminate other foods placed on the cutting board.

Acrylic, glass, marble, plastic or solid wood? You choose. Just follow these guidelines:

- If possible use two cutting boards: one strictly to cut raw meat, poultry and seafood; the other for ready-to-eat foods, such as breads and vegetables. Don't confuse them.
- Wash boards thoroughly in hot, soapy water after each use or place in dishwasher.
- Discard old cutting boards that have cracks, crevices, and excessive knife scars.

3. Cook to Proper Temperatures

Harmful bacteria are destroyed when food is cooked to proper temperatures. Buy a meat thermometer and use it!

Safe Cooking Temperatures

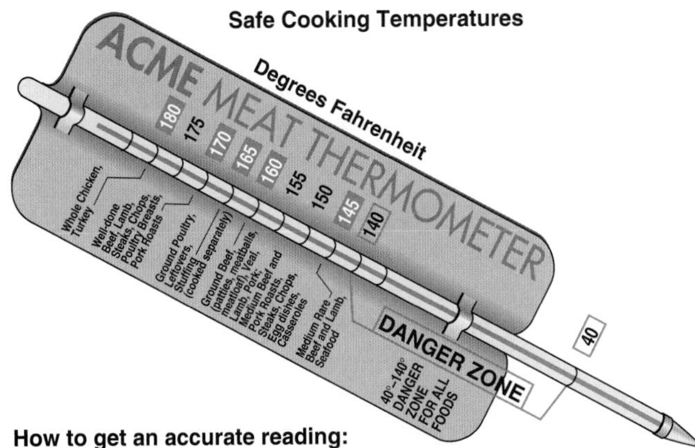

How to get an accurate reading:

- Red meats, roasts, steaks, chops, and poultry pieces: insert in center of the thickest part, away from bone, fat, and gristle.
- Poultry (whole bird): insert in inner thigh area near the breast, but not touching bone.
- Ground meat and poultry: place in thickest area of meatloaf or patty; with thin patties, insert sideways reaching the very center with the stem.
- Egg dishes and casseroles: insert in center or thickest area of the dish.
- Fish: cook until opaque and flakes easily with a fork.

4. Refrigerate Promptly Below 40°F

Refrigerate foods quickly to slow the growth of bacteria and prevent foodborne illness. Leftover foods from a meal should not stay out of refrigeration longer than 2 hours. In hot weather (80°F or above), this time is reduced to 1 hour. Set your refrigerator below 40°F. This will keep perishable foods out of what's called the "danger zone"—40°F or above. Keep a refrigerator thermometer inside your refrigerator at all times.

General Guidelines for Leftovers

Perishable food	Keeps up to
Cooked fresh vegetables	3-4 days
Cooked pasta	3-5 days
Cooked rice	1 week
Deli counter meats	5 days
Greens	1-2 days
Meat	
Ham, cooked and sliced	3-4 days
Hot dogs, opened	1 week
Lunch meats, prepackaged, opened	3-5 days
Cooked beef, pork, poultry, fish, meat casseroles	3-4 days
Cooked patties and nuggets, gravy and broth	1-2 days
Seafood, cooked	2 days
Soups and stews	3-4 days
Stuffing	1-2 days

When in doubt, throw it out!

Source: Adapted from The American Dietetic Association and the ConAgra Foundation's *Home Food Safety . . . It's In Your Hands* program, 1999. For more information, visit www.homefoodsafety.org, or call 800-366-1655 to receive a free home food safety brochure.

Safe Handling Instructions

THIS PRODUCT WAS PREPARED FROM INSPECTED AND PASSED MEAT AND/OR POULTRY. SOME FOOD PRODUCTS MAY CONTAIN BACTERIA THAT CAN CAUSE ILLNESS IF THE PRODUCT IS MISHANDLED OR COOKED IMPROPERLY. FOR YOUR PROTECTION, FOLLOW THESE SAFE HANDLING INSTRUCTIONS.

KEEP REFRIGERATED OR FROZEN.
THAW IN REFRIGERATOR OR MICROWAVE.

KEEP RAW MEAT AND POULTRY SEPARATE FROM OTHER FOODS.
WASH WORKING SURFACES (INCLUDING CUTTING BOARDS),
UTENSILS, AND HANDS AFTER TOUCHING RAW MEAT OR POULTRY.

COOK THOROUGHLY.

KEEP HOT FOODS HOT. REFRIGERATE LEFTOVERS IMMEDIATELY OR DISCARD.

The U.S. Department of Agriculture requires that all fresh meat and poultry products carry this label as a reminder to handle the products carefully.

illness in people who consume meat from infected animals. As a result of investigating these outbreaks, agricultural officials in the United Kingdom now prohibit the inclusion of mammalian meat and bonemeal in feed for all food-producing animals, since these tissues are suspected of transmitting the disease. As a result, the rate of newly reported cases of the disease is decreasing. Mad cow disease poses little or no concern to consumers in the United States, however, since the USDA has banned the import of cattle from Great Britain and other countries affected by the disease since 1989. USDA also bans the use of most mammalian protein tissues in the manufacture of animal feed given to ruminant animals (for example, cows, sheep, and goats).[13]

When it comes to picnics, choose foods that last without refrigeration, such as fresh fruits and vegetables, breads and crackers, and canned spreads and cheeses that can be opened and used on the spot. Aged cheeses, such as cheddar and Swiss, do well for an hour or two, but they should be kept in an ice chest for longer periods. Contrary to popular belief, salads made with mayonnaise pose no greater risk than foods prepared with any other type of dressing. Whether a dish contains mayonnaise or not, chill it well before, during, and after the picnic. As for burgers, chicken breasts, and other foods intended for grilling, don't partially cook them ahead of time and then throw them on the fire later. Partial cooking may not kill all the bacteria present, and because half-cooked items may not be heated thoroughly later, chances of bacterial contamination run high. Partial cooking is safe only if you take, say, a burger directly from the microwave oven to place on the grill.[14]

Seafood also should be handled with care, especially fish intended to be eaten raw or only lightly steamed. The foodborne infections that lurk in normal-appearing seafood can be much worse than those of spoilage: worms, parasites, severe viral intestinal disorders, and hepatitis. Raw fish dishes such as sushi and sashimi are safe for most healthy people to eat if they are prepared with fresh fish that has been commercially frozen at temperatures lower than most home freezers can attain; freezing kills any parasites that might be present. However, eating raw or undercooked oysters, clams, mussels, and whole scallops is especially risky. Such types of seafood sometimes carry a strain of bacteria known as *Vibrios* that can multiply even during refrigeration. While these bacteria are destroyed by thorough cooking, they can thrive in raw shellfish and cause serious illness. In fact, sometimes the bacteria cause a deadly blood poisoning that can kill a person within a day or two.[15] Because of the risk of contamination, the hazards of eating raw or undercooked seafood need to be weighed carefully, especially by vulnerable people, including those with liver disease, diabetes, gastrointestinal disorders, HIV infection, and other diseases that may compromise the body's ability to defend itself against food poisoning. The Savvy Diner feature that follows gives some tips for ensuring a safe catch by carefully buying, storing, and cooking seafood.

In general, remember that fresh food smells fresh. Any food that carries an off odor should not be eaten because the smell is probably the result of bacterial wastes and indicates that the number of bacteria in the food is dangerous. To be sure, not all types of food poisoning are detectable by odor, but if a food smells bad, chances are high that it is spoiled. Refer to the Savvy Diner feature for recommendations for preventing foodborne illness. See also the Food Safety Scorecard to help you rate your food safety knowledge.

Home Food Safety

The tradition of tea drinking has been popular for more than 2,000 years in China. Contaminated water may have been one of the reasons. A hot drink made with boiled water was less likely to cause digestive problems than plain water.[16]

In general:

- Buy only those foods stored below the frost line in store freezers.
- Do not buy or use items that appear to have been opened or tampered with.
- When running errands, make the grocery store your last stop. When you get home, refrigerate perishables immediately.
- Follow label instructions for storing and preparing packaged and frozen foods.
- Immediately refrigerate cooked foods that are not to be served right away. Use shallow, not deep, containers—the foods will cool faster.
- Maintain a clean, dry kitchen that is free of flies and insects. Wash or replace dirty sponges and towels; clean up food spills and crumb-filled crevices. Use hot, soapy water for countertops as well as for dirty dishes. Hot water and soap will immobilize bacteria and wash them away; cold water will not.

TO PREVENT ILLNESS FROM *CAMPYLOBACTER, SALMONELLA,* AND *STAPHYLOCOCCUS AUREUS* VARIETIES:

- Avoid cross-contamination: Wash hands with hot, soapy water and wash thoroughly before reusing utensils, cutting boards, or countertops that have been in contact with raw meats, poultry, or eggs.
- Thaw meats or poultry in the refrigerator, not at room temperature. If you must hasten the thawing, use cool running water or a microwave oven.
- Cook stuffing separately or stuff poultry just prior to cooking. Use a meat thermometer to avoid under-cooking. Insert the thermometer between the thigh and the body of a turkey or in the thickest part of other meats, making sure the tip is not in contact with bone. Cook to the temperature shown for that meat on the thermometer.
- Refrigerate leftovers promptly and heat them thoroughly to at least 140°F before serving.
- Use clean eggs with intact shells and cook eggs before eating them (soft-boiled for seven minutes, poached for five, or fried for three minutes on each side).
- Keep susceptible foods at temperatures colder or hotter than room temperature. Keep hot foods at 140°F or more. Keep cold foods at 40°F or less.
- Mix foods with utensils, not hands; keep hands and utensils away from mouth, nose, and hair.
- Do not prepare food if you have a skin infection or infectious disease. Anyone, though, may be a carrier of bacteria and should avoid coughing or sneezing over food.

TO PREVENT POISONING FROM *CLOSTRIDIUM BOTULINUM:*

- Before canning anything, seek professional advice. The U.S. Department of Agriculture Extension Service provides such information free of charge.
- Throw out food with off odors. (An off odor, however, is not necessarily detectable in a food containing toxins.)
- Do not even taste food that is suspect.
- Discard food from cans that leak or bulge. Dispose of the food in a manner that will protect other people and animals from its accidental use.

Continued

TO PREVENT ILLNESS FROM *ESCHERICHIA COLI:*

- Cook all meat and poultry to 160°F—until the flesh is no longer pink and all the juices run clear.
- Avoid cross-contamination: Wash hands, utensils, cutting boards, or countertops that have come into contact with raw meat or poultry with hot, soapy water.
- Order hamburgers and other beef items well-done when eating out. Return any undercooked food.
- Do not drink raw milk or potentially contaminated water.
- Wash fruits and vegetables thoroughly before eating.
- Steer clear of fresh unpasteurized apple cider, particularly if you're serving it to a high-risk person such as a child, the elderly, or anyone with a compromised immune system. As an alternative, buy pasteurized cider or heat fresh cider to a temperature of 160°F (a slow simmer) before serving.

To Ensure a Safe Catch: Seafood Safety Tips

BUYING

- Choose fresh seafood that smells like a "fresh ocean breeze," not unpleasantly "fishy."
- Choose fresh fish steaks and fillets that are moist, with no drying or browning around the edges.
- Buy seafood only from reputable dealers. You can't know what you're buying from the back of a pickup truck. It may not have undergone FDA or state inspection.
- Cook fish within two days of purchase.

STORING

- Keep fresh fish in the coldest part of your refrigerator—usually under the freezer or in the meat drawer—until ready to cook and serve.
- Store fresh fish in your refrigerator in the same wrapper it had in the store.
- Store canned fish in a clean, covered glass or plastic container in the refrigerator after opening.
- Refrigerate smoked, pickled, vacuum-packed, and modified-atmosphere-packed fish products.
- Keep cooked and raw seafood separate. It's not safe to put cooked seafood back in the original container used for raw seafood or to store raw and cooked seafood together.

COOKING

- For fin fish (baked, broiled, poached, fried, or stewed): allow 10 minutes cooking time for each inch of thickness. Turn the fish over halfway through the cooking time unless it is less than a half inch thick. Add 5 minutes to the total cooking time if the fish is wrapped in foil or cooked in a sauce. Properly cooked fish will flake easily with a fork and should be opaque and firm but not translucent.
- For molluscan shellfish: Steamed—cook 4 to 9 minutes from the start of steaming.

FOOD SAFETY SCORECARD

Improper storage of food not only increases the risk of food poisoning but also almost always results in a loss of nutrients and good taste. The following quiz is designed to measure your food safety savvy.

1. If you're packing a picnic, it's okay for the cold foods to be at room temperature prior to packing as long as they are placed in a cooler with ice or ice packs. *True or False?*
2. Foods prepared with mayonnaise—macaroni salad, potato salad and cole slaw—are common sources of food poisoning. *True or False?*
3. Raw ground meat or poultry can be stored in the refrigerator, but should be used within:
   a. one to two days.
   b. two to three days.
   c. three to four days.
   d. one week.
4. Signs that canned foods may be contaminated include:
   a. bulging.
   b. leaking.
   c. spurting of liquid when opened.
   d. all of the above.
5. Which of the following foods have been linked to food poisoning?
   a. cooked rice
   b. apple cider
   c. shellfish
   d. all of the above
6. The best place in the refrigerator to store milk is in the door. *True or False?*
7. When bringing home groceries from the market, it's a good idea to rewrap meat and poultry before placing them in the refrigerator or freezer. *True or False?*
8. Fresh fish should smell "fishy." *True or False?*
9. The best way to handle green-skinned potatoes is to:
   a. throw them away.
   b. soak them in cold water.
   c. peel the skin and remove some of the flesh prior to cooking.
   d. remove the green section after cooking.
10. Canned foods can be stored in the pantry indefinitely. *True or False?*
11. You can tell a food is contaminated by the way it looks, smells, or tastes. *True or False?*
12. The maximum time perishable foods can be kept at room temperature is:
    a. one-half hour.
    b. one hour.
    c. two hours.
    d. 24 hours.
13. It's okay to thaw frozen ham on the kitchen counter since salted and smoked meats are free of bacteria. *True or False?*
14. You can reduce your exposure to chemical contaminants in whole fish by proper cleaning. *True or False?*
15. Cloudy liquid around packaged hot dogs indicates bacteria have started growing. *True or False?*

Continued

FOOD SAFETY—*Continued*

SCORECARD

ANSWERS

1. False. Be sure food is cold or frozen before placing it in a cooler. This minimizes the chance of microbial growth. Use ice packs between food items. Frozen juice boxes can also be used and enjoyed later in the day after they have thawed.
2. False. Adding mayonnaise to food does not increase the chance of food poisoning. Most store-bought mayonnaise contains ingredients (vinegar, lemon juice and salt) that actually slow bacterial growth.
3. (a) Ground meat and ground poultry are more perishable than other meats. They should be refrigerated and cooked (or frozen) within one to two days of purchase.
4. (d) Never buy or use products with bulging lids, leaking cans or cracked jars. All are warning signs that a product may be contaminated with the bacteria that cause deadly botulism.
5. (d) While the most common offenders are poultry, meat, and shellfish, any food can cause food poisoning, including cooked rice, if mishandled.
6. False. The refrigerator door does not stay as cold as the rest of the refrigerator. Highly perishable foods like milk and eggs should be stored on an inside shelf. Use door for condiments.
7. False. Leave meat and poultry in store wrapping. The less you handle it, the better. This is especially true for ground meat and ground poultry.
8. False. Fresh fish should smell like a fresh sea breeze. If it smells "fishy," don't buy it.
9. (c) Green-skinned potatoes contain a natural toxin called solanine, which develops when potatoes are exposed to sunlight. It imparts a bitter taste and may cause stomach upset if eaten in large quantities. Before cooking, simply remove the green area and about one-half inch of flesh around it.
10. Canned foods have an extended but not infinite shelf life. Remember to place newly purchased cans behind older ones. To be safe, throw out beans and high-acid foods (for example, pineapple, tomatoes) after one year and other canned foods after two years. Use canned fish within six months.
11. False. Food spoilage may leave tell-tale signs such as changes in looks, smell or taste, but contaminated food does not. Remember this rule of thumb: "When in doubt, throw it out."
12. (c) The longer a perishable food is kept at room temperature, the greater the likelihood that bacteria will multiply to dangerous levels. Food kept unrefrigerated for more than two hours is a prime target for bacterial growth.
13. False. Many people think that salted or smoked meats are immune to bacterial contamination. That's not so, especially since many manufacturers have been gradually lowering the salt content of cured meats. With any meat or poultry, the safest way to thaw it is in the refrigerator overnight.
14. True. By removing the skin and trimming the fatty tissue along the back and belly, where chemicals tend to concentrate, you can reduce your exposure to contaminants.
15. True. Although hot dogs are processed to last longer than other meat products, *Listeria monocytogenes* can grow even under refrigeration. If you notice cloudy liquid, discard the franks. Freeze hot dogs if you don't plan to use them within two weeks. If opened, use within a week.

HOW DID YOU SCORE?

Count up the number of questions you answered correctly.

12–15 Congratulations! You're quite the food safety scholar.

8–11 You're fairly savvy when it comes to food safety, but don't push your luck. Brush up on the questions you had trouble with.

7 or below You're a likely target for food poisoning. Mend your ways; it's never too late.

Source: Adapted with permission from A. Schepers, What's Your Food Safety I.Q.?, *Environmental Nutrition,* 22 (1999): 1, 6.

PESTICIDES AND OTHER CHEMICAL CONTAMINANTS

Food producers and food processors exert major efforts to maintain a safe food supply. Some risk of consuming undesirable substances, or **contaminants,** however, is unavoidable. Our industrial society's reliance on chemical processes means that foods may become contaminated by a variety of chemicals introduced into the environment. In addition, agricultural techniques that necessitate the use of pesticides affect the food supply. The following section examines some of the major chemical players in the food supply and looks at some ways that scientists hope to reduce them.

CONTAMINANTS potentially dangerous substances, such as lead, that can accidentally get into foods.

Chemical Agents

Some of the problem industrial chemicals prevalent in the environment and food supply include **organic halogens,** such as polychlorinated biphenyl (PCB) and polybrominated biphenyl (PBB), and **heavy metals,** such as lead and mercury.

ORGANIC HALOGENS compounds that contain one or more of a class of atoms called halogens, including fluorine, chlorine, iodine, or bromine.

HEAVY METALS any of a number of mineral ions, such as mercury and lead, so named because of their relatively high atomic weight. Many heavy metals are poisonous.

Fortunately, episodes of direct, excessive chemical contamination such as chemical spills are few and far between. But when a particular area *does* suddenly become contaminated, the effects can be far-reaching. In 1973, for example, half a ton of PBB was accidentally mixed into some livestock feed that was distributed throughout the state of Michigan. Millions of animals ingested excess PBB, and so did the people who ate the beef that came from the tainted livestock. Dairy farmers recognized the seriousness of the accident when their cows began going dry, aborting their calves, and developing abnormal growths on their hooves. More than 30,000 cattle, sheep, and swine and more than a million chickens died as a result of PBB. Unfortunately, the problem didn't stop with the animals. By 1982, an estimated 97 percent of Michigan's residents had been contaminated with PBB, with side effects that included nervous system aberrations and alterations in the liver and immune systems.[17]

The Michigan incident is just one example of chemical contamination. The number of contaminants and the amount of information available about them are far beyond the scope of this text, but the principles that apply to all contaminants can be illustrated by giving details about just a few. A list of several chemical contaminants of particular concern in foods is presented in Table 11-3.

When considering the risks of chemical contamination, remember that even though a substance is toxic, the hazard it poses tends to be small because chemical levels are usually carefully controlled. When a chemical spill or other accident occurs, the risk can suddenly soar. That's why scientists differentiate between **toxicity** (a property of all substances) and **hazard** (the likelihood of a substance's actually causing harm). All substances are potentially toxic, but they are hazardous only if consumed in large enough quantities. In other words, the dose makes the poison. Thus, a chemical that is present in foods in minuscule amounts does not pose a significant hazard until for some reason it becomes concentrated in the food in excess, toxic amounts.

TOXICITY the ability of a substance to harm living organisms. All substances are toxic if present in high enough concentrations.

HAZARD state of danger; used to refer to any circumstance in which harm is possible.

Most experts agree that chemicals in foods pose a small hazard; the chief concern is accidental gross contamination. In some instances, however, chemical contamination can be subtle and insidious. The problem of chronic low-level lead poisoning is a prime example. In the United States, lead poisoning ranks as one of the most common childhood environmental health problems, affecting some two million children under the age of six.[18]

Many of the chemicals that contaminate foods are the waste products of industry.

Lead usually does not poison a person all at once; rather, low levels build up gradually in the soft tissues of the kidneys, bone marrow, liver, and brain. Over time, the accumulated lead can cause such health problems as diminished intelligence and impaired development. Pregnant women and young children are particularly vulnerable to the effects of lead because their bodies absorb high levels of calcium to meet their growth needs but the body cannot distinguish between

TABLE 11-3 EXAMPLES OF CONTAMINANTS IN FOODS

Name and Description	Sources	Toxic Effects	Typical Route to Food Chain
Cadmium (heavy metal)	Used in industrial processes, including electroplating, plastics, batteries, alloys, pigments, smelters, and burning fuels; present in cigarette smoke.	No immediately detectable symptoms; slowly and irreversibly damages kidneys and liver.	Enters air in smokestack emissions, settles on ground and is absorbed into plants, consumed by farm animals, and eaten in meat and produce by people. Sewage sludge and fertilizers leave large amounts in the soil; runoff contaminates shellfish.
Lead (heavy metal)	Lead crystal, improperly manufactured and old ceramic ware, paint, old plumbing, and leaded gasoline.	Displaces calcium, iron, zinc, and other minerals from their sites of action in the nervous system, bone marrow, kidneys, and liver, causing failure to function; harms fetuses, infants, and children, who easily absorb and retain lead; causes breakage of red blood cells (anemia) and interferes with the immune response.	Air pollution, leaded gasoline, water pipes, improperly manufactured and old ceramic ware, and lead crystal.
Mercury (heavy metal)	Widely dispersed in gases from the earth's crust; local high concentrations from industry, electrical equipment, paints, and agriculture.	Poisons the nervous system, especially in fetuses.	Inorganic mercury released into waterways by industry and acid rain is converted to methylmercury by bacteria and ingested by fish (tuna, swordfish, and others).
Polychlorinated biphenyl (PCB) (organic compound)	No natural source; produced for use in electrical equipment.	Causes long-lasting skin eruptions, eye irritation, growth retardation in children of exposed mothers, anorexia, fatigue, and other effects.	Discarded electrical equipment, accidental industrial leakage, or reuse of PCB containers for food.

Some antique ceramic and crystal items are best left on the shelf. The older the piece, the higher the chances that it leaches lead.

lead and calcium. As a result, they readily absorb lead. If they are calcium or iron deficient, as many women and children are, their bodies take in even more lead.

Historically, lead entered the food and water supply largely through leaded gasoline exhaust, which contaminates rainfall, which in turn pervades crop soil and water supplies. Lead contamination of the water supply has been and continues to be a source of concern (refer to the discussion on water in Chapter 7). In addition, lead solder used to seal the seams of cans was long a source of the contaminant. Fortunately, during the past two decades, the use of leaded gasoline has been greatly reduced, and lead-soldered cans have been eliminated. Today the levels of lead in foods and beverages are the lowest in history.[19]

Nevertheless, new sources of lead contamination have surfaced during recent years. When it comes to lead exposure via food and beverages, scientists have identified ceramic hollowware, lead crystal ware, and foil capsules on wine bottles as potential sources. Lead can leach from the glaze of ceramic bowls, mugs, and pitchers that have not been properly formulated or fired, especially when the food or beverage inside is hot and acidic, such as coffee, tea, or tomato soup. Wine and other alcoholic beverages also promote the leaching of lead, so experts advise against storing alcohol in pitchers and other containers made with lead crystal. In addition, it's a good idea to wipe with a damp cloth the rims of wine bottles with foil capsules, which have been shown to leach lead.

Many manufacturers have modified or are in the process of modifying their practices to reduce the potential for lead contamination. For help on lowering your own risk, see Table 11-4.

Although chemical contamination is not the greatest hazard posed by the food supply, there are still many unknowns. No one knows what level the contaminants accumulating in the environment may be reaching and threatening the health of the planet and its people. Another unknown factor is the interaction among contaminants; a substance that poses no threat by itself may combine with other chemicals lurking in the environment to form potent carcinogens. Still another unknown is the effect of time. Many potentially hazardous substances have been used for only a short time. What are the effects of prolonged exposure to them?

Despite the uncertainties, keep in mind that more often than not, healthful eating habits and overall good health protect against the toxicity of food contaminants and other environmental pollutants. Eating a wide variety of food ensures an adequate supply of essential nutrients and minimizes exposure to potential environmental contaminants.

Pesticide Residues

Pesticides are substances used to prevent, destroy, or repel harmful pests, including insects, spiders, bacteria, weeds, molds and mildews, rodents, and other living things. Unlike many other chemicals present in the food supply, pesticides are intended specifically to poison living things. The trouble with pesticides, of course, is that they can inadvertently harm wildlife, people, and other species.

PESTICIDES chemicals applied intentionally to plants, including foods, to prevent or eliminate pest damage. Pests include all living organisms that destroy or spoil foods: bacteria, molds and fungi, insects, and rats and other rodents, to name a few.

When farmers first began using pesticides, they chose potent ones—chemicals designed to keep on killing for as long as they remained in the soil. Unfortunately, after years of widespread pesticide use, some unsettling side effects of the chemicals surfaced. They washed from the soil into lakes, rivers, and oceans,

TABLE 11-4 GETTING THE LEAD OUT

To protect against overexposure to lead, use the following guidelines:

- Do not store fruit juices or other acidic foods in ceramic containers.
- Do not store beverages in lead crystal containers.
- Use antique or collectible housewares for food or beverages only on occasion.
- Do not eat or drink from items that show a dusty or chalky gray residue on the glaze after they are washed.
- Do not feed babies from crystal bottles.
- During pregnancy, avoid frequent use of ceramic mugs for hot beverages such as coffee or tea. In addition, do not use lead crystal ware daily.
- Before pouring from a wine bottle sealed with a foil capsule, wipe the rim of the bottle with a cloth dampened with water or lemon juice.
- Run cold water for several minutes before using it for drinking or cooking.
- If you would like to purchase a home test kit for lead leaching from ceramic ware, contact your local FDA office, listed in the phone directory.
- For a listing of lead-safe ceramic ware, write to the Environmental Defense Fund, P.O. Box 96969, Washington, D.C. 20090-6969.
- For additional questions about lead, call the National Lead Information Center at 800-532-3394. Callers may request an information packet in English or Spanish.

Source: Adapted from J. E. Foulke, Lead threat lessens, but mugs pose problem, *FDA Consumer*, April 1993, pp. 19–23.

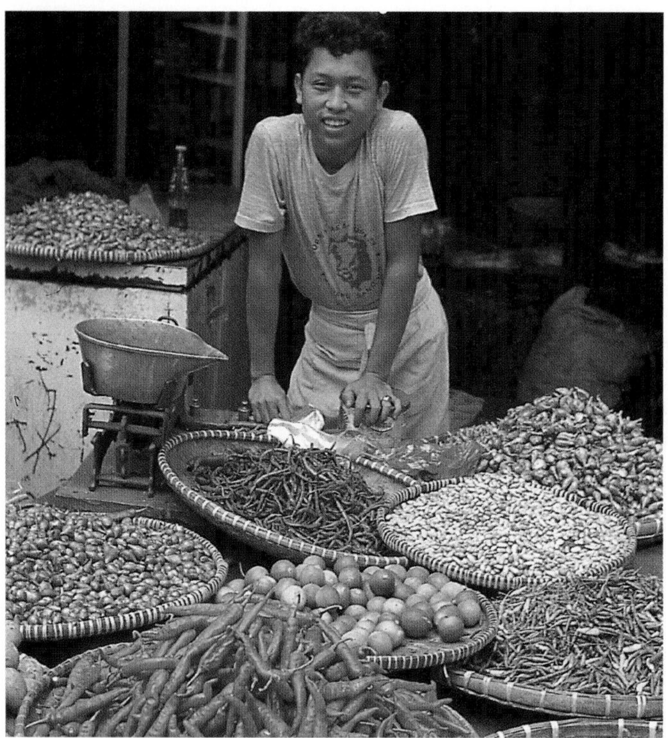

Foods imported from other countries may harbor residues of pesticides that have been banned for use in the United States.

contaminating water supplies; they poisoned farm workers, who breathed the chemicals day in and day out; they endangered many species of wildlife, disrupting the animals' abilities to reproduce and wiping out entire populations; and, most disturbing, they would not go away.

For example, in the 1960s, scientists observed that a popular pesticide called DDT had begun accumulating in the body fat of animals. DDT threatened the survival of the American eagle by weakening eggshells to the point of collapse, killing the developing chicks inside. It also appeared in big fish, carnivorous animals, people, and human breast milk. Finally, after years of widespread agricultural use, the United States banned DDT. However, the pesticide still lingers in the environment. Many foreign countries continue to use DDT, including countries from which the United States buys produce. Regrettably, despite the U.S. ban on DDT, U.S. companies are still allowed to sell it (as well as other banned pesticides) to countries where DDT use remains legal. This practice may come back to haunt us, however, when we buy imported produce that has been exposed to DDT. In addition, the DDT situation illustrates the need to consider environmental issues from a global perspective; residues of chemicals banned in one country can travel the entire world not only through the imported and exported foods but also through wind, rain, and waterways.

DDT taught us another lesson about pesticides: A contaminant that builds up in the body carries the greatest potential risk. In the body, a contaminant that is quickly broken down to some harmless compound poses the least risk to health. Likewise, if the body can easily and rapidly excrete a pesticide residue, the body may not be harmed by it. But if the residue enters the body and stays there, all the while interacting with the body's cells and systems, it may wreak havoc. Additional doses piled on top of the first ones compound the damaging effects. Moreover, when a substance resists breakdown either inside the body (by the body's own enzymes) or outside (by microorganisms) and furthermore accumulates from one species to the next, it builds up in the food chain (see Figure 11-2). DDT causes all these problems, which is why it is so deadly.

This is not to say that pesticides have no place in the farming community. Farmers would face daunting obstacles to growing crops without them. Careful use of the chemicals often boosts crop yields, which in turn helps to keep the price of fruits and vegetables down and the availability of a wide variety of produce high. Still, as the DDT experience illustrates, pesticides must be evaluated scrupulously and used judiciously.

The ideal pesticide destroys the target pest and then breaks down quickly to other products that pose minimal hazard to people and other animals. Scientists have made many strides in developing relatively innocuous pesticides over the years, and the search for better, safer pesticides continues. What's more, since the introduction of pesticides decades ago, many national and international agencies have adopted strict **regulations** for pesticide use.

In the United States, the Environmental Protection Agency (EPA) determines whether a particular chemical may be used on U.S. crops. In deciding whether to approve a pesticide, the EPA scrutinizes dozens of studies that assess the substance's possible effects on people, wildlife, fish, or plants as well as its potential to cause such problems as cancer and birth defects. To do so, the EPA examines the estimated amount of a pesticide residue that a person might be exposed to during the course of a 70-year lifetime and the **risk** of harm that

REGULATION a legal mandate that must be obeyed. Failure to follow a regulation brings about serious legal consequences.

RISK the harm a substance may confer. Scientists estimate risk by assessing the amount of a chemical that each person in a population might consume over time (also called *exposure*) and by considering how toxic the substance might be (*toxicity*).

risk = exposure × toxicity

exposure = amount of substance in food × amount of food eaten

FIGURE 11-2
HOW A FOOD CHAIN WORKS

A person who eats fish regularly may consume about 100 pounds of it in a year. These fish will, in turn, have eaten a few tons of small plant-eating fish during their lifetime. The little plant eaters will have ingested several tons of plants. If the plants have been contaminated with toxic chemicals, the bodies of the small fish that eat them will contain high concentrations of the chemicals; the larger fish that eat the little fish will harbor even higher amounts of the chemicals; and so on through the food chain. If none of the chemicals are lost along the way, the person at the end of the food chain ultimately eats the same amount of chemical contaminants that was contained in the original *several tons* of plants.

amount might pose. If the risk appears unacceptably high, the pesticide will be ruled out. All things considered, the process often requires years of research and costs millions of dollars.[20]

Once the EPA approves a pesticide for use, it sets forth safety standards. For those pesticides it allows, it decides which crops may be treated with it and how much may be applied. It also establishes a **tolerance**—that is, the maximum amount of a pesticide residue allowed in or on a food. The EPA also sets forth a **reference dose** for the pesticide. This represents the amount of a chemical that could be consumed daily without posing any health risk. To calculate the reference dose, scientists use animal studies to estimate the maximum amount of the chemical that a person could take in daily without suffering harm. They then take a fraction of this amount, usually 1/100, to ensure an extra **margin of safety.** In other words, the reference dose is 1/100 of the maximum amount of the substance that appears to be safe. Scientists factor in a large margin of safety as a precautionary measure to help ensure that even highly vulnerable people won't be harmed by the substance in question (see Figure 11-3 on page 398).[21]

After the EPA approves a pesticide for use, the FDA, in its ongoing monitoring program, begins to check for residues of it. Inspectors collect samples from packers, shippers, and other food handlers and then test for pesticide residues. If a food contains residues that exceed the EPA's tolerance limits, the FDA can seize the entire shipment and press criminal charges. In addition, the FDA conducts studies to determine whether pesticide residues from a variety of foods together fall below reference dose levels. Such cross-checks help to ensure that farmers are applying pesticides properly. Meanwhile, what can you do to protect yourself against unacceptably high levels of pesticides? The following Nutrition Action feature offers a perspective on that question.

(text continued on page 398)

TOLERANCE the maximum amount of a particular substance allowed on food.

REFERENCE DOSE the estimated amount of a chemical that could be consumed daily without causing harmful effects.

MARGIN OF SAFETY from a food safety standpoint, the margin is a zone between the maximum amount of a substance that appears to be safe and the amount allowed in the food supply.

NUTRITION ACTION

Should You Buy Organically Grown Produce?

ORGANICALLY GROWN FOODS crops or livestock grown and processed according to USDA regulations concerning use of pesticides, herbicides, fungicides, fertilizers, preservatives, other synthetic chemicals, growth hormones, antibiotics, or other drugs (see www.ams.usda.gov/nop).

The USDA organic seal is meant to provide information to consumers and will help organic farmers and ranchers further expand their already growing markets.

Although **organically grown foods** have been available for more than 40 years, marketers carved out a small but strong niche in the last two decades: "organically grown" produce that commands a premium price. Once sold only in health food stores, organic fruits, vegetables, meats, and other foods can now be found in most mainstream markets and represent 1 percent of food sales nationwide. Sales of organic food products have grown from $78 million in 1980 to about $6 billion today, and sales of organically grown crops are predicted to increase fourfold during the current decade. Whereas most sectors of agriculture are losing farmers, the number of organic farmers is expected to grow by 12 percent per year.

As a result of this growth the USDA has moved to create a national reference standard to define what is and what is not organic. Only crops grown and processed according to the USDA regulations concerning the use of fertilizers, herbicides, pesticides, fungicides, and other synthetic chemical agents can claim to have been grown organically. In March 2000, the USDA introduced a proposal which places strict regulations on organic farming. The new proposal mandates certification of organic products by USDA-approved state or private regulating authorities and sets strict labeling standards for processed organic foods. To gain the USDA organic seal on their labels, raw products must be 100 percent organic, and processed foods must contain 95 percent organic ingredients. If a food contains between 50 and 95 percent organic contents, the label can read "product made with organic ingredients." Additionally, irradiated or genetically engineered products cannot be labeled as organic.[22]

Consumers who opt for organics often believe that the expensive produce is especially healthful because it has been grown without the use of pesticides and contains no pesticide residues.[23] But organically grown produce is not always what people think it is, for a number of reasons. For one, organic growers often use pesticides naturally found in the environment, such as sulfur, nicotine, and copper, so "organically grown" doesn't always mean grown without pesticides.[24]

Similarly, the concept of "pesticide-free" produce is misleading because it suggests that produce that's not free of pesticides is harmful. Consider that today scientists can use high-tech laboratory techniques to detect infinitesimal amounts of chemical residues on produce. Table 11-5 gives some comparisons to help put the common measurements for pesticide residues into perspective. As the table suggests, whether a food contains any chemical residues or has been exposed to pesticides at some time in the past is not necessarily the right question to ask about its effect on your health. The real issue is whether the amount of residue that remains in the food at the time you eat it is harmful. If a food has been treated with a pesticide and the chemical has since evaporated, changed into a nontoxic compound, or been diluted below the point at which it can do any harm, the food may not be inferior to an untreated food. In fact, it may be nutritionally superior because it has not been weakened by attacks of pests.

None of this is to say that buying organically grown produce is a waste of money. If you can afford it and you're certain that it comes from a reputable farmer who is following sound organic growing techniques, organically grown

TABLE 11-5 PESTICIDES IN PERSPECTIVE

When you hear about pesticide residues in food, the amounts measured are either parts per million (ppm), parts per billion (ppb), or parts per trillion (ppt). The following comparisons show just how tiny these amounts are.

1 ppm:	1 ppb:	1 ppt:
1 gram of residue in 1 million grams of food	1 gram of residue in 1 billion grams of food	1 gram of residue in 1 trillion grams of food
1 inch in 16 miles	1 inch in 16,000 miles	1 inch in 16 million miles
1 minute in 2 years	1 second in 32 years	1 second in 32,000 years
1 cent in $10,000	1 cent in $10 million	1 square foot on floor tile the size of Indiana

Source: Adapted from International Food Information Council, *Pesticides and Food Safety* (Washington, D.C.: International Food Information Council, 1995), p. 2.

food casts a vote in favor of agricultural techniques that promote the well-being of the environment and farming communities over the long run. Reputable organic farmers typically do use fewer pesticides than conventional farmers because they use "eco-friendly" techniques such as crop rotation to help keep pests under control. This helps to protect the soil as well as the groundwater and the farm workers themselves against chemical contamination.

Still, many farmers who aren't meeting "organic" standards per se are using other techniques to keep pesticide use to a minimum. For example, more and more farmers have adopted a system called **integrated pest management**—a technique by which farmers cut back chemical use by combining strategies such as crop rotation, genetic engineering (discussed at the end of the chapter), and biological controls such as the release of a predator insect on a crop to get rid of another pest. What's more, no major health organization advocates eating only organically grown produce. The biggest health hazard, they say, is letting worries about pesticide residues stand in the way of eating a fruit- and vegetable-rich diet. To keep pesticide residues from all types of produce to a minimum, use the produce-handling tips in Table 11-6.

INTEGRATED PEST MANAGEMENT the use of biological controls, crop rotation, genetic engineering, and other tactics to reduce chemical use in the growing of crops.

If you grow your own produce, do so with care. Pesticides and disinfectants used on home gardens rank as a major source of chemical residue exposure to consumers. For information on safe use, storage, and disposal of home and garden pesticides, contact your local Cooperative Extension Service or the EPA's national pesticide hotline: 800-858-PEST.

TABLE 11-6 HANDLING PRODUCE PROPERLY*

To keep pesticide residues in fruits and vegetables to a minimum, use the following tips:

- Rinse produce thoroughly with water and scrub with a vegetable brush. When present, most pesticide residues reside on the surface of a product. Consider peeling produce to which wax has been applied; some wax may contain residual from antifungal chemicals.
- Do not use soap to wash produce. Soap was not designed to be used on food and may leave behind undesirable residues of its own.
- Discard the outer leaves of lettuce, cabbage, and other leafy vegetables. Rinse the interior leaves thoroughly to remove dirt and debris.
- Use a knife to peel an orange or grapefruit; do not bite into the peel.
- Eat a variety of fruits and vegetables. Farmers use different chemicals for different crops, so eating a wide variety of produce helps to cut down on exposure to any particular pesticide residue.

*To reduce pesticide residue intake from meats, trim fat from meat and remove the skin from poultry and fish (pesticide residues concentrate in the animal's fat tissue). Discard fats and oils in broths and pan drippings.

Source: Adapted from C. F. Chaisson and coauthors, *Pesticides in Food: A Guide for Professionals* (Chicago: The American Dietetic Association, 1991), p. 18.

FIGURE 11-3
MARGIN OF SAFETY

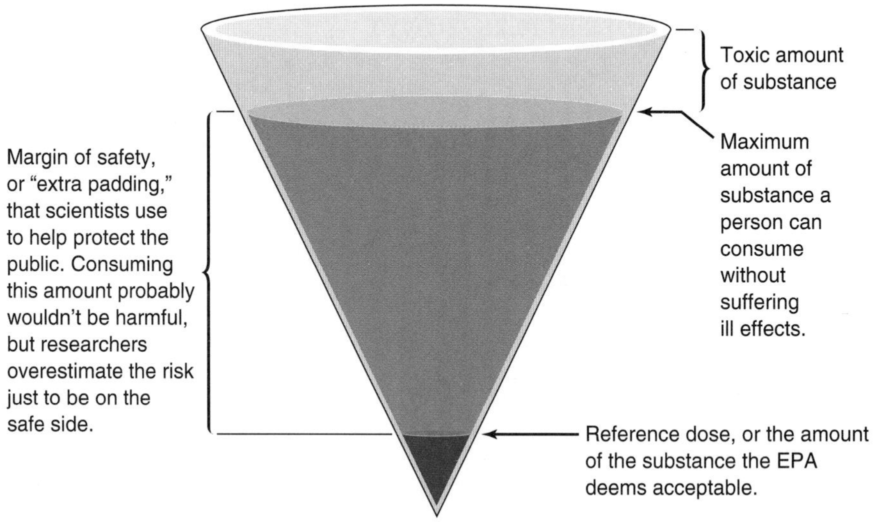

Toxic amount of substance

Maximum amount of substance a person can consume without suffering ill effects.

Margin of safety, or "extra padding," that scientists use to help protect the public. Consuming this amount probably wouldn't be harmful, but researchers overestimate the risk just to be on the safe side.

Reference dose, or the amount of the substance the EPA deems acceptable.

While EPA and FDA regulations go a long way in protecting the public from harmful pesticide residues, problems with the monitoring system still exist. As explained earlier, other countries may use pesticides banned in the United States, and imported foods might not be tested for the presence of those pesticides. In addition, consider that in 1993 the National Academy of Sciences issued the results of a large-scale, five-year study on the risk of pesticide residues in the diets of infants and children. Although it concluded that the U.S. food supply is safe for children, it called for changes in the methods used to assess health risks from pesticides to better account for differences between adults and children.[25]

Despite the uncertainties, however, major health organizations, including the American Academy of Pediatrics, the American Cancer Society, and the American Medical Association, agree that the health risks posed by pesticide residues are minimal and that the health risks of *not* eating fruits and vegetables for fear of consuming pesticide residues far outweigh the slight risk linked with those substances.

FOOD ADDITIVES

FOOD ADDITIVE any substance added to food, including substances used in the production, processing, treatment, packaging, transportation, or storage of food.

INTENTIONAL FOOD ADDITIVES substances intentionally added to food. Examples include nutrients, colors, spices, and herbs.

INCIDENTAL FOOD ADDITIVES (or indirect additives) substances that accidentally get into food as a result of contact with it during growing, processing, packaging, storing, or some other stage before the food is consumed.

From a safety standpoint, **food additives** rank among the *least* hazardous substances in food, although consumers tend to rank them high on their list of food-related risks. The great majority of food additives enhance the color, flavor, texture, or stability of foods or even improve the nutritional value of certain items, as shown in Table 11-7. Many additives are common substances such as vitamins, herbs, and spices, deliberately added to foods and called direct or **intentional additives.**

On the other hand, **incidental additives,** such as packaging materials or processing chemicals, get into the food by accident during processing. The federal government regulates all types of food additives and requires that food processors perform tests to determine whether additives are present in safe levels.

Manufacturers must go through a lengthy, costly process to get FDA approval for a new food additive. The manufacturer must conduct extensive research to show that the additive in question does what it is supposed to do, that it can be detected and measured in foods to which it has been added, and that it is safe in the amounts in which it will be used.

As with pesticide residues (described earlier), many food additives are allowed only in amounts that ensure a wide margin of safety. Most additives that pose any potential risk are allowed in foods only at levels 1/100 of those at which the

TABLE 11-7 MAJOR USES OF FOOD ADDITIVES

Function of Additive	Examples	Foods in Which Often Used
Impart or maintain consistency	Stabilizers, thickeners, anticaking agents, and emulsifiers including alginates, lecithin, mono- and di-glycerides, glycerine, pectin, guar gum	Baked goods, cake mixes, salad dressings, ice cream, processed cheese, table salt, chocolate
Improve or maintain nutritional value	Vitamins A and D, thiamin, niacin, folic acid, ascorbic acid (vitamin C), calcium citrate, zinc oxide, iron, iodine	Flour, bread, biscuits, breakfast cereals, pasta, margarine, milk, iodized salt, juices
Maintain palatability and wholesomeness	Ascorbic acid, butylated hydroxyanisole (BHA), butylated hydroxytoluene (BHT), benzoates, sulfites	Bread, crackers, frozen and dried fruit, margarine, lard, potato chips, cake mixes, meat
Produce light texture; control acidity or alkalinity	Yeast, sodium bicarbonate, citric acid, lactic acid, phosphoric acid	Cakes, cookies, quick breads, crackers, butter, soft drinks
Enhance flavor or impart desired color	Cloves, ginger, fructose, aspartame, MSG, FD&C Red No. 40, caramel, turmeric	Spice cake, gingerbread, soft drinks, yogurt, soup, candy, cheese, jams, gum

Source: Adapted from *Food Additives* (Rockville, MD; Washington, D.C.: Food and Drug Administration in cooperation with International Food Information Council, 1992), pp. 7–8.

risk is still known to be zero. Even nutrient food additives are subject to the margin of safety concept. Consider that while iodine has long been added to salt to prevent iodine deficiency, the amounts added have been controlled because excess amounts of the mineral can be deadly.

Food additives extend the shelf life of many commonly eaten foods.

The GRAS List and the Delaney Clause

Attention to two aspects of food law will help you to better understand the issues surrounding the use of additives in foods.

Substances Generally Recognized As Safe (GRAS). For the first half of the twentieth century, scientists evaluated the safety of food additives using a simple approach: An added substance was either "safe" and therefore permitted for use in foods or "poisonous and deleterious" and therefore banned. As the study of toxic agents advanced, scientists realized that eventually they would be able to show that virtually every substance poses a health hazard if the dose is large enough. Consequently, they recognized that simply classifying an additive as "safe" or "poisonous" failed to be an effective means of evaluation.

To get around the problem, Congress set forth the bill that would later become the Food Additives Amendment of 1958. As the members of Congress debated the bill, a question arose as to how to deal with the additives already in use. Congress decided that a "safe" substance in use prior to 1958 would be deemed a "Substance Generally Recognized as Safe" (a GRAS substance) and be put on the **GRAS list.** Substances not in use before that time would be classified as food additives and subject to regulation under the Food Additives Amendment.

GRAS (GENERALLY RECOGNIZED AS SAFE) LIST a list of ingredients, established by the FDA, that had long been in use and were believed safe. The list is subject to revision as new facts become known.

With the establishment of the amendment, the FDA put hundreds of substances on the GRAS list. Everything from vegetable oils, salt, pepper, sugar, caffeine, vinegar, and baking powder to meat, poultry, eggs, milk, seafood, cereals, fruit, and vegetables were—and still are—classified as GRAS substances.

The GRAS list came under scrutiny in 1969, however, when safety questions arose about the GRAS substance cyclamate, an artificial sweetener. After reviewing hundreds of studies, the FDA decided to ban cyclamate from use in foods and beverages. As a result of this incident, President Richard Nixon ordered a reevaluation of the safety of all substances on the GRAS list. The FDA conducted a sweeping review and removed about 300 substances from the list. Today, the more than 400 substances classified as GRAS are continually subject to reexamination as new facts and concerns arise.

Delaney Clause. Another piece of legislation came about during the debate regarding the Food Additives Amendment of 1958. At that time, James J. Delaney (U.S. Representative to New York) sponsored a provision to the bill that forbid the approval of any additive found to cause cancer in animals no matter how small the dose. The rationale for the provision stemmed from the widely held belief that it was possible to completely eliminate cancer-causing agents from the food supply. Congress voted to add the **Delaney clause** into the Food Additives Amendment of 1958, despite considerable debate.

DELANEY CLAUSE a provision in the 1958 Food Additives Amendment that prohibits manufacturers from using any substance that is known to cause cancer in animals or humans at any dose level.

In recent years, the Delaney clause often put regulatory agencies in a legal bind because it is virtually impossible to eliminate potential cancer-causing agents from the food supply. Consider that many substances, even those found naturally in a number of foods, can cause cancer when given to animals in large enough amounts. In fact, critics charged that the Delaney clause encouraged the use of studies in which animals were fed substances in doses hundreds of thousands of times greater than the dose a person could consume via food. The results of such studies can be meaningless, given that the dose makes the poison, as pointed out earlier in the chapter. In addition, as scientists continued to develop better techniques for detecting chemical residues in foods, the task of completely eliminating minuscule amounts became even more complex and unrealistic. As a result, the Food Quality Protection Act of 1996 eliminated the Delaney Clause from law.

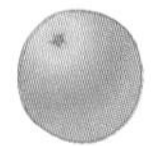

NEW TECHNOLOGIES ON THE HORIZON

Back in the 1860s, a French scientist named Louis Pasteur came up with a radical, newfangled process by which the disease-producing microorganisms in a food could be destroyed by exposing it to heat. Dubbed pasteurization, after Dr. Pasteur, the process marked a major breakthrough in the science of food safety. At the time, however, Dr. Pasteur's discovery met with widespread fear and opposition. In fact, it wasn't accepted as a vital public health measure until 1909, when the city of Chicago set forth the first U.S. law requiring pasteurization of milk.

Today, of course, few consumers even question the wisdom of drinking pasteurized milk. Still, according to many experts, new technologies such as irradiation have prompted a public outcry similar to the turn-of-the-century resistance to pasteurization.[26] As public health organizations strive to feed a fast-growing world population while ensuring a safe food supply, debate about new ways of doing so is sure to heat up. The following sections explore some of the controversial food technologies under consideration as alternatives to help improve the safety of our food supply and to maintain the nutritional value of the foods available in the marketplace.

Irradiation

IRRADIATION the process of exposing a substance to low doses of radiation, using gamma rays, X-rays, or electricity (electron beams) to kill insects, bacteria, and other potentially harmful microorganisms.

Irradiation ranks as one of the foremost technologies earmarked by food safety experts for increased future use in the United States. The process involves exposing food to low doses of radiation, which destroys insects and several types of bacteria, including *Salmonella*. Contrary to popular belief, irradiation does not make a food radioactive. Rather, the radiation rays pass through food, leaving behind no radioactive residues.

Aside from the misunderstanding that irradiation makes food radioactive, many of the criticisms about the process center around substances called **unique**

radiolytic products—compounds that are not present in food naturally or after conventional processing. Critics charge that these products may pose health hazards to consumers who eat irradiated food. According to a comprehensive report from the World Health Organization, however, concern about unique radiolytic products is "probably unfounded," because many of the substances found in irradiated foods are similar, if not identical, to compounds found in other foods. Even those that appear to be unique may not have been detected in conventionally processed foods because those items haven't undergone the same kind of extensive testing as have irradiated foods.[27]

UNIQUE RADIOLYTIC PRODUCTS substances unique to irradiated food and apparently created during the process of irradiation.

The FDA requires that the labels of all irradiated foods carry this internationally known radura symbol for irradiation. The circle in the middle represents an energy source; the five breaks in the outer circle symbolize rays generated by the energy source; and the two petals signify food. For unpackaged irradiated meat, the statement and logo must be displayed at the point of sale for consumers. The labeling requirements do not pertain to foodservice establishments such as restaurants.

Another common concern about irradiation is that it might alter the nutritional value of a food. Again, the World Health Organization, along with numerous other public health agencies, points out that nutrient losses, if any, are insignificant.

To be sure, that's not to say that irradiation poses no risks at all. One of the most troublesome is that the radioactive materials used in the irradiation process may put workers and communities at undue risk. However, as with any technology, the risk of irradiation must be weighed carefully against the benefits and risks of alternative technologies. Consider spices, which typically harbor high levels of pathogens. Often, manufacturers douse them with a toxic, explosive chemical called ethylene oxide to rid them of bacteria. Yet because of its toxic, explosive nature, ethylene oxide puts workers at risk, may pollute the air, and may leave behind residue on the spices. In fact, the gas, which was once used to sterilize medical supplies, has been deemed so dangerous that most manufacturers now use irradiation instead. (Cotton swabs, tampons, teething rings, and a number of other consumer goods are also sterilized via irradiation.)[28] Recently, USDA authorized irradiation of meat already inspected by the agency and approved as safe for consumption. As a result, food companies now have the option of using irradiation on raw meat and meat products (for example, ground beef, frozen hamburger patties, or frozen poultry).

The example of this alternative technology highlights the complexity of the issues surrounding irradiation. Although there are no easy answers about whether this or any other new technology will be best over the long run, it's crucial that both consumers and scientists look at all the angles before forming opinions one way or another. The World Health Organization, the Food and Agriculture Organization of the United Nations, the Food and Drug Administration, and numerous other agencies encourage the use of irradiation in the fight against foodborne disease and food loss, and some 35 countries have approved it for use. According to WHO, each year spoilage, insect infestation, and the like lead to losses of as much as 50 percent of the world's food supply, losses that could be eliminated with irradiation. What's more, the process may help prevent deaths resulting from foodborne illness. In the United States, for example, where incidence of *Salmonella* on poultry continues to rise and strains of bacteria such as *E. coli* O157:H7 keep increasing, irradiation might help to prevent thousands of deaths and illness caused by foodborne disease. Clearly, as the pressures to feed the world safely continue to mount, the public and scientific community will continue to examine the pros and cons of irradiation.

Genetic Engineering

In May 1994, a new breed of tomato made headlines. Called the Flavr Savr, the product garnered national repute as the first food created via **genetic engineering** to hit the market. It wasn't so much the product itself, which is simply a slow-ripening tomato, that caused such a stir, but rather the opening of the regulatory door that can pave the way to a whole new crop of genetically engineered products. What are these high-tech foods, and will we see more of them in the future?

GENETIC ENGINEERING the process of altering the genes of a plant in an effort to create a new plant with different traits. This process of recombining genes is also known as recombinant DNA; a form of biotechnology.

Just as a person's genes determine the person's hair color, eye color, and other characteristics, a plant's genes dictate the plant's structure, resistance to spoilage, and other qualities. With genetic engineering, scientists can alter a plant's genes

Farmers bred corn as we know it today from this wild corn.

in an effort to make a particular trait more desirable. To create the Flavr Savr tomato, scientists identified the gene that causes ripe tomatoes to soften and rot. They reversed the gene—or turned it backwards, so to speak—and then inserted the reversed gene back into tomato plants. The backward gene in the new tomatoes suppresses, or "turns off," the rotting gene. As a result, the Flavr Savr tomato stays ripe longer than regular tomatoes (see Figure 11-4). This allows growers more time to ship it to the consumer without refrigeration.[29]

Although the idea of altering a plant's genes to create a different version sounds new, scientists have been using a cruder version of the concept for centuries. In the 1500s, farmers crossed, say, a good food crop with another plant resistant to disease to form a hybrid that contained traits of both plants. Many of the fruits and vegetables we enjoy today were produced by this sort of crossbreeding. For example, corn as we know it came from breeding corn with a much harder outer shell on the kernel; kiwi came from a small, hard berry; and nectarines came from modified peaches.[30] What makes genetic engineering different, however, is the speed and precision it affords scientists. Whereas crossbreeding typically takes more than a decade, genetic engineering can yield new products in about five years.

TRANSGENETICS the process of transferring genes from one species to another unrelated species.

BIOTECHNOLOGY the science that alters the composition and characteristics of biological systems, organisms, or foods by manipulating their genetic makeup.

Another distinction is that genetic engineering allows scientists to "mix and match" genes from different species—say, mixing a gene from a fish with the genes of a vegetable. This type of research—called **transgenetics**—raises many unanswered ethical questions, especially among vegetarians and other people who fear that inserting an animal gene into a fruit or vegetable makes a food that is on some level both vegetable and animal. The possibilities created by this technology are seemingly endless. It has been suggested that ultimately, the world will obtain most of its food, fuel, fiber, and some of its pharmaceuticals from genetically altered vegetation and trees.[31] Major companies are spending billions of dollars annually on **biotechnology** as researchers continue to develop new applications for genetically engineered products.

The U.S. Department of Agriculture reports that, since 1987, about 40 new genetically engineered agricultural products have completed all the federal regulatory requirements and are sold commercially. Recently released government data on acreage of biotechnology-derived crops indicates that genetically engineered soybean, cotton, and corn account for 20 percent to 44 percent of total acreage planted in 1998.[32] The majority of these plants have been engineered with genes from bacteria that destroy pests or protect plants against herbicides. The resistant plants created through genetic engineering lower the farmers' dependence on the use of pesticides and weed killers while increasing overall production.

Genetic engineering is used to boost the nutritional value of foods. Advances in biotechnology can increase levels of natural health-enhancing substances in plants knowns as *phytochemicals*. These compounds, many of which are found in soybeans, tomatoes, garlic and other plant foods, have been shown to protect against cardiovascular disease, cancer, and free radical damage to tissues, as discussed earlier in Chapter 6. One such compound, beta-carotene, belongs to a class of compounds known as carotenoids. In 1998, the USDA scientists released three new tomato breeding lines that contain about 10 to 25 times more beta-carotene than typical tomatoes.[33] Researchers are close to succeeding at manipulating a gene that will improve the quality of the proteins made by sweet potato, an important food crop in poorer countries where high-quality protein foods are at a shortage. Early study results show the protein content of the genetically engineered potatoes increased as much as fivefold.[34]

FIGURE 11-4
CREATING THE FLAVR SAVR TOMATO

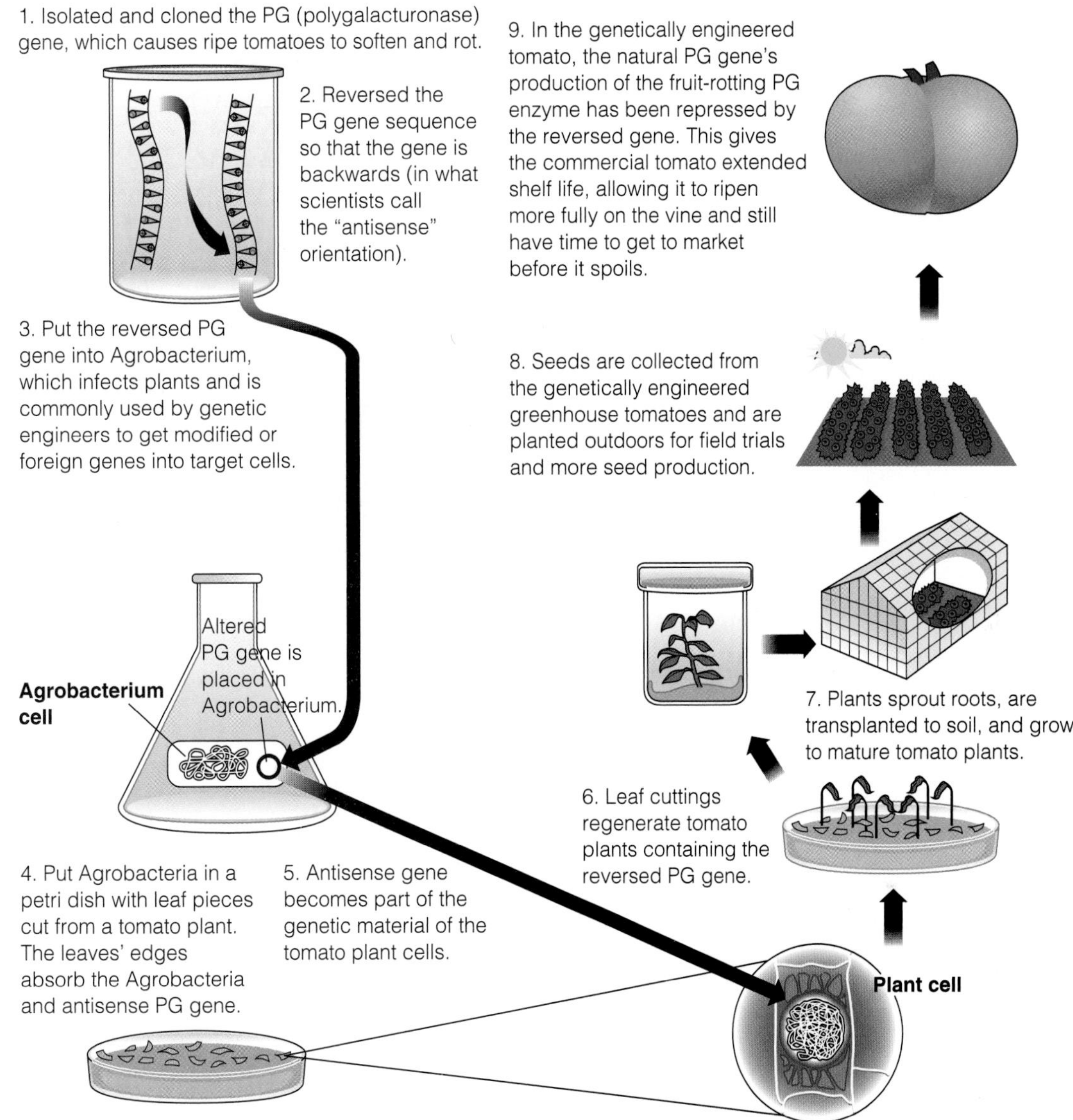

1. Isolated and cloned the PG (polygalacturonase) gene, which causes ripe tomatoes to soften and rot.
2. Reversed the PG gene sequence so that the gene is backwards (in what scientists call the "antisense" orientation).
3. Put the reversed PG gene into Agrobacterium, which infects plants and is commonly used by genetic engineers to get modified or foreign genes into target cells.
Agrobacterium cell
Altered PG gene is placed in Agrobacterium.
4. Put Agrobacteria in a petri dish with leaf pieces cut from a tomato plant. The leaves' edges absorb the Agrobacteria and antisense PG gene.
5. Antisense gene becomes part of the genetic material of the tomato plant cells.
Plant cell
6. Leaf cuttings regenerate tomato plants containing the reversed PG gene.
7. Plants sprout roots, are transplanted to soil, and grow to mature tomato plants.
8. Seeds are collected from the genetically engineered greenhouse tomatoes and are planted outdoors for field trials and more seed production.
9. In the genetically engineered tomato, the natural PG gene's production of the fruit-rotting PG enzyme has been repressed by the reversed gene. This gives the commercial tomato extended shelf life, allowing it to ripen more fully on the vine and still have time to get to market before it spoils.

Source: Reprinted with permission from Calgene's recipe for genetically engineered tomatoes, *FDA Consumer,* April 1995, p. 9.

In development are plants that will produce edible vaccines. This could be of value in developing countries where some environmental conditions can make conventional vaccine administration impractical. Also being investigated is an edible vaccine to protect against diarrhea, a major cause of infant mortality in developing countries.[35]

The introduction of genetically engineered food has not been without problems. For some, the concern is the potential impact of an organism that would not have been created under normal conditions. The long-term safety of the environment as the genetically-engineered plants multiply and mutate cannot be known. There is fear that naturally occurring cross-pollination between genetically engineered plants with nearby weeds may spread traits from plants to weeds. The so-called "superweed" would be resistant to insects and herbicides.

People with food allergies express concern that new varieties of food produced by transgenetics may introduce allergens not found in the food before it was

"Our turnips look like carrots, our potatoes taste like apples, our green peas are blue . . . stuff like that."

altered. Indeed, researchers have shown that allergens can be transferred through bioengineering, as in the case where an allergic protein showed up in soybeans that had been genetically altered with proteins from Brazil nuts.[36]

Another widely debated issue surrounding genetic engineering involves labeling. Many consumer groups have called for across-the-board labels on genetically engineered foods. But according to the FDA, with some exceptions, the food must carry distinct consequences to consumers who eat it before it must bear special labeling. When genes from peanuts and other foods known to be common causes of allergies are put into a food, the label must indicate that the food contains an allergen unless the manufacturer can prove that the item's potential to cause allergies has not been transferred via the gene. In addition, a food that has been genetically engineered to significantly change, say, its fiber content or nutrient composition must bear a label that states the nature of the change.[37] A number of petitions currently calling for labeling of foods with genetically engineered ingredients may set in motion the steps leading to FDA approval of food labeling for these foods in the future.[38]

Despite the concerns, many scientists are hopeful that careful use of genetic engineering will confer long-term benefits. For instance, the development of insect- and disease-resistant plants may allow farmers to grow crops with fewer chemicals. The possibilities are enormous, and each new product considered for entry into the marketplace will require careful scrutiny.

Spotlight
DOMESTIC AND WORLD HUNGER

This book has focused on the problems of *overnutrition*—obesity, heart disease, cancer, and others—diseases of economically developed nations. People in developing nations as well as people in the less privileged parts of developed nations suffer from problems of **undernutrition,** which is characterized by chronic debilitating hunger and malnutrition. These conditions are most visible in times of **famine,** but they are widespread and persistent even when famine does not occur.

All people need food. Regardless of our race, religion, gender, or nationality, our bodies experience similarly the effects of hunger and its companion **malnutrition**—listlessness, weakness, failure to thrive, stunted growth, mental retardation, muscle wastage, scurvy, pellagra, beriberi, anemia, rickets, osteoporosis, goiter, tooth decay, blindness, and a host of other effects, including death.[39] Apathy and shortened attention span are two of a number of behavioral symptoms often mistaken for laziness, lack of intelligence, or mental illness in undernourished people.

How are hunger and poverty related?

The phenomenon of *hunger* is today being discussed in terms of **food security** or **food insecurity.** Food security is defined as access by all people at all times to enough food for an active, healthy life and at a minimum includes the following: (1) the ready availability of nutritionally adequate and safe foods and (2) the ability to acquire personally acceptable foods in a socially acceptable way.[40] Food insecurity was once viewed as a problem of overpopulation and inadequate food production, but now many people recognize it as a problem of **poverty.** Food is *available* but is not *accessible* to the poor, who have neither land nor money. Poverty is much more than an economic condition and exists for many reasons, including overpopulation, greed, unemployment, and the lack of productive resources such as land, tools, and credit. Consequently, if we are to provide adequate nutrition for all the earth's hungry people, we must transform the economic, political, and social structures that both limit food production, distribution, and consumption and create a gap between rich and poor.

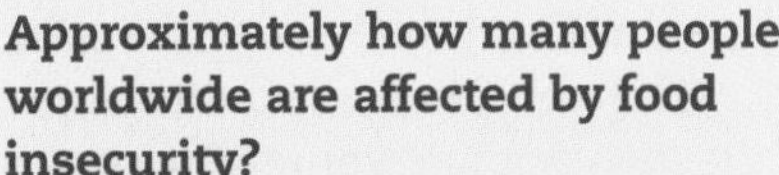

Approximately how many people worldwide are affected by food insecurity?

The Food and Agriculture Organization (FAO) estimates that of the more than 6 billion people in the world, at least 790 million—one in seven people worldwide—suffer from chronic, severe undernutrition, consuming too little food each day to meet even minimum energy requirements.[41] Protein energy malnutrition (PEM) contributes to almost 13 million childhood deaths yearly.[42] Worldwide, three micronutrient deficiencies are of particular concern: vitamin A deficiency, the world's most common cause of preventable child blindness and vision impairment; iron deficiency anemia; and iodine deficiency, causing high levels of goiter and child retardation:[43]

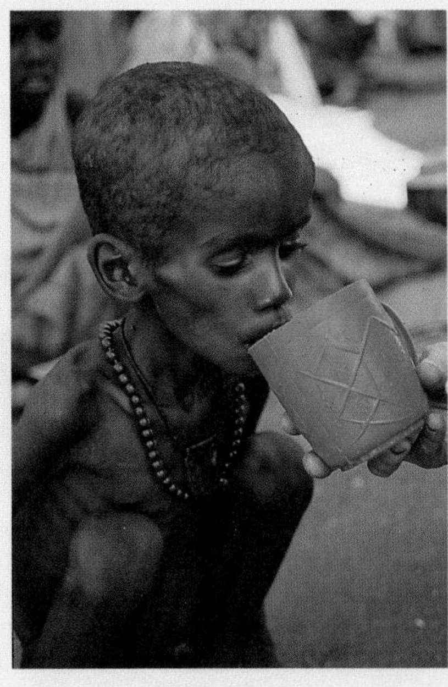

Feeding the hungry—in Somalia.

- *Vitamin A deficiency.* Some 13 million children below six years of age have **xerophthalmia.** Of these, an estimated 500,000 children become partially or totally blinded, as a result of an insufficiency of vitamin A in the diet. Vitamin A deficiency is also associated with other forms of malnutrition, infection, diarrhea, and a high rate of mortality.
- *Iron deficiency.* Iron deficiency anemia is estimated to affect some 2.1 billion people. In infancy and early childhood, iron deficiency is associated with decreased cognitive abilities and resistance to disease.
- *Iodine deficiency.* Iodine deficiency, the major preventable cause of mental retardation worldwide, is a risk factor for both physical and mental retardation in about one billion people. About 655 million people worldwide—especially in mountainous regions—are estimated to have goiter, and over

Infants can be the first to show the signs of undernutrition due to their high nutrient needs. No famine, no flood, no earthquake, no war has ever claimed the lives of 250,000 children in a single week. Yet malnutrition and disease claim that number of child victims every week.

three million suffer overt cretinism.

Worldwide, about 40,000 to 50,000 people die each day as a result of undernutrition. Millions of children die each year from the diseases of poverty: parasitic and infectious diseases such as dysentery, whooping cough, measles, tuberculosis, cholera, and malaria. These diseases interact with poor nutrition to form a vicious cycle in which the outcome for many is death.[44]

When does the risk for undernutrition run high?

Undernutrition runs high when nutrient needs are high, as in times of rapid growth. If family food is limited, pregnant and lactating women, infants, and children are the first to show the signs of undernutrition. Effects of food insecurity can be devastating to this group of the population.

During pregnancy, healthy women in developed countries gain an average of about 27 pounds, while low-income women show a weight gain of only 11 to 15 pounds. Babies of these women tend to have low birthweights (less than 5½ pounds). A low-birthweight baby has a greater than normal chance of having physical and mental birth defects, of contracting diseases, and of dying early in life. Low birthweight contributes to more than half of the deaths worldwide of children under five years of age. UNICEF refers to the **under-5 mortality rate** (U5MR) as the single best indicator of children's overall health and well-being.[45] UNICEF argues that the U5MR reflects a country's overall resources directed at children:

> *. . . the U5MR reflects the nutritional health and the health knowledge of mothers; the level of immunization and use of* ***oral rehydration therapy;*** *the availability of maternal and health services (including prenatal care); income and food availability in the family; the availability of clean water and safe sanitation; and the overall safety of the child's environment.*[46]

Until the middle of the twentieth century in most of the developing countries, babies were breastfed for their first year of life, with supplements of other milk and cereal gruel added to their diets after the first several months. Today, the percentage of infants who are exclusively breastfed to the age of four months has dropped to below 10 percent.[47] A number of factors contributed to this unfortunate decline, including the aggressive promotion and sale of infant formula to new mothers; the encouragement by health-care practitioners for mothers to bottle feed (with free samples sent home from the hospital after delivery of the newborn); and the global pattern of urbanization and accompanying loss of cultural ties supporting breastfeeding, combined with more women working outside the home.[48] Overall, WHO estimates that more than a million children's lives could be saved each year if all mothers gave their babies nothing but breast milk for the first four to six months of life.[49]

Breastfeeding permits infants in many developing countries to achieve weight and height gains equal to those of children in developed countries until about six months of age. In fact, replacing breast milk with infant formula in environments and economic circumstances that make it impossible to feed formula safely may lead to infant undernutrition. In the absence of sterilization and refrigeration, formula in bottles is an ideal breeding ground for bacteria. Feeding infants formula made with contaminated water often causes infections leading to diarrhea, dehydration, and inability to absorb nutrients.

Feeding the hungry—in the United States.

Even if infants are protected by breastfeeding at first, they must eventually be weaned. The weaning period is one of the most dangerous periods for children in developing countries. Newly weaned infants often receive nutrient-poor diluted cereals or starchy root crops, and the infants' foods are often prepared with contaminated water, making infection almost inevitable.

What is the status of food insecurity in the United States?

In the late 1990s, more than one in ten households experienced food insecurity. This represents more than 34 million people. Food insecurity rates are higher than average in female-headed households, households with children, especially black and Hispanic households, and households in inner-city areas.[50] A recent analysis of 26 state hunger surveys found consistent results across all surveys and evidence for several broad conclusions:[51]

- Food insecurity has become a chronic problem in the United States.
- Food insecurity is not due to food shortages. Hunger results from unequal distribution of economic resources—poverty.
- People who lack access to a variety of resources—not just food—are most at risk of hunger. When income is inadequate to meet the costs of housing, utilities, health care, and other fixed expenses, these items compete with and may take precedence over food.
- Private charity cannot solve the food insecurity problem. Voluntary activities are limited in expertise, time, and resources and are likely to require government support in order to continue.

Who are the hungry in the United States?

Through the late 1960s and 1970s, hunger was evident among the chronic poor: migrant workers, Native Americans, southern blacks, unemployed minorities, and some of the elderly as well as the newly unemployed blue-collar workers during the 1970s. Now, despite the strong economy of the 1990s, hunger is reaching into other segments of the population without regard for age, marital status, previous employment or successes, family ties, or efforts to change the situation. The millions who experience hunger today in the United States include the young, some of the elderly, the homeless, low-income women, ethnic minorities, and the new "working" poor—displaced farm families, and blue-collar and white-collar workers forced into minimum wage jobs.

What are some of the causes of food insecurity in the United States?

The most compelling single reason is poverty. Poverty and food insecurity are interdependent. Nutrition surveys investigating people's nutritional health in the United States have demonstrated consistently that the lower a family's income, the less adequate the family's nutrition status.

Although poverty is the major cause of food insecurity in the United States, other problems contribute as well, including alcoholism and chronic substance abuse, mental illness, loneliness, isolation, depression, and despair; the reluctance of people to accept what they perceive as "welfare" or "charity"; delays in receiving public assistance benefits; an increase in the number of single mothers without the means to care for their children; poor management of limited family financial resources; health problems of old age; lack of nutritional adequacy and balance in the food available to hungry people; lack of access to assistance programs; insufficient community food resources for the hungry; and insufficient community transportation systems to deliver food to hungry people who have no transportation.

As a result of the evidence accumulated during the 1960s and 1970s showing that food insecurity was a problem in the United States, the problems of poverty and hunger became national priorities. Old programs were revised and new programs were developed in an attempt to prevent malnutrition in those people found to be at greatest risk. The Food Stamp Program was expanded to serve more people. School lunch and breakfast programs were enlarged to support children nutritionally while they learned. Feeding programs to reach senior citizens were started. To provide food and nutrition education during the years when nutrition has the most crucial impact on growth, development, and future health, a supplemental food and nutrition program (WIC Program) was established for pregnant and breastfeeding women, infants, and children who were of low income and were nutritionally at risk. The result of these efforts was that food insecurity diminished as a serious problem for this country.[52] Now, however, food insecurity is increasing as a result of rising poverty and cuts in government aid.

What are people doing to help reduce problems of food insecurity?

The terms **food recovery** and **gleaning** refer to programs that collect excess wholesome food for delivery to hungry people.* Approximately 96 billion pounds—or over one-quarter of the 356 billion pounds of food produced in this country for human consumption—are lost at the retail and food service levels. In an effort to reduce food wastage, the 1997 National Summit on Food Recovery and Gleaning set a goal of providing an additional 500 million pounds of food a year to feeding organizations.

The Food Recovery and Gleaning Initiative is intended to supplement existing federal nutrition assistance programs. Currently, USDA is working with the National Restaurant Association to produce a food recovery handbook for its members, enabling schools to donate excess food from the School Lunch Program, encouraging airlines to donate unserved meals, working with the Department of Transportation to develop a comprehensive way to transport recovered foods, facilitating the donation of excess food from the Department of Defense, and providing technical assistance to community-based groups who seek to help.

Additionally, concerned citizens are working through community programs and churches to provide meals to the hungry. **Second Harvest,** the nation's largest supplier of surplus food, distributed over 900 million pounds of food to nearly 200 **food banks** and some 50,000 agencies for direct distribution around the nation in 1998.[53]

A community food bank can benefit hundreds of thousands of individuals through its partnerships with industry and the community:

- Industry donors contact the food bank when they have products to contribute: goods that are mislabeled, in damaged packages, underweight, close to expiration, or in oversupply.
- The food bank picks up large-scale donations, cases to trailer-loads, from the plant, distribution center, public warehouse, or retailer.
- At the food bank, nonprofit, charitable groups select products to stock their emergency food pantries, soup kitchens, shelters, child care centers, senior citizen programs, rehabilitation centers, and summer camps.
- Donated food reaches people in need: children and the elderly, the unemployed and low-wage earners, the homeless and frail homebound, the disabled and ill.

However, even the dramatic increases in the number of food banks, **food pantries, soup kitchens, prepared and perishable food programs,** and other emergency food assistance programs across the nation cannot keep pace with the growth in the number of hungry people seeking food assistance. Each day's worth of meals lasts only for that day, leaving the problem of poverty unsolved, as before. Moreover, one out of every five needy people is not even receiving meals. These people are left to scavenge garbage, to steal food or money to buy food, or to continue to starve.

Exactly how many people in the United States are homeless, and who are the homeless?

Estimates of the number of homeless vary and range from 600,000 to 3 million. In addition, nearly 3 million people spend more than 70 percent of their income on rent and are at risk of becoming homeless.[54] The U.S. Conference of Mayors surveyed 26 major cities to assess the status of hunger and homelessness in the urban United States during 1999. Lack of food, inadequate diets, poor nutritional status, and nutrition-related health problems—stunted growth, failure to thrive, low-birthweight babies, infant mortality, anemia, and compromised immune systems—are common among homeless persons.[55]

The lack of affordable housing leads the list of causes of homelessness identified by the mayors. Without adequate low-income housing, many poor people are forced to choose between shelter and food. Other causes of homelessness include unemployment, underemployment, poverty, inadequate public assistance benefit levels, the high cost of health care, substance

*The Good Samaritan Food Donation Act of 1996 encourages the donation of food and grocery products to nonprofit organizations such as homeless shelters, soup kitchens, and churches for distribution to needy individuals. The law provides uniform national protection to citizens, businesses, and nonprofit groups that in good faith donate, recover, and distribute excess food. The law limits the liability of donors to instances of gross negligence or intentional misconduct.

abuse and lack of needed services, and mental illness and the lack of needed services.

How are farm families affected by hunger?
Changes in the domestic economy are adverse not only on the receiving end but also on the producing end. U.S. farmers today lack significant control over what products they produce, what prices they must pay for supplies, and what prices they receive in return for their commodities. Just before 1980, the USDA urged farmers to increase corn and soybean production for export. To comply, farmers borrowed heavily to expand their production capabilities. Since that time, the costs to farmers for seed, fertilizer, equipment, and loans have steadily risen while crop prices have declined. Today, thousands of U.S. farmers are hungry, frustrated, and at the point of desperation concerning their debt.[56]

The number of hungry farm families is not known, but agencies that provide aid to the rural poor say the demand for food assistance is increasing. Ironically, farm families do not generally grow fruits, vegetables, and other crops and animals to feed themselves. With modern practices aimed at efficiency, most farmers raise two or three crops—for example, feed corn, sorghum, and wheat—and buy most of the food they eat themselves from the grocery store. As a result, when crop prices drop, farmers struggle to survive under the sagging prices and realize no significant profits. Eventually, the farmers go out of business entirely from lack of profits, just as would happen in any other type of business in the United States.

How does world hunger differ from hunger in the United States?
World hunger is more extreme than domestic hunger. In fact, most people would find it hard to imagine the severity of poverty in the developing world:

> *Many hundreds of millions of people in the poorest countries are preoccupied solely with survival and elementary needs. For them, work is frequently not available, or pay is low, and conditions barely tolerable. Homes are constructed of impermanent materials and have neither piped water nor sanitation. Electricity is a luxury. Health services are thinly spread, and in rural areas only rarely within walking distance. Permanent insecurity is the condition of the poor . . . in the wealthy countries, ordinary men and women face genuine economic problems. . . . But they rarely face anything resembling the total deprivation found in the poor countries.*[57]

World hunger is a problem of supply and demand, of inappropriate technology, of environmental abuse, of demographic distribution, of unequal access to resources, of extremes in dietary patterns, and of unjust economic systems. Oftentimes, people who are poor are powerless to change their situation because they have less access to vital resources such as education, training, food, and health services.

How are international trade and debt connected to hunger?
Over the years, developing countries have seen the prices of imported fuels and manufactured items rise much faster than the prices they receive for their export goods (such as bananas, coffee, and various raw materials) on the international market. The combination of high import costs with low export profits often pushes a developing country into accelerating international debt that sometimes leads to bankruptcy.

Debt and trade are closely related to the progress a country can make toward achieving an adequate diet for its people. As import prices increase relative to export prices, more of a country's total money base moves abroad to pay for the imports. With more and more of its money abroad, the country is forced to borrow money, usually at high interest rates, to continue functioning at home. Many of its financial resources must then go to pay the interest on the borrowed money, thus draining the economy further. Creditor nations may not demand much, or any, capital back, but they do require that interest be paid each year, and the interest can consume most of a country's gross national product. Large and growing debts can slow or halt a nation's attempt to deal effectively with its problems of local food insecurity. As more and more of its financial resources are being used to pay off interest on the country's trade debts, less and less money is available to deal with food insecurity at home. Each year, the debt crisis worsens and leads to further problems with hunger.

What about the role of multinational corporations in this issue?
Typically, large landowners and **multinational corporations** hire indigenous people for below-subsistence wages to work in the fertile farmlands growing crops to be exported for profit, leaving little fertile land for the local farmers to use to grow food. The local people work hard cultivating cash crops for others, not food crops for themselves. The money they earn is not even enough to buy the products they help to produce. They do not adequately share in the profits realized from the marketing of products grown with their labor. The results: imported foods—bananas, beef, cocoa, coconuts, coffee, pineapples, sugar, tea, winter tomatoes, and others—fill the grocery stores of developed countries, while the poor who labored to grow these foods have less food and fewer resources than when they farmed the land for their own use. Additional cropland is diverted for nonfood, cash crops—tobacco, rubber, cotton, and other agricultural products. These practices have also had an adverse

MINIGLOSSARY

APPROPRIATE TECHNOLOGY a technology that utilizes locally abundant resources in preference to locally scarce resources. Developing countries usually have a large labor force and little capital; the appropriate technology would therefore be labor intensive.

FAMINE widespread lack of access to food caused by natural disasters, political factors, or war; characterized by a large number of deaths due to starvation and malnutrition.

FOOD BANKS nonprofit community organizations that collect surplus commodities from the government and edible but often unmarketable foods from private industry for use by nonprofit charities, institutions, and feeding programs at nominal cost.

FOOD INSECURITY the inability to acquire or consume an adequate quality or sufficient quantity of food in socially acceptable ways, or the uncertainty that one will be able to do so.

FOOD PANTRIES centers usually attached to existing nonprofit agencies that distribute bags or boxes of groceries to people experiencing food emergencies. Foods distributed by pantries are prepared and consumed elsewhere. A referral or proof of need is often required. There are roughly two food pantries to every soup kitchen.

FOOD RECOVERY such activities as salvaging perishable produce from grocery stores and wholesale food markets; rescuing surplus prepared food from restaurants, corporate cafeterias, and caterers; and collecting non-perishable, canned or boxed processed food from manufacturers, supermarkets, or people's homes. The items recovered are donated to hungry people.

FOOD SECURITY access by all people at all times to enough food for an active and healthy life. Food security has two aspects: ensuring that adequate food supplies are available and ensuring that households whose members suffer from undernutrition have the ability to acquire food, either by producing it themselves or by being able to purchase it.

GLEANING the harvesting of excess food from farms, orchards, and packing houses to feed the hungry.

GOBI an acronym formed from the elements of UNICEF's Child Survival campaign—**G**rowth charts, **O**ral rehydration therapy, **B**reast milk, and **I**mmunization.

MALNUTRITION the impairment of health resulting from a relative deficiency or excess of food energy and specific nutrients necessary for health.

MULTINATIONAL CORPORATIONS international companies with direct investments and/or operative facilities in more than one country. U.S. oil and food companies are examples.

ORAL REHYDRATION THERAPY (ORT) the treatment of dehydration (usually due to diarrhea caused by infectious disease) with an oral solution; ORT as developed by UNICEF is intended to enable a mother to mix a simple solution for her child from substances that she has at home.

POVERTY the state of having too little money to meet minimum needs for food, clothing, and shelter. The U.S. Department of Health and Human Services defines the poverty level in the United States as an annual income of $16,450 for a family of four.

PREPARED AND PERISHABLE FOOD PROGRAMS (PPFPs) nonprofit programs that help to feed people in need by linking sources of unused, unserved cooked and fresh food—such as caterers, restaurants, hotel kitchens, and cafeterias—with social service agencies that serve meals to people who would otherwise go hungry. *Foodchain* is the national network of over 125 community-based PPFPs in 41 states and Canada.

SECOND HARVEST a national food banking network to which the majority of food banks belong.

SOUP KITCHEN small feeding operations attached to existing organizations such as churches, civic groups, or nonprofit agencies that serve prepared meals that are consumed on site. Soup kitchens generally do not require clients to prove need or show identification.

UNDER-5-MORTALITY RATE (U5MR) the number of children who die before the age of five for every 1,000 live births.

UNDERNUTRITION (also called **HUNGER**) as used in this discussion, a term that describes the domestic and world food problem of a continuous lack of the food energy and nutrients necessary to achieve and maintain health and protection from disease.

UNICEF the United Nations International Children's Emergency Fund, now referred to as the United Nations Children's Fund.

XEROPHTHALMIA (ZEER-ahf-THALL-me-uh) a severe form of eye disease in which the cornea hardens and may cause blindness. The problem results from vitamin A deficiency.

effect on the financial status of many U.S. farmers. The foreign cash crops often undersell the same U.S.-grown produce. The U.S. farmer cannot compete against these lower-priced imported foods and may be forced out of business.

Besides diverting acreage away from the traditional staples of the local diet, some multinational corporations may also contribute to hunger as a result of their marketing techniques. Their advertisements lead many consumers with limited incomes to associate such products as cola beverages, cigarettes, infant formulas, and snack foods with good health and prosperity. Such promotions are tragically inappropriate for these people. A poor family's nutrition status suffers when its tight budget is pinched further by the purchase of such unnecessary goods.

How does overpopulation fit into the picture?

The current world population is approximately 6 billion, and for the year 2050, the projected United Nations figure is approximately 9 billion. The earth may not be able to adequately support this many people. The world's present population is certainly of concern, as is the projected increase in that population. As important as the population question is, it is only one cause of the world food problem. Poverty seems to be at the root of both problems—hunger and overpopulation.

Three major factors affect population growth: birthrates, death rates, and standards of living. Low-income countries have high birthrates, high death rates, and low standards of living.

When people's standard of living rises, giving them better access to health care, family planning, and education, the death rate falls. In time, the birth rate also falls. As the standard of living continues to improve, the family earns sufficient income to risk having smaller num-

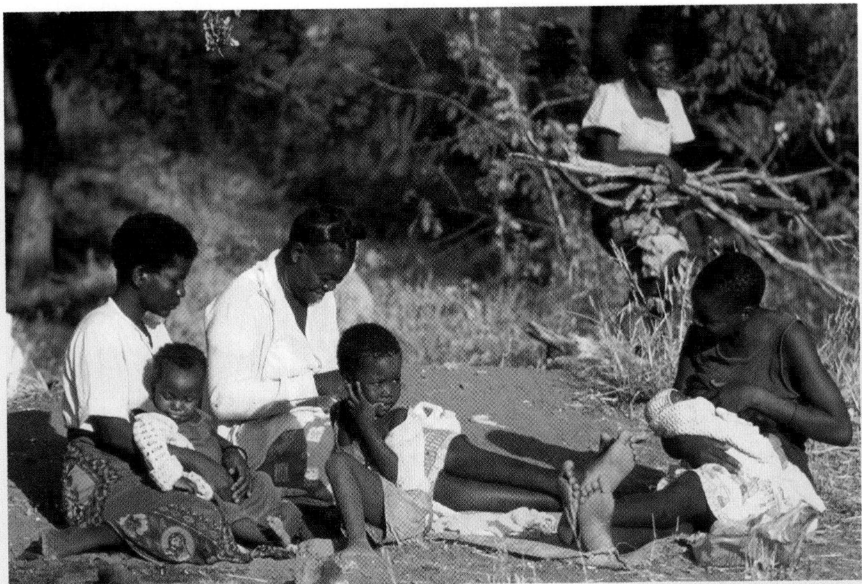

Vulnerability comes with age: Infants and children are disproportionately vulnerable to hunger. Breastfeeding offers the infant a safe and adequate supply of nutrients.

bers of children. A family depends on its children to cultivate the land, secure food and water, and provide for the adults in their old age. Under conditions of ongoing poverty, parents will choose to have many children to ensure that some will survive to adulthood. Children represent the "social security" of the poor. Improvements in economic status help relieve the need for this "insurance" and so help reduce the birth rate. The relationships between the infant mortality rate and the population growth rate reveal that hunger and poverty reflect both the level of national development and the people's sense of security.

As the world's population continues to grow, it threatens the world's capacity to produce adequate food. The activity of billions of human beings on the earth's limited surface is seriously and adversely affecting our planet: wiping out many of the varieties of plant life, heating up our climate, using up our freshwater supplies, and destroying the protective ozone layer that shields life from the sun's damaging rays—in short, overstraining the earth's ability to support life. Population control is one of the most pressing needs of this time in history.

Is there a better way to distribute world resources?

Land reform—giving people a meaningful opportunity to produce food for local consumption for example—can combine with population control to increase everyone's assets. Poor nations must be allowed to increase their agricultural productivity. Much is involved, but to put it simply, poor nations must gain greater access to five things simultaneously: land, capital, water, technology, and knowledge.[58] Equally important, each nation must adopt the political priority of improving the conditions of all its people. International food aid may be required temporarily during the development period, but eventually this aid will become less and less necessary.

Governments can learn from recent history the importance of developing local agricultural technology. A major effort made in the 1960s and 1970s—the green revolution—demonstrated the potential for increased grain production in Asia. It was an effort to bring the agricultural technology of the industrial world to the developing countries, but the high-yielding strains of wheat and rice that were selected required irrigation, chemical fertilizers, and pesticides—all costly and beyond the economic means of too many of the farmers in the developing world.

Instead of transplanting industrial technology into the developing countries, small, efficient farms and local structures for marketing, credit, transportation, food storage, and agricultural education should be developed. International research centers need to examine the conditions of tropical countries and orient their research toward **appropriate technology**—labor-intensive rather than energy-intensive agricultural methods. For example, labor-intensive technology, such as the use of manual grinders for grains, is appropriate in some places because it makes the best use of human, financial, and natural resources. A manual grinder can process 20 pounds of grain per hour, replacing the mortar and pestle, which in the same time can pound a maximum of only 3 pounds.[59] The specific technology that is appropriate for use varies from situation to situation.

The cornerstone of true development was best expressed decades ago by Mahatma Gandhi: "Whenever you are in doubt . . . apply the following test. Recall the face of the poorest and the weakest man whom you may have seen, and ask yourself if the step you contemplate is going to be of any use to him. Will he gain anything by it? Will it restore him to a control over his own life and destiny?"[60]

Environmental concerns must be taken more seriously as well. As important as the amount of land available for crop production is the condition of the soil and the availability of water. Soil erosion is now accelerating on every continent at a rate that threatens the world's ability to continue feeding itself.[61] Erosion of soil has always occurred; it is a natural process. But in the past, it has been compensated for by processes that build the soil up, such as the growth of trees.

Where forest has already been converted to farmland and there are no trees, farmers should alternate soil-devouring crops with soil-building crops, a practice known as crop rotation. When farmers must choose whether to make three times as much money planting corn year after year or to rotate crops and possibly go bankrupt, many choose the short-term profits. Ruin may not follow immediately, but it will follow.[62]

Is there hope for a world without hunger for women and children? Women make up 50 percent of the world's population. Any solution to the problems of poverty and hunger is incomplete and even hopeless if it fails to address the role of women in developing countries, for women and their children represent the majority of those living in poverty.

In many countries, over 90 percent of the population live in rural areas. The life of a woman living in rural poverty is oppressive. Typically, these women not only are the primary food producers but also are responsible for child care and food preparation. Often they have to work as harvesters on other people's lands as well. Husbands are frequently required to be absent from their homes, not by choice but because the changing global economy has forced many men to leave home in search of wages.

Women play a vital role in the nutrition of their nation's people. Their nutrition during pregnancy and lactation determines the future health of their children. If women are weakened by malnutrition themselves or are ignorant about how to feed their families, the consequences ripple outward to affect many other individuals. The importance of the role women play in these countries is increasingly appreciated, and many countries now offer development programs with women in mind.

Seven basic strategies are at the heart of women's programs:

- Removing barriers to financial credit.
- Providing access to time-saving technologies.
- Providing appropriate training to promote self-reliance.
- Teaching management and marketing skills.
- Making health and day-care services available.
- Forming women's support groups.[63]
- Providing information and technology to promote planned pregnancies.

The recognition of women's needs by some development organizations is an encouraging trend in the efforts to contend with the world hunger crisis.

There is hopeful news for children in developing countries—the group that is most strongly affected by poverty, malnutrition, and food insecurity and its relationship to the environment.[64] **GOBI,** a child survival plan set forth by UNICEF, has made outstanding progress in cutting the number of hunger-related child deaths. GOBI is an acronym formed from four simple, but profoundly important, elements of UNICEF's Child Survival campaign: **g**rowth charts, **o**ral rehydration therapy (ORT), **b**reast milk, and **i**mmunization.

A mother can learn to weigh her child every month and chart the child's growth on a specially designed paper growth chart. She can learn to detect for herself the early stages of hidden malnutrition that can leave a child irreparably retarded in mind and body. Then at least she can know she needs to take steps to remedy the malnutrition—if she can.

The importance of oral rehydration therapy (ORT) is that most children who die of malnutrition do not starve to death—they die because their health has been compromised by dehydration from infections causing diarrhea. Until recently, there was no easy way of stopping the infection-diarrhea cycle and saving their lives. Now, the spread of ORT is preventing an estimated one million dehydration deaths each year.[65] Oral rehydration therapy is the administration of a simple solution that mothers can make up themselves, using locally available ingredients, which increases a body's ability to absorb fluids 25-fold.[66] International development groups also provide mothers with packets of premeasured salt and sugar to be mixed with water in rural and urban areas. A safe and sanitary supply of drinking water is a prerequisite for the success of the ORT program.

The promotion of breastfeeding among mothers in developing countries has many benefits. Breast milk is hygienic, is readily available, is nutritionally sound, and provides infants with immunologic protection specific for their environment. In the developing world, the advantages of breastfeeding over formula feeding can mean the difference between life and death.

Immunizations (the I of GOBI) could prevent most of the five million deaths each year from measles, diphtheria, tetanus, whooping cough, poliomyelitis, and tuberculosis. Adequate protein nutrition is necessary, however, for vaccinations to be useful so that the protein in the vaccine itself is not used by the body as a source of protein. The immunization achievements of the last two decades are credited with the prevention of approximately three million deaths a year as well as the protection of many millions more from disease, malnutrition, blindness, deafness, and polio.[67]

The first World Summit for Children in history was convened by UNICEF in September of 1990, bringing together representatives of 159 nations for the purpose of making a renewed commitment to ending child deaths and child malnutrition. Significantly, *nutrition* was mentioned for the first time in world history as an internationally recognized human right.[68] An immediate result of this summit has been an increase in the number of governments actively adopting the child survival strategies of UNICEF—universal immunization, oral rehydration therapy, a massive effort to promote breastfeeding as the ideal food for at least the first four to six months of an infant's life, an attack on malnutrition involving nutrition surveillance focusing on growth monitoring and weighing of infants at least once every month for the first 18 months of the child's life, and nutrition and literacy education that will empower women in developing countries and lead to a reduction in nutrition-related diseases among vulnerable children.[69]

What can I do to help alleviate hunger problems?

The problems of hunger can appear so great that they sometimes seem approachable only by way of worldwide political decisions. Indeed, the members of the International Conference on Nutrition stressed that worldwide efforts to overcome hunger and malnutrition and to foster self-reliant development must be intensive. To this end, many individuals and groups are working to improve the chances of the future well-being of the world and its people through a number of national and international organizations.*

Solutions to the hunger problem depend on people's willingness to take action and to work together.[70] Regardless of the type and level of involvement a person chooses, each person can make a difference. Individual people can do any of following:

- Assist in government and community programs as volunteers.
- Help to develop means of informing low-income people of food-related services and programs for which they are eligible.
- Help to increase the accessibility of existing programs and services to those who need them.
- Document the needs that exist in their own communities.
- Join with others in the community who have similar interests.
- Establish a program for collecting and distributing nutritious foods that would otherwise have been wasted.
- Follow current hunger legislation, call and write legislators about hunger issues, lobby to draw political attention to the need for more job opportunities and a higher minimum wage.

Individuals can also help change the world through the personal choices they make each day. Our choices have an impact on the way the rest of the world's people live and die. Our nation, with 6 percent of the world's population, consumes about 40 percent of the world's food and energy resources. People in affluent nations have the freedom and means to choose their lifestyles. We can find ways to reduce our consumption of the world's nonrenewable resources and use only what is absolutely required.

Choosing a diet at the level of necessity rather than excess would reduce the resource demands made by our industrial agriculture. In fact, those who study the future are convinced that the hope of the world lies in everyone's adopting a simple lifestyle. As one such person put it, "the widespread simplification of life is vital to the well-being of the entire human family."[71] Personal lifestyles do matter, for a society is nothing more than the sum of its individuals. As we go, so goes our world.

*Organizations and groups working to end world hunger are listed in Appendix A.

Nutrition on the Web

nutrition.wadsworth.com	Go to the *Personal Nutrition* site to check for the latest updates to chapter topics or to access links to related Web sites.
www.homefoodsafety.org	Home food safety tips plus an interactive quiz to test your food safety knowledge, and useful links to related sites.
www.foodsafety.gov	The National Food Safety Database site provides food safety information, daily news stories, and many useful links.
www.fda.gov/search.html	Search for information on food safety topics.
ificinfo.health.org	Information on irradiation, biotechnology, and other food safety and nutrition issues.
www.safefood.org	Information on safe food handling practices.
www.nal.usda.gov/fnic/foodborne/foodborn.htm	Information from the USDA Foodborne Illness Education Center.
www.cic.info@pueblo.gsa.gov	Resources from the Consumer Information Center.
www.fightbac.org	Information from the Partnership for Food Safety Education.
www.cdc.gov	Food safety information and tips for international travelers.
www.cfsan.fda.gov	Information on food safety from the Center for Food Safety and Applied Nutrition.
www.healthfinder.gov/searchoptions/topicsaz/htm	Use this search engine to locate home food safety information for handling meat, poultry, and seafood.
www.epa.gov	Information from the Environmental Protection Agency.
www.ams.usda.gov/nop	Information on organic foods.
www.fsis.usda.gov	Information on food safety in the marketplace.
www.iastate.edu	Click on search and look for information on food irradiation.
www.aphis.usda.gov/biotechnology	Information on food biotechnology from the USDA.
www.who.org	Resources on world hunger, poverty, and overpopulation.
ww.brown.edu/Departments/World_Hunger_Program/index.html	Information and links to research, hunger advocacy, policies, and education.
www.secondharvest.org	Information about food distribution, community food banks, and food security topics.
www.frac.org	Information on U.S. food security issues and food assistance programs from the Food Research and Action Center.
www.thehungersite.com	You can click a button that says "Donate free food," and a food donation is made to the United Nations World Food Program.
www.wfp.org	A resource for information about world hunger from the United Nations World Food Programme.
www.worldhungeryear.org	Information about food insecurity in the United States.
www.fns.usda.gov/fns	Information on the U.S. food assistance food recovery programs.
www.fao.org/sd	Information about sustainable development provided by FAO.
www.oxfamamerica.org	Information regarding global hunger and poverty.
www.bread.org	Advocacy information on domestic and world hunger issues.
www.unicefusa.org	Information on child survival programs worldwide.

Check Yourself . . .

1. Name four of the top ten causes of microbial food contamination.
2. Identify three things you can do to reduce your risk of eating a food contaminated with *E. coli* 0157:H7.
3. Explain cross-contamination.
4. List five ways you can reduce the chances of consuming excess lead.
5. Name three functions of food additives.
6. Describe how overpopulation fits into the world hunger picture.
7. Explain how a food chain works.
8. Identify some of the pros and cons of buying organically grown produce.
9. Describe the Delaney clause and its limitations.
10. List a benefit and a concern associated with genetic engineering.

Answers to Check Yourself questions are found in Appendix G.

Appendix A

Nutrition Resources

People interested in nutrition often want to know where they can find reliable nutrition information. No matter where you live, there are several sources you can turn to.

The National Center for Nutrition and Dietetics (NCND) maintains a Consumer Nutrition Hot Line staffed by registered dietitians.

- Call 800-366-1655 to listen to nutrition messages recorded by a registered dietitian (RD) in English or Spanish, or to receive a referral to an RD in your area for individual or group nutrition counseling.

Local Resources of Reliable Nutrition Information

- The registered dietitians at your local hospitals or in private practice
- Local college and university nutrition instructors or professors
- State or County Extension Service home economists
- Registered dietitians in city, county, or state agencies
- Nutrition or health-related journals found in college or university libraries:

  American Family Physician
  American Journal of Clinical Nutrition
  Diabetes Forecast
  FDA Consumer
  Geriatrics
  Journal of the American Dietetic Association
  Journal of the American Medical Association
  Journal of Food Technology
  Journal of Nutrition Education
  Journal of Nutrition for the Elderly
  Medicine and Science in Sports and Exercise
  New England Journal of Medicine
  Nutrition Research
  Nutrition Reviews
  Nutrition Today

Postgraduate Medicine
Science
Science News
Scientific American

- Search for toll-free numbers for health information at www.healthfinder.gov

Health and Nutrition-Related Newsletters

American Institute for Cancer Research
www.aicr.org

Center for Science in the Public Interest (CSPI) Nutrition Action Healthletter
202-332-9110
www.cspinet.org

Community Nutrition Institute
(CNI) Nutrition Week
202-776-0595

Consumer Reports on Health
800-234-2188
www.consumerreports.org

Dairy Council Digest
www.nationaldairycouncil.com

Environmental Nutrition
800-829-5384

Harvard Medical School Health Letter
800-829-9045
www.hms.harvard.edu/news/index.html

Mayo Clinic Health Letter
800-333-9037
http://mayohealth.org

National Council for Reliable Health Information
www.ncrhi.org

Nutrition and the M.D.
800-638-6423
www.lww.com

Tufts University Health and Nutrition Letter
800-274-7581
www.healthletter.tufts.edu

University of California at Berkeley Wellness Letter
800-829-9080

Other Sources of Nutrition Information

You can request a free publication list from many of these sources

U.S. Government

Administration on Aging
330 Independence Avenue SW
Washington, DC 20201
202-619-0724
www.aoa.dhhs.gov

Centers for Disease Control and Prevention (CDC)
1600 Clifton Road NE
Atlanta, GA 30333
404-639-3311
www.cdc.gov

Federal Trade Commission (FTC)
Public Reference Branch
202-326-2222
www.ftc.gov

Food and Drug Administration (FDA)
Office of Consumer Affairs, HFE 1
Room 16-85
5600 Fishers Lane
Rockville, MD 20857
301-443-1544
www.fda.gov

FDA Center for Food Safety and Applied Nutrition Outreach and Information Center
200 C Street SW
Washington, DC 20204
888-SAFEFOOD or 888-723-3366
vm.cfsan.fda.gov

FDA Consumer Information Line
888-463-6332

FDA Office of Food Labeling, HFS 150
200 C Street SW
Washington, DC 20204
202-205-4561
www.cfsan.fda.gov

Food and Nutrition Information Center
National Agricultural Library, Room 304
10301 Baltimore Avenue
Beltsville, MD 20705-2351
301-504-5719
www.nal.usda.gov/fnic

National Aging Information Center
330 Independence Avenue SW
Washington, DC 20201
202-619-7501
www.aoa.dhhs.gov/naic

National Health Information Center (NHIC)
Office of Disease Prevention and Health Promotion
800-336-4797
nhic-nt.health.org

National Heart, Lung, and Blood Institute
Information Center
P.O. Box 30105
Bethesda, MD 20824-0105
301-592-8573
www.nhlbi.nih.gov/nhlbi/nhlbi.htm

National Institutes of Health (NIH)
9000 Rockville Pike
Bethesda, MD 20892
301-496-2433
www.nih.gov

National Institute on Aging
Public Information Office
31 Center Drive, MSC 2292
Bethesda, MD 20892
301-496-1752
www.nih.gov/nia

U.S. Department of Agriculture (USDA)
14th Street SW and Independence Avenue
Washington, DC 20250
202-720-2791
www.fns.usda.gov/fncs

USDA Center for Nutrition Policy and Promotion
1120 20th Street NW, Suite 200
North Lobby
Washington, DC 20036
202-208-2417
www.usda.gov/cnpp

USDA Food Safety and Inspection Service
Food Safety Education Office, Room 1180-S
Washington, DC 20250
202-690-0351
www.fsis.usda.gov

USDA Meat and Poultry Hotline
800-535-4555

U.S. EPA Safe Drinking Water Hotline
800-426-4791

U.S. Department of Education (DOE)
Accreditation Agency Evaluation Branch
1990 K Street NW
Washington, DC 20202
202-219-7011

U.S. Department of Health and Human Services
200 Independence Avenue SW
Washington, DC 20201
202-619-0257
www.os.dhhs.gov

U.S. Environmental Protection Agency (EPA)
401 Main Street SW
Washington, DC 20460
202-260-2090
www.epa.gov

U.S. Public Health Service
Assistant Secretary of Health
Humphrey Building, Room 725-H
200 Independence Avenue SW
Washington, DC 20201
202-690-7694

U.S. Government Printing Office
Superintendent of Documents
Washington, DC 20402
202-512-1071
www.access.gpo.gov/su_docs

Canadian Government

Bureau of Nutritional Sciences Food Directorate
Health Protection Branch
3-West
Sir Frederick Banting Research Centre, 2203A
Tunney's Pasture
Ottawa, Ontario K1A 0L2
www.hc-sc.gc.ca

Canadian Food Inspection Agency
Agriculture and Agri-Food Canada
59 Camelot Drive
Nepean, Ontario K1A 0Y9
613-225-CFIA or 613-225-2342
www.agr.ca

Nutrition & Healthy Eating Unit
Strategies and Systems for Health Directorate, 1917C
17th Floor—Jeanne Mance Bldg.
Tunney's Pasture
Ottawa, Ontario K1A 1B4
www.hc-sc.gc.ca

Nutrition Specialist
Health Support Services
Indian and Northern Health Services Directorate
Medical Services Branch
20th Floor—Jeanne Mance Bldg., 1920B
Tunney's Pasture
Ottawa, Ontario K1A 0L3
www.hc-sc.gc.ca

Professional Nutrition Organizations

American Society for Nutritional Sciences
9650 Rockville Pike
Bethesda, MD 20814
301-530-7050
www.nutrition.org

American Dietetic Association (ADA)
216 West Jackson Boulevard, Suite 800
Chicago, IL 60606-6995
800-877-1600; 312-899-0040
www.eatright.org

ADA, Consumer Nutrition Hotline
800-366-1655

American Society for Clinical Nutrition
9650 Rockville Pike
Bethesda, MD 20814-3998
301-530-7110; fax 301-571-1863
www.faseb.org/ascn

Dietitians of Canada
480 University Avenue, Suite 604
Toronto, Ontario M5G 1V2, Canada
416-596-0857; fax 416-596-0603
www.dietitians.ca

Human Nutrition Institute (INACG)
1126 Sixteenth Street NW
Washington, DC 20036
202-659-0789
www.ilsi.org

National Academy of Sciences/ National Research Council (NAS/NRC)
2101 Constitution Avenue, NW
Washington, DC 20418
202-334-2000
www.nas.edu

National Institute of Nutrition
265 Carling Avenue, Suite 302
Ottawa, Ontario K1S 2E1
613-235-3355; fax 613-235-7032
www.nin.ca

Society for Nutrition Education
1001 Connecticut Ave. NW,
Suite 528

Washington, DC 20036
202-452-8534
www.sne.org

International Agencies

Food and Agriculture Organization of the United Nations (FAO)
2175 K Street, Suite 300
Washington, DC 20437
202-653-2400
www.fao.org

International Food Information Council Foundation
1100 Connecticut Avenue NW, Suite 430
Washington, DC 20036
202-296-6540
ificinfo.health.org

UNICEF
3 United Nations Plaza
New York, NY 10017
212-326-7000
www.unicef.org

World Health Organization (WHO)
Regional Office
525 23rd Street NW
Washington, DC 20037
202-974-3000
www.who.org

Consumer and Advocacy Groups

Center for Science in the Public Interest
1875 Connecticut Avenue NW, Suite 300
Washington, DC 20009-5728
202-332-9110; fax 202-265-4954
www.cspinet.org

Children's Defense Fund
25 E Street, NW
Washington, DC 20001
202-628-8787
www.childrensdefense.org

Community Nutrition Institute
910 17th Street, NW
Suite 800
Washington, DC 20006
202-776-0595

Consumer Information Center
Pueblo, CO 81009
888-8 PUEBLO or 888-878-3256
www.pueblo.gsa.gov

Consumers Union of US Inc.
101 Truman Avenue
Yonkers, NY 10703-1057
914-378-2000
www.consumersunion.org

Food Research and Action Center
1875 Connecticut Avenue, NW
Washington, DC 20009
202-986-2200
www.frac.org

National Council for Reliable Health Information (formerly known as National Council Against Health Fraud, Inc. (NCAHF))
Independence, MO
816-228-4595
www.ncrhi.org

Professional and Service Organizations

Al-Anon Family Group Headquarters, Inc.
1600 Corporate Landing Parkway
Virginia Beach, VA 23454-5617
800-356-9996
www.al-anon.alateen.org

Alateen
1600 Corporate Landing Parkway
Virginia Beach, VA 23454-5617
800-356-9996
www.al-anon.alateen.org

Alcohol & Drug Abuse Information Line
800-252-6465

Alcoholics Anonymous (AA)
General Service Office
475 Riverside Drive
New York, NY 10115
212-870-3400
www.aa.org

Alliance for Food and Fiber
Food Safety Hotline:
800-266-0200

Alzheimer's Disease Education and Referral Center
P. O. Box 8250
Silver Spring, MD 20907-8250
800-438-4380
www.alzheimers.org

Alzheimer's Disease Information and Referral Service
919 North Michigan Avenue, Suite 1000
Chicago, IL 60611
800-272-3900
www.alz.org

Alzheimer Society of Canada
www.alzheimer.ca

American Academy of Allergy, Asthma, and Immunology
611 East Wells Street
Milwaukee, WI 53202
414-272-6071
www.aaaai.org

American Academy of Pediatrics
141 Northwest Point Boulevard
Elk Grove Village, IL 60007
847-228-5005
www.aap.org

American Anorexia & Bulimia Association, Inc.
165 West 46th Street #1108
New York, NY 10036
212-575-6200
www.aabainc.org

American Association of Family and Consumer Sciences
1555 King Street
Alexandria, VA 22314
703-706-4600
www.aafcs.org

American Association of Retired Persons
601 E Street NW
Washington, DC 20049
202-434-2277
www.aarp.org

American Cancer Society
National Home Office
1599 Clifton Road NE
Atlanta, GA 30329-4251
800-ACS-2345 or 800-227-2345
www.cancer.org

American College of Sports Medicine
P.O. Box 1440
Indianapolis, IN 46206-1440
317-637-9200
www.acsm.org

American Council on Exercise (ACE)
5820 Oberlin Drive, Suite 102
San Diego, CA 92121
800-529-8227
www.acefitness.org

American Council on Science and Health
1995 Broadway, 2nd Floor
New York, NY 10023
212-362-7044
www.acsh.org

American Dental Association
211 East Chicago Avenue
Chicago, IL 60611
312-440-2500
www.ada.org

American Diabetes Association
1660 Duke Street
Alexandria, VA 22314
800-232-3472 or 703-549-1500
www.diabetes.org

American Heart Association
7272 Greenville Avenue
Dallas, TX 75231
214-373-6300
800-275-0448
www.americanheart.org

American Institute for Cancer Research
1759 R Street, NW
Washington, DC 20009
800-843-8114 or 202-328-7744
www.aicr.org

American Medical Association
515 North State Street
Chicago, IL 60610

312-464-5000
www.ama-assn.org

American Public Health Association (APHA)
800 I Street
Washington, DC 20001
202-777-2742
www.apha.org

American Public Welfare Association
(Food Stamp Program Administrators)
810 First Street, NE
Washington, D.C. 20002
202-682-0100

American Red Cross
National Headquarters
8111 Gatehouse Road
Falls Church, VA 22042
703-206-7180
www.redcross.org

American School Food Service Association
(Child Nutrition Program Personnel)
1600 Duke Street, 7th Floor
Alexandria, VA 22314
703-739-3900
www.asfsa.org

American SIDS (Sudden Infant Death) Institute Information Line:
800-232-SIDS

Anorexia Nervosa and Related Eating Disorders (ANRED)
P.O. Box 5102
Eugene, OR 97405
541-344-1144
www.anred.com

Arthritis Foundation Information Line:
800-283-7800

Association of Birth Defect Children, Inc.
930 Woodcock Road, Suite 225
Orlando, FL 32803
407-245-7035
www.birthdefects.org

Asthma and Allergy Foundation of America
Information Line:
800-727-8462

Canadian Cancer Society
www.cancer.ca

Canadian Diabetes Association
15 Toronto Street, Suite 800
Toronto, ON M5C 2E3
800-BANTING or 800-226-8464
416-363-3373
www.diabetes.ca

Canadian Heart and Stroke Foundation
www.hsf.ca

Canadian Public Health Association
400-1565 Carling Avenue
Ottawa, Ontario K1Z 8R1
613-725-3769
www.cpha.ca

The Food Allergy Network
10400 Eaton Place, Suite 107
Fairfax, VA 22030-2208
800-929-4040 or 703-691-3179
www.foodallergy.org

Internet Health Resources
www.ihr.com

La Leche International, Inc.
1400 N. Meacham Road
Schaumburg, IL 60173
847-519-7730
www.lalecheleague.org

March of Dimes Birth Defects Foundation
1275 Mamaroneck Avenue
White Plains, NY 10605
914-428-7100
www.modimes.org

Narcotics Anonymous (NA)
P.O. Box 9999
Van Nuys, CA 91409
818-773-9999
www.wsoinc.com

National AIDS Hotline (CDC)
800-342-AIDS (English)
800-344-SIDA (Spanish)
800-2437-TTY (Deaf)
900-820-2437

National Association for Sickle Cell Disease Information Line:
800-421-8453

National Association of Anorexia Nervosa and Associated Disorders, Inc. (ANAD)
P.O. Box 7
Highland Park, IL 60035
847-831-3438
www.anad.org

National Association of WIC Directors
P.O. Box 53355
Washington, DC 20009
202-232-5492

National Cancer Institute
Office of Cancer Communications
Building 31, Room 10824
Bethesda, MD 20892
800-4-CANCER or 800-422-6237
www.nci.nih.gov

National Center for Education in Maternal and Child Health
2000 15th St. North, Suite 701
Arlington, VA 22201-2617
703-524-7802
www.ncemch.org

National Child Abuse Hotline
800-422-4453

National Council on Alcoholism and Drug Dependence (NCADD)
12 West 21st Street
New York, NY 10010
800-NCA-CALL or 800-622-2255
212-206-6770
www.ncadd.org

National Diabetes Information Clearinghouse
1 Information Way
Bethesda, MD 20892-3560
301-654-3327
www.niddk.nih.gov

National Digestive Disease Information Clearinghouse (NDDIC)
2 Information Way
Bethesda, MD 20892-3570
301-654-3810
www.niddk.nih.gov

National Lead Information Center
800-LEAD-FYI or 800-532-3394
800-424-LEAD or 800-424-5323

National Osteoporosis Foundation
1150 17th Street NW, Suite 500
Washington, DC 20036
202-223-2226
www.nof.org

National Pesticide Telecommunications Network (NPTN)
Oregon State University
333 Weniger Hall
Corvallis, OR 97331-6502
541-737-6091
www.ace.orst.edu/info/nptn

Nutrition Screening Initiative
1010 Wisconsin Avenue, NW
Suite 800
Washington, DC 20007
202-625-1662

Overeaters Anonymous (OA)
6075 Zenith Court NE
Rio Rancho, NM 87124
505-891-2664
www.overeatersanonymous.org

President's Council on Physical Fitness and Sports
Humphrey Building, Room 738 H
200 Independence Avenue SW
Washington, DC 20201
202-690-9000
www.indiana.edu/~preschal/council.html

Shape Up America!
6707 Democracy Boulevard, Suite 306
Bethesda, MD 20817
301-493-5368
www.shapeup.org

Weight Watchers International, Inc.
Consumer Affairs Department/IN
175 Crossways Park West
Woodbury, NY 11797

516-390-1400
www.weightwatchers.com

Trade Organizations

American Egg Board
1460 Renaissance Drive
Park Ridge, IL 60068
708-296-7043

American Meat Insitute
1700 North Moore Street
Suite 1600
Arlington, VA 22209
703-841-2400

Beech-Nut Nutrition Corporation
P.O. 618
St. Louis, MO 63188-0618
800-523-6633
www.beechnut.com

Borden Inc.
180 East Broad Street
Columbus, OH 43215
800-426-7336

Campbell Soup Company
Consumer Response Center
Campbell Place, Box 26B
Camden, NJ 08103-1701
800-257-8443
www.campbellssoup.com

General Mills, Inc.
Number One General Mills Boulevard
Minneapolis, MN 55426
800-328-6787
www.generalmills.com

Hoffmann-LaRoche, Inc.
340 Kingsland Street
Nutley, NJ 07110
973-235-5000

Kellogg Company
P.O. Box 3599
Battle Creek, MI 49016-3599
616-961-2000
www.kelloggs.com

Kraft Foods
Consumer Response and Information Center
One Kraft Court
Glenview, IL 60025
800-323-0768
www.kraftfoods.com

Mead Johnson Nutritionals
2400 West Lloyd Expressway
Evansville, IN 47721
800-247-7893
www.meadjohnson.com

Nabisco Consumer Affairs
100 DeForest Avenue
East Hanover, NJ 07936
800-NABISCO or 800-932-7800
www.nabisco.com

National Cattlemen's Beef Association
444 North Michigan Avenue
Chicago, IL 60611
312-467-5520

National Dairy Council
10255 West Higgins Road, Suite 900
Rosemont, IL 60018-5616
847-803-2000
www.dairyinfo.com

NutraSweet/KELCO
P.O. Box 2986
Chicago, IL 60654-0986
800-321-7254
www.equal.com

Pillsbury Company
Consumer Relations
P.O. Box 550
Minneapolis, MN 55440
800-767-4466
www.pillsbury.com

The Potato Board
7555 E. Hampton Avenue
Suite 412
Denver, CO 80231
303-369-7783

Procter and Gamble Company
One Procter and Gamble Plaza
Cincinnati, OH 45202
513-983-1100
www.pg.com

Ross Laboratories
625 Cleveland Avenue
Columbus, OH 43215
800-227-5767
www.ross.com

Soy Protein Council
1255 23rd Street NW
Washington, DC 20037
202-467-6610
www.spcouncil.org

Sunkist Growers
Citrus Hotline
P.O. Box 7888
Van Nuys, CA 91409-7888
Attention MS 236
800-CITRUS-5 or 800-248-7875
www.sunkist.com

United Fresh Fruit and Vegetable Association
727 North Washington Street
Alexandria, VA 22314
703-836-3410

United Soybean Board
P.O. Box 419200
St. Louis, MO 63141
www.talksoy.com

USA Rice Federation
4301 North Fairfax Drive, Suite 305
Arlington, VA 22203
Phone: 703-351-8161
www.usarice.com

Organizations Concerned with Hunger and Poverty

Bread for the World
1100 Wayne Avenue, Suite 1000
Silver Spring, MD 20910
301-608-2400
www.bread.org

Center on Hunger, Poverty and Nutrition Policy
Tufts University School of Nutrition
11 Curtis Avenue
Medford, MA 02155
617-627-3956
www.tufts.edu/nutrition/centeronhunger

Catholic Relief Services
209 West Fayette Street
Baltimore, MD 21201-3443
410-625-2220
800-235-2772
www.catholicrelief.org

Freedom from Hunger
P.O. Box 2000
1644 DaVinci Court
Davis, CA 95617
530-758-6200
www.freefromhunger.org

Church World Service/CROP
P.O. Box 968
Elkhart, IN 46515
219-264-3102
800-456-1310
www.churchworldservice.org

Oxfam America
26 West Street
Boston, MA 02111
617-482-1211
www.oxfamamerica.org

Second Harvest and Foodchain
116 South Michigan Avenue
Room 4
Chicago, IL 60603
312-263-2303
www.secondharvest.org

U.S. National Committee for World Food Day
1001 22nd Street, NW
Washington, DC 20437
202-653-2404

Worldwatch Institute
1776 Massachusetts Avenue NW Suite 800
Washington, DC 20036
202-452-1999
www.worldwatch.org

Appendix B

An Introduction to the Human Body

The brief anatomy lesson that follows is a lesson in "anatomy for nutrition's sake" to review the body systems and terminology referred to in this book. To make the body's design understandable, the first few paragraphs are devoted to the essential needs of the cells and the body's mechanisms that ensure that they are met.

 THE CELLS

The body is composed of millions of cells, and not one of them knows anything about food. While you get hungry for meat, milk, or bread, each cell of your body sits in its place waiting until the nutrients it needs pass by. Each of the body's **cells** is a self-contained, living entity (Figure B-1), although each depends on the rest of the body to supply its needs. Each cell keeps itself alive just as its single-celled ancestors did, living alone in the ocean 3 billion years ago, by taking up the substances it needs from the surrounding fluid and releasing the wastes it produces into that fluid.

The body cells' most basic need, always, is for energy fuel and the oxygen with which to burn it. Next, they need water, the environment in which they live. Then they need building blocks to maintain themselves—especially the materials they can't make for themselves. These building blocks—the **essential nutrients**—must be supplied preformed from food. These are among the limitations of our

FIGURE B-1
A TYPICAL CELL
(Simplified Diagram)

A membrane encloses each cell's contents.

A separate inner membrane encloses the cell's nucleus.

Inside the nucleus is the hereditary material, which contains the genes. The genes control the inheritance of the cell's characteristics and its day to day workings. They are faithfully copied each time the cell duplicates itself.

On these membranes, instructions from the genes are translated into proteins that perform functions in the body.

Many other structures are present. This is a mitochondrion, a structure that takes in nutrients and releases energy from them.

These finger-like projections are typical of cells that absorb nutrients in the intestines.

heredity from which there is no appeal, and they underlie the first principle of diet planning. Whatever foods we choose, they must provide energy, water, and the essential nutrients. In a sense, the body is only a system organized to provide for these needs of its cells.

In the human body every cell works in cooperation with every other to support the whole. The cell's **genes** determine the nature of that work. Each gene is a blueprint that directs the making of a piece of protein machinery—most often an **enzyme**—that helps to do the cell's work. Each cell contains a complete set of genes, but different ones are active in different types of cells. For example, in some intestinal cells, the genes for making digestive enzymes are active; in some of the body's fat cells, the genes for making enzymes that make and break down fat are active.

Cells are organized into tissues that perform specialized tasks governed by the genes that are active in them. For example, some cells are joined together to form muscle tissue, which can contract. Tissues also are organized in sets to form whole organs. In the heart organ, for example, muscle tissues, nerve tissues, connective tissues, and other types all work together to pump blood. Some jobs around the body require that several related organs cooperate to perform them. The organs that join together to work on a function are parts of a body system. For example, the heart, lungs, and blood vessels all work to deliver oxygen and nutrients to the body tissues as parts of the cardiovascular system. The next few sections present some body systems with special significance to nutrition.

THE BODY FLUIDS

Every cell of the body needs a continuous supply of water, oxygen, energy, and building materials. The body fluids supply these necessities, bathing the outside of all the cells (see Figure B-2). Every cell continuously uses up oxygen (producing carbon dioxide) and nutrients (producing waste products). The body fluids are the transport canals for these materials, carrying oxygen and nutrients to the cells and carbon dioxide and waste away from them. These fluids must circulate to pick up fresh supplies and deliver the wastes to points of disposal.

The fluids that bathe the cells and circulate around the body are the extracellular fluids, the **blood** and **lymph** (Figure B-3). Blood travels within the **arteries, veins,** and **capillaries,** as well as within the heart's chambers (Figure B-4). Lymph is derived from the blood in the capillaries; it squeezes out across their walls and circulates around the cells, permitting exchange of materials. Some of the lymph returns to the blood farther along the capillaries, and the rest travels around the body by way of its own vessels, eventually returning to the bloodstream elsewhere.

FIGURE B-2
ONE CELL AND ITS ASSOCIATED FLUIDS

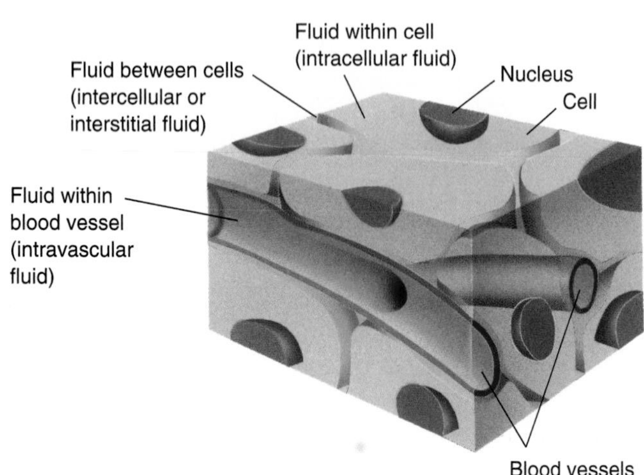

THE CIRCULATORY SYSTEM

As the blood, pumped by the heart, travels through the circulatory system, it picks up and delivers materials as needed. Its routing ensures that all cells will be served. Oxygen is picked up and carbon dioxide is released in the **lungs,** and all blood that circulates to the lungs is returned to the heart. From there, it must go to the other body tissues. Thus all tissues receive freshly oxygenated blood.

As it passes the digestive system, the blood delivers oxygen to the cells there and picks up nutrients from the **intestine** for distribution elsewhere. All blood leaving the digestive system must go next to the **liver,** which has the special task of chemically altering the absorbed mate-

FIGURE B-3
HOW THE BODY FLUIDS CIRCULATE AROUND CELLS

The upper left-hand box shows a tiny portion of tissue with blood flowing through its network of capillaries (greatly enlarged). The bottom right-hand box illustrates the movement of the extracellular fluid.

Blood enters tissues by way of artery.
Blood collects into veins for return to heart.
Blood circulates among cells by way of capillaries.
Fluid filters out of blood capillaries whose walls are made of cells with small spaces between them.
Exchange of materials takes place between cell fluid and extracellular fluid.
Fluid may flow back into capillary or into lymph vessel. Lymph enters the bloodstream later through a large lymphatic vessel that empties into a large vein.
In from air
Oxygen
Blood vessel
Carbon dioxide
Lungs
Out of body
Wastes
Blood vessel
Kidneys

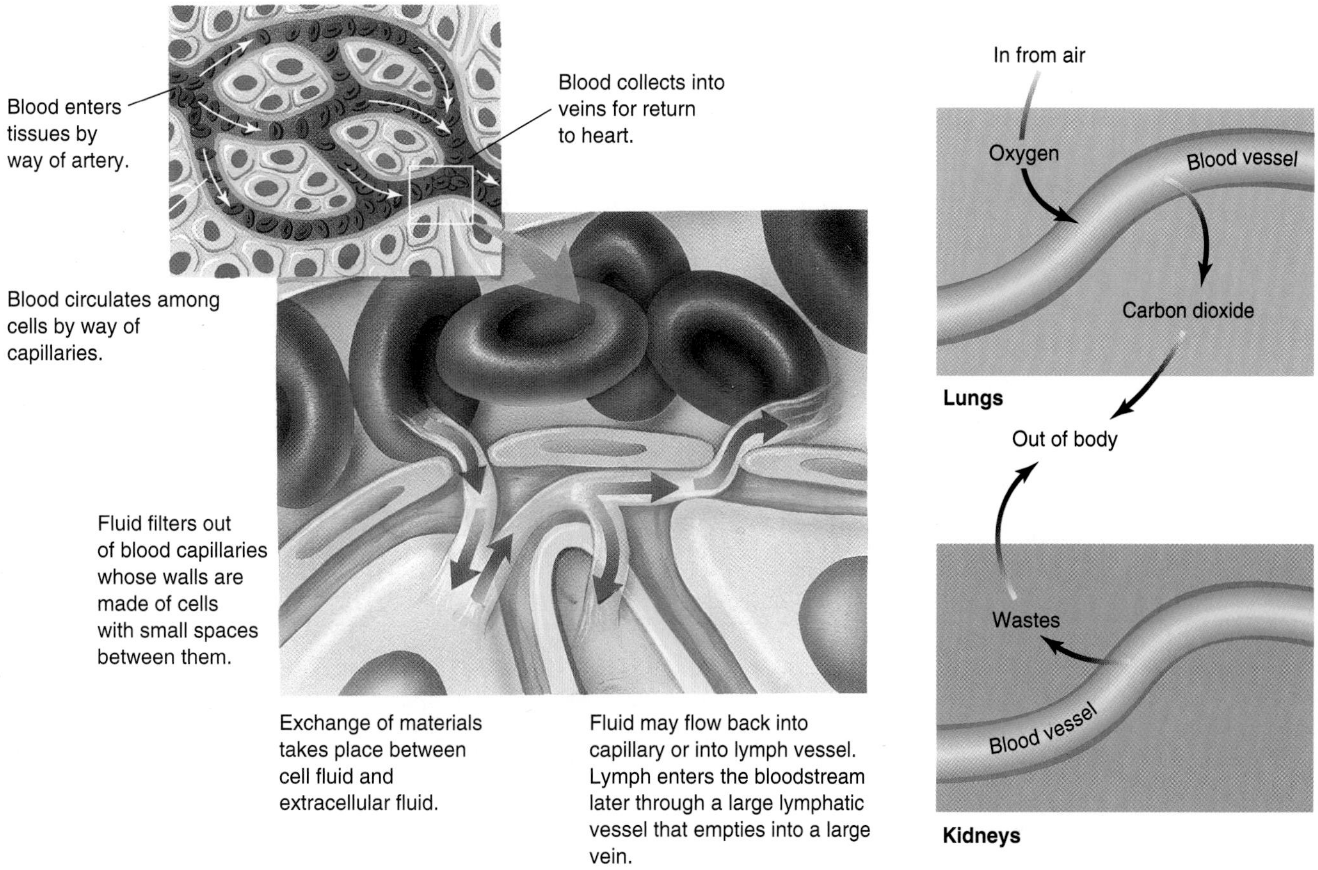

rials to make them better suited for use by other tissues. Then, in passing through the **kidneys,** the blood is cleansed of its wastes.

As it flows through the skin, the blood is cooled by radiating heat to the surroundings, helping to maintain the temperature of the body's internal organs. Fluid leaving the blood as lymph may ultimately evaporate from the lungs and skin or be used to make body secretions, such as digestive juices, which will be used within the body for various purposes. On its return to the heart, the blood has delivered most of its oxygen and picked up carbon dioxide from the body cells. Its next stop is the lungs once again, to release its carbon dioxide and replenish its oxygen.

In summary, the routing of the blood is as shown in Figure B-4:

- Heart to body to heart to lungs to heart (repeat).

The portion of the blood that flows by the digestive tract travels from:

- Heart to digestive tract to liver to heart.

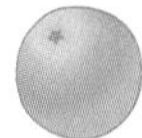

THE IMMUNE SYSTEM

Many of the body's cells cooperate to maintain its defenses against infection. The skin presents a physical barrier, and the body's cavities (lungs, digestive tract, and others) are lined with membranes that resist penetration by invading

FIGURE B-4
THE CARDIOVASCULAR SYSTEM

Blood leaves right side of heart, picks up oxygen in lungs, and returns to left side of heart. Blood leaves left side of heart, goes to the head, or to the digestive tract and then to the liver, or to the lower body, and then returns to right side of heart.

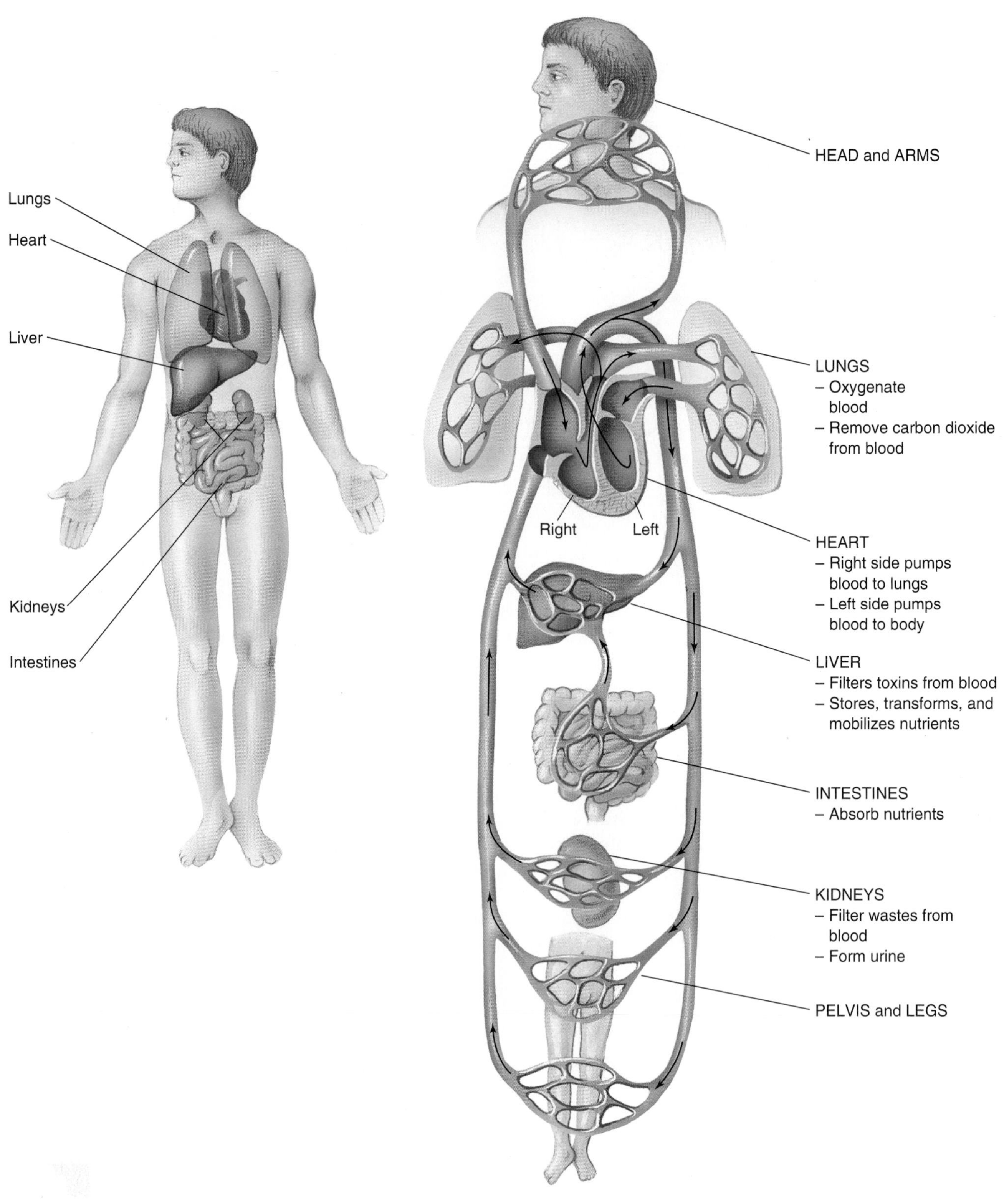

microbes or unwanted substances. The body's linings are easily damaged by nutrient deficiencies, and clinicians inspect both the skin and the inside of the mouth to detect signs of malnutrition. (The chapters on protein, vitamins, and minerals present details of the signs of deficiencies.)

When a wound or infection penetrates these first lines of defense (the skin and linings), the lymph and blood present internal defenses: cells and proteins that can inactivate, remove, or destroy microbes and foreign substances. Special cells are able to recognize the chemical structures of some foreign materials and to remember them for a time so that they can quickly mobilize their defenses when they see them again. This ability confers **immunity** against many diseases that you have previously fought and conquered. Some immune cells produce proteins that act as ammunition (**antibodies**) designed to destroy specific targets (**antigens**), and still other cells can gobble up and digest the invaders.

Immune system components reside in tissues all over the body—in the linings of the bones, in the digestive tract, in the blood vessels, in the lymph glands, and in glands of their own. They are in constant flux, being made and dismantled rapidly, and their maintenance requires a continuous supply of nutrients. A deficiency or an overdose of any nutrient is likely to affect the immune system adversely, and a deficiency of nutrients early in an infant's development can weaken that individual's immune defenses against infection for years.

THE HORMONAL AND NERVOUS SYSTEMS

The blood also carries messages, chemical signals from one system of cells to another, that communicate the changing needs of the living system. These chemical messages, or **hormones,** are secreted and released into the blood by the **endocrine** glands. For example, when the **pancreas** (a gland) experiences a too-high concentration of glucose in the blood, it releases **insulin** (a hormone). Insulin stimulates the liver, muscles, and fat cells to remove glucose from the blood and put it away. When the blood glucose level falls too low, the pancreas secretes another hormone, glucagon. The liver responds by releasing glucose into the blood once again.

More about the blood glucose level–Chapter 3.

Glands and hormones abound in the body, each gland a detector system to monitor a condition in the body that needs regulation and each hormone a messenger to stimulate certain tissues to take appropriate action. Examples of the working of these hormones appear throughout this book.

The body's other major communication system is, of course, the nervous system. With the brain and spinal cord as central controllers, the system receives and integrates messages from sensory receptors all over the body—sight, hearing, touch, smell, taste, and others—which all communicate to the brain the state of both the outer and inner worlds, including the availability of food and the need to eat. The system then returns instructions to the muscles and glands, telling them what to do.

The nervous system's part in hunger regulation is coordinated by the brain. The sensations of hunger and appetite are experienced in the **cortex** of the brain, the thinking, outer layer. However, much of the brain's regulatory work goes on in the deep brain centers, without the person's (or the cortex's) awareness. An organ there, the **hypothalamus** (Figure B-5), monitors many body conditions, including the availability of nutrients and water.

THE EXCRETORY SYSTEM

To dispose of waste, the kidneys straddle the circulatory system and filter each pass of the blood (see Figure B-6). Waste materials removed with water are collected as urine in tubes that deliver them to the urinary bladder, which is periodically emptied. Thus the blood is purified continuously throughout the day, and

FIGURE B-5
THE BRAIN'S HYPOTHALAMUS AND CORTEX

The hypothalamus monitors the body's conditions and sends signals to the brain's thinking portion, the cortex, which decides on actions.

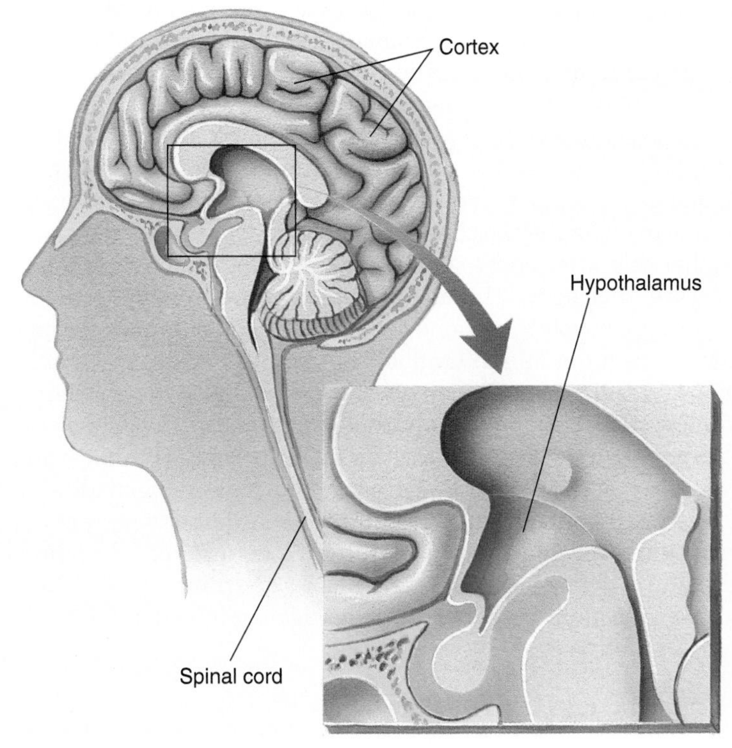

dissolved minerals are excreted as necessary (including sodium, to keep blood pressure from rising too high). As you might expect, the kidneys' work is regulated by hormones secreted by glands responsive to conditions in the blood (such as the sodium concentration).

FIGURE B-6
THE EXCRETORY SYSTEM

1. Blood enters kidney by way of arteries and disperses into capillaries.
2. Kidney filters waste from blood and sends it as urine to the bladder.
3. Bladder periodically eliminates urine.

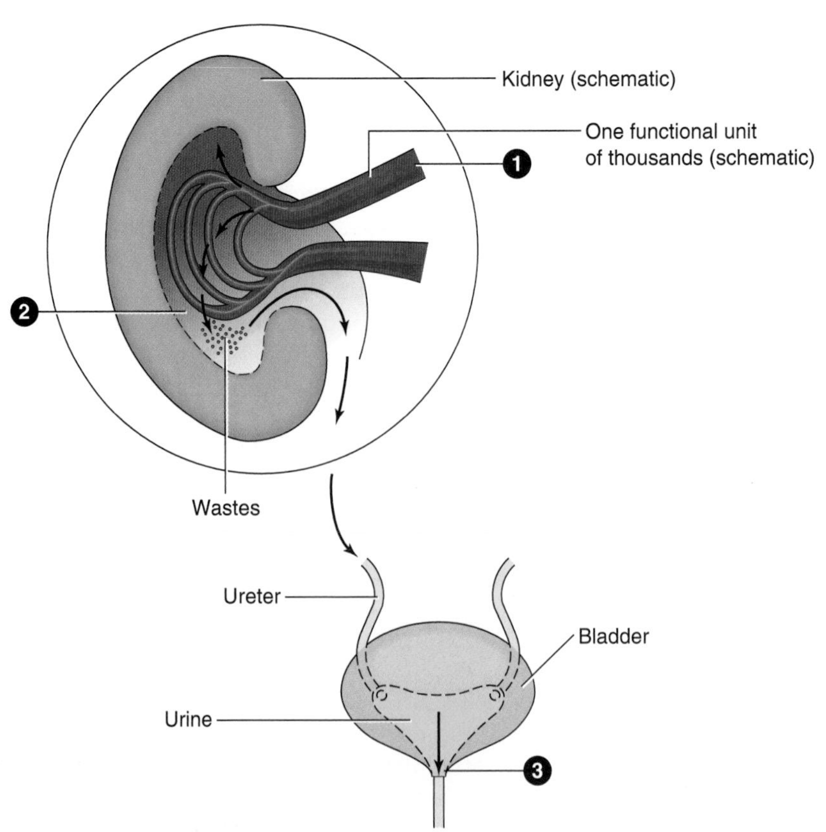

TEMPERATURE REGULATION

All the body's cells obtain energy by breaking down the nutrients—carbohydrate, fat, and to some extent protein—and one of the ways this energy is released is as heat. The heat is lost to the air through the skin surface. Temperature regulation involves speeding up or slowing down cellular heat production (**metabolism**) and increasing or decreasing heat loss through the skin. Specialized nerve cells in an area of the brain called the hypothalamus serve as a thermostat, measuring the temperature of the blood. These cells signal other cells, near the body surface, to respond appropriately. When the body is too hot, blood vessels immediately under the skin dilate, allowing warm blood to flow near the surface, where its heat can radiate away. The sweat glands also are activated to secrete warm fluid onto the skin surface, where its heat can be lost by evaporation. When the body is cold, these mechanisms shut down and shivering is triggered, generating heat.

By means of these systems of transportation, communication, waste disposal, and heat regulation, the cells of the multicellular human animal cooperate to provide one another with a circulating bath of warm, clean, nutritive fluid whose composition is finely regulated to meet their needs.

THE DIGESTIVE SYSTEM

You may eat meals only two or three times a day, but your body's cells need their nutrients 24 hours a day. Providing the needed nutrients requires the cooperation of millions of specialized cells. When the body's cells are deprived of fuel, certain nerve cells in the brain (the hypothalamus) detect this condition and generate nerve impulses that signal hunger to the conscious part of the brain, the cortex. They also stimulate the stomach to intensify its contractions, creating hunger pangs. Becoming conscious of hunger, then, you eat, delivering a complex mixture of chewed and swallowed food to the intestinal tract.

Many of the cells lining the intestinal tract secrete powerful juices and enzymes to disintegrate nutrients (especially carbohydrate and protein) into their component parts. Two organs outside the digestive tract—the liver with its associated gallbladder and the pancreas—also contribute digestive juices through a common duct into the small intestine. The presence of these digestive juices and enzymes requires that still other cells specialize in protecting the digestive system. They secrete a thick, viscous substance known as **mucus,** or the **mucous membrane,** which coats the intestinal tract lining and ensures that it will not itself be digested.

The process of digestion is diagrammed in Figure B-7 (see page A-14). The first part, the mouth, is designed for physically breaking down foods. The teeth cut off a bite-size portion and then, aided by the tongue, grind it finely enough to be mixed with saliva and swallowed. The esophagus carries the mixture to the stomach. The stomach is supplied with several sets of muscles to mix and grind it further and secretes acid and enzymes that will begin to break it apart chemically.

During the preparatory stage, as the complex carbohydrate known as starch is released from a food (such as bread), an enzyme present in the saliva starts to break it down chemically to smaller units. But this action is stopped when the carbohydrate units reach the stomach, because glands in the stomach wall exude hydrochloric acid. The salivary enzyme that breaks up starch is digested in the stomach, together with other proteins. Further dismantling of carbohydrate occurs after it leaves the stomach.

Fats and oils, taken as part of such complex foods as meats or nuts or in relatively pure form as butter or oil, are not much affected until after leaving the stomach.

Proteins are eaten as part of such foods as meat, milk, and legumes. Although no chemical action on them takes place in the mouth, chewing and mixing protein with saliva is an important part of preparing it for the chemical action that begins in the stomach. There, enzymes and hydrochloric acid break apart the

Digestive tract secretions:

Salivary glands:
Saliva.
Salivary amylase (enzyme that breaks down starch).

Stomach (gastric) glands:
Gastric juice.
Hydrochloric acid (uncoils protein).
Gastric protease (enzyme that breaks down protein).
Mucus (thick coating that protects the stomach wall from these secretions).

Intestinal cells:
Enzymes (break down carbohydrate and protein).
Mucus (thin coating that protects the intestinal wall).

Liver and gallbladder:
Bile (emulsifier that separates fat into small particles enzymes can attack).

Pancreas:
Bicarbonate (neutralizes acid fluid from stomach so intestinal and pancreatic enzymes can work on its contents).
Enzymes (break down carbohydrate, fat, and protein).

FIGURE B-7
THE DIGESTIVE SYSTEM

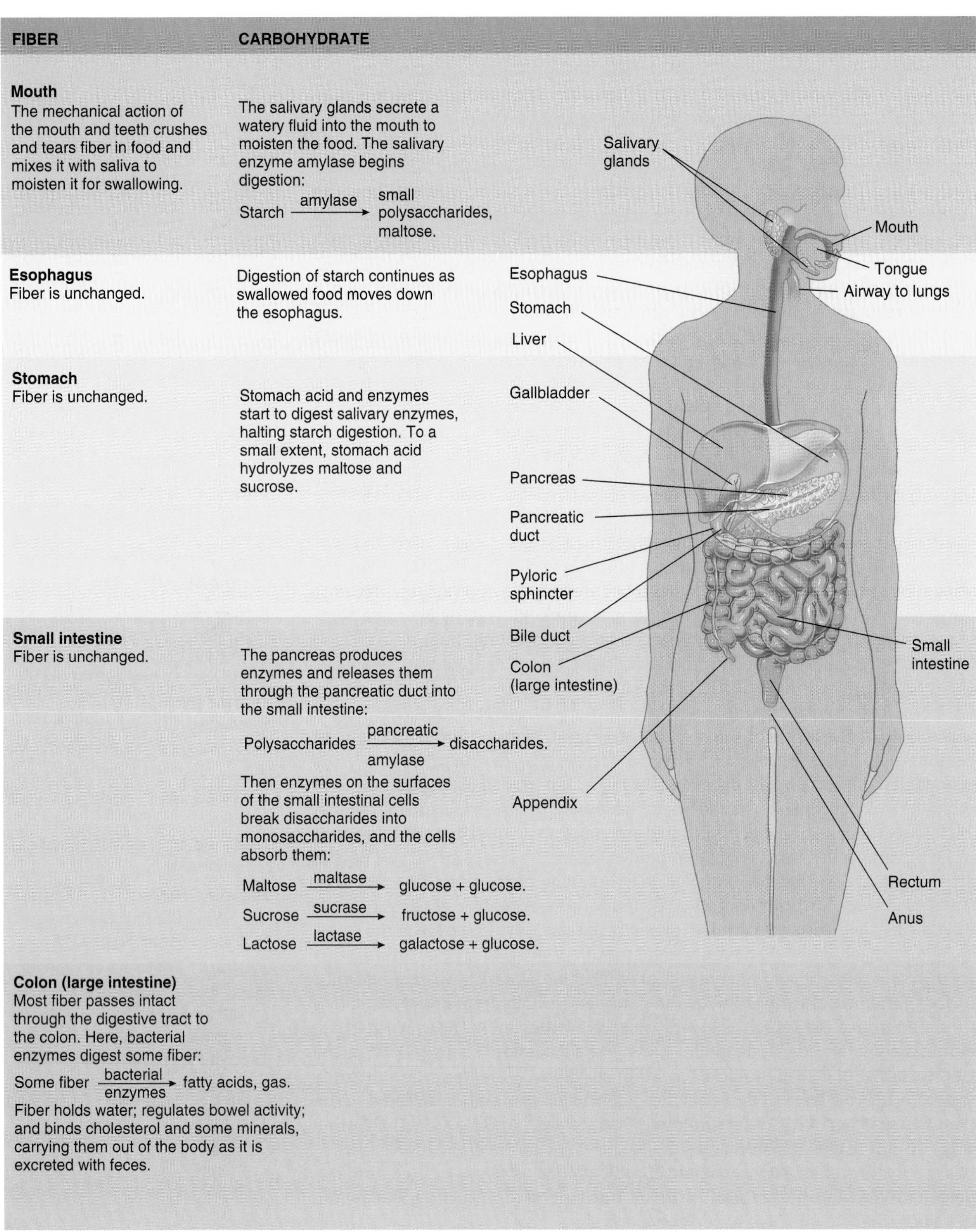

FIBER
CARBOHYDRATE
Mouth
The mechanical action of the mouth and teeth crushes and tears fiber in food and mixes it with saliva to moisten it for swallowing.
The salivary glands secrete a watery fluid into the mouth to moisten the food. The salivary enzyme amylase begins digestion:
Starch —amylase→ small polysaccharides, maltose.
Esophagus
Fiber is unchanged.
Digestion of starch continues as swallowed food moves down the esophagus.
Stomach
Fiber is unchanged.
Stomach acid and enzymes start to digest salivary enzymes, halting starch digestion. To a small extent, stomach acid hydrolyzes maltose and sucrose.
Small intestine
Fiber is unchanged.
The pancreas produces enzymes and releases them through the pancreatic duct into the small intestine:
Polysaccharides —pancreatic amylase→ disaccharides.
Then enzymes on the surfaces of the small intestinal cells break disaccharides into monosaccharides, and the cells absorb them:
Maltose —maltase→ glucose + glucose.
Sucrose —sucrase→ fructose + glucose.
Lactose —lactase→ galactose + glucose.
Colon (large intestine)
Most fiber passes intact through the digestive tract to the colon. Here, bacterial enzymes digest some fiber:
Some fiber —bacterial enzymes→ fatty acids, gas.
Fiber holds water; regulates bowel activity; and binds cholesterol and some minerals, carrying them out of the body as it is excreted with feces.
Salivary glands
Mouth
Tongue
Airway to lungs
Esophagus
Stomach
Liver
Gallbladder
Pancreas
Pancreatic duct
Pyloric sphincter
Bile duct
Colon (large intestine)
Small intestine
Appendix
Rectum
Anus

FIGURE B-7
THE DIGESTIVE SYSTEM—*continued*

	Fat	Protein	Vitamins	Minerals and Water
Mouth	Glands in the base of the tongue secrete a fat-digesting enzyme known as lingual lipase. Some hard fats begin to melt as they reach body temperature.	In the mouth, chewing crushes and softens protein-rich foods and mixes them with saliva to be swallowed.	No action.	The salivary glands add water to disperse and carry food.
Esophagus	Fat is unchanged.	No action.	No action.	No action.
Stomach	The degree of hydrolysis is slight for most fats but may be appreciable for milk fats. The stomach's churning action mixes fat with water and acid. A gastric enzyme accesses and hydrolyzes a small percentage of fat.	Stomach acid works to uncoil protein strands and activate stomach enzymes. Then the enzymes break the strands into smaller fragments: $\text{Protein} \xrightarrow[\text{HCl}]{\text{pepsin}} \text{smaller polypeptides}$	Water-soluble vitamins need little action by the digestive organs except absorption in the small intestine. However, vitamin B_{12} requires "intrinsic factor" produced by the stomach in order to be absorbed.	The stomach secretes enough watery fluid to turn a moist, chewed mass of swallowed food into a liquid. Stomach acid acts on iron to make it more absorbable. Vitamin C and a factor in meat also increase iron absorption.
Small intestine	The liver secretes bile; the gallbladder stores it and releases it through the common bile duct into the small intestine when fat arrives there. The bile emulsifies the fat, making it ready for enzyme action. The pancreas produces fat-digesting enzymes and releases them through the common bile duct into the small intestine. These enzymes split triglycerides into monoglycerides, free fatty acids, and glycerol, which are absorbed.	In the small intestine, the fragments of protein are split into free amino acids, dipeptides, and tripeptides with the help of enzymes from the pancreas and small intestine. Enzymes on the surface of the small intestinal cells break these peptides into amino acids, and they are absorbed through the cells into the blood. The large intestine carries any undigested protein residue out of the body. Normally, practically all the protein is digested and absorbed.	Bile emulsifies fat-soluble vitamins and aids in their absorption with other fats. Water-soluble vitamins are absorbed.	The small intestine, pancreas, and liver add enough fluid so that approximately 2 gallons are secreted into the intestine in a day. Many minerals are absorbed. Vitamin D aids in the absorption of calcium.
Colon	Some fat and cholesterol, trapped in fiber, exit in feces.		Bacteria produce vitamin K, which is absorbed.	More minerals and most of the water are absorbed.

large, complex protein molecules into smaller pieces known as peptides and finally into dipeptides, tripeptides, and amino acids.

The complicated chemical dismantling that takes place beyond the stomach requires that only small amounts be processed at one time. To accomplish this, the **pylorus,** a circular muscle surrounding the lower end of the stomach, controls the exit of the contents, allowing only a little at a time to be squirted forcefully into the small intestine. Gradually the stomach empties itself by means of these powerful squirts.

The small intestine is "the" organ of digestion and absorption; it finishes the job the mouth and stomach have started. It is actually about 20 feet long, but it is called small because its diameter is small compared with that of the large intestine. Its contents must touch its walls to make contact with the secretions and to be absorbed at the proper places. At the end of the small intestine, a circular muscle (similar in function to the pylorus at the end of the stomach) controls the flow of the contents going into the large intestine (colon).

The small intestine works with the precision of a laboratory chemist. As the thoroughly liquefied and partially digested nutrient mixture arrives there, hormonal messages tell the gallbladder to send its **emulsifier, bile,** in amounts matched to the amount of fat present. Other hormones notify the pancreas to release **bicarbonate** in amounts precisely adjusted to neutralize the stomach acid as well as enzymes of the appropriate kinds and quantities to continue dismantling whatever large molecules remain. Such messages also keep the strong muscles imbedded in the walls of the intestine contracting, in a squeezing activity called **peristalsis,** so that the contents will be pressed along to the next region. Peristalsis is stimulated by the presence of roughage or fiber and is quieted by the presence of fat, which requires a longer time for digestion.

Meanwhile, as the pancreatic and intestinal enzymes act on the bonds that hold the large nutrients together, smaller and smaller units make their appearance in the intestinal fluids. Finally, units that cells can use—glucose, glycerol, fatty acids, and amino acids, among others—are released.

Once the digestive system has broken food down to its nutrient components, it must deliver them to the rest of the body. The cells of the intestinal lining absorb nutrients from the mixture within the intestine and deposit them in the blood and lymph. Every molecule of nutrient must traverse one of these cells if it is to enter the body fluids. The cells are selective: they can recognize the nutrients needed by the body. The cells are also extraordinarily efficient: they absorb enough nutrients to nourish all the body's other cells.

The intestinal tract lining is composed of a single sheet of cells, and the sheet pokes out into millions of finger-shaped projections (**villi**). Each villus has its own capillary network and a lymph vessel so that as nutrients move across the cells, they can immediately mingle into the body fluids. On every villus every cell has a brushlike covering of tiny hairs (**microvilli**) that can trap the nutrient particles. Figure B-8 provides a close look at these details.

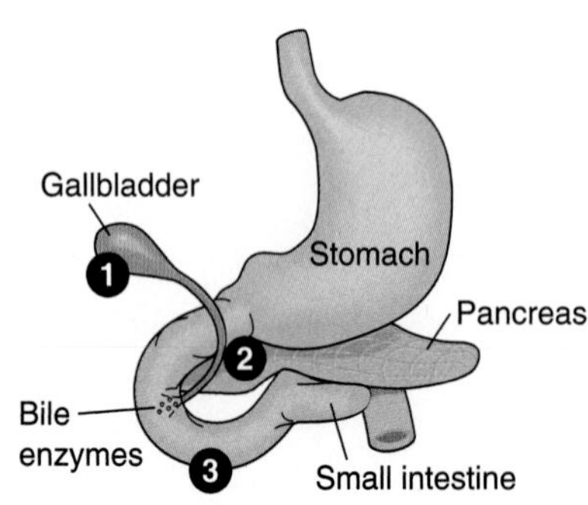

Small intestine–details

1. The gallbladder sends bile into the small intestine by way of a duct.
2. The pancreas sends enzymes (and bicarbonate).
3. The small intestine also secretes enzymes.

The small intestine's lining, villi and all, is wrinkled into thousands of folds, so that its absorbing surface is enormous. If the folds, and the villi that cover them, were spread out flat, the total area would equal a third of a football field in size. The billions of cells of that surface, although they weigh only 4 to 5 pounds, absorb enough nutrients in a few hours a day to nourish the other 150 or so pounds of body tissues.

Nutrients released early in the digestive process, such as simple sugars, and those requiring no special handling, such as the water-soluble vitamins, are absorbed high in the small intestine; nutrients that are released more slowly are absorbed further down. The lymphatic and circulatory systems then take over the job of transporting them to the cell consumers. The lymph at first carries most of the products of fat digestion and the fat-soluble vitamins, later delivering them to the blood. The blood carries the products of carbohydrate and protein digestion, the water-soluble vitamins, and the minerals. By the time the

FIGURE B-8
DETAILS OF THE LINING OF THE SMALL INTESTINE

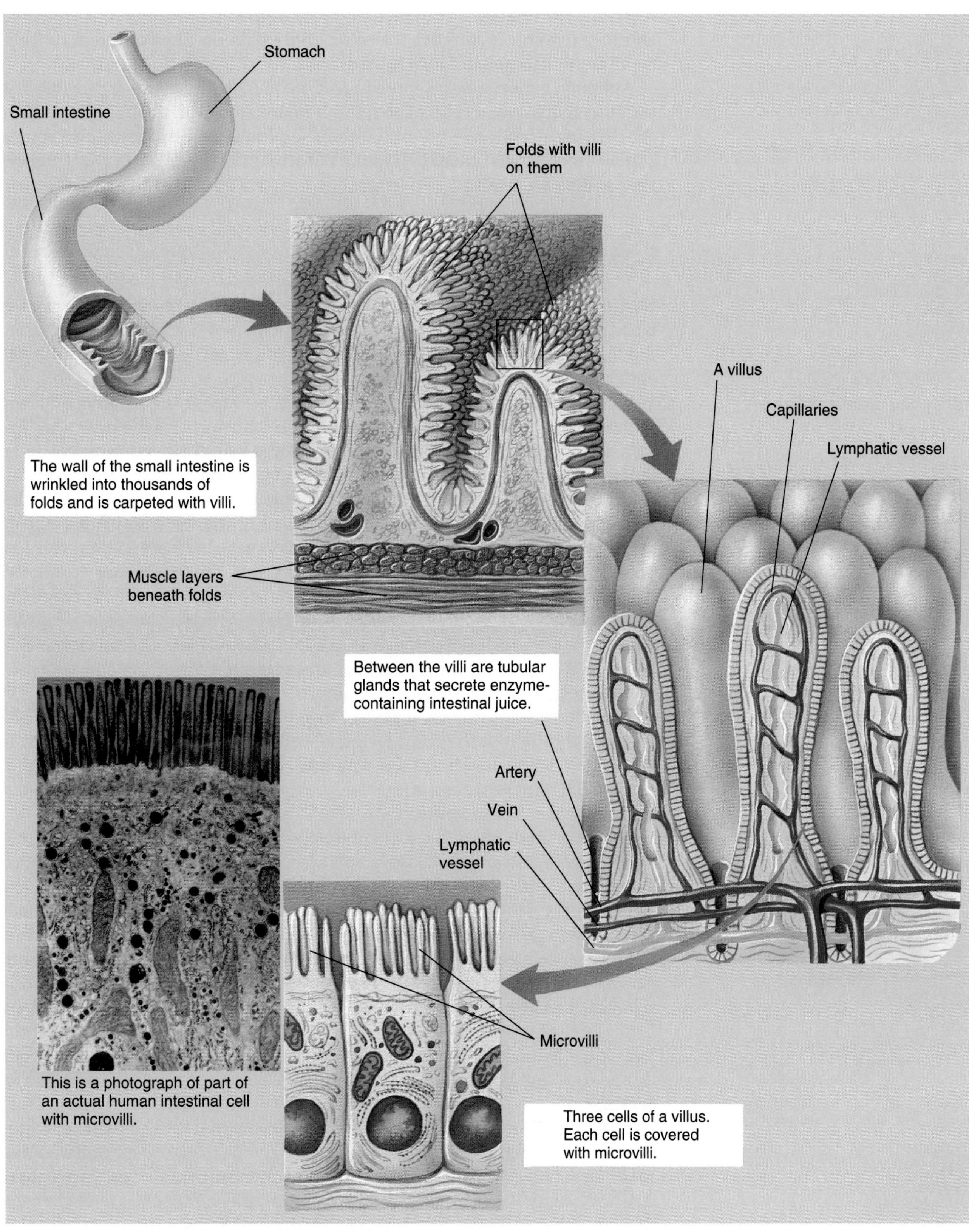
Stomach
Small intestine
Folds with villi on them
The wall of the small intestine is wrinkled into thousands of folds and is carpeted with villi.
Muscle layers beneath folds
A villus
Capillaries
Lymphatic vessel
Between the villi are tubular glands that secrete enzyme-containing intestinal juice.
Artery
Vein
Lymphatic vessel
Microvilli
This is a photograph of part of an actual human intestinal cell with microvilli.
Three cells of a villus. Each cell is covered with microvilli.

remaining mixture reaches the end of the small intestine, little is left but water, indigestible residue (mostly fiber), and dissolved minerals. The cells lining the colon are specialized for absorbing these minerals and retrieving the water for recycling. The final waste product, the feces, a smooth paste of a consistency suitable for excretion, is stored in the colon until excretion. Such a system can adjust to whatever mixture of foods is presented.

Although a meal may be eaten in half an hour, the nutrients it provides reach the body fluids over a span of about four hours. However, as already mentioned, the cells of the body need their nutrients around the clock. Providing a constant supply requires that there be systems of storage and release to meet the cells' needs between meals.

STORAGE SYSTEMS

Nutrients leave the digestive system by way of both circulatory systems—the blood and the lymph. The blood carries products of carbohydrate and protein digestion and some of the smaller fats; the lymph carries the larger fats in packages called chylomicrons (see Chapter 4).

More about liver glycogen—Chapter 3.
More about lipoproteins—Chapter 4.
More about fasting—Chapter 8.
More about protein deficiency—Chapter 5.

All nutrients leaving the digestive system by way of the blood are collected in thousands of capillaries in the membrane that supports the intestine. These converge into veins and then into a single large vein. This vein conveys its contents to the liver and there breaks up once again into a vast network of capillaries that weave among the liver cells, allowing them access to the newly arriving nutrients. The liver cells process these nutrients. They convert the sugars from carbohydrate mostly into the body's sugar, glucose; and if there is a surplus, they store some as glycogen and convert the remainder to fat. They reassemble fatty acids and glycerol from fat into larger fats and package them with protein for transport to other parts of the body. As for the amino acids from protein, the liver cells alter these as needed, making glucose from some if necessary and fat from others if there is an excess, or converting one amino acid into another to use in making proteins.

The nutrients leaving the digestive tract by way of the lymph as chylomicrons circulate throughout the body, giving all cells the opportunity to withdraw fats from them. Some also find their way into the blood and circulate through the liver, which removes them, alters their components, and releases new products, including other lipoproteins.

The new products of liver metabolism—glucose, fat packaged with protein (lipoproteins), and amino acids—are released into the bloodstream again and circulated to all other cells of the body. Surplus fat is then removed by cells specialized for its storage; these fat cells are located in deposits all over the body.

The liver's glycogen provides a reserve supply of the body's sugar, glucose, and thus can sustain cell activities if the intervals between meals become so long that glucose absorbed from ingested foods is used up. When the body is depending solely on liver glycogen, however, the supply is used up within three to six hours. Similarly, the fat cells store reserves of fat, the body's other principal energy nutrient. Unlike the liver, however the fat cells have virtually infinite storage capacity and can continue to supply fat for days, weeks, or even months when no food is eaten.

These storage systems for glucose and fat ensure that the cells will not go without energy nutrients even if the body is hungry for food, except under extreme conditions. Body stores also exist for many other nutrients, each with a characteristic capacity. For example, the third energy nutrient, protein, is held in an available pool (the amino acids in the liver and blood) that is rather rapidly depleted during protein deficiency. The liver and fat cells store many vitamins, and the bones provide reserves of calcium, sodium, and other minerals that can be drawn on to keep the blood levels constant and to meet cellular demands.

METABOLISM: BREAKING DOWN NUTRIENTS FOR ENERGY

The breaking down of body compounds is known as **catabolism.** These reactions usually release energy and are represented, wherever possible, by "down" arrows in chemical diagrams (see Figure B-9). Glycogen can be broken down to glucose, triglycerides to fatty acids and glycerol, and protein to amino acids. When the body needs energy, it breaks any or all of the four basic units—glucose, fatty acids, glycerol, and amino acids—into even smaller units. When the body does not require energy, the end-products of digestion (glucose, amino acids, glycerol, and fatty acids) are used to build body compounds in a process called **anabolism** (see Figure B-9). Anabolic reactions involve the conversion of glucose to glycogen or fat, the conversion of amino acids to body proteins or fat, and the synthesis of body fat from glycerol and fatty acids. Catabolism and anabolism are examples of **energy metabolism.**

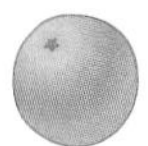

OTHER SYSTEMS

In addition to the systems described above, the body has many more: the bones, the muscles, the nerves, the lungs, the reproductive organs, and others. All of these cooperate so that each cell can carry on its own life. Each assures, through hormonal or nerve-mediated messages, that its needs will be met by the others, and each contributes to the welfare of the whole by doing the work it is specialized for.

Of the millions of cells in the body, only a small percentage comprise the cortex of the brain, in which the conscious mind resides. These receive messages from other cells when they require you to "become conscious" of a need for decision and action. In modern life the need may be as complex as, for example, to notice that you feel anxious and to decide to consult an advisor, or it may be such

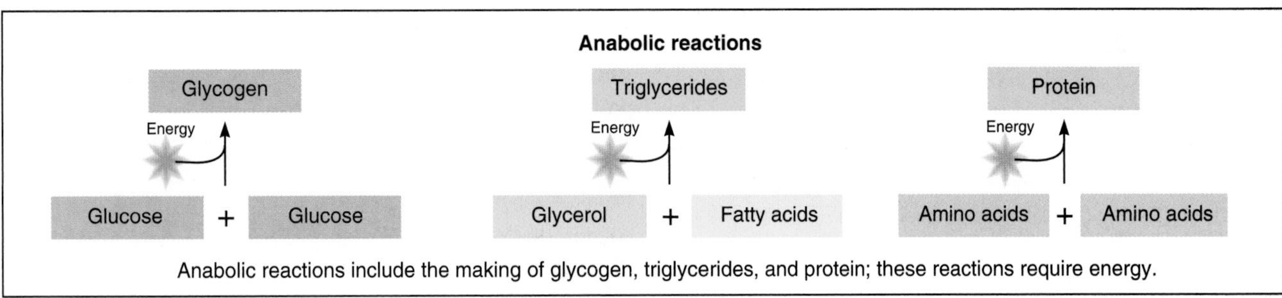

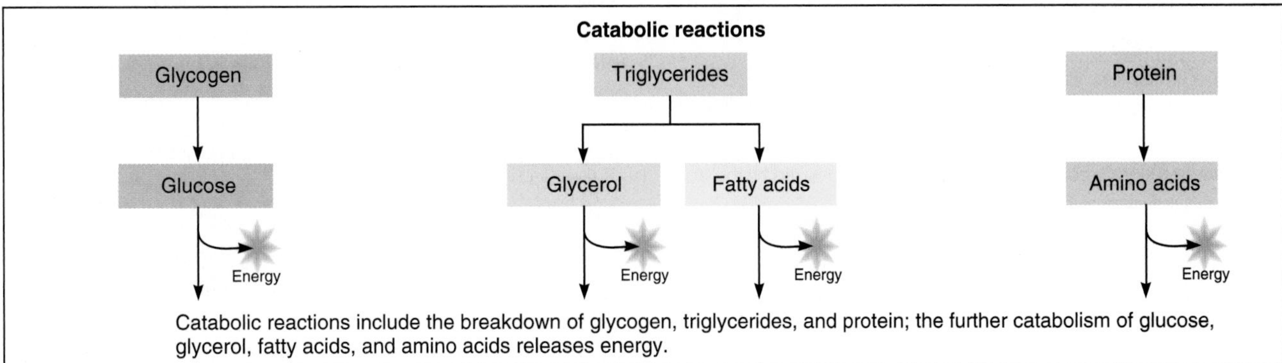

FIGURE B-9
REACTIONS OF ENERGY METABOLISM COMPARED

Source: Reprinted with permission from E. N. Whitney and coauthors, *Nutrition for Health and Health Care* (St. Paul, MN: West Publishing, 1995) p. 120.

a "simple" need as "I'm tired, I think I'll go to bed," or "I'm hungry, I guess I'd better eat."

Most of the body's work is done automatically and is finely regulated to achieve a state of well-being. But when your cortex does become involved, you would do well to "listen" to your body and to cultivate an understanding and appreciation of its needs. Then when you make decisions you will act to promote your health.

MINIGLOSSARY

ANABOLISM (ann-ABB-o-lism) reactions in which small molecules are put together to build larger ones. Anabolic reactions consume energy.
ana = up

ANTIBODIES proteins made by the immune system, expressly designed to combine with and to inactivate specific antigens.

ANTIGENS microbes or substances that are foreign to the body.

ARTERIES blood vessels that carry blood containing fresh oxygen supplies from the heart to the tissues.

BICARBONATE a chemical that neutralizes acid; a secretion of the pancreas.

BILE a compound made from cholesterol by the liver, stored in the gallbladder, and secreted into the small intestine. It emulsifies lipids to ready them for enzymatic digestion.

BLOOD the fluid of the circulatory system—water, red and white blood cells and other formed particles, proteins, nutrients, oxygen, and other constituents.

CAPILLARIES minute, weblike blood vessels that connect arteries to veins and permit transfer of materials between blood and tissues.

CATABOLISM (ca-TAB-o-lism) reactions in which large molecules are broken down to smaller ones. Catabolic reactions usually release energy.
kata = down

CELLS the smallest units in which independent life can exist. All living things are single cells or organisms made of cells.

CORTEX an outer covering; in the brain, that part in which conscious thought takes place.

EMULSIFIER (ee-MULL-sih-fire) a compound with both water-soluble and fat-soluble portions that can attract lipids into water to form an emulsion.

ENDOCRINE (EN-doh-crin) a term to describe a gland secreting or a hormone being secreted into the blood.
endo = into

ENERGY METABOLISM all the reactions by which the body obtains and spends the energy from food or body stores.

ENZYME a protein catalyst. A catalyst is a compound that facilitates (speeds up the rate of) a chemical reaction without itself being altered in the process.

ESSENTIAL NUTRIENTS compounds that can't be synthesized by the body in amounts sufficient to meet physiological needs.

GENE a unit of a cell's inheritance, made of a chemical, DNA, that is copied faithfully so that every time the cell divides, both its offspring get identical copies. Genes direct the cells' machinery to make the proteins that form each cell's structures and to do its work.

HORMONE a chemical messenger, secreted by one organ (a gland) in response to a condition in the body, that acts on another organ or organs to change that condition.

HYPOTHALAMUS (high-poh-THALL-uh-mus) a part of the brain that senses a variety of conditions in the blood, such as temperature, salt content, glucose content, and others, and signals other parts of the brain or body to change those conditions when necessary.

IMMUNITY the ability to successfully resist a disease, conferred on the body by way of the immune system's memory of previous exposure to that disease and its ability to mount a specific defense promptly and swiftly.

INSULIN a hormone from the pancreas that helps glucose get into cells.

INTESTINE a long, tubular organ of digestion and the site of nutrient absorption.

KIDNEYS the organs that filter the blood to remove waste material and forward it to the bladder for excretion.

LIVER the large, lobed organ that lies under the ribs and filters the blood, removing, processing, and readying for redistribution many of its materials.

LUNGS the organs of gas exchange. Blood circulating through the lungs releases its carbon dioxide and picks up fresh oxygen to carry to the tissues.

LYMPH (LIMF) the fluid outside the circulatory system that bathes the cells, derived from the blood by being pressed through the capillary walls; similar to the blood in composition but without red blood cells.

METABOLISM (meh-TAB-o-lism) total of all chemical reactions that go on in living cells.

MICROBES bacteria, viruses, or other organisms invisible to the naked eye; some cause disease.

MICROVILLI (MY-croh-VILL-ee, MY-croh-VILL-eye) tiny hairlike projections on each cell of the intestinal tract lining that can trap nutrient particles and translocate them into the cells (singular: **microvillus**).

MUCUS (MYOO-cus) a thick, slippery coating of the intestinal tract lining (and other body linings) that protects the cells from exposure to digestive juices. The adjective form is *mucous,* and the coating is often called the mucous membrane.

PANCREAS a gland that secretes the endocrine hormone insulin and also produces the exocrine secretions that aid digestion in the small intestine. (An *exocrine* secretion is one that is expelled through a duct into a body cavity or onto the surface of the skin; *exo* means "out." See also *endocrine*.)

PERISTALSIS (perri-STALL-sis) the wavelike squeezing motions of the stomach and intestines that push their contents along.

PYLORUS (pye-LORE-us) muscle that regulates the opening of the bottom of the stomach.

VEINS blood vessels that carry used blood from the tissues back to the heart.

VILLI (VILL-ee, VILL-eye) fingerlike projections of the sheet of cells that line the GI tract; the villi make the surface area much greater than it would otherwise be (singular: **villus**).

Canadian Dietary Guidelines and Recommendations

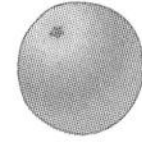

CANADA'S GUIDELINES FOR HEALTHY EATING

Canada's Guidelines for Healthy Eating were developed by the Communications/ Implementation Committee as the key nutrition messages to be communicated to healthy Canadians over two years of age. The guidelines encourage people to

- Enjoy a variety of foods.
- Emphasize cereals, breads, other grain products, vegetables, and fruits.
- Choose lower-fat dairy products, leaner meats, and foods prepared with little or no fat.
- Achieve and maintain a healthy body weight by enjoying regular physical activity and healthy eating.
- Limit salt, alcohol, and caffeine.

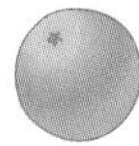

CANADA'S RECOMMENDED NUTRIENT INTAKES (RNI)

A major revision of the nutrient recommendations is underway in the United States and Canada. The Dietary Reference Intakes (DRI) reports are replacing both the 1989 RDA in the United States and the 1990 RNI in Canada. The new DRI are presented on the inside front cover. The RNI for vitamins, minerals, protein, and energy that do not yet have new DRI values are presented here in Tables C-1 and C-2.

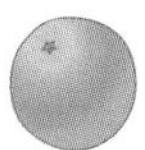

CANADA'S FOOD GUIDE TO HEALTHY EATING

Canada's Food Guide to Healthy Eating gives detailed information for selecting foods to meet *Canada's Guidelines for Healthy Eating* (see page A-24). The *Food Guide* was designed to meet the nutritional needs of all Canadians four years of age and older and takes a total diet approach, rather than emphasizing a single food, meal, or day's meals and snacks. The rainbow side of the *Food Guide* shows the four food groups with pictorial examples of foods in each group. The bar side shows the number of servings recommended for each group.

TABLE C-1 RECOMMENDED NUTRIENT INTAKES FOR CANADIANS, 1990

Age	Sex	Weight (kg)	Protein (g/day)[a]	Vitamins	Minerals		
				Vitamin A (RE/day)[b]	Iron (mg/day)	Iodine (mg/day)	Zinc (mg/day)
Infants (months)							
0–4	Both	6	12[c]	400	0.3[d]	30	2[e]
5–12	Both	9	12	400	7	40	3
Children (years)							
1	Both	11	13	400	6	55	4
2–3	Both	14	16	400	6	65	4
4–6	Both	18	19	500	8	85	5
7–9	M	25	26	700	8	110	7
	F	25	26	700	8	95	7
10–12	M	34	34	800	8	125	9
	F	36	36	800	8	110	9
13–15	M	50	49	900	10	160	12
	F	48	46	800	13	160	9
16–18	M	62	58	1000	10	160	12
	F	53	47	800	12	160	9
Adults (years)							
19–24	M	71	61	1000	9	160	12
	F	58	50	800	13	160	9
25–49	M	74	64	1000	9	160	12
	F	59	51	800	13[f]	160	9
50–74	M	73	63	1000	9	160	12
	F	63	54	800	8	160	9
75+	M	69	59	1000	9	160	12
	F	64	55	800	8	160	9
Pregnancy (additional amount needed)							
1st trimester			5	0	0	25	6
2nd trimester			20	0	5	25	6
3rd trimester			24	0	10	25	6
Lactation (additional amount needed)			20	400	0	50	6

Note: Recommended intakes of energy and of certain nutrients are not listed in this table because of the nature of the variables upon which they are based. The figures for energy are estimates of average requirements for expected patterns of activity (see Table C-2). For nutrients not shown, the following amounts are recommended based on at least 2000 calories per day and body weights as given: thiamin, 0.4 milligram per 1000 calories (0.48 milligram/5000 kilojoules); riboflavin, 0.5 milligram per 1000 calories (0.6 milligram/5000 kilojoules); niacin, 7.2 niacin equivalents per 1000 calories (8.6 niacin equivalents/5000 kilojoules); vitamin B_6, 15 micrograms, as pyridoxine, per gram of protein. Recommended intakes during periods of growth are taken as appropriate for individuals representative of the midpoint in each age group. All recommended intakes are designed to cover individual variations in essentially all of a healthy population subsisting upon a variety of common foods available in Canada.

Source: Health and Welfare Canada, *Nutrition Recommendations: The Report of the Scientific Review Committee* (Ottawa: Canadian Government Publishing Centre, 1990), Table 20, p. 204.

[a]The primary units are expressed per kilogram of body weight. The figures shown here are examples.

[b]One retinol equivalent (RE) corresponds to the biological activity of 1 microgram of retinol, 6 micrograms of beta-carotene, or 12 micrograms of other carotenes.

[c]The assumption is made that the protein is from breast milk or has the same biological value as breast milk and that, between 3 and 9 months, adjustment for the quality of the protein is made.

[d]Based on the assumption that breast milk is the source of iron.

[e]Based on the assumption that breast milk is the source of zinc.

[f]After menopause, the recommended intake is 8 milligrams per day.

FOOD LABELS

This section defines the terms found on food labels in Canada (see Table C-3). Figure C-1 on page A-26 illustrates what is found on a sample food label.

1 cup oatmeal
1/2 C ~~L~~ Nonfat
1 C coffee 1 T 1/2 n 1/2
1 tea sugar

1 W.W. Br
1 T P.B
1 T raspberry jam

3 dates 8 almonds

1 orange

3 chix ~~[illegible]~~ tacos

green tea

TABLE C-2 AVERAGE ENERGY REQUIREMENTS FOR CANADIANS

Age	Sex	Average Height (cm)	Average Weight (kg)	Requirements* (cal/kg)†	(MJ/kg)†	(cal/day)	(MJ/day)	(cal/cm)	(MJ/cm)
Infants (months)									
0–2	Both	55	4.5	120–100	0.50–0.42	500	2.0	9	0.04
3–5	Both	63	7.0	100–95	0.42–0.40	700	2.8	11	0.05
6–8	Both	69	8.5	95–97	0.40–0.41	800	3.4	11.5	0.05
9–11	Both	73	9.5	97–99	0.41	950	3.8	12.5	0.05
Children and adults (years)									
1	Both	82	11	101	0.42	1,100	4.8	13.5	0.06
2–3	Both	95	14	94	0.39	1,300	5.6	13.5	0.06
4–6	Both	107	18	100	0.42	1,800	7.6	17	0.07
7–9	M	126	25	88	0.37	2,200	9.2	17.5	0.07
	F	125	25	76	0.32	1,900	8.0	15	0.06
10–12	M	141	34	73	0.30	2,500	10.4	17.5	0.07
	F	143	36	61	0.25	2,200	9.2	15.5	0.06
13–15	M	159	50	57	0.24	2,800	12.0	17.5	0.07
	F	157	48	46	0.19	2,200	9.2	14	0.06
16–18	M	172	62	51	0.21	3,200	13.2	18.5	0.08
	F	160	53	40	0.17	2,100	8.8	13	0.05
19–24	M	175	71	42	0.18	3,000	12.6		
	F	160	58	36	0.15	2,100	8.8		
25–49	M	172	74	36	0.15	2,700	11.3		
	F	160	59	32	0.13	1,900	8.0		
50–74	M	170	73	31	0.13	2,300	9.7		
	F	158	63	29	0.12	1,800	7.6		
75+	M	168	69	29	0.12	2,000	8.4		
	F	155	64	23	0.10	1,500	6.3		

*Requirements can be expected to vary within a range of ±30%.
†First and last figures are averages at the beginning and end of the three-month period.

Source: Health and Welfare Canada. *Nutrition Recommendations: The Report of the Scientific Review Committee* (Ottawa: Canadian Government Publishing Centre, 1990), Tables 5 and 6, pp. 25, 27. Used with permission.

TABLE C-3 TERMS ON FOOD LABELS

calorie reduced 50% or fewer calories than the regular version.

light term may be used to describe anything (for example, light in colour, texture, flavour, taste, or calories); read the label to find out what is "light" about the product.

low calorie calorie-reduced and no more than 15 calories per serving.

low cholesterol no more than 3 mg of cholesterol per 100 g of the food and low in saturated fat; *does not* always mean low in total fat.

low fat no more than 3 g of fat per serving; *does not* always mean low in calories.

lower fat at least 25% less fat than the comparison food.

carbohydrate reduced not more than 50% of the carbohydrate found in the regular version.

source of dietary fibre a product that provides 2–4 g of fibre.

high source of dietary fibre a product that provides 4–6 g of fibre.

very high source of fibre a product that provides 6 g (or more) of fibre.

sugar free low in carbohydrates and calories; an "extra" food in the exchange system.

Healthy Canada

Health and Welfare Canada — Santé et Bien-étre social Canada

CANADA'S

Food Guide

TO HEALTHY EATING

Enjoy a variety of foods from each group every day.

Choose lower-fat foods more often.

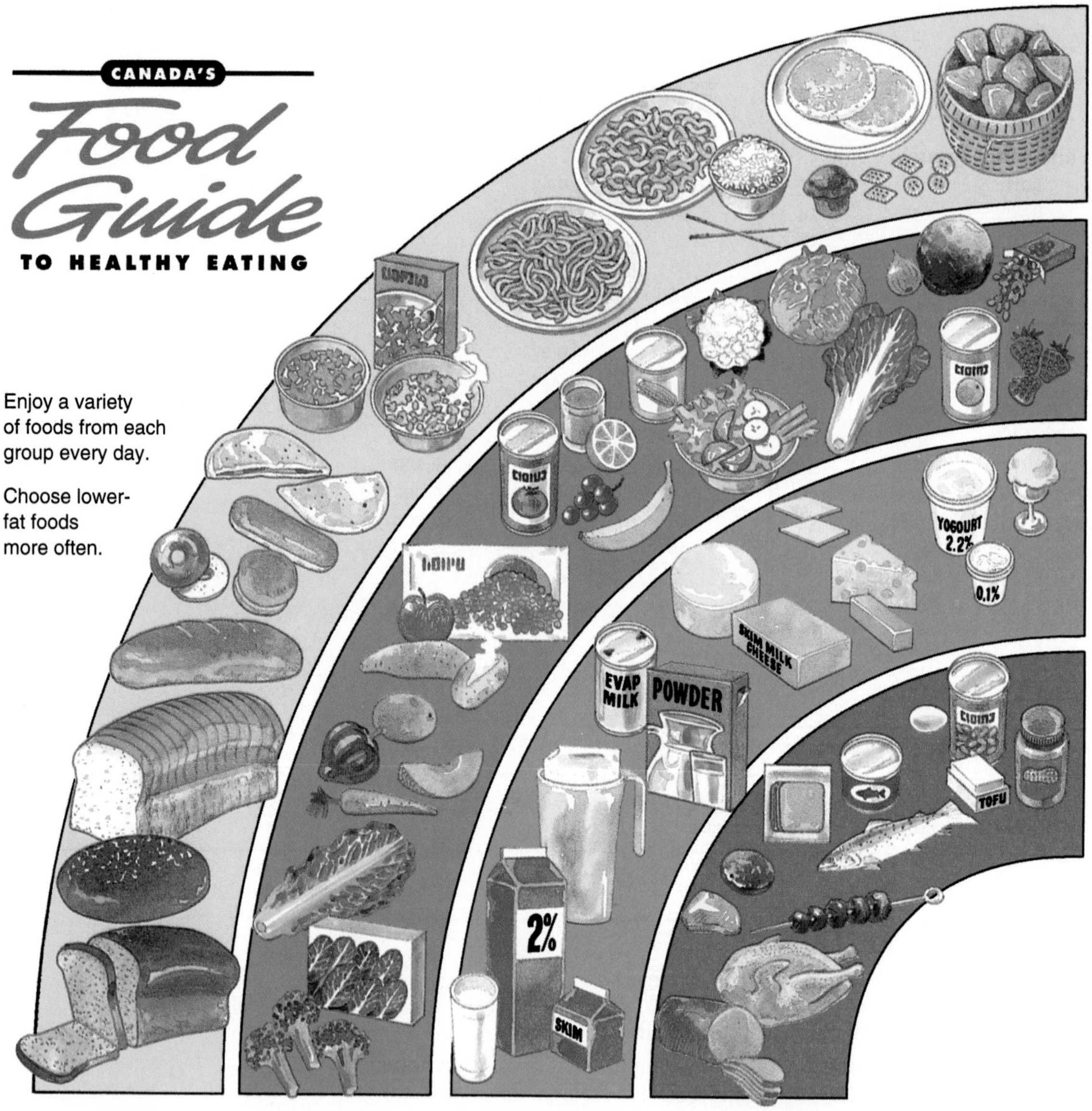

Grain Products
Choose whole grain and enriched products more often.

Vegetables & Fruit
Choose dark green and orange vegetables and orange fruit more often.

Milk Products
Choose lower-fat milk products more often.

Meat & Alternatives
Choose leaner meats, poultry and fish, as well as dried peas, beans and lentils more often.

CANADA'S
Food Guide
TO HEALTHY EATING
FOR PEOPLE FOUR YEARS AND OVER

Different People Need Different Amounts of Food

The amount of food you need every day from the 4 food groups and other foods depends on your age, body size, activity level, whether you are male or female and if you are pregnant or breast-feeding. That's why the Food Guide gives a lower and higher number of servings for each food group. For example, young children can choose the lower number of servings, while male teenagers can go to the higher number. Most other people can choose servings somewhere in between.

Grain Products
5–12
SERVINGS PER DAY

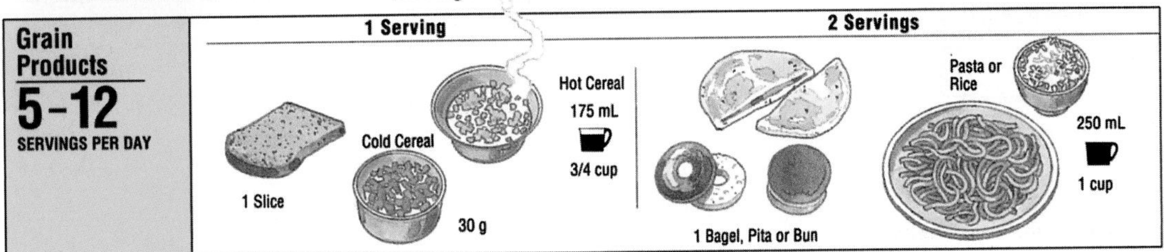

Vegetables & Fruit
5–10
SERVINGS PER DAY

Milk Products
SERVINGS PER DAY
Children 4–9 years: 2–3
Youth 10–16 years: 3–4
Adults: 2–4
Pregnant & Breast-feeding Women: 3–4

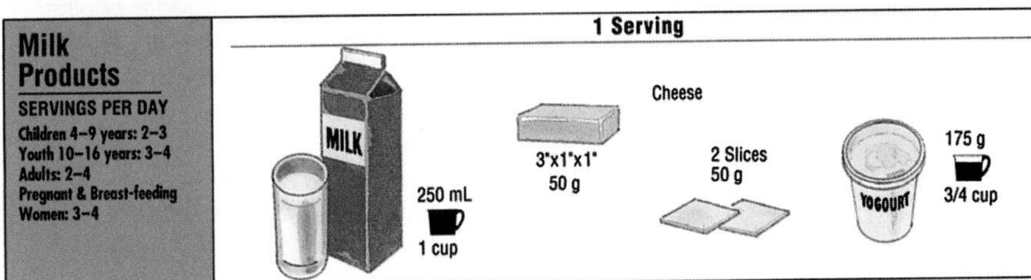

Meat & Alternatives
2–3
SERVINGS PER DAY

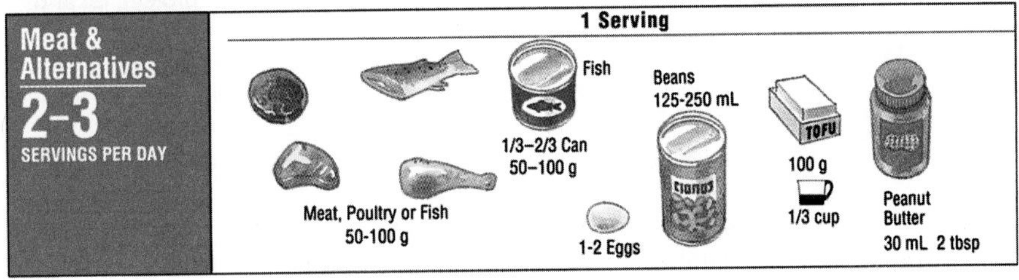

Other Foods

Taste and enjoyment can also come from other foods and beverages that are not part of the 4 food groups. Some of these foods are higher in fat or Calories, so use these foods in moderation.

Enjoy eating well, being active and feeling good about yourself. That's VITALITE.

ISBN 0-662-19648-1

FIGURE C-1
EXAMPLE OF A FOOD LABEL

OUR COMMITMENT TO QUALITY

Kellogg's is committed to providing foods of outstanding quality and freshness. If this product in any way falls below the high standards you've come to expect from Kellogg's, please send your comments and both top flaps to:
Consumer Affairs
KELLOGG CANADA INC.
Etobicoke, Ontario M9W 5P2

IF IT DOESN'T SAY Kellogg's* ON THE BOX, IT'S NOT Kellogg's* IN THE BOX.
SI LE NOM Kellogg's* N'EST PAS SUR LA BOÎTE, CE N'EST PAS Kellogg's* DANS LA BOÎTE.

- HIGH IN FIBRE
- LOW IN FAT
- PRESERVATIVE FREE
- SOURCE ÉLEVÉE DE FIBRES
- FAIBLE EN MATIÈRES GRASSES
- SANS AGENT DE CONSERVATION

NUTRITION INFORMATION
APPORT NUTRITIONNEL

	Per 40 g serving cereal (175 mL, ¾ cup) / Par ration de 40 g de céréale (175 mL, ¾ tasse)	Per 40 g serving cereal with 125 mL Partly Skimmed Milk (2%) / Par ration de 40 g de céréale avec 125 mL de lait partiellement écrémé (2,0 %)	
ENERGY	130Cal	195Cal	ÉNERGIE
	540kJ	810kJ	
PROTEIN	3.0g	7.3g	PROTÉINES
FAT	0.4g	2.9g	MATIÈRES GRASSES
CARBOHYDRATE	32g	38g	GLUCIDES
SUGARS*	11g	18g	*SUCRES
STARCH	16g	16g	AMIDON
DIETARY FIBRE	4.6g	4.6g	FIBRES ALIMENTAIRES
SODIUM	235mg	300mg	SODIUM
POTASSIUM	240mg	440mg	POTASSIUM

% of Recommended Daily Intake
% de l'apport quotidien conseillé

VITAMIN A	0%	7%	VITAMINE A
VITAMIN D	0%	23%	VITAMINE D
VITAMIN B1	62%	66%	VITAMINE B1
VITAMIN B2	3%	16%	VITAMINE B2
NIACIN	13%	18%	NIACINE
VITAMIN B6	13%	16%	VITAMINE B6
FOLACIN	11%	14%	FOLACINE
VITAMIN B12	0%	25%	VITAMINE B12
PANTOTHENATE	9%	15%	PANTOTHÉNATE
CALCIUM	1%	15%	CALCIUM
PHOSPHORUS	12%	23%	PHOSPHORE
MAGNESIUM	20%	27%	MAGNÉSIUM
IRON	38%	39%	FER
ZINC	16%	22%	ZINC

*Approximately half of the sugars occur naturally in the raisins.
Environ la moitié des sucres se retrouvent à l'état naturel dans les fruits.

Canadian Diabetes Association Food Choice Values: 40 g (175 mL, ¾ cup) cereal. Système des choix d'aliments de l'Association canadienne du diabète : 40 g (175 mL, ¾ tasse) céréale = 1 [■] + ½ [◆] + ½ [✱] choices/choix

INGREDIENTS / INGRÉDIENTS

WHOLE WHEAT, RAISINS (COATED WITH SUGAR, HYDROGENATED VEGETABLE OIL), WHEAT BRAN, SUGAR/GLUCOSE-FRUCTOSE, SALT, MALT (CORN FLOUR, MALTED BARLEY), VITAMINS (THIAMIN HYDROCHLORIDE, PYRIDOXINE HYDROCHLORIDE, FOLIC ACID, d-CALCIUM PANTOTHENATE), MINERALS (IRON, ZINC OXIDE).

BLÉ ENTIER, RAISINS SECS (ENROBÉS DE SUCRE, D'HUILE VÉGÉTALE HYDROGÉNÉE), SON DE BLÉ, SUCRE/GLUCOSE-FRUCTOSE, SEL, MALT (FARINE DE MAÏS, ORGE MALTÉ), VITAMINES (CHLORHYDRATE DE THIAMINE, CHLORHYDRATE DE PYRIDOXINE, ACIDE FOLIQUE, PANTOTHÉNATE DE d-CALCIUM), MINÉRAUX (FER, OXYDE DE ZINC).

Made by / Produit par
KELLOGG CANADA INC.
ETOBICOKE, ONTARIO
CANADA M9W 5P2
*Registered trademark of /
*Marque déposée de
KELLOGG CANADA INC. © 1994
00094

WHAT YOU WILL FIND ON A LABEL:

Nutrition Claims

- in Canada, it is optional for a company to decide to use claims,
- when claims appear on a label, they must follow government laws

Nutrition Information

- gives detailed nutrition facts about the product, including serving size and core list
- does not have to appear by law on food products in Canada
- refers to the food as packaged, so if you add milk, eggs or other food, the nutritional content of the food you eat can be very different

Serving Size

- the amount of food for which the information is given
- check the serving size: the serving size on the label may not be the same as the serving size you would actually eat (for example, the serving size of cereal may be ¾ cup, much smaller than your regular serving

Core List

- the energy (in Calories and kilojoules), grams of protein, fat and carbohydrate for each serving
- some products break down fat into monounsaturates, polyunsaturates, saturates, and cholesterol (to find out what these mean, look at the Fats & Oils section)
- carbohydrates may include the amount of sugars, starch and fibre, or may list these items separately

Sodium and Potassium (in milligrams)

Vitamins and Minerals (as percent of your recommended daily intake)

Canadian Diabetes Association Food Choice Values and Symbols

- the Values and Symbols are tools to help you fit the food into your meal plan, they are not an endorsement by CDA
- it is up to the food company to decide if they want their foods analyzed and assigned symbols
- when they are on a label, they have been assigned by a dietitian working for CDA, so you can be sure the information is correct

Ingredients

- must be found on all food labels by law
- ingredients are listed in decreasing order by weight, so what you see first is what you get the most of

Appendix D

Aids to Calculation

Many mathematical problems have been worked out for you as examples at appropriate places in the text. This appendix aims to help with the use of the metric system and with those problems not fully explained elsewhere.

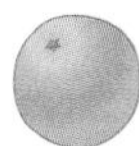

CONVERSION FACTORS*

Conversion factors are useful mathematical tools in everyday calculations, like the ones encountered in the study of nutrition. A conversion factor is a fraction in which the numerator (top) and the denominator (bottom) express the same quantity in different units. For example, 2.2 pounds and 1 kilogram are equivalent; they express the same weight. The conversion factor used to change pounds to kilograms or vice versa is:

$$\frac{2.2\text{ lb}}{1\text{ kg}} \quad \text{or} \quad \frac{1\text{ kg}}{2.2\text{ lb}}$$

Because either of these factors equals 1, a measurement can be multiplied by the factor without changing the value of the measurement. Thus its units can be changed.

The correct factor to use in a problem is the one with the unit you are seeking in the numerator (top) of the fraction. Following are three examples of problems commonly encountered in nutrition study; they illustrate the usefulness of conversion factors.

Example 1

Convert ¼ cup to an approximate number of milliliters for use in a recipe.

1. The conversion factor is:

$$\frac{1\text{ c}}{250\text{ ml}} \quad \text{or} \quad \frac{250\text{ ml}}{1\text{ c}}$$

2. Multiply ¼ cup by the factor:

$$\tfrac{1}{4}\,\cancel{\text{c}} \times \frac{250\text{ ml}}{1\,\cancel{\text{c}}} = 62.5\text{ ml, or about 60 ml}$$

*For a listing of specific conversion factors, see Table D-1 on page A-30.

Example 2

Convert the weight of 130 pounds to kilograms.

1. Choose the conversion factor in which the unit you are seeking is on top:

$$\frac{1\text{ kg}}{2.2\text{ lb}}$$

2. Multiply 130 pounds by the factor:

$$130\text{ lb} \times \frac{1\text{ kg}}{2.2\text{ lb}} = \frac{130\text{ kg}}{2.2} = 59\text{ kg (rounded off to nearest whole number)}$$

Example 3

How many grams of saturated fat are contained in a 3-ounce hamburger? A 4-ounce hamburger contains 7 grams of saturated fat.

1. You are seeking grams of saturated fat; therefore, the conversion factor is:

$$\frac{7\text{ g saturated fat}}{4\text{ oz hamburger}}$$

2. Multiply 3 ounces of hamburger by the conversion factor:

$$3\text{ oz hamburger} \times \frac{7\text{ g saturated fat}}{4\text{ oz hamburger}} = \frac{3 \times 7\text{ g}}{4} = \frac{21}{4}$$

$$= 5\text{ g saturated fat (rounded off to nearest whole number)}$$

PERCENTAGES

A percentage is a comparison between a number of items (perhaps your intake of calories) and a standard number (perhaps the number of calories recommended for your age and sex—your energy RDA). The standard number is the number you divide by. The answer you get after the division must be multiplied by 100 to be stated as a percentage (*percent* means "per 100").

Example 4

What percentage of the 1989 RDA for energy is your calorie intake?

1. Find your energy RDA (inside front cover). We'll use 2,100 calories to demonstrate.
2. Total your calorie intake for a day—for example, 1,200 calories.
3. Divide your calorie intake by the RDA calories:

   1,200 cal (your intake) ÷ 2,100 cal (RDA) = 0.571
4. Multiply your answer by 100 to state it as a percentage:

   0.571 × 100 = 57.1 = 57% (rounded off to the nearest whole number).

In some problems in nutrition, the percentage may be more than 100. For example, suppose your daily intake of vitamin A is 3,200 RE and your RDA (male) is 1,000 RE. Your intake as a percentage of the RDA is more than 100 percent (that is, you consume more than 100 percent of your vitamin A RDA). The following calculations show your vitamin A intake as a percentage of the RDA:

$$3{,}200 \div 1{,}000 = 3.2$$

$$3.2 \times 100 = 320\%\text{ of RDA}$$

Sometimes the comparison is between a part of a whole (for example, your calories from protein) and the total amount (your total calories). In this case, the total number is the one you divide by as shown in Example 5.

Example 5

What percentages of your total calories for the day come from protein, fat, and carbohydrate?

1. Using Appendix H and your diet record, find the total grams of protein, fat, and carbohydrate you consumed—for example, 60 grams protein, 80 grams fat, and 285 grams carbohydrate.
2. Multiply the number of grams by the number of calories from 1 gram of each energy nutrient (conversion factors):

$$60 \text{ g protein} \times \frac{4 \text{ cal}}{1 \text{ g protein}} = 240 \text{ cal}$$

$$80 \text{ g fat} \times \frac{9 \text{ cal}}{1 \text{ g fat}} = 720 \text{ cal}$$

$$285 \text{ g carbohydrate} \times \frac{4 \text{ cal}}{1 \text{ g carbohydrate}} = 1{,}140 \text{ cal}$$

$$240 + 720 + 1{,}140 = 2{,}100 \text{ cal}$$

3. Find the percentage of total calories from each energy nutrient (see Example 4):

$$\text{Protein: } 240 \div 2{,}100 = 0.114$$
$$0.114 \times 100 = 11.4 = 11\% \text{ of calories.}$$

$$\text{Fat: } 720 \div 2{,}100 = 0.343$$
$$0.343 \times 100 = 34.3 = 34\% \text{ of calories.}$$

$$\text{Carbohydrate: } 1{,}140 \div 2{,}100 = 0.543$$
$$0.543 \times 100 = 54.3 = 54\% \text{ of calories.}$$

$$11\% + 34\% + 54\% = 99\% \text{ of calories (total)}$$

The percentages total 99 percent rather than 100 percent because a little was lost from each number in rounding off. Either 99 or 101 is a reasonable total.

NUTRIENT UNITS

International Units (IU)

To convert IU (International Units) found on supplement labels to the units used in the DRI or RDA tables:

- RE:[a] from animal sources, divide by 3.33; and from vegetables and fruits, divide by 10.
- μg vitamin D: divide by 40 or multiply by 0.025.
- mg α-T:[b] multiply by 0.67, if the supplement is labeled "natural" (d-α-tocopherol); multiply by 0.45 if labeled dl-α-tocopherol.
- mg α-TE:[c] divide by 1.5. To convert α-TE to α-T, multiply by 0.8.

[a]Retinol equivalents (vitamin A).
[b]Alpha-tocopherol (vitamin E).
[c]Alpha-tocopherol equivalents (vitamin E).

Sodium

To convert milligrams of sodium to grams of salt:

mg sodium ÷ 400 = g of salt

The reverse is also true:

g salt × 400 = mg sodium

Folate

To convert micrograms of synthetic folate in supplements and enriched foods to Dietary Folate Equivalents (µg DFE):

µg synthetic folate × 1.7 = µg DFE

For naturally occurring folate, assign each microgram folate a value of 1 µg DFE:

µg folate = µg DFE

TABLE D-1 CONVERSION FACTORS

Length	Volume	Temperature	
1 inch = 2.54 cm	1 tbsp = 3 tsp or 15 ml	Steam — 100°C	212°F — Steam
1 ft = 30.48 cm	1 oz = 2 tbsp or 30 ml	Body temperature — 37°C	98.6°F — Body temperature
1 m = 39.37 inches	½ c = about 125 ml	Ice — 0°C	32°F — Ice
	1 c = 16 tbsp or about 250 ml	Centigrade	Fahrenheit
	1 qt = 4 c or .95 liter	$t_F = 9/5\ t_c + 32$	
	1 qt = 32 oz (fluid)	$t_c = 5/9\ (t_F - 32)$	
	1 l = 1.06 qt or 1000 ml		
	1 gal = 16 c or 4 qt or 3.79 liter		

Weight		Energy
1 oz = about 30 g (28.35 g)	1 tsp any powder or liquid = about 5 g or 5 ml	1 cal = 4.2 kJ = 0.004 mJ
1 lb = about 454 g	1 tbsp any powder or liquid = about 15 g or 15 ml	1 mJ = 240 cal
1 kg = 1000 g	½ c any vegetable, fruit, or fluid weighs about 100 g	1 kJ = 0.24 cal
1 kg = 2.2 lb		1 g carbohydrate = 4 cal = 17 kJ
1 g = 1000 mg		1 g fat = 9 cal = 37 kJ
1 mg = 1000 µg		1 g protein = 4 cal = 17 kJ
1 µg = 1/1000 mg		1 g alcohol = 7 cal = 29 kJ

TABLE D-2 BODY WEIGHTS: QUICK ESTIMATION OF DESIRABLE BODY WEIGHT*

Men	Women
For 5 ft, consider 106 lb a reasonable weight.	For 5 ft, consider 100 lb a reasonable weight.
For each inch over 5 ft, add 6 lb.	For each inch over 5 ft, add 5 lb.
Subtract 6 lb for each inch under 5 ft. Add 10% for a large-framed individual; subtract 10% for a small-framed individual.	Subtract 5 lb for each inch under 5 ft. Add 10% for a large-framed individual; subtract 10% for a small-framed individual.
Example: A man 5 ft 8 inches tall (medium frame) would start at 106 lb, add 48, and arrive at a reasonable weight of 154 lb.	For each year under 25 (down to 18), subtract 1 lb.
	Example: A woman 21 years old, 5 ft 4 inches tall (medium-frame), would start at 100 lb, add 20, and subtract 4, arriving at a reasonable weight of 116 lb.

*To determine frame size, see Table D-3.

TABLE D-3 FRAME SIZE

To make a simple approximation of your frame size:

Extend your arm and bend the forearm upward at a 90-degree angle. Keep the fingers straight and turn the inside of your wrist away from your body. Place the thumb and index finger of your other hand on the two prominent bones on either side of your elbow. Measure the space between your fingers against a ruler or a tape measure.* Compare the measurements with the following standards. (These standards represent the elbow measurements for medium-framed men and women of various heights. Measurements smaller than those listed indicate you have a small frame, and larger measurements indicate a large frame.†)

Men		Women	
Height in 1-in. Heels	*Elbow Breadth*	*Height in 1-in. Heels*	*Elbow Breadth*
5 ft 2 in. to 5 ft 3 in.	2½ to 2⅞ in.	4 ft 10 in. to 4 ft 11 in.	2¼ to 2½ in.
5 ft 4 in. to 5 ft 7 in.	2⅝ to 2⅞ in.	5 ft 0 in. to 5 ft 3 in.	2¼ to 2½ in.
5 ft 8 in. to 5 ft 11 in.	2¾ to 3 in.	5 ft 4 in. to 5 ft 7 in.	2⅜ to 2⅝ in.
6 ft 0 in. to 6 ft 3 in.	2¾ to 3⅛ in.	5 ft 8 in. to 5 ft 11 in.	2⅜ to 3⅝ in.
6 ft 4 in. and over	2⅞ to 3¼ in.	6 ft 0 in. and over	2½ to 3¾ in.

*For the most accurate measurement, have your health-care provider measure your elbow breadth with a caliper.

†A simple estimate of frame size can be derived by circling your wrist with the thumb and middle finger of your other hand. If the thumb and finger do not meet, you most likely have a large frame. If the thumb and finger just meet, you most likely have a medium frame, and if the fingers overlap greatly, you may have a small frame.

Appendix E

Food Exchange Systems

Chapter 2 introduced dietary guidelines, food group plans, and the exchange system. This appendix provides details of the U.S. and Canadian exchange systems.

THE U.S. EXCHANGE SYSTEM

The U.S. exchange system divides the foods suitable for use in planning a healthy diet into seven lists—the starch, fruit, milk, other carbohydrates, vegetable, meat and meat substitutes, and fat lists.* These lists are shown in Tables E-1 through E-7. Following these lists are three other sets of foods: free foods, combination foods, and fast foods (Tables E-8, E-9, and E-10).

*The U.S. Exchange System presented here is based on material in *Exchange Lists for Meal Planning*, 1995, prepared by committees of the American Diabetes Association and the American Dietetic Association, with permission of both organizations.

TABLE E-1 THE U.S. EXCHANGE SYSTEM: STARCH LIST

15 g carbohydrate, 3 g protein, 0–1 g fat, 80 cal
Cereals, grains, pasta, breads, crackers, snacks, starchy vegetables, and cooked dried beans, peas, and lentils are starches. In general, one starch is:

- ½ cup of cereal, grain, pasta, or starchy vegetable
- 1 ounce of a bread product, such as 1 slice of bread
- ¾ to 1 ounce of most snack foods (some snack foods may also have added fat)

Amount	Food
Bread	
½ (1 oz)	Bagel
2 slices (1½ oz)	Bread, reduced-calorie
1 slice (1 oz)	Bread, white, whole-wheat, pumpernickel, rye
2 (⅔ oz)	Bread sticks, crisp, 4 in. long × ½ in.
½	English muffin
½ (1 oz)	Hot dog or hamburger bun
½	Pita, 6 in. across
1 slice (1 oz)	Raisin bread, unfrosted
1 (1 oz)	Roll, plain, small
1	Tortilla, corn, 6 in. across
1	Tortilla, flour, 7–8 in. across
1	Waffle, 4½ in. square, reduced-fat
Cereals and Grains	
½ cup	Bran cereals
½ cup	Bulgur
½ cup	Cereals
¾ cup	Cereals, unsweetened, ready-to-eat
3 tbsp	Cornmeal (dry)
⅓ cup	Couscous
3 tbsp	Flour (dry)
¼ cup	Granola, low-fat
¼ cup	Grape-Nuts
½ cup	Grits
½ cup	Kasha
¼ cup	Millet
¼ cup	Muesli
½ cup	Oats
½ cup	Pasta
1½ cups	Puffed cereal
½ cup	Rice milk
⅓ cup	Rice, white or brown
½ cup	Shredded Wheat
½ cup	Sugar-frosted cereal
3 tbsp	Wheat germ
Starchy Vegetables	
⅓ cup	Baked beans
½ cup	Corn
1 (5 oz)	Corn on cob, medium
1 cup	Mixed vegetables with corn, peas, or pasta
½ cup	Peas, green
½ cup	Plantain
1 small (3 oz)	Potato, baked or boiled
½ cup	Potato, mashed
1 cup	Squash, winter (acorn, butternut)
½ cup	Yam, sweet potato, plain
Crackers and Snacks	
8	Animal crackers
3	Graham crackers, 2½ in. square
¾ oz	Matzoh
4 slices	Melba toast
24	Oyster crackers
3 cups	Popcorn (popped, no fat added or low-fat microwave)
¾ oz	Pretzels
2	Rice cakes, 4 in. across
6	Saltine-type crackers
15–20 (¾ oz)	Snack chips, fat-free (tortilla, potato)
2–5 (¾ oz)	Whole-wheat crackers, no fat added
Dried Beans. Peas, and Lentils (Count as 1 starch exchange, plus 1 very lean meat exchange)	
½ cup	Beans and peas (garbanzo, pinto, kidney, white, split, black-eyed)
⅔ cup	Lima beans
½ cup	Lentils
3 tbsp	Miso
Starchy Foods Prepared With Fat (Count as 1 starch exchange, plus 1 fat exchange)	
1	Biscuit, 2½ in. across
½ cup	Chow mein noodles
1 (2 oz)	Corn bread, 2 in. cube
6	Crackers, round butter type
1 cup	Croutons
16–25 (3 oz)	French-fried potatoes
¼ cup	Granola
1 (1½ oz)	Muffin, small
2	Pancakes, 4 in. across
3 cups	Popcorn, microwave
3	Sandwich crackers, cheese or peanut butter filling
⅓ cup	Stuffing, bread (prepared)
2	Taco shell, 6 in. across
1	Waffle, 4½ in. square
4–6 (1 oz)	Whole-wheat crackers, fat added

= 400 mg or more of sodium per serving.

TABLE E-2 U.S. EXCHANGE SYSTEM: FRUIT LIST

15 g carbohydrates, 60 cal

Fresh, frozen, canned, and dried fruits and fruit juices are on this list. In general, one fruit exchange is:

- 1 small to medium fresh fruit
- ½ cup of canned or fresh fruit or fruit juice
- ¼ cup of dried fruit

Amount	Food
Fruit	
1 (4 oz)	Apple, unpeeled, small
½ cup	Applesauce, unsweetened
4 rings	Apples, dried
4 whole (5½ oz)	Apricots, fresh
8 halves	Apricots, dried
½ cup	Apricots, canned
1 (4 oz)	Banana, small
¾ cup	Blackberries
¾ cup	Blueberries
⅓ melon (11 oz) or 1 cup cubes	Cantaloupe, small
12 (3 oz)	Cherries, sweet, fresh
½ cup	Cherries, sweet, canned
3	Dates
½ large or 2 medium (3½ oz)	Figs, fresh
1½	Figs, dried
½ cup	Fruit cocktail
½ (11 oz)	Grapefruit, large
¾ cup	Grapefruit sections, canned
17 (3 oz)	Grapes, small
1 slice (10 oz) or 1 cup cubes	Honeydew melon
1 (3½ oz)	Kiwi
¾ cup	Mandarin oranges, canned
½ fruit (5½ oz) or ½ cup	Mango, small
1 (5 oz)	Nectarine, small
1 (6½ oz)	Orange, small
½ fruit (8 oz) or 1 cup cubes	Papaya
1 (6 oz)	Peach, medium, fresh
½ cup	Peaches, canned
½ (4 oz)	Pear, large, fresh
½ cup	Pears, canned
¾ cup	Pineapple, fresh
½ cup	Pineapple, canned
2 (5 oz)	Plums, small
½ cup	Plums, canned
3	Prunes, dried
2 tbsp	Raisins
1 cup	Raspberries
1¼ cup whole berries	Strawberries
2 (8 oz)	Tangerines, small
slice (13½ oz) or 1¼ cup cubes	Watermelon
Fruit Juice	
½ cup	Apple juice/cider
⅓ cup	Cranberry juice cocktail
1 cup	Cranberry juice cocktail, reduced-calorie
⅓ cup	Fruit juice blend, 100% juice
⅓ cup	Grape juice
½ cup	Grapefruit juice
½ cup	Orange juice
½ cup	Pineapple juice
⅓ cup	Prune juice

TABLE E-3 U.S. EXCHANGE SYSTEM: MILK LIST

Nonfat and very lowfat milk = 12 g carbohydrate, 8 g protein, 0–3 g fat, 90 cal; low-fat milk = 12 g carbohydrate, 8 g protein, 5 g fat, 120 cal; whole milk = 12 g carbohydrate, 8 g protein, 8 g fat, 150 cal.

Amount	Food
Nonfat and Very Low-Fat Milk	
1 c	Nonfat milk
1 c	½% milk
1 c	1% milk
⅓ c	Dry nonfat milk
½ c	Evaporated nonfat milk
1 c	Nonfat or low-fat buttermilk
¾ c	Plain nonfat yogurt
1 c	Nonfat or low-fat fruit-flavored yogurt sweetened with aspartame or with a nonnutritive sweetener
Low-Fat Milk	
1 c	2% milk
¾ c	Plain low-fat yogurt
1 c	Sweet acidophilus milk
Whole Milk	
1 c	Whole milk
½ c	Evaporated whole milk
1 c	Goat's milk
1 c	Kefir

TABLE E-4 U.S. EXCHANGE SYSTEM: OTHER CARBOHYDRATES LIST

15 grams carbohydrate, or 1 starch, or 1 fruit, or 1 milk.
You can substitute food choices from this list for a starch, fruit, or milk choice on your meal plan. Some choices will also count as one or more fat choices.

Amount	Food	Exchanges Per Serving
1/12th cake	Angel food cake, unfrosted	2 carbohydrates
2 in. square	Brownie, small, unfrosted	1 carbohydrate, 1 fat
2 in. square	Cake, unfrosted	1 carbohydrate, 1 fat
2 in. square	Cake, frosted	2 carbohydrates, 1 fat
2 small	Cookie, fat-free	1 carbohydrate
2 small	Cookie or sandwich cookie with creme filling	1 carbohydrate, 1 fat
1 small	Cupcake, frosted	2 carbohydrates, 1 fat
¼ cup	Cranberry sauce, jellied	2 carbohydrates
1 medium (1½ oz)	Doughnut, plain cake	1½ carbohydrates, 2 fats
3¾ in. across (2 oz)	Doughnut, glazed	2 carbohydrates, 2 fats
1 bar (3 oz)	Fruit juice bars, frozen, 100% juice	1 carbohydrate
1 roll (¾ oz)	Fruit snacks, chewy (pureed fruit concentrate)	1 carbohydrate
1 tbsp	Fruit spreads, 100% fruit	1 carbohydrate
½ cup	Gelatin, regular	1 carbohydrate
3	Gingersnaps	1 carbohydrate
1 bar	Granola bar	1 carbohydrate, 1 fat
1 bar	Granola bar, fat-free	2 carbohydrates
⅓ cup	Hummus	1 carbohydrate, 1 fat
½ cup	Ice cream	1 carbohydrate, 2 fats
½ cup	Ice cream, light	1 carbohydrate, 1 fat
½ cup	Ice cream, fat-free, no sugar added	1 carbohydrate
1 tbsp	Jam or jelly, regular	1 carbohydrate
1 cup	Milk, chocolate, whole	2 carbohydrates, 1 fat
⅙ pie	Pie, fruit, 2 crusts	3 carbohydrates, 2 fats
⅛ pie	Pie, pumpkin or custard	1 carbohydrate, 2 fats
12–18 (1 oz)	Potato chips	1 carbohydrate, 2 fats
½ cup	Pudding, regular (made with low-fat milk)	2 carbohydrates
½ cup	Pudding, sugar-free (made with low-fat milk)	1 carbohydrate
¼ cup	Salad dressing, fat-free	1 carbohydrate
½ cup	Sherbet, sorbet	2 carbohydrates
½ cup	Spaghetti or pasta sauce, canned	1 carbohydrate, 1 fat
1 (2½ oz)	Sweet roll or Danish	2½ carbohydrates, 2 fats
2 tbsp	Syrup, light	1 carbohydrate
1 tbsp	Syrup, regular	1 carbohydrate
¼ cup	Syrup, regular	4 carbohydrates
6–12 (1 oz)	Tortilla chips	1 carbohydrate, 2 fats
⅓ cup	Yogurt, frozen, low-fat, fat-free	1 carbohydrate, 0–1 fat
½ cup	Yogurt, frozen, fat-free, no sugar added	1 carbohydrate
1 cup	Yogurt, low-fat with fruit	3 carbohydrates, 0–1 fat
5	Vanilla wafers	1 carbohydrate, 1 fat

= 400 mg or more sodium per exchange.

TABLE E-5 U.S. EXCHANGE SYSTEM: VEGETABLE LIST

5 g carbohydrate, 2 g protein, 25 cal. All portion sizes, except as otherwise noted, are ½ c of any cooked vegetable or vegetable juice, 1 c of any raw vegetable.

Artichoke
Artichoke hearts
Asparagus
Beans (green, wax, Italian)
Bean sprouts
Beets
Broccoli
Brussels sprouts
Cabbage
Carrots
Cauliflower
Celery
Cucumber
Eggplant
Green onions or scallions
Greens (collard, kale, mustard, turnip)
Kohlrabi
Leeks
Mixed vegetables (without corn, peas, or pasta)*
Mushrooms
Okra
Onions
Pea pods
Peppers (all varieties)
Radishes
Salad greens (endive, escarole, lettuce, romaine, spinach)
Sauerkraut
Spinach
Summer squash
Tomato
Tomatoes, canned
Tomato sauce
Tomato/vegetable juice
Turnips
Water chestnuts
Watercress

*Starchy vegetables such as corn, peas, and potatoes are found on the Starch List.
= 400 mg or more sodium per exchange.

TABLE E-6 U.S. EXCHANGE SYSTEM: MEAT/MEAT SUBSTITUTES

Very-lean meat = 7 g protein, 0-1 g fat, 35 cal; lean meat = 7 g protein, 3 g fat, 55 cal; medium-fat meat = 7 g protein, 5 g fat, 75 cal; high-fat meat = 7 g protein, 8 g fat, 100 cal.

Meat and meat substitutes that contain both protein and fat are on this list. In general, one meat exchange is:

- 1 oz meat, fish, poultry, or cheese
- ½ cup dried beans, cooked

Based on the amount of fat they contain, meats are divided into very lean, lean, medium-fat, and high-fat lists.

Amount	Food
Very Lean Meat and Substitutes List	
1 oz	*Poultry:* Chicken or turkey (white meat, no skin), Cornish hen (no skin)
1 oz	*Fish:* Fresh or frozen cod, flounder, haddock, halibut, trout; tuna, fresh or canned in water
1 oz	*Shellfish:* Clams, crab, lobster, scallops, shrimp, imitation shellfish
1 oz	*Game:* Duck or pheasant (no skin), venison, buffalo, ostrich
	Cheese with 1 gram or less fat per ounce:
¼ cup	Nonfat or low-fat cottage cheese
1 oz	Fat-free cheese
1 oz	*Other:* Processed sandwich meats with 1 gram or less fat per ounce, such as deli thin, shaved meats, chipped beef , turkey ham
2	Egg whites
¼ cup	Egg substitutes, plain
1 oz	Hot dogs with 1 gram or less fat per ounce
1 oz	Kidney (high in cholesterol)
1 oz	Sausage with 1 gram or less fat per ounce
Count as one very lean meat and one starch exchange:	
½ cup	Dried beans, peas, lentils (cooked)
Lean Meat and Substitutes List	
1 oz	*Beef:* USDA Select or Choice grades of lean beef trimmed of fat, such as round, sirloin, and flank steak; tenderloin; roast (rib, chuck, rump); steak (T-bone, porterhouse, cubed), ground round
1 oz	*Pork:* Lean pork, such as fresh ham; canned, cured, or boiled ham; Canadian bacon ; tenderloin, center loin chop
1 oz	*Lamb:* Roast, chop, leg
1 oz	*Veal:* Lean chop, roast
1 oz	*Poultry:* Chicken, turkey (dark meat, no skin), chicken white meat (with skin), domestic duck or goose (well-drained of fat, no skin)
	Fish:
1 oz	Herring (uncreamed or smoked)

Continued

TABLE E-6 U.S. EXCHANGE SYSTEM: MEAT/MEAT SUBSTITUTES—*Continued*

Amount	Food
Lean Meat and Substitutes List—*Continued*	
6 medium	Oysters
1 oz	Salmon (fresh or canned), catfish
2 medium	Sardines (canned)
1 oz	Tuna (canned in oil, drained)
1 oz	*Game:* Goose (no skin), rabbit
	Cheese:
¼ cup	4.5%-fat cottage cheese
2 tbsp	Grated Parmesan
1 oz	Cheese with 3 grams or less fat per ounce
	Other:
1½ oz	Hot dots with 3 grams or less fat per ounce (salt shaker symbol)
1 oz	Processed sandwich meat with 3 grams or less fat per ounce, such as turkey pastrami or kielbasa
1 oz	Liver, heart (high in cholesterol)
Medium-Fat Meat and Substitutes List	
1 oz	*Beef:* Most beef products fall into this category (ground beef, meatloaf, corned beef, short ribs, Prime grades of meat trimmed of fat, such as prime rib)
1 oz	*Pork:* Top loin, chop, Boston butt, cutlet
1 oz	*Lamb:* Rib roast, ground
1 oz	*Veal:* Cutlet (ground or cubed, unbreaded)
1 oz	*Poultry:* Chicken dark meat (with skin), ground turkey or ground chicken, fried chicken (with skin)
1 oz	*Fish:* Any fried fish product
	Cheese: With 5 grams or less fat per ounce
1 oz	Feta
1 oz	Mozzarella
¼ cup	Ricotta
	Other:
1	Egg (high in cholesterol, limit to 3 per week)
1 oz	Sausage with 5 grams or less fat per ounce
1 cup	Soy milk
¼ cup	Tempeh
4 oz or ½ cup	Tofu
High-Fat Meat and Substitutes List	
1 oz	*Pork:* Spareribs, ground pork, pork sausage
1 oz	*Cheese:* All regular cheeses, such as American (salt shaker symbol), cheddar, Monterey Jack, Swiss
1 oz	*Other:* Processed sandwich meats with 8 grams or less fat per ounce, such as bologna, pimento loaf, salami
1 oz	Sausage, such as bratwurst, Italian, knockwurst, Polish, smoked
1 (10/lb)	Hot dog (turkey or chicken) (salt shaker symbol)
3 slices (20 slices/lb)	Bacon
Count as one high-fat meat plus one fat exchange:	
1 (10/lb)	Hot dog (beef, pork, or combination) (salt shaker symbol)
2 tbsp	Peanut butter (contains unsaturated fat)

(salt shaker symbol) = 400 mg or more sodium per exchange.

TABLE E-7 U.S. EXCHANGE SYSTEM: FAT LIST

5 g fat, 45 cal.

Amount	Food
Monounsaturated Fats List	
⅛ (1 oz)	Avocado, medium
1 tsp	Oil (canola, olive, peanut)
8 large	Olives: ripe (black)
10 large	green, stuffed (salt shaker symbol)
	Nuts:
6 nuts	almonds, cashews
6 nuts	mixed (50% peanuts)
10 nuts	peanuts
4 halves	pecans
2 tsp	Peanut butter, smooth or crunchy
1 tbsp	Sesame seeds
2 tsp	Tahini paste
Polyunsaturated Fats List	
1 tsp	Margarine: stick, tub, or squeeze
1 tbsp	lower-fat (30% to 50% vegetable oil)
1 tsp	Mayonnaise: regular
1 tbsp	reduced-fat
4 halves	Nuts, walnuts, English
1 tsp	Oil (corn, safflower, soybean)
1 tbsp	Salad dressing: regular (salt shaker symbol)
2 tbsp	reduced-fat
2 tsp	Miracle Whip Salad Dressing®: regular
1 tbsp	reduced-fat
1 tbsp	Seeds: pumpkin, sunflower

Continued

TABLE E-7 U.S. EXCHANGE SYSTEM: FAT LIST—*Continued*

5 g fat, 45 cal.

Amount	Food	Amount	Food
Saturated Fats List*		2 tbsp	Coconut, sweetened, shredded
1 slice (20 slices/lb)	Bacon, cooked	2 tbsp	Cream, half and half
1 tsp	Bacon, grease	1 tbsp (½ oz)	Cream cheese: regular
1 tsp	Butter: stick	2 tbsp (1 oz)	reduced-fat
2 tsp	whipped		Fatback or salt pork, see below†
1 tbsp	reduced-fat	1 tsp	Shortening or lard
2 tbsp (½ oz)	Chitterlings, boiled	2 tbsp	Sour cream: regular
		3 tbsp	reduced-fat

*Saturated fats can raise blood cholesterol levels.

†Use a piece 1 in. × 1 in. × ¼ in. if you plan to eat the fatback cooked with vegetables. Use a piece 2 in. × 1 in. × ½ in. when eating only the vegetables with the fatback removed.

= 400 mg or more sodium per exchange.

TABLE E-8 U.S. EXCHANGE SYSTEM: FREE FOODS

A free food is any food or drink that contains less then 20 cal/serving. People with diabetes are advised to eat as much as they want of those items that have no serving size specified. They may eat two or three servings per day of those items that have a specific serving size. It is suggested that they spread the servings out through the day. Foods listed without a serving size can be eaten as often as you like.

Amount	Food
Fat-free Or Reduced-fat Foods	
1 tbsp	Cream cheese, fat-free
1 tbsp	Creamers, nondairy, liquid
2 tsp	Creamers, nondairy, powdered
1 tbsp	Mayonnaise, fat-free
1 tsp	Mayonnaise, reduced-fat
4 tbsp	Margarine, fat-free
1 tsp	Margarine, reduced-fat
1 tbsp	Miracle Whip®, nonfat
1 tsp	Miracle Whip®, reduced-fat
1 tbsp	Salad dressing, fat-free
2 tbsp	Salad dressing, fat-free, Italian
¼ cup	Salsa
1 tbsp	Sour cream, fat-free, reduced-fat
2 tbsp	Whipped topping, regular or light
Sugar-free Or Low-sugar Foods	
1 candy	Candy, hard, sugar-free
	Gelatin dessert, sugar-free
	Gelatin, unflavored
	Gum, sugar-free
2 tsp	Jam or jelly, low-sugar or light
	Sugar substitutes*
2 tbsp	Syrup, sugar-free
Drinks	
	Bouillon, broth, consommé
	Bouillon or broth, low-sodium
	Carbonated or mineral water
1 tbsp	Cocoa powder, unsweetened
	Coffee
	Club soda
	Diet soft drinks, sugar-free
	Drink mixes, sugar-free
	Tea
	Tonic water, sugar-free
Condiments	
1 tbsp	Catsup
	Horseradish
	Lemon juice
	Lime juice
	Mustard
1½ large	Pickles, dill
	Soy sauce, regular or light
1 tbsp	Taco sauce
	Vinegar
Seasonings	
	Flavoring extracts
	Garlic
	Herbs, fresh or dried
	Pimento
	Spices
	Tabasco® or hot pepper sauce
	Wine, used in cooking
	Worcestershire sauce

*Sugar substitutes, alternatives, or replacements that are approved by the Food and Drug Administration (FDA) are safe to use.

= 400 mg or more of sodium per choice.

TABLE E-9 U.S. EXCHANGE SYSTEM: COMBINATION FOODS

Much of the food we eat is mixed together in various combinations. These combination foods do not fit into any one exchange list. It can be quite hard to tell what is in a certain casserole dish or baked food item. This is a list of average values for some typical combination foods. This list will help you fit these foods into your meal plan. Ask your dietitian for information about any other foods you'd like to eat.

Amount	Food	Exchanges Per Serving
Entrees		
1 cup (8 oz)	Tuna noodle casserole, lasagna, spaghetti with meatballs, chili with beans, macaroni and cheese	2 carbohydrates, 2 medium-fat meats 1 carbohydrate, 2 lean meats 2 carbohydrates, 2 medium-fat meats, 1 fat
2 cups (16 oz)	Chow mein (without noodles or rice)	2 carbohydrates, 2 medium-fat meats, 2 fats
¼ of 10 in. (5 oz)	Pizza, cheese, thin crust	2 carbohydrates, 1 medium-fat meat, 4 fats
¼ of 10 in. (5 oz)	Pizza, meat topping, thin crust	
1 (7 oz)	Pot pie	
Frozen entrees		
1 (11 oz)	Salisbury steak with gravy, mashed potato	2 carbohydrates, 3 medium-fat meats, 3–4 fats
1 (11 oz)	Turkey with gravy, mashed potato, dressing	2 carbohydrates, 2 medium-fat meats, 2 fats
1 (8 oz)	Entree with less than 300 calories	2 carbohydrates, 3 lean meats
Soups		
1 cup (8 oz)	Bean	1 carbohydrate, 1 very lean meat
1 cup (8 oz)	Cream (made with water)	1 carbohydrate, 1 fat
½ cup (4 oz)	Split pea (made with water)	1 carbohydrate
1 cup (8 oz)	Tomato (made with water)	1 carbohydrate
1 cup (8 oz)	Vegetable beef, chicken noodle, or other broth-type	1 carbohydrate

= 400 mg or more sodium per exchange.

TABLE E-10 U.S. EXCHANGE SYSTEM: FAST FOODS*

Amount	Food	Exchanges Per Serving
2	Burritos with beef	4 carbohydrates, 2 medium-fat meats, 2 fats
6	Chicken nuggets	1 carbohydrate, 2 medium-fat meats, 1 fat
1 each	Chicken breast and wing, breaded and fried	1 carbohydrate, 4 medium-fat meats, 2 fats
1	Fish sandwich/tartar sauce	3 carbohydrates, 1 medium-fat meat, 3 fats
20–25	French fries, thin	2 carbohydrates, 2 fats
1	Hamburger, regular	2 carbohydrates, 2 medium-fat meats
1	Hamburger, large	2 carbohydrates, 3 medium-fat meats, 1 fat
1	Hot dog with bun	1 carbohydrate, 1 high-fat meat, 1 fat
1	Individual pan pizza	5 carbohydrates, 3 medium-fat meats, 3 fats
1 medium	Soft-serve cone	2 carbohydrates, 1 fat
1 sub (6 in.)	Submarine sandwich	3 carbohydrates, 1 vegetable, 2 medium-fat meats, 1 fat
1 (6 oz)	Taco, hard shell	2 carbohydrates, 2 medium-fat meats, 2 fats
1 (3 oz)	Taco, soft shell	1 carbohydrate, 1 medium-fat meat, 1 fat

*Ask at your fast-food restaurant for nutrition information about your favorite fast foods.

= 400 mg or more of sodium per serving.

 THE CANADIAN EXCHANGE SYSTEM

The *Good Health Eating Guide* is the Canadian exchange system of meal planning.* It contains several features similar to those of the U.S. exchange system including the following:

- Foods are divided into groups according to carbohydrate, protein, and fat content.
- Foods are interchangeable within a group.
- Most foods are eaten in measured amounts.
- An energy (calorie) value is given for each food group.

Tables E-11 through E-18 present the Canadian exchange system.

TABLE E-11 CANADIAN EXCHANGE SYSTEM: STARCH FOODS GROUP

15 g carbohydrate [starch], 2 g protein, 290 kJ (68 cal).

Food	Measure	Mass (Weight)
Breads		
Bagel	½	30 g
Bread crumbs	50 mL (¼ cup)	30 g
Bread cubes	250 mL (1 cup)	30 g
Bread sticks	2	20 g
Brewis, cooked	50 mL (¼ cup)	45 g
Chapati	1	20 g
Cookies, plain	2	20 g
English muffin, crumpet	½	30 g
Flour	40 mL (2½ tbsp)	20 g
Hamburger bun	½	30 g
Hot dog bun	½	30 g
Kaiser roll	½	30 g
Matzo, 15 cm	1	20 g
Melba toast, rectangular	4	15 g
Melba toast, rounds	7	15 g
Pita, 20 cm diameter (8″ diameter)	¼	30 g
Pita, 15 cm diameter (6″ diameter)	½	30 g
Plain roll	1 small	30 g
Pretzels	7	20 g
Raisin bread	1 slice	30 g
Rice cakes	2	30 g
Roti	1	20 g
Rusks	2	20 g
Rye, coarse or pumpernickel	½ slice	30 g
Soda crackers	6	20 g
Tortilla, corn (taco shell)	1	30 g
Tortilla, flour	1	30 g
White (French & Italian)	1 slice	25 g
Whole-wheat, cracked-wheat, rye, white-enriched	1 slice	30 g
Starch Vegetables		
Beans, peas (dried), cooked	125 mL (½ cup)	80 g
Breadfruit	1 slice	75 g
Corn, canned whole kernel	125 mL (½ cup)	85 g
Corn-on-the-cob	½ medium cob	140 g
Cornstarch	30 mL (2 tbsp)	15 g
Plantain	⅓ small	50 g
Popcorn, air-popped unbuttered	750 mL (3 cups)	20 g
Potatoes, whole (with or without skin)	½ medium	95 g
Yam, sweet potatoes (with or without skin)	½	75 g

Continued

*The tables for the Canadian exchange system are taken from *Good Health Eating Guide* (Toronto: Canadian Diabetes Association, 1994) and are used with the association's permission.

TABLE E-11 CANADIAN EXCHANGE SYSTEM: STARCH FOODS GROUP—*Continued*

Each of the following measured foods equals more than 1 Starch choice:

Food	Measure	Food Choices	Mass (Weight)
Bran Flakes	150 mL (⅔ cup)	1 Starch + ½ Sugars	24 g
Croissant, small	1 small	1 Starch + 1½ Fats & Oils	35 g
large	½ large	1 Starch + 1½ Fats & Oils	30 g
Corn, canned creamed	12 mL (½ cup)	1 Starch + ½ Fruits & Vegetables	113 g
Potato chips	15 chips	1 Starch + 2 Fats & Oils	30 g
Tortilla chips (nachos)	13 chips	1 Starch + 1½ Fats & Oils	20 g
Corn chips	30 chips	1 Starch + 2 Fats & Oils	30 g
Cheese twists	30 chips	1 Starch + 1½ Fats & Oils	30 g
Cheese puffs	27 chips	1 Starch + 2 Fats & Oils	30 g
Tea Biscuit	1	1 Starch + 2 Fats & Oils	30 g
Pancake, homemade using 50 mL (¼ cup) batter (6-in diameter)	1 medium	1½ Starch + 1 Fats & Oils	50 g
Potatoes, French fried (homemade or frozen)	10 regular size	1 Starch & 1 Fats & Oils	35 g
Soup, canned* (prepared with equal volume of water)	250 mL (1 cup)	1 Starch	260 g
Waffle, packaged	1	1 Starch + 1 Fats & Oils	35 g

Food	Measure	Mass (Weight)
Cereals		
Bran flakes, 100% Bran	125 mL (½ cup)	30 g
Cooked cereals, cooked	125 mL (½ cup)	125 g
dry	30 mL (2 tbsp)	20 g
Cornmeal, cooked	125 mL (½ cup)	125 g
dry	30 mL (2 tbsp)	20 g
Ready-to-eat unsweetened cereal	125 mL (½ cup)	20 g
Shredded wheat biscuit, rectangular or round	1	20 g
Shredded wheat, bite size	125 mL (½ cup)	20 g
Wheat germ	75 mL (⅓ cup)	30 g
Cornflakes	175 mL (⅔ cup)	20 g
Rice Krispies	175 mL (⅔ cup)	20 g
Cheerios	200 mL (¾ cup)	20 g
Muffets	1 muffet	20 g
Puffed rice	300 mL (1¼ cup)	15 g
Puffed wheat	425 mL (1⅔ cup)	20 g

Food	Measure	Mass (Weight)
Grains		
Barley, cooked	125 mL (½ cup)	120 g
dry	30 mL (2 tbsp)	20 g
Bulgar, kasha,		
cooked moist	125 mL (½ cup)	70 g
cooked crumbly	75 mL (⅓ cup)	40 g
dry	30 mL (2 tbsp)	20 g
Rice, brown & white (short & long grain)	125 mL (½ cup)	70 g
Rice, wild	75 mL (⅓ cup)	70 g
Tapioca, pearl and granulated quick cooking dry	30 mL (2 tbsp)	15 g
Couscous, cooked moist	125 mL (½ cup)	70 g
dry	30 mL (2 tbsp)	20 g
Quinoa, cooked moist	125 mL (½ cup)	70 g
dry	30 mL (2 tbsp)	20 g
Pasta		
Macaroni, cooked	125 mL (½ cup)	70 g
Noodles, cooked	125 mL (½ cup)	80 g
Spaghetti, cooked	125 mL (½ cup)	70 g

*Soup can vary according to brand and type. Check the label for Food Choice Values and Symbols or the core nutrient listing.

TABLE E-12 CANADIAN EXCHANGE SYSTEM: FRUITS AND VEGETABLES GROUP

10 g carbohydrate, 1 g protein, 190 kJ, (44 cal)

Food	Measure	Mass (Weight)
Fruits (fresh, frozen, without sugar, canned in water)		
Apple, raw		
(with or without skin)	½ medium	75 g
sauce unsweetened	125 mL (½ cup)	120 g
sweetened	*(see Combined Food Choices)*	
Apple butter	20 mL (4 tsp)	20 g
Apricot, raw	2 medium	115 g
canned in water	4 halves plus 30 mL (2 tbsp) liquid	110 g
Bake-apple (cloudberries), raw	125 mL (½ cup)	120 g
Banana, with peel	½ small	75 g
peeled	½ small	50 g
Berries (blackberries, blueberries, boysenberries, huckleberries, loganberries, raspberries)		
raw	125 mL (½ cup)	70 g
canned, in water	125 mL (½ cup), includes 30 mL (2 tbsp) liquid	100 g
Cantaloupe,		
wedge with rind	¼	240 g
cubed or diced	250 mL (1 cup)	160 g
Cherries,		
raw with pits	10	75 g
raw without pits	10	70 g
canned in water with pits	75 mL (⅓ cup), includes 30 mL (2 tbsp) liquid	90 g
canned, in water, without pits	75 mL (⅓ cup), includes 30 mL (2 tbsp) liquid	85 g
Crabapple, raw	1 small	55 g
Cranberries, raw	250 mL (1 cup)	100 g
Figs, raw	1 medium	50 g
canned in water	3 medium plus 30 mL (2 tbsp) liquid	100 g
Foxberries, raw	250 mL (1 cup)	100 g
Fruit cocktail, canned in water	125 mL (½ cup) includes 30 mL (2 tbsp) liquid	120 g
Fruit, mixed cut-up	125 mL (½ cup)	120 g
Gooseberries, raw	250 mL (1 cup)	150 g
canned in water	250 mL (1 cup), includes 30 mL (2 tbsp) liquid	230 g
Grapefruit, raw with rind	½ small	185 g
raw sectioned	125 mL (½ cup)	100 g
canned in water	125 mL (½ cup), includes 30 mL (2 tbsp) liquid	120 g
Grapes, raw slip skin	125 mL (½ cup)	75 g
raw seedless	125 mL (½ cup)	75 g
canned in water	75 mL (⅓ cup), includes 30 mL (2 tbsp) liquid	115 g
Honeydew melon,		
raw with rind	⅒	225 g
cubed or diced	250 mL (1 cup)	170 g
Guava, raw	½	50 g
Kiwi, raw with skin	2	155 g
Kumquats, raw	3	60 g
Loquats, raw	8	130 g
Lychee fruit, raw	8	120 g
Mandarin orange,		
raw with rind	1	135 g
raw sectioned	125 mL (½ cup)	100 g
canned in water	125 mL (½ cup), includes 30 mL (2 tbsp) liquid	100 g
Mango, raw without skin and seed		65 g
diced	75 mL (⅓ cup)	65 g
Nectarine	½ medium	75 g
Orange, raw with rind	1 small	130 g
raw sectioned	125 mL (½ cup)	95 g
Papaya, raw with skin and seeds	¼ medium	150 g
raw without skin and seeds	¼ medium	100 g
cubed or diced	125 mL (½ cup)	100 g
Peaches, raw with seed and skin	1 large	100 g
raw sliced or diced	125 mL (½ cup)	100 g
canned in water, halves or slices	125 mL (½ cup) includes 30 mL (2 tbsp) liquid	120 g
Pear, raw with skin and core	½	90 g
raw without skin and core	½	85 g
halves canned in water	1 half plus 30 mL (2 tbsp liquid)	60 g 90 g

Continued

TABLE E-12 CANADIAN EXCHANGE SYSTEM: FRUITS AND VEGETABLES GROUP—*Continued*

10 g carbohydrate, 1 g protein, 190 kJ, (44 cal)

Food	Measure	Mass (Weight)
Fruits (fresh, frozen, without sugar, canned in water)		
Persimmons,		
raw native	1	30 g
raw Japanese	¼	50 g
Pineapple, raw	1 slice	75 g
raw diced	125 mL (½ cup)	75 g
sliced canned in water	2 slices plus 15 mL (1 tbsp) liquid	100 g
diced canned in water	125 mL (½ cup) includes 30 mL (2 tbsp) liquid	100 g
sliced canned in juice	1 slice, plus 15 mL (1 tbsp) liquid	55 g
diced canned in juice	75 mL (⅓ cup) includes 15 mL (1 tbsp) liquid	55 g
Plum, raw	2 small	60 g
Damson	6	65 g
Japanese	1	70 g
canned in water	3 plus 30 mL (2 tbsp) liquid	100 g
canned in apple juice	2 plus 30 mL (2 tbsp) liquid	70 g
Pomegranate, raw	½	140 g
Strawberries, raw	250 mL (1 cup)	150 g
frozen/canned in water	250 mL (1 cup) includes 30 mL (2 tbsp) liquid	240 g
Rhubarb	250 mL (1 cup)	150 g
Tangelo, raw	1	205 g
Tangerine, raw		
medium-sized	1	115 g
raw sectioned	125 mL (½ cup)	100 g
Watermelon, raw with		
rind	1 wedge	310 g
cubed or diced	250 mL (1 cup)	160 g
Dried Fruit		
Apple	5 pieces	15 g
Apricot	4 halves	15 g
Banana flakes	30 mL (2 tbsp)	15 g
Currants	30 mL (2 tbsp)	15 g
Dates, without pits	2	15 g
Peach	½	15 g
Pear	½	15 g
Prunes, raw with pits	2	15 g
raw without pits	2	10 g
stewed no liquid	2	20 g
stewed with liquid	2 plus 15 mL (1 tbsp) liquid	35 g
Raisins	30 mL (2 tbsp)	15 g
Juices (no sugar added or unsweetened)		
Apricot, grape, guava, mango, prune	50 mL (¼ cup)	55 g
Apple, carrot, papaya, pear, pineapple, pomegranate	75 mL (⅓ cup)	80 g
Cranberry (*see Sugars section*)		
Clamato (*see Sugars section*)		
Grapefruit, loganberry, orange, raspberry, tangelo, tangerine	125 mL (½ cup)	130 g
Tomato, tomato-based mixed vegetables	250 mL (1 cup)	255 g
Vegetables (fresh, frozen or canned)		
Artichokes, French, globe	2 small	50 g
Beets, diced or sliced	125 mL (½ cup)	85 g
Carrots, diced		
cooked or uncooked	125 mL (½ cup)	75 g
Chestnuts, fresh	5	20 g
Parsnips, mashed	125 mL (½ cup)	80 g
Peas, fresh or frozen	125 mL (½ cup)	80 g
canned	75 mL (⅓ cup)	55 g
Pumpkin, mashed	125 mL (½ cup)	45 g
Rutabagas, mashed	125 mL (½ cup)	85 g
Sauerkraut	250 mL (1 cup)	235 g
Snowpeas	250 mL (1 cup)	135 g
Squash, yellow or winter mashed	125 mL (½ cup)	115 g
Succotash	75 mL (⅓ cup)	55 g
Tomatoes, canned	250 mL (1 cup)	240 g
Tomato paste	50 mL (¼ cup)	55 g
Tomato sauce*	75 mL (⅓ cup)	100 g
Turnip, mashed	125 mL (½ cup)	115 g
Vegetables, mixed	125 mL (½ cup)	90 g
Water chestnuts	8 medium	50 g

*Tomato sauce varies according to brand name. Check the label or discuss with your dietitian.

TABLE E-13 CANADIAN EXCHANGE SYSTEM: MILK GROUP

Type of Milk	Carbohydrate	Protein	Fat	Energy
Nonfat	6 g	4 g	0 g	170 kJ (40 cal)
1%	6 g	4 g	1 g	206 kJ (49 cal)
2%	6 g	4 g	2 g	244 kJ (58 cal)
Whole	6 g	4 g	4 g	319 kJ (76 cal)

Food	Measure	Mass (Weight)
Buttermilk	125 mL (½ c)	125 g
Evaporated milk	50 mL (¼ c)	50 g
Milk	125 mL (½ c)	125 g
Powdered milk, regular	30 mL (2 tbsp)	15 g
Instant	50 mL (¼ c)	15 g
Plain yogurt	125 mL (½ c)	125 g

Each of the following measured foods equals more than 1 Milk choice:

Food	Food Choices	Measure	Mass (Weight)
Milkshake	1 Milk + 3 Sugars + ½ Protein	250 mL (1 cup)	300 g
Chocolate Milk 2%	2 Milk 2% + 1 Sugars	250 mL (1 cup)	300 g
Frozen Yogurt	1 Milk + 1 Sugars	125 mL (½ cup)	125 g

TABLE E-14 CANADIAN EXCHANGE SYSTEM: SUGARS

10 g carbohydrate, 167 kJ (40 cal)

Food	Measure	Mass (Weight)
Beverages:		
Condensed milk	15 ml (1 tbsp)	
*Flavoured fruit crystals	75 mL (⅓ cup)	
*Iced tea mixes	75 mL (⅓ cup)	
Regular soft drinks	125 mL (½ cup)	
*Sweet drink mixes	75 mL (⅓ cup)	
Tonic water	125 mL (½ cup)	
Miscellaneous:		
Bubble gum (large square)	1 piece	5 g
Cranberry cocktail	75 mL (⅓ cup)	80 g
Cranberry cocktail, light	350 mL (1⅓ cup)	260 g
Cranberry sauce	30 mL (2 tbsp)	
Hard candy mints	2	5 g
Honey, molasses corn & cane syrup	10 mL (2 tsp)	15 g
Jelly beans	4	10 g
Licorice	1 short stick	10 g
Marshmallows	2 large	15 g
Popsicle	1 stick (½ popsicle)	
Powdered gelatin mix (Jello®) (reconstituted)	50 mL (¼ cup)	
Regular jam, jelly, marmalade	15 mL (1 tbsp)	
Sugar, white, brown, icing, maple	10 mL (2 tsp)	10 g
Sweet pickles	2 small	100 g
Sweet relish	30 mL (2 tbsp)	

Each of the following measured foods equal more than 1 Sugars choice:

Food	Choices	Measures	Mass (Weight)
Brownie	1 Sugars + 1 Fats & Oils	1	20 g
Clamato juice	1½ Sugars	175 mL (⅔ cup)	
Fruit salad, light syrup	1 Sugars + 1 Fruits & Vegetables	125 mL (½ cup)	130 g
Aero® bar	2½ Sugars + 2½ Fats & Oils	1 bar	43 g
Smarties®	4½ Sugars + 2 Fats & Oils	1 box	60 g
Sherbet	3 Sugars + ½ Fats & Oils	125 mL (½ cup)	95 g

*These have been made with water.

TABLE E-15 CANADIAN EXCHANGE SYSTEM: PROTEIN FOODS GROUP

7 g protein, 3 g fat, 230 kJ (55 cal)

Food	Measure	Mass (Weight)
Cheese		
Low fat cheese, about 7% milk fat (M.F.)	1 slice	30 g
Cottage cheese, 2% M.F. or less	50 mL (¼ cup)	55 g
Ricotta, about 7% M.F.	50 mL (¼ cup)	60 g
Fish		
Anchovy	(*see Extras*)	
Canned, drained e.g. tuna packed in water, mackerel salmon	50 mL (¼ cup) (⅓ of 6.5 oz can)	30 g
Cod tongues, cheeks	75 mL (⅓ cup)	50 g
Fillet or steak, e.g. Boston blue, cod, flounder, haddock, halibut, mackerel, orange roughy, perch, pickerel, pike, salmon, shad, snapper, sole, swordfish, trout, tuna, whitefish	1 piece	30 g
Herring	⅓ fish	30 g
Sardines, smelts	2 medium or 3 small	30 g
Squid, octopus	50 mL (¼ cup)	40 g
Shellfish		
Clams, mussels, oysters, scallops, snails	3 medium	30 g
Crab, lobster flaked	50 mL (¼ cup)	30 g
Shrimp:		
fresh	5 large	30 g
frozen	10 medium	30 g
canned	18 small	30 g
dry pack	50 mL (¼ cup)	30 g
Meat and Poultry (e.g. beef, chicken, goat, ham, lamb, pork, turkey, veal, wild game)		
Back, peameal bacon	3 thin slices	30 g
Chop	½ chop, with bone	40 g
Minced or ground, lean or extra-lean	30 mL (2 tbsp)	30 g
Sliced, lean	1 slice	30 g
Steak, lean	1 piece	30 g
Organ Meats		
Heart, liver	1 slice	30 g
Kidney, sweet breads, chopped	50 mL (¼ cup)	30 g
Tongue	1 slice	30 g
Tripe	5 pieces	60 g
Soyabean		
Bean curd or tofu	½ block	70 g
Eggs		
Egg in shell, raw or cooked	1 medium	50 g
Egg without shell, cooked or poached in water	1 medium	45 g
Egg, scrambled	50 mL (¼ cup)	55 g

Each of the following measured foods equal more than 1 protein choice:

Food	Choices	Measures	Mass (Weight)
Cheese	1 Protein + 1 Fats & Oils	1 piece	25 g
Cheese, coarsely grated, e.g. cheddar	1 Protein + 1 Fats & Oils	50 mL (¼ cup)	25 g
Cheese, dry, finely grated, e.g. parmesan	1 Protein + 1 Fats & Oils	45 mL	15 g
Cheese, ricotta, high fat	1 Protein + 1 Fats & Oils	50 mL (¼ cup)	55 g
Eel	1 Protein + 1 Fats & Oils	1 slice	50 g
Bologna	1 Protein + 1 Fats & Oils	1 slice	20 g
Canned luncheon meat	1 Protein + 1 Fats & Oils	1 slice	20 g
Corned beef, fresh	1 Protein + 1 Fats & Oils	1 slice	25 g
Corned beef, canned	1 Protein + 1 Fats & Oils	1 slice	25 g
Ground beef, medium fat	1 Protein + 1 Fats & Oils	30 mL (2 tbsp)	25 g
Meat spreads, canned	1 Protein + 1 Fats & Oils	45 mL	35 g
Mutton chop	1 Protein + 1 Fats & Oils	½ chop (with bone)	35 g
Pate	(*see Fats & Oils group*)		
Sausage, pork link	1 Protein + 1 Fats & Oils	1 link	25 g
Sausage, garlic Polish or knockwurst	1 Protein + 1 Fats & Oils	1 slice	50 g
Summer sausage or salami	1 Protein + 1 Fats & Oils	1 slice	40 g
Spareribs or shortribs, with bone	1 Protein + 1 Fats & Oils	1 large	65 g
Stewing beef	1 Protein + 1 Fats & Oils	1 cube	25 g
Weiner, hot dog	1 Protein + 1 Fats & Oils	½ medium	25 g
Peanut butter	1 Protein + 1 Fats & Oils	15 mL (1 tbsp)	15 g

TABLE E-16 CANADIAN EXCHANGE SYSTEM: FATS AND OILS GROUP

5 g fat, 190 kJ (45 cal)

Food	Measure	Mass (Weight)	Food	Measure	Mass (Weight)
Avocado	⅛	30 g	Walnuts	4 halves	10 g
Bacon, side crisp*	1 slice	5 g	Pumpkin and Squash Seeds	20 mL (4 tsp)	10 g
Butter*	5 mL (1 tsp)	5 g	Sesame Seeds	15 mL (1 tbsp)	10 g
Cheese spread	15 mL (1 tbsp)	15 g	Sunflower Seeds,		
Coconut,			shelled	15 mL (1 tbsp)	10 g
fresh*	45 mL (3 tbsp)	15 g	in shell	45 mL (3 tbsp)	15 g
dried*	15 mL (1 tbsp)	10 g	Oil, cooking and salad	5 mL (1 tsp)	5 g
Cream,			Olives,		
half-and-half (cereal) 10%*	30 mL (2 tbsp)	30 g	green	10	45 g
light (coffee) 20%*	15 mL (1 tbsp)	15 g	black (ripe)	7	57 g
whipping 32–37%*	15 mL (1 tbsp)	15 g	Pate, liverwurst, meat spreads	15 mL (1 tbsp)	15 g
Cream cheese*	15 mL (1 tbsp)	15 g	Salad dressing:		
Gravy*	30 mL (2 tbsp)	30 g	blue cheese, French	10 mL (2 tsp)	10 g
Lard*	5 mL (1 tsp)	5 g	Italian, mayonnaise, Thousand Island	5 mL (1 tsp)	5 g
Margarine	5 mL (1 tsp)	5 g	Salad dressing:		
Nuts, shelled:			low-calorie	30 mL (2 tbsp)	30 g
Almonds	8 nuts	5 g	Salt pork, raw or cooked*	5 mL (1 tsp)	5 g
Brazil Nuts	2 nuts	10 g	Sesame oil	5 mL (1 tsp)	5 g
Cashews	5 nuts	10 g	Sour cream, 12% M.F.	30 mL (2 tbsp)	30 g
Filberts, Hazelnuts	5 nuts	10 g	7% M.F.	60 mL (4 tbsp)	60 g
Macadamia	3 nuts	5 g	Shortening*	5 mL (1 tsp)	
Peanuts	10 nuts	10 g			
Pecans	5 halves	5 g			
Pignolias, Pine nuts	25 mL (5 tsp)	10 g			
Pistachios					
shelled	20 nuts	10 g			
in shell	20 nuts	20 g			

*These items contain higher amounts of saturated fat.

TABLE E-17 CANADIAN EXCHANGE SYSTEM: EXTRAS

Extra Vegetables

Use less than 125 mL (½ cup) for all vegetables listed, larger quantities may need to be counted as Fruits & Vegetables choice.

Artichokes	Cabbage	Kohlrabi	Rhubarb
Asparagus	Cauliflower	Leeks	Sauerkraut
Bamboo shoots	Celery	Lettuce	Shallots
Beans, string, green or yellow	Chard	Mushroom	Spinach
Bean sprouts, mung or soya	Cucumber	Okra	Sprouts, alfalfa, radish, etc.
Bitter melon (balsam pear)	Eggplant	Onions, green, mature	Tomato wedges
Bok choy	Endive	Parsley	Watercress
Broccoli	Fiddleheads	Peppers, green, red, yellow	Zucchini
Brussels sprouts	Greens, beet, dandelion, etc.	Radish	
	Kale	Rapini	

Condiments

2.5 g carbohydrate, 60 kJ (1 cal), limited to amount indicated

Food	Measure	Food	Measure
Anchovies	2 fillets	Dietetic fruit spreads	5 mL (1 tsp)
Barbecue sauce	15 mL (1 tbsp)	Maraschino cherries	1
Bran, natural	30 mL (2 tbsp)	Nondairy coffee whitener	5 mL (1 tsp)
Brewer's yeast	5 mL (1 tsp)	Nuts, chopped pieces	5 mL (1 tsp)
Carob powder	5 mL (1 tsp)	Pickles, unsweetened, dill	2
Catsup	5 mL (1 tsp)	sour mixed	11
Chili sauce	5 mL (1 tsp)	Sugar substitutes, granular	5 mL (1 tsp) (3 to 4 packages)
Cocoa powder	5 mL (1 tsp)		
Cranberry sauce, unsweetened	15 mL (1 tbsp)	Whipped toppings	15 mL (1 tbsp)

May be used without measuring

Free Foods

Artificial sweetener, such as cyclamate or aspartame	Dulse	Marjoram, cinnamon, etc.	Sugar-free Crystal drink
Baking powder, soda	Flavouring and extracts	Mineral water	Tea, clear
Bouillon or clear broth	Garlic	Mustard	Vinegar
Bouillon from cube, powder or liquid	Gelatin, unsweetened	Parsley	Water
Chow Chow, unsweetened	Ginger root	Pimentos	Worcestershire sauce
Coffee, clear	Herbal teas, unsweetened	Soda water, Club soda	Salt, pepper, thyme
Consomme	Horseradish, uncreamed	Soya sauce	
	Lemon juice/lemon wedge	Sugar-free jelly powder	
	Lime juice/lime wedge	Sugar-free soft drink	

TABLE E-18 CANADIAN EXCHANGE SYSTEM: COMBINED FOOD CHOICES

Food	Choices	Measures	Mass (Weight)
Angel food cake	½ Starch + 2½ Sugars	1/12 cake	50 g
Apple crisp (Apples, raisins, oatmeal, flour, brown sugar, butter)	½ Starch + 1½ Fruits & Vegetables + 1 Sugars + 1–2 Fats & Oils	125 mL (½ cup)	
Applesauce, sweetened	1 Fruits & Vegetables + 1 Sugars	125 mL (½ cup)	
Beans and pork in tomato sauce	1 Starch + ½ Fruits & Vegetables + ½ Sugars + 1 Protein	125 mL (½ cup)	135 g
Beef burrito	2 Starch + 3 Protein + 3 Fats & Oils		110 g
Brownie	1 Sugars + 1 Fats & Oils	1 brownie	20 g
Cabbage rolls* (Lean ground pork or lean ground beef, rice, cabbage, sauerkraut)	1 Starch + 2 Protein	3 rolls	310 g
Caesar salad (Romaine lettuce, dressing, sprinkled with croutons and parmesan cheese)	2–4 Fats & Oils	20 mL dressing (4 tsp)	
Cheesecake	½ Starch + 2 Sugars + ½ Protein + 5 Fats & Oils	1 piece	80 g
Chicken fingers (Bread crumbs, chicken breasts, oil)	1 Starch + 2 Protein + 2 Fats & Oils	6 Small	100 g
Chicken and snow pea Oriental (Boneless chicken, snow peas, rice)	2 Starch + ½ Fruits & Vegetables + 3 Protein + 1 Fats & Oil	500 mL (2 cups)	
Chili (Lean ground beef, kidney beans, tomatoes)	1½ Starch + ½ Fruits & Vegetables + 3½ Protein	300 mL (1¼ cup)	325 g
Chips			
Potato chips	1 Starch + 2 Fats & Oils	15 chips	30 g
Corn chips	1 Starch + 2 Fats & Oils	30 chips	30 g
Tortilla chips	1 Starch + 1½ Fats & Oils	13 chips	
Cheese twists	1 Starch + 1½ Fats & Oils	30 chips	30 g
Chocolate bar			
Aero®	2½ Sugars + 2½ Fats & Oils	bar	43 g
Smarties®	4½ Sugars + 2 Fats & Oils	package	60 g
Chocolate cake (without icing) (Flour, sugar, cocoa, oil)	1 Starch + 2 Sugars + 3 Fats & Oils	1/10 of an 8-in. pan	
Chocolate devil's food cake (without icing)	2 Starch + 2 Sugars + 3 Fats & Oils	1/12 of a 9-in. pan	
Chocolate milk	2 Milk 2% + 1 Sugars	250 mL (1 cup)	300 g
Clubhouse (tripledecker) sandwich (Cold meat, cheese, tomato, hold the mayo)	3 Starch + 3 Protein + 4 Fats & Oils		
Cookies			
chocolate chip	½ Starch + ½ Sugars + 1½ Fats & Oils	2 cookies	22 g
oatmeal	1 Starch + 1 Sugars + 1 Fats & Oils	2 cookies	40 g
Donut (Chocolate glazed, yeast, batter)	1 Starch + 1½ Sugars + 2 Fats & Oils	1 donut	65 g
Egg roll	1 Starch + ½ Protein + 1 Fats & Oils		75 g

*If eaten with sauce, add ½ Fruits & Vegetables choice.

Continued

TABLE E-18 CANADIAN EXCHANGE SYSTEM: COMBINED FOOD CHOICES—*Continued*

Food	Choices	Measures	Mass (Weight)
Four bean salad (Green beans, wax beans, lima beans, kidney beans, oil & vinegar)	1 Starch + ½ Protein + 1 Fats & Oils	125 mL (½ cup)	
French toast	1 Starch + ½ Protein + 2 Fats & Oils	1 slice	65 g
Fruit in heavy syrup (Fruit, sugar)	1 Fruits & Vegetables + 1½ Sugars	125 mL (½ cup)	
Granola bar (Oatmeal, chocolate chips, sugar)	½ Starch + 1 Sugars + 1–2 Fats & Oils		30 g
Granola cereal (Harvest crunch without fruit)	1 Starch + 1 Sugars + 2 Fats & Oils	125 mL (½ cup)	45 g
Hamburger	2 Starch + 3 Protein + 2 Fats & Oils	Junior burger	
Ice cream and cone, plain flavour			
Ice cream	½ Milk + 2–3 Sugars + 1–2 Fats & Oils		100 g
Cone	½ Sugars		4 g
Lasagna (Pasta, meat, cheese, tomato) 13″ × 9″ Pan;			
regular cheese	1 Starch + 1 Fruits & Vegetables + 3 Protein + 2 Fats & Oils	3″ × 4″ piece	
low-fat cheese	1 Starch + 1 Fruits & Vegetables + 3 Protein	3″ × 4″ piece	
Legumes			
Dried beans (kidney, navy, pinto, fava, chick-peas)	2 Starch + 1 Protein	250 mL (1 cup)	180 g
Dried peas	2 Starch + 1 Protein	250 mL (1 cup)	210 g
Lentils	2 Starch + 1 Protein	250 mL (1 cup)	210 g
Macaroni and cheese (Macaroni, cheese, flour, milk, margarine)	2 Starch + 2 Protein + 2 Fats & Oils	250 mL (1 cup)	210 g
Minestrone soup (Lean ground beef, potatoes, carrots, kidney beans, macaroni)	1½ Starch + ½ Fruits & Vegetables + ½ Fats & Oils	250 mL (1 cup)	
Muffin (All bran, flour, raisin, sugar)	1 Starch + ½ Sugars + 1 Fats & Oils	1 small muffin	45 g
Nuts (*dry or roasted without any oil added*)			
Almonds, dried sliced	½ Protein + 2 Fats & Oils	50 mL (¼ cup)	22 g
Brazil nuts, dried unblanched	½ Protein + 2½ Fats & Oils	5 large nuts	23 g
Cashew nuts, dry roasted	½ Starch + ½ Protein + 2 Fats & Oils	50 mL (¼ cup)	28 g
Filbert hazelnut, dry	½ Protein + 3½ Fats & Oils	50 mL (¼ cup)	30 g
Macadamia nuts, dried	½ Protein + 4 Fats & Oils	50 mL (¼ cup)	28 g
Peanuts, raw	1 Protein + 2 Fats & Oils	50 mL (¼ cup)	30 g
Pecans, dry roasted	½ Fruits & Vegetables + 3 Fats & Oils	50 mL (¼ cup)	22 g
Pine nuts, pignolia dried	1 Protein + 3 Fats & Oils	50 mL (¼ cup)	34 g
Pistachio nuts, dried	½ Fruits & Vegetables + ½ Protein + 2½ Fats & Oils	50 mL (¼ cup)	27 g
Pumpkin seeds, roasted	2 Protein + 2½ Fats & Oils	50 mL (¼ cup)	47 g
Sesame seeds, whole dried	½ Fruits & Vegetables + ½ Protein + 2½ Fats & Oils	50 mL (¼ cup)	30 g
Sunflower kernel, dried	½ Protein + 1½ Fats & Oils	50 mL (¼ cup)	17 g
Walnuts, dried chopped	½ Protein + 3 Fats & Oils	¼ cup (50 mL)	26 g
Perogies (Potato, cheese, dough)	2 Starch + 1 Protein + 1 Fats & Oils	3 perogies	

Continued

TABLE E-18 CANADIAN EXCHANGE SYSTEM: COMBINED FOOD CHOICES—*Continued*

Food	Choices	Measures	Mass (Weight)
Pie, fruit	1 Starch + 1 Fruits & Vegetables + 2 Sugars + 3 Fats & Oils	1 piece	120 g
Pizza, cheese (thin crust) (⅛ of a 12″)	1 Starch + 1 Protein + 1 Fats & Oils	1 slice	50 g
Pork stir fry (Boneless pork, snow peas, peppers, mushrooms)	½ to 1 Fruits & Vegetables + 3 Protein	200 mL (¾ cup)	
Potato salad (Potatoes, onions, mayonnaise, celery)	1 Starch + 1 Fats & Oils	125 mL (½ cup)	130 g
Potatoes, scalloped (Milk, potato, onions)	2 Starch + 1 Milk + 1–2 Fats & Oils	200 mL (¾ cup)	210 g
Pudding, bread or rice (Rice or bread)	1 Starch + 1 Sugar + 1 Fats & Oils	125 mL (½ cup)	
Pudding, vanilla (Milk, sugar)	1 Milk + 2 Sugars	125 mL (½ cup)	
Raisin bran cereal	1 Starch + ½ Fruits & Vegetables + ½ Sugars	175 mL (⅔ cup)	40 g
Rice Krispie squares	½ Starch + 1½ Sugars + ½ Fats & Oils	1 square	30 g
Shepherd's pie (Potatoes, lean ground beef, frozen mixed vegetables)	2 Starch + 1 Fruits & Vegetables + 3 Protein	325 mL (1⅓ cup)	
Sherbet, orange	3 Sugars + ½ Fats & Oils	125 mL (½ cup)	
Spaghetti and meat sauce (With 175 mL (⅔ cup) meat sauce)	2 Starch + 1 Fruits & Vegetables + 2 Protein + 3 Fats & Oils	250 mL (1 cup)	
Stew (Lean stewing beef, carrots, peas, potatoes)	2 Starch + 2 Fruits & Vegetables + 3 Protein + ½ Fats & Oils	200 mL (¾ cup)	
Sundae (Ice cream, chocolate syrup)	4 Sugars + 3 Fats & Oils	125 mL (½ cup)	
Tuna casserole (Noodles, tuna, egg, milk, cheese, bread crumbs)	1 Starch + 2 Protein + ½ Fats & Oils	125 mL (½ cup)	
Yogurt, fruit bottom (Milk, fruit, sugar)	1 Fruits & Vegetables + 1 Milk + 1 Sugars	125 mL (½ cup)	125 g
	½ Fruits & Vegetables + 1½ Milk + 1½ Sugars	175 mL (⅔ cup)	175 g
Yogurt, frozen	1 Milk + 1 Sugars	125 mL (½ cup)	125 g

Appendix F

Chapter Notes

Chapter 1 Notes

1. H. A. Guthrie, *Introductory Nutrition* (St. Louis, Mo.: The C. V. Mosby Company, 1983), pp. 2–4.
2. Federal Trade Commission News, General Nutrition Inc. Agrees to Pay $2.4 Million . . . , April 28, 1994.
3. S. H. Short, Health quackery: Our role as professionals, *Journal of the American Dietetic Association* 94 (1994): 607–608; Position of the American Dietetic Association: Identifying food and nutrition misinformation, *Journal of the American Dietetic Association* 95 (1995): 705–707.
4. S. H. Short, 1994; R. M. Philen and coauthors, Survey of advertising for nutritional supplements in health and bodybuilding magazines, *Journal of the American Medical Association* 268 (1992): 1008–1011.
5. Centers for Disease Control and Prevention, *National Vital Statistics Report* 47 (1999).
6. J. M. McGinnis and W. H. Foege, Actual causes of death in the United States, *Journal of the American Medical Association* 270 (1993): 2207–2212.
7. L. Breslow and N. Breslow, Health practices and disability: Some evidence from Alameda County, *Preventive Medicine* 22 (1993): 86–95.
8. J. M. McGinnis and W. H. Foege, 1993.
9. U.S. Department of Health and Human Services, Public Health Service, *Healthy People 2000: National Health Promotion and Disease Prevention Objectives* (Washington, D.C.: U.S. Government Printing Office, 1990), pp. 93–94.
10. National Center for Health Statistics (NCHS), *Healthy People 2000 Review, 1997* (Hyattsville, MD: Public Health Service, 1997), data obtained from the NCHS Web site at www.cdc.gov/nchswww.
11. S. C. Parks and coauthors, President's page: Challenging the future—Changing consumer eating habits create new opportunities in commercial foodservice, *Journal of the American Dietetic Association* 94 (1994): 908.
12. Princeton Survey Research Associates, *Shopping for Health* (Washington, D.C.: Food Marketing Institute, 1992), p. 12.
13. T. A. Pearson and coauthors, Does a cholesterol-lowering diet cost more? American Heart Association 66th Scientific Sessions Abstract, November 1993.
14. Portions of this discussion were adapted from M. A. Boyle and D. H. Morris, *Community Nutrition in Action: An Entrepreneurial Approach* (St. Paul: West Publishing Company, 1994), pp. 240–246.
15. B. W. Hickman and coauthors, Nutrition claims in advertising: A study of four women's magazines, *Journal of Nutrition Education* 25 (1993): 227–235.
16. K. Kotz and M. Story, Food advertisements during children's Saturday morning television programming: Are they consistent with dietary recommendations?, *Journal of the American Dietetic Association* 94 (1994): 1296–1300; G. Pazzaglia Sylvester and coauthors, Children's television and nutrition: Friends or foes?, *Nutrition Today* 30 (1995): 6–15.
17. American Dietetic Association, *New Survey Tracks Americans' Behaviors, Attitudes, and Knowledge of Nutrition and Health Issues* (Chicago: American Dietetic Association, 1997).
18. S. A. Oliveria and coauthors, Parent-child relationships in nutrient intake: The Framingham Children's Study, *American Journal of Clinical Nutrition* 56 (1992): 593–598.
19. J. MacClancy, *Consuming Culture: Why You Eat What You Eat* (New York: Henry Holt and Company, 1992), p. 38.
20. W. H. Glinsmann and G. K. Beauchamp, Babies need sugars in moderation, *Pediatric Basics* 69 (1994): 19–21.
21. Remarks of R. Wyden in *Deception and Fraud in the Diet Industry—Part 1: Hearing Before the House of Representatives, Subcommittee on Regulation, Business Opportunities, and Energy, Committee on Small Business* (Washington, D.C.: U.S. Government Printing Office, March 26, 1990), p. 1.
22. World Cancer Research Fund and American Institute for Cancer Research, *Food, Nutrition, and the Prevention of Cancer: A Global Perspective* (Washington, D.C.: American Institute for Cancer Research, 1997).
23. C. S. Fuchs and coauthors, Dietary fiber and the risk of colorectal cancer and adenoma in women, *New England Journal of Medicine* 340 (1999): 169–176.
24. World Cancer Research Fund, 1997.
25. D. R. Jacobs and coauthors, Whole-grain intake may reduce the risk of ischemic heart disease in post-menopausal women: The Iowa Women's Health Study, *American Journal of Clinical Nutrition* 68 (1998): 248–257.
26. The discussion about finding credible sources of information on the Internet is adapted from M. Boyle and D. Morris, *Community Nutrition in Action: An Entrepreneurial Approach* (Belmont, CA: Wadsworth Publishing Co., 1999), pp. 274–275.
27. R. Harris, Evaluating Internet research sources on-line, available at www.sccu.edu/faculty/R_Harris/evalu8it.htm; and Milton's Web, available at milton.mse.jhu.edu:8001/research/education/net.html.
28. *FDA Consumer*, October 1989.
29. I. Milner, The color of quackery? Fingering the phony nutritionists of the Yellow Pages, *Nutrition Forum* 11 (1994): 19–22.

Chapter 2 Notes

1. A. K. Kant and coauthors, Dietary diversity and subsequent mortality in the First National Health and Nutrition Examination Survey Epidemiologic Follow-up Study, *American Journal of Clinical Nutrition* 57 (1993): 434–440.
2. S. L. Anderson, A look at the Japanese dietary guidelines, *Journal of the American Dietetic Association* 90 (1990): 1527.
3. Food and Nutrition Board, Institute of Medicine, *Dietary Reference Intakes for Calcium, Phosphorus, Magnesium, Vitamin D, and Fluoride* (Washington, D.C.: National Academy Press), 1997, pp. S-1–S-13.
4. Personal communication with the Snack Food Association, Alexandria, Virginia, 1998.
5. American Medical Association and American Dietetic Association, *Targets for Adolescent Health: Nutrition and Physical Fitness* (Chicago: American Medical Association, 1991), p. 2.
6. American Heart Association, *Nutritious Nibbles* (Dallas: American Heart Association, 1984), p. 5.

7. U.S. Department of Agriculture, *Making Bag Lunches, Snacks, and Desserts Using the Dietary Guidelines,* USDA HNIS Home and Garden Bulletin No. 232-9 (Washington, D.C.: U.S. Government Printing Office), pp. 2–25.
8. L. S. Sims, A special issue (food labeling reform) deserves a special issue (of *Nutrition Today*)! *Nutrition Today,* September/ October 1993, p. 4.
9. Food and Drug Administration, FDA Backgrounder: The New Food Label, April 1994.
10. FDA Talk Paper, FDA Allows Whole Oat Foods to Make Health Claim on Reducing the Risk of Heart Disease, January 21, 1997.
11. HHS, News, U.S. Department of Health and Human Services, Folic Acid to Fortify U.S. Food Products to Prevent Birth Defects, February 29, 1996.
12. C. J. Geiger, Health claims: History, current regulatory status, and consumer research, *Journal of the American Dietetic Association* 98 (1998): 1312–1322.
13. FDA approves health claim labeling for foods containing soy protein, *Journal of the American Dietetic Association* 100 (2000): 292.
14. A. E. Sloan, The explosion of multi-cultural cuisine, *Food Technology* 48 (1994): 74–75.
15. J. F. Mariani, *Dictionary of American Food and Drink* (New York: Hearst Books, 1994), p. xvi.
16. S. J. Algert, E. Brzenzinski, and T. H. Ellison, Mexican American food practices, customs, and holidays, *Ethnic and Regional Food Practices, A Series* (Chicago and Alexandria, VA: The American Dietetic Association and American Diabetes Association, 1998), pp. 1–9, 23–26.
17. K. M. Ma, Chinese American food practices, customs, and holidays, *Ethnic and Regional Food Practices, A Series* (Chicago and Alexandria, VA: The American Dietetic Association, 1998), pp. 1–10, 27–31.
18. T. C. Campbell and J. Chen, Diet and chronic degenerative diseases: A summary of results from an ecologic study in rural China, in *Western Diseases* (Totowa, NJ: Humana Press, 1994), pp. 67–118.
19. Should Americans be eating more Chinese food? *The Tufts University Diet & Nutrition Letter* 8 (1990): 5.
20. M. Nestle, Mediterranean diets: Historical and research overview, *The American Journal of Clinical Nutrition* 61 (1995): 1313S–1320S.
21. Mariani, 1994, p. 295.
22. P. G. Kittler and K. Sucher, *Food and Culture in America* (New York: Van Nostrand Reinhold, 1989), pp. 182–194.
23. Personal communication with the NPD Group's National Eating Trends Service, Park Ridge, Illinois, 1994.
24. C. Higgins and coauthors, Jewish food practices, customs, and holidays, *Ethnic and Regional Food Practices, A Series* (Chicago and Alexandria, VA: The American Dietetic Association and the American Diabetes Association, 1998), pp. 1–7, 18–20.
25. J. Herman Viola and C. Margolis, *Seeds of Change* (Washington, D.C.: Smithsonian Institution Press, 1991), as cited in American Dietetic Association, *Food Folklore: Tales and Truths About What We Eat* (Minneapolis, MN: Chronimed Publishing, 1999).

Chapter 3 Notes

1. M. Lee and S. D. Krasinski, Human adult-onset lactase decline: An update, *Nutrition Reviews* 56 (1998): 1–8; F. L. Suarez and S. D. Savaiano, Diet, genetics and lactose intolerance, *Food Technology* 51 (1997): 74–76.
2. F. L. Suarez and coauthors, Tolerance to the daily ingestion of two cups of milk by individuals claiming lactose intolerance, *American Journal of Clinical Nutrition* 65 (1997): 1502–1506; L. D. McBean and G. D. Miller, Allaying fears and fallacies about lactose intolerance, *Journal of the American Dietetic Association* 98 (1998): 671–676.
3. M. Woodward and A. R. Walker, Sugar consumption and dental caries: Evidence from 90 countries, *British Dental Journal* 176 (1994): 297–302.
4. J. H. Shaw, Causes and control of dental caries, *New England Journal of Medicine* 317 (1987): 996–1004.
5. S. Kashket and coauthors, Lack of correlation between food retention on the human dentition and consumer perception of food stickiness, *Journal of Dental Research* 70 (1991): 1314–1319.
6. American Dental Association, *Diet & Dental Health* (Chicago: American Dental Association, 1993), p. 7.
7. M. E. Jensen, Responses of interproximal plaque pH to snack foods and effect of chewing sorbitol-containing gum, *Journal of the American Dental Association* 113 (1986): 262–266.
8. A. S. Papas and coauthors, Dietary models for root caries, *American Journal of Clinical Nutrition* 61 (1995): 417S–422S; American Dental Association, p. 7.
9. C. O. Enwonwu, Interface of malnutrition and periodontal diseases, *American Journal of Clinical Nutrition* 61 (1995): 430S–436S.
10. J. O. Alvarez, Nutrition, tooth development, and dental caries, *American Journal of Clinical Nutrition* 61 (1995): 410S–416S.
11. American Dental Association, p. 2.
12. MRCA Information Services, *Eating in America, Edition II (Eat II)* (Chicago: National Livestock and Meat Board, 1994), pp. 1–35.
13. E. Lehner and J. Lehner, *Folklore and Odysseys of Food and Medicinal Plants* (New York: Tudor Publishing Co., 1962) and R. Tannahill, *Food in History* (New York: Crown Trade Paperbacks, 1988), as cited in *Food Folklore: Tales and Truths about What We Eat* (Minneapolis, MN: Chronimed Publishing, 1999).
14. K. R. Westerterp, Food quotient, respiratory quotient, and energy balance, *American Journal of Clinical Nutrition* 57 (1993): 759S–765S; R. L. Atkinson, Role of diet in obesity treatment, an address presented at the North American Association for the Study of Obesity and Emory University School of Medicine conference Obesity Update: Pathophysiology, Clinical Consequences, and Therapeutic Options, Atlanta, Georgia, August 1992.
15. Position of the American Dietetic Association: Health implications of dietary fiber, *Journal of the American Dietetic Association* 97 (1997): 1157–1159.
16. E. Giovannucci and coauthors, Relationship of diet to risk of colorectal adenoma in men, *Journal of the National Cancer Institute* 84 (1992): 91–98; J. L. Freudenheim and coauthors, Risks associated with source of fiber and fiber components in cancer of the colon and rectum, *Cancer Research* 50 (1990): 3295–3300.
17. S. R. Glore and coauthors, Soluble fiber and serum lipids: A literature review, *Journal of the American Dietetic Association* 94 (1994): 425–436; M. A. Eastwood, The physiological effects of dietary fiber: An update, *Annual Review of Nutrition,* 12 (1992): 19–35.
18. D. J. Jenkins and coauthors, Effect on blood lipids of very high intakes of fiber in diets low in saturated fat and cholesterol, *New England Journal of Medicine* 329 (1993): 21–26; C. M. Ripsin and coauthors, Oat products and lipid lowering: A meta-analysis, *Journal of the American Medical Association* 267 (1992): 3317–3325.
19. J. Salmeron and coauthors, Dietary fiber, glycemic load, and risk of non-insulin-dependent diabetes mellitus in women, *Journal of the American Medical Association* 277 (1997): 472–477.
20. American Dietetic Association, Position of the American Dietetic Association: Health implications of dietary fiber, 1997.
21. F. Q. Nuttall, Dietary fiber in the management of diabetes, *Diabetes* 42 (1993): 503–508.
22. T. M. Wolever and coauthors, The glycemic index: Methodology and clinical implications, *American Journal of Clinical Nutrition* 54 (1991): 846–854.
23. The Expert Committee on the Diagnosis and Classification of Diabetes Mellitus, Report on the Expert Committee on the Diagnosis and Classification of Diabetes Mellitus, *Diabetes Care* 20 (1997): 1183–1197; Position Statement of the American Diabetes Association: Translation of the diabetes nutrition recommendations for health care institutions, *Diabetes Care* 21 (1998): S66–S68.
24. R. Butler and coauthors, Type 2 diabetes: Causes, complications, and new screening recommendations, *Geriatrics* 53 (1998): 47–54; C. Leontos and coauthors, National Diabetes Education Program: Opportunities and challenges, *Journal of the American Dietetic Association* 98 (1998): 73–75.
25. R. A. Hayward and coauthors, Starting insulin therapy in patients with Type 2 diabetes, *Journal of the American Medical Association* 278 (1997): 1663–1669.
26. Position Statement of American Diabetes Association, 1994.
27. Personal communication with Marketing Data Enterprises, Inc., Valley Stream, New York, 1998.
28. S. Cohen and coauthors, Saccharin and urothelial proliferation: A threshold phenomenon, *Journal of the American Societies for Experimental Biology* 6 (1992): A-1594.
29. Position of the American Dietetic Association: Use of nutritive and nonnutritive sweeteners, *Journal of the American Dietetic Association* 98 (1998): 580–585.
30. Council on Scientific Affairs, Saccharin: Review of safety issues, *Journal of the American Medical Association* 254 (1985): 2622–2644.
31. S. S. Schiffman and coauthors, Aspartame and susceptibility to headache, *New England Journal of Medicine* 317 (1987): 1181–1185.
32. Council on Scientific Affairs, Aspartame: Review of safety issues, *Journal of the American Medical Association* 254 (1985):400–402; American Council on Science and Health, *Low-Calorie Sweeteners* (New York: American Council on Science and Health, 1993), pp. 7–13; Position of the American Dietetic Association: Use of nutritive and nonnutritive sweeteners, *Journal of the American Dietetic Association* 98 (1998): 580–587.
33. J. E. Blundell and A. J. Hill, Paradoxical effects of an intense sweetener (aspartame) on appetite, *The Lancet* 1 (1986): 1092–1093.
34. B. J. Rolls, Effects of intense sweeteners on hunger, food intake, and body weight: *A review, American Journal of Clinical Nutrition* 53 (1991): 872–878; A. Drewnowski, Comparing the effects of aspartame and sucrose on

motivational ratings, taste preferences, and energy intakes in humans, *American Journal of Clinical Nutrition* 59 (1994): 338–345.

35. M. J. Franz and coauthors, Nutrition principles for the management of diabetes and related complications, *Diabetes Care* 17 (1994): 495–496.

Chapter 4 Notes

1. C. D. Berdanier, ω-3 Fatty acids: A Panacea? *Nutrition Today*, 29 (1994): 28–32.
2. K. K. Carroll, Biological effects of fish oil in relation to chronic diseases, *Lipids* 21 (1986): 731–732.
3. D. Kromhout and coauthors, The inverse relation between fish consumption and 20-year mortality from coronary heart disease, *New England Journal of Medicine* 312 (1985): 1205–1209; C. M. Albert and coauthors, Fish consumption and risk of sudden cardiac death, *Journal of the American Medical Association* 279 (1998): 23–28.
4. R. K. Chandra and coauthors, Decreased systemis thromboxane A_2 biosynthesis in normal human subjects fed a salmon-rich diet, *American Journal of Clinical Nutrition* 60 (1994): 369–373.
5. A. P. Simopoulos, Omega-3 fatty acids in health and disease and in growth and development, *American Journal of Clinical Nutrition* 54 (1991): 438–463; S. B. Eaton and coauthors, Dietary intake of long-chain polyunsaturated fatty acids during the paleolithic, *World Reviews Nutrition & Dietetics* 83 (1998): 12–23; H. Gerster, N–3 fish oil polyunsaturated fatty acids and bleeding, *Journal of Nutrition and Environmental Medicine* 5 (1995): 281–296.
6. B. E. Phillipson and coauthors, Reduction of plasma lipids, lipoproteins, and apoproteins by dietary fish oil in patients with hypertriglyceridemia, *New England Journal of Medicine* 312 (1985): 1210–1216; A. P. Simopoulos, ω-3 Fatty acids in health and disease and in growth and development, *American Journal of Clinical Nutrition* 54 (1991): 438–463; and W. S. Harris and F. Muzio, Fish oil reduces postprandial triglyceride concentration without accelerating lipid-emulsion removal rates, *American Journal of Clinical Nutrition* 58 (1993): 68–74.
7. W. C. Willett and A. Ascherio, Trans fatty acids: Are the effects only marginal? *American Journal of Public Health* 84 (1994): 722–724; A. Ascherio and coauthors, *Trans*-fatty acids and coronary heart disease, *New England Journal of Medicine* 340 (1999): 1994–1998.
8. J. T. Judd and coauthors, Dietary *trans* fatty acids: Effects on plasma lipids and lipoproteins of healthy men and women, *American Journal of Clinical Nutrition* 59 (1994): 861–868; R. Troisi and coauthors, Trans fatty acid intake in relation to serum lipid concentrations in adult men, *American Journal of Clinical Nutrition* 56 (1992): 1019–1024; A. H. Lichtenstein and coauthors, Hydrogenation impairs the hypolipidemic effect of corn oil in humans, *Arteriosclerosis and Thrombosis* 13 (1993): 154–161.
9. W. C. Willett and coauthors, Intake of *trans* fatty acids and risk of coronary heart disease among women, *The Lancet* 341 (1993): 581–585.
10. T. Byers, Hardened fats, hardened arteries, *New England Journal of Medicine* 337 (1997): 1545, A. H. Lichtenstein, Trans fatty acids, plasma lipid levels, and risk of developing cardiovascular disease: A statement for healthcare professionals from the American Heart Association, *Circulation* 95 (1997): 2588–2590.
11. W. C. Willett and A. Ascherio, 1994.
12. Community Nutrition Institute, Coming soon to a nutrition panel near you: Trans fatty acids information, *Nutrition Week*, November 19, 1999, pp. 1–2.
13. Nutrition Committee Advisory, American Heart Association, Trans fatty acids, May 17, 1994.
14. H. Gylling and T. A. Miettinen, Efficacy of plant stanol ester in lowering cholesterol in postmenopausal women and patients with diabetes, in T. T. Nguyen, ed., *New Developments in the Dietary Management of High Cholesterol* (Minneapolis, MN: McGraw-Hill Companies, 1998), pp. 39–43.
15. Expert Panel on Detection, Evaluation, and Treatment of High Blood Cholesterol in Adults, Summary of the Second Report of the National Cholesterol Education Program (NCEP) Expert Panel on Detection, Evaluation, and Treatment of High Blood Cholesterol in Adults (Adult Treatment Panel), *Journal of the American Medical Association* 269 (1993): 3014–3023.
16. G. Wolf, The role of oxidized low-density lipoprotein in the activation of peroxisome proliferator-activated receptor (gamma): Implications for atherosclerosis, *Nutrition Reviews* 57 (1999): 88–91; R. Ross, Atherosclerosis: An inflammatory disease, *New England Journal of Medicine* 340 (1999): 115–126.
17. S. L. Connor, The healthy heart: Challenges and opportunities for dietetics professionals in the 21st century, *Journal of the American Dietetic Association* 99 (1999): 164–165.
18. Adult Treatment Panel II, *Journal of the American Medical Association* 269 (1993): 3015–3023.
19. A. H. Lichtenstein and coauthors, Effects of canola, corn, and olive oils on fasting and postprandial plasma lipoproteins in humans as part of a National Cholesterol Education Program step 2 diet, *Arteriosclerosis and Thrombosis* 13 (1993): 1533–1542; J. M. Hodgson and coauthors, Can linoleic acid contribute to coronary artery disease? *American Journal of Clinical Nutrition* 58 (1993): 228–234; E. S. Sarkkinen and coauthors, Long-term effects of three fat-modified diets in hypercholesterolemic subjects, *Atherosclerosis* 105 (1994): 9–23.
20. C. V. Felton and coauthors, Dietary polyunsaturated fatty acids and composition of human aortic plaques, *The Lancet* 344 (1994): 1195–1196.
21. M. D. Lorgeril and coauthors, Mediterranean alpha-linolenic acid-rich diet in secondary prevention of coronary heart disease, *The Lancet* 343 (1994): 1454–1459.
22. H. Blackburn, Co-investigator, Seven Countries Study, as quoted in D. Schardt, B. Liebman, and S. Schmidt, Going Mediterranean, *Nutrition Action Health Letter* 21 (1994): 1–5.
23. E. B. Rimm and coauthors, Vitamin E consumption and the risk of coronary heart disease in men, *New England Journal of Medicine* 328 (1993): 1450–1456; M. J. Stampfer and coauthors, Vitamin E consumption and the risk of coronary heart disease in women, *New England Journal of Medicine* 328 (1993): 1444–1449; P. Knekt, Antioxidant vitamin intake and coronary mortality in a longitudinal population study, *American Journal of Epidemiology* 139 (1994): 1180–1189.
24. I. Jialal and S. M. Grundy, The effect of dietary supplementation with alpha tocopherol on the oxidative modification of low-density lipoprotein, *Journal of Lipid Research* 6 (1992): 899–906; J. A. Simon, Vitamin C and cardiovascular disease, *Journal of the American College of Nutrition* 11 (1992): 107–125; D. L. Morris, S. B. Kritchevsky, and C. E. Davis, Serum carotenoids and coronary heart disease: The lipid research clinics coronary primary prevention trial and follow-up study, *Journal of the American Medical Association* 272 (1994): 1439–1441.
25. D. Kritchevsky, Antioxidant vitamins in the prevention of cardiovascular disease, *Nutrition Today* 27 (1992): 30–33; D. Steinberg, Antioxidant vitamins and coronary heart disease, *New England Journal of Medicine* 328 (1993): 1487–1489; C. H. Hennekens, J. E. Buring, and R. Peto, Antioxidant vitamins: Benefits not yet proved, *New England Journal of Medicine* 330 (1994): 1080–1081.
26. M. H. Read, Dietary fat practices: Age, gender, nutrition knowledge, *Topics in Clinical Nutrition* 13 (1998): 53–60.
27. Personal communication with Consumer Affairs, Entenmann's, Bay Shore, NY.
28. Personal communication with Media relations, McDonald's Corporation, Oak Brook, IL.
29. Personal communication with Consumer Affairs, Kraft, Glenview, IL.
30. G. E. Ruoff, Reducing fat intake with fat substitutes, *American Family Physician* 43 (April 1991): 1235–1242.
31. *Simplesse All Natural Fat Substitute: A Scientific Overview*, (Deerfield, IL.: The Simplesse Company, 1991), pp. 3–10.
32. U.S. Department of Health and Human Services, HHS News, FDA Approves Fat Substitute, Olestra, January 24, 1996.
33. Position paper of the American Dietetic Association: Fat replacers, *Journal of the American Dietetic Association* 98 (1998): 463–468; C. C. Akoh, Fat replacers, 52 (1998): 47–56.
34. W. Root Food (New York: Simon & Schuster, Inc., 1986), as cited in American Dietetic Association, *Food Folklore: Tales and Truths About What We Eat* (Minneapolis, MN: Chronimed Publishing Company, 1999).
35. R. H. Eckel and R. M. Krauss, American Heart Association call to action: Obesity as a major risk factor for coronary heart disease, *Circulation* 97 (1998): 2099–2100.
36. F. B. Hu and coauthors, Dietary fat intake and the risk of coronary heart disease in women, *New England Journal of Medicine* 337 (1997): 1491–1499.
37. D. J. Jenkins and coauthors, Effect on blood lipids of very high intakes of fiber in diets low in saturated fat and cholesterol, *New England Journal of Medicine* 329 (1993): 21–26.
38. D. L. Sprecher and coauthors, Efficacy of psyllium in reducing serum cholesterol levels in hypercholesterolemic patients on high or low-fat diets, *Annals of Internal Medicine* 119 (1993): 545–554.
39. Expert Panel on Detection, Evaluation, and Treatment of High Blood Cholesterol Levels in Adults, *Journal of the American Medical Association* 269 (1993): 3015–3023.
40. R. R. Wing and coauthors, Change in waist-hip ratio with weight loss and its association with change in cardiovascular risk factors, *American Journal of Clinical Nutrition* 55 (1992): 1086–1092.
41. W. S. Harris and F. Muzio, Fish oil reduces postprandial triglyceride concentrations without accelerating lipid-emulsion removal rates, *American Journal of Clinical Nutrition* 58 (1993): 68–74; R. Vandongen, Effects on blood pressure of ω-3 fats in subjects at increased risk of cardiovascular disease, *Hypertension* 22 (1993): 371–379.

42. G. S. Omenn and coauthors, Preventing coronary heart disease: B vitamins and homocysteine, *Circulation* 97 (1998): 421–424; O. Nygard and coauthors, Major lifestyle determinants of plasma total homocysteine distribution: The Hordaland Homocysteine Study, *American Journal of Clinical Nutrition* 67 (1998): 263–270.
43. P. M. Ridker and coauthors, Association of moderate alcohol consumption and plasma concentration of endogenous tissue-type plasminogen activator, *Journal of the American Medical Association* 272 (1994): 929–933.
44. J. M. Gaziano and coauthors, Moderate alcohol intake, increased levels of high-density lipoprotein and its subfractions, and decreased risk of myocardial infarction, *New England Journal of Medicine* 329 (1993): 1829–1834; G. D. Friedman and A. L. Klatsky, Is alcohol good for your heart, *New England Journal of Medicine* 329 (1993): 1882–1883.
45. S. L. Englebardt, Eat, drink, go back to work, *American Health* June 1994, p. 90.
46. T. A. Pearson and P. Terry, What to advise patients about drinking alcohol: The clinician's conundrum, *Journal of the American Medical Association* 272 (1994): 967–968.
47. K. Breithaupt and coauthors, Protective effect of chronic garlic intake on elastic properties of the aorta in the elderly, *Circulation* 96 (1997): 2649–2655; A. Orekhov and J. Grunwald, Effects of garlic on atherosclerosis, *Nutrition* 13 (1997): 656–663; and J. Isaacsohm and coauthors, Garlic powder and plasma lipids and lipoproteins, *Archives of Internal Medicine* 158 (1998): 1189–1194.
48. L. Kushi and coauthors, Dietary antioxidant vitamins and death from coronary heart disease in postmenopausal women, *New England Journal of Medicine* 334 (1996): 1156–1162.
49. U.S. Department of Health and Human Services, Report of the Expert Panel on Blood Cholesterol Levels in Children and Adolescents, 1991; S. S. Gidding, The rationale for lowering serum cholesterol levels in American children, *American Journal of Diseases in Children* 147 (1993): 386–392; J. P. Stong and coauthors, Prevalence and extent of atherosclerosis in adolescents and young adults, *Journal of the American Medical Association* 281 (1999): 727–735.

Chapter 5 Notes

1. D. E. Matthews, The proteins and amino acids, in M. E. Shils, J. A. Olson, and M. Shike, eds., *Modern Nutrition in Health and Disease*, 9th ed. (Baltimore, MD: Williams and Wilkins, 1999), pp. 11–48.
2. W. J. Visek, Arginine needs, physiological state and usual diets: A reevaluation, *Journal of Nutrition* 116 (1986): 36–46.
3. D. W. Wilmore, Glutamine and the gut, *Gastroenterology* 107 (1994): 1885–1886.
4. Personal communication with Gerald Gleich, M.D., Department of Immunology, Mayo Clinic, Rochester, MN, 1992.
5. Personal communication with EMS Hotline (800-367-2829), 1995.
6. EMS Hotline, 1995; E. A. Belongia and coauthors, An investigation of the cause of eosinophilia-myaligia syndrome associated with tryptophan use, *New England Journal of Medicine* 323 (1990): 357–365.
7. C. Ballentine, The essential guide to amino acids, *FDA Consumer*, September 1985, pp. 23–25.
8. S. A. Anderson and D. J. Raiten, eds., *Safety of Amino Acids Used as Dietary Supplements* (Bethesda, MD: Life Sciences Research Office, Federation of American Societies for Experimental Biology, 1992), pp. v–x, 193–196, 213–215.
9. C. Ballentine, 1985.
10. A. Brezezinski, Melatonin in humans, *New England Journal of Medicine* 336 (1997): 186–95.
11. P. J. Rasch and coauthors, Protein dietary supplementation and physical performance, *Medicine and Science in Sports* 1 (1969): 195–199.
12. V. Lambert, Using "smart" drugs and drinks may not be smart, *FDA Consumer*, April 1993, pp. 24–26.
13. N. J. Smith and B. Worthington-Roberts, *Food for Sport* (Palo Alto, Calif.: Bull Publishing Company, 1989), p. 21.
14. R. K. Chandra and coauthors, Nutrition and immunity: Lessons from the past and new insights into the future, *American Journal of Clinical Nutrition* 53 (1991): 1087–1101; S. A. Shikora and coauthors, Nutrition and immunology: Clinician's approach, in R. A. Forse, ed., *Diet, Nutrition, and Immunity* (Boca Raton, FL: CRC Press, 1994), pp. 9–22.
15. V. R. Young, Soy protein in relation to human protein and amino acid nutrition, *Journal of the American Dietetic Association* 91 (1991): 828–835.
16. D. G. Schroeder and R. Martorell, Enhancing child survival by preventing malnutrition, *American Journal of Clinical Nutrition* 65 (1997): 1080–1081.
17. B. Torun and F. Chew, Protein-energy malnutrition, in M. E. Shils, J. A. Olson, and M. Shike, eds., *Modern Nutrition in Health and Disease* (Baltimore, MD: Williams and Wilkins, 1999), pp. 963–988.
18. M. Nestle and S. Guttmacher, Hunger in the United States: Rationale, methods, and policy implications of state hunger surveys, *Journal of Nutrition Education* 24 (1992): 18S–22S.
19. M. B. Zemel, Calcium utilization: Effect of varying level and source of dietary protein, *American Journal of Clinical Nutrition* 48 (1988): 880–883.
20. American Institute for Cancer Research, *Stopping Cancer Before It Starts*, (New York: Golden Books, 1999), pp. 214–217.
21. I. Chalmers, *The Great Food Almanac* (San Francisco: Collins Publishers, 1994).
22. C. L. Vecchia, Vegetable consumption and risk of chronic disease, *Epidemiology* 9 (1998): 208–210.
23. J. Carol and J. Sobal, Model of the process of adopting vegetarian diets: Health vegetarians and ethical vegetarians, *Journal of Nutrition Education* 30 (1998): 196–202; D. Farley, More people trying vegetarian diets, *FDA Consumer* October 1995, pp. 10–14.
24. E. Haddad, Meeting the RDAs with a vegetarian diet, *Topics in Clinical Nutrition* 10 (1995): 7–16.
25. Position of the American Dietetic Association: Vegetarian Diets, *Journal of the American Dietetic Association* 97 (1997): 1317–1321.
26. C. Lamberg-Allardt and coauthors, Low serum 25-hydroxyvitamin D concentrations and secondary hyperparathyroidism in middle-aged white strict vegetarians, *American Journal of Clinical Nutrition* 58 (1993): 684–689.
27. J. R. Hunt and coauthors, Zinc absorption, mineral balance, and blood lipids in women consuming controlled lacto-ovo-vegetarian and omnivorous diets for 8 weeks, *American Journal of Clinical Nutrition* 67 (1998): 421–430.
28. S. I. Barr and coauthors, Spinal bone mineral density in premenopausal vegetarian and nonvegetarian women: Cross-sectional and prospective comparisons, *Journal of the American Dietetic Association* 98 (1998): 760–765.
29. T. A. B. Sanders, Vegetarian diets and children, *Pediatric Clinics of North America* 42 (1995): 955.
30. K. Burke, The use of soy foods in a vegetarian diet, *Topics in Clinical Nutrition* 10 (1995): 37–43.
31. T. J. Key and coauthors, Dietary habits and mortality in 11,000 vegetarians and health-conscious people: Results of a 17 year follow up, *British Medical Journal* 313 (1996): 775–779.
32. P. Walter, Effects of vegetarian diets on aging and longevity, *Nutrition Reviews* 55 (1997): S61–S65; J. Swarner, The vegetarian diet and cancer prevention, *Topics in Clinical Nutrition* 10 (1995): 17–21; and C. Whitten, Vegetarian diets and ischemic heart disease, *Topics in Clinical Nutrition* 10 (1995): 27–33.
33. Soy and health: Discovering the health benefits of isoflavones fact sheet, United Soybean Board, 1999.
34. *U.S. Soyfoods Directory: Alphabetical List of Soyfoods* (Indiana: Indiana Soybean Board, 1998).
35. American Heart Association, 1997 Heart and Stroke Statistical Update, Dallas, TX, 1997.
36. K. K. Carrol, Review of clinical studies on cholesterol-lowering response to soy protein, *Journal of the American Dietetic Association* 91 (1991): 820–827.
37. J. R. Crouse and coauthors, A randomized trial comparing the effect of casein with that of soy protein containing varying amounts of isoflavones on plasma concentrations of lipids and lipoproteins, *Archives of Internal Medicine* 159 (1999): 2070–2076.
38. J. W. Anderson and coauthors, Meta-analysis of the effects of soy protein intake on serum lipids, *New England Journal of Medicine* 333 (1995): 276–282
39. R. M. Bakhit and coauthors, Intake of 25 g of soybean protein with or without fiber alters plasma lipids in men with elevated cholesterol concentrations, *Journal of Nutrition* 124 (1994): 213–222; and S. M. Potter and coauthors, Depression of plasma cholesterol in men by consumption of baked products containing soy protein, *American Journal of Clinical Nutrition* 58 (1993) 501–506.
40. S. M. Potter and coauthors, Soy protein and isoflavones: Their effects on blood lipids and bone density in postmenopausal women, *American Journal of Clinical Nutrition* 68 (1998): 1375S–1379S; D. L. Alekel and coauthors, Isoflavone-rich soy protein isolate exerts significant bone sparing in the lumbar spine in perimenopausal women, Third International Symposium on the Role of Soy in Preventing and Treating Chronic Disease, Washington, D.C., October 31–November 4, 1999.
41. National Osteoporosis Foundation, Fast Facts on Osteoporosis, (Washington, D.C.: National Osteoporosis Foundation 1997).
42. M. Lock, Menopause in cultural context, *Experimental Gerontology* 29 (1994): 307–317; American Heart Association, 1997.
43. A. L. Murkies and coauthors, Dietary flour supplementation decreases postmenopausal hot flushes: Effect of soy and wheat, *Maturitas* 221 (1995): 189–195.

44. D. C. Knight and J. A. Eden. A review of the clinical effects of phytoestrogens, *Obstetrics and Gynecology* 87 (1996): 897–904; D. H. Upmalis, F. C. Cone, and C. Lamia, Effect on biochemical parameters of an oral soy extract used in the treatment of vasomotor symptoms in menopausal women, Third International Symposium on the Role of Soy in Preventing and Treating Chronic Disease, Washington, D.C., October 31–November 4, 1999.

45. American Cancer Society, Cancer Facts & Figures 1999. (New York: ACS, Inc., 1999).

46. American Cancer Society, 1999.

47. L. N. Kolonel. Variability in diet and its relation to risk in ethnic and migrant groups, *Basic Life Sci.* 43 (1988): 129–135.

48. A. Nomura and coauthors, Breast cancer and diet among the Japanese in Hawaii, *American Journal of Clinical Nutrition* 31 (1978) 2020–2025.

49. R. K. Severson and coauthors, A prospective study of demographics, diet, and prostrate cancer among men of Japanese ancestry in Hawaii, *Cancer Research* 49 (1989): 1857–1860.

50. M. Messina and S. Barnes. The role of soy products in reducing risk of cancer, *Journal of the National Cancer Institute* 83 (1991): 541–546; D. Li and coauthors, Soybean isoflavones reduce experimental metastasis in mice, *Journal of Nutrition* 129 (1999): 1075–1078; J. R. Zhou and coauthors, Soybean components inhibit orthotopic growth of human prostate cancer cells in SCID mice, Third International Symposium on the Role of Soy in Preventing and Treating Chronic Disease, Washington, D.C., October 31–November 4, 1999.

Chapter 6 Notes

1. J. Jaramillo-Arango, The conquest of nutritional diseases (vitamins), in *The British Contribution to Medicine* (Edinburgh: E. & S. Livingstone, Ltd., 1953), pp. 140–162.

2. R. Hill and coauthors, The discovery of vitamins, in *The Chemistry of Life* (Cambridge: Cambridge University Press, 1970), pp. 156–170.

3. J. M. McKenney and coauthors, A comparison of the efficacy and toxic effects of sustained- vs. immediate-release niacin in hypercholesterolemic patients, *Journal of the American Medical Association* 271 (1994): 672–677; L. Lasagna, Over-the-counter niacin, *Journal of the American Medical Association* 271 (1994): 709–710; M. T. Behme, Nicotinamide and diabetes prevention, *Nutrition Reviews* 53 (1995): 137–139.

4. K. Steinmetz and J. Potter, Vegetables, fruit, and cancer prevention: A review, *Journal of the American Dietetic Association* 96 (1996): 1027–1039; A. Kwasniewska and coauthors, Folate deficiency and cervical intraepithelial neoplasia, *European Journal of Gynaecological Oncology* 18 (1997): 526–530.

5. A. Bendich, Folic acid and prevention of neural tube birth defects: critical assessment of FDA proposals to increase folic acid intakes, *Journal of Nutrition Education* 26 (1994): 294–299.

6. C. W. Suitor and L. B. Bailey, Dietary folate equivalents: Interpretation and application, *Journal of the American Dietetic Association* 100 (2000): 88–94.

7. M. Nestle, Folate fortification and neural tube defects: policy implications, *Journal of Nutrition Education* 26 (1994): 287–293.

8. K. M. Riggs and coauthors, Relations of vitamin B_{12}, vitamin B_6, folate, and homocysteine to cognitive performance in the Normative Aging Study, *American Journal of Clinical Nutrition* 63 (1996): 306–314.

9. G. S. Omenn and coauthors, Preventing coronary heart disease: B vitamins and homocysteine, *Circulation* 97 (1998): 421–424; O. Nygard and coauthors, Major lifestyle determinants of plasma total homocysteine distribution: The Hordaland Homocysteine Study, *American Journal of Clinical Nutrition* 67 (1998): 263–270.

10. J. Selhub and coauthors, Association between plasma homocysteine concentrations and extracranial carotid-artery stenosis, *New England Journal of Medicine* 332 (1995): 286–291; M. J. Stampfer and M. R. Malinow, Can lowering homocysteine levels reduce cardiovascular risk? *New England Journal of Medicine* 332 (1995): 328–329; J. Selhub and coauthors, Vitamin status and intake as primary determinants of homocysteinemia in an elderly population, *Journal of the American Medical Association* 270 (1993): 2693–2698.

11. H. I. Morrison and coauthors, Serum folate and risk of fatal coronary heart disease, *Journal of the American Medical Association* 275 (1996): 1893–1896; C. J. Boushey and coauthors, Quantitative assessment of plasma homocysteine as a risk factor for vascular disease, *Journal of the American Medical Association* 274 (1995): 1049–1057.

12. M. K. Berman and coauthors, Vitamin B_6 in premenstrual syndrome, *Journal of the American Dietetic Association* 90 (1990): 859–860.

13. P. M. Suter and coauthors, Reversal of protein-bound vitamin B_{12} malabsorption with antibiotics in atrophic gastritis, *Gastroenterology* 101 (1991): 1039–1045.

14. L. C. Pauling, *Vitamin C and the Common Cold* (San Francisco: W. H. Freeman, 1970).

15. L. A. Levin, Opthalmology, *Journal of the American Medical Association* 273 (1995): 1703–1705; R. C. Rose and coauthors, Ocular oxidants and antioxidant protection, *Proceedings of the Society for Experimental Biology and Medicine* 217 (1998): 397–407; B. Lyle and coauthors, serum carotenoids and tocopherols and incidence of age-related nuclear cataract, *American Journal of Clinical Nutrition* 69 (1999): 272–277; M. C. Leske, Antioxidant vitamins and nuclear opacities: The Longitudinal study of cataract, *Opthalmology* 105 (1998): 831–836.

16. M. F. Hollick and coauthors, The vitamin D content of fortified milk and infant formula, *New England Journal of Medicine* 326 (1992): 1213–1215.

17. National Osteoporosis Foundation, *The Osteoporosis Report* 9 (Winter 1993): 3.

18. W. Mertz, A balanced approach to nutrition for health: the need for biologically essential minerals and vitamins, *Journal of the American Dietetic Association* 94 (1994): 1259–1262.

19. P. S. Connolly, Treatment of nocturnal leg cramps, *Archives of Internal Medicine* 152 (1992): 1877–1880.

20. L. Mosca and coauthors, Antioxidant nutrient supplementation reduces the susceptibility of low density lipoprotein to oxidation in patients with coronary artery disease, *Journal of the American College of Cardiology* 30 (1997): 392–399; M. N. Diaz and coauthors, Antioxidants and atherosclerotic heart disease, *New England Journal of Medicine* 337 (1997): 408–416.

21. M. Stampfer and coauthors, Vitamin E consumption and the risk of coronary disease in women, *New England Journal of Medicine* 328 (1993): 1444–1449; E. Rimm and coauthors, Vitamin E consumption and the risk of coronary disease in men, *New England Journal of Medicine* 328 (1993): 1450–1456.

22. D. Feskanich and coauthors, Vitamin K intake and hip fractures in women: A prospective study, *American Journal of Clinical Nutrition* 69 (1999): 74–79; C. Vermeer, M. H. J. Knapen, and L. J. Schurgers, Vitamin K and metabolic bone disease, *Journal of Clinical Pathology* 51 (1998): 424–426; and P. Weber, The role of vitamins in the prevention of osteoporosis: A brief status report, *International Journal for Vitamin and Nutrition Research,* 69 (1999): 194–197.

23. J. Hirsh and V. Fuster, Guide to anticoagulant therapy, part 2: oral anticoagulants, *Circulation* 89 (1994): 1473.

24. A. L. Eldridge and E. T. Sheehan, Food supplement use and related beliefs: Survey of community college students, *Journal of Nutrition Education* 26 (1994): 259–265; A. Sarubin, Dietary Supplements: Industry statistics and government regulations, presented at the American Dietetic Association's Annual Meeting and Exhibition, Atlanta, GA, October, 1999.

25. Waverly Root, *Food* (New York: Simon & Schuster, Inc., 1986).

26. S. J. VanGarde and M. Woodburn, *Food Preservation and Safety* (Ames, Iowa: Iowa State University Press, 1994), p. 109.

27. J. Bailey, *Keeping Food Fresh* (New York: Harper & Row, 1989), p. 18.

28. VanGarde and Woodburn, p. 110.

29. M. A. McCarthy and R. H. Matthews, *Conserving Nutrients in Foods* Administrative Report No. 384 (Hyattsville, MD: Nutrition Monitoring Division, Human Nutrition Information Service, U.S. Department of Agriculture, 1988), p. 3.

30. VanGarde and Woodburn, pp. 94–98.

31. VanGarde and Woodburn, pp. 123–124.

32. N. R. Farnsworth and coauthors, Medicinal plants in therapy, *Bulletin of the World Health Organization* 63 (1985): 965–1170.

33. N. R. Farnsworth, The role of medicinal plants in drug development, chapter in Natural Products and Drug Development, P. Krogsgaard-Larsen, S. Brogger Christensen and H. Kofod, eds., Proceedings of Alfred Benzon Symposium 20, (Munksgaard: Copenhagen, 1984) pp. 17–30.

34. Natural Foods Merchandiser market reports for 1991, 1992, 1993, and 1994 (Boulder, CO: New Hope Communication).

35. P. Brevort, The economics of botanicals: The U.S. experience. Presentation at the NIH/OAM Conference on Botanicals: A Role in U.S. Health Care? (Washington, D.C., December 16, 1994).

36. G. B. Mahady, Herbal remedies: The promise scrutinized by science, Program for Collaborative Research in the Pharmaceutical Sciences, College of Pharmacy, University of Illinois at Chicago, April 1999.

37. Parts of this discussion are adapted from Functional foods: Opening the door to better health, *Food Insight,* November/December 1995.

38. W. H. S. Jones, ed., *Hippocrates* (1932): 351.

39. C. M. Hasler, Functional foods: The western perspective, *Nutrition Reviews* 54 (1996): 6–9.

40. Position of the American Dietetic Association: Functional Foods, *Journal of the American Dietetic Association* 99 (1999): 1278–1285.

41. A. B. Caragay, Cancer-preventive foods and ingredients, *Food Technology* 42 (1992): 65–68; A. S. Bloch, Phytochemicals and functional foods for cancer risk reduction, *Topics in Clinical Nutrition* 15 (2000): 24–28.

42. Anonymous, The nutraceutical initiative: A proposal for economic and regulatory reform (New York: The Foundation for Innovation in Medicine, 1991).
43. P. R. Thomas and R. Earl, eds., Enhancing the food supply, in *Opportunities in the Nutrition and Food Sciences* (Washington, D.C.: National Academy Press, 1994), pp. 98–142.
44. J. W. Anderson and coauthors, Meta-analysis of the effects of soy protein intake on serum lipids, *New England Journal of Medicine* 333 (1995): 276–282; M. Messina and S. Barnes, The role of soy products in reducing risk of cancer, *Journal of the National Cancer Institute* 83 (1991): 541–546; B. H. Arjmandi and coauthors, Dietary soybean protein prevents bone loss in an ovariectomized rat model of osteoporosis, *Journal of Nutrition* 126 (1996): 162–167; P. Albertazzi and coauthors, The effect of dietary soy supplementation on hot flushes, *Obstetrics and Gynecology* 91 (1997): 6–11.
45. *Food Insight*, International Food Information Council, Washington, D.C., May/June 1998.
46. *Food Insight*, 1998.

Chapter 7 Notes

1. C. M. McCay, Anorganic substances, in *Notes on the History of Nutrition Research*, ed. F. Verzar (Vienna: Hans Huber Publishers, 1973), pp. 156–184.
2. National Research Council, *Recommended Dietary Allowances—10th Edition* (Washington, D.C.: National Academy Press, 1989), pp. 248–249.
3. U.S. Environmental Protection Agency, Office of Water, *Is Your Drinking Water Safe?* (Washington, D.C.: U.S. Environmental Protection Agency, 1989).
4. U.S. Environmental Protection Agency and Centers for Disease Control and Prevention, Guidance for people with severely weakened immune systems, June 15, 1995.
5. U.S. Environmental Protection Agency Office of Water, *Lead and Your Drinking Water* (Washington, D.C.: U.S. Environmental Protection Agency, 1987).
6. Bottled water, *Food Industry Newsletter*, Vol. 28, April 12, 1999, pp. 1–2.
7. U.S. General Accounting Office, *Food Safety and Quality: Stronger FDA Standards and Oversight Needed for Bottled Water* (Washington, D.C.: U.S. General Accounting Office, 1991), p. 17.
8. V. Lambert, Bottled water: New trends, new rules, *FDA Consumer*, June 1993, pp. 9–11.
9. J. Stannard and coauthors, Fluoride content of some bottled waters and recommendation for fluoride supplementation, *Journal of Pedodontics* 14 (1990): 103–107.
10. Committee on Dietary Reference Intakes, *Dietary Reference Intakes for Calcium, Phosphorus, Magnesium, Vitamin D, and Fluoride*, (Washington, D.C.: National Academy Press, 1997).
11. R. P. Heaney and coauthors, Food factors influencing calcium availability, in *Nutritional Aspects of Osteoporosis*, eds. P. Burckharat and R. P. Heaney. Proceedings of the 2nd International Symposium on Osteoporosis, Lausanne, Switzerland, May 1994 (New York: Raven Press, 1995).
12. National Digestive Diseases Information Clearinghouse, *Lactose Intolerance*, NIH Publication No. 94-2751, April 1994; L. N. Aurisicchio and C. S. Pitchumoni, Lactose intolerance: Recognizing the link between diet and discomfort, *Postgraduate Medicine* 95 (1994): 113–116, 119–120; F. L. Suarez and coauthors, Lactose maldigestion is not an impediment to the intake of 1500 milligrams calcium daily as dairy products, *American Journal of Clinical Nutrition* 68 (1998): 1118–1122.
13. G. Wyshak and R. E. Frisch, Carbonated beverages, dietary calcium, the dietary calcium-phosphorus ratio, and bone fractures in girls and boys, *Journal of Adolescent Health* 15 (1994): 210–215; Z. Haurel and coauthors, Adolescents and calcium: What they do and do not know and how much they consume, *Journal of Adolescent Health* 2 (1998): 225–228.
14. E. Rubenowitz and coauthors, Magnesium in drinking water and death from myocardial infarction, *American Journal of Epidemiology* 143 (1996): 456–462.
15. G. G. Krishna and S. C. Kapoor, Potassium depletion exacerbates essential hypertension, *Annals of Internal Medicine* 115 (1991): 77–82.
16. O. Ophir and coauthors, Low blood pressure in vegetarians: The possible role of potassium, *American Journal of Clinical Nutrition* 37 (1983): 755–762.
17. D. Farley, High blood pressure: Controlling the silent killer, *FDA Consumer*, December 1991, pp. 28–33. M. H. Alderman and coauthors, Dietary sodium intake and mortality: The National Health and Nutrition Examination Survey, *Lancet* 351 (1998): 781–785.
18. S. A. Corrigan and coauthors, Weight reduction in the prevention and treatment of hypertension: A review of representative clinical trials, *American Journal of Health Promotion* 5 (1991): 208–214. P. K. Whelton and coauthors, Sodium reduction and weight loss in the treatment of hypertension in older persons, *Journal of the American Medical Association* 279 (1998): 839–846.
19. National High Blood Pressure Education Program Working Group, National High Blood Pressure Education Program Working Group Report on Primary Prevention of Hypertension, *Archives of Internal Medicine* 153 (1993): 186–208.
20. The Joint National Committee on Detection, Evaluation, and Treatment of High Blood Pressure, The Fifth Report of the Joint National Committee on Detection, Evaluation, and Treatment of High Blood Pressure, *Archives of Internal Medicine* 153 (1993): 154–183.
21. Joint National Committee, 1993.
22. National Research Council, *Diet and Health: Implications for Reducing Chronic Disease Risk, Executive Summary* (Washington, D.C.: National Academy Press, 1989), pp. 14–15.
23. B. M. Massie, To combat hypertension, increase activity, *The Physician and Sportsmedicine* 20 (1992): 89–111.
24. R. Stamler and coauthors, Primary prevention of hypertension by nutritional-hygienic means, *Journal of the American Medical Association* 262 (1989): 1801–1807.
25. T. Kotchen and D. McCarron, Dietary electrolytes and blood pressure: A statement for healthcare professionals from the American Heart Association, *Circulation* 98 (1998): 613–617; T. A. Kotchen and J. M. Kotchen, Dietary sodium and blood pressure: Interactions with other nutrients, *American Journal of Clinical Nutrition* 65 (1997): 708S–711S.
26. P. Suter, Potassium and hypertension, *Nutrition Reviews* 56 (1998): 151–153.
27. C. G. Osborne and coauthors, Evidence for the relationship of calcium to blood pressure, *Nutrition Reviews* 54 (1996): 365–381; D. A. McCarron and coauthors, Dietary calcium and blood pressure: Modifying factors in specific populations, American Journal of Clinical Nutrition (Supplement) 54 (1991): 215–219.
28. F. C. Luft, Dietary sodium, potassium and chloride intake and arterial hypertension, *Nutrition Today*, May/June 1989, pp. 11–12.
29. Joint National Committee, 1993.
30. The DASH clinical trial—*New England Journal of Medicine* 336 (1997): 1117.
31. M. Toussaint-Samat, *A History of Food* (Cambridge, MA: Blackwell Publishers, 1994), as cited in American Dietetic Association, *Food Folklore* (Minneapolis, MN: Chronimed Publishing, 1999), p. 18.
32. A. C. Looker and coauthors, Prevalence of iron deficiency in the United States, *Journal of the American Medical Association* 277 (1997): 973–976.
33. L. S. Stephenson, Possible new developments in community control of iron-deficiency anemia, *Nutrition Reviews* 53 (1995): 23–30.
34. E. R. Monsen, Iron nutrition and absorption: Dietary factors which impact iron bioavailablity, *Journal of the American Dietetic Association* 88 (1988): 786–790.
35. E. V. M. Borigato and F. E. Martinez, Iron nutritional status is improved in Brazilian preterm infants fed food cooked in iron pots, *The Journal of Nutrition* 128 (1998): 855–859.
36. A. Ascherio and coauthors, Dietary iron intake and risk of coronary disease among men, *Circulation* 89 (1994): 969–974; W. R. Proulx and C. M. Weaver, Ironing out heart disease, *Nutrition Today* 30 (1995): 16–23.
37. J. T. Salonen and coauthors, High stored iron levels are associated with excess risk of myocardial infarction in Eastern Finnish men, *Circulation* 86 (1992): 803–811.
38. R. Bahl and coauthors, Plasma zinc as a predictor of diarrheal and respiratory morbidity in children in an urban slum setting, *American Journal of Clinical Nutrition* 68 (1998): 414S–417S; P. C. Lira and coauthors, Effect of zinc supplementation on the morbidity, immune function, and growth of low-birthweight infants in northeast Brazil, *American Journal of Clinical Nutrition* 68 (1998): 418S–424S; and N. W. Solomons, Mild human zinc deficiency produces an imbalance between cell-mediated and humoral immunity, *Nutrition Reviews* 56 (1998): 27–32.
39. Zinc lozenges reduce the duration of common cold symptoms, *Nutrition Reviews* 55 (1997): 82–88.
40. A. S. Prasad, Discovery of human zinc deficiency and studies in an experimental human model, *American Journal of Clinical Nutrition* 53 (1991): 403–412.
41. J. C. King and C. L. Keen, Zinc, in M. E. Shils and J. A. Olson, eds., *Modern Nutrition in Health and Disease* (Baltimore, MD: Williams and Wilkins, 1999), pp. 223–240.
42. J. A. Pennington, A review of iodine toxicity reports, *Journal of the American Dietetic Association* 90 (1990): 1571–1581.
43. K. Lee and coauthors, Too much versus too little: The implications of current iodine intake in the United States, *Nutrition Reviews* 57 (1999): 177–181.
44. J. G. Hollowell and coauthors, Iodine nutrition in the United States. Trends and public health implications: Iodine excretion data from National Health and Nutrition Examination Surveys I and III, *Journal of Clinical Endocrinology and Metabolism* 83 (1998): 3401–3408.
45. Position of the American Dietetic Association: The impact of fluoride on

dental health, *Journal of the American Dietetic Association* 94 (1994): 1428–1431.

46. American Academy of Pediatrics, Fluoride supplementation for children, *Pediatrics* 95 (1995): 777; S. Levy and coauthors, Infants' fluoride ingestion from water, supplements, and dentifrice, *Journal of the American Dental Association* 126 (1995): 1625–1632.

47. B. J. Stoecker, Chromium, in *Modern Nutrition in Health and Disease* (Baltimore, MD: Williams and Wilkins, 1999), pp. 277–283.

48. D. H. Holben and A. M. Smith, The diverse role of selenium within selenoproteins: A review, *Journal of the American Dietetic Association* 99 (1999): 836–843; K. Yoshizawa and coauthors, Study of prediagnostic selenium level in toenails and the risk of advanced prostate cancer, *Journal of the National Cancer Institute* 90 (1998): 1219–1224; L. E. Clark and coauthors, Effects of selenium supplementation for cancer prevention in patients with carcinoma of the skin, *Journal of the American Medical Association* 276 (1996): 1957–1963.

49. F. H. Nielsen, Facts and fallacies about boron, *Nutrition Today*, May/June 1992, pp. 6–12, S. Meacham and coauthors, Effect of boron supplementation on blood and urinary calcium, magnesium, and phosphorus, and urinary boron in athletic and sedentary women, *American Journal of Clinical Nutrition* 61 (1995): 341–345.

50. The opening vignettes are from P. Ola and E. D'Aulaire, The health risk women can no longer ignore, *Reader's Digest*, August 1994, 91–95.

51. L. W. Turner and coauthors, Osteoporotic fracture among older U.S. women, *Journal of Aging and Health* 10 (1998):372–391; E. Siris and coauthors, Design of NORA, the National Osteoporosis Risk Assessment Program. A longitudinal U.S. Registry of postmenopausal women, *Osteoporosis International* 8 (1998): S62–S69.

52. T. A. Ricci and coauthors, Calcium supplementation suppresses bone turnover during weight reduction in postmenopausal women, *Journal of Bone and Mineral Research* 13 (1998): 1045–1050.

53. National Center for Injury Prevention and Control, Falls and hip fractures among the elderly, *Unintentional Injury Fact Sheet*, 1999 (available from www.cdc.gov).

54. V. Matkovic, Nutrition, genetics, and skeletal development, *Journal of the American College of Nutrition* 15 (1996): 556–569; G. M. Chan, K. Hoffman, and M. McMurray, Effects of dairy products on bone and body composition in pubertal girls *Journal of Pediatrics* 128 (1995): 551–556.

55. J. B. Anderson and P. A. Rondano, Peak bone mass development of females: Can young adult women improve their peak bone mass? *Journal of the American College of Nutrition* 15 (1996): 570–574.

56. R. L. Smith and coauthors, Prevention of postmenopausal osteoporosis: A comparative study of exercise, calcium supplementation, and hormone replacement therapy, *New England Journal of Medicine* 325 (1991): 1189–1195.

57. F. H. Anderson, Osteoporosis in men, *International Journal of Clinical Practice* 52 (1998): 176–180.

58. M. L. Rencken and coauthors, Bone density at multiple skeletal sites in amenorrheic athletes, *Journal of the American Medical Association* 276 (1996): 238–240.

59. H. Shibasaki and coauthors, The importance of body weight history in the occurrence and recovery of osteoporosis in patients with anorexia nervosa: Evaluation by dual X-ray absorptiometry and bone metabolic markers, *European Journal of Endocrinology* 139 (1998): 276–283.

60. C. J. Strange, Boning up on osteoporosis, *FDA Consumer* Reprint, August 1997.

61. E. L. Smith and C. Gilligan, Effects of inactivity and exercise on bone, *Physician and Sportsmedicine* 15 (1997): 91–102.

62. M. N. Hadley and S. V. Reddy, Smoking and the human vertebral column: A review of the impact of cigarette use on vertebral bone metabolism and spinal fusion, *Neurosurgery* 41 (1997): 116–124; C. W. Slemenda, Cigarettes and the skeleton, *New England Journal of Medicine* 330 (1994): 430–431.

63. D. Teegarden and coauthors, Dietary calcium, protein, and phosphorus are related to bone mineral density and content in young women, *American Journal of Clinical Nutrition* 68 (1998): 749–754; D. Teegarden and coauthors, Higher childhood and adolescent milk intakes are linked to higher calcium intakes in adulthood, *American Journal of Clinical Nutrition* 69 (1999): 1014–1017; R. P. Heaney and coauthors, Dietary changes favorably affect bone remodeling in older adults, *Journal of the American Dietetic Association* 99 (1999): 1228–1233.

64. D. Neumark-Sztainer and coauthors, Correlates of inadequate consumption of dairy products among adolescents, *Journal of Nutrition Education* 29 (1997): 12–20; L. Harnack, J. Stang, and M. Story, Soft drink consumption among U.S. children and adolescents: Nutritional consequences, *Journal of the American Dietetic Association* 99 (1999): 436–441.

65. P. Weber, The role of vitamins in the prevention of osteoporosis: A brief status report, *International Journal for Vitamin and Nutrition Research* 69 (1999): 194–197; G. Ferland, Vitamin K-dependent proteins: An update, *Nutrition Reviews* 56 (1998): 223–230.

66. U.S. Barzel and L. K. Massey, Excess dietary protein can adversely affect bone, *Journal of Nutrition* 128 (1998): 1051–1053; R. P. Heaney, Excess dietary protein may not adversely affect bone, *Journal of Nutrition* 128 (1998):1054–1057; S. J. Whiting and B. Lemke, Excess retinol intake may explain the high incidence of osteoporosis in northern Europe, *Nutrition Reviews* 57 (1999): 192–195; F. Ginty, A. Flynn, and K. D. Cashman, The effect of dietary sodium intake on biochemical markers of bone metabolism in young, *British Journal of Nutrition* 79 (1998): 343–350.

67. D. Feskanich and coauthors, Vitamin K intake and hip fractures in women: A prospective study, *American Journal of Clinical Nutrition* 69 (1999): 74–79.

68. Potassium, magnesium, and fruit and vegetable intakes linked to greater bone mineral density in the elderly, *American Journal of Clinical Nutrition* 69 (1999): 727–736.

69. B. Dawson-Hughes and coauthors, Effect of calcium and vitamin D supplements on bone density in men and women 65 years of age or older, *New England Journal of Medicine* 337 (1997): 670–675; B. Dawson-Hughes, Calcium, vitamin D, and risk of osteoporosis in adults: Essential information for the clinician, *Nutrition in Clinical Care* 1 (1998): 63–70; K. M. Chiu, Efficacy of calcium supplements on bone mass in postmenopausal women, *The Journal of Gerontology* 54 (1999): 275–280.

70. J. R. Saltzman, Nutritionally significant changes in gastrointestinal functioning with aging, *Nutrition in Clinical Care* 1 (1998): 20–29.

71. C. C. Collins and M. A. Summa, Clinical significance of lead content in dietary calcium supplements, *Nutrition in Clinical Care* 1 (1998): 156–159.

72. L. Pachucki-Hyde, Cutting your risk of osteoporosis, *Diabetes Self-Management*, May/June, 1998, pp. 36–43; E. Barrett-Connor, Hormone replacement therapy, *British Medical Journal* 317 (1998): 457–461; R. Lindsay, The role of estrogen in the prevention of osteoporosis, *Endocrinology and Metabolism Clinics of North America* 27 (1998): 399–405; M. McClung and coauthors, Alendronate prevents postmenopausal bone loss in women without osteoporosis, *Annals of Internal Medicine* 128 (1998): 253–261.

73. J. W. Nieves and coauthors, Calcium potentiates the effect of estrogen and calcitonin on bone mass: Review and analysis, *American Journal of Clinical Nutrition* 67 (1998): 18–24.

74. P. Taxel, Osteoporosis: Detection, prevention, and treatment in primary care, *Geriatrics* 53 (1998):22–40.

75. Top ten advances of 1993, *Harvard Health Letter*, 19(5) (1994): 7.

76. R. D. Lewis and C. M. Modlesky, Nutrition, physical activity, and bone health in women, *International Journal of Sports Nutrition* 8 (1998): 250–284; T. V. Nguyen and coauthors, Bone loss, physical activity, and weight change in elderly women: The Dubbo Osteoporosis Epidemiology Study *Journal of Bone Mineral Research* 13 (1998): 1458–1467.

77. R. L. Prince and coauthors, Prevention of postmenopausal osteoporosis: A comparative study of exercise, calcium supplementation, and hormone-replacement therapy, *New England Journal of Medicine* 325 (1991): 1189–1195.

78. N. K. Henderson and coauthors, The roles of exercise and fall risk reduction in the prevention of osteoporosis, *Endocrinology and Metabolism Clinics of North America* 27 (1998): 369–387.

Chapter 8 Notes

1. World Health Organization, as quoted in J. M. Rippe and coauthors, Public policy statement on obesity and health, *Nutrition in Clinical Care* 1 (1998): 34–47.

2. J. Koplan, U.S. obesity rates climb to epidemic proportions, *Journal of the American Medical Association* 280 (1999): 34–37.

3. National Heart, Lung, and Blood Institute Expert Panel, National Institutes of Health, *Clinical Guidelines on the Identification, Evaluation, and Treatment of Overweight and Obesity in Adults* (Washington, D.C.: Government Printing Office, 1998).

4. J. R. Rippe and coauthors, Public policy statement on obesity and health, *Nutrition in Clinical Care* 1 (1998): 34–47.

5. F. X. Pi-Sunyer, Health implications of obesity, *American Journal of Clinical Nutrition* 53 (1991): 1595S–1603S; I-Min Lee and coauthors, Body weight and mortality: A 27-year follow-up of middle-aged men, *Journal of the American Medical Association* 270 (1993): 2823–2828.

6. D. Festi and coauthors, Gallbladder motility and gallstone formation in obese patients following very low-calorie diets: Use it (fat) to lose it (well), *International Journal of Obesity and Related Metabolic Disorders* 22 (1998): 592–600.

7. Federal Trade Commission and Food and Drug Administration, *The Facts About Weight Loss Products and Programs*, DHHS Publication No. (FDA) 92-1189, 1992.

8. S. L. Gortmaker and coauthors, Social and economic consequences of overweight in adolescence and young adulthood, *New England Journal of Medicine* 329 (1993): 1008–1012; J. A. Cassell, Social anthropology and nutrition: A different look at obesity in America, *Journal of the American Dietetic Association* 95 (1995): 424–427.
9. R. Roubenoff and coauthors, Predicting body fatness: The body mass index versus estimation by bioelectrical impedance, *American Journal of Public Health* 85 (1995): 726–728.
10. P. Bjorntorp, Obesity, *Lancet* 350 (1997): 423–426.
11. Position of the American Dietetic Association, Weight management, *Journal of the American Dietetic Association* 97 (1997): 71–74.
12. The National Heart, Lung, and Blood Institute Expert Panel on the Identification, Evaluation, and Treatment of Overweight and Obesity in Adults, Executive summary of the clinical guidelines on the identification, evaluation and treatment of overweight and obesity in adults, *Journal of the American Dietetic Association* 98 (1998): 1178–1181.
13. R. J. Kuczmarski and coauthors, Varying body mass index cutoff points to describe overweight prevalence among U.S. adults: NHANES III (1988–1994), *Obesity Research* 5 (1997): 542–545.
14. National Heart, Lung, and Blood Institute expert panel, 1998.
15. L. VanHorn and coauthors, The dietitian's role in developing and implementing the first federal obesity guidelines, *Journal of the American Dietetic Association* 98 (1998): 1115–1117.
16. G. Blackburn, Effect of degree of weight loss on health benefits, *Obesity Research* 3 (1995): 211S–216S.
17. E. T. Poehlman and E. S. Horton, Energy needs: Assessment and requirements in humans, in M. E. Shils, J. A. Olson, M. Shike, and A. C. Ross, eds., *Modern Nutrition in Health and Disease* (Baltimore, MD: Williams and Wilkins, 1999), pp. 95–104.
18. P. A. Lachance, Human obesity, *Food Technology* 48 (1994): 127–138.
19. J. M. Friedman, Defective gene linked to obesity, *Nature*, December 1, 1994.
20. C. S. Mantzoros, The role of leptin in human obesity and disease: A review of the current evidence, *Annals of Internal Medicine* 130 (1999): 671–680; R. V. Considine, Serum immunoreactive-leptin concentrations in normal-weight and obese humans, *New England Journal of Medicine* 334 (1996): 292–295; J. M. Friedman, Leptin, leptin receptors, and the control of body weight, *Nutrition Reviews* 56 (1998): S38–S41.
21. M. A. McCrory and coauthors, Dietary variety within food groups: Association with energy intake and body fatness in men and women, *American Journal of Clinical Nutrition* 69 (1999): 440–447.
22. W. Dietz, Factors associated with childhood obesity, *Nutrition* 7(4) (1991): 290–291; R. C. Whitaker and coauthors, Predicting obesity in young adulthood from childhood and parental obesity, *New England Journal of Medicine* 337 (1997): 869–873; M. C. Bellizzi and W. H. Dietz, Workshop on childhood obesity: Summary of the discussion, *American Journal of Clinical Nutrition* 70 (1999): S173–S175.
23. F. Xavier Pi-Sunyer, Obesity, in M. E. Shils, J. A. Olson, M. Shike, and A. C. Ross, eds., *Modern Nutrition in Health and Disease* (Baltimore, M.D.: Williams and Wilkins, 1999), 1395–1418.
24. A. J. Stunkard and coauthors, A twin study of human obesity, *Journal of the American Medical Association* 256 (1986): 51–54; A. J. Stunkard and coauthors, An adoption study of human obesity, *New England Journal of Medicine* 314 (1986): 193–198; A. J. Stunkard and coauthors, The body-mass index of twins who have been reared apart, *New England Journal of Medicine* 322 (1990): 1483–1487; C. Bouchard and L. Perusse, Genetics of obesity, *Annual Review of Nutrition* 13 (1993): 337–354.
25. C. Bouchard and coauthors, The response to long-term overfeeding in identical twins, *New England Journal of Medicine* 322 (1990): 1477–1482.
26. Food and Nutrition Board, Institute of Medicine, P. R. Thomas, ed., *Weighing the Options: Criteria for Evaluating Weight-Management Programs* (Washington, D.C.: National Academy Press, 1995), pp. 1–25; J. M. Rippe and coauthors, Obesity as a chronic disease: Modern medical and lifestyle management, *Journal of the American Dietetic Association* 98 (1998): S9–S15.
27. K. D. Brownell and T. A. Wadden, Etiology and treatment of obesity: Understanding a serious, prevalent, and refractory disorder, *Journal of Consulting and Clinical Psychology* 60 (1992): 505–517.
28. B. J. Rolls and D. J. Shide, The influence of dietary fat on food intake and body weight, *Nutrition Reviews* 50 (1992): 283–290.
29. J. E. Blundell and J. J. Macdiarmid, Fat as a risk factor for overconsumption: Satiation, satiety, and patterns of eating, *Journal of the American Dietetic Association* 97 (1997): S63–S69.
30. American College of Sports Medicine and The American Dietetic Association, *Questioning 40/30/30: A Guide to Understanding Nutrition Advice* (Chicago: IL: American Dietetic Association, 1999); C. A. Titchenal and coauthors, Macronutrient composition of the *Zone Diet* based on computer analysis, *Medicine & Science in Sports & Exercise* 29 (1997): S126–S130.
31. The guidelines are adapted from AHA Medical/Scientific Statement, *American Heart Association Guidelines for Weight Management Programs for Healthy Adults* (Dallas, TX: American Heart Association, 1998).
32. J. Beedoe and coauthors, A review of low-calorie and very-low-calorie diet plans and possible metabolic consequences, *Topics in Clinical Nutrition* 6(1) (1990): 68–83.
33. National Task Force on the Prevention and Treatment of Obesity, Very low calorie diets, *Journal of the American Medical Association* 270 (1993): 967–974.
34. J. M. Rippe, The obesity epidemic: Challenges and opportunities, *Journal of the American Dietetic Association* 98 (1998): 85S.
35. F. M. Berg, Drug treatment for obesity, *Obesity and Health* 7 (1993): 8–12.
36. H. M. Connolly and coauthors, Valvular heart disease associated with fenfluramine-phentermine, *New England Journal of Medicine* 337 (1997): 581–588.
37. L. J. Aronne, Modern medical management of obesity: The role of pharmaceutical management, *Journal of the American Dietetic Association* 98 (1998): S23–S26; M. W. Schwartz and R. J. Seeley, The new biology of weight regulation, *Journal of the American Dietetic Association* 97 (1997): 54–58.
38. J. Stevens and coauthors, Effect of psyllium gum and wheat bran on spontaneous energy intake, *American Journal of Clinical Nutrition* 46 (1987): 812–817.
39. F. M. Berg, Chromium picolinate: Scam of the hour, *Obesity and Health* 7 (1993): 54–55.
40. P. Kurtzweil, Dieter's brews make tea time a dangerous affair, *FDA Consumer*, July/August 1997, pp. 6–11.
41. W. G. Van Gemert and coauthors, Quality of life assessment of morbidly obese patients: Effect of weight-reducing surgery, *American Journal of Clinical Nutrition* 67 (1998): 197–201; National Institute of Diabetes and Digestive and Kidney Diseases, *Gastric Surgery for Severe Obesity* (Bethesda, MD: National Institutes of Health, 1996), pp. 1–6.
42. S. M. Grundy, Multifactorial causation of obesity: Implications for prevention, *American Journal of Clinical Nutrition* 67 (1998): 563S–572S; National Task Force on the Prevention and Treatment of Obesity, Towards prevention of obesity, *Obesity Research* 2 (1994): 571–584.
43. P. Thomas, 1995.
44. F. X. Pi-Sunyer, The fattening of America, *Journal of the American Medical Association* 272 (1994): 238.
45. J. P. Foreyt and G. K. Goodrick, *Living Without Dieting* (New York: Warner Books, 1992), p. 28.
46. Ibid., 30.
47. K. D. Brownell, *The LEARN Program for Weight Control* (Dallas, TX: American Health Publishing Company, 1994), pp. 102–103.
48. Foreyt and Goodrick, pp. 43–58.
49. M. Shah and coauthors, Comparison of a low-fat, ad libitum complex carbohydrate diet with a low energy diet in moderately obese women, *American Journal of Clinical Nutrition* 59 (1994): 980–984; S. M. Shick and coauthors, Persons successful at long-term weight loss and maintenance continue to consume a low-energy, low-fat diet, *Journal of the American Dietetic Association* 98 (1998): 408–413.
50. M. T. McGuire and coauthors, Long-term maintenance of weight loss: Do people who lose weight through various weight loss methods use different methods to maintain their weight? *International Journal of Obesity and Related Metabolic Disorders* 22 (1998): 572–577; M. Senekal, A multidimensional weight-management program for women, *Journal of the American Dietetic Association* 99 (1999): 1257–1264; M. L. Klem and coauthors, A descriptive study of individuals successful at long-term maintenance of substantial weight loss, *American Journal of Clinical Nutrition* 66 (1997): 239–246.
51. K. Brownell, Yo-yo dieting, in *Nutrition 91/92*, ed. C. C. Cook-Fuller with S. Barrett (Guilford, CT: The Dushkin Publishing Group, 1991), pp. 132–134.
52. R. E. Anderson and coauthors, Changes in bone mineral content in obese dieting women, *Metabolism: Clinical and Experimental* 46 (1997): 857–861; L. Salamone and coauthors, *American Journal of Clinical Nutrition*, July 1999.
53. C. E. Ross, Overweight and depression, *Journal of Health and Social Behavior* 35 (1994): 63–78; M. E. Lean and coauthors, Impairment of health and quality of life in people with large waist circumference, *The Lancet* 351 (1998): 853–856.
54. National Task Force on the Prevention and Treatment of Obesity, Weight cycling, *Journal of the American Medical Association* 275 (1994): 1196–1202.
55. J. M. Rippe and S. Hess, The role of physical activity in the prevention and management of obesity, *Journal of the American Dietetic Association* 98 (1998): 31S–38S; P. D. Wood, Clinical applications of diet and physical activity in weight loss, *Nutrition Reviews* 54 (1998): S131–S135.
56. I. Chalmers, *The Great Food Almanac* (San Francisco: Collins Publishers, 1994).
57. J. O. Prochaska, *Changing for Good* (New York: William Morrow and Company, 1994), p. 47.
58. Ibid., 38–50.

59. H. Steiner and J. Lock, Anorexia nervosa and bulimia nervosa in children and adolescents: A review of the past ten years, *Journal of the American Academy of Child & Adolescent Psychiatry* 37 (1998): 352–357; G. C. Patton and coauthors, Onset of adolescent eating disorders: Population-based cohort study over 3 years, *British Journal of Medicine* 318 (1999): 765–768.
60. D. Neumark-Sztainer, Excessive weight preoccupation: Normative but not harmless, *Nutrition Today* 30 (1995): 68–74.
61. A. E. Becker and coauthors, Eating disorders, *New England Journal of Medicine* 340 (1999): 1092–1098; H. W. Hoek, The distribution of eating disorders, in K. D. Brownell and C. G. Fairburn, eds., *Eating Disorders and Obesity* (New York: Guilford Press, 1995), 207–211.
62. Task force on DSM-IV, 307.50 Eating Disorders Not Otherwise Specified, *DSM-IV Draft Criteria* (Washington, D.C.: American Psychiatric Association, 1993), P: 2.
63. C. W. Baker and K. D. Brownell, Binge eating disorder: Identification and management, *Nutrition in Clinical Care* 2 (1999): 344–353.
64. N. I. Hahn and M. M. Woolsey, When food becomes a cry for help, *Journal of the American Dietetic Association* 98 (1998): 395–398.
65. C. L. Rock, Nutritional and medical assessment and management of eating disorders, *Nutrition in Clinical Care* 2 (1999): 332–343.
66. T. Pryor and W. Wiederman, Personality features and expressed concerns of adolescents with eating disorders, *Adolescence* 33 (1998): 291–298.
67. W. Kaye and coauthors, Serotonin neuronal function and selective serotonin reuptake inhibitor treatment in anorexia and bulimia nervosa, *Biological Psychiatry* 44 (1998): 825–838.
68. A. Gila and coauthors, Subjective body-image dimensions in normal and anorexic adolescents, *British Journal of Medical Psychology* 71 (1998): 175–184.
69. A. Gila, 1998.
70. F. Klapper and coauthors, Psychiatric management of eating disorders, *Nutrition in Clinical Care* 2 (1999): 354–360.
71. D. Williamson, *Assessment of Eating Disorders: Obesity, Anorexia, and Bulimia Nervosa*, (New York: Pergamon Press, 1990); Practice guidelines for eating disorders, *American Journal of Psychiatry* 150 (1993): 212–218.
72. D. W. Reiff and K. K. L. Reiff, Position of the American Dietetic Association: Nutrition intervention in the treatment of anorexia nervosa, bulimia nervosa, and binge eating, *Journal of the American Dietetic Association* 94 (1994): 902–907; Position statement reaffirmed, 1999.
73. Klapper, 1999.
74. The guidelines listed are from M. Herrin, Dartmouth College Nutrition Education Program; for more information, contact Eating Disorders Awareness and Prevention, Inc. (800-931-2237, www.edap.org).

Chapter 9 Notes

1. S. N. Blair and coauthors, Changes in physical fitness and all-cause mortality: a prospective study of healthy and unhealthy men, *Journal of the American Medical Association* 273 (1995): 1093–1098.
2. Centers for Disease Control and Prevention and U.S. Department of Health and Human Services, *Physical Activity and Health: A Report of the Surgeon General*, (Washington, D.C.: U.S. Department of Health and Human Services, 1996).
3. The idea of positive addiction and this list of prerequisites for it originate with the psychologist W. Glasser, *Positive Addiction* (New York: Harper and Row, 1976), p. 93.
4. S. N. Blair, Diet and activity: The synergistic merger, *Nutrition Today* 30 (1995): 108–112.
5. The margin list is from Surgeon General's Report on Physical Activity and Health, 1996; adapted from B. E. Ainsworth and coauthors, Compendium of physical activities: Classification of energy costs of human physical activities, *Medicine and Science in Sports and Exercise* 25 (1993): 71–80.
6. M. A. Fiatarone and coauthors, Exercise training and nutritional supplementation for physical frailty in very elderly people, *New England Journal of Medicine* 330 (1994): 1769–1775.
7. American College of Sports Medicine, position stand: The recommended quantity and quality of exercise for developing and maintaining cardiorespiratory and muscular fitness, and flexibility in healthy adults, *Medicine and Science in Sports and Exercise* 30 (1998): 975–991.
8. M. Caldwell and coauthors, Weight-training increases fat-free mass and strength in untrained young women, *Journal of the American Dietetic Association*, 98 (1998): 414–418.
9. C. E. Broeder, The effects of either high-intensity resistance or endurance training on resting metabolic rate, *American Journal of Clinical Nutrition* 55 (1992): 802–810.
10. American College of Sports Medicine, 1998.
11. American College of Sports Medicine, *ACSM's Guidelines for Exercise Training and Prescription*, 5th ed. (Philadelphia: Williams and Wilkins, 1995), pp. 3–11.
12. F. Katch and W. McArdle, *Introduction to Nutrition, Exercise, and Health* (Philadelphia: Lea and Febiger, 1993) p. 329.
13. E. Hultman and coauthors, Work and exercise, in *Modern Nutrition in Health and Disease*, 9th ed., M. E. Shils, J. A. Olson, and M. Shike, eds. (Baltimore, MD: Williams and Wilkins, 1999), pp. 761–782.
14. A. N. Bosch, S. C. Dennis, and T. D. Noakes, Influence of carbohydrate loading on fuel substrate turnover and oxidation during prolonged exercise, *Journal of Applied Physiology* 74 (1993): 1921–1927; L. M. Burke and coauthors, Muscle glycogen storage after prolonged exercise: Effect of frequency of carbohydrate feedings, *American Journal of Clinical Nutrition* 64 (1996): 115–119.
15. J. A. Romijn and coauthors, Regulation of endogenous fat and carbohydrate metabolism in relation to exercise intensity and duration, *American Journal of Physiology* 265 (1993): E380–E391.
16. G. A. Brooks and J. Mercier, Balance of carbohydrate and lipid utilization during exercise: The "crossover" concept, *Journal of Applied Physiology* 76 (1994): 2253–2261.
17. D. C. Nieman, *Fitness and Sports Medicine: An Introduction* (Palo Alto, Calif.: Bull Publishing Company, 1990), pp. 246–258.
18. Position of the American Dietetic Association and the Canadian Dietetic Association, Nutrition for physical fitness and athletic performance for adults, *Journal of the American Dietetic Association* 93 (1993): 691–695.
19. A. Grandjean, What are the protein requirements for athletes? *Food and Nutrition News* 65 (2), March/April 1993, p. 11; M. E. Houston, Protein and amino acid needs of athletes, *Nutrition Today*, September/October 1992, pp. 36–39.
20. C. M. Cumming and coauthors, Recreational runners' beliefs and practices concerning water intake, *Journal of Nutrition Education* 26 (1994): 195–197.
21. American College of Sports Medicine, Position stand on exercise and fluid replacement, *Medicine and Science in Sports and Exercise* 28 (1996): i–vii; M. H. Williams, *Nutrition for Health, Fitness, and Sport*, 5th ed. (Boston: McGraw-Hill, 1999), p. 300; M. H. Williams, *Nutrition for Health, Fitness, and Sport*, 5th ed. (Boston: McGraw-Hill, 1999), p. 300. C. V. Gisolfi, Fluid balance for optimal performance, *Nutrition Reviews* 54 (1996): S159–S168.
22. M. Millard-Stafford, Fluid replacement during exercise in the heat, *Sports Medicine* 13 (1992): 223–233.
23. C. V. Gisolfi, Fluid balance for optimal performance, *Nutrition Reviews* 54 (1996): S159–S168.
24. R. J. Maughn, Fluid and electrolyte loss and replacement in exercise, in C. Williams and J. T. Devlin, eds., *Foods, Nutrition and Sports Performance: An International Scientific Consensus*. (London, E & FN Spon, 1992) pp. 19–33.
25. K. B. Wheeler and A. M. Cameron, Plasma volume: The hidden key to performance, *American Fitness Quarterly* (April 1990): 24–26.
26. Wheeler and Cameron, 1990.
27. Katch and McArdle, 1993.
28. M. Meydani and coauthors, Protective effect of vitamin E on exercise-induced oxidative damage in young and older adults, *American Journal of Physiology* 264 (1993).
29. L. Bucci, *Nutrients as Ergogenic Aids for Sports and Exercise* (Boca Raton, FL: CRC Press, 1993), pp. 1–161; M. Kaminski and R. Boal, An effect of ascorbic acid on delayed-onset muscle soreness, *Pain* 50 (1992): 317.
30. L. M. Weight, P. Jacobs, and T. D. Noakes, Dietary iron deficiency and sports anemia, *British Journal of Nutrition* 68 (1992): 253–260.
31. Nutrition and physical performance, in Nieman, 1990, pp. 221–268.
32. Y. I. Shu and J. D. Haas, Iron depletion without anemia and physical performance in young women, *American Journal of Clinical Nutrition* 66 (1997): 334–341; S. P. Bourque and coauthors, Twelve weeks of endurance exercise training does not affect iron status in women, *Journal of the American Dietetic Association* 97 (1997): 1116–1121.
33. K. Beals and M. M. Manore, The prevalence and consequence of subclinical eating disorders in female athletes, *International Journal of Sports Nutrition* 4 (1994): 175–195.
34. T. H. Murray, The ethics of drugs and sports, in *Drugs and Performance in Sports*, R. H. Strauss, ed. (Philadelphia: W. B. Saunders Company, 1987), pp. 11–21.
35. L. E. Armstong and C. M. Maresh, Vitamin and mineral supplements as nutritional aids to exercise performance and health, *Nutrition Reviews* 54 (1996): S149–S158; E. A. Applegate and L. E. Grivetti, Search for the competitive edge: A history of dietary fads and supplements, *Journal of Nutrition* 127 (1997): 869S–873S.
36. Position of the American Dietetic Association and Canadian Dietetic Association: Nutrition for physical fitness and athletic performance for adults, *Journal of the American Dietetic Association* 93 (1993): 691–696; L. Bucci, *Nutrients as Ergogenic Aids for Sports and Exercise* (Boca Raton, FL: CRC Press, 1993), pp. 1–161.
37. S. Barrett, Don't buy phony ergogenic aids, *Nutrition Forum* May/June 17 (1997): 19–21, 24.

38. G. Mirkin, Can bee pollen benefit health? *Journal of the American Medical Association* 262 (1989): 1854.
39. L. S. Walker and coauthors, Chromium picolinate and body composition and muscular performance in wrestlers, *Medcine and Science in Sports and Exercise* 30 (1998): 1730–1737; K. E. Grant and coauthors, Chromium and exercise training: Effect on obese women, *Medicine and Science in Sports and Exercise* 29 (1997): 992–998.
40. E. A. Applegate, 1997.
41. Food and Drug Administration, *Anabolic Steroids: Losing at Winning,* DHHS Publication No. (FDA) 88-3171; K. L. Ropp, No-win situation for athletes, *FDA Consumer,* December 1992, pp. 8–12; National Academy of Sports Medicine Policy Statement and Position Paper: Anabolic androgenic steroids, growth hormones, stimulants, ergogenics, and drug use in sports, in *Death in the Locker Room II: Drugs and Sports,* B. Goldman and R. Klatz (Chicago: Elite Sports Medicine Publications, 1992), pp. 328–373.
42. C. E. Yesalis and coauthors, Anabolic steroid use in the United States, *Journal of the American Medical Association* 270 (1993): 1217–1221.
43. Committee on Sports Medicine and Fitness, Adolescents and anabolic steroids: A subject review, *Pediatrics* 99 (1997): 443–447.
44. Nieman, 1990.
45. E. Lehner and J. Lehner, *Folklore & Odysseys of Food and Medicinal Plants* (New York: Tudor Publishing Co., 1962), as cited in American Dietetic Association, *Food Folklore: Tales and Truths About What We Eat* (Minneapolis, MN: Chronimed Publishing, 1999).
46. W. S. Holt, Jr., Nutrition and athletes, *American Family Physician* 47 (1993): 1757–1764.
47. A. R. Coggan and S. C. Swanson, Nutritional manipulations before and during endurance exercise, *Medicine and Science in Sports and Exercise* 24 (1992): S331–S335.
48. Substance abuse and Mental Health Services Administration (SAMHSA), Prevalence of past-month alcohol, cigarette, and other drug use, U.S. population, age 12 and older, 1996, *Substance Abuse and Mental Health Statistics Source Book* (Washington, D.C.: Government Printing Office, 1998), pp. 1–31.
49. Committee on Substance Abuse, Alcohol use and abuse: A pediatric concern, *Pediatrics* (1995): 439–442.
50. K. A. Douglas and coauthors, Results from the 1995 National College Health Risk Behavior Survey, *Journal of College Health,* September 1997, pp. 55–66.
51. C. L. Hart and coauthors, Alcohol consumption and mortality from all causes, coronary heart disease, and stroke: Results from a prospective cohort study of Scottish men with 21 years of follow up, *British Medical Journal* 318 (1999): 1725–1729.
52. S. Liu and coauthors, Prevalence of alcohol-impaired driving: Results from a national self-reported survey of health behaviors, *Journal of the American Medical Association* 277 (1997): 122–125.
53. Alcohol-related traffic fatalities, *FDA Consumer,* March 1993, p. 26; L. Archer, B. F. Grant, and D. A. Dawson, What if Americans drank less? The potential effect on the prevalence of alcohol abuse and dependence, *American Journal of Public Health* 85 (1995): 61–66.
54. I. Rossow and A. Amundsen, Alcohol abuse and mortality: A 40-year prospective study of Norwegian conscripts, *Social Science and Medicine* 44 (1997): 261–265; M. J. Thun and coauthors, Alcohol consumption and mortality among middle-aged and elderly U.S. adults, *New England Journal of Medicine* 337 (1997): 1705–1714.
55. Eighth Special Report to the U.S. Congress on Alcohol and Health, 1993.
56. R. A. Dietrich, Genetics of alcoholism: How do we find the answers and what do we do then? *Alcohol and Alcoholism* 25 (1990): 571–572.
57. P. Avogaro, Alcohol—A risk or protective factor in aging, in *Sedentary Life and Nutrition,* F. Fabris, L. Pernigotti, and E. Farrario, eds. (New York: Raven Press, Ltd., 1990), pp. 163–172.
58. R. Doll and coauthors, Mortality in relation to consumption of alcohol: 13 years' observations on male British doctors, *British Medical Journal* 309 (1994): 911–918; P. M. Ridker and coauthors, Association of moderate alcohol consumption and plasma concentration of endogenous tissue-type plasminogen activator, *Journal of the American Medical Association* 272 (1994): 929–933; R. Locher and coauthors, Ethanol suppresses smooth muscle cell proliferation in the postprandial state: A new antiatherosclerotic mechanism of alcohol? *American Journal of Clinical Nutrition* 67 (1998): 338–341.
59. M. Nestle, Alcohol guidelines for chronic disease prevention: From prohibition to moderation, *Nutrition Today* 32 (1997): 86–88.
60. C. A. Camargo and coauthors, Prospective study of moderate alcohol consumption and mortality in U.S. male physicians, *Archives of Internal Medicine* 137 (1997): 79–85; C. S. Fuchs and coauthors, Alcohol consumption and mortality among women, *New England Journal of Medicine* 332 (1995): 1245–1250.
61. T. A. Pearson and P. Terry, What to advise patients about drinking alcohol: The clinician's conundrum, *Journal of the American Medical Association* 272 (1994): 967–968.
62. Summary of the Report of a Working Group of the Royal Colleges of Physicians, Psychiatrists, and General Practitioners, Alcohol and the heart in perspective: Sensible limits reaffirmed, *Journal of the Royal College of Physicians of London* 29 (1995): 266–271.
63. J. M. Gaziano and coauthors, Moderate alcohol intake, increased levels of high-density lipoprotein and its subfractions, and decreased risk of myocardial infarction, *New England Journal of Medicine* 329 (1993): 1829–1834; and G. D. Friedman and A. L. Klatsky, Is alcohol good for your heart, *New England Journal of Medicine* 329 (1993): 1882–1883.
64. S. A. Smith-Warner and coauthors, Alcohol and breast cancer in women, *Journal of the American Medical Association* 279 (1998): 535–540.
65. H. K. Seitz and C. M. Oneta, Gastrointestinal alcohol dehydrogenase, *Nutrition Reviews* 56 (1998): 52–60.
66. M. Frezza and coauthors, High blood alcohol levels in women—The role of decreased gastric alcohol dehydrogenase activity and first-pass metabolism, *New England Journal of Medicine* 322 (1990): 95–99.
67. P. M. Suter and coauthors, Effects of alcohol on energy metabolism and body weight regulation: Is alcohol a risk factor for obesity? *Nutrition Reviews* 55 (1997): 157–171.
68. J. J. B. Anderson and B. R. Switzer, Effects of alcohol on nutritional status: Part I—Minerals, *Internal Medicine* 8 (1987): 69, 73–75, 79, 81–82, 85.
69. S. N. Mattson and coauthors, Heavy prenatal alcohol exposure with or without physical features of fetal alcohol syndrome leads to IQ deficits, *Journal of Pediatrics* 131 (1997): 718–721.
70. The questions are taken from the CAGE questionnaire to screen for alcohol abuse, in M. Carethers, Health promotion in the elderly, *American Family Physician* 45 (1992): 2253.

Chapter 10 Notes

1. L. Breslow and N. Breslow, Health practices and disability: Some evidence from Alameda County, *Preventive Medicine* 22 (1993): 86–95; A. J. Vita and coauthors, Aging, health risks, and cumulative disability, *New England Journal of Medicine,* 338 (1998):1035–1041.
2. B. Worthington-Roberts, The role of maternal nutrition in the prevention of birth defects, *Journal of the American Dietetic Association* 97 (1997): 184–185.
3. Food and Nutrition Board, Subcommittee on the Tenth Edition of the RDAs, *Recommended Dietary Allowances,* 10th ed. (Washington, D.C.: National Academy Press, 1989); Institute of Medicine, National Academy of Sciences, Food and Nutrition Board, *Nutrition During Pregnancy* (Washington, D.C.: National Academy Press, 1990), p. 17.
4. M. S. Bergman, Improving birth outcomes with nutrition intervention, *Topics in Clinical Nutrition* 13 (1997): 74–79.
5. T. O. Scholl and coauthors, Dietary and serum folate: Their influence on the outcome of pregnancy, *American Journal of Clinical Nutrition* 63 (1996): 520–525; R. J. Hine, What practitioners need to know about folic acid, *Journal of the American Dietetic Association* 96 (1996): 451–452; and P. Kurtzweil, How folate can help prevent birth defects, *FDA Consumer,* April 1997.
6. A. Kloeblen, Folate knowledge, intake from fortified grain products, and periconceptional supplementation patterns of a sample of low-income pregnant women according to the Health Belief Model, *Journal of the American Dietetic Association* 99 (1999): 33–38; G. Vozenilek, What they don't know could hurt them: Increasing public awareness of folic acid and neural tube defects, *Journal of the American Dietetic Association* 99 (1999): 20–22.
7. Y. Firth and coauthors, Estimation of individual intakes of folate in women of childbearing age with and without simulation of folic acid fortification, *Journal of the American Dietetic Association* 98 (1998): 985–988; A. A. Yates, Dietary Reference Intakes: The new basis for recommendations for calcium and related nutrients, B vitamins, and choline, *Journal of the American Dietetic Association* 98 (1998): 699–706.
8. B. Worthington-Roberts and R. M. Pitkin, Women's nutrition for optimal reproductive health, in *Call to Action: Better Nutrition for Mothers, Children, and Families,* ed. C. Sharbaugh (Washington, D.C.: National Center for Education in Maternal and Child Health, 1991), p. 124.
9. Institute of Medicine, 1990; C. W. Suitor and J. D. Gardner, *Journal of the American Dietetic Association* 90 (1990): 268.
10. M. Bonati, S. Nannini, and A. Addis, Vitamin A supplementation during pregnancy in developed countries, *Lancet* 345 (1995): 736–737; *New England Journal of Medicine,* November 23, 1995.
11. Institute of Medicine, 1990.

12. U.S. Department of Health and Human Services, Public Health Service, *Healthy People 2000: National Health Promotion and Disease Prevention Objectives* (Washington, D.C.: U.S. Government Printing Office, 1990).
13. C. M. Olson, Promoting positive nutritional practices during pregnancy and lactation, *American Journal of Clinical Nutrition* 59 (1994): 525S–531S; Predictors of poor maternal weight gain from baseline anthropometric, psychosocial, and demographic information in a Hispanic population, *Journal of the American Dietetic Association* 97 (1997): 1264–1268.
14. Institute of Medicine, 1990.
15. A. J. Rainville, Pica practices of pregnant women are associated with lower maternal hemoglobin level at delivery, *Journal of the American Dietetic Association* 98 (1998): 293–296.
16. A. Nehlig and G. Debry, Potential teratogenic and neurodevelopmental consequences of coffee and caffeine exposure: A review of human and animal data, *Neurotoxicology and Teratology* 16 (1994): 531–543.
17. Food and Nutrition Board, *Nutrition During Pregnancy* (Washington, D.C.: National Academy Press, 1990), pp. 397–399.
18. X. Ou Shu and coauthors, Maternal smoking, alcohol drinking, caffeine consumption, and fetal growth: Results from a prospective study, *Epidemiology* 6 (1995): 115–120; R. B. Ness and coauthors, Cocaine and tobacco use and the risk of spontaneous abortion, *New England Journal of Medicine* 340 (1999): 333–339.
19. U.S. Department of Health and Human Services, *Healthy People 2000*, 1990.
20. M. G. Bulterys, S. Greenland, and J. F. Kraus, Chronic fetal hypoxia and sudden infant death syndrome: Interaction between maternal smoking and low hematocrit during pregnancy, *Pediatrics* 86 (1990): 535–540; E. A. Mitchell and coauthors, Smoking and sudden infant death syndrome, *Pediatrics* 91 (1993): 893–896.
21. B. Worthington-Roberts and R. M. Pitkin, Women's nutrition for optimal reproductive health, pp. 113–136.
22. Ibid., 129.
23. Update: Trends in fetal alcohol syndrome—United States, 1979–1993, *Morbidity and Mortality Weekly Reports* 44 (1995): 249–251; American Academy of Pediatrics, Fetal alcohol syndrome and fetal alcohol effects, *Pediatrics* 91 (1993): 1004–1006.
24. N. I. Hahn and M. Erick, Battling morning (noon and night) sickness: New approaches for treating an age-old problem, *Journal of the American Dietetic Association* 94 (1994): 147–148.
25. J. C. King and J. Weininger, Pregnancy and lactation, in *Present Knowledge in Nutrition*, 6th ed., ed., M. L. Brown (Washington, D.C.: Nutrition Foundation, 1990), pp. 314–319.
26. D. R. Hollingsworth, *Pregnancy, Diabetes, and Birth*, 2nd ed. (Baltimore: Williams and Wilkins, 1992).
27. American Diabetes Association, Position statement: Gestational Diabetes Mellitus, *Diabetes Care* 21 (1998): S60–S61.
28. Position of the American Dietetic Association: Nutrition care for pregnant adolescents, *Journal of the American Dietetic Association* 94 (1994): 449–450.
29. M. Story and I. Alton, Nutrition issues, and adolescent pregnancy, *Nutrition Today* 30 (1995): 142–151.
30. M. L. Hediger and coauthors, Rate and amount of weight gain during adolescent pregnancy: Associations with maternal weight-for-height and birth weight, *American Journal of Clinical Nutrition* 52 (1990): 793–799.
31. Position of the American Dietetic Association: Nutrition care for pregnant adolescents, 1994.
32. R. Tannahil, *Food in History* (New York: Crown Publishers, 1988), p. 275.
33. R. Urgert and coauthors, Effects of cafestol and kahweol from coffee grounds on serum lipids and serum liver enzymes in humans, *American Journal of Clinical Nutrition* 61 (1995): 149–154.
34. A. Leviton and E. N. Allred, Correlates of decaffeinated coffee choice, *Epidemiology* 5 (1994): 537–540.
35. E. C. Strain and coauthors, Caffeine dependence syndrome, *Journal of the American Medical Association* 272 (1994): 1043–1048.
36. S. G. Oei and coauthors, Fetal arrhythmia caused by excessive intake of caffeine by pregnant women, *British Medical Journal* 298 (1989): 1075–1076.
37. C. Infante-Rivard and coauthors, Fetal loss associated with caffeine intake before and during pregnancy, *Journal of the American Medical Association* 270 (1993): 2940–2943; B. Eskenazi, Caffeine during pregnancy: Grounds for concern? *Journal of the American Medical Association* 270 (1993): 2973–2974.
38. N. L. Benowitz and coauthors, Persistent increase in caffeine concentrations in people who stop smoking, *British Medical Journal* 298 (1989): 1075–1076.
39. American Dietetic Association, Position of the American Dietetic Association: Promotion of breastfeeding, *Journal of the American Dietetic Association* 97 (1997): 662–665; M. A. Murtaugh, Optimal breastfeeding duration, *Journal of the American Dietetic Association* 97 (1997): 1252–1254.
40. Institute of Medicine, National Academy of Sciences, Food and Nutrition Board, *Nutrition During Lactation* (Washington, D.C.: National Academy Press, 1991), pp. 1–19.
41. American Academy of Pediatrics, Work Group on Breastfeeding, Breastfeeding and the use of human milk, Pediatrics 100 (1997): 1035–1039.
42. A. N. Eden, Infant feeding: Putting recommendations from the American Academy of Pediatrics Work Group on Breastfeeding into practice, *Nutrition in Clinical Care* 1 (1998): 89–91.
43. U.S. Department of Health and Human Services, *Healthy People 2000*, pp. 379–380.
44. Community Nutrition Institute, AIDS changes the debate on breastfeeding infants, *Nutrition Week* 27 (1997): 2–3; R. F. Black, Transmission of HIV-1 in the breastfeeding process, *Journal of the American Dietetic Association* 96 (1996): 267–274; R. D. Semba and M. C. Neville, Breastfeeding, mastitis, and HIV transmission: Nutritional implications, *Nutrition Reviews* 57 (1999): 146–153.
45. American Academy of Pediatrics, Committee on Nutrition, *Pediatric Nutrition Handbook*, 4th ed., ed. R. E. Kleinman (Elk Grove Village, IL.: American Academy of Pediatrics, 1998).
46. C. M. Trahms and J. Powell, Maternal and infant nutrition: An inquiry into the issues, *Topics in Clinical Nutrition* 11 (1996): 10–17; D. B. Johnson, Nutrition in infancy: Evolving views on recommendations, *Nutrition Today* 32 (1997): 63–68; J. S. Forsyth and coauthors, Relation of infant diet to childhood health: Seven year follow-up of cohort of children in Dundee infant feeding study, *British Medical Journal* January 3, 1998.
47. E. E. Ziegler, S. J. Fomon, and S. E. Nelson, *Journal of Pediatrics* 116 (1990): 11.
48. M. Irigoyen and coauthors, *Pediatrics* 88 (1991): 320.
49. Food and Nutrition Board, National Research Council, *Recommended Dietary Allowances*, 10th ed. (Washington, D.C.: National Academy Press, 1989), pp. 24–38.
50. L. L. Birch and coauthors, The variability of young children's energy intake, *New England Journal of Medicine* 324 (1991): 232–235; M. Sigman-Grant, Feeding Preschoolers: Balancing nutritional and developmental needs, *Nutrition Today* 27 (1992): 13–17.
51. E. Satter, *How to Get Your Kids to Eat . . . But Not Too Much* (Palo Alto, Calif.: Bull Publishing Company, 1987), pp. 13–28.
52. Position of the American Dietetic Association: Dietary guidance for healthy children aged 2 to 11 years, *Journal of the American Dietetic Association* 99 (1999): 93–101; M. F. Picciano, L. D. McBean, and V. A. Stallings, How to grow a healthy child: A conference report, *Nutrition Today* 34 (1999): 6–14.
53. R. Yip and coauthors, Pediatric nutrition surveillance system—United States, 1980–1991, *Morbidity and Mortality Weekly Report*, November 27, 1992, pp. 1–24.
54. P. C. DuRousseau and coauthors, Children in foster care: Are they at nutritional risk? *Journal of the American Dietetic Association* 91 (1991): 83–85; M. L. Taylor and S. A. Koblinsky, Dietary intake and growth status of young homeless children, *Journal of the American Dietetic Association* 93 (1993): 464–466.
55. ADA Reports, Position of the American Dietetic Association: Domestic food and nutrition security, *Journal of the American Dietetic Association* 98 (1998): 337–342.
56. T. B. Van Itallie, Predicting obesity in children, *Nutrition Reviews* 56 (1998): 154–155.
57. G. S. Berenson and coauthors, Association between multiple cardiovascular risk factors and atherosclerosis in children and young adults, *New England Journal of Medicine*, 338 (1998): 1650–1656; L. Van Horn, Primary prevention of cardiovascular disease starts in childhood, *Journal of the American Dietetic Association* 100 (2000): 41–43.
58. N. F. Krebs and S. L. Johnson, Guidelines for healthy children: Promoting eating, moving, and common sense, *Journal of the American Dietetic Association* 100 (2000): 37–39; R. E. Andersen and coauthors, Relationship of physical activity and television watching with body weight and level of fatness among children: Results from the Third National Health and Nutrition Examination Survey, *Journal of the American Medical Association* 279 (1998): 938–942.
59. G. P. Sylvester and coauthors, Children's television and nutrition: Friend or foes? *Nutrition Today* 30 (1995): 6–10; K. Kotz and coauthors, Food advertisements during children's Saturday morning television programming, *Journal of the American Dietetic Association* 94 (1994): 1296–1299.
60. A. Leung and coauthors, Children and television, *American Family Physician* 50 (1994): 909–913.
61. M. Golan, Parents as the exclusive agents of change in the treatment of childhood obesity, *The American Journal of Clinical Nutrition* 67 (1998): 1130–1135; T. Ramers, Maternal acceptability of a dietary intervention designed to lower children's intake of saturated fat and cholesterol: The

Dietary Intervention Study in Children (DISC), *Journal of the American Dietetic Association* 98 (1998): 31–34; S. C. Wilkins and coauthors, Family functioning is related to overweight in children, *Journal of the American Dietetic Association* 98 (1998): 572–574; and C. C. Francis and coauthors, Body composition, dietary intake, and energy expenditure in nonobese, prepubertal children of obese and nonobese biological mothers, *Journal of the American Dietetic Association* 99 (1999): 58–65.

62. A. C. Looker and coauthors, Prevalence of iron deficiency in the United States, *Journal of the American Medical Association* 277 (1997): 973–976.

63. S. S. Morey, American Academy of Pediatrics releases report on cholesterol levels in children and adolescents, *American Family Physician* 57 (1998): 2266–2268.

64. National Cholesterol Education Program, *Report of the Expert Panel on Blood Cholesterol Levels in Children and Adolescents* (Washington, D.C.: U.S. Department of Health and Human Services, Public Health Service, National Institutes of Health, NIH Publication No. 91-2732, 1991), pp. 1–22; F. E. Thompson and B. A. Dennison, Dietary sources of fats and cholesterol in U.S. children aged 2 through 5 years, *American Journal of Public Health* 84 (1994): 799–806.

65. L. E. Underwood, Normal adolescent growth and development, *Nutrition Today* 26 (1991): 11–16; G. B. Forbes, Body composition of adolescent girls, *Nutrition Today* 26 (1991): 17–20.

66. M. Story and coauthors, Adolescent nutrition: Trends and critical issues for the 1990s, in *Call to Action: Better Nutrition for Mothers, Children, and Families* ed. C. O. Sharbaugh (Washington, D.C.: National Center for Education in Maternal and Child Health, 1991), pp. 169–189.

67. D. Neumark-Sztainer and coauthors, Factors influencing food choices of adolescents: Findings from focus-group discussions with adolescents, *Journal of the American Dietetic Association* 99 (1999): 929–934, 937.

68. K. K. Christoffel and A. Ariza, The epidemiology of overweight in children: Relevance for clinical care, *Pediatrics* 101 (1998): 103; R. P. Troiano and K. M. Flegal, Overweight children and adolescents: Description, epidemiology, and demographics, *Pediatrics* 101 (1998): 497S–502S.

69. M. E. Shaw, Adolescent breakfast skipping: An Australian study, *Adolescence* 33 (1998): 851–861; B. Levine, Childhood obesity associated with the excessive consumption of soft drinks and fruit juices, *Topics in Clinical Nutrition* 13 (1997); 69–73; M. Story and D. Neumark-Sztainer, Competitive foods in schools: Issues, trends, and future directions, *Topics in Clinical Nutrition* 15 (1999): 37–46.

70. A. Rosenbloom and coauthors, Emerging epidemic of Type 2 Diabetes in youth, *Diabetes Care* 22 (1999): 345–351.

71. L. Brabin and B. J. Brabin, The cost of successful adolescent growth and development in girls in relation to iron and vitamin A status, *American Journal of Clinical Nutrition* 55 (1992): 955–958.

72. V. Matkovic, Diet, genetics, and peak bone mass of adolescent girls, *Nutrition Today* 26 (1991): 21–24.

73. National Cholesterol Education Program, *Report of the Expert Panel on Blood Cholesterol Levels in Children,* 1991.

74. U.S. Department of Health and Human Services, Public Health Service, *Surgeon General's Report on Nutrition and Health* (Washington, D.C.: U.S. Government Printing Office, DHHS Pub. No. 88-50210, 1988), pp. 345–380.

75. U.S. Department of Health and Human Services, *Healthy People 2000*, pp. 571–578.

76. B. Hunter, Food allergies: No trivia matter, *Consumers' Research Magazine* 82 (1999): 2.

77. Food allergies: How worried should you be? *Tufts University Health & Nutrition Letter* 17 (1999): 2.

78. Food Allergy Network, *Information About Food Allergies,* 1999.

79. H. Sampson, Food Allergy, *Journal of the American Medical Association* 278 (1997): 22–26.

80. S. Sicherer, Manifestations of food allergy: Evaluation and management *American Family Physician* 59 (1999): 2–8.

81. S. Gottlieb, Scientists develop vaccine strategy for peanut allergy, *British Medical Journal* 318 (1999): 71–88.

82. R. A. Thompson and R. T. Sherman, *Helping Athletes with Eating Disorders* (Champaign, IL: Human Kinetics Publishers, 1993).

83. U.S. Census Bureau, Sixty-five plus in the United States, statistical brief, http://www.census.gov/ftp/pub/socdemo/www/agebrief.html, June 6, 1999: 1–7.

84. K. G. Manton and E. Stallard, Longevity in the United States: Age and sex-specific evidence on life span limits from mortality patterns 1960–1990, *Journal of Gerontology* 51A (1996): B362–B375.

85. Ibid.

86. The discussion of demographic trends is adapted from B. Senauer, E. Asp, and J. Kinsey, *Food Trends and the Changing Consumer* (St. Paul: Eagan Press, 1991), pp. 199–213.

87. H. Kerschener, Productive aging: A quality of life agenda, *Journal of the American Dietetic Association* 98 (1998): 1445–1448; G. W. Auld and coauthors, Reported adoption of dietary fat and fiber recommendations among consumers, *Journal of the American Dietetic Association* 100 (2000): 52–58.

88. AMA Conference Proceedings, Unified Dietary Recommendations, Summary of a Scientific Conference on Preventive Nutrition: Pediatrics to Geriatrics, *Circulation* 100 (1999): 450–455.

89. Institute of Medicine, *The Second Fifty Years, Promoting Health and Preventing Disability* (Washington, D.C.: National Academy Press, 1992), pp. 1–21.

90. Institute of Medicine, *Extending Life, Enhancing Life: A National Research Agenda on Aging* (Washington, D.C.: National Academy Press, 1991) pp. 1–39.

91. J. E. Kerstetter, B. A. Holthausen, and P. A. Fitz, Malnutrition in the institutionalized older adult, *Journal of the American Dietetic Association* 92 (1992): 1109–1116.

92. J. T. Dwyer, J. J. Gallo, and W. Reichel, Assessing nutritional status in elderly patients, *American Family Physician* 47 (1993): 613–620.

93. J. Blumberg, Nutrient requirements of the healthy elderly—Should there be specific RDAs? *Nutrition Reviews* 52 (1994): S15–S18.

94. B. Dawson-Hughes and coauthors, Effect of calcium and vitamin D supplementation on bone density in men and women 65 years of age or older, *New England Journal of Medicine,* 337 (1997): 670–676.

95. R. Chernoff, Thirst and fluid requirements, *Nutrition Reviews* 52 (1994): S3–S5.

96. A. Sarubin, *The Health Professional's Guide to Popular Dietary Supplements* (Chicago, IL: The American Dietetic Association, 2000). F. Tripp, The use of dieteary supplements in the elderly: Current issues and recommendations, *Journal of the American Dietetic Association* 97 (1997): S181–S183; M. Freeman and coauthors, Cognitive, behavioral, and environmental correlates of nutrient supplement use among independently living older adults, *Journal of Nutrition for the Elderly* 17 (1998): 19–37; J. Howard and coauthors, Investigating relationships between nutrition knowledge, attitudes, and beliefs and dietary adequacy of the elderly, *Journal of Nutrition for the Elderly* 17 (1998): 38–51.

97. K. Shaw, Healthy Aging, *British Medical Journal* 315 (1997): 7115–7116.

98. Institute of Medicine, *Extending Life*, pp. 1–39.

99. Kerstetter, Holthausen, and Fitz, Malnutrition, p. 1113.

100. Community Nutrition Institute, *Nutrition Week*, April 30, 1993, p. 3.

101. *Nutrition Screening Initiative Survey* (Washington, D.C.: Peter D. Hart Research Associates, February 1990; President's page: The Nutrition Screening Initiative—An emerging force in public policy, *Journal of the American Dietetic Association* 93 (1993): 822.

102. M. A. Hess, President's page: ADA as an advocate for older Americans, *Journal of the American Dietetic Association* 91 (1991): 847–849.

103. J. V. White and coauthors, Nutrition Screening Initiative: Development and implementation of the public awareness checklist and screening tools, *Journal of the American Dietetic Association* 92 (1992): 163–167.

104. J. Weinberg, Psychologic implications of the nutritional needs of the elderly, *Journal of the American Dietetic Association* 60 (1972): 293–296.

105. P. G. Kittler and K. Sucher, *Food and Culture in America: A Nutrition Handbook*, 2nd ed. (Belmont, CA: Wadsworth Publishing Co., 1998).

106. D. L. Edwards and coauthors, Home-delivered meals benefit the diabetic elderly, *Journal of the American Dietetic Association* 93 (1993): 585–587; M. R. Neyman and coauthors, Effect of participation in congregate-site meal programs on nutritional status of the healthy elderly, *Journal of the American Dietetic Association* 96 (1996): 475; E. Fogler-Levitt and coauthors, Utilization of home-delivered meals by recipients 75 years of age or older, *Journal of the American Dietetic Association* 95 (1995): 552.

107. The list of advantages is adapted from D. A. Roe, *Geriatric Nutrition*, 3rd ed. (Englewood Cliffs, N.J.: Prentice-Hall, 1992) pp. 1–9.

108. World Cancer Research Fund and American Institute for Cancer Research, *Food, Nutrition, and the Prevention of Cancer: A Global Perspective* (Washington, D.C.: American Institute for Cancer Research, 1997), pp. 506–507; T. Sugimura, Cancer prevention: Past, present, and future, *Mutation Research* 402 (1998): 7–14; M. J. Hill, Nutrition and human cancer, *Annals of the New York Academy of Sciences* 883 (1997): 68–78; D. M. DeMarini, Dietary interventions of human carcinogenesis, *Mutation Research* 400 (1998): 457–465.

109. The American Institute for Cancer Research, *Stopping Cancer Before It Starts* (New York: Golden Books, 1999).

110. E. Giovannucci and B. Goldin, The role of fat, fatty acids, and total energy intake in the etiology of human colon cancer, *American Journal of Clinical Nutrition* 66 (1997): S1564–S1571; D. Y. Kim and coauthors, Stimulatory effect of high-fat diets on colon cell proliferation depend on the type of dietary fat and site of the colon, *Nutrition and Cancer* 30 (1998): 118–123.

111. E. Giovannucci and coauthors, Intake of fat, meat, and fiber in relation to risk of colon cancer in men, *Cancer Research* 54 (1994): 2390–2397.

112. C. S. Fuchs and coauthors, Dietary fiber and the risk of colorectal cancer and adenoma in women, *New England Journal of Medicine* 340 (1999): 169–176.
113. W. C. Willett and coauthors, Relation of meat, fat, and fiber intake to the risk of colon cancer in a prospective study among women, *New England Journal of Medicine* 323 (1990): 1664–1672.
114. B. S. Reddy, Role of dietary fiber in colon cancer: An overview, *American Journal of Medicine* 106 (1999): S16–S19; M. C. Jansen and coauthors, Dietary fiber and plant foods in relation to colorectal cancer mortality: The Seven Countries Study, *International Journal of Cancer* 81 (1999): 174–179.
115. M. Ferraroni and coauthors, Selected micronutrient intake and the risk of colorectal cancer, *British Journal of Cancer* 70 (1994): 1150–1155; P. R. Holt and coauthors, Modulation of abnormal colonic epithelial cell proliferation and differentiation by low-fat dairy foods, *Journal of the American Medical Association* 280 (1998): 1074–1079.
116. K. K. Carroll, Dietary fats and cancer, *American Journal of Clinical Nutrition* 53 (1991): 1064S–1067S; C. W. Welsch, Dietary fat, calories, and mammary gland tumorigenesis, In *Advances in Experimental Medicine and Biology* (New York: Plenum Press, 1992), pp. 203–221.
117. R. L. Prentice and coauthors, Dietary fat and breast cancer: A quantitative assessment of the epidemiological literature and a discussion of methodological issues, *Cancer Research* 49 (1989): 3147–3156; P. Toniolo and coauthors, Consumption of meat, animal products, protein, and fat and risk of breast cancer: A prospective cohort study in New York, *Epidemiology* 5 (1994): 391–397.
118. K. Katsouyanni and coauthors, The association of fat and other macronutrients with breast cancer: A case-controlled study from Greece, *British Journal of Cancer* 70 (1994): 537–541.
119. Xiu-Ying Qi and coauthors, The association between breast cancer and diet and other factors, *Asia Pacific Journal of Public Health* 7 (1994): 98–104.
120. M. D. Holmes and coauthors, Association of dietary intake of fat and fatty acids with risk of breast cancer, *Journal of the American Medical Association* 281 (1999): 914–920; D. J. Hunter and coauthors, Cohort studies of fat intake and the risk of breast cancer—A pooled analysis, *New England Journal of Medicine* 334 (1996): 356–361; D. Kritchevsky, Caloric restriction and experimental mammary carcinogenesis, *Breast Cancer Research and Treatment* 46 (1997): 161–167.
121. American Institute for Cancer Research, *Diet and Health Recommendations for Cancer Prevention* (Washington, D.C.: American Institute for Cancer Research, 1998); American Cancer Society Nutrition Advisory Committee, Most frequently asked questions about diet and cancer, *Nutrition Today* May/June 1997, pp. 125–127.
122. *Stopping Cancer Before It Starts*, 1999; B. J. Caan and coauthors, Body size and the risk of colon cancer in a large case-control study, *International Journal of Obesity and Related Metabolic Disorders* 22 (1998): 178–184; L. C. Yong and coauthors, Prospective study of relative weight and risk of breast cancer: The breast cancer detection demonstration project follow-up study, 1979 to 1987–1989, *American Journal of Epidemiology* 143 (1996): 985–995; Z. Huang and coauthors, Dual effects of weight and weight gain on breast cancer risk, *Journal of the American Medical Association* 278 (1997): 1407–1411.
123. K. A. Steinmetz and J. D. Potter, Vegetables, fruit, and cancer prevention: A review, *Journal of the American Dietetic Association* 96 (1996): 1027–1039.
124. G. Block, Vitamin C and cancer prevention: The epidemiologic evidence, *American Journal of Clinical Nutrition* 53 (1991): 270S–282S.
125. W. J. Craig, Phytochemicals: Guardians of our health, *Journal of the American Dietetic Association* 97 (1997): S199–S204; Z. Djuric and coauthors, Oxidative DNA damage levels in blood from women at high risk for breast cancer are associated with dietary intakes of meats, vegetables, and fruits, *Journal of the American Dietetic Association* 98 (1998): 524–528.
126. Steinmetz, 1996.
127. L. C. Clark and coauthors, Reducing cancer risk with selenium, *Journal of the American Medical Association* 276 (1995): 1957–1963.
128. P. Knekt and coauthors, Vitamin E and cancer prevention, *American Journal of Clinical Nutrition* 53 (1991): 283S–286S.
129. J. A. Baron and coauthors, Calcium supplements for the prevention of colorectal adenomas, *New England Journal of Medicine* 340 (1999): 101–107.
130. M. A. Rogers and coauthors, Consumption of nitrate, nitrite, and nitrosodimethylamine and the risk of upper aerodigestive tract cancer, *Cancer Epidemiology, Biomarkers, and Prevention* 4 (1995): 29–36; L. E. Hansson and coauthors, Nutrients and gastric cancer risk. A population-based case-control study in Sweden, *International Journal of Cancer* 57 (1994): 638–644.
131. M. Messina and S. Barnes, The role of soy products in reducing risk of cancer, *Journal of the National Cancer Institute* 83 (1991): 541–546.
132. S. A. Smith-Warner and coauthors, Alcohol and breast cancer in women: A pooled analysis of cohort studies, *Journal of the American Medical Association* 279 (1998): 535–540.
133. G. Howe and coauthors, The association between alcohol and breast cancer risk: Evidence from the combined analysis of six dietary case-control studies, *International Journal of Cancer* 47 (1991): 707–710; L. Holmberg and coauthors, Diet and breast cancer risk: Results from a population-based, case-control study in Sweden, *Archives of Internal Medicine* 154 (1994): 1805–1811; J. L. Freudenheim and coauthors, Lifetime alcohol consumption and risk of breast cancer, *Nutrition and Cancer* 23 (1995): 1–11.
134. E. Giovannucci, Tomatoes, tomato-based products, lycopene, and cancer: Review of the epidemiologic literature, *Journal of the National Cancer Institute* 91 (1998): 317–331.
135. A. Bloch and C. A. Thompson, Position of the American Dietetic Association: Phytochemicals and functional foods, *Journal of the American Dietetic Association* 95 (1995): 493–496; Position of the American Dietetic Association: Functional foods, *Journal of the American Dietetic Association* 99 (1999): 1278–1285; A. S. Bloch, Phytochemicals and functional foods for cancer risk reduction, *Topics in Clinical Nutrition* 15 (2000): 24–28.
136. J. Childers and coauthors, Chemoprevention of cervical cancer with folic acid: A Phase III Southwest Oncology Group Intergroup Study, *Cancer Epidemiology, Biomarkers, and Prevention* 4 (1995): 155–159.
137. American Institute for Cancer Research, *Diet and Health Recommendations for Cancer Prevention* (Washington, D.C.: American Institute for Cancer Research, 1998).
138. T. Byers, Dietary trends in the United States: Relevance to cancer prevention, *Cancer* 72 (1993): 1015–1018.
139. J. Dwyer, Diet and nutritional strategies for cancer risk reduction: Focus on the 21st century, *Cancer* 72 (1993): 1024–1031.
140. Committee on Comparative Toxicity of Naturally Occurring Carcinogens, *Carcinogens and Anticarcinogens in the Human Diet* (Washington, D.C.: National Academy Press, 1996).

Chapter 11 Notes

1. Updates: New estimates on foodborne diseases, *FDA Consumer* January/February 2000, p. 3; Science-based, unified approach needed to safeguard the nation's food supply. *Public Health Reports* 113 (1998): 482–483.
2. Position of the American Dietetic Association: Food and water safety, *Journal of the American Dietetic Association* 97 (1997): 1048–1053.
3. A. Hingley, Campylobacter: Low-profile bug is food poisoning leader, *FDA Consumer* September/October 1999, pp. 14–17.
4. Get rid of the garlic, FDA says, *FDA Consumer*, June 1989, p. 2.
5. B. P. Bell and coauthors, A multistate outbreak of *Escherichia coli* o157:H7-associated bloody diarrhea and hemolytic uremic syndrome from hamburgers: The Washington experience, *Journal of the American Medical Association* 272 (1994): 1349–1353; U.S. Department of Agriculture Food Safety and Inspection Service, *E. coli* o157:H7 at a glance, *Food News for Consumers* 10 (1993), p. 5.
6. R. E. Besser and coauthors, An outbreak of diarrhea and hemolytic uremic syndrome from *Escherichia coli* o157:H7 in fresh-pressed apple cider, *Journal of the American Medical Association* 269 (1993): 2217–2219.
7. S. C. Witt, *Biotechnology, Microbes and the Environment* (San Francisco: Center for Science Information, 1990), pp. 182–183.
8. E. Hilton and coauthors, Ingestion of yogurt containing *lactobacillus acidophilus* as prophylaxis for candidal vaginitis, *Annals of Internal Medicine* 116 (1992): 353–357; K. Gupta and coauthors, Inverse association of H2o2-producing lactobaccili and vaginal *Escherichia coli* colonization in women with recurrent urinary tract infections, *Journal of Infectious Diseases* 178 (1998): 446–450; Y. Delneste, A. Dounet-Hughes, and E. J. Schiffin, Functional foods: Mechanisms of action on immunocompetent cells, *Nutrition Reviews* 56 (1998): 593–598.
9. R. S. Horowitz and coauthors, Jin Bu Huan toxicity in children—Colorado 1993, *Morbidity and Mortality Weekly Report* 42 (1993): 633–635.
10. U.S. Department of Health and Human Services, Public Health Service, FDA warns consumers against Nature's Nutrition Formula One, statement issued February 28, 1995.
11. R. S. Koff, Herbal hepatotoxicity: Revisiting a dangerous alternative, *Journal of the American Medical Association* 273 (1995): 502.
12. N. D. Vietmeyer, The preposterous puffer, *National Geographic* 166 (1984): 260–270.
13. Bovine spongiform encephalopathy: "Mad cow disease," *Nutrition Reviews* 54 (1996): 208–210; Bovine spongiform encephalopathy, February 2000 update; www.foodsafety.gov.
14. U.S. Department of Agriculture Office of Public Affairs, News Division, "Grill" the USDA Hotline experts about safe summer cooking, Release No. 0483.94, 1994.
15. Department of Health and Human Services, Public Health Service, U.S.

Food and Drug Administration, *Getting Hooked on Seafood Safety*, DHHS Publication No. (FDA) 93-2266, May 1993.

16. J. Trager, *The Food Chronology* (New York: Henry Holt and Company, 1995), as cited in *Food Folklore:Tales and Truths About What We Eat* (Minneapolis, MN.: Chronimed Publishing, 1999).

17. 97% of Michigan population contaminated by 1973 spills, *Tallahassee Democrat*, April 16, 1982.

18. U.S. Public Health Service, Screening for lead exposure in children, *American Family Physician* 51 (1995): 139–143; America's children: Key national indicators of well-being, 1998, pp. 1–3 of "Special Features," available online at www.childstats.gov.

19. J. E. Foulke, Lead threat lessens, but mugs pose problem, *FDA Consumer*, April 1993, pp. 19–23.

20. International Food Information Council, *Pesticides and Food Safety* (Washington, D.C.: International Food Information Council Foundation, 1995), p. 2.

21. Ibid., 2–4; C. F. Chaisson and coauthors, *Pesticides in Food: A Guide for Professionals* (Chicago: The American Dietetic Association, 1991), pp. 4–8.

22. U.S. Department of Agriculture, Organic labeling claims, USDA Press Release, December 1999 (available from www.usda.gov/news/releases/1999/); Community Nutrition Institute, Proposed organic rule eliminates GE ingredients, irradiation, sewer sludge, *Nutrition Week* 10 (2000): 1–2.

23. Council for Agricultural Science and Technology (CAST), *Public Perceptions of Agrichemicals* (Ames, Iowa: CAST Task Force Report No. 123, 1995), pp. 17–21.

24. C. F. Chaisson and coauthors, *Pesticides in Food: A Guide for Professionals* (Chicago: The American Dietetic Association, 1991), p. 16.

25. National Academy of Sciences, National Research Council, *Pesticides in the Diets of Infants and Children* (Washington, D.C.: National Academy Press, 1993); "Provocative" report issued on use of pesticides, *Journal of the American Medical Association* 275 (1996): 899.

26. World Health Organization, *Safety and Nutritional Adequacy of Irradiated Food* (Geneva, Switzerland: World Health Organization, 1994), pp. 4–5; WHO Technical Report Series 890, *High-Dose Irradiation: Wholesomeness of Food Irradiated with Doses Above 10kGy* (Geneva: Switzerland: World Health Organization, 1999).

27. Position of the American Dietetic Association: Food irradiation, *Journal of the American Dietetic Association* 100 (2000): 246–253.

28. Position of the American Dietetic Association: Food irradiation, 2000; Community Nutrtion Institute, Irradiation of meat begins amidst failure of safe food safety policy, *Nutrition Week* 8 (2000): 1–7.

29. J. Henkel, Genetic engineering: Fast forwarding to the future, *FDA Consumer*, April 1995, pp. 6–11.

30. H. I. Miller, Foods of the future: The new biotechnology and FDA regulation, *Journal of the American Medical Association* 269 (1993): 910–912.

31. P. Abelson, A third technological revolution, *Science* 279 (1998): 5359; D.D. Stadler, A. F. Reeder, and C. H. Strohbehn, Application of genetic engineering to foods, *Topics in Clinical Nutrition* 14 (1999): 39–50; C. McCullum, The new biotechnology era: Issues for nutrition education and policy, *Journal of Nutrition Education* 29 (1997): 116–119.

32. United States Department of Agriculture, Agricultural Research Service, *USDA and Biotechnology*, www.usda.gov.

33. T. Weaver, *USDA releases new tomatoes with increased beta-carotene*, (Washington, D.C.: USDA Agricultural Research Service, 1998).

34. A. Moffaat, Toting up the harvest of transgenic plants, *Science* 282 (1998): 5397.

35. Ibid.

36. J. A. Nordlee and coauthors, Identification of a Brazil-nut allergen in transgenetic soybeans, *New England Journal of Medicine* 334 (1996): 688–692.

37. L. Thompson, Are bioengineered foods safe? *FDA Consumer* 34 (2000): 18–23.

38. Community Nutrition Institute, Consumers Union calls for GE testing, labeling, *Nutrition Week* 34 (1999): 1–2; C. Silva and R. Leonard, U.S. prepares to OK labels on GE foods, *Nutrition Week* 45 (1999): 1–2; Community Nutrition Institute, Biotech debate heats up in Boston: Strong grassroots movement forming, *Nutrition Week* 30 (2000): 1–3.

39. G. G. Graham, Starvation in the modern world, *New England Journal of Medicine* 328 (1993): 1058–1061.

40. G. Bickel, M. Andrews and S. Carlson, The magnitude of hunger: In a new national measure of food security, *Topics in Clinical Nutrition* 13 (1998): 15–30; A. Kendall and E. Kennedy, Position of the American Dietetic Association: Domestic food and nutrition security, *Journal of the American Dietetic Association* 98 (1998): 337–342; W. L. Hamilton and coauthors, *Household Food Security in the United States*, (Washington, D.C.: U.S. Department of Agriculture Food and Consumer Service, 1997); the household food security report is available via www.usda.fov/fcs/measure.htm.

41. *1999 World Population Data Sheet*, (Washington, D.C.: Population Reference Bureau, 1999); International Food Policy Research Institute, *Feeding the World, Preventing Poverty, and Protecting the Earth: A 2020 Vision* (Washington, D.C.: IFPRI, 1996); United Nations Food and Agriculture Organization (FAO) *The State of Food and Agriculture*, 1998 (available from www.fao.org).

42. UNICEF, *The State of the World's Children* 1999, (New York: Oxford University Press, 1998); L. Brown, *State of the World* 2000 (Washington, D.C.: Worldwatch Institute, 2000); www.worldwatch.org.

43. Position of the American Dietetic Association: World Hunger, *Journal of the American Dietetic Association* 95 (1995): 1160–1162.

44. Community Nutrition Institute, Malnutrition accounts for over half of child deaths worldwide, says UNICEF, *Nutrition Week* 28 (1998): 1–2.

45. Y. W. Bradshaw and coauthors, Borrowing against the future: children and third world indebtedness, *Social Forces* 71 (1993): 629–656.

46. Ibid.

47. L. Robertson, Breastfeeding practices in maternity wards in Swaziland, *Journal of Nutrition Education* 23 (1991): 284–287.

48. UNICEF, *The State of the World's Children*, 1999.

49. A. N. J. Malik and W. A. M. Cutting, Breastfeeding: The Baby Friendly Initiative, *British Medical Journal* 316 (1998): 1548–1549; American Academy of Pediatrics Work Group on Breastfeeding, Breastfeeding and the use of human milk, *Pediatrics* 100 (1997): 1035–1039.

50. USDA Center for Nutrition Policy and Promotion, Could there be hunger in America?, *Nutrition Insight*, September, 1998.

51. Position of the American Dietetic Association: Domestic food and nutrition security, 1998; B. W. Klein, Food security and hunger measures: Promising future for state and local household surveys, *Family Economics and Nutrition Review* 9 (1996): 31–37.

52. T. A. Fox, Food security: The federal commitment and the role of The American Dietetic Association, *Topics in Clinical Nutrition* 13 (1998): 5–14.

53. Community Nutrition Institute, Nearly 26 million people turned to food banks in 1997, Second Harvest says, *Nutrition Week*, March 1998; Community Nutrition Institute, Cities and charities report an increase demand for emergency food services, *Nutrition Week*, December 1999; B. O. Daponte, Food pantry use among low-income households in Allegheny County, Pennsylvania, *Journal of Nutrition Education* 30 (1998): 50–57.

54. The U.S. Conference of Mayors, 1999.

55. J. L. Wiecha, J. T. Dwyer, and M. Dunn-Strohecker, Nutrition and health services needs among the homeless, *Public Health Reports* 106(4) (1991): 364–374.

56. USDA National Agricultural Statistics Service, *Farms and Land in Farms*, 1998 (Washington, D.C.: USDA Interagency Agricultural Projections Committee, 1999).

57. P. Uvin, The state of world hunger, *Nutrition Reviews* 52 (1994): 151–161.

58. G. Gardner, *Shrinking Fields: Cropland Loss in a World of Eight Billion* (Washington, D.C.: Worldwatch Institute, 1996); M. W. Rosegrant, *Water Resources in the Twenty-First Century*, Discussion Paper 20, (Washington, D.C.: International Food Policy Research Institute, 1997); Bread for the World Institute, *Hunger 1998: Hunger in a Global Economy* (Silver Spring, MD: Bread for the World Institute, 1997); Food and Agriculture Organization of the United Nations, *The Sixth World Food Survey* (Rome: FAO Statistical Analysis Service, 1996).

59. E. Velempini and K. D. Travers, Accessibility of nutritious African foods for an adequate diet in Bulawayo, Zimbabwe, *Journal of Nutrition Education* 29 (1997): 120–127; W. Fawzi and coauthors, A prospective study of malnutrition in relation to child mortality in the Sudan, *American Journal of Clinical Nutrition* (1997): 1062; S. Dalton, An education and research opportunity in Nepal: Dietetics in a developing country, *Topics in Clinical Nutrition* 11 (1996): 39–46.

60. A. Durning, Life on the Brink, *World Watch Papers* 3(1990): 29.

61. L. R. Brown, *Who Will Feed China? Wake Up Call For a Small Planet*, (New York: W. W. Norton, 1995).

62. B. Stutz, The landscape of hunger, *Audubon*, March-April 1993, pp. 54–57; Newsbreaks: Effects of environmental degradation on nutrition, *Nutrition Today*, March/April 1992, p. 4.

63. Oxfam America, *Facts for Action: Women Creating a New World*, no. 3 (Boston: Oxfam America, 1991), pp. 2–3.

64. S. Lewis, Food security, environment, poverty, and the world's children, *Journal of Nutrition Education* 24 (1992): 35–55.

65. UNICEF, *The State of the World's Children*, 1999.

66. Position of the American Dietetic Association: World Hunger, 1995.

67. UNICEF, 1999.

68. S. Lewis, 1992.

69. S. Lewis, 1992, p. 55.

70. J. Poppendieck, *Sweet Charity? Emergency Food and the End of Entitlement* (New York: Viking Penguin Putnam, Inc., 1998).

71. D. Elgin, *Voluntary Simplicity: Toward a Way of Life That Is Outwardly Simple, Inwardly Rich* (New York: William Morrow, 1981), p. 25.

Appendix G

Answers to Check Yourself Questions

Chapter 1

1. Overnutrition contributes to heart disease, cancer, stroke, diabetes, and hypertension. **2.** As defined in the chapter, a *nutritionist* is someone who claims to be capable of advising people about their diets; some nutritionists are registered dietitians, whereas others are self-described experts whose training is questionable. In contrast, a *registered dietitian* is a nutrition professional who has graduated from a college program in dietetics approved by the American Dietetic Association, has completed an internship program or the equivalent, has passed a national registration examination, and maintains competencies in the field through regular continuing education.
3. Four of the many factors that influence our eating habits are personal preference, cultural traditions, economic considerations, and advertising (other factors include availability, social and psychological factors, and personal beliefs). **4.** Health fraud or quackery is conscious deceit practiced for profit (for example, the promotion of an unproven product or therapy). **5.** To trim fat and calories from fast-food meals, you can select an English muffin or bagel with jelly instead of a doughnut or danish; order sandwiches without mayonnaise or special sauces; and order a side salad with your burger instead of French fries. You can also order low-fat milk instead of a milkshake; use less sour cream and guacamole on nachos and tacos; and choose frozen yogurt instead of ice cream. **6.** Three red flags that can help you spot a quack are (1) promoter claims that the medical establishment (or government) is against him or her and won't accept the new "alternative" treatment, (2) the promoter uses testimonials and anecdotes from satisfied customers to support claims, and (3) the promoter uses a computer-scored questionnaire for diagnosing "nutrient deficiencies." Other red flags: The promoter claims that the product will make weight loss easy; the promoter promises that the product is made with a "secret formula," available only through this one company; or the treatment is only available through the back pages of magazines, over the phone, or by mail-order ads in the form of news stories or infomercials. **7.** Many of our eating habits arise from the traditions, belief systems, technologies, values, and norms of the culture in which we live. **8.** Three factors other than diet that influence longevity include avoiding excess alcohol, not smoking, and maintaining desirable weight. Other factors include exercising regularly and sleeping seven to eight hours a night. **9.** The Food and Drug Administration holds the authority to prosecute companies that display false nutrition information on product labels or enclosures, and the Federal Trade Commission can prosecute manufacturers who make fraudulent or misleading statements in their advertisements. **10.** The government initiative is called *Healthy People 2010: National Health Promotion and Disease Prevention Objectives*.

Chapter 2

1. Characteristics of a healthful diet can include any of the following: nutrient adequacy, balance, calorie control, moderation, and variety. **2.** Nutrient density refers to foods that are rich in nutrients (protein, vitamins, minerals) but relatively low in calories and fat. **3.** The Food Guide Pyramid recommends 6 to 11 servings from the *Bread, Cereal, Rice and Pasta Group,* 3 to 5 servings from the *Vegetable Group,* 2 to 4 servings from the *Fruit Group,* 2 to 3 servings from the *Milk, Yogurt, and Cheese Group,* and 2 to 3 servings from the *Meat, Poultry, Fish, Dry Beans, Eggs, and Nuts Group.* **4.** Nutrients that must appear on virtually all food labels include fat (total fat, saturated fat, cholesterol), carbohydrate (total carbohydrate, simple sugars, dietary fiber), sodium, protein, vitamins A and C, calcium, and iron. **5.** The % Daily Value on food labels can be used to check how the foods you eat fit into a healthful diet in terms of their fats, carbohydrates, fiber, and sodium. For example, if a label shows that a serving of a food contains 3 grams of fat (and a Daily Value for fat of 5 percent), this means that the food supplies 5 percent of the total fat that a person eating a 2,000 calorie diet should consume. **6.** The ten Dietary guidelines are: (1) Aim for a healthy weight; (2) Be physically active each day; (3) Let the Pyramid guide your food choices; (4) Choose a variety of grains daily, especially whole grains; (5) Choose a variety of fruits and vegetables daily; (6) Keep food safe to eat; (7) Choose a diet that is low in saturated fat and cholesterol and moderate in total fat; (8) Choose beverages and foods that limit your intake of sugars; (9) Choose and prepare foods with less salt; and (10) If you drink alcoholic beverages, do so in moderation. **7.** The RDA for calories offers only a very rough estimate of individual calorie needs, which can vary tremendously due to factors such as a person's ratio of muscle to fat and activity level. **8.** Examples of relatively low-fat snacks include fruit juices, low-fat yogurt, fresh fruits and vegetables, plain popcorn, pretzels, whole-grain crackers, and low-fat cheeses. **9.** Items in the ingredients list are listed in descending order by weight. **10.** The three calorie-yielding nutrients are carbohydrate, fat, and protein.

Chapter 3

1. The simple carbohydrates include glucose, fructose, galactose, maltose, sucrose, and lactose. These sugars are found as naturally occurring sugars in fresh fruits and vegetables (glucose and fructose), in milk and milk products (lactose), and in concentrated form in honey, corn syrup, table sugar, cakes, cookies, candy, soda pop, and other foods with added sugars. **2.** Complex carbohydrates include starch and fiber. Starch is found mostly as grains—as in wheat, rice, and corn products; fiber is found in plant foods such as whole grains, fruits, legumes, and vegetables. **3.** The primary role of carbohydrates in the body is to provide energy, and for certain cells (brain and nervous system) carbohydrates are the preferred energy source. Carbohydrate-rich foods are usually less expensive than protein as an energy source, and unlike fat—the other energy nutrient—carbohydrates are not associated with chronic diseases such as heart disease and cancer. **4.** The *Dietary Guidelines for Americans* recommend that we (1) choose a variety of grains daily, especially whole grains; (2) choose a variety of fruits and vegetables daily; and (3) choose beverages and foods that limit our intake of sugars. **5.** Sugar-containing foods can be incorporated into a carefully designed eating plan for the person with diabetes, since blood glucose levels are affected by a number of dietary factors, including the total amounts of carbohydrate (complex and simple), fiber, and fat consumed in a meal. **6.** Enrichment of refined grain products makes them comparable to whole-grain products with respect to five nutrients (thiamin, riboflavin, niacin, folate, and iron). However, since enrichment does not add back other important nutrients (for example, magnesium, vitamin B_6, zinc, and chromium) to refined products, whole-grain products are preferred choices over enriched products. **7.** Some of the health effects of fiber include (1) fiber serves as an aid to weight control by providing satiety; high-fiber foods are typically lower in calories than high-fat, low-fiber foods; (2) fiber provides bulk in the large intestine and promotes regularity; (3) fiber speeds transit time through the colon and may protect against colon cancer; (4) fiber may stabilize blood sugar levels by delaying glucose absorption; and (5) certain fibers may lower blood cholesterol. **8.** To increase the fiber content of your diet, you can choose whole-grain breads and cereals; choose whole fruits instead of juice; choose fiber-rich snacks such as popcorn, whole-grain crackers, fresh fruits and vegetables, and dried fruits; and add legumes to salads, soups, stews, and pasta dishes. **9.** Carbohydrate digestion begins in the mouth, where saliva is mixed with the food and an enzyme begins to split starch into smaller polysaccharides and maltose. Enzymes in the small intestine continue digestive action by splitting the polysaccharides into disaccharides. Enzymes on the surface of the intestinal cells finally split the disaccharides into monosaccharides and release them to be absorbed across the intestinal wall into the blood. **10.** The body maintains normal blood glucose concentrations by releasing the hormone insulin from the pancreas in response to high blood glucose concentrations. Insulin stimulates body cells to absorb the excess glucose. The body responds to low blood concentrations of glucose by releasing the hormone glucagon from the pancreas. Glucagon stimulates the liver to release its stored glucose into the blood.

Chapter 4

1. Fat in the diet provides calories, satiety, fat-soluble vitamins, aroma, and flavor. **2.** Fat in the body serves as a concentrated energy reserve, nourishes the skin and hair, provides the major components of cell membranes, insulates the body from extremes of body temperature, and cushions the vital organs to protect them from shock. **3.** (a) Examples of food sources of highly saturated fats are beef tallow, butter, coconut and palm oil, and lard. (b) Examples of highly monounsaturated fats include avocados, canola and olive oils, and peanuts. (c) Examples of highly polyunsaturated fats include corn and safflower oils, fish, almonds, and walnuts. (d) Examples of food sources of trans fatty acids include hydrogenated products, such as margarine, peanut butter, shortenings, commercial frying fats, baked goods, and any other foods listing "partially hydrogenated vegetable oil" on their label. (e) Examples of cholesterol-containing foods include animal-derived products such as liver, meat, egg yolks, and whole milk products. (f) Foods rich in omega-3 fatty acids include fish, shellfish, and flax seed; other sources are canola oil, leafy green vegetables, walnuts, wheat germ, and soy. **4.** Hydrogenation adds hydrogen to unsaturated fat to make it more solid and more resistant to chemical damage. **5.** *Chylomicrons* serve as a means of transporting newly digested fat from the intestine through the lymph and blood to body cells; *very-low-density lipoproteins (VLDL)* carry fats packaged or made by the liver to various tissues in the body; *low-density lipoproteins (LDL)* carry cholesterol (much of it synthesized in the liver) to body cells; *high-density lipoproteins (HDL)* carry cholesterol in the blood back to the liver for recycling or disposal. **6.** A high LDL level or a low HDL level are associated with increased risk of heart disease; a high HDL level is associated with a decreased risk of heart disease. **7.** Some hard fats begin to melt as they reach body temperature in the mouth. The stomach's churning action mixes fat with water and acid. The small intestine is the primary site for fat digestion. Bile emulsifies fat, which can then be broken down by pancreatic enzymes to fatty acids, glycerol, and monoglycerides. **8.** The current recommendations for fat in the diet are (1) eat no more than 30 percent of calories as fat; (2) eat no more than 8 to 10 percent of calories as saturated fat; (3) eat no more than 10 percent of calories as polyunsaturated fat; (4) eat 10 percent to 15 percent of calories as monounsaturated fat; and (5) limit daily cholesterol intake to no more than 300 milligrams. **9.** The leading risk factors for heart disease include high blood cholesterol (especially high LDL), cigarette smoking, high blood pressure, and obesity. **10.** People can raise their HDL levels by losing weight if they are overweight and by following a regular exercise routine.

Chapter 5

1. Nitrogen is the element that appears exclusively in protein. **2.** Protein synthesis will be halted if an essential amino acid is missing from the diet. **3.** The quality of a dietary protein depends on both its assortment of essential amino acids relative to human needs and on its digestibility. **4.** We are advised to consume about 12 percent of our calories from protein. **5.** The proteins in the body are used for growth and maintenance, enzyme action, hormones, antibodies, fluid and acid-base balance, transportation, body structures, and energy. **6.** Some of the risks associated with using amino acid supplements include irritability, insomnia, and impaired growth. **7.** Complementary proteins are two or more food proteins whose amino acid assortments complement each other in such a way that the essential amino acids limited in or missing from each other are supplied by the others. **8.** Yes, a vegetarian diet can meet protein needs when varied plant-based protein sources are included in the diet on a daily basis. **9.** Without careful planning the vegan diet may be limited in iron, vitamin D, calcium, riboflavin, vitamin B_{12}, and high-quality protein. **10.** Research findings suggest that soyfoods may play a role in lowering blood cholesterol level and risk for conditions such as heart disease, certain forms of cancer, and osteoporosis.

Chapter 6

1. Thiamin, riboflavin, niacin, biotin, and pantothenic acid are needed as coenzymes in energy metabolism; vitamin B_6 is used in protein metabolism; folate and vitamin B_{12} are needed for red blood cell formation and new cell synthesis; vitamin B_{12} also helps to maintain nerve cells; vitamin C acts as an antioxidant and is needed for the synthesis of collagen. **2.** Beta-carotene, a precursor of vitamin A, is an orange pigment found in plants that is converted into active vitamin A inside the body; preformed vitamin A is vitamin A already in its active form. **3.** People following very-low-calorie diets, strict vegetarians, women who are pregnant or breastfeeding, among others, may need a multivitamin/mineral supplement. **4.** It is always best to cook vegetables in the least amount of water and for the shortest period of time possible. More of the water-soluble vitamins will be lost by boiling the vegetables in water than by steaming them or cooking them in a microwave oven. **5.** People need not obtain all their vitamin D from food because the body can synthesize it with the help of sunlight. **6.** Retinal, one of the active forms of vitamin A, is synthesized from dietary vitamin A and functions as a portion of the visual pigments in the eyes. These pigments serve to transform light into nerve impulses that are interpreted by the brain as visual images. **7.** Women of childbearing age are advised to consume generous amounts of folate before and during the first few weeks of pregnancy to reduce the risk of producing a baby with neural tube defects. **8.** Phytochemicals are chemicals found in plants that are not nutrients but that appear to help fight diseases such as cancer. **9.** Health experts advise everyone to eat at least five servings of fruits and vegetables a day in order to obtain the maximum benefits from the vitamins, minerals, antioxidants, and phytochemicals available in produce. Possible benefits include an enhanced immune system and reduced risk for many chronic conditions such as heart disease, high blood pressure, certain types of cancer, and age-related vision loss. **10.** Look for a supplement that meets high standards (USP standards) for manufacturing. Look for a bottle or package with an expiration date. Also, choose a supplement that contains both vitamins and minerals in amounts no more than 100 to 150 percent of the recommended amount for each.

Chapter 7

1. Water in the body serves many functions including transporting nutrients to cells, acting as a shock absorber in joints and around the spinal cord, and helping to maintain body temperature. **2.** Chloride, the major negatively charged ion of the fluids outside the cells, helps maintain acid-base balance in the blood and is part of hydrochloric acid in the stomach. As the body's chief positively charged ion inside the cells, potassium plays a major role in maintaining water balance and cell integrity, and it is critical to maintaining the heartbeat. Magnesium is a part of the body's protein-making machinery, is necessary for the release of energy, and it helps to relax muscles after contraction. Sulfur helps strands of protein to assume and hold a particular shape, thus enabling them to perform their specific roles. **3.** Dietary factors that may adversely affect blood pressure include high salt intakes and low potassium or calcium intakes. Lifestyle changes recommended for reducing blood pressure include achieving and maintaining a healthful weight; adopting a low-fat eating plan rich in fruits, vegetables, legumes, and low-fat dairy products; limiting the use of salt in the diet; using alcohol only in moderation; and getting regular exercise. **4.** Major minerals are essential mineral nutrients found in the body in amounts greater than 5 grams; trace minerals are essential mineral nutrients found in the human body in amounts less than 5 grams. **5.** The bone minerals are calcium, phosphorus, magnesium, and fluoride. **6.** Iodine functions as part of the thyroid hormones, which regulate body temperature, metabolic rate, reproduction, and growth. Iodine deficiency can cause goiter (enlargement of the thyroid gland) and cretinism (severe mental and physical retardation of an infant exposed to a deficiency during pregnancy). **7.** Vitamin C and the MFP factor enhance absorption of iron from a meal. Phytic acid, tannic acid, and fiber can interfere with or inhibit the absorption of iron. **8.** Deficiency symptoms of zinc in children include growth retardation, delayed sexual development, poor appetite, and decreased taste sensitivity; zinc deficiency in adults results in poor taste perception and impaired wound healing. **9.** Osteoporosis refers to a loss of bone mass to the point that the skeleton is unable to sustain ordinary strains and leads to increased risk of bone fractures. Risk factors for osteoporosis include advanced age, female gender, estrogen deficiency, family history of osteoporosis, petite body build, sedentary lifestyle, smoking, excess alcohol use, and being of British, Northern European, Chinese, Japanese, or Mexican-American ancestry. **10.** Calcium supplements may be needed by those who either cannot or will not adjust their diets to get enough calcium from food. If using a calcium supplement, it is better absorbed in doses of 500 milligrams or less. Avoid supplements derived from dolomite, oyster shells, or bonemeal, as some have been found to be contaminated with lead.

Chapter 8

1. A primary factor influencing basal metabolic rate is body composition. The more lean tissue in a body, the higher the metabolic rate. **2.** The skinfold test gives a fair approximation of total body fat. **3.** Risk factors associated with obesity include arthritis, certain types of cancer, heart disease, decreased longevity, diabetes, gallbladder and liver disease, hypertension, respiratory problems, and varicose veins. **4.** It is likely that the most important single contributor to the obesity problem in this country is underactivity. **5.** A successful weight-loss program includes healthful eating habits, exercise, behavior modification, and a weight-maintenance component. **6.** A weekly weight loss of one to two pounds (or less) is recommended. **7.** Behavior modification techniques include setting small, achievable goals; keeping a record of behavior patterns; planning periodic rewards; and evaluating progress on a regular basis. **8.** Central obesity refers to excess fat on the abdomen and around the trunk and is associated with a greater risk for developing diabetes, hypertension, and heart disease. **9.** The measure of a successful weight-loss program is keeping off the lost weight. **10.** A person with anorexia nervosa has an intense fear of gaining weight, a refusal to maintain body weight at or above a minimal normal weight for age and height, a disturbance in self perception of weight status, and amenorrhea. The person with bulimia nervosa experiences repeated episodes of uncontrolled binge eating followed by recurrent compensatory behaviors such as self-induced vomiting, misuse of laxatives and diuretics, or excessive exercise.

Chapter 9

1. The four components of fitness are strength, flexibility, muscle endurance, and cardiovascular endurance. Cardiovascular endurance has top priority. **2.** The benefits of exercise include, among others, improved sleep, enhanced immunity, improved self-confidence, increased caloric expenditure, stronger bones, and lowered risk for conditions such as heart disease. **3.** Aerobic exercise (running, swimming, rollerblading) requires oxygen and uses fat as the main source of energy. Anaerobic exercise (sprinting) burns stored glycogen as fuel without relying on a source of oxygen. **4.** Target heart rate is the heartbeat rate that will achieve a cardiovascular conditioning effect for a given person—fast enough to push the heart but not so fast as to strain it. To determine your target heart rate, subtract your age from 220 and then multiply this number by 60 percent and 85 percent to find lower and upper limits. **5.** The recommended prescription for cardiovascular fitness is that you exercise within your target heart rate range at least three times per week for 20 to 60 minutes. **6.** The training effect is the effect of regular exercise on the cardiovascular system—including improvements in heart, lung, and muscle function and increased blood volume. **7.** Anabolic steroids are synthetic male hormones that appear to help build muscle. There are numerous side effects including acne, liver abnormalities, temporary infertility, roid rages, dizziness, hypertension, heart disease, stroke, and stunted growth. **8.** The American Dietetic Association recommends that athletes consume 1 to 1.5 grams of protein per kilogram of healthy body weight. **9.** Water is always an adequate fluid-replacement beverage for exercise lasting less than one hour; properly balanced sports drinks may also be considered by the endurance athlete. **10.** The eating plan should consist mostly of whole, minimally processed foods low in fat, and high in complex carbohydrates, vitamins, and minerals.

Chapter 10

1. During pregnancy, a woman has increased needs for nearly all nutrients; a 30 milligram supplement of iron is recommended. **2.** Birthweight is the most potent indicator of the infant's future health status. **3.** Practices to avoid during pregnancy include maternal use of nonfood substances, excess caffeine, low-calorie diets, megadoses of certain vitamins, tobacco, alcohol, and illicit drugs. The practice of pica can lead to malnutrition or intestinal blockage; excess caffeine can cause birth defects in animals and lower birthweights in humans; any diet causing ketosis can deprive the fetal brain of glucose and lead to congenital deformity; excess intakes of vitamins A and D can cause fetal malformations; smoking restricts the blood supply to the fetus and can lead to low birthweight, stunted growth, and retarded development; the use of alcohol or other drugs can lead to irreversible brain damage and physical and mental retardation in the infant. **4.** Three advantages of breastfeeding are immunological protection, the receipt of *Lactobacillus bifidus* (favors growth of friendly intestinal bacteria), and protection from allergy development. **5.** The baby should receive breast milk or infant formula through the first year of life. At 4 to 6 months of age, juices and fruits that contain vitamin C are among the first foods introduced to a baby's diet. Iron-fortified rice cereal, followed by other cereals are also introduced at this time as a source of iron for the baby. Mashed vegetables and fruits, infant breads, and crackers can be introduced from 6 to 8 months; soft protein foods, toast, teething crackers, soft-cooked vegetables, and fruit are usually introduced between 8 and 10 months; whole eggs and whole milk can be introduced at 1 year of age. **6.** To prevent nutrient deficiencies, all children should consume adequate amounts of fruits, vegetables, legumes, and enriched or whole-grain breads, cereals, and snacks. Calcium- and iron-rich food selections should be encouraged. To help prevent obesity, children age 2 and older should be encouraged to consume a diet containing no more than 30 percent of calories from fat daily and should learn to enjoy a physically active lifestyle. To prevent dental caries, children should be encouraged to practice good dental hygiene from an early age and to avoid snacking between meals on carbohydrate-rich snacks. **7.** Three nutrition-related problems common among teenagers are undernutrition, low calcium intakes, and eating disorders. **8.** Physiological changes associated with aging include sensory impairments, altered endocrine, gastrointestinal, and cardiovascular functions, oral problems, and changes in the skeletal system. **9.** Calorie needs decline with age; nutrient requirements generally remain the same as one ages, but some evidence suggests an increased need for certain vitamins. **10.** Nutrients thought to be protective of cancer formation include beta-carotene, fiber, vitamin C, vitamin E, selenium, and calcium. High fat intakes are associated with increased risk of cancer formation.

Chapter 11

1. The leading causes of microbial food contamination include: *botulinum, Campylobacter jejuni, Listeria monocytogenes, Clostridium perfringens, Salmonella, Escherichia coli, Shigella, Staphylococcus aureus,* and *Vibrio vulnificus* bacteria, as well as Hepatitis A virus. **2.** To reduce your risk of eating a food contaminated with *E. coli*, be sure to cook all meat and poultry to 160 degrees Fahrenheit, avoid cross-contamination, and do not drink raw milk. **3.** Cross contamination refers to the accidental transfer of bacteria from one food to another that occurs, for example, by chopping vegetables on the same cutting board used to skin poultry. **4.** To reduce your chances of consuming excess lead, do not store foods or beverages in lead crystal containers; run cold water for several minutes before using it for drinking or cooking; and don't store fruit juices or other acidic foods in ceramic containers. **5.** Food additives function to enhance flavor, impart color, and improve nutritional value of foods (other functions include improving texture or stability of foods). **6.** As the world's population continues to grow, it threatens the world's capacity to produce adequate food. However, children represent the "social security" of the poor. It takes improved economic status to help reduce birthrate in developing countries. **7.** If the food eaten by one species has been contaminated by chemicals, it will accumulate in the food chain from one species to the next. **8.** A benefit of buying organic foods is that they are grown without pesticides, herbicides, fungicides, or drugs. The downside to buying organic foods is their higher price. **9.** The Delaney Clause prohibits manufacturers from using any substance that is known to cause cancer in animals or humans at any dose level. The Delaney Clause is limited because it is virtually impossible to eliminate potential cancer-causing agents from the food supply, and it overlooks the principle that it is the dose that matters in cancer formation. **10.** A positive side of genetic engineering is that it can boost the nutritional value of foods. A negative is that the long-term safety of genetic engineering for the environment is unknown.

Appendix H

Table of Food Composition

This edition of the table of food composition contains more complete values for several nutrients than any comparable table.* These include dietary fiber; saturated, monounsaturated, and polyunsaturated fat; vitamin B_6; vitamin E; folate; magnesium; and zinc. The table includes a wide variety of foods from all food groups and is updated yearly to reflect current food patterns. For example, this edition includes many new nonfat items; several new ethnic items such as adzuki beans, tahitian taro, and gai choy chinese mustard; and a new selection of vegetarian foods.

Sources of Data

To achieve a complete and reliable listing of nutrients for all the foods, over 1,200 sources of information are researched. Government sources are the primary base for all data for most foods. In addition to USDA data (from Release 12 and surveys), provisional USDA information—both published and unpublished—is included.

Even with all the government sources available, however, some nutrient values are still missing; and as the USDA updates various data, it sometimes reports conflicting values for the same items. To fill in the missing values and resolve discrepancies, other reliable sources of information are used. These sources include journal articles, food composition tables from Canada and England, information from other nutrient data banks and publications, unpublished scientific data, and manufacturers' data.

The data for brand foods are listed as provided by the food manufacturers and the food chain restaurants. This information changes often because recipes and formulations are modified to meet consumer preferences, and the data are usually limited to those nutrients required for food labels. To provide more complete information, values for several nutrients are often estimated based on known values for major ingredients.

Accuracy

The energy and nutrients in recipes and combination foods vary widely, depending on the ingredients. The amounts of various fatty acids and cholesterol are influenced by the type of fat used (the specific type of oil, vegetable shortening, butter, margarine, etc.).

Estimates of nutrient amounts for foods and nutrients include all possible adjustments in the interest of accuracy. When multiple values are reported for a nutrient, the numbers are averaged and weighted with consideration of the original number of analyses in the separate sources. Whenever water percentages are available, estimates of nutrient amounts are adjusted for water content. When no water is given, water percentage is assumed to be that shown in the table. Whenever a reported weight appeared inconsistent, many kitchen tests were made, and the average weight of the typical product was given as tested.

When estimates of nutrient amounts in cooked foods are derived from reported amounts in raw foods, published retention factors are applied. Data for combination foods are modified to include newer data for major ingredients.

*This food composition table has been prepared for West-Wadsworth Publishing Company and is copyrighted by ESHA Research in Salem, Oregon—the developer and publisher of the Food Processor®, Genesis® R&D, and the Computer Chef® nutrition software systems. The major sources for the data are from the USDA, supplemented by more than 1200 additional sources of information. Because the list of references is so extensive, it is not provided here, but is available from the publisher.

Considerable effort has been made to report the most accurate data available. The table is revised annually, and the authors welcome any suggestions or comments for future editions.

Average Values

It is important to know that many different nutrient values can be reported for foods, even by reliable sources. Many factors influence the amounts of nutrients in foods, including the mineral content of the soil, the method of processing, genetics, the diet of the animal or the fertilizer of the plant, the season of the year, methods of analysis, the difference in moisture content of the samples analyzed, the length and method of storage, and methods of cooking the food. The mineral content of water also varies according to the source.

Although each nutrient is presented as a single number, each number is actually an average of a range of data. More detailed reports from the USDA, for example, indicate the number of samples and the standard deviation of the data. One can also find different reported values for foods as older data are replaced with newer data from more recent analyses using newer analytical techniques. Therefore, nutrient data should be viewed and used only as a guide, a close approximation of nutrient content.

Dietary Fiber

There can be many different reported values for dietary fiber in foods because information depends on the type of analytical technique used. The fiber data in this table are primarily from the USDA/ARS Human Nutrition Information Service in Hyattsville, Maryland; Composition of Foods by Southgate and Paul (England); and many journal articles.

Vitamin A

Vitamin A is reported in retinol equivalents (RE). The amount of vitamin A can vary by the season of the year and the maturity of the plant. Reported values in both dairy products and plants are higher in summer and early fall than in winter. The values reported here represent year-round averages. The organ meats of all animal products (liver especially) contain large amounts of vitamin A, which vary widely, depending on the background of the animal. The vitamin is also present in very small amounts in regular meat and is often reported as a trace.

Vitamin E

Vitamin E is actually a combination of various forms of this nutrient, and the measure of alpha-tocopherol equivalents (α-TE) summarizes the activity of the various types of tocopherols and tocotrienols into one measure.

Fats

Total fats, as well as the breakdown of total fats to saturated, monounsaturated, and polyunsaturated fats, are listed in the table. The fatty acids seldom add up to the total due to rounding and to other fatty acid components that are not included in these basic categories, such as *trans*-fatty acids and glycerol. *Trans*-fatty acids can comprise a large share of the total fat in margarine and shortening (hydrogenated oils) and in any foods that include them as ingredients.

Enrichment-Fortification

The mandatory enrichment values for foods are presented as appropriate, including the new values for folate enrichment folacin in grain products.

Niacin

Niacin values are for preformed niacin and do not include additional niacin that may form in the body from the conversion of tryptophan.

Using the Table

The items in this table have been organized into several categories, which are listed at the head of each right-hand page. As the key shows, each group has been color-coded to make it easier to find individual items.

In an effort to conserve space, the following abbreviations have been used in the food descriptions and nutrient breakdowns:

- diam = diameter
- ea = each
- enr = enriched
- f/ = from
- frzn = frozen
- g = grams
- liq = liquid
- pce = piece
- pkg = package
- w/ = with
- w/o = without
- t = trace
- o = zero (no nutrient value)
- blank space = information not available

Table H-1

Food Composition

(Computer code number is for West Diet Analysis program) (For purposes of calculations, use "0" for t, <1, <.1, <.01, etc.)

Computer Code Number	Food Description	Measure	Wt (g)	H_2O (%)	Ener (cal)	Prot (g)	Carb (g)	Dietary Fiber (g)	Fat (g)	Fat Breakdown (g) Sat	Mono	Poly
	BEVERAGES											
	Alcoholic:											
	Beer:											
1	Regular (12 fl oz)	1½ c	356	92	146	1	13	1	0	0	0	0
2	Light (12 fl oz)	1½ c	354	95	99	1	5	0	0	0	0	0
1506	Nonalcoholic (12 fl oz)	1½ c	360	98	32	1	5	0	0	0	0	0
	Gin, rum, vodka, whiskey:											
3	80 proof	1½ fl oz	42	67	97	0	0	0	0	0	0	0
4	86 proof	1½ fl oz	42	64	105	0	<1	0	0	0	0	0
5	90 proof	1½ fl oz	42	62	110	0	0	0	0	0	0	0
	Liqueur:											
1359	Coffee liqueur, 53 proof	1½ fl oz	52	31	175	<1	24	0	<1	.1	t	.1
1360	Coffee & cream liqueur, 34 proof	1½ fl oz	47	46	154	1	10	0	7	4.5	2.1	.3
1361	Crème de menthe, 72 proof	1½ fl oz	50	28	186	0	21	0	<1	t	t	.1
	Wine, 4 fl oz:											
6	Dessert, sweet	½ c	118	72	181	<1	14	0	0	0	0	0
7	Red	½ c	118	88	85	<1	2	0	0	0	0	0
8	Rosé	½ c	118	89	84	<1	2	0	0	0	0	0
9	White medium	½ c	118	90	80	<1	1	0	0	0	0	0
1592	Nonalcoholic	1 c	232	98	14	1	3	0	0	0	0	0
1593	Nonalcoholic light	1 c	232	98	14	1	3	0	0	0	0	0
1409	Wine cooler, bottle (12 fl oz)	1½ c	340	90	169	<1	20	<1	<1	t	t	t
1595	Wine cooler, cup	1 c	227	90	113	<1	13	<1	<1	t	t	t
	Carbonated:											
10	Club soda (12 fl oz)	1½ c	355	100	0	0	0	0	0	0	0	0
11	Cola beverage (12 fl oz)	1½ c	372	89	153	0	39	0	0	0	0	0
12	Diet cola w/aspartame (12 fl oz)	1½ c	355	100	4	<1	<1	0	0	0	0	0
13	Diet soda pop w/saccharin (12 fl oz)	1½ c	355	100	0	0	<1	0	0	0	0	0
14	Ginger ale (12 fl oz)	1½ c	366	91	124	0	32	0	0	0	0	0
15	Grape soda (12 fl oz)	1½ c	372	89	160	0	42	0	0	0	0	0
16	Lemon-lime (12 fl oz)	1½ c	368	89	147	0	38	0	0	0	0	0
17	Orange (12 fl oz)	1½ c	372	88	179	0	46	0	0	0	0	0
18	Pepper-type soda (12 fl oz)	1½ c	368	89	151	0	38	0	<1	.1	0	0
19	Root beer (12 fl oz)	1½ c	370	89	152	0	39	0	0	0	0	0
20	Coffee, brewed	1 c	237	99	5	<1	1	0	<1	t	0	t
21	Coffee, prepared from instant	1 c	238	99	5	<1	1	0	<1	t	0	t
	Fruit drinks, noncarbonated:											
22	Fruit punch drink, canned	1 c	248	88	117	0	29	<1	<1	t	t	t
1358	Gatorade	1 c	241	93	60	0	15	0	0	0	0	0
23	Grape drink, canned	1 c	250	87	125	<1	32	<1	0	0	0	0
1304	Koolade sweetened with sugar	1 c	262	90	97	0	25	0	<1	t	t	t
1356	Koolade sweetened with nutrasweet	1 c	240	95	43	0	11	0	0	0	0	0
26	Lemonade,frzn concentrate (6-oz can)	¾ c	219	52	396	1	103	1	<1	.1	t	.1
27	Lemonade, from concentrate	1 c	248	89	99	<1	26	<1	<1	t	t	t
28	Limeade, frzn concentrate (6-oz can)	¾ c	218	50	408	<1	108	1	<1	t	t	.1
29	Limeade, from concentrate	1 c	247	89	101	0	27	<1	<1	t	t	t
24	Pineapple grapefruit, canned	1 c	250	88	118	<1	29	<1	<1	t	t	.1
25	Pineapple orange, canned	1 c	250	87	125	3	29	<1	0	0	0	0
	Fruit and vegetable juices: see Fruit and Vegetable sections											
	Ultra Slim Fast, ready to drink, can:											
30411	Chocolate Royale	1 ea	350	84	220	10	38	5	3	1	1	.5
30415	French Vanilla	1 ea	350	84	220	10	38	5	3	1	1.5	.5
30413	Strawberries n' cream	1 ea	350	83	220	10	42	5	3	1	1.5	.5
1357	Water, bottled: Perrier (6½ fl oz)	1 ea	192	100	0	0	0	0	0	0	0	0
1594	Water, bottled: Tonic water	1½ c	366	91	124	0	32	0	0	0	0	0
	Tea:											
30	Brewed, regular	1 c	237	100	2	0	1	0	<1	t	0	t
1662	Brewed, herbal	1 c	237	100	2	0	<1	0	<1	t	t	t
32	From instant, sweetened	1 c	259	91	88	<1	22	0	<1	t	t	t
31	From instant, unsweetened	1 c	237	100	2	<1	<1	0	0	0	0	0

PAGE KEY: A–4 = Beverages A–6 = Dairy A–10 = Eggs A–10 = Fat/Oil A–14 = Fruit A–20 = Bakery A–28 = Grain A–32 = Fish A–34 = Meats A–38 = Poultry A–40 = Sausage A–42 = Mixed/Fast A–46 = Nuts/Seeds A–50 = Sweets A–52 = Vegetables/Legumes A–62 = Vegetarian Foods A–64 = Misc A–66 = Soups/Sauces A–68 = Fast A–84 = Convenience A–88 = Baby foods

Chol (mg)	Calc (mg)	Iron (mg)	Magn (mg)	Pota (mg)	Sodi (mg)	Zinc (mg)	VT-A (RE)	Thia (mg)	VT-E (a-TE)	Ribo (mg)	Niac (mg)	V-B6 (mg)	Fola (μg)	VT-C (mg)
0	18	.11	21	89	18	.07	0	.02	0	.09	1.61	.18	21	0
0	18	.14	18	64	11	.11	0	.03	0	.11	1.39	.12	14	0
0	25	.04	32	90	18	.04	0	.02	0	.09	1.63	.18	22	0
0	0	.02	0	1	<1	.02	0	<.01	0	<.01	<.01	0	0	0
0	0	.02	0	1	<1	.02	0	<.01	0	<.01	<.01	0	0	0
0	0	.02	0	1	<1	.02	0	<.01	0	<.01	<.01	0	0	0
0	1	.03	2	16	4	.02	0	<.01	0	.01	.07	0	0	0
7	8	.06	1	15	43	.07	20	0	.12	.03	.04	.01	0	0
0	0	.03	0	0	2	.02	0	0	0	0	<.01	0	0	0
0	9	.28	11	109	11	.08	0	.02	0	.02	.25	0	<1	0
0	9	.51	15	132	6	.11	0	.01	0	.03	.1	.04	2	0
0	9	.45	12	117	6	.07	0	<.01	0	.02	.09	.03	1	0
0	11	.38	12	94	6	.08	0	<.01	0	.01	.08	.02	<1	0
0	21	.93	23	204	16	.19	0	0	0	.02	.23	.05	2	0
0	21	.93	23	204	16	.19	0	0	0	.02	.23	.05	2	0
0	19	.92	18	152	29	.2	<1	.02	.02	.02	.15	.04	4	6
0	13	.61	12	102	19	.13	<1	.01	.02	.02	.1	.03	3	4
0	18	.04	4	7	75	.35	0	0	0	0	0	0	0	0
0	11	.11	4	4	15	.04	0	0	0	0	0	0	0	0
0	14	.11	4	0	21	.28	0	.02	0	.08	0	0	0	0
0	14	.14	4	7	57	.18	0	0	0	0	0	0	0	0
0	11	.66	4	4	26	.18	0	0	0	0	0	0	0	0
0	11	.3	4	4	56	.26	0	0	0	0	0	0	0	0
0	7	.26	4	4	40	.18	0	0	0	0	.05	0	0	0
0	19	.22	4	7	45	.37	0	0	0	0	0	0	0	0
0	11	.15	0	4	37	.15	0	0	0	0	0	0	0	0
0	18	.18	4	4	48	.26	0	0	0	0	0	0	0	0
0	5	.12	12	128	5	.05	0	0	0	0	.53	0	<1	0
0	7	.12	10	86	7	.07	0	0	0	<.01	.67	0	0	0
0	20	.52	5	62	55	.3	3	0	0	.06	.05	0	3	73
0	0	.12	2	26	96	.05	0	.01	0	0	0	0	0	0
0	7	.25	10	87	2	.07	<1	.02	0	.02	.25	.05	2	40
0	42	.13	3	3	37	.08	0	0	0	<.01	<.01	0	<1	31
0	17	.65	5	50	50	.26	2	.02	0	.05	.05	0	5	77
0	15	1.58	11	147	9	.17	22	.06	0	.21	.16	.05	22	39
0	7	.4	5	37	7	.1	5	.01	0	.05	.04	.01	5	10
0	11	.22	9	129	0	.09	0	.02	0	.02	.22	0	9	26
0	7	.07	2	32	5	.05	0	<.01	0	<.01	.05	0	2	7
0	17	.77	15	153	35	.15	9	.07	0	.04	.67	.1	26	115
0	12	.67	15	115	7	.15	13	.07	0	.05	.52	.12	27	56
5	400	2.7	140	530	220	2.24	525	.52	7	.59	7	.7	120	21
5	400	2.8	140	450	460	2.1	525	.52	7	.59	7	.7	120	21
5	400	2.7	140	450	460	2.24	525	.52	7	.59	7	.7	120	21
0	27	0	0	0	2	0	0	0	0	0	0	0	0	0
0	4	.04	0	0	15	.37	0	0	0	0	0	0	0	0
0	0	.05	7	88	7	.05	0	0	0	.03	0	0	12	0
0	5	.19	2	21	2	.09	0	.02	0	.01	0	0	1	0
0	5	.05	5	49	8	.08	0	0	0	.05	.09	<.01	10	0
0	5	.05	5	47	7	.07	0	0	0	<.01	.09	<.01	1	0

Table H-1
Food Composition

(Computer code number is for West Diet Analysis program) (For purposes of calculations, use "0" for t, <1, <.1, <.01, etc.)

Computer Code Number	Food Description	Measure	Wt (g)	H_2O (%)	Ener (cal)	Prot (g)	Carb (g)	Dietary Fiber (g)	Fat (g)	Fat Breakdown (g) Sat	Mono	Poly
DAIRY												
	Butter: see Fats and Oils, #158,159,160											
	Cheese, natural:											
33	Blue	1 oz	28	42	99	6	1	0	8	5.2	2.2	.2
34	Brick	1 oz	28	41	104	6	1	0	8	5.3	2.4	.2
35	Brie	1 oz	28	48	93	6	<1	0	8	4.9	2.2	.2
36	Camembert	1 oz	28	52	84	6	<1	0	7	4.3	2	.2
37	Cheddar:	1 oz	28	37	113	7	<1	0	9	5.9	2.6	.3
38	1" cube	1 ea	17	37	68	4	<1	0	6	3.6	1.6	.2
39	Shredded	1 c	113	37	455	28	1	0	37	24	10.6	1.1
1406	Low fat, low sodium	1 oz	28	65	48	7	1	0	2	1.2	.6	.1
	Cottage:											
984	Low sodium, low fat	1 c	225	83	162	28	6	0	2	1.4	.7	.1
40	Creamed, large curd	1 c	225	79	232	28	6	0	10	6.4	2.9	.3
41	Creamed, small curd	1 c	210	79	216	26	6	0	9	6	2.7	.3
42	With fruit	1 c	226	72	280	22	30	0	8	4.9	2.2	.2
43	Low fat 2%	1 c	226	79	203	31	8	0	4	2.8	1.2	.1
44	Low fat 1%	1 c	226	82	164	28	6	0	2	1.5	.7	.1
46	Cream	1 tbs	15	54	52	1	<1	0	5	3.3	1.5	.2
983	low fat	1 tbs	15	64	35	2	1	0	3	1.7	.9	.1
47	Edam	1 oz	28	42	100	7	<1	0	8	4.9	2.3	.2
48	Feta	1 oz	28	55	74	4	1	0	6	4.2	1.3	.2
49	Gouda	1 oz	28	41	100	7	1	0	8	4.9	2.2	.2
50	Gruyère	1 oz	28	33	116	8	<1	0	9	5.3	2.8	.5
51	Gorgonzola	1 oz	28	43	97	6	1	0	8	5		
1676	Limburger	1 oz	28	48	92	6	<1	0	8	4.7	2.4	.1
53	Monterey Jack	1 oz	28	41	104	7	<1	0	8	5.3	2.4	.3
54	Mozzarella, whole milk	1 oz	28	54	79	5	1	0	6	3.7	1.8	.2
55	Mozzarella, part-skim milk, low moisture	1 oz	28	49	78	8	1	0	5	3	1.4	.1
56	Muenster	1 oz	28	42	103	7	<1	0	8	5.3	2.4	.2
2422	Neufchatel	1 oz	28	62	73	3	1	0	7	4.1	1.9	.2
1399	Nonfat cheese (Kraft Singles)	1 oz	28	61	44	6	4	0	0	0	0	0
59	Parmesan, grated:	1 oz	28	18	128	12	1	0	8	5.5	2.4	.2
57	Cup, not pressed down	1 c	100	18	456	42	4	0	30	19.7	8.7	.7
58	Tablespoon	1 tbs	6	18	27	2	<1	0	2	1.2	.5	t
60	Provolone	1 oz	28	41	98	7	1	0	7	4.9	2.1	.2
61	Ricotta, whole milk	1 c	246	72	428	28	7	0	32	20.4	8.9	.9
62	Ricotta, part-skim milk	1 c	246	74	339	28	13	0	19	12.1	5.7	.6
63	Romano	1 oz	28	31	108	9	1	0	8	4.8	2.2	.2
64	Swiss	1 oz	28	37	105	8	1	0	8	5	2	.3
976	low fat	1 oz	28	60	50	8	1	0	1	.9	.4	t
	Pasteurized processed cheese products:											
65	American	1 oz	28	39	105	6	<1	0	9	5.5	2.5	.3
66	Swiss	1 oz	28	42	93	7	1	0	7	4.5	2	.2
67	American cheese food, jar	½ c	57	43	187	11	4	0	14	9	4.1	.4
68	American cheese spread	1 tbs	15	48	43	2	1	0	3	2	.9	.1
982	Velveeta cheese spread, low fat, low sodium, slice	1 pce	34	62	61	9	1	0	2	1.5	.7	.1
	Cream, sweet:											
69	Half & half (cream & milk)	1 c	242	81	315	7	10	0	28	17.3	8	1
70	Tablespoon	1 tbs	15	81	19	<1	1	0	2	1.1	.5	.1
71	Light, coffee or table:	1 c	240	74	468	6	9	0	46	28.8	13.4	1.7
72	Tablespoon	1 tbs	15	74	29	<1	1	0	3	1.8	.8	.1
73	Light whipping cream, liquid:	1 c	239	63	698	5	7	0	74	46.1	21.7	2.1
74	Tablespoon	1 tbs	15	63	44	<1	<1	0	5	2.9	1.4	.1
75	Heavy whipping cream, liquid:	1 c	238	58	821	5	7	0	88	54.7	25.5	3.3
76	Tablespoon	1 tbs	15	58	52	<1	<1	0	6	3.4	1.6	.2
77	Whipped cream, pressurized:	1 c	60	61	154	2	7	0	13	8.3	3.8	.5
78	Tablespoon	1 tbs	4	61	10	<1	<1	0	1	.6	.3	t
79	Cream, sour, cultured:	1 c	230	71	492	7	10	0	48	29.9	13.9	1.8
80	Tablespoon	1 tbs	14	71	30	<1	1	0	3	1.8	.8	.1

Chol (mg)	Calc (mg)	Iron (mg)	Magn (mg)	Pota (mg)	Sodi (mg)	Zinc (mg)	VT-A (RE)	Thia (mg)	VT-E (a-TE)	Ribo (mg)	Niac (mg)	V-B6 (mg)	Fola (µg)	VT-C (mg)
21	148	.09	6	72	391	.74	64	.01	.18	.11	.29	.05	10	0
26	189	.12	7	38	157	.73	85	<.01	.14	.1	.03	.02	6	0
28	51	.14	6	43	176	.67	51	.02	.18	.15	.11	.07	18	0
20	109	.09	6	52	236	.67	71	.01	.18	.14	.18	.06	17	0
29	202	.19	8	28	174	.87	85	.01	.1	.1	.02	.02	5	0
18	123	.12	5	17	106	.53	51	<.01	.06	.06	.01	.01	3	0
119	815	.77	31	111	702	3.51	342	.03	.41	.42	.09	.08	21	0
6	197	.2	8	31	6	.86	17	.01	.05	.01	.02	.02	5	0
9	137	.31	11	194	29	.85	25	.04	.25	.36	.29	.16	27	0
33	135	.31	12	190	911	.83	108	.05	.27	.37	.28	.15	27	0
31	126	.29	11	177	851	.78	101	.04	.26	.34	.26	.14	26	0
25	108	.25	9	151	915	.65	81	.04	.21	.29	.23	.12	22	0
19	155	.36	14	217	918	.95	45	.05	.13	.42	.32	.17	30	0
10	138	.32	12	193	918	.86	25	.05	.25	.37	.29	.15	28	0
16	12	.18	1	18	44	.08	66	<.01	.14	.03	.01	.01	2	0
8	17	.25	1	25	44	.11	33	<.01	.07	.04	.02	.01	3	0
25	205	.12	8	53	270	1.05	71	.01	.21	.11	.02	.02	5	0
25	138	.18	5	17	312	.81	36	.04	.01	.24	.28	.12	9	0
32	196	.07	8	34	229	1.09	49	.01	.1	.09	.02	.02	6	0
31	283	.05	10	23	94	1.09	84	.02	.1	.08	.03	.02	3	0
30	170	.18			280		43							0
25	139	.04	6	36	224	.59	88	.02	.18	.14	.04	.02	16	0
25	209	.2	8	23	150	.84	71	<.01	.09	.11	.03	.02	5	0
22	145	.05	5	19	104	.62	67	<.01	.1	.07	.02	.02	2	0
15	205	.07	7	26	148	.88	53	.01	.13	.1	.03	.02	3	0
27	201	.11	8	37	176	.79	88	<.01	.13	.09	.03	.02	3	0
21	21	.08	2	32	112	.15	74	<.01	.26	.05	.03	.01	3	0
4	221	0		81	427		126		0	.1				0
22	385	.27	14	30	521	.89	48	.01	.22	.11	.09	.03	2	0
79	1375	.95	51	107	1861	3.19	173	.04	.8	.39	.31	.1	8	0
5	82	.06	3	6	112	.19	10	<.01	.05	.02	.02	.01	<1	0
19	212	.15	8	39	245	.9	74	<.01	.1	.09	.04	.02	3	0
124	509	.93	28	258	207	2.85	330	.03	.86	.48	.26	.11	30	0
76	669	1.08	36	308	308	3.3	278	.05	.53	.45	.19	.05	32	0
29	298	.22	11	24	336	.72	39	.01	.2	.1	.02	.02	2	0
26	269	.05	10	31	73	1.09	71	.01	.14	.1	.03	.02	2	0
10	269	.05	10	31	73	1.09	18	.01	.05	.1	.02	.02	2	0
26	172	.11	6	45	400	.84	81	.01	.13	.1	.02	.02	2	0
24	216	.17	8	60	384	1.01	64	<.01	.19	.08	.01	.01	2	0
36	327	.48	17	159	678	1.7	125	.02	.4	.25	.08	.08	4	0
8	84	.05	4	36	202	.39	28	.01	.11	.06	.02	.02	1	0
12	233	.15	8	61	2	1.13	22	.01	.17	.13	.03	.03	3	0
89	254	.17	25	315	98	1.23	259	.08	.27	.36	.19	.09	6	2
6	16	.01	2	19	6	.08	16	<.01	.02	.02	.01	.01	<1	<1
159	231	.1	21	293	95	.65	437	.08	.36	.35	.14	.08	6	2
10	14	.01	1	18	6	.04	27	<.01	.02	.02	.01	<.01	<1	<1
265	166	.07	17	231	82	.6	705	.06	1.43	.3	.1	.07	9	1
17	10	<.01	1	14	5	.04	44	<.01	.09	.02	.01	<.01	1	<1
326	154	.07	17	179	89	.55	1001	.05	1.5	.26	.09	.06	9	1
21	10	<.01	1	11	6	.03	63	<.01	.09	.02	.01	<.01	1	<1
46	61	.03	6	88	78	.22	124	.02	.36	.04	.04	.02	2	0
3	4	<.01		6	5	.01	8	<.01	.02	<.01	<.01	<.01	<1	0
102	267	.14	26	331	123	.62	449	.08	1.3	.34	.15	.04	25	2
6	16	.01	2	20	7	.04	27	<.01	.08	.02	.01	<.01	2	<1

Table H-1

Food Composition

(Computer code number is for West Diet Analysis program) (For purposes of calculations, use "0" for t, <1, <.1, <.01, etc.)

Computer Code Number	Food Description	Measure	Wt (g)	H_2O (%)	Ener (cal)	Prot (g)	Carb (g)	Dietary Fiber (g)	Fat (g)	Fat Breakdown (g) Sat	Mono	Poly
	DAIRY—Continued											
	Cream products—imitation and part dairy:											
81	Coffee whitener, frozen or liquid	1 tbs	15	77	20	<1	2	0	1	1.4	t	0
82	Coffee whitener, powdered	1 tsp	2	2	11	<1	1	0	1	.6	t	0
83	Dessert topping, frozen, nondairy:	1 c	75	50	239	1	17	0	19	16.4	1.2	.4
84	Tablespoon	1 tbs	5	50	16	<1	1	0	1	1.1	.1	t
85	Dessert topping, mix with whole milk:	1 c	80	67	151	3	13	0	10	8.6	.7	.2
86	Tablespoon	1 tbs	5	67	9	<1	1	0	1	.5	t	t
88	Dessert topping, pressurized	1 c	70	60	185	1	11	0	16	13.3	1.3	.2
87	Tablespoon	1 tbs	4	60	11	<1	1	0	1	.8	.1	t
91	Sour cream, imitation:	1 c	230	71	478	6	15	0	45	40.9	1.3	.1
92	Tablespoon	1 tbs	14	71	29	<1	1	0	3	2.5	.1	t
89	Sour dressing, part dairy:	1 c	235	75	418	8	11	0	39	31.3	4.6	1.1
90	Tablespoon	1 tbs	15	75	27	1	1	0	2	2	.3	.1
	Milk, fluid:											
93	Whole milk	1 c	244	88	150	8	11	0	8	5.1	2.7	.3
94	2% reduced-fat milk	1 c	244	89	121	8	12	0	5	2.9	1.3	.2
95	2% milk solids added	1 c	245	89	125	9	12	0	5	2.9	1.4	.2
96	1% lowfat milk	1 c	244	90	102	8	12	0	3	1.6	.7	.1
97	1% milk solids added	1 c	245	90	104	9	12	0	2	1.5	.7	.1
98	Nonfat milk, vitamin A added	1 c	245	91	85	8	12	0	<1	.3	.1	t
99	Nonfat milk solids added	1 c	245	90	90	9	12	0	1	.4	.2	t
100	Buttermilk, skim	1 c	245	90	99	8	12	0	2	1.3	.6	.1
	Milk, canned:											
101	Sweetened condensed	1 c	306	27	982	24	166	0	27	16.8	7.4	1
103	Evaporated, nonfat	1 c	256	79	199	19	29	0	1	.3	.2	t
	Milk, dried:											
104	Buttermilk, sweet	1 c	120	3	464	41	59	0	7	4.3	2	.3
105	Instant, nonfat, vit A added (makes 1 qt)	1 ea	91	4	326	32	47	0	1	.4	.2	t
106	Instant nonfat, vit A added	1 c	68	4	243	24	35	0	<1	.3	.1	t
107	Goat milk	1 c	244	87	168	9	11	0	10	6.5	2.7	.4
108	Kefir	1 c	233	88	149	8	11	0	8			
	Milk beverages and powdered mixes:											
	Chocolate:											
109	Whole	1 c	250	82	209	8	26	2	8	5.3	2.5	.3
110	2% fat	1 c	250	84	179	8	26	1	5	3.1	1.5	.2
111	1% fat	1 c	250	84	158	9	26	1	2	1.5	.7	.1
	Chocolate-flavored beverages:											
112	Powder containing nonfat dry milk:	1 oz	28	1	101	3	22	<1	1	.7	.4	t
113	Prepared with water	1 c	275	86	138	4	30	3	2	.9	.5	t
114	Powder without nonfat dry milk:	1 oz	28	1	98	1	25	2	1	.5	.3	t
115	Prepared with whole milk	1 c	266	81	226	9	31	1	9	5.5	2.6	.3
116	Eggnog, commercial	1 c	254	74	343	10	34	0	19	11.3	5.7	.9
974	Eggnog, 2% reduced-fat	1 c	254	85	189	12	17	0	8	3.7	2.7	.7
1027	Instant Breakfast, envelope,powder only:	1 ea	37	7	131	7	24	<1	1	.3	.1	t
1028	Prepared with whole milk	1 c	281	77	280	15	36	<1	9	5.4	2.5	.3
1029	Prepared with 2% milk	1 c	281	78	252	15	36	<1	5	3.3	1.5	.2
1283	Prepared with 1% milk	1 c	281	79	233	15	36	<1	3	1.9	.9	.1
1284	Prepared with nonfat milk	1 c	282	80	216	16	36	<1	1	.7	.3	t
117	Malted milk, chocolate, powder:	3 tsp	21	1	79	1	18	<1	1	.5	.2	.1
118	Prepared with whole milk	1 c	265	81	228	9	30	<1	9	5.5	2.6	.4
1661	Ovaltine with whole milk	1 c	265	81	225	9	29	<1	9	5.5	2.5	.4
119	Malted mix powder, natural:	3 tsp	21	2	87	2	16	<1	2	.9	.4	.3
120	Prepared with whole milk	1 c	265	81	236	10	27	<1	10	6	2.8	.6
121	Milk shakes, chocolate	1 c	166	71	211	6	34	1	6	3.8	1.8	.2
122	Milk shakes, vanilla	1 c	166	75	184	6	30	1	5	3.1	1.4	.2
	Milk desserts:											
134	Custard, baked	1 c	282	79	296	14	30	0	13	6.6	4.3	1
1548	Low-fat frozen dessert bars	1 ea	81	72	88	2	19	0	1	.2	.1	.4
	Ice cream, vanilla (about 10% fat):											
123	Hardened: ½ gallon	1 ea	1064	61	2138	37	251	0	117	72.4	33.8	4.4
124	Cup	1 c	132	61	265	5	31	0	14	9	4.2	.5
126	Soft serve	1 c	172	60	370	7	38	0	22	12.9	6	.8

PAGE KEY: A–4 = Beverages A–6 = Dairy A–10 = Eggs A–10 = Fat/Oil A–14 = Fruit A–20 = Bakery A–28 = Grain A–32 = Fish A–34 = Meats A–38 = Poultry A–40 = Sausage A–42 = Mixed/Fast A–46 = Nuts/Seeds A–50 = Sweets A–52 = Vegetables/Legumes A–62 = Vegetarian Foods A–64 = Misc A–66 = Soups/Sauces A–68 = Fast A–84 = Convenience A–88 = Baby foods

Chol (mg)	Calc (mg)	Iron (mg)	Magn (mg)	Pota (mg)	Sodi (mg)	Zinc (mg)	VT-A (RE)	Thia (mg)	VT-E (a-TE)	Ribo (mg)	Niac (mg)	V-B6 (mg)	Fola (μg)	VT-C (mg)
0	1	<.01		29	12	<.01	1	0	.24	0	0	0	0	0
0		.02		16	4	.01	<1	0	<.01	<.01	0	0	0	0
0	5	.09	1	14	19	.02	64	0	.14	0	0	0	0	0
0		.01		1	1	<.01	4	0	.01	0	0	0	0	0
8	72	.03	8	121	53	.22	39	.02	.11	.09	.05	.02	3	1
<1	5	<.01		8	3	.01	2	<.01	.01	.01	<.01	<.01	<1	<1
0	4	.01	1	13	43	.01	33	0	.12	0	0	0	0	0
0		<.01		1	2	0	2	0	.01	0	0	0	0	0
0	6	.9	15	370	235	2.71	0	0	.34	0	0	0	0	0
0		.05	1	22	14	.16	0	0	.02	0	0	0	0	0
13	266	.07	23	381	113	.87	5	.09	.29	.38	.17	.04	28	2
1	17	<.01	1	24	7	.06	<1	.01	.02	.02	.01	<.01	2	<1
33	290	.12	33	371	120	.93	76	.09	.24	.39	.2	.1	12	2
18	298	.12	33	376	122	.95	139	.09	.17	.4	.21	.1	12	2
18	314	.12	35	397	128	.98	140	.1	.17	.42	.22	.11	13	2
10	300	.12	34	381	123	.95	144	.09	.1	.41	.21	.1	12	2
10	314	.12	35	397	128	.98	145	.1	.1	.42	.22	.11	13	2
4	301	.1	28	407	126	.98	149	.09	.1	.34	.22	.1	13	2
5	316	.12	35	419	130	1	149	.1	.1	.43	.22	.11	13	2
9	284	.12	27	370	257	1.03	20	.08	.15	.38	.14	.08	12	2
104	869	.58	79	1135	389	2.88	248	.27	.65	1.27	.64	.16	34	8
9	742	.74	69	850	294	2.3	300	.11	.01	.79	.44	.14	22	3
83	1420	.36	132	1910	620	4.82	65	.47	.48	1.9	1.05	.41	57	7
17	1119	.28	106	1551	500	4.01	646	.38	.02	1.58	.81	.31	45	5
12	836	.21	80	1159	373	3	483	.28	.01	1.18	.61	.23	34	4
28	327	.12	34	498	122	.73	137	.12	.22	.34	.68	.11	1	3
		.3	33	373	107									
30	280	.6	32	418	149	1.03	72	.09	.23	.4	.31	.1	12	2
17	285	.6	33	423	151	1.03	143	.09	.13	.41	.31	.1	12	2
7	288	.6	33	425	152	1.03	148	.09	.06	.41	.32	.1	12	2
1	91	.33	23	199	141	.41	1	.03	.04	.16	.16	.03	0	<1
3	129	.47	33	270	198	.6	1	.04	.06	.21	.22	.04	0	1
0	10	.88	27	165	59	.43	1	.01	.11	.04	.14	<.01	2	<1
32	301	.8	53	497	165	1.28	77	.1	.21	.43	.32	.1	12	2
149	330	.51	47	419	138	1.17	203	.09	.58	.48	.27	.13	2	4
194	269	.71	32	367	155	1.26	197	.11	1.01	.55	.21	.15	30	2
4	105	4.74	84	350	142	3.16	554	.31	5.31	.07	5.27	.42	105	28
38	396	4.86	117	719	262	4.09	630	.41	5.51	.47	5.46	.52	118	31
23	401	4.86	118	726	264	4.12	693	.41	5.41	.48	5.46	.53	118	31
14	406	4.86	118	731	266	4.12	698	.41	5.36	.48	5.47	.53	118	31
9	407	4.83	112	755	268	4.14	703	.4	5.3	.42	5.47	.52	118	31
1	13	.48	15	130	53	.17	4	.04	.08	.04	.42	.03	4	<1
34	305	.61	48	498	172	1.09	79	.13	.26	.44	.62	.13	16	3
34	384	3.76	53	620	244	1.17	901	.73	.32	1.26	10.9	1.02	32	34
4	63	.15	19	159	104	.21	18	.11	.08	.19	1.1	.09	10	1
37	355	.26	53	530	223	1.14	95	.2	.32	.59	1.31	.19	22	3
22	188	.51	28	332	161	.68	38	.1	.11	.41	.27	.08	6	1
18	203	.15	20	289	136	.6	53	.07	.1	.3	.31	.09	5	1
245	316	.85	39	431	217	1.49	169	.09	.68	.64	.24	.14	28	1
1	82	.07	10	111	47	.26	38	.03	.07	.11	.06	.03	3	1
468	1361	.96	149	2117	851	7.34	1244	.44	0	2.55	1.23	.51	53	6
58	169	.12	18	263	106	.91	154	.05	0	.32	.15	.06	7	1
157	225	.36	21	304	105	.89	265	.08	.64	.31	.16	.08	15	1

Table H-1

Food Composition

(Computer code number is for West Diet Analysis program) (For purposes of calculations, use "0" for t, <1, <.1, <.01, etc.)

Computer Code Number	Food Description	Measure	Wt (g)	H_2O (%)	Ener (cal)	Prot (g)	Carb (g)	Dietary Fiber (g)	Fat (g)	Fat Breakdown (g) Sat	Mono	Poly
	DAIRY—Continued											
	Ice cream, rich vanilla (16% fat):											
127	Hardened: ½ gallon	1 ea	1188	60	2554	49	264	0	154	88.9	41.5	5.5
128	Cup	1 c	148	57	357	5	33	0	24	14.8	6.9	.9
1724	Ben & Jerry's	½ c	108		230	4	21	0	17	10		
	Ice milk, vanilla (about 4% fat):											
129	Hardened: ½ gallon	1 ea	1048	68	1456	40	238	0	45	27.7	12.9	1.7
130	Cup	1 c	132	68	183	5	30	0	6	3.5	1.6	.2
131	Soft serve (about 3% fat)	1 c	176	70	222	9	38	0	5	2.9	1.3	.2
	Pudding, canned (5 oz can = .55 cup):											
135	Chocolate	1 ea	142	69	189	4	32	1	6	1	2.4	2
136	Tapioca	1 ea	142	74	169	3	27	<1	5	.9	2.3	1.9
137	Vanilla	1 ea	142	71	185	3	31	<1	5	.8	2.2	1.9
	Puddings, dry mix with whole milk:											
138	Chocolate, instant	1 c	294	74	326	9	55	3	9	5.4	2.7	.5
139	Chocolate, regular, cooked	1 c	284	74	315	9	51	3	10	5.9	2.8	.4
140	Rice, cooked	1 c	288	72	351	9	60	<1	8	5.1	2.4	.3
141	Tapioca, cooked	1 c	282	74	321	8	55	0	8	5.1	2.3	.3
142	Vanilla, instant	1 c	284	73	324	8	56	0	8	4.9	2.4	.4
143	Vanilla, regular, cooked	1 c	280	75	311	8	52	0	8	5.1	2.4	.4
	Sherbet (2% fat):											
132	½ gallon	1 ea	1542	66	2127	17	469	8	31	17.9	8.3	1.2
133	Cup	1 c	198	66	273	2	60	1	4	2.3	1.1	.2
144	Soy milk	1 c	245	93	81	7	4	3	5	.7	1	2.6
2301	Soy milk, fortified, fat free	1 c	240	88	110	6	22	1	0	0	0	0
	Yogurt, frozen, low-fat											
1584	Cup	1 c	144	65	229	6	35	0	8	4.9	2.3	.3
1512	Scoop	1 ea	79	74	78	4	15	0	<1	.1	t	t
	Yogurt, lowfat:											
1172	Fruit added with low-calorie sweetener	1 c	241	86	122	12	19	1	<1	.2	.1	t
145	Fruit added	1 c	245	74	250	11	47	<1	3	1.7	.7	.1
146	Plain	1 c	245	85	155	13	17	0	4	2.4	1	.1
147	Vanilla or coffee flavor	1 c	245	79	209	13	34	0	3	2	.8	.1
148	Yogurt, made with nonfat milk	1 c	245	85	137	15	19	0	<1	.3	.1	t
149	Yogurt, made with whole milk	1 c	245	88	150	9	11	0	8	5.1	2.2	.2
	EGGS											
	Raw, large:											
150	Whole, without shell	1 ea	50	75	74	6	1	0	5	1.5	1.9	.7
151	White	1 ea	33	88	16	3	<1	0	0	0	0	0
152	Yolk	1 ea	17	49	61	3	<1	0	5	1.6	2	.7
	Cooked:											
153	Fried in margarine	1 ea	46	69	91	6	1	0	7	1.9	2.8	1.3
154	Hard-cooked, shell removed	1 ea	50	75	77	6	1	0	5	1.6	2	.7
155	Hard-cooked, chopped	1 c	136	75	211	17	2	0	14	4.4	5.5	1.9
156	Poached, no added salt	1 ea	50	75	74	6	1	0	5	1.5	1.9	.7
157	Scrambled with milk & margarine	1 ea	61	73	101	7	1	0	7	2.2	2.9	1.3
1681	Egg substitute, liquid:	½ c	126	83	106	16	1	0	4	.8	1.1	2
1254	Egg Beaters, Fleischmann's	½ c	122		60	12	2	0	0	0	0	0
1262	Egg substitute, liquid, prepared	½ c	105	80	100	14	1	0	4	.8	1.1	1.9
	FATS and OILS											
158	Butter: Stick	½ c	114	16	817	1	<1	0	92	57.7	27.4	3.4
159	Tablespoon:	1 tbs	14	16	100	<1	<1	0	11	7.1	3.4	.4
8025	Unsalted	1 tbs	14	18	100	<1	<1	0	11	7.1	3.4	.4
160	Pat (about 1 tsp)	1 ea	5	16	36	<1	<1	0	4	2.5	1.2	.2
1682	Whipped	1 tsp	3	16	21	<1	<1	0	2	1.5	.7	.1
	Fats, cooking:											
1363	Bacon fat	1 tbs	14		125	0	0	0	14	6.3	5.9	1.1
1362	Beef fat/tallow	1 c	205	0	1849	0	0	0	205	103	87.3	8.2
1364	Chicken fat	1 c	205		1845	0	0	0	205	61.1	91.6	42.8
161	Vegetable shortening:	1 c	205	0	1812	0	0	0	205	52.1	91.2	53.5
162	Tablespoon	1 tbs	13	0	115	0	0	0	13	3.3	5.8	3.4

PAGE KEY: A–4 = Beverages A–6 = Dairy A–10 = Eggs A–10 = Fat/Oil A–14 = Fruit A–20 = Bakery A–28 = Grain A–32 = Fish A–34 = Meats A–38 = Poultry A–40 = Sausage A–42 = Mixed/Fast A–46 = Nuts/Seeds A–50 = Sweets A–52 = Vegetables/Legumes A–62 = Vegetarian Foods A–64 = Misc A–66 = Soups/Sauces A–68 = Fast A–84 = Convenience A–88 = Baby foods

Chol (mg)	Calc (mg)	Iron (mg)	Magn (mg)	Pota (mg)	Sodi (mg)	Zinc (mg)	VT-A (RE)	Thia (mg)	VT-E (a-TE)	Ribo (mg)	Niac (mg)	V-B6 (mg)	Fola (μg)	VT-C (mg)
1081	1556	2.49	143	2102	725	6.18	1829	.58	4.4	2.16	1.13	.57	107	9
90	173	.07	16	235	83	.59	272	.06	0	.24	.12	.06	7	1
95	150	.36			55		214		0					0
147	1456	1.05	157	2211	891	4.61	493	.61	0	2.78	.94	.68	63	8
18	183	.13	20	279	112	.58	62	.08	0	.35	.12	.09	8	1
21	276	.11	25	389	123	.93	51	.09	0	.35	.21	.08	11	2
4	128	.72	30	256	183	.6	16	.04	.18	.22	.49	.04	4	3
1	119	.33	11	148	168	.38	0	.03	.13	.14	.44	.14	4	1
10	125	.18	11	160	192	.35	9	.03	.18	.2	.36	.02	0	0
32	300	.85	53	488	835	1.23	62	.1	.18	.41	.28	.11	12	3
34	315	1.02	43	463	293	1.28	74	.09	.17	.49	.29	.1	11	2
32	297	1.09	37	372	314	1.09	58	.22	.17	.4	1.28	.1	11	2
34	293	.17	34	372	341	.96	76	.08	.23	.4	.21	.11	11	2
31	287	.2	34	364	812	.94	71	.09	.18	.39	.21	.1	11	2
34	300	.14	36	381	448	.98	76	.08	.17	.4	.21	.09	11	2
77	833	2.16	123	1480	709	7.4	216	.39	.88	1.05	1.48	.52	77	66
10	107	.28	16	190	91	.95	28	.05	.11	.13	.19	.07	10	9
0	10	1.42	47	345	29	.56	7	.39	.02	.17	.36	.1	4	0
0	400	1.44		20	60		0	.07		.1	3			0
3	206	.43	20	304	125	.6	82	.05	.07	.32	.41	.11	9	1
1	137	.07	13	175	53	.67	1	.03	<.01	.16	.08	.04	8	1
3	369	.61	41	550	139	1.83	6	.1	.17	.45	.5	.11	32	26
10	372	.17	36	478	143	1.81	27	.09	.07	.44	.23	.1	23	2
15	448	.2	43	573	172	2.18	39	.11	.1	.52	.28	.12	27	2
12	419	.17	40	537	161	2.03	32	.1	.08	.49	.26	.11	26	2
4	488	.22	47	625	187	2.38	5	.12	.01	.57	.3	.13	30	2
31	296	.12	28	380	114	1.45	73	.07	.22	.35	.18	.08	18	1
213	24	.72	5	60	63	.55	95	.03	.52	.25	.04	.07	23	0
0	2	.01	4	47	54	<.01	0	<.01	0	.15	.03	<.01	1	0
218	23	.6	2	16	7	.53	99	.03	.54	.11	<.01	.07	25	0
211	25	.72	5	61	162	.55	114	.03	.75	.24	.03	.07	17	0
212	25	.59	5	63	62	.52	84	.03	.52	.26	.03	.06	22	0
577	68	1.62	14	171	169	1.43	228	.09	1.43	.7	.09	.16	60	0
212	24	.72	5	60	140	.55	95	.02	.52	.21	.03	.06	17	0
215	43	.73	7	84	171	.61	119	.03	.8	.27	.05	.07	18	<1
1	67	2.65	11	416	223	1.64	272	.14	.61	.38	.14	<.01	19	0
0	80	2.16		170	200		80		.59					
1	63	2.51	10	394	211	1.55	258	.11	.58	.34	.12	<.01	13	0
250	27	.18	2	30	942	.06	860	.01	1.8	.04	.05	<.01	3	0
31	3	.02		4	116	.01	106	<.01	.22	<.01	.01	0	<1	0
31	3	.02		4	2	.01	106	<.01	.22	<.01	.01	0	<1	0
11	1	.01		1	41	<.01	38	0	.08	<.01	<.01	0	<1	0
7	1	<.01		1	25	<.01	23	0	.05	<.01	<.01	0	<1	0
14		0			76	<.01	0	0	.31	0	0	0	0	0
223	0	0	0		<1	0	0	0	3.08	0	0	0	0	0
174	0	0	0	0	0	0	0	0	5.54	0	0	0	0	0
0	0	0	0	0	0	0	0	0	17	0	0	0	0	0
0	0	0	0	0	0	0	0	0	1.08	0	0	0	0	0

Table H-1

Food Composition (Computer code number is for West Diet Analysis program) (For purposes of calculations, use "0" for t, <1, <.1, <.01, etc.)

Computer Code Number	Food Description	Measure	Wt (g)	H_2O (%)	Ener (cal)	Prot (g)	Carb (g)	Dietary Fiber (g)	Fat (g)	Fat Breakdown (g) Sat	Mono	Poly
	FATS and OILS—Continued											
163	Lard:	1 c	205	0	1849	0	0	0	205	81.1	87	28.3
164	Tablespoon	1 tbs	13	0	117	0	0	0	13	5.1	5.5	1.8
	Margarine:											
165	Imitation (about 40% fat), soft:	1 c	232	58	800	1	1	0	90	17.9	36.4	32
166	Tablespoon	1 tbs	14	58	48	<1	<1	0	5	1.1	2.2	1.9
167	Regular, hard (about 80% fat):	½ c	114	16	820	1	1	0	92	18	40.8	29
168	Tablespoon	1 tbs	14	16	101	<1	<1	0	11	2.2	5	3.6
169	Pat	1 ea	5	16	36	<1	<1	0	4	.8	1.8	1.3
170	Regular, soft (about 80% fat):	1 c	227	16	1625	2	1	0	183	31.3	64.7	78.5
171	Tablespoon	1 tbs	14	16	100	<1	<1	0	11	1.9	4	4.8
2056	Saffola, unsalted	1 tbs	14	20	100	0	0	0	11	2	3	4.5
2057	Saffola, reduced fat	1 tbs	14	37	60	0	0	0	8	1.3	2.7	4.4
172	Spread (about 60% fat), hard:	1 c	227	37	1225	1	0	0	138	32	59	41.1
173	Tablespoon	1 tbs	14	37	76	<1	0	0	9	2	3.6	2.5
174	Pat	1 ea	5	37	27	<1	0	0	3	.7	1.2	1
175	Spread (about 60% fat), soft:	1 c	227	37	1225	1	0	0	138	29.3	71.5	31.3
176	Tablespoon	1 tbs	14	37	76	<1	0	0	9	1.8	4.4	1.9
2160	Touch of Butter (47% fat)	1 tbs	14		60	0	0	0	7	1.5	3.1	1.5
	Oils:											
1585	Canola:	1 c	218	0	1927	0	0	0	218	15.5	128	64.5
1586	Tablespoon	1 tbs	14	0	124	0	0	0	14	1	8.2	4.1
177	Corn:	1 c	218	0	1927	0	0	0	218	29.4	54.1	131
178	Tablespoon	1 tbs	14	0	124	0	0	0	14	1.9	3.5	8.4
179	Olive:	1 c	216	0	1909	0	0	0	216	29.4	159	21.3
180	Tablespoon	1 tbs	14	0	124	0	0	0	14	1.9	10.3	1.4
1683	Olive, extra virgin	1 tbs	14		126	0	0	0	14	2	10.8	1.3
181	Peanut:	1 c	216	0	1909	0	0	0	216	40	99.8	71.3
182	Tablespoon	1 tbs	14	0	124	0	0	0	14	2.6	6.5	4.6
183	Safflower:	1 c	218	0	1927	0	0	0	218	19.8	26.4	162
184	Tablespoon	1 tbs	14	0	124	0	0	0	14	1.3	1.7	10.4
185	Soybean:	1 c	218	0	1927	0	0	0	218	32	50.8	126
186	Tablespoon	1 tbs	14	0	124	0	0	0	14	2.1	3.3	8.1
187	Soybean/cottonseed:	1 c	218	0	1927	0	0	0	218	39.5	64.3	105
188	Tablespoon	1 tbs	14	0	124	0	0	0	14	2.5	4.1	6.7
189	Sunflower:	1 c	218	0	1927	0	0	0	218	25.3	42.5	143
190	Tablespoon	1 tbs	14	0	124	0	0	0	14	1.6	2.7	9.2
	Salad dressings/sandwich spreads:											
191	Blue cheese, regular	1 tbs	15	32	76	1	1	0	8	1.5	1.8	4.2
1040	Low calorie	1 tbs	15	79	15	1	<1	<1	1	.2	.5	.4
1684	Caesar's	1 tbs	12	36	56	1	<1	<1	5	1	3.7	.5
192	French, regular	1 tbs	16	38	69	<1	3	0	7	1.5	1.3	3.5
193	Low calorie	1 tbs	16	71	21	<1	3	0	1	.1	.2	.5
194	Italian, regular	1 tbs	15	40	70	<1	2	0	7	1	1.7	4.2
195	Low calorie	1 tbs	15	84	16	<1	1	<1	1	.2	.3	.9
	Kraft, Deliciously Right:											
2150	1000 Island	1 tbs	16		35	0	4	0	2	.5		
2153	Bacon & tomato	1 tbs	16		31	1	2	0	3	.5		
2154	Cucumber ranch	1 tbs	16		31	0	1	0	3	.5		
2151	French	1 tbs	16		25	0	3	0	1	.2		
2152	Ranch	1 tbs	16		52	0	3	0	5	.8		
199	Mayo type, regular	1 tbs	15	40	58	<1	4	0	5	.7	1.3	2.7
1030	Low calorie	1 tbs	14	54	36	<1	3	0	3	.4	.7	1.4
	Mayonnaise:											
197	Imitation, low calorie	1 tbs	15	63	35	<1	2	0	3	.5	.7	1.6
196	Regular (soybean)	1 tbs	14	17	100	<1	<1	0	11	1.7	3.2	5.8
1488	Regular, low calorie, low sodium	1 tbs	14	63	32	<1	2	0	3	.5	.6	1.4
1493	Regular, low calorie	1 tbs	15	63	35	<1	2	0	3	.5	.7	1.6
198	Ranch, regular	1 tbs	15	39	80	0	<1	0	8	1.2		
2251	Low calorie	1 tbs	14	69	30	<1	1	0	2	.5		
1685	Russian	1 tbs	15	34	74	<1	2	0	8	1.1	1.8	4.4
1502	Salad dressing, low calorie, oil free	1 tbs	15	88	4	<1	1	<1	<1	0	0	0

Chol (mg)	Calc (mg)	Iron (mg)	Magn (mg)	Pota (mg)	Sodi (mg)	Zinc (mg)	VT-A (RE)	Thia (mg)	VT-E (a-TE)	Ribo (mg)	Niac (mg)	V-B6 (mg)	Fola (μg)	VT-C (mg)
195		0			<1	.23	0	0	2.46	0	0	0	0	0
12		0			<1	.01	0	0	.16	0	0	0	0	0
0	41	0	4	59	2227	0	1853	.01	5.41	.05	.03	.01	2	<1
0	2	0		4	134	0	112	<.01	.33	<.01	<.01	<.01	<1	<1
0	34	.07	3	48	1075	0	911	.01	14.6	.04	.03	.01	1	<1
0	4	.01		6	132	0	112	<.01	1.79	<.01	<.01	<.01	<1	<1
0	1	<.01		2	47	0	40	<.01	.64	<.01	<.01	0	<1	<1
0	60	0	5	86	2447	0	1813	.02	27.2	.07	.04	.02	2	<1
0	4	0		5	151	0	112	<.01	1.68	<.01	<.01	<.01	<1	<1
	0	0			0		51							0
	0	0			115		51							0
0	47	0	4	68	2256	0	1813	.02	11.4	.06	.04	.01	2	<1
0	3	0		4	139	0	112	<.01	.7	<.01	<.01	<.01	<1	<1
0	1	0		1	50	0	40	0	.25	<.01	<.01	0	<1	<1
0	47	0	4	68	2256	0	1813	.02	20.5	.06	.04	.01	2	<1
0	3	0		4	139	0	112	<.01	1.26	<.01	<.01	<.01	<1	<1
0	0	0		0	110		100		1.27					0
0	0	0	0	0	0	0	0	0	45.8	0	0	0	0	0
0	0	0	0	0	0	0	0	0	2.94	0	0	0	0	0
0	0	0	0	0	0	0	0	0	46	0	0	0	0	0
0	0	0	0	0	0	0	0	0	2.95	0	0	0	0	0
0		.82		0	<1	.13	0	0	26.8	0	0	0	0	0
0		.05		0	<1	.01	0	0	1.74	0	0	0	0	0
									1.74					
0		.06			<1	.02	0	0	27.9	0	0	0	0	0
0		<.01			<1	<.01	0	0	1.81	0	0	0	0	0
0	0	0	0	0	0	0	0	0	94	0	0	0	0	0
0	0	0	0	0	0	0	0	0	6.03	0	0	0	0	0
0		.04		0	0	0	0	0	39.7	0	0	0	0	0
0		<.01		0	0	0	0	0	2.55	0	0	0	0	0
0	0	0	0	0	0	0	0	0	61.5	0	0	0	0	0
0	0	0	0	0	0	0	0	0	3.95	0	0	0	0	0
0	0	0	0	0	0	0	0	0	110	0	0	0	0	0
0	0	0	0	0	0	0	0	0	7.08	0	0	0	0	0
3	12	.03	0	6	164	.04	10	<.01	1.4	.01	.01	.01	1	<1
<1	13	.07	1	1	180	.04	<1	<.01	.13	.01	.01	<.01	<1	<1
12	23	.2	3	21	207	.13	7	<.01	.72	.02	.5	.01	2	1
0	2	.06	0	13	219	.01	21	<.01	1.35	<.01	0	<.01	1	0
0	2	.06	0	13	126	.03	21	0	.19	0	0	0	0	0
0	1	.03		2	118	.02	4	<.01	1.56	<.01	0	<.01	1	0
1		.03	0	2	118	.02	0	0	.22	0	0	0	0	0
2	0	0		27	160		0		.19					0
1	0	0		21	155		0		.75					0
0	0	0		10	232		0		.73					0
0	0	0		7	130		50		.42					0
0	0	0		5	165		0		1.31					0
4	2	.03		1	107	.03	13	<.01	.6	<.01	<.01	<.01	1	0
4	2	.03		1	99	.02	9	<.01	.6	<.01	0	<.01	1	0
4		0		1	75	.02	0	0	.96	0	0	0	0	0
8	3	.07		5	79	.02	12	0	1.65	0	<.01	.08	1	0
3	0	0	0	1	15	.01	1	0	.53	<.01	0	0	<1	0
4		0		1	75	.02	0	0	.96	0	0	0	0	0
5	0	0			105		0							0
5	5	0			120		0		.7					0
3	3	.09		24	130	.06	31	.01	1.53	.01	.09	<.01	2	1
0	1	.04	2	7	256	<.01	<1	0	0	0	<.01	<.01	<1	<1

Table H-1

Food Composition

(Computer code number is for West Diet Analysis program) (For purposes of calculations, use "0" for t, <1, <.1, <.01, etc.)

Computer Code Number	Food Description	Measure	Wt (g)	H_2O (%)	Ener (cal)	Prot (g)	Carb (g)	Dietary Fiber (g)	Fat (g)	Fat Breakdown (g) Sat	Mono	Poly
	FATS and OILS—Continued											
	Salad dressing, no cholesterol											
1605	Miracle Whip	1 tbs	15	57	48	0	2	0	4	1.1	1.1	2.1
203	Salad dressing, from recipe, cooked	1 tbs	16	69	25	1	2	0	2	.5	.6	.3
200	Tartar sauce, regular	1 tbs	14	34	74	<1	1	<1	8	1.5	2.6	4.1
1503	Low calorie	1 tbs	14	63	31	<1	2	<1	2	.4	.6	1.3
201	Thousand island, regular	1 tbs	16	46	60	<1	2	0	6	1	1.3	3.2
202	Low calorie	1 tbs	15	69	24	<1	2	<1	2	.2	.4	.9
204	Vinegar and oil	1 tbs	16	47	72	0	<1	0	8	1.5	2.4	3.9
	Wishbone:											
2180	Creamy Italian, lite	1 tbs	15		26	<1	2		2	.4	.9	.7
2166	Italian, lite	1 tbs	16	90	6	0	1		<1	0	.2	.1
8427	Ranch, lite	1 tbs	15	56	50	0	2	0	4	.7		
	FRUITS and FRUIT JUICES											
	Apples:											
	Fresh, raw, with peel:											
205	2¾" diam (about 3 per lb w/cores)	1 ea	138	84	81	<1	21	4	<1	.1	t	.1
206	3¼" diam (about 2 per lb w/cores)	1 ea	212	84	125	<1	32	6	1	.1	t	.2
207	Raw, peeled slices	1 c	110	84	63	<1	16	2	<1	.1	t	.1
208	Dried, sulfured	10 ea	64	32	156	1	42	6	<1	t	t	.1
209	Apple juice, bottled or canned	1 c	248	88	117	<1	29	<1	<1	t	t	.1
210	Applesauce, sweetened	1 c	255	80	194	<1	51	3	<1	.1	t	.1
211	Applesauce, unsweetened	1 c	244	88	105	<1	28	3	<1	t	t	t
	Apricots:											
212	Raw, w/o pits (about 12 per lb w/pits)	3 ea	105	86	50	1	12	3	<1	t	.2	.1
	Canned (fruit and liquid):											
213	Heavy syrup	1 c	240	78	199	1	52	4	<1	t	.1	t
214	Halves	3 ea	120	78	100	1	26	2	<1	t	t	t
215	Juice pack	1 c	244	87	117	2	30	4	<1	t	t	t
216	Halves	3 ea	108	87	52	1	13	2	<1	t	t	t
217	Dried, halves	10 ea	35	31	83	1	22	3	<1	t	.1	t
218	Dried, cooked, unsweetened, w/liquid	1 c	250	76	213	3	55	8	<1	t	.2	.1
219	Apricot nectar, canned	1 c	251	85	141	1	36	2	<1	t	.1	t
	Avocados, raw, edible part only:											
220	California	1 ea	173	73	306	4	12	8	30	4.5	19.5	3.5
221	Florida	1 ea	304	80	340	5	27	16	27	5.3	14.8	4.5
222	Mashed, fresh, average	1 c	230	74	370	5	17	11	35	5.6	22.1	4.5
	Bananas, raw, without peel:											
223	Whole, 8¾" long (175g w/peel)	1 ea	118	74	109	1	28	3	1	.2	t	.1
224	Slices	1 c	150	74	138	2	35	4	1	.3	.1	.1
1285	Bananas, dehydrated slices	½ c	50	3	173	2	44	4	1	.3	.1	.2
225	Blackberries, raw	1 c	144	86	75	1	18	8	1	t	.1	.3
	Blueberries:											
226	Fresh	1 c	145	85	81	1	20	4	1	t	.1	.2
227	Frozen, sweetened	10 oz	284	77	230	1	62	6	<1	t	.1	.2
228	Frozen, thawed	1 c	230	77	186	1	51	5	<1	t	t	.1
	Cherries:											
229	Sour, red pitted, canned water pack	1 c	244	90	88	2	22	3	<1	.1	.1	.1
230	Sweet, red pitted, raw	10 ea	68	81	49	1	11	2	1	.1	.2	.2
231	Cranberry juice cocktail,vitamin C added	1 c	253	85	144	0	36	<1	<1	t	t	.1
1411	Cranberry juice, low calorie	1 c	237	95	45	0	11	<1	0	0	0	0
232	Cranberry-apple juice,vitamin C added	1 c	245	83	164	<1	42	<1	0	0	0	0
233	Cranberry sauce, canned, strained	1 c	277	61	418	1	108	3	<1	t	.1	.2
234	Dates, whole, without pits	10 ea	83	22	228	2	61	6	<1	.2	.1	t
235	Dates, chopped	1 c	178	22	490	4	131	13	1	.3	.3	.1
236	Figs, dried	10 ea	190	28	485	6	124	18	2	.4	.5	1.1
	Fruit cocktail, canned, fruit and liq:											
237	Heavy syrup pack	1 c	248	80	181	1	47	2	<1	t	t	.1
238	Juice pack	1 c	237	87	109	1	28	2	<1	t	t	t
	Grapefruit:											
	Raw 3¾" diam (half w/rind = 241g)											
239	Pink/red, half fruit, edible part	1 ea	123	91	37	1	9	2	<1	t	t	t
240	White, half fruit, edible part	1 ea	118	90	39	1	10	1	<1	t	t	t
241	Canned sections with light syrup	1 c	254	84	152	1	39	1	<1	t	t	.1

PAGE KEY: A–4 = Beverages A–6 = Dairy A–10 = Eggs A–10 = Fat/Oil A–14 = Fruit A–20 = Bakery A–28 = Grain A–32 = Fish A–34 = Meats A–38 = Poultry A–40 = Sausage A–42 = Mixed/Fast A–46 = Nuts/Seeds A–50 = Sweets A–52 = Vegetables/Legumes A–62 = Vegetarian Foods A–64 = Misc A–66 = Soups/Sauces A–68 = Fast A–84 = Convenience A–88 = Baby foods

Chol (mg)	Calc (mg)	Iron (mg)	Magn (mg)	Pota (mg)	Sodi (mg)	Zinc (mg)	VT-A (RE)	Thia (mg)	VT-E (a-TE)	Ribo (mg)	Niac (mg)	V-B6 (mg)	Fola (μg)	VT-C (mg)
0	0	<.01	0	0	102	0	2	0	.64	0	0	0	0	0
9	13	.08	1	19	117	.06	20	.01	.3	.02	.04	<.01	1	<1
7	3	.13		11	99	.02	9	<.01	2.24	<.01	0	<.01	1	<1
3	2	.09		5	83	.02	2	0	.83	<.01	.01	<.01	<1	<1
4	2	.1		18	112	.02	15	<.01	.18	<.01	<.01	<.01	1	0
2	2	.09		17	150	.02	14	<.01	.18	<.01	0	<.01	1	0
0	0	0	0	1	<1	0	0	0	1.41	0	0	0	0	0
<1	0	0			148		0	0	.56	0	0			0
0	1	0			255		2	0	.24	0	0			<1
2	0	0			120		0							0
0	10	.25	7	159	0	.05	7	.02	.44	.02	.11	.07	4	8
0	15	.38	11	244	0	.08	11	.04	.68	.03	.16	.1	6	12
0	4	.08	3	124	0	.04	4	.02	.09	.01	.1	.05	<1	4
0	9	.9	10	288	56	.13	4	0	.35	.1	.59	.08	0	2
0	17	.92	7	295	7	.07	<1	.05	.02	.04	.25	.07	<1	2
0	10	.89	8	156	8	.1	3	.03	.03	.07	.48	.07	2	4
0	7	.29	7	183	5	.07	7	.03	.02	.06	.46	.06	1	3
0	15	.57	8	311	1	.27	274	.03	.93	.04	.63	.06	9	10
0	22	.72	17	336	10	.26	295	.05	2.14	.05	.9	.13	4	7
0	11	.36	8	168	5	.13	148	.02	1.07	.03	.45	.06	2	4
0	29	.73	24	403	10	.27	412	.04	2.17	.05	.84	.13	4	12
0	13	.32	11	178	4	.12	183	.02	.96	.02	.37	.06	2	5
0	16	1.65	16	482	3	.26	253	<.01	.52	.05	1.05	.05	4	1
0	40	4.18	42	1222	7	.65	590	.01	1.25	.07	2.36	.28	0	4
0	18	.95	13	286	8	.23	331	.02	.2	.03	.65	.05	3	2
0	19	2.04	71	1096	21	.73	106	.19	2.32	.21	3.32	.48	113	14
0	33	1.61	103	1483	15	1.28	185	.33	2.37	.37	5.84	.85	162	24
0	25	2.35	90	1377	23	.97	140	.25	3.08	.28	4.42	.64	142	18
0	7	.37	34	467	1	.19	9	.05	.32	.12	.64	.68	22	11
0	9	.46	43	594	1	.24	12	.07	.4	.15	.81	.87	29	14
0	11	.57	54	746	1	.3	15	.09	0	.12	1.4	.22	7	3
0	46	.82	29	282	0	.39	23	.04	1.02	.06	.58	.08	49	30
0	9	.25	7	129	9	.16	14	.07	1.45	.07	.52	.05	9	19
0	17	1.11	6	170	3	.17	11	.06	2.02	.15	.72	.17	19	3
0	14	.9	5	138	2	.14	9	.05	1.63	.12	.58	.14	15	2
0	27	3.34	15	239	17	.17	183	.04	.32	.1	.43	.11	19	5
0	10	.26	7	152	0	.04	14	.03	.09	.04	.27	.02	3	5
0	8	.38	5	45	5	.18	1	.02	0	.02	.09	.05	<1	90
0	21	.09	5	52	7	.05	1	.02	0	.02	.08	.04	<1	76
0	17	.15	5	66	5	.1	1	.01	0	.05	.15	.05	<1	78
0	11	.61	8	72	80	.14	6	.04	.28	.06	.28	.04	3	6
0	27	.95	29	541	2	.24	4	.07	.08	.08	1.83	.16	10	0
0	57	2.05	62	1160	5	.52	9	.16	.18	.18	3.92	.34	22	0
0	274	4.24	112	1352	21	.97	25	.13	9.5	.17	1.32	.43	14	2
0	15	.72	12	218	15	.2	50	.04	.72	.05	.93	.12	6	5
0	19	.5	17	225	9	.21	73	.03	.47	.04	.95	.12	6	6
0	13	.15	10	159	0	.09	32	.04	.31	.02	.23	.05	15	47
0	14	.07	11	175	0	.08	1	.04	.29	.02	.32	.05	12	39
0	36	1.02	25	328	5	.2	0	.1	.63	.05	.62	.05	22	54

Table H-1

Food Composition

(Computer code number is for West Diet Analysis program) (For purposes of calculations, use "0" for t, <1, <.1, <.01, etc.)

Computer Code Number	Food Description	Measure	Wt (g)	H_2O (%)	Ener (cal)	Prot (g)	Carb (g)	Dietary Fiber (g)	Fat (g)	Fat Breakdown (g) Sat	Mono	Poly
	FRUITS and FRUIT JUICES											
	Grapefruit juice:											
242	Fresh, white, raw	1 c	247	90	96	1	23	<1	<1	t	t	.1
243	Canned, unsweetened	1 c	247	90	94	1	22	<1	<1	t	t	.1
244	Sweetened	1 c	250	87	115	1	28	<1	<1	t	t	.1
	Frozen concentrate, unsweetened:											
245	Undiluted, 6-fl-oz can	¾ c	207	62	302	4	72	1	1	.1	.1	.2
246	Diluted with 3 cans water	1 c	247	89	101	1	24	<1	<1	t	t	.1
	Grapes, raw European (adherent skin):											
247	Thompson seedless	10 ea	50	81	35	<1	9	<1	<1	.1	t	.1
248	Tokay/Emperor, seeded types	10 ea	50	81	35	<1	9	<1	<1	.1	t	.1
	Grape juice:											
249	Bottled or canned	1 c	253	84	154	1	38	<1	<1	.1	t	.1
	Frozen concentrate, sweetened:											
250	Undiluted, 6-fl-oz can, vit C added	¾ c	216	54	387	1	96	1	1	.2	t	.2
251	Diluted with 3 cans water, vit C added	1 c	250	87	128	<1	32	<1	<1	.1	t	.1
1410	Low calorie	1 c	253	84	154	1	38	<1	<1	.1	t	.1
252	Kiwi fruit, raw, peeled (88 g with peel)	1 ea	76	83	46	1	11	3	<1	t	t	.2
253	Lemons, raw, without peel and seeds (about 4 per lb whole)	1 ea	58	89	17	1	5	2	<1	t	t	.1
	Lemon juice:											
254	Fresh:	1 c	244	91	61	1	21	1	0	0	0	0
255	Tablespoon	1 tbs	15	91	4	<1	1	<1	0	0	0	0
256	Canned or bottled, unsweetened:	1 c	244	92	51	1	16	1	1	.1	t	.2
257	Tablespoon	1 tbs	15	92	3	<1	1	<1	<1	t	t	t
258	Frozen, single strength, unsweetened:	1 c	244	92	54	1	16	1	1	.1	t	.2
2298	Tablespoon	1 tbs	15	92	3	<1	1	<1	<1	t	t	t
	Lime juice:											
260	Fresh:	1 c	246	90	66	1	22	1	<1	t	t	.1
261	Tablespoon	1 tbs	15	90	4	<1	1	<1	<1	t	t	t
262	Canned or bottled, unsweetened	1 c	246	92	52	1	16	1	1	.1	.1	.2
263	Mangos, raw, edible part (300 g w/skin & seeds)	1 ea	207	82	135	1	35	4	1	.1	.2	.1
	Melons, raw, without rind and contents:											
264	Cantaloupe, 5" diam (2⅓ lb whole with refuse), orange flesh	½ ea	276	90	97	2	23	2	1	.2	t	.3
265	Honeydew, 6½" diam (5¼ lb whole with refuse), slice = ⅒ melon	1 pce	160	90	56	1	15	1	<1	t	t	.1
266	Nectarines, raw, w/o pits, 2¼" diam	1 ea	136	86	67	1	16	2	1	.1	.2	.3
	Oranges, raw:											
267	Whole w/o peel and seeds, 2⅝" diam (180 g with peel and seeds)	1 ea	131	87	62	1	15	3	<1	t	t	t
268	Sections, without membranes	1 c	180	87	85	2	21	4	<1	t	t	t
	Orange juice:											
269	Fresh, all varieties	1 c	248	88	112	2	26	<1	<1	.1	.1	.1
270	Canned, unsweetened	1 c	249	89	105	1	24	<1	<1	t	.1	.1
271	Chilled	1 c	249	88	110	2	25	<1	1	.1	.1	.2
	Frozen concentrate:											
272	Undiluted (6-oz can)	¾ c	213	58	339	5	81	2	<1	.1	.1	.1
273	Diluted w/3 parts water by volume	1 c	249	88	112	2	27	<1	<1	t	t	t
1345	Orange juice, from dry crystals	1 c	248	88	114	0	29	0	0	0	0	0
274	Orange and grapefruit juice, canned	1 c	247	89	106	1	25	<1	<1	t	t	t
	Papayas, raw:											
275	½" slices	1 c	140	89	55	1	14	3	<1	.1	.1	t
276	Whole, 3½" diam by 5⅛" w/o seeds and skin (1 lb w/refuse)	1 ea	304	89	119	2	30	5	<1	.1	.1	.1
1031	Papaya nectar, canned	1 c	250	85	143	<1	36	1	<1	.1	.1	.1
	Peaches:											
277	Raw, whole, 2½" diam, peeled, pitted (about 4 per lb whole)	1 ea	98	88	42	1	11	2	<1	t	t	t
278	Raw, sliced	1 c	170	88	73	1	19	3	<1	t	.1	.1
	Canned, fruit and liquid:											
279	Heavy syrup pack:	1 c	262	79	194	1	52	3	<1	t	.1	.1
280	Half	1 ea	98	79	72	<1	19	1	<1	t	t	t

Chol (mg)	Calc (mg)	Iron (mg)	Magn (mg)	Pota (mg)	Sodi (mg)	Zinc (mg)	VT-A (RE)	Thia (mg)	VT-E (a-TE)	Ribo (mg)	Niac (mg)	V-B6 (mg)	Fola (μg)	VT-C (mg)
0	22	.49	30	400	2	.12	2	.1	.12	.05	.49	.11	25	94
0	17	.49	25	378	2	.22	2	.1	.12	.05	.57	.05	26	72
0	20	.9	25	405	5	.15	0	.1	.12	.06	.8	.05	26	67
0	56	1.01	79	1001	6	.37	6	.3	.37	.16	1.6	.32	26	248
0	20	.35	27	336	2	.12	2	.1	.12	.05	.54	.11	9	83
0	5	.13	3	92	1	.02	3	.05	.35	.03	.15	.05	2	5
0	5	.13	3	92	1	.02	3	.05	.35	.03	.15	.05	2	5
0	23	.61	25	334	8	.13	3	.07	0	.09	.66	.16	7	<1
0	28	.78	32	160	15	.28	6	.11	.38	.2	.93	.32	9	179
0	10	.25	10	52	5	.1	2	.04	.12	.06	.31	.1	3	60
0	23	.61	25	334	8	.13	3	.07	0	.09	.66	.16	7	<1
0	20	.31	23	252	4	.13	14	.01	.85	.04	.38	.07	29	74
0	15	.35	5	80	1	.03	2	.02	.14	.01	.06	.05	6	31
0	17	.07	15	303	2	.12	5	.07	.22	.02	.24	.12	31	112
0	1	<.01	1	19	<1	.01	<1	<.01	.01	<.01	.01	.01	2	7
0	27	.32	19	249	51	.15	5	.1	.22	.02	.48	.1	25	60
0	2	.02	1	15	3	.01	<1	.01	.01	<.01	.03	.01	2	4
0	19	.29	19	217	2	.12	2	.14	.22	.03	.33	.15	23	77
0	1	.02	1	13	<1	.01	<1	.01	.01	<.01	.02	.01	1	5
0	22	.07	15	268	2	.15	2	.05	.22	.02	.25	.11	20	72
0	1	<.01	1	16	<1	.01	<1	<.01	.01	<.01	.01	.01	1	4
0	29	.57	17	185	39	.15	5	.08	.17	.01	.4	.07	19	16
0	21	.27	19	323	4	.08	805	.12	2.32	.12	1.21	.28	29	57
0	30	.58	30	853	25	.44	889	.1	.41	.06	1.58	.32	47	116
0	10	.11	11	434	16	.11	6	.12	.24	.03	.96	.09	10	40
0	7	.2	11	288	0	.12	101	.02	1.21	.06	1.35	.03	5	7
0	52	.13	13	237	0	.09	27	.11	.31	.05	.37	.08	40	70
0	72	.18	18	326	0	.13	38	.16	.43	.07	.51	.11	54	96
0	27	.5	27	496	2	.12	50	.22	.22	.07	.99	.1	75	124
0	20	1.1	27	436	5	.17	45	.15	.22	.07	.78	.22	45	86
0	25	.42	27	473	2	.1	20	.28	.47	.05	.7	.13	45	82
0	68	.75	72	1435	6	.38	60	.6	.68	.14	1.53	.33	330	294
0	22	.25	25	473	2	.12	20	.2	.47	.04	.5	.11	109	97
0	62	.2	2	50	12	.1	551	<.01	0	.04	0	0	143	121
0	20	1.14	25	390	7	.17	30	.14	.17	.07	.83	.06	35	72
0	34	.14	14	360	4	.1	39	.04	1.57	.04	.47	.03	53	86
0	73	.3	30	781	9	.21	85	.08	3.4	.1	1.03	.06	116	188
0	25	.85	7	77	12	.37	27	.01	.05	.01	.37	.02	5	7
0	5	.11	7	193	0	.14	53	.02	.69	.04	.97	.02	3	6
0	8	.19	12	335	0	.24	92	.03	1.19	.07	1.68	.03	6	11
0	8	.71	13	241	16	.24	86	.03	2.33	.06	1.61	.05	8	7
0	3	.26	5	90	6	.09	32	.01	.87	.02	.6	.02	3	3

Table H-1

Food Composition (Computer code number is for West Diet Analysis program) (For purposes of calculations, use "0" for t, <1, <.1, <.01, etc.)

Computer Code Number	Food Description	Measure	Wt (g)	H_2O (%)	Ener (cal)	Prot (g)	Carb (g)	Dietary Fiber (g)	Fat (g)	Fat Breakdown (g) Sat	Mono	Poly
	FRUITS and FRUIT JUICES—Continued											
281	Juice pack:	1 c	248	87	109	2	29	3	<1	t	t	t
282	Half	1 ea	98	87	43	1	11	1	<1	t	t	t
283	Dried, uncooked	10 ea	130	32	311	5	80	11	1	.1	.4	.5
284	Dried, cooked, fruit and liquid	1 c	258	78	199	3	51	7	1	.1	.2	.3
	Frozen, slice, sweetened:											
285	10-oz package,vitamin C added	1 ea	284	75	267	2	68	5	<1	t	.1	.2
286	Cup, thawed measure, vitamin C added	1 c	250	75	235	2	60	4	<1	t	.1	.2
1032	Peach nectar, canned	1 c	249	86	134	1	35	1	<1	t	t	t
	Pears:											
	Fresh, with skin, cored:											
287	Bartlett, 2½" diam (about 2½ per lb)	1 ea	166	84	98	1	25	4	1	t	.1	.2
288	Bosc, 2⅛" diam (about 3 per lb)	1 ea	139	84	82	1	21	3	1	t	.1	.1
289	D'Anjou, 3" diam (about 2 per lb)	1 ea	209	84	123	1	32	5	1	t	.2	.2
	Canned, fruit and liquid:											
290	Heavy syrup pack:	1 c	266	80	197	1	51	4	<1	t	.1	.1
291	Half	1 ea	76	80	56	<1	15	1	<1	t	t	t
292	Juice pack:	1 c	248	86	124	1	32	4	<1	t	t	t
293	Half	1 ea	76	86	38	<1	10	1	<1	t	t	t
294	Dried halves	10 ea	175	27	459	3	122	13	1	.1	.2	.3
1033	Pear nectar, canned	1 c	250	84	150	<1	39	1	<1	t	t	t
	Pineapple:											
295	Fresh chunks, diced	1 c	155	86	76	1	19	2	1	t	.1	.2
	Canned, fruit and liquid:											
	Heavy syrup pack:											
296	Crushed, chunks, tidbits	½ c	127	79	99	<1	26	1	<1	t	t	.1
297	Slices	1 ea	49	79	38	<1	10	<1	<1	t	t	t
298	Juice pack, crushed, chunks, tidbits	1 c	250	83	150	1	39	2	<1	t	t	.1
299	Juice pack, slices	1 ea	47	83	28	<1	7	<1	<1	t	t	t
300	Pineapple juice, canned, unsweetened	1 c	250	85	140	1	34	<1	<1	t	t	.1
	Plantains, yellow flesh, without peel:											
301	Raw slices (whole=179 g w/o peel)	1 c	148	65	181	2	47	3	1	.2	t	.1
302	Cooked, boiled, sliced	1 c	154	67	179	1	48	4	<1	.1	t	.1
	Plums:											
303	Fresh, medium, 2⅛" diam	1 ea	66	85	36	1	9	1	<1	t	.3	.1
304	Fresh, small, 1½" diam	1 ea	28	85	15	<1	4	<1	<1	t	.1	t
	Canned, purple, with liquid:											
305	Heavy syrup pack:	1 c	258	76	230	1	60	3	<1	t	.2	.1
306	Plums	3 ea	138	76	123	<1	32	1	<1	t	.1	t
307	Juice pack:	1 c	252	84	146	1	38	3	<1	t	t	t
308	Plums	3 ea	138	84	80	1	21	1	<1	t	t	t
1698	Pomegranate, fresh	1 ea	154	81	105	1	26	1	<1	.1	.1	.1
	Prunes, dried, pitted:											
309	Uncooked (10 = 97 g w/pits, 84 g w/o pits)	10 ea	84	32	201	2	53	6	<1	t	.3	.1
310	Cooked, unsweetened, fruit & liq (250 g w/pits)	1 c	248	70	265	3	70	16	1	t	.4	.1
311	Prune juice, bottled or canned	1 c	256	81	182	2	45	3	<1	t	.1	t
	Raisins, seedless:											
312	Cup, not pressed down	1 c	145	15	435	5	115	6	1	.2	t	.2
313	One packet, ½ oz	½ oz	14	15	42	<1	11	1	<1	t	t	t
	Raspberries:											
314	Fresh	1 c	123	87	60	1	14	8	1	t	.1	.4
315	Frozen, sweetened	10 oz	284	73	293	2	74	12	<1	t	t	.3
316	Cup, thawed measure	1 c	250	73	258	2	65	11	<1	t	t	.2
317	Rhubarb, cooked, added sugar	1 c	240	68	278	1	75	5	<1	t	t	.1
	Strawberries:											
318	Fresh, whole, capped	1 c	144	92	43	1	10	3	1	t	.1	.3
	Frozen, sliced, sweetened:											
319	10-oz container	10 oz	284	73	273	2	74	5	<1	t	.1	.2
320	Cup, thawed measure	1 c	255	73	245	1	66	5	<1	t	t	.2
	Tangerines, without peel and seeds:											
321	Fresh (2⅜" whole) 116 g w/refuse	1 ea	84	88	37	1	9	2	<1	t	t	t
322	Canned, light syrup, fruit and liquid	1 c	252	83	154	1	41	2	<1	t	t	t

PAGE KEY: A–4 = Beverages A–6 = Dairy A–10 = Eggs A–10 = Fat/Oil A–14 = Fruit A–20 = Bakery A–28 = Grain A–32 = Fish A–34 = Meats A–38 = Poultry A–40 = Sausage A–42 = Mixed/Fast A–46 = Nuts/Seeds A–50 = Sweets A–52 = Vegetables/Legumes A–62 = Vegetarian Foods A–64 = Misc A–66 = Soups/Sauces A–68 = Fast A–84 = Convenience A–88 = Baby foods

Chol (mg)	Calc (mg)	Iron (mg)	Magn (mg)	Pota (mg)	Sodi (mg)	Zinc (mg)	VT-A (RE)	Thia (mg)	VT-E (a-TE)	Ribo (mg)	Niac (mg)	V-B6 (mg)	Fola (μg)	VT-C (mg)
0	15	.67	17	317	10	.27	94	.02	3.72	.04	1.44	.05	8	9
0	6	.26	7	125	4	.11	37	.01	1.47	.02	.57	.02	3	4
0	36	5.28	55	1294	9	.74	281	<.01	0	.28	5.69	.09	<1	6
0	23	3.38	33	826	5	.46	52	.01	0	.05	3.92	.1	<1	10
0	9	1.05	14	369	17	.14	79	.04	2.53	.1	1.85	.05	9	268
0	7	.92	12	325	15	.12	70	.03	2.23	.09	1.63	.04	8	236
0	12	.47	10	100	17	.2	65	.01	.2	.03	.72	.02	3	13
0	18	.41	10	208	0	.2	3	.03	.83	.07	.17	.03	12	7
0	15	.35	8	174	0	.17	3	.03	.69	.06	.14	.02	10	6
0	23	.52	12	261	0	.25	4	.04	1.05	.08	.21	.04	15	8
0	13	.58	11	173	13	.21	0	.03	1.33	.06	.64	.04	3	3
0	4	.17	3	49	4	.06	0	.01	.38	.02	.18	.01	1	1
0	22	.72	17	238	10	.22	2	.03	1.24	.03	.5	.03	3	4
0	7	.22	5	73	3	.07	1	.01	.38	.01	.15	.01	1	1
0	59	3.68	58	933	10	.68	1	.01	0	.25	2.4	.13	0	12
0	12	.65	7	32	10	.17	<1	<.01	.25	.03	.32	.03	3	3
0	11	.57	22	175	2	.12	3	.14	.15	.06	.65	.13	16	24
0	18	.48	20	132	1	.15	1	.11	.13	.03	.36	.09	6	9
0	7	.19	8	51	<1	.06	<1	.04	.05	.01	.14	.04	2	4
0	35	.7	35	305	2	.25	10	.24	.25	.05	.71	.18	12	24
0	7	.13	7	57	<1	.05	2	.04	.05	.01	.13	.03	2	4
0	42	.65	32	335	2	.27	1	.14	.05	.05	.64	.24	58	27
0	4	.89	55	739	6	.21	167	.08	.4	.08	1.02	.44	33	27
0	3	.89	49	716	8	.2	140	.07	.22	.08	1.16	.37	40	17
0	3	.07	5	114	0	.07	21	.03	.4	.06	.33	.05	1	6
0	1	.03	2	48	0	.03	9	.01	.17	.03	.14	.02	1	3
0	23	2.17	13	235	49	.18	67	.04	1.81	.1	.75	.07	6	1
0	12	1.16	7	126	26	.1	36	.02	.97	.05	.4	.04	3	1
0	25	.86	20	388	3	.28	255	.06	1.76	.15	1.19	.07	7	7
0	14	.47	11	213	1	.15	139	.03	.97	.08	.65	.04	4	4
0	5	.46	5	399	5	.18	0	.05	.85	.05	.46	.16	9	9
0	43	2.08	38	626	3	.44	167	.07	1.22	.14	1.65	.22	3	3
0	57	2.75	50	828	5	.59	77	.06	<.01	.25	1.79	.54	<1	7
0	31	3.02	36	707	10	.54	1	.04	.03	.18	2.01	.56	1	10
0	71	3.02	48	1088	17	.39	1	.23	1.02	.13	1.19	.36	5	5
0	7	.29	5	105	2	.04	<1	.02	.1	.01	.11	.03	<1	<1
0	27	.7	22	187	0	.57	16	.04	.55	.11	1.11	.07	32	31
0	43	1.85	37	324	3	.51	17	.05	1.28	.13	.65	.1	74	47
0	37	1.63	32	285	2	.45	15	.05	1.13	.11	.57	.08	65	41
0	348	.5	29	230	2	.19	17	.04	.48	.05	.48	.05	13	8
0	20	.55	14	239	1	.19	4	.03	.2	.09	.33	.08	25	82
0	31	1.68	20	278	9	.17	6	.04	.4	.14	1.14	.08	42	118
0	28	1.5	18	250	8	.15	5	.04	.36	.13	1.02	.08	38	106
0	12	.08	10	132	1	.2	77	.09	.2	.02	.13	.06	17	26
0	18	.93	20	197	15	.6	212	.13	.86	.11	1.12	.11	12	50

Table H-1

Food Composition (Computer code number is for West Diet Analysis program) (For purposes of calculations, use "0" for t, <1, <.1, <.01, etc.)

Computer Code Number	Food Description	Measure	Wt (g)	H_2O (%)	Ener (cal)	Prot (g)	Carb (g)	Dietary Fiber (g)	Fat (g)	Fat Breakdown (g) Sat	Mono	Poly
	FRUITS and FRUIT JUICES—Continued											
323	Tangerine juice, canned, sweetened	1 c	249	87	125	1	30	<1	<1	t	t	.1
	Watermelon, raw, without rind and seeds:											
324	Piece, 1/16 wedge	1 pce	286	91	91	2	20	1	1	.1	.3	.4
325	Diced	1 c	152	91	49	1	11	1	1	.1	.2	.2
	BAKED GOODS: BREADS, CAKES, COOKIES, CRACKERS, PIES											
326	Bagels, plain, enriched, 3½" diam.	1 ea	71	33	195	7	38	2	1	.2	.1	.5
1663	Bagel, oat bran	1 ea	71	33	181	8	38	3	1	.1	.2	.3
	Biscuits:											
327	From home recipe	1 ea	60	29	212	4	27	1	10	2.6	4.2	2.5
328	From mix	1 ea	57	29	191	4	28	1	7	1.6	2.4	2.5
329	From refrigerated dough	1 ea	74	27	276	4	34	1	13	8.7	3.4	.5
330	Bread crumbs, dry, grated (see # 364, 365 for soft crumbs)	1 c	108	6	427	13	78	3	6	1.4	2.3	1.7
2087	Bread sticks, brown & serve	1 ea	57	34	150	7	28	1	1	.5	.5	.5
	Breads:											
331	Boston brown, canned, 3¼" slice	1 pce	45	47	88	2	19	2	1	.1	.1	.3
332	Cracked wheat (¼ cracked-wheat & ¾ enr wheat flour): 1-lb loaf	1 ea	454	36	1180	39	225	25	18	4.2	8.6	3.1
333	Slice (18 per loaf)	1 pce	25	36	65	2	12	1	1	.2	.5	.2
334	Slice, toasted	1 pce	23	30	65	2	12	1	1	.2	.5	.2
335	French/Vienna, enriched: 1-lb loaf	1 ea	454	34	1243	40	236	14	14	2.9	5.5	3.1
337	Slice, 4¾ x 4 x ½"	1 pce	25	34	68	2	13	1	1	.2	.3	.2
336	French, slice, 5 x 2½"	1 pce	25	34	68	2	13	1	1	.2	.3	.2
	French toast: see Mixed Dishes, and Fast Foods, #691											
2083	Honey wheatberry	1 pce	38	38	100	3	18	2	1	0	.5	
338	Italian, enriched: 1-lb loaf	1 ea	454	36	1230	40	227	12	16	3.9	3.7	6.3
339	Slice, 4½ x 3¼ x ¾"	1 pce	30	36	81	3	15	1	1	.3	.2	.4
340	Mixed grain, enriched: 1-lb loaf	1 ea	454	38	1135	45	211	29	17	3.7	6.9	4.2
341	Slice (18 per loaf)	1 pce	26	38	65	3	12	2	1	.2	.4	.2
342	Slice, toasted	1 pce	24	32	65	3	12	2	1	.2	.4	.2
343	Oatmeal, enriched: 1-lb loaf	1 ea	454	37	1221	38	220	18	20	3.2	7.2	7.7
344	Slice (18 per loaf)	1 pce	27	37	73	2	13	1	1	.2	.4	.5
345	Slice, toasted	1 pce	25	31	73	2	13	1	1	.2	.4	.5
346	Pita pocket bread, enr, 6½" round	1 ea	60	32	165	5	33	1	1	.1	.1	.3
	Pumpernickel (⅔ rye & ⅓ enr wheat flr):											
347	1-lb loaf	1 ea	454	38	1135	40	216	29	14	2	4.2	5.6
348	Slice, 5 x 4 x ⅜"	1 pce	26	38	65	2	12	2	1	.1	.2	.3
349	Slice, toasted	1 pce	29	32	80	3	15	2	1	.1	.3	.4
350	Raisin, enriched: 1-lb loaf	1 ea	454	34	1243	36	237	19	20	4.9	10.5	3.1
351	Slice (18 per loaf)	1 pce	26	34	71	2	14	1	1	.3	.6	.2
352	Slice, toasted	1 pce	24	28	71	2	14	1	1	.3	.6	.2
353	Rye, light (⅓ rye & ⅔ enr wheat flr): 1-lb loaf	1 ea	454	37	1175	39	219	26	15	2.9	6	3.6
354	Slice, 4¾ x 3¾ x 7/16"	1 pce	32	37	83	3	15	2	1	.2	.4	.3
355	Slice, toasted	1 pce	24	31	68	2	13	2	1	.2	.3	.2
356	Wheat (enr wheat & whole-wheat flour): 1-lb loaf	1 ea	454	37	1160	43	213	25	19	3.9	7.3	4.5
357	Slice (18 per loaf)	1 pce	25	37	65	2	12	1	1	.2	.4	.2
358	Slice, toasted	1 pce	23	32	65	2	12	1	1	.2	.4	.2
359	White, enriched: 1-lb loaf	1 ea	454	35	1293	36	225	9	26	5.4	5.9	12.6
360	Slice	1 pce	42	35	120	3	21	1	2	.5	.5	1.2
361	Slice, toasted	1 pce	38	29	119	3	21	1	2	.5	.5	1.2
366	Whole-wheat: 1-lb loaf	1 ea	454	38	1116	44	209	31	19	4.2	7.6	4.5
367	Slice (16 per loaf)	1 pce	28	38	69	3	13	2	1	.3	.5	.3
368	Slice, toasted	1 pce	25	30	69	3	13	2	1	.3	.5	.3
	Bread stuffing, prepared from mix:											
369	Dry type	1 c	200	65	356	6	43	6	17	3.5	7.6	5.2
370	Moist type, with egg and margarine	1 c	232	65	390	9	51	5	17	3.4	7.4	4.9

Chol (mg)	Calc (mg)	Iron (mg)	Magn (mg)	Pota (mg)	Sodi (mg)	Zinc (mg)	VT-A (RE)	Thia (mg)	VT-E (a-TE)	Ribo (mg)	Niac (mg)	V-B6 (mg)	Fola (μg)	VT-C (mg)
0	45	.5	20	443	2	.07	105	.15	.22	.05	.25	.08	11	55
0	23	.49	31	332	6	.2	106	.23	.43	.06	.57	.41	6	27
0	12	.26	17	176	3	.11	56	.12	.23	.03	.3	.22	3	15
0	52	2.53	21	72	379	.62	0	.38	.02	.22	3.24	.04	62	0
0	9	2.19	40	145	360	1.48	<1	.23	.17	.24	2.1	.14	57	<1
2	141	1.74	11	73	348	.32	14	.21	1.45	.19	1.77	.02	37	<1
2	105	1.17	14	107	544	.35	15	.2	.23	.2	1.72	.04	3	<1
5	89	1.64	9	87	584	.29	24	.27	.44	.18	1.63	.03	6	0
0	245	6.61	50	239	931	1.32	<1	.83	.95	.47	7.4	.11	118	0
0	60	2.7			290		0	.22		.1	1.6			0
<1	31	.94	28	143	284	.22	5	.01	.13	.05	.5	.04	5	0
														0
0	195	12.8	236	804	2442	5.63	0	1.63	2.56	1.09	16.7	1.38	277	
0	11	.7	13	44	135	.31	0	.09	.14	.06	.92	.08	15	0
0	11	.7	13	44	135	.31	0	.07	.14	.05	.83	.07	7	0
0	341	11.5	123	513	2764	3.95	0	2.36	1.07	1.49	21.6	.19	431	0
0	19	.63	7	28	152	.22	0	.13	.06	.08	1.19	.01	24	0
0	19	.63	7	28	152	.22	0	.13	.06	.08	1.19	.01	24	0
0	20	.72			200		0	.12	.24	.07	.8			0
0	354	13.3	123	499	2651	3.9	0	2.15	1.26	1.33	19.9	.22	431	0
0	23	.88	8	33	175	.26	0	.14	.08	.09	1.31	.01	28	0
0	413	15.8	241	926	2210	5.77	0	1.85	2.79	1.55	19.8	1.51	363	1
0	24	.9	14	53	127	.33	0	.11	.16	.09	1.14	.09	21	<1
0	24	.9	14	53	127	.33	0	.08	.16	.08	1.02	.08	16	<1
0	300	12.3	168	645	2719	4.63	9	1.81	1.56	1.09	14.3	.31	281	2
0	18	.73	10	38	162	.27	1	.11	.09	.06	.85	.02	17	<1
0	18	.73	10	38	163	.28	<1	.09	.09	.06	.77	.02	13	<1
0	52	1.57	16	72	322	.5	0	.36	.02	.2	2.78	.02	57	0
0	309	13	245	944	3046	6.72	0	1.48	2.3	1.38	14	.57	363	0
0	18	.75	14	54	174	.38	0	.08	.13	.08	.8	.03	21	0
0	21	.91	17	66	214	.47	0	.08	.17	.09	.89	.04	20	0
0	300	13.2	118	1030	1770	3.27	1	1.54	3.44	1.81	15.8	.31	395	2
0	17	.75	7	59	101	.19	<1	.09	.2	.1	.9	.02	23	<1
0	17	.76	7	59	102	.19	<1	.07	.2	.09	.81	.02	18	<1
0	331	12.8	182	754	2996	5.18	2	1.97	2.51	1.52	17.3	.34	390	1
0	23	.91	13	53	211	.36	<1	.14	.18	.11	1.22	.02	27	<1
0	19	.74	10	44	174	.3	<1	.09	.15	.08	.9	.02	17	<1
0	572	15.8	209	627	2447	4.77	0	2.09	3	1.45	20.5	.49	204	0
0	26	.83	11	50	133	.26	0	.1	.14	.07	1.03	.02	19	0
0	26	.83	11	50	132	.26	0	.08	.14	.06	.93	.02	15	0
14	259	13.5	86	663	1629	2.91	100	1.84	4.95	1.74	16.3	.23	413	1
1	24	1.25	8	61	151	.27	9	.17	.46	.16	1.51	.02	38	<1
1	24	1.24	8	61	150	.27	8	.13	.46	.14	1.35	.02	12	<1
0	327	15	390	1144	2392	8.81	0	1.59	4.72	.93	17.4	.81	227	0
0	20	.92	24	71	148	.54	0	.1	.29	.06	1.08	.05	14	0
0	20	.93	24	71	148	.54	0	.08	.23	.05	.97	.04	9	0
0	64	2.18	24	148	1086	.56	162	.27	2.8	.21	2.96	.08	202	0
0	148	3.8	35	304	1069	.74	160	.39	2.78	.33	3.69	.12	39	4

Table H-1

Food Composition (Computer code number is for West Diet Analysis program) (For purposes of calculations, use "0" for t, <1, <.1, <.01, etc.)

Computer Code Number	Food Description	Measure	Wt (g)	H_2O (%)	Ener (cal)	Prot (g)	Carb (g)	Dietary Fiber (g)	Fat (g)	Fat Breakdown (g) Sat	Mono	Poly
	BAKED GOODS: BREADS, CAKES, COOKIES, CRACKERS, PIES—Continued											
	Cakes, prepared from mixes using enriched flour and veg shortening, w/frostings made from margarine:											
	Angel food:											
371	Whole cake, 9 ¾" diam tube	1 ea	340	33	877	20	197	5	3	.4	.2	1.2
372	Piece, 1⁄12 of cake	1 pce	28	33	72	2	16	<1	<1	t	t	.1
373	Boston cream pie, ⅛ of cake	1 pce	123	45	310	3	53	2	10	3.1	5.4	1.2
	Coffee cake:											
374	Whole cake, 7¾ x 5⅝ x 1¼"	1 ea	336	30	1068	18	177	4	32	6.3	13	10.7
375	Piece, ⅙ of cake	1 pce	56	30	178	3	30	1	5	1	2.2	1.8
	Devil's food, chocolate frosting:											
376	Whole cake, 2 layer, 8 or 9" diam	1 ea	1021	23	3747	42	557	29	167	47.9	91.9	19.5
377	Piece, 1⁄16 of cake	1 pce	64	23	235	3	35	2	10	3	5.8	1.2
378	Cupcake, 2½" diam	1 ea	42	23	154	2	23	1	7	2	3.8	.8
	Gingerbread:											
379	Whole cake, 8" square	1 ea	603	33	1863	24	306	7	61	15.8	34	8.1
380	Piece, 1⁄9 of cake	1 pce	67	33	207	3	34	1	7	1.8	3.8	.9
	Yellow, chocolate frosting, 2 layer:											
381	Whole cake, 8 or 9" in diam	1 ea	1024	22	3880	39	567	18	178	49	99	21.4
382	Piece, 1⁄16 of cake	1 pce	64	22	243	2	35	1	11	3.1	6.2	1.3
	Cakes from home recipes w/enr flour:											
	Carrot cake, made with veg oil, cream cheese frosting:											
383	Whole, 9 x 13" cake	1 ea	1776	21	7743	82	838	21	469	86.8	116	242
384	Piece, 1⁄16 of cake, 2¼ x 3¼" slice	1 pce	111	21	484	5	52	1	29	5.4	7.2	15.1
	Fruitcake, dark:											
385	Whole cake, 7½"diam tube, 2¼"high	1 ea	1376	25	4458	40	848	51	125	15.4	57.4	44.6
386	Piece, 1⁄32 of cake, ⅔" arc	1 pce	43	25	139	1	26	2	4	.5	1.8	1.4
	Sheet, plain, made w/veg shortening, no frosting:											
387	Whole cake, 9" square	1 ea	774	24	2817	35	433	3	108	29.9	51.5	25.5
388	Piece, 1⁄9 of cake	1 pce	86	24	313	4	48	<1	12	3.3	5.7	2.8
	Sheet, plain, made w/margarine, uncooked white frosting:											
389	Whole cake, 9" square	1 ea	576	22	2148	20	339	2	83	13.8	35.5	29.5
390	Piece, 1⁄9 of cake	1 pce	64	22	239	2	38	<1	9	1.5	3.9	3.3
	Cakes, commerical:											
	Cheesecake:											
401	Whole cake, 9" diam	1 ea	960	46	3081	53	245	4	216	111	74.4	13.2
402	Piece, 1⁄12 of cake	1 pce	80	46	257	4	20	<1	18	9.2	6.2	1.1
	Pound cake:											
393	Loaf, 8½ x 3½ x 3"	1 ea	340	25	1319	19	166	2	68	38.1	19	3.7
394	Slice, 1⁄17 of loaf, 2" slice	1 pce	28	25	109	2	14	<1	6	3.1	1.6	.3
	Snack cakes:											
395	Chocolate w/creme filling,Ding Dong	1 ea	50	20	188	2	30	<1	7	1.6	2.7	2.1
396	Sponge cake w/creme filling,Twinkie	1 ea	43	20	157	1	27	<1	5	1.2	1.9	1.5
1677	Sponge cake, 1⁄12 of 12" cake	1 pce	38	30	110	2	23	<1	1	.3	.4	.2
	White, white frosting, 2 layer:											
397	Whole cake, 8 or 9" diam	1 ea	1136	20	4260	37	716	11	153	68.2	60.1	15.4
398	Piece, 1⁄16 of cake	1 pce	71	20	266	2	45	1	10	4.3	3.8	1
	Yellow, chocolate frosting, 2 layer:											
399	Whole cake, 8 or 9" in diam	1 ea	1024	22	3880	39	567	18	178	49	99	21.4
400	Piece, 1⁄16 of cake	1 pce	64	22	243	2	35	1	11	3.1	6.2	1.3
1332	Bagel chips	5 pce	70	3	298	6	52	6	7	1.2	2	3.4
2225	Bagel chips, onion garlic, toasted	1 oz	28		193	5	31	3	8	1.7	5.2	0
1035	Cheese puffs/Cheetos	1 c	20	1	111	2	11	<1	7	1.3	4.1	1
	Cookies made with enriched flour:											
	Brownies with nuts:											
403	Commercial w/frosting, 1½ x 1¾ x ⅞"	1 ea	61	14	247	3	39	1	10	2.6	5.1	1.6
1902	Fat free fudge, Entenmann's	1 pce	40	24	110	2	27	1	0	0	0	0

PAGE KEY: A–4 = Beverages A–6 = Dairy A–10 = Eggs A–10 = Fat/Oil A–14 = Fruit A–20 = Bakery A–28 = Grain A–32 = Fish A–34 = Meats A–38 = Poultry A–40 = Sausage A–42 = Mixed/Fast A–46 = Nuts/Seeds A–50 = Sweets A–52 = Vegetables/Legumes A–62 = Vegetarian Foods A–64 = Misc A–66 = Soups/Sauces A–68 = Fast A–84 = Convenience A–88 = Baby foods

Chol (mg)	Calc (mg)	Iron (mg)	Magn (mg)	Pota (mg)	Sodi (mg)	Zinc (mg)	VT-A (RE)	Thia (mg)	VT-E (a-TE)	Ribo (mg)	Niac (mg)	V-B6 (mg)	Fola (µg)	VT-C (mg)
0	476	1.77	41	316	2546	.24	0	.35	.34	1.67	3	.1	119	0
0	39	.15	3	26	210	.02	0	.03	.03	.14	.25	.01	10	0
45	28	.47	7	48	177	.2	28	.5	1.3	.33	.23	.03	18	<1
165	457	4.8	60	376	1414	1.51	134	.56	5.58	.59	5.11	.17	228	1
27	76	.8	10	63	236	.25	22	.09	.93	.1	.85	.03	38	<1
470	439	22.5	347	2042	3410	7.04	286	.28	17.3	1.36	5.89	.32	174	1
29	27	1.41	22	128	214	.44	18	.02	1.08	.08	.37	.02	11	<1
19	18	.92	14	84	140	.29	12	.01	.71	.06	.24	.01	7	<1
211	416	20	96	1453	2761	2.47	96	1.14	8.26	1.12	9.41	.23	60	1
23	46	2.22	11	161	307	.27	11	.13	.92	.12	1.05	.02	7	<1
563	379	21.3	307	1822	3450	6.35	276	1.23	27.6	1.61	12.8	.3	225	1
35	24	1.33	19	114	216	.4	17	.08	1.73	.1	.8	.02	14	<1
959	444	22.2	320	1989	4368	8.7	6819	2.42	74.9	2.77	17.9	1.35	213	19
60	28	1.39	20	124	273	.54	426	.15	4.68	.17	1.12	.08	13	1
69	454	28.5	220	2105	3715	3.72	261	.69	42.9	1.36	10.9	.63	261	5
2	14	.89	7	66	116	.12	8	.02	1.34	.04	.34	.02	8	<1
503	495	11.7	108	611	2322	2.74	372	1.24	11	1.39	10.1	.26	54	2
56	55	1.3	12	68	258	.3	41	.14	1.22	.15	1.12	.03	6	<1
323	357	6.16	35	305	1981	1.44	109	.58	10.9	.4	2.88	.2	156	1
36	40	.68	4	34	220	.16	12	.06	1.22	.04	.32	.02	17	<1
528	490	6.05	106	864	1987	4.9	1545	.27	10.1	1.85	1.87	.5	173	6
44	41	.5	9	72	166	.41	129	.02	.84	.15	.16	.04	14	<1
751	119	4.69	37	405	1353	1.56	530	.47	2.24	.78	4.45	.12	139	<1
62	10	.39	3	33	111	.13	44	.04	.18	.06	.37	.01	11	<1
8	36	1.68	20	61	213	.28	2	.11	1.01	.15	1.22	.01	14	<1
7	19	.55	3	39	157	.13	2	.07	.83	.06	.52	.01	12	<1
39	27	1.03	4	38	93	.19	17	.09	.17	.1	.73	.02	15	0
91	545	9.09	60	659	2658	1.76	368	1.14	20.4	1.48	10.2	.16	64	1
6	34	.57	4	41	166	.11	23	.07	1.28	.09	.64	.01	4	<1
563	379	21.3	307	1822	3450	6.35	276	1.23	27.6	1.61	12.8	.3	225	1
35	24	1.33	19	114	216	.4	17	.08	1.73	.1	.8	.02	14	<1
0	9	1.38	41	167	419	.9	0	.13	.46	.12	1.57	.19	58	0
0	0	2.52			490		0	.39	<.01	.24	3.5			0
1	12	.47	4	33	210	.08	7	.05	1.02	.07	.65	.03	24	<1
10	18	1.37	19	91	190	.44	12	.16	1.3	.13	1.05	.02	13	<1
0	0	1.08		90	140		0		.01					0

Table H-1

Food Composition

(Computer code number is for West Diet Analysis program) (For purposes of calculations, use "0" for t, <1, <.1, <.01, etc.)

Computer Code Number	Food Description	Measure	Wt (g)	H_2O (%)	Ener (cal)	Prot (g)	Carb (g)	Dietary Fiber (g)	Fat (g)	Fat Breakdown (g) Sat	Mono	Poly
	BAKED GOODS: BREADS, CAKES, COOKIES, CRACKERS, PIES—Continued											
	Chocolate chip cookies:											
405	Commercial, 2¼" diam	4 ea	60	12	275	2	35	2	15	4.5	7.8	1.6
406	Home recipe, 2¼" diam	4 ea	64	6	312	4	37	2	18	5.2	6.7	5.4
407	From refrigerated dough, 2¼" diam	4 ea	64	13	284	3	39	1	13	4.5	6.5	1.3
408	Fig bars	4 ea	64	16	223	2	45	3	5	.9	2.6	.8
2052	Fruit bar, no fat	1 ea	28		90	2	21	0	0	0	0	0
2162	Fudge, fat free, Snackwell	1 ea	16	14	53	1	12	<1	<1	.1	.1	t
409	Oatmeal raisin, 2⅝" diam	4 ea	60	6	261	4	41	2	10	1.9	4.1	3
410	Peanut butter, home recipe, 2⅝"diam	4 ea	80	6	380	7	47	2	19	3.5	8.7	5.8
411	Sandwich-type, all	4 ea	40	2	189	2	28	1	8	1.7	4.7	1.1
412	Shortbread, commercial, small	4 ea	32	4	161	2	21	1	8	2	4.3	1
413	Shortbread, home recipe, large	2 ea	22	3	120	1	12	<1	7	4.5	2.1	.3
414	Sugar from refrigerated dough, 2" diam	4 ea	48	5	232	2	31	<1	11	2.8	6.2	1.4
1874	Vanilla sandwich, Snackwell's	2 ea	26	4	109	1	21	1	2	.5	.8	.2
415	Vanilla wafers	10 ea	40	5	176	2	29	1	6	1.4	2.4	1.5
416	Corn chips	1 c	26	1	140	2	15	1	9	1.2	2.5	4.3
	Crackers (enriched):											
417	Cheese	10 ea	10	3	50	1	6	<1	3	.9	.9	.5
418	Cheese with peanut butter	4 ea	28	4	135	4	16	1	6	1.4	3.4	1.2
	Fat free, enriched:											
2161	Cracked pepper, Snackwell	1 ea	15	2	60	2	12	<1	<1	.1	t	.1
2159	Wheat, Snackwell	7 ea	15	1	60	2	12	1	<1	.1	.1	.1
2075	Whole wheat, herb seasoned	5 ea	14	5	50	2	11	2	0	0	0	0
2077	Whole wheat, onion	5 ea	14	4	50	2	11	2	0	0	0	0
419	Graham, enriched	2 ea	14	4	59	1	11	<1	1	.4	.7	.2
420	Melba toast, plain, enriched	1 pce	5	5	19	1	4	<1	<1	t	t	.1
1514	Rice cakes, unsalted, enriched	2 ea	18	6	70	1	15	1	<1	.1	.2	.2
421	Rye wafer, whole grain	2 ea	22	5	73	2	18	5	<1	t	t	.1
422	Saltine-enriched	4 ea	12	4	52	1	9	<1	1	.3	.8	.2
1971	Saltine, unsalted tops, enriched	2 ea	6		25	1	4	0	<1	0	0	0
423	Snack-type, round like Ritz, enriched	3 ea	9	3	45	1	5	<1	2	.4	1	.7
424	Wheat, thin, enriched	4 ea	8	3	38	1	5	<1	2	.7	.8	.2
425	Whole-wheat wafers	2 ea	8	3	35	1	5	1	1	.2	.8	.2
426	Croissants, 4½ x 4 x 1¾"	1 ea	57	23	231	5	26	1	12	6.7	3.2	.7
1699	Croutons, seasoned	½ c	20	4	93	2	13	1	4	1	1.9	.5
	Danish pastry:											
427	Packaged ring, plain, 12 oz	1 ea	340	21	1349	19	181	1	65	13.5	40.9	6.4
428	Round piece, plain, 4¼" diam, 1" high	1 ea	88	21	349	5	47	<1	17	3.5	10.6	1.6
429	Ounce, plain	1 oz	28	21	111	2	15	<1	5	1.1	3.4	.5
430	Round piece with fruit	1 ea	94	29	335	5	45		16	3.3	10.1	1.6
	Desserts, 3 x 3" piece:											
1348	Apple crisp	1 pce	78	61	127	1	25	1	3	.6	1.2	.9
1353	Apple cobbler	1 pce	104	57	199	2	35	2	6	1.2	2.8	2
1349	Cherry crisp	1 pce	138	77	146	2	24	1	5	.9	2.5	1.8
1352	Cherry cobbler	1 pce	129	66	198	2	34	1	6	1.2	2.8	1.9
1350	Peach crisp	1 pce	139	75	155	2	27	2	5	.9	2.5	1.7
1351	Peach cobbler	1 pce	130	64	204	2	36	2	6	1.2	2.8	1.9
	Doughnuts:											
431	Cake type, plain, 3¼" diam	1 ea	47	21	198	2	23	1	11	1.8	4.5	3.8
432	Yeast-leavened, glazed, 3¾" diam	1 ea	60	25	242	4	27	1	14	3.5	7.8	1.7
	English muffins:											
433	Plain, enriched	1 ea	57	42	134	4	26	2	1	.2	.2	.5
434	Toasted	1 ea	52	37	133	4	26	2	1	.1	.2	.5
1504	Whole wheat	1 ea	66	46	134	6	27	4	1	.2	.3	.6
1414	Granola bar, soft	1 ea	28	6	124	2	19	1	5	2	1.1	1.5
1415	Granola bar, hard	1 ea	25	4	118	3	16	1	5	.6	1.1	3
1985	Granola bar, fat free, all flavors	1 ea	42	10	140	2	35	3	0	0	0	0
	Muffins, 2½" diam, 1½" high:											
	From home recipe:											
435	Blueberry	1 ea	57	39	165	4	23	1	6	1.4	1.6	3.1
436	Bran, wheat	1 ea	57	35	164	4	24	4	7	1.5	1.8	3.6
437	Cornmeal	1 ea	57	32	183	4	25	2	7	1.6	1.8	3.5

Chol (mg)	Calc (mg)	Iron (mg)	Magn (mg)	Pota (mg)	Sodi (mg)	Zinc (mg)	VT-A (RE)	Thia (mg)	VT-E (a-TE)	Ribo (mg)	Niac (mg)	V-B6 (mg)	Fola (μg)	VT-C (mg)
0	9	1.45	21	56	196	.28	1	.07	1.74	.12	.97	.1	23	0
20	25	1.57	35	143	231	.59	105	.12	1.86	.11	.87	.05	21	<1
15	16	1.44	15	115	134	.32	11	.12	1.31	.12	1.27	.03	36	0
0	41	1.86	17	132	224	.25	3	.1	.45	.14	1.2	.05	17	<1
0	0	.36			95		0		.01					0
0	3	.29	5	26	71	.08	<1	.02	<.01	.02	.26	<.01		0
20	60	1.59	25	143	323	.52	98	.15	1.5	.1	.76	.04	18	<1
25	31	1.78	31	185	414	.66	125	.18	3.04	.17	2.81	.07	44	<1
0	10	1.55	18	70	242	.32	<1	.03	1.21	.07	.83	.01	17	0
6	11	.88	5	32	146	.17	4	.11	.98	.1	1.07	.01	19	0
20	4	.58	3	15	102	.09	67	.08	.18	.06	.64	<.01	2	0
15	43	.88	4	78	225	.13	5	.09	1.54	.06	1.16	.01	25	0
<1	17	.61	5	28	95	.16	<1	.05		.07	.69	.01		0
23	19	.95	6	39	125	.14	7	.11	.54	.13	1.24	.03	20	0
0	33	.34	20	37	164	.33	2	.01	.35	.04	.31	.06	5	0
1	15	.48	4	14	99	.11	3	.06	.1	.04	.47	.05	8	0
1	22	.82	16	69	278	.3	10	.11	1.24	.1	1.83	.42	25	0
<1	26	.73	4	19	148	.14	<1	.05		.06	.78	.01		<1
<1	28	.58	7	43	169	.21	<1	.04		.07	.73	.02		0
0	0				80		100							2
0	0	0			80		100							2
0	3	.52	4	19	85	.11	0	.03	.27	.04	.58	.01	8	0
0	5	.18	3	10	41	.1	0	.02	.01	.01	.21	<.01	6	0
0	2	.27	24	52	5	.54	1	.01	.02	.03	1.41	.03	4	0
0	9	1.31	27	109	175	.62	<1	.09	.44	.06	.35	.06	3	<1
0	14	.65	3	15	156	.09	0	.07	.2	.05	.63	<.01	15	0
0		.36		5	50				.1					
0	11	.32	2	12	76	.06	0	.04	.4	.03	.36	<.01	7	0
2	3	.28	5	16	70	.13	<1	.04	.02	.03	.34	.01	1	0
0	4	.25	8	24	53	.17	0	.02	.31	.01	.36	.01	3	0
43	21	1.16	9	67	424	.43	78	.22	.24	.14	1.25	.03	35	<1
1	19	.56	8	36	248	.19	1	.1	.32	.08	.93	.02	18	0
105	143	6.94	54	371	1261	1.87	20	.99	3.06	.75	8.5	.2	211	10
27	37	1.8	14	96	326	.48	5	.25	.79	.19	2.2	.05	55	3
9	12	.57	4	30	104	.15	2	.08	.25	.06	.7	.02	17	1
19	22	1.4	14	110	333	.48	24	.29	.85	.21	1.8	.06	31	2
0	22	.58	5	76	142	.12	24	.07		.06	.6	.03	4	2
1	21	.79	6	106	288	.16	76	.1	1.11	.09	.74	.04	3	<1
0	26	2.14	11	154	74	.15	150	.06	.93	.08	.6	.06	11	3
1	28	1.81	9	133	294	.2	135	.1	1.01	.11	.85	.05	9	2
0	20	.89	12	189	70	.19	108	.06	1.13	.05	1.06	.03	6	5
1	24	.91	10	159	291	.23	105	.09	1.16	.09	1.19	.03	6	3
17	21	.92	9	60	257	.26	8	.1	1.63	.11	.87	.03	22	<1
4	26	1.22	13	65	205	.46	6	.22	1.75	.13	1.71	.03	26	0
0	99	1.43	12	75	264	.4	0	.25	.07	.16	2.21	.02	46	<1
0	98	1.41	11	74	262	.39	0	.2	.07	.14	1.98	.02	38	<1
0	175	1.62	47	139	420	1.06	0	.2	.46	.09	2.25	.11	28	0
<1	29	.72	21	91	78	.42	0	.08	.34	.05	.14	.03	7	0
0	15	.74	24	84	73	.51	4	.07	.33	.03	.39	.02	6	<1
0	0	3.6			5		100							0
22	107	1.29	9	69	251	.31	16	.15	1.03	.16	1.26	.02	7	1
20	106	2.39	44	181	335	1.57	136	.19	1.31	.25	2.29	.18	30	4
26	147	1.49	13	82	333	.35	23	.17	1.08	.18	1.36	.05	10	<1

Table H-1

Food Composition (Computer code number is for West Diet Analysis program) (For purposes of calculations, use "0" for t, <1, <.1, <.01, etc.)

Computer Code Number	Food Description	Measure	Wt (g)	H_2O (%)	Ener (cal)	Prot (g)	Carb (g)	Dietary Fiber (g)	Fat (g)	Fat Breakdown (g) Sat	Mono	Poly
	BAKED GOODS: BREADS, CAKES, COOKIES, CRACKERS, PIES—Continued											
	From commercial mix:											
438	Blueberry	1 ea	50	36	150	3	24	1	4	.7	1.8	1.5
439	Bran, wheat	1 ea	50	35	138	3	23	2	5	1.2	2.3	.7
440	Cornmeal	1 ea	50	30	161	4	25	1	5	1.4	2.6	.6
1864	Nabisco Newtons, fat free, all flavors	1 ea	23		69	1	16		0	0	0	0
	Pancakes, 4" diam:											
441	Buckwheat, from mix w/ egg and milk	1 ea	30	54	62	2	8	1	2	.6	.6	.8
442	Plain, from home recipe	1 ea	38	53	86	2	11	1	4	.8	.9	1.7
443	Plain, from mix; egg, milk, oil added	1 ea	38	53	74	2	14	<1	1	.2	.3	.3
1468	Pan dulce, sweet roll w/topping	1 ea	79	21	291	5	48	1	9	2	3.9	2.7
	Piecrust,with enriched flour, vegetable shortening, baked:											
444	Home recipe, 9" shell	1 ea	180	10	949	12	85	3	62	15.5	27.4	16.4
	From mix:											
445	Piecrust for 2-crust pie	1 ea	320	10	1686	21	152	5	111	27.6	48.6	29.2
446	1 pie shell	1 ea	160	11	802	11	81	3	49	12.3	27.7	6.2
	Pies, 9" diam; pie crust made with vegetable shortening, enriched flour:											
447	Apple: Whole pie	1 ea	1000	52	2370	19	340	16	110	21.1	59.4	20.9
448	Piece, ⅙ of pie	1 pce	167	52	396	3	57	3	18	3.5	9.9	3.5
449	Banana cream: Whole pie	1 ea	1152	48	3098	51	379	8	157	43.3	65.9	38
450	Piece, ⅙ of pie	1 pce	192	48	516	8	63	1	26	7.2	11	6.3
451	Blueberry: Whole pie	1 ea	1176	51	2881	32	394	16	140	34.3	60.2	36.2
452	Piece, ⅙ of pie	1 pce	196	51	480	5	66	3	23	5.7	10	6
453	Cherry: Whole pie	1 ea	1140	46	3078	32	439	17	139	34.1	60.5	37.1
454	Piece, ⅙ of pie	1 pce	240	46	648	7	92	4	29	7.2	12.7	7.8
455	Chocolate cream: Whole pie	1 ea	1194	63	2150	49	281	12	97	35.5	38.2	18.6
456	Piece, ⅙ of pie	1 pce	199	63	358	8	47	2	16	5.9	6.4	3.1
457	Custard: Whole pie	1 ea	630	61	1323	35	131	10	73	17.5	36.3	12.1
458	Piece, ⅙ of pie	1 pce	105	61	221	6	22	2	12	2.9	6	2
459	Lemon meringue: Whole pie	1 ea	678	42	1817	10	320	8	59	10.6	24.6	19.6
460	Piece, ⅙ of pie	1 pce	113	42	303	2	53	1	10	1.8	4.1	3.3
461	Peach: Whole pie	1 ea	1111	45	2994	26	443	16	130	31.1	55.7	37.4
462	Piece, ⅙ of pie	1 pce	139	45	375	3	55	2	16	3.9	7	4.7
463	Pecan: Whole pie	1 ea	678	19	2712	27	388	24	125	25.5	73.2	20.1
464	Piece, ⅙ of pie	1 pce	113	19	452	5	65	4	21	4.2	12.2	3.4
465	Pumpkin: Whole pie	1 ea	654	58	1373	25	179	18	62	13.2	32.8	10.5
466	Piece, ⅙ of pie	1 pce	109	58	229	4	30	3	10	2.2	5.5	1.7
467	Pies, fried, commercial: Apple	1 ea	85	40	266	2	33	1	14	6.5	5.8	1.2
468	Pies, fried, commercial: Cherry	1 ea	128	38	404	4	54	3	21	3.1	9.5	6.9
	Pretzels, made with enriched flour:											
469	Thin sticks, 2¼" long	1 oz	28	3	107	3	22	1	1	.2	.4	.3
470	Dutch twists	10 pce	60	3	229	5	47	2	2	.4	.8	.7
471	Thin twists, 3¼ x 2¼ x ¼"	10 pce	60	3	229	5	47	2	2	.4	.8	.7
	Rolls & buns, enriched, commercial:											
472	Cloverleaf rolls, 2½" diam, 2" high	1 ea	28	32	84	2	14	1	2	.5	1	.3
473	Hot dog buns	1 ea	43	34	123	4	22	1	2	.5	1.1	.4
474	Hamburger buns	1 ea	43	34	123	4	22	1	2	.5	1.1	.4
475	Hard roll, white, 3¾" diam, 2" high	1 ea	57	31	167	6	30	1	2	.3	.6	1
476	Submarine rolls/hoagies, 11¼ x 3 x 2½"	1 ea	135	31	392	12	75	4	4	.9	1.3	1.4
	Rolls & buns, enriched, home recipe:											
477	Dinner rolls 2½" diam, 2" high	1 ea	35	29	112	3	19	1	3	.7	1.1	.7
	Sports/fitness bar:											
2043	Forza energy bar	1 ea	70	18	231	10	45	4	1			
2042	Power bar	1 ea	65		230	10	45	3	2			
2041	Tiger sports bar	1 ea	65	17	229	11	40	4	2			
478	Toaster pastries, fortified (Poptarts)	1 ea	52	12	204	2	37	1	5	.8	2.1	2
2132	Toaster strudel pastry—cream cheese	1 ea	54	32	188	3	24	<1	9	2.7		
2134	Toaster strudel pastry—french toast	1 ea	54	32	188	3	24	<1	9	2.9		

PAGE KEY: A–4 = Beverages A–6 = Dairy A–10 = Eggs A–10 = Fat/Oil A–14 = Fruit A–20 = Bakery A–28 = Grain A–32 = Fish A–34 = Meats A–38 = Poultry A–40 = Sausage A–42 = Mixed/Fast A–46 = Nuts/Seeds A–50 = Sweets A–52 = Vegetables/Legumes A–62 = Vegetarian Foods A–64 = Misc A–66 = Soups/Sauces A–68 = Fast A–84 = Convenience A–88 = Baby foods

Chol (mg)	Calc (mg)	Iron (mg)	Magn (mg)	Pota (mg)	Sodi (mg)	Zinc (mg)	VT-A (RE)	Thia (mg)	VT-E (a-TE)	Ribo (mg)	Niac (mg)	V-B6 (mg)	Fola (µg)	VT-C (mg)
23	12	.56	5	39	219	.19	11	.07	.7	.16	1.12	.04	5	<1
34	16	1.27	28	73	234	.57	15	.1	.75	.12	1.44	.09	8	0
31	37	.97	10	65	398	.32	22	.12	.75	.14	1.05	.05	5	<1
					77									
20	77	.56	17	70	160	.35	20	.05	.62	.08	.4	.04	5	<1
22	83	.68	6	50	167	.21	20	.08	.36	.11	.6	.02	14	<1
5	48	.59	8	66	239	.15	3	.08	.32	.08	.65	.03	3	<1
26	13	1.82	10	57	140	.35	87	.23	1.35	.21	1.98	.04	22	<1
0	18	5.2	25	121	976	.79	0	.7	9.94	.5	5.96	.04	121	0
0	32	9.25	45	214	1734	1.41	0	1.25	17.7	.89	10.6	.08	214	0
0	96	3.44	24	99	1166	.62	0	.48	8.83	.3	3.79	.09	19	0
0	110	4.5	70	650	2660	1.6	300	.28	16.5	.27	2.63	.38	220	32
0	18	.75	12	109	444	.27	50	.05	2.76	.04	.44	.06	37	5
588	864	12	184	1900	2764	5.53	806	1.6	16.9	2.38	12.1	1.53	311	18
98	144	2	31	317	461	.92	134	.27	2.82	.4	2.02	.25	52	3
0	82	14.5	94	588	2175	2.35	47	1.8	24.7	1.55	14	.4	270	8
0	14	2.41	16	98	363	.39	8	.3	4.12	.26	2.33	.07	45	1
0	114	21.1	103	878	2177	2.28	547	1.69	21.7	1.43	14.6	.39	308	11
0	24	4.44	22	185	458	.48	115	.35	4.56	.3	3.07	.08	65	2
109	1028	8.84	170	1705	2085	4.93	235	1.03	11.4	2.43	7.34	.37	59	6
18	171	1.47	28	284	348	.82	39	.17	1.9	.41	1.22	.06	10	1
208	504	3.65	69	668	1512	3.28	315	.25	7.5	1.31	1.84	.3	126	2
35	84	.61	12	111	252	.55	52	.04	1.25	.22	.31	.05	21	<1
305	380	4.14	102	603	990	3.32	353	.42	9.7	1.42	4.4	.2	88	22
51	63	.69	17	101	165	.55	59	.07	1.62	.24	.73	.03	15	4
0	59	12.1	79	1047	2025	1.88	386	1.41	26.2	1.19	15.3	.2	58	589
0	7	1.52	10	131	253	.23	48	.18	3.28	.15	1.91	.02	7	74
217	115	7.05	122	502	2874	3.86	319	.62	17.2	.83	1.69	.14	183	7
36	19	1.18	20	84	479	.64	53	.1	2.86	.14	.28	.02	30	1
131	392	5.17	98	1007	1844	2.94	3139	.36	10.5	1	1.22	.37	131	10
22	65	.86	16	168	307	.49	523	.06	1.75	.17	.2	.06	22	2
13	13	.88	8	51	325	.17	33	.1	.37	.08	.98	.03	4	1
0	28	1.56	13	83	479	.29	22	.18	.55	.14	1.83	.04	23	2
0	10	1.21	10	41	480	.24	0	.13	.06	.17	1.47	.03	48	0
0	22	2.59	21	88	1029	.51	0	.28	.13	.37	3.15	.07	103	0
0	22	2.59	21	88	1029	.51	0	.28	.13	.37	3.15	.07	103	0
<1	33	.88	6	37	146	.22	0	.14	.22	.09	1.13	.01	27	<1
0	60	1.36	9	61	241	.27	0	.21	.2	.13	1.69	.02	41	0
0	60	1.36	9	61	241	.27	0	.21	.2	.13	1.69	.02	41	0
0	54	1.87	15	62	310	.54	0	.27	.1	.19	2.42	.03	54	0
0	122	3.78	27	122	783	.85	0	.54	.1	.33	4.47	.05	40	0
13	21	1.04	7	53	145	.24	28	.14	.35	.14	1.21	.02	15	<1
0	300	6.3	160	220	65	5.25		1.5	20	1.7	20	2	400	60
0	300	5.4	140	150	110	5.25		1.5		1.7	20	2	400	60
	349	4.49	140	279	100		50	1.5	19.9	1.69	19.9	1.99	399	60
0	13	1.81	9	58	218	.34	149	.15	.97	.19	2.05	.2	34	<1
12	12	.97			217		17		1					0
12	12	.97			217		17		1					0

Table H-1
Food Composition

(Computer code number is for West Diet Analysis program) (For purposes of calculations, use "0" for t, <1, <.1, <.01, etc.)

Computer Code Number	Food Description	Measure	Wt (g)	H_2O (%)	Ener (cal)	Prot (g)	Carb (g)	Dietary Fiber (g)	Fat (g)	Fat Breakdown (g) Sat	Mono	Poly
	BAKED GOODS: BREADS, CAKES, COOKIES, CRACKERS, PIES—Continued											
	Tortilla chips:											
1271	Plain	10 pce	18	2	90	1	11	1	5	.9	2.8	.7
1036	Nacho flavor	1 c	26	2	129	2	16	1	7	1.3	3.9	.9
1037	Taco flavor	1 pce	18	2	86	1	11	1	4	.8	2.6	.6
	Tortillas:											
479	Corn, enriched, 6" diam	1 ea	26	44	58	2	12	1	1	.1	.2	.3
480	Flour, 8" diam	1 ea	49	27	159	4	27	2	3	.6	1.4	1.4
1301	Flour, 10" diam	1 ea	72	27	234	6	40	2	5	.8	2.1	2
481	Taco shells	1 ea	14	4	63	1	9	1	3	.4	1.5	.6
	Waffles, 7" diam:											
482	From home recipe	1 ea	75	42	218	6	25	1	11	2.2	2.6	5.1
483	From mix, egg/milk added	1 ea	75	42	218	5	26	1	10	1.7	2.7	5.2
1510	Whole grain, prepared from frozen	1 ea	39	43	107	4	13	1	5	1.6	1.9	.9
	GRAIN PRODUCTS: CEREAL, FLOUR, GRAIN, PASTA and NOODLES, POPCORN											
484	Barley, pearled, dry, uncooked	1 c	200	10	704	20	155	31	2	.5	.3	1.1
485	Barley, pearled, cooked	1 c	157	69	193	4	44	6	1	.1	.1	.3
2009	Breakfast bars, fat free, all flavors	1 ea	38	25	110	2	26	3	0	0	0	0
	Breakfast bar, Snackwell:											
2165	Apple-cinnamon	1 ea	37	16	119	1	29	1	<1	.1	t	.1
2164	Blueberry	1 ea	37	16	121	1	29	1	<1	t	t	.1
2163	Strawberry	1 ea	37	16	120	1	29	1	<1	t	t	.1
	Breakfast cereals, hot, cooked w/o salt added:											
	Corn grits (hominy) enriched:											
486	Regular/quick prep w/o salt, yellow:	1 c	242	85	145	3	31	<1	<1	.1	.1	.2
487	Instant, prepared from packet, white	1 ea	137	82	89	2	21	1	<1	t	t	.1
	Cream of wheat:											
488	Regular, quick, instant	1 c	239	87	129	4	27	1	<1	.1	.1	.3
489	Mix and eat, plain, packet	1 ea	142	82	102	3	21	<1	<1	t	t	.2
1664	Farina cereal, cooked w/o salt	1 c	233	88	117	3	25	3	<1	t	t	.1
490	Malt-O-Meal, cooked w/o salt	1 c	240	88	122	4	26	1	<1	.1	.1	t
494	Maypo	1 c	216	83	153	5	29	5	2	.4	.7	.8
	Oatmeal or rolled oats:											
491	Regular, quick, instant,nonfortified cooked w/o salt	1 c	234	85	145	6	25	4	2	.4	.7	.9
	Instant, fortified:											
492	Plain, from packet	½ c	118	85	70	4	12	2	1	.2	.4	.4
493	Flavored, from packet	½ c	109	76	106	3	21	2	1	.2	.5	.5
	Breakfast cereals, ready to eat:											
495	All-Bran	1 c	62	3	160	8	46	20	2	.4	.4	1.3
1306	Alpha Bits	1 c	28	1	110	2	24	1	1	.1	.2	.2
1307	Apple Jacks	1 c	33	3	120	2	30	1	<1	.1	.1	.2
1308	Bran Buds	1 c	90	3	240	8	72	36	2	.4	.4	1.4
1305	Bran Chex	1 c	49	2	156	5	39	8	1	.2	.3	.7
1309	Honey BucWheat Crisp	1 c	38	5	147	4	31	3	1	.2	.3	.6
1310	C.W. Post, plain	1 c	97	2	421	9	73	7	13	1.7	6	4.7
1311	C.W. Post, with raisins	1 c	103	4	446	9	74	14	15	11	1.7	1.4
496	Cap'n Crunch	1 c	37	2	147	2	32	1	2	.5	.4	.3
1312	Cap'n Crunchberries	1 c	35	2	140	2	30	1	2	.5	.3	.3
1313	Cap'n Crunch, peanut butter	1 c	35	2	146	3	28	1	3	.7	1.1	.7
497	Cheerios	1 c	23	3	84	2	17	2	1	.3	.5	.2
1314	Cocoa Krispies	1 c	41	2	159	3	36	1	1	.7	.2	.2
1316	Cocoa Pebbles	1 c	32	2	131	1	27	1	2	1.1	.4	.1
1315	Corn Bran	1 c	36	3	120	2	30	6	1	.3	.3	.4
1317	Corn Chex	1 c	28	2	110	2	25	<1	<1	t	t	t
498	Corn Flakes, Kellogg's	1 c	28	3	100	2	24	1	<1	.1	t	.1
499	Corn Flakes, Post Toasties	1 c	24	3	93	2	21	1	<1	t	t	t
1340	Corn Pops	1 c	31	3	120	1	28	<1	<1	.1	.1	t
1318	Cracklin' Oat Bran	1 c	65	4	252	5	48	8	8	3.4	3.8	.9
1038	Crispy Wheat `N Raisins	1 c	43	7	150	3	35	3	1	.1	.1	.2

PAGE KEY: A–4 = Beverages A–6 = Dairy A–10 = Eggs A–10 = Fat/Oil A–14 = Fruit A–20 = Bakery A–28 = Grain A–32 = Fish A–34 = Meats A–38 = Poultry A–40 = Sausage A–42 = Mixed/Fast A–46 = Nuts/Seeds A–50 = Sweets A–52 = Vegetables/Legumes A–62 = Vegetarian Foods A–64 = Misc A–66 = Soups/Sauces A–68 = Fast A–84 = Convenience A–88 = Baby foods

Chol (mg)	Calc (mg)	Iron (mg)	Magn (mg)	Pota (mg)	Sodi (mg)	Zinc (mg)	VT-A (RE)	Thia (mg)	VT-E (a-TE)	Ribo (mg)	Niac (mg)	V-B6 (mg)	Fola (μg)	VT-C (mg)
0	28	.27	16	35	95	.27	4	.01	.24	.03	.23	.05	2	0
1	38	.37	21	56	184	.31	11	.03	.35	.05	.37	.07	4	<1
1	28	.36	16	39	142	.23	16	.04	.24	.04	.36	.05	4	<1
0	45	.36	17	40	42	.24	6	.03	.04	.02	.39	.06	30	0
0	61	1.62	13	64	234	.35	0	.26	.62	.14	1.75	.02	60	0
0	90	2.38	19	94	344	.51	0	.38	.91	.21	2.57	.04	89	0
0	35	.36	15	34	25	.19	6	.04	.59	.02	.24	.04	4	0
52	191	1.73	14	119	383	.51	49	.2	1.73	.26	1.55	.04	34	<1
38	93	1.22	15	134	458	.35	19	.15	1.5	.19	1.23	.07	9	<1
39	84	.69	15	91	150	.45	25	.08	.53	.13	.75	.04	7	<1
0	58	5	158	560	18	4.26	4	.38	.26	.23	9.2	.52	46	0
0	17	2.09	34	146	5	1.29	2	.13	.08	.1	3.23	.18	25	0
0	20	.72			25		20							1
<1	17	5	6	68	103	3.88	260	.39		.44	5.2	.52		<1
<1	14	4.83	5	43	107	3.85	260	.39		.44	5.2	.52		<1
<1	14	4.82	6	47	102	3.83	260	.39		.44	5.2	.52		2
0	0	1.55	10	53	0	.17	14	.24	.12	.14	1.96	.06	75	0
0	8	8.19	11	38	289	.21	0	.15	.03	.08	1.38	.05	47	0
0	50	10.3	12	45	139	.33	0	.24	.03	0	1.43	.03	108	0
0	20	8.09	7	38	241	.24	125	.43	.02	.28	4.97	.57	101	0
0	5	1.17	5	30	0	.16	0	.19	.03	.12	1.28	.02	54	0
0	5	9.6	5	31	2	.17	0	.48	.03	.24	5.76	.02	5	0
0	112	7.56	45	190	233	1.34	633	.65	1.51	.65	8.42	.86	9	26
0	19	1.59	56	131	2	1.15	5	.26	.23	.05	.3	.05	9	0
0	109	4.2	28	66	190	.58	302	.35	.14	.19	3.65	.49	100	0
0	112	4.45	34	91	169	.66	306	.35	.14	.25	3.92	.51	100	<1
0	200	9	280	620	560	7.5	450	.75	1.14	.85	10	1	186	30
0	8	2.66	16	54	178	1.48	371	.36	.02	.42	4.93	.5	99	0
0	0	4.5	8	35	150	3.75	225	.38	.05	.43	5	.5	116	15
0	60	13.5	240	809	599	11.3	676	1.17	1.42	1.26	15	1.53	270	45
0	29	14	69	216	345	6.47	11	.64	.56	.26	8.62	.88	173	26
0	54	10.9	43	142	361	.68	913	.9	8.99	1.03	12.1	1.88	11	36
<1	47	15.4	67	198	167	1.64	1284	1.26	.68	1.46	17.1	1.75	342	0
<1	50	16.4	74	261	161	1.64	1363	1.34	.72	1.55	18.1	1.85	364	0
0	7	6.18	13	47	286	5.14	5	.51	.18	.58	6.85	.68	137	0
0	9	6.06	13	49	256	5.39	6	.5	.25	.57	6.72	.67	135	<1
0	3	5.85	24	80	264	4.87	5	.49	.19	.55	6.48	.65	130	0
0	42	6.21	25	68	218	2.88	288	.29	.16	.33	3.84	.38	77	11
0	0	2.38	11	79	278	1.97	298	.49	.19	.57	6.6	.66	123	20
0	5	2.02	13	53	180	1.7	424	.42	.04	.48	5.63	.58	113	0
0	27	10.1	19	75	338	5	5	.1	.19	.56	6.66	.67	134	0
0	3	8.01	4	23	306	.1	14	.36	.07	.07	4.93	.5	99	15
0	0	8.68	3	25	300	.17	225	.36	.03	.43	5	.5	99	15
0	1	.63	4	28	252	.07	318	.31	.06	.36	4.22	.43	85	0
0	0	1.86	2	25	120	1.55	225	.4	.03	.43	5.18	.5	109	15
0	26	2.41	79	305	226	1.95	299	.5	.43	.56	6.63	.66	181	20
0	54	3.52	33	180	223	.85	293	.29	.45	.33	3.91	.39	78	0

Table H-1

Food Composition

(Computer code number is for West Diet Analysis program) (For purposes of calculations, use "0" for t, <1, <.1, <.01, etc.)

Computer Code Number	Food Description	Measure	Wt (g)	H_2O (%)	Ener (cal)	Prot (g)	Carb (g)	Dietary Fiber (g)	Fat (g)	Fat Breakdown (g) Sat	Mono	Poly
	GRAIN PRODUCTS: CEREAL, FLOUR, GRAIN, PASTA and NOODLES, POPCORN—Continued											
1319	Fortified Oat Flakes	1 c	48	3	180	8	36	1	1	.2	.3	.4
500	40% Bran Flakes, Kellogg's	1 c	39	4	121	4	32	7	1	.2	.2	.5
501	40% Bran Flakes, Post	1 c	47	3	152	5	37	9	1	.1	.1	.4
502	Froot Loops	1 c	32	2	120	2	28	1	1	.4	.2	.3
518	Frosted Flakes	1 c	41	3	159	2	37	1	<1	.1	t	.1
1320	Frosted Mini-Wheats	1 c	51	5	170	5	41	5	1	.2	.1	.6
1321	Frosted Rice Krispies	1 c	35	2	135	2	32	<1	<1	.1	.1	.1
1324	Fruit & Fibre w/dates	1 c	57	9	193	5	43	8	3	.4	1.3	1
1322	Fruity Pebbles	1 c	32	3	130	1	28	<1	2	1.4	.1	.1
503	Golden Grahams	1 c	39	3	150	2	33	1	1	.2	.4	.2
504	Granola, homemade	½ c	61	5	285	9	32	6	15	2.9	4.8	6.5
505	Granola, low fat	½ c	47	3	181	5	38	3	3	0		
1670	Granola, low fat, commercial	½ c	45	5	165	4	35	2	2	.6	.7	.9
505	Grape Nuts	½ c	55	3	196	7	45	5	<1	t	t	.1
1326	Grape Nuts Flakes	1 c	39	3	144	4	32	4	1	.6	.1	.2
1665	Heartland Natural with raisins	1 c	110	5	468	11	76	6	16	4	4.2	6.2
1327	Honey & Nut Corn Flakes	1 c	37	2	148	3	31	1	2	.3	.7	.6
506	Honey Nut Cheerios	1 c	33	2	126	3	27	2	1	.3	.5	.2
1328	HoneyBran	1 c	35	2	119	3	29	4	1	.3	.1	.3
1329	HoneyComb	1 c	22	1	86	1	20	1	<1	.2	.1	.1
1330	King Vitaman	1 c	21	2	81	2	18	1	1	.2	.3	.2
1039	Kix	1 c	19	2	72	1	16	1	<1	.1	.1	t
1331	Life	1 c	44	4	167	4	35	3	2	.3	.6	.8
507	Lucky Charms	1 c	32	2	124	2	27	1	1	.2	.4	.2
1323	Mueslix Five Grain	1 c	82	5	279	7	63	7	3	.5	1	1.2
508	Nature Valley Granola	1 c	113	4	510	12	74	7	20	2.6	13.3	3.8
1666	Nutri Grain Almond Raisin	1 c	40	6	147	3	31	3	2	.1	1	1.2
1336	100% Bran	1 c	66	3	178	8	48	19	3	.6	.6	1.9
509	100% Natural cereal, plain	1 c	104	2	462	11	71	8	17	7.5	7.5	2.3
1337	100% Natural with apples & cinnamon	1 c	104	2	477	11	70	7	20	15.5	1.8	1.3
1338	100% Natural with raisins & dates	1 c	110	3	496	12	72	7	20	13.6	3.7	1.7
510	Product 19	1 c	33	4	110	2	28	1	<1	t	.2	.2
1339	Quisp	1 c	30	3	121	2	25	1	2	.5	.4	.2
511	Raisin Bran, Kellogg's	1 c	61	9	200	6	47	8	1	.1	.1	.4
512	Raisin Bran, Post	1 c	56	9	172	5	42	8	1	.2	.1	.5
1667	Raisin Squares	1 c	71	9	241	6	55	7	2	.2	.2	.6
1041	Rice Chex	1 c	33	3	130	2	29	1	<1	t	t	t
513	Rice Krispies, Kellogg's	1 c	28	2	111	2	25	<1	<1	t	t	t
514	Rice, puffed	1 c	14	4	54	1	12	<1	<1	t	t	t
515	Shredded Wheat	1 c	43	5	154	5	35	4	1	.1	.1	.4
516	Special K	1 c	31	3	110	6	22	1	<1	t	t	.2
517	Super Golden Crisp	1 c	33	1	123	2	30	<1	<1	.1	.1	.1
519	Honey Smacks	1 c	36	3	133	3	32	1	1	.4	.1	.3
1341	Tasteeos	1 c	24	2	94	3	19	3	1	.2	.2	.2
1342	Team	1 c	42	4	164	3	36	1	1	.1	.2	.3
520	Total, wheat, with added calcium	1 c	40	3	140	4	32	4	1	.2	.2	.1
521	Trix	1 c	28	2	114	1	24	1	2	.4	.9	.3
1344	Wheat Chex	1 c	46	2	169	5	38	4	1	.2	.1	.5
1043	Wheat cereal, puffed, fortified	1 c	12	4	44	2	9	1	<1	t	t	.1
522	Wheaties	1 c	29	3	106	3	23	2	1	.2	.2	.1
523	Buckwheat flour, dark	1 c	120	11	402	15	85	12	4	.8	1.1	1.1
525	Buckwheat, whole grain, dry	1 c	170	10	583	23	122	17	6	1.3	1.8	1.8
526	Bulgar, dry, uncooked	1 c	140	9	479	17	106	26	2	.3	.2	.8
527	Bulgar, cooked	1 c	182	78	151	6	34	8	<1	.1	.1	.2
	Cornmeal:											
528	Whole-ground, unbolted, dry	1 c	122	10	442	10	94	9	4	.6	1.2	2
530	Degermed, enriched, dry	1 c	138	12	505	12	107	10	2	.3	.6	1
38041	Degermed, enriched, baked	1 c	138	12	505	12	107	10	2	.3	.6	1
	Macaroni, cooked:											
532	Enriched	1 c	140	66	197	7	40	2	1	.1	.1	.4
533	Whole wheat	1 c	140	67	174	7	37	4	1	.1	.1	.3

Chol (mg)	Calc (mg)	Iron (mg)	Magn (mg)	Pota (mg)	Sodi (mg)	Zinc (mg)	VT-A (RE)	Thia (mg)	VT-E (a-TE)	Ribo (mg)	Niac (mg)	V-B6 (mg)	Fola (μg)	VT-C (mg)
0	68	13.7	58	228	220	2.54	636	.62	.34	.72	8.45	.86	169	0
0	0	10.9	81	229	309	5.03	505	.51	7.22	.58	6.71	.66	138	20
0	21	13.4	102	251	431	2.49	622	.61	.54	.7	8.27	.85	166	0
0	0	4.51	8	35	150	3.75	225	.37	.12	.43	5	.5	96	15
0	0	6.15	4	26	264	.2	298	.5	.05	.56	6.61	.66	123	20
0	0	15	60	170	0	1.5	0	.37	.46	.42	4.64	.5	102	0
0	0	2.42	8	27	256	.42	303	.49	.03	.56	6.72	.66	140	20
0	30	10.1	81	335	270	3.02	725	.75	1.32	.85	10.1	1	201	0
0	4	2.02	9	24	178	1.7	424	.42	.03	.48	5.63	.58	113	0
0	19	5.85	12	69	357	4.88	293	.49	.29	.55	6.51	.65	130	19
0	49	2.56	109	328	15	2.48	2	.45	7.87	.17	1.25	.19	52	1
0		2.71	36	143	90	5.64	226	.56	7.57	.64	7.52	.75	151	
0	15	1.35	30	127	101	2.84	169	.27	4.03	.31	3.74	.36	90	0
0	5	15.7	37	184	382	1.21	728	.71	.14	.82	9.68	.99	194	0
0	16	11.2	43	136	220	.78	516	.51	.1	.58	6.86	.7	138	0
0	66	4.02	141	415	226	2.83	7	.32	.77	.14	1.54	.2	44	1
0	0	3.03	3	40	249	.2	152	.26	.09	.3	3.37	.33	74	10
0	22	4.95	32	94	285	4.13	248	.41	.34	.47	5.51	.55	110	16
0	16	5.57	46	151	202	.9	463	.45	.81	.52	6.16	.63	23	19
0	4	2.09	7	25	124	1.17	291	.29	.09	.33	3.87	.4	78	0
0	3	5.92	18	58	176	2.65	212	.26	1.42	.3	3.53	.35	71	8
0	28	5.13	6	26	167	2.38	238	.24	.05	.27	3.17	.32	63	9
0	134	12.3	43	109	240	5.5	2	.55	.22	.62	7.35	.73	147	0
0	35	4.8	21	58	217	4	240	.4	.14	.45	5.34	.53	107	16
0	38	8.94	82	369	107	7.46	747	.75	8.94	.84	9.84	.99	197	1
0	85	3.53	107	375	183	2.27	0	.35	7.97	.12	1.25	.16	17	0
0	122	1.14	13	147	139	3.06	0	.32	4.38	.35	4.08	.41	80	0
0	46	8.12	312	652	457	5.74	0	1.58	1.53	1.78	20.9	2.11	47	63
1	100	3.11	109	457	28	2.5	1	.36	1.19	.17	1.84	.19	26	<1
0	157	2.89	72	514	52	2	6	.33	.73	.57	1.87	.11	17	1
0	160	3.12	124	538	47	2.11	7	.31	.77	.65	2.09	.16	45	0
0	0	19.8	18	55	308	16.5	248	1.65	24.4	1.88	22	2.21	429	66
0	6	5.1	15	40	216	4.26	4	.42	.15	.48	5.67	.56	113	0
0	40	4.5	80	350	390	3.75	225	.37	.56	.43	5	.5	122	0
0	26	8.9	95	345	365	2.97	741	.73	1.3	.84	9.86	1.01	198	0
0	0	21.7	54	335	4	1.99	0	.5	.38	.57	6.67	.64	142	0
0	5	9.44	8	38	276	.45	2	.43	.04	.01	5.81	.59	116	17
0	5	.7	12	27	206	.46	371	.52	.03	.59	6.92	.69	138	15
0	1	.41	4	16	1	.15	0	.06	.01	.01	.87	0	1	0
0	16	1.81	57	155	4	1.42	0	.11	.23	.12	2.26	.11	21	0
0	0	8.4	16	55	250	3.75	225	.53	.08	.59	7.01	.71	93	15
0	7	2.08	20	48	51	1.75	437	.43	.12	.49	5.81	.59	116	0
0	0	2.4	21	53	67	.4	300	.5	.18	.58	6.66	.68	133	20
0	11	6.86	26	71	183	.69	318	.31	.17	.36	4.22	.43	85	13
0	6	12	12	71	260	.58	556	.55	.1	.63	7.39	.76	7	22
0	344	24	43	129	265	20	500	2	31.3	2.27	26.8	2.67	533	80
0	30	4.2	3	16	184	3.5	210	.35	.56	.4	4.68	.47	93	14
0	18	13.2	58	173	308	1.23	0	.6	.17	.17	8.1	.83	162	24
0	3	.56	16	44	1	.37	<1	.05	.08	.03	1.43	.02	4	0
0	53	7.83	31	101	215	.68	218	.36	.36	.41	4.84	.48	97	14
0	49	4.87	301	692	13	3.74	0	.5	1.24	.23	7.38	.7	65	0
0	31	3.74	393	782	2	4.08	0	.17	1.75	.72	11.9	.36	51	0
0	49	3.44	230	574	24	2.7	0	.32	.22	.16	7.15	.48	38	0
0	18	1.75	58	124	9	1.04	0	.1	.05	.05	1.82	.15	33	0
0	7	4.21	155	350	43	2.22	57	.47	.82	.24	4.43	.37	31	0
0	7	5.7	55	224	4	.99	57	.99	.45	.56	6.94	.35	258	0
0	7	5.7	55	224	4	.99	57	.79	.5	.5	6.25	.32	181	0
	Calc		Magn		Sodi			Thia	VT-E	Ribo	Niac	V-B6	Fola	VT-C
0	10	1.96	25	43	1	.74	0	.29	.04	.14	2.34	.05	98	0
0	21	1.48	42	62	4	1.13	0	.15	.14	.06	.99	.11	7	0

Table H-1

Food Composition

(Computer code number is for West Diet Analysis program) (For purposes of calculations, use "0" for t, <1, <.1, <.01, etc.)

Computer Code Number	Food Description	Measure	Wt (g)	H_2O (%)	Ener (cal)	Prot (g)	Carb (g)	Dietary Fiber (g)	Fat (g)	Fat Breakdown (g) Sat	Mono	Poly
	GRAIN PRODUCTS: CEREAL, FLOUR, GRAIN, PASTA and NOODLES, POPCORN—Continued											
534	Vegetable, enriched	1 c	134	68	172	6	36	2	<1	t	t	.1
535	Millet, cooked	1 c	240	71	286	8	57	3	2	.4	.4	1.2
	Noodles (see also Pasta and Spaghetti):											
1507	Cellophane noodles, cooked	1 c	190	79	160	<1	39	<1	<1	t	t	t
1995	Cellophane noodles, dry	1 c	140	13	491	<1	121	1	<1	t	t	t
537	Chow mein, dry	1 c	45	1	237	4	26	2	14	2	3.5	7.8
536	Egg noodles, cooked, enriched	1 c	160	69	213	8	40	2	2	.5	.7	.7
538	Spinach noodles, dry	3½ oz	100	8	372	13	75	11	2	.2	.2	.6
1343	Oat bran, dry	¼ c	24	7	59	4	16	4	2	.3	.6	.7
	Pasta, cooked:											
1418	Fresh	2 oz	57	69	75	3	14	1	1	.1	.1	.2
1417	Linguini/Rotini	1 c	140	66	197	7	40	4	1	.1	.1	.4
	Popcorn:											
539	Air popped, plain	1 c	8	4	31	1	6	1	<1	t	.1	.2
1042	Microwaved, low fat, low sodium	1 c	6	3	25	1	4	1	1	.1	.2	.3
540	Popped in vegetable oil/salted	1 c	11	3	55	1	6	1	3	.5	.9	1.5
541	Sugar-syrup coated	1 c	35	3	151	1	28	2	4	1.3	1	1.6
	Rice:											
542	Brown rice, cooked	1 c	195	73	216	5	45	4	2	.4	.6	.6
2215	Mexican rice, cooked	1 c	226		820	16	180	6	30	4	1	1
2216	Spanish rice, cooked	1 c	246	85	130	3	28	2	1			
	White, enriched, all types:											
543	Regular/long grain, dry	1 c	185	12	675	13	148	2	1	.3	.4	.3
544	Regular/long grain, cooked	1 c	158	68	205	4	45	1	<1	.1	.1	.1
545	Instant, prepared without salt	1 c	165	76	162	3	35	1	<1	.1	.1	.1
	Parboiled/converted rice:											
546	Raw, dry	1 c	185	10	686	13	151	3	1	.3	.3	.3
547	Cooked	1 c	175	72	200	4	43	1	<1	.1	.1	.1
1486	Sticky rice (glutinous), cooked	1 c	174	77	169	4	37	2	<1	.1	.1	.1
548	Wild rice, cooked	1 c	164	74	166	7	35	3	1	.1	.1	.4
1700	Rice and pasta (Rice-a-Roni), cooked	1 c	202	72	246	5	43	1	6	1.1	2.3	1.9
549	Rye flour, medium	1 c	102	10	361	10	79	15	2	.2	.2	.8
1044	Soy flour, low-fat	1 c	88	3	325	45	30	9	6	.9	1.3	3.3
	Spaghetti pasta:											
550	Without salt, enriched	1 c	140	66	197	7	40	4	1	.1	.1	.4
551	With salt, enriched	1 c	140	66	197	7	40	2	1	.1	.1	.4
552	Whole-wheat spaghetti, cooked	1 c	140	67	174	7	37	6	1	.1	.1	.3
1302	Tapioca-pearl, dry	1 c	152	11	544	<1	135	1	<1	t	t	t
553	Wheat bran, crude	1 c	58	10	125	9	37	25	2	.4	.4	1.3
554	Wheat germ, raw	1 c	115	11	414	27	60	15	11	1.9	1.6	6.9
555	Wheat germ, toasted	1 c	113	6	432	33	56	15	12	2.1	1.7	7.5
1669	Wheat germ, with brown sugar & honey	1 c	113	3	420	30	66	11	9	1.5	1.2	5.5
556	Rolled wheat, cooked	1 c	240	84	149	5	33	4	1	.1	.1	.5
557	Whole-grain wheat, cooked	1 c	150	86	84	4	20	3	<1	.1	.1	.2
	Wheat flour (unbleached):											
	All-purpose white flour, enriched:											
558	Sifted	1 c	115	12	419	12	88	3	1	.2	.1	.5
559	Unsifted	1 c	125	12	455	13	95	3	1	.2	.1	.5
560	Cake or pastry, enriched, sifted	1 c	96	12	348	8	75	2	1	.1	.1	.4
561	Self-rising, enriched, unsifted	1 c	125	11	443	12	93	3	1	.2	.1	.5
562	Whole wheat, from hard wheats	1 c	120	10	407	16	87	15	2	.4	.3	.9
	MEATS: FISH and SHELLFISH											
1045	Bass, baked or broiled	4 oz	113	69	165	27	0	0	5	1.4	1.5	2.3
1046	Bluefish, baked or broiled	4 oz	113	63	180	29	0	0	6	1.4	2	2.7
1686	Catfish, breaded/flour fried	4 oz	113	49	325	21	14	1	20	5	9	4.7
	Clams:											
563	Raw meat only	1 ea	145	82	107	19	4	0	1	.3	.4	.7
564	Canned, drained	1 c	160	64	237	41	8	0	3	.7	.9	1.5
1290	Steamed, meat only	10 ea	95	64	141	24	5	0	2	.4	.5	.9

PAGE KEY: A–4 = Beverages A–6 = Dairy A–10 = Eggs A–10 = Fat/Oil A–14 = Fruit A–20 = Bakery A–28 = Grain A–32 = Fish A–34 = Meats A–38 = Poultry A–40 = Sausage A–42 = Mixed/Fast A–46 = Nuts/Seeds A–50 = Sweets A–52 = Vegetables/Legumes A–62 = Vegetarian Foods A–64 = Misc A–66 = Soups/Sauces A–68 = Fast A–84 = Convenience A–88 = Baby foods

Chol (mg)	Calc (mg)	Iron (mg)	Magn (mg)	Pota (mg)	Sodi (mg)	Zinc (mg)	VT-A (RE)	Thia (mg)	VT-E (a-TE)	Ribo (mg)	Niac (mg)	V-B6 (mg)	Fola (µg)	VT-C (mg)
0	15	.66	25	41	8	.59	7	.15	.05	.08	1.43	.03	87	0
0	7	1.51	106	149	5	2.18	0	.25	.43	.2	3.19	.26	46	0
0	14	1	3	5	9	.23	0	.07	.06	0	.09	.02	1	0
0	35	3.04	4	14	14	.57	0	.21	.18	0	.28	.07	3	0
0	9	2.13	23	54	198	.63	4	.26	.07	.19	2.68	.05	40	0
53	19	2.54	30	45	11	.99	10	.3	.08	.13	2.38	.06	102	0
0	58	2.13	174	376	36	2.76	46	.37	.04	.2	4.55	.32	48	0
0	14	1.3	56	136	1	.75	0	.28	.41	.05	.22	.04	12	0
19	3	.65	10	14	3	.32	3	.12	.09	.09	.56	.02	36	0
0	10	1.96	25	43	1	.74	0	.29	.08	.14	2.34	.05	98	0
0	1	.21	10	24	<1	.27	2	.02	.01	.02	.15	.02	2	0
0	1	.14	9	14	29	.23	1	.02	.06	.01	.12	.01	1	0
0	1	.31	12	25	97	.29	2	.01	.03	.01	.17	.02	2	<1
2	15	.61	12	38	72	.2	3	.02	.42	.02	.77	.01	1	0
0	19	.82	84	84	10	1.23	0	.19	.53	.05	2.98	.28	8	0
0	300	9			2700		120							96
0		.72			1340									
0	52	7.97	46	213	9	2.02	0	1.07	.24	.09	7.75	.3	427	0
0	16	1.9	19	55	2	.77	0	.26	.08	.02	2.34	.15	92	0
0	13	1.04	8	7	5	.4	0	.12	.08	.08	1.45	.02	68	0
0	111	6.59	57	222	9	1.78	0	1.1	.24	.13	6.72	.65	427	0
0	33	1.98	21	65	5	.54	0	.44	.09	.03	2.45	.03	87	0
0	3	.24	9	17	9	.71	0	.03	.07	.02	.5	.04	2	0
0	5	.98	52	166	5	2.2	0	.08	.38	.14	2.12	.22	43	0
2	16	1.9	24	85	1147	.57	0	.25	.27	.16	3.6	.2	89	<1
0	24	2.16	76	347	3	2.03	0	.29	1.36	.12	1.76	.27	19	0
0	165	5.27	202	2261	16	1.04	4	.33	.17	.25	1.9	.46	361	0
0	10	1.96	25	43	1	.74	0	.29	.08	.14	2.34	.05	98	0
0	10	1.96	25	43	140	.74	0	.29	.38	.14	2.34	.05	98	0
0	21	1.48	42	62	4	1.13	0	.15	.07	.06	.99	.11	7	0
0	30	2.4	2	17	2	.18	0	.01	0	0	0	.01	6	0
0	42	6.15	354	686	1	4.22	0	.3	1.35	.33	7.89	.75	46	0
0	45	7.2	275	1025	14	14.1	0	2.16	20.7	.57	7.83	1.5	323	0
0	51	10.3	362	1070	5	18.9	0	1.89	20.5	.93	6.32	1.11	398	7
0	56	9.1	307	1089	12	15.7	11	1.51	24.9	.78	5.34	.56	376	0
0	17	1.49	53	170	0	1.15	0	.17	.48	.12	2.14	.17	26	0
0	9	.88	35	99	1	.73	0	.12	.3	.03	1.5	.08	12	0
0	17	5.34	25	123	2	.8	0	.9	.07	.57	6.79	.05	177	0
0	19	5.8	27	134	2	.87	0	.98	.07	.62	7.38	.05	193	0
0	13	7.03	15	101	2	.59	0	.86	.06	.41	6.52	.03	148	0
0	423	5.84	24	155	1587	.77	0	.84	.07	.52	7.29	.06	193	0
0	41	4.66	166	486	6	3.52	0	.54	1.48	.26	7.64	.41	53	0
98	116	2.16	43	515	102	.94	40	.1	.84	.1	1.72	.16	19	2
86	10	.7	47	539	87	1.18	156	.08	.71	.11	8.19	.52	2	0
92	41	1.44	34	376	598	1.05	33	.4	2.48	.18	3.37	.21	19	1
49	67	20.3	13	455	81	1.99	131	.12	1.45	.31	2.57	.09	23	19
107	147	44.8	29	1004	179	4.37	274	.24	3.04	.68	5.36	.18	46	35
64	87	26.6	17	597	106	2.59	162	.14	1.86	.4	3.18	.1	27	21

Table H-1

Food Composition

(Computer code number is for West Diet Analysis program) (For purposes of calculations, use "0" for t, <1, <.1, <.01, etc.)

Computer Code Number	Food Description	Measure	Wt (g)	H_2O (%)	Ener (cal)	Prot (g)	Carb (g)	Dietary Fiber (g)	Fat (g)	Fat Breakdown (g) Sat	Mono	Poly
	MEATS: FISH and SHELLFISH—Continued											
	Cod:											
565	Baked	4 oz	113	76	119	26	0	0	1	.2	.1	.5
566	Batter fried	4 oz	113	67	196	20	8	<1	9	2.2	3.6	2.6
567	Poached, no added fat	4 oz	113	77	116	25	0	0	1	.2	.1	.3
	Crab, meat only:											
1048	Blue crab, cooked	1 c	118	77	120	24	0	0	2	.3	.3	.8
1049	Dungeness crab, cooked	1 c	118	73	130	26	1	0	1	.2	.3	.5
568	Blue crab, canned	1 c	135	76	134	28	0	0	2	.4	.3	.6
1587	Crab, imitation, from surimi	4 oz	113	74	115	14	11	0	1	.3	.2	.8
569	Fish sticks, breaded pollock	2 ea	56	46	152	9	13	<1	7	1.8	2.8	1.8
572	Flounder/sole, baked	4 oz	113	73	132	27	0	0	2	.5	.4	.9
1599	Grouper, baked or broiled	4 oz	113	73	133	28	0	0	1	.4	.4	.6
573	Haddock, breaded, fried	4 oz	113	55	264	22	14	1	13	3.2	5.4	3.3
1050	Haddock, smoked	4 oz	113	71	131	28	0	0	1	.3	.3	.5
	Halibut:											
17291	Baked	4 oz	113	72	158	30	0	0	3	.7	1	1.4
1051	Smoked	4 oz	113	64	203	34			4	.6	1.2	1.5
1054	Raw	4 oz	113	78	124	23	0	0	3	.7	.8	1.1
575	Herring, pickled	4 oz	113	55	296	16	11	0	20	4.4	11	4.8
1052	Lobster meat, cooked w/moist heat	1 c	145	76	142	30	2	0	1	.2	.2	.5
1687	Ocean perch, baked/broiled	4 oz	113	73	137	27	0	0	2	.4	1	.7
576	Ocean perch, breaded/fried	4 oz	113	59	249	22	9	<1	13	3.2	5.7	3.4
1056	Octopus, raw	4 oz	113	80	93	17	2	0	1	.3	.2	.3
	Oysters:											
577	Raw, Eastern	1 c	248	85	169	17	10	0	6	2	.9	2.8
578	Raw, Pacific	1 c	248	82	201	23	12	0	6	1.3	.9	2.2
	Cooked:											
579	Eastern, breaded, fried, medium	5 ea	73	65	144	6	8	<1	9	2.7	1.7	4.6
580	Western, simmered	5 ea	125	64	204	24	12	0	6	1.9	.9	2.7
581	Pollock, baked, broiled, or poached	4 oz	113	74	128	27	0	0	1	.3	.2	.6
	Salmon:											
582	Canned pink, solids and liquid	4 oz	113	69	157	22	0	0	7	1.7	2.1	2.3
583	Broiled or baked	4 oz	113	62	244	31	0	0	12	2.2	6	2.7
584	Smoked	4 oz	113	72	132	21	0	0	5	1.2	2.3	1.1
585	Atlantic sardines, canned, drained, 2 = 24 g	4 oz	113	60	235	28	0	0	13	1.9	4.4	6.4
586	Scallops, breaded, cooked from frozen	6 ea	93	58	200	17	9	<1	10	2	2.5	5.3
1588	Scallops, imitation, from surimi	4 oz	113	74	112	14	12	0	1	.1	.1	.3
1688	Scallops, steamed/boiled	½ c	60	76	64	10	1	0	2	.3	.7	.6
	Shrimp:											
587	Cooked, boiled, 2 large = 11g	16 ea	88	77	87	18	0	0	1	.2	.2	.5
588	Canned, drained	½ c	64	73	77	15	1	0	1	.3	.3	.7
589	Fried, 2 large = 15 g, breaded	12 ea	90	53	218	19	10	<1	11	2	3	5.9
1057	Raw, large, about 7g each	14 ea	98	76	104	20	1	0	2	.3	.3	1
1589	Shrimp, imitation, from surimi	4 oz	113	75	114	14	10	0	2	.3	.2	1
1053	Snapper, baked or broiled	4 oz	113	70	145	30	0	0	2	.4	.4	.7
1060	Squid, fried in flour	4 oz	113	64	198	20	9	0	8	2.1	3.1	2.4
1590	Surimi	4 oz	113	76	112	17	8	0	1	.2	.2	.6
1058	Swordfish, raw	4 oz	113	76	137	22	0	0	5	1.3	1.7	1
1059	Swordfish, baked or broiled	4 oz	113	69	175	29	0	0	6	1.7	2.2	1.3
590	Trout, baked or broiled	4 oz	113	70	170	26	0	0	7	1.8	2	2.2
	Tuna, light, canned, drained solids:											
591	Oil pack	1 c	145	60	287	42	0	0	12	2.2	4.3	4.2
592	Water pack	1 c	154	74	179	39	0	0	1	.4	.2	.5
1061	Bluefin tuna, fresh	4 oz	113	68	163	26	0	0	6	1.4	1.8	1.6
	MEATS: BEEF, LAMB, PORK and others											
	BEEF, cooked, trimmed to ½" outer fat:											
	Braised, simmered, pot roasted:											
	Relatively fat, choice chuck blade:											
593	Lean and fat, piece 2½ x 2½ x ¾"	4 oz	113	47	393	30	0	0	29	13	14.8	1.2
594	Lean only	4 oz	113	55	297	35	0	0	16	7.3	8.2	.7

PAGE KEY: A–4 = Beverages A–6 = Dairy A–10 = Eggs A–10 = Fat/Oil A–14 = Fruit A–20 = Bakery A–28 = Grain A–32 = Fish A–34 = Meats A–38 = Poultry A–40 = Sausage A–42 = Mixed/Fast A–46 = Nuts/Seeds A–50 = Sweets A–52 = Vegetables/Legumes A–62 = Vegetarian Foods A–64 = Misc A–66 = Soups/Sauces A–68 = Fast A–84 = Convenience A–88 = Baby foods

Chol (mg)	Calc (mg)	Iron (mg)	Magn (mg)	Pota (mg)	Sodi (mg)	Zinc (mg)	VT-A (RE)	Thia (mg)	VT-E (a-TE)	Ribo (mg)	Niac (mg)	V-B6 (mg)	Fola (μg)	VT-C (mg)
62	16	.55	47	276	88	.65	16	.1	.39	.09	2.84	.32	9	1
64	43	.92	36	443	124	.62	17	.13	.92	.13	2.58	.23	10	1
61	23	.54	41	496	69	.63	14	.09	.32	.08	2.48	.28	8	1
118	123	1.07	39	382	329	4.98	2	.12	1.18	.06	3.89	.21	60	4
90	70	.51	68	481	446	6.45	37	.07	1.33	.24	4.27	.2	50	4
120	136	1.13	53	505	450	5.43	2	.11	1.35	.11	1.85	.2	57	4
23	15	.44	49	102	950	.37	23	.04	.12	.03	.2	.03	2	0
63	11	.41	14	146	326	.37	17	.07	.77	.1	1.19	.03	10	0
77	20	.38	65	389	119	.71	12	.09	2.6	.13	2.46	.27	10	0
53	24	1.29	42	537	60	.58	56	.09	.71	.01	.43	.4	11	0
96	63	1.92	46	345	523	.59	33	.08	1.56	.14	4.49	.28	19	<1
87	55	1.58	61	469	862	.56	25	.05	.56	.05	5.73	.45	17	0
46	68	1.21	121	651	78	.6	61	.08	1.23	.1	8.05	.45	16	0
59	87	1.56	154	833	2260	.78	86	.11	1.11	.14	10.8	.64	22	0
36	53	.95	94	509	61	.47	53	.07	.96	.08	6.61	.39	14	0
15	87	1.38	9	78	983	.6	292	.04	1.81	.16	3.73	.19	3	0
104	88	.57	51	510	551	4.23	38	.01	2.1	.1	1.55	.11	16	0
61	155	1.33	44	396	108	.69	16	.15	1.84	.15	2.76	.3	12	1
71	136	1.57	38	323	431	.67	23	.14	2.41	.18	2.68	.24	15	1
54	60	5.99	34	396	260	1.9	51	.03	1.36	.04	2.37	.41	18	6
131	112	16.5	117	387	523	225	74	.25	1.98	.24	3.42	.15	25	9
124	20	12.7	55	417	263	41.2	201	.17	2.11	.58	4.98	.12	25	20
59	45	5.07	42	178	304	63.6	66	.11	1.66	.15	1.2	.05	23	3
125	20	11.5	55	378	265	41.5	183	.16	2.21	.55	4.53	.11	19	16
108	7	.32	82	437	131	.68	26	.08	.32	.09	1.86	.08	4	0
62	241	.95	38	368	626	1.04	19	.03	1.53	.21	7.39	.34	17	0
98	8	.62	35	424	75	.58	71	.24	1.42	.19	7.54	.25	6	0
26	12	.96	20	198	886	.35	29	.03	1.53	.11	5.33	.31	2	0
160	432	3.3	44	449	571	1.48	76	.09	.34	.26	5.93	.19	13	0
57	39	.76	55	310	432	.99	20	.04	1.77	.1	1.4	.13	34	2
25	9	.35	49	116	898	.37	23	.01	.12	.02	.35	.03	2	0
19	15	.15	32	168	246	.55	27	.01	.81	.04	.6	.08	7	1
172	34	2.72	30	160	197	1.37	58	.03	.66	.03	2.28	.11	3	2
111	38	1.75	26	134	108	.81	11	.02	.59	.02	1.77	.07	1	1
159	60	1.13	36	203	310	1.24	50	.12	1.35	.12	2.76	.09	7	1
149	51	2.36	36	181	145	1.09	53	.03	.8	.03	2.5	.1	3	2
41	21	.68	49	101	797	.37	23	.03	.12	.04	.19	.03	2	0
53	45	.27	42	590	64	.5	40	.06	.71	<.01	.39	.52	7	2
294	44	1.14	43	315	346	1.97	12	.06	2.09	.52	2.94	.07	16	5
34	10	.29	49	127	162	.37	23	.02	.28	.02	.25	.03	2	0
44	5	.91	30	325	102	1.3	41	.04	.56	.11	10.9	.37	2	1
56	7	1.18	38	417	130	1.66	46	.05	.71	.13	13.3	.43	3	1
78	97	.43	35	506	63	.58	17	.17	.57	.11	6.52	.39	21	2
26	19	2.02	45	300	513	1.31	33	.05	1.74	.17	18	.16	8	0
46	17	2.36	42	365	521	1.19	26	.05	.82	.11	20.5	.54	6	0
43	9	1.15	56	285	44	.68	740	.27	1.13	.28	9.77	.51	2	0
112	11	3.45	21	275	67	7.57	0	.08	.26	.27	3.54	.32	10	0
120	15	4.16	26	297	80	11.6	0	.09	.16	.32	3.02	.33	7	0

Table H-1

Food Composition

(Computer code number is for West Diet Analysis program) (For purposes of calculations, use "0" for t, <1, <.1, <.01, etc.)

Computer Code Number	Food Description	Measure	Wt (g)	H_2O (%)	Ener (cal)	Prot (g)	Carb (g)	Dietary Fiber (g)	Fat (g)	Fat Breakdown (g) Sat	Mono	Poly
	MEATS: BEEF, LAMB, PORK and others—Continued											
	Relatively lean, like choice round:											
595	Lean and fat, pce 4⅛ x 2½ x ¾"	4 oz	113	52	311	32	0	0	19	8.5	9.7	.8
596	Lean only	4 oz	113	57	249	36	0	0	11	4.8	5.4	.5
	Ground beef, broiled, patty 3 x ⅝":											
597	Extra lean, about 16% fat	4 oz	113	54	299	32	0	0	18	8	9	.8
598	Lean, 21% fat	4 oz	113	53	316	32	0	0	20	8.9	10.1	.8
	Roasts, oven cooked, no added liquid:											
	Relatively fat, prime rib:											
601	Lean and fat, piece 4⅛ x 2¼ x ½"	4 oz	113	46	425	25	0	0	35	15.8	17.9	1.5
602	Lean only	4 oz	113	58	271	31	0	0	16	7	7.9	.7
	Relatively lean, choice round:											
603	Lean and fat, piece 2½ x 2½ x ¾"	4 oz	113	59	272	30	0	0	16	7.1	8.1	.7
604	Lean only	4 oz	113	65	198	33	0	0	6	2.9	3.3	.3
1701	Steak, rib, broiled, lean	4 oz	113	58	250	32	0	0	13	5.7	6.4	.5
	Steak, broiled, relatively lean,											
606	choice sirloin, lean only	4 oz	113	62	228	34	0	0	9	4	4.6	.4
	Steak, broiled, relatively fat,											
	choice T-bone:											
1063	Lean and fat	4 oz	113	52	349	26	0	0	26	11.8	13.3	1.1
1064	Lean only	4 oz	113	61	232	30	0	0	11	5.1	5.8	.5
	Variety meats:											
1086	Brains, panfried	4 oz	113	71	221	14	0	0	18	6.7	7	3.9
599	Heart, simmered	4 oz	113	64	198	32	<1	0	6	2.9	1.5	1.5
600	Liver, fried	4 oz	113	56	245	30	9	0	9	3.1	1.8	1.9
1062	Tongue, cooked	4 oz	113	56	320	25	<1	0	23	10.1	10.7	.9
607	Beef, canned, corned	4 oz	113	58	283	31	0	0	17	7.5	8.5	.7
608	Beef, dried, cured	1 oz	28	56	46	8	<1	0	1	.5	.5	.1
	LAMB, domestic, cooked:											
	Chop, arm, braised (5.6 oz raw w/bone):											
609	Lean and fat	1 ea	70	44	242	21	0	0	17	7.8	7.3	1.4
610	Lean only	1 ea	55	49	153	19	0	0	8	3.6	3.4	.6
	Chop, loin, broiled (4.2 oz raw w/bone):											
611	Lean and fat	1 ea	64	52	202	16	0	0	15	6.8	6.4	1.2
612	Lean only	1 ea	46	61	99	14	0	0	4	2.1	1.9	.4
1067	Cutlet, avg of lean cuts, cooked	4 oz	113	54	330	28	0	0	23	10.9	10.2	1.9
	Leg, roasted, 3 oz = 4⅛ x 2¼ x ½":											
613	Lean and fat	4 oz	113	57	292	29	0	0	19	8.7	8.1	1.6
614	Lean only	4 oz	113	64	216	32	0	0	9	4.1	3.8	.7
615	Rib, roasted, lean and fat	4 oz	113	48	406	24	0	0	34	15.7	14.7	2.8
616	Rib, roasted, lean only	4 oz	113	60	262	30	0	0	15	7	6.6	1.2
1065	Shoulder, roasted, lean and fat	4 oz	113	56	312	25	0	0	23	10.5	9.8	1.9
1066	Shoulder, roasted, lean only	4 oz	113	63	231	28	0	0	12	5.7	5.3	1.1
	Variety meats:											
1069	Brains, panfried	4 oz	113	76	164	14	0	0	11	4.4	3.7	1.9
1068	Heart, braised	4 oz	113	64	209	28	2	0	9	3.9	2.7	1.1
1070	Sweetbreads, cooked	4 oz	113	60	264	26	0	0	17	8.1	6.5	1.4
1071	Tongue, cooked	4 oz	113	58	311	24	0	0	23	8.8	11.6	1.4
	PORK, cured, cooked (see also Sausages and Lunch Meats)											
617	Bacon, medium slices	3 pce	19	13	109	6	<1	0	9	3.3	4.5	1.1
1087	Breakfast strips, cooked	2 pce	23	27	106	7	<1	0	8	2.9	3.8	1.3
618	Canadian-style bacon	2 pce	47	62	87	11	1	0	4	1.3	1.9	.4
	Ham, roasted:											
619	Lean and fat, 2 pces 4⅛ x 2¼ x ¼"	4 oz	113	64	201	25	0	0	10	3.4	5	1.7
620	Lean only	4 oz	113	68	164	24	2	0	6	2.1	3	.6
621	Ham, canned, roasted, 8% fat	4 oz	113	69	154	24	1	0	6	1.8	2.8	.5
	PORK, fresh, cooked:											
	Chops, loin (cut 3 per lb with bone):											
1291	Braised, lean and fat	1 ea	89	58	213	24	0	0	12	4.5	5.4	1
1292	Lean only	1 ea	80	61	163	23	0	0	7	2.7	3.3	.6
622	Broiled, lean and fat	1 ea	82	58	197	23	0	0	11	3.9	4.8	.8
623	Broiled, lean only	1 ea	74	61	149	22	0	0	6	2.2	2.7	.4

Chol (mg)	Calc (mg)	Iron (mg)	Magn (mg)	Pota (mg)	Sodi (mg)	Zinc (mg)	VT-A (RE)	Thia (mg)	VT-E (a-TE)	Ribo (mg)	Niac (mg)	V-B6 (mg)	Fola (μg)	VT-C (mg)
108	7	3.53	25	319	56	5.55	0	.08	.21	.27	4.21	.37	11	0
108	6	3.91	28	348	58	6.19	0	.08	.2	.29	4.61	.41	12	0
112	10	3.13	28	417	93	7.27	0	.08	.2	.36	6.61	.36	12	0
114	14	2.77	27	394	101	7.01	0	.07	.23	.27	6.75	.34	12	0
96	12	2.61	21	334	71	5.92	0	.08	.27	.19	3.8	.26	8	0
91	11	2.95	28	425	84	7.84	0	.09	.14	.24	4.64	.34	9	0
81	7	2.07	27	406	67	4.87	0	.09	.23	.18	3.92	.4	7	0
78	6	2.2	30	446	70	5.36	0	.1	.12	.19	4.24	.43	8	0
90	15	2.9	30	445	78	7.9	0	.11	.16	.25	5.42	.45	9	0
101	12	3.8	36	455	75	7.37	0	.15	.16	.33	4.84	.51	11	0
76	9	3.06	26	363	72	5.03	0	.1	.24	.24	4.46	.37	8	0
67	7	3.58	32	427	80	6	0	.12	.16	.28	5.23	.44	9	0
2254	10	2.51	17	400	179	1.53	0	.15	2.37	.29	4.27	.44	7	4
218	7	8.49	28	263	71	3.54	0	.16	.81	1.74	4.6	.24	2	2
545	12	7.1	26	411	120	6.16	12123	.24	.72	4.68	16.3	1.62	249	26
121	8	3.83	19	203	68	5.42	0	.03	.4	.4	2.43	.18	6	1
97	14	2.35	16	154	1136	4.03	0	.02	.17	.17	2.75	.15	10	0
12	2	1.26	9	124	972	1.47	0	.02	.04	.06	1.53	.1	3	0
84	17	1.67	18	214	50	4.26	0	.05	.1	.17	4.66	.08	13	0
67	14	1.49	16	186	42	4.02	0	.04	.1	.15	3.48	.07	12	0
64	13	1.16	15	209	49	2.23	0	.06	.08	.16	4.54	.08	11	0
44	9	.92	13	173	39	1.9	0	.05	.07	.13	3.15	.07	11	0
110	12	2.26	25	340	77	4.67	0	.12	.15	.32	7.48	.16	19	0
105	12	2.24	27	354	75	4.97	0	.11	.17	.3	7.45	.17	23	0
101	9	2.4	29	382	77	5.58	0	.12	.2	.33	7.16	.19	26	0
110	25	1.81	23	306	82	3.94	0	.1	.11	.24	7.63	.12	17	0
99	24	2	26	356	91	5.05	0	.1	.17	.26	6.96	.17	25	0
104	23	2.23	26	284	75	5.91	0	.1	.16	.27	6.95	.15	24	0
98	21	2.41	28	299	77	6.83	0	.1	.2	.29	6.51	.17	28	0
2308	14	1.9	16	232	151	1.54	0	.12	1.73	.27	2.79	.12	6	14
281	16	6.24	27	212	71	4.16	0	.19	.79	1.34	4.93	.34	2	8
452	14	2.4	21	329	59	3.03	0	.02	.78	.24	2.89	.06	15	23
214	11	2.97	18	179	76	3.38	0	.09	.36	.47	4.17	.19	3	8
16	2	.31	5	92	303	.62	0	.13	.1	.05	1.39	.05	1	0
24	3	.45	6	107	483	.85	0	.17	.08	.08	1.75	.08	1	0
27	5	.38	10	183	727	.8	0	.39	.15	.09	3.25	.21	2	0
67	9	1.51	25	462	1695	2.79	0	.82	.45	.37	6.95	.35	3	0
60	9	1.67	16	324	1359	3.25	0	.85	.29	.23	4.54	.45	3	0
34	7	1.04	24	393	1282	2.52	0	1.18	.29	.28	5.53	.51	6	0
71	19	.95	17	333	43	2.12	2	.56	.3	.23	3.93	.33	3	1
63	14	.9	16	310	40	1.98	2	.53	.3	.21	3.67	.31	3	<1
67	27	.66	20	294	48	1.85	2	.88	.27	.24	4.3	.35	5	<1
61	23	.63	20	278	44	1.76	2	.85	.31	.23	4.1	.35	4	<1

Table H-1

Food Composition

(Computer code number is for West Diet Analysis program) (For purposes of calculations, use "0" for t, <1, <.1, <.01, etc.)

Computer Code Number	Food Description	Measure	Wt (g)	H_2O (%)	Ener (cal)	Prot (g)	Carb (g)	Dietary Fiber (g)	Fat (g)	Fat Breakdown (g) Sat	Mono	Poly
	MEATS: BEEF, LAMB, PORK and others—Continued											
624	Panfried, lean and fat	1 ea	78	53	216	23	0	0	13	4.7	5.5	1.5
625	Panfried, lean only	1 ea	63	59	152	16	0	0	10	3.2	3.9	1.2
626	Leg, roasted, lean and fat	4 oz	113	55	308	30	0	0	20	7.3	8.9	1.9
627	Leg, roasted, lean only	4 oz	113	61	233	35	0	0	9	3.2	4.3	.9
628	Rib, roasted, lean and fat	4 oz	113	56	288	31	0	0	17	6.7	7.9	1.4
629	Rib, roasted, lean only	4 oz	113	59	252	32	0	0	13	4.9	5.9	1
630	Shoulder, braised, lean and fat	4 oz	113	48	372	32	0	0	26	9.6	11.8	2.6
631	Shoulder, braised, lean only	4 oz	113	54	280	36	0	0	14	4.7	6.5	1.3
1088	Spareribs, cooked, yield from 1 lb raw with bone	4 oz	113	40	449	33	0	0	34	12.5	15.3	3.1
1095	Rabbit, roasted (1 cup meat = 140 g)	4 oz	113	61	223	33	0	0	9	4	2	3
	VEAL, cooked:											
632	Cutlet, braised or broiled, 4⅛ x 2¼ x ½"	4 oz	113	52	321	34	0	0	19	7.6	7.6	1.3
633	Rib roasted, lean, 2 pieces 4⅛ x 2¼ x ¼"	4 oz	113	60	258	27	0	0	16	6.1	6.1	1.1
634	Liver, panfried	4 oz	113	67	186	24	3	0	8	3.3	1.9	2.2
1096	Venison (deer meat), roasted	4 oz	113	65	179	34	0	0	4	1.4	1	.7
	MEATS: POULTRY and POULTRY PRODUCTS											
	CHICKEN, cooked:											
	Fried, batter dipped:											
635	Breast	1 ea	280	52	728	69	25	1	37	9.9	15.3	8.6
636	Drumstick	1 ea	72	53	193	16	6	<1	11	3	4.7	2.7
637	Thigh	1 ea	86	51	238	19	8	<1	14	3.8	5.9	3.3
638	Wing	1 ea	49	46	159	10	5	<1	11	2.9	4.5	2.5
	Fried, flour coated:											
639	Breast	1 ea	196	57	435	62	3	<1	17	4.9	7.1	3.8
1212	Breast, without skin	1 ea	86	60	161	29	<1	<1	4	1.1	1.5	.9
640	Drumstick	1 ea	49	57	120	13	1	<1	7	1.8	2.7	1.6
641	Thigh	1 ea	62	54	162	17	2	<1	9	2.5	3.7	2.1
1099	Thigh, without skin	1 ea	52	59	113	15	1	<1	5	1.4	2	1.3
642	Wing	1 ea	32	49	103	8	1	<1	7	1.9	2.9	1.6
	Roasted:											
643	All types of meat	1 c	140	64	266	40	0	0	10	2.9	3.8	2.4
644	Dark meat	1 c	140	63	287	38	0	0	14	3.7	5.2	3.2
645	Light meat	1 c	140	65	242	43	0	0	6	1.8	2.2	1.4
646	Breast, without skin	1 ea	172	65	284	53	0	0	6	1.8	2.2	1.3
647	Drumstick, without skin	1 ea	44	67	76	12	0	0	2	.7	.8	.6
1703	Leg, without skin	1 ea	95	65	181	26	0	0	8	2.2	2.9	1.9
648	Thigh	1 ea	62	59	153	16	0	0	10	2.7	3.9	2.1
1100	Thigh, without skin	1 ea	52	63	109	13	0	0	6	1.6	2.2	1.3
649	Stewed, all types	1 c	140	67	248	38	0	0	9	2.6	3.5	2.2
656	Canned, boneless chicken	4 oz	113	69	186	25	0	0	9	2.5	3.6	2
1102	Gizzards, simmered	1 c	145	67	222	39	2	0	5	1.5	1.3	1.5
1101	Hearts, simmered	1 c	145	65	268	38	<1	0	11	3.3	2.9	3.3
2300	Liver, simmered: Ounce	3 oz	85	68	133	21	1	0	5	1.6	1.1	.8
1098	Liver, simmered: Piece = 20 g	6 ea	120	68	188	29	1	0	7	2.2	1.6	1.1
	DUCK, roasted:											
1293	Meat with skin, about 2.7 cups	½ ea	382	52	1287	73	0	0	108	36.9	49.3	13.9
651	Meat only, about 1.5 cups	½ ea	221	64	444	52	0	0	25	9.2	8.2	3.2
	GOOSE, domesticated, roasted:											
1294	Meat only, about 4.2 cups	½ ea	591	57	1406	173	0	0	75	23.6	40.2	10.9
1295	Meat with skin, about 5.5 cups	½ ea	774	52	2360	195	0	0	170	53.2	80.5	24.8
	TURKEY:											
	Roasted, meat only:											
652	Dark meat	4 oz	113	63	211	33	0	0	8	2.7	1.8	2.5
653	Light meat	4 oz	113	66	177	34	0	0	4	1.2	.6	1
654	All types, chopped or diced	1 c	140	65	238	42	0	0	7	2.3	1.5	2
1103	Ground, cooked	4 oz	113	59	266	31	0	0	15	4.1	5.5	3.6
1106	Gizzard, cooked	2 ea	134	65	218	39	1	0	5	1.5	1	1.5
1107	Heart, cooked	4 ea	64	64	113	17	1	0	4	1.1	.8	1.1
1108	Liver, cooked	1 ea	75	66	127	18	3	0	4	1.4	1.1	.8

PAGE KEY: A–4 = Beverages A–6 = Dairy A–10 = Eggs A–10 = Fat/Oil A–14 = Fruit A–20 = Bakery A–28 = Grain A–32 = Fish A–34 = Meats A–38 = Poultry A–40 = Sausage A–42 = Mixed/Fast A–46 = Nuts/Seeds A–50 = Sweets A–52 = Vegetables/Legumes A–62 = Vegetarian Foods A–64 = Misc A–66 = Soups/Sauces A–68 = Fast A–84 = Convenience A–88 = Baby foods

Chol (mg)	Calc (mg)	Iron (mg)	Magn (mg)	Pota (mg)	Sodi (mg)	Zinc (mg)	VT-A (RE)	Thia (mg)	VT-E (a-TE)	Ribo (mg)	Niac (mg)	V-B6 (mg)	Fola (μg)	VT-C (mg)
72	21	.71	23	332	62	1.8	2	.89	.32	.24	4.37	.37	5	1
52	14	.67	16	230	49	2.44	1	.46	.3	.23	2.8	.26	3	<1
106	16	1.14	25	398	68	3.34	3	.72	.34	.35	5.16	.45	11	<1
108	8	1.29	33	442	73	3.4	3	.91	.46	.4	5.56	.38	3	<1
82	32	1.06	24	476	52	2.33	2	.82	.41	.34	6.92	.37	3	<1
80	29	1.11	25	494	53	2.41	2	.86	.55	.36	7.25	.38	3	<1
123	20	1.82	21	417	99	4.72	3	.61	.5	.35	5.89	.4	5	<1
129	9	2.2	25	458	115	5.62	3	.68	.58	.41	6.71	.46	6	<1
137	53	2.09	27	362	105	5.2	3	.46	.52	.43	6.19	.4	5	0
93	21	2.57	24	433	53	2.57	0	.1	.96	.24	9.53	.53	12	0
133	32	1.23	27	316	90	4.1	0	.04	.45	.34	10.2	.29	16	0
124	12	1.1	25	333	104	4.62	0	.06	.4	.3	7.89	.28	15	0
634	8	2.96	21	232	60	10.8	9095	.15	.42	2.19	9.58	.55	858	35
127	8	5.05	27	379	61	3.11	0	.2	.28	.68	7.58	.42	5	0
238	56	3.5	67	563	770	2.66	56	.32	2.97	.41	29.4	1.2	42	0
62	12	.97	14	134	194	1.68	19	.08	.88	.15	3.67	.19	13	0
80	15	1.25	18	165	248	1.75	25	.1	1.05	.19	4.92	.22	16	0
39	10	.63	8	68	157	.68	17	.05	.52	.07	2.58	.15	9	0
174	31	2.33	59	508	149	2.16	29	.16	1.12	.26	26.9	1.14	12	0
78	14	.98	27	237	68	.93	6	.07	.36	.11	12.7	.55	3	0
44	6	.66	11	112	44	1.42	12	.04	.41	.11	2.96	.17	5	0
60	9	.92	15	147	55	1.56	18	.06	.52	.15	4.31	.2	7	0
53	7	.76	13	135	49	1.45	11	.05	.3	.13	3.7	.2	5	0
26	5	.4	6	57	25	.56	12	.02	.18	.04	2.14	.13	2	0
125	21	1.69	35	340	120	2.94	22	.1	.58	.25	12.8	.66	8	0
130	21	1.86	32	336	130	3.92	31	.1	.81	.32	9.17	.5	11	0
119	21	1.48	38	346	108	1.72	13	.09	.37	.16	17.4	.84	6	0
146	26	1.79	50	440	127	1.72	10	.12	.66	.2	23.6	1.03	7	0
41	5	.57	11	108	42	1.4	8	.03	.25	.1	2.68	.17	4	0
89	11	1.24	23	230	86	2.72	18	.07	.55	.22	6	.35	8	0
58	7	.83	14	138	52	1.46	30	.04	.35	.13	3.95	.19	4	0
49	6	.68	12	124	46	1.34	10	.04	.3	.12	3.4	.18	4	0
116	20	1.64	29	252	98	2.79	21	.07	.42	.23	8.57	.36	8	0
70	16	1.79	14	156	568	1.59	38	.02	.24	.15	7.15	.4	5	2
281	14	6.02	29	260	97	6.35	81	.04	2.29	.35	5.77	.17	77	2
351	28	13.1	29	191	70	10.6	13	.1	2.32	1.07	4.06	.46	116	3
536	12	7.2	18	119	43	3.69	4176	.13	1.45	1.49	3.78	.49	655	13
757	17	10.2	25	168	61	5.21	5895	.18	2.04	2.1	5.34	.7	924	19
321	42	10.3	61	779	225	7.11	241	.66	2.5	1.03	18.5	.69	23	0
197	26	5.97	44	557	144	5.75	51	.57	1.55	1.04	11.3	.55	22	0
567	83	17	148	2293	449	18.7	71	.54	9.16	2.3	24.1	2.78	71	0
704	101	21.9	170	2546	542	20.3	163	.6	13.5	2.5	32.3	2.86	15	0
96	36	2.63	27	328	89	5.04	0	.07	.94	.28	4.12	.41	10	0
78	21	1.53	32	345	72	2.31	0	.07	.12	.15	7.73	.61	7	0
106	35	2.49	36	417	98	4.34	0	.09	.59	.25	7.62	.64	10	0
115	28	2.18	27	305	121	3.23	0	.06	.45	.19	5.45	.44	8	0
311	20	7.29	25	283	72	5.57	74	.04	.27	.44	4.11	.16	70	2
145	8	4.41	14	117	35	3.37	5	.04	.13	.56	2.08	.2	51	1
470	8	5.85	11	146	48	2.32	2805	.04	2.41	1.07	4.46	.39	500	1

Table H-1

Food Composition

(Computer code number is for West Diet Analysis program) (For purposes of calculations, use "0" for t, <1, <.1, <.01, etc.)

Computer Code Number	Food Description	Measure	Wt (g)	H_2O (%)	Ener (cal)	Prot (g)	Carb (g)	Dietary Fiber (g)	Fat (g)	Fat Breakdown (g) Sat	Mono	Poly
	MEATS: POULTRY and POULTRY PRODUCTS—Continued											
	POULTRY FOOD PRODUCTS (see also items in Sausages & Lunchmeats section):											
1567	Chicken patty, breaded, cooked	1 ea	75	49	213	12	11	<1	13	4.1	6.4	1.6
659	Turkey and gravy, frozen package	3 oz	85	85	57	5	4	<1	2	.8	.8	.4
	Turkey breast, Louis Rich:											
1104	Barbecued	2 oz	56	72	58	12	2	0	<1	.2	.2	.1
1943	Hickory smoked	1 pce	80		80	16	2	0	1	0		
1947	Honey roasted	1 pce	80		80	16	3	0	1	.5		
1945	Oven roasted	1 pce	80		70	16		0	1	0		
661	Turkey patty, breaded, fried	2 oz	57	50	161	8	9	<1	10	2.7	4.3	2.7
662	Turkey, frozen, roasted, seasoned	4 oz	113	68	175	24	3	0	7	2.1	1.4	1.9
1704	Turkey roll, light meat	1 pce	28	72	41	5	<1	0	2	.6	.7	.5
	MEATS: SAUSAGES and LUNCHMEATS (see also Poultry Food Products)											
1072	Beerwurst/beer salami, beef	1 oz	28	53	92	3	<1	0	8	3.6	3.9	.3
1074	Beerwurst/beer salami, pork	1 oz	28	61	67	4	1	0	5	1.8	2.5	.7
1075	Berliner sausage	1 oz	28	61	64	4	1	0	5	1.7	2.2	.4
	Bologna:											
1297	Beef	1 pce	23	55	72	3	<1	0	7	2.8	3.2	.3
2115	Beef, light, Oscar Mayer	1 pce	28	65	56	3	2	0	4	1.6	2	.1
663	Beef & pork	1 pce	28	54	88	3	1	0	8	3	3.7	.7
2155	Healthy Favorites	1 pce	23		22	3	1	0	<1	0		
1298	Pork	1 pce	23	61	57	4	<1	0	5	1.6	2.2	.5
2114	Regular, light, Oscar Mayer	1 pce	28	65	56	3	2	0	4	1.6	2	.4
664	Turkey	1 pce	28	65	56	4	<1	0	4	1.4	1.3	1.2
1970	Turkey, Louis Rich	1 pce	28	67	57	3	<1	0	5	1.5	1.8	1.3
665	Braunschweiger sausage	2 pce	57	48	205	8	2	0	18	6.2	8.5	2.1
1073	Bratwurst, link	1 ea	70	51	226	10	2	0	19	6.9	9.3	2
666	Brown & serve sausage links, cooked	2 ea	26	45	102	4	1	0	10	3.4	4.4	1
1089	Cheesefurter/cheese smokie	2 ea	86	52	281	12	1	0	25	9	11.8	2.6
2157	Chicken breast, Healthy Favorites	4 pce	52		40	9	1	0	0	0	0	0
1556	Chorizo, pork & beef	1 ea	60	32	273	15	1	0	23	8.6	11	2.1
1090	Corned beef loaf, jellied	1 pce	28	69	43	6	0	0	2	.7	.7	.1
	Frankfurters:											
1077	Beef, large link, 8/package	1 ea	57	55	180	7	1	0	16	6.9	7.9	.8
1078	Beef and pork, large link, 8/package	1 ea	57	54	182	6	1	0	17	6.2	8	1.6
667	Beef and pork, small link, 10/pkg	1 ea	45	54	144	5	1	0	13	4.9	6.3	1.2
668	Turkey frankfurter, 10/package	1 ea	45	63	102	6	1	0	8	2.7	2.5	2.2
1968	Turkey/chicken frank 8/pkg	1 ea	43		80	6	1	0	6	2		
	Ham:											
669	Ham lunchmeat, canned, 3 x 2 x ½"	1 pce	21	52	70	3	<1	0	6	2.3	3	.7
670	Chopped ham, packaged	2 pce	42	64	96	7	0	0	7	2.4	3.4	.9
2156	Honey ham, Healthy Favorites	4 pce	52	73	55	9	2	0	1	.4	.8	.1
2113	Oscar Mayer lower sodium ham	1 pce	21	73	23	3	1	0	1	.3	.4	.1
673	Turkey ham lunchmeat	2 pce	57	71	73	11	<1	0	3	1	.7	.9
1091	Kielbasa sausage	1 pce	26	54	81	3	1	0	7	2.6	3.4	.8
1092	Knockwurst sausage, link	1 ea	68	55	209	8	1	0	19	6.9	8.7	2
1093	Mortadella lunchmeat	2 pce	30	52	93	5	1	0	8	2.8	3.4	.9
1097	Olive loaf lunchmeat	2 pce	57	58	134	7	5	<1	9	3.3	4.5	1.1
1952	Turkey breast, fat free	1 pce	28	77	22	4	1	0	<1	.1	.1	t
1080	Turkey pastrami	2 pce	57	71	80	10	1	0	4	1	1.2	.9
1969	Turkey salami	1 pce	28	72	41	4	<1	0	3	.9	1	.8
1081	Pepperoni sausage	2 pce	11	27	55	2	<1	0	5	1.8	2.3	.5
1094	Pickle & pimento loaf	2 pce	57	57	149	7	3	<1	12	4.5	5.5	1.5
1082	Polish sausage	1 oz	28	53	91	4	<1	0	8	2.9	3.8	.9
674	Pork sausage, cooked, link, small	2 ea	26	45	96	5	<1	0	8	2.8	4.1	.8
1079	Pork sausage, cooked, patty	4 oz	113	45	417	22	1	0	35	12.1	17.7	3.3
675	Salami, pork and beef	2 pce	57	60	143	8	1	0	11	4.6	5.2	1.1
677	Salami, pork and beef, dry	3 pce	30	35	125	7	1	0	10	3.7	5.1	1
676	Salami, turkey	2 pce	57	66	112	9	<1	0	8	2.3	2.6	2
	Sandwich spreads:											
1300	Ham salad spread	2 tbs	30	63	65	3	3	0	5	1.5	2.2	.8
678	Pork and beef	2 tbs	30	60	70	2	4	<1	5	1.8	2.3	.8
1296	Chicken/turkey	2 tbs	26	66	52	3	2	0	4	.9	.8	1.6

Chol (mg)	Calc (mg)	Iron (mg)	Magn (mg)	Pota (mg)	Sodi (mg)	Zinc (mg)	VT-A (RE)	Thia (mg)	VT-E (a-TE)	Ribo (mg)	Niac (mg)	V-B6 (mg)	Fola (µg)	VT-C (mg)
45	12	.94	15	185	399	.78	22	.07	1.46	.1	5.04	.23	8	<1
15	12	.79	7	52	471	.59	11	.02	.3	.11	1.53	.08	3	0
25	14	.62	16	175	599	.59	0	.02		.06	5.35	.22	2	0
35	0	.72			1060		0							0
35	0	.72			940		0							0
35	0				910		0							0
35	8	1.25	9	157	456	.82	6	.06	1.36	.11	1.31	.11	16	0
60	6	1.84	25	337	768	2.87	0	.05	.43	.18	7.09	.3	6	0
12	11	.36	4	70	137	.44	0	.02	.04	.06	1.96	.09	1	0
17	3	.42	3	49	288	.68	0	.02	.05	.03	.95	.05	1	0
16	2	.21	4	71	347	.48	0	.15	.06	.05	.91	.1	1	0
13	3	.32	4	79	363	.69	0	.11	.06	.06	.87	.06	1	0
13	3	.38	3	36	226	.5	0	.01	.04	.02	.55	.03	1	0
13	4	.34	4	44	314	.53	0							0
15	3	.42	3	50	285	.54	0	.05	.06	.04	.72	.05	1	0
7		.18			255									
14	3	.18	3	65	272	.47	0	.12	.06	.04	.9	.06	1	0
15	14	.39	5	46	312	.45	0							0
28	23	.43	4	56	246	.49	0	.01	.15	.05	.99	.06	2	0
22	34	.45	5	51	242	.57	0	.01		.05	1.08	.05		0
89	5	5.34	6	113	652	1.6	2405	.14	.2	.87	4.77	.19	25	0
44	34	.72	11	197	778	1.47	0	.17	.19	.16	2.31	.09	3	0
16	2	.62	4	70	248	.3	0	.21	.06	.09	.96	.06	1	0
58	50	.93	11	177	931	1.94	33	.21	.27	.14	2.49	.11	3	0
25		.72			620									
53	5	.95	11	239	741	2.05	0	.38	.13	.18	3.08	.32	1	0
13	3	.57	3	28	267	1.15	0	0	.05	.03	.49	.03	2	0
35	11	.81	2	95	585	1.24	0	.03	.11	.06	1.38	.07	2	0
28	6	.66	6	95	638	1.05	0	.11	.14	.07	1.5	.07	2	0
22	5	.52	4	75	504	.83	0	.09	.11	.05	1.18	.06	2	0
48	48	.83	6	81	642	1.4	0	.02	.28	.08	1.86	.1	4	0
40	60	1.08			480		0							0
13	1	.15	2	45	271	.31	0	.08	.05	.04	.66	.04	1	<1
21	3	.35	7	134	576	.81	0	.26	.11	.09	1.63	.15	<1	0
24	6	.7	18	144	635	1.02	0							0
9	1	.3	5	197	174	.42	0							0
32	6	1.57	9	185	568	1.68	0	.03	.36	.14	2.01	.14	3	0
17	11	.38	4	70	280	.52	0	.06	.06	.06	.75	.05	1	0
39	7	.62	7	135	687	1.13	0	.23	.39	.09	1.86	.12	1	0
17	5	.42	3	49	374	.63	0	.04	.07	.05	.8	.04	1	0
22	62	.31	11	169	846	.79	34	.17	.14	.15	1.05	.13	1	0
9	3	.34	8	59	387	.24	0							0
31	5	.95	8	148	596	1.23	0	.03	.12	.14	2.01	.15	3	0
21	11	.35	6	61	281	.65	0							0
9	1	.15	2	38	224	.27	0	.03	.02	.03	.55	.03	<1	0
21	54	.58	10	194	792	.8	4	.17	.14	.14	1.17	.11	3	0
20	3	.4	4	66	245	.54	0	.14	.06	.04	.96	.05	1	<1
22	8	.33	4	94	336	.65	0	.19	.07	.07	1.18	.09	1	<1
94	36	1.42	19	408	1462	2.84	0	.84	.29	.29	5.11	.37	2	2
37	7	1.52	9	113	607	1.22	0	.14	.12	.21	2.02	.12	1	0
24	2	.45	5	113	558	.97	0	.18	.08	.09	1.46	.15	1	0
47	11	.92	9	139	572	1.03	0	.04	.32	.1	2.01	.14	2	0
11	2	.18	3	45	274	.33	0	.13	.52	.04	.63	.04	<1	0
11	4	.24	2	33	304	.31	3	.05	.52	.04	.52	.04	1	0
8	3	.16	3	48	98	.27	11	.01	.57	.02	.43	.03	1	<1

Table H-1

Food Composition

(Computer code number is for West Diet Analysis program) (For purposes of calculations, use "0" for t, <1, <.1, <.01, etc.)

Computer Code Number	Food Description	Measure	Wt (g)	H_2O (%)	Ener (cal)	Prot (g)	Carb (g)	Dietary Fiber (g)	Fat (g)	Fat Breakdown (g) Sat	Mono	Poly
	MEATS: SAUSAGES and LUNCHMEATS—Continued											
1084	Smoked link sausage, beef and pork	1 ea	68	52	228	9	1	0	21	7.2	9.7	2.2
1083	Smoked link sausage, pork	1 ea	68	39	265	15	1	0	22	7.7	9.9	2.6
1085	Summer sausage	2 pce	46	51	154	7	<1	0	14	5.5	6	.6
1076	Turkey breakfast sausage	1 pce	28	60	64	6	0	0	5	1.6	1.8	1.2
679	Vienna sausage, canned	2 ea	32	60	89	3	1	0	8	3	4	.5
	MIXED DISHES and FAST FOODS											
	MIXED DISHES:											
1445	Almond Chicken	1 c	242	77	275	20	18	4	14	2	5.3	5.8
1981	Baked beans, fat free, honey	½ c	120	73	110	7	24	7	0	0	0	0
1454	Bean cake	1 ea	32	23	130	2	16	1	7	1	2.9	2.6
680	Beef stew w/ vegetables, homemade	1 c	245	82	218	16	15	2	10	4.9	4.5	.5
1109	Beef stew w/ vegetables, canned	1 c	245	82	194	14	17	2	8	2.4	3.1	.3
1116	Beef, macaroni, tomato sauce casserole	1 c	226	76	255	16	26	2	10	3.8	4.1	.5
2295	Beef fajita	1 ea	223	63	409	17	46	4	17	5.1	7.6	3.9
1265	Beef flauta	1 ea	113	49	360	16	13	2	27	4.9	11.6	9.1
681	Beef pot pie, homemade	1 pce	210	55	517	21	39	3	30	8.4	14.7	7.3
1898	Broccoli, batter fried	1 c	85	74	123	3	9	2	9	1.3	2.2	4.9
1462	Buffalo wings/spicy chicken wings	2 pce	32	53	98	8	<1	<1	7	1.8	2.8	1.6
1675	Carrot raisin salad	½ c	88	58	204	1	21	2	14	2	3.9	7.3
2248	Cheeseburger deluxe	1 ea	219	52	563	28	38		33	15	12.6	2
682	Chicken à la king, homemade	1 c	245	68	468	27	12	1	34	12.7	14.3	6.2
683	Chicken & noodles, homemade	1 c	240	71	367	22	26	2	18	5.9	7.1	3.5
684	Chicken chow mein, canned	1 c	250	89	95	6	18	2	1	0	.1	.8
685	Chicken chow mein, homemade	1 c	250	78	255	31	10	1	10	2.4	4.3	3.1
1266	Chicken fajita	1 ea	223	61	405	22	50	4	13	2.4	6	3.5
1264	Chicken flauta	1 ea	113	52	343	14	13	2	27	4.3	11.1	9.6
686	Chicken pot pie, homemade (⅓)	1 pce	232	57	545	23	42	3	31	10.9	14.5	5.8
1672	Chili con carne	½ c	127	77	128	12	11	2	4	1.7	1.7	.3
1112	Chicken salad with celery	½ c	78	53	268	11	1	<1	25	3.1	4.5	15.8
1382	Chicken teriyaki, breast	1 ea	128	67	176	26	7	<1	4	.9	1	.9
687	Chili with beans, canned	1 c	256	75	287	15	30	11	14	6	6	.9
1479	Chinese pastry	1 oz	28	46	67	1	13	<1	1	.2	.4	.8
688	Chop suey with beef & pork	1 c	220	63	425	22	31	3	24	5	8.6	9.3
690	Coleslaw	1 c	132	74	195	2	17	2	15	2.1	3.2	8.5
689	Corn pudding	1 c	250	76	273	11	32	4	13	6.4	4.3	1.8
1110	Corned beef hash, canned	1 c	220	67	398	19	23	1	25	11.9	10.9	.9
1255	Deviled egg (½ egg + filling)	1 ea	31	69	62	4	<1	0	5	1.2	1.7	1.5
	Egg foo yung patty:											
1467	Meatless	1 ea	86	78	113	6	3	1	8	1.9	3.3	2.1
1458	With beef	1 ea	86	74	129	9	3	<1	9	2.2	3.2	2.4
1465	With chicken	1 ea	86	74	130	9	4	<1	9	2.1	3.1	2.5
1602	Egg roll, meatless	1 ea	64	70	102	3	10	1	6	1.2	2.5	1.6
1550	Egg roll, with meat	1 ea	64	66	114	5	9	1	6	1.5	2.7	1.5
1113	Egg salad	1 c	183	57	586	17	3	0	56	10.6	17.4	24.2
691	French toast w/wheat bread, homemade	1 pce	65	54	151	5	16	<1	7	2	3	1.7
1355	Green pepper, stuffed	1 ea	172	75	229	11	20	2	11	5	4.9	.5
1487	Hot & sour soup (Chinese)	1 c	244	88	133	12	5	<1	6	2	2.5	1
2242	Hamburger deluxe	1 ea	110	49	279	13	27		13	4.1	5.3	2.6
1997	Hummous/hummus	¼ c	62	65	106	3	12	3	5	.8	2.2	2
	Lasagna:											
1346	With meat, homemade	1 pce	245	67	382	22	39	3	15	7.7	5	.9
1111	Without meat, homemade	1 pce	218	69	298	15	39	3	9	5.4	2.4	.6
1117	Frozen entree	1 ea	340	75	390	24	42	4	14	6.7	5.5	.8
1606	Lo mein, meatless	1 c	200	82	134	6	27	3	1	.1	.1	.3
1607	Lo mein, with meat	1 c	200	70	285	17	31	2	10	1.9	2.9	4.5
692	Macaroni & cheese, canned	1 c	240	80	228	9	26	1	10	4.2	3.1	1.4
693	Macaroni & cheese, homemade	1 c	200	58	430	17	40	1	22	8.9	8.8	3.6
1115	Macaroni salad, no cheese	1 c	177	60	461	5	28	2	37	4	6	25.5
1120	Meat loaf, beef	1 pce	87	63	182	16	4	<1	11	4.4	4.7	.5
1119	Meat loaf, beef and pork (⅓)	1 pce	87	60	205	15	4	<1	14	5.2	6.3	.9
1303	Moussaka (lamb & eggplant)	1 c	250	82	237	16	13	4	13	4.6	5.4	1.9

PAGE KEY: A–4 = Beverages A–6 = Dairy A–10 = Eggs A–10 = Fat/Oil A–14 = Fruit A–20 = Bakery A–28 = Grain A–32 = Fish A–34 = Meats A–38 = Poultry A–40 = Sausage A–42 = Mixed/Fast A–46 = Nuts/Seeds A–50 = Sweets A–52 = Vegetables/Legumes A–62 = Vegetarian Foods A–64 = Misc A–66 = Soups/Sauces A–68 = Fast A–84 = Convenience A–88 = Baby foods

Chol (mg)	Calc (mg)	Iron (mg)	Magn (mg)	Pota (mg)	Sodi (mg)	Zinc (mg)	VT-A (RE)	Thia (mg)	VT-E (a-TE)	Ribo (mg)	Niac (mg)	V-B6 (mg)	Fola (μg)	VT-C (mg)
48	7	.99	8	129	643	1.43	0	.18	.15	.12	2.2	.12	1	0
46	20	.79	13	228	1020	1.92	0	.48	.17	.17	3.08	.24	3	1
34	6	1.17	6	125	571	1.18	0	.07	.1	.15	1.98	.12	1	0
23	5	.51	6	75	188	.96	0	.03	.14	.08	1.4	.08	1	0
17	3	.28	2	32	305	.51	0	.03	.07	.03	.51	.04	1	0
35	81	2	59	551	615	1.54	75	.08	2.64	.19	8.59	.42	31	10
0	40	2.7			135		450							12
0	3	.67	6	57	55	.16	0	.07	1.14	.05	.55	.02	9	0
64	29	2.94	40	613	292	5.29	568	.15	.49	.17	4.66	.28	37	17
34	29	2.21	39	426	1006	4.24	262	.07	.34	.12	2.45	.2	31	7
39	26	2.7	40	522	862	3.14	97	.22	.57	.22	4.31	.29	20	14
26	76	3.69	38	427	850	2.38	52	.46	2.08	.3	4.73	.32	25	29
45	50	2.15	29	292	187	4.18	15	.07	4	.15	2.13	.25	10	14
44	29	3.78	6	334	596	3.17	519	.29	3.78	.29	4.83	.24	29	6
16	67	.94	20	242	62	.38	102	.08	2.1	.13	.75	.11	43	53
26	5	.4	6	59	61	.56	17	.01	.23	.04	2.06	.13	1	<1
10	26	.75	14	317	118	.19	1452	.08	5.03	.05	.64	.22	9	5
88	206	4.66	44	445	1108	4.6	129	.39	1.18	.46	7.38	.28	81	8
186	127	2.45	20	404	760	1.8	272	.1	.98	.42	5.39	.23	11	12
96	26	2.16	26	149	600	1.53	10	.05		.17	4.32	.19	10	0
7	45	1.25	14	418	725	1.3	28	.05	.05	.1	1	.09	12	12
77	57	2.5	28	473	718	2.12	50	.07	.75	.22	4.25	.41	19	10
41	83	3.7	51	532	439	1.77	55	.48	2.04	.37	6.64	.35	41	22
37	52	.97	27	243	189	1.18	21	.05	4.06	.1	3.21	.22	8	14
72	70	3.02	25	343	594	2	735	.32	3.25	.32	4.87	.46	29	5
67	34	2.6	23	347	505	1.79	84	.06	.81	.57	1.24	.16	23	1
48	16	.62	11	138	201	.79	31	.03	6.27	.07	3.28	.34	8	1
80	27	1.75	36	309	1866	1.94	16	.08	.35	.2	8.69	.46	13	3
43	120	8.78	115	934	1336	5.12	87	.12	1.88	.27	.92	.34	58	4
0	8	.51	6	28	3	.18	<1	.05	.25	<.01	.41	.02	1	0
46	39	4.16	54	515	818	3.52	134	.36	1.82	.37	5.63	.44	44	20
7	45	.96	12	236	356	.26	66	.05	5.28	.04	.11	.14	51	11
250	100	1.4	37	403	138	1.25	90	1.03	.52	.32	2.47	.29	63	7
73	29	4.4	36	440	1188	3.3	0	.02	.48	.2	4.62	.43	20	0
121	15	.35	3	36	94	.3	49	.02	.86	.14	.02	.05	13	0
184	31	1.04	12	118	310	.7	86	.04	1.57	.25	.44	.09	30	5
180	26	1.11	11	145	184	1.16	92	.05	1.79	.24	.74	.15	22	3
182	27	.86	12	144	187	.81	95	.05	1.87	.25	.96	.13	22	3
30	12	.76	9	98	306	.25	15	.08	.81	.1	.81	.05	12	3
37	13	.78	10	124	304	.46	14	.16	.78	.13	1.31	.1	9	2
574	74	1.8	13	180	665	1.42	260	.08	8.87	.66	.09	.46	62	0
76	64	1.09	11	86	311	.44	81	.13	.31	.21	1.06	.05	15	<1
34	16	1.77	20	233	201	2.28	44	.15	.75	.1	2.74	.3	17	55
23	29	1.83	27	351	1562	1.17	2	.19	.12	.22	4.58	.15	12	1
26	63	2.63	22	227	504	2.06	9	.23	.82	.2	3.69	.12	52	2
0	31	.97	18	108	151	.68	1	.06	.62	.03	.25	.25	37	5
56	258	3.22	50	461	745	3.25	158	.23	1.15	.33	3.97	.21	19	15
31	252	2.5	44	375	714	1.77	156	.22	1.07	.27	2.49	.17	17	15
55	263	3.44	64	752	823	3.7	248	.29	3.45	.39	5.07	.32	28	41
0	47	2.06	33	389	623	.92	130	.23	.35	.24	2.83	.19	49	12
30	25	2.11	39	246	276	1.63	6	.37	1.51	.24	4.25	.28	41	8
24	199	.96	31	139	730	1.2	73	.12	.14	.24	.96	.02	8	<1
42	362	1.8	37	240	1086	1.2	234	.2	.12	.4	1.8	.05	10	1
27	31	1.56	20	170	352	.53	44	.18	10.3	.1	1.43	.33	20	4
84	29	1.61	14	187	145	3.23	23	.05	.31	.22	2.61	.15	11	1
84	33	1.42	14	213	381	2.68	23	.19	.32	.22	2.68	.17	10	1
97	68	1.79	40	557	432	2.56	105	.15	.81	.31	4.14	.23	45	6

Table H-1

Food Composition

(Computer code number is for West Diet Analysis program) (For purposes of calculations, use "0" for t, <1, <.1, <.01, etc.)

Computer Code Number	Food Description	Measure	Wt (g)	H_2O (%)	Ener (cal)	Prot (g)	Carb (g)	Dietary Fiber (g)	Fat (g)	Fat Breakdown (g) Sat	Mono	Poly
	MIXED DISHES and FAST FOODS—Continued											
1899	Mushrooms, batter fried	5 ea	70	66	148	2	8	1	12	2.1	3	6.4
715	Potato salad with mayonnaise and eggs	½ c	125	76	179	3	14	2	10	1.8	3.1	4.7
1674	Pizza, combination, 1⁄12 of 12″ round	1 pce	79	48	184	13	21		5	1.5	2.5	.9
1673	Pizza, pepperoni, 1⁄12 of 12″ round	1 pce	71	46	181	10	20		7	2.2	3.1	1.2
694	Quiche Lorraine ⅛ of 8″ quiche	1 pce	176	54	508	20	20	1	39	17.6	13.8	4.9
1449	Ramen noodles, cooked	1 c	227	82	156	6	29	3	2	.4	.5	.5
1671	Ravioli, meat	½ c	125	68	194	10	18	1	9	2.9	3.6	1
1597	Fried rice (meatless)	1 c	166	68	264	5	34	1	12	1.7	3	6.3
2142	Roast beef hash	½ c	117	66	230	9	11	1	16	7	5.8	3.2
	Spaghetti (enriched) in tomato sauce											
	With cheese:											
695	Canned	1 c	250	80	190	5	38	2	1	0	.4	.5
696	Home recipe	1 c	250	77	260	9	37	2	9	2	5.4	1.2
	With meatballs:											
697	Canned	1 c	250	78	258	12	28	6	10	2.1	3.9	3.9
698	Home recipe	1 c	248	70	332	19	39	8	12	3.3	6.3	2.2
716	Spinach soufflé	1 c	136	74	219	11	3	3	19	9.5	5.7	2.2
1553	Sweet & sour pork	1 c	226	77	231	15	25	1	8	2.2	2.9	2.4
1263	Sweet & sour chicken breast	1 ea	131	79	117	8	15	1	3	.6	.8	1.5
1515	Three bean salad	1 c	150	82	139	4	13	3	8	1.2	1.9	4.9
717	Tuna salad	1 c	205	63	383	33	19	0	19	3.2	5.9	8.4
1121	Tuna noodle casserole, homemade	1 c	202	75	237	17	25	1	7	1.9	1.5	3.2
1270	Waldorf salad	1 c	137	58	408	4	13	2	40	4.1	7.3	27
	FAST FOODS and SANDWICHES (see end of this appendix for additional Fast Foods)											
699	Burrito, beef & bean	1 ea	116	52	255	11	33	3	9	4.2	3.5	.6
700	Burrito, bean	1 ea	109	52	225	7	36	4	7	3.5	2.4	.6
2106	Burrito, chicken con queso	1 ea	306	76	280	12	53	5	6	1.5		
701	Cheeseburger with bun, regular	1 ea	154	55	359	18	28		20	9.2	7.2	1.5
702	Cheeseburger with bun, 4-oz patty	1 ea	166	51	417	21	35		21	8.7	7.8	2.7
703	Chicken patty sandwich	1 ea	182	47	515	24	39	1	29	8.5	10.4	8.4
704	Corndog	1 ea	175	47	460	17	56		19	5.2	9.1	3.5
1922	Corndog, chicken	1 ea	113	52	271	13	26		13			
705	Enchilada	1 ea	163	63	319	10	28		19	10.6	6.3	.8
706	English muffin with egg, cheese, bacon	1 ea	146	49	383	20	31	1	20	9	6.8	2.1
	Fish sandwich:											
707	Regular, with cheese	1 ea	183	45	523	21	48	<1	28	8.1	8.9	9.4
708	Large, no cheese	1 ea	158	47	431	17	41	<1	23	5.2	7.7	8.2
709	Hamburger with bun, regular	1 ea	107	45	275	14	33	1	10	3.5	3.7	1.8
710	Hamburger with bun, 4-oz patty	1 ea	215	50	576	32	39		32	12	14.1	2.8
711	Hotdog/frankfurter with bun	1 ea	98	54	242	10	18		14	5.1	6.8	1.7
	Lunchables:											
2129	Bologna & American cheese	1 ea	128		450	18	19	0	34	15		
2130	Ham & cheese	1 ea	128		320	22	19	0	17	8		
2117	Honey ham & Amer. w/choc pudding	1 ea	176		390	18	34	<1	20	9		
2118	Honey turkey & cheddar w/Jello	1 ea	163		320	17	27	<1	16	9		
2131	Pepperoni & American cheese	1 ea	128		480	20	19	0	36	17		
2125	Salami & American cheese	1 ea	128		430	18	18	0	32	15		
2127	Turkey & cheddar cheese	1 ea	128		360	20	20	1	22	11		
712	Pizza, cheese, ⅛ of 15″ round	1 pce	63	48	140	8	20	1	3	1.5	1	.5
	SANDWICHES:											
	Avocado, chesse, tomato & lettuce:											
1276	On white bread, firm	1 ea	210	58	478	15	41	5	29	8.8	11.3	7.2
1278	On part whole wheat	1 ea	201	59	444	14	34	7	29	8.6	11.4	7.3
1277	On whole wheat	1 ea	214	58	468	16	40	8	30	8.7	11.6	7.5
	Bacon, lettuce & tomato sandwich:											
1137	On white bread, soft	1 ea	124	53	308	10	28	2	18	4.5	6.1	6.1
1139	On part whole wheat	1 ea	124	54	303	10	26	3	17	4.3	6.2	6.1
1138	On whole wheat	1 ea	137	53	328	12	32	4	18	4.4	6.4	6.3

PAGE KEY: A–4 = Beverages A–6 = Dairy A–10 = Eggs A–10 = Fat/Oil A–14 = Fruit A–20 = Bakery A–28 = Grain A–32 = Fish A–34 = Meats A–38 = Poultry A–40 = Sausage A–42 = Mixed/Fast A–46 = Nuts/Seeds A–50 = Sweets A–52 = Vegetables/Legumes A–62 = Vegetarian Foods A–64 = Misc A–66 = Soups/Sauces A–68 = Fast A–84 = Convenience A–88 = Baby foods

Chol (mg)	Calc (mg)	Iron (mg)	Magn (mg)	Pota (mg)	Sodi (mg)	Zinc (mg)	VT-A (RE)	Thia (mg)	VT-E (a-TE)	Ribo (mg)	Niac (mg)	V-B6 (mg)	Fola (μg)	VT-C (mg)
14	54	.76	8	180	121	.42	10	.07	.92	.22	1.65	.05	8	1
85	24	.81	19	318	661	.39	41	.1	2.33	.07	1.11	.18	8	12
20	101	1.53	18	179	382	1.11	101	.21		.17	1.96	.09	32	2
14	65	.94	9	153	267	.52	55	.13		.23	3.05	.06	37	2
205	201	1.9	27	271	549	1.66	243	.23	1.91	.44	4.71	.19	17	3
38	20	1.89	24	51	1349	.76	204	.22	.09	.1	1.75	.06	9	<1
84	32	2.03	20	259	619	1.67	94	.15	1.52	.22	2.95	.14	14	11
42	30	1.84	24	134	286	.89	21	.21	2.46	.11	2.25	.15	22	4
40	10	.9	22	362	695	2.99	0	.09		.12	2.33	.3	12	0
7	40	2.75	21	303	955	1.12	120	.35	2.13	.27	4.5	.13	6	10
7	80	2.25	26	408	955	1.3	140	.25	2.75	.17	2.25	.2	8	12
22	52	3.25	20	245	1220	2.39	100	.15	1.5	.17	2.25	.12	5	5
74	124	3.72	40	665	1009	2.45	159	.25	1.64	.3	3.97	.2	10	22
184	230	1.35	38	201	763	1.29	675	.09	1.22	.3	.48	.12	80	3
38	28	1.36	34	390	1219	1.46	28	.55	.62	.21	3.6	.41	10	20
23	16	.79	21	187	732	.66	20	.06	.39	.08	3.06	.18	6	12
0	35	1.42	25	224	514	.54	23	.07	1.96	.09	.4	.04	53	4
27	35	2.05	39	365	824	1.15	55	.06	1.95	.14	13.7	.17	16	5
41	34	2.3	30	182	772	1.2	13	.18	1.18	.15	7.78	.2	10	1
21	43	.88	39	270	236	.63	39	.1	8.67	.05	.36	.36	27	6
24	53	2.46	42	329	670	1.93	32	.27	.7	.42	2.71	.19	58	1
2	57	2.27	44	328	495	.76	16	.32	.87	.3	2.04	.15	44	1
10	40	.72			600		40							15
52	182	2.65	26	229	976	2.62	71	.32	1.34	.23	6.38	.15	65	2
60	171	3.42	30	335	1050	3.49	65	.35		.28	8.05	.18	61	2
60	60	4.68	35	353	957	1.87	31	.33	.55	.24	6.81	.2	100	9
79	102	6.18	17	263	973	1.31	37	.28	.7	.7	4.17	.09	103	0
64					668									
44	324	1.32	50	240	784	2.51	186	.08	1.47	.42	1.91	.39	65	1
234	207	3.29	34	213	784	1.81	158	.48	.6	.53	3.93	.16	47	1
68	185	3.5	37	353	939	1.17	97	.46	1.83	.42	4.23	.11	91	3
55	84	2.61	33	340	615	.99	30	.33	.87	.22	3.4	.11	85	3
43	51	2.46	22	215	564	2.05	13	.26	.43	.32	4.7	.13	52	3
103	92	5.55	45	527	742	5.81	4	.34	1.61	.41	6.73	.37	84	1
44	23	2.31	13	143	670	1.98	0	.23	.27	.27	3.65	.05	48	<1
85	300	2.7			1620		60							0
60	300	1.8			1770		80							
55	250	2.7			1540		40							
50	20	6			1360		80							
95	250	2.7			1840		60							
80	250	2.7			1740		60							
70	300	1.8			1650		60							
9	117	.58	16	110	336	.81	74	.18		.16	2.48	.04	35	1
34	294	3.06	54	581	550	1.71	140	.37	4.55	.39	3.77	.32	80	11
31	291	3.1	67	617	525	1.91	140	.35	4.55	.39	3.94	.35	80	11
31	281	3.53	102	679	593	2.68	140	.36	4.17	.37	4.29	.42	92	11
21	52	1.99	20	233	590	.96	31	.35	2.34	.19	3.2	.14	34	12
20	63	2.27	33	283	604	1.19	31	.36	2.7	.21	3.62	.17	37	12
20	55	2.68	64	342	670	1.9	31	.37	2.36	.2	3.97	.24	48	12

Table H-1

Food Composition (Computer code number is for West Diet Analysis program) (For purposes of calculations, use "0" for t, <1, <.1, <.01, etc.)

Computer Code Number	Food Description	Measure	Wt (g)	H_2O (%)	Ener (cal)	Prot (g)	Carb (g)	Dietary Fiber (g)	Fat (g)	Fat Breakdown (g) Sat	Mono	Poly
	MIXED DISHES and FAST FOODS—Continued											
	Cheese, grilled:											
1140	On white bread, soft	1 ea	119	37	400	18	30	1	24	13.2	7.5	2
1142	On part whole wheat	1 ea	119	37	396	18	28	3	24	13	7.6	2.1
1141	On whole wheat	1 ea	132	38	420	20	33	4	24	13.1	7.8	2.2
1596	Chicken fillet	1 ea	182	47	515	24	39	1	29	8.5	10.4	8.4
	Chicken salad:											
1143	On white bread, soft	1 ea	110	40	369	11	31	1	23	3.7	6.1	12.1
1145	On part whole wheat	1 ea	110	41	364	11	29	4	23	3.5	6.1	12.1
1144	On whole wheat	1 ea	123	41	387	13	34	5	23	3.6	6.3	12.2
1146	Corned beef & swiss on rye	1 ea	156	49	420	28	22	<1	26	9.4	7.4	6.3
	Egg salad:											
1147	On white bread, soft	1 ea	117	43	380	10	31	1	25	4.4	6.8	12
1149	On part whole wheat	1 ea	116	43	374	10	29	3	25	4.1	6.8	12
1148	On whole wheat	1 ea	130	43	400	12	35	5	25	4.3	7.1	12.2
	Ham:											
1279	On rye bread	1 ea	150	60	283	22	21	<1	13	2.4	3.8	6
1151	On white bread, soft	1 ea	157	55	334	22	30	1	14	3	4.4	5.9
1153	On part whole wheat	1 ea	156	55	328	22	28	3	14	2.7	4.4	6
1152	On whole wheat	1 ea	169	54	352	24	34	4	15	2.9	4.7	6.1
	Ham & cheese:											
1280	On white bread, soft	1 ea	157	49	403	23	30	1	22	8.4	5.9	6.1
1282	On part whole wheat	1 ea	156	50	397	23	28	3	21	8.1	5.9	6.2
1281	On whole wheat	1 ea	170	49	424	24	34	4	22	8.3	6.2	6.4
1150	Ham & swiss on rye	1 ea	150	54	339	22	22	<1	19	6.5	5.1	6
	Ham salad:											
1154	On white bread, soft	1 ea	131	47	362	11	37	1	20	4.8	6.7	7.4
1156	On part whole wheat	1 ea	131	48	357	11	35	3	20	4.5	6.8	7.5
1155	On whole wheat	1 ea	144	47	380	12	40	4	20	4.7	7	7.6
1157	Patty melt: Ground beef & cheese on rye	1 ea	182	46	561	37	22	3	37	13.2	11.7	8.4
	Peanut butter & jelly:											
1158	On white bread, soft	1 ea	101	26	351	12	47	3	15	3.1	6.7	3.9
1160	On part whole wheat	1 ea	101	27	346	12	45	5	15	2.9	6.7	4
1159	On whole wheat	1 ea	114	28	370	13	50	6	15	3	7	4.1
1161	Reuben, grilled: Corned beef, swiss cheese, sauerkraut on rye	1 ea	239	64	462	28	25	2	29	9.9	9.5	7.1
	Roast beef:											
713	On a bun	1 ea	139	49	346	21	33		14	3.6	6.8	1.7
1162	On white bread, soft	1 ea	157	46	404	29	34	1	17	3.4	4.2	8.2
1164	On part whole wheat	1 ea	156	47	398	29	32	3	17	3.2	4.3	8.3
1163	On whole wheat	1 ea	169	46	422	31	38	4	17	3.3	4.5	8.4
	Tuna salad:											
1165	On white bread, soft	1 ea	122	46	327	14	35	2	15	2.5	3.8	7.9
1167	On part whole wheat	1 ea	122	47	322	14	33	4	15	2.2	3.8	8
1166	On whole wheat	1 ea	135	46	346	16	39	5	15	2.3	4.1	8.1
	Turkey:											
1168	On white bread, soft	1 ea	156	54	346	24	29	1	15	2.4	3.2	8.3
1170	On part whole wheat	1 ea	155	54	338	24	27	3	14	2.1	3.2	8.3
1169	On whole wheat	1 ea	169	53	365	26	33	4	15	2.3	3.5	8.5
	Turkey ham:											
1272	On rye bread	1 ea	150	60	280	21	20	<1	14	2.5	2.8	6.9
1273	On white bread, soft	1 ea	156	55	331	21	29	1	14	3	3.4	6.8
1275	On part whole wheat	1 ea	156	56	326	21	28	3	14	3	4.2	5.6
1274	On whole wheat	1 ea	169	55	350	23	33	4	15	2.9	3.7	7
714	Taco	1 ea	171	58	369	21	27		20	11.4	6.6	1
	Tostada:											
1114	With refried beans	1 ea	144	66	223	10	26	7	10	5.4	3	.7
1118	With beans & beef	1 ea	225	70	333	16	30	4	17	11.5	3.5	.6
1354	With beans & chicken	1 ea	156	68	248	19	18	3	11	5.3	3.9	1.6
	NUTS, SEEDS, and PRODUCTS											
	Almonds:											
1365	Dry roasted, salted	1 c	138	3	810	22	33	19	71	6.7	46.2	14.9

Chol (mg)	Calc (mg)	Iron (mg)	Magn (mg)	Pota (mg)	Sodi (mg)	Zinc (mg)	VT-A (RE)	Thia (mg)	VT-E (a-TE)	Ribo (mg)	Niac (mg)	V-B6 (mg)	Fola (μg)	VT-C (mg)
55	399	1.81	25	154	1143	2.05	212	.24	1.13	.34	1.91	.06	24	<1
54	412	2.12	39	209	1160	2.31	212	.25	1.53	.36	2.39	.1	28	<1
54	402	2.54	73	271	1226	3.08	212	.26	1.14	.35	2.74	.17	40	<1
60	60	4.68	35	353	957	1.87	31	.33	.55	.24	6.81	.2	100	9
32	60	2.04	18	139	460	.8	24	.25	6.16	.18	3.68	.25	26	1
31	73	2.35	33	195	475	1.06	24	.26	6.57	.2	4.16	.29	30	1
30	63	2.79	68	259	543	1.85	24	.27	6.14	.19	4.5	.36	42	1
82	268	3.12	28	225	1392	3.65	81	.19	2.59	.33	2.72	.17	19	1
157	71	2.18	16	113	526	.74	76	.26	4.52	.31	1.98	.2	37	0
155	83	2.47	31	169	539	1	75	.27	4.91	.33	2.45	.23	41	0
155	73	2.92	66	234	611	1.8	76	.28	4.53	.32	2.82	.3	53	0
47	48	2.3	26	364	1566	2.11	8	.99	2.36	.31	5.45	.47	15	23
47	60	2.39	29	368	1619	2.04	8	1.02	2.34	.33	5.99	.47	24	22
45	71	2.68	43	421	1630	2.29	8	1.03	2.73	.35	6.45	.5	27	22
45	62	3.1	77	483	1696	3.05	8	1.04	2.34	.33	6.78	.57	39	22
61	232	2.28	30	315	1620	2.34	90	.76	2.53	.36	4.64	.36	25	15
59	244	2.58	45	368	1630	2.59	90	.77	2.93	.39	5.09	.39	28	15
59	236	3.02	79	432	1707	3.37	90	.78	2.56	.37	5.47	.46	40	15
57	258	2.25	29	344	1602	2.59	79	.72	2.52	.36	4.06	.35	16	15
30	56	2.07	19	159	921	1.06	8	.51	3.29	.22	3.25	.17	22	4
29	69	2.38	33	216	936	1.32	8	.52	3.69	.24	3.74	.21	26	4
29	59	2.81	69	279	1001	2.11	8	.53	3.29	.23	4.09	.28	38	4
113	222	4.19	36	391	701	7.11	123	.25	3.5	.46	6.14	.35	25	<1
2	60	2.25	56	245	293	1.07	<1	.27	.12	.17	5.33	.13	40	<1
0	72	2.55	70	299	308	1.32	<1	.28	.51	.19	5.8	.17	44	<1
0	63	2.97	104	361	375	2.09	<1	.29	.14	.17	6.14	.24	56	<1
80	288	4.24	38	361	1949	3.73	130	.21	4.46	.34	2.79	.27	38	13
51	54	4.23	31	316	792	3.39	21	.37	.19	.31	5.87	.26	57	2
45	60	3.98	28	432	1595	3.78	12	.29	3.3	.3	6.39	.39	30	12
43	72	4.27	42	485	1607	4.02	12	.31	3.69	.32	6.84	.42	34	12
43	62	4.7	77	547	1672	4.79	12	.31	3.31	.31	7.18	.49	45	12
14	60	2.24	22	161	567	.67	22	.25	2.72	.18	5.53	.11	25	1
13	73	2.55	37	217	582	.93	22	.26	3.12	.2	6	.16	29	1
12	63	2.98	72	280	649	1.72	22	.27	2.71	.19	6.34	.23	41	1
45	56	2	29	302	1585	1.33	12	.26	3.46	.23	8.97	.4	24	0
43	68	2.28	43	354	1589	1.57	11	.27	3.82	.25	9.38	.44	28	0
43	59	2.73	77	418	1665	2.35	12	.28	3.47	.23	9.77	.51	39	0
55	51	4.06	25	342	1185	3	8	.22	2.8	.33	4.3	.29	17	<1
55	62	4.09	28	346	1248	2.9	8	.27	2.75	.35	4.87	.28	25	0
53	74	4.37	42	400	1262	3.15	8	.28	3.57	.37	5.34	.32	29	0
53	65	4.81	76	462	1329	3.91	8	.29	2.76	.35	5.68	.39	41	0
56	221	2.41	70	474	802	3.93	147	.15	1.88	.44	3.21	.24	68	2
30	210	1.89	59	403	543	1.9	85	.1	1.15	.33	1.32	.16	43	1
74	189	2.45	67	491	871	3.17	173	.09	1.8	.49	2.86	.25	85	4
53	168	1.79	47	365	433	2.28	86	.11	1.87	.2	4.52	.32	53	3
0	389	5.24	420	1062	1076	6.76	0	.18	7.66	.83	3.89	.1	88	1

Table H-1

Food Composition (Computer code number is for West Diet Analysis program) (For purposes of calculations, use "0" for t, <1, <.1, <.01, etc.)

Computer Code Number	Food Description	Measure	Wt (g)	H_2O (%)	Ener (cal)	Prot (g)	Carb (g)	Dietary Fiber (g)	Fat (g)	Fat Breakdown (g) Sat	Mono	Poly
	NUTS, SEEDS, and PRODUCTS—Continued											
	Almonds:											
718	Slivered, packed, unsalted	1 c	108	4	636	22	22	12	56	5.3	36.6	11.9
719	Whole, dried, unsalted	1 c	142	4	836	28	29	15	74	7	48.1	15.6
720	Ounce	1 oz	28	4	165	6	6	3	15	1.4	9.5	3.1
721	Almond butter:	1 tbs	16	1	101	2	3	1	9	.9	6.1	2
4572	Salted	1 tbs	16	1	101	2	3	1	9	.9	6.1	2
722	Brazil nuts, dry (about 7)	1 c	140	3	918	20	18	8	93	22.7	32.2	33.7
	Cashew nuts, dry roasted:											
723	Salted:	1 c	137	2	786	21	45	4	64	12.8	37.4	10.7
724	Ounce	1 oz	28	2	161	4	9	1	13	2.6	7.6	2.2
4621	Unsalted:	1 c	137	2	786	21	45	4	64	12.8	37.4	10.7
4621	Ounce	1 oz	28	2	161	4	9	1	13	2.6	7.6	2.2
725	Oil roasted:	1 c	130	4	749	23	37	5	63	12.6	36.9	10.6
726	Ounce	1 oz	28	4	161	5	8	1	13	2.7	7.9	2.3
4622	Unsalted:	1 c	130	4	749	21	37	5	63	12.6	36.9	10.6
4622	Ounce	1 oz	28	4	161	5	8	1	13	2.7	7.9	2.3
727	Cashew butter, unsalted	1 tbs	16	3	94	3	4	<1	8	1.6	4.7	1.3
4662	Cashew butter, salted	1 tbs	16	3	94	3	4	<1	8	1.6	4.7	1.3
728	Chestnuts, European, roasted (1 cup = approx 17 kernels)	1 c	143	40	350	5	76	7	3	.6	1.1	1.2
	Coconut, raw:											
729	Piece 2 x 2 x ½"	1 pce	45	47	159	2	7	4	15	13.5	.6	.2
730	Shredded/grated, unpacked	½ c	40	47	142	1	6	4	13	12	.6	.1
	Coconut, dried, shredded/grated:											
731	Unsweetened	1 c	78	3	515	6	19	13	50	45.1	2.1	.6
732	Sweetened	1 c	93	13	466	3	44	4	33	29.6	1.4	.4
733	Filberts/hazelnuts, chopped:	1 c	135	5	853	18	21	8	84	6.2	66.3	8.1
734	Ounce	1 oz	28	5	177	4	4	2	17	1.3	13.7	1.7
735	Macadamias, oil roasted, salted:	1 c	134	2	962	10	17	12	103	15.4	80.9	1.8
736	Ounce	1 oz	28	2	201	2	4	3	21	3.2	16.9	.4
1368	Macadamias, oil roasted, unsalted	1 c	134	2	962	10	17	12	103	15.4	80.9	1.8
	Mixed nuts:											
737	Dry roasted, salted	1 c	137	2	814	24	35	12	71	9.4	43	14.8
738	Oil roasted, salted	1 c	142	2	876	24	30	13	80	12.4	45	18.9
1369	Oil roasted, unsalted	1 c	142	2	876	27	30	14	80	12.4	45	18.9
	Peanuts:											
739	Oil roasted, salted	1 c	144	2	837	38	27	13	71	9.8	35.3	22.5
740	Ounce	1 oz	28	2	163	7	5	3	14	1.9	6.9	4.4
1370	Oil roasted, unsalted	1 c	144	2	837	38	27	10	71	9.8	35.3	22.5
741	Dried, salted	1 c	146	2	854	35	31	12	73	10.1	36.1	22.9
742	Ounce	1 oz	28	2	164	7	6	2	14	1.9	6.9	4.4
743	Peanut butter:	½ c	128	1	759	33	25	8	65	14.3	31.1	17.7
1371	Tablespoon	2 tbs	32	1	190	8	6	2	16	3.6	7.8	4.4
744	Pecan halves, dried, unsalted:	1 c	108	5	720	9	20	8	73	5.8	45.6	18
745	Ounce	1 oz	28	5	187	2	5	2	19	1.5	11.8	4.7
1372	Pecan halves, dry roasted, salted	¼ c	28	1	185	2	6	3	18	1.4	11.3	4.5
746	Pine nuts/piñons, dried	1 oz	28	6	176	3	5	3	17	2.6	6.4	7.2
747	Pistachios, dried, shelled	1 oz	28	4	162	6	7	3	14	1.8	9.2	2
1373	Pistachios,dry roasted,salted,shelled	1 c	128	2	776	19	35	14	68	8.8	45.7	10.2
748	Pumpkin kernels, dried, unsalted	1 oz	28	7	151	7	5	1	13	2.4	4	5.8
1374	Pumpkin kernels, roasted, salted	1 c	227	7	1184	75	30	9	96	18.1	29.7	43.6
749	Sesame seeds, hulled, dried	¼ c	38	5	223	10	4	3	21	2.9	7.9	9.1
	Sunflower seed kernels:											
750	Dry	¼c	36	5	205	8	7	4	18	1.9	3.4	11.8
751	Oil roasted	¼c	34	3	209	7	5	2	20	2	3.7	12.9
752	Tahini (sesame butter)	1 tbs	15	3	91	3	3	1	8	1.2	3.2	3.7
1334	Trail mix w/chocolate chips	1 c	146	7	707	21	66	8	47	9.3	19.8	16.5
753	Black walnuts, chopped:	1 c	125	4	759	31	15	6	71	4.8	15.9	46.9
754	Ounce	1 oz	28	4	170	7	3	1	16	1.1	3.6	10.5
755	English walnuts, chopped:	1 c	120	4	770	17	22	6	74	7.2	17	46.9
756	Ounce	1 oz	28	4	180	4	5	1	17	1.7	4	10.9

PAGE KEY: A–4 = Beverages A–6 = Dairy A–10 = Eggs A–10 = Fat/Oil A–14 = Fruit A–20 = Bakery A–28 = Grain A–32 = Fish A–34 = Meats A–38 = Poultry A–40 = Sausage A–42 = Mixed/Fast A–46 = Nuts/Seeds A–50 = Sweets A–52 = Vegetables/Legumes A–62 = Vegetarian Foods A–64 = Misc A–66 = Soups/Sauces A–68 = Fast A–84 = Convenience A–88 = Baby foods

Chol (mg)	Calc (mg)	Iron (mg)	Magn (mg)	Pota (mg)	Sodi (mg)	Zinc (mg)	VT-A (RE)	Thia (mg)	VT-E (a-TE)	Ribo (mg)	Niac (mg)	V-B6 (mg)	Fola (μg)	VT-C (mg)
0	287	3.95	320	791	12	3.15	0	.23	25.9	.84	3.63	.12	63	1
0	378	5.2	420	1039	16	4.15	0	.3	34.1	1.11	4.77	.16	83	1
0	74	1.02	83	205	3	.82	0	.06	6.72	.22	.94	.03	16	<1
0	43	.59	48	121	2	.49	0	.02	3.25	.1	.46	.01	10	<1
0	43	.59	48	121	72	.49	0	.02	3.25	.1	.46	.01	10	<1
0	246	4.76	315	840	3	6.43	0	1.4	10.6	.17	2.27	.35	6	1
0	62	8.22	356	774	877	7.67	0	.27	.78	.27	1.92	.35	95	0
0	13	1.68	73	158	179	1.57	0	.06	.16	.06	.39	.07	19	0
0	62	8.22	356	774	22	7.67	0	.27	.78	.27	1.92	.35	95	0
0	13	1.68	73	158	4	1.57	0	.06	.16	.06	.39	.07	19	0
0	53	5.33	332	689	814	6.18	0	.55	2.03	.23	2.34	.32	88	0
0	11	1.15	71	148	175	1.33	0	.12	.44	.05	.5	.07	19	0
0	53	5.33	332	689	22	6.18	0	.55	2.03	.23	2.34	.32	88	0
0	11	1.15	71	148	5	1.33	0	.12	.44	.05	.5	.07	19	0
0	7	.8	41	87	2	.83	0	.05	.25	.03	.26	.04	11	0
0	7	.8	41	87	98	.83	0	.05	.25	.03	.26	.04	11	0
0	41	1.3	47	847	3	.81	3	.35	1.72	.25	1.92	.71	100	37
0	6	1.09	14	160	9	.49	0	.03	.33	.01	.24	.02	12	1
0	6	.97	13	142	8	.44	0	.03	.29	.01	.22	.02	11	1
0	20	2.59	70	424	29	1.57	0	.05	1.05	.08	.47	.23	7	1
0	14	1.79	46	313	244	1.69	0	.03	1.26	.02	.44	.25	8	1
0	254	4.41	385	601	4	3.24	9	.67	32.3	.15	1.54	.83	97	1
0	53	.92	80	125	1	.67	2	.14	6.69	.03	.32	.17	20	<1
0	60	2.41	157	441	348	1.47	1	.28	.55	.15	2.71	.26	21	0
0	13	.5	33	92	73	.31	<1	.06	.11	.03	.57	.05	4	0
0	60	2.41	157	441	9	1.47	1	.28	.55	.15	2.71	.26	21	0
0	96	5.07	308	818	917	5.21	1	.27	8.22	.27	6.44	.41	69	1
0	153	4.56	334	825	926	7.21	3	.71	8.52	.31	7.19	.34	118	1
0	153	4.56	334	825	16	7.21	3	.71	8.52	.31	7.19	.34	118	1
0	127	2.64	266	982	624	9.55	0	.36	10.7	.16	20.6	.37	181	0
0	25	.51	52	191	121	1.86	0	.07	2.07	.03	4	.07	35	0
0	127	2.64	266	982	9	9.55	0	.36	10.7	.16	20.6	.37	181	0
0	79	3.3	257	961	1186	4.83	0	.64	10.8	.14	19.7	.37	212	0
0	15	.63	49	184	228	.93	0	.12	2.07	.03	3.78	.07	41	0
0	49	2.36	204	856	598	3.74	0	.11	12.8	.13	17.2	.58	95	0
0	12	.59	51	214	149	.93	0	.03	3.2	.03	4.29	.14	24	0
0	39	2.3	138	423	1	5.91	14	.92	3.35	.14	.96	.2	42	2
0	10	.6	36	110	<1	1.53	4	.24	.87	.04	.25	.05	11	1
0	10	.61	37	104	218	1.59	4	.09	.84	.03	.26	.05	11	1
0	2	.86	65	176	20	1.2	1	.35	.98	.06	1.22	.03	16	1
0	38	1.9	44	306	2	.37	6	.23	1.46	.05	.3	.07	16	2
0	90	4.06	166	1241	998	1.74	31	.54	8.26	.31	1.8	.33	76	9
0	12	4.2	150	226	5	2.09	11	.06	.28	.09	.49	.06	16	1
0	98	33.8	1212	1829	1305	16.9	86	.48	2.27	.72	3.95	.2	130	4
0	50	2.96	132	155	15	3.91	3	.27	.86	.03	1.78	.05	36	0
0	42	2.44	127	248	1	1.82	2	.82	18.1	.09	1.62	.28	82	<1
0	19	2.28	43	164	1	1.77	2	.11	17.1	.09	1.4	.27	80	<1
0	21	.95	53	69	<1	1.58	1	.24	.34	.02	.85	.02	15	0
6	159	4.95	235	946	177	4.58	7	.6	15.6	.33	6.44	.38	95	2
0	72	3.84	253	655	1	4.28	37	.27	3.28	.14	.86	.69	82	4
0	16	.86	57	147	<1	.96	8	.06	.73	.03	.19	.15	18	1
0	113	2.93	203	602	12	3.28	14	.46	3.14	.18	1.25	.67	79	4
0	26	.68	47	141	3	.76	3	.11	.73	.04	.29	.16	18	1

Table H-1
Food Composition

(Computer code number is for West Diet Analysis program) (For purposes of calculations, use "0" for t, <1, <.1, <.01, etc.)

Computer Code Number	Food Description	Measure	Wt (g)	H_2O (%)	Ener (cal)	Prot (g)	Carb (g)	Dietary Fiber (g)	Fat (g)	Fat Breakdown (g) Sat	Mono	Poly
	SWEETENERS and SWEETS (see also Dairy [milk desserts] and Baked Goods)											
757	Apple butter	2 tbs	36	52	66	<1	17	<1	<1	t	t	t
1124	Butterscotch topping	2 tbs	41	32	103	1	27	<1	<1	t	t	0
1125	Caramel topping	2 tbs	41	32	103	1	27	<1	<1	t	t	0
	Cake frosting, creamy vanilla:											
1127	Canned	2 tbs	39	13	163	<1	27	<1	7	1.9	3.4	.9
1123	From mix	2 tbs	39	12	165	<1	28	<1	6	1.3	2.6	2.2
	Cake frosting, lite:											
2061	Milk chocolate	1 tbs	16	18	58	<1	11	<1	1	.4		
2062	Vanilla	1 tbs	16	15	60	0	12	<1	1	.4		
	Candy:											
1128	Almond Joy candy bar	1 oz	28	10	131	1	16	1	7	4.8	1.8	.4
2069	Butterscotch morsels	¼ c	43	1	246	0	31	0	12	12.5	0	0
758	Caramel, plain or chocolate	1 pce	10	8	38	<1	8	<1	1	.7	.1	t
1961	Chewing gum, sugarless	1 pce	3		6	0	2		0	0	0	0
	Chocolate (see also #784, 785, 971):											
	Milk chocolate:											
759	Plain	1 oz	28	1	144	2	17	1	9	5.2	2.8	.3
760	With almonds	1 oz	28	1	147	3	15	2	10	4.8	3.8	.6
761	With peanuts	1 oz	28	1	155	5	11	2	11	3.4	5.1	2.5
762	With rice cereal	1 oz	28	2	139	2	18	1	7	4.4	2.4	.2
763	Semisweet chocolate chips	1 c	168	1	805	7	106	10	50	29.9	16.8	1.6
764	Sweet dark chocolate (candy bar)	1 ea	41	1	226	2	25	2	13	8.3	4.6	.4
765	Fondant candy, uncoated (mints, candy corn, other)	1 pce	16	7	57	0	15	0	0	0	0	0
1697	Fruit Roll-Up (small)	1 ea	14	11	49	<1	12	<1	<1	.1	.2	.1
766	Fudge, chocolate	1 pce	17	10	65	<1	13	<1	1	.9	.4	.1
767	Gumdrops	1 c	182	1	703	0	180	0	0	0	0	0
768	Hard candy, all flavors	1 pce	6	1	22	0	6	0	<1	0	0	0
769	Jellybeans	10 pce	11	6	40	0	10	0	<1	t	t	t
1134	M&M's plain chocolate candy	10 pce	7	2	34	<1	5	<1	1	.9	.5	t
1135	M&M's peanut chocolate candy	10 pce	20	2	103	2	12	1	5	2.1	2.2	.8
1130	Mars almond bar	1 ea	50	4	234	4	31	1	11	2.7	5.5	2.8
1129	Milky Way candy bar	1 ea	60	6	254	3	43	1	10	4.7	3.6	.4
1708	Milk chocolate-coated peanuts	1 c	149	2	773	19	74	7	50	21.8	19.4	6.4
1709	Peanut brittle, recipe	1 c	147	2	666	11	102	3	28	7.4	12.5	6.9
1132	Reese's peanut butter cup	2 ea	50	3	271	5	27	2	16	5.5	6.5	2.8
1133	Skor English toffee candy bar	1 ea	39	3	217	2	22	1	13	8.5	4.3	.5
1131	Snickers candy bar (2.2oz)	1 ea	62	5	297	5	37	2	15	5.6	6.5	3
1482	Fruit juice bar (2.5 fl oz)	1 ea	77	78	63	1	16	0	<1	t	0	t
771	Gelatin dessert/Jello, prepared	½ c	135	85	80	2	19	0	0	0	0	0
1702	SugarFree	½ c	117	98	8	1	1	0	0	0	0	0
772	Honey:	1 c	339	17	1030	1	279	1	0	0	0	0
773	Tablespoon	1 tbs	21	17	64	<1	17	<1	0	0	0	0
774	Jams or preserves:	1 tbs	20	29	54	<1	14	<1	<1	0	t	t
775	Packet	1 ea	14	34	34	<1	9	<1	<1	t	t	0
776	Jellies:	1 tbs	19	28	51	<1	13	<1	<1	t	t	t
777	Packet	1 ea	14	28	38	<1	10	<1	<1	t	t	t
1136	Marmalade	1 tbs	20	33	49	<1	13	<1	0	0	0	0
770	Marshmallows	1 ea	7	16	22	<1	6	<1	<1	t	t	t
1126	Marshmallow creme topping	2 tbs	38	18	118	1	30	<1	<1	t	t	t
778	Popsicle/ice pops	1 ea	128	80	92	0	24	0	0	0	0	0
	Sugars:											
779	Brown sugar	1 c	220	2	827	0	214	0	0	0	0	0
780	White sugar, granulated:	1 c	200		774	0	200	0	0	0	0	0
781	Tablespoon	1 tbs	12		46	0	12	0	0	0	0	0
782	Packet	1 ea	6		23	0	6	0	0	0	0	0
783	White sugar, powdered, sifted	1 c	100		389	0	99	0	<1	t	t	t
	Sweeteners:											
1711	Equal, packet	1 ea	1	12	4	<1	1	0	<1	0	t	t
1712	Sweet 'N Low, packet	1 ea	1		0	0	1	0	0	0	0	0

Chol (mg)	Calc (mg)	Iron (mg)	Magn (mg)	Pota (mg)	Sodi (mg)	Zinc (mg)	VT-A (RE)	Thia (mg)	VT-E (a-TE)	Ribo (mg)	Niac (mg)	V-B6 (mg)	Fola (μg)	VT-C (mg)
0	2	.05	1	33	0	.02	0	<.01	.01	<.01	.03	.01	<1	1
<1	22	.08	3	34	143	.08	11	<.01	0	.04	.02	.01	1	<1
<1	22	.08	3	34	143	.08	11	<.01	0	.04	.02	.01	1	<1
0	1	.04		14	35	0	88	0	.79	<.01	<.01	0	0	0
0	4	.09	1	9	87	.04	42	.01	.79	.01	.13	<.01	0	0
0	1	.24			40		<1							0
0		.02			29		0							0
1	17	.39	18	69	41	.22	1	.01	.63	.04	.13	.02		<1
0	0	0		80	46		0	.03		.04	.03			0
1	14	.01	2	21	24	.04	1	<.01	.05	.02	.02	<.01	<1	<1
				0	0									
6	53	.39	17	108	23	.39	15	.02	.35	.08	.09	.01	2	<1
5	63	.46	25	124	21	.37	4	.02	.53	.12	.21	.01	3	<1
3	32	.52	34	150	11	.68	6	.08	1.3	.05	2.12	.04	23	0
5	48	.21	14	96	41	.31	3	.02	.35	.08	.13	.02	3	<1
0	54	5.26	193	613	18	2.72	3	.09	2	.15	.72	.06	5	0
<1	11	.98	47	139	3	.61	1	.01	.41	.1	.27	.02	1	0
0		.01		3	6	.01	<1	0	0	<.01	0	0	0	0
0	4	.14	3	41	9	.03	2	.01	.04	<.01	.01	.04	1	1
2	7	.08	4	17	10	.07	8	<.01	.02	.01	.02	<.01	<1	<1
0	5	.73	2	9	80	0	0	0	0	<.01	<.01	0	0	0
0		.02			2	<.01	0	0	0	0	0	0	0	0
0		.12		4	3	.01	0	0	0	0	0	0	0	0
1	7	.08	3	19	4	.07	4	<.01	.06	.01	.02	<.01	<1	<1
2	20	.23	12	69	10	.27	5	.03	.43	.04	.41	.02	7	<1
4	84	.55	36	163	85	.55	22	.02	.3	.16	.47	.03	9	<1
8	78	.46	20	145	144	.43	34	.02	.39	.13	.21	.03	6	1
13	155	1.95	134	748	61	2.8	0	.17	3.8	.26	6.33	.31	12	0
19	44	2.03	73	306	664	1.43	69	.28	2.41	.08	5.15	.15	103	0
2	39	.6	42	176	159	.7	9	.02	.66	.1	1.99	.04	27	<1
20	51	.02	13	93	108	.3	27	.01	.53	.13	.03	.01		<1
8	58	.47	42	209	165	.88	24	.13	.95	.1	2.26	.07	25	<1
0	4	.15	3	41	3	.04	2	.01	0	.01	.12	.02	5	7
0	3	.04	1	1	57	.04	0	0	0	<.01	<.01	<.01	0	0
0	2	.01	1	0	56	.03	0	0	0	<.01	<.01	<.01	0	0
0	20	1.42	7	176	14	.75	0	0	0	.13	.41	.08	7	2
0	1	.09		11	1	.05	0	0	0	.01	.02	<.01	<1	<1
0	4	.2	1	18	2	.01	<1	<.01	.02	.01	.04	<.01	2	<1
0	3	.07	1	11	6	.01	<1	0	0	<.01	<.01	<.01	5	1
0	2	.04	1	12	7	.01	<1	0	0	<.01	.01	<.01	<1	<1
0	1	.03	1	9	5	.01	<1	0	0	<.01	<.01	<.01	<1	<1
0	8	.03		7	11	.01	1	<.01	0	<.01	.01	<.01	7	1
0		.02			3	<.01	<1	0	0	0	<.01	0	<1	0
0	1	.08	1	2	17	.01	<1	0	0	0	.03	<.01	<1	0
0	0	0	1	5	15	.03	0	0	0	0	0	0	0	0
0	187	4.2	64	761	86	.4	0	.02	0	.01	.18	.06	2	0
0	2	.12	0	4	2	.06	0	0	0	.04	0	0	0	0
0		.01	0		<1	<.01	0	0	0	<.01	0	0	0	0
0		<.01	0		<1	<.01	0	0	0	<.01	0	0	0	0
0	1	.06	0	2	1	.03	0	0	0	0	0	0	0	0
0		<.01			<1	0	0	0	0	0	0	0	0	0
0	0	0			0		0							0

Table H-1

Food Composition

(Computer code number is for West Diet Analysis program) (For purposes of calculations, use "0" for t, <1, <.1, <.01, etc.)

Computer Code Number	Food Description	Measure	Wt (g)	H_2O (%)	Ener (cal)	Prot (g)	Carb (g)	Dietary Fiber (g)	Fat (g)	Fat Breakdown (g) Sat	Mono	Poly
	SWEETENERS and SWEETS—Continued											
	Syrups, chocolate:											
785	Hot fudge type	2 tbs	43	22	149	2	27	1	4	2.4	1.6	1.4
784	Thin type	2 tbs	38	29	93	1	25	1	<1	.3	.2	t
786	Molasses, blackstrap	2 tbs	41	29	96	0	25	0	0	0	0	0
1710	Light cane syrup	2 tbs	41	24	103	0	27	0	0	0	0	0
787	Pancake table syrup (corn and maple)	2 tbs	40	24	115	0	30	0	0	0	0	0
	VEGETABLES and LEGUMES											
788	Alfalfa sprouts	1 c	33	91	10	1	1	1	<1	t	.1	t
1815	Amaranth leaves, raw, chopped	1 c	28	92	7	1	1	<1	<1	t	t	t
1816	Amaranth leaves, raw, each	1 ea	14	92	4	<1	1	<1	<1	t	t	t
1817	Amaranth leaves, cooked	1 c	132	91	28	3	5	2	<1	.1	.1	.1
1987	Arugula, raw, chopped	½ c	10	92	2	<1	<1	<1	<1	t	t	t
789	Artichokes, cooked globe (300 g with refuse)	1 ea	120	84	60	4	13	6	<1	t	t	.1
1177	Artichoke hearts, cooked from frozen	1 c	168	86	76	5	15	8	1	.2	t	.4
1176	Artichoke hearts, marinated	1 c	130	81	128	3	10	6	10	1.5	2.3	5.9
2021	Artichoke hearts, in water	½ c	100	91	37	2	6	0	0	0	0	0
	Asparagus, green, cooked:											
	From fresh:											
790	Cuts and tips	½ c	90	92	22	2	4	1	<1	.1	t	.1
791	Spears, ½" diam at base	4 ea	60	92	14	2	3	1	<1	t	t	.1
	From frozen:											
792	Cuts and tips	½ c	90	91	25	3	4	1	<1	.1	t	.2
793	Spears, ½" diam at base	4 ea	60	91	17	2	3	1	<1	.1	t	.1
794	Canned, spears, ½" diam at base	4 ea	72	94	14	2	2	1	<1	.1	t	.2
795	Bamboo shoots, canned, drained slices	1 c	131	94	25	2	4	2	1	.1	t	.2
1795	Bamboo shoots, raw slices	1 c	151	91	41	4	8	3	<1	.1	t	.2
1798	Bamboo shoots, cooked slices	1 c	120	96	14	2	2	1	<1	.1	t	.1
	Beans (see also alphabetical listing this section):											
1990	Adzuki beans, cooked	½ c	115	66	147	9	28	1	<1	t	t	t
796	Black beans, cooked	½ c	86	66	114	8	20	7	<1	.1	t	.2
	Canned beans (white/navy):											
803	With pork and tomato sauce	½ c	127	73	124	7	25	6	1	.5	.6	.2
804	With sweet sauce	½ c	130	71	144	7	27	7	2	.7	.8	.2
805	With frankfurters	½ c	130	69	185	9	20	9	9	3.1	3.7	1.1
	Lima beans:											
797	Thick seeded (Fordhooks), cooked from frozen	½ c	85	73	85	5	16	5	<1	.1	t	.1
798	Thin seeded (Baby), cooked from frozen	½ c	90	72	94	6	18	5	<1	.1	t	.1
799	Cooked from dry, drained	½ c	94	70	108	7	20	7	<1	.1	t	.2
1998	Red Mexican, cooked f/dry	½ c	112	70	126	8	23	9	<1	.1	.1	.2
	Snap bean/green string beans cuts and french style:											
800	Cooked from fresh	½ c	63	89	22	1	5	2	<1	t	t	.1
801	Cooked from frozen	½ c	68	91	19	1	4	2	<1	t	t	.1
802	Canned, drained	½ c	68	93	14	1	3	1	<1	t	t	t
1713	Snap bean, yellow, cooked f/fresh	½ c	63	89	22	1	5	2	<1	t	t	.1
	Bean sprouts (mung):											
806	Raw	½ c	52	90	16	2	3	1	<1	t	t	t
807	Cooked, stir fried	½ c	62	84	31	3	7	1	<1	t	t	t
808	Cooked, boiled, drained	½ c	62	93	13	1	3	<1	<1	t	t	t
1788	Canned, drained	½ c	63	96	8	1	1	<1	<1	t	t	t
	Beets, cooked from fresh:											
809	Sliced or diced	½ c	85	87	37	1	8	2	<1	t	t	.1
810	Whole beets, 2" diam	2 ea	100	87	44	2	10	2	<1	t	t	.1
	Beets, canned:											
811	Sliced or diced	½ c	79	91	24	1	6	1	<1	t	t	t
812	Pickled slices	½ c	114	82	74	1	19	2	<1	t	t	t
813	Beet greens, cooked, drained	½ c	72	89	19	2	4	2	<1	t	t	.1

PAGE KEY: A–4 = Beverages A–6 = Dairy A–10 = Eggs A–10 = Fat/Oil A–14 = Fruit A–20 = Bakery A–28 = Grain A–32 = Fish A–34 = Meats A–38 = Poultry A–40 = Sausage A–42 = Mixed/Fast A–46 = Nuts/Seeds A–50 = Sweets A–52 = Vegetables/Legumes A–62 = Vegetarian Foods A–64 = Misc A–66 = Soups/Sauces A–68 = Fast A–84 = Convenience A–88 = Baby foods

Chol (mg)	Calc (mg)	Iron (mg)	Magn (mg)	Pota (mg)	Sodi (mg)	Zinc (mg)	VT-A (RE)	Thia (mg)	VT-E (a-TE)	Ribo (mg)	Niac (mg)	V-B6 (mg)	Fola (μg)	VT-C (mg)
5	43	.52	21	92	56	.34	9	.01	0	.09	.09	.01	2	<1
0	5	5.17	25	183	58	.28	494	<.01	.01	.31	12.8	.01	2	<1
0	353	7.18	88	1021	23	.41	0	.01	0	.02	.44	.29	<1	0
0	68	1.76	100	376	6	.12	0	.03	0	.02	.08	.27	0	0
0		.04	1	1	33	.02	0	<.01	0	<.01	.01	0	0	0
0	11	.32	9	26	2	.3	5	.02	—	.04	.16	.01	12	3
0	60	.65	15	171	6	.25	82	.01	.22	.04	.18	.05	24	12
0	30	.32	8	85	3	.13	41	<.01	.11	.02	.09	.03	12	6
0	276	2.98	73	846	28	1.16	366	.03	.66	.18	.74	.23	75	54
0	16	.15	5	37	3	.05	24	<.01	.04	.01	.03	.01	10	1
0	54	1.55	72	425	114	.59	22	.08	.23	.08	1.2	.13	61	12
0	35	.94	52	444	89	.6	27	.1	.32	.26	1.54	.15	200	8
0	30	1.24	37	335	688	.41	21	.05	1.43	.13	1.06	.11	114	40
0	0	1.35		0	250		12							4
0	18	.66	9	144	10	.38	49	.11	.34	.11	.97	.11	131	10
0	12	.44	6	96	7	.25	32	.07	.23	.08	.65	.07	88	6
0	21	.58	12	196	4	.5	74	.06	1.13	.09	.94	.02	122	22
0	14	.38	8	131	2	.34	49	.04	.75	.06	.62	.01	81	15
0	11	1.32	7	124	207	.29	38	.04	.31	.07	.69	.08	69	13
0	10	.42	5	105	9	.85	1	.03	.5	.03	.18	.18	4	1
0	20	.75	5	805	6	1.66	3	.23	1.51	.11	.91	.36	11	6
0	14	.29	4	640	5	.56	0	.02	.8	.06	.36	.12	3	1
0	32	2.3	60	612	9	2.04	1	.13	.11	.07	.82	.11	139	0
0	23	1.81	60	305	1	.96	1	.21	.07	.05	.43	.06	128	0
9	71	4.17	44	381	559	7.44	15	.07	.69	.06	.63	.09	29	4
9	79	2.16	44	346	437	1.95	14	.06	.7	.08	.46	.11	49	4
8	62	2.25	36	306	559	2.43	19	.07	.61	.07	1.17	.06	39	3
0	19	1.16	29	347	45	.37	16	.06	.25	.05	.91	.1	18	11
0	25	1.76	50	370	26	.49	15	.06	.58	.05	.69	.1	14	5
0	16	2.25	40	478	2	.89	0	.15	.17	.05	.4	.15	78	0
0	42	1.86	48	369	240	.87	<1	.13	.08	.07	.37	.11	94	2
0	29	.81	16	188	2	.23	42	.05	.09	.06	.39	.03	21	6
0	33	.6	16	86	6	.33	27	.02	.09	.06	.26	.04	16	3
0	18	.61	9	74	178	.2	24	.01	.09	.04	.14	.02	22	3
0	29	.81	16	188	2	.23	5	.05	.18	.06	.39	.03	21	6
0	7	.47	11	77	3	.21	1	.04	.02	.06	.39	.05	32	7
0	8	1.18	20	136	6	.56	2	.09	.01	.11	.74	.08	43	10
0	7	.4	9	63	6	.29	1	.03	.01	.06	.51	.03	18	7
0	9	.27	6	17	88	.18	1	.02	.01	.04	.14	.02	6	<1
0	14	.67	20	259	65	.3	3	.02	.25	.03	.28	.06	68	3
0	16	.79	23	305	77	.35	4	.03	.3	.04	.33	.07	80	4
0	12	1.44	13	117	153	.17	1	.01	.24	.03	.12	.04	24	3
0	12	.47	17	169	301	.3	1	.01	.15	.05	.29	.06	30	3
0	82	1.37	49	654	174	.36	367	.08	.22	.21	.36	.09	10	18

Table H-1

Food Composition (Computer code number is for West Diet Analysis program) (For purposes of calculations, use "0" for t, <1, <.1, <.01, etc.)

Computer Code Number	Food Description	Measure	Wt (g)	H_2O (%)	Ener (cal)	Prot (g)	Carb (g)	Dietary Fiber (g)	Fat (g)	Fat Breakdown (g) Sat	Mono	Poly
	VEGETABLES and LEGUMES—Continued											
	Broccoli, raw:											
817	Chopped	½ c	44	91	12	1	2	1	<1	t	t	.1
818	Spears	1 ea	31	91	9	1	2	1	<1	t	t	.1
	Broccoli, cooked from fresh:											
819	Spears	1 ea	180	91	50	5	9	5	1	.1	t	.3
820	Chopped	½ c	78	91	22	2	4	2	<1	t	t	.1
	Broccoli, cooked from frozen:											
821	Spear, small piece	½ c	92	91	26	3	5	3	<1	t	t	.1
822	Chopped	½ c	92	91	26	3	5	3	<1	t	t	.1
1603	Broccoflower, steamed	½ c	78	90	25	2	5	2	<1	t	t	.1
823	Brussels sprouts, cooked from fresh	½ c	78	87	30	2	7	2	<1	.1	t	.2
824	Brussels sprouts, cooked from frozen	½ c	78	87	33	3	6	3	<1	.1	t	.2
	Cabbage, common varieties:											
825	Raw, shredded or chopped	1 c	70	92	17	1	4	2	<1	t	t	.1
826	Cooked, drained	1 c	150	94	33	2	7	3	1	.1	t	.3
	Cabbage, Chinese:											
1178	Bok choy, raw, shredded	1 c	70	95	9	1	2	1	<1	t	t	.1
827	Bok choy, cooked, drained	1 c	170	96	20	3	3	3	<1	t	t	.1
1937	Kim chee style	1 c	150	92	31	2	6	2	<1	t	t	.2
828	Pe Tsai, raw, chopped	1 c	76	94	12	1	2	2	<1	t	t	.1
1796	Pe Tsai, cooked	1 c	119	95	17	2	3	3	<1	t	t	.1
	Cabbage, red, coarsely chopped:											
829	Raw	1 c	89	92	24	1	5	2	<1	t	t	.1
830	Cooked, drained	1 c	150	94	31	2	7	3	<1	t	t	.1
831	Cabbage, savoy, coarsely chopped, raw	1 c	70	91	19	1	4	2	<1	t	t	t
1785	Cabbage, savoy, cooked	1 c	145	92	35	3	8	4	<1	t	t	.1
1896	Capers	1 ea	5	86		<1	<1	<1	<1			
	Carrots, raw:											
832	Whole, 7½ x 1⅛"	1 ea	72	88	31	1	7	2	<1	t	t	.1
833	Grated	½ c	55	88	24	1	6	2	<1	t	t	t
	Carrots, cooked, sliced, drained:											
834	From raw	½ c	78	87	35	1	8	3	<1	t	t	.1
835	From frozen	½ c	73	90	26	1	6	3	<1	t	t	t
836	Carrots, canned, sliced, drained	½ c	73	93	17	<1	4	1	<1	t	t	.1
837	Carrot juice, canned	1 c	236	89	94	2	22	2	<1	.1	t	.2
	Cauliflower, flowerets:											
838	Raw	½ c	50	92	12	1	3	1	<1	t	t	t
839	Cooked from fresh, drained	½ c	62	93	14	1	3	2	<1	t	t	.1
840	Cooked, from frozen, drained	½ c	90	94	17	1	3	2	<1	t	t	.1
	Celery, pascal type, raw:											
841	Large outer stalk, 8 x 1½"(root end)	1 ea	40	95	6	<1	1	1	<1	t	t	t
842	Diced	1 c	120	95	19	1	4	2	<1	t	t	.1
1789	Celeriac/celery root, cooked	1 c	155	92	39	1	9	2	<1	.1	.1	.2
1179	Chard, swiss, raw, chopped	1 c	36	93	7	1	1	1	<1	t	t	t
1180	Chard, swiss, cooked	1 c	175	93	35	3	7	4	<1	t	t	t
1855	Chayote fruit, raw	1 ea	203	94	39	2	9	3	<1	.1	t	.1
1856	Chayote fruit, cooked	1 c	160	93	38	1	8	4	1	.2	.1	.3
	Chickpeas (see Garbanzo Beans #854)											
	Collards, cooked, drained:											
843	From raw	½ c	95	92	25	2	5	3	<1	t	t	.1
844	From frozen	½ c	85	88	31	3	6	3	<1	.1	t	.2
	Corn, yellow, cooked, drained:											
845	From raw, on cob, 5" long	1 ea	77	73	72	2	17	2	1	.1	.2	.3
846	From frozen, on cob, 3½" long	1 ea	63	73	59	2	14	2	<1	.1	.1	.2
847	Kernels, cooked from frozen	½ c	82	77	66	2	16	2	<1	.1	.1	.2
	Corn, canned:											
848	Cream style	½ c	128	79	92	2	23	2	1	.1	.2	.3
849	Whole kernel, vacuum pack	½ c	105	77	83	3	20	2	1	.1	.2	.2
	Cowpeas (see Black-eyed peas #814-816)											
850	Cucumber slices with peel	7 pce	28	96	4	<1	1	<1	<1	t	t	t
1948	Cucumber, kim chee style	1 c	150	91	31	2	7	2	<1	t	0	.1

Chol (mg)	Calc (mg)	Iron (mg)	Magn (mg)	Pota (mg)	Sodi (mg)	Zinc (mg)	VT-A (RE)	Thia (mg)	VT-E (a-TE)	Ribo (mg)	Niac (mg)	V-B6 (mg)	Fola (μg)	VT-C (mg)
0	21	.39	11	143	12	.18	68	.03	.73	.05	.28	.07	31	41
0	15	.27	8	101	8	.12	48	.02	.51	.04	.2	.05	22	29
0	83	1.51	43	526	47	.68	250	.1	3.04	.2	1.03	.26	90	134
0	36	.65	19	228	20	.3	108	.04	1.32	.09	.45	.11	39	58
0	47	.56	18	166	22	.28	174	.05	.95	.07	.42	.12	28	37
0	47	.56	18	166	22	.28	174	.05	1.52	.07	.42	.12	52	37
0	25	.55	16	251	18	.39	5	.06	.23	.07	.59	.14	38	49
0	28	.94	16	247	16	.26	56	.08	.66	.06	.47	.14	47	48
0	19	.58	19	254	18	.28	46	.08	.45	.09	.42	.22	79	36
0	33	.41	10	172	13	.13	9	.03	.07	.03	.21	.07	30	22
0	46	.25	12	146	12	.13	19	.09	.16	.08	.42	.17	30	30
0	73	.56	13	176	45	.13	210	.03	.08	.05	.35	.14	46	31
0	158	1.77	19	631	58	.29	437	.05	.2	.11	.73	.28	69	44
0	145	1.28	27	375	995	.35	426	.07	.24	.1	.75	.34	88	80
0	58	.24	10	181	7	.17	91	.03	.09	.04	.3	.18	60	20
0	38	.36	12	268	11	.21	115	.05	.14	.05	.59	.21	63	19
0	45	.44	13	183	10	.19	4	.04	.09	.03	.27	.19	18	51
0	55	.52	16	210	12	.22	4	.05	.18	.03	.3	.21	19	52
0	24	.28	20	161	20	.19	70	.05	.07	.02	.21	.13	56	22
0	43	.55	35	267	35	.33	129	.07	.15	.03	.03	.22	67	25
0	2	.05			105		1							0
0	19	.36	11	233	25	.14	2025	.07	.33	.04	.67	.11	10	7
0	15	.27	8	178	19	.11	1547	.05	.25	.03	.51	.08	8	5
0	24	.48	10	177	51	.23	1914	.03	.33	.04	.39	.19	11	2
0	20	.34	7	115	43	.17	1292	.02	.31	.03	.32	.09	8	2
0	18	.47	6	131	177	.19	1005	.01	.31	.02	.4	.08	7	2
0	57	1.09	33	689	68	.42	2584	.22	.02	.13	.91	.51	9	20
0	11	.22	7	152	15	.14	1	.03	.02	.03	.26	.11	28	23
0	10	.2	6	88	9	.11	1	.03	.02	.03	.25	.11	27	27
0	15	.37	8	125	16	.12	2	.03	.04	.05	.28	.08	37	28
0	16	.16	4	115	35	.05	5	.02	.14	.02	.13	.03	11	3
0	48	.48	13	344	104	.16	16	.05	.43	.05	.39	.1	34	8
0	40	.67	19	268	95	.31	0	.04	.31	.06	.66	.16	5	6
0	18	.65	29	136	77	.13	119	.01	.68	.03	.14	.04	5	11
0	102	3.96	151	961	313	.58	550	.06	3.31	.15	.63	.15	15	31
0	34	.69	24	254	4	1.5	12	.05	.24	.06	.95	.15	189	16
0	21	.35	19	277	2	.5	8	.04	.19	.06	.67	.19	29	13
0	113	.44	16	247	9	.4	297	.04	.84	.1	.55	.12	88	17
0	179	.95	25	213	42	.23	508	.04	.42	.1	.54	.1	65	22
0	2	.47	22	193	3	.48	16	.13	.07	.05	1.17	.17	23	4
0	2	.38	18	158	3	.4	13	.11	.06	.04	.96	.14	19	3
0	3	.29	16	121	4	.33	18	.07	.07	.06	1.07	.11	25	3
0	4	.49	22	172	365	.68	13	.03	.11	.07	1.23	.08	57	6
0	5	.44	24	195	286	.48	25	.04	.09	.08	1.23	.06	52	9
0	4	.07	3	40	1	.06	6	.01	.02	.01	.06	.01	4	1
0	13	7.23	12	176	1531	.76	49	.04	.24	.04	.69	.16	34	5

Table H-1

Food Composition (Computer code number is for West Diet Analysis program) (For purposes of calculations, use "0" for t, <1, <.1, <.01, etc.)

Computer Code Number	Food Description	Measure	Wt (g)	H_2O (%)	Ener (cal)	Prot (g)	Carb (g)	Dietary Fiber (g)	Fat (g)	Fat Breakdown (g) Sat	Mono	Poly
	VEGETABLES and LEGUMES—Continued											
	Dandelion Greens:											
851	Raw	1 c	55	86	25	1	5	2	<1	.1	t	.2
852	Chopped, cooked, drained	1 c	105	90	35	2	7	3	1	.2	t	.3
853	Eggplant, cooked	1 c	99	92	28	1	7	2	<1	t	t	.1
1714	Endive, fresh, chopped	1 c	50	94	8	1	2	2	<1	t	t	t
856	Escarole/curly endive, chopped	1 c	50	94	8	1	2	2	<1	t	t	t
854	Garbanzo beans (chickpeas), cooked	1 c	164	60	269	14	45	12	4	.4	1	1.9
1939	Grape leaf, raw:	1 ea	3	73	3	<1	1	<1	<1	t	t	t
7914	Cup	1 c	14	73	13	1	2	2	<1	t	t	.1
855	Great northern beans, cooked	1 c	177	69	209	15	37	12	1	.2	t	.3
857	Jerusalem artichoke, raw slices	1 c	150	78	114	3	26	2	<1	0	t	t
1794	Jicama	1 c	120	90	46	1	11	6	<1	t	t	t
	Kale, cooked, drained:											
858	From raw	1 c	130	91	36	2	7	3	1	.1	t	.3
859	From frozen	1 c	130	90	39	4	7	3	1	.1	t	.3
860	Kidney beans, canned	1 c	256	77	218	13	40	16	1	.1	.1	.5
1181	Kohlrabi, raw slices	1 c	135	91	36	2	8	5	<1	t	t	.1
861	Kohlrabi, cooked	1 c	165	90	48	3	11	2	<1	t	t	.1
1183	Leeks, raw, chopped	1 c	89	83	54	1	13	2	<1	t	t	.1
1182	Leeks, cooked, chopped	1 c	104	91	32	1	8	1	<1	t	t	.1
862	Lentils, cooked from dry	1 c	198	70	230	18	40	16	1	.1	.1	.3
1288	Lentils, sprouted, stir-fried	1 c	124	69	125	11	26	5	1	.1	.1	.2
1289	Lentils, sprouted, raw	1 c	77	67	82	7	17	3	<1	t	.1	.2
	Lettuce:											
	Butterhead/Boston types:											
863	Head, 5" diameter	¼ ea	41	96	5	1	1	<1	<1	t	t	t
864	Leaves, inner or outer	4 ea	30	96	4	<1	1	<1	<1	t	t	t
	Iceberg/crisphead:											
867	Chopped or shredded	1 c	55	96	7	1	1	1	<1	t	t	.1
865	Head, 6" diameter	1 ea	539	96	65	5	11	8	1	.1	t	.5
866	Wedge, ¼ head	1 ea	135	96	16	1	3	2	<1	t	t	.1
868	Looseleaf, chopped	½ c	28	94	5	<1	1	1	<1	t	t	t
869	Romaine, chopped	½ c	28	95	4	<1	1	<1	<1	t	t	t
870	Romaine, inner leaf	3 pce	30	95	4	<1	1	1	<1	t	t	t
1930	Luffa, cooked (Chinese okra)	1 c	178	89	57	3	13	6	<1	.1	t	.1
	Mushrooms:											
871	Raw, sliced	½ c	35	92	9	1	2	<1	<1	t	t	.1
872	Cooked from fresh, pieces	½ c	78	91	21	2	4	2	<1	t	t	.1
1962	Stir fried, shitake slices	½ c	73	83	40	1	10	2	<1	t	t	t
873	Canned, drained	½ c	78	91	19	1	4	2	<1	t	t	.1
1951	Mushroom caps, pickled	8 ea	47	92	11	1	2	1	<1	t	t	.1
	Mustard greens:											
874	Cooked from raw	½ c	70	94	10	2	1	1	<1	t	.1	t
875	Cooked from frozen	½ c	75	94	14	2	2	2	<1	t	.1	t
876	Navy beans, cooked from dry	1 c	182	63	258	16	48	12	1	.3	.1	.4
	Okra, cooked:											
877	From fresh pods	8 ea	85	90	27	2	6	2	<1	t	t	t
878	From frozen slices	1 c	184	91	51	4	11	5	1	.1	.1	.1
1236	Batter fried from fresh	1 c	92	69	175	3	11	2	13	2.1	3.4	7.1
1930	Chinese, (Luffa), cooked	1 c	178	89	57	3	13	6	<1	.1	t	.1
	Onions:											
879	Raw, chopped	½ c	80	90	30	1	7	1	<1	t	t	t
880	Raw, sliced	½ c	58	90	22	1	5	1	<1	t	t	t
881	Cooked, drained, chopped	½ c	105	88	46	1	11	1	<1	t	t	.1
882	Dehydrated flakes	¼ c	14	4	49	1	12	1	<1	t	t	t
1934	Onions, pearl, cooked	½ c	93	87	41	1	9	1	<1	t	t	.1
883	Spring/green onions, bulb and top, chopped	½ c	50	90	16	1	4	1	<1	t	t	t
884	Onion rings, breaded, heated f/frozen	2 ea	20	28	81	1	8	<1	5	1.7	2.2	1
1917	Palm hearts, cooked slices	1 c	146	69	150	4	39	2	<1	.1	.1	t
885	Parsley, raw, chopped	½ c	30	88	11	1	2	1	<1	t	.1	t
888	Parsnips, sliced, cooked	½ c	78	78	63	1	15	3	<1	t	.1	t

Chol (mg)	Calc (mg)	Iron (mg)	Magn (mg)	Pota (mg)	Sodi (mg)	Zinc (mg)	VT-A (RE)	Thia (mg)	VT-E (a-TE)	Ribo (mg)	Niac (mg)	V-B6 (mg)	Fola (μg)	VT-C (mg)
0	103	1.71	20	218	42	.23	770	.1	1.38	.14	.44	.14	15	19
0	147	1.89	25	244	46	.29	1228	.14	2.63	.18	.54	.17	13	19
0	6	.35	13	246	3	.15	6	.07	.03	.02	.59	.08	14	1
0	26	.41	7	157	11	.39	103	.04	.22	.04	.2	.01	71	3
0	26	.41	7	157	11	.39	103	.04	.22	.04	.2	.01	71	3
0	80	4.74	79	477	11	2.51	5	.19	.57	.1	.86	.23	282	2
0	11	.08	3	8	<1	.02	81	<.01	.06	.01	.07	.01	2	<1
0	51	.37	13	38	1	.09	378	.01	.28	.05	.33	.06	12	2
0	120	3.77	88	692	4	1.56	<1	.28	.53	.1	1.21	.21	181	2
0	21	5.1	25	644	6	.18	3	.3	.28	.09	1.95	.12	20	6
0	14	.72	14	180	5	.19	2	.02	5.48	.03	.24	.05	14	24
0	94	1.17	23	296	30	.31	962	.07	1.11	.09	.65	.18	17	53
0	179	1.22	23	417	19	.23	826	.06	.23	.15	.87	.11	19	33
0	61	3.23	72	658	873	1.41	0	.27	.13	.22	1.17	.06	130	3
0	32	.54	26	473	27	.04	5	.07	.65	.03	.54	.2	22	84
0	41	.66	31	561	35	.51	7	.07	2.76	.03	.64	.25	20	89
0	52	1.87	25	160	18	.11	9	.05	.82	.03	.36	.21	57	11
0	31	1.14	15	90	10	.06	5	.03	.63	.02	.21	.12	25	4
0	38	6.59	71	731	4	2.51	2	.33	.22	.14	2.1	.35	358	3
0	17	3.84	43	352	12	1.98	5	.27	.11	.11	1.49	.2	83	16
0	19	2.47	28	248	8	1.16	4	.18	.07	.1	.87	.15	77	13
0	13	.12	5	105	2	.07	40	.02	.18	.02	.12	.02	30	3
0	10	.09	4	77	1	.05	29	.02	.13	.02	.09	.01	22	2
0	10	.27	5	87	5	.12	18	.02	.15	.02	.1	.02	31	2
0	102	2.7	48	852	48	1.19	178	.25	1.51	.16	1.01	.22	302	21
0	26	.67	12	213	12	.3	45	.06	.38	.04	.25	.05	76	5
0	19	.39	3	74	3	.08	53	.01	.12	.02	.11	.01	14	5
0	10	.31	2	81	2	.07	73	.03	.12	.03	.14	.01	38	7
0	11	.33	2	87	2	.07	78	.03	.13	.03	.15	.01	41	7
0	112	.8	101	570	420	.97	103	.23	1.22	.1	1.54	.33	81	29
0	2	.43	3	130	1	.26	0	.04	.04	.16	1.44	.03	7	1
0	5	1.36	9	278	2	.68	0	.06	.09	.23	3.48	.07	14	3
0	2	.32	10	85	3	.97	0	.03	.09	.12	1.1	.12	15	<1
0	9	.62	12	101	332	.56	0	.07	.09	.02	1.24	.05	10	0
0	2	.5	5	139	95	.28	0	.03	.05	.16	1.42	.03	6	1
0	52	.49	10	141	11	.08	212	.03	1.41	.04	.3	.07	51	18
0	76	.84	10	104	19	.15	335	.03	1.31	.04	.19	.08	52	10
0	127	4.51	107	670	2	1.93	<1	.37	.73	.11	.97	.3	255	2
0	54	.38	48	274	4	.47	49	.11	.59	.05	.74	.16	39	14
0	177	1.23	94	431	6	1.14	94	.18	1.27	.23	1.44	.09	269	22
15	104	.77	37	214	137	.5	43	.13	3.08	.1	.75	.13	37	10
0	112	.8	101	570	420	.97	103	.23	1.22	.1	1.54	.33	81	29
0	16	.18	8	126	2	.15	0	.03	.1	.02	.12	.09	15	5
0	12	.13	6	91	2	.11	0	.02	.07	.01	.09	.07	11	4
0	23	.25	12	174	3	.22	0	.04	.14	.02	.17	.13	16	5
0	36	.22	13	227	3	.26	0	.07	.19	.01	.14	.22	23	10
0	21	.22	10	154	218	.19	0	.04	.12	.02	.15	.12	14	5
0	36	.74	10	138	8	.19	19	.03	.06	.04	.26	.03	32	9
0	6	.34	4	26	75	.08	5	.06	.14	.03	.72	.01	13	<1
0	26	2.47	15	2636	20	5.45	10	.07	.73	.25	1.25	1.06	30	10
0	41	1.86	15	166	17	.32	156	.03	.54	.03	.39	.03	46	40
0	29	.45	23	286	8	.2	0	.06	.78	.04	.56	.07	45	10

Table H-1

Food Composition

(Computer code number is for West Diet Analysis program) (For purposes of calculations, use "0" for t, <1, <.1, <.01, etc.)

Computer Code Number	Food Description	Measure	Wt (g)	H_2O (%)	Ener (cal)	Prot (g)	Carb (g)	Dietary Fiber (g)	Fat (g)	Fat Breakdown (g) Sat	Mono	Poly
	VEGETABLES and LEGUMES—Continued											
	Peas:											
	Black-eyed, cooked:											
814	From dry, drained	½ c	86	70	100	7	18	6	<1	.1	t	.2
815	From fresh, drained	½ c	82	75	79	3	17	4	<1	.1	t	.1
816	From frozen, drained	½ c	85	66	112	7	20	5	1	.1	.1	.2
889	Edible pod peas, cooked	½ c	80	89	34	3	6	2	<1	t	t	.1
890	Green, canned, drained:	½ c	85	82	59	4	11	3	<1	.1	t	.1
5267	Unsalted	½ c	124	86	66	4	12	4	<1	.1	t	.2
891	Green, cooked from frozen	½ c	80	79	62	4	11	4	<1	t	t	.1
1786	Snow peas, raw	½ c	49	89	21	1	4	1	<1	t	t	t
1787	Snow peas, raw	10 ea	34	89	14	1	3	1	<1	t	t	t
892	Split, green, cooked from dry	½ c	98	69	116	8	21	8	<1	.1	.1	.2
1187	Peas & carrots, cooked from frozen	½ c	80	86	38	2	8	2	<1	.1	t	.2
1186	Peas & carrots, canned w/liquid	½ c	128	88	49	3	11	3	<1	.1	t	.2
	Peppers, hot:											
893	Hot green chili, canned	½ c	68	92	14	1	3	1	<1	t	t	t
894	Hot green chili, raw	1 ea	45	88	18	1	4	1	<1	t	t	t
1715	Hot red chili, raw, diced	1 tbs	9	88	4	<1	1	<1	<1	t	t	t
1988	Jalapeno, raw	1 ea	45	90	11	<1	2		<1			
895	Jalapeno, chopped, canned	½ c	68	89	18	1	3	2	1	.1	t	.3
1918	Jalapeno wheels, in brine (Ortega)	2 tbs	29		10	0	2		0	0	0	0
	Peppers, sweet, green:											
896	Whole pod (90 g with refuse), raw	1 ea	74	92	20	1	5	1	<1	t	t	.1
897	Cooked, chopped (1 pod cooked = 73g)	½ c	68	92	19	1	5	1	<1	t	t	.1
	Peppers, sweet, red:											
1286	Raw, chopped	½ c	75	92	20	1	5	1	<1	t	t	.1
1807	Raw, each	1 ea	74	92	20	1	5	1	<1	t	t	.1
1287	Cooked, chopped	½ c	68	92	19	1	5	1	<1	t	t	.1
	Peppers, sweet, yellow:											
1872	Raw, large	1 ea	186	92	50	2	12	2	<1	.1	t	.2
1873	Strips	10 pce	52	92	14	1	3	<1	<1	t	t	.1
898	Pinto beans, cooked from dry	½ c	85	64	116	7	22	7	<1	.1	.1	.2
1191	Poi, two finger	½ c	120	72	134	<1	33	<1	<1	t	t	.1
	Potatoes:											
	Baked in oven, 4¾"x2⅓" diam											
899	With skin	1 ea	202	71	220	5	51	5	<1	.1	t	.1
900	Flesh only	1 ea	156	75	145	3	34	2	<1	t	t	.1
901	Skin only	1 ea	58	47	115	2	27	5	<1	t	t	t
	Baked in microwave, 4¾"x 2⅓"dm:											
902	With skin	1 ea	202	72	212	5	49	5	<1	.1	t	.1
903	Flesh only	1 ea	156	74	156	3	36	2	<1	t	t	.1
904	Skin only	1 ea	58	63	77	3	17	3	<1	t	t	t
	Boiled, about 2½" diam:											
905	Peeled after boiling	1 ea	136	77	118	3	27	2	<1	t	t	.1
906	Peeled before boiling	1 ea	135	77	116	2	27	2	<1	t	t	.1
	French fried, strips 2–3½" long:											
907	Oven heated	10 ea	50	35	167	2	20	2	9	3	5.7	.7
908	Fried in vegetable oil	10 ea	50	38	158	2	20	2	8	1.9	4.7	.7
1188	Fried in veg and animal oil	10 ea	50	38	158	2	20	2	8	1.9	4.7	.7
909	Hashed browns from frozen	1 c	156	56	340	5	44	3	18	7	8	2.1
	Mashed:											
910	Home recipe with whole milk	½ c	105	78	81	2	18	2	1	.4	.2	.1
911	Home recipe with milk and marg	½ c	105	76	111	2	17	2	4	1.1	1.9	1.3
912	Prepared from flakes; water, milk, margarine, salt added	½ c	110	76	124	2	16	3	6	1.6	2.5	1.7
	Potato products, prepared:											
	Au gratin:											
913	From dry mix	½ c	123	79	114	3	16	1	5	3.5	1.5	.2
914	From home recipe, using butter	½ c	122	74	161	7	14	2	9	4.8	3.2	1.3
	Scalloped:											
915	From dry mix	½ c	122	79	113	3	16	1	5	3.2	1.5	.2
916	From home recipe, using butter	½ c	123	81	106	4	13	2	5	1.7	1.7	.9

PAGE KEY: A–4 = Beverages A–6 = Dairy A–10 = Eggs A–10 = Fat/Oil A–14 = Fruit A–20 = Bakery A–28 = Grain A–32 = Fish A–34 = Meats A–38 = Poultry A–40 = Sausage A–42 = Mixed/Fast A–46 = Nuts/Seeds A–50 = Sweets A–52 = Vegetables/Legumes A–62 = Vegetarian Foods A–64 = Misc A–66 = Soups/Sauces A–68 = Fast A–84 = Convenience A–88 = Baby foods

Chol (mg)	Calc (mg)	Iron (mg)	Magn (mg)	Pota (mg)	Sodi (mg)	Zinc (mg)	VT-A (RE)	Thia (mg)	VT-E (a-TE)	Ribo (mg)	Niac (mg)	V-B6 (mg)	Fola (µg)	VT-C (mg)
0	21	2.16	46	239	3	1.11	2	.17	.24	.05	.43	.09	179	<1
0	105	.92	43	343	3	.84	65	.08	.18	.12	1.15	.05	104	2
0	20	1.8	42	319	4	1.21	7	.22	.33	.05	.62	.08	120	2
0	34	1.58	21	192	3	.3	10	.1	.31	.06	.43	.11	23	38
0	17	.81	14	147	214	.6	65	.1	.32	.07	.62	.05	38	8
0	22	1.26	21	124	11	.87	47	.14	.47	.09	1.04	.08	35	12
0	19	1.26	23	134	70	.75	54	.23	.14	.08	1.18	.09	47	8
0	21	1.02	12	98	2	.13	7	.07	.19	.04	.29	.08	20	29
0	15	.71	8	68	1	.09	5	.05	.13	.03	.2	.05	14	20
0	14	1.26	35	355	2	.98	1	.19	.38	.05	.87	.05	64	<1
0	18	.75	13	126	54	.36	621	.18	.26	.05	.92	.07	21	6
0	29	.96	18	128	333	.74	739	.09	.24	.07	.74	.11	23	8
0	5	.34	10	127	798	.12	41	.01	.47	.03	.54	.1	7	46
0	8	.54	11	153	3	.13	35	.04	.31	.04	.43	.12	10	109
0	2	.11	2	31	1	.03	97	.01	.06	.01	.09	.02	2	22
				2	2		30		.37					53
0	16	1.28	10	131	1136	.23	116	.03	.47	.03	.27	.13	10	7
0				55	390		10		.2					21
0	7	.34	7	131	1	.09	47	.05	.51	.02	.38	.18	16	66
0	6	.31	7	113	1	.08	40	.04	.47	.02	.32	.16	11	51
0	7	.34	7	133	1	.09	428	.05	.52	.02	.38	.19	16	143
0	7	.34	7	131	1	.09	422	.05	.51	.02	.38	.18	16	141
0	6	.31	7	113	1	.08	256	.04	.47	.02	.32	.16	11	116
0	20	.86	22	394	4	.32	45	.05	1.28	.05	1.66	.31	48	342
0	6	.24	6	110	1	.09	12	.01	.36	.01	.46	.09	13	96
0	41	2.22	47	398	2	.92	<1	.16	.8	.08	.34	.13	146	2
0	19	1.06	29	220	14	.26	2	.16	.22	.05	1.32	.33	26	5
0	20	2.75	54	844	16	.65	0	.22	.1	.07	3.33	.7	22	26
0	8	.55	39	610	8	.45	0	.16	.06	.03	2.18	.47	14	20
0	20	4.08	25	332	12	.28	0	.07	.02	.06	1.78	.36	12	8
0	22	2.5	54	903	16	.73	0	.24	.1	.06	3.45	.69	24	30
0	8	.64	39	641	11	.51	0	.2	.06	.04	2.54	.5	19	24
0	27	3.45	21	377	9	.3	0	.04	.02	.04	1.29	.28	10	9
0	7	.42	30	515	5	.41	0	.14	.07	.03	1.96	.41	14	18
0	11	.42	27	443	7	.36	0	.13	.07	.03	1.77	.36	12	10
0	6	.83	11	270	307	.2	0	.04	.25	.02	1.34	.11	11	3
0	9	.38	17	366	108	.19	0	.09	.25	.01	1.63	.12	14	5
6	9	.38	17	366	108	.19	0	.09	.25	.01	1.63	.12	14	5
0	23	2.36	26	680	53	.5	0	.17	.3	.03	3.78	.2	10	10
2	27	.28	19	314	318	.3	6	.09	.05	.04	1.18	.24	9	7
2	27	.27	19	303	310	.28	21	.09	.31	.04	1.13	.23	8	6
4	54	.24	20	256	365	.2	23	.12	.77	.05	.74	.01	8	11
18	102	.39	18	269	540	.29	38	.02	1.48	.1	1.15	.05	8	4
18	145	.78	24	483	528	.84	46	.08	.64	.14	1.21	.21	13	12
13	44	.46	17	248	416	.3	26	.02	.18	.07	1.26	.05	12	4
7	70	.7	23	465	412	.49	23	.08	.4	.11	1.29	.22	13	13

Table H-1

Food Composition

(Computer code number is for West Diet Analysis program) (For purposes of calculations, use "0" for t, <1, <.1, <.01, etc.)

Computer Code Number	Food Description	Measure	Wt (g)	H_2O (%)	Ener (cal)	Prot (g)	Carb (g)	Dietary Fiber (g)	Fat (g)	Fat Breakdown (g) Sat	Mono	Poly
	VEGETABLES and LEGUMES—Continued											
	Potato Salad (see Mixed Dishes #715)											
1192	Potato puffs, cooked from frozen	½ c	64	53	142	2	19	2	7	3.3	2.8	.5
918	Pumpkin, cooked from fresh, mashed	½ c	123	94	25	1	6	1	<1	t	t	t
919	Pumpkin, canned	½ c	123	90	42	1	10	4	<1	.2	t	t
1891	Radicchio, raw, shredded	½ c	20	93	5	<1	1	<1	<1	t	t	t
1894	Radicchio, raw, leaf	10 ea	80	93	18	1	4	1	<1	t	t	.1
920	Red radishes	10 ea	45	95	8	<1	2	1	<1	t	t	t
1793	Daikon radishes (Chinese) raw	½ c	44	95	8	<1	2	1	<1	t	t	t
921	Refried beans, canned	½ c	126	76	118	7	19	7	2	.6	.7	.2
1375	Rutabaga, cooked cubes	½ c	85	89	33	1	7	2	<1	t	t	.1
922	Sauerkraut, canned with liquid	½ c	118	92	22	1	5	3	<1	t	t	.1
923	Seaweed, kelp, raw	½ c	40	82	17	1	4	1	<1	.1	t	t
924	Seaweed, spirulina, dried	½ c	8	5	23	5	2	<1	1	.2	.1	.2
1866	Shallots, raw, chopped	1 tbs	10	80	7	<1	2	<1	<1	t	t	t
1557	Snow peas, stir-fried	½ c	83	89	35	2	6	2	<1	t	t	.1
925	Soybeans, cooked from dry	½ c	86	63	149	15	9	5	8	1.1	1.7	4.4
1996	Soybeans, dry roasted	½ c	86	1	387	34	28	7	19	2.7	4.1	10.6
	Soybean products:											
926	Miso	½ c	138	41	284	17	39	7	8	1.2	1.9	4.7
	Soy milk (see #144 and #2301 under Dairy)											
	Tofu (soybean curd):											
7540	Extra firm, silken	½ c	126	88	69	9	3	<1	2	.4	.4	1.3
7542	Firm, silken	½ c	126	87	77	9	3	<1	3	.5	.7	1.9
927	Regular	½ c	124	87	76	8	2	<1	5	.7	1	2.6
7541	Soft, silken	½ c	124	89	68	6	4	<1	3	.4	.6	1.9
	Spinach:											
928	Raw, chopped	½ c	28	92	6	1	1	1	<1	t	t	t
929	Cooked, from fresh, drained	½ c	90	91	21	3	3	2	<1	t	t	.1
930	Cooked from frozen (leaf)	½ c	95	90	27	3	5	3	<1	t	t	.1
931	Canned, drained solids:	½ c	107	92	25	3	4	3	1	.1	t	.2
5149	Unsalted	½ c	107	92	25	3	4	3	1	.1	t	.2
	Spinach soufflé (see Mixed Dishes)											
	Squash, summer varieties,cooked w/skin:											
932	Varieties averaged	½ c	90	94	18	1	4	1	<1	.1	t	.1
933	Crookneck	½ c	90	94	18	1	4	1	<1	.1	t	.1
934	Zucchini	½ c	90	95	14	1	4	1	<1	t	t	t
	Squash, winter varieties, cooked:											
	Average of all varieties, baked:											
935	Mashed	1 c	245	89	96	2	21	7	2	.3	.1	.6
936	Cubes	1 c	205	89	80	2	18	6	1	.3	.1	.5
937	Acorn, baked, mashed	½ c	123	83	69	1	18	5	<1	t	t	.1
1218	Acorn, boiled, mashed	½ c	122	90	41	1	11	3	<1	t	t	t
	Butternut squash:											
938	Baked cubes	½ c	103	88	41	1	11	3	<1	t	t	t
1219	Baked, mashed	½ c	103	88	41	1	11	3	<1	t	t	t
1193	Cooked from frozen	½ c	120	88	47	1	12	3	<1	t	t	t
1194	Hubbard, baked, mashed	½ c	120	85	60	3	13	3	1	.2	.1	.3
1195	Hubbard, boiled, mashed	½ c	118	91	35	2	8	3	<1	.1	t	.2
1196	Spaghetti, baked or boiled	½ c	77	92	22	<1	5	1	<1	t	t	.1
1189	Succotash, cooked from frozen	½ c	85	74	79	4	17	3	1	.1	.1	.4
	Sweet potatoes:											
939	Baked in skin, peeled, 5 x 2″ diam	1 ea	114	73	117	2	28	3	<1	t	t	.1
940	Boiled without skin, 5 x 2″ diam	1 ea	151	73	159	2	37	3	<1	.1	t	.2
941	Candied, 2½ x 2″	1 pce	105	67	144	1	29	3	3	1.4	.7	.2
	Canned:											
942	Solid pack	½ c	128	74	129	3	30	2	<1	.1	t	.1
943	Vacuum pack, mashed	½ c	127	76	116	2	27	2	<1	.1	t	.1
944	Vacuum pack, 3¾ x 1″	2 pce	80	76	73	1	17	1	<1	t	t	.1
1940	Taro shoots, cooked slices	1 c	140	95	20	1	4	1	<1	t	t	t
1941	Taro, tahitian, cooked slices	1 c	137	86	60	6	9	1	1	.2	.1	.4
	Tomatillos:											
1877	Raw, each	1 ea	34	92	11	<1	2	1	<1	t	.1	.1
1875	Raw, chopped	1 c	132	92	42	1	8	3	1	.2	.2	.6

Chol (mg)	Calc (mg)	Iron (mg)	Magn (mg)	Pota (mg)	Sodi (mg)	Zinc (mg)	VT-A (RE)	Thia (mg)	VT-E (a-TE)	Ribo (mg)	Niac (mg)	V-B6 (mg)	Fola (μg)	VT-C (mg)
0	19	1	12	243	477	.19	1	.12	.03	.05	1.38	.15	11	4
0	18	.7	11	283	1	.28	1330	.04	1.3	.1	.51	.05	10	6
0	32	1.71	28	253	6	.21	2713	.03	1.3	.07	.45	.07	15	5
0	4	.11	3	60	4	.12	1	<.01	.45	.01	.05	.01	12	2
0	15	.45	10	242	18	.5	2	.01	1.81	.02	.2	.05	48	6
0	9	.13	4	104	11	.13	<1	<.01	0	.02	.13	.03	12	10
0	12	.18	7	100	9	.07	0	.01	0	.01	.09	.02	12	10
10	44	2.09	42	336	377	1.47	0	.03	.39	.02	.4	.18	14	8
0	41	.45	20	277	17	.3	48	.07	.13	.03	.61	.09	13	16
0	35	1.73	15	201	780	.22	2	.02	.12	.03	.17	.15	28	17
0	67	1.14	48	36	93	.49	5	.02	.35	.06	.19	<.01	72	1
0	10	2.28	16	109	84	.16	5	.19	.4	.29	1.02	.03	8	1
0	4	.12	2	33	1	.04	125	.01	.01	<.01	.02	.03	3	1
0	36	1.73	20	166	3	.22	11	.11	.32	.06	.47	.13	28	42
0	88	4.42	74	443	1	.99	1	.13	1.68	.24	.34	.2	46	1
0	120	3.4	196	1173	2	4.1	2	.37	3.96	.65	.91	.19	176	4
0	91	3.78	58	226	5032	4.58	12	.13	.01	.34	1.19	.3	45	0
0	39	1.5	34	195	80	.76	0	.1	.18	.04	.3	.01		0
0	41	1.3	34	244	45	.77	0	.13	.24	.05	.31	.01		0
0	138	1.38	33	149	10	.79	1	.06	.01	.05	.66	.06	55	<1
0	38	1.02	36	223	6	.64	0	.12	.25	.05	.37	.01		0
0	28	.76	22	156	22	.15	188	.02	.53	.05	.2	.05	54	8
0	122	3.21	78	419	63	.68	737	.09	.86	.21	.44	.22	131	9
0	139	1.44	66	283	82	.66	739	.06	.91	.16	.4	.14	103	12
0	136	2.46	81	370	29	.49	939	.02	1.39	.15	.41	.11	105	15
0	136	2.46	81	370	29	.49	939	.02	1.39	.15	.41	.11	105	15
0	24	.32	22	173	1	.35	26	.04	.11	.04	.46	.06	18	5
0	24	.32	22	173	1	.35	26	.04	.11	.04	.46	.08	18	5
0	12	.31	20	228	3	.16	22	.04	.11	.04	.38	.07	15	4
0	34	.81	20	1070	2	.64	872	.21	.29	.06	1.72	.18	69	23
0	29	.68	16	896	2	.53	730	.17	.25	.05	1.44	.15	57	20
0	54	1.14	53	538	5	.21	53	.2	.15	.02	1.08	.24	23	13
0	32	.68	32	321	4	.13	32	.12	.15	.01	.65	.14	14	8
0	42	.62	30	293	4	.13	721	.07	.17	.02	1	.13	20	16
0	42	.62	30	293	4	.13	721	.07	.17	.02	1	.13	20	16
0	23	.7	11	160	2	.14	401	.06	.16	.05	.56	.08	20	4
0	20	.56	26	430	10	.18	725	.09	.14	.06	.67	.21	19	11
0	12	.33	15	253	6	.12	473	.05	.14	.03	.39	.12	11	8
0	16	.26	8	90	14	.15	8	.03	.09	.02	.62	.08	6	3
0	13	.76	20	225	38	.38	20	.06	.31	.06	1.11	.08	28	5
0	32	.51	23	397	11	.33	2487	.08	.32	.14	.69	.27	26	28
0	32	.85	15	278	20	.41	2574	.08	.42	.21	.97	.37	17	26
8	27	1.19	12	198	73	.16	440	.02	3.99	.04	.41	.04	12	7
0	38	1.7	31	269	96	.27	1936	.03	.35	.11	1.22	.3	14	7
0	28	1.13	28	396	67	.23	1013	.05	.32	.07	.94	.24	21	33
0	18	.71	18	250	42	.14	638	.03	.2	.05	.59	.15	13	21
0	20	.57	11	482	3	.76	7	.05	1.4	.07	1.13	.16	4	26
0	204	2.14	70	854	74	.14	241	.06	3.7	.27	.66	.16	10	52
0	2	.21	7	91	<1	.07	4	.01	.13	.01	.63	.02	2	4
0	9	.82	26	354	1	.29	14	.06	.5	.05	2.44	.07	9	15

Table H-1

Food Composition

(Computer code number is for West Diet Analysis program) (For purposes of calculations, use "0" for t, <1, <.1, <.01, etc.)

Computer Code Number	Food Description	Measure	Wt (g)	H_2O (%)	Ener (cal)	Prot (g)	Carb (g)	Dietary Fiber (g)	Fat (g)	Fat Breakdown (g) Sat	Mono	Poly
	VEGETABLES and LEGUMES—Continued											
	Tomatoes:											
945	Raw, whole, 2 ⅗" diam	1 ea	123	94	26	1	6	1	<1	.1	.1	.2
946	Raw, chopped	1 c	180	94	38	2	8	2	1	.1	.1	.2
947	Cooked from raw	1 c	240	92	65	3	14	2	1	.1	.2	.4
948	Canned, solids and liquid:	1 c	240	94	46	2	10	2	<1	t	t	.1
5741	Unsalted	1 c	240	94	46	2	10	2	<1	t	t	.1
1879	Tomatoes, sundried:	1 c	54	15	139	8	30	7	2	.2	.3	.6
1881	Pieces	10 pce	20	15	52	3	11	2	1	.1	.1	.2
1885	Oil pack, drained	10 pce	30	54	64	2	7	2	4	.6	2.6	.6
2020	Tomato, raw	1 ea	123	94	26	1	6	1	<1	.1	.1	.2
949	Tomato juice, canned:	1 c	243	94	41	2	10	1	<1	t	t	.1
5397	Unsalted	1 c	243	94	41	2	10	2	<1	t	t	.1
	Tomato products, canned:											
950	Paste, no added salt	1 c	262	74	215	10	51	11	1	.2	.2	.6
951	Puree, no added salt	1 c	250	87	100	4	24	5	<1	.1	.1	.2
952	Sauce, with salt	1 c	245	89	73	3	18	3	<1	.1	.1	.2
953	Turnips, cubes, cooked from fresh	1 c	156	94	33	1	8	3	<1	t	t	.1
	Turnip greens, cooked:											
954	From fresh, leaves and stems	1 c	144	93	29	2	6	5	<1	.1	t	.1
955	From frozen, chopped	1 c	164	90	49	6	8	6	1	.2	t	.3
956	Vegetable juice cocktail, canned	1 c	242	93	46	2	11	2	<1	t	t	.1
	Vegetables, mixed:											
957	Canned, drained	½ c	81	87	38	2	7	2	<1	t	t	.1
958	Frozen, cooked, drained	½ c	91	83	54	3	12	4	<1	t	t	.1
1818	Water chestnuts, Chinese, raw	½ c	62	73	60	1	15	2	<1	t	t	t
	Water chestnuts, canned:											
959	Slices	½ c	70	86	35	1	9	2	<1	t	t	t
960	Whole	4 ea	28	86	14	<1	3	1	<1	t	t	t
1190	Watercress, fresh, chopped	½ c	17	95	2	<1	<1	<1	<1	t	t	t
	VEGETARIAN FOODS:											
7509	Bacon strips, meatless	3 ea	15	49	46	2	1	<1	4	.7	1.1	2.3
1511	Baked beans, canned	½ c	127	73	118	6	26	6	1	.1	t	.2
7526	Bakon crumbles	¼ c	7	16	28	2	1	<1	2			
7548	Chicken, breaded, fried, meatless	1 pce	57	70	97	6	3	3	7	1	2.9	2.5
7547	Chicken slices, meatless	2 ea	60	59	132	10	4	3	8	1.3	2	4.4
7557	Chili w/meat substitute	½ c	107	64	141	19	15	4	2	.3	.6	.9
7549	Fish stick, meatless	2 ea	57	45	165	13	5	3	10	1.6	2.5	5.4
7550	Frankfurter, meatless	1 ea	51	58	102	10	4	2	5	.8	1.2	2.7
7504	GardenBurger, patty	1 ea	45	53	87	5	13	3	1	.4	.3	.7
7505	GardenSausage, patty	1 ea	35	15	117	4	22	5	1	.7	.3	t
7551	Luncheon slice, meatless	1 sl	67	46	188	17	6	3	11	1.7	2.6	5.6
7560	Meatloaf, meatless	1 ea	71	58	142	15	6	3	6	1	1.5	3.3
1171	Nuteena	1 ea	55	58	162	6	6	2	13	5.1	5.8	1.7
7556	Pot pie, meatless	1 ea	227	59	524	15	41	5	34	9.5	12.6	9.8
7554	Soyburger, patty	1 ea	71	58	142	15	6	3	6	1	1.5	3.3
7562	Soyburger w/cheese, patty	1 ea	135	50	316	21	29	4	13	4.2	3.9	3.7
7564	Tempeh	1 c	166	55	330	31	28	9	13	1.9	2.9	7.2
7670	Vegan burger, patty	1 ea	78	71	75	11	6	4	<1	.1	.3	.2
	Vegetarian foods, Green Giant:											
7677	Breakfast links	3 ea	68	65	114	12	5	4	5	.7	1.2	3.1
7676	Breakfast patties	2 ea	57	65	95	10	5	3	4	.6	1	2.6
	Burger, harvest, patty:											
7673	Italian	1 ea	90	65	139	17	8	5	4	1.4	.3	.4
7674	Original	1 ea	90	65	137	18	8	5	4	1.3	.1	.4
7675	Southwestern	1 ea	90	65	135	16	9	5	4	1.4	.2	.4
	Vegetarian foods, Loma Linda											
7727	Chik nuggets, frozen	5 pce	85	47	245	12	13	5	16	2.5	4	8.8
7753	Chik-fried, frozen	1 pce	57	51	178	11	1	1	15	1.9	3.7	8.7
7744	Franks, big, canned	1 ea	51	59	110	10	2	2	7	1.1	1.7	3.8
7747	Linketts, canned	1 ea	35	60	72	7	1	1	4	.7	1.2	2.5
1173	Redi-burger, patty	1 ea	85	59	172	16	5	4	10	1.5	2.4	5.8
7755	Swiss stake w/gravy, canned	1 pce	92	71	120	9	8	4	6	.8	1.5	3.3

Chol (mg)	Calc (mg)	Iron (mg)	Magn (mg)	Pota (mg)	Sodi (mg)	Zinc (mg)	VT-A (RE)	Thia (mg)	VT-E (a-TE)	Ribo (mg)	Niac (mg)	V-B6 (mg)	Fola (µg)	VT-C (mg)
0	6	.55	13	273	11	.11	76	.07	.47	.06	.77	.1	18	23
0	9	.81	20	400	16	.16	112	.11	.68	.09	1.13	.14	27	34
0	14	1.34	34	670	26	.26	178	.17	.91	.14	1.8	.23	31	55
0	72	1.32	29	530	355	.38	144	.11	.77	.07	1.76	.22	19	34
0	72	1.32	29	545	24	.38	144	.11	.91	.07	1.76	.22	19	34
0	59	4.91	105	1850	1131	1.07	47	.28	<.01	.26	4.89	.18	37	21
0	22	1.82	39	685	419	.4	17	.11	<.01	.1	1.81	.07	14	8
0	14	.8	24	470	80	.23	39	.06	.16	.11	1.09	.1	7	31
0	6	.55	13	273	11	.11	76	.07	.47	.06	.77	.1	18	23
0	22	1.41	27	535	877	.34	136	.11	2.21	.07	1.64	.27	48	44
0	22	1.41	27	535	24	.34	136	.11	2.21	.07	1.64	.27	48	44
0	92	5.08	134	2454	231	2.1	639	.41	11.3	.5	8.44	1	59	111
0	42	3.1	60	1065	85	.55	320	.18	6.3	.13	4.3	.38	27	26
0	34	1.89	47	909	1482	.61	240	.16	3.43	.14	2.82	.38	23	32
0	34	.34	12	211	78	.31	0	.04	.05	.04	.47	.1	14	18
0	197	1.15	32	292	42	.2	792	.06	2.48	.1	.59	.26	170	39
0	249	3.18	43	367	25	.67	1308	.09	4.79	.12	.77	.11	65	36
0	27	1.02	27	467	653	.48	283	.1	.77	.07	1.76	.34	51	67
0	22	.85	13	236	121	.33	944	.04	.49	.04	.47	.06	19	4
0	23	.75	20	154	32	.45	389	.06	.33	.11	.77	.07	17	3
0	7	.04	14	362	9	.31	0	.09	.74	.12	.62	.2	10	2
0	3	.61	3	83	6	.27	0	.01	.35	.02	.25	.11	4	1
0	1	.24	1	33	2	.11	0	<.01	.14	.01	.1	.04	2	<1
0	20	.03	4	56	7	.02	80	.01	.17	.02	.03	.02	2	7
0	3	.36	3	25	220	.06	1	.66	1.04	.07	1.13	.07	6	0
0	63	.37	41	376	504	1.78	22	.19	.67	.08	.54	.17	30	4
	8	.44	11	120	172	.25	0	.06		.02	.12	.02	7	0
0	13	.97	7	171	228	.37	0	.4	1.11	.27	2.68	.28	32	0
0	21	.78	10	198	474	.42	0	.66	1.61	.24	3.18	.42	46	0
0	53	4.24	36	362	527	1.26	78	.12	1.25	.07	1.21	.15	82	16
0	54	1.14	13	342	279	.8	0	.63	2.25	.51	6.84	.85	58	0
0	17	.92	9	76	219	.61	0	.56	.98	.61	8.16	.5	40	0
0	36	1.35		129	112		18	.05		.09		.06		<1
0	181	.33		307	78		3	.08		.2		.13		<1
0	27	1.54	15	188	576	1.07	0	.64	2.01	.37	7.37	.74	67	0
0	21	1.49	13	128	391	1.28	0	.64	1.23	.43	7.1	.85	55	0
0	9	.27	33	166	119	.46	0	.1		.35	1.04	.45	49	0
20	66	2.9	31	331	538	1.05	729	.65	4	.4	4.47	.31	40	10
0	21	1.49	13	128	391	1.28	0	.64	1.23	.43	7.1	.85	55	0
13	146	2.71	26	211	931	1.97	45	.77	1.43	.55	8.14	.86	69	1
0	154	3.75	116	609	10	3	115	.22	.03	.18	7.69	.5	86	0
0	79	2.66	15	398	351	.69	0	.23	.01	.51	3.78	.18	225	0
0	65	1.84			340	4.56	0	.18		.09	.27	.18		0
0	54	2			285	3.82	0	.15		.07	2.28	.15		0
0	74	2.61			374	6.93	3	.28		.14	4.05	.28		0
0	76	2.7			378	7.2	0	.29		.14	4.32	.29		0
0	71	2.52			371	6.66	3	.27		.13	4.05	.27		0
2	40	1.4		153	709	.43	0	.67		.3	2.89	.45		0
4	2	.63		76	503	.2	0	.98		.46	2.1	.35		0
2	8	.77		51	243	.89	0	.26		.46	1.98	.14		0
1	4	.39		29	160	.46	0	.13		.22	.64	.29		0
1	12	1.06	16	121	455	1.11	0	.14		.3	1.9	.51	21	0
2	24	.31		225	433	.41	0	1.25		.65	5.41	1		0

Table H-1

Food Composition

(Computer code number is for West Diet Analysis program) (For purposes of calculations, use "0" for t, <1, <.1, <.01, etc.)

Computer Code Number	Food Description	Measure	Wt (g)	H_2O (%)	Ener (cal)	Prot (g)	Carb (g)	Dietary Fiber (g)	Fat (g)	Fat Breakdown (g) Sat	Mono	Poly
	VEGETARIAN FOODS:—Continued											
1174	Vege-Burger, patty	1 ea	55	71	66	10	2	2	2	.4	.6	.5
	Vegetarian foods, Morningstar Farms:											
7672	Better-n-burgers, svg	1 ea	78	71	75	11	6	4	<1	.1	.3	.2
7766	Better-n-eggs	¼ c	57	88	23	5	<1	0	<1	.1	.1	.1
57436	Breakfast links	2 pce	45	60	63	8	2	2	2	.5	.7	1.3
7752	Breakfast strips	2 pce	16	43	56	2	2	<1	4	.7	1.1	2.6
7725	Burger crumbles, svg	1 ea	55	60	116	11	3	3	6	1.6	2.3	2.5
7726	Burger, spicy black bean	1 ea	78	60	113	11	15	5	1	.2	.3	.4
7665	Chik pattie	1 ea	71	51	177	7	15	2	10	1.3	2.6	5.9
7724	Frank, deli	1 ea	45	52	109	10	3	3	7	1	2.1	3.5
7722	Garden vege pattie	1 ea	67	60	104	11	9	4	4	.5	1.1	2.2
7746	Grillers	1 ea	64	55	139	14	5	3	7	1.7	2.2	3
7664	Prime pattie	1 ea	64	64	94	16	4	3	2	.2	.4	.6
	Vegetarian foods, Worthington:											
7634	Beef style, meatless, frzn	3 pce	55	58	113	9	4	3	7	1.2	2.7	2.6
7732	Burger, meatless, patty	¼ c	55	71	60	9	2	1	2	.3	.5	1.1
1846	Chik slices, canned	2 pce	60	78	62	6	1	1	4	.6	.9	2.3
1833	Chili, canned	½ c	106	73	136	9	10	4	7	1.1	1.7	4.1
1835	Choplets, slices, canned	2 pce	92	72	93	17	3	2	2	.9	.3	.3
7608	Corned beef style, meatless, frzn	4 pce	57	55	138	10	5	2	9	1.9	4.1	3.1
1831	Country stew, canned	1 c	240	81	208	13	20	5	9	1.6	2.3	4.8
7632	Egg roll, meatless, frzn	1 ea	85	53	181	6	20	2	8	1.7	4.5	2.3
1838	Numete, slices, canned	1 pce	55	58	132	6	5	3	10	2.4	4.4	2.7
1839	Prime stakes, slices, canned	1 pce	92	71	136	9	4	4	9	1.4	2.9	4.9
1840	Protose, slices, canned	1 pce	55	53	131	13	5	3	7	1	3	2.4
7606	Roast, dinner, meatless, frzn	1 ea	85	63	180	12	5	3	12	2.2	5	5.2
1842	Saucette links, canned	1 pce	38	62	86	6	1		6	1.1	1.6	3.8
1844	Savory slices, canned	1 pce	28	66	48	3	2	1	3	1.2	1.3	.6
7735	Stakelets, frzn	1 pce	71	58	145	12	6	2	8	1.4	2.7	3.9
1847	Turkee slices, canned	1 pce	33	64	68	5	1	1	5	.8	1.9	2.1
	MISCELLANEOUS											
	Baking powders for home use:											
	Sodium aluminum sulfate:											
962	With monocalcium phosphate monohydrate	1 tsp	5	2	6	<1	2	0	0	0	0	0
963	With monocalcium phosphate monohydrate, calcium sulfate	1 tsp	5	5	3	0	1	<1	0	0	0	0
964	Straight phosphate	1 tsp	5	4	3	<1	1	<1	0	0	0	0
965	Low sodium	1 tsp	5	6	5	<1	2	<1	<1	t	0	t
1204	Baking soda	1 tsp	5		0	0	0	0	0	0	0	0
966	Basil, dried	1 tbs	5	6	13	1	3	2	<1	t	t	.1
2068	Cajun seasoning	1 tsp	3	5	6	<1	1	<1	<1			
961	Carob flour	1 c	103	4	185	5	92	41	1	.1	.2	.2
967	Catsup:	¼ c	61	67	63	1	17	1	<1	t	t	.1
968	Tablespoon	1 tbs	15	67	16	<1	4	<1	<1	t	t	t
1200	Cayenne/red pepper	1 tbs	5	8	16	1	3	1	1	.2	.1	.4
969	Celery seed	1 tsp	2	6	8	<1	1	<1	<1	t	.3	.1
1203	Chili powder:	1 tbs	8	8	25	1	4	3	1	.2	.3	.6
970	Teaspoon	1 tsp	3	8	9	<1	2	1	<1	.1	.1	.2
	Chocolate:											
971	Baking, unsweetened, square	1 oz	28	1	146	3	8	4	15	9.1	5.2	.5
	For other chocolate items, see Sweeteners & Sweets											
972	Cilantro/coriander, fresh	1 tbs	1	93		<1	<1	<1	<1	0	t	0
2287	Cinnamon	1 tsp	2	10	5	<1	2	1	<1	t	t	t
1197	Cornstarch	1 tbs	8	8	30	<1	7	<1	<1	t	t	t
2239	Curry powder	1 tsp	2	10	6	<1	1	1	<1	t	.1	.1
1202	Dill weed, dried	1 tbs	3	7	8	1	2	<1	<1	t	.1	t
975	Garlic cloves	1 ea	3	59	4	<1	1	<1	<1	t	0	t
2238	Garlic powder	1 tsp	3	6	10	<1	2	<1	<1	t	t	t
977	Gelatin, dry, unsweetened: Envelope	1 ea	7	13	23	6	0	0	<1	t	t	t
978	Ginger root, slices, raw	2 pce	5	82	3	<1	1	<1	<1	t	t	t

Chol (mg)	Calc (mg)	Iron (mg)	Magn (mg)	Pota (mg)	Sodi (mg)	Zinc (mg)	VT-A (RE)	Thia (mg)	VT-E (a-TE)	Ribo (mg)	Niac (mg)	V-B6 (mg)	Fola (μg)	VT-C (mg)
0	8	.5	12	30	114	.58	0	.2		.25	.78	.31	15	0
0	79	2.66	15	398	351	.69	0	.23	.01	.51	3.78	.18	225	0
2	7	.63		68	90	.51	64	.01		.26	0	.11		0
1	15	2.14	16	59	338	.36	0	6.95		.22	5.19	.33	12	0
<1	7	.27		15	220	.05	0	.75		.04	.6	.07		0
0	40	3.2	1	89	238	.82	0	4.96	.34	.18	1.49	.27		0
1	56	1.84	44	269	499	.93	14	8.03	.36	.14	0	.21		0
1	11	1.02		163	536	.31	0	2.15		.16	1.51	.14		0
2	16	.26	4	50	524	.38	0	.14	1.26	.02	0	.01		0
1	34	.72	29	180	382	.59	20	6.47	.98	.1	0	0	29	0
2	43	1.16		127	256	.49	0	11.7		.24	2.99	.37		0
1	46	2.14		142	247	.74	0	.51		.25	.92	.41		2
0	4	2.63		44	624	.22	0	.89		.34	6.46	.56		0
0	4	1.73		25	269	.38	0	.13		.1	1.96	.24		0
1	9	.73		111	257	.26	0	.06		.05	.37	.08		0
0	20	1.49		195	523	.57	0	.02		.03	1.04	.31		0
0	6	.37		40	500	.65	0	.05		.05	0	.05		0
1	6	1.17		58	524	.26	0	10.6		.07	1.36	.3		0
2	51	5.09		270	826	1.03	216	1.85		.29	4.22	.86		0
1	15	.57		96	384	.31	0	1.22		.19	0	.03		0
0	10	1.12		155	272	.56	0	.08		.06	.54	.2		0
2	12	.38		82	445	.38	0	.12		.13	1.98	.38		0
<1	1	1.84		50	283	.7	0	.18		.13	1.34	.24		0
2	36	2.87		38	566	.64	0	2.13		.25	6.02	.6		0
1	9	1.15		25	205	.26	0	.59		.08	.09	.13		0
<1		.47		14	179	.08	0	.08		.06	.48	.1		0
2	49	.99		95	484	.5	0	1.51		.12	3.1	.26		0
1	3	.47		16	203	.11	0	1.13		.05	.39	.09		0
0	97	0		7	547	0	0	0	0	0	0	0	0	0
0	294	.55	1	1	530	<.01	0	0	0	0	0	0	0	0
0	368	.56	2		395	<.01	0	0	0	0	0	0	0	0
0	217	.41	1	505	4	.04	0	0	<.01	0	0	0	0	0
0	0	0	0	0	1368	0	0	0	0	0	0	0	0	0
0	106	2.1	21	172	2	.29	47	.01	.08	.02	.35	.06	14	3
				29	474									
0	358	3.03	56	852	36	.95	1	.05	.65	.47	1.96	.38	30	<1
0	12	.43	13	293	723	.14	62	.05	.9	.04	.84	.11	9	9
0	3	.1	3	72	178	.03	15	.01	.22	.01	.21	.03	2	2
0	7	.39	8	101	1	.12	208	.02	.24	.05	.43	.1	5	4
0	35	.9	9	28	3	.14	<1	.01	.02	.01	.06	.01	<1	<1
0	22	1.14	14	153	81	.22	279	.03	.08	.06	.63	.15	8	5
0	8	.43	5	57	30	.08	105	.01	.03	.02	.24	.06	3	2
0	21	1.77	87	233	4	1.12	3	.02	.34	.05	.31	.03	2	0
0	1	.02		5	<1	<.01	3	<.01	.02	<.01	.01	<.01	<1	<1
0	25	.76	1	10	1	.04	1	<.01	0	<.01	.03	<.01	1	1
0		.04			1	<.01	0	0	0	0	0	0	0	0
0	10	.59	5	31	1	.08	2	<.01	.01	.01	.07	.01	3	<1
0	53	1.46	13	99	6	.1	18	.01		.01	.08	.04		1
0	5	.05	1	12	1	.03	0	.01	0	<.01	.02	.04	<1	1
0	2	.08	2	33	1	.08	0	.01	0	<.01	.02	.08	<1	1
0	4	.08	2	1	14	.01	0	<.01	0	.02	.01	0	2	0
0	1	.02	2	21	1	.02	0	<.01	.01	<.01	.03	.01	1	<1

Table H-1

Food Composition

(Computer code number is for West Diet Analysis program) (For purposes of calculations, use "0" for t, <1, <.1, <.01, etc.)

Computer Code Number	Food Description	Measure	Wt (g)	H_2O (%)	Ener (cal)	Prot (g)	Carb (g)	Dietary Fiber (g)	Fat (g)	Fat Breakdown (g) Sat	Mono	Poly
	MISCELLANEOUS—Continued											
1198	Horseradish, prepared	1 tbs	15	85	7	<1	2	<1	<1	t	t	.1
1997	Hummous/hummus	1 c	246	65	421	12	50	12	21	3.1	8.8	7.8
1909	Mustard, country dijon	1 tsp	5		5	<1	<1	0	0	0	0	0
2019	Mustard, gai choy Chinese	1 tbs	16	94	3	<1	1		<1			
979	Mustard, prepared (1 packet = 1 tsp)	1 tsp	5	80	4	<1	<1	<1	<1	t	.2	t
	Miso (see #926 under Vegetables and Legumes, Soybean products)											
980	Olives, green	5 ea	20	78	23	<1	<1	<1	3	.3	1.9	.2
981	Olives, ripe, pitted	5 ea	22	80	25	<1	1	1	2	.3	1.7	.2
26008	Onion powder	1 tsp	2	5	7	<1	2	<1	<1	t	t	t
2237	Oregano, ground	1 tsp	2	7	6	<1	1	1	<1	.1	t	.1
2236	Paprika	1 tsp	2	10	6	<1	1	<1	<1	t	t	.2
887	Parsley, freeze dried	¼ c	1	2	3	<1	<1	<1	<1	t	t	t
	Parsley, fresh (see #885 and #886)											
985	Pepper, black	1 tsp	2	10	5	<1	1	1	<1	t	t	t
	Pickles:											
986	Dill, medium, 3¾ x 1¼" diam	1 ea	65	92	12	<1	3	1	<1	t	t	t
987	Fresh pack, slices, 1½" diam x ¼"	2 pce	15	79	11	<1	3	<1	<1	0	0	t
988	Sweet, medium	1 ea	35	65	41	<1	11	<1	<1	t	t	t
989	Pickle relish, sweet	1 tbs	15	63	21	<1	5	<1	<1	t	t	t
	Popcorn (see Grain Products #539-541)											
917	Potato chips:	10 pce	20	2	107	1	11	1	7	2.2	2	2.4
44076	Unsalted	1 oz	28	2	150	2	15	1	10	3.1	2.8	3.4
1201	Sage, ground	1 tsp	1	8	3	<1	1	<1	<1	.1	t	t
1347	Salsa, from recipe	1 tbs	15	93	3	<1	1	<1	<1	t	t	t
2218	Salsa, pico de gallo, medium	1 tbs	15	92	2	0	1	<1	0	0	0	0
990	Salt	1 tsp	6		0	0	0	0	0	0	0	0
	Salt Substitutes:											
1205	Morton, salt substitute	1 tsp	6			0	<1		0	0	0	0
1207	Morton, light salt	1 tsp	6			0	<1		0	0	0	0
2067	Seasoned salt, no MSG	1 tsp	5	5	4	<1	1	<1	<1			
991	Vinegar, cider	½ c	120	94	17	0	7	0	0	0	0	0
2172	Balsamic	1 tbs	15	64	21	0	4	0	0	0	0	0
2176	Malt	1 tbs	15	90	5	0	<1	0	0	0	0	0
2182	Tarragon	1 tbs	15	95	3	0	<1	0	0	0	0	0
2181	White wine	1 tbs	15	89	5	0	<1	0	0	0	0	0
	Yeast:											
992	Baker's, dry, active, package	1 ea	7	8	21	3	3	1	<1	t	.2	t
993	Brewer's, dry	1 tbs	8	5	23	3	3	3	<1	t	t	0
	SOUPS, SAUCES, and GRAVIES											
	SOUPS, canned, condensed:											
	Unprepared, condensed:											
1210	Cream of celery	1 c	251	85	181	3	18	2	11	2.8	2.6	5
1215	Cream of chicken	1 c	251	82	233	7	18	<1	15	4.2	6.5	3
1216	Cream of mushroom	1 c	251	81	259	4	19	1	19	5.1	3.6	8.9
1220	Onion	1 c	246	86	113	8	16	2	3	.5	1.5	1.3
	Prepared w/equal volume of whole milk:											
994	Clam chowder, New England	1 c	248	85	164	9	17	1	7	2.9	2.3	1.1
1209	Cream of celery	1 c	248	86	164	6	14	1	10	3.9	2.5	2.6
995	Cream of chicken	1 c	248	85	191	8	15	<1	11	4.6	4.5	1.6
996	Cream of mushroom	1 c	248	85	203	6	15	<1	14	5.1	3	4.6
1214	Cream of potato	1 c	248	87	149	6	17	<1	6	3.8	1.7	.6
1213	Oyster stew	1 c	245	89	135	6	10	0	8	5	2.1	.3
997	Tomato	1 c	248	85	161	6	22	3	6	2.9	1.6	1.1
	Prepared with equal volume of water:											
998	Bean with bacon	1 c	253	84	172	8	23	9	6	1.5	2.2	1.8
999	Beef broth/bouillon/consommé	1 c	240	98	17	3	<1	0	1	.3	.2	t
1000	Beef noodle	1 c	244	92	83	5	9	1	3	1.1	1.2	.5
1001	Chicken noodle	1 c	241	92	75	4	9	1	2	.7	1.1	.6
1002	Chicken rice	1 c	241	94	60	4	7	1	2	.5	.9	.4
1208	Chili beef	1 c	250	85	170	7	21	9	7	3.3	2.8	.3
1003	Clam chowder, Manhattan	1 c	244	92	78	2	12	1	2	.4	.4	1.3

PAGE KEY: A–4 = Beverages A–6 = Dairy A–10 = Eggs A–10 = Fat/Oil A–14 = Fruit A–20 = Bakery A–28 = Grain A–32 = Fish A–34 = Meats A–38 = Poultry A–40 = Sausage A–42 = Mixed/Fast A–46 = Nuts/Seeds A–50 = Sweets A–52 = Vegetables/Legumes A–62 = Vegetarian Foods A–64 = Misc A–66 = Soups/Sauces A–68 = Fast A–84 = Convenience A–88 = Baby foods

Chol (mg)	Calc (mg)	Iron (mg)	Magn (mg)	Pota (mg)	Sodi (mg)	Zinc (mg)	VT-A (RE)	Thia (mg)	VT-E (a-TE)	Ribo (mg)	Niac (mg)	V-B6 (mg)	Fola (μg)	VT-C (mg)
0	8	.06	4	37	47	.12	<1	<.01	<.01	<.01	.06	.01	9	4
0	123	3.86	71	428	600	2.71	5	.23	2.46	.13	1.01	.98	146	19
0				10	120									
0	4	.1	2	6	63	.03	0	0	.09	0	0	<.01	0	0
0	12	.32	4	11	480	.01	6	0	.6	0	0	<.01	<1	0
0	19	.73	1	2	192	.05	9	<.01	.66	0	.01	<.01	0	<1
0	7	.05	2	19	1	.05	0	.01	<.01	<.01	.01	.03	3	<1
0	31	.88	5	33	<1	.09	14	.01	.03	.01	.12	.02	5	1
0	4	.47	4	47	1	.08	121	.01	.01	.03	.31	.04	2	1
0	2	.54	4	63	4	.06	63	.01	.06	.02	.1	.01	15	1
0	9	.58	4	25	1	.03	<1	<.01	.02	<.01	.02	.01	<1	<1
0	6	.34	7	75	833	.09	21	.01	.1	.02	.04	.01	1	1
0	5	.27	1	30	101	0	2	0	.02	<.01	0	<.01	0	1
0	1	.21	1	11	329	.03	5	<.01	.06	.01	.06	<.01	<1	<1
0	3	.12	1	30	107	.01	1	0	.02	<.01	0	0	0	1
0	5	.33	13	255	119	.22	0	.03	.98	.04	.77	.13	9	6
0	7	.46	19	357	2	.3	0	.05	1.37	.05	1.07	.18	13	9
0	16	.28	4	11	<1	.05	6	.01	.02	<.01	.06	.01	3	<1
0	1	.06	1	24	58	.02	22	.01	.04	<.01	.06	.01	2	5
0					130									
0	1	.02			2325	.01	0	0	0	0	0	0	0	0
	33			3018	<1									
	2		4	1560	1170									
				15	1542									
0	7	.72	26	120	1	0	0	0	0	0	0	0	0	0
	2	.07		10	3		<1	.07		.07	.07			<1
	2	.07		13	4		<1	.07		.07	.07			1
		.07		2	1		<1	.07		.07	.07			<1
	1	.07		12	1		<1	.07		.07	.07			<1
0	4	1.16	7	140	3	.45	<1	.16	.01	.38	2.79	.11	164	<1
0	17	1.38	18	151	10	.63	0	1.25		.34	3.03	.4	313	0
28	80	1.26	13	246	1900	.3	60	.06	.38	.1	.66	.02	5	<1
20	68	1.2	5	176	1972	1.26	113	.06	.33	.12	1.64	.03	3	<1
3	65	1.05	10	168	1736	1.19	0	.06	2.61	.17	1.62	.02	8	2
0	54	1.35	5	138	2115	1.23	0	.07	.57	.05	1.21	.1	30	2
22	186	1.49	22	300	992	.8	40	.07	.15	.24	1.03	.13	10	3
32	186	.69	22	310	1009	.2	67	.07	.97	.25	.44	.06	8	1
27	181	.67	17	273	1046	.67	94	.07	.24	.26	.92	.07	8	1
20	179	.59	20	270	918	.64	37	.08	1.34	.28	.91	.06	10	2
22	166	.55	17	322	1061	.67	67	.08	.1	.24	.64	.09	9	1
32	167	1.05	20	235	1041	10.3	44	.07	.49	.23	.34	.06	10	4
17	159	1.81	22	449	744	.29	109	.13	2.6	.25	1.52	.16	21	68
3	81	2.05	45	402	951	1.03	89	.09	.08	.03	.57	.04	32	2
0	14	.41	5	130	782	0	0	<.01	0	.05	1.87	.02	5	0
5	15	1.1	5	100	952	1.54	63	.07	<.01	.06	1.07	.04	19	<1
7	17	.77	5	55	1106	.39	72	.05	.07	.06	1.39	.03	22	<1
7	17	.75	0	101	815	.26	65	.02	.05	.02	1.13	.02	1	<1
12	42	2.13	30	525	1035	1.4	150	.06	.17	.07	1.07	.16	17	4
2	27	1.63	12	188	578	.98	98	.03	.73	.04	.82	.1	10	4

Table H-1

Food Composition

(Computer code number is for West Diet Analysis program) (For purposes of calculations, use "0" for t, <1, <.1, <.01, etc.)

Computer Code Number	Food Description	Measure	Wt (g)	H_2O (%)	Ener (cal)	Prot (g)	Carb (g)	Dietary Fiber (g)	Fat (g)	Fat Breakdown (g) Sat	Mono	Poly
	SOUPS, SAUCES, and GRAVIES—Continued											
1004	Cream of chicken	1 c	244	91	117	3	9	<1	7	2.1	3.3	1.5
1005	Cream of mushroom	1 c	244	90	129	2	9	<1	9	2.4	1.7	4.2
1006	Minestrone	1 c	241	91	82	4	11	1	3	.6	.7	1.1
1211	Onion	1 c	241	93	58	4	8	1	2	.3	.7	.7
1007	Split pea & ham	1 c	253	82	190	10	28	2	4	1.8	1.8	.6
1008	Tomato	1 c	244	90	85	2	17	<1	2	.4	.4	1
1009	Vegetable beef	1 c	244	92	78	6	10	<1	2	.9	.8	.1
1010	Vegetarian vegetable	1 c	241	92	72	2	12	<1	2	.3	.8	.7
	Ready to serve:											
1707	Chunky chicken soup	1 c	251	84	178	13	17	2	7	2	3	1.4
	SOUPS, dehydrated:											
	Prepared with water:											
1299	Beef broth/bouillon	1 c	244	97	19	1	2	0	1	.3	.3	t
1376	Chicken broth	1 c	244	97	22	1	1	0	1	.3	.4	.4
1013	Chicken noodle	1 c	252	94	53	3	7	1	1	.3	.5	.4
1122	Cream of chicken	1 c	261	91	107	2	13	<1	5	3.4	1.2	.4
1014	Onion	1 c	246	96	27	1	5	1	1	.1	.3	.1
1217	Split pea	1 c	255	87	125	7	21	3	1	.4	.7	.3
1015	Tomato vegetable	1 c	253	93	56	2	10	<1	1	.4	.3	.1
	Unprepared, dry products:											
1011	Beef bouillon, packet	1 ea	6	3	14	1	1	0	1	.3	.2	t
1012	Onion soup, packet	1 ea	39	4	115	5	21	4	2	.5	1.4	.3
	SAUCES											
	From dry mixes, prepared with milk:											
1016	Cheese sauce	1 c	279	77	307	17	23	1	17	9.3	5.3	1.6
1017	Hollandaise	1 c	259	84	240	5	14	<1	20	11.6	5.9	.9
1018	White sauce	1 c	264	81	240	10	21	<1	13	6.4	4.7	1.7
	From home recipe:											
1206	Lowfat cheese sauce	¼ c	61	73	85	6	4	<1	5	2.1	1.9	.9
1019	White sauce, medium	¼ c	72	77	102	2	6	<1	8	2.3	3.2	2
	Ready to serve:											
2202	Alfredo sauce, reduced fat	¼ c	69		170	5	16	0	10	6		
1020	Barbeque sauce	1 tbs	16	81	12	<1	2	<1	<1	t	.1	.1
1706	Chili sauce, tomato base	1 tbs	17	68	18	<1	4	<1	<1	t	t	t
2126	Creole sauce	¼ c	62	89	25	1	4	1	1	.1	.2	.3
2124	Hoisin sauce	1 tbs	17	47	35	<1	7	0	1	0		
2199	Pesto sauce	2 tbs	16		83	2	1	0	8	1.8	5.4	.7
1021	Soy sauce	1 tbs	16	71	8	1	1	<1	<1	t	t	t
2123	Szechuan sauce	1 tbs	16	71	21	<1	3	<1	1	.1	.3	.4
1380	Teriyaki sauce	1 tbs	18	68	15	1	3	<1	0	0	0	0
	Spaghetti sauce, canned:											
1377	Plain	1 c	249	75	271	5	40	8	12	1.7	6.1	3.3
1378	With meat	1 c	250	74	300	8	37	8	14	2.7	7	3.2
1379	With mushrooms	½ c	123	84	108	2	13	1	3	.4	1.5	.8
	GRAVIES											
	Canned:											
1022	Beef	1 c	233	87	123	9	11	1	5	2.7	2.2	.2
1023	Chicken	1 c	238	85	188	5	13	1	14	3.4	6.1	3.6
1024	Mushroom	1 c	238	89	119	3	13	1	6	1	2.8	2.4
1025	From dry mix, brown	1 c	258	92	75	2	13	<1	2	.8	.7	.1
1026	From dry mix, chicken	1 c	260	91	83	3	14	<1	2	.5	.9	.4
	FAST FOOD RESTAURANTS											
	ARBY'S											
1402	Bac'n cheddar deluxe	1 ea	231	59	512	21	39	<1	31	8.7	12.7	10.1
	Roast beef sandwiches:											
1403	Regular	1 ea	155	47	383	22	35	1	18	7	8	3.5
1404	Junior	1 ea	89	48	233	11	23	<1	11	4.1	5.2	2.5
1405	Super	1 ea	254	58	552	24	54	1	28	7.6	12.2	8.4
1407	Beef 'n cheddar	1 ea	194	50	508	25	43		26	7.7	12	6.8
1408	Chicken breast sandwich	1 ea	204	52	445	22	52	1	22	3	9.7	10.1
1412	Ham'n cheese sandwich	1 ea	169	54	355	25	34	<1	14	5.1	5.8	3.8
1726	Italian sub sandwich	1 ea	297	58	671	34	47		39	12.8	15.7	8.5

PAGE KEY: A–4 = Beverages A–6 = Dairy A–10 = Eggs A–10 = Fat/Oil A–14 = Fruit A–20 = Bakery A–28 = Grain A–32 = Fish A–34 = Meats A–38 = Poultry A–40 = Sausage A–42 = Mixed/Fast A–46 = Nuts/Seeds A–50 = Sweets A–52 = Vegetables/Legumes A–62 = Vegetarian Foods A–64 = Misc A–66 = Soups/Sauces A–68 = Fast A–84 = Convenience A–88 = Baby foods

Chol (mg)	Calc (mg)	Iron (mg)	Magn (mg)	Pota (mg)	Sodi (mg)	Zinc (mg)	VT-A (RE)	Thia (mg)	VT-E (a-TE)	Ribo (mg)	Niac (mg)	V-B6 (mg)	Fola (μg)	VT-C (mg)
10	34	.61	2	88	986	.63	56	.03	.2	.06	.82	.02	2	<1
2	46	.51	5	100	881	.59	0	.05	1.24	.09	.72	.01	5	1
2	34	.92	7	313	911	.73	234	.05	.07	.04	.94	.1	36	1
0	26	.67	2	67	1053	.61	0	.03	.29	.02	.6	.05	15	1
8	23	2.28	48	400	1006	1.32	45	.15	.15	.08	1.47	.07	3	2
0	12	1.76	7	264	695	.24	68	.09	2.49	.05	1.42	.11	15	66
5	17	1.12	5	173	791	1.54	190	.04	.32	.05	1.03	.08	10	2
0	22	1.08	7	210	822	.46	301	.05	.79	.05	.92	.05	11	1
30	25	1.73	8	176	889	1	131	.08	.18	.17	4.42	.05	5	1
0	10	.02	7	37	1361	.07	<1	<.01	.02	.02	.36	0	0	0
0	15	.07	5	24	1483	.01	12	.01	.02	.03	.19	0	2	0
3	33	.5	8	30	1282	.2	5	.07	.02	.06	.88	.01	18	<1
3	76	.26	5	214	1184	1.57	123	.1	.15	.2	2.61	.05	5	1
0	12	.15	5	64	849	.06	<1	.03	.1	.06	.48	0	1	<1
3	20	.94	43	224	1147	.56	5	.21	.13	.14	1.26	.05	39	0
0	8	.63	20	104	1146	.17	20	.06	.81	.05	.79	.05	10	6
1	4	.06	3	27	1018	0	<1	<.01	.01	.01	.27	.01	2	0
2	55	.58	25	260	3493	.23	1	.11	.42	.24	1.99	.04	6	1
53	569	.28	47	552	1565	.97	117	.15	.33	.56	.32	.14	13	2
52	124	.9	8	124	1564	.7	220	.04	.26	.18	.06	.5	22	<1
34	425	.26	264	444	797	.55	92	.08	1.58	.45	.53	.07	16	3
11	166	.25	10	100	389	.73	58	.03	.55	.14	.16	.03	4	<1
8	75	.21	9	100	82	.26	89	.05	.98	.12	.28	.03	4	1
30	150	0	8	80	600		80	0		.1	0			0
0	3	.14	3	28	130	.03	14	<.01	.18	<.01	.14	.01	1	1
0	3	.14	2	63	227	.05	24	.01	.05	.01	.27	.02	1	3
0	35	.31	9	187	339	.1	24	.03	.61	.02	.53	.07	9	0
0	0	0			250		0							0
4	64	.09	6	15	129	.29	39	.01		.03	0	.02	<1	0
0	3	.32	5	29	914	.06	0	.01	0	.02	.54	.03	2	0
0	2	.12	2	13	218	.02	10	<.01	.07	<.01	.1	.01	1	<1
0	4	.31	11	40	690	.02	0	<.01	0	.01	.23	.02	4	0
0	70	1.62	60	956	1235	.52	306	.14	4.98	.15	3.76	.88	54	28
15	68	1.94	60	952	1179	1.37	577	.13	5.91	.17	4.51	.87	52	26
0	15	1	15	332	494	.34	241	.08	1.35	.08	.93	.16	12	9
7	14	1.63	5	189	1304	2.33	0	.07	.15	.08	1.54	.02	5	0
5	48	1.12	5	259	1373	1.9	264	.04	.37	.1	1.05	.02	5	0
0	17	1.57	5	252	1356	1.67	0	.08	.19	.15	1.6	.05	29	0
3	67	.23	10	57	1075	.31	0	.04	.05	.08	.81	0	0	0
3	39	.26	10	62	1133	.32	0	.05	.05	.15	.78	.03	3	3
38	110	4.32		491	1094	3	40	.34		.46	9.6			11
43	60	4.86	16	422	936	3.75	0	.28		.48	11	.2	14	1
22	40	2.7	8	201	519	1.5		.18		.25	6.6	.1	7	
43	90	6.48	25	533	1174	3.75	30	.39		.58	12.4	.3	21	9
52	150	6.12		321	1166	3		.42		.63	9.8			1
45	60	2.88	30	330	1019	.15		.22		.54	9	.38	18	5
55	170	2.7	31	382	1400	.9	40	.82		.37	7.8	.31	26	24
69	410	4.32		565	2062		100	.91		.49	8.2			11

Table H-1

Food Composition (Computer code number is for West Diet Analysis program) (For purposes of calculations, use "0" for t, <1, <.1, <.01, etc.)

Computer Code Number	Food Description	Measure	Wt (g)	H_2O (%)	Ener (cal)	Prot (g)	Carb (g)	Dietary Fiber (g)	Fat (g)	Fat Breakdown (g) Sat	Mono	Poly
	FAST FOOD RESTAURANTS											
	ARBY'S—Continued											
1413	Turkey sandwich, deluxe	1 ea	195	69	260	20	33	<1	6	1.6	2.3	2.4
1680	Turkey sub sandwich	1 ea	277	62	486	33	46		19	5.3	6	7
	Milkshakes:											
1419	Chocolate	1 ea	340	74	451	10	76	<1	12	2.8	7	1.7
1420	Jamocha	1 ea	326	75	368	9	59	0	10	2.5	6.4	1.6
1421	Vanilla	1 ea	312	77	330	10	46	0	11	3.9	5.3	2.3
1728	Salad, roast chicken	1 ea	400	88	204	24	12		7	3.3	.9	.9
1729	Sports drink, Upper Ten	1 ea	358	88	169	0	42		0	0	0	0
	Source: Arby's											
	BURGER KING											
1423	Croissant sandwich, egg, sausage & cheese	1 ea	176	46	600	22	25	1	46	16		
	Whopper sandwiches:											
1425	Whopper	1 ea	270	58	640	27	45	3	39	11		
1426	Whopper with cheese	1 ea	294	57	730	33	46	3	46	16		
	Sandwiches:											
1629	BK broiler chicken sandwich	1 ea	248	59	550	30	41	2	29	6		
1432	Cheeseburger	1 ea	138	48	380	23	28	1	19	9		
1434	Chicken sandwich	1 ea	229	45	710	26	54	2	43	9		
1427	Double beef	1 ea	351	57	870	46	45	3	56	19		
1428	Double beef & cheese	1 ea	375	56	960	52	46	3	63	24		
1433	Double cheeseburger with bacon	1 ea	218	48	640	44	28	1	39	18		
1431	Hamburger	1 ea	126	48	330	20	28	1	15	6		
1437	Ocean catch fish fillet	1 ea	255	51	700	26	56	3	41	6		
1435	Chicken tenders	1 ea	88	50	230	16	14	2	12	3		
1439	French fries (salted)	1 svg	116	40	370	5	43	3	20	5		
1630	French toast sticks	1 svg	141	33	500	4	60	1	27	7		
1440	Onion rings	1 svg	124	51	310	4	41	6	14	2	8	4
1441	Milk shakes, chocolate	1 ea	284	75	320	9	54	3	7	4		
1442	Milk shakes, vanilla	1 ea	284	75	300	9	53	1	6	4		
1443	Fried apple pie	1 ea	113	47	300	3	39	2	15	3		
	Source: Burger King Corporation											
	CHICK-FIL-A											
	Sandwiches:											
69153	Chargrilled chicken	1 ea	150	54	280	27	36	1	3	1		
69152	Chicken	1 ea	167	61	290	24	29	1	9	2		
69155	Chicken salad	1 ea	167	55	320	25	42	1	5	2		
69154	Chicken salad club	1 ea	232	62	390	33	38	2	12	5		
	Salads:											
52139	Carrot and raisin	1 ea	76	53	150	5	28	2	2	0		
52136	Chicken plate	1 ea	468	85	290	21	40	6	5	0		
52134	Chicken garden, charbroiled	1 ea	397	89	170	26	10	5	3	1		
52135	Chick-n-strips	1 ea	451	86	290	32	21	5	9	2		
52138	Cole slaw	1 ea	79	70	130	6	11	1	6	1		
52137	Tossed salad	1 ea	130	85	70	5	13	1	0	0	0	0
15263	Chicken nuggets, svg	1 ea	110	51	290	28	12	60	12	3		
15262	Chicken-n-strips, svg	1 ea	119	59	230	29	10	0	8	2		
50885	Hearty breast of chicken soup, svg	1 ea	215	86	110	16	10	1	1	0		
7973	Waffle potato fries, svg	1 ea	85	28	290	1	49	0	10	4		
46489	Cheesecake, svg	1 ea	88	52	270	13	7	0	21	9		
49134	Fudge nut brownie, svg	1 ea	88	8	416	12	49	0	19	3.6		
20601	Icedream, svg	1 ea	127	74	140	11	16	0	4	1		
48214	Lemon pie, svg	1 ea	99	56	280	1	19	0	22	6		
	Source: Chick-Fil-A											

Chol (mg)	Calc (mg)	Iron (mg)	Magn (mg)	Pota (mg)	Sodi (mg)	Zinc (mg)	VT-A (RE)	Thia (mg)	VT-E (a-TE)	Ribo (mg)	Niac (mg)	V-B6 (mg)	Fola (μg)	VT-C (mg)
33	130	3.42	30	353	1262	1.5	40	.08		.41	15.4	.52	20	12
51	400	4.68		500	2033		20	13.2		.54	18.8			
36	250	.72	48	410	341	1.5	60	.12		.68	.8	.14	14	5
35	250	2.7	36	525	262	1.5	60	.12		.68	.8	.14	14	2
32	300	2.7	36	686	281	1.5	60	.12		.68	4	.14	37	2
43	170	1.98		877	508		485	.33		.54	5.6			51
0				0	40									
260	150	3.6			1140		80							0
90	80	4.5			870		100	.33		.41	7	.35		9
115	250	4.5			1350		150	.34		.48	7	.33		9
80	60	5.4			480		60							6
65	100	2.7			770		60							0
60	100	3.6			1400		0							0
170	80	7.2			940		100	.34		.56	10			9
195	250	7.2			1420		150	.35		.63	10			9
145	200	4.5			1240		80	.31		.42	6			0
55	40	1.8			530		20	.28		.31	4.89			0
90	60	2.7			980		20							1
35	0	.72			530		0							0
0	0	1.08			240		0							4
0	60	2.7			490		0							0
0	100	1.44			810		0							0
20	200	1.8			230		60	.13		.55	.13			0
20	300	0			230		60	.11		.57	.13			4
0	0	1.44			230		0							6
40					640									
50					870									
10					810									
70					980									
6					650									
35					570									
25					650									
20					430									
15					430									
0					0									
14					770									
20					380									
45					760									
5					960									
10					510									
36					773									
40					240									
5					550									

Table H-1

Food Composition (Computer code number is for West Diet Analysis program) (For purposes of calculations, use "0" for t, <1, <.1, <.01, etc.)

Computer Code Number	Food Description	Measure	Wt (g)	H_2O (%)	Ener (cal)	Prot (g)	Carb (g)	Dietary Fiber (g)	Fat (g)	Fat Breakdown (g) Sat	Mono	Poly
	FAST FOOD RESTAURANTS—Continued											
	DAIRY QUEEN											
	Ice cream cones:											
1446	Small vanilla	1 ea	142	63	230	6	38	0	7	4.5		
1447	Regular vanilla	1 ea	213	64	350	8	57	0	10	7		
1448	Large vanilla	1 ea	253	65	410	10	65	0	12	8		
1450	Chocolate dipped	1 ea	234	59	510	9	63	1	25	13		
1453	Chocolate sundae	1 ea	241	62	410	8	73	0	10	6		
1455	Banana split	1 ea	369	67	510	8	96	3	12	8		
1456	Peanut buster parfait	1 ea	305	51	730	16	99	2	31	17		
1457	Hot fudge brownie delight	1 ea	305	52	710	11	102	1	29	14	12	2
1459	Buster bar	1 ea	149	45	450	10	41	2	28	12		
1645	Breeze, strawberry, regular	1 ea	383	70	460	13	99	1	1	1	0	0
1460	Dilly bar	1 ea	85	55	210	3	21	0	13	7	3	3
1461	DQ ice cream sandwich	1 ea	61	46	150	3	24	1	5	2		
1463	Milk shakes, regular	1 ea	397	71	520	12	88	<1	14	8	2	2
1464	Milk shakes, large	1 ea	461	71	600	13	101	<1	16	10	2	2
1466	Milk shakes, malted	1 ea	418	68	610	13	106	<1	14	8	2	2
1470	Misty slush, small	1 ea	454	88	220	0	56	0	0	0	0	0
2250	Starkiss	1 ea	85	75	80	0	21	0	0	0	0	0
	Yogurt:											
1641	Yogurt cone, regular	1 ea	213	66	280	9	59	0	1	.5		
1643	Yogurt sundae, strawberry	1 ea	255	69	300	9	66	1	<1	.5	0	0
	Sandwiches:											
1481	Cheeseburger, double	1 ea	219	55	540	35	30	2	31	16		
1480	Cheeseburger, single	1 ea	152	55	340	20	29	2	17	8		
1474	Chicken	1 ea	191	56	430	24	37	2	20	4		
1647	Chicken fillet, grilled	1 ea	184	64	310	24	30	3	10	2.5		
1475	Fish fillet sandwich	1 ea	170	57	370	16	39	2	16	3.5		
1476	Fish fillet with cheese	1 ea	184	56	420	19	40	2	21	6	7	8
1477	Hamburger, single	1 ea	128	56	269	16	27	2	11	4.6	5.6	.9
1478	Hamburger, double	1 ea	212	62	440	30	29	2	22	10		
	Hotdog:											
1483	Regular	1 ea	99	57	240	9	19	1	14	5		
1484	With cheese	1 ea	113	55	290	12	20	1	18	8	8	2
1485	With chili	1 ea	128	61	280	12	21	2	16	6		
1489	French fries, small	1 ea	71	39	210	3	29	3	10	2	5	3
1490	French fries, large	1 ea	128	40	390	5	52	6	18	4	8	6
1491	Onion rings	1 ea	85	46	240	4	29	2	12	2.5		
	Source: International Dairy Queen											
	HARDEES											
1734	Frisco burger hamburger	1 ea	242		760	36	43		50	18		
1736	Frisco grilled chicken salad	1 ea	278		120	18	2		4	1		
1737	Peach shake	1 ea	345		390	10	77		4	3		
	Source: Hardees											
	JACK IN THE BOX											
	Breakfast items:											
1492	Breakfast Jack sandwich	1 ea	121	49	300	18	30	0	12	5	5	2.5
1494	Sausage crescent	1 ea	156	39	580	22	28	0	43	16		
1495	Supreme crescent	1 ea	153	40	530	23	34	0	33	10	18.9	7.8
1496	Pancake platter	1 ea	231	45	610	15	87	0	22	9	7.6	3.5
1497	Scrambled egg platter	1 ea	213	52	560	18	50	0	32	9	16.6	4.4
	Sandwiches:											
1654	Bacon cheeseburger	1 ea	242	49	710	35	41	0	45	15	15.7	8.7
1499	Cheeseburger	1 ea	110	41	330	16	32	0	15	6	5.9	2.3
1739	Chicken caesar pita sandwich	1 ea	237	59	520	27	44	4	26	6		
1655	Chicken sandwich	1 ea	160	52	400	20	38	0	18	4		
1656	Chicken sandwich, sourdough ranch	1 ea	225		490	29	45	1	21	6		
1505	Chicken supreme	1 ea	245	55	620	25	48	0	36	11	14.8	11.4
1583	Double cheeseburger	1 ea	152	44	450	24	35	0	24	12	11.6	3.1

PAGE KEY: A–4 = Beverages A–6 = Dairy A–10 = Eggs A–10 = Fat/Oil A–14 = Fruit A–20 = Bakery A–28 = Grain A–32 = Fish A–34 = Meats A–38 = Poultry A–40 = Sausage A–42 = Mixed/Fast A–46 = Nuts/Seeds A–50 = Sweets A–52 = Vegetables/Legumes A–62 = Vegetarian Foods A–64 = Misc A–66 = Soups/Sauces A–68 = Fast A–84 = Convenience A–88 = Baby foods

Chol (mg)	Calc (mg)	Iron (mg)	Magn (mg)	Pota (mg)	Sodi (mg)	Zinc (mg)	VT-A (RE)	Thia (mg)	VT-E (a-TE)	Ribo (mg)	Niac (mg)	V-B6 (mg)	Fola (μg)	VT-C (mg)
20	200	1.08		250	115		122	.05		.28				
30	300	1.8		390	170		150	.09		.38	.16	.13		2
40	350	1.8		451	200		200	.11		.4	.2			2
30	300	1.8		435	200		150	.09		.38	.16	.13		2
30	250	1.44		394	210		150	.08		.35	.4	.19		0
30	250	1.8		860	180		200	.15		.25	.4	.2		15
35	300	1.8		660	400		150	.15		.51	3	.22		1
35	300	5.4		510	340		80	.15		.68	.3	.18		1
15	150	1.08		400	280		80	.09		.17	3	.08		0
10	450	2.7		530	270		0	.13		.73				9
10	100	.36		170	75		60	.03		.14		.06		0
5	60	.72		105	115		40	.03		.25	.4	.05		0
45	400	1.44		570	230		80	.12		.59	.8	.19		<1
50	450	1.44		660	260		200	.15		.68	.8			<1
45	400	1.44		570	230		80	.12		.59	.8	.19		<1
0	0	0			20		0							0
0	0	0			10		0							0
5	300	1.8		285	170		0	.09		.38				2
5	300	1.8		352	180		0	.09		.49				6
115	250	4.5		426	1130		150	.29		.49	6.78			4
55	150	3.6		263	850		60	.29		.33	3.89			4
55	40	1.8		350	760		0	.37		.34	11			0
50	200	2.7		330	1040		0	.3		1.02	12			0
45	40	1.8		280	630		0	.3		.22	3			0
60	100	1.8		290	850		80	.3		.25	5			0
42	56	2.5		234	584		37	.27		.23	3.6			3
90	60	4.5		444	680		60	.32		.45	7.49			6
25	60	1.8		170	730		20	.22		.14	2			4
40	150	1.8		180	950		60	.22		.17	2			4
35	60	1.8		262	870		80	.23		.14	3			4
0	0	.72		430	115		0	.09		.03	2			5
0	0	1.44		780	200		0	.15		.07	3			9
0	0	1.08		90	135		0	.09		.05	.4			0
70					1280									
60					520									
25					290									
185	200	2.7		220	890		80	.47		.41	3			9
185	150	2.7		260	1010		100	.6		.51	4.6			0
210	150	3.6		270	930		150	.65		.54	4.2			12
100	100	1.8		310	890		80	.03		.85	7			6
380	150	4.5		450	1060		150			.66	5			9
110	250	5.4		540	1240		80	.24		.48	8.8	.39		9
35	200	2.7		200	510		60	.23		.23	3			1
55	250	2.7		490	1050		80							2
45	150	1.8		180	1290		40							0
65	150	1.8		340	1060									0
75	200	2.7		190	1520		100	.39		.32	11			2
75	250	3.6		320	900		100	.15		.34	6			0

Table H-1

Food Composition

(Computer code number is for West Diet Analysis program) (For purposes of calculations, use "0" for t, <1, <.1, <.01, etc.)

Computer Code Number	Food Description	Measure	Wt (g)	H_2O (%)	Ener (cal)	Prot (g)	Carb (g)	Dietary Fiber (g)	Fat (g)	Fat Breakdown (g) Sat	Mono	Poly
	FAST FOOD RESTAURANTS—Continued											
	JACK IN THE BOX—Continued											
1651	Grilled sourdough burger	1 ea	223	48	670	32	39	0	43	16	17.8	7.9
1498	Hamburger	1 ea	97	42	280	13	31	0	11	4	4.9	2
1500	Jumbo Jack burger	1 ea	229	55	560	26	41	0	32	10	13	8
1501	Jumbo Jack burger with cheese	1 ea	242	55	610	29	41	0	36	12	15	9
1740	Monterey roast beef sandwich	1 ea	238	57	540	30	40	3	30	9		
1508	Tacos, regular	1 ea	78	57	190	7	15	2	11	4		
1509	Tacos, super	1 ea	126	59	280	12	22	3	17	6		
	Teriyaki bowl:											
1679	Beef	1 ea	440	62	640	28	124	7	3	1		
1668	Chicken	1 ea	440	62	580	28	115	6	1			
1516	French fries	1 ea	109	38	350	4	45	4	17	4		
1517	Hash browns	1 ea	57	53	160	1	14	1	11	2.5	6.8	.3
1518	Onion rings	1 ea	103	34	380	5	38	0	23	6	15.2	.9
	Milkshakes:											
1519	Chocolate	1 ea	322	72	390	9	74	0	6	3.5	2.1	
1520	Strawberry	1 ea	298	74	330	9	60	0	7	4	2	
1521	Vanilla	1 ea	304	73	350	9	62	0	7	4	1.8	
1522	Apple turnover	1 ea	110	34	350	3	48	0	19	4	10.6	1.5
	Source: Jack in the Box Restaurant, Inc											
	KENTUCKY FRIED CHICKEN											
	Rotisserie gold:											
1472	Dark qtr, no skin	1 ea	117	66	217	27	0	0	12	3.5		
1473	Dark qtr, w/skin	1 ea	146	62	333	30	1		24	6.6		
1513	White qtr with wing, w/skin	1 ea	176	65	335	40	1		19	5.4		
1525	White qtr with wing, no skin	1 ea	117	63	199	37	0	0	6	1.7		
	Original Recipe:											
1253	Center breast	1 ea	103	52	260	25	9	<1	14	3.8	7.8	2
1251	Side breast	1 ea	83	47	245	18	9	<1	15	4.2	8.8	2.2
1250	Drumstick	1 ea	57	54	152	14	3	<1	8	2.2	4.1	1.3
1252	Thigh	1 ea	95	49	287	18	8	<1	21	5.3	9.4	3.1
1249	Wing	1 ea	53	45	172	12	5	<1	12	3	6	1.8
	Hot & spicy:											
1451	Center breast	1 ea	125	48	360	28	13		22	5		
1452	Side breast	1 ea	120	43	400	22	16		28	6		
1430	Thigh	1 ea	119	47	370	24	10		27	6		
1471	Wing	1 ea	61	38	220	14	5		16	4		
	Extra crispy recipe:											
1261	Center breast	1 ea	118	48	330	26	14	<1	20	4.8	10.8	2.1
1259	Side breast	1 ea	116	40	400	21	19	<1	27	5.5	12.9	2.3
1258	Drumstick	1 ea	65	49	190	14	6	<1	12	3.4	7.7	1.7
1260	Thigh	1 ea	109	43	380	23	7	<1	30	7.7	16	4.2
1257	Wing	1 ea	59	34	240	13	8	<1	17	4.4	10.7	2.5
1390	Baked beans	½ c	167	70	200	8	36	6	3	1.5	.7	.4
1526	Breadstick	1 ea	33	30	110	3	17	0	3	0		
1388	Buttermilk biscuit	1 ea	65	28	234	5	28	<1	13	3.4	6.2	2.3
1391	Chicken little sandwich	1 ea	47	35	169	6	14	<1	10	2	4.7	3.4
1269	Coleslaw	1 svg	90	75	114	1	13	<1	6	1	1.7	3.4
1527	Cornbread	1 ea	56	26	228	3	25	1	13	2		
1268	Corn-on-the-cob	1 ea	151	70	222	4	27	8	12	2	1	1.5
1429	Chicken, hot wings	1 svg	135	38	471	27	18		33			
1386	Kentucky fries	1 svg	77	42	228	3	26	3	12	3.2		
1381	Kentucky nuggets	6 ea	95	48	284	16	15	<1	18	4		
1534	Macaroni & cheese	1 svg	114	71	162	7	15	0	8	3		
1387	Mashed potatoes & gravy	1 svg	120	80	103	1	16	<1	5	.4	.5	.2
1530	Pasta salad	1 svg	108	78	135	2	14	1	8	1		
1389	Potato salad	½ c	188	74	271	5	27	3	16	3	4.2	7.3
1383	Potato wedges	1 svg	92	59	192	3	25	3	9	3		
1535	Red beans & rice	1 svg	112	76	114	4	18	3	3	1		
1529	Vegetable medley salad	1 ea	114	77	126	1	21	3	4	1		
	Source: Kentucky Fried Chicken Corp											

PAGE KEY: A–4 = Beverages A–6 = Dairy A–10 = Eggs A–10 = Fat/Oil A–14 = Fruit A–20 = Bakery A–28 = Grain A–32 = Fish A–34 = Meats A–38 = Poultry A–40 = Sausage A–42 = Mixed/Fast A–46 = Nuts/Seeds A–50 = Sweets A–52 = Vegetables/Legumes A–62 = Vegetarian Foods A–64 = Misc A–66 = Soups/Sauces A–68 = Fast A–84 = Convenience A–88 = Baby foods

Chol (mg)	Calc (mg)	Iron (mg)	Magn (mg)	Pota (mg)	Sodi (mg)	Zinc (mg)	VT-A (RE)	Thia (mg)	VT-E (a-TE)	Ribo (mg)	Niac (mg)	V-B6 (mg)	Fola (μg)	VT-C (mg)
110	200	4.5		510	1140		150	.65		.48	8	.33		6
25	100	2.7		190	430		20	.15		.26	2			1
65	100	4.5		450	700		40	.36		.29	1.8			6
80	200	5.4		460	780		60	.36		.44	1.6			6
75	300	3.6		500	1270		80							5
20	100	1.08	35	240	410	1.2	0	.07		.17	1	.13		0
30	150	1.8	45	370	720	1.8	0	.12		.08	1.4	.18		2
25	150	4.5		430	930		1000							6
30	100	1.8		380	1220		1100							9
0	0	1.08		690	190		0	.18		.03	3.8			24
0	0	.36		190	310		0	.05			1			6
0	20	1.8		130	450		0	.29		.17	2.6			2
25	300	.72		680	210		0	.15		.6	.4			0
30	300	0		550	180		0	.15		.43	.4			0
30	300	0		570	180		0	.15		.34	.4			0
0	0	1.8		80	460		0	.2		.12	1.8			9
128	10	.18			772		15							1
163	10	.18			980		15							1
157	10	.18			1104		15							1
97	10	.18			667		15							1
92	30	.72			609		15	.09		.17	11.5			
78	68	1.2			604		15	.06		.13	6.9			
75	21	1.1			269		15	.05		.12	3.2			
112	40	1.08			591		31	.08		.3	5.5			
59	30	.54			383		15	.03		.08	3.7			
80	20	.72			750		15							6
80	40	1.08			850		15							6
100	20	1.08			670		15							6
65	20	.72			440		30							
75	33	.8			740		15	.11		.13	13.1			
75	20	.72			710		15	.09		.1	8.5			
65	20	.36			310		30	.06		.12	3.7			
90	49	1.2			520		30	.1		.21	6.5			
65	20	.36			320		30			.04	.06	3.3		
5	61	2.17	44	348	812	1.96	38	.09		.06	.76	.11	49	3
0	30	.18			15		0							0
3	43	1.92			565		28	.26		.2	2.77			
18	23	1.7			331		5	.16		.12	2.2			
5	30	.36			177		32	.03		.03	.2			27
42	60	.72			194		10							
0	0	.36			76		20	.14		.11	1.8			2
150	40	3.24			1230		15							6
4	11	.98			535		0							0
66	2	.1			865		15	.02		.02	1	.05		<1
16	120	.72			531		190							0
<1	20	.4			388		15			.04	1.2			
1	20	1.08			663		110							7
16	16	3.25	23	385	636	.44	120	.1		.03	.9	.29	11	
3					428		0							
4	10	.72			315									
0	20	.36			240		375							5

Table H-1

Food Composition

(Computer code number is for West Diet Analysis program) (For purposes of calculations, use "0" for t, <1, <.1, <.01, etc.)

Computer Code Number	Food Description	Measure	Wt (g)	H_2O (%)	Ener (cal)	Prot (g)	Carb (g)	Dietary Fiber (g)	Fat (g)	Fat Breakdown (g) Sat	Mono	Poly
	FAST FOOD RESTAURANTS—Continued											
	LONG JOHN SILVER'S											
1528	Chicken plank dinner, 3 piece	1 ea	399	56	890	32	101		44	9.5	24.8	9.4
1531	Clam chowder	1 ea	198	86	140	11	10	1	6	1.8	2.5	1.7
1532	Clam dinner	1 ea	361	46	990	24	114		52	10.9	31.4	9.9
	Fish, batter fried:											
1523	Fish & fryes (fries), 3 piece	1 ea	384	54	980	31	92		50	11.3	28.4	9.7
1524	Fish & fryes, 2 piece	1 ea	261	54	610	27	52		37	7.9	23.5	5.3
2240	Fish and lemon crumb dinner, 3 piece	1 ea	493	71	610	39	86		13	2.2	3.9	5.3
2241	Fish and lemon crumb dinner, 2 piece	1 ea	334	77	330	24	46		5	.9	1.6	1.2
1533	Fish & chicken dinner	1 ea	431	55	950	36	102		49	10.6	28.8	9.5
1537	Shrimp dinner, batter fried	1 ea	331	54	840	18	88		47	9.7	27.2	9.1
	Salads:											
1541	Cole slaw	1 ea	98	70	140	1	20	1	6	1	1.5	3.5
1539	Ocean chef salad	1 ea	234	89	110	12	13	2	1	.4	.4	.2
1540	Seafood salad	1 ea	278	79	380	15	12	2	31	5.1	8.2	17.5
1542	Fryes (fries) serving	1 ea	85	43	250	3	28	1	15	2.5	7.4	5.1
1543	Hush puppies	1 ea	24	40	70	2	10	<1	2	.4	1.3	.2
	Source: Long John Silver's, Lexington KY											
	McDONALD'S											
	Sandwiches:											
1221	Big mac	1 ea	216	53	510	25	46	3	26	9.3	7.5	4.1
1226	Cheeseburger	1 ea	122	46	318	15	36	2	13	5.6	3.8	1.1
1224	Filet-o-fish	1 ea	145	49	364	14	41	1	16	3.7	3.8	5.6
1225	Hamburger	1 ea	108	49	266	12	35	2	9	3.2	2.8	.9
1444	McChicken	1 ea	189	52	491	17	42	2	29	5.4	8.5	10.2
1591	McLean deluxe	1 ea	214	64	345	23	37	2	12	4.4	3.6	1.2
1438	McLean deluxe with cheese	1 ea	228	63	398	26	38	2	16	6.8	4.6	1.3
1222	Quarter-pounder	1 ea	171	52	415	23	36	2	20	7.8	6.7	1.3
1223	Quarter-pounder with cheese	1 ea	199	50	520	28	37	2	29	12.6	8.7	1.6
1227	French fries, small serving	1 ea	68	40	207	3	26	2	10	1.7	3.1	2.5
1228	Chicken McNuggets	4 pce	73	51	198	12	10	0	12	2.5	3.7	2.4
	Sauces (packet):											
1229	Hot mustard	1 ea	30	60	63	1	7	1	4	.5	1.1	2
1230	Barbecue	1 ea	32	58	53	<1	12	<1	<1	.1	.1	.2
1231	Sweet & sour	1 ea	32	57	55	<1	12	<1	<1	.1	.1	.3
	Low-fat (frozen yogurt) milk shakes:											
1232	Chocolate	1 ea	295		348	13	62	1	6	3.5	.1	.7
1233	Strawberry	1 ea	294		343	12	63	<1	5	3.4	.1	.6
1234	Vanilla	1 ea	293		308	12	54	<1	5	3.3	.1	.6
	Low-fat (frozen yogurt) sundaes:											
1237	Hot caramel	1 ea	182	56	307	7	62	1	3	2	.3	1
1235	Hot fudge	1 ea	179	60	293	8	53	2	5	4.7	.1	.4
1267	Strawberry	1 ea	178	65	239	6	51	1	1	.7	.1	.2
1238	Vanilla	1 ea	90	68	118	4	24	<1	1	.5	.2	t
1241	Cookies, McDonaldland	1 ea	56	3	258	4	41	1	9	1.7	6.4	.8
1242	Cookies, chocolaty chip	1 ea	56	3	282	3	36	1	14	3.9	4.4	1
1240	Muffin, apple bran, fat-free	1 ea	75	39	182	4	40	1	1	.2	.1	.3
1239	Pie, apple	1 ea	84	35	289	3	37	1	14	3.7	4.5	.3
	Breakfast items:											
1243	English muffin with spread	1 ea	63	33	189	5	30	2	6	2.4	1.5	1.3
1244	Egg McMuffin	1 ea	137	57	289	17	27	1	13	.7	4.5	1.6
1245	Hotcakes with marg & syrup	1 ea	222	44	557	8	100	2	14	2.4	4.6	5.8
1246	Scrambled eggs	1 ea	102	73	170	13	1	0	12	3.6	5.3	1.7
1247	Pork sausage	1 ea	43	45	173	6	<1	0	16	5.5	6.4	2.1
1248	Hashbrown potatoes	1 ea	53	55	130	1	13	1	8	1.3	2.3	1.9
1392	Sausage McMuffin	1 ea	112	42	361	13	26	1	23	8.3	8.2	2.8
1393	Sausage McMuffin with egg	1 ea	163	52	443	19	27	1	29	10	10.7	3.6
1394	Biscuit with biscuit spread	1 ea	76	32	260	4	32	1	13	3.8	3.7	.8

PAGE KEY: A–4 = Beverages A–6 = Dairy A–10 = Eggs A–10 = Fat/Oil A–14 = Fruit A–20 = Bakery A–28 = Grain A–32 = Fish A–34 = Meats A–38 = Poultry A–40 = Sausage A–42 = Mixed/Fast A–46 = Nuts/Seeds A–50 = Sweets A–52 = Vegetables/Legumes A–62 = Vegetarian Foods A–64 = Misc A–66 = Soups/Sauces A–68 = Fast A–84 = Convenience A–88 = Baby foods

Chol (mg)	Calc (mg)	Iron (mg)	Magn (mg)	Pota (mg)	Sodi (mg)	Zinc (mg)	VT-A (RE)	Thia (mg)	VT-E (a-TE)	Ribo (mg)	Niac (mg)	V-B6 (mg)	Fola (μg)	VT-C (mg)
55	200	4.5		1170	2000	3	40	.52		.51	16			9
20	200	1.8		380	590	.6	150	.09		.25	2			
75	200	4.5		910	1830	3	40	.75		.42	12			12
70	200	4.5		1120	1530	3	40	.45		.42	8			15
60	40	1.8		900	1480	1.2		.37		.34	8			9
125	200	5.4		990	1420	2.25	700	.75		.59	24			6
75	80	1.8		440	640	.9	1000	.3		.25	14			18
75	200	4.5		1280	2090	3	40	.6		.59	14			9
100	200	3.6		840	1630	3	40	.45		.42	9			9
15	60	.72		190	260	.6	40	.06		.07	2			
40	100	3.6		95	730	.3	500	.12		.14	3			21
55	150	4.5		130	980	.9	200	.15		.25	3			21
0	200	.72		370	500	.3	0	.09			1.6			6
	40	.72		65	25	.3		.06		.03	.8			
76	202	4.32	46	456	932	4.81	66	.49	1.01	.44	6.08	.25	49	3
42	134	2.73	27	281	766	2.62	64	.33	.46	.31	3.81	.15	24	2
37	123	1.85	32	266	708	.7	21	.32	1.52	.23	2.58	.07	30	0
28	126	2.73	24	260	531	2.25	22	.33	.23	.26	3.81	.14	21	2
52	128	2.5	32	319	797	1.06	29	.91	6.16	.24	7.74	.38	37	1
59	131	4.29	40	537	811	4.9	74	.42	.63	.34	7.16	.29	44	8
73	139	4.29	43	559	1046	5.26	115	.42	.85	.39	7.16	.3	47	8
70	127	4.33	33	405	692	4.66	33	.39	.36	.32	6.78	.24	27	3
97	143	4.5			1160		115	.39	.81	.43	6.78	.26	33	3
0	9	.53	26	469	135	.32	0	.05	.83	0	1.94	.24	26	8
42	9	.65	17	210	353	.69	0	.08	.96	.11	5.15	.21		0
3	7	.78		29	85		4	.01		.01	.15			0
0	4	0		51	277		0	.01		.01	.17			4
0	2	.16		7	158		74	0		.01	.08			0
24	372	1.04		543	241		46	.12		.51	.4	.1		3
24	366	.29		542	170		46	.12		.51	.4	.11		3
24	360	.29		533	193		45	.12		.51	.31			3
7	246	.15		344	197		18	.09		.34	.27			2
5	258	.59		441	190		7	.09		.34	.29			2
5	221	.25		325	115		6	.06		.34	.25			2
3	132	.23		175	84		4	<.01		.01	.23			1
0	10	1.73	11	62	267	.38	0	.24	.99	.16	2.01	.03		0
3	28	1.78	24	142	229	.4	0	.14	.92	.16	1.48			0
0	34	1.29	13	77	215	.33	0	.14	0	.14	1.32	.03	5	1
0	17	1.23	7	69	221	.23		.19	1.5	.12	1.55	.04	9	27
13	103	1.59	13	69	386	.42	33	.25	.13	.31	2.61	.04	57	1
234	151	2.44	24	199	730	1.56	100	.49	.85	.45	3.33	.15	33	2
11	108	1.98	27	285	746	.53	119	.24	1.2	.26	1.86	.09	<1	<1
424	50	1.19	10	126	143	1.06	168	.07	.92	.51	.06	.12	44	0
33	7	.5	7	102	292	.78	0	.18	.26	.06	1.7	.09		0
0	7	.3	11	213	332	.15	0	.08	.58	.02	.9	.08	8	3
46	132	2.07	22	191	751	1.51	48	.56	.66	.27	3.76	.14	16	0
257	156	2.8	26	251	821	2.07	117	.59	1.11	.49	3.79	.19	33	0
0	68	1.85	9	105	836	.3	2	.29	.81	.23	2.23	.03	5	0

Table H-1

Food Composition (Computer code number is for West Diet Analysis program) (For purposes of calculations, use "0" for t, <1, <.1, <.01, etc.)

Computer Code Number	Food Description	Measure	Wt (g)	H_2O (%)	Ener (cal)	Prot (g)	Carb (g)	Dietary Fiber (g)	Fat (g)	Fat Breakdown (g) Sat	Mono	Poly
	FAST FOOD RESTAURANTS—Continued											
	McDONALD'S—Continued											
1395	Biscuit with sausage	1 ea	119	37	433	10	32	1	29	8.6	10.1	2.8
1396	Biscuit with sausage & egg	1 ea	170	48	518	16	33	1	35	10.5	12.7	3.7
1397	Biscuit with bacon, egg, cheese	1 ea	152	46	450	17	33	1	27	8.7	8.9	2.3
	Salads:											
1398	Chef salad	1 ea	313	86	206	19	9	3	11	4.2	3	1.2
1400	Garden salad	1 ea	234	92	84	6	7	3	4	1.1	1.4	.7
1401	Chunky chicken salad	1 ea	296	87	164	23	8	3	5	1.3	1.6	1
	Source: McDonald's Corporation											
	PIZZA HUT											
	Pan pizza:											
1657	Cheese	2 pce	216	51	522	24	56	4	22	10	6.8	3.4
1658	Pepperoni	2 pce	208	49	531	22	56	4	24	8	9.9	3.7
1659	Supreme	2 pce	273	56	622	30	56	6	30	12	12	4.2
1660	Super supreme	2 pce	286	56	645	30	56	6	34	12		
	Thin 'n crispy pizza:											
1649	Cheese	2 pce	174	52	411	22	42	4	16	8	4.4	2.3
1623	Pepperoni	2 pce	168	48	431	22	42	2	20	8		
1622	Supreme	2 pce	232	57	514	28	42	4	26	10		
1620	Super supreme	2 pce	247	57	541	28	44	4	28	12		
	Hand tossed pizza:											
1619	Cheese	2 pce	216	53	470	26	58	4	14	7.9		
1618	Pepperoni	2 pce	208	51	477	24	58	4	16	8		
1648	Supreme	2 pce	273	56	568	32	60	6	24	10		
1617	Super supreme	2 pce	286	57	591	32	60	6	26	10		
	Personal pan pizza:											
1610	Pepperoni	1 ea	255	50	637	27	69	5	28	10	11.8	4.5
1609	Supreme	1 ea	327	57	721	33	70	6	34	12	14.7	5.6
	Source: Pizza Hut											
	SUBWAY											
	Deli style sandwich:											
69104	Bologna	1 ea	171	64	292	10	38	2	12	4		
69102	Ham	1 ea	171	69	234	11	37	2	4	1		
69103	Roast beef	1 ea	180	69	245	13	38	2	4	1		
69105	Seafood and crab:	1 ea	178	66	298	12	37	2	11	2		
69106	With light mayo	1 ea	178	68	256	12	37	2	7	2		
69108	Tuna:	1 ea	178		354	11	37	2	18	3		
69107	With light mayo	1 ea	178	67	279	11	38	2	9	2		
69101	Turkey	1 ea	180	69	235	12	38	2	4	1		
	Sandwiches, 6 inch:											
	B.L.T.:											
69135	On white bread	1 ea	191	67	311	14	38	3	10	3		
69136	On wheat bread	1 ea	198	65	327	14	44	3	10	3		
	Chicken taco sub:											
69131	On white bread	1 ea	286	70	421	24	43	3	16	5		
69132	On wheat bread	1 ea	293	69	436	25	49	4	16	5		
	Club :											
69117	On white bread	1 ea	246	73	297	21	40	3	5	1		
69118	On wheat bread	1 ea	253	71	312	21	46	3	5	1		
	Cold cut trio:											
69113	On white bread	1 ea	246	71	362	19	39	3	13	4		
69114	On wheat bread	1 ea	253	68	378	20	46	3	13	4		
	Ham:											
69115	On white bread	1 ea	232	73	287	18	39	3	5	1		
69115	On wheat bread	1 ea	239	71	302	19	45	3	5	1		

PAGE KEY: A–4 = Beverages A–6 = Dairy A–10 = Eggs A–10 = Fat/Oil A–14 = Fruit A–20 = Bakery A–28 = Grain A–32 = Fish A–34 = Meats A–38 = Poultry A–40 = Sausage A–42 = Mixed/Fast A–46 = Nuts/Seeds A–50 = Sweets A–52 = Vegetables/Legumes A–62 = Vegetarian Foods A–64 = Misc A–66 = Soups/Sauces A–68 = Fast A–84 = Convenience A–88 = Baby foods

Chol (mg)	Calc (mg)	Iron (mg)	Magn (mg)	Pota (mg)	Sodi (mg)	Zinc (mg)	VT-A (RE)	Thia (mg)	VT-E (a-TE)	Ribo (mg)	Niac (mg)	V-B6 (mg)	Fola (µg)	VT-C (mg)
33	75	2.35	15	207	1128	1.08	2	.48	1.07	.29	3.93	.12	5	0
245	100	2.95	20	271	1199	1.61	59	.51	1.53	.55	3.96	.18	27	0
238	103	2.6	20	245	1315	1.64	99	.39	1.49	.57	3.32	.13	30	0
179	157	1.81	40	605	727	2.16	1179	.33	1.45	.37	4.32	.36	100	22
139	52	1.34	24	407	61	.73	1114	.12	.95	.24	.65	.16	96	22
76	54	1.62	44	673	318	1.52	1973	.51	1.28	.21	8.46	.52	83	30
50	288	3	63	337	1002	4.32	211	.6		.64	5.48	.18		7
48	206	3.21	55	399	1140	4.14	190	.62		.48	5.31	.16	0	8
60	234	4.6	81	620	1529	6	195	.86		.84	6.4	.33		11
68	236	4.39	80	592	1649	5.99	201	.83		.73	7.13			12
50	291	2.06	56	307	1070	4.23	217	.46		.46	5.65	.18		6
50	208	2.2	51	330	1255	4.02	199	.48		.49	5.97			7
62	238	3.6	79	631	1591	5.4	197	.7		.57	6.27			12
70	238	3.41	73	563	1762	5.47	208	.72		.53	6.57			9
50	284	3	71	388	1242	4.6	198	.48		.48	5.3			10
48	202	3.21	84	610	1380	6.01	187	.72		.56	7.59			13
60	232	4.6	87	589	1769	5.48	192	.82		.66	8.45			14
68	232	4.39	89	607	1889	5.65	198	.84		.68	8.71			14
55	250	4	60	406	1338	3.8	233	.56		.66	8.16	.2		10
66	276	5.19	74	603	1757	4.69	240	.73		.82	9.91	.4		14
20	39	3			744		113							14
14	24	3			773		113							14
13	23	3			638		113							14
17	24	3			544		113							14
16	24	3			556		118							14
18	26	3			557		116							14
16	26	3			583		126							14
12	26	3			944		113							14
16	27	3			945		120							15
16	33	3			957		120							15
52	118	4			1264		209							18
52	124	4			1275		209							18
26	29	4			1341		120							15
26	35	4			1352		120							15
64	49	4			1401		130							16
64	55	4			1412		130							16
28	28	3			1308		120							15
28	35	3			1319		120							15

Table H-1

Food Composition

(Computer code number is for West Diet Analysis program) (For purposes of calculations, use "0" for t, <1, <.1, <.01, etc.)

Computer Code Number	Food Description	Measure	Wt (g)	H_2O (%)	Ener (cal)	Prot (g)	Carb (g)	Dietary Fiber (g)	Fat (g)	Fat Breakdown (g) Sat	Mono	Poly
	FAST FOOD RESTAURANTS—Continued											
	SUBWAY—Continued											
	Italian B.M.T.											
69139	On white bread	1 ea	246	66	445	21	39	3	21	8		
69140	On wheat bread	1 ea	253	64	460	21	45	3	22	7		
	Meatball:											
69129	On white bread	1 ea	260	70	404	18	44	3	16	6		
69130	On wheat bread	1 ea	267	67	419	19	51	3	16	6		
	Melt with turkey, ham, bacon, cheese:											
69127	On white bread	1 ea	251	70	366	22	40	3	12	5		
69128	On wheat bread	1 ea	258	68	382	23	46	3	12	5		
	Pizza sub:											
69133	On white bread	1 ea	250	66	448	19	41	3	22	9		
69134	On wheat bread	1 ea	257	65	464	19	48	3	22	9		
	Roast beef:											
69121	On white bread	1 ea	232	72	288	19	39	3	5	1		
69122	On wheat bread	1 ea	239	70	303	20	45	3	5	1		
	Roasted chicken breast:											
69125	On white bread	1 ea	246	70	332	26	41	3	6	1		
69126	On wheat bread	1 ea	253	68	348	27	47	3	6	1		
	Seafood and crab:											
69145	On white bread:	1 ea	246	69	415	19	38	3	19	3		
69147	With light mayo	1 ea	246	72	332	19	39	3	10	2		
69146	On wheat bread:	1 ea	253	67	430	20	44	3	19	3		
69148	With light mayo	1 ea	253	70	347	20	45	3	10	2		
	Spicy italian:											
69123	On white bread	1 ea	232	64	467	20	38	3	24	9		
69124	On wheat bread	1 ea	239	62	482	21	44	3	25	9		
	Steak and cheese:											
69119	On white bread	1 ea	257	68	383	29	41	3	10	6		
69120	On wheat bread	1 ea	264	67	398	30	47	3	10	6		
	Tuna:											
69141	On white bread:	1 ea	246	62	527	18	38	3	32	5		
69143	With light mayo	1 ea	246	70	376	18	39	3	15	2		
69142	On wheat bread:	1 ea	253	62	542	19	44	3	32	5		
69144	With light mayo	1 ea	253	68	391	19	46	3	15	2		
	Turkey:											
69111	On white bread	1 ea	232	73	273	17	40	3	4	1		
69112	On wheat bread	1 ea	239	71	289	18	46	3	4	1		
	Turkey breast and ham:											
69137	On white bread	1 ea	232	73	280	18	39	3	5	1		
69138	On wheat bread	1 ea	239	71	295	18	46	3	5	1		
	Veggie delite:											
69109	On white bread	1 ea	175	71	222	9	38	3	3	0		
69110	On wheat bread	1 ea	182	69	237	9	44	3	3	0		
	Salads:											
52128	B.L.T.	1 ea	276	91	140	7	10	2	8	3		
52124	B.M.T., classic Italian	1 ea	331	86	274	14	11	1	20	7		
52127	Chicken taco	1 ea	370	87	250	18	15	2	14	5		
52115	Club	1 ea	331	91	126	14	12	1	3	1		
52120	Cold cut trio	1 ea	330	89	191	13	11	1	11	3		
52123	Ham	1 ea	316	91	116	12	11	1	3	1		
52129	Meatball	1 ea	345	88	233	12	16	2	14	5		
52131	Melt	1 ea	336	88	195	16	12	1	10	4		
52121	Pizza	1 ea	335	86	277	12	13	2	20	8		
52126	Roast beef	1 ea	316	92	117	12	11	1	3	1		
52119	Roasted chicken breast	1 ea	331	89	162	20	13	1	4	1		
52117	Seafood and crab:	1 5	331	88	244	13	10	2	17	3		
52116	With light mayo	1 5	331	90	161	13	11	2	8	1		
52130	Steak and cheese	1 ea	342	87	212	22	13	1	8	5		
52122	Tuna:	1 ea	331	84	356	12	10	1	30	5		
52118	With light mayo	1 ea	331	89	205	12	11	1	13	2		
52114	Turkey breast	1 ea	316	92	102	11	12	1	2	1		
52125	With ham	1 ea	316	92	109	11	11	1	3	1		

PAGE KEY: A–4 = Beverages A–6 = Dairy A–10 = Eggs A–10 = Fat/Oil A–14 = Fruit A–20 = Bakery A–28 = Grain A–32 = Fish A–34 = Meats A–38 = Poultry A–40 = Sausage A–42 = Mixed/Fast A–46 = Nuts/Seeds A–50 = Sweets A–52 = Vegetables/Legumes A–62 = Vegetarian Foods A–64 = Misc A–66 = Soups/Sauces A–68 = Fast A–84 = Convenience A–88 = Baby foods

Chol (mg)	Calc (mg)	Iron (mg)	Magn (mg)	Pota (mg)	Sodi (mg)	Zinc (mg)	VT-A (RE)	Thia (mg)	VT-E (a-TE)	Ribo (mg)	Niac (mg)	V-B6 (mg)	Fola (μg)	VT-C (mg)
56	44	4			1652		151							15
56	50	4			1664		151							15
33	32	4			1035		142							16
33	39	4			1046		142							16
42	93	4			1735		155							15
42	100	3			1746		156							15
50	103	4			1609		238							16
50	110	3			1621		238							16
20	25	4			928		120							15
20	32	3			939		120							15
48	35	3			967		123							15
48	42	3			978		123							15
34	28	3			849		121							15
32	28	3			873		131							15
34	34	3			860		121							15
32	34	3			884		131							15
57	40	4			1592		169							15
57	47	4			1604		169							15
70	88	5			1106		175							18
70	95	5			1117		176							18
36	32	3			875		125							15
32	32	3			928		146							15
36	38	3			886		126							15
32	38	3			940		146							15
19	30	4			1391		120							15
19	37	3			1403		120							15
24	29	3			1350		120							15
24	36	3			1361		120							15
0	25	3			582		120							15
0	32	3			593		120							15
16	24	1			672		273							32
56	41	2			1379		303							32
52	115	3			990		361							35
26	26	2			1067		273							32
64	46	2			1127		282							33
28	25	2			1034		273							32
33	30	2			761		295							33
42	90	2			1461		308							32
50	100	2			1336		390							33
20	23	2			654		273							32
48	32	2			693		276							32
34	25	2			575		273							32
32	25	2			599		284							32
70	86	3			832		328							35
36	29	2			601		278							32
32	29	2			654		298							32
19	28	2			1117		273							32
24	27	2			1076		273							32

Table H-1

Food Composition

(Computer code number is for West Diet Analysis program) (For purposes of calculations, use "0" for t, <1, <.1, <.01, etc.)

Computer Code Number	Food Description	Measure	Wt (g)	H_2O (%)	Ener (cal)	Prot (g)	Carb (g)	Dietary Fiber (g)	Fat (g)	Fat Breakdown (g) Sat	Mono	Poly
	FAST FOOD RESTAURANTS—Continued											
	SUBWAY—Continued											
52113	Veggie delite	1 ea	260	94	51	2	10	1	1	0		
	Cookies:											
47662	Brazil nut and chocolate chip	1 ea	48	12	229	3	27	1	12	3.5		
47655	Chocolate chip:	1 ea	48	14	209	2	29	1	10	3.5		
47658	With M&M's	1 ea	48	14	209	2	29	1	10	3		
47659	Chocolate chunk	1 ea	48	14	209	2	29	1	10	3.5		
47656	Oatmeal raisin	1 ea	48	15	199	3	29	1	8	2		
47657	Peanut butter	1 ea	48	13	219	3	26	1	12	2.5		
47660	Sugar	1 ea	48	11	229	2	28	0	12	3		
47661	White chip macademia	1 ea	48	12	229	2	28	1	12	2.5		
	Source: Subway International											
	TACO BELL											
	Breakfast burrito:											
1601	Bacon breakfast burrito	1 ea	99	48	291	11	23		17	4		
1627	Country breakfast burrito	1 ea	113	55	220	8	26	2	14	5		
1626	Fiesta breakfast burrito	1 ea	92	44	280	9	25	2	16	6		
1625	Grande breakfast burrito	1 ea	177	56	420	13	43	3	22	7		
1604	Sausage breakfast burrito	1 ea	106	49	303	11	23		19	6		
	Burritos:											
1544	Bean with red sauce	1 ea	198	58	380	13	55	13	12	4		
1545	Beef with red sauce	1 ea	198	57	432	22	42	4	19	8	6.7	.7
1546	Beef & bean with red sauce	1 ea	198	57	412	17	50	5	16	6	6.1	2.1
1569	Big beef supreme	1 ea	298	64	520	24	54	11	23	10		
1552	Chicken burrito	1 ea	171	58	345	17	41		13	5		
1547	Supreme with red sauce	1 ea	248	64	428	16	50	10	18	7.8		
1571	7 layer burrito	1 ea	234	61	438	13	55	11	19	5.8		
1538	Chilito	1 ea	156	49	391	17	41		18	9		
1549	Chilito, steak	1 ea	257	62	496	26	47		23	10		
	Tacos:											
1551	Taco	1 ea	78	58	180	9	12	3	10	4		
1554	Soft taco	1 ea	99	63	242	10	13	3	11	4.4		
1536	Soft taco supreme	1 ea	128	64	234	11	21	3	13	6.3		
1568	Soft taco, chicken	1 ea	128	63	212	15	22	2	7	2.6		
1572	Soft taco, steak	1 ea	100	63	180	12	16	2	8	1.9		
1555	Tostada with red sauce	1 ea	156	67	264	9	27	11	13	4.4		
1558	Mexican pizza	1 ea	223	53	578	21	43	8	35	10.1		
1559	Taco salad with salsa	1 ea	585	71	923	33	70	17	56	16.3		
1560	Nachos, regular	1 ea	106	40	343	5	36	3	19	4.3		
1561	Nachos, bellgrande	1 ea	287	51	708	19	77	16	36	10.1		
1562	Pintos & cheese with red sauce	1 ea	128	68	203	10	19	11	10	4.3		
1563	Taco sauce, packet	1 ea	9	94	2	<1	<1	<1	<1	0	0	0
1564	Salsa	1 ea	10	28	27	1	6		<1	0	0	0
1565	Cinnamon twists	1 ea	35	6	175	1	24	0	7	0		
1628	Caramel roll	1 ea	85	19	353	6	46		16	4		
	Source: Taco Bell Corporation											
	WENDY'S											
	Hamburgers:											
1566	Single on white bun, no toppings	1 ea	133	44	360	24	31	2	16	6		
1570	Cheeseburger, bacon	1 ea	166	55	380	20	34	2	19	7	10.3	1.4
1730	Chicken sandwich, grilled	1 ea	189	62	310	27	35	2	8	1.5		
	Baked potatoes:											
1573	Plain	1 ea	284	71	310	7	71	7	0	0	0	0
1574	With bacon & cheese	1 ea	380	69	530	17	78	7	18	4	10.7	3.3
1575	With broccoli & cheese	1 ea	411	74	470	9	80	9	14	2.5	8	2.5
1576	With cheese	1 ea	383	68	570	14	78	7	23	8	9.2	4.8

Chol (mg)	Calc (mg)	Iron (mg)	Magn (mg)	Pota (mg)	Sodi (mg)	Zinc (mg)	VT-A (RE)	Thia (mg)	VT-E (a-TE)	Ribo (mg)	Niac (mg)	V-B6 (mg)	Fola (μg)	VT-C (mg)
0	23	1			308		136							32
10	32	1.99			115		0							0
10	16	1.99			139		0							0
15	16	1			139		0							0
10	16	1			139		0							0
15	32	1			159		0							0
0	16	1			179		0							0
20	0	.72			179		0							0
10	16	1			139		0							0
181	80	1.8			652		310							
195	80	1.08			690		250							0
25	80	.72			580		150							0
205	100	1.8			1050		500							0
183	80	1.8			661		320							
10	150	2.7		495	1100		450	.04		2.02	1.98	.31		0
57	160	3.96		380	1303		530	.4		2.14	3.44	.32		1
32	170	3.78	50	442	1221	2.67	450	.49		.41	3.09	.59	38	1
55	150	2.7			1520		600							5
57	140	2.52			854		440							1
34	146	8.75	48	410	1196		486	.39		2.04	2.81	.34		5
21	165	2.98			1058		248							5
47	300	3.06			980		950							
78	200	2.7			1313		970							2
25	80	1.08		159	330		100	.05		.14	1.2	.12		0
27	88	1.19		211	363		110	.42		.24	2.95	1.08		0
31	90	1.62			532		135							3
37	85	1.52			571		63							1
19	62	1.13			797		31							0
13	132	1.59		401	573		441	.05		.17	.63	.26		1
46	253	3.65	80	408	1054	5.37	405	.32		.33	2.96	1.12	60	5
65	326	6.84		1048	1931	1.67	1736	.51		.76	4.8	.56	10	26
5	107	.77		160	610	1.68	64	.17		.16	.68	.19	10	0
32	184	3.31		674	1205		138	.1		.34	2.17			3
16	160	1.92	110	384	693	2.17	267	.05		.15	.43	.21	68	0
0	0	.07		9	75		30	0			.02			<1
0	50	.6		376	709		168	.02		.14	0			10
0	0	.45		27	238		50	.1		.04	.71	.04		0
15	60	1.44			312		330							4
65	110	4.14		296	580		0	.43		.38	6.71			0
60	170	3.42	38	375	850	5.9	80	.3		.31	6.43	.26	28	6
65	100	2.7			790		40							6
0	30	3.78	75	1187	25	.74	0	.31	.14	.12	4.3	.8	31	36
20	180	4.32	87	1498	1390	2.75	100	.24		.19	5.04	.94	36	36
5	210	4.5	93	1745	470	.97	350	.34		.29	4.5	.97	74	72
30	380	4.14	85	1510	640	.67	200	.25		.28	3.6	.88	36	36

Table H-1

Food Composition (Computer code number is for West Diet Analysis program) (For purposes of calculations, use "0" for t, <1, <.1, <.01, etc.)

Computer Code Number	Food Description	Measure	Wt (g)	H_2O (%)	Ener (cal)	Prot (g)	Carb (g)	Dietary Fiber (g)	Fat (g)	Fat Breakdown (g) Sat	Mono	Poly
	FAST FOOD RESTAURANTS—Continued											
	WENDY'S—Continued											
1577	With chili & cheese	1 ea	439	69	630	20	83	9	24	9		
1578	With sour cream & chives	1 ea	314	71	380	8	74	8	6	4		
1579	Chili	1 ea	227	81	210	15	21	5	7	2.5		
1582	Chocolate chip cookies	1 ea	57	6	270	3	36	1	13	6		
1580	French fries	1 ea	130	41	390	5	50	5	19	3	11.9	2.4
1581	Frosty dairy dessert	1 ea	227	68	330	8	56	0	8	5		
	Source: Wendy's International											
	CONVENIENCE FOODS and MEALS											
	BUDGET GOURMET											
1695	Chicken cacciatore	1 ea	312	80	300	20	27		13			
1692	Linguini & shrimp	1 ea	284	77	330	15	33		15			
1691	Scallops & shrimp	1 ea	326	79	320	16	43		9			
2245	Seafood newburg	1 ea	284	74	350	17	43		12			
1693	Sirloin tips with country gravy	1 ea	284	80	310	16	21		18			
1694	Sweet & sour chicken with rice	1 ea	284	72	350	18	53		7			
1689	Teriyaki chicken	1 ea	340	77	360	20	44		12			
1690	Veal parmigiana	1 ea	340	75	440	26	39		20			
1696	Yankee pot roast	1 ea	312	77	380	27	22		21			
	Source: The All American Gourmet Co.											
	HAAGEN DAZS											
1755	Ice cream bar, vanilla almond	1 ea	107		371	6	26		27	14.1	10	3
	Sorbet:											
1758	Lemon	½ c	113		140	0	35		0	0	0	0
1760	Orange	½ c	113		140	0	36		0	0	0	0
1759	Raspberry	½ c	113		110	0	27		0	0	0	0
	Yogurt, frozen:											
1753	Chocolate	½ c	98		171	8	26		4	2	2	0
1754	Strawberry	½ c	98		171	6	27		4	2	2	0
	Yogurt extra, frozen:											
1752	Brownie nut	½ c	101		220	8	29		9	4	4	1
1751	Raspberry rendezvous	½ c	101		132	4	26		2	1	1	0
	Source: Pillsbury											
	HEALTHY CHOICE											
	Entrees:											
2112	Fish, lemon pepper	1 ea	303	78	290	14	47	7	5	1		
1624	Lasagna	1 ea	383	76	390	26	60	9	5	2		
2111	Meatloaf, traditional	1 ea	340	79	320	16	46	7	8	4		
2104	Zucchini lasagna	1 ea	396	80	329	20	58	11	1	1		
2110	Dinner, pasta shells marinara	1 ea	340	74	360	25	59	5	3	1.5		
	Low-fat ice cream:											
973	Brownie	½ c	71	60	120	3	22	2	2	1	.3	.7
259	Butter pecan	½ c	71	60	120	3	22	1	2	1	.3	.7
650	Chocolate chip	½ c	71	62	120	3	21	<1	2	1	1	0
1608	Cookie & cream	½ c	71	62	120	3	21	<1	2	1.5	.5	0
650	Chocolate chip	½ c	71	62	120	3	21	<1	2	1	1	0
45	Rocky road	½ c	71	53	140	3	28	2	1	1	.5	0
1621	Vanilla	½ c	71	66	100	3	18	1	2	.5	1.5	0
391	Vanilla fudge	½ c	71	62	120	3	21	1	2	1.5		
	Source: ConAgra Frozen Foods, Omaha, NE											

Chol (mg)	Calc (mg)	Iron (mg)	Magn (mg)	Pota (mg)	Sodi (mg)	Zinc (mg)	VT-A (RE)	Thia (mg)	VT-E (a-TE)	Ribo (mg)	Niac (mg)	V-B6 (mg)	Fola (µg)	VT-C (mg)
40	330	5.04	122	1745	770	4.15	200	.33		.29	4.5	.99	55	36
15	80	4.32	71	1438	40	.91	300	.23		.14	3.04	.8	32	48
30	80	2.9		501	800		80	.11		.15	2.66			4
30	10	1.8	13	89	120	.41	0	.05		.06	.36	.03	5	0
0	20	1.08	55	845	120	.62	0	.18		.04	3.6	.33	40	6
35	310	1.08	46	544	200	.97	150	.11		.47	.32	.13	17	0
60	150	1.8			810		40	.23		.51	5			21
75	10	3.6			1250		1000	.3		.17	3			2
70	150	.72			690		150			.26	3			12
70	100	.72			660		40	.23		.26	2			
40	60	.36			570		150	.15		.17	4	.28		2
40	60	.72			640		80	.12		.34	3			2
55	80	1.4			610		300	.15		.34	6			12
165	30	4.5			1160		1000	.45		.6	6			6
70	150	1.8			690		600	.15		.43	7			6
90	161	.38		221	85		161			.18				
0				30	20									7
0				80	20									20
0				60	15									7
40	147	.71		241	45		20			.17				
50	147			141	45		20	.03		.17				5
55	152	.73		250	60		20			.14				
20	81			97	25		0			.1				5
25	20	1.08			360		100							30
15	150	3.6		500	550		100	.3		.26	2			6
35	40	1.8			460		150							54
10	199	2.69			309		249							0
25	400	1.8			390		100							4
2	80	0		268	55		40							0
2	100	0		211	60		40							0
2	100	0		240	50		40							0
2	100			254	90		60	.03		.15				2
2	100	0		240	50		40							0
2	100	0		168	60		40	.03		.15				0
5	100			254	50		60	.05		.22				2
2	100	0		296	50		40							0

Table H-1

Food Composition

(Computer code number is for West Diet Analysis program) (For purposes of calculations, use "0" for t, <1, <.1, <.01, etc.)

Computer Code Number	Food Description	Measure	Wt (g)	H_2O (%)	Ener (cal)	Prot (g)	Carb (g)	Dietary Fiber (g)	Fat (g)	Fat Breakdown (g) Sat	Mono	Poly
	CONVENIENCE FOODS and MEALS—Continued											
	HEALTH VALLEY											
	Soups, fat-free:											
2001	Beef broth, no salt added	1 c	240	98	18	5	0	0	0	0	0	0
2073	Beef broth, w/salt	1 c	240	98	30	5	2	0	0	0	0	0
2016	Black bean & vegetable	1 c	240	85	110	11	24	12	0	0	0	0
2017	Chicken broth	1 c	240	97	30	6	0	0	0	0	0	0
2018	14 garden vegetable	1 c	240	90	80	6	17	4	0	0	0	0
2015	Lentil & carrot	1 c	240	85	90	10	25	14	0	0	0	0
2014	Split pea & carrot	1 c	240	89	110	8	17	4	0	0	0	0
2013	Tomato vegetable	1 c	240	90	80	6	17	5	0	0	0	0
	Source: Health Valley											
	LA CHOY											
2100	Egg rolls, mini, chicken	1 svg	106	53	220	8	35	3	6	1.5		
2099	Egg rolls, mini, shrimp	1 svg	106	56	210	7	35	3	4	1		
	Source: Beatrice/Hunt Wesson											
	LEAN CUISINE											
	Dinners:											
1639	Baked cheese ravioli	1 ea	241	77	250	12	32	4	8	3	2	1
1632	Chicken chow mein	1 ea	255	81	210	13	28	2	5	1	2	1
1633	Lasagna	1 ea	291	79	270	19	34	5	6	2.5	1.5	.5
1634	Macaroni & cheese	1 ea	255	78	270	13	39	2	7	3.5	1.5	.5
1631	Spaghetti w/meatballs	1 ea	269	74	290	17	40	4	7	2	3	1.5
	Pizza:											
1636	French bread sausage pizza	1 ea	170	53	420	19	41	4	20	5	13.9	1.1
	Source: Stouffer's Foods Corp, Solon, OH											
	TASTE ADVENTURE SOUPS											
1905	Black bean	1 c	242		139	6	28	6	1			
1904	Curry lentil	1 c	241		138	6	30	5	1			
1906	Lentil chili	1 c	242		181	11	33	6	1			
1903	Split pea	1 c	244		140	5	27	5	1			
	Source: Taste Adventure Soups											
	WEIGHT WATCHERS											
	Cheese, fat-free slices:											
1978	Cheddar, sharp	2 pce	21	65	30	5	2	0	0	0	0	0
1980	Swiss	2 pce	21	65	30	5	2	0	0	0	0	0
1977	White	2 pce	21	65	30	5	2	0	0	0	0	0
1979	Yellow	2 pce	21	65	30	5	2	0	0	0	0	0
	Dinners:											
2029	Chicken chow mein	1 ea	255	81	200	12	34	3	2	.5		
1646	Oven fried fish	1 ea	218	78	230	15	25	2	8	2.5	5	2
1972	Margarine, reduced fat	1 tbs	14	49	59	0	0	0	7	1.5		
	Pizza:											
1653	Cheese	1 ea	163	48	390	23	49	6	12	4	3	1
1650	Deluxe combination pizza	1 ea	186	56	380	23	47	6	11	3.5	5	2
1652	Pepperoni pizza	1 ea	158	48	390	23	46	4	12	4	5	2
	Desserts:											
1644	Chocolate brownie	1 ea	182	75	190	6	35	4	4	1	2	1
2024	Chocolate eclair	1 ea	60	45	151	3	24	2	5	1.5		
2247	Chocolate mousse	1 ea	78	44	190	6	33	3	4	1.5		
1642	Strawberry cheesecake	1 ea	111	62	180	7	28	2	5	2	1	2

PAGE KEY: A–4 = Beverages A–6 = Dairy A–10 = Eggs A–10 = Fat/Oil A–14 = Fruit A–20 = Bakery A–28 = Grain A–32 = Fish A–34 = Meats A–38 = Poultry A–40 = Sausage A–42 = Mixed/Fast A–46 = Nuts/Seeds A–50 = Sweets A–52 = Vegetables/Legumes A–62 = Vegetarian Foods A–64 = Misc A–66 = Soups/Sauces A–68 = Fast A–84 = Convenience A–88 = Baby foods

Chol (mg)	Calc (mg)	Iron (mg)	Magn (mg)	Pota (mg)	Sodi (mg)	Zinc (mg)	VT-A (RE)	Thia (mg)	VT-E (a-TE)	Ribo (mg)	Niac (mg)	V-B6 (mg)	Fola (μg)	VT-C (mg)
0				196	74						.98			
0	0	0		196	160		0				.98			5
0	40	3.6		676	280		2000	.34		.11	1.35	.22	135	9
0	20	1.8		147	170		0			.03	2.45			1
0	40	1.8		406	250		2000	.26		.08	2.25	.18	27	15
0	60	5.4		439	220		2000	.1		.16	5.63	.45	27	2
0	40	5.4		439	230		2000	.1		.16	5.63	.45		9
0	40	5.4		609	240		2000	.1		.08	2.25	.13	<1	9
5	20	1.44			460		20							0
5	20	1.44			510		20							0
55	200	1.08	42	400	500	1.5	150	.06		.25	1.2	.2	48	6
35	20	.36	30	300	510	1.1	20	.15		.17	5			6
25	150	1.8	44	620	560	2.9	100	.15		.25	3	.32		12
20	250	.72		170	550		20	.12		.25	1.2			0
30	100	2.7	47	480	520	2.5	80	.15		.25	3	.2		4
35	250	2.7	39	340	900	2.2	80	.45		.51	5	.07		6
				650	565									
				467	584									
				650	448									
				484	591									
0	99	0		64	306		56							0
0	99	0		74	276		56							0
0	99	0		64	306		56							0
0	99	0		64	306		56							0
25	40	.72		360	430		300							36
25	20	1.44		370	450		40	.09		.14	1.6			0
0	0	0		5	128		49							0
35	700	1.8		290	590		80	.3		.51	3	.06		6
40	500	3.6		370	550		150	.3		.51	3	.2		5
45	450	1.8		320	650		80	.23		.51	3			5
5	80	1.08		230	160		0	.06		.03	.2	.03		0
0	40	0		65	151		0							0
5	60	1.8		320	150		0							0
15	80	.36		115	230		40	.06		.07	1.6			2

Table H-1

Food Composition

(Computer code number is for West Diet Analysis program) (For purposes of calculations, use "0" for t, <1, <.1, <.01, etc.)

Computer Code Number	Food Description	Measure	Wt (g)	H_2O (%)	Ener (cal)	Prot (g)	Carb (g)	Dietary Fiber (g)	Fat (g)	Fat Breakdown (g) Sat	Mono	Poly
	CONVENIENCE FOODS and MEALS—Continued											
	WEIGHT WATCHERS—CONTINUED											
2027	Triple chocolate cheesecake	1 ea	89	52	199	7	32	1	5	2.5		
	Source: Weight Watchers											
	SWEET SUCCESS:											
	Drinks, prepared:											
1776	Chocolate chip	1 c	265	81	180	15	30	6	3	1.6		
1777	Chocolate fudge	1 c	265	81	180	15	30	6	2			
1774	Chocolate mocha	1 c	265	81	180	15	30	6	1	1		
1778	Milk chocolate	1 c	265	81	180	15	30	6	2	1		
1775	Vanilla	1 c	265	81	180	15	33	6	1	.6		
	Drinks, ready to drink:											
2147	Chocolate mint	1 c	297	82	187	11	36	6	3	0		
2148	Strawberry	1 c	265	82	167	10	32	5	3	0		
	Shakes:											
1771	Chocolate almond	1 c	250	82	158	9	30	5	2	0	2.1	.2
1773	Chocolate fudge	1 c	250	82	158	9	30	5	2	0	2.1	.2
1768	Chocolate mocha	1 c	250	82	158	9	30	5	2	0	.6	1.8
1769	Chocolate raspberry truffle	1 c	250	82	158	9	30	5	2	0	2.2	.2
1770	Vanilla creme	1 c	250	82	158	9	30	5	2	0	2.1	.3
	Snack bars:											
1767	Chocolate brownie	1 ea	33	9	120	2	23	3	4	2	.5	.6
1766	Chocolate chip	1 ea	33	9	120	2	23	3	4	2	.4	.5
1921	Oatmeal raisin	1 ea	33	9	120	2	23	3	4	2		
1765	Peanut butter	1 ea	33	9	120	2	23	3	4	2	.6	.6
	Source: Foodway National Inc, Boise, ID											
	BABY FOODS											
1720	Apple juice	½ c	125	88	59	0	15	<1	<1	t	t	t
1721	Applesauce, strained	1 tbs	16	89	7	<1	2	<1	<1	t	t	t
1716	Carrots, strained	1 tbs	14	92	4	<1	1	<1	<1	t	t	t
1718	Cereal, mixed, milk added	1 tbs	15	75	17	1	2	<1	1	.3		
1719	Cereal, rice, milk added	1 tbs	15	75	17	<1	3	<1	1	.3		
1723	Chicken and noodles, strained	1 tbs	16	88	8	<1	1	<1	<1	.1	.1	t
1722	Peas, strained	1 tbs	15	87	6	1	1	<1	<1	t	t	t
1717	Teething biscuits	1 ea	11	6	43	1	8	<1	<1	.2	.2	.1

Chol (mg)	Calc (mg)	Iron (mg)	Magn (mg)	Pota (mg)	Sodi (mg)	Zinc (mg)	VT-A (RE)	Thia (mg)	VT-E (a-TE)	Ribo (mg)	Niac (mg)	V-B6 (mg)	Fola (μg)	VT-C (mg)
10	80	1.08		169	199		0							0
6	500	6.3	140	600	288	5.25	350	.52	7.05	.59	7	.7	140	21
6	500	6.3	140	750	336	5.25	350	.52	7.05	.59	7	.7	140	21
6	500	6.3	140	800	336	5.25	350	.52	7.05	.59	7	.7	140	21
6	500	6.3	140	750	336	5.25	350	.52	7.05	.59	7	.7	140	21
6	500	6.3	140	830	312	5.25	250	.52	7.05	.59	7	.7	140	21
6	470	5.94	131	526	226	5.05	329	.5	6.56	.56	6.53	.65	131	20
5	419	5.3	117	310	175	4.51	294	.45	5.86	.5	5.83	.58	117	17
5	396	5	110	443	190	4.25	277	.42	5.53	.47	5.5	.55	110	16
5	396	5	110	443	175	4.25	277	.42	5.53	.47	5.5	.55	110	16
5	396	5	110	403	175	4.25	277	.42	5.53	.47	5.5	.55	110	16
5	383	5	110	443	175	4.25	277	.42	5.53	.47	5.5	.55	110	16
5	396	5	110	293	175	4.25	277	.42	5.53	.47	5.5	.55	110	16
3	150	2.71	60	140	45	.59	150	.22	3.01	.25	3	.3	60	9
3	150	2.71	60	110	40	.59	150	.22	3.01	.25	3	.3	60	9
3	150	2.71	60		30	.59	150	.22	3.01	.25	3	.3	60	9
3	150	2.71	60	125	35	.59	150	.22	3.01	.25	3	.3	60	9
0	5	.71	4	114	4	.04	2	.01	.75	.02	.1	.04	<1	72
0	1	.03		11	<1	<.01	<1	<.01	.1	<.01	.01	<.01	<1	6
0	3	.05	1	27	5	.02	160	<.01	.07	.01	.06	.01	2	1
2	33	1.56	4	30	7	.11	4	.06		.09	.87	.01	2	<1
2	36	1.83	7	28	7	.1	4	.07		.07	.78	.02	1	<1
3	4	.07	1	6	3	.05	18	<.01	.04	.01	.07	<.01	2	<1
0	3	.14	2	17	1	.05	8	.01	.08	.01	.15	.01	4	1
0	29	.39	4	35	40	.1	1	.03	.05	.06	.48	.01	5	1

Glossary

Accreditation approval; in the case of hospitals or university departments, approval by a professional organization of the educational program offered. There are phony accrediting agencies; the genuine ones are listed in a directory called *Accredited Institutions of Postsecondary Education.*

Acesulfame-K (AY-see-sul-fame) a derivative of acetoacetic acid approved for use in the United States in 1988. Since it is not metabolized by the body, acesulfame K does not contribute calories and is excreted from the body unchanged. It is currently approved for use in more than 70 countries and found in more than 100 international products, including chewing gum, gelatins, nondairy creamers, powdered drink mixes, and puddings.

Acetaldehyde (ass-et-AL-duh-hide) a substance to which drinking alcohol (ethanol) is metabolized.

Acid-Base Balance equilibrium between acid and base concentrations in the body fluids.

Acidosis (a-sih-DOSE-sis) blood acidity above normal, indicating excess acid.

Acids compounds that release hydrogens in a watery solution; acids have a low pH.

Acquired Immune Deficiency Syndrome (AIDS) an immune system disorder caused by the human immunodeficiency virus (HIV). Its attack on the individual's immune cells (T-cells) results in a decreased ability to fight foreign organisms, thus increasing the individual's susceptibility to a variety of opportunistic infections. AIDS is transmitted to a person through direct contact of the person's body fluids with contaminated body fluids. It is most often transmitted through sexual intercourse, contaminated needles, contaminated blood products, or from mother to infant during pregnancy or lactation.

Activities of Daily Living (ADL) include bathing, dressing, grooming, transferring from bed to chair, going to the bathroom, and feeding oneself.

Additives substances that are added to foods, but are not normally consumed by themselves as foods.

Adequacy characterizes a diet that provides all of the essential nutrients, fiber, and energy (calories) in amounts sufficient to maintain health.

Adequate Intake (AI) the estimated amount of a nutrient that should be consumed when sufficient scientific evidence is not available to calculate an EAR and RDA.

Adipose Tissue the body's fat tissue, consisting of masses of fat-storing cells and blood vessels to nourish them.

Adverse Reaction an unusual response to food, including food allergies and food intolerances.

Aerobic requiring oxygen.

Aflatoxin a poisonous toxin produced by molds.

Agave a plant with spiny-margined leaves and flowers.

Age-Related Macular Degeneration oxidative damage to the central portion of the eye—called the macula—that allows you to focus and see details clearly (peripheral vision remains unimpaired).

Alcohol Dehydrogenase a liver enzyme that converts ethanol to acetaldehyde. The MEOS also oxidizes alcohol.

Alcoholism a dependency on alcohol marked by compulsive uncontrollable drinking with negative effects on physical health, family relationships, and social health.

Alkalosis (al-kah-LOH-sis) blood alkalinity above normal.

Allergy an immune reaction to a foreign substance, such as a component of food; also called *hypersensitivity*.

Allyl Sulfides compounds in garlic that may help lower blood cholesterol levels and protect against some types of cancer.

Alternative Sweeteners nutritive (calorie-containing) sweeteners such as fructose, sorbitol, mannitol, and xylitol.

Amaranth a golden-colored grain.

Amenorrhea the absence or cessation of menstruation.

Amine (a-MEEN) **Group** the nitrogen-containing portion of an amino acid.

Amino (a-MEEN-o) **Acids** building blocks of protein; each is a compound with an amine group at one end, an acid group at the other, and a distinctive side chain.

Amniotic Sac the "bag of waters" in the uterus in which the fetus floats.

Anabolic Steroids synthetic male hormones with a chemical structure similar to that of cholesterol; such hormones have wide-ranging effects on body functioning.

Anaerobic not requiring oxygen.

Anaphylaxis (an-ah-fa-LAX-is) a potentially fatal whole-body allergic reaction to an offending substance causing reduced oxygen supply to the heart and other body tissues. Symptoms include difficulty breathing, low blood pressure, pale skin, a weak, rapid pulse, and loss of consciousness.

Anemia any condition in which the blood is unable to deliver sufficient oxygen to the cells of the body. Examples include a shortage or abnormality of the red blood cells. Many nutrient deficiencies and diseases can cause anemia.

Anencephaly (an-en-SEFF-ah-lee) a severe neural tube defect in which the brain fails to form; anencephaly leads to death soon after birth.

Anorexia Nervosa literally "nervous lack of appetite," a disorder (usually seen in teenage girls) involving self-starvation to the extreme.

Antibodies large proteins of the blood and body fluids, produced by one type of immune cell in response to invasion of the body by unfamiliar molecules (mostly foreign proteins). Antibodies inactivate the foreign substances and so protect the body. The foreign substances are called antigens.

Antigen a substance foreign to the body that elicits the formation of antibodies or an inflammation reaction from immune system cells.

Antioxidant (anti-OX-ih-dant) a compound that protects other compounds from oxygen by itself reacting with oxygen; it helps to prevent damage done to the body as a result of chemical reactions that involve the use of oxygen.

Antipromoters compounds in foods that act in several ways to oppose the formation of cancer.

Appendicitis inflammation and/or infection of the appendix, a sac protruding from the large intestine.

Appetite the psychological desire to find and eat food, experienced as a pleasant sensation, often in the absence of hunger.

Appropriate Technology a technology that utilizes locally abundant resources in preference to locally scarce resources. Developing countries usually have a large labor force and little capital; the appropriate technology would therefore be labor intensive.

Arousal as used in this context, heightened activity of certain brain centers associated with excitement and anxiety.

Artesian Water or **Artesian Well Water** water drawn from a well that taps a confined water-bearing rock or rock formation.

Artificial Sweeteners nonnutritive sugar replacements such as acesulfame-K, aspartame, saccharin, and sucralose.

Ascorbic Acid one of the active forms of vitamin C (the other is *dehydroascorbic acid*); an antioxidant nutrient.

Aspartame a dipeptide containing the amino acids aspartic acid and phenylalanine and used in the United States and Canada since 1981. Although it is digested as protein and supplies calories, it is so sweet that only small amounts, which contribute negligible calories, are needed to sweeten foods. Thus, it is classified as a nonnutritive sweetener. Often sold under the trade name NutraSweet, aspartame is blended with lactose and an anticaking agent and sold commercially as Equal.

Atherosclerosis (ATH-er-oh-scler-OH-sis) a type of cardiovascular disease; the most common kind of hardening of the arteries characterized by the formation of fatty deposits, or plaques, in their inner walls.

Athletic Amenorrhea cessation of menstruation associated with strenuous athletic training.

Atrophic Gastritis an age-related condition characterized by the stomach's inability to produce enough acid, which in turn leads to vitamin B_{12} deficiencies.

Atrophy a decrease in size (for example, of a muscle) in response to disuse.

Avidin a protein found in raw egg whites that can bind biotin and inhibit its absorption.

Bake to cook in an oven surrounded by heat.

Balance a feature of a diet that provides a number of types of foods in balance with one another, such that foods rich in one nutrient do not crowd out of the diet foods that are rich in another nutrient.

Balanced Meal a meal containing sufficient but not excessive amounts of foods from each of the food groups and therefore sufficient but not excessive amounts of carbohydrates, fat, protein, vitamins, and minerals.

Balance Study a laboratory study in which a person is fed a controlled diet and the intake and excretion of a nutrient are measured.

Ballistic Stretches stretches characterized by short, choppy, sometimes painful movements that often pull connective tissues beyond their elastic limits.

Basal Metabolic Rate (BMR) the rate at which the body spends energy to support its basal metabolism. The BMR accounts for the largest component of a person's daily energy (calorie) needs.

Basal Metabolism the sum total of all the chemical activities of the cells necessary to sustain life, exclusive of voluntary activities—that is, the ongoing activities of the cells when the body is at rest.

Bases compounds that accept hydrogens from solutions; bases have a high pH.

Behavior Modification a process developed by psychologists for helping people make lasting behavior changes.

Benzocaine an anesthetic found in gum or candy form that numbs the taste buds and reduces the desire for food.

Beriberi the thiamin deficiency disease, characterized by irregular heartbeat, paralysis, and extreme wasting of muscle tissue.

Beta-Carotene an orange pigment found in plants that is converted into vitamin A inside the body. Beta-carotene is also an antioxidant.

Bialy a flat breakfast roll that is softer than a bagel.

Bicarbonate a common alkaline chemical; a secretion of the pancreas; also, the active ingredient of baking soda.

Bifidus (BIFF-id-us) **Factor** a factor in colostrum and breast milk that favors the growth in the infant's intestinal tract of the "friendly" bacteria *Lactobacillus bifidus* so that other, less desirable intestinal inhabitants will not flourish.

Bile a mixture of compounds, including cholesterol, made by the liver, stored in the gallbladder, and secreted into the small intestine. Bile emulsifies lipids to ready them for enzymatic digestion and helps transport them into the intestinal wall cells.

Binders in foods, chemical compounds that can combine with nutrients (especially minerals) to form complexes the body cannot absorb. Examples of such binders are **phytic** (FIGHT-ic) **acid** and **oxalic** (ox-AL-ic) **acid.**

Bing thin pancakes.

Binge Drinker a person who drinks 4 or more drinks in a short period.

Binge-Eating Disorder an eating disorder characterized by uncontrolled chronic episodes of overeating (compulsive overeating) without other symptoms of eating disorders. Typically, the episodes of binge eating occur at least twice a week on average for a period of six months or more.

Bioelectrical Impedance estimation of body fat content made by measuring how quickly electrical current is conducted through the body.

Biological Value (BV) a measure of protein quality, assessed by determining how well a given food or food mixture supports nitrogen retention.

Biotechnology the science that manipulates biological systems or organisms to modify their products or components or create new products.'

Bisphosphonates drugs that decrease the risk of fractures by acting on the bone-dismantling cells (osteoclasts) and inhibiting their resorption of bone tissue; an example is alendronate (Fosamax).

Black, Cuban, or Turtle Beans medium-size black-skinned ovals that have a rich, sweet taste. They are best served in Mexican and Latin American dishes or thick soups and stews.

Black-Eyed Peas small and oval shaped, creamy white legumes with a black spot. They have a vegetable flavor with mealy texture. Use in salads with rice and greens.

Blood Lipid Profile a test that determines the amounts and kinds of lipids in the blood, normally as part of a diagnosis for cardiovascular disease risk.

Body Composition the proportion of muscle, bone, fat, and other tissue that make up a person's total body weight.

Body Mass Index an index of a person's weight in relation to height which correlates with total body fat content; calculated by dividing the weight of a person by the square of the person's height.

Bok Choy a vegetable with broad, white or greenish-white stalks and dark green leaves; also called Chinese chard.

Bolillo a roll-like bread often used instead of tortillas or to make sandwiches.

Bone Density a measure of bone strength that reflects the degree of bone mineralization. The bone density test compares your bone density to that of a healthy young adult.

Bone Density Tests use a dual beam of low-level X-rays to take a snapshot of bone density in the spine, wrist, and hip. Simpler, less precise tests use ultrasound to measure bone density in the wrist or heel; these tests can be done in a physician's office and may help identify individuals in need of more precise testing.

Braise to cook by browning in fat and then simmering in a covered container with a little liquid.

Bran the fibrous protective covering of a whole grain and the chief source of fiber in grain (removed during refining).

Broil to cook quickly over or under a direct source of intense heat, allowing fats to drip away.

Buffers compounds that help keep a solution's acidity (amount of acid) or alkalinity (amount of base) constant.

Bulimia Nervosa, Bulimarexia (byoo-LEE-me-uh, byoo-lee-ma-REX-ee-uh) binge eating (literally, "eating like an ox"), combined with an intense fear of becoming fat and sometimes followed by self-induced vomiting or the taking of laxatives.

Burritos warm flour tortillas stuffed with a mixture of egg, meat, beans, and/or avocado.

Caffeine a type of compound, called a methylxanthine, found in coffee beans, cola nuts, cocoa beans, and tea leaves. A central nervous system stimulant, caffeine's effects include increasing the heart rate, boosting urine production, and raising the metabolic rate.

Caffeine Dependence Syndrome dependence on caffeine characterized by at least three of the four following criteria: withdrawal symptoms such as headache and fatigue; caffeine consumption despite knowledge that it may be causing harm; repeated, unsuccessful attempts to cut back on caffeine; and tolerance to caffeine.

Calcitonin a hormone used as a drug to decrease the rate of bone loss in osteoporosis; administered as a nasal spray (Miacalcin) or by injection, calcitonin works by inhibiting the bone resorption activity of osteoclasts.

Calorie the unit used to measure energy. Technically, when we see the term *calorie* on food labels or talk about the amount of calories our bodies need, we are referring to *kilocalories (kcal)*—the amount of heat required to raise the temperature of one kilogram of water one degree Celsius. Use of the term *kilocalorie*, however, tends to be reserved for laboratories and technical journals. Throughout this book, we will use the term *calorie* rather than kilocalorie (*calor* means heat).

Calorie Control control of consumption of energy (calories); a feature of a sound diet plan.

Cancer a disease in which cells multiply out of control and disrupt normal functioning of one or more organs.

Carbohydrate Loading a regimen of intense exercise, followed by eating a high-carbohydrate diet, that allows muscles to temporarily store more glycogen than their normal capacity; also called *glycogen loading*.

Carbohydrates compounds made of single sugars or multiples of them and composed of carbon, hydrogen, and oxygen atoms.

Carcinogen a cancer-causing substance.

Cardiovascular Conditioning or **Training Effect** the effect of regular exercise on the cardiovascular system—including improvements in heart, lung, and muscle function and increased blood volume.

Cardiovascular Disease (CVD) disease of the heart and blood vessels. The two most common forms of CVD are atherosclerosis and hypertension.

Carotenoids a group of pigments (yellow, orange, and red) found in plant foods. The most prevalent carotenoids in the diet are beta-carotene (a precursor of vitamin A), alpha-carotene, lycopene, lutein, zeaxanthin, and beta-cryptoxanthin. Carotenoids have a variety of effects in the body including possible antioxidant activity and enhancement of the immune system.

Carrageenan a seaweed derivative used by food manufacturers to add "body" to numerous products, including ice cream, frozen yogurt, and salad dressings.

Cassava a starchy root that is never eaten raw because it must be cooked to eliminate its bitter smell.

Cataracts thickening of the lens of the eye that can lead to blindness.

Cellophane Noodles thin, translucent noodles made from mung beans.

Central Obesity excess fat on the abdomen and around the trunk. Peripheral obesity is excess fat on the arms, thighs, hips, and buttocks.

Challah an egg-containing yeast bread, often braided, and served on the Sabbath and holidays.

Cherimoya a fruit with a rough green outer skin and sherbetlike flesh.

Chilaquiles tortilla casserole often made with eggs or meat.

Chiles Rellenos roasted mild green chili pepper stuffed with cheese, dipped in egg batter, and fried.

Chinese Broccoli a green leafy vegetable often stir-fried; also called Chinese kale.

Chitterlings (chitlins) pig intestine.

Chlorophyll the green pigment of plants that traps energy from sunlight and uses this energy in photosynthesis (the synthesis of carbohydrate by green plants).

Cholesterol (koh-LESS-ter-all) one of the sterols, manufactured in the body for a variety of purposes and also found in animal-derived foods.

Choline a nonessential nutrient used to make the phospholipid lecithin and other molecules.

Chorizo spicy beef or pork sausage.

Choy Sum a bright-green vegetable commonly stir-fried; also called field mustard or Chinese flowering cabbage.

Chylomicron (KIGH-loh-MY-cron) a type of lipoprotein that transports newly digested fat from the intestine through lymph and blood.

Cirrhosis (seer-OH-sis) advanced liver disease, often associated with alcoholism, in which liver cells have died and hardened and have permanently lost their function.

Club Soda artificially carbonated water containing added salts and minerals.

Coenzymes enzyme helpers; small molecules that interact with enzymes and enable them to do their work. Many coenzymes are made from water-soluble vitamins.

Cofactor a mineral element that, like a coenzyme, works with an enzyme to facilitate a chemical reaction.

Collagen the characteristic protein of connective tissue, including scars, ligaments, tendons, and the underlying matrix on which bones and teeth are built.

Colon Cancer cancer of the large intestine (colon), the terminal portion of the digestive tract (see Appendix B).

Colostrum (co-LAHS-trum) a milklike secretion from the breast, rich in protective factors, present during the first day or so after delivery and before milk appears.

Complementary Proteins two or more food proteins whose amino acid assortments complement each other in such a way that the essential amino acids limited in or missing from each are supplied by the others.

Complete Proteins proteins containing all the essential amino acids in the right proportion relative to need. The *quality* of a food protein is judged by the proportions of essential amino acids that it contains relative to our needs. Animal proteins are the highest in quality.

Complex Carbohydrates long chains of sugars (glucose) arranged as starch or fiber. Also called polysaccharides.

Constipation hardness and dryness of bowel movements associated with discomfort in passing them.

Contaminants potentially dangerous substances, such as lead, that can accidentally get into foods.

Contamination Iron iron found in foods as the result of contamination by inorganic iron salts from iron cookware, iron-containing soils, and the like.

Control Group a group of individuals with matching characteristics to the group being treated in an intervention study who receive a sham treatment or no treatment at all.

Correlation a simultaneous change in two factors, such as a decrease in blood pressure with regular aerobic activity (a direct or positive correlation) or the decrease in incidence of bone fractures with increasing calcium intakes (an inverse or negative correlation).

Correspondence School a school from which courses can be taken and degrees granted by mail. Schools that are accredited offer respectable courses and degrees.

Cortical Bone the dense outer ivorylike layer of bone that provides an exterior shell over trabecular bone.

Cretinism (CREE-tin-ism) severe mental and physical retardation of an infant caused by iodine deficiency during pregnancy.

Cross-Contamination the inadvertent transfer of bacteria from one food to another that occurs, for instance, by chopping vegetables on the same cutting board used to skin poultry.

Cross-Reaction the reaction of one antigen with antibodies developed against another antigen.

Cruciferous Vegetables vegetables with cross-shaped blossoms; examples include broccoli, Brussels sprouts, cauliflower, cabbage, rutabagas, and turnips.

Culture knowledge, beliefs, customs, laws, morals, art, and literature acquired by members of a society and passed along to succeeding generations.

Daily Values the amount of fat, sodium, fiber, and other nutrients health experts say should make up a healthful diet. The % Daily Values that appear on food labels tell you the percentage of a nutrient that a serving of the food contributes to a healthful diet.

Degenerative Disease chronic disease characterized by deterioration of body organs as a result of misuse and neglect; poor eating habits, smoking, lack of exercise, and other lifestyle habits often contribute to degenerative diseases, including heart disease, cancer, osteoporosis, and diabetes.

Delaney Clause a provision in the 1958 Food Additives Amendment that prohibited manufacturers from using any substance that was known to cause cancer in animals or humans at any dose level.

Denaturation the change in shape of a protein brought about by heat, alcohol, acids, bases, or other agents. Many well-known poisons are salts of heavy metals such as mercury and silver; these salts alter the structure of proteins wherever they touch them.

Dental Caries decay of the teeth, or cavities.

Dental Plaque a colorless film, consisting of bacteria and their by-products, that is constantly forming on the teeth.

Designer Estrogens (Selective Estrogen Receptor Modulators—SERMs) drugs that act on *estrogen receptors* in osteoblasts to promote an increase in bone mass; an example is raloxifene (Evista). Unlike estrogen, SERMs have little effect on reproductive tissues of the breast or uterus. *Estrogen receptors* are cellular molecules that bind to estrogen, selective estrogen receptor modulators, or phytoestrogens and deliver these compounds to the nucleus of the cell.

Designer Foods foods "fortified" with phytochemicals or plants bred to contain high levels of phytochemicals; also known as "future foods."

DEXA Bone Scan a method to measure bone density that uses small amounts of x-ray radiation. DEXA stands for dual energy x-ray absorptiometry.

Diabetes (dye-uh-BEET-eez) a disorder (technically termed *diabetes mellitus*) characterized by elevated blood glucose and insufficiency or relative ineffectiveness of insulin, which renders a person unable to regulate the blood glucose level normally.

Diastolic Pressure the second figure in a blood pressure reading, which reflects the arterial pressure when the heart is between beats.

Dietary Folate Equivalent a unit of measure that mathematically equalizes the difference in absorption between less absorbable food folate and highly absorbable synthetic folate added to enriched foods and found in supplements (see Appendix D).

Dietary Reference Intakes (DRI) a set of four lists of reference values for nutrients that can be used for planning and assessing diets for healthy populations. DRI are guides for meeting the daily nutritional needs of virtually all healthy people in a specific age and gender group and include the Recommended Dietary Allowances (RDA), Estimated Average Requirements (EAR), Adequate Intakes (AI), and Tolerable Upper Intake Levels (UL).

Dim Sum steamed or fried dumplings stuffed with pork, shrimp, beef, sweet paste, or preserves and steamed or fried.

Dipeptides (dye-PEP-tides) protein fragments two amino acids long. A peptide is a strand of amino acids.

Diploma Mill a correspondence school that grinds out degrees—sometimes worth no more than the cost of the paper they are printed on—the way a grain mill grinds out flour.

Disability any restriction on or impairment in performing an activity in the manner or within the range considered normal for a human being.

Disaccharides pairs of single sugars linked together (*di* means two).

Disordered Eating eating food as an outlet for emotional stress rather than in response to internal physiological cues.

Diuretics (dye-you-RET-ics) medications causing increased water excretion.

Diverticulosis (dye-ver-tic-you-LOCE-iss) outpocketings of weakened areas of the intestinal wall, like blowouts in a tire, that can rupture, causing dangerous infections.

Drink a dose of any alcoholic beverage that delivers one-half ounce of pure ethanol: 5 ounces of wine, 12 ounces of beer, 1.5 ounces of hard liquor (whiskey, gin, rum, or vodka).

Drugs substances that can modify one or more of the body's functions.

Dysentery (DISS-en-terry) an infection of the digestive tract that causes diarrhea.

Eating Disorder general term for several conditions (anorexia nervosa, bulimia nervosa, binge-eating disorder) that exhibit an excessive preoccupation with body weight, a fear of body fatness, and a distorted body image.

Eclampsia a severe extension of preeclampsia characterized by convulsions.

Edema (eh-DEEM-uh) swelling of body tissue caused by leakage of fluid from the blood vessels, seen in (among other conditions) protein deficiency.

Electrolytes compounds that partially dissociate in water to form ions; examples are sodium, potassium, and chloride.

Empty-Calorie Foods a phrase used to indicate that a food supplies calories but negligible nutrients.

Emulsifier a substance that mixes with both fat and water and can break fat globules into small droplets, thereby suspending fat in water.

Endosperm the bulk of the edible part of a grain; contains starch grains embedded in a protein matrix.

Endurance the ability to sustain an effort for a long time. One type, muscle endurance, is the ability of a muscle to contract repeatedly within a given time without becoming exhausted. Another type, cardiovascular endurance, is the ability of the cardiovascular system to sustain effort over a period of time.

Energy the capacity to do work, such as moving or heating something.

Enriched refers to a process by which the B vitamins thiamin, riboflavin, niacin, folic acid, and the mineral iron are added to refined grains and grain products at levels specified by law. Enrichment refers only to refined grain products such as wheat-flour products, cornmeal, grits, and polished rice.

Enterotoxin a toxic compound, produced by microorganisms, that harms the gastrointestinal tract.

Enzymes protein catalysts. A catalyst facilitates a chemical reaction without itself being altered in the process.

Eosinophilia-Myalgia (ee-o-sin-o-FIL-ia my-AL-jia) Syndrome (EMS) a disease characterized primarily by a high level of eosinophils, a type of white blood cell, as well as myalgia—that is, muscle pain and weakness.

EPA, DHA eicosapentaenoic (EYE-cossa-PENTA-ee-NO-ick) acid, docosahexaenoic (DOE-cossa-HEXA-ee-NO-ick) acid; omega-3 fatty acids made from linolenic acid in the tissues of fish.

Epidemiological Study a study of a population to search for possible correlations between nutrition factors and health patterns over time.

Epithelial (ep-ih-THEE-lee-ul) **Tissue** those cells that form the outer surface of the body and line the body cavities and the principal passageways leading to the exterior. Examples include the cornea, digestive tract lining, respiratory tract lining, and skin. The epithelial cells produce mucus to protect these tissues from bacteria and other potentially harmful substances. Without this mucus, infections become more likely.

Ergogenic Aids anything that helps to increase the capacity to work or exercise.

Essential Amino Acids amino acids that cannot be synthesized by the body or that cannot be synthesized in amounts sufficient to meet physiological need.

Essential Fatty Acid a fatty acid that cannot be synthesized in the body in amounts sufficient to meet physiological need; linoleic acid and linolenic acid are essential fatty acids.

Essential Nutrients nutrients that must be obtained from food because the body cannot make them for itself.

Estimated Average Requirement (EAR) the average daily intake that meets the estimated nutrient needs of half of the individuals of a specific age and gender.

Estrogen a major female hormone—important in connection with nutrition because it maintains calcium balance and because its secretion abruptly declines at menopause.

Estrogen Replacement Therapy (ERT) administration of estrogen to replace the natural hormone that declines with menopause. Since ERT may increase the risk of uterine and breast cancer, *hormone replacement therapy*—the administration of a combination of estrogen with the hormone, progesterone—is often used. The combination of hormones lowers the risk of cancer.

Ethnic Cuisine the traditional foods eaten by the people of a particular culture.

Euphoria (you-FORE-ee-uh) a feeling of great well-being that people often seek through the use of drugs such as alcohol.

Exchange Lists lists of foods with portion sizes specified; the foods on a single list are similar with respect to nutrient and calorie content and so can be mixed and matched in the diet (see Appendix E).

Exercise Stress Test a test that monitors heart function during exercise to detect abnormalities that may not show up under ordinary conditions; exercise physiologists and trained physicians or health-care professionals can administer the test.

Experimental Group the participants in a study who receive the real treatment or intervention under investigation.

External Cue Theory the theory that some people eat in response to such external factors as the presence of food or the time of day rather than to such internal factors as hunger.

Famine widespread lack of access to food caused by natural disasters, political factors, or war; characterized by a large number of deaths due to starvation and malnutrition.

Fat Cell Theory states that during the growing years, fat cells respond to overfeeding by producing additional fat cells (**hyperplastic obesity**); the number of fat cells eventually becomes fixed, and overfeeding from this point on causes the body to enlarge existing fat cells (**hypertrophic obesity**). Hypertropic obesity is the more common type and is usually seen in adults.

Fats lipids that are solid at room temperature.

Fatty Acids basic units of fat composed of chains of carbon atoms with an acid group at one end and hydrogen atoms attached all along their length.

Fatty Liver an early stage of liver disease seen in several conditions (kwashiorkor, alcoholic liver disease), characterized by accumulation of fat in the liver cells.

Female Athlete Triad a condition characterized by disordered eating, lack of menstrual periods, and low bone density.

Fetal Alcohol Effect (FAE) abnormalities from prenatal alcohol exposure, not sufficient for a diagnosis of fetal alcohol syndrome, but physically or mentally impairing to the child.

Fetal Alcohol Syndrome (FAS) the cluster of symptoms seen in an infant or child whose mother consumed excess alcohol during pregnancy, including retarded growth, impaired development of the central nervous system, and facial malformations.

Fibers the indigestible residues of food, composed mostly of polysaccharides. The term *dietary fiber* refers to the fiber that resists human digestive enzymes. The best known of the fibers are **cellulose, hemicellulose, pectin,** and **gums.**

First Amendment the amendment to the U.S. Constitution that guarantees freedom of the press.

Fitness the body's ability to meet physical demands, composed of four components: flexibility, strength, muscle endurance, and cardiovascular endurance.

Flexibility the ability to bend or extend without injury; flexibility depends on the elasticity of the muscles, tendons, and ligaments and on the condition of the joints.

Fluid Balance distribution of fluid among body compartments.

Fluorosis (floor-OH-sis) discoloration of the teeth from ingestion of too much fluoride during tooth development.

Foam Cells cells from the immune system containing scavenged oxidized LDL-cholesterol that are thought to initiate arterial plaque formation.

Food Additive any substance added to food, including substances used in the production, processing, treatment, packaging, transportation, or storage of food.

Food Allergen a substance in food—usually a protein—that is seen by the body as harmful and causes the immune system to mount an allergic reaction.

Food Allergy an adverse reaction to an otherwise harmless substance that involves the body's immune system.

Food Aversion a strong desire to avoid a particular food.

Food Banks nonprofit community organizations that collect surplus commodities from the government and edible but often unmarketable foods from private industry for use by nonprofit charities, institutions, and feeding programs at nominal cost.

Food Composition Tables tables that list the nutrient profile of commonly eaten foods.

Food Group Plan a diet-planning tool, such as the Food Guide Pyramid, that groups foods according to similar origin and nutrient content and then specifies the number of foods from each group that a person should eat.

Food Insecurity the inability to acquire or consume an adequate quality or sufficient quantity of food in socially acceptable ways, or the uncertainty that one will be able to do so.

Food Intolerance a general term for any adverse reaction to a food or food component that does not involve the body's immune system.

Food Intoxication illness caused by eating food that contains a harmful toxin.

Food Pantries centers usually attached to existing nonprofit agencies that distribute bags or boxes of groceries to people experiencing food emergencies. Foods distributed by pantries are prepared and consumed elsewhere. A referral or proof of need is often required. There are roughly two food pantries to every soup kitchen.

Food Recovery such activities as salvaging perishable produce from grocery stores and wholesale food markets; rescuing surplus prepared food from restaurants, corporate cafeterias, and caterers; and collecting non-perishable, canned or boxed processed food from manufacturers, supermarkets, or people's homes. The items recovered are donated to hungry people.

Food Security access by all people at all times to enough food for an active and healthy life. Food security has two aspects: ensuring that adequate food supplies are available and ensuring that households whose members suffer from undernutrition have the ability to acquire food, either by producing it themselves or by being able to purchase it.

Foodborne Illness or **Food Poisoning** illness occurring as a result of eating food contaminated with disease-producing microorganisms, such as bacteria, viruses, or parasites, or toxic substances such as environmental pollutants.

Foodborne Infection illness caused by eating a food containing bacteria or other microorganisms capable of growing and thriving in a person's tissues.

Fortified Foods foods to which nutrients have been added. Typically, commonly eaten foods are chosen for fortification with added nutrients to help prevent a deficiency of a nutrient (iodized salt, milk with vitamin D) or to reduce the risk of chronic disease (juices with added calcium).

Frame Size the size of a person's bones and musculature. A person with a large frame can weigh more than one the same height with a small frame without increased risks.

Free Radicals highly toxic compounds created in the body as a result of chemical reactions that involve oxygen. Environmental pollutants such as cigarette smoke and ozone also prompt the formation of free radicals.

From a Community Water System or From a Municipal Source statement that must appear on bottles containing water derived from a municipal water supply. The phrase must conspicuously precede or follow the name of the brand.

Fructose (FROOK-toce) fruit sugar—the sweetest of the single sugars. Another single sugar, **galactose** (ga-LACK-toce), occurs bonded to glucose in the sugar of milk.

Functional Food a general term for foods that provide an *additional* physiological or psychological benefit beyond that of meeting basic nutritional needs. Also called *medical foods.*

Fusion Cuisine a term used to describe food that combines the elements of two or more cuisines—say, European and Oriental—to create a new one.

Galactose a monosaccharide; part of the disaccharide, lactose (milk sugar).

Garbanzo Beans or **Chick-Peas** large, round, and tan colored legumes. They have a nutty flavor and crunchy texture. Use in soups and stews and puréed for dips.

Gastroplasty surgery on the stomach (also called stomach stapling) that reduces its volume to less than 2 ounces (the size of a shot glass) to prevent overeating.

Gelfilte Fish a chopped fish mixture often made with pike and whitefish as well as matzoh crumbs, eggs, and seasonings.

Genetic Engineering the process of altering the genes of a plant in an effort to create a new plant with different traits. This process of recombining genes is also known as recombinant DNA; a form of biotechnology.

Germ the nutrient-rich and fat-dense inner part of a whole grain (removed during refining).

Gestational Diabetes the appearance of abnormal glucose tolerance during pregnancy, with a return to normal following pregnancy.

Gleaning the harvesting of excess food from farms, orchards, and packing houses to feed the hungry.

Glucagon (GLUE-cuh-gon) a hormone released by the pancreas that signals the liver to release glucose into the bloodstream.

Glucose (GLOO-koce) the building block of carbohydrate; a single sugar used in both plant and animal tissues as quick-energy.

Glutinous Rice short-grained, opaque, white rice that turns sticky when cooked.

Glycemic Effect a measure of the extent to which a food raises the blood glucose level and elicits an insulin response as compared with pure glucose; also referred to as the *glycemic index* of a food.

Glycerol (GLISS-er-all) an organic compound, three carbons long, that serves as the backbone for triglycerides.

Glycogen (GLY-co-gen) a polysaccharide composed of chains of glucose, manufactured in the body and stored in liver and muscle. As a storage form of glucose, liver glycogen can be broken down by the liver to maintain a constant blood glucose level when carbohydrate intake is inadequate.

GOBI an acronym formed from the elements of UNICEF's Child Survival campaign—**G**rowth charts, **O**ral rehydration therapy, **B**reast milk, and **I**mmunization.

Goiter (GOY-ter) enlargement of the thyroid gland caused by iodine deficiency.

GRAS (Generally Recognized as Safe) List a list of ingredients, established by the FDA, that had long been in use and were believed safe. The list is subject to revision as new facts become known.

Grazing eating small amounts of food at intervals throughout the day rather than—or in addition to—eating regular meals.

Great Northern Beans medium white and kidney shaped beans. Enjoy the delicate flavor and firm texture in salads, soups, and main dishes.

Green Soybeans (Edamame) these large soybeans are harvested when the beans are still green and sweet and can be served as a snack or a main vegetable dish, after boiling in water for 15 to 20 minutes. They are high in protein and fiber and contain no cholesterol. Edamame is more often found in Asian and natural food stores, shelled, or still in the pod.

Grits coarsely ground cornmeal.

Ground Water water that comes from an underground body of water that does not come into contact with any surface water.

Guava a sweet juicy fruit with green or yellow skin and red or yellow flesh.

Hard Water water with a high concentration of minerals such as calcium and magnesium.

Hazard state of danger; used to refer to any circumstance in which harm is possible.

HDL (high-density lipoprotein) carries cholesterol in the blood back to the liver for recycling or disposal.

Health Claim a statement on the food label linking the nutritional profile of a food to a reduced risk of a particular disease, such as osteoporosis or cancer. Manufacturers must adhere to strict government guidelines when making such claims.

Health Fraud conscious deceit practiced for profit, such as the promotion of a false or an unproven product or therapy.

Heat Stroke an acute and dangerous reaction to heat buildup in the body, requiring emergency medical attention; also called *sun stroke.*

Heavy Metals any of a number of mineral ions, such as mercury and lead, so named because of their relatively high atomic weight. Many heavy metals are poisonous.

Heme (HEEM) **Iron** the iron-holding part of the hemoglobin protein, found in meat, fish, and poultry. About 40 percent of the iron in meat, fish, and poultry is bound into heme. Meat, fish, and poultry also contain a factor (MFP factor) other than heme that promotes the absorption of iron, even of the iron from other foods eaten at the same time as the meat.

Hemoglobin (HEEM-oh-globe-in) the oxygen-carrying protein of the blood; found in the red blood cells.

Hemorrhoids (HEM-or-oids) swollen, hardened (varicose) veins in the rectum, usually caused by the pressure resulting from constipation.

Histamine a substance released by cells of the immune system during an allergic reaction to an antigen, causing inflammation, itching, hives, dilation of blood vessels, and a drop in blood pressure.

Hominy hulled, dried corn kernels with certain parts removed.

Homocysteine (ho-mo-SIS-teen) a chemical that appears to be toxic to the blood vessels of the heart. High blood levels of homocysteine have been associated with low blood levels of vitamin B_{12}, vitamin B_6, and folate.

Hormones chemical messengers. Hormones are secreted by a variety of glands in the body in response to altered conditions. Each affects one or more target tissues or organs and elicits specific responses to restore normal conditions.

Hunger the physiological drive to find and eat food, experienced as an unpleasant sensation.

Husk the outer, inedible covering of a grain.

Hydrogenation (high-droh-gen-AY-shun) the process of adding hydrogen to unsaturated fat to make it more solid and more resistant to chemical change.

Hydrolyzed Vegetable Protein (HVP) a protein obtained from any vegetable, including soybeans. The protein is broken down into amino acids by a chemical process called acid hydrolysis. HVP is a flavor enhancer that can be used in soups, broths, sauces, gravies, flavoring and spice blends, canned and frozen vegetables, and meats and poultry.

Hydrostatic Weighing or **Underwater Weighing** a weighing method in which the less a person weighs under water compared to the person's out-of-water weight, the greater the proportion of body fat (fat is less dense or more buoyant than lean tissue).

Hyperglycemia an abnormally high blood glucose concentration, often a symptom of diabetes.

Hypertension sustained high blood pressure.

Hypertrophy an increase in size in response to use.

Hypoglycemia (HIGH-po-gligh-SEEM-ee-uh) an abnormally low blood glucose concentration—below 60 to 70 mg/100 ml.

Hypothalamus (high-poh-THALL-ah-mus) a part of the brain that senses a variety of conditions in the blood, such as temperature, salt content, and glucose content, and then signals other parts of the brain or body to change those conditions when necessary.

Immunity specific disease resistance derived from the immune system's memory of prior exposure to specific disease agents and its ability to mount a swift response against them.

Incidental Food Additives (or indirect additives) substances that accidentally get into food as a result of contact with it during growing, processing, packaging, storing, or some other stage before the food is consumed.

Incomplete Protein a protein lacking or low in one or more of the essential amino acids.

Ingredients List a listing of the ingredients in a food, with items listed in descending order of predominance by weight. All food labels are required to bear an ingredients list.

Inorganic being or composed of matter other than plant or animal.

Insoluble Fiber includes the fiber types called cellulose, hemicellulose, and lignin; insoluble fibers do not dissolve in water.

Insulin a hormone secreted by the pancreas in response to high blood glucose levels; it assists cells in drawing glucose from the blood.

Insulin Resistance Syndrome a combination of four risk factors—diabetes, obesity, high blood pressure, and high blood cholesterol—that increase a person's risk of developing cardiovascular disease; also called *Syndrome X*.

Integrated Pest Management the use of biological controls, crop rotation, genetic engineering, and other tactics to reduce chemical use in the growing of crops.

Intentional Food Additives substances intentionally added to food. Examples include nutrients, colors, spices, and herbs.

Internet a network of millions of computers connected to universities, government agencies, and commercial and nonprofit organizations around the world and linked together to form a "mega-network."

Intervention Study a population study examining the effects of a treatment on experimental subjects compared to a control group.

Intestinal Flora the normal bacterial inhabitants of the digestive tract.

Intrinsic Factor a compound made in the stomach that is necessary for the body's absorption of vitamin B_{12}.

Ions (EYE-ons) electrically charged particles, such as sodium (positively charged) and chloride (negatively charged).

Iron-Deficiency Anemia a reduction of the number and size of red blood cells and a loss of their color because of iron deficiency.

Iron Overload a condition in which the body contains more iron than it needs or can handle; excess iron is toxic and can damage the liver. The most common cause of iron overload is the genetic disorder hemochromatosis.

Irradiation the process of exposing a substance to low doses of radiation, using gamma rays, X-rays, or electricity (electron beams) to kill insects, bacteria, and other potentially harmful microorganisms.

Isoflavones compounds found in many fruits, vegetables, and soy-based foods that are thought to play a role in fighting breast cancer by blocking the action of the hormone estrogen.

Jicama a crisp, bean root vegetable that is tan outside and white inside and is always eaten raw; jicama is as popular in Mexico as the potato is in the United States.

Jujube Chinese date.

Kasha cracked buckwheat, barley, millet, or wheat that is served as a cooked cereal or potato substitute.

Kashrut biblical ordinances regarding which foods are fit to eat.

Ketosis abnormal amounts of ketone bodies in the blood and urine; ketone bodies are produced from the incomplete breakdown of fat when glucose is unavailable for the brain and nerve cells.

Kidney Beans large, red, and kidney shaped beans (the white variety is called cannellini). They have a bland taste and soft texture but tough skins. Use in chili, bean stews, and Mexican dishes for red; Italian dishes for white.

Knish a potato pastry filled with ground meat, potato, or kasha.

Kosher fit, proper, or in accordance with religious law.

Kwashiorkor (kwash-ee-OR-core) a deficiency disease caused by inadequate protein in the presence of adequate food energy.

L-tryptophan an essential amino acid that has been sold in tablets, capsules, and powders as a dietary supplement.

Lactoferrin (lak-toe-FERR-in) a factor in breast milk that binds and helps absorb iron and keeps it from supporting the growth of the infant's intestinal bacteria.

Lactose a double sugar composed of glucose and galactose; commonly known as milk sugar.

Lactose Intolerance inability to digest lactose as a result of a lack of the necessary enzyme lactase. Symptoms include nausea, abdominal pain, diarrhea, or excessive gas that occurs anywhere from 15 minutes to a couple of hours after consuming milk or milk products.

LDL (low-density lipoprotein) carries cholesterol (much of it synthesized in the liver) to body cells. A high blood cholesterol level usually reflects high LDL.

Lecithin (LESS-ih-thin) a phospholipid, a major constituent of cell membranes, manufactured by the liver, and also found in many foods.

Legumes (leg-GYOOMS) plants of the bean and pea family having roots with nodules that contain bacteria that can trap nitrogen from the air in the soil and make it into compounds that become part of the seed. The seeds are rich in high-quality protein compared with those of most other plant foods.

Lentils these legumes are small, flat, and round. Usually brown colored, lentils also can be green, pink, or red. They have a mild taste with firm texture. Best used when combined with grains or vegetables in salads, soups, or stews.

Leptin an appetite-suppressing hormone produced in the fat cells that communicates information about the body's fat content to the brain (*leptos* means slender).

Life Expectancy the average number of years lived by people in a given society.

Life Span the maximum number of years of life attainable by a member of a given species.

Lifestyle Diseases conditions that may be aggravated by modern lifestyles that include too little exercise, poor diets, and excessive drinking and smoking. Lifestyle diseases are also referred to as diseases of affluence.

Lima or **Butter Beans** limas are soft and mealy in texture. They are flat, oval shaped, and white tinged with green. The smaller variety has a milder taste. Use in soups and stews.

Limiting Amino Acid a term given to the essential amino acid in shortest supply (relative to the body's need) in a food protein; it therefore *limits* the body's ability to make its own proteins.

Linoleic (lin-oh-LAY-ic) **Acid, Linolenic** (lin-oh-LEN-ic) **Acid** polyunsaturated fatty acids, essential for human beings.

Lipids a family of compounds that includes triglycerides (fats and oils), phospholipids (lecithin), and sterols (cholesterol).

Lipoprotein Lipase (LPL) an enzyme located on the surfaces of fat cells that enables the cell to convert blood triglycerides into fatty acids and glycerol to be pulled into the cell for reassembly and storage as body fat.

Lipoproteins (LIP-oh-PRO-teens) clusters of lipids associated with protein that serve as transport vehicles for lipids in blood and lymph. The four main types of lipoproteins are chylomicrons, VLDL, LDL and HDL.

Liposuction a type of surgery (also called lipectomy) that vacuums out fat cells that have accumulated, typically in the buttocks and thighs. If the person continues to eat more calories than are expended through physical activity, fat will return to the fat cells that remain in those regions.

Litchi small, round fruits with orange-red skin and opaque, white flesh; also called litchee or lychee.

Longan a small, round fruit with smooth brown skin and clear pulp.

Low Birthweight (LBW) a birthweight of 5½ lb (2,500 g) or less, used as a predictor of poor health in the newborn and as a probable indicator of poor nutrition status of the mother during and/or before pregnancy. Normal birthweight for a full-term baby is 6½ to 8¾ lb (about 3,000 to 4,000 g). LBW infants are of two different types. Some are premature (they are born early). Others have suffered growth failure in the uterus; they may or may not be born early, but they are small.

Lox smoked salmon.

Lymph (LIMF) the body fluid that transports the products of fat digestion toward the heart and eventually drains back into the bloodstream; lymph consists of the same components as blood with the exception of red blood cells.

Macular Degeneration a progressive loss of function of the part of the retina (macula) that is necessary for focused vision; often leads to blindness.

Mad Cow Disease (Bovine Spongiform Encephalopathy or BSE) a rare and fatal degenerative disease first diagnosed in 1986 in cattle in the United Kingdom. The bovine disease may be passed to humans who eat the meat of infected animals and may lead to death due to brain and nerve damage.

Major Mineral an essential mineral nutrient found in the human body in amounts greater than 5 grams.

Malnutrition the impairment of health resulting from a relative deficiency or excess of food energy and specific nutrients necessary for health.

Maltose a double sugar composed of two glucose units.

Mantou steamed bread.

Marasmus (ma-RAZ-mus) an energy-deficiency disease; starvation.

Margin of Safety from a food safety standpoint, the margin is a zone between the maximum amount of a substance that appears to be safe and the amount allowed in the food supply.

Matzoh a crackerlike bread eaten most often at Passover.

Meat Alternatives meat alternatives made from soybeans contain soy protein or tofu and other ingredients mixed together to simulate various kinds of meat. These meat alternatives are sold as frozen, canned, or dried foods.

Meat Replacements textured vegetable protein products formulated to look and taste like meat, fish, or poultry. Many of these are designed to match the known nutrient contents of animal protein foods.

Megadose a dose of ten or more times the amount normally recommended. An overdose is an amount high enough to cause toxicity symptoms. Megadoses taken over a long period often result in an overdose.

Menopause the time of life at which a woman's menstrual cycle ceases, usually at about 45 to 50 years of age.

MEOS (Microsomal Ethanol-Oxidizing System) a system of enzymes in the liver that oxidize not only alcohol but also several classes of drugs. (The **microsomes** are tiny particles of membranes with associated enzymes that can be collected from broken-up cells.)

Metabolism the sum of all physical and chemical reactions taking place in living cells; includes reactions in which the cells derive energy (ATP) from the energy nutrients, and reactions in which the cells synthesize body compounds (glycogen, body fat, body proteins).

Milk Allergy the most common food allergy; caused by the protein in raw milk. Milk allergy is sometimes overcome by cooking the milk to denature the protein; it is sometimes alleviated by an abstinence from and a gradual reintroduction to milk.

Milk Anemia iron-deficiency anemia caused by drinking so much milk that iron-rich foods are displaced from the diet.

Mineral Water water that is drawn from an underground source and that contains at least 250 parts per million of dissolved solids.

Minerals small, naturally occurring, inorganic, chemical elements; the minerals serve as structural components and in many vital processes in the body.

Miso a rich, salty condiment that characterizes the essence of Japanese cooking. The Japanese make miso soup and use miso to flavor a variety of foods. A smooth paste, miso is made from soybeans and a grain such as rice, plus salt and a mold culture, and then aged in cedar vats for one to three years.

Moderation the attribute of a diet that provides no unwanted constituent in excess.

Monoglyceride (mon-oh-GLISS-er-ide) a glycerol molecule with one fatty acid attached to it. A *diglyceride* is a glycerol molecule with two fatty acids attached to it.

Monosaccharide a single sugar (*mono* means one).

Monounsaturated Fatty Acid a fatty acid containing one point of unsaturation, found mostly in vegetable oils such as olive, canola, and peanut.

Multinational Corporations international companies with direct investments and/or operative facilities in more than one country. U.S. oil and food companies are examples.

Myoglobin the oxygen carrying protein of the muscles (*myo* means muscle).

Narcotic (nar-KOT-ic) any drug that dulls the senses, induces sleep, and becomes addictive with prolonged use.

Neural Tube Defects include any of a number of birth defects in the orderly formation of the neural tube during early gestation. Both the brain and the spinal cord develop from the neural tube; defects result in various central nervous system disorders. The two main types are *spina bifida* (incomplete closure of the bony casing around the spinal cord) and *anencephaly* (a partially or completely missing brain).

Neurotoxin a poisonous compound that disrupts the nervous system.

Neurotransmitters chemicals that are released at the end of a nerve cell when a nerve impulse arrives there; they diffuse across the gap to the next nerve cell and alter the membrane of that second cell to either excite it or inhibit it.

Niacin Equivalents (NEs) the amount of niacin present in food, including the niacin that can theoretically be made from tryptophan contained in the food.

Night Blindness slow recovery of vision following flashes of bright light at night; an early symptom of vitamin A deficiency.

Nitrogen Balance the amount of nitrogen consumed compared with the amount of nitrogen excreted in a given time period.

Nondairy Soy Frozen Dessert Nondairy frozen desserts are made from soy milk or soy yogurt. Soy ice cream is one of the most popular desserts made from soybeans and can be found in many grocery stores.

Nonheme Iron the iron found in plant foods.

Nursing Bottle Syndrome (also called baby bottle tooth decay) decay of all the upper and sometimes the back lower teeth that occurs in infants given carbohydrate-containing liquids when they sleep. The syndrome can also develop in babies given bottles of liquid to carry around and sip all day.

Nutraceuticals a term without any legal or scientific meaning, but used to refer to foods, nutrients, phytochemicals, or dietary supplements believed to have medicinal attributes. The term is sometimes used to extend the credibility of known nutrients to a wide group of substances which have, to date, scientifically unproven effects.

Nutrient Content Claims claims such as "low-fat" and "low-calorie" used on food labels to help consumers who don't want to scrutinize the Nutrition Facts panel get an idea of a food's nutritional profile. These claims must adhere to specific definitions set forth by the Food and Drug Administration.

Nutrient Dense refers to a food that supplies large amounts of nutrients relative to the number of calories it contains. The higher the level of nutrients and the fewer the number of calories, the more nutrient dense the food is.

Nutrients substances obtained from food and used in the body to promote growth, maintenance, and repair.

Nutrition Facts Panel a detailed breakdown of the nutritional content of a serving of a food that must appear on virtually all packaged foods sold in the United States.

Nutrition the study of foods, their nutrients and other chemical components, their actions and interactions in the body, and their influence on health and disease.

Nutritional Yeast a fortified food supplement containing B vitamins, iron, and protein that can be used to improve the quality of a vegetarian diet.

Nutritionist a person who claims to be capable of advising people about their diets. Some nutritionists are registered dietitians, whereas others are self-described experts whose training is questionable.

Obesity body weight high enough above normal weight to constitute a health hazard; conventionally defined as weight 20 percent or more above the desirable weight for height, or a BMI of 30.0 or greater.

Oils lipids that are liquid at normal room temperature.

Olestra an artificial fat derived from vegetable oils and sugar combined in such a way that the body cannot break them down. Sold under the brand name Olean®, olestra does not contribute calories to food. It can, however, prevent absorption of some nutrients. Thus, the FDA requires all products made with it to bear this warning: "This Product Contains Olestra. Olestra may cause abdominal cramping and loose stools. Olestra inhibits the absorption of some vitamins and nutrients. Vitamins A, D, E, and K have been added."

Omega the last letter—the far end—of the Greek alphabet (ω), used by chemists to refer to the position of the end-most double bond in a fatty acid. The **omega-6** fatty acids have their end-most double bonds after the sixth carbon in the chain; the **omega-3** acids, after the third.

Opportunistic Infections infections produced by organisms that do not affect people whose immune systems are working normally. An example is the unusual form of pneumonia caused by *Pneumocystis carinii*, often seen in individuals with AIDS.

Oral Rehydration Therapy (ORT) the treatment of dehydration (usually due to diarrhea caused by infectious disease) with an oral solution; ORT as developed by UNICEF is intended to enable a mother to mix a simple solution for her child from substances that she has at home.

Organic of, related to, or containing carbon compounds.

Organic Halogens compounds that contain one or more of a class of atoms called halogens, including fluorine, chlorine, iodine, or bromine.

Organically Grown Foods crops or livestock grown and processed according to USDA regulations concerning use of pesticides, herbicides, fungicides, fertilizers, preservatives, other synthetic chemicals, growth hormones, antibiotics, or other drugs (see www.ams.usda.gov/nop); does not include foods that have been genetically engineered or treated with irradiation.

Oriental Radish large, cylindrically shaped vegetables with smooth skin; also called daikon.

Osteoblast a bone-building cell; responsible for formation of bone.

Osteoclast a bone-destroying cell; responsible for resorption and removal of bone.

Osteomalacia (os-tee-o-mal-AY-shuh) the disease resulting from vitamin D deficiency in adults. (Its counterpart in children is called rickets.) Osteomalacia is characterized by bowed legs and a curved spine.

Osteoporosis (OSS-tee-oh-pore-OH-sis) also known as adult bone loss; a disease in which the bones become porous and fragile.

Overload an extra physical demand placed on the body. A principle of training is that for a body system to improve, its workload must be increased by increments over time.

Overnutrition calorie or nutrient overconsumption severe enough to cause disease or increased risk of disease; a form of malnutrition.

Overweight conventionally defined as weight between 10 percent and 20 percent above the desirable weight for height, or a body mass index (BMI) of 25.0 through 29.9.

Oxidation interaction of a compound with oxygen; a damaging effect by a chemically reactive and unstable form of oxygen.

Oxidized LDL-Cholesterol (o-LDL) the cholesterol in LDLs that is attacked by reactive oxygen molecules inside the walls of the arteries; o-LDL is taken up by scavenger cells and deposited in plaque.

Pasteurization the process of sterilizing food via heat treatment.

Peak Bone Mass the highest bone density achieved for an individual—accumulated over the first three decades of life; typically occurs by 30 years of age. After age 30, bone resorption slowly begins to exceed bone formation.

Pellagra (pell-AY-gra) niacin deficiency characterized by diarrhea, inflammation of the skin, and, in severe cases, mental disorders and death.

Peptide Bond a bond that connects one amino acid with another.

Periodontal Disease inflammation or degeneration of the tissues that surround and support the teeth.

Peristalsis the wavelike muscular squeezing of the esophagus, stomach, and small intestine that pushes their food contents along.

Pesticides chemicals applied intentionally to plants, including foods, to prevent or eliminate pest damage. Pests include all living organisms that destroy or spoil foods: bacteria, molds and fungi, insects, and rats and other rodents, to name a few.

pH the concentration of hydrogen ions. The lower the pH, the stronger the acid; pH 2 is a strong acid; pH 7 is neutral; and a pH above 7 is alkaline.

Phenylketonuria an inborn error of metabolism, detectable at birth, in which the body lacks the enzyme needed to convert the amino acid phenylalanine to the amino acid tyrosine. As a result, derivatives of phenylalanine accumulate in the blood and tissues, where they can cause severe damage, including mental retardation.

Phenylpropanolamine Hydrochloride (PPA) a stimulant of the central nervous system available in over–the–counter weight-loss products used to suppress appetite.

Phospholipids (FOSS-foh-LIP-ids) one of the three main classes of lipids; a lipid similar to a triglyceride but containing phosphorus.

Photosynthesis a process in which plants use the green pigment chlorophyll to trap the energy of the sun and produce glucose from carbon dioxide and water.

Phytochemicals (FIGH-toe-CHEM-icals) physiologically active compounds found in plants that are not nutrients but that appear to help promote health and reduce risk for cancer, heart disease, and other conditions.

Pica the craving of nonfood items such as clay, ice, and laundry starch. Pica does not appear to be limited to any particular geographic area, race, sex, culture, or social status.

Pigment a molecule capable of absorbing certain wavelengths of light. Pigments in the eye permit us to perceive different colors.

Pinto Beans medium oval legumes that are mottled beige and brown with an earthy flavor. They are most often used in Mexican dishes, such as refried beans, stews, or dips.

Placebo a sham or neutral treatment given to a control group; an inert, harmless "treatment" that the group's members cannot recognize as different from the real thing, which minimizes the chance that a result of the treatment will appear to have occurred due to the **placebo effect**—the healing effect that the belief in the treatment, rather than the treatment itself, often has.

Placebo Effect an improvement in a person's sense of well-being or physical health in response to the use of a placebo (a substance having no medicinal properties or medicinal effects).

Placenta (pla-SEN-tuh) the organ inside the uterus in which the mother's and fetus's circulatory systems intertwine and in which exchange of materials between maternal and fetal blood takes place. The fetus receives nutrients and oxygen across the placenta; the mother's blood picks up carbon dioxide and other waste materials to be excreted via her lungs and kidneys.

Plantain a greenish, starchy banana; because it is starchy even when ripe, it is never eaten raw and is usually pan-fried.

Poach to cook foods (fish, an egg without its shell, etc.) in liquid such as water, wine, juice, or bouillon near the boiling point.

Point of Unsaturation a site in a molecule where the bonding is such that additional hydrogen atoms can easily be added.

Polypharmacy the taking of three or more medications regularly; occurs in one-third of those over 65 years.

Polyunsaturated Fatty Acid (sometimes abbreviated PUFA) a fatty acid in which two or more points of unsaturation occur, found in nuts and vegetable oils such as safflower, sunflower, and soybean, and in fatty fish.

Postnatal after birth.

Poverty the state of having too little money to meet minimum needs for food, clothing, and shelter. The U.S. Department of Health and Human Services defines the poverty level in the United States as an annual income of $16,450 for a family of four.

Precursor a compound that can be converted into another compound. For example, beta-carotene is a precursor of vitamin A.

Preeclampsia a condition characterized by hypertension, fluid retention, and protein in the urine.

Preformed Vitamin A vitamin A in its active form.

Pregnancy-Induced Hypertension (PIH) high blood pressure that develops during the second half of pregnancy.

Premenstrual Syndrome (PMS) a cluster of physical, emotional, and psychological symptoms that some women experience seven to ten days before menstruating. Symptoms can include acne, anxiety, food cravings (especially for sweets), back pain, breast tenderness, cramps, depression, fatigue, headaches, irritability, moodiness, water retention, and weight gain. Because a clear-cut treatment for the symptoms of PMS has not been identified, women who suffer from the problem rank as prime targets for unproved nutritional remedies for the condition.

Prenatal prior to birth.

Prepared and Perishable Food Programs (PPFPS) nonprofit programs that help to feed people in need by linking sources of unused, unserved cooked and fresh food—such as caterers, restaurants, hotel kitchens, and cafeterias—with social service agencies that serve meals to people who would otherwise go hungry.

Problem Drinker (or alcohol abuser) a person who experiences psychological, social, family, employment, or school problems because of alcohol. Problem drinkers often binge drink and turn to alcohol when facing problems or making decisions.

Pro-Oxidant a compound that stimulates free radical damage.

Protein Digestibility-Corrected Amino Acid Score (PDCAAS) a measure of protein quality; the PDCAAS takes into account both the amino acid balance of a food and its digestibility.

Protein Quality a measure of the essential amino acid content of a protein relative to the essential amino acid needs of the body.

Protein Synthesis the process by which cells assemble amino acids into proteins. Each individual is unique because of minute differences in the ways his or her body proteins are made. The instructions for making every protein in a person's body are transmitted in the genetic information the person receives at conception.

Protein-Energy Malnutrition (PEM), also called **Protein-Calorie Malnutrition (PCM)** the world's most widespread malnutrition problem, including both kwashiorkor and marasmus as well as the states in which they overlap.

Protein-Sparing a description of the effect of carbohydrate and fat, which, by being available to yield energy, allow amino acids to be used to build body proteins.

Proteins compounds—composed of atoms of carbon, hydrogen, oxygen, and nitrogen—arranged as strands of amino acids. Some amino acids also contain atoms of sulfur.

Purified Water (also know as Demineralized Water, Distilled Water, Deionized Water, or Reverse Osmosis Water) water from which all the minerals have been removed, thereby eliminating the possibility that the minerals might corrode, say, a steam iron.

Purslane leafy vegetable that can be used in salads or cooked like spinach.

Quackery fraud. A quack is a person who practices health fraud.

Queso Blanco, Fresco, and **Mexicano** soft white cheeses made of part-skim milk.

Recommended Dietary Allowances (RDA) daily dietary intake levels sufficient to meet the nutrient requirements of approximately 98 percent of healthy people.

Recommended Nutrient Intakes (RNI) nutrition guidelines set forth by the Canadian government for the Canadian people; similar to the RDA used in the United States. (See Appendix C for a full listing of the RNI).

Red Beans this versatile bean is a medium-size, dark red oval. The taste and texture are similar to kidney beans. Use in soups and stews, and serve with rice.

Reference Daily Intakes (RDI) a table devised in 1968 listing one suggested daily intake for vitamins, minerals, and protein. The RDI are not intended as nutritional goals that everyone should meet. Rather, they are for use on food labels to help people get an idea of the amount of nutrients a serving of the product contributes to the diet. The RDI were known as the U.S. RDA (the U.S. Recommended Daily Allowances) until the food labels were revamped in 1993.

Reference Dose the estimated amount of a chemical that could be consumed daily without causing harmful effects.

Reference Protein egg white protein, the standard with which other proteins are compared to determine protein quality.

Refined refers to the process by which the coarse parts of food products are removed. For example, the refining of wheat into flour involves removing three of the four parts of the kernel—the chaff, the bran, and the germ—leaving only the endosperm.

Registered Dietitian (RD) a professional who has graduated from a program of dietetics accredited by the Commission on Accreditation for Dietetics Education (CADE) of the American Dietetic Association (ADA), has served an internship program or the equivalent to gain practical skills, has passed a registration examination, and maintains competencies through continuing education. Some states require licensing for dietitians; that is, they have legislation in place obligating anyone who wants to use the title "dietitian" to receive permission by passing a state examination. Other states do not require dietitians to be licensed. RD (the abbreviation for registered dietitian) is often used to refer to such a professional in the same way M.D. designates a medical doctor.

Regulation a legal mandate that must be obeyed. Failure to follow a regulation brings about serious legal consequences.

Requirement the minimum amount of a nutrient that will prevent the development of deficiency symptoms. Requirements differ from the RDA and AI, which include a substantial margin of safety to cover the requirements of different individuals.

Retina (RET-in-uh) the paper-thin layer of light-sensitive cells lining the back of the inside of the eye.

Retinal (RET-in-al) one of the active forms of vitamin A that functions in the pigments of the eye. Other active forms of vitamin A include retinol.

Retinol one of the active forms of vitamin A.

Retinol Equivalents (RE) a measure of the amount of retinol the body will derive from a food containing preformed vitamin A or beta-carotene. Note that some tables list vitamin A in terms of *International Units (IU)*. See Appendix D for methods of converting from one measure to another.

Rice Sticks flat, opaque, wide noodles made from rice flour.

Rice Vermicelli thin, white noodles made from rice flour.

Rickets a disease that occurs in children as a result of vitamin D deficiency and that is characterized by abnormal growth of bone, which in turn leads to bowed legs and an outward-bowed chest.

Risk the harm a substance may confer. Scientists estimate risk by assessing the amount of a chemical that each person in a population might consume over time (also called *exposure*) and by considering how toxic the substance might be (*toxicity*).

Saccharin a zero-calorie sweetener discovered in 1879 and used in the United States since the turn of the century. A possible link to bladder cancer has led to saccharin's being banned as a food additive in Canada, although it is available there as a tabletop sweetener; it is used with a warning label in the United States and is the sweetening agent in Sweet 'N Low.

Salt a pair of charged mineral particles, such as sodium (Na+) and chloride (Cl–), that associate together. In water, they dissociate and help to carry electric current—that is, they become electrolytes.

Salt Sensitive the tendency for blood pressure to rise in proportion to salt consumption that certain people seem to have from birth.

Satiety the feeling of fullness or satisfaction that people feel following a meal.

Saturated Fatty Acid a fatty acid carrying the maximum possible number of hydrogen atoms (having no points of unsaturation). Saturated fats are found in animal foods like meat, poultry, and full-fat dairy products, and in tropical oils such as palm and coconut.

Sauté (saw-TAY, the French word for stir-fry) to cook in a pan using little fat; foods are stirred frequently to prevent sticking.

Schmaltz chicken fat.

Scurvy the vitamin C deficiency disease characterized by bleeding gums, tooth loss, and even death in severe cases.

Second Harvest a national food banking network to which the majority of food banks belong.

Seltzer tap water injected with carbon dioxide and containing no added salts.

Serving the amount of food a person might eat, similar to a helping.

Set-Point Theory the theory that the body tends to maintain a certain weight by adjusting hunger, appetite, and food energy intake on the one hand and metabolism (energy output) on the other so that a person's conscious efforts to alter weight may be foiled.

Simple Carbohydrates (sugars) the single sugars (monosaccharides) and the pairs of sugars (disaccharides) linked together.

Simplesse® the trade name for a protein-based, low-calorie artificial fat, approved by the FDA for use in foods such as frozen desserts; cannot be used for frying or baking.

Skinfold Test a clinical test of body fatness in which the thickness of a fold of skin on the back of the arm (the triceps), below the shoulder blade (subscapular), or in other areas is measured with an instrument called a caliper. Obesity is defined by triceps skinfold thickness equal to or greater than 18–19 mm in adult men or 25–26 mm in women.

Social Group a group of people, such as a family, who depend on one another and share a set of norms, beliefs, values, and behaviors.

Soft Water water containing a high sodium concentration.

Soluble Fiber includes the fiber types called pectin, gums, mucilages, some hemicelluloses,and algal substances (e.g., carageenan); soluble fibers either dissolve or swell when placed in water. Psyllium seed husk is an ingredient in certain cereals and bulk-forming laxatives and contains both soluble and insoluble properties.

Sopa rice or pasta that is fried and cooked in consomme.

Soul Food a term coined in the mid-1960s to promote ethnic pride and solidarity among African-Americans.

Soup Kitchen small feeding operations attached to existing organizations such as churches, civic groups, or nonprofit agencies that serve prepared meals that are consumed on site. Soup kitchens generally do not require clients to prove need or show identification.

Soy Cheese soy cheese is made from soy milk. Its creamy texture makes it an easy substitute for sour cream or cream cheese.

Soy Flour soy flour is made from roasted soybeans ground into a fine powder. Soy flour gives a protein boost to recipes. Soy flour is gluten-free so yeast-raised breads made with soy flour are more dense in texture.

Soy Milk, Soy Beverages soybeans that are soaked, ground fine, and strained, produce a fluid called soybean milk. Soy milk is an excellent source of high quality protein and B vitamins.

Soy Nuts roasted soy nuts are whole soybeans that have been soaked in water and then baked until browned.

Soy Protein, Texturized texturized soy protein usually refers to products made from texturized soy flour. Texturized soy flour is made by running defatted soy flour through an extrusion cooker, which allows for many different forms and sizes. When hydrated, it has a chewy texture. It is widely used as a meat extender.

Soy Yogurt soy yogurt is made from soy milk. Its creamy texture makes it an easy substitute for sour cream or cream cheese.

Soy-Nut Butter made from roasted, whole soy nuts, which are then crushed and blended with soy oil and other ingredients, soy-nut butter has a slightly nutty taste, significantly less fat than peanut butter, and provides many other nutritional benefits.

Soybeans as soybeans mature in the pod they ripen into a hard, dry bean. Most soybeans are yellow. However, there are also brown and black varieties.

Sparkling Bottled Water water whose carbon dioxide (the ingredient that makes soda pop bubbly) is naturally present. That is, carbonation is not added from an outside source.

Split Peas green or yellow, these small halved peas supply an earthy flavor with mealy texture. They are best used in soups and with rice or grains.

Sports Anemia a temporary condition of low blood hemoglobin level, associated with the early stage of athletic training.

Spring Water water derived from an underground formation from which water flows naturally to the surface of the earth and to which minerals have not been added or taken away. It may be collected either at the spring itself or through a hole tapping the underground formation feeding the spring.

Stanol Esters members of the sterol class of lipids, derived from plants, and capable of reducing blood cholesterol levels when eaten as part of a low-fat diet.

Staple Food a food used frequently or daily in the diet—for example, potatoes (in Ireland) or rice (in Asia).

Staple Grain a grain used frequently or daily in the diet—for example, corn (in Mexico) or rice (in Asia).

Starch a plant polysaccharide composed of hundreds of glucose molecules, digestible by human beings.

Static Stretches stretches that lengthen tissues without injury; characterized by long-lasting, painless, pleasurable stretches.

Steam to cook foods suspended over boiling water.

Sterols (STEER-alls) one of the three main classes of lipids; a lipid with a structure similar to that of cholesterol.

Strength the ability of muscles to work against resistance.

Stress Fracture bone damage or breakage caused by stress on bone surfaces during exercise.

Sucralose a nonnutritive sweetener approved in 1998 as a tabletop sweetener and for use in a variety of desserts, confections, and nonalcoholic beverages. Sucralose is a noncaloric, heat-stable sweetener derived from a chlorinated form of sugar. Although sucralose is made from sugar, the body does not recognize it as sugar, and the sucralose molecule is excreted in the urine essentially unchanged.

Sucrose (SOO-crose) a double sugar composed of glucose and fructose.

Sugar alcohols, (maltitol, mannitol, sorbitol, isomalt, xylitol) can be derived from fruits or commercially produced from dextrose; absorbed more slowly and metabolized differently than other sugars in the human body. The sugar alcohols are not readily used by ordinary mouth bacteria and therefore are associated with less cavity formation. Although the sugar alcohols are used as sugar substitutes, they do add calories (about 1.5 to 3 calories per gram) to a food product. They are found in a wide variety of chewing gums, candies, and dietetic foods. Sorbitol and mannitol can have a laxative effect in some people.

Systolic Pressure the first figure in a blood pressure reading, which reflects arterial pressure caused by the contraction of the heart's left ventricle.

Tannins compounds found in tea and coffee that bind iron.

Target Heart Rate the heartbeat rate that will achieve a cardiovascular conditioning effect for a given person—fast enough to push the heart but not so fast as to strain it.

Taro a starchy vegetable with brown, hairy skin and a pink-purple interior.

Tempeh tempeh, a traditional Indonesian food, is a chunky, tender soybean cake. Whole soybeans, sometimes mixed with another grain such as rice or millet, are fermented into a rich cake of soybeans with a smoky nutty flavor.

Tocopherol a type of alcohol; the active form of vitamin E is alpha-tocopherol.

Tofu and Tofu Products tofu, also known as soybean curd, is a soft cheeselike food made by curdling fresh hot soy milk with a coagulant. Tofu is a bland product that easily absorbs the flavors of other ingredients with which it is cooked. Tofu is rich in high-quality protein and B vitamins and is low in sodium.

Tolerable Upper Intake Level (UL) the maximum amount of a nutrient that is unlikely to pose risk of harm in healthy people. The UL exceeds the RDA and is not intended to be an amount that people should regularly consume.

Tolerance the maximum amount of a particular substance allowed on food.

Tonic Water artificially carbonated water with added sugar and/or high-fructose corn syrup, sodium, and quinine.

Toxicants poisons, that is, agents that cause physical harm or death when present in large amounts.

Toxicity the ability of a substance to harm living organisms. All substances are toxic if present in high enough concentrations.

Trabecular (tra-BECK-you-lar) **Bone** the lacy inner network of calcium-containing crystals—spongelike in appearance—that supports the bone's structure.

Trace Mineral an essential mineral nutrient found in the human body in amounts less than 5 grams.

Trans Fatty Acid a type of fatty acid created when an unsaturated fat is hydrogenated. Found primarily in margarines, shortenings, commercial frying fats, and baked goods, trans fatty acids have been implicated in research as culprits in heart disease.

Transgenetics the process of transferring genes from one species to another unrelated species.

Triglycerides (try-GLISS-er-ides) the major class of dietary lipids, including fats and oils. A triglyceride is made up of three units known as fatty acids and one unit called glycerol.

Trimester one-third of the normal duration of pregnancy; the first trimester is 0 to 13 weeks, the second is 13 to 26 weeks, and the third trimester is 26 to 40 weeks.

Tripeptides (try-PEP-tides) protein fragments three amino acids long.

Under-5-Mortality Rate (U5MR) the number of children who die before the age of five for every 1,000 live births.

Undernutrition severe underconsumption of calories or nutrients leading to disease or increased susceptibility to disease; a form of malnutrition.

Underwater Weighing (hydrostatic weighing) a measure of density and volume; the less a person weighs underwater compared to the person's out-of-water weight, the greater the proportion of body fat (fat is less dense or more buoyant than lean tissue).

Underweight 10 percent or more below the desirable weight for height, or a BMI of less than 18.5.

UNICEF the United Nations International Children's Emergency Fund, now referred to as the United Nations Children's Fund.

Unique Radiolytic Products substances unique to irradiated food and apparently created during the process of irradiation.

Unsaturated Fatty Acid a fatty acid with one or more points of unsaturation. Unsaturated fats are found in foods from both plant and animal sources. Unsaturated fatty acids are further divided into monounsaturated fatty acids and polyunsaturated fatty acids.

Unspecified Eating Disorders some people suffer from unspecified eating disorders; that is, they exhibit some but not all of the criteria for specific eating disorders.

Urea (yoo-REE-uh) the principal nitrogen-excretion product of metabolism, generated mostly by the removal of amine groups from unneeded amino acids or from those amino acids being sacrificed to a need for energy.

Variety a feature of a diet in which different foods are used for the same purposes on different occasions—the opposite of *monotony.*

Vegan (VEE-gun) a person who eats only plant foods.

Vegetarian people who exclude animal flesh from their diet and in some cases other animal products such as milk, cheese, and eggs.

Vitamins organic, or carbon containing, essential nutrients vital to life and needed in minute amounts.

VLDL (very-low-density lipoprotein) carries fats packaged or made by the liver to various tissues in the body.

Waist Circumference a measure used to assess a person's abdominal (visceral) fat; excess fat in the abdomen increases a person's risk for obesity-related health problems.

Well Water water derived from a rock formation by way of a hole bored, drilled, or otherwise constructed in the ground.

White Navy Beans these beans are small, white ovals and are best used in soups and stews and as baked beans.

Whole Food a food that is altered as little as possible from the plant or animal tissue from which it was taken—such as beets, milk, oats, potatoes, or apples.

Whole Grain refers to a grain that is milled in its entirety (all but the husk), not refined. Whole grains include wheat, corn, rice, rye, oats, amaranth, barley, buckwheat, sorghum, and millet; two others—bulgur and couscous—are processed from wheat grains.

World Wide Web the portion of the Internet that contains sites hosted by government agencies, industry, academic institutions, health organizations, and other groups or individuals. Web sites vary in sophistication from simple displays of text to pages with sophisticated graphics, videos, and sounds.

Xerophthalmia (ZEER-ahf-THALL-me-uh) a severe form of eye disease in which the cornea hardens and may cause blindness. The problem results from vitamin A deficiency.

Yard-Long Beans thin, tender string beans that grow to as long as 18 inches.

Yo-Yo Dieting the practice of losing weight and then regaining it, only to lose it and regain it again.

Zapote an apple-size fruit with green skin and black flesh.

Index

A

Accreditation, 24
Accuracy of Internet information, checking, 22
Accutane (isotretinoin), 326
Acesulfame K (Sunette/Sweet One), 93, 96
Acetaldehyde, 313, 315
Acid-base balance, 141, 142–43, 254
Acidosis, 143
Acids, 143
Aconite, 191
Acquired immune deficiency syndrome (AIDS), 142
Activities of daily living (ADL), 351
Activity pyramid, 286
Addiction, alcohol. *See* Alcoholism
Additives, food, 105, 398–400
Adequacy, dietary, 28, 29
Adequate Intake (AI), 33–34, 35
Adolescence, 342–48
  nutrients needed during, 343
  nutrition-related problems in, 343–48
  pregnancy in, 327
Adulthood, nutrition in, 348–63
  dietary recommendations for health, 350
  factors affecting nutrition status, 357–59
  nutritional needs and intakes, 351–54
  nutrition-related problems in older adults, 354
  preparing for old age, 363–64
  sources of nutritional assistance, 359–63
Adverse reaction, 345
Advertising, food choices and, 15–16
Aerobic exercise, 269, 293–94
Aerobic metabolism, 292–93
Aflatoxins, 382
African-American food, 59–60, 61, 62–63
Age-related macular degeneration, 180, 181
Aging
  biologic function changes from ages 30–70, 357
  demographic trends and, 348–50
  healthy, 321
  nutritional needs and, 351–54
  nutrition status and, 351
  scorecard, 349
Aging well pyramid, 364
Agricultural technology, 411
AIDS (acquired immune deficiency syndrome), 142
Alcohol, 9, 31, 311–16
  athletic performance and, 306
  benefits of, 312
  binge drinking, 311, 315
  brain effects of, 314, 315
  breastfeeding and, 332
  cancer and, 312, 369–70
  consumption in U.S., 311
  Dietary Guidelines on, 41
  effect on body, 311
  fitness and, 311–16
  heart disease and, 130, 312
  how the body handles, 312–13
  hypertension and, 218
  interaction with other drugs, 314
  moderate drinking, 9, 130, 312, 316
  nutrition and, 311–16
  osteoporosis and, 231
  during pregnancy, 315, 326
Alcohol abuse, 311
Alcohol dehydrogenase, 312, 314, 315, 316
Alcoholism
  defined, 315
  diagnostic signs of, 316
  distinguishing features of, 311
  nutritional problems of, 313–14
Alitame (Novasweet), 93
Alkalosis, 143
Allergens, food, 345
Allergies, food, 335, 337, 345–47, 403–4
  milk allergy, 210, 211
Allicin, 190, 193
All-or-nothing attitude, 262
Allyl sulfides, 194
Alpha-lactalbumin, 331
Alpha-linolenic acid, 105
Alternative sweeteners, 96
Amebiasis, 381
Amenorrhea, 231, 302
American Academy of Family Physicians, 358
American Academy of Pediatrics, 326, 331, 333, 337
American Cancer Society, 36, 368
American College of Sports Medicine, 285, 287–89, 290, 292, 306
American Dental Association (ADA), 74
American Diabetes Association, 91, 96
  Web site, 92
American Dietetic Association, 24, 94, 95, 228, 298, 306, 327, 358
American Heart Association, guidelines of, 36
American Institute for Cancer Research, 20, 370
American Institute of Nutrition, 4
American Medical Association, 52, 94, 116
Amine group, 136, 167
Amino acids, 136–40. *See also* Protein(s)
  absorption of, 144
  defined, 136
  as ergogenic aids, 304, 305–6
  essential, 136, 137, 144–46, 155–56
  limiting, 145
  nonessential, 136, 144
  protein synthesis from, 140
  supplements, 137–39, 305–6
Anabolic steroids, 304, 305
Anaerobic metabolism, 292–93
Anaphylaxis, 346
Anemia, 172, 175, 183

Anemia, *continued*
 defined, 172
 iron-deficiency, 223, 342, 344
 sports, 302
Anencephaly, 173
Anorexia nervosa, 275–78, 279, 344, 348
Antibodies, 141, 345
 in breast milk, 330–31
 defined, 142
Anticarcinogens, 161
Antigens, 142
Antioxidants, 198
 as additive to oils, 105
 aging and need for, 352
 cancer and, 369
 carotenoids as, 177, 178, 180, 181, 194, 370, 402
 defined, 105, 176
 phytochemicals as, 178, 193–95, 370, 402
 recovery from exercise/performance and, 301
 reducing blood cholesterol levels and, 116
 vitamin C as, 169, 176–78, 355
 vitamin E as, 169, 177, 178, 183–84, 355, 369
Appendicitis, 82
Appetite, 14, 249, 339
Appetite suppressants, 256, 257
Appropriate technology, 410, 411
Arginine, 136, 138
Aristotle, 73
Arousal, 250
Arsenic, 228
Artificial sweeteners, 93–96
Ascorbic acid. *See* Vitamin C
Aspartame (NutraSweet/Equal), 93–96
Atherosclerosis, 82, 109, 112–13, 114, 125–31, 342, 358
Athletes. *See also* Exercise(s); Fitness
 amino acid supplements used by, 138–39, 305–6
 anabolic steroids used by, 304, 305
 caffeine and alcohol effects on performance, 306
 nutritional supplements and, 303–6
 pregame meal for, 308–10
Athletic amenorrhea, 231, 302
Atrophic gastritis, 175
Atrophy, 287
Attitude
 eating attitudes quiz, 280
 weight loss and, 262
Availability of food, 14

B

Baby boom, 348
Bacteria, foodborne illnesses caused by, 378–82
Balance, dietary, 28, 29
Ballistic stretches, 289
Basal metabolic rate (BMR), 246
Basal metabolism, 246, 263
Bases, 143
 acid-base balance, 141, 142–43, 254
Beecher, Gary, 197
Bee pollen, 303, 304
Behavior, obesity and, 249–51
Behavior modification, 272–74
Belladonna, 191
Benecol, 197, 198
Benzocaine, 257
Beriberi, 170
Beta-carotene, 181, 194, 355, 402
 cancer and, 368, 369, 370
 defined, 180
Beverages. *See also* Alcohol; Fluid balance; Water
 fluid-replacement drinks, 299–300
 nutrient density of selected, 30
BHA and BHT additives, 105
Bifidus factor, 330–31
Bile, 109, 110
Binders, 209
Binge drinking, 311, 315
Binge-eating disorder, 261, 275, 277, 278
Bioelectrical impedance, 242, 243, 244
Bioengineering. *See* Genetic engineering
Biological value (BV), 146
Biotechnology, 402
Biotin, 168, 179
Birth defects, 173, 321, 326
Birthrate, standard of living and, 410–11
Bisphosphonates, 234, 235
Black cohosh, 190
Blood cholesterol level, 101, 112–17
 heart disease and high, 112–13, 126–29
 lowering, 113–17
 standards for, 129
Blood-clotting system, vitamin K and, 184
Blood glucose level, 84-85, 87–89
 hunger and, 250
 hypoglycemia and, 89–90
 maintaining, 87–89
 soluble fiber and, 83
Blood lipid profile, 101
Blood pressure, diet and, 218–21. *See also* Hypertension
Blue cohosh, 191
Body, introduction to, A-7–A-20
Body build, osteoporosis and, 231
Body composition
 metabolic rate and, 246
 weight loss and, 265–66
Body image, eating disorders and, 275, 276, 277, 278
Body mass, lean, 146, 265, 267
Body mass index (BMI), 243–44, 245
Bone(s)
 bone-healthy lifestyle scorecard, 236
 cortical, 230, 235
 exercise and, 302
 fluoride for, 227–28
 nutrients for, 209, 232
 peak bone mass, 231, 232, 235
 remodeling during life of, 231, 232
 trabecular, 230, 232, 235
Bone density, 235, 266
Bone density test, 234
Bone fractures, 229, 230
Bone loss. *See* Osteoporosis
Borage, 191
Boron, 228
Bottled water, 205–6
Botulism, 379–80
Bovine spongiform encephalopathy (BSE), 384–86
Brain, alcohol's effects on, 314, 315
Bran, 80
Breads, 80–81
Breast cancer, 312, 367–68
Breastfeeding, 330–32
 avoidance of herbal products during, 383
 benefits of, 330–31
 in developing countries, 406, 413
 food allergies and, 346
 food guide for, 324
 nutrition and, 330
 vegetarian diet and, 156
Brillat-Savarin, Anthelme, 2, 200
Broom (herb), 191
Bruche, Hilde, 278
Buddhism, dietary practices in, 17
Buffers, 143
Bulimia nervosa, 238, 275, 276, 277–78, 279, 344
Butter, margarine vs., 106–8
B vitamins
 B_6, 168, 174–75
 B_{12}, 168, 175
 biotin, 168, 179
 exercise-supporting functions of, 301
 folate, 168, 172–74, 321, 370
 homocysteine converted by, 129–30, 174
 niacin, 168, 171–72
 pantothenic acid, 168, 179
 riboflavin, 168, 171, 172
 thiamin, 168, 170–71
 vegetarian diet and, 156

C

Cadmium, 392
Caffeine, 328–29
 athletic performance and, 306
 breastfeeding and, 332
 content in drinks and foods, 329
 defined, 328
 as ergogenic aid, 304
 pregnancy and, 325, 328–29
Caffeine dependence syndrome, 328
Calcitonin, 230*n*, 234, 235
Calcium, 29, 32, 33, 206, 208–12
 adolescent intakes of, 344
 aging and need for, 352
 cancer and, 369
 as cofactor, 208
 consumption scorecard, 213
 foods containing, 323
 hormones regulating blood level of, 230
 hypertension and, 219–20
 in milk and milk substitutes, 209–12
 osteoporosis and, 33, 34, 209, 231–32
 during pregnancy, 322–23
 recommended intakes of, 9, 210, 233
 sources of, 209–12
 supplements, 232–34
 vegetarian diet and, 157–58
Calorie control, 28, 29
Calories
 in alcohol, 313

cancer and, 368
defined, 31
in fast food, 11, 12
for older adults, 351
during pregnancy, 324
RDA for, 35
spent during various activities, 268
tips for cutting back on, 264
very-low-calorie diets (VLCDs), 256
Campbell, T. Colin, 57
Campylobacteriosis, 378, 380
Campylobacter jejuni, 378, 380, 387
Canada, A-21–A-26
Cancer, 365–72
antipromoters and cellular repair, 367
breast, 312, 367–68
colon, 19–20, 82, 367
diet and, 365–72
factors decreasing/increasing risk of, 371
fiber and, 19–20, 369
initiation of, 365, 366
lifestyle and, 365
malnutrition secondary to, 358
phytochemicals to prevent, 194, 195
promotion of, 365, 366
recommendations for preventing, 370–72
saccharin and, 94
skin, 369
soy and, 161–62
Caprenin, 118
Capsaicin, 190
Capsicum, 190
Carbohydrate loading, 296
Carbohydrates, 9, 67–97
blood glucose level and, maintaining, 87–89
calorie value of, 31
cancer and, 368
categories and sources of, 68, 69
complex, 68, 69, 72–84
consumption scorecard, 86
contribution to body stores, 251, 252–53
Daily Value for, 46
diabetes and, 89, 90–92
digestion of, 84–87
for energy, 31, 68, 310
fiber, 68, 69, 81–84
hypoglycemia and, 89–90
low-carbohydrate diet, 255–56, 325
as nutrient, 31
primary role of, 68
selection tips for, 77–79
simple, 68, 69–72
in sports beverage, 299–300
starch, 68, 69, 72–81
sugar alternatives, 93–96
Cardiovascular conditioning (training effect), 293–94
Cardiovascular disease (CVD), 101. *See also* Heart disease
Cardiovascular endurance, 290
Caries, dental, 73, 344
Carnitine, 304
Carotenoids, 177, 178, 180, 181, 194, 370, 402
Carrageenan, 118
CARS Checklist, 21–22
Carver, George Washington, 318
Casal, Gaspar, 166–67
Cells, A-7–A-8
Centers for Disease Control and Prevention, 94, 137, 204
Central obesity, 242, 243
Chaff (husk), 80
Chain length, fatty acid, 102
Chaparral, 191
Chemical agents in foods, 391–93
Chewing gum, sugar-free, 96
Childhood, early and middle, 337–42
growth and nutrient needs in, 337–39
nutrition-related problems of, 341–42
other factors influencing nutrition in, 340–41
strategies to foster healthful eating habits
in, 339
Children. *See also* Adolescence; Infancy
Dietary Guidelines on fats for, 131
hunger and, 405–7, 410–11, 412–13
protein-energy malnutrition in, 147–48, 405
rickets in, 166, 167, 182
Chinese foods, 56–57, 61, 62
Chloride, 206, 214, 215, 217, 220
Chlorophyll, 181
Cholecalciferol. *See* Vitamin D
Cholesterol, 100, 101, 111
in adolescence, 344
blood level of, 101, 112–17, 126–29
in childhood, 342
coffee drinking and, 328
content of selected foods, 109
Daily Value for, 46
Dietary Guidelines on, 129
fiber and, 82–83, 127
functions of, 109
garlic and, 131
good (HDL) vs. bad (LDL), 112–13
heart disease and, 112–13, 126–29
lowering level of, 113–17, 127, 171
oxidized LDL-, 112–13, 114
soy in diet and, 160
stanol esters and, 108
trans fatty acids and, 107–8
Choline, 139
Chondroitin sulfate, 356
Chromium, 207, 228
Chromium picolinate, 260, 303–5
Chylomicrons, 110–11
Ciguatera fish poisoning, 383
Cirrhosis, 313, 314, 315
Cis fatty acids, 106
Clostridium botulinum, 379–80, 383, 387
Clostridium perfringens, 380
Cobalamin (Vitamin B_{12}), 168, 175, 228
Cobalt, 207, 228
Coenzyme Q10, 304, 355
Coenzymes, 167
water-soluble vitamins as, 167–79
Cofactor, 208
Coffee, effects of, 328–29. *See also* Caffeine
Cold remedies, 178, 225
Collagen, 176
Colon cancer, 19–20, 82, 367
Colostrum, 330–31
Comfrey, 191
Complementary proteins, 145
Complete proteins, 145, 146
Complex carbohydrates, 68, 69, 72–84
breads, 80–81
defined, 68
fiber, 68, 69, 81–84
starch, 68, 69, 72–81
whole foods, 76–80, 81
Congregate meal program, 360–62
Consensus of published studies, credibility and, 21
Constipation, 82
Contaminants in food, 391–98
chemical agents, 391–93
defined, 391
examples of, 392
interaction among, 393
organically grown foods and, 396–97
pesticide residues, 393–98
Contamination iron, 224
Control group, 20, 24
Cooking
fats in, 122–23
home food safety and, 387–88
partial, 386
proteins in, 154
recipe modifications, 124
temperatures, 372, 384, 385
terms, 123
Copper, 207, 228
Correlations, 20, 24
Correspondence schools, 23, 24
Cortical bone, 230, 235
Cravings during pregnancy, 325, 326
Creatine, 304
Credibility of nutrition information, checking, 20–21, 22
Cretinism, 226, 406
Crop rotation, 412
Crossbreeding, 402
Cross-contamination, 384
Cross-reaction, 346
Cruciferous vegetables, 368
Cryptosporidium, 204
Culture, food choices and, 17
CyberDiet Web site, 14
Cyclamate, 93, 399–400
Cysteine, 136*n*

D

Daidzein, 194
Daily Values (DV), 34, 46–49
Dairy Council of California Web site, 213
DASH-Dietary Approaches to Stop Hypertension diet, 220–21
DDT, 394
Death, leading causes of, 6, 350
Deficiency diseases, 6, 166–69
Degenerative diseases, 6, 14
Dehydroepiandrosterone (DHEA), 304, 355–56
Delaney, James J., 400
Delaney clause, 400
Dementia, 354, 358
Demographic trends, 348–50
Denaturation of proteins, 140

Dental caries, 73, 344
Dental health, 73–74, 358
Dental plaque, 73
Dependence, 311, 328
Depression, malnutrition secondary to, 358
Designer estrogens, 234, 235
Designer foods, 196
Desserts, alternatives to sweet, 71–72
Deutsch, R.M., 98, 134
Developing countries
 breastfeeding in, 406, 413
 role of women in, 412–13
Dexfenfluramine (Redux), 257
Dextrins, 75
DHEA (dehydroepiandrosterone), 304, 355–56
Diabetes, 89, 90–92
 blood glucose and, 89, 90–91
 defined, 82, 89
 determining risk for, 91
 gestational, 327
 malnutrition secondary to, 358
 obesity and, 240
 soy and, 162
 sugar substitutes and, 95–96
 Type 1 and 2, 90–91
 Web site on, 92
Diary, food, 272–73
Diet
 avoiding using word, 261–62
 blood pressure and, 218–21
 cancer and, 365–72
 elimination, 347
 heart disease and, 125–31
 Mediterranean, 58, 115–16, 130
 nature of shoppers' concerns about, 5
 physical endurance and, 295–96, 297
 varied, importance of, 372
 vegetarian, 17, 155–62, 369
 weight loss. *See* Weight-loss diets
Diet aids, unproven or fraudulent, 138, 255
Dietary Folate Equivalent (DFE), 172*n*
Dietary Goals for the United States, 36
Dietary Guidelines for Americans, 36, 37
Dietary Reference Intakes (DRI), 33–35
Dietary Supplement Health Education Act (DSHEA, 1994), 192
Diet Center, 258
Dieting. *See* Weight loss
Diet planning, 38–52. *See also* Nutrients
 checking your own diet, 53
 exchange lists and, 51–52, A-32–A-50
 food composition tables and, 52
 food group plans, 38–42
 food labels and, 42–51
 principles of, 28–30
Dietition, registered, 24
Digestion
 of carbohydrates, 84–87
 of lipids, 110–11
 of proteins, 140, 144, 145
Dim sum, 57
Dining out, weight control and, 270–71
Dipeptides, 144
Diploma mills, 23, 24
Disability, 354, 357
Disaccharides, 69–70
Disease(s). *See also specific diseases*
 associated with lack of dietary fiber, 82–83
 breastfeeding and communicable, 332
 deficiency, 6, 166–69
 degenerative, 6, 14
 foodborne illnesses, 376–83
 functional foods and prevention of, 196–97
 lifestyle, 38, 80, 285
 malnutrition secondary to, 358
 nutrient/disease relationships, 49–51
 nutrition-responsive and -unresponsive, 351
 reducing risk of, 350
Disordered eating, 275, 279
Diuretics, 217, 257
Diverticulosis, 82
DRI (Dietary Reference Intakes), 33–35
Drink (alcoholic), defined, 312, 315
Drinking. *See* Alcohol
Drug(s)
 alcohol as, 311
 alcohol interaction with, 314
 breastfeeding and, 332
 defined, 315
 polypharmacy, 354
 pregnancy and, 326
 weight loss and, 255–60
Dwarfism, 225
Dysentery, 147

E

EAR (estimated average requirement), 33, 34, 35
Eating attitudes quiz, 280
Eating behavior, 249–51. *See also* Weight control
Eating disorders, 231, 275–80
 among adolescents, 344–48
 athletic amenorrhea and, 302
 defined, 279
 diagnosis of, 276
 disordered eating vs., 275
 prevention of, 278–79
 treatment of, 277–78
 unspecified, 275, 279
 warning signs of, 278, 279
Eating habits, 18, 262, 339
 breaking old, 272–74
Echinacea, 190
Eclampsia, 327
Edema, 147
Eicosanoids, 103*n*
80/20 rule, 29
Elderly. *See* Older adults
Elderly Nutrition Program, 360
Electrolytes, 214–17
Elimination diet, 347
Emerson, R.W., 282
Empty-calorie foods, 70
Emulsifiers, 105, 109
Endosperm, 80
End-stage kidney disease, 358
Endurance, 290
 diet and physical, 295–96, 297
Energy
 carbohydrates for, 31, 68, 310
 defined, 31
 for exercise, 292–94
 fat as reserve of, 87, 100
 metabolsim, A-19
 proteins as, 31, 141, 143–44
Energy balance
 basal metabolic rate and, 246
 physical activity and, 246–47
 total energy needs, 247
 weight control and, 244–47
Energy output scorecard, 248
Energy-yielding nutrients, 31
Enriched foods, 80, 81
Enrichment Act (1942), 80
Entamoeba histolytica, 381
Enterotoxin, 377, 379
Environmental concerns, 411–12
Environmental Protection Agency (EPA), 203–4, 394–95
Enzymes, 70, 85, 141
 coenzymes, 167–79
Eosinophilia-myalgia syndrome (EMS), 137
Ephedra (ma huang), 191, 260, 383
Epidemiological study, 20, 24
Epithelial tissue, 180
Ergogenic aids, 303–6
Erosion, soil, 411–12
Escherichia coli 0157:H7, 380, 381–82, 388, 401
Essential amino acids, 136, 137, 144
 protein quality of foods and, 144–46
 in vegetarian diet, 155–56
Essential fatty acids, 102, 103
Essential nutrients, 30–31
Estimated average requirement (EAR), 33, 34, 35
Estrogen, 229, 231, 235, 266
 designer, 234, 235
Estrogen replacement therapy (ERT), 234, 235
Ethics, transgenetics and, 402
Ethnic cuisines, 17, 54–63
Ethylene oxide, 401
Euphoria, 311, 315
Exchange lists, 51–52, A-32–A-50
Exercise(s). *See also* Fitness
 activity pyramid, 286
 aerobic, 269, 293–94
 anaerobic, 292–93
 bones and, 302
 calories expended in, 268
 for cardiovascular endurance, 290
 energy for, 292–94
 examples of moderate, 285
 fats used in, 295, 296–97
 for flexibility, 289–90
 fluid needs and, 298–300
 fuels for, 294–97
 glucose use during, 294–96
 habits, 262
 HDL levels and, 129
 minerals and, 301–2
 obesity and lack of, 250–51
 for older adults, 351
 osteoporosis prevention and, 235
 physical activity scorecard, 291
 physical conditioning, 286–87
 reasons to, 284–87

spot-reducing, 267
strength training, 287–89
stretching, 289–90
target heart rate during, 294
vitamins and, 301
weight loss and, 266–69
Exercise stress test, 290
Experimental group, 20, 24
External cue theory, 249

F

Family, food choices and, 16
Famine, 405, 410
Farm families, hunger among, 409
Fast food, 11–13, 14, 271
Fasting, 246, 252, 253–54
Fat(s)
absorption of, 111
calorie value of, 31
cancer and, 368
characteristics of, in foods, 104–5
consumption scorecard, 121
contribution to body stores, 251, 252–53
controlling dietary, 117–32
cooking and, 122–23
Daily Value for, 46
defined, 100
digestion of, 111
effects on blood lipids, 115
energy intake from, determining, 31
energy storage in, 87, 100
in fast food, 11, 12
food labels and content of, 47, 122
functions in body, 100–101
functions in food, 101
how body handles, 110–11
for infants and children, 342
as nutrient, 31, 101
recipe modifications, 124
saturated vs. unsaturated, 102–3, 105
substitutes, 118–20
terminology of, 101–2
tips for cutting back on, 131, 264
total daily allowance, determining, 120
Fat, body
body weight vs., 241–42
contribution of excess energy nutrients to, 251–53
dietary recommendations for, 9
distribution of, 242–43, 244
functions of, 100–101
measuring, 242
stress and accumulation of, 250
use for energy during exercise, 295, 296–97
Fat cell, 101
Fat cell theory, 249
Fat-soluble vitamins, 32, 169, 179–85
Fatty acids
cis, 106
defined, 102
essential, 102, 103
omega-6 vs. omega-3, 102, 103–4, 105
points of unsaturation in, 103, 104–5
saturated, 102
trans, 106–8
unsaturated, 103
Fatty liver, 313, 315
FDA. *See* Food and Drug Administration (FDA)
Feasting, 252
Federal Trade Commission (FTC), 4, 22
Feeding formula, 332–33, 406
Female athlete triad, 302
Fenfluramine (Pondimin), 257
Fen-phen, 257
herbal, 260
Fetal alcohol effect, 326
Fetal alcohol syndrome (FAS), 315, 326
Feverfew, 190
Fiber, 9, 68, 69, 81–84
aging and need for, 352
blood cholesterol and, 82–83, 127
cancer and, 19–20, 369
daily intake of, 84
Daily Value for, 46
defined, 81
in Food Guide Pyramid, 85
health and, 81–84
insoluble, 82, 83
iron absorption and, 224
soluble, 82–83
tracking intake of, 79
Fiber pills, 257
First Amendment, 21, 22, 24
Fisher, M.F.K., 164
Fish oils, 103–4
Fitness, 282–317
alcohol and, 311–16
bones and, 302
components of, 287–90
defined, 284
energy for exercise, 292–94
exercise guidelines for, 285
fluid needs and exercise, 298–300
food for, 308–10
fuels for exercise, 294–97
getting started on lifetime, 290–92
physical activity scorecard, 291
protein needs for, 297–98
quackery scorecard, 307
reasons to exercise, 284–87
supplements and, 303–6
vitamins and minerals for exercise, 300–302
Five a day plus scorecard, 188
Flavonoids, 194
Flavr Savr tomato, 401–2, 403
Flexibility, 289–90
Flossing, 74
Fluid balance, 143. *See also* Beverages; Water
fluid needs and exercise, 298–300
proteins and, 142, 143
Fluid-replacement drinks, 299–300
Fluoridation in U.S., 227
Fluoride, 74, 207, 227–28, 333
Fluorosis, 227
Foam cells, 112–13, 114
Folate, 168, 172–74, 370
requirements during pregnancy, 172–73, 321
vitamin B_{12} and, 175
Food additives, 105, 398–400
Food Additives Amendment (1958), 399, 400
Food allergens, 345
Food allergies, 335, 337, 345–47, 403–4
milk allergy, 210, 211
Food Allergy Network, 346
Food and Agriculture Organization (FAO), 35–36, 401, 405
Food and Drug Administration (FDA)
approval of food additives, 398
aspartame and, 94
bottled water standards, 205, 206
on caffeine consumption during pregnancy, 325
folate fortification and, 173
food labels and, 42, 197
functional foods and, 198
health fraud and, 22
irradiation and, 401
medicinal herbs and, 192
pesticide monitoring by, 395
supplements and, 137
trans fatty acids and, 107
Food aversions, 345
Food banks, 360, 408, 410
Foodborne illnesses (food poisoning), 376–83
causes of, 377
defined, 376
microbial agents, 378–82
natural toxins, 382–83
types of, 377–78
Foodborne infections, 377–78
Food chain, pesticides in, 394, 395
Food choices, factors affecting, 14–18
Food composition tables, 52, A-68–A-155
Food diary, keeping, 272–73
Food group plans, 38–42
Food Guide Pyramid, 38
actual consumption pyramid, 76
for adults over age 70, 353
African-American, 59
carbohydrates in, 75, 77–79
Chinese-American, 56
DASH diet, 221
fats in, 117
fiber in, 85
Jewish-American, 60
Mediterranean, 58
Mexican-American, 55
mineral sources in, 208
protein in, 150
sample balanced weight-loss diets using, 264
vegetarian, 156
vitamin sources in, 170
for young children, 338
Food insecurity, 405–9, 410
Food intolerance, 345–46
Food intoxications, 377
Food labels. *See* Labels, food
Food pantries, 410
Food Quality Protection Act (1996), 400
Food recovery, 408, 410
Food Recovery and Gleaning Initiative, 408
Food safety. *See* Safety, food
Food security, 405, 410
Food Stamp program, 359–60, 408
Food supply, availability of, 14

Forearm fracture, bone loss and, 230
Formula, infant, 332–33, 406
Fortified foods, 196
  calcium-fortified, 210, 211
  defined, 45, 80
  folate-fortified, 321*n*
Foundation for Innovation in Medicine, 196
Four Food Group Plan, 38
Fractures, bone, 229, 230
  stress fractures, 302
Frame size, A-31
Fraud, health, 4–5, 19, 22–23, 138–39
Free radicals, 116, 177, 178
Frequency of activity, 287
Friends, food choices and, 16–17
Fructose, 69, 96
Fruit
  nutrients in, 71
  simple sugars in, 70, 71
  storage and preparation of, 186–87
FTC (Federal Trade Commission), 4, 22
Fuels for exercise, 294–97
Functional foods, 196–98
Funk, Casimir, 167
Fusion cuisine, 54

G

Galactose, 69
Gandhi, Mahatma, 411
Garlic, 131, 190, 193, 196
Gastritis, atrophic, 175
Gastrointestinal disorders, malnutrition secondary to, 358
Gastroplasty, 260–61
Gender
  body composition and, 246
  HDL-cholesterol levels and, 129
  osteoporosis and, 231
General Nutrition, Inc. (GNC), 4
Genetic engineering, 347, 401–4
Genetics vs. environment, obesity and, 247–51
Genistein, 194
Germander, 191
Germ (wheat), 80
Gestational diabetes, 327
Giardia lamblia, 381
Giardiasis, 381
Ginger, 190
Ginkgo biloba, 190, 356
Ginseng, 304
Gleaning, 408, 410
Gleich, Gerald, 137
Glinsman, Walter, 198
Glucagon, 89
Glucosamine, 356
Glucose
  blood glucose level, 83, 84–85, 87–90, 250
  conversion of carbohydrates to, 84–85
  defined, 69
  soyfoods and control of, 162
  use during exercise, 294–96
Glutamine, 136*n*
Gluten intolerance, 345
Glycemic effect, 89*n*
Glycemic index, 89*n*
Glycerol, 102
Glycogen, 87, 88, 252, 253, 306
  use during exercise, 294–96
Glycogen-loading technique, 296
GOBI, 412–13
Goiter, 6, 226, 405–6
Good Samaritan Food Donation Act (1996), 408
Government, media misinformation and, 21
Grains, 80–81
  staple, 72–75
GRAS (Generally Recognized as Safe) list, 399–400
Grazing, 39–40
Green revolution, 411
Ground meat, bacterial contamination of, 384
Growth
  in early and middle childhood, 337–39
  in infancy, 331
  protein and, 140–41
Growth charts, 412–13

H

Habits, eating, 18, 262, 339
  breaking old, 272–74
Hardening of arteries. *See* Atherosclerosis
Hard water, 203
Hazard, 391
HDL. *See* High-density lipoproteins (HDL)
Headaches, aspartame and, 94
Health
  complex carbohydrates and, 76–80
  dental, 73–74, 358
  fiber and, 81–84
  lipids and, 111–17
  phytochemicals and, 194, 195
  protein and, 146–54
  sugar and, 70, 73–74
  vegetarian diet and, 158
Health claims, 48, 49–51
  legitimate, 197–98
Health fraud, 4–5, 19
  claims about amino acids and, 138–39
  detecting, 22–23
Health Management Resources, 258
Health (nutrient content claim), 50
Health promotion, nutrition and, 5–10
Healthy People 2000: National Health Promotion and Disease Prevention Objectives, 9–10, 331
Healthy People 2010 Nutrition-Related Objectives for the Nation, 10, 242
Heaney, Robert, 235
Heart
  exercise for, 293–94
  health scorecard, 128
Heart disease
  alcohol and, 130, 312
  antioxidants and, 116
  cholesterol and, 112–13, 126–29
  diet and, 125–31
  homocysteine and, 129–30, 173–74
  obesity and, 240–41
  phytochemicals and, 194, 195
  prevention of, 130–31, 194, 195
  risk factors linked to, 113, 125–26
  soy and, 160
  vitamin E and, 183
Heart rate, maximum (MHR), 294
Heart rate, target, 294
Heat exhaustion, 299
Heat stroke, 299
Heavy metals, 391–93
Heme iron, 224
Hemoglobin, 223
Hemorrhoids, 82
Hepatitis A virus, 381
Herbicides, 402
Herbs
  medicinal, 189–92
  natural toxicants in, 382–83
  supplements for weight loss, 260
High blood pressure. *See* Hypertension
High-density lipoproteins (HDL), 111, 113, 126–27, 129, 160
High-protein, low-carbohydrate diets, 254–55
High-risk individuals for heart disease, 129
Hinduism, dietary practices in, 17
Hip fracture, bone loss and, 229, 230
Hippocrates, 4, 166, 196
Hispanic food, 54–55, 63
Histamine, 345–46
Histidine, 136
HMB (beta-hydroxy-beta-methylbutyrate), 304
Home-Delivered Meals Program, 360–62
Home food safety, 384, 385, 387–88
Homeless in U.S., number of, 408–9
Homocysteine, 129–30, 173–74
Honey, bacteria in, 379
Hormones, 141
  age-related decline in, 231
  blood calcium level regulation by, 230
  blood glucose level and, 88–89
  changes during exercise in, 298
  defined, 141
  hunger and, 250
  metabolic rate and, 246
  thyroid, 226
Human growth hormone (GH), 138, 356
Human immunodeficiency virus (HIV), breastfeeding and, 332
Hunger, 249–50
  children and, 405–7, 410–11, 412–13
  defined, 14, 249, 410
  among farm families, 409
  individuals' efforts to alleviate, 413
  poverty and, 405, 407
  in U.S., 407–9
  worldwide, 405–13
Husk (chaff), 80
Hydration, schedule of, 299
Hydrogenation, 105
Hydrolyzed vegetable protein (HVP), 160
Hydrostatic weighing, 242, 243
Hyperglycemia, 90
Hyperplastic obesity, 249
Hypertension
  alcohol and, 218
  atherosclerosis and, 125
  calcium and, 219–20

chloride and, 220
DASH diet for, 220–21
defined, 215
diagnosis of, 219
malnutrition secondary to, 358
obesity and, 218, 240
in pregnancy, 326–27
sodium and, 215, 218–19
Hypertrophic obesity, 249
Hypertrophy, 286
Hypoglycemia, 89–90
Hypothalamus, eating behavior and, 250

I

Immune system, 345, 370, A-9, A-11
Immunity, 142
Immunity factors in colostrum and breast milk, 330–31
Immunizations, 413
Immunoglobulin E (IgE), 345
Imported foods, contaminants of, 398
Imported produce, contaminants on, 394
Incidental food additives, 398
Income, food choices and, 15. *See also* Poverty
Incomplete proteins, 145
Indoles, 194, 370
Infancy, 330–37, 342
breastfeeding in, 156, 324, 330–32, 346, 383, 406, 413
feeding formula in, 332–33, 406
food allergic reactions in, 335, 337, 346
foods in, 334–36
growth and development during, 331
nutrition-related problems in, 336–37
supplements during, 333
undernutrition in, 406–7
vitamin K deficiency in newborns, 184
Infant formula, 332–33, 406
Infant mortality rate, 324, 411
Infections
foodborne, 377–78
opportunistic, 142
Ingredients list on labels, 43, 45
Inorganic compounds, 202. *See also* Minerals
Insoluble fiber, 82, 83
Insulin, 88, 90–91, 140
Integrated pest management, 397
Intensity of activity, 287
Intentional food additives, 398
International Academy of Nutrition Consultants, 23
International Conference on Nutrition, 413
International cuisines. *See* Ethnic cuisines
Internet, 21–22, 24
Intervention study, 20, 24
Intestinal bypass surgery, 260
Intestinal flora, 184
Intoxications, food, 377
Intrinsic factor, 175
Iodine, 207, 226–27
Iodine deficiency, 405–6
Ions, 214
Iron, 29, 157, 207, 223–24
aging and need for, 352
contamination, 224
exercise-related functions of, 301–2
foods containing, 322
heme, 224
for infants, 334
pregnancy and, 321–22
ways to enhance absorption of, 224
Iron deficiency, 337, 405
Iron-deficiency anemia, 223, 342, 344
Iron overload, 224
Irradiation, 400–401
Islam, dietary practices in, 17
Isoflavones, 159–61, 194, 195
Isomalt, 96
Isothiocyanates, 194
Isotretinoin (Accutane), 326
Italian food, 57–59

J

Japanese food, 62
Jenny Craig, 258
"Jewish" foods, 60–61
Jin Bu Huan, 382
Johnson, Ben, 305
Joint National Committee on Detection, Evaluation, and Treatment of High Blood Pressure, 218
Journal, peer-reviewed, 20
Judaism, dietary practices in, 17

K

Kashrut, laws of, 61
Kava, 190
Ketosis, 89, 147, 254
Kombucha tea, 191
Kosher food, 17, 61
Kwashiorkor, 147, 148

L

Labels, food, 42–51
Daily Values on, 46–49
fats on, 47, 122
food allergies and reading, 347
for genetically engineered foods, 404
health claims on, 48, 49–51
ingredients list on, 43, 45
legal requirements for, 42–43
names of sugars in, 79
nutrient content claims on, 49, 50
Nutrition Facts panel, 43–45
sodium in, 222
types of, 46
Lactase, 70
Lactation. *See* Breastfeeding
Lactic acid, 292
Lactobacillus bifidus, 330–31
Lactoferrin, 331
Lactose, 69, 70, 212
Lactose intolerance, 70, 210–12, 345
Land reform, 411
Large intestine, digestion in. *See* Digestion
Later life, nutrition during, 348–63
L-cysteine, 139
LDL. *See* Low-density lipoproteins (LDL)
Lead poisoning, 391–93
from drinking water, 204–5, 392
Lean body mass, 146, 265, 267
Lecithin, 109
Legumes, 75
defined, 150
miniglossary of, 151
soybeans and soyfoods, 159–62, 197, 369
Leptin, 248
Life cycle, nutrition through, 318–73
aging scorecard, 349
early and middle childhood, 337–42
infancy, 330–37
later life, 348–63
meals for one, 361
pregnancy, 320–30
preparing for old age, 363–64
teenage years, 342–48
Life expectancy, 348–50
Life span, maximum, 348
Lifestyle, effect of, 7, 8, 285, 328, 365, 413
Lifestyle diseases, 38, 80, 285
Light (nutrient content claim), 50
Lignan, 194
Limiting amino acids, 145
Limonene, 194
Lind, James, 166
Linoleic acid (omega-6), 102, 103, 105, 331
Linolenic acid (omega-3), 102, 103–4, 105, 115
Lipase inhibitors, 257
Lipids, 98–133. *See also* Fat(s); Oil(s)
defined, 100
digestion of, 110–11
effects of types of fat on blood, 115
functions of, 100–101
health and, 111–17
types of, 101–2, 109
Lipoprotein lipase (LPL), 249
Lipoproteins, 110–11
defined, 111
high-density (HDL), 111, 113, 126–27, 129, 160
low-density (LDL), 111, 112–13, 127, 129, 160
very-low-density (VLDL), 111
Liposuction, 261
Listeria monocytogenes, 380
Listeriosis, 380
Liver
alcohol processing by, 312–13
digestion of carbohydrates and, 87
fatty, 313, 315
glucose storage as glycogen in, 294–96
Lobelia, 191
Longevity game scorecard, 8
"Lookist" attitude, 262
Low birthweight (LBW), 324, 326, 406
Low-carbohydrate diet, 255, 325
Low-density lipoproteins (LDL), 111, 112–13, 127, 129, 160
Low-risk individuals for heart disease, 129
L-tryptophan, 137
Lutein, 194
Lycopene, 194, 196, 370

Lymph, 110
Lysine, 138

M

Macular degeneration, age-related, 180, 181
Mad cow disease, 384–86
Magnesium, 206, 214
Ma huang (Ephedra), 191, 260, 383
Major minerals, 206, 207
Malnutrition
  defined, 6, 410
  effects of hunger, 405
  growth charts and recognition of, 412–13
  iron deficiency and, 223
  metabolic rate and, 246
  in older adults, 354, 358
  postnatal, 320
  prenatal, 320
  protein energy (PEM), 147–48, 405
  secondary to disease or physiologic state, 358
Maltitol, 96
Maltose, 69
Manganese, 207, 228
Mannitol, 96
Marasmus, 147–48
Margarine vs. butter, 106–8
Margin of safety, 395, 398–99
Mariani, John, 54
Mast cells, 345
Maternal weight gain, 324–25
Maturation, teen nutrition and, 342–48
Maximum heart rate (MHR), 294
Maximum life span, 348
Meals/recipes
  modifying recipes, 124
  for one, 361
  for one-year-old, 336
  planning, weight loss and, 262–66
  pregame meal, 308–10
  for snacks, 39–40
  translating dietary recommendations into actual, 130–31
  vegetarian, 158
Meat alternatives/replacements, 156, 160
Meat thermometer, 385
Media, food choices and, 15–16, 19–21
Medicinal herbs, 189–92
Mediterranean diet, 58, 115–16, 130
Megadoses, 174–75
Melatonin, 138, 356
Menopause, 129, 161, 229, 231, 235
Menstruation, athletic amenorrhea and cessation of, 302
Mental activity, energy needed for, 247
MEOS (microsomal ethanol-oxidizing system), 314, 315
Mercury, 392
Meridia (sibutramine), 257
Metabolism
  aerobic, 292–93
  alcohol, 312–13
  anaerobic, 292–93
  basal, 246, 263
  energy, A-19
  lean body mass and, 267
Mexican food, 54–55, 61
Microbial food agents, 378–82
Milk, 209–10
  for infants, 330–33, 336
  substitutes, 210–12
Milk allergy, 210, 211
Milk thistle, 190
Milne, A.A., 67
Minerals, 207–37. *See also specific mineral*
  aging and need for, 352
  cancer and, 368–69
  defined, 31, 202
  as ergogenic aid, 305
  exercise-related functions of, 301–2
  on food labels, 45, 48
  guide to, 206–7
  major, 206, 207
  as nutrients, 31–32
  supplements, 185
  trace, 207, 223–28
  in vegetarian diet, 157–58
Misinformation in news, 20–21
Miso, 160
Moderate alcohol intake, 9, 130, 312, 316
Moderation, dietary, 28, 29
Molds, food contamination by, 382
Molybdenum, 207, 228
Monoglycerides, 110
Monosaccharides, 69
Monoterpenes, 194
Monounsaturated fat, 105, 114–15
Monounsaturated fatty acid, 102, 103
Morning sickness, 326
Multinational corporations, hunger and, 409–10
Muscle endurance, 290
Muscle protein buildup after exercise, 298
Muscles
  delivery of oxygen by heart and lungs to, 293
  energy used by, 292–93
  glucose storage as glycogen in, 294–96
  unbalanced development of, 302

N

Narcotic, 314, 315
National Academy of Sciences (NAS), 33, 177, 398
  Institute of Medicine of, 196
National Association of Anorexia Nervosa and Associated Disorders (ANAD), 277
National Cancer Institute (NCI), 94, 365, 368
  National 5 a Day for Better Health program, 370–72
National Center for Complementary and Alternative Medicine (NCCAM), 192
National Clearinghouse for Alcohol and Drug Information, 316
National Council for Reliable Health Information (NCRHI), 20–21, 23
National Council on Aging, 358
National Dental Research Survey, 344
National Institute of Dental Health, 228
National Institute on Alcohol Abuse and Alcoholism, 311
National Nutritional Foods Association, 192
National Osteoporosis Foundation, 234
National Research Council, 218, 228
National Restaurant Association, 54, 408
National Summit on Food Recovery and Gleaning, 408
Natural toxins, 382–83
Nervous system, source of energy for, 253, 254
Neural tube defects, 173, 321
Neurotoxins, 377
News stories, critiquing nutrition, 20–21
Niacin, 168, 171–72
Niacin equivalents (NEs), 171
Nickel, 228
Night blindness, 180
Nitrosamines, 369
Nixon, Richard, 400
Nonessential amino acids, 136, 144
Nonessential nutrients, 31
Nonheme iron, 224
Norris, K., 238
Novasweet (alitame), 93
Nursing bottle syndrome, 74
Nursing homes, 362–63
Nutraceuticals, 196
NutraSweet/Equal (aspartame), 93–96
Nutrient content claims, 49, 50
Nutrient density, 29, 30, 41
Nutrient/disease relationships, health claims about, 49–51
Nutrient excesses, 36
Nutrients, 30–38
  for adolescents, 343
  amounts eaten daily, 32
  for bone growth and maintenance, 209, 232
  in breads, 80, 81
  challenge of dietary guidelines, 36–38
  classes of, 30, 31–32
  defined, 30
  Dietary Reference Intakes (DRI), 33–35
  in early and middle childhood, 337–39
  energy-yielding, 31
  essential and nonessential, 30–31
  fat as, 31, 101
  in food labels, 42–51
  for infants, 334–36
  on Nutrition Facts panel, 44
  for older adults, 351–54
  for pregnancy, 320–24
  recommendations for, 32–36
  in sugar, 71
  water as most essential, 32, 202–6
Nutri/System, 258
Nutrition
  alcohol and, 311–16
  defined, 4
  factors affecting food choices and, 14–18
  health promotion and, 5–10
  recognition of discipline, 4
  resources, A-1–A-6
  separating fact from fiction about, 19–24
Nutritional yeast, 156
Nutrition Facts panel, 43–45

Nutritionist, 23, 24
Nutrition Labeling and Education Act (1990), 42–43
Nutrition Recommendations for Canadians, 36, A-21–A-26
Nutrition Screening Initiative, 358

O

Obesity
 among adolescents, 344
 behavior and, 249–51
 causes of, 247–51
 central, 242, 243
 in children, 341–42
 defined, 82, 240
 fiber and, health benefits of, 83
 genetics vs. environment and, 247–51
 high dietary protein and, 148–49
 hyperplastic, 249
 hypertension and, 218, 240
 hypertrophic, 249
 measures of, 241–42
 obese vs. healthy self, 262, 263
 physical risks of, 240–41, 242
 rising trend toward, 240, 242
 social and economic handicap from, 241
 surgical procedures for, 260–61
 weight gain during pregnancy and, 324
ob gene, 248
Oil(s). *See also* Fat(s)
 defined, 100
 fish, 103–4
 olive oil, 57–58, 115
 preventing spoilage of, 105
 saturated and unsaturated compared, 105
 vegetable, 106, 108
Older adults. *See also* Aging
 demographics and, 348–50
 factors affecting nutrition status of, 357–59
 nutritional needs and intakes, 351–54
 nutrition-related problems of, 354
 preparation for old age, 363–64
 sources for nutritional assistance, 359–63
O-LDL (Oxidized LDL-cholesterol), 112–13, 114
Oleic acid (omega-9), 102
Olestra, 118, 119, 120
Olive oil, 57–58, 115
Omega-3 (linolenic acid) fatty acid, 102, 103–4, 105, 115
Omega-6 (linoleic acid) fatty acid, 102, 103–4, 105, 331
Opportunistic infections, 142
Optifast, 259
Oral disease, 358
Oral rehydration therapy (ORT), 410, 413
Organically grown food, 396–97
Organic halogens, 391, 392
Osteoarthritis, 358
Osteoblasts, 231, 235
Osteoclasts, 231, 235
Osteomalacia, 182
Osteoporosis, 34, 229–35
 bone fractures and, 229, 230
 calcium and, 33, 34, 209, 231–32
 excess protein in diet and, 148
 malnutrition secondary to, 358
 prevention of, 231–32, 235
 soy and, 160–61
Overload, 286–87
Overnutrition, 6. *See also* Obesity
Overpopulation, 410–11
Overweight, 10, 240. *See also* Obesity
Oxalic acid, 209
Oxidized LDL-cholesterol (o-LDL), 112-13, 114

P

Palmitic acid, 102
Pancreas, 87, 88–89
Pantothenic acid, 168, 179
Paralytic shellfish poisoning, 383
Parasites, 381
Parathyroid hormones, 230*n*
Parthenolide, 190
Partially hydrogenated vegetable oil, 106, 108
Pasteur, Louis, 374, 400
Pasteurization, 376, 400
Pauling, Linus, 178
PBB (polybrominated biphenyl), 391
PCB (polychlorinated biphenyl), 391, 392
Peak bone mass, 231, 232, 235
Pectin, 127
Peer-reviewed journal, 20
Pellagra, 6, 166–67, 171
Pennyroyal, 191
Pepsin, 145
Peptide bond, 140
Percent fat free, 50, 51
Perfringens food poisoning, 380
Periodontal disease, 74
Pesticides, 393–98, 402
pH, 143
Phenols, 194
Phentermine, 257
Phenylalanine, 139
Phenylketonuria (PKU), 94, 96
Phenylpropanolamine (PPA) hydrochloride, 257
Phosphate salt, 304
Phospholipids, 101, 109, 111, 212
Phosphorus, 206, 212
Photosynthesis, 69
Physical activity. *See also* Exercise(s); Fitness
 energy balance and, 246–47
 obesity and lack of, 250–51
 scorecard, 291
Physical conditioning, 286–87
Phytic acid, 209, 224
Phytochemicals, 178, 193–95, 197, 370, 372, 402
Phytoestrogens, 159, 161, 194
Pica, 325
Pigment, 180, 195
PKU (Phenylketonuria), 94, 96
Placebo, 20, 24, 94, 96
Placebo effect, 20, 305
Placenta, 320
Plaque
 arterial, 125
 dental, 73
Plasma volume, exercise and, 298, 299
Point of unsaturation, 103, 104–5
Poisoning, food. *See* Foodborne illnesses
Poke root, 191
Polybrominated biphenyl (PBB), 391
Polychlorinated biphenyl (PCB), 391, 392
Polypharmacy, 354
Polyphenols, 372
Polysaccharides, 69, 81
Polyunsaturated fat, 105, 114, 115
Polyunsaturated fatty acid, 102, 103
Pondimin (fenfluramine), 257
Population
 aging of, 348–50
 hunger and overpopulation, 410–11
Postnatal period, malnutrition in, 320
Potassium, 206, 214, 215–17, 219
Poverty
 defined, 410
 hunger and, 405, 407
 rural, 409
 undernutrition among adolescents in, 343
Precursor, 180
Preeclampsia, 327
Preformed vitamin A, 180
Pregame meal, 308–10
Pregnancy, 320–30
 adolescent, 327
 alcohol consumption during, 315, 326
 avoidance of herbal products during, 383
 caffeine intake during, 325, 328–29
 common nutrition-related problems in, 326–27
 cravings during, 325, 326
 folate required during, 172–73, 321
 food guide for, 324
 iodine requirements during, 226
 maternal weight gain, 324–25
 nutritional needs in, 320–24
 practices to avoid in, 325–26
 readiness scorecard, 323
 trimesters in, 321, 325
 undernutrition during, 406
 vegetarian diet during, 156
 zinc requirements during, 225
Pregnancy-induced hypertension, 327
Premature baby, breastfeeding, 332
Premenstrual syndrome (PMS), 174
Prenatal period, malnutrition in, 320
Preparation of food, safety in, 383–88. *See also* Cooking
Prepared and perishable food programs (PPFPs), 410
Prices, food, 15
Prick skin test (PST), 347
Probiotics, 196–97, 382
Problem drinkers, 311, 315
Processed foods, 215, 216
Produce-handling tips, 397
Protein(s), 134–63
 acid-base balance and, 141, 142–43
 aging and need for, 352

Protein(s), *continued*
- antibodies, 141, 142, 330–31, 345
- calorie value of, 31
- cancer and, 368
- complementary, 145
- complete, 145, 146
- consumption scorecard, 152
- contribution to body stores, 251–52
- cooking, 154
- defined, 136
- denaturation of, 140
- digestion of, 140, 144, 145
- as energy, 31, 141, 143–44
- enzymes as, 141
- excessive, 148–49
- for fitness, 297–98
- fluid balance and, 142, 143
- functions of body, 140–44
- growth and maintenance and, 140–41
- health and, 146–54
- hormones and, 141
- how body handles, 144, 145
- hydrolyzed vegetable (HVP), 160
- incomplete, 145
- as nutrient, 31
- pregnancy and, 321
- protein quality of foods, 144–46
- recommended intakes of, 146
- reference, 146, 148
- scorecard, 152
- shopping for, 153
- as source of life's variety, 140
- sources of, 149–51, 153–54
- structure of, 136
- transport, 141, 143
- in vegetarian diet, 155–56, 157
- vitamin B_6 and metabolism of, 174

Protein deficiency, 142
Protein digestibility-corrected amino acid score (PDCAAS), 146*n*
Protein energy malnutrition (PEM), 147–48, 405
Protein quality, 146
Protein-sparing, 144
Protein supplements, 137–39, 148
Protein synthesis, 140
Protozoa, 381
Psyllium, 191
Puffer poisoning, 383
Pulse, how to take your, 294
Pyridoxine (Vitamin B_6), 168, 174–75
Pyruvate, 304

Q

Quackery, 4–5
- detecting, 22–23
- fitness quackery scorecard, 307

R

Race and ethnicity, osteoporosis and, 231
Radioallergosorbent test (RAST), 347
Randomized, controlled study, 20
Raw meats, safety in preparing, 384, 385
Recommended Dietary Allowances (RDA), 33, 34, 35
Redux (dexfenfluramine), 257
Reference dose, 395, 398
Reference protein, 146, 148
Refined foods, 80
Refrigeration of foods, 384, 385
Registered dietitian (RD), 24, 259
Regulation(s)
- of herbal supplements, 192
- of organic farming, 396
- for pesticide use, 394

Religion, food choices and, 17
Requirement, 33
Research, credibility of nutrition, 20–21
Resveratrol, 198
Retina, 180
Retinal, 180
Retinol, 181
Retinol equivalents (RE), 181
Retirement centers, 362–63
Riboflavin, 168, 171, 172
Rickets, 166, 167, 182
Risk, 394–95
Root vegetables, 75
Roughage. *See* Fiber

S

Saccharin (Sweet'N Low), 93, 94, 96
Safety, food, 374–404
- common mistakes in, 383
- food additives and, 105, 398–400
- foodborne illnesses (food poisoning), 376–83
- in home, 384, 385, 387–88
- microbial food agents and, 378–82
- natural toxins and, 382–83
- new technologies for, 400–404
- organically grown produce and, 396–97
- pesticides and other chemical contaminants, 391–98, 402
- rank of areas of concern, 378
- scorecard, 389–90
- seafood, 383, 386, 388
- in storage and preparation, 383–88

Safety, margin of, 395, 398–99
St. John's wort, 191, 260
Salatrim, 118
Saliva, 73, 85
Salivary glands, digestion and, 87
Salmonella bacteria, 378, 380, 387, 401
Salmonellosis, 378–79, 380
Salt. *See also* Sodium
- defined, 215
- iodized, 226, 227
- seasoning foods without excess, 222

Salt sensitivity, 218
Salt substitutes, 222
Salt tablets, 300
Saponins, 194
Sassafras, 191
Satiety, 101, 249–50
Saturated fat, 46, 101, 105, 114–15, 118
- unsaturated vs., 102–3, 105

Saturated fatty acid, 102
Saturation, degree of, 102
Saw palmetto, 191
School lunch and breakfast programs, 408
School lunches, 340
Scombroid fish poisoning, 383
Scullcap, 191
Scurvy, 166
Seafood, safety tips for, 383, 386, 388
Second Harvest, 408, 410
Sedentary lifestyle, diseases associated with, 285
Selenium, 178, 207, 228, 369
Self, obese vs. healthy, 262, 263
Serotonin, 257
Servings, 38, 41, 42, 43
Set-point theory, 247
Seventh-Day Adventists, 17, 365–67
Shapedown Pediatric Obesity Program, 259
Shigellosis, 380, 381–82
Shopping
- for carbohydrates, 77–78
- checking for fat content, 122
- for proteins, 153
- for reduced-sodium diet, 222

Sibutramine (Meridia), 257
Silicon, 228
Silymarin, 190
Simple carbohydrates, 68, 69–72. *See also* Sugar(s)
Simplesse, 119–20
Skin cancer, 369
Skinfold test, 242, 243
Small intestine, digestion in. *See* Digestion
"Smart drinks," 139
Smoking
- breastfeeding and, 332
- caffeine metabolism and, 329
- malnutrition secondary to, 358
- osteoporosis and, 231
- pregnancy and, 325–26
- vitamin C consumption and, 177

Snacking, grazer's guide to, 39–40
Social groups, food choices and, 16–17
Sodium, 32, 206, 214, 215
- Daily Value for, 46, 216
- dietary recommendations for, 9
- in fast food, 11, 12
- in fluid-replacement drinks, 300
- hypertension and, 217, 218–19
- reduced-sodium diet, 222
- sources of, 216
- in whole unprocessed vs. processed foods, 215, 216

Sodium bicarbonate, 304
Sodium-potassium pump, 143
Soft water, 203
Soil erosion, 411–12
Solanine, 382
Soluble fibers, 82–83
Solution, The, 259
Sorbitol, 96
Soul food, 59–60
Soup kitchen, 410
Soyfoods, benefits of, 159–62, 197, 369
Spices, sweet, 72
Spina bifida, 173

Spinal vertebrae fracture, bone loss
and, 230
Splenda (sucralose), 93, 96
Sports, 293. *See also* Exercise(s); Fitness
Sports anemia, 302
Sports drinks, 299–300
Spot-reducing exercise, 267
Standard of living, birthrate and, 410–11
Stanol esters, 108, 197
Staphylococcal food poisoning, 381
Staphylococcus aureus, 377, 379, 381, 387
Staple grain, 72–75
Starch, 68, 69, 72–81
  in breads, 80–81
  defined, 72
  sources of, 72–75
  in whole foods, 76–80
Static stretches, 289
Steroids, anabolic, 304, 305
Sterols, 101. *See also* Cholesterol
Stimulant drugs, weight loss
and, 256
Stomach, digestion in. *See* Digestion
Storage of food
  safety in, 383–88
  vitamin preservation and, 186–87
Strength and strength training, 287–89
Stress, eating behavior and, 250
Stress fracture, 302
Stretching exercises, 289–90
Stroke, 358
Subjects in research study, 21
Substances generally recognized as safe
(GRAS) list, 399–400
Sucralose (Splenda), 93, 96
Sucrose, 69
Sudden infant death syndrome
(SIDS), 326
Sugar(s), 9, 68, 69–72
  alternatives to, 93–96
  disaccharides, 69–70
  health and, 70, 73–74
  keeping sweetness in diet, 70–72
  monosaccharides, 69
  nutrients in, 71
  in selected foods, 72
Sugar alcohols, 96
Sugar-free chewing gum, 96
Sulfur, 206, 214
Sulphoraphane, 194
Sunette/Sweet One (acesulfame K), 93, 96
Supplements
  amino acid, 137–39
  anti-aging, claims for, 355–56
  calcium, 232–34
  fitness and, 303–6
  foods vs., 372
  guidelines for choosing, 185
  herbal, 189–92, 260
  during infancy, 333
  mineral, 185
  for older adults, 353–54
  potassium, 217
  protein, 137–39, 148
  vitamin, 185
Surgery, weight loss and, 260–61
Sweat, 298–99
Sweeteners
  alternative, 96
  artificial, 93–96
Sweetness in diet, keeping, 70–72
Sweet'N Low (saccharin), 93, 94, 96
Sweet spices, 72

T

Take Control, 197
Tannins, 224
Target heart rate, 294
Tastes in food, 17–18
Tea, benefits from drinking, 372
Technology(ies)
  agricultural, 411
  appropriate, 410, 411
  food safety, 400–404
Teenage years. *See* Adolescence
Teeth
  dental caries in, 73, 344
  dental health and, 73–74, 358
  fluoride for, 74, 227–28
  tooth decay, 70, 73, 95
Television, obesity in children and, 342
10-calorie rule, 263
Testosterone, 289
Tetrodotoxin, 383
Thermometer, meat, 385
Thiamin, 168, 170–71
Thyroid gland, basal metabolic rate
and, 246
Thyroid hormones, 226
Time for activity, 287
Tin, 228
Title III legislation, 360
Tofu and tofu products, 160
Tolerable Upper Intake Level (UL), 34, 35
Tolerance
  alcohol, 311
  pesticide, 395
Tooth decay. *See* Teeth
TOPS (Take Off Pounds Sensibly), 259
Toxicants, natural food, 382–83
Toxicity, 391
  iron, 224
  from vitamins, 168–69
Trabecular bone, 230, 232, 235
Trace minerals, 207, 223–28
Training effect, 293–94
Trans fat, effect on blood lipids, 115
Trans fatty acids, 106–8
Transgenetics, 402, 403–4
Transient hypertension of pregnancy,
326–27
Transport proteins, 141, 143
Trichinella spiralis, 381
Trichinosis, 381
Triglycerides, 101, 102
Trimesters in pregnancy, 321, 325
Tripeptides, 144
Tryptophan, 171
Tubers, 75
Tufts University *Nutrition Navigator* Web
site, 22
Twin studies of obesity, 249
Tyramine, 345
Tyrosine, 136*n*

U

UL (Tolerable Upper Intake Level), 34, 35
Under-5-mortality rate (U5MR), 406, 410
Undernutrition, 341, 405–13
  among adolescents, 343
  deaths worldwide from, 406
  defined, 15, 410
  in infancy, 406–7
Underwater weighing, 242, 243
Underweight, defined, 240
UNICEF, 406, 410, 412, 413
Unique radiolytic products, 400–401
U.S. Conference of Mayors, 408
U.S. Department of Agriculture (USDA),
402, 408
  USDA organic seal, 396
U.S. Department of Health and Human Ser-
vices, 9
Unsaturated fatty acid, 103
  cis vs. trans, 106–8
Unsaturation, point of, 103, 104–5
Unspecified eating disorder, 275, 279
Urea, 144

V

Vaccines, edible, 403
Valerian, 191
Vanadium, 228
Variety, dietary, 28–30
Vegetable oil, partially hydrogenated,
106, 108
Vegetables
  cruciferous, 368
  storage and preparation of, 186–87
Vegetarian diet, 155–62
  cancer and, 369
  easy-to-prepare meals and snacks, 158
  health benefits of, 158
  minerals in, 157–58
  proteins in, 155–56, 157
  religion and, 17
  soy in, 159–62
  types of, 155
  vitamins in, 156–57
Very-low-calorie diets (VLCD), 256
Very-low-density lipoprotein (VLDL), 111
Vibrio bacteria, infection with, 378, 381, 386
Viruses, 381
Vision, vitamin A and, 179–80
Vitamin(s), 164–99
  aging and need for, 352
  B vitamins, 129–30, 156, 170–75, 179,
212, 301
  cancer and, 368–69
  classes of, 32
  deficiency diseases, 166–67
  defined, 31, 167
  discovery of, 4, 166, 167
  as ergogenic aid, 305
  exercise-related functions of, 301
  fat-soluble, 32, 169, 179–85
  five a day plus scorecard and, 188
  on food labels, 45, 48
  guide to, 168–69
  as nutrients, 31–32

Vitamin(s), *continued*
during pregnancy, 323
preservation of, 186–87
sources in Food Guide Pyramid, 170
supplements, choosing, 185
in vegetarian diet, 156–57
water-soluble, 32, 167–79
Vitamin A, 19, 169, 179–81
Vitamin A deficiency, 405
Vitamin B_6 (pyridoxine), 168, 174–75
Vitamin B_{12} (cobalamin), 168, 175, 228
Vitamin C and the Common Cold (Pauling), 178
Vitamin C (ascorbic acid), 166, 169, 176–79, 355, 368
Vitamin D (cholecalciferol), 109, 156–57, 169, 181–83, 230*n*, 231, 333
Vitamin E, 169, 177, 178, 183–84, 355, 369
Vitamin K, 169, 184
VLDL (very-low-density lipoprotein), 111

W

Waist circumference, 244
Water
balance in body, 203
bottled, 205–6
exercise and, 298–300
fluoridated, 227
functions in body, 202, 203
hard vs. soft, 203
for infants, 334
lead in drinking, 204–5, 392
as nutrient, 32, 202–6
for older adults, 352, 353
safety of drinking, 203–5
sources in diet, 203
Water-soluble vitamins, 32, 167–79
Weaning period, undernutrition in, 407
Web sites on nutrition, 25, 64, 97, 132, 151, 163, 199, 237, 281, 362, 373, 414
CyberDiet, 14
Dairy Council of California, 213
Intelihealth, 52
Mayo Clinic, 116
on diabetes, 92
on eating disorders, 280
for fitness, 310
Produce OASIS web site, 179
for weight loss, 266
Weight control, 238-81. *See also* Weight gain; Weight loss
breaking old habits and, 272–74
diabetes risk and, 91
dining out and, 270–71
eating disorders and, 275–80
energy balance and, 244–47
energy output scorecard and, 248
fiber and, 82
guidelines for evaluating programs, 255
healthful weight, defining, 241–44
healthy weight scorecard, 245
weight for height standards, 241
Weight cycling, 266
Weight gain, 251–53. *See also* Obesity
maternal, 324–25
risk of breast cancer and, 368
strategies for, 269
yo-yo effect and, 265–66
Weight loss, 253–61
artificial sweeteners and, 94–95
attitude and, 262
behavior modification and, 272–74
complex carbohydrates and, 80
diet aids and, 138, 255
drugs and, 255–60
eating disorders and, 275–80
exercise and, 266–69
fasting and, 246, 253–54
meal planning and, 262–66
plateau in, 265
profile of successful dieters, 265
strategies for, 261–69, 274
surgery and, 260–61
yo-yo effect in, 265–66
Weight-loss diets
comparing programs, 258–59
guidelines for evaluating programs, 255
high-protein, low-carbohydrate, 254–55
sample, using Food Guide Pyramid, 264
very-low-calorie (VLCD), 256
Weight Watchers, 259
Wheat bran, 83
Wheat kernel, main parts of, 80
Whole foods, 76–80, 81
Whole grain foods, 80, 81
WIC (Women's, Infant's, and Children's) program, 327, 408
Withdrawal symptoms, quitting caffeine and, 329
Women
iron requirements of, 223
role of, in developing countries, 412–13
World Health Organization (WHO), 189, 240, 401
FAO/WHO recommendations, 35–36
World resources, distribution of, 411–12
World Summit for Children (1990), 413
World view, food choices and, 18
World Wide Web, 22, 24. *See also* Web sites on nutrition
Wormwood, 191
Wrist fracture, bone loss and, 230

X

Xerophthalmia, 405, 410
Xylitol, 96

Y

Yeast, nutritional, 156
Yo-yo effect, 265–66

Z

Zeaxanthin, 194
Zinc, 148, 157, 207, 225–26, 322

Photo Credits

Chapter opening and other custom photos: Scott Hirko/Hespenheide Design

Chapter 1 5 © 2000 J. Share/Stone; **11** J. Gerard Smith/Photo Researchers, Inc.; **14** Dale Durfee/Tony Stone Images; **16** Blair Seitz/Photo-Researchers, Inc.; **18** © Bob Daemmrich/Stock Boston, Inc.; **21** B. Daemmrich/The Image Works; **23** Courtesy of Marilyn Herbert

Chapter 2 29 Felicia Martinez/PhotoEdit; **31** (top) John Kelly/Tony Stone Images; Felicia Martinez/PhotoEdit; **34** Bob Thomas/Tony Stone Images; **36** Felicia Martinez/PhotoEdit; **39** Key Sanders/Tony Stone Images; **41** (left) Polara Studios; Photo Researchers, Inc.; **44** Ziggy Kaluzny/Tony Stone Images; **49** Charles Winters/Photo Researchers, Inc.; **54** Michael Newman/PhotoEdit; **57** Tony Freeman/PhotoEdit; Felicia Martinez/PhotoEdit; **59** Bonnie Kamin/PhotoEdit; **60** Bill Aron/PhotoEdit

Chapter 3 70 Felicia Martinez/PhotoEdit; **71** Tony Freeman/Photo Edit; **73** Mark Harwood/Tony Stone Images; **75** (left) © 2000 PhotoDisc; Polara Studios; **81** Juan-Pablo Lira/Image Bank; **82** Felicia Martinez/PhotoEdit; **84** Mark Antman/The Image Works; **89** Polara Studios

Chapter 4 100 Lori Adamski Peek/Tony Stone Images; **104** (top) Quest Photographic, Inc.; Michael Newman/PhotoEdit; **108** Don and Pat Valenti/Tony Stone Images; **113** Felicia Martinez/PhotoEdit; **115, 118, 121** Quest Photographic, Inc.; **123** (top) Thomas Del Brase/Tony Stone Images; (center) James Jackson/Tony Stone Images; (bottom) David Frazier/Tony Stone Images; **125** Courtesy of ICI, Pharmaceuticals Division, Cheshire UK; **131** Polara Studios

Chapter 5 137 © 2000 PhotoDisc; **138** (top) Jean Marc Barey, Vandystadt/Photo Researchers, Inc.; Felicia Martinez/PhotoEdit; **140** Stephen Frisch/Stock Boston, Inc.; **144** (top) © 2000 PhotoDisc; Polara Studios; **148** Photo Library, United Nations Food and Agricultural Organization/Rome; **149** Polara Studios; **153** © 2000 PhotoDisc; **154** © Japack Company/CORBIS; **155** Polara Studios; **157** (3) Tony Freeman/PhotoEdit; (4) Felicia Martinez/PhotoEdit

Chapter 6 166 Thomas Braise/Tony Stone Images; **167** Biophoto Association/Photo Researchers, Inc.; **171, 172** Quest Photographic, Inc.; **173** (top) Quest Photographic, Inc.; NMSB/Custom Medical Stock Photo; **174, 175, 176, 177, 180** Quest Photographic, Inc.; **182** (top) Quest Photographic, Inc.; David Young-Wolff/PhotoEdit; **183, 184** Quest Photographic, Inc.; **187** (top) David R. Frazier/Photo Library; (center) Quest Photographic, Inc.; (bottom) Felicia Martinez/PhotoEdit; **189** (top) Esbin-Anderson; Jessica Wecker/Courtesy of Celestial Seasoning Herbs; **193** Zane Williams/Tony Stone Images

Chapter 7 202 Camera M.D. Studios; **205** © The Stock Market/Michael A. Keller 2000; **209** Quest Photographic, Inc.; **210** California Milk Processors Board and Manitoba Milk Producers; **212, 214, 216, 217** Quest Photographic, Inc.; **218** (top) David Young-Wolff/PhotoEdit; © Charles Feil/Stock Boston, Inc.; **223** Quest Photographic, Inc.; **224** (top)Polara Studios; Michael Newman/PhotoEdit; **225** (top)Courtesy of H. Sanstead, University of Texas at Galveston; Quest Photographic, Inc.; **227** © Lester V. Bergman/CORBIS; **229** Larry Mulvehill/Photo Researchers, Inc.; **230** With permission from Dempster et al., J. Bone Min. Res: I 15–21, 1986; **232** Courtesy of Gjon Mills; **233** Michael Newman/PhotoEdit; **236** © 2000 David Madison/Stone

Chapter 8 240 Tony Freeman/PhotoEdit; **243** (top three) David Young-Wolff/PhotoEdit; (bottom left) Alan Oddie/PhotoEdit; (bottom right) B. Daemmrich/Stock Boston, Inc.; **247** David Madison/Tony Stone Images; **251** © 2000 PhotoDisc; **267** David Young-Wolff/PhotoEdit; **269** Polara Studios; **270** Jeff Greenberg/PhotoEdit; **272** Philip and Karen Smith/Tony Stone Images; **275** Tony Freeman/PhotoEdit; **277** Michael Newman/PhotoEdit; Felicia Martinez/PhotoEdit

Chapter 9 284 Richard Hutchings/PhotoEdit; **286** Tony Freeman/PhotoEdit; **290** Elena Rooraid/PhotoEdit; **292** (top) Esbin-Anderson; Tim Davis/Photo Researchers, Inc.; **294** Michael Newman/PhotoEdit; **295** Tony Freeman/PhotoEdit; **296** A. Kubacsi, Explorer/Photo Researchers, Inc.; **297** Tony Freeman/PhotoEdit; **298** David Young-Wolff/PhotoEdit; **300** Michael Newman/PhotoEdit; **302** Tony Freeman/PhotoEdit; **303** E. M. Wallop/The Stock Market; **307** © E. Weber/Visuals Unlimited; **309** Felicia Martinez/PhotoEdit; **315** Aaron Haupt/Stock Boston, Inc.

Chapter 10 **320** Petit-Format, Nestles/Photo Researchers, Inc.; **321** © Lennart Nilsson/Albert Bonniers Förlag AB from *A Child is Born*, Dell Publishing Co.; **325** David Sams/Stock Boston, Inc.; **326** Jones, K. L.; Smith, D. W.; Ulleland, C. N; and Steissaguth, A. P. *The Lancet*, June 9, 1973; pp. 1267–1271; **328** Dion Ogust/The Image Works; **331** Reprinted with permission from J. Brown, *Nutrition Now* (St. Paul, MN: West Publishing); **332** Myrleen Ferguson/PhotoEdit; **333** J. Gerard Smith/Photo Researchers, Inc.; **335, 336** Robert Brenner/PhotoEdit; **337** Anthony Vannelli; **341** Jeff Greenberg/Photo Researchers, Inc.; **342** Michael Newman/PhotoEdit; **343** Marilyn Ferguson Cate/PhotoEdit; **344** Mary Kate Denny/PhotoEdit; **351** Michael Newman/PhotoEdit; Bill Bachmann/PhotoEdit; **354** Tom Ives; **368** Karl Weidmann/Photo Researchers, Inc.; **370** Polara Studios; **372** © The Stock Market/Roy Morsch 2000

Chapter 11 **377** © 2000 PhotoDisc; **379** © 2000 Martin Chaffer/ Stone; **384** Felicia Martinez/PhotoEdit; **385** David Young-Wolff/PhotoEdit; **391** Phil Borden/PhotoEdit; **392** David Young-Wolff/PhotoEdit; **394** © George Loun/Visuals Unlimited; **396** (top) Nancy Richmond/The Image Works; James Wilson/Woodfin Camp & Associates; **399** Polara Studios; **402** Antonio Montaner/Smithsonian; **405** Bruce Brander/Photo Researchers, Inc.; **406** © Louise Gubb/The Image Works; **407** UPI/Corbis-Bettman; **411** © Louise Gubb/The Image Works

Median Heights and Weights and Recommended Energy Intakes (United States)

Age (years)	Weight		Height		Average Energy Allowance	
	kg	lb	cm	inches	cal per kg	cal per day[a]
Infants						
0.0–0.5	6	13	60	24	108	650
0.5–1.0	9	20	71	28	98	850
Children						
1–3	13	29	90	35	102	1,300
4–6	20	44	112	44	90	1,800
7–10	28	62	132	52	70	2,000
Males						
11–14	45	99	157	62	55	2,500
15–18	66	145	176	69	45	3,000
19–24	72	160	177	70	40	2,900
25–50	79	174	176	70	37	2,900
51+	77	170	173	68	30	2,300
Females						
11–14	46	101	157	62	47	2,200
15–18	55	120	163	64	40	2,200
19–24	58	128	164	65	38	2,200
25–50	63	138	163	64	36	2,200
51+	65	143	160	63	30	1,900
Pregnant (2nd and 3rd trimesters)						+ 300
Lactating						+ 500

[a]Average energy allowances have been rounded.

Source: Recommended Dietary Allowances, © 1989 by the National Academy of Sciences, National Academy Press, Washington, D.C.

Recommended BMI Cutoff Values for Adolescents

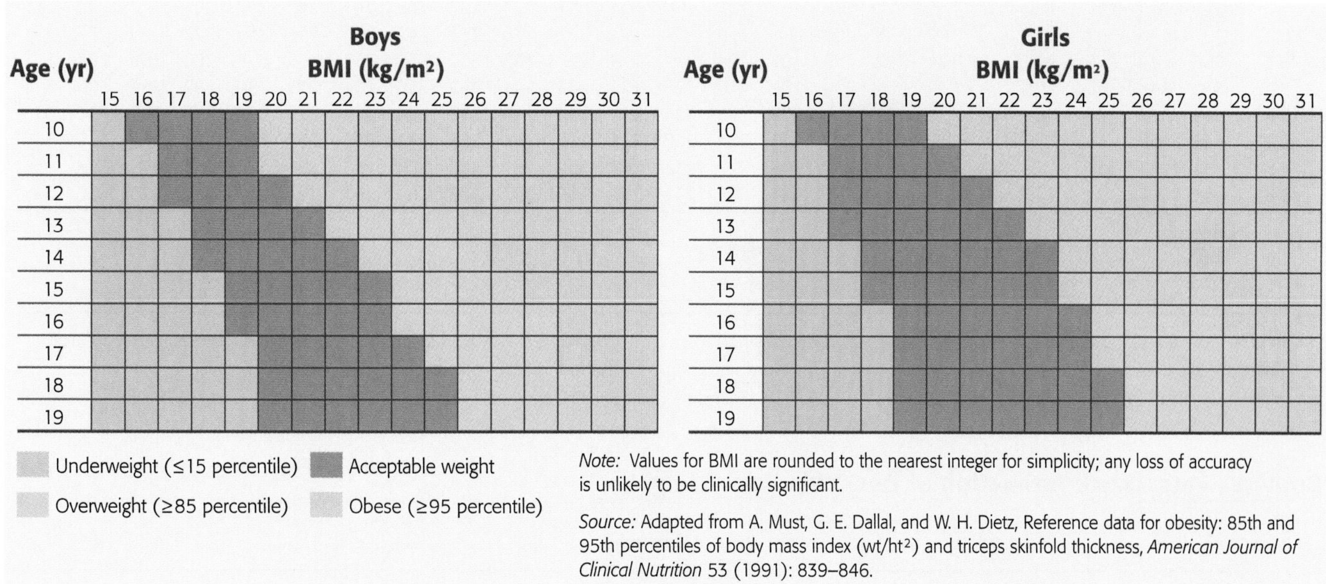

Note: Values for BMI are rounded to the nearest integer for simplicity; any loss of accuracy is unlikely to be clinically significant.

Source: Adapted from A. Must, G. E. Dallal, and W. H. Dietz, Reference data for obesity: 85th and 95th percentiles of body mass index (wt/ht^2) and triceps skinfold thickness, *American Journal of Clinical Nutrition* 53 (1991): 839–846.

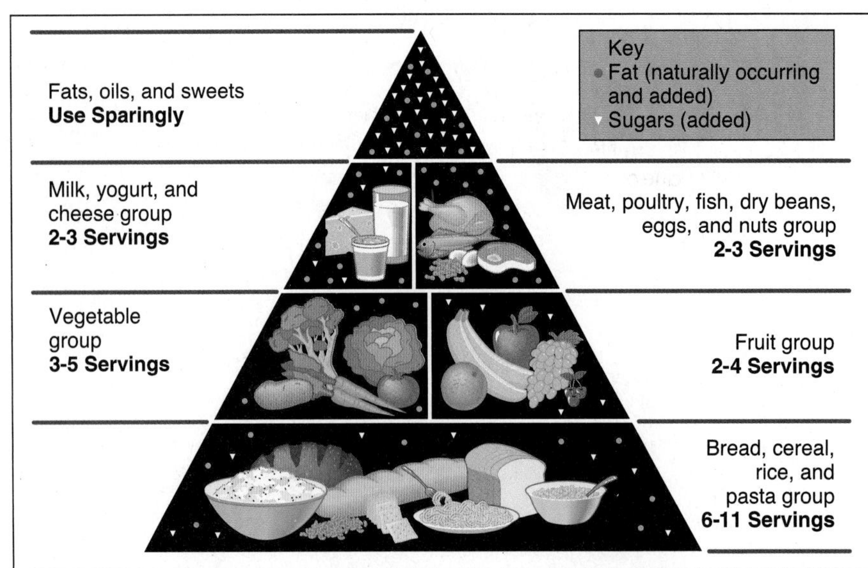

Food Guide Pyramid

A Guide to Daily Food Choices

The breadth of the base shows that grains (breads, cereals, rice, and pasta) deserve most emphasis in the diet. The tip is smallest: use fats, oils, and sweets sparingly.

Source: USDA, 1992

Body Mass Index (BMI)

Height	18	19	20	21	22	23	24	25	26	27	28	29	30	31	32	33	34	35	36	37	38	39	40
	Body Weight (pounds)																						
4′10″	86	91	96	100	105	110	115	119	124	129	134	138	143	148	153	158	162	167	172	177	181	186	191
4′11″	89	94	99	104	109	114	119	124	128	133	138	143	148	153	158	163	168	173	178	183	188	193	198
5′0″	92	97	102	107	112	118	123	128	133	138	143	148	153	158	163	168	174	179	184	189	194	199	204
5′1″	95	100	106	111	116	122	127	132	137	143	148	153	158	164	169	174	180	185	190	195	201	206	211
5′2″	98	104	109	115	120	126	131	136	142	147	153	158	164	169	175	180	186	191	196	202	207	213	218
5′3″	102	107	113	118	124	130	135	141	146	152	158	163	169	175	180	186	191	197	203	208	214	220	225
5′4″	105	110	116	122	128	134	140	145	151	157	163	169	174	180	186	192	197	204	209	215	221	227	232
5′5″	108	114	120	126	132	138	144	150	156	162	168	174	180	186	192	198	204	210	216	222	228	234	240
5′6″	112	118	124	130	136	142	148	155	161	167	173	179	186	192	198	204	210	216	223	229	235	241	247
5′7″	115	121	127	134	140	146	153	159	166	172	178	185	191	198	204	211	217	223	230	236	242	249	255
5′8″	118	125	131	138	144	151	158	164	171	177	184	190	197	203	210	216	223	230	236	243	249	256	262
5′9″	122	128	135	142	149	155	162	169	176	182	189	196	203	209	216	223	230	236	243	250	257	263	270
5′10″	126	132	139	146	153	160	167	174	181	188	195	202	209	216	222	229	236	243	250	257	264	271	278
5′11″	129	136	143	150	157	165	172	179	186	193	200	208	215	222	229	236	243	250	257	265	272	279	286
6′0″	132	140	147	154	162	169	177	184	191	199	206	213	221	228	235	242	250	258	265	272	279	287	294
6′1″	136	144	151	159	166	174	182	189	197	204	212	219	227	235	242	250	257	265	272	280	288	295	302
6′2″	141	148	155	163	171	179	186	194	202	210	218	225	233	241	249	256	264	272	280	287	295	303	311
6′3″	144	152	160	168	176	184	192	200	208	216	224	232	240	248	256	264	272	279	287	295	303	311	319
6′4″	148	156	164	172	180	189	197	205	213	221	230	238	246	254	263	271	279	287	295	304	312	320	328
6′5″	151	160	168	176	185	193	202	210	218	227	235	244	252	261	269	277	286	294	303	311	319	328	336
6′6″	155	164	172	181	190	198	207	216	224	233	241	250	259	267	276	284	293	302	310	319	328	336	345
	Underweight (<18.5)	Healthy Weight (18.5–24.9)						Overweight (25–29.9)					Obese (≥ 30)										

Find your height along the left-hand column and look across the row until you find the number that is closest to your weight. The number at the top of that column identifies your BMI.

Chapter 8 describes how BMI correlates with disease risks. The area shaded in blue represents healthy weight ranges.

Source: National Heart, Lung, and Blood Institute, 1998.

Body Weights: Quick Estimation of Desirable Body Weight

Men	Women
For 5 ft, consider 106 lb a reasonable weight. For each inch over 5 ft, add 6 lb. Subtract 6 lb for each inch under 5 ft. Add 10% for a large-framed individual; subtract 10% for a small-framed individual. (See Appendix D, Table D-3: Frame Sizes.) *Example:* A man 5 ft 8 inches tall (medium frame) would start at 106 lb, add 48, and arrive at a reasonable weight of 154 lb.	For 5 ft, consider 100 lb a reasonable weight. For each inch over 5 ft, add 5 lb. Subtract 5 lb for each inch under 5 ft. Add 10% for a large-framed individual; subtract 10% for a small-framed individual. For each year under 25 (down to 18), subtract 1 lb. *Example:* A woman 21 years old, 5 ft 4 inches tall (medium frame), would start at 100 lb, add 20, and subtract 4, arriving at a reasonable weight of 116 lb.